ELECTROMAGNETIC FIELDS FOR ENGINEERS

Master the fundamentals of undergraduate electromagnetics with this concise and accessible textbook, linking theoretical principles to real-world engineering applications. Lightning, nuclear fusion, superconductors: over 80 real-world TechNote case studies throughout the book show how key electromagnetic principles work in a wide variety of natural effects and man-made devices. Learn in confidence: over 170 annotated step-by-step examples, with illustrated field patterns, aid student visualization of key physical principles, and help to build a solid foundation for future study. Dive deeper: SideNotes provide detailed proofs and context without distracting from core learning, and carefully designed appendices provide additional mathematical assistance when needed. Make progress: over 375 end-of-chapter homework problems to assess and extend student understanding. Flexible instructor support: start your students off with statics, or dive straight into dynamics, with this versatile full-color textbook for a one- or two-semester course, supported by lecture slides, instructor solutions, MATLAB animations, and PowerPoint and jpeg figures.

DANIEL S. ELLIOTT is Professor of Electrical and Computer Engineering, and Physics and Astronomy, at Purdue University, where he has over 39 years of experience teaching undergraduate and graduate courses in electromagnetics, lasers, optics, circuits, and introductory physics. He is a Member of the IEEE and a Fellow of the American Physical Society and Optica.

"This textbook takes a unique approach to teaching electromagnetic theory, bridging theoretical principles to real-world engineering applications. It provides well-crafted case studies, which the author uses to demonstrate both natural and man-made electromagnetics phenomena. Throughout the text, students and researchers have the opportunity to examine the beauty of electromagnetics and understand its pivotal role in shaping the technological landscape of today and tomorrow."

Julio Urbina, *The Pennsylvania State University*

"Daniel Elliott's new textbook provides a comprehensive dive into electromagnetic fields. While covering the traditional topics with care, it is also infused with interesting snippets in the form of 'TechNotes' that apply the concepts to all sorts of modern day-to-day electrical devices and applications, and 'Biographical Notes' on the many historical figures who created this discipline. I'm finding it hard to put down!"

Andrew F. Peterson, *Georgia Institute of Technology*

Electromagnetic Fields for Engineers

Daniel S. Elliott
Purdue University, Indiana

CAMBRIDGE
UNIVERSITY PRESS

CAMBRIDGE
UNIVERSITY PRESS

Shaftesbury Road, Cambridge CB2 8EA, United Kingdom

One Liberty Plaza, 20th Floor, New York, NY 10006, USA

477 Williamstown Road, Port Melbourne, VIC 3207, Australia

314–321, 3rd Floor, Plot 3, Splendor Forum, Jasola District Centre, New Delhi – 110025, India

103 Penang Road, #05–06/07, Visioncrest Commercial, Singapore 238467

Cambridge University Press is part of Cambridge University Press & Assessment,
a department of the University of Cambridge.

We share the University's mission to contribute to society through the pursuit of
education, learning and research at the highest international levels of excellence.

www.cambridge.org
Information on this title: www.cambridge.org/highereducation/isbn/9781009309448

DOI: 10.1017/9781009309431

First published 2025

Printed in the United Kingdom by CPI Group Ltd, Croydon, CR0 4YY

A catalogue record for this publication is available from the British Library

Library of Congress Cataloging-in-Publication Data
Names: Elliott, D. S., 1953– author.
Title: Electromagnetic fields for engineers / D. S. Elliott, The Elmore Family School of Electrical and Computer
     Engineering and the Department of Physics and Astronomy, Purdue University.
Description: First edition. | New York : Cambridge University Press, [2024] | Includes bibliographical
     references and index.
Identifiers: LCCN 2023045653 (print) | LCCN 2023045654 (ebook) | ISBN 9781009309448 (hardback) |
     ISBN 9781009309431 (epub)
Subjects: LCSH: Electromagnetic fields–Textbooks.
Classification: LCC TK454.4.E5 E44 2024 (print) | LCC TK454.4.E5 (ebook) | DDC 530.14/1–dc23/eng/20240304
LC record available at https://lccn.loc.gov/2023045653
LC ebook record available at https://lccn.loc.gov/2023045654

ISBN 978-1-009-30944-8 Hardback

Additional resources for this publication at www.cambridge.org/elliott.

Contents

Detailed Contents

Preface

This textbook on electromagnetic fields originated from lecture notes that I have developed and used over the years at Purdue University. I have taught this course frequently, and in writing this text I have attempted to use what I have learned in the classroom. Specifically, I have found that there are several stumbling blocks often encountered by students, and I have tried to address these directly in the text. The study of electromagnetics is heavily mathematical, and we can make little headway in this field if we cannot calculate the forces, field quantities, current, and charge distributions that are so important to the phenomena that we observe. At the same time, however, it is important to develop a physical understanding of these effects. Without this picture, the many nuances and details become overwhelming. I have attempted to present a balanced discussion, including both the mathematical detail and physical picture in my treatment of this topic.

I have included in this text numerous examples with solutions, technical notes (labeled "TechNotes"), "SideNotes," and biographical sketches of several of the historical figures who played critical roles in developing the guiding principles of electromagnetics (labeled "Biographical Notes"). In developing example problems, I devote a significant effort to how to set up the calculations. I presume that the student has an understanding of differential and integral calculus, and is familiar with vector notations and operations. At the same time, I have found over the years that a review of these topics is beneficial, or even necessary, for most students. That is, the typical student has used these tools previously, but is perhaps a little rusty, and a reminder right up front is very helpful. Therefore, I use the first few class meetings of the semester to review these math tools, which are contained in Appendices A–E, before we start on electrostatics. In working examples, I often take the approach of repeating the guiding principles as we apply them to a specific example. Perhaps I have overdone this, but I have found that repetition helps drive these ideas home, and cuts down on the guesswork that many students employ as they try to put the pieces together. The TechNotes that I include help relate many of the concepts of electromagnetics to devices or techniques that depend on them. Detailed understanding of these applications often requires a much deeper knowledge than we can hope to achieve in this course, but seeing these applications should help make the material more relevant and meaningful. In the biographical sketches I have tried to place the contributions of historical figures, such as Coulomb, Cavendish, Ampere, Maxwell, and Hertz, in the context of their times, and to show how the contributions of each have added to our collective knowledge. Finally, there are many proofs and theorems that are detailed and tedious, and perhaps a bit tangential to the main discussion. For completeness, I didn't want to totally ignore these details, but at the same time I didn't want to be distracted from the main discussion. I have set these discussions aside in SideNotes, which can be found in many of the chapters. Readers who choose to skip these should still come away with the major points of the discussion. But for those who want to dive a little more deeply into the details, these SideNotes should provide the necessary rigor.

Courses on electromagnetics at various universities follow many different styles and formats. In some electrical engineering schools, the course is taught in a single semester; in others, two semesters; and in others still, one required semester followed by a second optional course for those who want to go deeper into the material. Many courses start with electrostatics and magnetostatics, in order to establish the fundamental principles of static fields, and develop the complete set of Maxwell's Equations one step at a time. That is the approach we take at my institution, and the outline that I have followed for this text. After statics, we discuss uniform plane waves and transmission lines, but not waveguides or antennas. These topics are covered in a second optional course. A rough course outline for our first-semester required course is as follows:

Weeks	Topics
1–3	Math review, electrostatics
4	Current, materials
5–6	Poisson's and Laplace's Equations, resistance, capacitance
7–8	Magnetostatics
9–10	Magnetic dipole, materials, energy, inductance
11	Faraday's Law, displacement current, Maxwell's Equations
12	Wave equation, wave propagation
13	Power, Poynting vector
14	Reflection
15	Transmission lines

Notice the review of mathematical tools in the first week of our course. We take this opportunity to review vectors, coordinate systems, differential operations (gradient, divergence, and curl) in the different coordinate systems, and the Divergence Theorem and Stokes' Theorem. There are different philosophies on this as well. Often, these topics are reviewed as they are needed throughout the course. For this reason, these mathematical review topics have been placed in Appendices A–E at the end of the text. I have also included discussions of Taylor's Expansions (Appendix H), complex numbers (Appendix I), and hyperbolic trigonometric functions (Appendix J). These could be treated as part of a mathematics review at the beginning of the course, but we typically touch on these topics as needed.

As an alternative to the statics-first approach, many electromagnetics courses are built on the premise that students have seen static fields before, and they dive right in with the full set of Maxwell's Equations and their applications. This leaves more time for the study of waves. This course is similar to our second-semester optional course, whose outline is as follows:

Weeks	Topics
1–3	Time-varying fields and Maxwell's Equations
4–6	Plane electromagnetic waves
7–9	Theory and application of transmission lines
10–12	Waveguides and cavity resonators
13–15	Antennas and radiating systems

This text can be used with this course outline by simply skipping the chapters on electrostatics and magnetostatics, or touching on them briefly, as needed.

In either format (statics first, or straight into dynamics), many courses start with an in-depth discussion of transmission lines. This textbook can be used for courses that follow this format by simply starting immediately with Chapter 9, excluding Section 9.2, which connects electric and magnetic fields to potentials and currents. In the remainder of Chapter 9, I discuss wave and pulse propagation on transmission lines in terms of the potential and current, which can easily be accessed by students before mastering electric and magnetic fields.

In order to make the textbook as appealing to as many different electrical engineering curricula as possible, I have included many topics that some courses may use, but others will choose to omit. It was not my intent that each topic be covered by any particular course. That would clearly be more material than could reasonably be absorbed by the students. Rather, I included many topics that would give the instructors options to include in their course, or topics that students might refer back to in their professional lives as needed.

I have many people to thank for their direct and indirect contributions to this text: numerous colleagues whom I have worked with throughout my career and have helped me develop my techniques and passion for teaching; professors who taught me so many years ago; and students who I have worked with in more recent years. The adage that we learn the most when we teach it to others is a constant truth. To Professors Kevin Webb and Thomas Roth, thank you for your input. A particular debt of gratitude is owed to Professor David Janes, who carefully read through the entire manuscript and offered many excellent suggestions for improvements. Jonah Quirk, Gary Hudson, Austin Beidelman, and Harper Westphal helped with figures or photographs, and Lindsay Elliott was a constant source of suggestions, advice, and comments. But none have been so impactful on me as my family, who have over the years kept me grounded on what is really important, and provided significance to my daily routine. To them, I am forever in debt.

To the Student

Applications of electromagnetics surround us every day. They have become an integral part of our routines, so much so that we often take them for granted. Electric and magnetic fields allow us to communicate (cell phones, wireless internet connections, radios, etc.), distribute power (transmission lines, waveguides, unguided waves, etc.), entertain (CDs and DVDs, TV remotes), and help keep us healthy (magnetic resonance imaging [MRI], X-rays, etc.). Freely propagating waves and guided waves are important in energy transfer, communication systems, and high-frequency electronic systems. We all enjoy our high-speed computer systems, but perhaps do not fully appreciate the importance of propagation effects that must be properly treated as bit rates go up. By this time in your education as an engineering student, you have already encountered basic components of electrical circuits, such as resistors, capacitors, inductors, transformers, etc. Each and every one of these devices depends on electric and/or magnetic fields to function, and we must understand their properties to extend their uses to new purposes. Scientifically, electromagnetic theory helped form our understanding of modern physical sciences, playing an important role in the development of the theory of relativity, integration of optics and radio waves, and the basis for the structure of matter at the atomic level. In this text, we will develop the theory of electric and magnetic fields, and explore their uses in elementary devices.

We have gained an understanding of the properties of electric and magnetic fields over the past 200 years through observations of the forces that are exerted between charged bodies (for electric forces) and between magnetic systems, such as current-carrying conductors and permanent magnets. These field properties can be summarized through a set of four equations known collectively as **Maxwell's Equations**. In this textbook we will develop these equations and explore uses of each to solve different types of problems. Maxwell's Equations, which were developed over the years to describe concisely the qualitative and quantitative empirical observations of electric and magnetic phenomena, are individually known as Gauss' Law, Faraday's Law, Ampere's Law, and a fourth equation that suffers the ignominy of having no name, but which describes the fact that there are no magnetic charges. We will develop each of these equations in this text, starting with static (non-time-varying) fields, and adding time-varying effects later. For static fields, we will see that electric effects and magnetic effects can be treated separately. Thus we start with electrostatic effects, governed by **Gauss' Law** and a limited form of **Faraday's Law**, which together describe the electric field produced by electric charges. We will define the electric potential, explore the energy stored in electric systems, and the capacitance of electrical conductors to hold charge (i.e. their "capacity" to store charge). We will move on to magnetostatic systems, which obey a limited form of **Ampere's Law** and the fourth, unnamed Maxwell Equation. In this study, we will define a magnetic potential, explore the energy stored in magnetic systems, and examine the inductance of current-carrying conductors. These studies of static systems will start us on the way to an understanding of time-varying phenomena, essential for the operation of transformers, and for propagation and generation of electromagnetic waves. To make the transition from static fields to dynamic fields, we must complete Faraday's Law and Ampere's Law. The discovery by key scientists of the nineteenth century that a common set of laws that govern electric and magnetic forces also lead to a description of propagation of waves was a major advance in our scientific understanding of fields and technological development of electromagnetic effects.

TechNote 1 Electromagnetics in the Cinema

Electric and magnetic effects have long been featured in popular media. While they do make for exciting entertainment, they often lack much connection with reality. A few examples are illustrated in Fig. 1. Magneto, a major antagonist in the *X-Men* movies and shown in Fig. 1(a), is well known for his ability to control magnetic fields, and through that to exert huge forces on anything containing metal. (Note that Magneto twists and moves any metal, not only ferromagnetic metals – that is, those containing iron, cobalt, or nickel.) In Fig. 1(b), Luke Skywalker is armed with his light saber. Again, light sabers stir our imaginations but are, unfortunately, totally unrealistic. Except in unusual conditions, light does not interact with light, so when two light sabers clash they should just pass through one another. In addition, light emitted by the shaft of the saber cannot simply end 1.5 m from the hilt, but rather continue propagating off to infinity. Finally, in Fig. 1(c), Zeus is shown holding two bolts of lightning, which he is prepared to hurl at any foe. Hopefully, in studying this text, you will gain a better appreciation for what is realistic and practical behavior of electric and magnetic fields, and still be able to appreciate these superheroes in the cinema.

(a) (b) (c)

FIGURE 1 Electromagnetics in the cinema. (a) Magneto in *X-Men: First Class.* (b) Luke Skywalker wielding a light saber. (c) Zeus holding thunderbolts.

In teaching this course over the years, we have, unfortunately, seen that some students seem to get lost in the details, and fail to see the bigger picture of electric and magnetic fields. We cannot avoid the details, as it is only through mastering these that we can develop a true understanding. We will, however, do our best to help guide you, and help avoid pitfalls and sandtraps. To this end, we find it useful if we can maintain a global view of each topic as we treat it. For example, it may be helpful to keep the following summary of electromagnetic fields in mind:

1. Electric charges exert forces on each other. A more useful picture of this force is that electrical charges produce electric fields, and these fields exert forces on other charges. We will examine this quantitatively using Coulomb's Law and Gauss' Law.
2. Moving charges exert forces on each other. We will describe this effect in terms of the magnetic field created by a moving charge, or current, and the force exerted by the

magnetic field on another moving charge. This magnetic field is described through the Biot–Savart Law and Ampere's Law.

3. Charge is conserved. Charges can be moved around, or redistributed, through electrical currents, but we can neither destroy nor create them.

4. Since electric and magnetic fields exert forces on charges and currents, we understand that charges and currents can be redistributed when they are free to do so.

5. Time-varying magnetic fields produce an electric field. This is known as, and quantified through, Faraday's Law.

6. Time-varying electric fields produce a magnetic field. This is known as, and quantified through, Ampere's Law.

7. Through the previous two principles, we can understand propagation of energy through electromagnetic waves. These waves can be guided by conductors, but can also propagate without conductors as freely propagating waves.

SI Units

We will use the International System of units, abbreviated SI from the French Système International d'Unités, throughout our discussion of electromagnetic field theory. The seven basic quantities – length (meter, m), mass (kilogram, kg), time (second, s), amount of substance (mole), current (ampere, A), temperature (kelvin, K), and luminous intensity (candela, cd) – are used to define all measurable quantities. These units have been agreed upon by international commission, and are periodically revisited and redefined to improve upon their utility. Standards for each quantity have been developed over the years, as we have learned how to make measurements of higher precision of our physical world, and learned of new methods to improve their universality, accessibility, and precise transfer.

The meter (m), the standard unit of length, was originally defined in terms of dimensions of the Earth; specifically, the meter was defined as 10^{-7} times the distance from the equator to the North Pole. As a more practical measure, the meter was later defined as the distance between two marks inscribed on a platinum iridium bar maintained under controlled conditions in a laboratory located in Sèvres, France (near Paris). More recently, the standard of length and time have been unified through the speed of light, c; the meter is defined as the distance traveled by light through vacuum in $(299\,792\,458)^{-1}$ s.

The unit of mass, the kilogram (kg), is approximately the mass of one liter $(10^{-3}\ \mathrm{m}^3)$ of water at standard temperature and pressure (STP). Since 1889, a standard mass in the form of a block of platinum–iridium has been maintained at the International Bureau of Weights and Measures in Sèvres. Effective May 20, 2019, the kilogram has been defined electronically using a device known as a Kibble balance, along with a new fixed definition of the Planck constant, $h = 6.626\,070\,15 \times 10^{-34}$ Js. (The Planck constant is well known in quantum physics as the constant that relates the energy of a quantum of light, a photon, to its frequency. We will not have any further need of the Planck constant in our study of classical electromagnetics.)

The second (s) is the unit of time, a unit that is shared with the British system of units. Originally defined in terms of a fraction of a day, the second is now defined using atomic measures. Specifically, the second is defined as the time for $9\,192\,631\,770$ oscillations of a microwave field corresponding to the transition between two components of the ground state of atomic cesium. This transition frequency is one of the most precisely measured quantities in history, and together with a similar transition in atomic rubidium, forms the basis for "atomic clocks." These remarkable instruments are essential components of our global positioning system (GPS), which we all enjoy when driving cross country, for example.

The SI unit for the amount of matter is called the mole. A mole consists of $6.022\,140\,76 \times 10^{23}$ particles, and the Avogadro constant, N_A, is defined as $6.022\,140\,76 \times 10^{23}$ units/mole, where a unit is an atom, molecule, ion, electron, or any other particle. The abbreviation for mole is mol.

With these standards for length, mass, and time, we are in a position to start relating other measures to these primary standards. For example, the velocity of an object is the rate at which its position changes. How do we determine the velocity of an object? Well, there are many ways to do this, but fundamentally we can measure the distance that an object moves, and how much time it takes to move that distance, and the velocity of this object is the ratio of these two quantities. So we now can see a connection back to a measurement of two of our standard metrics – length and time – and we determine that $[\mathbf{u}] = $ m/s. (**Note of notation:** We will use square brackets to denote the units of a quantity, and bold type face to denote a vector quantity. For example, "$[\mathbf{u}] = $ m/s" represents "The units for the vector velocity is meters per second." Vectors are quantities that are represented not only by a magnitude, but also by a direction. For the example given, the magnitude of the velocity of an object is the speed at which it moves, and the direction of the vector is the direction in which the object is moving. See Appendix A for a detailed review of scalar and vector definitions and vector arithmetic.) Similarly, acceleration is the rate at which the velocity changes, $\mathbf{a} = \Delta\mathbf{u}/\Delta t$. Therefore $[\mathbf{a}] = $ m/s^2. As we remarked earlier, electric and magnetic fields can apply forces on charged bodies, and by Newton's Second Law we know that the net force applied to an object will cause it to accelerate, $\mathbf{F} = m\mathbf{a}$, where m is the mass of the object. Therefore, $[\mathbf{F}] = $ kg m/s^2, a unit known as the newton (N). And as we apply a force to an object and move it over a distance from one position $\mathbf{r}_1$ to a new position $\mathbf{r}_2$, we must do work on that object, and therefore will change its energy by the amount $\Delta W = -\int_{\mathbf{r}_1}^{\mathbf{r}_2} \mathbf{F} \cdot d\mathbf{r}$. (The notation in this expression may not be familiar to all readers. If not, do not be alarmed. We will define it and explore its properties in Appendix C.) The unit of energy is $[W] = $ N m, which is known as the joule (J).

While we can apply the preceding standards in a broad arena of systems, electrical as well as non-electrical, the ampere (A), our unit of electrical current, is clearly electrical in nature. Prior to 2019, the ampere was defined in terms of a mechanical force; specifically, two long, straight wires separated by 1 m carrying oppositely directed currents of magnitude 1 A feel a repulsive force per unit length of the wires of 0.2 μN/m. Since 2019 the ampere has been defined as 1 coulomb (C) per second, where the fundamental charge of a proton is $e = 1.602\,176\,634 \times 10^{-19}$ C. While the volt is not a fundamental unit, it is important in electromagnetics. The volt is the unit of electric potential of an object (to be defined and discussed in more detail later), derived from the joule and coulomb. Specifically, V = J/C.

We will not make use of kelvin or candela in our study of electromagnetics, and so make no further comment here on their definitions.

Example 1 Equivalent Units

You will recall that the force exerted by a compressed or extended spring is $\mathbf{F} = -k(x - x_0)\,\mathbf{a}_x$, where k is the spring constant, x is the length of the spring, x_0 is the equilibrium length of the spring, and $\mathbf{a}_x$ is the unit vector along the axis of the spring. Find the units of the spring constant k, reducing it to a combination of the seven basic units.

Solution:
We can rearrange the spring force equation to write $k = F/(x - x_0)$. Note that we have dropped the vector notation, as only the magnitudes are needed to solve for units. We know that $[F] = $ N = kg m/s^2, and $[(x - x_0)] = $ m, so $[k] = $ N/m = kg/s^2.

We will examine resistors, capacitors, and inductors, elementary electrical devices that we first encountered in our study of linear circuits, in more detail during our study of static fields, and visit them again when deriving the properties of transmission lines. The units of resistance R, capacitance C, and inductance L in various equivalent forms, are:

- $[R]$ = ohm (Ω) = V/A
- $[C]$ = farad (F) = A s/V = C/V = C^2/J = J/V^2
- $[L]$ = henry (H) = V s/A = J/A^2

These unit conversions, and many others, are included in Appendix L for your convenience. We will, of course, return to these later in the course as we treat them each in more detail, but for now we will be satisfied to know that we can, and should as a rule, trace these units back to the primary seven fundamental constants, and not lose sight of their significance. In fact, we can often detect errors in our derivations or problem solutions if we only keep a careful eye on the units of quantities we are determining.

PROBLEMS

P-1 As we will discuss in a later chapter, the magnitude of the electric field in vacuum due to a point charge q is $E = \dfrac{1}{4\pi\varepsilon_0}\dfrac{q}{r^2}$, where r is the distance from the point charge. ε_0 is known as the permittivity of vacuum, and has units F/m. Show that the unit for the electric field is V/m.

P-2 As we will discuss in a later chapter, the magnitude of the magnetic flux density in vacuum due to an elemental current I of length $d\ell$ is $B = \dfrac{\mu_0}{4\pi}\dfrac{Id\ell}{r^2}$, where r is the distance from the current element. μ_0 is known as the permeability of vacuum, and has units H/m. (a) Show that the tesla, the unit for the magnetic flux density, is equivalent to V s/m^2. (b) Show that the tesla is also equivalent to N/(A m).

Modifying Prefixes

The values of the physical quantities that we will discuss in this textbook range over many orders of magnitude, and carrying 15 zeros for a measurement of time, for example, gets tedious, cumbersome, and prone to errors. For this reason, we often use Latin or Greek prefixes to modify the units that we have introduced previously. In Table 1 we list the common prefixes, their definitions, and common examples in which they might be used. There are, of course, many prefixes that we have not included in Table 1, but these will suffice for our purposes, and we will limit our usage in this text to those you see listed here. Note that all of these prefixes, with one exception, are integer powers of 1 000. The exception is *centi*, which is commonly used only for length (i.e. centimeters). The examples that we have listed in this table include units of time (fs, ps, ns, μs, and ms), length (nm, μm, mm, and cm), mass (kg), capacitance (pF, μF), charge (pC, μC), power (MW–TW), and frequency (MHz, GHz, THz). Use of these prefixes can simplify our discussions, and aid in our computations. I have watched many students writing out 0.000 000 005 s, for example, rather than the much more concise 5 ns, and then struggle to keep the correct number of zeros in their calculation. Besides being streamlined, using the prefixes significantly decreases your chances of making errors. They are commonly used by professionals in technical conversation, and there really is no substitute for learning their meaning and usage. So we encourage you to become familiar with these units right now, and use them throughout our study of electric and magnetic fields. You will be rewarded many times over.

Table 1

Common prefixes that modify units of measures.			
Prefix	**Abbrev.**	**Definition**	**Common examples**
femto-	f	10^{-15}	fs
pico-	p	10^{-12}	ps, pF, pC
nano-	n	10^{-9}	ns, nm
micro-	μ	10^{-6}	μs, μF, μm, μC
milli-	m	10^{-3}	ms, mm
centi-	c	10^{-2}	cm
kilo-	k	10^{3}	km, kg, kHz
mega-	M	10^{6}	MW, MHz
giga-	G	10^{9}	GW, GHz
tera-	T	10^{12}	TW, THz

Example 2 Equivalent Units

Using the definitions of the prefixes listed in Table 1, and the units discussed here, write the equivalent units of: (a) pC/μF, and (b) mW ns.

Solution:

(a) pC/μF = 10^{-12} C/ 10^{-6} F = 10^{-6} V = μV.

(b) mW ns = $(10^{-3}$ W$)(10^{-9}$ s$)$ = 10^{-12} J = pJ.

PROBLEMS

P-1 Write the equivalent units for kN km.

P-2 Write the equivalent units for mJ/ns.

CHAPTER 1
Electrostatics I

KEY WORDS

point charge; Coulomb force; inverse square law; permittivity; electric field; Coulomb's Law; source point; field point; charge density (linear, surface, and volume); Gauss' Law; conservative force; Gaussian surface; electric potential; equipotential surfaces; electric dipole moment; conductor; conductivity; dielectric material; polarization; semiconductor; tangential and normal field components; displacement field; linear, isotropic, homogeneous medium; electric susceptibility; relative permittivity; dielectric strength.

Electrostatic fields form the basis for such everyday applications as air cleaners, photocopiers, spark plugs, and corona discharges, to name a few. In this chapter we will explore several techniques that can be used to determine the electric field produced by a single electric charge, or a distribution of electric charges, including two fundamental relations known as Coulomb's Law and Gauss' Law. Gauss' Law is one of two differential properties obeyed by electric fields that we will explore. We will show that the electric field can be determined from an electric potential, and explore methods of determining the electric potential. Using the electric potential can often be a simpler approach for finding the electric field. We will then examine the influence of materials, including conductors and dielectrics (or insulators), on electric fields. This will require us to re-examine Gauss' Law to include material effects. We will also develop a set of relations, known as boundary conditions, that relate the fields on the two sides of an interface between two materials. Our studies begin with empirical observations of the force between two charged bodies.

1.1 Coulomb's Law

Every body, whether large (macroscopic) or small (microscopic), has a property called its electric charge. At the atomic level, all matter consists of electrons, neutrons, and protons. Electrons, residing in clouds centered on the nucleus, are very low in mass and can be very mobile. The primary carrier of electric charge, therefore, is usually the electron, whose charge is measured to be $q_e = -e$, where $e = 1.602\,176\,634 \times 10^{-19}$C is called the fundamental or elementary charge. The central nucleus contains most of the mass of an atom, and consists of electrically neutral neutrons and positively charged protons. To the accuracy that we can measure the charge of the proton, q_p, it is precisely equal in magnitude but opposite in sign to that of the electron. An atom or larger body is said to be electrically neutral when it contains the same number of electrons and protons, such that its total charge is equal to zero.

Our starting point to understand electric fields comes from empirical observations of the forces between two charged bodies. For simplicity, we treat these bodies as very small in comparison to the distance d between the bodies. These are called **point charges**. The force between these charges can be attractive (toward one another) or repulsive (directly away from one another), but it is *always* parallel to the line joining the two charged bodies. Repulsive forces are illustrated in Fig. 1.1, in which $\mathbf{F}_{12}$ represents the force exerted on charge q_1 by charge q_2, and $\mathbf{F}_{21}$ represents the force exerted on charge q_2 by charge q_1. As is shown, the forces $\mathbf{F}_{12}$ and $\mathbf{F}_{21}$ are always equal in magnitude but opposite in direction to one another. (This is, of course, required by Newton's Third Law.) Laboratory measurements show that the magnitude of these forces decreases as the distance d between the two charged bodies increases, following a $1/d^2$ dependence. This **Coulomb force**, as the electric force between charged bodies is often called, is therefore said to be an example of an **inverse square law** force. (The measurement by Charles-Augustin de Coulomb of the $1/d^2$ dependence of the electric force between two point charges is described in Biographical Note 1.1.) Another common inverse square law force is gravity, in which the gravitational attraction between two masses is observed to vary as $1/d^2$ as well. Unlike the gravitational force, which is always attractive, the electrostatic force can be attractive or repulsive, depending on the signs of the two charges. Opposite signs are assigned to the charges of bodies that attract one another, and the same sign to the charges of bodies that repel each other.

The electric force between the two charges, which we have described above with words, can be written concisely using vector notation. In Fig. 1.1, the charge and position of charge 1 is q_1 and $\mathbf{r}_1$, and that of charge 2 is q_2 and $\mathbf{r}_2$. The force exerted on charge q_1 by charge q_2 is described by

$$\mathbf{F}_{12} = \frac{1}{4\pi\varepsilon_0} \frac{q_1\,q_2}{|\mathbf{r}_1 - \mathbf{r}_2|^2}\,\mathbf{a}_{12},$$

$$(1.1)$$

where $\mathbf{a}_{12}$ is the unit vector pointing in the direction from $\mathbf{r}_2$ to $\mathbf{r}_1$:

$$\mathbf{a}_{12} = \frac{\mathbf{r}_1 - \mathbf{r}_2}{|\mathbf{r}_1 - \mathbf{r}_2|}.$$

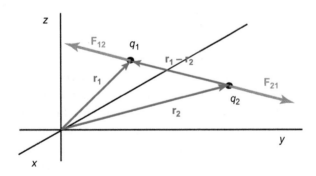

FIGURE 1.1 The force **F** between two point charges varies as $1/d^2$, where $d = |\mathbf{r}_1 - \mathbf{r}_2|$ is the distance between the charges.

Note the following three properties of the force $\mathbf{F}_{12}$ contained in Eq. (1.1):

1. $\mathbf{F}_{12}$ decreases as the inverse of the square of the distance $d = |\mathbf{r}_1 - \mathbf{r}_2|$.
2. $\mathbf{F}_{12}$ points *away* from point charge q_2 if $q_1 q_2 > 0$, and *toward* q_2 if $q_1 q_2 < 0$.
3. The magnitude of $\mathbf{F}_{12}$ scales as the product of the charges $q_1 q_2$.

The factor $1/4\pi\varepsilon_0 = 8.988 \times 10^9$ N m^2/C^2 is valid in the SI system of units. (This proportionality factor does take different forms in other systems of units, so don't be dismayed if you encounter other expressions for the Coulomb force in texts that employ other unit systems.) ε_0 is known as the **permittivity of free space**, or the **permittivity of vacuum**, and has a numerical value of $\varepsilon_0 = 8.854 \times 10^{-12}$ C^2/N m^2, or using the unit conversions from Appendix L, $\varepsilon_0 = 8.854$ pF/m. The force $\mathbf{F}_{21}$ that is exerted on charge q_2 by charge q_1 is, by Newton's Third Law, exactly equal but opposite to $\mathbf{F}_{12}$; that is, $\mathbf{F}_{21} = -\mathbf{F}_{12}$.

Biographical Note 1.1 Charles-Augustin de Coulomb (1736–1806)

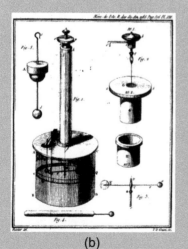

(a)

(b)

FIGURE 1.2 (a) A portrait of Charles-Augustin de Coulomb. (b) A sketch of the torsion balance used by Coulomb in his measurements.

Charles de Coulomb, pictured in Fig. 1.2(a), was a French military officer and engineer who oversaw numerous construction projects for the French army during his career. His contribution to our understanding of electric forces was through his precise measurements of the force between two charged bodies; specifically, the dependence of this force on the inverse of the square of the distance between the charges. He carried out his measurements using a torsion balance consisting of a thin needle suspended from a thread of silver, copper, or silk. (See Fig. 1.2(b) for a sketch of his torsion balance.) A pith ball (a non-conducting sphere) that could be charged was skewered onto one end of the needle, while a counter-balancing weight was added to the opposite end, such that the needle lay in a horizontal plane. Upon approaching the charged pith ball on the torsion balance with a second, similarly charged pith ball, the torsion balance was repelled, as shown in Fig. 1.3(a). Equilibrium was achieved when the electrostatic repulsive force was precisely balanced by the torsional force of the balance. Precise calibration of the restoring torque of the balance and the distance between the charged pith balls led Coulomb to the inverse square law of the force. Similar measurements of oppositely charged bodies were problematic, however, in that these bodies attract one another, and upon approaching one body with another they tend to make contact and redistribute the charges. Coulomb therefore devised a different technique to show that the inverse square law is valid for oppositely charged bodies, as shown in Fig. 1.3(b). In this case, he mounted the second charged body at a distance from the axis that was greater than the radius of the torsion balance, and set the balance into oscillation in a horizontal plane. Just like the period of oscillation of a pendulum swinging back and forth under the influence of gravity depends on the gravitational force ($T = 2\pi\sqrt{L/g}$, where L is the length of the pendulum and $g = 9.8$ m/s^2 is the acceleration due to gravity on Earth), Coulomb used the period of the oscillation of the torsional balance to determine the force of electrostatic attraction between the two charged bodies. Once again, he was able to show that the force decreased as the inverse of the square of the distance. The unit of charge is named the coulomb in his honor.

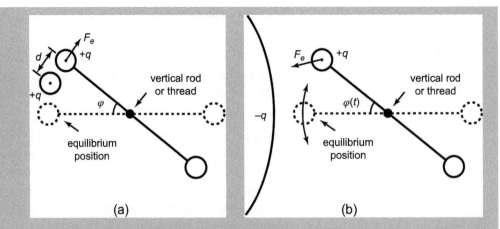

FIGURE 1.3 A torsion balance. In (a), the force between like charges is determined through the position of the balance arm. At this position, the torque caused by the repulsive electrostatic force balances the torque caused by twisting the central rod or thread. In (b), the attractive force between oppositely charged bodies is determined through the period of oscillation of the balance.

Detailed measurements showing the inverse square law nature of the electrostatic force between two point charges were also carried out by Henry Cavendish. His work is summarized in Biographical Note 1.2.

Biographical Note 1.2 Henry Cavendish (1731–1810)

A British natural philosopher, chemist, and physicist, Henry Cavendish carried out a wide variety of high-precision scientific investigations. Of primary relevance to the topic of this text were his precise measurements in the early 1770s of the electrostatic force between two charged bodies using a torsion balance, showing an inverse square law dependence. Unfortunately, Cavendish did not publish these results; rather, James Clerk Maxwell, nearly a century later, studied Cavendish's notes and published these in 1879. Coulomb, who published his studies of the inverse square law in three reports starting in 1785, is generally credited with the relation now known as Coulomb's Law.

In other studies, Cavendish studied the conductivity of solutions of electrolytes, in which he found a relation consistent with Ohm's Law. It is remarkable that these studies were carried out before ammeters were available for measuring current. Instead, Cavendish used his body as part of the circuit, and estimated the electrical current from the shock sensation that he felt. Again, he did not publish any report of these investigations, and so Georg Ohm is credited with the first report of this relation. In other electrically related studies, Cavendish proposed such ideas as electric potential, capacitance, the dielectric constant of insulators, and the capacitance of a parallel plate capacitor.

Cavendish also studied important topics in chemistry and earth science, in which he showed that burning hydrogen gas produced water, found that hydrogen gas was much less dense than air, and measured the density of the Earth.

While our description of the "force at a distance" between the two charges described above is correct and valid in the context in which we have described it, it is not as complete as we wish it to be. A more general, convenient, and physically appropriate approach will be to discuss this force $\mathbf{F}_{12}$ in terms of the **electric field** $\mathbf{E}(\mathbf{r})$ produced by charge q_2, which in turn exerts the force

$$\mathbf{F}_{12} = q_1 \mathbf{E}(\mathbf{r}) \tag{1.2}$$

on the charge q_1. This may seem like a trivial change, but remember that we are only considering static fields at this time, and when we include time-varying effects later, and specifically generation of propagating electromagnetic waves, this field description will become essential. From Eq. (1.1), the electric field at location $\mathbf{r}$ produced by a single charge q_2 at location $\mathbf{r}'$ is

$$\mathbf{E}(\mathbf{r}) = \frac{1}{4\pi\varepsilon_0} \frac{q_2}{|\mathbf{r} - \mathbf{r}'|^2} \mathbf{a}_{\mathbf{rr}'}. \tag{1.3}$$

This expression, known as **Coulomb's Law**, is a simple expression for the electric field generated by a single point charge. The unit vector $\mathbf{a}_{\mathbf{rr}'}$ in this expression points from the location $\mathbf{r}'$ of the charge to the location $\mathbf{r}$ at which we determine the field,

$$\mathbf{a}_{\mathbf{rr}'} = \frac{\mathbf{r} - \mathbf{r}'}{|\mathbf{r} - \mathbf{r}'|}. \tag{1.4}$$

At any location $\mathbf{r}$ surrounding the point charge q_2, this electric field $\mathbf{E}(\mathbf{r})$ points radially away from the charge when $q_2 > 0$, or radially inward for $q_2 < 0$. There are two critical features in the previous statement. That is, (1) the electric field due to a point charge is in the *radial* direction, and (2) electric field lines *originate* on positive charges and *terminate* on negative charges. Furthermore, as advertised, the force on a charge q_1 at any location is $\mathbf{F}_{12} = q_1 \mathbf{E}(\mathbf{r})$, in perfect agreement with the force given in Eq. (1.1).

As a note on notation, we will customarily use primed coordinates (e.g. $\mathbf{r}'$) to denote the location of the charge, or more generally the **source** (when we later discuss magnetic fields produced by currents, the primed coordinates $\mathbf{r}'$ will indicate a segment of the current), and unprimed coordinates (e.g. $\mathbf{r}$) to denote the location of the point at which we are determining the field. The former is called the **source point**, the latter the **field point** or **observation point**.

Example 1.1 Coulomb's Law, Point Charge

Determine the electric field $\mathbf{E}(\mathbf{r})$ at point A in Fig. 1.4 due to an electron at point B. Express $\mathbf{E}(\mathbf{r})$ in (a) magnitude and direction form, as well as (b) in terms of its x-, y-, and z-components.

FIGURE 1.4 The electron at point B produces an electric field in the surrounding space, which we compute at point A in Example 1.1.

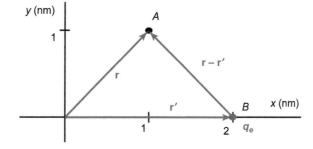

Solution:
Eq. (1.3) gives the electric field $\mathbf{E}(\mathbf{r})$ due to a single charge. Our first task is to define the source point (the location of the electron, point B) that generates the field, and the field point (the location at which we determine $\mathbf{E}(\mathbf{r})$, point A). We read both of these points

from Fig. 1.4, and find

$$\mathbf{r}' = 2\,\text{nm} \cdot \mathbf{a}_x + 0 \cdot \mathbf{a}_y + 0 \cdot \mathbf{a}_z = (2, 0, 0)\,\text{nm}$$

and

$$\mathbf{r} = 1\,\text{nm} \cdot \mathbf{a}_x + 1\,\text{nm} \cdot \mathbf{a}_y + 0 \cdot \mathbf{a}_z = (1, 1, 0)\,\text{nm}.$$

(Note the use of nanometers as the unit of length here. "Nano" of course is the prefix for 10^{-9}, and it is convenient to express length scales at the atomic level using nanometers. At the end of our solution we will write our result using the SI unit meter.) We can determine the relative position vector

$$\mathbf{r} - \mathbf{r}' = (1, 1, 0)\,\text{nm} - (2, 0, 0)\,\text{nm} = (-1, 1, 0)\,\text{nm},$$

its magnitude

$$|\mathbf{r} - \mathbf{r}'| = \sqrt{(-1\,\text{nm})^2 + (1\,\text{nm})^2 + (0\,\text{nm})^2} = \sqrt{2}\,\text{nm},$$

and the unit vector pointing from B to A

$$\mathbf{a}_{BA} = \frac{\mathbf{r} - \mathbf{r}'}{|\mathbf{r} - \mathbf{r}'|} = \frac{(-1, 1, 0)\,\text{nm}}{\sqrt{2}\,\text{nm}} = \frac{1}{\sqrt{2}}(-1, 1, 0). \tag{1.5}$$

Since the charged body that generates the field is an electron, the charge q_2 is $q_2 = -e = -1.602 \times 10^{-19}$ C. Now we are ready to use Eq. (1.3) to find the magnitude of the electric field at $\mathbf{r}$:

$$\begin{aligned}
|\mathbf{E}| &= \frac{1}{4\pi\varepsilon_0} \frac{|q_2|}{|\mathbf{r} - \mathbf{r}'|^2} \\
&= (8.988 \times 10^9\,\text{N m}^2/\text{C}^2) \times \frac{(1.602 \times 10^{-19}\,\text{C})}{(\sqrt{2}\,\text{nm})^2} \left(\frac{\text{J}}{\text{Nm}}\right)\left(\frac{\text{VC}}{\text{J}}\right)\left(\frac{\text{nm}}{10^{-9}\,\text{m}}\right)^2 \\
&= 7.2 \times 10^8\,\text{V/m}.
\end{aligned}$$

The extra terms inside the parentheses in the second line of this equation are conversion factors for the units. (Refer to Appendix L for these and other useful unit conversions.) Using this form of $|\mathbf{E}|$ and the unit vector $\mathbf{a}_{BA}$ from Eq. (1.5), the electric field vector is

$$\mathbf{E} = -|\mathbf{E}|\,\mathbf{a}_{BA} = -7.2 \times 10^8\,\text{V/m} \times \frac{1}{\sqrt{2}}(-1, 1, 0),$$

where we include the minus sign since q_2 is negative.

To write $\mathbf{E}$ in terms of its x-, y-, and z-components, we multiply through, which yields

$$\mathbf{E} = -5.9 \times 10^8\,\text{V/m} \times (-1, 1, 0).$$

Thus the electric field components are $E_x = 5.9 \times 10^8$ V/m, $E_y = -5.9 \times 10^8$ V/m, and $E_z = 0$. Note that this electric field (at position $\mathbf{r}$) points directly toward the charge (at position $\mathbf{r}'$), as expected for a negative charge.

1.1.1 Ensemble of Discrete Charges

When the physical arrangement contains more than one charge, the total electric field generated by the ensemble of charges can be determined by applying the principle of **superposition**. That is, the total electric field $\mathbf{E}$ generated by the ensemble is simply the vector sum of the fields due to each charge individually. Let the location and charge of each of the point charges be $\mathbf{r}'_i$ and q_i, respectively, where $i = 1 \rightarrow N$ is an index that labels each of the N charges. Then the total field $\mathbf{E}$ at a location $\mathbf{r}$ is

$$\mathbf{E}(\mathbf{r}) = \sum_{i=1}^{N} \mathbf{E}_i(\mathbf{r}) = \frac{1}{4\pi\varepsilon_0} \sum_{i=1}^{N} \frac{q_i}{|\mathbf{r} - \mathbf{r}'_i|^2} \, \mathbf{a}_{\mathbf{rr}'_i}.$$

The unit vector $\mathbf{a}_{\mathbf{rr}'_i}$ in this expression points from charge i to the field point, and differs, of course, for each of the charges. An alternate form for this equation, in which the unit vectors are replaced using Eq. (1.4), is

$$\mathbf{E}(\mathbf{r}) = \frac{1}{4\pi\varepsilon_0} \sum_{i=1}^{N} \frac{q_i(\mathbf{r} - \mathbf{r}'_i)}{|\mathbf{r} - \mathbf{r}'_i|^3}. \tag{1.6}$$

These expressions are perfectly equivalent. Do not be fooled into thinking that the field of Eq. (1.6) decreases inversely with the *cube* of the distance from source point to field point, as you must not ignore the factor $\mathbf{r} - \mathbf{r}'_i$ in the numerator.

Example 1.2 Coulomb's Law, Discrete Charges

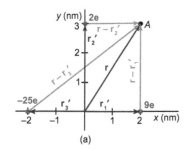

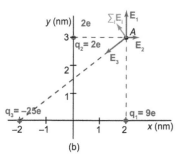

FIGURE 1.5 Three point charges generate an electric field $\mathbf{E}$ at point A. In (a), the location of the three charges and point A, given in terms of the position vectors $\mathbf{r}'_1$, $\mathbf{r}'_2$, $\mathbf{r}'_3$, and $\mathbf{r}$ (the dark blue arrows), are shown. The relative position vectors pointing from the source points to the field point A (light blue arrows) are also shown. In (b), the red arrows represent the electric fields $\mathbf{E}_1$, $\mathbf{E}_2$, and $\mathbf{E}_3$ due to the three individual charges, while the light blue arrow is the net field $\sum_i \mathbf{E}_i$.

Determine the electric field $\mathbf{E}(\mathbf{r})$ at point A in Fig. 1.5 due to the three charges shown. Express $\mathbf{E}(\mathbf{r})$ (a) in its magnitude and direction form, as well as (b) in terms of its x-, y-, and z-components.

Solution:

Eq. (1.6) gives the electric field $\mathbf{E}(\mathbf{r})$ due to a finite number of point charges. Start by defining the coordinates $\mathbf{r}'_i$ of each charge q_i. From Fig. 1.5, the source points are

$$\mathbf{r}'_1 = (2, 0, 0) \text{ nm},$$
$$\mathbf{r}'_2 = (0, 3, 0) \text{ nm, and}$$
$$\mathbf{r}'_3 = (-2, 0, 0) \text{ nm}.$$

The field location is $\mathbf{r} = (2, 3, 0)$ nm, so the three relative position vectors are

$$\mathbf{r} - \mathbf{r}'_1 = (0, 3, 0) \text{ nm},$$
$$\mathbf{r} - \mathbf{r}'_2 = (2, 0, 0) \text{ nm, and}$$
$$\mathbf{r} - \mathbf{r}'_3 = (4, 3, 0) \text{ nm},$$

each pointing from source (charge i) to point A. The magnitude of each of these is determined using Eq. (A.2):

$$|\mathbf{r} - \mathbf{r}'_1| = \sqrt{0^2 + (3)^2 + 0^2} \text{ nm} = 3 \text{ nm},$$
$$|\mathbf{r} - \mathbf{r}'_2| = \sqrt{(2)^2 + 0^2 + 0^2} \text{ nm} = 2 \text{ nm, and}$$
$$|\mathbf{r} - \mathbf{r}'_3| = \sqrt{(4)^2 + (3)^2 + 0^2} \text{ nm} = 5 \text{ nm}.$$

The sum of the individual fields is then

$$\mathbf{E} = \frac{1}{4\pi\varepsilon_0} \left[\frac{(9e)(0, 3, 0)\text{nm}}{(3 \text{ nm})^3} + \frac{(2e)(2, 0, 0) \text{ nm}}{(2 \text{ nm})^3} + \frac{(-25e)(4, 3, 0) \text{ nm}}{(5 \text{ nm})^3} \right]$$

$$= \frac{e}{4\pi\varepsilon_0} \left[(0, 1, 0) + \left(\frac{1}{2}, 0, 0\right) - \left(\frac{4}{5}, \frac{3}{5}, 0\right) \right] \text{nm}^{-2}$$

$$= \frac{e}{4\pi\varepsilon_0} \left(-\frac{3}{10}, \frac{4}{10}, 0 \right) \text{nm}^{-2} = (-0.432, 0.576, 0) \times 10^9 \text{ V/m}.$$

The magnitude of this vector $|\mathbf{E}|$ is found using Eq. (A.2),

$$|\mathbf{E}| = 0.720 \times 10^9 \text{ V/m},$$

and the direction unit vector using Eq. (A.6),

$$\mathbf{a} = \frac{\mathbf{E}}{|\mathbf{E}|} = (-0.6, 0.8, 0).$$

1.1.2 Continuous Charge Distributions and Charge Density

Although charges are always discrete entities, we are often interested in effects produced by not a few individual charges, but rather a very large number of charges. Remember that the charge of an individual electron or proton is only $\pm 1.602 \times 10^{-19}$ C, which is an extremely small charge. In the macroscopic world, it is only when a very large number of charges are present that the fields they generate become significant. For example, consider a simple parallel plate capacitor that you probably learned about in your introductory physics and linear circuits classes. (We'll discuss capacitors in more detail later in this text.) For a capacitor with capacitance $C = 0.1$ μF charged to a voltage of $V = 10$ V, the charge stored on one of the conducting plates of the capacitor is $+Q$, and on the other is $-Q$, where $Q = CV = (10^{-7}$ F$)(10$ V$) = 10^{-6}$ C $= 1.0$ μC. While any quantity that is measured in units of *micro*-something may seem to be infinitesimally small, the number of excess electrons on the negatively charged conductor of the capacitor is quite large: $Q/(-q_e) = 10^{-6}$ C$/1.6 \times 10^{-19}$ C $= 6.2 \times 10^{12}$. (Similarly, the positively charged plate has a deficiency of this same number of electrons.) When this many electrons are present, the individual loses meaning. After all, what significance can one or two electrons have in a sea of 6.2 trillion? For this reason, we are usually free to treat the number of electrons, and therefore the charge Q, as a continuous, rather than a discrete, variable.

As an example, we first consider charge that is distributed along a straight or curved line, as if along a slender thread. Such a linear charge distribution is illustrated in Fig. 1.6. If the charge within a short segment of length $\Delta \ell_i$ of this line is Δq_i, then the **linear charge density** ρ_ℓ, that is, the charge per unit length, is defined as approximately

$$\rho_\ell(\mathbf{r}'_i) \approx \frac{\Delta q_i}{\Delta \ell_i}.$$

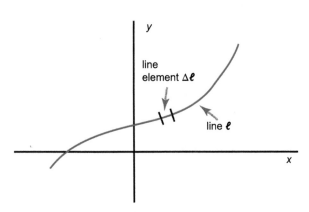

FIGURE 1.6 Linear charge of density $\rho_\ell(\mathbf{r}')$. The charge inside the element of length $\Delta \ell$ is $\rho_\ell(\mathbf{r}')\Delta \ell$.

The index i labels each individual element of the entire ensemble. This becomes more precise in the limit as the segment length $\Delta \ell_i$ becomes short,

$$\rho_\ell(\mathbf{r}'_i) = \lim_{\Delta \ell_i \to 0} \frac{\Delta q_i}{\Delta \ell_i} = \frac{dq}{d\ell}. \tag{1.7}$$

Note the notation $\rho_\ell(\mathbf{r}')$ here. In general, the linear charge density will not be uniform, but rather it will vary along the line. To express this variation, we write that ρ_ℓ is a function of $\mathbf{r}'$, shorthand notation for the source position. This notation is concise, general, and easily applied in any coordinate system.

To determine the total charge contained within a large interval of a linear distribution, we must divide the entire line ℓ into a large number of small elements, each of length $\Delta \ell_i'$ and containing charge $\Delta q_i = \rho_\ell(\mathbf{r}_i')\Delta \ell_i'$, and sum these bits of charge:

$$Q = \sum_i \Delta q_i = \lim_{\Delta \ell_i' \to 0} \sum_i \rho_\ell(\mathbf{r}_i')\Delta \ell_i' = \int_\ell \rho_\ell(\mathbf{r}')d\ell'. \tag{1.8}$$

Here the subscript ℓ on the integral sign indicates that the integration is to be carried out over the line ℓ. You must choose the proper form of the line element $d\ell$ according to the geometry of the problem.

We will soon see that any excess charge on a conductor resides on the exterior surface of that conductor. It is therefore of interest to define the **surface charge density**, $\rho_s(\mathbf{r}')$ as the amount of charge per unit area of a surface. Similar to the definition of the linear charge density in Eq. (1.7), the surface charge density $\rho_s(\mathbf{r}')$ is defined by dividing the entire surface s into a large number of small elemental areas, each of area Δs_i and containing a charge Δq_i. Then the average surface charge density in element i at location $\mathbf{r}'$ is

$$\rho_s(\mathbf{r}') = \lim_{\Delta s_i \to 0} \frac{\Delta q_i}{\Delta s_i},$$

and the total charge on the entire surface is

$$Q = \sum_i \Delta q_i = \lim_{\Delta s_i' \to 0} \sum_i \rho_s(\mathbf{r}_i')\Delta s_i' = \int_s \rho_s(\mathbf{r}')ds'. \tag{1.9}$$

The subscript s on the integral sign is a reminder that this is a surface integral.

For charges distributed within a volume (picture a three-dimensional cloud of charge), the distribution is defined in terms of the **volume charge density** $\rho_v(\mathbf{r}')$; that is, the charge per unit volume. This density is defined by dividing the entire volume v into small elements, each of volume Δv_i and containing a charge Δq_i. Then the average volume charge density in element i is $\Delta q_i/\Delta v_i$, and the volume charge density at location $\mathbf{r}'$ is

$$\rho_v(\mathbf{r}') = \lim_{\Delta v_i \to 0} \frac{\Delta q_i}{\Delta v_i}.$$

The total charge in the entire volume is

$$Q = \sum_i \Delta q_i = \int_v \rho_v(\mathbf{r}')dv'. \tag{1.10}$$

Similar to the notation for line or surface integrals in Eqs. (1.8) or (1.9), respectively, the subscript v on the integral sign is used for a volume integral.

We will now explore how we can determine the electric field $\mathbf{E}$ that is created by a distribution of charges. The development follows directly our previous discussion of the field due to a finite number of discrete charges, which we derived by asserting that $\mathbf{E}$ obeys superposition, and that the total field must be the sum of fields due to individual charges. We consider some region in space v in which there is a charge density $\rho_v(\mathbf{r}')$. (Note again the use of the primed notation on $\mathbf{r}'$ to indicate the location of the charge, or source.) We divide this volume into small elements, each of volume $\Delta v_i'$, containing a charge of $q_i = \Delta v_i' \rho_v(\mathbf{r}_i')$, where the elemental volume $\Delta v_i'$ at $\mathbf{r}'$ must be small enough that the charge density $\rho_v(\mathbf{r}_i')$ and the position vector $\mathbf{r}-\mathbf{r}'$ are essentially uniform within each. The index i labels each of the small volumes and charges. By applying Eq. (1.6), the electric field $\mathbf{E}(\mathbf{r})$ at location $\mathbf{r}$ can be found by the following summation:

$$\mathbf{E}(\mathbf{r}) = \frac{1}{4\pi\varepsilon_0} \sum_i \frac{q_i(\mathbf{r} - \mathbf{r}'_i)}{|\mathbf{r} - \mathbf{r}'_i|^3}$$

$$= \frac{1}{4\pi\varepsilon_0} \sum_i \frac{\left[\Delta v'_i \, \rho_v(\mathbf{r}'_i)\right] (\mathbf{r} - \mathbf{r}'_i)}{|\mathbf{r} - \mathbf{r}'_i|^3}.$$

In the limit as the volume elements become smaller, and summation becomes an integral,

$$\mathbf{E}(\mathbf{r}) = \frac{1}{4\pi\varepsilon_0} \lim_{\Delta v'_i \to 0} \sum_i \frac{\rho_v(\mathbf{r}'_i)(\mathbf{r} - \mathbf{r}'_i)}{|\mathbf{r} - \mathbf{r}'_i|^3} \, \Delta v'_i$$

or

$$\boxed{\mathbf{E}(\mathbf{r}) = \frac{1}{4\pi\varepsilon_0} \int_v \frac{\rho_v(\mathbf{r}')(\mathbf{r} - \mathbf{r}')}{|\mathbf{r} - \mathbf{r}'|^3} \, dv'.} \tag{1.11}$$

Recall that the subscript v on the integral symbol is simply to remind us that this is a volume integral. Also, $\rho_v(\mathbf{r}')$ in the numerator of this integrand is the volume charge density as a function of the location $\mathbf{r}'$, where $\mathbf{r}'$ is shorthand notation for the source position, representing x', y', and z' in rectangular coordinates; ρ', ϕ', and z' in cylindrical coordinates; or R', θ', and ϕ' in spherical coordinates. (See Appendix B for a review of rectangular, cylindrical, and spherical coordinates.) Do not make the all-too-common mistake of interpreting $\rho_v(\mathbf{r}')$ as ρ_v *times* $\mathbf{r}'$. It is, in fact, ρ_v *as a function of* $\mathbf{r}'$! Eq. (1.11) allows us to determine the vector electric field $\mathbf{E}(\mathbf{r})$ by integrating over the volume of the charge distribution. Recall that we derived this equation using only the observation that the force between charges obeys the inverse square law, Eq. (1.1), and the assertion that the electric field obeys superposition.

Example 1.3 Coulomb's Law, Spherical Charge

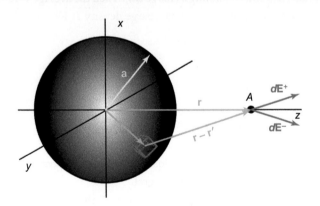

FIGURE 1.7 A sphere of radius a and uniform charge density ρ_v.

Determine the electric field $\mathbf{E}(\mathbf{r})$ outside the sphere of radius a and uniform charge density ρ_v shown in Fig. 1.7.

Solution:
Recall that the electric field lines originate on positive charges, so if the charge density ρ_v of this sphere is positive, then we expect that the resulting electric field $\mathbf{E}(\mathbf{r})$ points *away* from the sphere. Also, by the symmetry of the charged sphere, we expect that the electric field has only a radial component (i.e. that $\mathbf{E}(\mathbf{r})$ points *directly* away from the center of the sphere). For example, the electric field $\mathbf{E}(\mathbf{r})$ at a point on the z-axis has no component in the x- or y-directions. This can be proven by assuming the opposite, that a component E_x or E_y exists, and then rotating the entire geometry 180° about the z-axis. Then E_x would become $-E_x$, and E_y would become $-E_y$. But upon rotation, nothing about the charge distribution changed, since the sphere is uniformly charged. This implies that E_x and E_y are both zero, as we stated.

We will use Eq. (1.11) to find an explicit expression for $\mathbf{E}(\mathbf{r})$, but have several choices to make in getting started. First, we choose a coordinate system in which the origin is at the center of the sphere. Second, spherical coordinates are the most natural, convenient choice when working with a charged sphere. And finally, the analysis is simplified by choosing the field point $\mathbf{r}$ to be a point that lies on the z-axis. This point is labeled A in

Fig. 1.7. We are free to make this choice because of the symmetry of the sphere, and will generalize our result to other locations later.

The next step is to define the source point (the location of an element of charge within the sphere),

$$\mathbf{r}' = (x', y', z') = (R' \sin \theta' \cos \phi', R' \sin \theta' \sin \phi', R' \cos \theta'),$$

and the field point, or observation point, A,

$$\mathbf{r} = (0, 0, R),$$

where R is the distance between A and the center of the sphere. Notice that even when using spherical *coordinates* $(R, \theta, \text{and } \phi)$ to define the source and observation points, we write these position vectors using their rectangular *components*, x, y, and z. This is necessary, since only when vectors are expressed in rectangular components can they be simply added, subtracted, or integrated. The vector $\mathbf{r} - \mathbf{r}'$ is

$$\mathbf{r} - \mathbf{r}' = (-R' \sin \theta' \cos \phi', -R' \sin \theta' \sin \phi', R - R' \cos \theta'), \qquad (1.12)$$

and its magnitude is

$$|\mathbf{r} - \mathbf{r}'| = \left[(-R' \sin \theta' \cos \phi')^2 + (-R' \sin \theta' \sin \phi')^2 + (R - R' \cos \theta')^2 \right]^{1/2}.$$

After grouping the first two terms together and expanding the third term, $|\mathbf{r} - \mathbf{r}'|$ is

$$|\mathbf{r} - \mathbf{r}'| = \left[R'^2 \sin^2 \theta' \left(\cos^2 \phi' + \sin^2 \phi' \right) + \left(R^2 - 2RR' \cos \theta' + R'^2 \cos^2 \theta' \right) \right]^{1/2}.$$

Using $\cos^2 \phi' + \sin^2 \phi' = 1$ and $\cos^2 \theta' + \sin^2 \theta' = 1$, this becomes

$$|\mathbf{r} - \mathbf{r}'| = \left[R'^2 + R^2 - 2RR' \cos \theta' \right]^{1/2}. \qquad (1.13)$$

Using Eqs. (1.12) and (1.13) in Eq. (1.11), the electric field at point A is

$$\mathbf{E}(\mathbf{r}) = \frac{1}{4\pi\varepsilon_0} \int_v \frac{\rho_v(\mathbf{r}')(-R' \sin \theta' \cos \phi', -R' \sin \theta' \sin \phi', R - R' \cos \theta')}{\left[R'^2 + R^2 - 2RR' \cos \theta' \right]^{3/2}} dv'. \qquad (1.14)$$

A note of explanation of the notation is probably helpful here. In the numerator of the integrand, notice the x-, y-, and z-components of the vector $\mathbf{r} - \mathbf{r}'$ enclosed within parentheses. In the following, we must evaluate each of the components of $\mathbf{E}(\mathbf{r})$ separately, $E_x(\mathbf{r})$ using the x-component of $\mathbf{r} - \mathbf{r}'$, $E_y(\mathbf{r})$ using the y-component of $\mathbf{r} - \mathbf{r}'$, and similarly for $E_z(\mathbf{r})$. The notation used in Eq. (1.14) is simply the shorthand notation for this.

As reviewed in Appendix C, the volume element for integration in spherical coordinates is $dv' = R'^2 dR' \sin \theta' d\theta' d\phi'$, where the range of R' is 0 to a, the range of θ' is 0 to π, and for ϕ', 0 to 2π. Since we are told that the charge density is uniform inside the sphere, we can set $\rho_v(\mathbf{r}') = \rho_v$ (i.e. a constant, independent of position), and pull it outside the integral. We next examine the three components E_x, E_y, and E_z separately, as these involve evaluation of slightly different integrals. For E_x, Eq. (1.14) gives us

$$E_x = \frac{\rho_v}{4\pi\varepsilon_0} \int_0^{2\pi} \int_0^\pi \int_0^a \frac{-R' \sin \theta' \cos \phi'}{\left[R'^2 + R^2 - 2RR' \cos \theta' \right]^{3/2}} R'^2 dR' \sin \theta' d\theta' d\phi'.$$

ϕ' appears only in the numerator of this integrand, as $\cos \phi'$, so carrying out the integration in ϕ' first leads immediately to $E_x = 0$. Similarly, for E_y, Eq. (1.14) gives us

$$E_y = \frac{\rho_v}{4\pi\varepsilon_0} \int_0^{2\pi} \int_0^\pi \int_0^a \frac{-R' \sin \theta' \sin \phi'}{\left[R'^2 + R^2 - 2RR' \cos \theta' \right]^{3/2}} R'^2 dR' \sin \theta' d\theta' d\phi',$$

which upon integration in ϕ' yields $E_y = 0$. These results are as we predicted earlier. A second symmetry argument that leads us to the same conclusion is that, for each element within the sphere at the position given by (x', y', z'), there is an identical element at $(-x', -y', z')$. We show in Fig. 1.7 the elemental fields $d\mathbf{E}^+$ and $d\mathbf{E}^-$ due to these two elements within the spherical volume v. The magnitudes of $d\mathbf{E}^+$ and $d\mathbf{E}^-$ must be identical, since they result from volume elements that hold an identical charge and they are equidistant from point A. Due to the location of the source elements, however, the components dE_x^+ and dE_x^- at point A due to these two elements are equal in magnitude but opposite in sign, and therefore exactly cancel each other. Similarly, the components dE_y^+ and dE_y^- cancel one another. Therefore, by symmetry, only the component E_z survives. Of course, we arrived at the same conclusion by carrying out the integrations explicitly, but the symmetry argument is powerful and can save us a great deal of work.

We now direct our attention to solving for the E_z component. From Eq. (1.14), this field component is

$$E_z = \frac{\rho_v}{4\pi\varepsilon_0} \int_0^{2\pi} \int_0^{\pi} \int_0^a \frac{R - R'\cos\theta'}{\left[R'^2 + R^2 - 2RR'\cos\theta'\right]^{3/2}} R'^2 dR' \sin\theta' d\theta' d\phi'.$$

Integration in ϕ' is the easiest place to start, since the integrand is independent of ϕ', leading to

$$E_z = \frac{\rho_v}{2\varepsilon_0} \int_0^{\pi} \int_0^a \frac{R - R'\cos\theta'}{\left[R'^2 + R^2 - 2RR'\cos\theta'\right]^{3/2}} R'^2 dR' \sin\theta' d\theta'.$$

Next we evaluate the integral in θ'. We use the substitution $x = \cos\theta'$, and rely on the integral tables in Appendix G, specifically Eqs. (G.5) and (G.6). After several steps and simplifications, E_z is

$$E_z = \frac{\rho_v}{\varepsilon_0 R^2} \int_0^a R'^2 dR'.$$

At one stage leading to this equation, we were required to evaluate the absolute value $|R - R'|$. Be sure to choose the appropriate absolute value that is valid when $R > a$. (You will face a similar decision in Prob. P1-5 at the end of this chapter, when asked to find an expression for $\mathbf{E}(R)$ that is valid for $R < a$.) Finally, evaluation of the integral in R' results in

$$E_z(R) = \frac{\rho_v}{\varepsilon_0 R^2} \frac{a^3}{3}.$$

Since the total charge of the sphere is $Q = \int_v \rho_v dv' = \rho_v v$, where $v = \frac{4\pi}{3}a^3$ is the volume of the sphere, this reduces to

$$E_z(R) = \frac{1}{4\pi\varepsilon_0} \frac{Q}{R^2}.$$

Notice that this electric field $E_z(R)$ is identical in this region to the field created by a point charge of the same charge Q.

Finally, recall that at the beginning of this solution we chose the z-axis such that the point A lies along it. But the symmetry of the physical geometry dictates that the field $\mathbf{E}$ must be spherically symmetric, allowing us to write that the electric field $\mathbf{E}$ at *any* point outside the sphere is

$$\mathbf{E}(R) = \frac{1}{4\pi\varepsilon_0} \frac{Q}{R^2} \mathbf{a}_R.$$

The electric field resulting from a distribution of charge lying on a surface or along a line can be determined by following the same derivation that led to Eq. (1.11). For surface charges, with surface charge density given by $\rho_s(\mathbf{r}')$, the surface is divided into small elements of area $\Delta s_i'$, each containing a charge $\Delta s_i'\rho_s(\mathbf{r}_i')$. Then applying Eq. (1.6), the electric field can be written

$$\mathbf{E}(\mathbf{r}) = \sum_i \frac{1}{4\pi\varepsilon_0} \frac{\left[\Delta s_i'\,\rho_s(\mathbf{r}_i')\right](\mathbf{r} - \mathbf{r}_i')}{|\mathbf{r} - \mathbf{r}_i'|^3},$$

which, as the elemental areas $\Delta s_i'$ become small, becomes a surface integral. The resulting electric field is

$$\mathbf{E}(\mathbf{r}) = \frac{1}{4\pi\varepsilon_0} \int_s \frac{\rho_s(\mathbf{r}')(\mathbf{r} - \mathbf{r}')}{|\mathbf{r} - \mathbf{r}'|^3}\, ds'. \tag{1.15}$$

For a charge distributed along a line, with density $\rho_\ell(\mathbf{r}')$, a similar approach leads to

$$\mathbf{E}(\mathbf{r}) = \frac{1}{4\pi\varepsilon_0} \int_\ell \frac{\rho_\ell(\mathbf{r}')(\mathbf{r} - \mathbf{r}')}{|\mathbf{r} - \mathbf{r}'|^3}\, d\ell'. \tag{1.16}$$

The subscripts s and ℓ on the integral signs of Eqs. (1.15) and (1.16) remind us that the integration is to be carried out over the entire charged surface or line distribution. Remember also that $\rho_s(\mathbf{r}')$ and $\rho_\ell(\mathbf{r}')$ indicate the charge density as a function of $\mathbf{r}'$, not charge density times $\mathbf{r}'$.

Example 1.4 Coulomb's Law, Surface Charge

Determine the electric field $\mathbf{E}(\mathbf{r})$ at the point P on the z-axis that results from a surface charge density $\rho_s(\mathbf{r}') = \rho_{s,0}\cos\phi'$ on the outer wall of the cylinder of radius a and length L, as shown in Fig. 1.8.

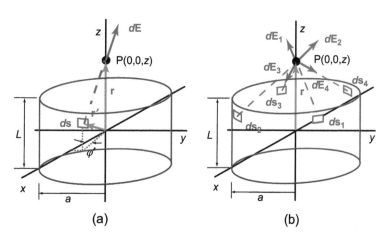

FIGURE 1.8 (a) A cylindrical shell of surface charge density $\rho_s(\mathbf{r}') = \rho_{s,0}\cos\phi'$. The radius of the cylinder is a, and it extends from $z = -L/2$ to $z = L/2$. (b) The contribution to the electric field $d\mathbf{E}$ from four symmetrically placed surface elements. Each of these contributions is of the same magnitude, but in different directions.

(a)   (b)

Solution:

This charge density is positive for $x' > 0$, and negative for $x' < 0$. Since electric field lines originate on positive charges, and terminate on negative charges, we expect that the electric field lines inside the cylinder point primarily in the $-x$-direction.

We will use Eq. (1.15) to find $\mathbf{E}(0,0,z)$. Referring to the diagram in Fig. 1.8(a), the location of a surface element on the cylinder wall, using the variables ϕ' and z', is

$$\mathbf{r}' = (x',y',z') = (a\cos\phi', a\sin\phi', z'),$$

and the location of the observation point P on the z-axis, a distance z from the origin, is

$$\mathbf{r} = (0, 0, z).$$

Note again that we express the position vectors $\mathbf{r}'$ and $\mathbf{r}$ in terms of their x-, y-, and z-components, using the cylindrical variables (ρ, ϕ, and z). The vector $\mathbf{r} - \mathbf{r}'$, which points from the source point to the observation point, is

$$\mathbf{r} - \mathbf{r}' = (-a\cos\phi', -a\sin\phi', (z - z')), \tag{1.17}$$

and its magnitude is

$$|\mathbf{r} - \mathbf{r}'| = \left[(-a\cos\phi')^2 + (-a\sin\phi')^2 + (z - z')^2\right]^{1/2} = \left[a^2 + (z - z')^2\right]^{1/2}. \tag{1.18}$$

Using Eqs. (1.17) and (1.18) in Eq. (1.15), the electric field is

$$\mathbf{E}(z) = \frac{1}{4\pi\varepsilon_0}\int_s \frac{\rho_s(\mathbf{r}')(-a\cos\phi', -a\sin\phi', (z - z'))}{[a^2 + (z - z')^2]^{3/2}}ds'. \tag{1.19}$$

We are given $\rho_s(\mathbf{r}') = \rho_{s,0}\cos\phi'$, and, on the outer wall of the cylinder, the surface elemental area is $ds' = a\,d\phi'\,dz'$. The range of ϕ' is 0 to 2π and the range of z' is $-L/2$ to $+L/2$.

Each of the three components of $\mathbf{E}$ must be evaluated individually. Before doing so, let us stop to reflect on the form that we expect the result to assume. The charge distribution, $\rho_s(\mathbf{r}') = \rho_{s,0}\cos\phi'$, is an even function in y', but odd in x'. Consider the four surface elements ds_1, ds_2, ds_3, and ds_4 shown in Fig. 1.8(b). These elements are placed symmetrically about the cylinder, such that ds_1 is located at (x', y', z'), ds_2 is located at $(x', -y', z')$, ds_3 is at $(-x', -y', z')$, and ds_4 is at $(-x', y', z')$. The magnitudes of the charge contained within each element are identical, although the sign of the charge is positive in elements ds_1 and ds_2 and negative in elements ds_3 and ds_4, and each element is the same distance from the point P on the axis at $\mathbf{r} = (0, 0, z)$. The magnitudes of the field elements $d\mathbf{E}_1$ through $d\mathbf{E}_4$ generated by the charge in surface elements ds_1 through ds_4, respectively, are therefore equal. Additionally, $d\mathbf{E}_1$ and $d\mathbf{E}_2$ point *away* from the surface elements ds_1 and ds_2, since the charges in these elements are positive, while $d\mathbf{E}_3$ and $d\mathbf{E}_4$ point *toward* elements ds_3 and ds_4. The z-components of $d\mathbf{E}_3$ and $d\mathbf{E}_4$ exactly balance the z-components of $d\mathbf{E}_1$ and $d\mathbf{E}_2$, and the y-components of $d\mathbf{E}_1$ and $d\mathbf{E}_3$ exactly balance the y-components of $d\mathbf{E}_2$ and $d\mathbf{E}_4$. We conclude, therefore, that the y- and z-components of the electric field at any point P along the z-axis must be identically zero. The x-components $dE_{1,x}$, $dE_{2,x}$, $dE_{3,x}$, and $dE_{4,x}$, on the other hand, are each negative, and the x-component of $\mathbf{E}$ therefore survives.

Let us now carry out the integration to see how our expectations fare, starting with E_y, which by Eq. (1.19) is of the form

$$E_y(z) = \frac{1}{4\pi\varepsilon_0}\int_{-L/2}^{+L/2}\int_0^{2\pi} \frac{\rho_{s,0}\cos\phi'\,(-a\sin\phi')}{[a^2 + (z - z')^2]^{3/2}}a\,d\phi'\,dz'.$$

Integrating first in ϕ' is simple, since ϕ' appears only in the numerator of the integrand as $\cos\phi'\sin\phi'$. This is most easily evaluated using trigonometry identities,

$$\int_0^{2\pi}\cos\phi'\sin\phi'\,d\phi' = \int_0^{2\pi}\frac{1}{2}\sin 2\phi'\,d\phi' = \frac{1}{2}\left(\frac{-\cos 2\phi'}{2}\right)\Bigg|_0^{2\pi} = 0.$$

Therefore, $E_y(z) = 0$, which is consistent with the conclusion based on the symmetry of the charge distribution that we reached previously. We next evaluate $E_z(z)$, which from Eq. (1.19) is

$$E_z(z) = \frac{1}{4\pi\varepsilon_0} \int_{-L/2}^{+L/2} \int_0^{2\pi} \frac{\rho_{s,0}\cos\phi'\,(z-z')}{[a^2 + (z-z')^2]^{3/2}} \, a \, d\phi' \, dz'.$$

Integration in ϕ' first is again the easier approach, and

$$\int_0^{2\pi} \cos\phi' \, d\phi' = \sin\phi' \Big|_0^{2\pi} = 0$$

leads to the result that $E_z(z) = 0$, also consistent with our symmetry argument result.

Finally, we evaluate $E_x(z)$,

$$E_x(z) = \frac{1}{4\pi\varepsilon_0} \int_{-L/2}^{+L/2} \int_0^{2\pi} \frac{\rho_{s,0}\cos\phi'\,(-a\cos\phi')}{[a^2 + (z-z')^2]^{3/2}} \, a \, d\phi' \, dz'.$$

The ϕ' integration is

$$\int_0^{2\pi} \cos^2\phi' \, d\phi' = \int_0^{2\pi} \frac{1}{2}(1 + \cos 2\phi') \, d\phi' = \frac{1}{2}\left(\phi' + \frac{\sin 2\phi'}{2}\right)\Big|_0^{2\pi} = \pi,$$

which reduces $E_x(z)$ to

$$E_x(z) = \frac{-\rho_{s,0}\, a^2}{4\varepsilon_0} \int_{-L/2}^{+L/2} \frac{dz'}{[a^2 + (z-z')^2]^{3/2}}.$$

Substituting $w = z' - z$ and using Eq. (G.12) leads to the result

$$E_x(z) = \frac{-\rho_{s,0}}{4\varepsilon_0}\left(\frac{(L/2 - z)}{\sqrt{a^2 + (L/2 - z)^2}} + \frac{L/2 + z}{\sqrt{a^2 + (L/2 + z)^2}}\right). \tag{1.20}$$

This function is plotted in Fig. 1.9 for four different cylinder lengths L. Its maximum value is at $z = 0$, where it is equal to

$$E_x(0) = \frac{-\rho_{s,0}L}{4\varepsilon_0\sqrt{a^2 + (L/2)^2}}.$$

Inside the cylinder, for $L > a$, $E_x(z)$ is relatively constant, decreasing to only

$$E_x(\pm L/2) = \frac{-\rho_{s,0}L}{4\varepsilon_0\sqrt{a^2 + L^2}}$$

at either end of the cylinder. As $L \to \infty$, $E_x(z)$ approaches $-\rho_{s,0}/2\varepsilon_0$.

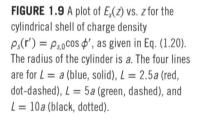

FIGURE 1.9 A plot of $E_x(z)$ vs. z for the cylindrical shell of charge density $\rho_s(\mathbf{r}') = \rho_{s,0}\cos\phi'$, as given in Eq. (1.20). The radius of the cylinder is a. The four lines are for $L = a$ (blue, solid), $L = 2.5a$ (red, dot-dashed), $L = 5a$ (green, dashed), and $L = 10a$ (black, dotted).

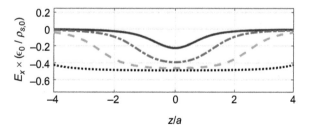

Now let's try finding the electric field due to a line of charge. This will be a uniform distribution of charge in the next example, since that makes the integration easier to carry out, but, of course, Eq. (1.16) is equally applicable for a non-uniform charge, as we will show in the subsequent example.

Example 1.5 Coulomb's Law, Linear Charge

Determine the electric field $\mathbf{E}(\mathbf{r})$ due to a straight, finite-length line charge of uniform linear charge density ρ_ℓ and length L.

Solution:
Let the line charge lie along the z-axis, extending from $z = -L/2$ to $z = +L/2$, as illustrated in Fig. 1.10(a). Before we start a rigorous solution, let's think how we expect the field lines to behave. We know that field lines always point away from positive charges, and they get weaker with increasing distance from the source. We presume $\rho_\ell > 0$. (If not, we need only invert all the field lines.) For locations close to the line charge, but not close to the ends, the electric field lines should lie primarily in the x–y plane, pointing away from the z-axis. We reach this conclusion by thinking about the electric field due to each of the individual segments of the line charge, and realizing that for these points there will be significant cancellation of the many E_z components. The radial components, on the other hand, all point away from the line, and so there is no cancellation here. At the other extreme, as we examine the field at large distances, then the fact that the charge is distributed along a line becomes unimportant, and we should expect that the field approaches that of a point charge of total charge $\rho_\ell L$. Let's proceed now with our calculation of the electric field, and see how our predictions fare.

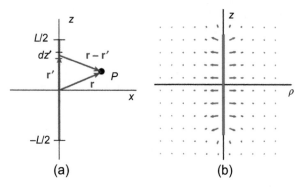

FIGURE 1.10 (a) A finite-length line charge of uniform linear charge density ρ_ℓ extending from $z = -L/2$ to $z = +L/2$. (b) A vector plot of the electric field for this finite-length line charge, as given in Eqs. (1.21) and (1.22).

We will determine the field at a position in the x–z plane. After solution of the problem, the result can be generalized to be valid at any location. Using Eq. (1.16), the electric field is

$$\mathbf{E}(\mathbf{r}) = \frac{1}{4\pi\varepsilon_0} \int_\ell \frac{\rho_\ell(\mathbf{r}')(\mathbf{r} - \mathbf{r}')}{|\mathbf{r} - \mathbf{r}'|^3} \, d\ell',$$

with $d\ell' = dz'$, $\mathbf{r} = (x, 0, z)$, and $\mathbf{r}' = (0, 0, z')$. The vector pointing from the source point to the field point P is $\mathbf{r} - \mathbf{r}' = (x, 0, z - z')$, and its magnitude is $|\mathbf{r} - \mathbf{r}'| = \sqrt{x^2 + (z - z')^2}$. Using these in Eq. (1.16), the electric field is

$$\mathbf{E}(\mathbf{r}) = \frac{\rho_\ell}{4\pi\varepsilon_0} \int_{-L/2}^{L/2} \frac{(x, 0, z - z')}{[x^2 + (z - z')^2]^{3/2}} \, dz'.$$

We evaluate the x- and z-components of $\mathbf{E}$ separately. The former is

$$E_x(x, 0, z) = \frac{x\rho_\ell}{4\pi\varepsilon_0} \int_{-L/2}^{L/2} \frac{dz'}{[x^2 + (z - z')^2]^{3/2}}.$$

Substituting $w = z' - z$, this becomes

$$E_x(x, 0, z) = \frac{x\rho_\ell}{4\pi\varepsilon_0} \int_{-L/2-z}^{L/2-z} \frac{dw}{[x^2 + w^2]^{3/2}}.$$

Using Eq. (G.12) from the integral tables in Appendix G, the result for $E_x(x, 0, z)$ is

$$E_x(x, 0, z) = \frac{\rho_\ell}{4\pi\varepsilon_0 \, x} \left[\frac{L/2 - z}{\sqrt{x^2 + (L/2 - z)^2}} + \frac{L/2 + z}{\sqrt{x^2 + (L/2 + z)^2}} \right].$$

To generalize this result for E_x to a point P not restricted to the x–z plane, recognize from the cylindrical symmetry of the line charge that this represents the solution for the ρ component, with ρ substituted for x:

$$E_\rho = \frac{\rho_\ell}{4\pi\varepsilon_0\,\rho}\left[\frac{L/2-z}{\sqrt{\rho^2+(L/2-z)^2}}+\frac{L/2+z}{\sqrt{\rho^2+(L/2+z)^2}}\right]. \tag{1.21}$$

The z-component of the electric field is

$$E_z(x,0,z) = \frac{\rho_\ell}{4\pi\varepsilon_0}\int_{-L/2}^{L/2}\frac{(z-z')\,dz'}{[x^2+(z-z')^2]^{3/2}}.$$

To solve this integral, we again substitute $w = z' - z$, and use Eq. (G.13), resulting in

$$E_z(x,0,z) = \frac{\rho_\ell}{4\pi\varepsilon_0}\left[\frac{1}{\sqrt{x^2+(L/2-z)^2}}-\frac{1}{\sqrt{x^2+(L/2+z)^2}}\right].$$

Writing this in a form that is valid for points not restricted to the x–z plane, we substitute $x \to \rho$, resulting in

$$E_z = \frac{\rho_\ell}{4\pi\varepsilon_0}\left[\frac{1}{\sqrt{\rho^2+(L/2-z)^2}}-\frac{1}{\sqrt{\rho^2+(L/2+z)^2}}\right]. \tag{1.22}$$

See Fig. 1.10(b) for a plot of the vector electric field, as expressed in Eqs. (1.21) and (1.22).

An important special case of this line charge is that of an infinitely long, uniform line charge. As $L \to \infty$, each of the two terms inside the square brackets of Eq. (1.21) tend to 1, and E_ρ becomes

$$E_\rho = \frac{\rho_\ell}{2\pi\varepsilon_0\,\rho}.$$

Conversely, the two terms inside the square brackets of Eq. (1.22) tend to 0, and E_z becomes

$$E_z = 0.$$

These results can be expressed together in vector form as

$$\boxed{\mathbf{E}_\rho = \frac{\rho_\ell}{2\pi\varepsilon_0\,\rho}\mathbf{a}_\rho.} \tag{1.23}$$

That is, the electric field due to an infinitely long uniform line charge is directed radially outward from the line charge, and decreases as $1/\rho$. We will make extensive use of this result later.

You will recall that, at the outset of this example, we made some predictions of this electric field in two limiting cases. Let's revisit those to see how we did, examining first the components E_ρ and E_z for a point close to the middle of the line charge. For simplicity, we will set $z = 0$. (The field components do vary somewhat away from $z = 0$, but not enough to warrant the additional complexity of these estimates.) At this plane, E_z vanishes exactly for any ρ, as can be seen directly from Eq. (1.22). Examination of Eq. (1.21) for E_ρ at $z = 0$ and $\rho \ll L/2$ shows that the term inside the square brackets approaches 2, and the electric field reduces to $\mathbf{E} \simeq \mathbf{a}_\rho\rho_\ell/2\pi\varepsilon_0\rho$. This matches Eq. (1.23), the electric field for an infinitely long line charge that we just discussed. This should make sense, since for a point close to the axis at $z = 0$, the distance to the end of the line charge is much, much larger than any other distance, and the length of the charge distribution is essentially infinite. Our second prediction was that at a long distance from the charge segment, the electric field should approach that of a point charge, of total charge $\rho_\ell L$. We will verify this at two specific points. First, examination of Eq. (1.21) for

E_ρ at $z = 0$ and $\rho \gg L/2$ shows that the expression inside the square brackets approaches L/ρ. Alternatively, we can examine E_z at $z \gg L/2$, $\rho = 0$, and find that in this limit the term inside the square brackets of Eq. (1.22) approaches L/z^2. Either of these limits validates our prediction that the electric field decreases as the inverse distance squared, and is proportional to the total charge of the line, $\rho_\ell L$. With a little more work, the same result (i.e. that E is proportional to the total charge $\rho_\ell L$, that it decreases as the square of the inverse of the distance from the charge, and that it points directly away from the origin) can be verified for the more general case of a distant point P not on the z-axis nor in the $z = 0$ plane. We leave this as a homework exercise (Prob. P1-11).

In the previous example, the charge density was uniform. Now let's tackle a similar problem, but one in which the charge density is not uniform.

Example 1.6 Non-uniform Linear Charge Distribution

A linear charge distribution lies along the z-axis, as shown in Fig. 1.11(a). From $z = 0$ to $z = L/2$, the charge density is $\rho_\ell(z) = \rho_{\ell,0}$, while from $z = 0$ to $z = -L/2$, the charge density is $\rho_\ell(z) = -\rho_{\ell,0}$. The charge density elsewhere is 0. Determine the field $\mathbf{E}(\mathbf{r})$ that results from this charge distribution.

FIGURE 1.11 (a) A finite-length line charge of non-uniform linear charge density $\rho_\ell(z)$. We will determine the electric field **E** at the point P in Example 1.6. (b) A vector plot of the electric field for a finite-length line charge of non-uniform linear charge density ρ_ℓ, as given in Eqs. (1.21) and (1.22).

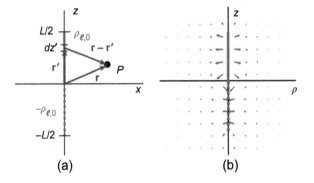

(a)   (b)

Solution:

We follow much the same process as that of the previous two examples. Since the charge is distributed along a line, we will use Eq. (1.16) to find the electric field $\mathbf{E}(\mathbf{r})$. We denote the location of the charge

$$\mathbf{r}' = (0, 0, z')$$

since it lies along the z-axis, and the observation point P is located at

$$\mathbf{r} = (x, 0, z).$$

In writing $\mathbf{r}$, we have again chosen to use the coordinate system in which the observation point $\mathbf{r}$ lies in the x–z plane. This simplifies our analysis a bit, and we are free to make this assignment since the $\mathbf{E}(\mathbf{r})$ field resulting from the charge density given, which is cylindrically symmetric about the z-axis, must display the same cylindrical symmetry. When we are finished with our solution, the result can be generalized to be valid for any observation point, not necessarily lying in this plane.

The vector $\mathbf{r} - \mathbf{r}'$, which points from the source point to the observation point, is

$$\mathbf{r} - \mathbf{r}' = (x, 0, z - z'), \tag{1.24}$$

and its magnitude is

$$|\mathbf{r} - \mathbf{r}'| = \left[(x)^2 + (z - z')^2\right]^{1/2}. \tag{1.25}$$

Using Eqs. (1.24) and (1.25) in Eq. (1.16), the electric field is

$$\mathbf{E}(\mathbf{r}) = \frac{1}{4\pi\varepsilon_0} \int_{-L/2}^{L/2} \frac{\rho_\ell(z')(x, 0, z - z')}{\left[x^2 + (z - z')^2\right]^{3/2}} \, dz'.$$

Breaking the integral into two parts, one for the region of charge density $-\rho_{\ell,0}$ and the other for the region of charge density $\rho_{\ell,0}$, this becomes

$$\mathbf{E}(\mathbf{r}) = \frac{\rho_{\ell,0}}{4\pi\varepsilon_0} \left[-\int_{-L/2}^{0} \frac{(x, 0, z - z')}{\left[x^2 + (z - z')^2\right]^{3/2}} \, dz' + \int_{0}^{L/2} \frac{(x, 0, z - z')}{\left[x^2 + (z - z')^2\right]^{3/2}} \, dz' \right].$$

As with the earlier examples, we evaluate the x- and z-components of $\mathbf{E}(\mathbf{r})$ separately. The x-component is

$$E_x(\mathbf{r}) = \frac{\rho_{\ell,0}}{4\pi\varepsilon_0} \left[-\int_{-L/2}^{0} \frac{x \, dz'}{\left[x^2 + (z - z')^2\right]^{3/2}} + \int_{0}^{L/2} \frac{x \, dz'}{\left[x^2 + (z - z')^2\right]^{3/2}} \right].$$

This integral is solved by substituting $w = z' - z$ and using the integral Eq. (G.12) from Appendix G. After making the transformation $x \to \rho$, as permitted due to the cylindrical symmetry of the line charge about the z-axis, the result is

$$E_\rho(\mathbf{r}) = \frac{\rho_{\ell,0}}{4\pi\varepsilon_0} \left(\frac{1}{\rho}\right) \left[\frac{2z}{\sqrt{\rho^2 + z^2}} + \frac{L/2 - z}{\sqrt{\rho^2 + (L/2 - z)^2}} - \frac{L/2 + z}{\sqrt{\rho^2 + (L/2 + z)^2}} \right]. \tag{1.26}$$

Next we evaluate $E_z(\mathbf{r})$,

$$E_z(\mathbf{r}) = \frac{\rho_{\ell,0}}{4\pi\varepsilon_0} \left[-\int_{-L/2}^{0} \frac{(z - z') \, dz'}{\left[x^2 + (z - z')^2\right]^{3/2}} + \int_{0}^{L/2} \frac{(z - z') \, dz'}{\left[x^2 + (z - z')^2\right]^{3/2}} \right].$$

Using the same substitution $w = z' - z$, this time with the integral Eq. (G.13) from Appendix G, we find, with x replaced by ρ,

$$E_z(\mathbf{r}) = \frac{\rho_{\ell,0}}{4\pi\varepsilon_0} \left[\frac{-2}{\sqrt{\rho^2 + z^2}} + \frac{1}{\sqrt{\rho^2 + (L/2 - z)^2}} + \frac{1}{\sqrt{\rho^2 + (L/2 + z)^2}} \right]. \tag{1.27}$$

A plot of this field pattern is shown in Fig. 1.11(b). Notice how the field lines originate on the positive charges, sweep around, and terminate on the negative charges. Notice also, since the net charge of this line is zero, there is significant cancellation among the contributions to the field, especially at large distances from the charges, and the magnitude of the field tends to drop away rather quickly.

1.1.3 Electric Field for Point, Line, and Uniform Sheet of Charge

To this point, we have applied Coulomb's Law in several different forms (Eqs. (1.3), (1.11), (1.15), and (1.16)) to find the electric field due to a variety of charge distributions. There are three cases that deserve special mention, however, due to their simplicity and their application as an approximation to other more complex problems. These cases are highlighted in this section.

The first case is the electric field due to a point charge Q, or equivalently, outside a uniform sphere of charge. As we showed in Eq. (1.3) (but simplified by placing the charge at the origin), this electric field,

$$\mathbf{E}(R) = \frac{1}{4\pi\varepsilon_0} \frac{Q}{R^2} \mathbf{a}_R,$$

points radially outward from the charge (inward if $Q < 0$), and decreases with increasing distance as R^{-2}.

We also treated an infinitely long uniform line charge lying along the z-axis in Example 1.5, and found that the electric field due to this charge is

$$\mathbf{E}_\rho = \frac{\rho_\ell}{2\pi\varepsilon_0 \, \rho}\mathbf{a}_\rho.$$

This field points radially outward, but in this case $\mathbf{E}$ points out from the z-axis. Also, the field magnitude decreases with increasing distance, but for the line charge it decreases as ρ^{-1}, a slower decrease than for $\mathbf{E}$ due to a point charge or a sphere of uniform charge.

Finally, in a case that we have not yet treated, but which we will encounter later (Prob. P1-14 and Example 1.9), the electric field $\mathbf{E}$ due to an infinite uniform sheet of charge of surface charge density ρ_s lying in the x–y plane is given simply as

$$\mathbf{E} = \frac{\rho_s}{2\varepsilon_0}\mathbf{a}_z$$

for $z > 0$, and

$$\mathbf{E} = -\frac{\rho_s}{2\varepsilon_0}\mathbf{a}_z$$

for $z < 0$. This field points away from the charged sheet on either side, but its magnitude does not depend on the distance from the sheet.

Note the relation between the position dependence for these three fields and the dimensionality of the charge distribution. For a uniform sheet of charge, which is a distribution of charge in two dimensions, the dependence of the electric field on position is z^0. For a uniform line charge, a charge distribution that extends in just one dimension, $|\mathbf{E}(\rho)|$ decreases as ρ^{-1}. And for a point charge, which is of zero dimensions, $|\mathbf{E}(R)|$ decreases as R^{-2}. Combining these results, a uniform charge distributed in n dimensions produces an electric field whose magnitude depends on position as d^{2-n}, where d is the appropriate position coordinate (z, ρ, or R) for the geometry.

1.1.4 Applications of Electrostatics

We close this section with several notes about common applications of electrostatics.

TechNote 1.1 Electrostatic Air Cleaners

Electrostatic air cleaners can be used to remove particles and dust from room air. As the air is drawn into the cleaner, the dust is ionized through its surface interactions with a mesh of polyurethane or polypropylene fibers, for example. Once charged, the dust and particulates are separated from the air with a number of oppositely charged conducting plates. Electrostatic air cleaners can be an effective means for filtering dust and particulates from the air, but they can also produce harmful ozone, and they require periodic cleaning.

Electrostatic spraying, used widely in industrial applications, is described in Tech-Note 1.2.

TechNote 1.2 Electrostatic Spraying

Electrostatic spraying can be used to uniformly and efficiently coat surfaces with paints, disinfectants, or other solutions. A charge is transferred to the liquid droplets as they are atomized in a nozzle by applying a high potential to the nozzle. The surface to be sprayed is grounded, or held at a potential opposite that of the nozzle. The charged liquid droplets are attracted to the target, accelerated by the electrostatic field between the nozzle and the target. This technique is characterized by uniform coatings, even on surfaces not directly in line with the sprayer, and efficient use of the paint or solution. Electrostatic spraying has been in use since the 1930s, and is in wide use in the automotive industry and other industrial settings.

Photocopiers, common in nearly every business setting, rely on the deposition of charges on a drum, and the attraction of the toner to these charges, to create images on paper, as described in TechNote 1.3.

TechNote 1.3 Copy Machines

Copy machines, or photocopiers, are in wide use in offices today. The copy machine was invented in 1938 by Chester Carlson (1906–1968), and patented in 1944. Looking back, it is remarkable to learn that Carlson had difficulty finding industrial interest in his invention. Ultimately, he and Battelle Memorial Institute formed a partnership with Haloid to develop the copy machine further, and the first copy machine, or Xerox machine, came to market in 1949. In 1961, Haloid changed its name to Xerox.

The copying process, illustrated in Fig. 1.12, makes use of electrostatic charges and a process known as photoconduction. A photoconducting material is one in which the electrical conductivity is enhanced after the material is illuminated with light. (We'll talk more about electrical conductivity later, but simply put it quantifies the ability of electric charges to move about within a material.) In the photocopying process the photoconductor is prepared by depositing a negative charge on its surface. A document is scanned with a bright light source, and its image is projected onto the photoconducting drum. In regions where the light is bright, the surface charge is conducted into the material, where it combines with surface charges on the drum. This creates a charged

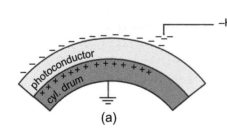
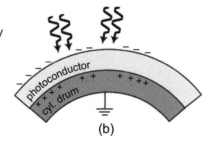
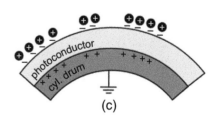

(a)   (b)   (c)

FIGURE 1.12 Representation of the photocopying process. In (a), the photoconductor is prepared by depositing a uniform charge on its surface. In (b), an image of the document to be reproduced is projected onto the photoconductor, removing the charges from those regions of the photoconductor which are illuminated. A toner (a fine, pigmented powder) that is deposited on the photoconductor adheres only to those regions that remain charged in (c). When a sheet of paper rolls with the drum, the toner is transferred to the paper, and when heated the toner melts and bonds to the paper.

image of the original document on the photoconductor surface. Upon application to the drum of a fine, pigmented, positively charged powder called toner, the toner adheres to the negatively charged regions of the photoconductor drum. When paper is brought into contact with the drum, the toner is transferred to the paper, and when heated the toner melts and bonds permanently to the paper.

In this section we have developed Coulomb's Law, in several different forms, which allows us to determine the electric field produced by a set of known discrete charges, or a known density of charge distributed throughout a volume, across a surface, or along a line. This can be a powerful technique, although, as you have probably already noticed, the integration can become challenging. Furthermore, we need to know the location of the charges in advance. Coulomb's Law must always be correct, of course, but we can only apply it to find the electric field when the magnitude and location of the charges are known. In many situations this condition is not satisfied. For example, as we will describe in more detail later, charges on conductors are free to move around, often in response to the electric field surrounding the conductor. In this case we don't know, at least initially, the location of the charges. As we will see, it will be necessary to have at our disposal other techniques that allow us to determine electric fields. For this reason, we must delve more deeply into the general properties of electric fields, as we will explore in the next section.

1.2 Differential Properties of $\mathbf{E}$

Now that we have examined the use of Coulomb's Law to determine the electric field $\mathbf{E}(\mathbf{r})$ at some point $\mathbf{r}$ due to discrete charges at $\mathbf{r}'_i$, or a continuous distribution of charges whose density is given by $\rho_v(\mathbf{r}')$, $\rho_s(\mathbf{r}')$, or $\rho_\ell(\mathbf{r}')$, we are ready to move on to explore some of the more general properties of electric fields. We start by developing a pair of differential relations that all static electric fields must satisfy, which will help us find solutions for $\mathbf{E}(\mathbf{r})$ in a wider range of situations.

The quantities that we seek are $\nabla \cdot \mathbf{E}(\mathbf{r})$ and $\nabla \times \mathbf{E}(\mathbf{r})$ throughout a region of space v. Knowledge of these two spatial differential operators will almost completely specify $\mathbf{E}(\mathbf{r})$ within that volume v. (See Appendix E.3 for a discussion of the Completeness Theorem.) When $\nabla \cdot \mathbf{E}(\mathbf{r})$ and $\nabla \times \mathbf{E}(\mathbf{r})$ are known, $\mathbf{E}(\mathbf{r})$ is uniquely determined save for an additive constant value, which we can usually find by other means. We cannot possibly underestimate the importance of these vector divergence and curl properties that we are about to introduce. As will be discussed later, these vector differential properties allow us to define a potential function for electric fields, they allow us to derive a set of boundary conditions that fields must obey at the interface between two different media, they allow us to describe propagation of waves (after we have introduced time-varying fields, of course), and much, much more. So while they may seem to be rather abstract concepts, especially at first, we want to be sure to master their meaning and learn how to apply them to a variety of problems.

1.2.1 Div $\mathbf{E}$

We start by considering the divergence of an electrostatic field $\mathbf{E}(\mathbf{r})$. For preparation, see Example D.5, in which it is shown that the divergence of a vector field of the form $\mathbf{K} = CR^{-2}\mathbf{a}_R$ is zero, a result that is valid everywhere except at the origin. Since the electric field due to a point charge is similar to the vector **K** ($\mathbf{E}(\mathbf{r})$ is also directed radially outward from the charge, and its magnitude decreases as $1/R^2$), $\mathbf{\nabla} \cdot \mathbf{E}$ for a point charge is identically zero (except, of course, at the location of the charge, where it is not defined). To explore the properties of $\mathbf{\nabla} \cdot \mathbf{E}$ further, consider the volume integral

$$\int_v \mathbf{\nabla} \cdot \mathbf{E} \, dv, \tag{1.28}$$

where $\mathbf{E}(\mathbf{r})$ is the electric field in the space surrounding a single point charge Q located at the origin, and the volume v is the interior of a sphere of radius a, centered at the origin. We start with a few comments. First, while we have chosen the volume to be a sphere centered on the charge, this turns out to be solely for our own convenience. We'll return to this point shortly. And second, a few of you may have had the initial impulse to declare, "Why, this integral must be zero, since we already know that $\mathbf{\nabla} \cdot \mathbf{E}$ for a point charge is zero." Not so fast, however, since we must not forget about the origin itself. Remember, our solution for $\mathbf{\nabla} \cdot \mathbf{E}$ was not valid at the origin. So if we don't know the value of $\mathbf{\nabla} \cdot \mathbf{E}$ at the origin, the volume integral that we seek is not calculated quite so directly. The solution to this dilemma can be found in the Divergence Theorem (see Appendix D.2 for a review of the Divergence Theorem), which states

$$\int_v \mathbf{\nabla} \cdot \mathbf{E} \, dv = \oint_s \mathbf{E}(\mathbf{r}) \cdot d\mathbf{s}.$$

The surface integral on the right-hand side is the surface integral over the closed surface s enclosing the volume v. This surface integral of the electric field is known as the **electric flux** and denoted Φ_E, in close analogy with the flow of currents through some cross-sectional area, as is reviewed in the discussion of vector surface integration in Appendix C.3.2. To evaluate this integral, we only need to know the electric field $\mathbf{E}$ at the surface of the sphere, but not at interior points within it. The surface element $d\mathbf{s}$ is an elemental patch on the surface of the sphere, whose vector points radially outward, $d\mathbf{s} = a^2 \sin\theta \, d\theta \, d\phi \, \mathbf{a}_R$. For a point charge, then,

$$\int_v \mathbf{\nabla} \cdot \mathbf{E} \, dv = \oint_s \mathbf{E}(\mathbf{r}) \cdot d\mathbf{s} = \int_{\phi=0}^{2\pi} \int_{\theta=0}^{\pi} \frac{1}{4\pi\varepsilon_0} \frac{Q}{a^2} \mathbf{a}_R \cdot a^2 \sin\theta \, d\theta \, d\phi \, \mathbf{a}_R.$$

Notice that by choosing to evaluate this surface integral on the surface of a sphere, the magnitude $|\mathbf{E}(\mathbf{r})|$ of the electric field is a constant, equal to $Q/4\pi\varepsilon_0 a^2$, where a, as a reminder, is the radius of the sphere. The factors a cancel one another, and $\mathbf{a}_R \cdot \mathbf{a}_R = 1$, so the electric flux is

$$\Phi_E = \oint_s \mathbf{E}(\mathbf{r}) \cdot d\mathbf{s} = \frac{Q}{4\pi\varepsilon_0} \int_0^{2\pi} \int_0^{\pi} \sin\theta \, d\theta \, d\phi = Q/\varepsilon_0, \tag{1.29}$$

where the integral in ϕ gives us 2π and the integral in θ is simply 2. Then the total flux of the electric field vector resulting from a point charge Q is simply equal to Q/ε_0.

It turns out that the shape or size of this volume v really doesn't make any difference at all, as long as the origin, or more specifically the charge Q, is located within it. We can demonstrate this by breaking the volume v up into two parts, say v_1 and v_2. We don't really care about the shape of these two parts, but let's agree that the charge Q is clearly in one of these, say v_1. Since the integral in Eq. (1.28) can be written

$$\int_{v_1} \mathbf{\nabla} \cdot \mathbf{E} \, dv \; + \int_{v_2} \mathbf{\nabla} \cdot \mathbf{E} \, dv, \tag{1.30}$$

and since the integrand $\mathbf{\nabla} \cdot \mathbf{E}$ is zero everywhere within the charge-free volume v_2, the second integral is zero, and only the first integral in Eq. (1.30) remains. But we didn't specify the shape or size of v_1, only that the charge Q was located within it. By this we conclude that the shape or size of v is immaterial in computing Eq. (1.28), and we are free to define v in whatever manner we choose.

These features then lead to an even more useful form of Eq. (1.29). That is, the electric flux through a surface surrounding multiple charges is

$$\Phi_E = \oint_s \mathbf{E}(\mathbf{r}) \cdot d\mathbf{s} = \oint_s \sum_i \mathbf{E}_i(\mathbf{r}) \cdot d\mathbf{s},$$

where $\mathbf{E}_i(\mathbf{r})$ is the field due to just the charge labeled i, and we have applied superposition to write that the total electric field is the sum of the fields due to the individual charges. Since the fields are well-behaved (i.e. continuous), we can interchange the order of summation and integration, so

$$\oint_s \mathbf{E}(\mathbf{r}) \cdot d\mathbf{s} = \sum_i \oint_s \mathbf{E}_i(\mathbf{r}) \cdot d\mathbf{s}.$$

But each surface integral of $\mathbf{E}_i(\mathbf{r})$ must yield Q_i/ε_0 by Eq. (1.29), so the flux is

$$\oint_s \mathbf{E}(\mathbf{r}) \cdot d\mathbf{s} = \sum_i Q_i/\varepsilon_0,$$

and since $\sum_i Q_i$ represents the total charge enclosed within the volume, Q_{enc},

$$\Phi_E = \oint_s \mathbf{E}(\mathbf{r}) \cdot d\mathbf{s} = Q_{enc}/\varepsilon_0. \qquad (1.31)$$

This relationship is extremely important in the study of electric fields, and is known as the integral form of **Gauss' Law**, in recognition of Carl Friedrich Gauss. (See Biographical Note 1.3.) Be aware that Gauss' Law in this form is not complete, in that we have considered only an ensemble of charges located in vacuum. We will modify this a bit later when we consider electric fields inside materials.

Biographical Note 1.3 Johann Carl Friedrich Gauss (1777–1855)

The work of Johann Carl Friedrich Gauss (pictured in Fig. 1.13), a German mathematician, influenced the development of algebra, astronomy, surveying, non-Euclidean geometry, magnetism, and many more fields. The impact of his work carries through to today, as we recognize with the Gaussian probability distribution, Gauss' Law, and the cgs unit for magnetic flux density, 1 gauss = 10^{-4} T.

FIGURE 1.13 Portrait of Carl Friedrich Gauss (1777–1855).

Finally, it is useful to convert the integral form of Gauss' Law into an alternative form, known as the differential form,

$$\boxed{\nabla \cdot \mathbf{E}(\mathbf{r}) = \rho_v/\varepsilon_0.}$$

(1.32)

Like its integral counterpart, Eq. (1.32) is valid only in vacuum, and we will modify it later when we include materials. The derivation of Eq. (1.32) is rather straightforward, and starts by considering Eq. (1.31) where the enclosed charge is constituted from a continuous volume distribution of charge, of density $\rho_v(\mathbf{r})$. Then the electric flux is

$$\int_v \nabla \cdot \mathbf{E}\, dv = \oint_s \mathbf{E}(\mathbf{r}) \cdot d\mathbf{s} = Q_{enc}/\varepsilon_0 = \frac{1}{\varepsilon_0} \int_v \rho_v(\mathbf{r})\, dv.$$

The first equality follows from the Divergence Theorem (Eq. (D.10)), the second from the integral form of Gauss' Law (Eq. (1.31)), and the third from the relationship between the charge density ρ_v in a volume v and the total charge Q_{enc} contained within that volume (Eq. (C.10)). Let us focus on the first and last terms in this expression, the two involving volume integrals. Since these are valid for *any* region in space (remember that the volume v is arbitrary), then the only way that the integrals can be equal to one another for all volumes v is if the integrands themselves are equal. This constitutes proof of the differential form of Gauss' Law (Eq. (1.32)).

1.2.2 Curl **E** for Static Fields

The second differential operator of the static electric field is curl **E**, or $\nabla \times \mathbf{E}$. Since the electric field for a point charge points radially outward, and contains no component that circulates about the source, we can immediately infer that, for a static electric field produced by a point charge, $\nabla \times \mathbf{E} = 0$. If this descriptive argument is not convincing, then we can use Eq. (D.13) to compute the curl of the electric field given in Eq. (1.3),

$$\nabla \times \mathbf{E}(\mathbf{r}) = \nabla \times \left(\frac{1}{4\pi\varepsilon_0} \frac{q}{|\mathbf{r} - \mathbf{r}'|^2}\, \mathbf{a}_{rr'} \right),$$

where we must note that the derivatives are taken with respect to the spatial coordinates $\mathbf{r}$ – that is, $x, y,$ and z, not the source coordinate $\mathbf{r}' = (x', y', z')$. Writing this out explicitly,

$$\nabla \times \mathbf{E}(\mathbf{r}) = \frac{q}{4\pi\varepsilon_0} \nabla \times \left(\frac{(x - x')\, \mathbf{a}_x + (y - y')\, \mathbf{a}_y + (z - z')\, \mathbf{a}_z}{\left[(x - x')^2 + (y - y')^2 + (z - z')^2 \right]^{3/2}} \right).$$

We consider one component at a time, and use the abbreviation $D = [(x - x')^2 + (y - y')^2 + (z - z')^2]^{1/2}$. Then,

$$(\nabla \times \mathbf{E}(\mathbf{r}))_x = \frac{q}{4\pi\varepsilon_0} \left[\frac{\partial}{\partial y} \frac{(z - z')}{D^3} - \frac{\partial}{\partial z} \frac{(y - y')}{D^3} \right]$$

$$= \frac{q}{4\pi\varepsilon_0} \left[(z - z') \frac{(-3/2)2(y - y')}{D^5} - (y - y') \frac{(-3/2)2(z - z')}{D^5} \right] = 0.$$

Similarly, $(\nabla \times \mathbf{E}(\mathbf{r}))_y$ and $(\nabla \times \mathbf{E}(\mathbf{r}))_z$ can be shown to be zero as well, completing the demonstration that $\nabla \times \mathbf{E}(\mathbf{r}) = 0$ for a point charge. This proof would have been much simpler to carry out in spherical coordinates, which we leave to the student as Prob. P1-22 at the end of this chapter.

If more than one charge is present, the resulting field is the superposition of field contributions from each individual charge $E_k(\mathbf{r})$, and

$$\nabla \times \mathbf{E}(\mathbf{r}) = \nabla \times \sum_k \mathbf{E}_k(\mathbf{r}) = \sum_k \nabla \times \mathbf{E}_k(\mathbf{r}) = 0.$$

For continuous charge distributions the proof follows the same argument, and we conclude that the static field $\mathbf{E}(\mathbf{r})$ resulting from *any* distribution of discrete and continuous charges must obey

$$\boxed{\nabla \times \mathbf{E} = 0,} \tag{1.33}$$

a result that is valid for all *static* electric fields. When we introduce time-varying fields later, we will need to modify this relation. Equation (1.33) is the beginning of what will become Faraday's Law, and the form given here is called the differential form. An equivalent, integral, form can be derived by considering the path integral of E around *any* closed path c and imposing Stokes' Theorem. We start by defining a closed path c, as shown in Fig. 1.14(a). The line integral of E around this path must be, by Stokes' Theorem,

$$\oint_c \mathbf{E} \cdot d\boldsymbol{\ell} = \int_s \nabla \times \mathbf{E} \cdot d\mathbf{s},$$

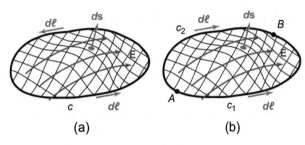

FIGURE 1.14 An illustration of the path integral of $\mathbf{E} \cdot d\ell$. In (a), the path c is a closed loop. In (b), paths c_1 and c_2 each lead from point A to point B.

where s is *any* surface bounded by c. But from Eq. (1.33), we know that $\nabla \times \mathbf{E}$ is zero everywhere, leading to the conclusion that

$$\boxed{\oint_c \mathbf{E} \cdot d\boldsymbol{\ell} = 0} \tag{1.34}$$

for any closed path c. Just like the differential form, Eq. (1.34) is valid only for static electric fields.

Before we conclude this section, let us write this path integral in a somewhat different form, with the goal of showing that the line integral of E between two points is independent of the path followed between those two points. Consider the two paths labeled c_1 and c_2 in Fig. 1.14(b), each leading from point A to point B. The difference between $\int \mathbf{E} \cdot d\ell$ for these two paths is given by

$$\Delta = \int_{c_1} \mathbf{E} \cdot d\boldsymbol{\ell} - \int_{c_2} \mathbf{E} \cdot d\boldsymbol{\ell}.$$

But the path integral of $\mathbf{E} \cdot d\ell$ along c_2 is just -1 times the path integral backwards along this path from B to A. We'll call this path $-c_2$, and the difference between the path integrals Δ becomes

$$\Delta = \int_{c_1} \mathbf{E} \cdot d\boldsymbol{\ell} + \int_{-c_2} \mathbf{E} \cdot d\boldsymbol{\ell}.$$

Finally, the path integral along c_1 plus the path integral along $-c_2$ must be 0, since this combined path forms a closed loop. We therefore conclude that the line integral $\int_\ell \mathbf{E} \cdot d\ell$ between two points is the same for any path between those end points. This is an important result, as it allows us to define an electric potential function, which will be introduced in Section 1.4. At that time, we will discuss the work that must be done in moving a charge along a path from point A to point B against an electrostatic force $q\mathbf{E}(\mathbf{r})$. In those terms, Eq. (1.34) tells us that the work is equal in magnitude but opposite in sign to the work required to move that same charge from point B to point A, whether by the same or a different path. This is a characteristic of a **conservative force**. Equivalently, for a conservative force, $\nabla \times \mathbf{F}(\mathbf{r}) = 0$. We will make extensive use of these notions later. Before we do that, let's explore Gauss' Law further.

1.3 Application of Gauss' Law in Vacuum

The integral form of Gauss' Law, as expressed in Eq. (1.31), provides us with a powerful alternative to Coulomb's Law for calculating the electric field in certain cases. While the number of these cases is somewhat limited, due to the high degree of symmetry required, these cases do tend to be important, and we will certainly appreciate the ease with which Gauss' Law can be applied. Since Gauss' Law in its integral form involves an integration over a closed surface, $\oint \mathbf{E} \cdot d\mathbf{s}$, it might seem backward to try to use this integral to learn anything about the integrand $\mathbf{E}$. This is where the importance of symmetry of the charge distribution, and careful selection of the surface over which we carry out the integration, which is called the **Gaussian surface**, become evident. A Gaussian surface is simply a surface over which we apply Gauss' Law. It is not necessarily connected with any physical boundary. As we will see in the following examples, the symmetry of the charge distribution can be used in order to argue that only certain components of the field exist, and that these can depend only on a limited set of spatial coordinates. We must then construct a Gaussian surface, composed of (1) surfaces over which the fields are uniform, (2) other surfaces over which $\mathbf{E} = 0$, and (3) still other surfaces over which $\mathbf{E}$ and $d\mathbf{s}$ are perpendicular. The second and third do not contribute to the total surface integral, and we can easily evaluate the first for $\mathbf{E}$ on those surfaces. Let us illustrate this technique further by working an example.

Example 1.7   Electric Field due to Spherical Charge

Consider a spherical distribution of charge of radius a. Let the volume charge density ρ_v be uniform throughout the sphere. Use Gauss' Law to determine the electric field at a distance R from the center of the sphere.

Solution:

We start by using the symmetry of the spherical charge distribution to determine the direction of $\mathbf{E}$, and any restrictions on its dependence on spatial coordinates. For this charge distribution, we expect that $\mathbf{E}$ at any point P must be directed radially (i.e. $\mathbf{E} = E_R \mathbf{a}_R$) and that E_R depends only on the distance R from the center of the sphere to the point P, but not on θ or ϕ. To show that $\mathbf{E} = E_R \mathbf{a}_R$, assume that there is a component of $\mathbf{E}$ in the θ or ϕ direction. We call it E_t, where the t stands for transverse. Now if we rotate the sphere about the axis that connects the center of the sphere with point P by 180°, E_t must be reversed. But the charge distribution is perfectly symmetric, so the charge distribution after the rotation is identical to the distribution before the rotation. Therefore E_t must be unchanged as well. The only conclusion is that E_t must be 0, and therefore the field is radial. To show that E_R depends only on R, we consider the field at two points, both a distance R from the center of the charged sphere, but at different coordinates θ and ϕ. Since the charge distribution is perfectly symmetric about its center, the field at these two points must be the same, and we conclude that E_R must be independent of θ and ϕ.

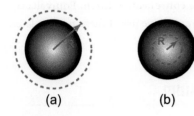

(a)   (b)

FIGURE 1.15 A sphere of radius a and charge density ρ_v. The dashed red lines indicate the Gaussian surface that we construct in order to determine $E_R(R)$ for (a) $R > a$, and (b) $R < a$.

Now we are ready to apply Gauss' Law. We start by constructing a Gaussian surface about the charge distribution. In Fig. 1.15, two examples of Gaussian surfaces are shown. Both are spherical, and are concentric with the spherical charge. In (a), the Gaussian surface is larger than the charged sphere, while in (b) it is smaller. The Gaussian surface shown in (a) is used to determine $\mathbf{E}$ at a radius $R > a$ and the Gaussian surface shown in (b) to determine $\mathbf{E}$ for $R < a$.

We begin by evaluating the electric flux $\oint_s \mathbf{E} \cdot d\mathbf{s}$ over the surface of the spherical Gaussian surface. This is valid for both cases, $R > a$ and $R < a$. The surface element for the outer surface of the sphere is $d\mathbf{s} = \mathbf{a}_R \, R^2 \sin\theta \, d\theta \, d\phi$, so

$$\Phi_E = \oint_s \mathbf{E} \cdot d\mathbf{s} = \int_{\phi=0}^{2\pi} \int_{\theta=0}^{\pi} (E_R(R)\,\mathbf{a}_R) \cdot \mathbf{a}_R \, R^2 \sin\theta \, d\theta \, d\phi = E_R \, R^2 \int_{\theta=0}^{\pi} \sin\theta \, d\theta \int_{\phi=0}^{2\pi} d\phi,$$

where we pulled E_R and R^2 outside of the integral since the integration variables are θ and ϕ, and E_R and R^2 are independent of these variables. Evaluating the integrals, the electric flux is

$$\Phi_E = \oint_s \mathbf{E} \cdot d\mathbf{s} = 4\pi E_R \, R^2. \tag{1.35}$$

Next we must determine the total charge contained within the Gaussian surface. This differs for the two cases, $R > a$ and $R < a$, so we treat them separately. For $R > a$ (i.e. outside the sphere),

$$Q_{enc} = \int_v \rho_v \, dv = \int_{\phi=0}^{2\pi} \int_{\theta=0}^{\pi} \int_{R=0}^{a} \rho_v R^2 dR \sin\theta \, d\theta \, d\phi,$$

where the limits of integration in R extend from 0 to a, since the charge density is zero for $R > a$. Carrying out the integration yields $Q_{enc} = (4\pi/3)a^3\rho_v$. This charge is not unexpected, since the charged sphere is completely inside the Gaussian surface, the volume of the charged sphere is $v = (4\pi/3)a^3$, and the total charge is $Q = \rho_v v$ when the charge density is uniform. Setting Q_{enc}/ε_0 equal to Eq. (1.35) and solving for E_R results in

$$E_R(R) = \frac{a^3 \rho_v}{3\varepsilon_0 R^2}.$$

Using $Q = (4\pi/3)a^3\rho_v$, we can also write $E_R(R)$ as

$$E_R(R) = \frac{Q}{4\pi\varepsilon_0 R^2},$$

which you will recognize as the radial electric field produced by a point charge of charge Q. In other words, the electric field outside the charged sphere is indistinguishable from the electric field due to a point charge of the same total charge Q.

For $R < a$, the evaluation is similar, but now the Gaussian surface is smaller than the charged sphere, so only a portion of the total charge Q is enclosed within the Gaussian surface,

$$Q_{enc} = \int_v \rho_v \, dv = \int_{\phi=0}^{2\pi} \int_{\theta=0}^{\pi} \int_{R'=0}^{R} \rho_v R'^2 dR' \sin\theta \, d\theta \, d\phi,$$

as reflected in the limit of integration for the R' integral. (Note that we have added a prime symbol to the R' variable to distinguish the variable of integration from the radius of the Gaussian surface.) This equation yields $Q_{enc} = (4\pi/3)R^3\rho_v$, which when set equal to Eq. (1.35) and solved for E_R results in

$$E_R(R) = \frac{\rho_v R}{3\varepsilon_0}.$$

So inside the spherical charge, the field strength grows linearly with radial distance R. The complete solution for $E_R(R)$ is then, gathering the results,

$$E_R(R) = \begin{cases} \rho_v R/3\varepsilon_0 & R < a \\ \rho_v a^3/3\varepsilon_0 R^2 & R > a \end{cases}. \tag{1.36}$$

This result is plotted in Fig. 1.16. Note that $E_R(R)$ is continuous everywhere, including at $R = a$.

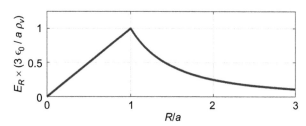

FIGURE 1.16 A plot of E_R versus R for the radial field produced by the charged sphere shown in Fig. 1.15, as given in Eq. (1.36).

Notice that, while we worked this problem for a uniform charge density ρ_v, this condition is not required for application of Gauss' Law. For a spherical distribution, we do require spherical symmetry (i.e. we require that a rotation about the origin leaves the charge distribution unchanged). So in order to successfully apply Gauss' Law to find the solution, the charge density cannot depend upon the angles θ or ϕ, but it could vary with R.

In the previous example, the integral $\oint_s \mathbf{E} \cdot d\mathbf{s}$ was evaluated in order to determine the integrand $\mathbf{E}$. To accomplish this, we had to invoke symmetry arguments to argue that only certain components of $\mathbf{E}$ are present (E_R), and that this component depends only on specific variables (R, but not θ or ϕ). In some cases we can use these same types of arguments to conclude that $\mathbf{E} \cdot d\mathbf{s}$ vanishes on some surfaces. The following example illustrates this.

Example 1.8 Electric Field due to Tube of Charge

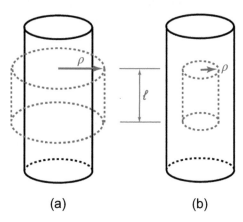

FIGURE 1.17 A long tube of radius a and surface charge density ρ_s. The dashed red lines indicate the Gaussian surface that we construct in order to determine $E_\rho(\rho)$ for (a) $\rho > a$, and (b) $\rho < a$.

(a)   (b)

Consider an infinitely long, hollow tube of radius a and uniform surface charge density ρ_s. Use Gauss' Law to determine the electric field at a distance ρ from the axis of the tube.

Solution:

Using a symmetry argument similar to that of the previous example, we assert that the electric field $\mathbf{E}$ produced by the infinitely long tube of charge must be directed radially outward (in the $\mathbf{a}_\rho$-direction), and that E_ρ must be independent of ϕ and z. We can therefore write $\mathbf{E} = \mathbf{a}_\rho E_\rho(\rho)$. Then we construct a Gaussian surface as shown in Fig. 1.17. The Gaussian surface of length ℓ and radius $\rho > a$, as shown in Fig. 1.17(a), is used to determine the field outside the tube, and the Gaussian surface of length ℓ and radius $\rho < a$, as shown in Fig. 1.17(b), to determine the field inside the tube. We start by evaluating $\oint_s \mathbf{E} \cdot d\mathbf{s}$ over the Gaussian surface. This part of our analysis is valid for $\rho > a$ as well as $\rho < a$. The Gaussian surface shown has three components: the top, the bottom, and the cylindrical wall. The surface elements for the top and bottom surfaces are $d\mathbf{s} = \mathbf{a}_z \rho \, d\rho \, d\phi$ and $d\mathbf{s} = -\mathbf{a}_z \rho \, d\rho \, d\phi$, respectively. In either case, $d\mathbf{s}$ is perpendicular to $\mathbf{E}$, so $\mathbf{E} \cdot d\mathbf{s} = \mathbf{a}_\rho E_\rho(\rho) \cdot (\pm \mathbf{a}_z \rho \, d\rho \, d\phi) = 0$. Only the surface flux through the cylindrical wall of the Gaussian surface contributes. For this surface, the surface element is $d\mathbf{s} = \mathbf{a}_\rho \rho \, d\phi \, dz$, and the electric flux becomes

$$\Phi_E = \oint_s \mathbf{E} \cdot d\mathbf{s} = \int_{z=0}^{\ell} \int_{\phi=0}^{2\pi} \left(\mathbf{a}_\rho E_\rho(\rho) \right) \cdot \mathbf{a}_\rho \rho \, d\phi \, dz$$

$$= E_\rho(\rho) \rho \int_{z=0}^{\ell} \int_{\phi=0}^{2\pi} d\phi \, dz,$$

where $E_\rho(\rho)$ and ρ can be pulled outside the integral since they are independent of ϕ and z. Thus the flux is

$$\Phi_E = \oint_s \mathbf{E} \cdot d\mathbf{s} = E_\rho(\rho) \, 2\pi\rho\ell. \tag{1.37}$$

Next we turn our attention to finding Q_{enc}, the charge enclosed within the Gaussian surface. For $\rho > a$, the enclosed charge is

$$Q_{enc} = \int_s \rho_s \, ds = \int_{z=0}^{\ell} \int_{\phi=0}^{2\pi} \rho_s a \, d\phi \, dz = \rho_s 2\pi a\ell,$$

where the integral is computed over the surface of the tube. For $\rho < a$, the Gaussian surface is inside the tube walls, and Q_{enc} is zero. Setting the surface flux in Eq. (1.37) equal to Q_{enc}/ε_0 and solving for E_ρ leads to

$$E_\rho(\rho) = \begin{cases} 0 & \rho < a \\ a\rho_s/\varepsilon_0\rho & \rho > a \end{cases}. \tag{1.38}$$

This result is plotted in Fig. 1.18. Notice that the electric field outside the cylindrical shell is identical to the electric field due to a line charge of linear charge density $\rho_\ell = 2\pi a\rho_s$. ($2\pi a$ is, of course, the circumference of the tube.)

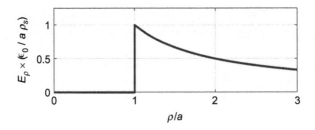

FIGURE 1.18 A plot of E_ρ versus ρ for the radial field given in Eq. (1.38), produced by the long tube of charge shown in Fig. 1.17.

We have one final example to work illustrating the application of Gauss' Law to determine $\mathbf{E}$. Before we do so, let's review the steps that we have followed, so that we can recognize each step in the following solution.

1. First, use the symmetry of the charge distribution in order to argue that only certain field components exist, and that these components can depend only on a limited set of variables.

2. Next, construct a Gaussian surface consisting of surfaces over which the field must be uniform, or others whose contribution to the total surface integral is zero (because $\mathbf{E} = 0$ or $\mathbf{E} \cdot d\mathbf{s} = 0$).

3. Finally, apply Gauss' Law in order to determine $\mathbf{E}$.

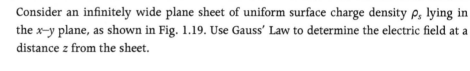

Example 1.9 Electric Field due to Infinite Plane Sheet of Charge

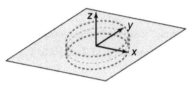

FIGURE 1.19 The blue surface is a large sheet of charge density ρ_s. The dashed red lines indicate the Gaussian surface constructed for the determination of $E_z(z)$, where z is the direction normal to the surface. The two flat faces of the Gaussian surface are of area Δs, and are parallel to the sheet of charge; one a distance z above the charge, the other the same distance z below.

Consider an infinitely wide plane sheet of uniform surface charge density ρ_s lying in the x–y plane, as shown in Fig. 1.19. Use Gauss' Law to determine the electric field at a distance z from the sheet.

Solution:

The solution for the electric field due to this infinite sheet of charge is similar to that of the previous two examples. For the infinite sheet of charge lying in the x–y plane, the field can point only in the z-direction, and the magnitude of E_z can depend only on z. These conclusions follow from the observation that the sheet is symmetric upon rotation about any axis parallel to the z-axis. If we presume that components E_x and E_y exist, and we rotate the plane 180° about this axis, then E_x upon rotation becomes $-E_x$, and E_y becomes $-E_y$. But the sheet of charge is unchanged; it still extends uniformly to infinity in the x- and y-directions. Therefore field components E_x and E_y must be zero, and only E_z remains. Similarly, if we consider two points equidistant from the plane, but at different positions x and y, then the field must be unchanged, since in either case the sheet of charge extends to infinity in either direction. Therefore E_z can depend only on the distance z, and we can write $\mathbf{E} = \mathbf{a}_z E_z(z)$. Finally, we expect $E_z(-z)$ to be equal to $-E_z(+z)$. In other words, $\mathbf{E}$ either points *toward* the surface in both regions (for $\rho_s < 0$), or it points *away* from the surface in both regions (for $\rho_s > 0$).

We construct the Gaussian surface shown in Fig. 1.19. This surface consists of two faces of area Δs, parallel to and a distance $\pm z$ from the charge sheet, and a side wall joining the two faces. The precise shape of the faces is not important. The elemental surface $d\mathbf{s}$ for the side wall points in a direction perpendicular to the z-direction, so $\mathbf{E} \cdot d\mathbf{s} = 0$ for this surface. Since the surface element $d\mathbf{s}$ always points outward, $d\mathbf{s} = \mathbf{a}_z ds$ for $z > 0$ and $d\mathbf{s} = -\mathbf{a}_z ds$ for $z < 0$. Therefore $\mathbf{E} \cdot d\mathbf{s}$ on the two faces has the same sign. The electric flux is

$$\Phi_E = \oint_s \mathbf{E} \cdot d\mathbf{s} = \int_s \mathbf{a}_z E_z(z) \cdot \mathbf{a}_z ds + \int_s \mathbf{a}_z E_z(-z) \cdot (-\mathbf{a}_z ds)$$
$$= E_z(z)\Delta s - E_z(-z)\Delta s$$
$$= 2E_z(z)\Delta s.$$

The charge enclosed within the Gaussian surface lies entirely in the surface, and is $Q_{enc} = \int_s \rho_s \, ds = \rho_s \, \Delta s$; so completing Gauss' Law, the electric field is

$$E_z(z) = \begin{cases} -\rho_s/2\varepsilon_0 & z < 0 \\ \rho_s/2\varepsilon_0 & z > 0 \end{cases} . \tag{1.39}$$

This result is plotted in Fig. 1.20. Notice that, aside from the sign difference on opposite sides of the sheet of charge, the electric field for an infinite sheet of charge is independent of distance z from the sheet.

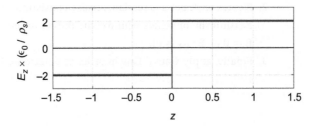

FIGURE 1.20 A plot of E_z versus z for the field produced by the infinite sheet of charge shown in Fig. 1.19, given by Eq. (1.39).

Gauss' Law can be used to argue that the electric field inside a hollow conducting enclosure, known as a Faraday Cage, must be zero. This is described in TechNote 1.4.

TechNote 1.4 Faraday Cage

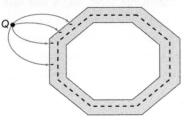

FIGURE 1.21 A point charge Q outside a conducting enclosure induces charges of the opposite sign on the exterior of the conductor, but no charges on the interior.

A Faraday cage, invented by Michael Faraday in 1836, is a conducting enclosure that isolates the interior region from electric fields. Electric fields outside the Faraday cage will induce a charge distributed on the exterior of the cage, with no effect whatsoever on the interior of the cage. For example, if an external, positively charged object approaches the cage from one side, as shown in Fig. 1.21, it induces a negative charge on the adjacent outside surface of the Faraday cage. The induced charge is drawn from a ground reservoir if the cage is grounded, or from the opposite outside surface if the cage is isolated from ground. No charge, however, appears on the inside surface of the enclosure. We can understand this effect by considering the Gaussian surface, shown as the dashed line in the figure. The Gaussian surface is constructed entirely inside the conductor, and is a three-dimensional figure completely enclosing the interior. We will argue in Section 1.5 that the electric field within a conductor must be zero, so that the total electric flux passing through this surface must also be zero. By Gauss' Law, the total charge inside the Gaussian surface is zero as well. That is, electric charge exists only on the outside surface of the conductor, not on the inside surface. With no charge on the inside surface, there can be no electric field inside the enclosure.

A Faraday cage can be constructed from a conducting screen, with the result that wavelengths of electromagnetic waves that are larger than the openings in the screen are blocked, while shorter wavelengths transmit to the interior. Thus, inside the Faraday cage, reception of your favorite radio station is not possible, but you would be able to clearly see outside, since the sub-micron wavelength of visible radiation is much smaller than the openings in the screen, while the long wavelength radio wave is not.

In this section we have applied the integral form of Gauss' Law in order to determine the electric field E. We can only do this in geometries which are sufficiently symmetric, as we must be able to argue that the electric field is uniform across the Gaussian surface that we construct, or that the electric field is zero or lies in the surface such that $\mathbf{E} \cdot d\mathbf{s}$ is zero. So the cases in which we can apply this are somewhat limited. On the other hand, these conditions are often satisfied, and when we can use Gauss' Law for the solution, the savings in effort over other techniques is considerable.

Before we leave this section, recall the trend that we discussed in Section 1.1 regarding the spatial dependence of E for the spherical, cylindrical, and planar geometries. The solutions that we derived in this section show the same behavior: E outside the spherical charge distribution, or a point charge, diminishes as R^{-2}, while for a cylindrical distribution E decreases as ρ^{-1}, and for the planar distribution as z^0 (i.e. independent of z). In other words, the symmetry of the charge determines the power law for how the electric field depends on distance.

1.4 Electric Potential

To this point we have discussed two different techniques for determining the electric field E produced by a set of discrete charges or a continuous distribution of charges: Coulomb's Law and Gauss' Law. In this section we develop an additional technique that involves a new scalar field quantity known as the **electric potential**. We show in Appendix E.2 (SideNote E.1) that an irrotational vector field (i.e. a vector field whose curl is zero) can always be written as the gradient of a scalar function. In addition, we showed in Section 1.2 that $\nabla \times \mathbf{E}$ for any static electric field E must be zero. Therefore, a scalar potential function V exists such that

$$\mathbf{E} = -\boldsymbol{\nabla} V. \tag{1.40}$$

For reasons that will become clear in a few minutes, this scalar field is called the electric potential. (We hope you will forgive the use of the symbol V for potential. This is common notation, but we must avoid confusion with volume, for which we have been using the lower case v. The distinction should also be clear from the context in which it is used, but if you are a student who searches for the formula with the correct symbols to plug into, instead of understanding the physics of the problem, this confusion will likely lead to difficulties.)

So we know that the potential V must exist, but we do not have any notion as to the physical significance of V, nor do we know how to determine this potential. Before exploring the answers to these issues, we address the minus sign in Eq. (1.40). This is standard notation, and since we only know that a scalar field whose gradient is E exists, we are free to call this scalar field $-V$. Including the minus sign in Eq. (1.40) in this way is consistent with the idea that electric field lines originate on positive charges and terminate on negative charges, and that the potential near positive charges is positive, and the potential near negative charges is negative.

Let us now investigate the physical significance of the potential V, which we will do by considering the work required to move a charge q against an electric field E along a path, as shown in Fig. 1.22. The work done in moving an object against a force in this manner is (see Eq. (C.1))

$$W = -\int_A^B \mathbf{F} \cdot d\boldsymbol{\ell},$$

the units of which are $[W] = \text{J}$, or joules. In the present discussion, the charge is moved against the force applied by an electric field E, which is of the form Eq. (1.2), $\mathbf{F} = q\mathbf{E}$. (Notice that in Fig. 1.22, the force F, at least at one point along the pathway, appears to be directed in somewhat the same direction as $d\boldsymbol{\ell}$, so that $\mathbf{F} \cdot d\boldsymbol{\ell} > 0$. If this holds true along most of the pathway, we would expect that the work $W < 0$ – that is, the charge moves to a position of lower potential energy! This situation is analogous to moving an object down a hillside, for which $d\boldsymbol{\ell}$ points down the hill. The gravitational force clearly

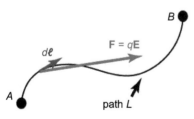

FIGURE 1.22 A path illustrating the calculation of the work required to move a charge from point A to point B.

points down the hill, and so here $\mathbf{F} \cdot d\boldsymbol{\ell} > 0$ as well. And you will recall that the potential energy of the object is greater at the top of the hill than at the bottom.) The work done in moving the charged object from A to B is

$$W = -q \int_A^B \mathbf{E} \cdot d\boldsymbol{\ell}, \tag{1.41}$$

or, using $\mathbf{E} = -\boldsymbol{\nabla}V$,

$$W = q \int_A^B \boldsymbol{\nabla}V \cdot d\boldsymbol{\ell}.$$

Now this is a very interesting result, which can be seen by expanding $\boldsymbol{\nabla}V$ and $d\boldsymbol{\ell}$, and explicitly determining the scalar product $\boldsymbol{\nabla}V \cdot d\boldsymbol{\ell}$. We work this in rectangular coordinates, although any system will do. Straight from the definition of the gradient of a scalar function (in rectangular coordinates), the gradient of the potential is

$$\boldsymbol{\nabla}V = \mathbf{a}_x \frac{\partial V}{\partial x} + \mathbf{a}_y \frac{\partial V}{\partial y} + \mathbf{a}_z \frac{\partial V}{\partial z},$$

and the path element $d\boldsymbol{\ell}$ is defined as $d\boldsymbol{\ell} = \mathbf{a}_x\, dx + \mathbf{a}_y\, dy + \mathbf{a}_z\, dz$. The dot product $\boldsymbol{\nabla}V \cdot d\boldsymbol{\ell}$ is then

$$\boldsymbol{\nabla}V \cdot d\boldsymbol{\ell} = \frac{\partial V}{\partial x}\, dx + \frac{\partial V}{\partial y}\, dy + \frac{\partial V}{\partial z}\, dz,$$

and the work done in moving the charged body is

$$W = q \int_A^B \left(\frac{\partial V}{\partial x}\, dx + \frac{\partial V}{\partial y}\, dy + \frac{\partial V}{\partial z}\, dz \right).$$

But since the potential is a function only of position (i.e. $V = V(x, y, z)$), the incremental change in potential can be written in terms of partial derivatives,

$$dV = (\partial V/\partial x)\, dx + (\partial V/\partial y)\, dy + (\partial V/\partial z)\, dz,$$

and the work is simply

$$W = q \int_A^B dV.$$

The integral of a perfect differential dV is simply the difference between the integrand evaluated at the two end points,

$$W = q\,(V_B - V_A). \tag{1.42}$$

This equation implies that the work done in moving the charged body from point A to point B depends only on the potential difference between those two end points, times the charge q on the body. The work does not depend on the path taken to move from A to B, only the potential V at the end points. We introduced this idea earlier (Section 1.2), a bit less rigorously, as a characteristic of a conservative force, which is a direct result of the irrotational nature of the static electric field.

Comparing Eqs. (1.41) and (1.42) gives us

$$\boxed{V_B - V_A = -\int_A^B \mathbf{E} \cdot d\boldsymbol{\ell},} \tag{1.43}$$

which allows us to determine the potential difference between two points A and B using the electric field $\mathbf{E}$ along any path joining those two points. Notice that the potential V at a point is always referenced to the potential at another point. For example, we often assign a ground plane where $V = 0$, or commonly we will assign a $V = 0$ reference as a point an infinite distance away.

Example 1.10 Potential Difference between Two Points Near an Infinite Line Charge

We showed in Example 1.5 that the electric field generated by an infinitely long line charge of uniform linear charge density ρ_ℓ is $\mathbf{E} = \mathbf{a}_\rho E_\rho(\rho)$, where

$$E_\rho(\rho) = \frac{\rho_\ell}{2\pi\varepsilon_0\,\rho}.$$

(See Eq. (1.23).) (a) Determine the potential V_B at position $\mathbf{r}_B = (0, 1, 0)$ cm relative to the potential V_A at position $\mathbf{r}_A = (2, 0, 0)$ cm, for a charge density $\rho_\ell = 2.0$ nC/m. (b) Use the result of part (a) to determine the work required to move a 1 μC charge from $\mathbf{r}_A$ to $\mathbf{r}_B$.

Solution:

(a) Equation (1.43) relates the potential difference $V_B - V_A$ to the path integral of the electric field. We must first choose the path over which to perform the integration. Remember that the potential difference between these two points does not depend on the path we choose, so we should choose a path that makes our calculation easy. Since $\mathbf{r}_A$ points along the x-axis, and $\mathbf{r}_B$ is along the y-axis, we choose the path inward on the x-axis to $(1, 0, 0)$ cm, which we'll label $\mathbf{r}_C$, and then follow a circular path of radius 1 cm from $\mathbf{r}_C$ to $\mathbf{r}_B$. Then,

$$V_B - V_A = -\int_A^B \mathbf{E} \cdot d\boldsymbol{\ell}$$
$$= -\int_A^C \mathbf{E} \cdot d\boldsymbol{\ell} - \int_C^B \mathbf{E} \cdot d\boldsymbol{\ell},$$

where we have broken the path integral into two parts; one for the segment from $\mathbf{r}_A$ to $\mathbf{r}_C$, and the second for the segment from $\mathbf{r}_C$ to $\mathbf{r}_B$.

For the first segment, the path is purely radial, and $d\boldsymbol{\ell} = \mathbf{a}_\rho d\rho$. Then

$$-\int_A^C \mathbf{E} \cdot d\boldsymbol{\ell} = -\int_{\rho_A}^{\rho_B} \frac{\rho_\ell}{2\pi\varepsilon_0\,\rho} \mathbf{a}_\rho \cdot \mathbf{a}_\rho d\rho$$
$$= -\frac{\rho_\ell}{2\pi\varepsilon_0} \ln(\rho)\Big|_{\rho_A}^{\rho_B} = \frac{\rho_\ell}{2\pi\varepsilon_0} \ln\left(\frac{\rho_A}{\rho_B}\right).$$

For the second segment, the potential difference is zero, since on this segment $d\boldsymbol{\ell} = \mathbf{a}_\phi \rho d\phi$, which is perpendicular to E. Therefore the potential difference $V_B - V_A$ is

$$V_B - V_A = \frac{\rho_\ell}{2\pi\varepsilon_0} \ln\left(\frac{\rho_A}{\rho_B}\right) = \frac{2.0 \text{ nC/m}}{2\pi(8.854 \text{ pF/m})} \ln(2) = 25 \text{ V}.$$

Note that the potential at $\mathbf{r}_B$ is +25 V greater than that at $\mathbf{r}_A$ since this point is closer to the positive line charge.

(b) To find the work required to move a 1 μC charge from $\mathbf{r}_A$ to $\mathbf{r}_B$, we use Eq. (1.42):

$$W = q(V_B - V_A) = (1 \text{ }\mu\text{C})(25 \text{ V}) = 25 \text{ }\mu\text{J}.$$

Since the potential V_B is greater than the potential V_A, we must do positive work to move the positive charge from $\mathbf{r}_A$ to $\mathbf{r}_B$.

We would be remiss if we didn't mention units of Eq. (1.42). If we move a 1 C charge (which is an enormous charge by most standards) across a 1 V potential difference, we must expend 1 J of work. The unit of potential is the volt, abbreviated V, and is equivalent to 1 J/C.

We next explore a little further the analogy between the electric potential, which is related to potential energy of a charge, and the gravitational potential. When determining the potential due to some distribution of charges, we often encounter surfaces over which the electric potential does not vary. (Later, when we discuss conductors, we will show that for static fields the potential of a conductor must be uniform, for example.) These surfaces are called **equipotential surfaces**. There is a useful analogy between these equipotential surfaces, or lines when we view a two-dimensional projection of the surfaces, and the contours on a contour map used for illustrating the elevation of some region of geographic interest. The contour lines indicate lines of constant elevation. When hiking along a path that follows a contour line, a hiker would stay on a level course, neither gaining nor losing elevation. Therefore their gravitational potential energy would remain constant. If she/he would turn by 90° to face the summit of the hill, and start walking in this direction, the hiker would be ascending the hillside at the maximum possible slope. This is the direction given by the gradient of the potential energy. A small spacing between the contours on a contour map indicates a slope that is very steep, while large spacing between contours indicates that the slope of the hillside is rather gentle. These observations about the contour lines and the gradient apply directly to equipotentials and their gradients as well.

Now that we have developed at least the beginning of an understanding of the physical significance of the electric potential, let's look at the electric potential for some specific charge distributions. We start, of course, with the point charge, and ask the following question: What is the electric potential at the point P, a distance R from a single point charge Q that is located at the origin? This is illustrated in Fig. 1.23. We already know the vector electric field due to a point charge,

$$\mathbf{E}(\mathbf{r}) = \frac{Q\,\mathbf{a}_R}{4\pi\varepsilon_0\,R^2},$$

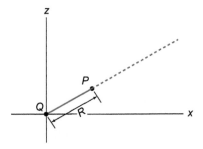

FIGURE 1.23 The dotted line illustrates the path over which $\mathbf{E} \cdot d\boldsymbol{\ell}$ is evaluated to determine the potential at point P due to the charge Q at the origin, relative to the potential $V = 0$ at $R = \infty$.

so using Eq. (1.43) we should be able to find the potential at point P. We need to choose a reference point for this determination, which we choose as $V = 0$ at $R = \infty$. We must also choose a path over which to carry out the integration. As we showed earlier, the potential $V(\mathbf{r})$ at a point does not depend on the path followed to get there, so we are free to choose the path that is most convenient for us. The simplest path is the radial path, coming directly along a straight line from ∞ to the point P a distance R from the charge Q. This path is shown as the dotted line in Fig. 1.23. Then the path element is $d\boldsymbol{\ell} = \mathbf{a}_R\,dR$, and

$$V(R) = -\int_{\infty}^{R} \left(\frac{Q\,\mathbf{a}_R}{4\pi\varepsilon_0\,R^2} \right) \cdot \mathbf{a}_R\,dR$$

$$= -\frac{Q}{4\pi\varepsilon_0} \int_{\infty}^{R} \frac{dR}{R^2} = -\frac{Q}{4\pi\varepsilon_0} \left(\frac{-1}{R} \right) \Bigg|_{\infty}^{R} = \frac{Q}{4\pi\varepsilon_0\,R}. \tag{1.44}$$

This potential for a point charge may appear to have a form that is similar to that of the electric field, but there are some very important distinctions, and we must not confuse them. First, the potential is a scalar, rather than a vector, in that it has no associated direction. And second, the potential drops off with increasing R as R^{-1}, rather than the R^{-2} dependence of $\mathbf{E}$.

Before moving on, we should write this potential in a slightly different form. In deriving Eq. (1.44) we placed the charge at the origin. More generally, the potential at position $\mathbf{r}$ due to a point charge at a location specified by $\mathbf{r}'$ is

$$V(\mathbf{r}) = \frac{Q}{4\pi\varepsilon_0 \, |\mathbf{r} - \mathbf{r}'|}. \tag{1.45}$$

In order to determine the potential resulting from a distribution of discrete charges, we can use the principle of superposition and simply sum the potentials due to the individual charges within the ensemble. We demonstrate this by applying superposition for the electric field $\mathbf{E}$. Applying Eq. (1.43), the potential due to N discrete charges is

$$V(\mathbf{r}) = -\int_A^B \mathbf{E} \cdot d\boldsymbol{\ell} = -\int_A^B \sum_i^N \mathbf{E}_i \cdot d\boldsymbol{\ell}.$$

Since the fields are well-behaved functions, the order of summation and integration is interchangeable, so

$$V(\mathbf{r}) = -\sum_i^N \int_A^B \mathbf{E}_i \cdot d\boldsymbol{\ell},$$

and since each integral in the sum is just $-V_i(\mathbf{r})$ due to just one of the charges i,

$$V(\mathbf{r}) = \sum_i^N V_i(\mathbf{r}),$$

indicating that the potential obeys superposition. For example, if we have N discrete point charges Q_i, where i is an index that labels the individual charges, with each individual charge located at a position specified by $\mathbf{r}_i'$, then the potential at position $\mathbf{r}$ given by Eq. (1.45) is easily extended as

$$V(\mathbf{r}) = \frac{1}{4\pi\varepsilon_0} \sum_{i=1}^N \frac{Q_i}{|\mathbf{r} - \mathbf{r}_i'|}. \tag{1.46}$$

Example 1.11 Potential due to Discrete Charges

Consider the three discrete charges described in Example 1.2 for the calculation of the electric field. Determine the potential $V(\mathbf{r})$ due to these charges, and from this potential, the electric field $\mathbf{E}(\mathbf{r})$ at the point $\mathbf{r} = (2, 3, 0)$.

Solution:
We apply Eq. (1.46) to find the potential at an arbitrary point (x, y, z). Since we will need to take the spatial derivative of the potential to find $\mathbf{E}$, we will evaluate $V(\mathbf{r})$ at an arbitrary location $\mathbf{r} = (x, y, z)$ first. Using the source points $\mathbf{r}_1' = (2, 0, 0)$, $\mathbf{r}_2' = (0, 3, 0)$, and $\mathbf{r}_3' = (-2, 0, 0)$, the potential $V(\mathbf{r})$ is

$$V(\mathbf{r}) = \frac{1}{4\pi\varepsilon_0}\left(\frac{9e}{\sqrt{(x-2)^2 + y^2 + z^2}} + \frac{2e}{\sqrt{x^2 + (y-3)^2 + z^2}} + \frac{-25e}{\sqrt{(x+2)^2 + y^2 + z^2}} \right).$$

The equipotential contours, that is lines of constant electric potential described by this function, are plotted in Fig. 1.24. Near the negative charge at $\mathbf{r}_3'$, the potential surface gets large and negative (blue and green contours), forming circles around the charge. Near the positive charges, the equipotential lines increase (positive potentials are yellow and red), and reach peaks at the charges. In either case, the shapes of the equipotentials become circular near the individual charges, centered on those charges, since the potential is dominated by that one charge.

FIGURE 1.24 The equipotential lines for the potential landscape produced by the three point charges of Example 1.2. The potentials for a few of the lines, in units of $e/4\pi\varepsilon_0\,nm$, are labeled in the figure. Potentials close to the positive charges are positive, while those close to the negative charges are negative. Not shown are additional contours of greater positive potential around charges q_1 and q_2, and additional contours of greater negative potential around the q_3 charge. The red arrows, which are perpendicular to the equipotential lines, indicate the electric field vectors, and are strongest in regions where the equipotentials are closely spaced.

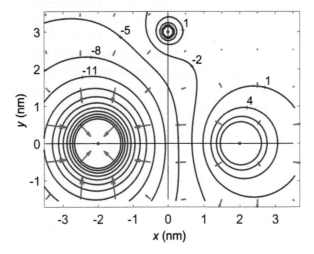

To find the electric field **E**, we must determine the gradient of the potential function,

$$\mathbf{E}(\mathbf{r}) = -\boldsymbol{\nabla}V(\mathbf{r}) = \frac{-e}{4\pi\varepsilon_0}\left\{9\left(-\frac{1}{2}\right)\frac{2(x-2)\mathbf{a}_x + 2y\mathbf{a}_y + 2z\mathbf{a}_z}{\left[(x-2)^2 + y^2 + z^2\right]^{3/2}}\right.$$
$$+ 2\left(-\frac{1}{2}\right)\frac{2x\mathbf{a}_x + 2(y-3)\mathbf{a}_y + 2z\mathbf{a}_z}{\left[x^2 + (y-3)^2 + z^2\right]^{3/2}}$$
$$\left. - 25\left(-\frac{1}{2}\right)\frac{2(x+2)\mathbf{a}_x + 2y\mathbf{a}_y + 2z\mathbf{a}_z}{\left[(x+2)^2 + y^2 + z^2\right]^{3/2}}\right\}.$$

This equation can be simplified only slightly by canceling factors of 2, so we won't bother to rewrite that here, but rather jump directly to the evaluation of **E** at $\mathbf{r} = (2, 3, 0)$, where we find

$$\mathbf{E}(2, 3, 0) = \frac{-e}{4\pi\varepsilon_0}\left\{-9\,\frac{3\mathbf{a}_y}{[3^2]^{3/2}} - 2\,\frac{2\mathbf{a}_x}{[2^2]^{3/2}} + 25\,\frac{4\mathbf{a}_x + 3\mathbf{a}_y}{[4^2 + 3^2]^{3/2}}\right\}$$
$$= \frac{-e}{4\pi\varepsilon_0}\left\{-\mathbf{a}_y - \frac{1}{2}\mathbf{a}_x + \frac{4\mathbf{a}_x + 3\mathbf{a}_y}{5}\right\}$$
$$= \frac{-e}{4\pi\varepsilon_0}\left\{\frac{3}{10}\mathbf{a}_x - \frac{4}{10}\mathbf{a}_y\right\}$$

This result is consistent with our result found in Example 1.2.

A few words of caution are in order at this point. We have been discussing how to determine the electric potential $V(\mathbf{r})$ due to a distribution of charges. If our ultimate goal is to use $V(\mathbf{r})$ to determine the electric field $\mathbf{E}(\mathbf{r})$ at a specific point by computing $-\boldsymbol{\nabla}V(\mathbf{r})$, we must be careful that we first find the potential $V(\mathbf{r})$ *as a function of* the spatial coordinates. If we were to, instead, evaluate the potential V at the specific location $\mathbf{r}$, this would not be sufficient, as we would not be able to find the gradient of this function. We have seen many students succumb to this temptation, and reach the conclusion that the electric field is zero because when one takes the derivative of the value of the potential at that point, rather than of the function $V(\mathbf{r})$, the result is, of course, zero. Also, do not make the mistake of concluding that if $V = 0$ at a point, then $\mathbf{E} = 0$ at that point as well. This is not correct. The electric field is related to how rapidly (in space) the potential V is changing, and the potential can be zero and still be strongly varying at a particular point. We can see examples of this along the $V = 0$ equipotential line, which runs between two of the blue–green lines in Fig. 1.24. The potential V is zero for any point on this contour, yet the electric field is not.

Now that we have found the potential $V(\mathbf{r})$ due to a single point charge, and then a number of discrete charges, we extend this to write the electric potential due to continuous distributions of charge, given by the volume charge density ρ_v, the surface charge density ρ_s, or the linear charge density ρ_ℓ. We will derive the first of these, but then assert that the derivation of the second and third follow along similar lines and present only the result. The development follows directly from our previous discussion of the potential due to a finite number of discrete charges. We consider some region in space v' in which there is a charge density $\rho_v(\mathbf{r}')$. This volume can be divided into many small elements, each of volume $\Delta v_i'$, containing a charge of $q_i = \Delta v_i' \, \rho_v(\mathbf{r}_i')$. By applying Eq. (1.46), the electric potential $V(\mathbf{r})$ at location $\mathbf{r}$ can be found by the following summation:

$$V(\mathbf{r}) = \sum_i \frac{1}{4\pi\varepsilon_0} \frac{q_i}{|\mathbf{r} - \mathbf{r}_i'|} = \sum_i \frac{1}{4\pi\varepsilon_0} \frac{\left[\Delta v_i' \, \rho_v(\mathbf{r}_i')\right]}{|\mathbf{r} - \mathbf{r}_i'|}.$$

In the limit as the volume elements become smaller, the potential is

$$V(\mathbf{r}) = \lim_{\Delta v_i' \to 0} \sum_i \frac{1}{4\pi\varepsilon_0} \frac{\rho_v(\mathbf{r}_i')}{|\mathbf{r} - \mathbf{r}_i'|} \, \Delta v_i'$$

or

$$V(\mathbf{r}) = \frac{1}{4\pi\varepsilon_0} \int_v \frac{\rho_v(\mathbf{r}')}{|\mathbf{r} - \mathbf{r}'|} \, dv'. \tag{1.47}$$

Thus we can determine the scalar electric potential by integrating over the volume of the charge distribution. The derivation for the potential due to a surface distribution of charge density ρ_s follows along similar lines, leading to

$$V(\mathbf{r}) = \frac{1}{4\pi\varepsilon_0} \int_s \frac{\rho_s(\mathbf{r}')}{|\mathbf{r} - \mathbf{r}'|} \, ds', \tag{1.48}$$

and for a linear distribution of density $\rho_\ell(\mathbf{r}')$,

$$V(\mathbf{r}) = \frac{1}{4\pi\varepsilon_0} \int_\ell \frac{\rho_\ell(\mathbf{r}')}{|\mathbf{r} - \mathbf{r}'|} \, d\ell'. \tag{1.49}$$

Example 1.12 Potential due to a Charged Sphere

A spherical shell having a uniform surface charge density ρ_s and radius a is shown in Fig. 1.25. Determine the potential $V(\mathbf{r})$ at point P (which may be inside or outside the sphere) due to this charged shell.

FIGURE 1.25 A spherical shell of uniform surface charge density ρ_s.

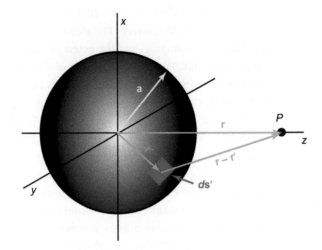

Solution:

Since the charge is a surface charge, we should use Eq. (1.48) to find the potential. For simplicity, let the shell be centered at the origin, and let the point P lie along the z-axis. We are allowed to make the latter statement since the shell is spherically symmetric, so that the potential can depend only on the distance R from the center of the sphere, $V(\mathbf{r}) = V(R)$. The coordinates of the point P are $\mathbf{r} = (0, 0, R)$. The location of an elemental patch on the surface of the sphere is given by $\mathbf{r'} = (x', y', z') = (a \sin \theta' \cos \phi', a \sin \theta' \sin \phi', a \cos \theta')$. The surface element on the spherical shell is $ds' = a^2 \sin \theta' \, d\theta' \, d\phi'$, and the distance $|\mathbf{r} - \mathbf{r'}|$ is

$$|\mathbf{r} - \mathbf{r'}| = \left[(a \sin \theta' \cos \phi')^2 + (a \sin \theta' \sin \phi')^2 + (R - a \cos \theta')^2 \right]^{1/2}.$$

Using $\cos^2 \theta' + \sin^2 \theta' = 1$ and $\cos^2 \phi' + \sin^2 \phi' = 1$, this reduces to

$$|\mathbf{r} - \mathbf{r'}| = \left[a^2 + R^2 - 2aR \cos \theta' \right]^{1/2}.$$

Inserting these into Eq. (1.48), the potential is

$$V(R) = \frac{1}{4\pi\varepsilon_0} \int_{\phi=0}^{2\pi} \int_{\theta=0}^{\pi} \frac{\rho_s}{[a^2 + R^2 - 2aR \cos \theta']^{1/2}} a^2 \sin \theta' \, d\theta' \, d\phi'$$

$$= \frac{1}{4\pi\varepsilon_0} \frac{q}{2} \int_{\theta=0}^{\pi} \frac{\sin \theta' \, d\theta'}{[a^2 + R^2 - 2aR \cos \theta']^{1/2}},$$

where $q = 4\pi a^2 \rho_s$ is the total charge of the sphere. (When the surface charge density is uniform, the total charge is simply the charge density times the surface area.) Substituting $x = \cos \theta'$ and using the integral Eq. (G.4) from Appendix G, the potential is

$$V(R) = \frac{-q}{8\pi\varepsilon_0 \, a \, R} \left[|a - R| - (a + R) \right] = \begin{cases} \dfrac{q}{4\pi\varepsilon_0 a} & R < a \\[2mm] \dfrac{q}{4\pi\varepsilon_0 R} & R > a \end{cases}. \qquad (1.50)$$

Note that the potential is constant inside the shell, and decreases as $1/R$ outside. This function is plotted in Fig. 1.26. We can find $\mathbf{E}$ for this potential by computing $-\nabla V$. The constant potential for $R < a$ gives $E = 0$ in that region. The $1/R$ dependence of the potential gives a $1/R^2$ dependence of $\mathbf{E}$.

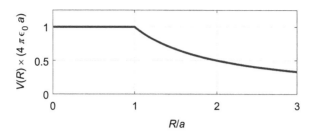

FIGURE 1.26 A plot of the potential $V(R)$, given in Eq. (1.50), resulting from a spherical shell of uniform charge density, as determined in Example 1.12.

Notice in this example that finding the potential $V(\mathbf{r})$ was relatively simple. This observation may not always hold true, but it is often the case that the integral that must be evaluated to find the potential is substantially easier than the integral for the electric field $\mathbf{E}(\mathbf{r})$. In addition, we must integrate only once, since the potential is a scalar, whereas the electric field, which is a vector, may require evaluation of three integrals, one for each spatial component of $\mathbf{E}(\mathbf{r})$. Of course, once the potential has been found, we still have the additional step of evaluating its gradient to find $\mathbf{E}(\mathbf{r})$. Even so, this can be the easier approach, and we will often employ it.

Example 1.13 Potential, Finite Line Charge

Determine the potential $V(\mathbf{r})$ due to a finite-length uniform line charge of linear charge density ρ_ℓ and of length L.

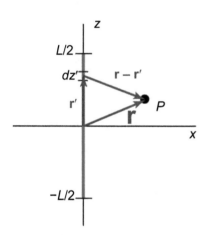

FIGURE 1.27 A finite-length line charge of uniform linear charge density ρ_ℓ and length L.

Solution:
The line charge, aligned with the z-axis and extending from $-L/2$ to $L/2$, and the field position $\mathbf{r}$, are shown in Fig. 1.27. The source point is $\mathbf{r}' = (0, 0, z')$, and for simplicity we let the field point P lie in the x–z plane at position $\mathbf{r} = (x, 0, z)$. We are free to make this choice because of the cylindrical symmetry of the charge distribution. Upon rotation about the z-axis, the potential $V(\mathbf{r})$ cannot change; that is, V does not depend on ϕ. Therefore the potential at $\mathbf{r} = (\rho, 0, z)$ (in cylindrical coordinates) must be the same as the potential at $\mathbf{r} = (x, 0, z)$ (in rectangular coordinates). The distance $|\mathbf{r} - \mathbf{r}'|$ is $[\rho^2 + (z - z')^2]^{1/2}$. Using Eq. (1.49), the potential at $\mathbf{r}$ is

$$V(\mathbf{r}) = V(\rho, z) = \frac{1}{4\pi\varepsilon_0} \int_{-L/2}^{L/2} \frac{\rho_\ell\, dz'}{[\rho^2 + (z - z')^2]^{1/2}}.$$

To evaluate this integral, we substitute $u = z' - z$, and use the integral Eq. (G.7) from Appendix G, resulting in

$$\begin{aligned}
V(\mathbf{r}) &= \frac{\rho_\ell}{4\pi\varepsilon_0} \int_{-L/2-z}^{L/2-z} \frac{du}{[\rho^2 + u^2]^{1/2}} \\
&= \frac{\rho_\ell}{4\pi\varepsilon_0} \ln\left(u + \sqrt{u^2 + \rho^2}\right) \Big|_{-L/2-z}^{L/2-z} \\
&= \frac{\rho_\ell}{4\pi\varepsilon_0} \ln\left(\frac{(L/2 - z) + \sqrt{(L/2 - z)^2 + \rho^2}}{-(L/2 + z) + \sqrt{(L/2 + z)^2 + \rho^2}}\right).
\end{aligned}$$

(1.51)

This solution is valid for any value of z.

This result can be evaluated at $z = 0$, and simplified a bit, to show that the potential at the midpoint of the linear charge is

$$V(\rho) = \frac{\rho_\ell}{2\pi\varepsilon_0} \ln\left(\frac{L/2 + \sqrt{(L/2)^2 + \rho^2}}{\rho} \right). \tag{1.52}$$

We previously determined the electric field due to an infinitely long line charge. Are these results consistent with one another? To compare these results, we should let $L \to \infty$, and determine the gradient of the potential (times -1). We do need to be careful, however, since as $L \to \infty$, the potential $V(\rho)$ of Eq. (1.52) diverges. (The natural logarithm diverges weakly as the argument gets large, but it does diverge.) The deleterious effect of this divergence can be avoided by taking the gradient of Eq. (1.52) first to find the electric field at the midpoint of a finite-length line charge, and then taking the limit of this field strength as $L \to \infty$. The result is

$$\mathbf{E}(\mathbf{r}) = \lim_{L\to\infty} \left(-\boldsymbol{\nabla} V(\mathbf{r}) \right)$$

$$= \lim_{L\to\infty} \left(-\frac{\rho_\ell}{2\pi\varepsilon_0} \frac{\partial}{\partial\rho} \left[\ln\left(L/2 + \sqrt{(L/2)^2 + \rho^2} \right) - \ln\rho \right] \mathbf{a}_\rho \right)$$

$$= -\frac{\rho_\ell}{2\pi\varepsilon_0} \lim_{L\to\infty} \left[\frac{1}{L/2 + \sqrt{(L/2)^2 + \rho^2}} \times \frac{1}{2} \times \frac{2\rho}{\sqrt{(L/2)^2 + \rho^2}} - \frac{1}{\rho} \right] \mathbf{a}_\rho.$$

As $L \to \infty$, the first term inside the square brackets vanishes, and the electric field is

$$\mathbf{E}(\mathbf{r}) = \frac{\rho_\ell}{2\pi\varepsilon_0\rho} \mathbf{a}_\rho,$$

agreeing with the result we derived using Gauss' Law or Coulomb's Law.

SideNote 1.1 Note on Integration

Before we move on, we note that there is sometimes confusion surrounding the integration leading to Eq. (1.51). That is, we might have chosen to divide the integral shown in Eq. (1.51) into two parts before evaluating it, one with limits of $-L/2 - z$ to 0, the other with limits of 0 and $L/2 - z$. Then we could change the limits of integration of the first integral to 0 to $L/2 + z$, which we can do since the integrand is an even function in the variable u, leading to the result

$$V(\mathbf{r}) = \frac{\rho_\ell}{4\pi\varepsilon_0} \left[\ln\left(\frac{(L/2 + z) + \sqrt{(L/2 + z)^2 + \rho^2}}{\rho} \right) \right.$$

$$\left. + \ln\left(\frac{(L/2 - z) + \sqrt{(L/2 - z)^2 + \rho^2}}{\rho} \right) \right]. \tag{1.53}$$

While this result appears to be quite different from our previous result, they are, in fact, the same. To show their equivalence, we start with Eq. (1.51), and multiply top and bottom by a common factor:

$$V(\mathbf{r}) = \frac{\rho_\ell}{4\pi\varepsilon_0} \ln\left(\frac{f_- + \sqrt{f_-^2 + \rho^2}}{-f_+ + \sqrt{f_+^2 + \rho^2}} \times \frac{f_+ + \sqrt{f_+^2 + \rho^2}}{f_+ + \sqrt{f_+^2 + \rho^2}} \right),$$

where we have substituted f_+ for $L/2 + z$ and f_- for $L/2 - z$. This reduces to

$$V(\mathbf{r}) = \frac{\rho_\ell}{4\pi\varepsilon_0} \ln\left(\frac{\left(f_- + \sqrt{f_-^2 + \rho^2}\right)\left(f_+ + \sqrt{f_+^2 + \rho^2}\right)}{-f_+^2 + \left[f_+^2 + \rho^2\right]} \right).$$

The denominator simplifies, giving us

$$V(\mathbf{r}) = \frac{\rho_\ell}{4\pi\varepsilon_0} \ln\left(\frac{\left(f_- + \sqrt{f_-^2 + \rho^2}\right)\left(f_+ + \sqrt{f_+^2 + \rho^2}\right)}{\rho^2} \right),$$

and we can rearrange the logarithm to write

$$V(\mathbf{r}) = \frac{\rho_\ell}{4\pi\varepsilon_0} \left[\ln\left(\frac{\left(f_- + \sqrt{f_-^2 + \rho^2}\right)}{\rho} \right) + \ln\left(\frac{\left(f_+ + \sqrt{f_+^2 + \rho^2}\right)}{\rho} \right) \right].$$

Comparison with Eq. (1.53) then shows agreement with Eq. (1.51).

1.4.1 Electric Dipole

While we're on the topic of the electric potential is a good time to introduce a simple configuration of a pair of charges, $+q$ and $-q$, separated by a distance d, known as an **electric dipole**. Even though its net charge is zero (by virtue of the positive and negative charges balancing one another), the separation of the charges allows the electric dipole to produce an electric field and an electric potential in the surrounding space.

Let us consider the potential $V(\mathbf{r})$ produced at a field point $\mathbf{r} = (x, y, z)$ by the electric dipole shown in Fig. 1.28. With the positive charge at $(0, 0, d/2)$ and the negative charge at $(0, 0, -d/2)$, the potential at a position (x, y, z) is

$$V(x, y, z) = \frac{q}{4\pi\varepsilon_0}\left(\frac{1}{\sqrt{x^2 + y^2 + (z - d/2)^2}} - \frac{1}{\sqrt{x^2 + y^2 + (z + d/2)^2}} \right). \quad (1.54)$$

The equipotential lines determined by this equation are plotted in Fig. 1.29. The equipotentials encircling the positive charge represent positive potentials, rising without limit as they approach $(0, 0, d/2)$. Similarly, negative equipotential lines encircle the negative charge, forming a valley in the potential surface. The plane at $z = 0$ is equidistant between the two charges, and has potential $V = 0$, as we should expect due to the asymmetry of the charges.

Let us now examine this potential function at distances far removed from the dipole. Specifically, we examine $V(x, y, z)$ at $R \gg d$, where $R = \sqrt{x^2 + y^2 + z^2}$. The result of this analysis will be very important as we consider electric fields in dielectric media, coming soon. Starting with Eq. (1.54), the potential is

$$V(x, y, z) \simeq \frac{q}{4\pi\varepsilon_0}\left(\frac{1}{\sqrt{x^2 + y^2 + z^2 - zd}} - \frac{1}{\sqrt{x^2 + y^2 + z^2 + zd}} \right),$$

where we have expanded the $(z \pm d/2)^2$ term in each denominator, and dropped the $d^2/4$ term. Dropping this term is valid since d^2 is much smaller than all other terms present. (While it may be tempting to use the same reasoning to also drop the term zd in each

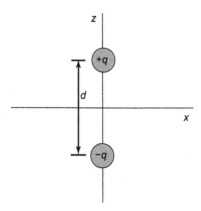

FIGURE 1.28 An electric dipole, consisting of two charges, one positive and one negative, separated by a distance d.

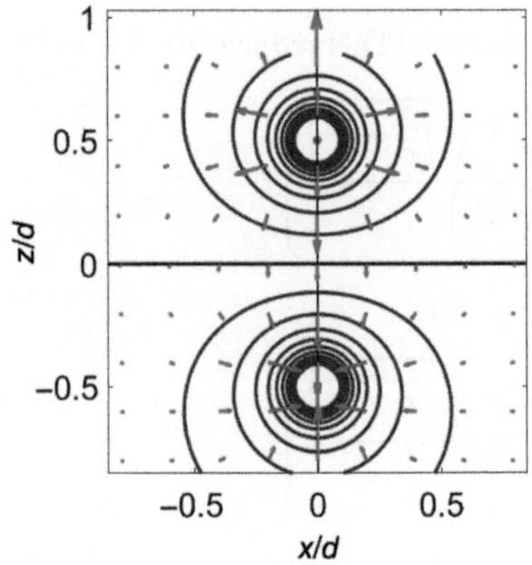

FIGURE 1.29 The equipotential lines due to the electric dipole shown in Fig. 1.28. The potential along the lines surrounding the positive charge are large and positive, while those around the negative charge are large and negative. The red arrows indicate the magnitude and direction of the electric field **E**. Notice that the electric field lines cross the equipotential lines at right angles, and that the electric field magnitude is greater where the equipotentials are closely spaced.

denominator, these terms will in fact lead to the most important term in the result for the potential, so we will keep them.) Next we pull out a factor R from each denominator,

$$V(x, y, z) \simeq \frac{q}{4\pi\varepsilon_0} \left(\frac{1}{R} \frac{1}{\sqrt{1 - zd/R^2}} - \frac{1}{R} \frac{1}{\sqrt{1 + zd/R^2}} \right).$$

Since $z = R\cos\theta$, we can write zd/R^2 as $d\cos\theta/R$, which must be much less than 1 since R is much larger than d. Then, using the Taylor expansion (Eq. (H.9))

$$\frac{1}{\sqrt{1 \pm \varepsilon}} \simeq 1 \mp \frac{1}{2}\varepsilon + \cdots$$

valid for small ε, the potential becomes

$$V(x, y, z) = \frac{q}{4\pi\varepsilon_0} \frac{1}{R} \left(\left[1 + \frac{1}{2} \frac{d\cos\theta}{R} + \cdots \right] - \left[1 - \frac{1}{2} \frac{d\cos\theta}{R} + \cdots \right] \right)$$
$$\approx \frac{qd\cos\theta}{4\pi\varepsilon_0 R^2}.$$

Now we see why it was important earlier to retain the terms zd, since these are the largest terms that remain in this potential. We are nearly finished, but let us first write this potential in a more general form. Recall that our analysis started by considering the dipole to be lined up along the z-axis, where θ is the angle between the z-axis and the field point vector **r**. More generally, the **electric dipole moment** vector is defined as

$$\mathbf{p} = q\mathbf{d}, \tag{1.55}$$

with **d** having magnitude d (the spacing between the $+q$ and $-q$ charges), and pointing from the negative charge to the positive charge. Using this notation, $qdR\cos\theta = \mathbf{p} \cdot \mathbf{r}$, and the potential at a position **r** due to an arbitrarily oriented electric dipole is

$$V(x, y, z) = \frac{\mathbf{p} \cdot \mathbf{r}}{4\pi\varepsilon_0 R^3} = \frac{\mathbf{p} \cdot \mathbf{a}_R}{4\pi\varepsilon_0 R^2}. \tag{1.56}$$

Notice the $1/R^2$ dependence of this potential, whereas the potential due to a point charge decreases as $1/R$. The potential of the dipole drops off faster with increasing R due to cancellation of the individual potentials of the $+q$ and $-q$ charges. As we get farther and farther away from the electric dipole, the small separation between the charges becomes relatively a smaller and smaller distance, and the dipole looks increasingly like a neutral object.

We can use the potential produced by an electric dipole to find the electric field by taking its gradient,

$$\mathbf{E} = -\boldsymbol{\nabla}V.$$

Using the spherical coordinate form of the gradient operator, Eq. (D.3), the field is

$$\mathbf{E} = \frac{p}{4\pi\varepsilon_0 R^3} \left(\mathbf{a}_R \, 2\cos\theta + \mathbf{a}_\theta \sin\theta \right). \tag{1.57}$$

This expression is valid only at locations far from the dipole (i.e. for $R \gg d$). See Fig. 1.29 for a plot of the electric field lines. At the edges of this diagram, where R is starting to become large, the field lines are starting to approach the form given by Eq. (1.57). The electric field lines originate on the positive charge of the dipole, bend around and down toward the negative charge, crossing the center line ($z = 0$) perpendicularly (this is purely the $\mathbf{a}_\theta$ direction), and point inward toward the negative charge.

In TechNote 1.5 a naturally occurring example of an electric dipole distribution, the electric eel, is described.

TechNote 1.5 Electric Eels

Electric eels are capable of producing large electric fields, which they use to locate and stun their prey, and for protection. The charging process takes place in electrocytes, cells that are capable of pumping positive sodium or potassium ions out of the cells attached to transport proteins. Each cell is capable of producing a potential difference of only 0.15 V, but when stacked one upon another in series, as in a voltaic cell, an eel can produce a potential difference as large as 860 V between their head and tail. After charging their body, they flex their body so that their head and tail are close to one another, with their prey in the space between (Fig. 1.30). The short spacing between head and tail increases the electric field in this region, and they can deliver a stunning shock to their prey. It is reported that electric eels can shock themselves if they're not careful.

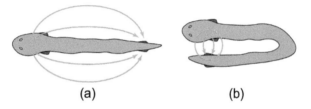

(a)   (b)

FIGURE 1.30 The electric field lines created by a fully charged eel, when straight (a) or folded (b).

1.5 Conductors and Dielectrics in Static Fields

LEARNING OBJECTIVES

After studying this section, you should be able to:
- describe the electrical characteristics of conductors and insulators;
- identify a conductor as an equipotential; and
- identify the location of excess charge on a conductor.

Up to now we have considered the electric field and its properties when the only objects existing in the whole universe are the charge (or perhaps a group of charges). We know from experience, however, that real physical systems include material objects, and we should reasonably expect that the behavior of these materials in response to electric fields will influence the field patterns that one might observe. In this section, we will start to explore some of these materials and their properties.

Broadly speaking, there exist three classes of materials that exhibit different electrical behaviors: conductors, dielectrics (or insulators), and semiconductors. In **conductors**, electrical charges can freely move in response to an applied electric field. Examples of conductors include metals, salt water, and ionized gases. We will characterize a material's ability to conduct an electrical current through a parameter called its **conductivity**, σ, which is the proportionality constant between the current density $\mathbf{J}_v$ induced in the material and $\mathbf{E}$, the electric field in that material; that is, $\mathbf{J}_v = \sigma\mathbf{E}$. We will discuss current density and conductivity in much more detail in Chapter 3, but for now be content to understand that the current density is the current per unit cross-sectional area of a conductor, and the conductivity of a good conductor is very large, such that even a small electric field can induce a sizable current density $\mathbf{J}_v$. There are several important consequences of the large conductivity of good conductors:

1. Under static conditions, the electric field **E** within the conductor must be zero (or at least very, very small). Otherwise, the electric field would induce currents (remember, $J_v = \sigma \mathbf{E}$!), with these currents representing a redistribution of the free charges, until we reach a condition in which **E** is zero.

2. Excess free charges can reside only at the surfaces of a conductor, since any net interior charge would produce an electric field, and this field would exert a force on neighboring free charges, moving them around until **E** becomes zero. Thus, the volume charge density ρ_v inside a conductor must be zero, and any unbalanced charges are driven to the surfaces.

3. Since the electric field in a conductor must be zero, we understand a conductor (especially a perfect conductor, for which the conductivity $\sigma \to \infty$) to be an equipotential. Since the potential difference between two points A and B is $-\int \mathbf{E} \cdot d\boldsymbol{\ell}$ (where the path integral is carried out on some path joining points A and B), and **E** must be zero everywhere inside a conductor, the potential at any two points in a conductor must be the same. Again, this is valid only in the static case.

4. As we will show in the next section, there are certain conditions that electric fields must satisfy at the interface between a conductor and a dielectric. These conditions are called boundary conditions, and these boundary conditions will have far-reaching consequences, even helping us to determine the amplitude of a reflected wave when an electromagnetic wave is incident upon an interface between two media.

The charges in **perfect dielectric materials**, on the other hand, are firmly attached to each other, and are not free to move around without constraint. That is not to say that there is no electric response in a dielectric material, as an electric field in the dielectric medium will pull the positive charges in one direction and the negative charges in the opposite direction. But the positive and negative charges, while displaced from one another by the electric field, remain bound to one another. We call this the **polarization** of the dielectric material, and we can apply our recent analysis of the potential due to a single electric dipole to understanding this material response.

Semiconductors, as their name implies, have properties intermediate to those of conductors and dielectrics. Semiconductors are extremely important as circuit components, and understanding the local fields and their influence on the electrical properties is critical, but this is an entire course within itself, and we will defer the behaviors and properties of semiconductor materials and devices to that other course.

1.6 Boundary Condition at a Conductor–Vacuum Interface

Application of Gauss' Law and $\nabla \times \mathbf{E} = 0$ at the interface between two materials allows us to derive important conditions that must be obeyed by electrostatic fields at these boundaries. In this section, we will explore these boundary conditions for electric fields at the interface between a conductor and vacuum.

When discussing boundary conditions that electric fields must obey at the interface between two media, we consider separately the component of the electric field that lies in the plane of the interface, known as the **tangential component**, and the component of the electric field that is perpendicular to the interface, known as the **normal component**. As we will show here, the tangential electric field component just outside a conductor must be identically zero, and the normal component must be equal to ρ_s/ε_0, where ρ_s is the surface charge density at the surface of the conductor. We will now prove these two electrostatic boundary conditions. To do so, recall the vector differential properties that we derived for electrostatic fields, namely $\nabla \times \mathbf{E} = 0$ and $\nabla \cdot \mathbf{E} = \rho_v/\varepsilon_0$. Actually, we will use the integral equivalents of these two laws:

$$\oint_c \mathbf{E} \cdot d\ell = 0,$$

for any closed path c, which we presented previously as Eq. (1.34), and Gauss' Law, Eq. (1.31),

$$\oint_s \mathbf{E} \cdot d\mathbf{s} = Q_{enc}/\varepsilon_0$$

for any closed surface s.

We start by considering the tangential component of the electric field at the surface of a conductor. Consider the conductor–vacuum interface shown in Fig. 1.31. In this figure, the boundary between a conductor (below the surface) and vacuum (above the surface) is shown. The path shown by the dashed lines is rectangular in shape, with the four sides labeled 1–4. Sides 1 and 3 are parallel (i.e. tangential) to the surface, with side 1 lying just below the surface (in the conductor), and side 3 just above the surface (in the vacuum). The length of these sides is Δw. The other two sides of this rectangular loop are of length Δh, but Δh is very small since sides 1 and 3 are very close to the surface. Therefore, $\Delta h \ll \Delta w$. Then applying Eq. (1.34) to this closed loop, the path integral reduces to just two terms,

$$\oint_c \mathbf{E} \cdot d\ell = E_{t,1}\Delta w - E_{t,3}\Delta w = 0,$$

where the subscript t represents the tangential component of E. Remember that the dot product designation in $\mathbf{E} \cdot d\ell$ picks out only the component of E that is parallel to $d\ell$. Thus for sides 1 and 3, only the tangential component E_t contributes to the path integral, while for sides 2 and 4, only the normal components contribute. But the contributions of the normal components are insignificant, since $\Delta h \ll \Delta w$. Notice also the signs of the two tangential terms. Since $d\ell$ represents an element of the closed path c, it is directed in opposite directions for sides 1 and 3, and $E_{t,1}\Delta w$ and $E_{t,3}\Delta w$ enter the sum with opposite signs. Since we know that $E_{t,1}$ must be zero (remember that this leg of the closed path is inside the conductor), the equation above tells us that $E_{t,3}$ must also be zero. Also, remember that the path we chose was quite arbitrary, forcing us to conclude that the tangential component of E just outside a conductor must *always* be zero – that is, E at the surface of a conductor must be normal to the surface.

We next explore the magnitude of the normal component of E just outside a conductor. To do this, we apply Gauss' Law, Eq. (1.31), to a closed surface such as that shown in Fig. 1.32. This closed surface has two flat faces of area Δs, one just below the interface (in the conductor) and the other just above the interface (in the vacuum). The side of this volume is of height Δh, which is exceedingly small, since the flat faces of the closed surface are very close to the interface. We quantify this as $\Delta h \ll \sqrt{\Delta s}$. Now we apply Eq. (1.31), where there are three contributions to the surface integral, one for each face, and one for the side wall. Of these three surfaces, only the flux through the flat face on the vacuum side is significant. (There is no flux through the bottom face since the electric field inside the conductor is zero, and the flux through the sides is negligible since that surface area is small, and since E is perpendicular to $d\mathbf{s}$ for the sides.) Therefore the total surface flux $\oint_s \mathbf{E} \cdot d\mathbf{s}$ is just $E_n\Delta s$, where E_n is the normal component of the electric field

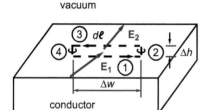

FIGURE 1.31 An interface between a conductor and vacuum. The dashed rectangle shows the path used to evaluate $\oint_c \mathbf{E} \cdot d\ell$ to establish the boundary condition for the tangential component of **E**.

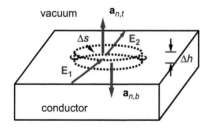

FIGURE 1.32 An interface between a conductor and vacuum. The dashed-line "pillbox" shows the surface used to evaluate $\oint_s \mathbf{E} \cdot d\mathbf{s}$ to establish boundary conditions for the normal component of **E**.

just outside the conductor. To complete the analysis of E_n, we must evaluate Q_{enc}, the charge enclosed within the volume bounded by the surface. As we wrote earlier, free charges in a conductor can only exist at the surface of the conductor. The volume charge density must be $\rho_v = 0$. Therefore, the charge enclosed within the closed surface must be $\rho_s \Delta s$, where ρ_s is the surface charge density in the surface of the conductor. Then by Eq. (1.31), the normal electric field component is $E_n = \rho_s/\varepsilon_0$. In summary, then, the electric field at the surface of the conductor must be normal to the conductor surface, and the magnitude of the field must be ρ_s/ε_0.

Example 1.14 Charge Inside a Conducting Sphere

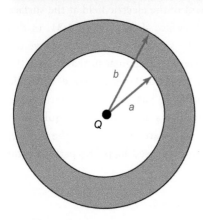

FIGURE 1.33 A hollow, thick-walled, spherical conductor, with a charge Q located at the center.

A charge Q is positioned at the center of a conducting, thick-walled, electrically neutral, spherical shell (inside radius a, outside radius b), as shown in Fig. 1.33. Determine the electric field $\mathbf{E(r)}$ and the potential $V(\mathbf{r})$ at all locations, and the surface charge density on the two surfaces of the conducting shell.

Solution:
We have applied Gauss' Law to find the electric field in similar problems previously, so we will abbreviate this part of our solution. In brief, because of the symmetry of the problem, the electric field must be radial, and E_R can depend only on the distance R, or $\mathbf{E(r)} = \mathbf{a}_R E_R(R)$. Using a spherical Gaussian surface of radius $R < a$, Gauss' Law leads to $\mathbf{E(r)} = Q\,\mathbf{a}_R/4\pi\varepsilon_0 R^2$. The electric field in the region $a < R < b$ (i.e. inside the conductor) must be zero. Finally, for $R > b$, we again apply Gauss' Law. Constructing a spherical Gaussian surface with $R > b$, the total charge enclosed within this sphere must be the charge at the center, Q, plus the charge on the surfaces of the conducting shell. But we were told that the shell is electrically neutral, so there is no net charge here, and $Q_{enc} = Q$. We conclude that $\mathbf{E(r)} = Q\,\mathbf{a}_R/4\pi\varepsilon_0 R^2$ for this region as well.

Now that we have determined the electric field $\mathbf{E(r)}$ inside $(R < a)$ and outside $(R > b)$ the sphere, we can use this result in order to determine the surface charge density on the two surfaces of the conductor. The normal field at the inner surface of the conductor is $|\mathbf{E}(a)| = Q/4\pi\varepsilon_0 a^2$. Since the radial field is pointed *outward* from the central charge, it is pointed *into* the conductor at the inner surface, so the surface charge density here is negative, and of magnitude $\rho_s(a) = -\varepsilon_0|\mathbf{E}(a)| = -Q/4\pi a^2$. At the outer surface of the conductor, the electric field is of magnitude $|\mathbf{E}(b)| = Q/4\pi\varepsilon_0 b^2$, and pointing away from the conductor. Therefore the surface charge density here is $\rho_s(b) = +\varepsilon_0|\mathbf{E}(b)| = Q/4\pi b^2$. Notice that when we integrate these surface charge densities over the entire inner and outer surfaces, we find that the total charge on the inner surface is $-Q$ and the total charge on the outer surface is $+Q$. Therefore the charge Q at the center induces an equal but opposite charge on the inner wall of the conductor. (Recall that electric fields terminate on negative charges.) Since the conductor is electrically neutral, the $-Q$ charge on the inner surface implies a $+Q$ charge on the outer surface, and this charge produces the electric field in the region $R > b$.

The final part of our solution is to find the electric potential $V(\mathbf{r})$. We use Eq. (1.43), with reference to $V = 0$ at $R = \infty$. For $R > b$,

$$V(R) = -\int_{\infty}^{R} \mathbf{E} \cdot d\boldsymbol{\ell} = \frac{Q}{4\pi\varepsilon_0 R}.$$

Inside the conductor the electric field is zero, so the potential in this region is constant,

$$V(R) = -\int_\infty^b \mathbf{E} \cdot d\ell - \int_b^R \mathbf{E} \cdot d\ell = \frac{Q}{4\pi\varepsilon_0 b}.$$

For $R < a$,

$$V(R) = -\int_\infty^b \mathbf{E} \cdot d\ell - \int_b^a \mathbf{E} \cdot d\ell - \int_a^R \mathbf{E} \cdot d\ell = \frac{Q}{4\pi\varepsilon_0}\left(\frac{1}{R} - \frac{1}{a} + \frac{1}{b}\right).$$

This potential is plotted in Fig. 1.34.

FIGURE 1.34 A plot of the electric potential for the thick-walled conductor of Example 1.14.

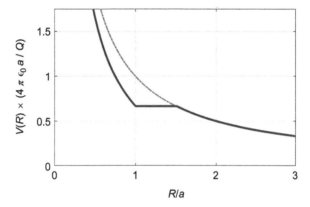

While in this example the interior charge was positioned at the center of the conducting sphere, Gauss' Law can be used to show that for a hollow conductor of *any* shape, and for charges located *anywhere* in the interior of this hollow conductor, the total surface charge induced on the interior wall of the conductor is always equal to, but opposite in sign from, the total charge inside the conductor. We leave the proof of this to Prob. P1-56 at the end of this chapter.

The operation of a high-voltage generator known as a van de Graaff generator, described in TechNote 1.6, is based on the principles of charge transport into a hollow conductor and the resulting induction of surface charges.

TechNote 1.6 van de Graaff Generator

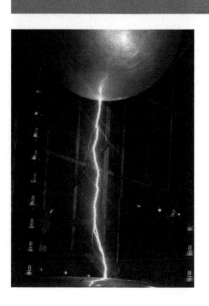

The van de Graaff generator, developed by Robert van de Graaff in 1929, is capable of generating extremely high electric potentials. The original purpose of this high-voltage source was for use in particle accelerators for physics experiments. They still find wide use today in nuclear research, nuclear medicine, and science education. A rotating drum at the base of the van de Graaff generator drives a nonconducting belt, which carries electrical charge into the interior of a conducting dome (Fig. 1.35(a)). The belt picks up the charge from a biased brush that is positioned in near-contact with the belt. When the brush is positively biased, it can attract negative charge from the belt, leaving the belt with a net positive charge. Inside the dome, the positive charges on the belt attract negative charges on a second brush, which are drawn off the brush to neutralize the belt. That leaves this brush with a positive charge, which is quickly conducted to the exterior of the dome, independent of how much charge is already present on the dome. Recall that the excess charge on a conductor is always located on the outside surface of that conductor.

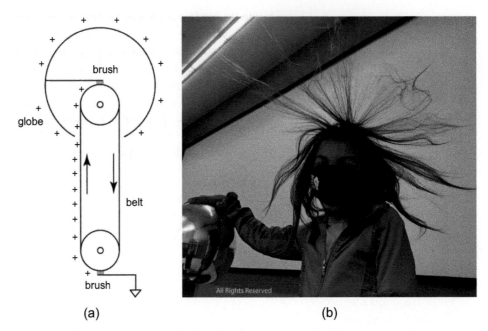

(a)   (b)

FIGURE 1.35 (a) A schematic diagram of a van de Graaff generator. Charge is carried by the belt into the interior of the spherical dome at the top. After this charge is transferred to a brush inside the dome, it is easily conducted to the dome, regardless of how much charge already resides on the dome. (b) This young experimental physicist (masked during the COVID-19 pandemic) is charged by a van de Graaff generator, demonstrating that like charges repel.

The total maximum charge that can accumulate on the dome is determined by primarily two factors: the radius of the spherical dome and the humidity of the surrounding air. The humidity is important since the dielectric strength of dry air is 30 kV/cm, but lower for humid air. (We'll discuss the dielectric strength, or breakdown field strength, in more detail in Section 1.10.) When the electric field outside the dome exceeds the dielectric strength of the air, the air molecules will ionize (i.e. they will be stripped of one or more electrons), and the air becomes conducting. On humid days, therefore, the charge on the dome will "leak off," or discharge into the air much more readily than on dry days. (For this same reason, you are more likely to shock yourself after walking over a wool carpet in the winter, when the air is dry, than in the summer, when the air is humid. Your body can accumulate a larger charge in the dry air.) The radius a of the sphere is important since the radial electric field just outside the dome surface is of amplitude $E_R(a) = Q/4\pi\varepsilon_0 a^2$, where Q is the total charge on the sphere. Since E_R is limited by the dielectric strength of the air, a larger radius a leads to a larger possible total charge Q. Also, since the potential of the sphere is $V_0 = Q/4\pi\varepsilon_0 a$, which we can write as $aE_R(a)$, we immediately see that a sphere of larger radius can be charged to a greater potential V_0. Let's add some numerical values. For a spherical dome of radius $a = 20$ cm, and for a dielectric strength of dry air $E_{max} = 30$ kV/cm, the maximum voltage of the dome is $V_{max} = aE_{max} = 600$ kV, and the charge on the dome is $Q = 4\pi\varepsilon_0 a V_{max} = 13.4$ µC. Potentials as high as ~5 MV are possible with larger-sized van de Graaff generators. In Fig. 1.35(b) we see a young lady charged up by keeping her hand in contact with an operating van de Graaff generator. The charges that her body accumulates repel one another, and therefore collect in her hair, causing it to stand on end.

As we have shown in this section, conductors are characterized by several important properties. Under static conditions, electric fields inside conductors must be zero, and the potential within a conductor is always uniform. Outside the conductor, the electric field must be perpendicular to the surface, and the surface charge density ρ_s must be equal to E_n/ε_0, where E_n is the normal component of the electric field. (We will have to modify this relation very soon when we consider dielectric materials. In its current form it is valid only at the interface between a conductor and vacuum.) Finally, we cannot have a net charge within the interior of a conductor; excess charges reside only on the surfaces.

1.7 Perfectly Insulating Dielectric Materials

In this section we discuss perfect dielectric materials – that is, materials that do not allow free currents – and their response to application of an electric field. Examples of dielectric materials include ceramics, rubber, and glass. In contrast to conductors, in which the charge carriers (typically the electrons) are unbound and free to move in response to an electric field, the charges in a dielectric material are tightly bound. Upon application of an electric field to a dielectric material, the field will pull the positive charges in one direction, and the negative charges in the opposite direction, but the charges remain bound and their displacement is small. This is referred to as *polarizing* the material, and these dipoles in turn produce a field, as we will see in the following.

In contrast to the response of a conductor, the positive and negative charges in a perfect dielectric are only shifted slightly in position, but are not free to move over a macroscopic distance. Two physical mechanisms can contribute to the electric polarization of a medium. In one, the molecules of a medium may themselves be permanent electric dipoles, which are randomly oriented in the absence of an electric field, but upon application of an electric field become partially aligned. Alternatively, other media consist of non-polar atoms or molecules, which become distorted, or polarized, when a field is applied. In other somewhat rare instances, we encounter materials that are polarized even in the absence of an applied electric field. Regardless of the physical origin of the dipole moments of the constituent particles, the **polarization field P** is defined as the product of the number density n and the average dipole moment $\langle\mathbf{p}\rangle$ of the electric dipoles in a medium:

$$\mathbf{P} = n\langle\mathbf{p}\rangle = \lim_{\Delta v' \to 0} \frac{1}{\Delta v'} \sum_{k=1}^{n\Delta v'} \mathbf{p}_k, \tag{1.58}$$

where $n\Delta v'$ is the number of dipoles in the volume $\Delta v'$, the index k labels each individual dipole inside the volume $\Delta v'$, and $\mathbf{p}_k$ is the vector dipole moment of the dipole k. Recall from Eq. (1.56) that the scalar potential field created by a single electric dipole located at the origin is

$$V(x, y, z) = \frac{\mathbf{p} \cdot \mathbf{a}_R}{4\pi\varepsilon_0 \, R^2}.$$

The net potential produced by all the dipoles contained within a small volume $\Delta v'$ is found by adding the potential of each of the individual dipoles

$$dV(x, y, z) = \sum_k \frac{\mathbf{p}_k \cdot (\mathbf{r} - \mathbf{r}'_k)}{4\pi\varepsilon_0 \, |\mathbf{r} - \mathbf{r}'_k|^3},$$

where $\mathbf{r}'_k$ represents the position of an individual dipole. As we let the elemental volume $\Delta v'$ become small, the differences between the various $\mathbf{r}'_k$ within $\Delta v'$ become insignificant, and we can replace each $\mathbf{r}'_k$ by $\mathbf{r}'$, without the subscript k. Then the potential becomes

$$dV = \frac{\left(\sum_k \mathbf{p}_k\right) \cdot (\mathbf{r} - \mathbf{r}')}{4\pi\varepsilon_0 \, |\mathbf{r} - \mathbf{r}'|^3}.$$

Using Eq. (1.58) to replace the sum over the individual dipoles in this expression with $\mathbf{P}(\mathbf{r}')\Delta v'$, the potential due to the dipoles in the small volume $\Delta v'$ is

$$dV = \frac{\mathbf{P}(\mathbf{r}')\Delta v' \cdot (\mathbf{r} - \mathbf{r}')}{4\pi\varepsilon_0 \, |\mathbf{r} - \mathbf{r}'|^3}.$$

Summing all these contributions due to the dipoles in the different volumes $\Delta v'$, in the limit as $\Delta v' \to 0$, the potential becomes

$$V(\mathbf{r}) = \int_{v'} \frac{\mathbf{P}(\mathbf{r}') \cdot (\mathbf{r} - \mathbf{r}')}{4\pi\varepsilon_0 \, |\mathbf{r} - \mathbf{r}'|^3} dv'. \tag{1.59}$$

We can simplify this a bit using Eq. (D.6),

$$\frac{\mathbf{r} - \mathbf{r}'}{|\mathbf{r} - \mathbf{r}'|^3} = \mathbf{\nabla}'\left(\frac{1}{|\mathbf{r} - \mathbf{r}'|}\right),$$

where, as a reminder, the prime on the Del operator indicates differentiation with respect to the primed coordinates (which designate the location of the dipoles). Inserting Eq. (D.6) into Eq. (1.59), the potential is

$$V(\mathbf{r}) = \frac{1}{4\pi\varepsilon_0} \int_{v'} \mathbf{P}(\mathbf{r}') \cdot \mathbf{\nabla}'\left(\frac{1}{|\mathbf{r} - \mathbf{r}'|}\right) dv'. \tag{1.60}$$

Now we can use the vector field equivalent of the product rule for differentiation, discussed in Appendix E (see Eq. (E.6) and Example E.1),

$$\mathbf{\nabla}' \cdot (f\mathbf{A}) = f\mathbf{\nabla}' \cdot \mathbf{A} + \mathbf{\nabla}'f \cdot \mathbf{A},$$

for any scalar field f and vector field $\mathbf{A}$. In this case, f represents $|\mathbf{r} - \mathbf{r}'|^{-1}$ and $\mathbf{A}$ represents $\mathbf{P}$, or

$$\mathbf{\nabla}' \cdot \left(\frac{\mathbf{P}(\mathbf{r}')}{|\mathbf{r} - \mathbf{r}'|}\right) = \frac{1}{|\mathbf{r} - \mathbf{r}'|} \mathbf{\nabla}' \cdot \mathbf{P}(\mathbf{r}') + \mathbf{P}(\mathbf{r}') \cdot \mathbf{\nabla}'\left(\frac{1}{|\mathbf{r} - \mathbf{r}'|}\right).$$

Using this in Eq. (1.60), the potential is

$$V(\mathbf{r}) = \frac{1}{4\pi\varepsilon_0}\left[\int_{v'} \mathbf{\nabla}' \cdot \left(\frac{\mathbf{P}(\mathbf{r}')}{|\mathbf{r} - \mathbf{r}'|}\right) dv' - \int_{v'} \frac{1}{|\mathbf{r} - \mathbf{r}'|} \mathbf{\nabla}' \cdot \mathbf{P}(\mathbf{r}') \, dv'\right].$$

Finally, we can apply the Divergence Theorem to write the first integral as a surface integral,

$$V(\mathbf{r}) = \frac{1}{4\pi\varepsilon_0} \oint_{s'} \frac{\mathbf{P}(\mathbf{r}') \cdot d\mathbf{s}'}{|\mathbf{r} - \mathbf{r}'|} - \frac{1}{4\pi\varepsilon_0} \int_{v'} \frac{\mathbf{\nabla}' \cdot \mathbf{P}(\mathbf{r}')}{|\mathbf{r} - \mathbf{r}'|} \, dv'. \tag{1.61}$$

Now this is in a form very similar to something we have seen before. Recall that the potential $V(\mathbf{r})$ due to a surface charge distribution, given previously in Eq. (1.48), is

$$V(\mathbf{r}) = \frac{1}{4\pi\varepsilon_0} \int_s \frac{\rho_s(\mathbf{r}')}{|\mathbf{r} - \mathbf{r}'|} \, ds',$$

while for a volume charge distribution the potential is

$$V(\mathbf{r}) = \frac{1}{4\pi\varepsilon_0} \int_v \frac{\rho_v(\mathbf{r}')}{|\mathbf{r} - \mathbf{r}'|}\, dv'.$$

Comparing the two integrals in Eq. (1.61) to these, we identify **effective charge densities** $\rho_{s,eff}(\mathbf{r}')$ and $\rho_{v,eff}(\mathbf{r}')$, defined as

$$\boxed{\rho_{s,eff}(\mathbf{r}) = \mathbf{P}(\mathbf{r}) \cdot \mathbf{a}_n,}$$
(1.62)

where $\mathbf{a}_n$ is the surface normal pointing outward from the dipolar volume, and

$$\boxed{\rho_{v,eff}(\mathbf{r}) = -\boldsymbol{\nabla} \cdot \mathbf{P}(\mathbf{r}).}$$
(1.63)

(Notice that we have dropped the primed notation on the position vector $\mathbf{r}$, since we no longer need to distinguish between source and field positions.) The potential, $V(\mathbf{r})$, generated by these effective charges is perfectly equivalent to the potential generated by free charges introduced earlier.

It is interesting to look a little more deeply into these effective charge densities in order to better appreciate their significance. Consider a medium containing a large number of electric dipoles, as shown in Fig. 1.36. For this example, all the dipoles are aligned in the same direction (upwards), with density increasing from bottom to top. Now let us "count" the effective charges distributed throughout the volume. In the interior of the medium, negative charges of one dipole pair are adjacent to positive charges of another. If the polarization of the medium were perfectly uniform, all these interior charges would be perfectly balanced, and no net effective charge in the interior of the dipolar distribution would remain. (Remember that the polarization is related to the density of dipoles in the medium.) In the case of varying dipole density, however, the net charges within the medium are not perfectly balanced, represented by $\boldsymbol{\nabla} \cdot \mathbf{P}(\mathbf{r})$. The unpaired dipole charges represent an effective charge density. In the figure, for example, there exists a net *negative* effective charge between the two horizontal dashed lines in the middle of the dipolar medium. This negative charge results from the dipoles slightly below the center of this region, which contribute a positive charge to this region, and the dipoles slightly above the center, which contribute a negative charge. Since the density of dipoles increases from bottom to top, the net charge between the two lines, or anywhere in the interior of this slab, is negative. We can also understand the effective *surface* charge density in a similar way. Again referring to Fig. 1.36, a number of negative charges of the dipoles at the bottom of the medium have no positive charges to balance them. Similarly at the top, we see an excess of unpaired positive charges. These unpaired charges represent an effective surface charge density, $\rho_{s,eff}(\mathbf{r})$.

FIGURE 1.36 A medium containing an inhomogeneous distribution of electric dipoles.

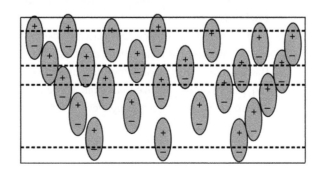

Example 1.15 Effective Charge Density

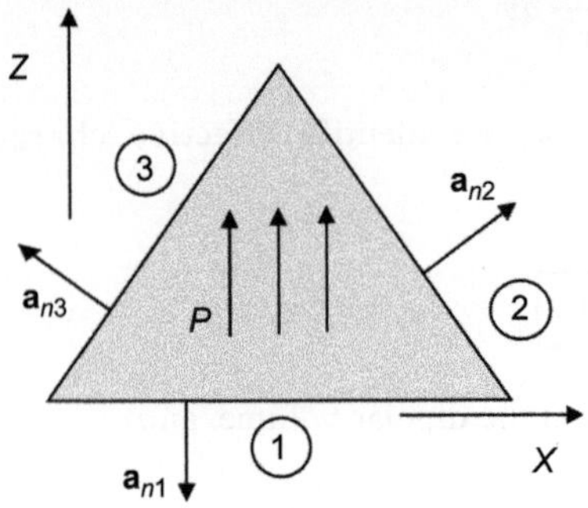

FIGURE 1.37 A triangular block of a polarized medium. The polarization **P** is uniform, and points in the z-direction.

Consider a slab of polarized dielectric material as shown in Fig. 1.37. The dielectric is in the shape of an equilateral triangle, of side L, and the polarization $\mathbf{P} = P_0\,\mathbf{a}_z$ is uniform within the triangle. (a) Determine the effective charge densities $\rho_{s,eff}(\mathbf{r})$ and $\rho_{v,eff}(\mathbf{r})$. (b) Under the approximation that the slab is thin (i.e. $t \ll L$, where t is the thickness of the slab), determine the resulting electric field **E** at the center of the equilateral triangle.

Solution:

(a) We must first determine the effective charge densities, for which we use Eqs. (1.62) and (1.63). For the effective surface charge densities we must identify the surface normal vector for the three sides. These are $\mathbf{a}_{n1} = -\mathbf{a}_z$, $\mathbf{a}_{n2} = \mathbf{a}_x \sin 60° + \mathbf{a}_z \cos 60° = (\sqrt{3}\mathbf{a}_x + \mathbf{a}_z)/2$, and $\mathbf{a}_{n3} = (-\sqrt{3}\mathbf{a}_x + \mathbf{a}_z)/2$. Using these we find that the effective surface charge densities on the three sides are $\rho_{s1,eff} = \mathbf{P} \cdot \mathbf{a}_{n1} = -P_0$, $\rho_{s2,eff} = \mathbf{P} \cdot \mathbf{a}_{n2} = P_0/2$, and $\rho_{s3,eff} = \mathbf{P} \cdot \mathbf{a}_{n3} = P_0/2$. Within the dielectric, the effective volume charge density is $\rho_{v,eff} = -\nabla \cdot \mathbf{P}(\mathbf{r})$, which, because the polarization is uniform, is zero.

(b) We use the effective charge densities determined in (a) to find the field at the center of the triangular slab. Since the thickness t of the slab is small, we can treat each side of the triangular slab as a linear charge of density $\rho_{\ell,i} = \rho_{si,eff}\,t$, for $i = 1$, 2, or 3. We can use the results of Example 1.5, in which we determined the electric field due to a line charge of length L. At the center of the line charge, a distance $\rho = L/2\sqrt{3}$ (the center of the equilateral triangle) from the line, the field is, using Eqs. (1.21) and (1.22), $\mathbf{E} = E_\rho \mathbf{a}_\rho$, where $E_\rho = 3\rho_\ell/(2\pi\varepsilon_0 L)$. Applying this to sides 1, 2, and 3, the electric fields due to the individual sides are

$$\mathbf{E}_1 = -3P_0 t/(2\pi\varepsilon_0 L)\mathbf{a}_z,$$
$$\mathbf{E}_2 = 3(P_0/2)t/(2\pi\varepsilon_0 L)(-\sqrt{3}\mathbf{a}_x - \mathbf{a}_z)/2,$$

and

$$\mathbf{E}_3 = 3(P_0/2)t/(2\pi\varepsilon_0 L)(\sqrt{3}\mathbf{a}_x - \mathbf{a}_z)/2.$$

The total field at the center is

$$\mathbf{E} = \mathbf{E}_1 + \mathbf{E}_2 + \mathbf{E}_3 = -9P_0 t/(4\pi\varepsilon_0 L)\mathbf{a}_z.$$

In Example 1.15 we chose the thickness of the slab t to be small so that we could use our previous result for the electric field produced by a finite-length line charge given by Eqs. (1.21) and (1.22). The restriction of small t was not necessary, and we could have found the resulting electric field with this technique even for large thickness t.

In summary, then, dielectric materials can be modeled as a distribution of microscopic electric dipoles. The electric field due to a polarized material is found by first determining the effective charges distributed through the volume (Eq. (1.63)) and/or over the surface (Eq. (1.62)) of the polarized material. These effective charges are then used in Coulomb's Law to determine the resulting electric field.

1.8 Gauss' Law Revisited

Now that we have learned how to treat polarized materials, and how to determine the electric field that these dipoles generate, we need to return to our previous discussion of the differential properties of the electrostatic field to determine how these laws must be modified to account for polarized dielectric media. Recall that earlier we derived Gauss' Law in its differential ($\nabla \cdot \mathbf{E} = \rho_v/\varepsilon_0$) and integral ($\oint \mathbf{E} \cdot d\mathbf{s} = Q_{enc}/\varepsilon_0$) forms, and what will become Faraday's Law, also in its differential ($\nabla \times \mathbf{E} = 0$) and integral ($\oint \mathbf{E} \cdot d\boldsymbol{\ell} = 0$) forms. The latter of these does not involve any electrical charges, so no accommodation of this law for the effective charges is required. Gauss' Law, however, certainly does involve charge, so let's explore this.

When we previously discussed Gauss' Law, we only considered the free charges of density $\rho_v(\mathbf{r})$, valid in a vacuum. Since then, we introduced the effective charge density of a polarized material, which is of the form $\rho_{v,eff}(\mathbf{r}) = -\nabla \cdot \mathbf{P}(\mathbf{r})$. (See Eq. (1.63)). Since the electric field $\mathbf{E}$ is the total field resulting from *all* electric charges, *free* as well as the *bound* charges, Gauss' Law must be modified to include the additional effect of these effective charges,

$$\nabla \cdot \mathbf{E} = \left(\rho_v + \rho_{v,eff}\right)/\varepsilon_0.$$

The electric field $\mathbf{E}$ generated by the effective charges is every bit as real as an electric field generated by free charges, and the electric force experienced by a test charge is due to the total electric field in that region in space, whether generated by free or bound (effective) charges. Using Eq. (1.63), the previous equation can be written as

$$\nabla \cdot \mathbf{E} = \rho_v/\varepsilon_0 - \nabla \cdot \mathbf{P}/\varepsilon_0,$$

and rearranging,

$$\nabla \cdot (\mathbf{E} + \mathbf{P}/\varepsilon_0) = \rho_v/\varepsilon_0.$$

Let us now define a new electric field quantity,

$$\boxed{\mathbf{D} \equiv \varepsilon_0\mathbf{E} + \mathbf{P},} \tag{1.64}$$

which is called the **displacement field**. The units of the displacement field are $[\mathbf{D}] = \text{C/m}^2$, which is consistent with $[\varepsilon_0\mathbf{E}] = (\text{F/m})(\text{V/m}) = \text{C/m}^2$. In terms of the displacement field $\mathbf{D}$, Gauss' Law becomes

$$\boxed{\nabla \cdot \mathbf{D} = \rho_v.} \tag{1.65}$$

In some limited cases we can associate $\mathbf{D}$ as the part of the total field $\mathbf{E}$ that is generated by the free charges. (This is not correct in a general case, so we should not take this picture too far. But it can be a handy way to attach an approximate picture to the various field quantities that we have discussed.)

Equation (1.65) represents the final form of Gauss' Law in differential form. There is, of course, an equivalent integral form, which can be derived as follows. The volume integral of $\nabla \cdot \mathbf{D}$ in some volume v must be, by Eq. (1.65) (i.e. Gauss' Law in its differential form).

$$\int_v \nabla \cdot \mathbf{D} \, dv = \int_v \rho_v \, dv.$$

The right-hand side is simply the total amount of free charge Q_{enc} enclosed within the volume v. The left-hand side may not look especially useful, until we apply the

Divergence Theorem (i.e. Eq. (D.10)) and write this volume integral as the surface integral of the displacement field,

$$\oint_s \mathbf{D} \cdot d\mathbf{s} = Q_{enc}. \qquad (1.66)$$

So we see that for any volume v, the surface flux of $\mathbf{D}$ through the surface must be equal to the total charge enclosed within the volume.

These revised forms of Gauss' Law, Eqs. (1.65) and (1.66), are valid in all media, including conductors and dielectrics (and vacuum, which is simply a special case of dielectric), and replace the earlier forms of Gauss' Law, Eqs. (1.31) and (1.32), which were valid only in vacuum. *You should adopt this form of Gauss' Law for all of your work*, and no longer use the earlier forms, which are not as generally valid.

In the following sections we will apply Gauss' Law to determine the boundary conditions that electric fields must satisfy at the interface between different materials (either conductor/dielectric or dielectric/dielectric), examine the properties of 'simple' (i.e. linear, isotropic, homogeneous) media, and apply Gauss' Law to determine the displacement field $\mathbf{D}$ and the electric field $\mathbf{E}$ in several symmetric geometries.

1.9 Boundary Conditions for Perfect Dielectrics

In Section 1.6 we derived the boundary conditions that an electrostatic field $\mathbf{E}$ must satisfy at the interface between a conductor and vacuum. In this section we will determine the boundary conditions that must be satisfied by electrostatic fields at the interface between two materials, either conductor/dielectric or dielectric/dielectric interfaces. As before, we will apply Gauss' Law (but now we use the complete form of Gauss' Law, Eq. (1.66)), and $\oint_c \mathbf{E} \cdot d\boldsymbol{\ell} = 0$ to relate tangential and normal components of fields on one side of the interface to those on the other. The dielectrics that we consider here are non-conducting.

We first consider the tangential field component – that is, the component of the field that lies in the plane of the interface. As before, we apply $\oint_c \mathbf{E} \cdot d\boldsymbol{\ell} = 0$ to the rectangular contour shown in Fig. 1.38. Unlike our previous treatment, however, we will not specify (initially) whether medium 1 (below the interface) and medium 2 (above the interface) are a conductor or dielectric material. Following the same argument as earlier, we derive

$$\oint_c \mathbf{E} \cdot d\boldsymbol{\ell} = E_{t,1}\Delta w - E_{t,3}\Delta w = 0.$$

We have again chosen sides 1 and 3 (the two sides of the rectangular path that are parallel to the surface) to be very close to the surface, so that $\Delta h \ll \Delta w$. The contributions of paths 2 and 4 are therefore insignificant, and we do not include them

FIGURE 1.38 An interface between two different media. The dashed rectangle shows the path used to evaluate $\oint_c \mathbf{E} \cdot d\boldsymbol{\ell}$ to establish the boundary condition for the tangential component of **E**.

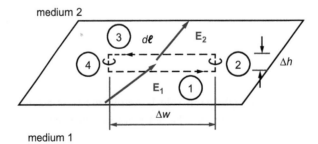

FIGURE 1.39 An interface between two media. The dashed-line "pillbox" shows the surface used to evaluate $\oint_s \mathbf{E} \cdot d\mathbf{s}$ to establish boundary conditions for the normal component of $\mathbf{E}$.

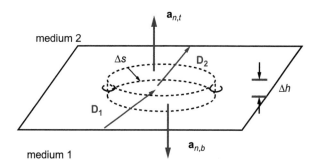

in this expression. We conclude therefore that, at the interface between two media, the tangential components of $\mathbf{E}$ must be continuous, or

$$E_{1,t} = E_{2,t}. \tag{1.67}$$

If one of these media is a conductor, then the field in that medium is zero, and $E_{1,t} = E_{2,t} = 0$, which, of course, is consistent with our previous conclusion.

Next we will apply Gauss' Law at the interface to see what we can learn about the normal field components. As before, we construct a pillbox shape as shown in Fig. 1.39. One face of this pillbox is just above the interface; the other is just below the interface. Since these faces are very close to the surface, the surface area of the sides of the pillbox is very small in comparison to the area of the two faces. Applying Gauss' Law, the displacement field flux is

$$\oint_s \mathbf{D} \cdot d\mathbf{s} = \mathbf{D}_1 \cdot \mathbf{a}_{n,b} \Delta s + \mathbf{D}_2 \cdot \mathbf{a}_{n,t} \Delta s = Q_{enc},$$

where $\mathbf{a}_{n,b}$ is the normal unit vector at the bottom of the pillbox, pointing downward, and $\mathbf{a}_{n,t}$ is the normal unit vector at the top of the pillbox, pointing upward. In order to standardize the notation in a form consistent with earlier notation, let us define $\mathbf{a}_{n,1}$, the unit vector pointing *outward* from medium 1 (and therefore *into* medium 2). In terms of this unit vector, $\mathbf{a}_{n,b} = -\mathbf{a}_{n,1}$ and $\mathbf{a}_{n,t} = +\mathbf{a}_{n,1}$. Then the previous equation becomes

$$(\mathbf{D}_2 - \mathbf{D}_1) \cdot \mathbf{a}_{n,1} \Delta s = Q_{enc}.$$

Now let us examine Q_{enc}. Recall that this term includes only free charges. Since the faces of the pillbox are very close to the interface between the two media, the only free charge that can be enclosed within the pillbox is the surface charge, and the magnitude of this charge is $Q_{enc} = \int \rho_s ds \approx \rho_s \Delta s$. Using this in the equation above, the elemental surface area Δs is common to each term and can be factored out, and the normal fields must satisfy

$$(\mathbf{D}_2 - \mathbf{D}_1) \cdot \mathbf{a}_{n,1} = \rho_s. \tag{1.68}$$

The dot product on the left-hand side picks out only the normal components of the displacement field, and this equation tells us that the normal component of $\mathbf{D}$ can be discontinuous, but only when there is a surface charge (free charges only) at the interface. For perfect dielectric media containing no free charges, D_n must be continuous at the interface. If one of the media is a conductor, there can be, of course, free charges. Since there is no field inside the conductor, the normal displacement field is $D_n = \rho_s$ – that is, the normal component of the displacement field just outside a conductor is equal to the surface charge density on the surface of the conductor. In the case of vacuum, this result is consistent with our previous result.

In summary, we have shown that the tangential component of **E** must be continuous at the interface between two media (Eq. (1.67)), and the discontinuity in the normal component of **D** must be equal to the surface charge density ρ_s at the interface (Eq. (1.68)). We will work an example problem that illustrates the application of these boundary conditions, but first we'll define a simple dielectric medium.

1.10 Simple Dielectric Media

We will, for the most part, be concerned only with what are called "simple" dielectric media. By simple, we mean that the medium is **linear, isotropic**, and **homogeneous**. *Linear* media are dielectric media in which the polarization field induced upon application of an external electric field is proportional to the amplitude of the field, written as

$$\mathbf{P} = \varepsilon_0 \chi_e \mathbf{E}, \tag{1.69}$$

where χ_e is called the **electric susceptibility** of the dielectric medium. For vacuum, $\chi_e = 0$, as there are no electric dipoles. (Recall that the polarization field is defined in terms of the density of electric dipoles and the average dipole moment.) In other dielectric media, such as air, water, ceramic, etc., the susceptibility will be a positive quantity that characterizes the polarization created in the medium by the electric field. (Most media exhibit nonlinear behavior under the influence of extremely strong electric fields, but we restrict our attention to smaller fields and strictly linear behavior.) In an *isotropic* medium, the magnitude of the polarization of the medium does not depend on the direction of the applied field **E**. The resulting **P** is therefore parallel to **E**. Isotropic media include air, glass, water, ceramic, or any other media which has no regular structure. Most (but not all) crystal materials are anisotropic, and the polarization of the material is different for application of **E** in different directions. These are important materials, but we must defer their discussion to another course. Finally, in a *homogeneous* medium, the polarization field is the same throughout the medium. The susceptibility is uniform throughout the medium, independent of location.

For simple media, Eq. (1.69) can be used in Eq. (1.64) to write the displacement field as

$$\mathbf{D} = \varepsilon_0 \mathbf{E} + \mathbf{P} = \varepsilon_0 \mathbf{E} + \varepsilon_0 \chi_e \mathbf{E},$$

and factoring out the common terms

$$\boxed{\mathbf{D} = \varepsilon_0 \left(1 + \chi_e\right) \mathbf{E}.} \tag{1.70}$$

The **permittivity**, or **absolute permittivity**, of the medium is defined as

$$\varepsilon = \varepsilon_0 \left(1 + \chi_e\right), \tag{1.71}$$

and the **relative permittivity**, or **dielectric constant**, as

$$\varepsilon_r = \frac{\varepsilon}{\varepsilon_0} = \left(1 + \chi_e\right). \tag{1.72}$$

The dielectric constants of several common materials are listed in Table 1.1. The dielectric constant and the permittivity contain the same information, of course, differing only by the factor ε_0, but be careful not to forget this factor when solving problems.

For simple (i.e. linear, isotropic, homogeneous) media, the displacement field **D** and the electric field **E** are linearly proportional, and parallel to one another, and Eq. (1.70) in simplified form is

$$\mathbf{D} = \varepsilon \mathbf{E}. \tag{1.73}$$

Table 1.1

Relative permittivity ε_r (or dielectric constant), defined in Eq. (1.72), of several common materials. The absolute permittivity is $\varepsilon = \varepsilon_r \varepsilon_0$, where $\varepsilon_0 = 8.854 \times 10^{-12}$ F/m.	
Material	ε_r
Vacuum	1.00000
Air	1.0006
Pyrex glass	4.7
Teflon (PTFE)	2.1
Polyethylene	2.25
Diamond	5.5–10
Undoped silicon	11.68
Quartz (SiO_2)	3.9
Mylar	3.1
Mica	3–6
Sapphire	8.9–11.1
Water	80.1
Methanol	30

Now that we have defined simple dielectric media, let's return to the topic of boundary conditions at an interface between two simple media, and work an example to illustrate the application of these conditions.

Example 1.16 Boundary Conditions at an Interface

Consider the boundary between two perfect (i.e. non-conducting) simple dielectric media defined by the surface $x = -\sqrt{3}y$, as shown in Fig. 1.40. The permittivities of the two media are $\varepsilon_1 = 2\varepsilon_0$ and $\varepsilon_2 = 3\varepsilon_0$. The electric field in medium 1 is $\mathbf{E}_1 = E_0 \mathbf{a}_x$. Determine the electric field $\mathbf{E}_2$ near the surface in medium 2.

FIGURE 1.40 The electric field $\mathbf{E}_1$ and $\mathbf{E}_2$ at the interface between two media. The interface is defined by $x = -\sqrt{3}y$, the electric field in medium 1 is $\mathbf{E}_1 = E_0 \mathbf{a}_x$, and the permittivity of the two media are $\varepsilon_1 = 2\varepsilon_0$ and $\varepsilon_2 = 3\varepsilon_0$.

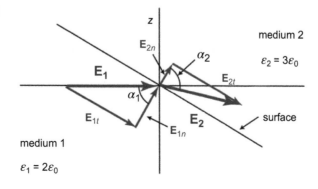

Solution:

We start by finding the unit normal vector for the surface, as described in Example D.2 of Appendix D.1,

$$\mathbf{a}_n = \frac{\boldsymbol{\nabla}\left(x + \sqrt{3}y\right)}{\left|\boldsymbol{\nabla}\left(x + \sqrt{3}y\right)\right|},$$

where $x + \sqrt{3}y = 0$ is the equation that defines the surface. The gradient in the numerator is $\boldsymbol{\nabla}\left(x + \sqrt{3}y\right) = \mathbf{a}_x + \sqrt{3}\mathbf{a}_y$, whose magnitude is $\left|\boldsymbol{\nabla}\left(x + \sqrt{3}y\right)\right| = \sqrt{1^2 + 3} = 2$, so the unit normal vector is

$$\mathbf{a}_n = \frac{1}{2}\left(\mathbf{a}_x + \sqrt{3}\mathbf{a}_y\right).$$

$\mathbf{a}_n$ can be used to break the electric field $\mathbf{E}_1$ into its normal and tangential components. E_{1n}, the magnitude of $|\mathbf{E}_{1n}|$, is

$$E_{1n} = \mathbf{a}_n \cdot \mathbf{E}_1 = \frac{1}{2}\left(\mathbf{a}_x + \sqrt{3}\mathbf{a}_y\right) \cdot E_0\mathbf{a}_x = \frac{1}{2}E_0,$$

and

$$\mathbf{E}_{1n} = E_{1n}\mathbf{a}_n = \frac{E_0}{4}\left(\mathbf{a}_x + \sqrt{3}\mathbf{a}_y\right).$$

To find the tangential component of $\mathbf{E}_1$, we use

$$\begin{aligned}\mathbf{E}_{1t} &= \mathbf{E}_1 - \mathbf{E}_{1n}\\ &= E_0\mathbf{a}_x - \frac{E_0}{4}\left(\mathbf{a}_x + \sqrt{3}\mathbf{a}_y\right)\\ &= \frac{E_0}{4}\left(3\mathbf{a}_x - \sqrt{3}\mathbf{a}_y\right).\end{aligned}$$

Now that we have the normal and tangential components of $\mathbf{E}_1$, we're ready to apply the boundary conditions that the electric field must satisfy, that is Eq. (1.67)

$$\mathbf{E}_{2t} = \mathbf{E}_{1t} = \frac{E_0}{4}\left(3\mathbf{a}_x - \sqrt{3}\mathbf{a}_y\right),$$

and Eq. (1.68)

$$\mathbf{D}_{2n} - \mathbf{D}_{1n} = \rho_s.$$

The surface charge density ρ_s must be zero since the dielectrics are non-conducting, so $\mathbf{D}_{2n} = \mathbf{D}_{1n}$. Then the normal electric field in medium 2 is

$$\mathbf{E}_{2n} = \frac{1}{\varepsilon_2}\mathbf{D}_{2n} = \frac{1}{\varepsilon_2}\mathbf{D}_{1n} = \frac{\varepsilon_1}{\varepsilon_2}\mathbf{E}_{1n} = \left(\frac{2\varepsilon_0}{3\varepsilon_0}\right)\frac{E_0}{4}\left(\mathbf{a}_x + \sqrt{3}\mathbf{a}_y\right) = \frac{E_0}{6}\left(\mathbf{a}_x + \sqrt{3}\mathbf{a}_y\right),$$

where Eq. (1.73) was used to relate $\mathbf{D}$ to $\mathbf{E}$. Finally, we combine $\mathbf{E}_{2t}$ and $\mathbf{E}_{2n}$ to get the total field $\mathbf{E}_2$,

$$\mathbf{E}_2 = \mathbf{E}_{2t} + \mathbf{E}_{2n} = \frac{E_0}{4}\left(3\mathbf{a}_x - \sqrt{3}\mathbf{a}_y\right) + \frac{E_0}{6}\left(\mathbf{a}_x + \sqrt{3}\mathbf{a}_y\right) = \frac{E_0}{12}\left(11\mathbf{a}_x - \sqrt{3}\mathbf{a}_y\right).$$

Note that the magnitude of the electric field in the second medium is $|\mathbf{E}_2| = 0.928\,E_0$. Thus, $|\mathbf{E}_2| < |\mathbf{E}_1|$, and the electric field $\mathbf{E}_2$ makes a greater angle with the surface normal than does $\mathbf{E}_1$, as can be shown by finding

$$\alpha_1 = \cos^{-1}\left(\frac{\mathbf{E}_1 \cdot \mathbf{a}_n}{E_0}\right) = \cos^{-1}\left(\frac{1}{2}\right) = 60°$$

and

$$\alpha_2 = \cos^{-1}\left(\frac{\mathbf{E}_2 \cdot \mathbf{a}_n}{|\mathbf{E}_2|}\right) = \cos^{-1}\left(\frac{2}{\sqrt{31}}\right) = 68.9°.$$

Table 1.2

Dielectric strength of several insulating materials.	
Material	**E_{max} (kV/cm)**
Air	12–30
Porcelain	16–110
Bakelite	130–150
Rubber	60–200
Paper	80
Teflon (PTFE)	400
Mica	1 500–2 200

While we are on this topic of dielectrics, we will also define a material parameter known as the dielectric strength of a material. For every medium, there is a maximum electric field that the material can withstand before it breaks down electrically – that is, before its electrons are ripped off the nuclei and the material starts to conduct. In a solid material this typically occurs in a very irreversible manner, and the dielectric medium will never be restored to its original condition. This maximum electric field is called the **dielectric strength** of the material. See Table 1.2 for the dielectric strength of several common materials.

These values are typically in the range of 10^7 V/m. Recall that in Example 1.1 we calculated the electric field strength due to an electron at distance of 10 Å, and found the result to be $\sim 7 \times 10^8$ V/m. This value is in the same neighborhood as the dielectric strength of materials, which can be understood in terms of an external field strength that is strong enough to overcome the binding field that holds the electrons in place within an atom.

Electrical breakdown is not an uncommon event, and you have likely experienced or observed some of the following examples previously. Our first example is the common spark plug, used in all internal combustion engines.

TechNote 1.7 Spark plug

Spark plugs are used in internal combustion engines to create the electrical spark that initiates the combustion of the gasoline/air mixture in the cylinder, the basis for operation of the engine. A spark plug, as shown in Fig. 1.41, has a central conductor, electrically isolated from the grounded body. When a high-voltage pulse is applied to the center conductor of the spark plug, the air in the gap between the conductors breaks down (is ionized), and starts conducting. This spark initiates combustion of the fuel, and the expanding gas in the cylinder forces the engine piston downward.

The first spark plug is sometimes attributed to Edmond Berger, an African American from New York, in 1839, although he did not file a patent claim. The first commercially successful internal combustion engine, which of course used spark plugs in its operation, was invented in 1860 by Belgian-born French engineer Étienne Lenoir.

FIGURE 1.41 A photograph of a spark plug.

A prime example of electrical discharges in nature is lightning, often accompanying serious rainstorms. Many buildings are fitted with lightning rods to protect against dangerous lightning strikes.

TechNote 1.8 Lightning and Lightning Rods

Lightning is a massive electrical discharge that results after a large accumulation of electrical charge in the clouds. Although the charging mechanism is not completely understood, it is known that it occurs in the central region of clouds, where the temperature is below freezing ($-25\,°C$ to $-15\,°C$), and where the clouds consist of a mix of small ice crystals, super-cooled liquid water (water below the freezing temperature which has not yet solidified), and larger hail particles known as graupel. Air currents in the clouds carry the small ice crystals and super-cooled liquid water upwards, while the larger, heavier graupel drop or remain suspended. As a result of collisions between the ice crystals and the graupel, charge is transferred from one to the other, and the ice crystals become positively charged, and the graupel becomes negatively charged. The net result is an accumulation of negative charge in the lower regions of the cloud, and positive charge at the top. The negative charge in the lower regions of the cloud induce a positive charge in the surface of the Earth, and a large electric field is established in the atmosphere between the cloud and the ground. When this electric field exceeds the dielectric strength (the breakdown field) of the air, the air ionizes and becomes more conductive. The result is a cloud-to-ground lightning strike. An average of 51 people are killed each year in the United States due to lightning strikes; an estimated 2 000 worldwide. Lightning can also strike from cloud to cloud. Heat lightning is lightning that is so distant that the accompanying thunder cannot be heard.

Lightning rods placed on the rooftops of buildings or other structures, as illustrated in Fig. 1.42, can help safeguard those structures from damage from lightning strikes. A lightning rod is a conducting rod, which when connected through a heavy wire to the ground, can provide a safe, low-resistance pathway for the huge currents in lightning strikes to be conducted to ground. Lightning rods were first introduced by Benjamin Franklin in the 1750s.

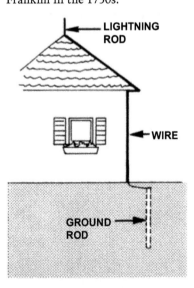

FIGURE 1.42 Lightning rod installed on a house.

The Tesla coil, first introduced by Nikola Tesla in the 1890s, is a high-frequency, high-voltage generator capable of producing large electrical discharges in the air. Tesla coils are described in TechNote 1.9.

TechNote 1.9 Tesla Coil

A Tesla coil is a doubly resonant circuit capable of producing an extremely high-voltage (50 kV to a few MV), low-current, high-frequency electrical discharge. The Tesla coil was invented in 1891 by Nikola Tesla, pictured in Fig. 1.43(a), an American electrical engineer (born in present-day Croatia) who immigrated to the United States in 1884. Tesla envisioned the Tesla coil as a means of transmitting power wirelessly over long distances.

The schematic shown in Fig. 1.43(b) is similar to Tesla's original design. The capacitor C1 and inductor L1 form a resonant circuit, which is charged through the secondary of the lines step-up transformer (labeled T) on the left. When the capacitor voltage reaches the breakdown voltage of the spark gap (SG), C1 discharges through L1, and high-frequency oscillation at the resonant frequency is initiated. L1 and L2 are weakly coupled co-axial coils, with the number of turns in L2 much larger than the number of turns in L1. The capacitance C2 is the stray capacitance of the top of the coil with surrounding grounded surfaces, and the stray capacitance between coils of the secondary. To optimize the operation of the Tesla coil, the resonant frequency of the primary is tuned to that of the secondary.

When in operation, the potential at the top of the coil is very high, oscillating at a frequency typically in the 50 kHz to 1 MHz range. See Fig. 1.43(c) for a photograph of the electrical arc sustained near this tip. The dome at the top of the coil in this photo allows the potential to reach higher levels before causing electrical breakdown of the surrounding air. Notice the glowing fluorescent tube to the right of the Tesla coil. The gas inside the long linear tube is ionized by the strong electric field generated by the Tesla coil. Modern day Tesla coils may use relatively simple transistor circuits which don't require careful tuning of the resonant frequencies to achieve high-voltage discharges at the top of the coil.

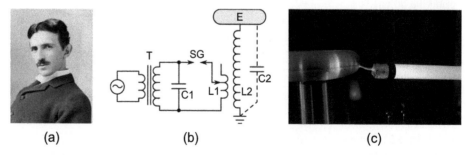

(a)   (b)   (c)

FIGURE 1.43 (a) Photograph of Nikola Tesla age 34, circa 1890. (b) Resonant driving circuit for a Tesla coil. (c) An operating Tesla coil. Note the fluorescent tube to the right of the Tesla coil. The strong electric field produced by the Tesla coil ionizes the gas inside the tube, causing the lamp to glow. (Photograph used with permission.)

The last example of electrical discharges highlighted here is that of corona discharges, commonly found near high-voltage transmission lines. These discharges are a major source of losses in the power grid.

TechNote 1.10 Corona Discharge

At points where the electric field outside a charged conductor exceeds the dielectric strength of the medium, the intense electric field can ionize that medium, causing the medium to become conductive. This is known as a corona discharge. Corona discharges in air near a high-tension utility line are shown. Corona discharges such as this are in fact one of the major sources of losses from the power grid. The dielectric strength of air is around 30 kV/cm. The electric field near sharp corners of a conductor can exceed this value, allowing for the discharge seen. In addition to power loss, corona discharges can also be responsible for ozone production, material degradation, and electromagnetic interference. On the other hand, there are several useful applications of coronas, such as in air purifiers, copy machines, and surface cleaning.

LEARNING OBJECTIVES

After studying this section, you should be able to:

- apply Gauss' Law to determine the displacement field in geometries of high symmetry;
- use the permittivity and **D** to find the electric field; and
- use **D** and **E** to find the polarization field.

1.11 Examples of $\mathbf{E}$, $\mathbf{D}$, and $\mathbf{P}$ in Dielectric Media

In the following examples we explore applications of Gauss' Law, Eq. (1.66), and the relationship between the displacement field **D**, the polarization field **P**, and the total electric field **E**.

Example 1.17 Charged Sphere with Dielectric Medium

A sphere of uniform charge density ρ_v, radius b, and permittivity ε_1 is shown in Fig. 1.44. The permittivity for $R > b$ is ε_2. Determine the displacement field $\mathbf{D(r)}$, the electric field $\mathbf{E(r)}$, and the polarization field $\mathbf{P(r)}$, for $R < b$ and $R > b$.

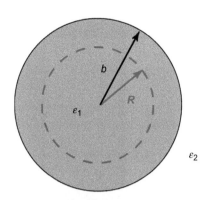

FIGURE 1.44 A sphere of uniform charge density ρ_v, radius b, and permittivity ε_1, considered in Example 1.17.

Solution:

We will use Gauss' Law in its now complete form Eq. (1.66). We start by constructing a Gaussian surface of radius R, concentric with the spherical charge. Using the symmetry of the spherical charge, we infer that $\mathbf{D}(\mathbf{r})$ must have only a radial component, and its magnitude must depend only on the radial coordinate R. On the Gaussian surface, then, $\mathbf{D}(\mathbf{r})$ is uniform, and

$$\oint_s \mathbf{D} \cdot d\mathbf{s} = D_R(R)\, 4\pi R^2 = Q_{enc}.$$

This result is valid for $R < b$ as well as for $R > b$. Next we need to evaluate the charge Q_{enc} enclosed within the Gaussian surface. For $R < b$, this is

$$Q_{enc} = \int_v \rho_v dv = \int_{\phi=0}^{2\pi} \int_{\theta=0}^{\pi} \int_{R'=0}^{R} \rho_v R'^2 dR' \sin\theta\, d\theta\, d\phi = \rho_v \left(\frac{4\pi}{3} R^3 \right).$$

Together, these two equations give us $D_R(R) = \rho_v R/3$ for $R < b$.

For $R > b$, the charge enclosed within the Gaussian surface is found using

$$Q_{enc} = \int_v \rho_v dv = \int_{\phi=0}^{2\pi} \int_{\theta=0}^{\pi} \int_{R'=0}^{b} \rho_v R'^2 dR' \sin\theta\, d\theta\, d\phi = \rho_v \left(\frac{4\pi}{3} b^3 \right).$$

Notice the upper limit of integration of the R' integral is b, since the charge only extends out to this radius. Setting this equal to the flux $\oint_s \mathbf{D} \cdot d\mathbf{s}$ and solving for $D_R(R)$ yields $D_R(R) = \rho_v b^3/3R^2$ for $R > b$. The results for the displacement field are summarized as

$$D_R(R) = \begin{cases} \rho_v R/3 & R < b \\ \rho_v b^3/3R^2 & R > b \end{cases}.$$

The displacement field $\mathbf{D}$, which is normal to the interface between the two regions (at $R = b$), is continuous at the interface, as required by the boundary condition Eq. (1.68), since there is no surface charge.

Using $\mathbf{E} = \mathbf{D}/\varepsilon$ the electric field is

$$E_R(R) = \begin{cases} \rho_v R/3\varepsilon_1 & R < b \\ \rho_v b^3/3\varepsilon_2 R^2 & R > b \end{cases}.$$

Finally, the polarization field $\mathbf{P}$ in each medium is determined using Eq. (1.64), with the results

$$P_R(R) = \begin{cases} \dfrac{\rho_v R}{3}\left(1 - \dfrac{\varepsilon_0}{\varepsilon_1}\right) & R < b \\[4mm] \dfrac{\rho_v b^3}{3R^2}\left(1 - \dfrac{\varepsilon_0}{\varepsilon_2}\right) & R > b \end{cases}$$

in the two regions.

Let's try a similar example, this time with a cylindrical geometry.

Example 1.18 Co-axial Conductors Spaced by a Dielectric Medium

Consider two co-axial conductors of length L, as shown in Fig. 1.45. The center conductor, of radius a, is held at potential V_0 relative to the grounded outer conductor (or shield), which is of radius b. The dielectric permittivity of the medium in the space between the conductors is ε. Determine the electric field $\mathbf{E(r)}$ in each region in terms of the potential V_0 for this configuration. (Ignore fringe field effects at the ends of the cylinder.)

Solution:

The field inside either of the conductors must be zero. On the outside (i.e. $\rho > b$), the electric field must be zero as well since the shield is grounded. All that remains is to determine the field in the region between the conductors. For this, we resort to Gauss' Law. Consider a cylindrical Gaussian surface of radius ρ and length h centered on the axis of the cylinder. By the cylindrical symmetry of the configuration, $\mathbf{D}$ must be directed radially, and its magnitude can depend only on the distance from the axis ρ. Then application of Gauss' Law to this surface yields

$$\oint_s \mathbf{D} \cdot d\mathbf{s} = D_\rho 2\pi\rho h = Q_{enc} = \rho_{s,a} 2\pi a h,$$

where $\rho_{s,a}$ is the surface charge density on the inner conductor. Solving this for D_ρ gives us $D_\rho = \rho_{s,a} a/\rho$. Since $\mathbf{E} = \mathbf{D}/\varepsilon$, the electric field is

$$E_\rho = \frac{\rho_{s,a} a}{\varepsilon \rho}.$$

This is the electric field that we seek, but here it is expressed in terms of the surface charge density $\rho_{s,a}$, rather than the potential V_0. We can determine V_0 from $\mathbf{E(r)}$ using

$$V_0 = -\int_b^a \mathbf{E(r)} \cdot d\boldsymbol{\ell},$$

where $d\boldsymbol{\ell}$ is an element along any path from the reference (the outer conductor) to the inner conductor. Using $\mathbf{E} = \mathbf{a}_\rho \rho_{s,a} a/\varepsilon\rho$ and $d\boldsymbol{\ell} = \mathbf{a}_\rho d\rho$, the potential is

$$V_0 = \frac{\rho_{s,a} a \ln(b/a)}{\varepsilon}.$$

Eliminating $\rho_{s,a}$, the electric field is

$$\mathbf{E} = \mathbf{a}_\rho \frac{V_0}{\ln(b/a)\rho}.$$

Notice that $\mathbf{E}$ decreases as $1/\rho$, and is independent of the permittivity ε of the dielectric.

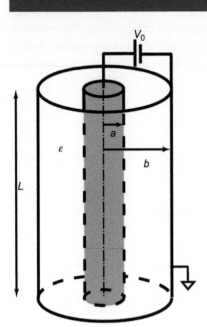

FIGURE 1.45 Two co-axial conductors of length L.

Example 1.19 Parallel Plate Conductors Filled with Two Dielectrics

A pair of parallel conducting plates is separated by two dielectric materials, as shown in Fig. 1.46. The bottom dielectric medium is air, of thickness d_1 and permittivity ε_0, while the upper material is of thickness d_2 and permittivity $\varepsilon_2 > \varepsilon_0$. The area of each plate is A, and the spacing between the plates is much less than $\sqrt{A}$. Determine the electric field $\mathbf{E}$, the displacement field $\mathbf{D}$, and the polarization field $\mathbf{P}$ in the space between the conducting plates.

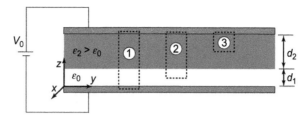

FIGURE 1.46 Parallel conducting plates separated by two dielectrics. The media between the plates are air, of thickness d_1 and permittivity ε_0, and another dielectric material of thickness d_2 and permittivity $\varepsilon_2 > \varepsilon_0$.

Solution:

Since we are told that the spacing between the plates is small, we can ignore the "fringe" fields at the edge of the capacitor, and approximate the fields **D** and **E** as uniform and directed in the $-\mathbf{a}_z$ direction. We apply Gauss' Law to determine **D** in the two dielectrics. As before, we choose the Gaussian surfaces based upon the symmetry of the elements in the configuration. Three different Gaussian surfaces, as shown in Fig. 1.46, are useful. The first allows us to relate the charge densities on the top and bottom conductors, the second will lead us to the displacement field **D** in the lower dielectric medium, and the third for a determination of **D** in the upper dielectric. In each case, the shape of the Gaussian surface in the x–y plane (not visible in our view from the side) is unimportant. We only need to specify Δs, the area of the top and bottom surfaces (lying parallel to the x–y plane). Since **D** is directed in the $-\mathbf{a}_z$ direction, and the surface $d\mathbf{s}$ for the side of the Gaussian surface points somewhere in the x–y plane (i.e. perpendicular to **D**), the sidewalls do not contribute to $\oint_s \mathbf{D} \cdot d\mathbf{s}$ in any case.

With the voltage source (power supply or battery) connected to the conductors, a charge collects on the inside surfaces of the top and bottom conducting plates. Let us label the charge on the top plate Q_t and the charge on the bottom plate Q_b, and the respective surface charge densities $\rho_{s,t} = Q_t/A$ and $\rho_{s,b} = Q_b/A$. Applying Eq. (1.66) to the Gaussian surface labeled #1, which has its top surface inside the top conductor and its bottom surface inside the bottom conductor, we find

$$\oint_s \mathbf{D} \cdot d\mathbf{s} = Q_{enc} = \rho_{s,t}\Delta s + \rho_{s,b}\Delta s.$$

Since the top and bottom surfaces lie inside the conductors, where the field must be zero, and since the sides are perpendicular to **D**, the flux $\oint_s \mathbf{D} \cdot d\mathbf{s}$ must be identically zero. This requires that $\rho_{s,t}$ and $\rho_{s,b}$ are equal but opposite. Let us label the surface charge on the top and bottom plates ρ_s and $-\rho_s$, respectively.

Next we consider the second Gaussian surface shown in Fig. 1.46, and use this surface to determine **D** within the lower dielectric medium between the two plates, identified as air. $\mathbf{D} \cdot d\mathbf{s}$ is zero on the top surface since this surface is inside the conductor, and zero on the sides since $d\mathbf{s}$ is perpendicular to **D**. The only surface with non-zero flux is the bottom surface, for which $\mathbf{D} = D_z\mathbf{a}_z$ and $d\mathbf{s} = -\mathbf{a}_z ds$, resulting in a flux of $\oint_s \mathbf{D} \cdot d\mathbf{s} = -D_z\Delta s$. The net charge contained within this Gaussian surface is $\rho_s \Delta s$, since the only charge is that which is on the bottom surface of the top conductor. From this we learn that D_z in the bottom dielectric is equal to $-\rho_s$. D_z is negative since the field lines point away from the positive charges in the upper plate (and toward the negative charges in the bottom plate).

Finally, we use the third Gaussian surface to determine **D** in the upper dielectric. The approach closely parallels that of the preceding paragraph, with the result that D_z in the top dielectric is $-\rho_s$ – that is, the same as we found in the lower dielectric. Therefore,

$$\mathbf{D} = -\mathbf{a}_z \rho_s$$

in either medium. Note that this result is consistent with the requirement that the normal component of $\mathbf{D}$ must be continuous at the surface between the two dielectrics.

The electric field $\mathbf{E}$ is different in the two dielectric media, since $\mathbf{E} = \mathbf{D}/\varepsilon$, and we write these as

$$\mathbf{E} = \begin{cases} -\mathbf{a}_z \dfrac{\rho_s}{\varepsilon_2} & d_1 < z < d_1 + d_2 \\ -\mathbf{a}_z \dfrac{\rho_s}{\varepsilon_0} & 0 < z < d_1 \end{cases}.$$

We can find the polarization field $\mathbf{P}$ using Eq. (1.64), $\mathbf{P} = \mathbf{D} - \varepsilon_0 \mathbf{E}$, with the result

$$\mathbf{P} = \begin{cases} -\mathbf{a}_z \rho_s \left(1 - \dfrac{\varepsilon_0}{\varepsilon_2}\right) & d_1 < z < d_1 + d_2 \\ 0 & 0 < z < d_1 \end{cases}.$$

Only in the upper dielectric is there a non-zero polarization field $\mathbf{P}$, since the bottom half is filled with air, whose density is sparse.

Let us examine the polarization density $\mathbf{P}$ in this last example, and the effective charge densities represented by it, a bit more. Our goal is to use this example to gain a physical picture of the charges and various fields. Using Eq. (1.62), we compute the effective surface charge densities $\rho_{s,eff} = \mathbf{P} \cdot \mathbf{a}_n$. At the top surface of the upper dielectric ($z = d_1 + d_2$), the unit normal vector $\mathbf{a}_n$ is $+\mathbf{a}_z$, and $\rho_{s,eff} = P_z = D_z - \varepsilon_0 E_z = -(1 - \varepsilon_0/\varepsilon_2)\rho_s$. At the bottom of the upper dielectric ($z = d_1$), $\mathbf{a}_n$ is $-\mathbf{a}_z$, and $\rho_{s,eff}$ here is $-P_z = (1 - \varepsilon_0/\varepsilon_2)\rho_s$. Since $\mathbf{P}$ is uniform in the upper dielectric, Eq. (1.63) tells us that $\rho_{v,eff}$ is zero. This is consistent with our picture that the positively charged upper conductor attracts the negative charges in the polarizable material, and repels the positive charges. Thus there is a net negative effective charge on the top surface of the dielectric material, and positive on the bottom. These effective surface charges on the dielectric material tend to mitigate the effect of the free surface charges on the conductor plate, such that the net electric field $\mathbf{E}$ is *reduced* in this material. Effectively, the net charge density at $z = d_1 + d_2$ is $\rho_s + \rho_{s,eff} = \rho_s \varepsilon_0/\varepsilon_2$. Recall the overly simplified interpretation of the fields $\mathbf{E}$, $\mathbf{D}$, and $\mathbf{P}$ that we introduced earlier, in which $\mathbf{D}$ is the field due to the free charges that reside on the conductor plates, $\mathbf{P}$ is the field due to the polarization effective charges, and $\mathbf{E}$ is the net field due to the free charges as well as the effective charges. In the present example, this model is valid. The electric force felt by a hypothetical test charge is due to the net field $\mathbf{E}$ — that is, the field due to the free charges and effective polarization charges combined.

Let's try another example.

Example 1.20 Polarized Cube

Consider the polarized cube shown in Fig. 1.47. The dimensions of the cube are $a \times a \times d$, in the x-, y-, and z-directions, respectively, and the polarization $\mathbf{P} = P_0 \mathbf{a}_z$ is uniform inside the cube, and zero elsewhere. (a) Determine the effective charge densities $\rho_{s,eff}(\mathbf{r})$ and $\rho_{v,eff}(\mathbf{r})$. (b) Use Coulomb's Law to write the integral that could be used to solve for the electric field $\mathbf{E}$ anywhere. (c) Solve this integral for $\mathbf{E}$ at a position along the z-axis. Plot $E_z(z)$ and $D_z(z)$ vs. z.

FIGURE 1.47 A polarized cube. The height is *d* and the sides are each *a*.

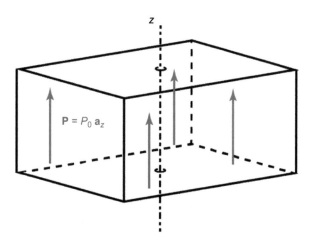

Solution:

(a) We start by finding the effective charges on the surfaces of the cube, and distributed throughout the volume of the cube. Using Eqs. (1.62) and (1.63), these charge densities are

$$\rho_{s,eff}(\mathbf{r}) = \mathbf{P} \cdot \mathbf{a}_z = P_0$$

on the top face ($z = d/2$), and

$$\rho_{s,eff}(\mathbf{r}) = \mathbf{P} \cdot (-\mathbf{a}_z) = -P_0$$

on the bottom face ($z = -d/2$). The other four faces are uncharged. The effective volume charge density is zero, since the polarization field **P** is uniform. This charge distribution should make sense to us, since the electric dipole moment vector **p** points from the negative charge of the dipole to the positive. The unmatched charges are therefore positive on the top face, and negative on the bottom.

(b) We can use these effective charges in Coulomb's Law to determine the electric field that they generate. For the general field point at $\mathbf{r} = (x, y, z)$, with effective charge located at $\mathbf{r}' = (x', y', \pm d/2)$, we use the form of Coulomb's Law valid for surface charges, Eq. (1.15), with $\rho_{s,eff}$ for ρ_s and $dx'\,dy'$ for the surface element,

$$
\mathbf{E}(\mathbf{r}) = \frac{1}{4\pi\varepsilon_0} \left\{ \int_{-a/2}^{a/2} \int_{-a/2}^{a/2} \frac{P_0\,(x-x', y-y', z-d/2)\,dx'\,dy'}{\left[(x-x')^2 + (y-y')^2 + (z-d/2)^2\right]^{3/2}} \right.
$$
$$
\left. - \int_{-a/2}^{a/2} \int_{-a/2}^{a/2} \frac{P_0\,(x-x', y-y', z+d/2)\,dx'\,dy'}{\left[(x-x')^2 + (y-y')^2 + (z+d/2)^2\right]^{3/2}} \right\}.
$$

The first integral gives us the field due to the positive effective charges on the top surface, while the second integral is for the negative effective charges on the bottom surface. For the general field point **r**, these integrals must be evaluated numerically. Along the z-axis, however, we can evaluate the integrals exactly, as we show next.

(c) By symmetry, the E_x and E_y components vanish for any point along the z-axis. This can be shown rigorously from the integrals above. The integrand for the x-component is odd in x', while the integrand for the y-component is odd in y'. Since these integrals are evaluated over an even interval ($-a/2$ to $+a/2$), the integrals are zero. Only the z-component remains to be evaluated. For this, we have two integrals to evaluate, one at $z' = +d/2$, the other at $z' = -d/2$, but these integrals are similar. Let's save ourselves some work by first evaluating the field due to a single square of uniform charge density.

For simplicity, we'll place this square at $z = 0$. After we have this result, we'll be able to easily modify it to find the electric field for our situation. The integral we will evaluate is

$$Y = \int_{-a/2}^{a/2} \int_{-a/2}^{a/2} \frac{z\, dx'\, dy'}{\left[(x - x')^2 + (y - y')^2 + z^2\right]^{3/2}}.$$

We use the integral Eq. (G.13) to evaluate the integral in dx', along with the observation that the integrand is even in x' and y', to write

$$Y = 4z \int_0^{a/2} \frac{dy'}{y'^2 + z^2} \frac{a/2}{\sqrt{(a/2)^2 + y'^2 + z^2}}.$$

Next we make the substitution

$$y' = \sqrt{(a/2)^2 + z^2}\, \tan u,$$

whose differential is

$$dy' = \sqrt{(a/2)^2 + z^2}\, \sec^2 u\, du.$$

Since $1 + \tan^2 u = \sec^2 u$, the integral simplifies to

$$Y = 2az \int_0^{u_{max}} \frac{\cos u\, du}{z^2 + (a/2)^2 \sin^2 u},$$

where the upper limit of the integral is $u_{max} = \tan^{-1}(1/\sqrt{1 + (2z/a)^2})$. With a final substitution $w = \sin u$, this becomes

$$Y = 2az \int_0^{w_{max}} \frac{dw}{z^2 + (a/2)^2 w^2},$$

where $w_{max} = \sin(u_{max}) = [2 + (2z/a)^2]^{-1/2}$. We evaluate this integral with the aid of integral Eq. (G.10) to find

$$Y = \sin^{-1}\left(\frac{a^2}{a^2 + (2z)^2}\right).$$

So now we can use this result for the integral Y to determine the field due to the two effective surface charges, $+P_0$ at $z = +d/2$ and $-P_0$ at $z = -d/2$. The electric field is

$$E(z) = \begin{cases} -E^+(z) + E^-(z) & z < -d/2 \\ -E^+(z) - E^-(z) & -d/2 < z < d/2, \\ E^+(z) - E^-(z) & z > d/2 \end{cases} \qquad (1.74)$$

where

$$E^+(z) = \frac{P_0}{\pi \varepsilon_0} \sin^{-1}\left(\frac{a^2}{a^2 + (2z - d)^2}\right)$$

and

$$E^-(z) = \frac{P_0}{\pi \varepsilon_0} \sin^{-1}\left(\frac{a^2}{a^2 + (2z + d)^2}\right).$$

$E(z)$ vs. z is plotted in Fig. 1.48(a).

We determine the displacement field $D_z(z)$ along the z-axis using Eq. (1.64), and plot $D_z(z)$ vs. z in Fig. 1.48(b). Notice that $D_z(z)$ is continuous at the two z-faces of the polarized cube, in agreement with the boundary condition for normal field components, Eq. (1.68). Note that $\rho_s = 0$ at these surfaces, since there are no free charges in the dielectric, only bound "effective" charges. The absence of free charges also has implications for our simplified model in which we identified $\mathbf{D}$ as the field due to free charges. This model clearly breaks down in this example, so we are reminded that this simple model is limited.

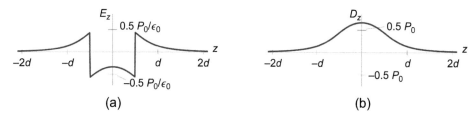

(a)

(b)

FIGURE 1.48 A plot of (a) the electric field $E_z(z)$ (Eq. (1.74)) and (b) the displacement field $D_z(z)$ along the z-axis of the polarized cube considered in Example 1.20.

Table 1.3

Collection of key formulas from Chapter 1.			
$\mathbf{F}_{12} = \dfrac{1}{4\pi\varepsilon_0} \dfrac{q_1\, q_2}{\|\mathbf{r}_1 - \mathbf{r}_2\|^2}\, \mathbf{a}_{12}$	(1.1)	$V(\mathbf{r}) = \dfrac{1}{4\pi\varepsilon_0} \displaystyle\sum_{i=1}^{N} \dfrac{Q_i}{\|\mathbf{r} - \mathbf{r}'_i\|}$	(1.46)
$\mathbf{E}(\mathbf{r}) = \dfrac{1}{4\pi\varepsilon_0} \dfrac{q_2}{\|\mathbf{r} - \mathbf{r}'\|^2}\, \mathbf{a}_{rr'}$	(1.3)	$V(\mathbf{r}) = \dfrac{1}{4\pi\varepsilon_0} \displaystyle\int_v \dfrac{\rho_v(\mathbf{r}')}{\|\mathbf{r} - \mathbf{r}'\|}\, dv$	(1.47)
$\mathbf{E}(\mathbf{r}) = \dfrac{1}{4\pi\varepsilon_0} \displaystyle\sum_{i=1}^{N} \dfrac{q_i(\mathbf{r} - \mathbf{r}'_i)}{\|\mathbf{r} - \mathbf{r}'_i\|^3}$	(1.6)	$V(\mathbf{r}) = \dfrac{1}{4\pi\varepsilon_0} \displaystyle\int_s \dfrac{\rho_s(\mathbf{r}')}{\|\mathbf{r} - \mathbf{r}'\|}\, ds'$	(1.48)
$\mathbf{E}(\mathbf{r}) = \dfrac{1}{4\pi\varepsilon_0} \displaystyle\int_v \dfrac{\rho_v(\mathbf{r}')(\mathbf{r} - \mathbf{r}')}{\|\mathbf{r} - \mathbf{r}'\|^3}\, dv$	(1.11)	$V(\mathbf{r}) = \dfrac{1}{4\pi\varepsilon_0} \displaystyle\int_\ell \dfrac{\rho_\ell(\mathbf{r}')}{\|\mathbf{r} - \mathbf{r}'\|}\, d\ell'$	(1.49)
$\mathbf{E}(\mathbf{r}) = \dfrac{1}{4\pi\varepsilon_0} \displaystyle\int_s \dfrac{\rho_s(\mathbf{r}')(\mathbf{r} - \mathbf{r}')}{\|\mathbf{r} - \mathbf{r}'\|^3}\, ds'$	(1.15)	$V(x, y, z) = \dfrac{\mathbf{p} \cdot \mathbf{r}}{4\pi\varepsilon_0\, R^3} = \dfrac{\mathbf{p} \cdot \mathbf{a}_R}{4\pi\varepsilon_0\, R^2}$	(1.56)
$\mathbf{E}(\mathbf{r}) = \dfrac{1}{4\pi\varepsilon_0} \displaystyle\int_\ell \dfrac{\rho_\ell(\mathbf{r}')(\mathbf{r} - \mathbf{r}')}{\|\mathbf{r} - \mathbf{r}'\|^3}\, d\ell'$	(1.16)	$\mathbf{E} = \dfrac{p}{4\pi\varepsilon_0\, R^3}\,(\mathbf{a}_R\, 2\cos\theta + \mathbf{a}_\theta \sin\theta)$	(1.57)
$\mathbf{E}_\rho = \dfrac{\rho_\ell}{2\pi\varepsilon_0\, \rho}\,\mathbf{a}_\rho$	(1.23)	$\rho_{s,eff}(\mathbf{r}) = \mathbf{P}(\mathbf{r}) \cdot \mathbf{a}_n$	(1.62)
$\Phi_E = \oint_s \mathbf{E}(\mathbf{r}) \cdot d\mathbf{s} = Q_{enc}/\varepsilon_0$	(1.31)	$\rho_{v,eff}(\mathbf{r}) = -\boldsymbol{\nabla} \cdot \mathbf{P}(\mathbf{r})$	(1.63)
$\boldsymbol{\nabla} \cdot \mathbf{E}(\mathbf{r}) = \rho_v/\varepsilon_0$	(1.32)	$\mathbf{D} \equiv \varepsilon_0\mathbf{E} + \mathbf{P}$	(1.64)
$\boldsymbol{\nabla} \times \mathbf{E} = 0$	(1.33)	$\boldsymbol{\nabla} \cdot \mathbf{D} = \rho_v$	(1.65)
$\oint_c \mathbf{E} \cdot d\ell = 0$	(1.34)	$\oint_s \mathbf{D} \cdot d\mathbf{s} = Q_{enc}$	(1.66)
$\mathbf{E} = -\boldsymbol{\nabla}V$	(1.40)	$E_{1,t} = E_{2,t}$	(1.67)
$V_B - V_A = -\int_A^B \mathbf{E} \cdot d\ell$	(1.43)	$(\mathbf{D}_2 - \mathbf{D}_1) \cdot \mathbf{a}_{n,1} = \rho_s$	(1.68)
$V(\mathbf{r}) = \dfrac{Q}{4\pi\varepsilon_0\, \|\mathbf{r} - \mathbf{r}'\|}$	(1.45)	$\mathbf{D} = \varepsilon_0\,(1 + \chi_e)\,\mathbf{E}$	(1.70)

These examples serve to illustrate several methods for determining the electric field in dielectric media. In the next chapter we turn to a discussion of the energy stored in an electric field, where we will make use of these new skills.

SUMMARY   It is remarkable how much we have been able to develop on the basis of just two key concepts: (1) empirical observations of the force between charged bodies; and (2) the notion that the electric field obeys simple superposition. The observation that electric forces between point charges are directed along the line joining the bodies, and that the magnitude varies as the inverse of the square of the distance between the bodies, allows us to invent the concept of the electric field, and to determine the electric field for any known distribution of charges. Based on the electric field for a point charge, and superposition, we were able to derive rather simple differential laws the electric fields must satisfy: Gauss' Law and $\nabla \times \mathbf{E}(\mathbf{r}) = 0$. The latter differential law also allowed us to define the electric potential, which we developed on the basis of the work required to move a charge along a path against an electric field $\mathbf{E}(\mathbf{r})$. We also introduced dielectric, or insulating, and conducting materials. In dielectrics the charges are bound, but polarizable, while in conductors the charge carriers (typically electrons) are free to move around. As a result, the electric field inside conductors must be zero (at least for static fields). It is interesting to note all the different techniques that we may find useful in determining the electric field in different problems. These include Coulomb's Law, Gauss' Law, and the electric potential (see Table 1.3).

PROBLEMS   **Coulomb's Law**

P1-1   Determine the vector force on the electron at the origin for a charge q at location **r** given by:

(a) $q = -e$ at $\mathbf{r} = (1, 1, 1)$ nm
(b) $q = -e$ at $\mathbf{r} = (1, 0, 1)$ nm
(c) $q = +e$ at $\mathbf{r} = (1, 0, 9)$ nm
(d) $q = +e$ at $\mathbf{r} = (0, 0, 10)$ nm

where $e = 1.602 \times 10^{-19}$ C is the fundamental unit of charge.

P1-2   Determine the electric field at the origin due to a charge:

(a) $q = -e$ at $\mathbf{r}' = (1, 1, 1)$ nm
(b) $q = -e$ at $\mathbf{r}' = (1, 0, 1)$ nm
(c) $q = +e$ at $\mathbf{r}' = (1, 0, 9)$ nm
(d) $q = +e$ at $\mathbf{r}' = (0, 0, 10)$ nm

where $e = 1.602 \times 10^{-19}$ C is the fundamental unit of charge.

P1-3   Determine the electric field at the position $\mathbf{r} = (1, 1, 0)$ nm due to a charge:

(a) $q = -e$ at $\mathbf{r}' = (1, 1, 1)$ nm
(b) $q = -e$ at $\mathbf{r}' = (1, 0, 1)$ nm
(c) $q = +e$ at $\mathbf{r}' = (1, 0, -2)$ nm
(d) $q = +e$ at $\mathbf{r}' = (0, 0, 1)$ nm

where $e = 1.602 \times 10^{-19}$ C is the fundamental unit of charge.

P1-4   Consider three point charges, labeled a, b, and c. Charge a is located at $\mathbf{r}'_a = (1, 0, 1)$ and has charge $q_a = -e$, where $e = 1.602 \times 10^{-19}$ C. Charges b and c are located at $\mathbf{r}'_b = (0, 1, 0)$ and $\mathbf{r}'_c = (3, 2, 0)$, and have charges $q_b = +e$ and $q_c = -2e$, respectively. Determine the electric field $\mathbf{E}(\mathbf{r})$ at:

(a) $\mathbf{r} = (1, 1, 1)$ nm
(b) $\mathbf{r} = (2, 0, 1)$ nm
(c) $\mathbf{r} = (1, 0, -2)$ nm
(d) $\mathbf{r} = (0, 0, 1)$ nm

P1-5 In Example 1.3 we used Coulomb's Law to determine the electric field $E(r)$ outside a sphere of uniform charge density. Carry out a similar analysis to determine the electric field inside that sphere of charge.

P1-6 Use Coulomb's Law to determine the electric field $E(r)$ (a) inside and (b) outside an infinitely long cylinder of uniform charge density ρ_v and radius a. (c) Compare your answer for (b) to the electric field generated by an infinitely long line charge of linear charge density ρ_ℓ. (d) Show that $|E(r)|$ is continuous at $\rho = a$.

P1-7 Use Coulomb's Law to determine the electric field $E(r)$ (a) inside and (b) outside an infinitely wide slab of uniform charge density ρ_v. The slab fills the space defined by $-d/2 < z < d/2$. (c) Compare your answer for (b) to the electric field generated by an infinitely wide sheet of surface charge density ρ_s. (d) Sketch a plot of $|E(r)|$ vs. z. (e) Show that $|E(r)|$ is continuous at $z = \pm d/2$.

P1-8 For each of the charge distributions shown in Fig. 1.49, indicate with an arrow the direction of the electric field E at each of the labeled points. (Write "0" if the electric field is zero at a location.)

FIGURE 1.49 Charge distributions for use in Prob. P1-8. In (a), q is positive. In (b), $\rho_\ell(y) = y$ C/m. In (c), ρ_ℓ is a positive constant.

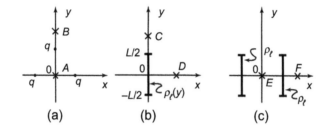

(a)   (b)   (c)

P1-9 Consider a spherical uniform charge density ρ_v between $a < R < b$. Determine the electric field E for the three regions $R < a$, $a < R < b$, and $R > b$.

P1-10 Set up the Coulomb Law integral that would allow us to evaluate the electric field E at position $(x, 0, z)$ produced by a non-uniform, spherical charge density $\rho_v = \rho_0 \cos\theta$ for $R < a$. Show that E_ϕ for this charge distribution must be zero.

P1-11 Consider the finite-length line charge described in Example 1.5. Use the results derived there for E_ρ and E_z, given by Eqs. (1.21) and (1.22), respectively, to show that for distances far from the line charge, that is $R \gg L$, the electric field reduces to the form expected for a point charge of total charge $Q = \rho_\ell L$. [Hint: You must convert these components of E, which here are given in cylindrical coordinates, into spherical components. Equations (B.14) are close enough for this task. Use $\cos\theta = z/R$, $\sin\theta = \rho/R$, and $R^2 = \rho^2 + z^2$, and keep the leading terms as R gets much larger than L. See Appendix H for useful Taylor expansions.]

P1-12 Determine the electric field E for the regions $R < a$ and $R > a$ produced by a spherical uniform surface charge density ρ_s of radius a. How is the discontinuity in E_R at $R = a$ related to ρ_s? (Discontinuities in E are addressed in Section 1.3.)

P1-13 Consider a spherical surface charge density $\rho_s = \rho_0 \cos\theta$ of radius a. Set up the Coulomb Law integral that would allow us to evaluate the electric field E at position $(x, 0, z)$. Show that E_ϕ for this charge distribution must be zero. Sketch the electric field lines inside and outside the sphere for $\rho_0 > 0$.

P1-14 Determine the electric field E due to an infinite sheet of charge of surface charge density ρ_s lying in the x–y plane using Coulomb's Law.

P1-15 Use Coulomb's Law to determine the electric field E along the z-axis of a round disc of radius a and surface charge density ρ_s lying in the x–y plane.

P1-16 Determine the electric field E along the z-axis due to a uniform rectangular surface distribution of charge, of sides $2a \times 2b$ and surface charge density ρ_s lying in the x–y plane. Examine your result for (a) small z and (b) large z. How do these limiting case results compare with the field due to an infinite sheet of uniform charge and a point charge, respectively?

P1-17 Determine the electric field E along the z-axis due to a round loop of uniform charge, of radius a and linear charge density ρ_ℓ lying in the x–y plane using Coulomb's Law.

P1-18 Use Coulomb's Law to determine the electric field E along the z-axis of a square loop of charge, of side $2a$ and linear charge density ρ_ℓ lying in the x–y plane.

Differential Properties of E

P1-19 Evaluate $\nabla \times E$ to determine which of the following vector fields are allowable electrostatic fields. Evaluate $\nabla \cdot E$ to find the charge density ρ_v associated with the field:

(a) $E_1 = E_0 \, (x\mathbf{a}_x + y\mathbf{a}_y + z\mathbf{a}_z)/[x^2 + y^2 + z^2]^{1/2}$
(b) $E_2 = E_0 \, (-y\mathbf{a}_x + x\mathbf{a}_y + z\mathbf{a}_z)/[x^2 + y^2 + z^2]^{1/2}$
(c) $E_3 = E_0 \, (x\mathbf{a}_x)$
(d) $E_4 = E_0 \, (x\mathbf{a}_x + y\mathbf{a}_y + z\mathbf{a}_z)/[x^2 + y^2 + z^2]^{3/2}$.

P1-20 Evaluate $\nabla \times E$ to determine which of the following vector fields are allowable electrostatic fields. Evaluate $\nabla \cdot E$ to find the charge density ρ_v associated with the field:

(a) $E_1 = E_0 \, (\rho\mathbf{a}_\rho + z\mathbf{a}_z)/[\rho^2 + z^2]^{1/2}$
(b) $E_2 = E_0 \, (\rho\mathbf{a}_\phi + z\mathbf{a}_z)/[\rho^2 + z^2]^{1/2}$
(c) $E_3 = E_0 \, (1/\rho)\mathbf{a}_\rho$
(d) $E_4 = E_0 \, \mathbf{a}_\rho$.

P1-21 Evaluate $\nabla \times E$ to determine which of the following vector fields are allowable electrostatic fields. Evaluate $\nabla \cdot E$ to find the charge density ρ_v associated with the field:

(a) $E_1 = E_0 \, \mathbf{a}_R/R^2$
(b) $E_2 = E_0 \, \mathbf{a}_R/R$
(c) $E_3 = E_0 \, \mathbf{a}_\phi/R^2$
(d) $E_4 = E_0 \, (\cos\theta\mathbf{a}_R - \sin\theta\mathbf{a}_\theta)/R^3$.

P1-22 In Section 1.2 we showed, using rectangular coordinates, that $\nabla \times E(\mathbf{r}) = 0$ for the electric field produced by a point charge. Prove this same result using spherical coordinates.

P1-23 In Example 1.5 we found the electric field due to a uniform line charge of finite length. Verify that the electric field $E(\mathbf{r})$ given by Eqs. (1.21) and (1.22) satisfies the requisite differential properties for electrostatic fields in a charge-free region, that is, $\nabla \cdot E(\mathbf{r}) = 0$ and $\nabla \times E(\mathbf{r}) = 0$.

P1-24 In Example 1.6 we found the electric field due to a non-uniform line charge of finite length. Verify that the electric field $E(\mathbf{r})$ given by Eqs. (1.26) and (1.27) satisfies the requisite differential properties for electrostatic fields in a charge-free region, that is, $\nabla \cdot E(\mathbf{r}) = 0$ and $\nabla \times E(\mathbf{r}) = 0$.

P1-25 In Prob. P1-5 we used Coulomb's Law to determine the electric field $E(\mathbf{r})$ inside a sphere of uniform charge density ρ_v. Verify that this electric field $E(\mathbf{r})$ satisfies the requisite differential properties for a charge-filled vacuum, that is, $\nabla \cdot E(\mathbf{r}) = \rho_v/\varepsilon_0$ and $\nabla \times E(\mathbf{r}) = 0$.

P1-26 In Prob. P1-6 we used Coulomb's Law to determine the electric field $\mathbf{E}(\mathbf{r})$ inside and outside a uniformly charged cylinder of charge density ρ_v. Verify that this electric field $\mathbf{E}(\mathbf{r})$ satisfies the requisite differential properties for a charge-filled vacuum, that is, $\boldsymbol{\nabla} \cdot \mathbf{E}(\mathbf{r}) = \rho_v/\varepsilon_0$ and $\boldsymbol{\nabla} \times \mathbf{E}(\mathbf{r}) = 0$.

P1-27 In Prob. P1-7 we used Coulomb's Law to determine the electric field $\mathbf{E}(\mathbf{r})$ produced by an infinitely wide thick slab of charge of uniform charge density ρ_v. Verify that this electric field $\mathbf{E}(\mathbf{r})$ satisfies the requisite differential properties for a charge-filled vacuum, that is, $\boldsymbol{\nabla} \cdot \mathbf{E}(\mathbf{r}) = \rho_v/\varepsilon_0$ and $\boldsymbol{\nabla} \times \mathbf{E}(\mathbf{r}) = 0$.

P1-28 In Prob. P1-9 we used Coulomb's Law to determine the electric field $\mathbf{E}(\mathbf{r})$ generated by a spherical uniform charge density ρ_v between $a < R < b$. Verify that this electric field satisfies the requisite differential properties, that is, $\boldsymbol{\nabla} \cdot \mathbf{E}(\mathbf{r}) = 0$ for $R < a$ and $R > b$, $\boldsymbol{\nabla} \cdot \mathbf{E}(\mathbf{r}) = \rho_v/\varepsilon_0$ for $a < R < b$, and $\boldsymbol{\nabla} \times \mathbf{E}(\mathbf{r}) = 0$ everywhere.

P1-29 In Prob. P1-12 we used Coulomb's Law to determine the electric field $\mathbf{E}(\mathbf{r})$ generated by a spherical uniform surface charge density ρ_s of radius a. Verify that this electric field satisfies the requisite differential properties in the regions $R < a$ and $R > a$.

Application of Gauss' Law in Vacuum

P1-30 The volume charge density of a spherical distribution of charge is ρ_v between $a < R < b$, and zero elsewhere. Use Gauss' Law to determine the electric field $\mathbf{E}$ for the three regions $R < a$, $a < R < b$, and $R > b$. Compare your results to those of Prob. P1-9.

P1-31 A spherical uniform surface charge density ρ_s has a radius a. Use Gauss' Law to determine the electric field $\mathbf{E}$ for the regions $R < a$ and $R > a$. Compare your results to those of Prob. P1-12.

P1-32 An infinitely long line charge of uniform linear charge density ρ_ℓ lies along the z-axis. Use Gauss' Law to determine the electric field $\mathbf{E}$. Compare your results to those of Example 1.5.

P1-33 Two infinitely long line charges lie along the z-direction: one of charge density ρ_ℓ positioned at $x = d/2$, the other of charge density $-\rho_\ell$ positioned at $x = -d/2$. Determine the electric field $\mathbf{E}$ at an arbitrary point $(x, y, 0)$.

P1-34 Consider an infinitely long cylinder of radius a and charge density $\rho_v(\rho) = \rho_0/\rho$ centered along the z-axis. Use Gauss' Law to determine the electric field $\mathbf{E}(\rho)$ for (a) $\rho < a$ and (b) $\rho > a$.

P1-35 The charge density of a long cylindrically symmetric distribution of charge is $\rho_v = 5/\rho$ (nC/m^2) for $\rho < 5$ m, and zero for $\rho > 5$ m. The permittivity is $\varepsilon = \varepsilon_0$ everywhere. (a) Determine the electric flux $\Phi_E = \int_s \mathbf{E} \cdot d\mathbf{s}$ through the surface defined by $\rho = a = 2$ m between $z = -h/2$ and $z = +h/2$, where $h = 2$ m. (b) Determine $\mathbf{E}$ at $\rho = 2$ m.

P1-36 Consider a thick slab of charge of uniform charge density ρ_v lying between $-d/2 < z < d/2$. The charge extends to $\pm\infty$ in the x- and y-directions. Use Gauss' Law to determine the electric field $\mathbf{E}(z)$ for $z < -d/2$, $-d/2 < z < +d/2$, and $z > d/2$. Compare your results to those of Prob. P1-7.

P1-37 Two thin infinitely wide parallel uniform sheets of charge, one of surface charge density $-\rho_s$ and the other $+\rho_s$, are located at $z = \pm d/2$, respectively. Use superposition to determine E_z vs. z for $z < -d/2$, $-d/2 < z < +d/2$, and $z > d/2$. Plot E_z vs. z.

P1-38 A thick slab of charge density $\rho_v(z) = \rho_0 z$ lies between $-d/2 < z < d/2$. The charge extends to $\pm\infty$ in the x- and y-directions. Formulate an argument based on Prob. P1-37 to show that the electric field is zero for $|z| > d/2$. Use Gauss' Law to determine the electric field $E(z)$ for $-d/2 < z < d/2$. Plot E_z vs. z. Where is the magnitude of this field the greatest?

P1-39 A thick slab of charge density $\rho_v(z) = \rho_0|z|$ lies between $-d/2 < z < d/2$. The charge extends to $\pm\infty$ in the x- and y-directions. Use Gauss' Law to determine the electric field $E(z)$ for $z < -d/2$, $-d/2 < z < +d/2$, and $z > d/2$. Plot E_z vs. z.

Electric Potential

P1-40 The electric field in a region is $E(r) = a_R\rho_0R^2/(4\varepsilon_0)$ for $R < a$, and $a_R\rho_0a^4/(4\varepsilon_0R^2)$ for $R > a$. (a) Determine the charge density $\rho_v(r)$ for $R < a$ and for $R > a$. (b) Determine the electric potential $V(r)$ for $R < a$ and for $R > a$. (Use $V = 0$ at $R = \infty$ as the reference.) (c) Determine the total charge Q of the distribution. (d) For the region $R > a$, compare $E(r)$ and $V(r)$ for this charge distribution to that for a point charge of the same total charge Q.

P1-41 The electric field in a region is $E(r) = \rho_0/\varepsilon_0(xa_x + za_z)$, where $\rho_0 = 1\ \mu C/m^3$. (a) Determine the charge density $\rho_v(r)$ in this space. (b) Determine the potential difference $V(r_b) - V(r_a)$ for $r_a = (1, 0, 1)$ m and $r_b = (-1, 1, 0)$ m. (c) Determine the work required to move a 1 μC charge from r_a to r_b.

P1-42 Consider a point charge $q_1 = -1$ nC located at $r' = (-1, 0, 0)$ nm, and a second point charge $q_2 = 9$ nC located at $r' = (2, 0, 0)$ nm. (a) What is the potential at the origin? Find the point (x, y, z) at which the electric field E is zero. (b) Decrease q_2 to 4 nC and repeat.

P1-43 For the three point charges described in Prob. P1-4, determine the electric potential $V(r)$ at an arbitrary point r produced by these three charges. Use this potential to determine the electric field $E(r)$ at:

(a) $r = (1, 1, 1)$ nm

(b) $r = (2, 0, 1)$ nm

(c) $r = (1, 0, -2)$ nm

(d) $r = (0, 0, 1)$ nm.

P1-44 The electric potential in a region of space is $V(x, y, z) = (x^2 - y^2 + 2z^2)$ V. (a) Determine $E(r)$ in this space. (b) Determine $\nabla \times E(r)$. (c) Determine the charge density ρ_v in this region. (d) Determine the work W required to move a $q = +1\ \mu$C charge from point A to point B along the path shown in Fig. 1.50. Coordinates are given in meters.

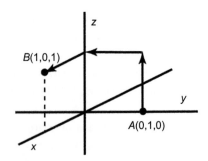

FIGURE 1.50 The path over which the charge is moved in Prob. P1-44.

P1-45 Consider the spherical uniform charge density ρ_v between $a < R < b$, as described in Prob. P1-9. Use Eq. (1.47) to determine the electric potential $V(r)$ for the three regions $R < a$, $a < R < b$, and $R > b$. Use $V(r)$ to determine the electric field $E(r)$.

P1-46 Consider the charge density $\rho_v = \rho_0 \cos\theta$ for $R < a$, and no charge for $R > a$, as described in Prob. P1-10. (a) Use Eq. (1.47) to write an integral expression for the electric potential $V(r)$ for this charge distribution. (b) Using the symmetry of the charge distribution, identify the point(s) for which the electric potential must be zero.

P1-47 Determine the electric potential $V(r)$ for a spherical uniform surface charge density ρ_s of radius a for the regions $R < a$ and $R > a$. Use $V(r)$ to determine the electric field $E(r)$. Compare your result with the result of Prob. P1-12.

P1-48 A spherical shell of radius a has a surface charge density $\rho_s = \rho_0 \cos\theta$, as described in Prob. P1-13. (a) Use Eq. (1.48) to write an integral expression for the electric potential $V(r)$ for this charge distribution valid for the regions $R < a$

and $R > a$. (b) Use the symmetry of the charge distribution to identify the point(s) for which the electric potential must be zero.

P1-49 Determine the electric potential $V(\mathbf{r})$ due to an infinite sheet of charge of surface charge density ρ_s lying in the x–y plane. Use $V(\mathbf{r})$ to determine the electric field $\mathbf{E}$. Compare your result with the result of Prob. P1-14. [In this problem, we encounter a pesky, weak divergence as we carry out the integral called for by Eq. (1.48). To counter this, remember that the potential is always defined relative to some reference location. Here we choose the reference plane at $z = 0$, and find $V(z) - V(0)$.]

P1-50 (a) Determine the electric potential $V(\mathbf{r})$ along the z-axis of a round disc of radius a and surface charge density ρ_s lying in the x–y plane, as described in Prob. P1-15. (b) Use $V(\mathbf{r})$ to determine the z-component of the electric field E_z. (c) Can the potential of part (a) be used to determine E_x or E_y along the z-axis? Why or why not? (d) Write an integral expression for the potential $V(\mathbf{r})$ at an arbitrary point (x, y, z).

P1-51 (a) Determine the electric potential $V(\mathbf{r})$ along the z-axis of a round loop of uniform charge, of radius a and linear charge density ρ_ℓ lying in the x–y plane and centered at the origin. (b) Use $V(\mathbf{r})$ to determine the z-component of the electric field E_z. (c) What does this result tell us about E_z at $z = 0$? Explain why this result makes sense. (d) Examine E_z in the limit $z \gg a$. How does this compare to the electric field due to a point charge? (e) Write an integral expression for the potential $V(\mathbf{r})$ at an arbitrary point (x, y, z).

P1-52 (a) Determine the electric potential $V(\mathbf{r})$ along the z-axis of a square loop of charge, of side $2a$ and linear charge density ρ_ℓ lying in the x–y plane and centered at the origin. (b) Use $V(\mathbf{r})$ to determine the z-component of the electric field E_z. (c) Write an integral expression for the potential $V(\mathbf{r})$ at an arbitrary point (x, y, z).

P1-53 (a) Determine the electric potential due to two infinite uniform sheets of charge, $+\rho_s$ at $z = +d/2$ and $-\rho_s$ at $z = d/2$. [Hint: To avoid a divergence in the integral, first find the potential due to two discs of finite radius a, then take the limit as $a \to \infty$.] (b) Use $V(\mathbf{r})$ to determine the electric field $\mathbf{E}$ between the sheets and outside the sheets.

Electric Dipole

P1-54 Derive Eq. (1.57) from Eq. (1.56).

P1-55 (a) Sketch the electric field lines for the electric dipole given by Eq. (1.57), at $\pi/6$ increments from $\theta = 0$ to $\theta = \pi$. (b) Determine the E_x and E_z components of the electric field of this dipole at each of these locations in the x–z plane.

Boundary Conditions of a Conductor–Vacuum Interface

P1-56 Use Gauss' Law to show that a charge Q placed inside a cavity contained within a conductor induces a total charge $-Q$ that is distributed over the inner surface of the conductor, even when the cavity is irregular in shape, and that this induced charge is arbitrarily close to the surface.

P1-57 In Fig. 1.51 we show a charge Q in the cavity inside the conductor ((a) and (b)), or outside the conductor (c). The conductor is shown as the shaded area, and is electrically neutral. (We say that a body is electrically neutral when the net charge on the body is zero. The positive and negative charges may be distributed differently, but when summed over the entire body they are balanced.) Sketch the electric field lines inside the cavity and outside the conductor. Show surface charges on the inner and outer surfaces of the conductor.

P1-58 In Fig. 1.51 we show a charge Q in the cavity inside the conductor ((a) and (b)), or outside the conductor (c). The conductor is shown as the shaded area, and is electrically grounded. (This conductor has zero potential, but can be charged.) Sketch the electric field lines inside the cavity and outside the conductor. Show surface charges on the conductor.

FIGURE 1.51 The charge Q is in the cavity inside the conductor in (a) and (b), and outside the conductor in (c). The conductor is shown as the shaded area. For use in Probs. P1-57 and P1-58.

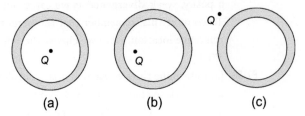

(a)   (b)   (c)

P1-59 In Fig. 1.52 we show a charge Q in the cavity inside the conductor, or outside the conductor. The conductor is shown as the shaded area, and is electrically neutral. Sketch the electric field lines inside the cavity and outside the conductor. Show surface charges on the conductor.

FIGURE 1.52 The charge Q is in the cavity inside the conductor, or outside the conductor. The conductor is shown as the shaded area. For use in Probs. P1-59 and P1-60.

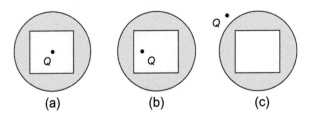

(a)   (b)   (c)

P1-60 In Fig. 1.52 we show a charge Q in the cavity inside the conductor, or outside the conductor. The conductor is shown as the shaded area, and is electrically grounded. Sketch the electric field lines inside the cavity and outside the conductor. Show surface charges on the conductor.

Perfectly Insulating Dielectric Materials

P1-61 A polarized medium is in the shape of a disc, as shown in Fig. 1.53(a). The polarization field inside the disc is $\mathbf{P}(\mathbf{r}) = P_0 \mathbf{a}_z$. Determine the effective surface and volume charge densities on the disc. Use these effective charges to determine the electric field $\mathbf{E}(\mathbf{r})$ for a point on the z-axis $\mathbf{r} = (0, 0, z)$ inside ($|z| < d/2$) or outside ($|z| > d/2$) the disc. Sketch the off-axis electric field lines.

FIGURE 1.53 Polarized media for Probs. P1-61 and P1-62. In (a), $\mathbf{P} = P(z)\mathbf{a}_z$. In (b), $\mathbf{P} = P(\rho)\mathbf{a}_\rho$.

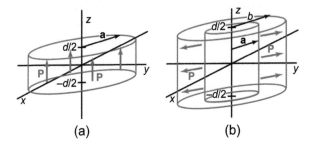

(a)   (b)

P1-62 A polarized medium is in the shape of a hollow cylinder, as shown in Fig. 1.53(b). The polarization field is $\mathbf{P}(\mathbf{r}) = P_0 \rho \mathbf{a}_\rho$. Determine the effective surface and volume charge densities. Assuming that the height d of the medium is large compared to its radius b (allowing us to ignore fringe effects at the edge), determine the electric field $\mathbf{E}$ in the interior of the polarized medium $a < \rho < b$.

P1-63 A polarized medium is in the shape of a disc with a hole, similar to that shown in Fig. 1.53(b). Unlike that figure, however, the polarization field inside the disc is $\mathbf{P}(\mathbf{r}) = P_0\mathbf{a}_z$. Determine the effective surface and volume charge densities on the disc. Use these effective charges to determine the electric field $\mathbf{E}(\mathbf{r})$ for a point on the z-axis, $\mathbf{r} = (0, 0, z)$ for any z.

P1-64 Consider a polarized medium in the shape of a sphere of radius a. The polarization field is $\mathbf{P}(\mathbf{r}) = P_0\mathbf{a}_z$. Determine the effective surface and volume charge densities.

Gauss' Law Revisited

P1-65 A charged conducting sphere of radius $a = 10$ cm is embedded within a dielectric medium. The electric field $\mathbf{E}$ outside the conductor is $\mathbf{E} = \mathbf{a}_R E_0/R^2$, where $E_0 = 50$ Vm. Determine the surface charge density on the sphere when the surrounding dielectric medium has a relative dielectric constant (a) $\varepsilon_r = 1$, and (b) $\varepsilon_r = 2$. (c) Determine the potential V_0 of the sphere relative to $V = 0$ at $R = \infty$ for both cases.

P1-66 Consider a pair of concentric spherical conductors. The inner sphere is of radius a and contains a charge Q; the larger sphere is grounded and of inner radius c. The space between the conductors is filled with two non-conducting, charge-free dielectrics, of permittivity ε_1 $(a < R < b)$ and ε_2 $(b < R < c)$. (a) Use Gauss' Law to determine the displacement field $\mathbf{D}$ in the space between the conductors. (b) Find the electric field $\mathbf{E}$ in the two dielectric media. (c) Determine the potential difference V_0 between the conductors. (d) Determine the surface charge densities ρ_s on the surfaces of the conductors. (e) Compare the total charge on the surface of the conductors at $R = c$ and $R = a$.

P1-67 A pair of co-axial cylindrical conductors have length L. The inner cylinder is of radius a and contains a charge Q; the larger cylinder is grounded and of inner radius $c \ll L$. The space between the conductors is filled with two non-conducting, charge-free dielectrics, of permittivity ε_1 $(a < \rho < b)$ and ε_2 $(b < \rho < c)$. (a) Use Gauss' Law to determine the displacement field $\mathbf{D}$ in the space between the conductors. (b) Find the electric field $\mathbf{E}$ in the two dielectric media. (c) Determine the potential difference V_0 between the conductors. (d) Determine the surface charge densities ρ_s on the surfaces of the conductors. (e) Compare the total charge on the surface of the conductors at $\rho = c$ and $\rho = a$.

P1-68 Determine the electric field $\mathbf{E}$ in the region surrounding a point charge Q embedded in a dielectric medium of permittivity $\varepsilon = \varepsilon_r\varepsilon_0$. Compare to $\mathbf{E}$ for a similar charge in vacuum.

Boundary Conditions between Two Perfect Dielectrics

P1-69 A dielectric medium of relative permittivity $\varepsilon_r = 2$ fills the space $z < 0$, as shown in Fig. 1.54. Above this surface is vacuum, $\varepsilon_r = 1$. The electric field $\mathbf{E}$ just above the interface between these two media is at an angle $\alpha_1 = 45°$ from the z-axis. Determine the angle α_2.

P1-70 Two perfect dielectric media fill the spaces $x < 0$ (medium 1) and $x > 0$ (medium 2). The relative permittivity for $x < 0$ is $\varepsilon_1 = 2$, while for $x > 0$ it is $\varepsilon_2 = 3$. The electric field $\mathbf{E}$ for $x = 0^-$ is $\mathbf{E}_1 = (4\mathbf{a}_x - 6\mathbf{a}_y)$ (V/m). Determine the electric field $\mathbf{E}_2$ at $x = 0^+$. (The superscript "$-$" or "$+$" indicates the negative or positive side of the location $x = 0$.) Find α_1 and α_2, the angles between the surface normal and $\mathbf{E}_1$ and $\mathbf{E}_2$, respectively.

P1-71 The plane interface between two perfect dielectric media is defined by $x = 2y$. The relative permittivity for $x < 2y$ is $\varepsilon_1 = 2$, while for $x > 2y$, it is $\varepsilon_2 = 3$. The

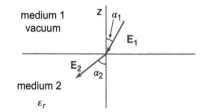

FIGURE 1.54 The interface between two dielectric media, for use in Prob. P1-69.

electric field $\mathbf{E}$ for $x = 2y^-$ is $\mathbf{E}_1 = (4\mathbf{a}_x - 6\mathbf{a}_y)$ (V/m). Determine the electric field $\mathbf{E}$ at $x = 2y^+$. (The superscript "−" or "+" indicates the negative or positive side of the plane $x = 2y$.) Find α_1 and α_2, the angles between the surface normal and $\mathbf{E}_1$ and $\mathbf{E}_2$, respectively.

P1-72 Two perfect dielectric media have relative permittivity $\varepsilon_1 = 2$ ($x < 0$) and $\varepsilon_2 = 3$ ($x > 0$). The displacement field $\mathbf{D}$ for $x = 0^-$ is $\mathbf{D}_1 = (4\mathbf{a}_x - 6\mathbf{a}_y)$ C/m^2. (a) Determine the electric field $\mathbf{E}$ at $x = 0^+$. (The superscript "−" or "+" indicates the negative or positive side of the location $x = 0$.) (b) Find the angle α_1 between $\mathbf{D}_1$ and the surface normal, and the angle α_2 between $\mathbf{D}_2$ and the surface normal.

P1-73 Consider the plane interface between two perfect dielectric media defined by $x = 2y$. The relative permittivity for $x < 2y$ is $\varepsilon_1 = 2$, while for $x > 2y$, it is $\varepsilon_2 = 3$. The displacement field $\mathbf{D}$ for $x = 2y^-$ is $\mathbf{D}_1 = (4\mathbf{a}_x - 6\mathbf{a}_y)$ (C/m^2). Determine the displacement field $\mathbf{D}$ at $x = 2y^+$. (The superscript "−" or "+" indicates the negative or positive side of the location $x = 2y$.) Find the angles between $\mathbf{D}_1$ and the surface normal α_1, and $\mathbf{D}_2$ and the surface normal α_2.

P1-74 Two media abut one another at the x–y plane. Their permittivities are $\varepsilon_1 = 4\varepsilon_0$ for $z > 0$ and ε_2 for $z < 0$. A surface charge density ρ_s lies in the $z = 0$ plane. The displacement fields in the two regions are $\mathbf{D}_1 = (10\mathbf{a}_x + 16\mathbf{a}_z)$ μC/m^2 above the interface, and $\mathbf{D}_2 = (5\mathbf{a}_x + 12\mathbf{a}_z)$ μC/m^2 below the interface. Determine ε_2 and ρ_s. Determine $\mathbf{E}$ and $\mathbf{P}$ in the two media.

Boundary Conditions for Perfect Dielectrics

P1-75 The $x = 0$ plane separates a perfect conductor ($x < 0$) and a dielectric medium ($x > 0$). The surface charge density in the surface of the conductor is $\rho_s = 5$ nC/m^2, and the permittivity of the dielectric medium is $\varepsilon = 3\varepsilon_0$. Determine $\mathbf{D}$, $\mathbf{E}$, and $\mathbf{P}$ in the dielectric medium close to the interface. Use $\mathbf{P}$ to determine the effective surface charge density $\rho_{s,eff}$ at $x = 0$ associated with the induced electric dipoles.

P1-76 Consider the plane interface between a perfect conductor ($x < 2y$) and a dielectric medium with permittivity $\varepsilon = 4\varepsilon_0$ ($x > 2y$). A surface charge of density $\rho_s = 2$ nC/m^2 lies at the conductor surface. Determine $\mathbf{D}$, $\mathbf{E}$, and $\mathbf{P}$ in the two media. Use $\mathbf{P}$ to determine the effective surface charge density $\rho_{s,eff}$ due to the electric dipoles.

Maximum Electric Field in Media

P1-77 A pair of planar conductors are spaced by $d = 0.1$ mm and have area $A = 1$ cm^2. The material between the plates is (a) air or (b) mica. Determine the maximum potential difference that can safely be applied to the conductors. [Hint: Use data given in Tables 1.1 and 1.2.]

P1-78 A conducting sphere of radius $a = 1.0$ cm is coated with a 0.2 cm thick layer of Teflon. A second conducting sphere of radius 1.5 cm is concentric with the first. Determine the maximum charge Q that can safely be stored on the inner conductor, and the maximum potential difference that can safely be applied between the spheres. [Hint: Use data given in Tables 1.1 and 1.2.]

P1-79 Determine the maximum voltage that can be applied to a van de Graaff generator of radius 1.0 m. How much charge is stored on the sphere? Assume that the dome of the van de Graaff is spherical and surrounded by dry air. [Hint: Use data given in Tables 1.1 and 1.2.]

CHAPTER 2
Electrostatics II

KEY WORDS

electrostatic energy; energy density; principle of virtual displacement; Poisson's Equation; Laplace's Equation; donor; hole; acceptor; junction; capacitance; capacitance per unit length; Method of Images; Uniqueness Theorem.

LEARNING OBJECTIVES

After studying this section, you should be able to:
- determine the electrostatic energy of a system, given the charge distribution and potential;
- determine the electrostatic energy density of a system, given the electric field and displacement field; and
- determine the electrostatic energy of a system from its energy density.

With the introduction in the previous chapter of the electric field and electric potential, and their properties in materials, we are now ready to examine the energy stored in electric fields, the electric forces that can be exerted on objects, and the capacitance between conductors. We cover these topics in this chapter. We also introduce additional methods that can be used for determining electric fields and potentials.

2.1 Electrostatic Energy

We now turn our attention to the energy stored within an electric field, and will show that the **electrostatic energy** of a system can be determined in terms of the electric field strength throughout the region. We will also introduce the density of the energy of the field, relevant, for example, to the energy transported by electromagnetic waves, as will be discussed later.

In the previous chapter we discussed the energy that we must supply to move a charge against an electric force. We can extend this idea to determine the electrostatic energy of a system of charges, which we equate to the energy required to construct the ensemble of charges. This should make sense to us, in light of the fact that electrostatic forces are conservative, and all the energy that we expend in assembling the charges can be recovered later if we let the charges return to their original locations. So let us now consider the work that we must do to bring in from infinity the first charge, then the second charge, followed by the third charge, and so forth until we have assembled the complete distribution of the charge. The first charge is, of course, easy. Since there are no other charges in the neighborhood to create a potential, the work that we must do to bring this first charge in is zero, $W_1 = 0$. For the second charge, however, we must work against the potential produced by the first charge, and the work in bringing charge q_2 in from infinity is $W_2 = q_2 V_2$, where V_2 is the potential at the final location of charge q_2 due to the first charge, or $V_2 = q_1/4\pi\varepsilon_0 R_{12}$, and R_{12} is the final distance between the charges q_1 and q_2. We can write this in an equivalent, but more symmetric, form as

$$W_2 = \frac{1}{2}\left[q_1 V_2(\mathbf{r}_1) + q_2 V_1(\mathbf{r}_2)\right],$$

where $V_2(\mathbf{r}_1)$ is the potential at the location of charge q_1 due to charge q_2, and $V_1(\mathbf{r}_2)$ is the potential at the location of charge q_2 due to charge q_1. Next, we bring in the third charge, q_3, requiring work $\Delta W = q_3 q_1/4\pi\varepsilon_0 R_{13} + q_3 q_2/4\pi\varepsilon_0 R_{23}$. Again we write this in the symmetric form

$$\Delta W = \frac{1}{2}\left[q_1 V_3(\mathbf{r}_1) + q_3 V_1(\mathbf{r}_3) + q_2 V_3(\mathbf{r}_2) + q_3 V_2(\mathbf{r}_3)\right].$$

Adding the work required to assemble the first three charges, the total work is

$$W_3 = W_2 + \Delta W$$
$$= \frac{1}{2}\left[q_1 V_2(\mathbf{r}_1) + q_2 V_1(\mathbf{r}_2) + q_1 V_3(\mathbf{r}_1) + q_3 V_1(\mathbf{r}_3) + q_2 V_3(\mathbf{r}_2) + q_3 V_2(\mathbf{r}_3)\right].$$

Regrouping the terms yields

$$W_3 = \frac{1}{2}\left[q_1 \left(V_2(\mathbf{r}_1) + V_3(\mathbf{r}_1)\right) + q_2 \left(V_1(\mathbf{r}_2) + V_3(\mathbf{r}_2)\right) + q_3 \left(V_1(\mathbf{r}_3) + V_2(\mathbf{r}_3)\right)\right].$$

Then, defining the potential at $\mathbf{r}_k$,

$$V(\mathbf{r}_k) = \frac{1}{4\pi\varepsilon_0} \sum_{j=1, j\neq k}^{N} \frac{q_j}{|\mathbf{r}_k - \mathbf{r}_j|}, \tag{2.1}$$

due to all the other charges q_j located at $\mathbf{r}_j$ $(j \neq k)$, where N is the number of charges in the ensemble ($N = 3$ to this point in the present case), the total work required to assemble the charges is

$$W_e = \frac{1}{2} \sum_{k=1}^{N} q_k\, V(\mathbf{r}_k). \tag{2.2}$$

To this point we have considered an ensemble of discrete charges. We now generalize this derivation to the case of a continuous distribution of charge. Dividing the entire volume v containing the charge into a set of smaller elemental volumes Δv_k, each containing a charge $\rho_v(\mathbf{r}_k)\Delta v_k$, the summation over the discrete charges of Eq. (2.2) becomes

$$W_e = \frac{1}{2} \sum_{k=1}^{N} \rho_v(\mathbf{r}_k)\Delta v_k\, V(\mathbf{r}_k).$$

In the limit as the elemental volumes Δv_k become infinitesimally small, the summation becomes an integral over the charge distribution of the form

$$W_e = \frac{1}{2} \int_v \rho_v(\mathbf{r}) V(\mathbf{r}) dv. \tag{2.3}$$

Let's work an example to illustrate this.

Example 2.1 Electrostatic Energy of a Uniform Spherical Charge Distribution

Consider the sphere of uniform charge density shown in Fig. 1.44, and used in Example 1.17. Determine the electrostatic energy of this configuration.

Solution:
We must evaluate the integral given in Eq. (2.3), for which the charge density ρ_v and the potential $V(R)$ are required. The former is given in the problem statement; for the latter,

we use the results for **E** for this charge distribution, which we derived using Gauss' Law in Example 1.17. These results were

$$E_R(R) = \begin{cases} \dfrac{\rho_v R}{3\varepsilon_1} & R < b \\ \dfrac{\rho_v b^3}{3\varepsilon_2 R^2} & R > b \end{cases}.$$

To determine $V(R)$, we use Eq. (1.43), with the reference potential $V = 0$ at $R = \infty$,

$$V(R) = -\int_\infty^R E_R(R) dR. \tag{2.4}$$

When evaluating this integral to find the potential outside the sphere ($R > b$), the integral is simply

$$V(R) = -\int_\infty^R \frac{\rho_v b^3 \, dR}{3\varepsilon_2 R^2} = \frac{\rho_v b^3}{3\varepsilon_2 R},$$

where we used the form of the electric field valid in the region $R > b$. To find the potential at a location inside the sphere ($R < b$), we have to work a little harder. We still make use of Eq. (2.4), but the integral must be broken into two parts, since E_R has a different form for $R > b$ and $R < b$:

$$V(R) = -\int_\infty^b \frac{\rho_v b^3 \, dR}{3\varepsilon_2 R^2} - \int_b^R \frac{\rho_v R \, dR}{3\varepsilon_1} = \frac{\rho_v b^2}{3\varepsilon_2} + \frac{\rho_v}{6\varepsilon_1}\left(b^2 - R^2\right).$$

Now that we have determined the potential $V(R)$ everywhere, we are ready to use Eq. (2.3) to determine the energy of the configuration:

$$\begin{aligned} W_e &= \frac{1}{2}\int_v V(R)\,\rho_v(R)\,dv \\ &= \frac{1}{2}\int_{\phi=0}^{2\pi}\int_{\theta=0}^{\pi}\int_{R=0}^{b}\left[\frac{\rho_v b^2}{3\varepsilon_2} + \frac{\rho_v}{6\varepsilon_1}\left(b^2 - R^2\right)\right]\rho_v R^2 dR \sin\theta \, d\theta \, d\phi \\ &= \frac{2\pi\rho_v^2 b^5}{9}\left[\frac{1}{5\varepsilon_1} + \frac{1}{\varepsilon_2}\right]. \end{aligned}$$

In the second line of the previous integral, notice the limit of integration of the R-integral. The volume integral of the first line is to be taken over all space. Since $\rho_v = 0$ for $R > b$, however, the upper bound in the R-integral in the second line is b.

An important modification of Eq. (2.3) is needed to treat the energy of systems that contain conductors. As discussed earlier, conductors are equipotential objects whose excess charge resides on their surfaces. In place of the volume integral of Eq. (2.3), we should use a surface integral of the form

$$W_e = \frac{1}{2}\sum_i \int_s \rho_s(\mathbf{r})V_i(\mathbf{r})ds = \frac{1}{2}\sum_k V_k Q_k, \tag{2.5}$$

where k is an index that labels the different conductors, and $Q_k = \int_s \rho_s(\mathbf{r})ds$ is the total charge on body k.

Example 2.2 Electrostatic Energy of a System with Conductors

Two planar parallel conductors of surface area S are separated by a distance d. The permittivity of the medium between the conductors is ε. A potential V_0 is applied to the top plate, while the bottom conductor is grounded. Use Eq. (2.5) to determine the energy stored by this configuration.

Solution:

In order to apply Eq. (2.5), the potential V of each conductor and the charge Q stored on each conductor are needed. The potentials are given, but we must work a little harder to find the charges. From previous treatments of similar parallel plate configurations, we know that the electric field $\mathbf{E}$ between the conductors is uniform, it points from the positive to the negative conductor (i.e. downward), and its magnitude is V_0/d. Furthermore, we know that the surface charge density on the inside surface of the top conductor is $\rho_s = |\mathbf{D}|$, where the displacement field between the plates is $\mathbf{D} = \varepsilon\mathbf{E}$. (The surface charge density on the bottom conductor is $-\rho_s$.) Since ρ_s is uniform across the area of the plate, the total charge is simply $Q = \rho_s S = \varepsilon V_0 S/d$.

Now let's apply Eq. (2.5) to determine the energy of this system. The bottom conductor is grounded, so only the charge-potential product of the top plate contributes to the summation, and the energy is

$$W_e = \frac{1}{2} V_0 Q = \frac{1}{2} V_0 \frac{\varepsilon V_0 S}{d} = \frac{1}{2}\left(\frac{\varepsilon S}{d}\right) V_0^2.$$

Some of you may recognize $\varepsilon S/d$ as the capacitance of the parallel plates, which will be covered in more detail in Section 2.5.

We would now like to develop the field energy in another form. While the expression given in Eq. (2.3) is certainly correct, it is often not the most convenient form to use, and there is a question of the proper interpretation of the integrand. We would like to place this in a form that uses the electric field directly. To accomplish this, we start with Eq. (2.3) and use the differential form of Gauss' Law, Eq. (1.65), to replace the charge density ρ_v,

$$W_e = \frac{1}{2} \int_v \rho_v(\mathbf{r}) V(\mathbf{r}) dv = \frac{1}{2} \int_v \mathbf{\nabla} \cdot \mathbf{D}\, V(\mathbf{r}) dv.$$

Now we can apply a vector differentiation product rule that we used earlier, given in Eq. (E.6) and repeated here:

$$\mathbf{\nabla} \cdot \left(f\mathbf{A}\right) = f\mathbf{\nabla} \cdot \mathbf{A} + \mathbf{\nabla}f \cdot \mathbf{A},$$

valid for any continuous scalar field f and vector field $\mathbf{A}$. In this case, with $\mathbf{A}$ as $\mathbf{D}$ and f as V, the energy becomes

$$W_e = \frac{1}{2} \int_v \left[\mathbf{\nabla} \cdot \left(\mathbf{D}(\mathbf{r})V(\mathbf{r})\right) - \mathbf{D}(\mathbf{r}) \cdot \mathbf{\nabla}V(\mathbf{r})\right] dv. \tag{2.6}$$

We apply the Divergence Theorem to the first integral,

$$\frac{1}{2} \int_v \mathbf{\nabla} \cdot \left(\mathbf{D}(\mathbf{r})V(\mathbf{r})\right) dv = \frac{1}{2} \int_s \mathbf{D}(\mathbf{r})V(\mathbf{r}) \cdot d\mathbf{s},$$

and recognize that this surface integral tends to zero as the surface becomes very large. (For a localized charge distribution, such as a point charge, or a group of charges, the displacement field decreases as R^{-2} and the potential decreases as R^{-1}, while the surface area increases only as R^2. Thus the integral decreases as R^{-1}, and we can safely disregard this integral by making our integration space large enough.) This leaves us with the second integral in Eq. (2.6) only; when we replace $\boldsymbol{\nabla}V(\mathbf{r})$ with $-\mathbf{E}(\mathbf{r})$ this gives us

$$W_e = \frac{1}{2}\int_v \mathbf{D}\cdot\mathbf{E}\,dv. \tag{2.7}$$

This expression for the energy stored in an electric field suggests further interpretation. That is, since the integration over all space of the quantity $(1/2)\mathbf{D}\cdot\mathbf{E}$ gives us the total energy stored by the field, we can interpret the integrand in this integral as the stored electrostatic energy *per unit volume*, termed the **energy density** of the field. That is, the energy density of an electric field is

$$w_e = \frac{1}{2}\mathbf{D}\cdot\mathbf{E}. \tag{2.8}$$

Example 2.3 Energy and Energy Density, Parallel Plate System

In Example 2.2 we used the potential and charge of each plate to find the energy stored in a configuration consisting of two planar parallel conductors of surface area S separated by a distance d. A potential V_0 was applied to the top plate relative to the bottom conductor. Use Eqs. (2.8) and (2.7) to determine the energy density w_e and total energy W_e stored by this same configuration.

Solution:
The electric field in the region between the conductors is $\mathbf{E} = -\mathbf{a}_z V_0/d$, where the direction $-\mathbf{a}_z$ points from the top plate to the bottom plate. The displacement field is $\mathbf{D} = \varepsilon\mathbf{E}$. Therefore, using Eq. (2.8), the energy density of the field between the plates is

$$w_e = \frac{1}{2}\mathbf{D}\cdot\mathbf{E} = \frac{1}{2}\frac{\varepsilon V_0^2}{d^2}.$$

The total energy is found by integrating the energy density over the volume between the plates. Since the energy density is uniform, the integral is

$$W_e = \frac{1}{2}\int_v w_e\,dv = \frac{1}{2}w_e S d = \frac{1}{2}\frac{\varepsilon V_0^2 S}{d},$$

in agreement with the result of Example 2.2.

Example 2.4 Energy Density and Energy

Consider again the charge distribution described in Example 2.1. Use Eqs. (2.8) and (2.7) to determine the energy density of the field and the total energy stored by the field.

Solution:
We have already determined $\mathbf{D}$ and $\mathbf{E}$ for this charge distribution, so we will borrow these results and move on from there. This gives an energy density in the two regions of

$$w_e(R) = \begin{cases} \dfrac{\rho_v^2 R^2}{18\varepsilon_1} & R < b \\[3mm] \dfrac{\rho_v^2 b^6}{18\varepsilon_2 R^4} & R > b \end{cases}.$$

To determine the total stored energy, we integrate the energy density over all space:

$$\begin{aligned} W_e = \int_v w_e \, dv &= \int_{\phi=0}^{2\pi} \int_{\theta=0}^{\pi} \int_{R=0}^{b} \frac{\rho_v^2 R^2}{18\varepsilon_1} R^2 dR \sin\theta \, d\theta \, d\phi \\ &\quad + \int_{\phi=0}^{2\pi} \int_{\theta=0}^{\pi} \int_{R=b}^{\infty} \frac{\rho_v^2 b^6}{18\varepsilon_2 R^4} R^2 dR \sin\theta \, d\theta \, d\phi \\ &= \frac{2\pi \rho_v^2 \, b^5}{9} \left[\frac{1}{5\varepsilon_1} + \frac{1}{\varepsilon_2} \right]. \end{aligned}$$

The first term represents the energy stored in the region $R < b$, while the second term is the energy stored in the region $R > b$. This total stored energy is consistent with the result of the previous example, but in that first result we could not separate the energy according to where it was stored.

Example 2.5 Electrostatic Energy, Co-axial Conductors

Two co-axial conductors of length L are shown in Fig. 1.45. The center conductor has radius a and is held at a potential V_0 relative to the outer conductor (or shield), which is of radius b. The dielectric permittivity of the medium in the space between the conductors is ε. Determine the electrostatic energy stored by this configuration. (Ignore fringe field effects at the ends of the cylinder.)

Solution:

We have already found, in Example 1.18, the displacement field $\mathbf{D}$ and the electric field $\mathbf{E}$ in the dielectric medium between the conductors. This gives us an energy density that varies with radius ρ as

$$w_e(\rho) = \frac{1}{2}\mathbf{D}\cdot\mathbf{E} = \frac{\rho_{s,a}^2 a^2}{2\varepsilon\rho^2}$$

and a total energy of

$$W_e = \int_v w_e(\rho) \, dv = \int_{z=0}^{L} \int_{\phi=0}^{2\pi} \int_{\rho=a}^{b} \frac{\rho_{s,a}^2 \, a^2}{2\varepsilon\rho^2} \rho d\rho \, d\phi \, dz = \pi L \frac{\rho_{s,a}^2 \, a^2}{\varepsilon} \ln(b/a).$$

Using the potential difference between the conductors,

$$V_0 = \frac{\rho_{s,a} \, a}{\varepsilon} \, \ln(b/a),$$

which we also derived in Example 1.18, the total energy stored in the electric field is

$$W_e = \frac{\pi\varepsilon L V_0^2}{\ln(b/a)}. \tag{2.9}$$

We will introduce the capacitance of electric conductors soon, and as we should expect, the electric energy W_e of a system is closely connected to its capacitance and potential. But first, let's look at another application of the electric energy of a system, that is, a method for determining the force felt by objects due to electric effects.

2.2 Electrostatic Force

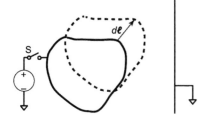

FIGURE 2.1 A charged conductor near a grounded plane. Upon displacement of the object on the left by $d\boldsymbol{\ell}$, the energy stored within the electric field in the surrounding space varies, and we can determine the force felt by the conductor through this change in the electrostatic energy.

Having explored the electrostatic energy of a system, we can now make use of this energy to determine the mechanical force applied to a dielectric or conducting object within the system. This determination is based on the **principle of virtual displacement**, which applies to a broad range of mechanical systems.

In applying this method, we consider a configuration such as that shown in Fig. 2.1. While we have illustrated this with a pair of conductors in this figure, the discussion is actually much more general, and we can apply these ideas to dielectric bodies as well. In formulating this approach, we must consider energy and work of three different varieties. These are: (1) the work done on the object by the electric force, dW_F; (2) the change in energy stored by the electric fields, dW_e; and (3) the work done by the sources, dW_s (but only if the sources are still connected to conductors within the system). Since energy must be conserved, we know that the work done by the sources must equal the sum of the electric work done on the object plus the increase in energy stored by the fields,

$$dW_s = dW_F + dW_e. \tag{2.10}$$

Now let's see what we know about each of these quantities. We start with the mechanical work done by the system (the charged bodies, the capacitor plates, etc.) in moving the object on the left in Fig. 2.1 by an incremental amount $d\boldsymbol{\ell}$. We denote the electrical force on the element – that is, the force that we want to determine – as $\mathbf{F}_e$. Then the mechanical work done on the object will be

$$dW_F = +\mathbf{F}_e \cdot d\boldsymbol{\ell}. \tag{2.11}$$

Note the + sign in this definition. If the electric force $\mathbf{F}_e$ is in the same direction as the displacement $d\boldsymbol{\ell}$, this work done on the object is positive, increasing the energy (perhaps potential or kinetic) of the object. Conversely, dW_F is negative if $\mathbf{F}_e$ opposes $d\boldsymbol{\ell}$. The inverse relation of Eq. (2.11) is

$$\mathbf{F}_e = \boldsymbol{\nabla} W_F, \tag{2.12}$$

giving us the force $\mathbf{F}_e$ from the incremental work dW_F. As a reminder, $\boldsymbol{\nabla}$ represents the gradient operation. In this context, we must determine the work done on the object per unit length of displacement. The direction of the gradient is the direction of maximum work done, which is the direction of the force $\mathbf{F}_e$. Second, we must consider the work dW_s done by any sources that are connected to the system. Note the source in Fig. 2.1. The position of the switch, whether opened or closed, is important here. If the switch is closed, then the source can do work on the system as charges flow from the source to the conductors (or the other way), and we must include this work in our analysis. If the switch is open, no current can flow, and the source does no work; that is, $dW_s = 0$. And finally, in moving the element, the electrostatic energy of the system changes by an amount that we will denote dW_e. We have already explored how to determine the electrostatic energy W_e of a system, and so we will use those skills to determine the *change* in electrostatic energy dW_e.

Now let's use Eqs. (2.10) and (2.12) to determine the force $\mathbf{F}_e$,

$$\mathbf{F}_e = \boldsymbol{\nabla} W_F = \boldsymbol{\nabla}(W_s - W_e), \tag{2.13}$$

and consider separately the two cases of switch closed or switch open. If the voltage source *does* remain connected to the conductor, then the potentials V of the conductor will remain constant (since it is held fixed by the sources). The charges on the conductors, however, may change by an amount dq, as charge is transferred from the source to the body. (We haven't discussed capacitance yet, but for those of you already familiar with this concept, the capacitance of the system depends on the geometry of the conductors and dielectric materials. If these bodies move, then the capacitance must change, and under constant potential conditions the charge stored by the capacitor must change as well.) The work done by the source is then $dW_s = V\,dq$. This is related to, but distinct from, the electrostatic energy of the system W_e. Recall from the last section that we determined the electrostatic energy of the system as the work required to assemble all the charges of the system, and showed, in its first form, that this energy is $W_e = 1/2 \sum_i q_i V_i$, where the index i labels the different bodies in the system. In this case, there is only the one conductor at potential V, and if we change the configuration of the system with the source connected, the potential of this body must remain constant, while charge dq is transferred from the source to the body. Therefore we find that dW_e is simply $(1/2)V\,dq$, or precisely one-half dW_s, the work done by the sources. From Eq. (2.13), we need $dW_s - dW_e$, which, with the sources connected to the system, reduces to simply dW_e. Therefore the electric force $\mathbf{F}_e$ felt by the object is

$$\boxed{\mathbf{F}_e = \boldsymbol{\nabla} W_e\big|_{constant\ V}\cdot} \qquad (2.14)$$

The "constant V" notation is a reminder that we are to determine this gradient under conditions where the potentials of the conductors remain fixed.

Conversely, to determine the force $\mathbf{F}_e$ with the source disconnected, the charge on the conductor must remain the same (the charge has no escape path), but the potential can vary. (It may at first seem strange that the potential can change, but in fact it can.) Obviously, the source can do no work on the system in this case, since it is disconnected from the system, so $dW_s = 0$ and Eq. (2.13) becomes

$$\boxed{\mathbf{F}_e = -\boldsymbol{\nabla} W_e\big|_{constant\ Q}\cdot} \qquad (2.15)$$

We can use either of these expressions to determine the electrostatic force on the body, since they will both provide the same result. It is usually a matter of convenience which form we choose to use. We will illustrate these ideas with an example.

Example 2.6 Electrostatic Force

Consider the pair of parallel conducting plates charged by the source to a potential V_0, as shown in Fig. 1.46. The medium in the lower part of the structure is air, of thickness d_1 and permittivity ε_0, while the medium in the upper part is of thickness d_2 and permittivity $\varepsilon_2 > \varepsilon_0$. The area of each plate is S, and the spacing between the plates is much less than $\sqrt{S}$. Determine the force on the lower plate. Is this force attractive or repulsive?

Solution:
We will work this two ways, with and without the source connected. Since we are told that the spacing between the plates is small, we can ignore fringe effects near the edges of the plates, and we know that the **D** field is uniform in the region between the plates, and in the $-\mathbf{a}_z$ direction.

With the source disconnected, the charge on the capacitor plates is fixed, $+Q$ on the upper plate and $-Q$ on the lower plate. By application of Gauss' Law, we showed in Example 1.19 that $\mathbf{D} = -\rho_s \mathbf{a}_z$ everywhere, where $\rho_s = Q/S$ is the surface charge density on the upper plate. The electric field $\mathbf{E}$ is different in the two dielectric media, since $\mathbf{E} = \mathbf{D}/\varepsilon$, and we write these as

$$\mathbf{E} = \begin{cases} -\mathbf{a}_z \dfrac{\rho_s}{\varepsilon_2} & d_1 < z < d_1 + d_2 \\[2mm] -\mathbf{a}_z \dfrac{\rho_s}{\varepsilon_0} & 0 < z < d_1 \end{cases}.$$

The electrostatic energy density in the two dielectric media is

$$w_e = \frac{1}{2}\mathbf{D} \cdot \mathbf{E} = \begin{cases} \dfrac{\rho_s^2}{2\varepsilon_2} & d_1 < z < d_1 + d_2 \\[2mm] \dfrac{\rho_s^2}{2\varepsilon_0} & 0 < z < d_1 \end{cases},$$

and the total energy stored is

$$W_e = \int_v w_e\, dv = \left(\frac{\rho_s^2}{2\varepsilon_2}\right) S d_2 + \left(\frac{\rho_s^2}{2\varepsilon_0}\right) S d_1. \tag{2.16}$$

This expression tells us the total energy stored in the system in terms of the surface charge density, which of course is related to the total charge Q on the top or bottom plate. If the bottom plate was displaced vertically by a distance Δz, while the charge was held fixed, then the energy of the system would change, and we can use Eq. (2.16) to determine this variation in the electric energy. By Eq. (2.15), the force on the bottom plate is

$$\mathbf{F}_e = -\boldsymbol{\nabla} W_e\big|_{\text{constant } Q} = -\mathbf{a}_z \frac{\partial}{\partial z} W_e,$$

where z is the position of the bottom plate. Since moving this plate upward (i.e. increasing z) causes the thickness d_1 to decrease by an equal amount, this implies that $\partial/\partial z = -\partial/\partial d_1$, and since all other terms (ρ_s, d_2, S, etc.) in Eq. (2.16) for the energy remain fixed as d_1 varies, the force becomes

$$\mathbf{F}_e = +\mathbf{a}_z \frac{\partial}{\partial d_1} W_e = \mathbf{a}_z \frac{\rho_s^2}{2\varepsilon_0} S. \tag{2.17}$$

The force is in the positive $\mathbf{a}_z$ direction, which means that it is attractive. The electrostatic force tends to decrease the air gap. This is good news, since otherwise we might have quite a problem with capacitors flying apart as soon as we apply a source to charge them.

We can also analyze this force on the lower plate when the source remains connected, so we turn our attention now to this case. This analysis follows similar lines, until we reach Eq. (2.16). In this case, the potential difference between the plates V_0 remains fixed, while ρ_s can vary as the thickness d_1 is varied, so we need to eliminate ρ_s from this expression, using in its place a term that includes V_0, the constant potential across the plates. We can do this using

$$V_0 = -\int_0^{d_1 + d_2} \mathbf{E} \cdot d\boldsymbol{\ell},$$

evaluated on any path from the bottom plate to the top plate. We choose a direct path in the $\mathbf{a}_z$ direction, using $d\boldsymbol{\ell} = \mathbf{a}_z\, dz$, to find

$$V_0 = (\rho_s/\varepsilon_0)\, d_1 + (\rho_s/\varepsilon_2)\, d_2. \tag{2.18}$$

Using this expression, we can write the electrostatic energy of the capacitor as

$$W_e = \frac{V_0^2 S}{2\left(d_1/\varepsilon_0 + d_2/\varepsilon_2\right)}.$$

In this expression, the thickness d_1 is the only variable. Everything else remains constant, and evaluating the force using Eq. (2.14) leads to

$$\mathbf{F}_e = \nabla W_e\big|_{constant\ V} = \frac{V_0^2 S\, \mathbf{a}_z}{2\varepsilon_0\left(d_1/\varepsilon_0 + d_2/\varepsilon_2\right)^2}.$$

As expected, the electric force $\mathbf{F}_e$ calculated this way is precisely the same as we found in Eq. (2.17).

While the solution presented above is correct, and in many cases is the simplest way to determine the force on an object, you might feel that it is not very intuitive. Most would agree that a simple picture of this force would have its benefits. Let's see if we can construct one. Actually, we use two distinct methods. The first is based on the result of Example 1.9 or Prob. P1-14, in which we showed that the electric field to either side of a uniform, infinite sheet of charge is $\pm\rho_s/2\varepsilon_0$. The parallel plates in this example are not infinitely wide, but they are closely spaced, so this electric field is valid everywhere except near the edges.

The net electric field felt by a bit of charge in the bottom plate due to the charge on the top plate, plus the effective charge on the two surfaces of the dielectric medium, is

$$\mathbf{E} = -\mathbf{a}_z\left(\frac{\rho_s}{2\varepsilon_0} - \frac{\rho_{s,eff}}{2\varepsilon_0} + \frac{\rho_{s,eff}}{2\varepsilon_0}\right) = -\mathbf{a}_z\frac{\rho_s}{2\varepsilon_0}.$$

(Note that this $\mathbf{E}$ is not the total field in the regions between the conductors, as we have not included the charges within the bottom plate.) This electric field exerts a force on the charges within one element of surface Δs in the bottom plate,

$$\Delta\mathbf{F} = \Delta q\mathbf{E} = -\rho_s\Delta s\mathbf{E} = +\mathbf{a}_z\frac{\rho_s^2\Delta s}{2\varepsilon_0}.$$

When the forces $\Delta\mathbf{F}$ for each of the elements Δs of the bottom plate are summed, the net force on the bottom plate is

$$\mathbf{F} = +\mathbf{a}_z\frac{\rho_s^2 S}{2\varepsilon_0},$$

in agreement with the result above.

Alternatively, we can determine the Coulomb force of Eq. (1.1) on the charges in a small elemental area Δs_{bot} of the bottom plate due to the charges in a small elemental area Δs_{top} of the upper plate, integrate over the entire area of the top plate to find the net force on the charges in Δs_{bot} of the bottom plate, and then integrate over the area of the bottom plate to find the total force on the bottom plate. This exercise is tedious, so we don't show the details here, but the result matches that of the earlier approaches.

In this section we have shown that, by quantifying the energy of a system of conductors and dielectrics, and the dependence of this energy on the position of an element, we can determine the electric force exerted on the element. The force, of course, can always be formulated in terms of the Coulomb force between the electric charges within the system, but it is often much simpler to determine using the approach outlined in this section. We will also apply this approach in Chapter 6 when we discuss forces and torques in magnetic systems.

2.3 Poisson's Equation and Laplace's Equation

In this section we will introduce another powerful tool for determination of the electric potential in a region of space. This is based on a pair of related differential equations, known as **Poisson's Equation** and **Laplace's Equation**, which govern the spatial variation of the scalar electric potential $V(\mathbf{r})$. These relations are named after Pierre-Simon Laplace and Siméon-Denis Poisson (see Biographical Notes 2.1 and 2.2).

Biographical Note 2.1 Pierre-Simon Laplace (1749–1827)

Pierre-Simon Laplace, portrayed in Fig. 2.2(a), was a French mathematician, astronomer, and physicist whose mathematical analysis of the physical world was wide-ranging and deep. Among problems that Laplace studied were the stability of the solar system, the dynamics of the ocean tides, potential theory, and probability theory. He developed the series later known as the Legendre polynomials, Laplace transforms, and the differential operator of relevance here, the Laplacian.

FIGURE 2.2 (a) Portrait of Pierre-Simon de Laplace (1749–1827).

Biographical Note 2.2 Siméon-Denis Poisson (1781–1840)

Siméon-Denis Poisson was a French mathematician and physicist known for his important contributions to definite integrals, electromagnetic theory, and probability. Poisson, shown in Fig. 2.2(b), studied under the direction of Laplace and Lagrange, and became a professor at École Polytechnique in 1802. The Poisson distribution is a statistical model of random events with applications in radioactivity and other random events.

FIGURE 2.2 (b) Lithographic portrait of Siméon-Denis Poisson (1781–1840).

To derive Poisson's Equation, let us start with Gauss' Law,

$$\nabla \cdot \mathbf{D} = \rho_v,$$

which is valid for all cases. In a simple, linear medium, we can write $\mathbf{D} = \varepsilon \mathbf{E}$, so Gauss' Law becomes

$$\nabla \cdot (\varepsilon \mathbf{E}) = \rho_v.$$

Using $\mathbf{E} = -\boldsymbol{\nabla}V$, this becomes

$$-\boldsymbol{\nabla} \cdot (\varepsilon\boldsymbol{\nabla}V) = \rho_v.$$

If the medium is homogeneous (i.e. the permittivity ε does not vary with position), then

$$-\varepsilon\,\nabla^2 V = \rho_v,$$

where ∇^2 is the Laplacian operator, $\boldsymbol{\nabla}\cdot\boldsymbol{\nabla}$. The Laplacian operator is a second-order spatial derivative. (See Appendix E.2 for an introduction to the Laplacian operator. You will find explicit expressions for ∇^2 in rectangular, cylindrical, and spherical coordinates in Eqs. (E.19)–(E.21) in that appendix.)

Minor rearrangement of the previous equation gives us

$$\boxed{\nabla^2 V = -\rho_v/\varepsilon,}\tag{2.19}$$

a relation known as **Poisson's Equation**.

At first glance, it may not seem like the potential $V(\mathbf{r})$ due to a charge density ρ_v as given by Eq. (1.47) can satisfy Poisson's Equation, but in fact it does so very nicely, as we show in SideNote 2.1.

SideNote 2.1 Proof of Solution to Poisson's Equation

We will show here that the potential given by Eq. (1.47) satisfies Poisson's Equation, Eq. (2.19). We start by finding the gradient of the potential $V(\mathbf{r})$, as given by Eq. (1.47):

$$\boldsymbol{\nabla}V(\mathbf{r}) = \frac{1}{4\pi\varepsilon_0}\boldsymbol{\nabla}\left(\int_v \frac{\rho_v(\mathbf{r}')}{|\mathbf{r}-\mathbf{r}'|}\,dv'\right).$$

As a reminder, the gradient $\boldsymbol{\nabla}$ includes derivatives with respect to the *unprimed* coordinates (the field location), while the integral is carried out over the *primed* coordinates (the charge locations). We can reverse the order of differentiation and integration to write this as

$$\boldsymbol{\nabla}V(\mathbf{r}) = \frac{1}{4\pi\varepsilon_0}\int_v \boldsymbol{\nabla}\left(\frac{1}{|\mathbf{r}-\mathbf{r}'|}\right)\rho_v(\mathbf{r}')\,dv',$$

where $\boldsymbol{\nabla}$ inside the integrand acts only on $|\mathbf{r}-\mathbf{r}'|^{-1}$ since this is the only term in the integrand in which $\mathbf{r}$ (the unprimed coordinate) appears. Now we use Eq. (D.5) to write this integral as

$$\boldsymbol{\nabla}V(\mathbf{r}) = \frac{1}{4\pi\varepsilon_0}\int_v \frac{-(\mathbf{r}-\mathbf{r}')}{|\mathbf{r}-\mathbf{r}'|^3}\rho_v(\mathbf{r}')\,dv'.$$

(We should recognize this as simply -1 times the electric field, so we're on the right path.)

Now we are ready to evaluate the divergence of $\boldsymbol{\nabla}V(\mathbf{r})$:

$$\nabla^2 V(\mathbf{r}) = \boldsymbol{\nabla}\cdot\boldsymbol{\nabla}V(\mathbf{r}) = \frac{1}{4\pi\varepsilon_0}\boldsymbol{\nabla}\cdot\left(\int_v \frac{-(\mathbf{r}-\mathbf{r}')}{|\mathbf{r}-\mathbf{r}'|^3}\rho_v(\mathbf{r}')\,dv'\right).$$

Again, since the derivatives used in the divergence $\boldsymbol{\nabla}$ are taken with respect to the *unprimed* coordinates, while the integral is carried out over the *primed* coordinates, we are allowed to pull the $\boldsymbol{\nabla}$ inside the integral, where it acts only on the $(\mathbf{r}-\mathbf{r}')/|\mathbf{r}-\mathbf{r}'|^3$ term in the integrand. We show in Example D.5 that the divergence

$$\mathbf{\nabla} \cdot \left(\frac{(\mathbf{r} - \mathbf{r}')}{|\mathbf{r} - \mathbf{r}'|^3} \right),$$

is equal to zero, except at $r - r' = 0$. At $\mathbf{r} = \mathbf{r}'$, however, this divergence becomes undefined, so we must treat this integral carefully near this point. Let's see what we can do. We want $\nabla^2 V(\mathbf{r})$, which we have in the form

$$\nabla^2 V(\mathbf{r}) = \frac{1}{4\pi\varepsilon_0} \int_v \mathbf{\nabla} \cdot \left(\frac{-(\mathbf{r} - \mathbf{r}')}{|\mathbf{r} - \mathbf{r}'|^3} \right) \rho_v(\mathbf{r}') \, dv'.$$

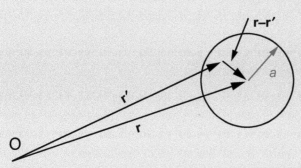

The volume integral is to be carried out over the entire charge distribution, but the integrand vanishes everywhere except at $\mathbf{r} = \mathbf{r}'$. So let's decrease the volume of integration to something more tractable, such as a small sphere of radius a, centered at the position $\mathbf{r}$, as shown in Fig. 2.3. The radius a is chosen to be small enough that the charge density $\rho_v(\mathbf{r}')$ is uniform over this volume, and equal to $\rho_v(\mathbf{r})$. We can therefore pull the charge density $\rho_v(\mathbf{r})$ outside of the integral, and write $\nabla^2 V(\mathbf{r})$ as

FIGURE 2.3 The spherical volume of radius a over which we integrate Eq. (2.20).

$$\nabla^2 V(\mathbf{r}) = \frac{\rho_v(\mathbf{r})}{4\pi\varepsilon_0} \int_v \mathbf{\nabla} \cdot \left(\frac{-(\mathbf{r} - \mathbf{r}')}{|\mathbf{r} - \mathbf{r}'|^3} \right) dv'. \qquad (2.20)$$

Next we notice a symmetry in the integrand upon interchange between the two position vectors $\mathbf{r}$ and $\mathbf{r}'$. Because of this symmetry, we are free to write $\nabla^2 V(\mathbf{r})$ as

$$\nabla^2 V(\mathbf{r}) = \frac{\rho_v(\mathbf{r})}{4\pi\varepsilon_0} \int_v \mathbf{\nabla}' \cdot \left(\frac{(\mathbf{r} - \mathbf{r}')}{|\mathbf{r} - \mathbf{r}'|^3} \right) dv',$$

where $\mathbf{\nabla}'$ indicates we are to take derivatives with respect to the *primed* spatial variables. We can apply the Divergence Theorem to change this volume integral into a surface integral

$$\nabla^2 V(\mathbf{r}) = \frac{\rho_v(\mathbf{r})}{4\pi\varepsilon_0} \int_s \left(\frac{(\mathbf{r} - \mathbf{r}')}{|\mathbf{r} - \mathbf{r}'|^3} \right) \cdot d\mathbf{s}',$$

where the surface integral is carried out over the surface of the sphere of radius a centered on the position $\mathbf{r}$. For this surface integral, the only $\mathbf{r}'$ of interest are position vectors that point to points on the surface of the sphere in Fig. 2.3. For these points, $\mathbf{r} - \mathbf{r}'$ is directed radially *inward*, its magnitude is a (the radius of the sphere), and the denominator in the integrand is a^3. The surface element $d\mathbf{s}'$ is directed radially *outward* from the center of the sphere, so

$$\nabla^2 V(\mathbf{r}) = \frac{\rho_v(\mathbf{r})}{4\pi\varepsilon_0} \int_s \left(\frac{-\mathbf{a}_{\mathbf{rr}'}}{a^2} \right) \cdot \mathbf{a}_{\mathbf{rr}'} \, ds'.$$

(We used the unit vector $\mathbf{a}_{\mathbf{rr}'}$ defined in Eq. (1.4).) Since the integrand is constant $(-a^{-2})$, the surface integral in this expression is simply the product of the integrand and the surface area of the sphere $(4\pi a^2)$, or -4π, and the final value of $\nabla^2 V(\mathbf{r})$ is

$$\nabla^2 V(\mathbf{r}) = \frac{\rho_v(\mathbf{r})}{4\pi\varepsilon_0}(-4\pi) = -\frac{\rho_v(\mathbf{r})}{\varepsilon_0},$$

as we set out to prove.

When the charge density $\rho_v = 0$, Poisson's Equation becomes

$$\boxed{\nabla^2 V = 0,} \qquad (2.21)$$

a relation known as **Laplace's Equation**. While Laplace's Equation is little more than a special case of Poisson's Equation, it is an important special case, and it is appropriate any time we have a charge-free region. We will spend a lot of time examining solutions to Laplace's Equation under different geometries.

The significance of Poisson's Equation cannot possibly be overstated. It opens for us a wide array of solutions, analytical as well as numerical, for the electric potential $V(\mathbf{r})$, and from the potential the electric field $\mathbf{E}(\mathbf{r})$. It may appear to be complicated, especially if you are unaccustomed to the Laplacian ∇^2 notation, but in fact it is quite a simple relation, and it leads us to solutions for the electric potential which are simple in form as well.

There are a few special properties of the potential that are a direct consequence of Laplace's Equation, valid in a charge-free region:

1. The potential $V(\mathbf{r})$ at any location $\mathbf{r}$ is equal to the average value of the potential at the locations surrounding $\mathbf{r}$. Specifically, if we construct a sphere around the point $\mathbf{r}$, then the potential at the center of the sphere is equal to the average of the potential over the surface of that sphere. (Recall that we have already demonstrated this for a charge-free region inside a conducting sphere.)
2. The potential $V(\mathbf{r})$ can have neither local maxima nor local minima.
3. If s is a closed surface, then the value of $V(\mathbf{r})$ for *any* point inside the surface is bounded by $V_{min} < V(\mathbf{r}) < V_{max}$, where V_{min} and V_{max} are the minimum and maximum values of the potential V on the surface s.

SideNote 2.2 Proof of General Properties of Solutions to Laplace's Equation

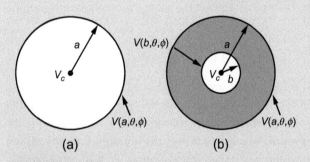

FIGURE 2.4 (a) A spherical surface of radius a in a charge-free region of space. The potential at the center of this sphere is equal to the average value of the potential on the surface of the sphere. (b) The spherical 'shell' region, over which the Divergence Theorem is applied in Eq. (2.24).

We'll start by proving the first of these statements. The second and third then follow easily. Let $V(\mathbf{r})$ represent the potential in some charge-free region of space. We construct a sphere of radius a within this region, and for convenience choose the origin as the center of the sphere. (This sphere is not necessarily a physical surface, merely a mathematical construct.) This sphere is illustrated in Fig. 2.4(a). We will show that the potential V_c at the center of the sphere is equal to the average value of the potential on the surface of the sphere. That is,

$$V_c = \frac{1}{4\pi a^2} \int_{s_a} V(a, \theta, \phi)ds.$$

Since we are considering a charge-free region, we know that Laplace's Equation $\nabla^2 V(\mathbf{r}) = 0$ is valid throughout the region. Next we define a second scalar field $f(\mathbf{r})$, where

$$f(\mathbf{r}) = 1/R. \tag{2.22}$$

Later in this proof, we will need $\nabla f(\mathbf{r})$ and $\nabla^2 f(\mathbf{r})$, so we compute them here at the outset. These are

$$\nabla f(\mathbf{r}) = \mathbf{a}_R \frac{\partial f}{\partial R} = -\mathbf{a}_R \frac{1}{R^2} \tag{2.23}$$

and

$$\nabla^2 f(\mathbf{r}) = \frac{1}{R^2} \frac{\partial}{\partial R} \left(R^2 \frac{\partial f}{\partial R} \right) = 0.$$

We see from this last relation that $f(\mathbf{r})$ satisfies Laplace's Equation, just as $V(\mathbf{r})$ does.

Now let's consider the function $V(\mathbf{r})\boldsymbol{\nabla} f(\mathbf{r}) - f(\mathbf{r})\boldsymbol{\nabla} V(\mathbf{r})$. As a product of a scalar field and a vector field, this function is a vector field, and by the Divergence Theorem the surface integral of its normal component is equal to the volume integral of its divergence:

$$\oint_s \left[V(\mathbf{r})\boldsymbol{\nabla} f(\mathbf{r}) - f(\mathbf{r})\boldsymbol{\nabla} V(\mathbf{r}) \right] \cdot d\mathbf{s} = \int_v \boldsymbol{\nabla} \cdot \left[V(\mathbf{r})\boldsymbol{\nabla} f(\mathbf{r}) - f(\mathbf{r})\boldsymbol{\nabla} V(\mathbf{r}) \right] dv, \qquad (2.24)$$

for any closed surface s, where v is the volume bounded by the surface s. We will apply this to the "shell" bounded by two concentric spherical surfaces, shown in Fig. 2.4(b). The outer sphere is of radius a, the inner of radius b. The former is the same sphere shown in Fig. 2.4(a). We label these two surfaces s_a and s_b, respectively.

We start by examining the integrand of the volume integral on the right, which we will show must be zero. We apply the product rule Eq. (E.6) to write

$$\boldsymbol{\nabla} \cdot \left[V(\mathbf{r})\boldsymbol{\nabla} f(\mathbf{r}) - f(\mathbf{r})\boldsymbol{\nabla} V(\mathbf{r}) \right]$$

$$= \left[\boldsymbol{\nabla} V(\mathbf{r}) \cdot \boldsymbol{\nabla} f(\mathbf{r}) + V(\mathbf{r})\nabla^2 f(\mathbf{r}) \right]$$

$$- \left[\boldsymbol{\nabla} f(\mathbf{r}) \cdot \boldsymbol{\nabla} V(\mathbf{r}) + f(\mathbf{r})\nabla^2 V(\mathbf{r}) \right]$$

$$= V(\mathbf{r})\nabla^2 f(\mathbf{r}) - f(\mathbf{r})\nabla^2 V(\mathbf{r}).$$

But $f(\mathbf{r})$ and $V(\mathbf{r})$ each satisfy Laplace's Equation, and so $\boldsymbol{\nabla} \cdot \left[V(\mathbf{r})\boldsymbol{\nabla} f(\mathbf{r}) - f(\mathbf{r}) \, \boldsymbol{\nabla} V(\mathbf{r}) \right]$ must be zero, as we claimed. Therefore the surface integral on the left-hand side of Eq. (2.24) must be zero as well,

$$\oint_s \left[V(\mathbf{r})\boldsymbol{\nabla} f(\mathbf{r}) - f(\mathbf{r})\boldsymbol{\nabla} V(\mathbf{r}) \right] \cdot d\mathbf{s} = 0.$$

Now let's break up the surface s into its two components s_a and s_b. The surface element $d\mathbf{s}$ for the surface s_a is $a^2 \sin\theta \, d\theta \, d\phi \, \mathbf{a}_R$, where $\mathbf{a}_R$ is the unit normal vector, pointing outward from the volume. On the inner surface s_b, the unit normal vector is $-\mathbf{a}_R$ (remember, on this surface pointing out of the volume means pointing in toward the center), so the surface element is $d\mathbf{s} = -b^2 \sin\theta \, d\theta \, d\phi \, \mathbf{a}_R$. We don't need the full detail here, so we'll write these two simply as $d\mathbf{s} = ds\,\mathbf{a}_R$ and $d\mathbf{s} = -ds\,\mathbf{a}_R$ for surfaces s_a and s_b, respectively. Using these elements, and using Eqs. (2.22) and (2.23), the surface integral becomes

$$\oint_s \left[V(\mathbf{r})\boldsymbol{\nabla} f(\mathbf{r}) - f(\mathbf{r})\boldsymbol{\nabla} V(\mathbf{r}) \right] \cdot d\mathbf{s}$$

$$= \int_{s_a} \left[V(\mathbf{r})\left(-\mathbf{a}_R \frac{1}{R^2} \right) - \frac{1}{R}\boldsymbol{\nabla} V(\mathbf{r}) \right]_{R=a} \cdot \mathbf{a}_R \, ds$$

$$- \int_{s_b} \left[V(\mathbf{r})\left(-\mathbf{a}_R \frac{1}{R^2} \right) - \frac{1}{R}\boldsymbol{\nabla} V(\mathbf{r}) \right]_{R=b} \cdot \mathbf{a}_R \, ds$$

$$= -\frac{1}{a^2} \int_{s_a} V(\mathbf{r})ds - \frac{1}{a} \int_{s_a} \boldsymbol{\nabla} V(\mathbf{r}) \cdot d\mathbf{s} + \frac{1}{b^2} \int_{s_b} V(\mathbf{r})ds + \frac{1}{b} \int_{s_b} \boldsymbol{\nabla} V(\mathbf{r}) \cdot d\mathbf{s},$$

which, as we already said, must be zero. By the Divergence Theorem, the two surface integrals of the gradient $\nabla V(\mathbf{r})$ vanish ($\int_s \nabla V(\mathbf{r}) \cdot d\mathbf{s} = \int_v \nabla^2 V(\mathbf{r}) dv = 0$ since $\nabla^2 V(\mathbf{r}) = 0$), so

$$-\frac{1}{a^2} \int_{S_a} V(\mathbf{r}) ds + \frac{1}{b^2} \int_{S_b} V(\mathbf{r}) ds = 0.$$

Now we let the radius $b \to 0$, such that the inner sphere becomes very small. Then the potential $V(\mathbf{r})$ on this surface is nearly uniform, and since the potential is a continuous function, must be equal to V_c, the potential at the central point:

$$\frac{1}{b^2} \int_{S_b} V(\mathbf{r}) ds \to \frac{1}{b^2} V_c \int_{S_b} ds = \frac{1}{b^2} V_c \left(4\pi b^2\right) = 4\pi V_c$$

and

$$V_c = \frac{1}{4\pi a^2} \int_{S_a} V(\mathbf{r}) ds,$$

as we set out to show.

The second property for any solution of Laplace's Equation that we listed above, that the potential $V(\mathbf{r})$ does not have local maxima or local minima, follows directly from this first property. Since the potential at a local maxima is greater than the potential at any adjacent point, this contradicts the notion that the potential at this point is the average of the potential at the surrounding points. Similarly for a local minima, for which the potential must be less than the potential at any surrounding point. Therefore, we conclude that the potential in a charge-free region cannot have any local maxima or minima. An alternative way of reaching the same conclusion is through the physical interpretation of the gradient of a function, which we present in Appendix D. That is, the gradient tells us the slope of a function, or how rapidly (in space) the function changes when we take a step Δx in the x-direction, Δy in the y-direction, and Δz in the z-direction. Recall that for a one-dimensional function $f(x)$, a local maximum in $f(x)$ is identified by the slope df/dx equal to zero, and the second derivative d^2f/dx^2 negative. Similarly, we identify a local minimum in $f(x)$ by the slope df/dx equal to zero, and the second derivative d^2f/dx^2 positive. The idea is the same here. For a scalar function $V(\mathbf{r})$, local maxima and minima occur where $\nabla V(\mathbf{r})$ is zero, and $\nabla^2 V(\mathbf{r})$ is negative (maxima) or positive (minima). But we know that $\nabla^2 V(\mathbf{r})$ must be zero in a charge-free region, and so this potential cannot have any local maxima or minima.

Finally, the third property listed above, that is, that the potential $V(\mathbf{r})$ at any location $\mathbf{r}$ inside a closed region is always between the maximum potential V_{max} and the minimum potential V_{min} on the bounding surface of the charge-free region, follows similar logic.

From these general properties of the potential in a charge-free region, we can summarize by saying that the static electric potential in a charge-free region is a rather well-behaved function. It cannot have any wild gyrations, extreme variations, or pockets of excess. It simply varies smoothly from one surface to another. These conclusions will be borne out by the simple examples that follow. This brings us, of course, to specific solutions of Laplace's and Poisson's Equations, since, in addition to the general properties of solutions to Laplace's Equations that we were able to define and prove, we can use these simple second-order differential equations to determine the potential distribution for specific cases. Of course, we must know the potential at the boundaries of the region, known as the boundary conditions, but once we know these we are able to solve these differential equations for an explicit determination of the potential.

Laplace's and Poisson's Equations can be solved for specific geometries numerically as well as analytically. Numerous software packages are available for solving for the potential, given a specified set of boundary conditions. We will not go into these in any detail, except to say that in problems of practical concern, in which there are several surfaces of irregular shape and various potentials surrounding a region, analytic techniques become impractical or impossible to define, and numerical solutions are the only game in town.

We will now work a couple of examples to illustrate the use of Laplace's Equation for determining the potential in specific geometries. We will limit our discussion at present to cases in which the potential depends only on a single variable. We call these one-dimensional solutions. Two- and three-dimensional solutions are of considerable interest, and typically of more practical importance, but we defer discussion of these problems to other texts.

Example 2.7 Laplace's Equation, Parallel Conductors with Single Dielectric Material

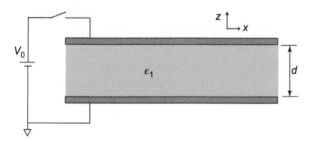

FIGURE 2.5 Two parallel conducting plates with a dielectric medium of permittivity ε_1 and thickness d in the space between. (For use in Example 2.7.)

Consider the simple pair of parallel conducting plates shown in Fig. 2.5. The region between the conductors is free of charge, and of permittivity ε. The spacing of the plates is d, the potential of the upper plate is V_0, and the lower plate is grounded. Under the approximation that d is much less than the size of the plate, determine the potential $V(z)$ in the region between the conducting plates.

Solution:

We start by considering the dimensionality of the solution. Since the plates are parallel to one another, and their size is large, moving to the left or right a bit from any location within the region between the plates shouldn't affect the potential, so we conclude that the potential is independent of x and y. This statement breaks down, of course, near the edges of the plates, but we are told that the size of the plates is much larger than their spacing, so we can safely ignore these small effects. Since the potential depends only on z, we will use the form of the Laplacian in rectangular coordinates, with $\partial V / \partial x$ and $\partial V / \partial y$ both equal to 0. Laplace's Equation reduces to

$$\frac{\partial^2 V(z)}{\partial z^2} = 0.$$

We solve this by direct integration. Integrating both sides once leads to

$$\frac{\partial V(z)}{\partial z} = C_1,$$

where C_1 is a constant of integration, which we must yet determine. Integrating a second time, we get

$$V(z) = C_1 z + C_2,$$

where C_2 is another constant of integration. So the solution to Laplace's Equation is a simple straight line function – that is, $V(z)$ varies linearly with z from plate to plate.

We must still find the constants C_1 and C_2, which we will do using the potentials of the plates. Evaluating $V(z) = C_1 z + C_2$ at $z = 0$, where the potential is 0, gives $C_2 = 0$, and evaluating $V(z) = C_1 z$ at $z = d$, where the potential is V_0, gives $C_1 = V_0/d$. Our final result is $V(z) = V_0 z/d$. We can check our solution by showing that $V(z)$ does indeed (1) satisfy Laplace's Equation $\partial^2 V(z)/\partial z^2 = 0$, and (2) that it has the correct values at the boundaries, $z = 0$ and $z = d$.

That first example may seem trivial, and it is, but it illustrates the basic concept of using Laplace's or Poisson's equation to determine the potential in a region of space. We determine the functional form of the potential, in a form that includes a couple of unknown constants. We then use any boundary conditions available in order to determine those constants. Let us now examine another example, one of a pair of parallel conducting plates, but in which a pair of dielectrics fill the space between the plates.

Example 2.8 Laplace's Equation, Parallel Conductors with Two Dielectric Materials

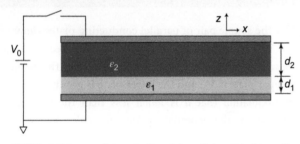

FIGURE 2.6 Two parallel conducting plates with two dielectric media, the first of thickness d_1 and permittivity ε_1, the second of thickness d_2 and permittivity ε_2. (For use in Example 2.8.)

Consider the simple pair of parallel conducting plates shown in Fig. 2.6. The region between the conductors is free of charge. Two dielectric slabs fill the space between the conductors, the bottom one of thickness d_1 and permittivity ε_1, and the second of thickness d_2 and permittivity ε_2. The potential of the upper plate is V_0, the lower plate is grounded, and the spacing between the plates is much less than their size. Determine the potential $V(z)$ in the region between the plates.

Solution:

As in the previous example, the potential must be independent of x and y, and Laplace's Equation reduces to

$$\frac{\partial^2 V(z)}{\partial z^2} = 0.$$

Using direct integration, we again obtain a general solution of the same form, valid within either the bottom dielectric or the top dielectric, except, of course, the constants of integration in these two regions are different. We use

$$V(z) = C_1 z + C_2$$

in the bottom dielectric (i.e. $0 < z < d_1$), and

$$V(z) = C_3 z + C_4$$

in the top dielectric (i.e. $d_1 < z < d_1 + d_2$). With four constants to determine, we will need to identify four boundary conditions. Two of these should be obvious. The potential at $z = 0$ is $V(z) = 0$, giving us $C_2 = 0$, and the potential at $z = d_1 + d_2$ is V_0, giving us $V_0 = C_3 (d_1 + d_2) + C_4$. For two more conditions we look at the boundary between the two dielectrics at $z = d_1$. First, we should expect that the potential is continuous at the boundary, or

$$V(d_1) = C_1 d_1 + C_2 = C_3 d_1 + C_4.$$

(The electric potential must always be continuous, since if this were not true, $\mathbf{E} = -\boldsymbol{\nabla}V$ would become infinite at the point of the discontinuity.) And finally, we can apply the boundary conditions that we determined in Section 1.5 for normal and tangential field components. In the present case, the electric fields are in the $-\mathbf{a}_z$ direction, and are therefore normal to the interface. The appropriate boundary condition then is that D_n is continuous across the boundary, or

$$-\varepsilon_1 \frac{\partial V}{\partial z}\bigg|_{z=d_1^-} = -\varepsilon_2 \frac{\partial V}{\partial z}\bigg|_{z=d_1^+},$$

which reduces to $\varepsilon_1 C_1 = \varepsilon_2 C_3$. (Remember that, for a perfect dielectric, the conductivity is zero and there are no free charges. Therefore $\rho_s = 0$ at the interface between the dielectrics.)

Solving these four equations for C_1, C_2, C_3, and C_4, we obtain

$$V(z) = \begin{cases} \dfrac{V_0\left[\varepsilon_1\left(z-d_1\right)+\varepsilon_2\, d_1\right]}{\left(\varepsilon_2\, d_1 + \varepsilon_1\, d_2\right)} & d_1 < z < (d_1 + d_2) \\[4mm] \dfrac{V_0\, \varepsilon_2\, z}{\left(\varepsilon_2\, d_1 + \varepsilon_1\, d_2\right)} & 0 < z < d_1 \end{cases} \tag{2.25}$$

Note that we can use this solution for the potential $V(z)$ to find the electric field $\mathbf{E}$ in the region between the plates. As a check of our solution, we can show (1) that $V(z)$ satisfies Laplace's Equation, and (2) that it satisfies the four conditions established during the course of our solution, that is, $V(z)$ at $z = 0$ and $z = (d_1 + d_2)$, that $V(z)$ is continuous at the interface between the dielectrics $(d_1 = d_2)$, and that $\mathbf{D}$ is the same in the two dielectrics. We leave this demonstration to Prob. P2-10.

Example 2.9 Laplace's Equation, Co-axial Conductors

In our final example of an application of Laplace's Equation, we consider the pair of parallel co-axial cylindrical conductors shown in Fig. 1.45. The region between the conductors has permittivity ε and is free of charge. The potential of the inner conductor relative to the grounded outer shield is V_0, and any fringe effects at the end of the cylinders are negligible. Determine the potential $V(\rho)$.

Solution:
By symmetry the potential must be independent of ϕ and z. (The independence of z is strictly true only when we are not close to the ends.) In cylindrical coordinates, when we retain only the ρ terms, Laplace's Equation reduces to

$$\frac{1}{\rho}\frac{\partial}{\partial \rho}\left(\rho\,\frac{\partial V(\rho)}{\partial \rho}\right) = 0,$$

where we used Eq. (E.20) for the Laplacian operator in cylindrical coordinates. We obtain a general solution using direct integration of the form

$$V(\rho) = C_1\, \ln \rho + C_2.$$

To determine the constants C_1 and C_2 we apply the boundary conditions $V(a) = V_0$ and $V(b) = 0$, which we solve to find

$$V(\rho) = V_0\, \frac{\ln\left(b/\rho\right)}{\ln\left(b/a\right)}.$$

The gradient of this potential yields the electric field **E** in the region between the cylindrical conductors. We can also verify that this potential $V(\rho)$ is V_0 at $\rho = a$ and zero at $\rho = b$.

In TechNotes 2.1 and 2.2 we use Poisson's Equation to describe the approximate properties of a semiconductor junction and of a battery.

TechNote 2.1 Semiconductor Junctions

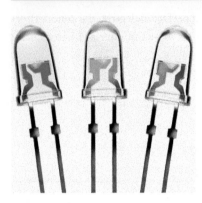

Semiconductor junctions, or p–n junctions, are the basis for solid-state devices such as diodes and bipolar transistors. In a semiconductor, the energy of electrons can lie in either of two bands, the valence band or the conduction band, but not in the band gap region that separates these bands. A common semiconductor material is silicon, for which the band gap energy at room temperature is 1.14 eV. The electronic structure of the silicon (Si) atom has four electrons in the outer shell, and at room temperature these electrons reside largely in the valence band. As such, they are not very mobile, and the conductivity of silicon is low, ~ 0.0017 S/m. (See Table 3.1. We'll talk more about conductivity in Chapter 3, but for now understand that semiconductors conduct better than insulators, but not as well as conductors.) By replacing some of the silicon atoms by atoms with five electrons in the outer shell, such as phosphorus (P), arsenic (As), or antimony (Sb), the material conductivity increases considerably as the "extra" electron in the outer orbital of these atoms starts to fill the conduction band. These atoms are called **donor** atoms, and the material is said to be n-type doped, as the dopant atoms contribute an extra electron, which has a negative charge, to the material. Alternatively, silicon can be doped by atoms with only three electrons in the outer shell, such as boron (B), aluminum (Al), or indium (In). These materials have one fewer electron than silicon in the outer shell, which leads to fewer electrons in the valence band. These unoccupied states in an ocean of electrons take on the properties of positively charged carriers, which are called **holes**. These holes are also mobile under the influence of an electric field, and so contribute to the conductivity of the material. These materials are said to be p-doped, due to the positive charge of the holes, and these atoms are called **acceptor** atoms.

In an n-doped semiconductor, the material is normally charge-neutral, as the negative charge of the extra electrons is offset by the positive charge of the dopant atoms. Similarly, in a p-doped semiconductor, the positive charge of the holes is balanced by the negative charge of the dopant atoms, and the material is charge-neutral. In a semi-conductor that is p-doped on one side and n-doped on the other, however, the situation is quite different (Fig. 2.7). In the absence of an externally applied electric field in the material, as would be the case when both materials are electrically neutral, the electrons in the n-doped region near the **junction** (the interface between the n-doped region and the p-doped region) are free to drift, or diffuse, from the n-doped region toward the p-doped region. Similarly, the holes in the p-doped region are free to drift toward the n-doped region. (Keep in mind that the electrons and holes are mobile, but the acceptor and donor impurity atoms are not. The electrons and holes are collectively called the charge carriers.) When electrons and holes are present in the same region of the semiconductor, they can recombine with one another. That is, the electrons in the conduction band can fill the holes in the valence band. As the density of electrons in the n-doped region and holes in the p-doped region decrease, these regions are no longer charge-neutral, since the charge of the dopants is no longer offset by the charge of the carriers. This region in which the carrier charge densities is reduced is known as the depletion region. The n-doped side of the depletion region becomes positively charged, and the p-doped side becomes negatively charged. This is shown pictorially in Fig. 2.7, in which the simplifying assumption that all of the carriers in the depletion zone recombine is made, and the carrier density becomes zero on both sides. The width of the depletion region on the p-doped side is labeled d_p, and that on the n-doped side d_n.

Even without application of an external potential to the semiconductor material, an electric field pointing from the positive charges of the n-doped side of the junction to the negative charges of the p-doped side is established in the depletion region, due to the net charge in the depletion region. This electric field inhibits additional diffusion of charge carriers into the depletion region, and the system reaches a steady state. The electric field in the depletion region leads to a potential difference between the two sides of the junction. Based on the direction of the electric field, the potential of the n-doped region is higher than that of the p-doped region.

We can approximate the electric field in the depletion region and the potential difference across the junction based upon the principles that we have just discussed. We'll use the following notation for the number densities of various charged entities: n_D, donor atoms (in the n-doped region); n_A, acceptor atoms (in the p-doped region); n_e, electrons (in the n-doped region); and n_h, holes (in the p-doped region). At any location in the semiconductor, the charge density is given by $\rho_v = e\left[(n_D - n_e) - (n_A - n_h)\right]$, where $e = 1.602 \times 10^{-19}$ C is the fundamental charge unit. We'll assume the simplified model shown for the charge distribution in Fig. 2.7. That is, in the n-doped and p-doped neutral regions the density of carriers is equal to the density of dopant impurity atoms, and the net charge density in these regions is zero. Furthermore, we'll assume that the carrier density is zero in the depletion region, so the charge density is $\rho_v = -en_A$ on the acceptor side, and $\rho_v = +en_D$ on the donor side. We then apply Poisson's Equation:

$$\nabla^2 V = -\rho_v/\varepsilon.$$

Since the charge density varies in only one dimension, the potential can also depend only on a single dimension, which we'll label the x-direction. Then Poisson's Equation becomes

$$\frac{d^2 V}{dx^2} = -\frac{\rho_v}{\varepsilon}.$$

We integrate both sides once to get

$$E_x(x) = -\frac{dV}{dx} = -\int_{-d_p}^{x} \frac{d^2 V}{dx^2}\, dx = \begin{cases} -en_A(x + d_p)/\varepsilon & -d_p < x < 0 \\ en_D(x - d_n)/\varepsilon & 0 < x < d_n \end{cases},$$

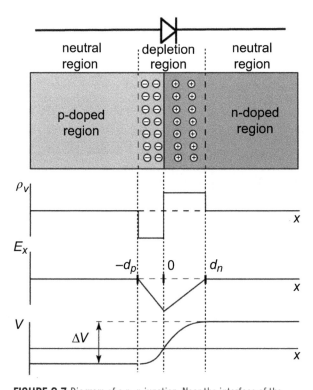

FIGURE 2.7 Diagram of a p–n junction. Near the interface of the n-doped region (on the right) and the p-doped region (on the left), known as the junction, electrons and holes recombine with one another, rendering this "depletion region" electrically charged: negative acceptor ions on the left, and positive donor ions on the right. This leads to an electric field in this region, and a potential difference across the junction. Plots of charge density, electric field, and electric potential over the junction region are shown beneath the junction diagram.

where we have used $E_x = 0$ in the neutral regions, along with the boundary conditions that E_x is continuous at $x = -d_p$ and $x = 0$. (The latter is valid when the permittivity ε is the same in each region and there are no surface charges at the boundaries.)

We integrate a second time to determine the electric potential in the depletion region relative to the p-doped region:

$$V(x) = \begin{cases} en_A(x + d_p)^2/2\varepsilon & -d_p < x < 0 \\ en_A d_p^2/2\varepsilon + en_D(d_n x - x^2/2)/\varepsilon & 0 < x < d_n \end{cases}.$$

Evaluating this potential at the edge of the depletion region, $x = d_n$, we find the potential of the n-doped neutral region relative to that of the p-doped neutral region,

$$\Delta V = q(n_A d_p^2 + n_D d_n^2)/2\varepsilon.$$

Defining the total width of the depletion region $d = d_p + d_n$, and using $d_p n_A = d_n n_D$, which states that the total number of holes and electrons that recombine in the depletion zone equal one another, we can show that $d_p = n_D d/(n_A + n_D)$ and $d_n = n_A d/(n_A + n_D)$. Replacing d_p and d_n in the expression for ΔV allows us to write

$$\Delta V = \frac{e}{2\varepsilon}\left(\frac{n_A n_D}{n_A + n_D}\right)d^2.$$

This expression can be inverted and solved for the width of the depletion region,

$$d = \sqrt{\frac{2\varepsilon}{e}\left(\frac{n_A + n_D}{n_A n_D}\right)\Delta V},$$

which shows the dependence of the width of the depletion region on the potential difference ΔV.

We set up this discussion to explore the potential difference across the junction and the depletion width when no external potential is applied to the junction. We can, however, extend this treatment to the case when an external potential is applied, and show that this junction material tends to conduct current in one direction, but not the reverse direction. Thus, the p–n junction functions as a solid-state diode. First consider applying a source with the positive terminal connected to the p-type material, and the negative terminal to the n-type material. This is known as the forward-biased condition. The positive potential at the p-type material repels the holes, forcing them into the depletion region. Recall that the depletion region on the p-doped side is negatively charged, so these holes tend to partially neutralize these charges. Similarly, the negative bias on the n-doped material repels the electrons on this side, forcing them toward the junction, where they neutralize some of the positive charges. Both of these effects tend to reduce the width d of the depletion region, and increase the current from the positive to the negative terminals. If we now reverse the leads of the source, and connect the negative lead to the p-type material and the positive to the n-type material, the charge carriers on both sides of the junction are drawn away from the junction, increasing its width and reducing the current. The junction is now reverse biased. Thus, the diode conducts current when forward biased, but conducts little current when reverse biased.

Let us go back and examine the approximation that we made in our analysis that all of the charge carriers within the depletion region had recombined, so that the charge density is $\rho_v = -en_A$ on the acceptor side and $\rho_v = +en_D$ on the donor side. This approximation made our analysis much simpler, but it clearly limits the validity of our result. To do this analysis properly, we should consider the effect of the electric field in the depletion zone on the diffusion of the carriers. That is, not only does the carrier density in the depletion zone affect the electric field in this region, but the electric field also affects the carrier density. We therefore would require a solution that considers both effects in a self-consistent solution. This is beyond the scope of our treatment here, and we defer this analysis to courses dealing with semiconductors.

Batteries use chemical energy to produce electric potential, and indirectly electrical current. The introduction in 1800 of the first voltaic cell by Alessandro Volta was critical to the subsequent discovery by Hans Christian Oersted that electrical currents produce magnetic fields (see Biographical Note 4.1) and many other advances. In TechNote 2.2, a very simple model of a battery is presented to illustrate an application of Poisson's Equation.

TechNote 2.2 Batteries

Batteries, in which chemical energy is used to create an electric potential, date back to the late eighteenth century. In 1780, Luigi Galvani (1737–1798), an Italian scientist, noted that the leg of a dissected frog, when suspended from an iron or brass hook, would twitch when the leg came into contact with a probe of a different metal. Galvani attributed the twitching of the frog's leg to electricity generated within the tissues of the frog's legs, which he called animal electricity. After learning of Galvani's work and his interpretation of the effect, Alexandro Volta (1745–1827), another noted Italian scientist, embarked upon further studies of the effect, and concluded that, rather than electricity generated within the frog's legs, an electric potential was produced at the points of contact between the electrodes and the frog's legs, and the legs were stimulated by the current passing through the tissue. Volta found that the stimulation varied for different metal probes, and described his findings to the president of the Royal Society of London in 1800. A multi-cell pile constructed by Volta, composed of alternating layers of zinc and silver, separated by cloth or paper soaked in a salt solution, is shown in Fig. 2.8(a). A diagram of a similar multi-element pile is shown in Fig. 2.8(b). Each element, commonly called a cell, consists of a zinc plate and a copper plate separated by an electrolyte. One element is stacked on top of another. When connected in series, as shown, the potential difference of this four-cell battery is four times the potential produced by a single cell.

We show a typical battery configuration in Fig. 2.8(c). In this diagram, we see a pair of electrodes, labeled anode and cathode, made of different types of metals. The region between the electrodes is filled with an electrolyte, a liquid or paste material consisting of high concentrations of positive and negative ions. These ions can easily diffuse within the electrolyte. Battery function depends on two different chemical reactions that take place at the surfaces of the electrodes. At the surface of the anode, molecules of the electrolyte undergo oxidation, a chemical process by which molecules in the electrolyte donate one or more electrons to the anode. In this reaction, the anode becomes negatively charged and neutral molecules in the electrolyte become positive ions, or negative ions are neutralized. At the cathode, the opposite type of reaction, called reduction, takes place. In reduction processes, the molecules in the electrolyte at the surface of the cathode accept electrons from the cathode. Thus, the cathode becomes positively charged and neutral molecules in the electrolyte become negatively charged, or positive ions are neutralized. When the anode is connected by a wire to an electrical element, such as a resistor, which is then connected to the cathode, excess electrons on the anode can flow through the wire and resistor to the cathode, where they reduce molecules in the electrolyte. Since electrons carry a negative charge, we generally describe this as a current flowing from the cathode through the electrical element to the anode.

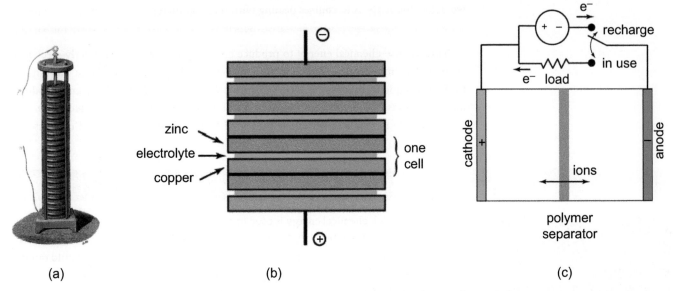

(a)

(b)

(c)

FIGURE 2.8 (a) Voltaic pile – the first electrical battery, created c. 1800 by Alessandro Volta. Each cell is composed of alternating layers of zinc and silver, separated by cloth or paper soaked in a salt solution. (b) A schematic of a simple voltaic pile. Each element consists of zinc and copper plates, separated by an electrolyte, such as saltwater brine or sulfuric acid. The zinc cathode is oxidized to form Zn^{2+}, which is added to the electrolyte. Hydrogen gas H_2, formed through reduction of H^+ ions from the electrolyte, is formed at the copper anode. (c) Schematic representation of a rechargeable battery, connected to a load for normal battery operation, or to a source for recharging.

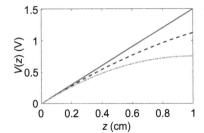

FIGURE 2.9 The potential $V(z)$ in the region between the anode and cathode for a simple model of a battery described in the text based on a parallel plate configuration. The straight, solid red line shows the potential when the electrolyte is perfectly neutralized. For the purple dashed line and the green dot-dashed line, charges in the electrolyte lead to a reduced potential difference between the electrodes. In each case, the charge density in the electrolyte is modeled as being uniform within the electrolyte.

Diffusion of the ions within the electrolyte is a critical process of the battery action. Without ionic diffusion, the electrolyte near the anode would quickly accumulate a large positive charge, and near the cathode a negative charge. An important function of the electrolyte is to allow free diffusion of the ions within the region between the electrodes. Batteries often contain a membrane made of a polymer, called the separator, as shown in Fig. 2.8(c), or an ionic conduit known as a salt bridge, which allows some ions, but not others, to pass to help neutralize the net charge density within the electrolyte. For example, the separator may preferentially allow negative ions to move toward the anode, where they can neutralize positive ions. To illustrate the importance of this diffusion process, we have plotted the potential $V(z)$ in the region between the anode (located at $z = 0$) and the cathode (located at $z = d = 1$ cm) in Fig. 2.9. For simplicity, we have chosen a simple parallel plate battery configuration, with $d = 1$ cm representing the spacing between the anode and cathode, and the potential of the anode being zero. These plots show solutions of Poisson's Equation, $\nabla^2 V = -\rho_v/\varepsilon$, for different values of the volume charge density ρ_v. In the parallel plate geometry, the potential $V(z)$ depends only on z, the linear coordinate from the anode to the cathode, and Poisson's Equation becomes $\partial^2 V(z)/\partial z^2 = -\rho_v/\varepsilon$. When ρ_v is constant (i.e. the charge is uniformly distributed in the region between the anode and cathode, which is a simplification that we use for this analysis), the solution for the potential is $V(z) = -\rho_v z^2/2\varepsilon + C_1 z + C_2$, where C_1 and C_2 are constants of integration. We determine these constants using the following boundary conditions at the anode: $V(0) = 0$ and $D_z(0) = \rho_s$, where ρ_s is the surface charge density of the anode. (Remember that ρ_s is negative at the anode, which indicates that $D_z(0)$ is also negative. This is expected, since **E** and **D** point from the positive cathode to the negative anode.) The result for the potential is $V(z) = -\rho_s z/\varepsilon - \rho_v z^2/2\varepsilon$. The red solid line in Fig. 2.9 corresponds to $\rho_v = 0$. (That is, the densities of the positive ions and negatives ions are equal to each other, and the electrolyte is electrically neutral.) We used $\rho_s = 4$ nC/m^2 for each case, which results in $V(d) = 1.5$ V when $\rho_v = 0$. For the other two lines in Fig. 2.9, $\rho_v = -\rho_s/2d$ (purple dashed line), or $\rho_v = -\rho_s/d$ (green dot-dashed line), leading to significant suppression of $V(d)$. Although our model is highly simplified, this example illustrates the importance of ion diffusion in the electrolyte to keep it electrically neutral.

Many types of batteries, but not all, are rechargeable, meaning that if a current passes through them in the opposite direction they can be restored to their initial state, or nearly so. This is illustrated in the model shown in Fig. 2.8(c). When the switch in the external circuit is flipped upward, the "charger" is integrated into the circuit. In this situation, the reduction and oxidation processes are reversed, as is the ion flow in the electrolyte.

2.4 A Numerical Solution of Laplace's Equation

In Section 2.3 we introduced Laplace's Equation, found analytic solutions to several simple one-dimensional geometries, and examined several general properties of solutions to this important equation. One such property is that the potential at one point is the average of the potentials of all the surrounding points. As is demonstrated in this section, this property can be used as a means of solving for the potential in a region numerically.

The first step of the numerical solution technique demonstrated here is to divide up the solution space (the region between a set of conductors, whose potentials are known) into a grid pattern consisting of a set of discrete points. An example is shown in Fig. 2.10. The three conductors have potentials of $V_0 = 80$ V, $-V_0$, and 0 V, as labeled. Our task is to determine the potential at each of the labeled grid points in the region between the conductors. Before we start, recall that we have previously worked several problems involving two parallel conductors, for which, in the limit of small spacing between the conductors, the edge effects near the ends of the conductors are minimal and the potential varies linearly from one conductor to the other. The two conductors on the left in this figure are parallel to one another, but the spacing is not small, especially in light of the proximity of the grounded conductor to the right. Therefore, it might be expected that the potential at the edge is more interesting than the simple, linearly varying potential asserted earlier for closely spaced plates.

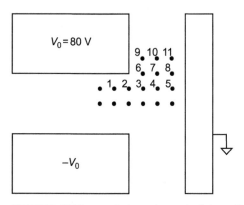

FIGURE 2.10 Three conductors, whose potentials are $V_0 = 80$ V, $-V_0$, and 0 V, as shown, surround a region in which the potential is unknown. The dots indicate the grid pattern for numerical calculation of the potentials.

You will notice that the grid pattern drawn between the conductors is not complete. First, the grid points extend only a short distance into the region between the conductors on the left. The reason for this is that farther to the left, the edge effects will fade, and the potential will vary uniformly from $+V_0$ to $-V_0$ from top to bottom. Extending the grid pattern will be left to Prob. P2-17. Second, only grid points in the top half of the solution space are plotted. This simplification is allowed due to the symmetry of the configuration, which tells us that the potential of any point along the horizontal mid-line must be zero, and that the potential at any point in the lower half is the negative of the potential of the mirror-image point in the top half. Therefore, our job is reduced by calculating the potential only in the upper half, and using $V = 0$ at the midpoint as a boundary condition. The following solution is found by guessing a set of potentials at each of the grid points, and then iteratively determining the potential at one grid point by averaging of the potentials of the surrounding grid points until convergence is obtained. (*Convergence* means that further iterations produce only a minimal variation in the result.)

Example 2.10 Numerical Solution of Laplace's Equation

Determine the potential at each of the labeled grid points in the region between the conductors shown in Fig. 2.10.

Solution:

The grid pattern has been decided for us. The next step in the solution is to guess at the potentials for the labeled grid points. Obviously, a better guess leads to more rapid convergence of the solution, but exact guesses are not required. The values we chose are listed in the following table, labeled as the first iteration.

Iteration	1	2	3	4	5	6
Grid Pt.						
1	40.00	40.00	40.00	38.75	39.22	39.11
2	40.00	40.00	35.00	36.88	36.45	36.14
3	40.00	20.00	28.75	26.56	25.45	24.88
4	30.00	25.00	18.75	16.62	15.62	15.11
5	20.00	11.25	8.83	7.82	7.33	7.09
6	10.00	55.00	50.62	48.73	47.76	47.26
7	60.00	35.00	31.09	29.20	28.24	27.75
8	20.00	16.56	14.67	13.72	13.24	13.00
9	60.00	58.75	57.27	56.38	55.90	55.66
10	40.00	38.44	36.78	35.86	35.38	35.14
11	20.00	18.75	17.86	17.40	17.16	17.03

Each potential in this table was found by averaging the potentials at the four nearest neighboring grid points. For example, grid point 1 is surrounded by the following four grid points: the top conductor (potential V_0); the unlabeled point to the left (potential $V_0/2$, as it is assumed that this point is far enough from the edge of the conductor that the potential varies smoothly from top to bottom); the grid point below (0 V, as this point is on the mid-line between the top and bottom conductors); and grid point 2 (potential V_2, the initial guess for the potential at this point). The average of these four potentials is

$$V_1(i) = \frac{V_0 + V_0/2 + 0 + V_2(i-1)}{4} = 40.00 \text{ V},$$

which is entered in the table as grid point 1, iteration 2. Notice the index i, which represents the iteration number. When calculating the ith value of V_1, we use the $(i-1)$th value of V_2.

For grid point 2, the process is similar. What is different is that, for the potential of the point to the left, V_1, we have already updated this potential, so this revised value is the value used in determining the average:

$$V_2(i) = \frac{V_0 + V_1(i) + 0 + V_3(i-1)}{4} = 40.00 \text{ V}.$$

This value is entered in the table as grid point 2, iteration 2.

Point-by-point, the average values are determined for all grid points, and listed in the column labeled iteration 2. One last example of an average value is

$$V_7(i) = \frac{V_{10}(i-1) + V_6(i) + V_4(i) + V_8(i-1)}{4} = 35.00 \text{ V.}$$

This value is calculated with the previous iteration value of V_{10} and V_8, which have not yet been updated, but with the new values for V_6 and V_4, which have been updated. The entire process is repeated for iteration 3, using the most recent potentials of each of the nearest neighbors, and so forth.

Notice that the initial guesses for grid points 6 and 7 were quite far off from the final solution. This was clearly an error in typing in the initial values. Still the solutions converged rather well, and variations of each potential after just the third or fourth iteration were minimal.

The values of the potentials can be improved in a couple of ways. A finer grid pattern is an obvious choice. In this technique, we approximate derivatives of the potential using differences between discrete points. Using grid points that are closer to one another makes this a better approximation. It also increases the effort needed to calculate all of the average potentials, however. Therefore, using a finer grid pattern in regions where the potential varies rapidly can help. Also helpful are techniques available to help achieve smoother convergence. In some cases the potential values may oscillate about a final value rather than smoothly approaching this value. This can increase the number of iterations needed to reach a final value. There are numerous, much more sophisticated, software packages available for numerically calculating potentials. These will not be covered here, but if you require precise potential plots these can be used to improve precision and decrease your programming time.

2.5 Capacitance

LEARNING OBJECTIVES

After studying this section, you should be able to:

- determine the capacitance of a pair of conductors using the electric field in the region between the capacitors;
- determine the capacitance of a pair of conductors using the electrostatic energy stored in the capacitor; and
- determine the capacitance per unit length of a pair of long, uniform parallel conductors.

For each of the systems of conductors that we have previously studied, we have found that the potential difference between the conductors is linearly proportional to the charge on the conductors. This is generally true (not limited to just the systems we selected), and it is useful to define the **capacitance** of a system of conductors as the total charge on a conducting body per unit potential difference,

$$C = \frac{Q}{V_1 - V_2},$$
(2.26)

where the charge on one conductor is $+Q$, the charge on the other $-Q$, and the potential difference between the conductors is $V_1 - V_2$. The capacitance depends only on the geometry of the system and material properties – that is, the permittivity of the dielectric in the region between the conductors – and is independent of the charge on the conductors or the potential. The name capacitance is a descriptive name, in that it describes the *capacity* of a conductor to hold charge. The unit of capacitance is the farad, named after Michael Faraday, usually abbreviated F. In terms of units that we have already studied, F = C/V. (We need to avoid confusion among symbols here. We use C to represent the capacitance of a system, as we just introduced, but C also stands for the unit of charge, the Coulomb. This is the standard notation.) You probably recall from your introductory physics classes that a simple parallel plate capacitor of plate area S, plate separation d, and permittivity ε is $C = \varepsilon S/d$. At least parts of this expression should make sense to us. For example, we know that the charge on the capacitor plates resides on the inside surfaces, so that as the plate area S increases, there is more surface area to hold charge, and we should therefore expect that the capacitance increases as well. Similarly, if the charge on the plates is held constant, the electric field is fixed and the potential difference between the plates increases with the plate separation d, so the capacitance should decrease with increasing d, making its placement in the denominator logical. Finally, the permittivity in the numerator tells us that, with increasing permittivity, the capacitor can store larger charge for a given potential difference. From our introduction to dielectric materials in Section 1.5, we know that the electric field in a dielectric material is reduced by the electric dipoles induced in the material, so that, for a given charge stored on the conductors, the potential difference between the conductors is reduced, leading to an increased capacitance. We'll explore this idea a bit more in the examples later in this section.

Let us now develop a more general concept of capacitance. This will allow us perhaps more insight into its significance, but also allow us to define and determine the capacitance of a wider set of conductor geometries. In particular, we are later headed toward a quantitative description of wave propagation on transmission lines, and one of the important parameters that we will use to describe this will be the **capacitance per unit length** of parallel conductors.

We consider two conducting bodies of arbitrary shape, as shown in Fig. 2.11, in which a charge $+Q$ is on the conductor on the left side and a charge $-Q$ on the other, and the dielectric medium between the conductors is a simple medium (i.e. linear, isotropic, and homogeneous), such that $\mathbf{D} = \varepsilon\mathbf{E}$ everywhere. Each conductor, of course, is an equipotential, and we denote the potential difference as $V_{12} = V_1 - V_2$. Since electric field lines originate at positive charges and terminate on negative charges, the electric field lines in this figure point from the left conductor to the right conductor. The charges on the two conductors reside on the surfaces. We will make use of the relation between the surface charge density and the displacement field $\mathbf{D}$, and the relation between the potential difference between the two bodies and the electric field $\mathbf{E}$. Constructing a Gaussian surface completely enclosing the left conductor (the dashed line in the figure), and applying Gauss' Law, the total charge on one plate is

$$Q = \oint_s \mathbf{D} \cdot d\mathbf{s}.$$

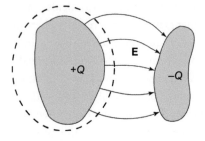

FIGURE 2.11 Two charged conductors. The capacitance $C = Q/V_{12}$ of a system of conductors is computed using the fields that surround the conductors.

Keep in mind that this can be applied for *any* Gaussian surface that encloses the conductor, so we are free to choose the surface that is most convenient for us. Similarly, we can determine the potential difference between the conductors in terms of the path integral of $\mathbf{E}$,

$$V_{12} = V_1 - V_2 = -\int_2^1 \mathbf{E} \cdot d\boldsymbol{\ell}.$$

Since the capacitance of these conductors is defined as the ratio of the charge Q to the potential difference, this quantity is

$$\boxed{C = \frac{Q}{V_{12}} = \frac{\oint_s \mathbf{D} \cdot ds}{-\int_2^1 \mathbf{E} \cdot d\boldsymbol{\ell}}.} \tag{2.27}$$

This expression is quite general, and it allows us, for any pair of conductors for which we can determine the $\mathbf{E}$ (and $\mathbf{D}$) field in the region between, to determine the capacitance.

An alternate method of determining the capacitance of a structure is based upon the electrostatic energy of the capacitor. In Section 2.1 we discussed the energy stored in the electric fields, either in terms of the charge density and potential (Eq. (2.3)) or in terms of **E** and **D** generated by the charges (Eq. (2.7)). In a typical capacitor, consisting of two conductors separated by a dielectric medium, the charges of the system are surface charges on the surfaces of the conductors, and Eq. (2.3) takes the form

$$W_e = \frac{1}{2}\int_{S_1} \rho_s(\mathbf{r})V_1 ds_1 + \frac{1}{2}\int_{S_2} \rho_s(\mathbf{r})V_2 ds_2,$$ (2.28)

where the index 1 or 2 labels the different conductor surfaces, each at potential V_1 or V_2, respectively. The potential of each surface is a constant, so

$$\int_{S_k} \rho_s(\mathbf{r})V_k ds_k = QV_k,$$

where the total charge on either conductor is $\pm Q = \int \rho_s ds_k$. Therefore the total energy of the capacitor is

$$W_e = \frac{1}{2}[QV_1 - QV_2] = \frac{1}{2}QV_{12}.$$

Using Eq. (2.27), this energy can be written as

$$W_e = \frac{1}{2}CV_{12}^2$$

or

$$W_e = \frac{Q^2}{2C}.$$

Thus the capacitance can be determined from the electrostatic energy as

$$\boxed{C = \frac{2W_e}{V_{12}^2} = \frac{Q^2}{2W_e}.}$$ (2.29)

Example 2.11 Capacitance, Parallel Conductors with Single Dielectric Material

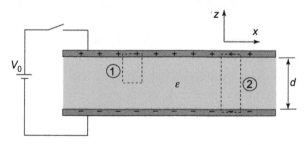

FIGURE 2.12 Two parallel conducting plates of area S with a dielectric medium of permittivity ε and thickness d in the space between. In Example 2.11 we will determine the capacitance of this configuration.

Determine the capacitance of the simple parallel plate capacitor shown in Fig 2.12. The plate area is S, the plate spacing is d, and the dielectric medium between the plates is of permittivity ε. Ignore fringe field effects at the edge of the plates.

Solution:

When the switch is closed, the capacitor is charged to a potential V_0, with charge Q residing on the upper plate and $-Q$ on the lower. Ignoring edge effects, the charge density on the upper plate is uniform, and equal to $\rho_s = Q/S$. To determine the displacement field **D** in the dielectric medium, we construct a Gaussian surface, labeled #1 in the figure. By the symmetry of the conductors, the x- and y-components of **D** must be zero, so $\mathbf{D} = \mathbf{a}_z D_z$. The flux is zero through the upper surface of the Gaussian surface (since this surface is inside the conductor, where the field must be zero) and through the sides (since $\mathbf{D} \perp \mathbf{a}_n$, the surface normal on these surfaces). So the displacement field flux is $\oint_s \mathbf{D} \cdot d\mathbf{s} = -D_z \Delta s$, where Δs is the area of the bottom face of the Gaussian surface. By Gauss' Law, $\oint_s \mathbf{D} \cdot d\mathbf{s}$ must be equal to the charge enclosed within the Gaussian surface. This charge resides on the inside surface of the conductor, and is equal to $\rho_s \Delta s$. Therefore D_z in the dielectric is equal to $-\rho_s$, and is independent of z. We have also constructed a larger Gaussian surface, labeled #2, with its upper face inside the top conductor and its lower face inside the bottom conductor, to show that the surface charge density on the inside surface of the lower conductor is equal to $-\rho_s$. (In words, the surface charge densities on the two conducting plates are equal but opposite.) Now that we know D_z, we use $E_z = D_z/\varepsilon$ everywhere in the region between the plates to determine the capacitance of the configuration. The total charge on the plates is $\pm Q$, where $Q = \rho_s S$. The potential difference between the plates is $V_0 = -\int \mathbf{E} \cdot d\boldsymbol{\ell}$, integrated along any path from the bottom plate to the top plate. But $\mathbf{E}$ is uniform, and choosing a straight-line path, $d\boldsymbol{\ell} = \mathbf{a}_z \, dz$, we find $V_0 = \rho_s d/\varepsilon$. These values of Q and V_0 give us $C = Q/V_0 = \varepsilon S/d$, the result that we recall from our earlier discussion.

Using the energy of the charged capacitor requires some of the same steps. The energy, using Eq. (2.7), is

$$W_e = \frac{1}{2} \int_v \mathbf{D} \cdot \mathbf{E} \, dv = \frac{1}{2} \left(\frac{\varepsilon V_0}{d} \right) \left(\frac{V_0}{d} \right) Sd = \frac{1}{2} V_0^2 \frac{\varepsilon S}{d}.$$

Then using Eq. (2.29), the capacitance is $C = \varepsilon S/d$, in agreement with the result in the previous paragraph.

Let's look a little more closely at the dielectric medium in this example. We'd like to understand its role in the capacitor, and how a material with a high permittivity increases the capacitance of the structure. To achieve this, we look to the polarization $\mathbf{P}$ induced in the material, and the effective charges at the surfaces. We showed previously in this solution that the displacement field D_z (remember, in this simple geometry all fields are in the z-direction) is equal to the surface charge density ρ_s in the top conductor, and $E_z = D_z/\varepsilon = \rho_s/\varepsilon$. By Eq. (1.64),

$$P_z = D_z - \varepsilon_0 E_z = \rho_s - \varepsilon_0 \rho_s/\varepsilon = \rho_s \left(1 - \varepsilon_r^{-1} \right).$$

Using P_z in Eq. (1.62) yields the effective charge densities, $-\rho_{s,eff}$ on the top surface of the dielectric and $+\rho_{s,eff}$ on the bottom, where $\rho_{s,eff} = \rho_s(1 - \varepsilon_r^{-1})$. So for small relative permittivity ε_r just greater than 1, these effective charges are rather small, much less than ρ_s, but as ε_r gets large the effective charge $\rho_{s,eff}$ can become a significant fraction of ρ_s. Since the effective charge is negative on the top side and positive on the bottom, the electric field resulting from these bound charges is opposite that due to free charges, and the total field in the region between the plates is reduced, relative to what it would be without the dielectric material. (In fact, we already saw this, in that $E_z = \rho_s/\varepsilon$, so a large permittivity leads to a small electric field E_z.) A decreased magnitude of the electric field tells us that the potential difference between the conductors is also reduced, and the capacitance increased. So we see that the large capacitance of a structure with a dielectric material with a large permittivity is a result of the large effective surface charge density on the dielectric.

Now lets look at a slightly more complex geometry, a parallel plate capacitor with two dielectrics.

Example 2.12 Capacitance, Parallel Conductors with Two Dielectric Materials

Determine the capacitance of the parallel plate capacitor with two dielectric materials shown in Fig 2.6. The plate area is S, the lower dielectric material is of thickness d_1 and permittivity ε_1, and the upper dielectric material is of thickness d_2 and permittivity ε_2. Ignore fringe field effects at the edge of the plates.

Solution:

We will first find, using Gauss' Law, $\mathbf{D}$ in the region between the conductor plates. Similar to the process described in Example 1.19, the displacement field is $D_z = -\rho_s$ for medium #1 or #2, where ρ_s is the surface charge density on the upper plate and the surface charge density on the lower plate is $-\rho_s$. From $\mathbf{D}$ we find $\mathbf{E} = E_z \mathbf{a}_z$, where

$$E_z = \begin{cases} D_z/\varepsilon_1 = -\rho_s/\varepsilon_1 & 0 < z < d_1 \\ D_z/\varepsilon_2 = -\rho_s/\varepsilon_2 & d_1 < z < d_1 + d_2 \end{cases}$$

Now with E_z in hand, we can find the potential difference $V_0 = -\int \mathbf{E} \cdot d\boldsymbol{\ell}$ along any path from the lower plate to the upper plate, and since the field is uniform within each of the two dielectrics, the integration yields

$$V_0 = \frac{\rho_s d_1}{\varepsilon_1} + \frac{\rho_s d_2}{\varepsilon_2}.$$

The capacitance is then

$$C = \frac{Q}{V_0} = \frac{S}{(d_1/\varepsilon_1 + d_2/\varepsilon_2)}.$$

Now let's examine a few geometries that differ from the parallel plates of the previous two examples, such that fields are no longer uniform. The first of these also differs in another important aspect. That is, we have only a single conductor. Still, capacitance is a useful concept, even in this case.

Example 2.13 Capacitance, Isolated Sphere

Find the capacitance of an isolated spherical conductor of radius a surrounded by a dielectric medium of permittivity ε.

Solution:

Our procedure will follow along the same line as the previous example, although the details differ. First, let's make sure we understand what the capacitance means, since it is a single-conductor system. If we apply a charge Q to the conducting sphere, this charge will, of course, reside on the outside surface of the conductor, and we can determine the electric field in the space surrounding the sphere using Gauss' Law or Coulomb's Law, and from this we can determine the potential V_0 of the sphere relative to $V = 0$ at $R = \infty$ by taking the line integral of $\mathbf{E}$. The capacitance of this isolated body is then the ratio of the charge Q on the conductor to its potential V_0. While we often think of a two-conductor system when we talk of capacitance, this is not necessary. Here we have a single conductor, whose potential (relative to infinity) is proportional to the charge it stores, and we can discuss its capacitance in precisely the same language as the more typical parallel plate element, for example.

We have treated a spherical charge distribution previously (although not a conductor), so the following procedure should be familiar. We construct a spherical Gaussian surface of radius $R > a$ that is concentric with the conductor, as shown in Fig. 1.15(a). Gauss' Law applied here gives $\oint \mathbf{D} \cdot d\mathbf{s} = D_R 4\pi R^2 = Q$, or $D_R = Q/4\pi R^2$. We divide by ε to get $E_R = Q/4\pi\varepsilon R^2$. The line integral of $\mathbf{E}$ is $V_0 = -\int \mathbf{E} \cdot d\boldsymbol{\ell}$ integrated from ∞ where $V = 0$ to a. This gives $V_0 = Q/4\pi\varepsilon a$, leading to a capacitance of $C = Q/V_0$ of $4\pi\varepsilon a$. For example, a sphere of radius $a = 20$ cm in air has a capacitance of 22 pF. This capacitance is consistent with the charge and potential we found for the van de Graaff generator example in TechNote 1.6.

Example 2.14 Capacitance of Two Co-axial Conductors

Find the capacitance of a pair of cylindrical co-axial conductors of radii a and b and length L, as first considered in Example 1.18. The space between the conductors is filled with a dielectric medium of permittivity ε. Ignore fringe field effects at the end.

Solution:

As we found in the previous example, the displacement field in the region between the conductors is radial, and of magnitude $D_\rho = \rho_{s,a} a/\rho$, where $\rho_{s,a}$ is the surface charge density on the inner cylindrical conductor. Since we can ignore fringe effects at the ends of the cylinders, the surface charge density is uniform, and is equal to $\rho_{s,a} = Q/2\pi aL$, where Q is the total charge on the inner conductor and $2\pi aL$ is the total surface area. Then we can write the displacement field as $D_\rho = Q/2\pi L\rho$. The electric field is $E_\rho = D_\rho/\varepsilon$, and the voltage of the inner conductor relative to the grounded outer shield is $V_0 = Q\ln(b/a)/2\pi\varepsilon L$. The capacitance is then $Q/V_0 = 2\pi\varepsilon L/\ln(b/a)$.

Let us make two special notes regarding capacitance: one related to a topic already discussed, the other related to a topic that will follow. First, we discussed in Section 2.1 the energy stored in an electrostatic field. For example, we calculated the energy stored in the field of a pair of co-axial conductors in Example 2.5, in which we found $W_e = \pi\varepsilon L V_0^2/\ln(b/a)$. (See Eq. (2.9).) From your experience with capacitors, you should recall that the energy stored in a capacitor of capacitance C charged to voltage V_0 is $W_c = (1/2)CV_0^2$. We should reasonably expect that these results are consistent, and indeed they are. We will ask you to verify this relation for specific geometries in Probs. P2-18–P2-20.

Second, an important topic that we will treat later is that of transmission lines, consisting of a pair of parallel, conducting wires (of various geometries) that can guide high-frequency electromagnetic waves from a source to a load. As an example, the signal for your television, whether you receive your service from a rooftop antenna, the cable service, or from your dish antenna (after reception from the satellite signal) is commonly carried by a co-axial transmission line. One of the parameters that we will use for describing wave propagation on a transmission line is its capacitance per unit length C'. From Example 2.14, we see that $C' = C/L$ for the co-axial line is

$$\boxed{C' = \frac{2\pi\varepsilon}{\ln(b/a)}.}$$
(2.30)

When we consider transmission lines in detail, we will also consider two additional important configurations of conductors. These are the pair of plane parallel conductors and two round parallel conductors, commonly called a twin-lead transmission line. We already derived the capacitance $C = \varepsilon S/d$ of a pair of parallel conductors of area S and spacing d. The capacitance per unit length C' for the plane parallel transmission line is simply

$$\boxed{C' = \frac{w\varepsilon}{d},}$$

(2.31)

where w is the width of the conductor plates. For the twin-lead configuration, the electric field pattern is a bit more complicated, and we are not ready yet to discuss C' for this transmission line. We will develop the tools necessary for this analysis in the next section, and include C' there.

In TechNotes 2.3–2.8, several examples of capacitances found in common technologies are described. The first example is a capacitive touch screen, common in many computers, smart phones, and other hand-held devices.

TechNote 2.3 Touch Screens

There are several forms of touch screens based on different technologies. Perhaps the most common are resistive and capacitive touch screens. Figure 2.13 shows a student doing his schoolwork on a computer with a touch screen.

A resistive touch screen is made with two conducting layers, the outer of which is flexible, and separated from the inner rigid layer by a number of non-conducting spacers. When this screen is touched, the outer flexible layer is deformed, making contact with the inner layer and allowing current to flow. This touchscreen is sensitive to pressure, but not to swiping or touching at different locations. The transparent conductive material is typically indium tin oxide (ITO).

In contrast, a capacitive touch screen is sensitive to swiping and multiple contacts. This screen is fabricated with a grid pattern of conducting pads. The capacitance between adjacent conducting pads changes with a touch of your finger, or some other conducting object, since your finger distorts the electric field lines in the space surrounding the conductors. The spatial resolution of these sensors can be very high, allowing imaging of finger prints, for example. As with the resistive touch screen, the transparent conductive material is typically ITO.

FIGURE 2.13 A young scholar doing his school work from home on a computer with a touch screen during the COVID-19 pandemic.

MEMS devices perform a number of functions, such as measurements of acceleration or orientation. One example of a MEMS element which illustrates capacitance is described in TechNote 2.4.

TechNote 2.4 MEMS, rf MEMS

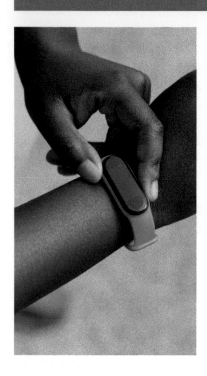

Microelectromechanical systems (MEMS) are small-scale mechanical structures actuated and/or read electrically to carry out specific tasks or sensing roles. Applications include uses in inkjet printers, accelerometers (for airbag deployment or stability control in automobiles, or in consumer electronics such as the Fitbit or Nintendo Wii), pressure sensors for tire air pressure monitoring, bio-MEMS devices in health sciences, ultrasound transducers, switches, and many more. A class of MEMS devices known as rf MEMS are used to switch and control radio frequency circuits, offering high isolation, low power dissipation, and low cost, size, and weight. MEMS are fabricated using standard semiconductor fabrication technologies.

As an example of a MEMS device, consider the accelerometer shown in Fig. 2.14. This micron-scale device consists of a proof mass supported by two sets of springs, one set at either end. Rotor fingers extending from either side of the proof mass interleave two sets of stator fingers, so-called since their positions are fixed. When the MEMS device is accelerated in the direction along the axis, the springs at the end of the proof mass in the center flex, and the distances between the rotor fingers and stator fingers varies. Thus, a measurement of the variation of the capacitances between these fingers provides a means of determining the acceleration of the device containing the MEMS element. Three accelerometers are needed to determine the acceleration in an arbitrary direction. Another example of a MEMS device measures the orientation of an instrument, such as your cell phone. A microlever in this sensor sags under the force of gravity, and the change in capacitance between this lever and a pair of stationary electrodes is processed to determine the orientation of the device.

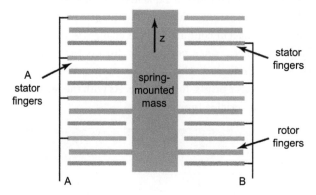

FIGURE 2.14 A diagram of a MEMS accelerometer. When this element is accelerated in a direction parallel to its axis (the z-direction), the position of the spring-mounted proof mass shifts. This decreases the spacing between half of the stationary stator fingers (B) and the rotor fingers (on the proof mass), and increases the spacing of the other half of the stator fingers (A). Thus, a measurement of these capacitances allows a determination of the acceleration of the device.

The condenser microphone converts acoustic waves into electrical impulses using the change in capacitance between a diaphragm (which is deformed by the acoustic wave) and a rigid backplate. (Condenser is synonymous with capacitor, especially in

British usage.) The most common form of condenser microphone today uses a permanent electric dipolar material, known as an electret. The condenser microphone is described in TechNote 2.5.

TechNote 2.5 Condenser Microphone

Microphones convert acoustic signals into electrical signals for amplification, transmission to distant locations, or recording. The two primary types of microphones are the dynamic and the condenser microphone. In either type of microphone, an incident acoustic wave causes a diaphragm in the microphone to vibrate. (A diaphragm is a thin, flexible, large-area disc supported at the periphery.) The manner in which the vibrations of the diaphragm are translated into an electrical signal differs for dynamic and condenser microphones. The principle of a dynamic microphone is the Faraday effect, so the discussion of that device is deferred to Chapter 5.

In a condenser microphone, shown schematically in Fig. 2.15(a), the vibrations of the diaphragm are sensed using the capacitance between this diaphragm and a backplate. As the capacitance changes, the charge stored on the capacitor changes, drawing current from the power source through the series resistor. The resistor voltage is amplified, forming the microphone output signal.

The most common form of condenser microphone makes use of an electret, which eliminates the need for a power source. A schematic of an electret condenser microphone is shown in Fig. 2.15(b). An electret is a material consisting of aligned molecules possessing a permanent electric dipole moment. A common electret material is poly-tetrafluoroethylene (PTFE). To fabricate an electret, the material is heated and an electric field applied to align the polar molecules. Upon cooling, the aligned dipoles are frozen into place. The first electret microphone was designed by Yaguchi in 1921 in Japan. In 1961, James West and Gerhard Sessler of Bell Labs invented the foil electret microphone using a thin Teflon foil, which made the electret microphone commercially feasible. Today more than 90% of microphones are foil electret microphones.

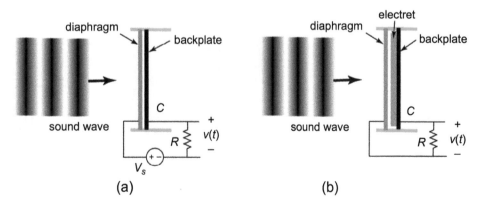

FIGURE 2.15 Schematic diagram of (a) a condenser microphone without an electret, and (b) a foil electret condenser microphone. Sound waves incident upon the diaphragm cause the diaphragm to vibrate, producing a varying potential across the capacitor. The electret condenser microphone does not require a voltage supply.

Variable capacitors have been in use for quite some time, as, for example, in old radio and television tuners. These devices are described in TechNote 2.6.

TechNote 2.6 Variable Capacitor

Resonant LC filters are used as tuners on televisions and radios. To select the radio station to listen to or the television channel to watch, the resonant frequency of the tuner must be changed, which requires using a variable capacitor (or, less commonly, a variable inductor). We show a variable capacitor from an old AM radio in Fig. 2.16(a). Half of the interleaved conducting plates are fixed, while the other half are rotatable. By turning the dial, the rotatable conducting plates are rotated into or away from the fixed plates, varying the capacitance. For the unit shown, the capacitance is 1.8 nF when the movable and fixed plates fully overlap, but only 45 pF when the rotatable plates are at the other extreme, such that only the edges of the plates overlap. When the inductor of the LC circuit has an inductance of $L \sim 50\ \mu$H, this range of capacitance is more than sufficient to cover the AM frequency range of 540–1 600 kHz.

It can be instructive to calculate the maximum capacitance of this capacitor. We base this estimate on the schematic of the capacitor shown in Fig. 2.16((b). As can be seen from the photograph in Fig. 2.16(a), the capacitor consists of two "clusters" of interleaved plates. In one cluster, $N_a = 9$ fixed plates are interleaved with $N_a + 1$ rotatable plates, while in the other, $N_b = 15$ fixed plates are interleaved with $N_b + 1$ rotatable plates. All of the moveable plates are connected to one another electrically, as are all of the fixed plates. Thus all of these capacitors are acting in parallel. But it is a bit more nuanced than this simple statement implies, in that each fixed plate is positioned between two rotatable plates. When a voltage V_0 is applied between the two sets of conductors, a surface charge ρ_s collects on both sides of the fixed plates, and a surface charge $-\rho_s$ on each face of the rotatable conductors that face a fixed plate. Therefore, if the area of the overlapping portion of the conductors is A, the total charge on the fixed conductors in the first cluster is $2N_a\rho_s A$, in the second cluster is $2N_b\rho_s A$, giving a combined charge $Q = 2(N_a + N_b)\rho_s A$. An equal but opposite charge accumulates on the rotatable conductor. Since $\rho_s = D_n = \varepsilon_0 E$, and $V_0 = Ed$, where d is the spacing between the conducting plates, the total capacitance is

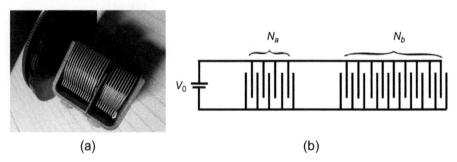

(a)

(b)

FIGURE 2.16 (a) A variable capacitor. Half of the interleaved conducting plates are fixed, while the other half are rotatable. The fixed plates are connected in parallel, as are the rotatable plates, increasing the capacitance by a factor of ∼50 over that of a single pair of conductors. (b) A schematic representation of the interleaved plates of the variable capacitor. The spacing between each set of adjacent conductors is d. When a potential is applied, a surface charge ρ_s accumulates on both sides of the fixed plates, and $-\rho_s$ on each rotatable surface facing a fixed plate.

$$C = \frac{Q}{V_0} = \frac{2(N_a + N_b)\varepsilon_0 EA}{Ed} = \frac{2(N_a + N_b)\varepsilon_0 A}{d}.$$

For the capacitor shown in the figure, $N_a = 9$, $N_b = 15$, $A \approx 6.2$ cm^2, and $d = 0.15 -$ 0.25 mm. (This capacitor is old and worn, and the spacing is not uniform between all plates.) Using these values, we estimate the capacitance of the assembly to be between 1.1 nF and 1.8 nF, in reasonable agreement with the measured maximum capacitance of 1.8 nF.

Capacitance measurements can be used to determine fluid pressures using a capacitance manometer, described in TechNote 2.7.

TechNote 2.7 Capacitance Manometer

The capacitance between conductors can be used to make a sensor to measure the pressure of a gas, using a device known as a capacitance manometer. Figure 2.17 shows a conducting diaphragm, in which the slightly curved orange arc separates two chambers containing gases at different pressures. Located on the electrode assembly are two additional conductors: a circular conductor in the center and a ring-shaped conductor concentric with the first. (In the cross-sectional slice shown in the figure, the ring appears at the top and bottom.) When the pressure of the gas in the chamber on the left changes, the diaphragm distends, changing the spacing between the diaphragm and the electrode assembly. This change in spacing between the conductors leads to a change in the capacitance δC between the conductors on the electrode assembly. The change δC between the central electrode and the diaphragm is greater than that of the ring, since the displacement of the diaphragm is greatest at the center. Thus, a measurement of the relative capacitance of these two capacitors can provide a means for measuring of the pressure difference between the left and right chambers. A variety of these devices is available commercially, with absolute or relative pressure readings from 10 mPa (0.1 mTorr) to many atmospheres.

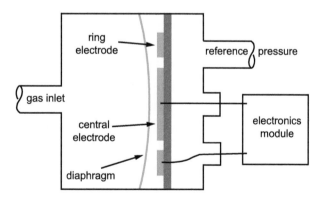

FIGURE 2.17 A schematic diagram of a capacitance manometer. A change in pressure of the gas at the inlet on the left distends the diaphragm, and changes the capacitance between the diaphragm and central and ring electrodes. Thus, the gas pressure at the inlet is measured through the capacitance values.

Supercapacitors use thin layers of electrolytic solutions to store large amounts of energy, especially for rapid charge–discharge cycles. These devices are described in TechNote 2.8.

TechNote 2.8 Supercapacitor

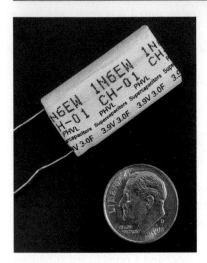

Supercapacitors, also called ultracapacitors, are electrical elements used to store electrical energy. Some of their features are similar to those of capacitors, while others are common with rechargeable batteries. Similar to both, they consist of two conducting electrodes. Unlike conventional capacitors, however, the region between the electrodes is filled not with a dielectric, but with an electrolytic solution containing positive and negative ions. Unlike batteries, however, there is no (or little) chemical activity at the electrodes. Instead, the electrodes attract the ions from the electrolyte to store large quantities of charge, with little potential difference between the electrodes.

Supercapacitors are often used in applications in which they are subjected to many rapid charge–discharge cycles. They don't degrade with multiple charging cycles as much as rechargeable batteries, but they are typically larger and more costly. For example, their energy density when charged is ~5% that of a lithium ion battery. In comparison to an electrolytic capacitor, however, the energy density per unit volume or mass of a supercapacitor is ~10–100 times greater. It achieves this high energy density even though the voltage of the supercapacitor is much lower. Note the chart in Fig. 2.18(a), which provides a qualitative comparison of the voltage vs. time for a rechargeable battery (in green) and of a supercapacitor (in orange). Similar to conventional capacitors, the voltage of a supercapacitor decreases as it discharges, while for a rechargeable battery the voltage remains relatively constant while discharging, until it is time to recharge the battery.

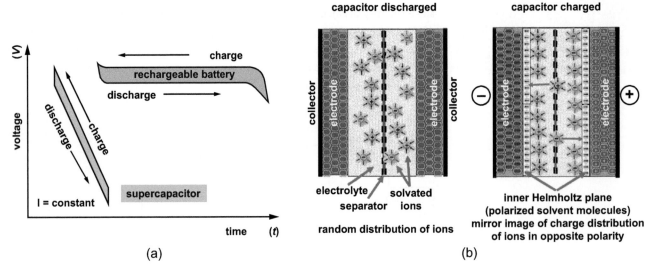

(a)   (b)

FIGURE 2.18 (a) A qualitative plot comparing the characteristics of a supercapacitor relative to a rechargeable battery. The voltage across a supercapacitor decreases as it discharges, similar to a conventional capacitor. For a battery, however, the potential remains relatively constant, until ultimately its energy is exhausted and it requires recharging. (b) Cross-sectional view of a supercapacitor. On the left, the supercapacitor is discharged, and the density of the positive and negative ions in the electrolyte is uniform. On the right, the supercapacitor is charged, and the positive ions of the electrolyte are attracted to the negative electrode, and the negative ions are attracted to the positive electrode.

Supercapacitors excel in their ability to deliver their stored energy quickly to a load, and to subsequently rapidly recharge. This is often quantified through their power density. This capability makes supercapacitors extremely useful for applications such as energy storage in regenerative braking systems in vehicles, which convert the kinetic energy of a moving vehicle to electrical energy stored in a battery or supercapacitor, which can then be used to power an electric motor to convert the energy back to kinetic energy of the vehicle. They also find application in wind turbines, which require a burst of energy to adjust the pitch of the turbine blades to optimize the performance of the turbine. Their ability to perform after multiple charge–discharge cycles is also a key advantage in turbines, since the supercapacitor is not easily accessible at the top of the tower, and therefore replacement is difficult.

Supercapacitors consist of two electrodes separated by a permeable separator, or membrane, as shown in Fig. 2.18(b). The space between the electrodes contains an electrolytic solution. An electrolyte is a solution of positive and negative ions in a solvent. When the supercapacitor is discharged, as shown in the left panel of Fig. 2.18(b), the positive and negative ions in the electrolyte are randomly distributed, with equal densities on both sides of the membrane. When the supercapacitor is charged, however, the negative electrode attracts the positive ions and repels the negative ions. Similarly, the positive electrode attracts the negative ions and repels the positive ions. The ions adhere to the electrode surface, and the two layers of charge (the free charges at the surface of the electrode and the opposite-charge ionic charges at the surface of the electrolyte, separated by a single mono-layer of solvent molecules) form what is called an electrical double layer (EDL). In an electrical double layer, no charge is exchanged between the electrodes and the electrolyte, and no chemical reactions proceed. The capacitance of EDL supercapacitors is typically in the range 0.1–100 F. Another charge-storing mechanism, called electrochemical pseudocapacitance, is also important in some types of supercapacitors. In pseudocapacitance the ions in the electrolyte undergo faradaic reduction or oxidation reactions at the surfaces of the electrodes, resulting in much higher capacitance, in the range 100–1 200 F.

We can understand the high capacitance of supercapacitors from the small effective spacing between the positive and negative charges. Recall that the capacitance of a conventional parallel plate capacitor is $C = \varepsilon A/d$, where ε is the permittivity of the dielectric between the charge layers, A is the area of the plates, and d is the spacing between the plates. In a conventional capacitor, d can be as small as 1 μm. In a supercapacitor, however, the effective spacing d is the distance between the charged layers in the electrical double layer, which is typically 0.3–0.8 nm. This spacing, which is about 10^3 times smaller than in a conventional capacitor, results in the extremely large capacitance of a supercapacitor.

Many different materials are used for the electrodes in a supercapacitor, but a common material for EDL supercapacitors is activated carbon, which has a very large surface area per unit volume. For electrochemical pseudocapacitance supercapacitors, a transition metal oxide such as MnO_2 or a noble metal oxide such as RuO_2 are common. A promising material, still in early stages of research, for storing charge in supercapacitors is based on porous graphene. Graphene is a two-dimensional layer of carbon atoms. (By two-dimensional, we mean that the carbon atoms form a single layer.)

LEARNING OBJECTIVES

After studying this section, you should be able to:

- sketch the electric field lines and equipotential lines between an arbitrary set of charged conductors;
- determine locations where the electric field is maximized; and
- estimate the capacitance of a configuration of conductors.

2.6 Sketching Field Lines and Equipotentials

In the previous chapter we introduced the electric field (based upon the electric force felt by a charged body in the vicinity of other charged objects) and the electric potential (based upon the potential energy of a charge near other charged objects), and we explored some of the properties of these fields and potentials. In this section we will introduce a graphical technique for drawing electric field lines and equipotential lines in the space between or surrounding one of more conductors. This technique can be helpful in solving problems, but also in building an intuitive picture in our minds for the behavior of electric fields and potentials.

We introduce this technique using the pair of co-axial conductors shown in Fig. 2.19(a). The inner conductor is positively charged, and the outer conductor is grounded. The electric field lines point radially outward, as shown in the figure. (We worked a similar problem in Example 1.8.) The dashed lines in this figure show the equipotential surfaces. The conductors are cylindrically symmetric, so the potential can depend only on the radial distance from the axis of the cylinders. The electric field lines and the equipotential lines, which are perpendicular to one another, divide the region between the conductors into smaller patches, one of which is shaded. Let's examine this quasi-square patch, expanded in (b), in a little more detail. The top and bottom of this patch are formed by the electric field lines, and the dashed equipotential lines form the left and right sides. The difference in potential between these two lines is

$$\Delta V = - \int_\ell \mathbf{E} \cdot d\boldsymbol{\ell} \approx E \Delta \ell.$$

(This approximation is valid when $\Delta \ell$ is chosen to be small enough that E is roughly constant over the interval.) Since no field lines pass through the top or bottom sides, and there is no charge within the region, the electric flux passing through the surface on the left must be equal to the electric flux passing through the right side,

$$\Phi_E = \int_s \mathbf{E} \cdot d\mathbf{s} \approx E \Delta w \Delta t.$$

Δt is the thickness into the page of the patch. The ratio of Φ_E to ΔV is therefore

$$\frac{\Phi_E}{\Delta V} = \frac{\Delta w}{\Delta \ell} \Delta t.$$

Therefore, as long as the potential difference ΔV between the equipotential lines and the electric flux Φ_E for the various blocks are chosen to be uniform, then the ratio of side dimensions $\Delta w / \Delta \ell$ is uniform as well. Inspection of Fig. 2.19(a) shows that for each of the patches between the conductors the ratio $\Delta w / \Delta \ell$ is roughly unity. As the radius

FIGURE 2.19 (a) Electric equipotential lines (dashed) and field lines (solid) for two co-axial conducting cylinders. (b) An expanded view of the shaded area of (a).

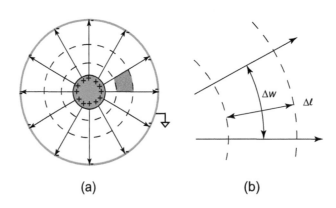

(a) (b)

increases, the electric field decreases, requiring a greater distance between equipotential lines. But the field lines spread out as well, keeping the area roughly square in shape. This is a general property, which can be applied to a graphical technique of mapping field lines and equipotential surfaces between conductors.

Let's apply this technique now to the configuration shown in Fig. 2.20, consisting of a pair of parallel conducting plates and a spherical conductor centered between the plates. The electric field lines and equipotentials in this geometry are sketched in the diagram. These would be much more difficult to calculate analytically than any that we have previously considered, but following the rules outlined above we can find a reasonable graphical solution. Note that the electric field lines cross the equipotential lines at right angles, and each of the patches are approximately square in dimension. (There are some odd-shaped patches as well, but as smaller patches are chosen these are minimized.) Where the equipotential lines are more closely spaced, the electric field lines are denser as well. The electric field lines meet the conductors, which are equipotential surfaces, at right angles, and the charge density ρ_s is greatest at positions where the electric field is greatest.

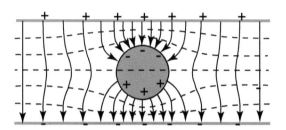

FIGURE 2.20 Electric equipotential lines (dashed) and field lines (solid) for a conducting sphere between two parallel conductors.

This graphical technique can also be used to show that the electric field just outside a sharp conductor can be very intense. For example, consider the two pairs of conductors shown in Fig. 2.21(a) and (b). In (a), the electric field lines and equipotential lines for a pair of spheres are shown, while in (b) one of the spheres has been replaced by a pointed conductor. In either case, the equipotential lines near the surfaces of conductors are nearly parallel to the conductor surface. Near a sharp corner in the conductor, therefore, where the equipotential lines bend rapidly as well, the electric field lines are more densely spaced. This region then experiences very large electric field strengths, and the surface charge density on the conductor near the sharp point is very large. This can be demonstrated dramatically with a van de Graaff generator. (See TechNote 1.6.) When a sharply pointed grounding rod is brought close to a charged van de Graaff, you can often see or hear a steady discharge between the grounding rod and the van de Graaff. The air near the pointed grounding rod is ionized by the strong field, and a steady stream of charges flows between the grounding rod and the van de Graaff. Due to the large electric field near the point, the potential of the van de Graaff is limited. If the grounding rod is replaced by a large grounded sphere, as shown in Fig. 2.21(a), the potential variations between the two conductors is much more uniform. As a result, the potential of the van de Graaff generator can reach much higher values before ionizing the air, and when that happens the discharge is often sudden and loud.

The graphical technique that we have introduced in this section can also give us a means of approximating the capacitance C between conductors, or the capacitance per

FIGURE 2.21 Electric equipotential lines (dashed) and field lines (solid) for (a) two conducting spheres, and (b) a conducting sphere and a pointed conductor.

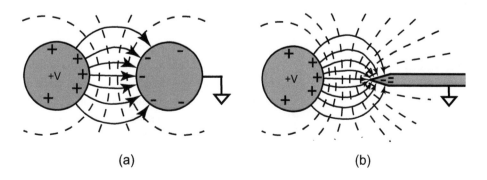

(a)   (b)

unit length C' between long parallel conductors. As an example, we return to the pair of co-axial conductors shown in Fig. 2.19(a). The capacitance per unit length C' is

$$C' = \frac{\rho_\ell}{N\Delta V},$$

where ρ_ℓ is the charge per unit length on the inner or outer conductor, N is the number of patches between the inner and outer conductors ($N = 3$ in Fig. 2.19(a)), and $N\Delta V$ is the potential difference between the conductors. But ρ_ℓ is equal to $\rho_s M\Delta w$, where M is the number of patches around the conductor ($M = 12$ in Fig. 2.19(a)), $M\Delta w$ is the circumference of the conductor, and the surface charge density ρ_s is equal to the normal displacement field D_n at the surface of the conductor, so

$$\rho_\ell = D_n M\Delta w = M\varepsilon\left(E_n\Delta w\right) = M\varepsilon\frac{\Phi_E}{\Delta t}.$$

Thus the capacitance per unit length C' between the conductors is

$$C' = \frac{\rho_\ell}{N\Delta V} = \frac{M\varepsilon\Phi_E/\Delta t}{N\Delta V} = \varepsilon\frac{M}{N}\frac{\Phi_E}{\Delta V\Delta t} = \varepsilon\frac{M}{N}\frac{\Delta w}{\Delta\ell}.$$

Since $\Delta w/\Delta\ell \approx 1$, C' is approximately

$$C' \approx \varepsilon\frac{M}{N}. \tag{2.32}$$

For the example of Fig. 2.19(a), $C' = \varepsilon\frac{12}{3} = 35\varepsilon_r$ pF/m, where ε_r is the relative permittivity of the dielectric. This value compares well with the analytic value of $C' = 39\varepsilon_r$ pF/m for a pair of co-axial conductors using Eq. (2.30) with measurements of a and b from the figure.

The graphical technique described in this section for sketching electric field lines and equipotentials is a useful tool for visualizing field patterns and charge densities. It can be especially helpful for geometries of conductors for which field patterns may be difficult to calculate analytically. We have used this technique to show that electric field strengths can be very large at the tip of sharply pointed conductors, and have illustrated a means of estimating the capacitance per unit length for pairs of conductors such as those in transmission lines. In the next section, we will describe another technique for determining electric fields and potentials near conductors.

2.7 Method of Images

As our final technique for determining the electric potential in a region of space, we introduce a technique known as the **Method of Images**. This technique is an extremely powerful method for determining the potential due to a discrete charge in the neighborhood of a conducting surface. We should expect that, since the charges within the conductor are free to move, the discrete charge will attract, or induce, charge of the opposite sign to the region in the conductor that is closest to the charge. But it is not known *a priori* what this surface charge distribution will be, and so it is difficult with the tools that we have developed to this point to determine the potential in the region outside the conductor. But we are not completely ignorant of any electrical conditions, since we do know the potential of the conductor. (Don't make the common mistake of thinking that the charge distribution in a grounded conductor is zero. This is absolutely not correct.) As the "Method of Images" name implies, we will solve for the potential in the region outside the conductor that is produced by this configuration by inventing a simpler distribution, consisting of the original charge plus one (or perhaps a few) "image" charges. We can show that these two configurations will produce precisely the same potential distribution.

"How do we know that the original charge plus conductor creates the same potential as the original charge plus the images?" you might ask. We will rely upon the **Uniqueness Theorem** to answer this important question. The Uniqueness Theorem tells us that if we have a potential function $V(\mathbf{r})$ that satisfies Poisson's Equation ($\nabla^2 V(\mathbf{r}) = -\rho_v/\varepsilon$), and that satisfies the given boundary conditions, then we have found the proper and complete solution for the potential $V(\mathbf{r})$ in this region. In other words, the solution is unique. This is important because there are many ways to find $V(\mathbf{r})$, including insightful guesses. Once we find one solution that works, the Uniqueness Theorem tells us that we are done.

SideNote 2.3 Proof of the Uniqueness Theorem

The proof of the Uniqueness Theorem is as follows. Let us assume that the Uniqueness Theorem is not correct; that is, that there are two solutions $V_1(\mathbf{r})$ and $V_2(\mathbf{r})$ that each satisfy Poisson's Equation over the volume v, and that satisfy the boundary conditions at the bounding surfaces s of the volume v. The Poisson's Equations for these two potentials are

$$\nabla^2 V_1(\mathbf{r}) = -\rho_v/\varepsilon$$

and

$$\nabla^2 V_2(\mathbf{r}) = -\rho_v/\varepsilon.$$

Subtracting one from the other leads to

$$\nabla^2 \left(V_2(\mathbf{r}) - V_1(\mathbf{r}) \right) = \nabla^2 \tilde{V}(\mathbf{r}) = 0,$$

where $\tilde{V}(\mathbf{r})$ represents the difference between the potentials $V_2(\mathbf{r}) - V_1(\mathbf{r})$. We will show that $\tilde{V}(\mathbf{r})$ must be identically zero throughout the volume v, telling us that $V_2(\mathbf{r})$ and $V_1(\mathbf{r})$ are in fact identical.

Let's take a look at the function $\nabla \cdot (\tilde{V}(\mathbf{r})\nabla \tilde{V}(\mathbf{r}))$. (It likely is not obvious why we choose this function. Bear with us and it will be made clear.) We use the product differentiation rule of Eq. (E.6), using $\tilde{V}$ and $\nabla \tilde{V}$ for Φ and $\mathbf{E}$, respectively, to write

$$\nabla \cdot (\tilde{V}(\mathbf{r})\nabla \tilde{V}(\mathbf{r})) = \nabla \tilde{V}(\mathbf{r}) \cdot \nabla \tilde{V}(\mathbf{r}) + \tilde{V}(\mathbf{r})\left(\nabla \cdot \nabla \tilde{V}(\mathbf{r}) \right).$$

But we already know that $\nabla \cdot \nabla \tilde{V}(\mathbf{r})$, which is the same as $\nabla^2 \tilde{V}(\mathbf{r})$, is zero. Also, $\nabla \tilde{V}(\mathbf{r}) \cdot \nabla \tilde{V}(\mathbf{r})$ can be written in the more compact notation $|\nabla \tilde{V}(\mathbf{r})|^2$, so this equation is

$$\nabla \cdot (\tilde{V}(\mathbf{r})\nabla \tilde{V}(\mathbf{r})) = |\nabla \tilde{V}(\mathbf{r})|^2.$$

Next we examine the volume integral of this relation over the entire volume v,

$$\int_v \nabla \cdot (\tilde{V}(\mathbf{r})\nabla \tilde{V}(\mathbf{r})) \, dv = \int_v |\nabla \tilde{V}(\mathbf{r})|^2 \, dv.$$

Applying the Divergence Theorem to the volume integral on the left side, this integral can be written as

$$\oint_s (\tilde{V}(\mathbf{r})\nabla \tilde{V}(\mathbf{r})) \cdot d\mathbf{s} = \int_v |\nabla \tilde{V}(\mathbf{r})|^2 \, dv.$$

The integrand of the surface integral on the left is the product of $\tilde{V}(\mathbf{r})$ and $\nabla \tilde{V}(\mathbf{r})$, but $\tilde{V}(\mathbf{r})$ must be zero everywhere on this surface, since both potentials

$V_1(\mathbf{r})$ and $V_2(\mathbf{r})$ satisfy the boundary conditions for our geometry. So we are left with

$$\int_v |\boldsymbol{\nabla} \tilde{V}(\mathbf{r})|^2 \, dv = 0.$$

Since the integrand $|\boldsymbol{\nabla} \tilde{V}(\mathbf{r})|^2$ of this integral must be zero or positive everywhere, and since its integral over the region v is zero, we can only conclude that $\boldsymbol{\nabla} \tilde{V}(\mathbf{r})$ is identically zero throughout the volume v. Recall that the gradient of any function tells us the spatial rate of change of the function. So if the gradient is zero, then the function is constant throughout that region. And since $\tilde{V}(\mathbf{r})$ is zero at the surface s, then $\tilde{V}(\mathbf{r})$ must be zero everywhere in the volume v. With this, we have proven that potentials $V_1(\mathbf{r})$ and $V_2(\mathbf{r})$ are identical, and the solution $V(\mathbf{r})$ is unique.

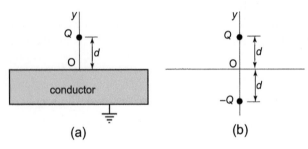

FIGURE 2.22 (a) A charge Q over a grounded conducting plane. (b) A charge Q and its image charge $-Q$.

If we accept the Uniqueness Theorem, we are ready to introduce the Method of Images, which we will do by example. Consider a point charge Q above an infinitely wide grounded plane, as shown in Fig. 2.22(a). The distance from the plane to the charge is d. The region above the plane is charge-free (aside from Q, of course), and the permittivity in this region is ε_0. Let us explore the use of this Method of Images to determine the potential in the region above the grounded plane. For concreteness, we define the normal to the ground plane as the y-axis, with the ground plane at $y = 0$ and the charge at $y = d$, and start by listing all that we know about the equipotential surfaces and the electric field lines for this configuration. For example, we know that the grounded plane is, obviously, an equipotential surface, at $V = 0$. Close to the charge Q, the potential should become very large, and the equipotentials in this region should circle around the charge Q. As we move far away from Q and from the grounded plane, we expect that the potential becomes very small, with $V \to 0$ as $y \to \infty$. We also know that the charge Q will attract negative charges to the surface of the grounded plane directly beneath the charge, distributed over some width, and that some of the electric field lines that originate at the positive charge Q will terminate on these negative surface charges. The electric field lines at $y = 0$ should be normal to the surface of the conductor.

Now let us examine the potential due to a different charge configuration – that is, that of the pair of discrete charges shown in Fig. 2.22(b). The charge $+Q$ is located at $y = d$ (the same location as the charge in the original problem) and the charge $-Q$ is located at $y = -d$. We call this the image charge. We recognize this pair of charges as an electric dipole, and we have already examined the potential due to this type of charge distribution. Our notation is a bit different from before, however, so we write out explicitly the potential at position (x, y, z), which we will call point P, as the sum of the potential due to each charge individually:

$$V(x, y, z) = \frac{Q}{4\pi\varepsilon_0} \left[\frac{1}{R_+} - \frac{1}{R_-} \right], \tag{2.33}$$

where $R_+ = [x^2 + (y - d)^2 + z^2]^{1/2}$ and $R_- = [x^2 + (y + d)^2 + z^2]^{1/2}$ are the distances from the $+Q$ and $-Q$ point charges, respectively, to the point P.

Let us now compare the potential at some specific locations for these two different cases. For any point on the plane defined by $y = 0$, the potential is zero in Fig. 2.22(a), because this coincides with the grounded plane. The potential at any point in this plane in Fig. 2.22(b) is also zero, since the distances R_+ and R_- are equal to one another. Also, for $y \to \infty$, the potential $V \to 0$ in both cases. So we see that the potential given in Eq. (2.33), which we wrote for the pair of point charges shown in Fig. 2.22(b), satisfies the boundary conditions (i.e. the potentials at $y = 0$ and $y \to \infty$) of the configuration in Fig. 2.22(a). Clearly the potential in Eq. (2.33) satisfies Laplace's Equation. By the Uniqueness Theorem, then, we have found the potential that we seek. The charge Q a distance d above a grounded conducting plane produces precisely the same potential in the region $y > 0$ as is produced by the pair of charges $+Q$ and $-Q$ positioned at $y = \pm d$.

Having successfully found $V(\mathbf{r})$, let us now proceed to find $\mathbf{E}(\mathbf{r})$ in this same region. We do this using $\mathbf{E} = -\boldsymbol{\nabla} V$, and find

$$\mathbf{E} = \frac{Q}{4\pi\varepsilon_0} \left[\frac{\mathbf{a}_x\, x + \mathbf{a}_y\,(y - d) + \mathbf{a}_z\, z}{R_+^3} - \frac{\mathbf{a}_x\, x + \mathbf{a}_y\,(y + d) + \mathbf{a}_z\, z}{R_-^3} \right].$$

This is, of course, the $\mathbf{E}$ field due to two point charges, $+Q$ located at $\mathbf{R}'_+ = (0, d, 0)$ and $-Q$ located at $\mathbf{R}'_- = (0, -d, 0)$.

Next, we can use this result for $\mathbf{E}$ to determine the surface charge density ρ_s in the $y = 0$ plane. Let us start by evaluating $\mathbf{E}$ at the $y = 0$ plane. Since $R_+ = R_- = [x^2 + d^2 + z^2]^{1/2}$, we get

$$\begin{aligned} \mathbf{E}(y = 0) &= \frac{Q}{4\pi\varepsilon_0} \left[\frac{\mathbf{a}_x\, x + \mathbf{a}_y\,(-d) + \mathbf{a}_z\, z}{[x^2 + d^2 + z^2]^{3/2}} - \frac{\mathbf{a}_x\, x + \mathbf{a}_y\, d + \mathbf{a}_z\, z}{[x^2 + d^2 + z^2]^{3/2}} \right] \\ &= \frac{Q}{4\pi\varepsilon_0} \frac{-\mathbf{a}_y\, 2d}{[x^2 + d^2 + z^2]^{3/2}}. \end{aligned}$$

Note that this field is normal to the surface, and pointing *into* the surface, as expected. By the boundary conditions that must be satisfied at any conducting surface, the charge density is $\rho_s = D_n$, so

$$\rho_s(x, 0, z) = D_y(y = 0) = \varepsilon_0\, \mathbf{E}(y = 0) \cdot \mathbf{a}_y = \frac{Q}{2\pi\varepsilon_0} \frac{-d}{[x^2 + d^2 + z^2]^{3/2}}.$$

This charge density has its maximum value at $x = 0$, $z = 0$ (directly beneath the charge Q), which should make sense. Integrating this charge density over the surface gives us the total charge induced in the grounded plane, and we find that charge to be precisely $-Q$, a result that perhaps we could have anticipated had we given this some forethought.

As a final note, it is interesting to realize that the charge $+Q$ experiences an attractive force pulling it toward the grounded plane. We already discussed that Q induces a negative charge in the conductor. These negative charges, in turn, create a potential that draws Q toward the plane. The magnitude of this force can be found by returning to our image charge, and we can immediately write $\mathbf{F} = -\mathbf{a}_y\, Q^2/[4\pi\varepsilon_0(2d)^2]$, where $2d$ is the distance between Q and its image $-Q$.

The Method of Images can be applied to a variety of other geometries. We will consider just one here, that of an infinitely long line charge of linear charge density ρ_ℓ parallel to a ground plane, as illustrated in Fig. 2.23(a). This example is related to a pair of parallel conducting wires, which is one form of transmission line that we consider in detail later in this text.

Let the line charge be a distance $y = d$ above the ground plane. Based upon our success with the discrete charge above the ground plane, a good guess is that the image charge is a line charge of linear charge density $-\rho_\ell$ positioned at $y = -d$. The original

FIGURE 2.23 (a) A line charge over a grounded conducting plane. (b) The line charge and its image.

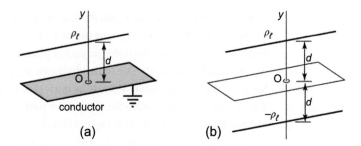

(a)   (b)

and the image line charges are shown in Fig. 2.23(b). Let's give this a try and see if we find a solution that satisfies the correct boundary conditions: $V = 0$ in the $y = 0$ plane, and $V \to 0$ as $y \to \infty$. We will choose the z-axis as the direction parallel to the line charge, and return to Eq. (1.52) for the potential due to a single line charge of length L. This potential has a weak divergence as the length L becomes large, but fortunately this inconvenience disappears when the potential due to the pair of line charges, $+\rho_\ell$ and its image $-\rho_\ell$, is considered. At some point P at $(x, y, 0)$, where the line charge extends from $z = -L/2$ to $z = +L/2$, the potential $V(\mathbf{r})$ is

$$V(x, y, z) = \frac{\rho_\ell}{2\pi\varepsilon_0} \left[\ln\left(\frac{L/2 + \sqrt{(L/2)^2 + \rho_+^2}}{\rho_+} \right) - \ln\left(\frac{L/2 + \sqrt{(L/2)^2 + \rho_-^2}}{\rho_-} \right) \right],$$

where $\rho_+ = [x^2 + (y-d)^2]^{1/2}$ and $\rho_- = [x^2 + (y+d)^2]^{1/2}$ are the radial distances from the ρ_ℓ line charge and the image $-\rho_\ell$ line charge to the point P. Combining the two terms gives

$$V(x, y, z) = \frac{\rho_\ell}{2\pi\varepsilon_0} \ln\left[\left(\frac{\rho_-}{\rho_+} \right) \left(\frac{L/2 + \sqrt{(L/2)^2 + \rho_+^2}}{L/2 + \sqrt{(L/2)^2 + \rho_-^2}} \right) \right].$$

Since the line charge is very long, $L \gg \rho_+$ and ρ_-, and the ratio in the previous equation

$$\frac{L/2 + \sqrt{(L/2)^2 + \rho_+^2}}{L/2 + \sqrt{(L/2)^2 + \rho_-^2}} \to 1.$$

Therefore, for the long line charge and its image, the potential is simply

$$V(\mathbf{r}) = \frac{\rho_\ell}{2\pi\varepsilon_0} \ln\left(\frac{\rho_-}{\rho_+} \right). \tag{2.34}$$

To show that this is the proper solution for the potential produced by the line charge above the grounded plane, we can now examine the value of this potential at the $y = 0$ plane and for $y \to \infty$, to see if indeed we match the known boundary conditions. At the $y = 0$ plane, the distances ρ_+ and ρ_- are equal, and since the $\ln(1) = 0$, this boundary condition is satisfied. Likewise, as $y \to \infty$, the ratio $\rho_-/\rho_+ \to 1$, and again the potential $V \to 0$. Since the potential given by Eq. (2.34), which we calculated for the original line charge ρ_ℓ at $y = d$ and an image line charge $-\rho_\ell$ at $y = -d$, satisfies the boundary conditions of a line charge ρ_ℓ at $y = d$ over a grounded plane, and it clearly satisfies Laplace's Equation, we are assured by the Uniqueness Theorem that this potential is the proper solution for our configuration, and our choice of the image at $y = -d$ with charge density $-\rho_\ell$ is correct.

As we hinted earlier, this configuration turns out to be much more important than might be evident to this point in our discussion. It pulls together many of the concepts that we have developed to this point in our study of electric fields. But also, and perhaps a more compelling motivation, this simple configuration of a line charge above a conducting plane, or equivalently, as we have now shown, the configuration of two parallel line charges, allows us an opening into an important class of transmission lines consisting of a pair of parallel conductors, often called a twin-lead transmission line. We will study waves guided by transmission lines later, but for now our interest is in the capacitance per unit length of this system. With that motivation, let us proceed. We start by examining the equipotential surfaces in the space above the conducting plane. Using the potential given by Eq. (2.34), an equipotential surface is defined by all points (x, y, z) such that the ratio of distances ρ_+/ρ_- is a constant. Let us label this ratio $\mathcal{R}$, where $\mathcal{R}$ must be positive, and less than 1 since $\rho_+ < \rho_-$ for $y > 0$. The potential of this surface is

$$V_{\mathcal{R}} = \frac{\rho_\ell}{2\pi\varepsilon_0} \ln\left(1/\mathcal{R}\right). \tag{2.35}$$

We show five equipotential surfaces in Fig. 2.24 for $\mathcal{R} = 1/8$, $1/4$, $3/8$, $1/2$, and $3/4$. These equipotential surfaces are precisely circular in shape, with radius

$$a = d\left(\frac{2\mathcal{R}}{1 - \mathcal{R}^2}\right), \tag{2.36}$$

centered at (x_0, y_0), where $x_0 = 0$ and

$$y_0 = d\left(\frac{1 + \mathcal{R}^2}{1 - \mathcal{R}^2}\right). \tag{2.37}$$

Up to this point, we have been discussing a simple line charge above a grounded plane, but now we see that the solution applies to a much larger scope of configurations. Since the equipotential surfaces are cylinders, the potential given by Eq. (2.34) will also apply to the case of a cylindrical conductor (i.e. a wire) lying above the ground plane, as shown in Fig. 2.25(a). When the position y_0, radius a, and potential V_0 of this wire precisely conform to those of one of the equipotential surfaces defined in Eqs. (2.35)–(2.37) for the line charge geometry, the potential in the region $y > 0$, not including the interior of the wire, will be given by Eq. (2.34) as well. How can we state this? Well, the potential at the surface of the wire is the same in both cases (one boundary condition), the potential at the ground plane is the same (a second boundary condition), and the potential at $y \to \infty$ is the same (the final boundary condition). Since the potentials match at all boundaries, we are assured by the Uniqueness Theorem that our solution for this new geometry – that is, i.e. the wire of radius a above the grounded plane – is the correct one.

In the equations above, we gave the potential $V(\mathbf{r})$ in terms of the linear charge density ρ_ℓ and the distance d. This is perhaps not the most convenient form for us, since for the configuration consisting of a wire above the plane we are likely to know the wire position

FIGURE 2.24 The equipotential surfaces (a) for a line charge of charge density ρ_ℓ over a grounded conducting plane, and (b) for a charged wire over a grounded conducting plane. When the wire of radius a in (b) coincides with one of the equipotential surfaces in (a), the equipotential surfaces outside the wire are identical.

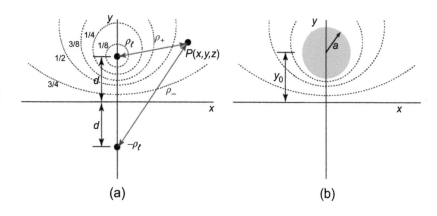

(a)   (b)

FIGURE 2.25 (a) A charged cylindrical conductor over a grounded conducting plane. (b) A charged cylindrical conductor and its image.

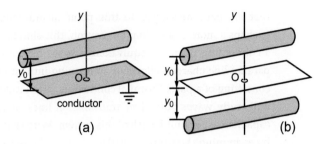

y_0, radius a, and potential V_0. It is useful, therefore, to solve for the equivalent ρ_ℓ and d in terms of y_0, a, and V_0. This involves a page of algebra, which we will not reproduce here, but the result is

$$d = \sqrt{y_0^2 - a^2}$$

and

$$\rho_\ell = \frac{2\pi\varepsilon_0 V_0}{\ln(1/\mathcal{R})},$$

where

$$\mathcal{R} = \frac{y_0}{a} - \sqrt{\left(\frac{y_0}{a}\right)^2 - 1},$$

or equivalently,

$$1/\mathcal{R} = \frac{y_0}{a} + \sqrt{\left(\frac{y_0}{a}\right)^2 - 1}.$$

After a bit more algebra involving hyperbolic cosines and such, ρ_ℓ can be written as

$$\rho_\ell = \frac{2\pi\varepsilon_0 V_0}{\cosh^{-1}(y_0/a)}. \tag{2.38}$$

(See Appendix J for a review and summary of properties of hyperbolic trigonometric functions.) The charge on the wire is located on its surface, of course, and we can determine the surface charge density on the wire by determining the displacement field **D** just outside the surface. Using the gradient of the potential, we find that **D** is normal to the wire (as we know it must be), and we use $\rho_s = D_n$ to find the surface charge density ρ_s. If we then integrate the ρ_s around the circumference of the wire, we find that the charge per unit length ρ_ℓ on the wire is as given in Eq. (2.38). We might have guessed at this result, but it is reassuring to work it out all the same. The capacitance per unit length of the conducting wire over the grounded plane is

$$C' = \frac{\rho_\ell}{V_0} = \frac{2\pi\varepsilon_0}{\cosh^{-1}(y_0/a)}. \tag{2.39}$$

Finally, we are ready to extend this analysis to that of two parallel conducting wires, as shown in Fig. 2.25(b). The potential $V(\mathbf{r})$ for any point exterior to the conductors is found immediately by realizing that this configuration, with one wire at position $y = +y_0$ and potential V_0 and the other wire at position $y = -y_0$ and potential $-V_0$, is precisely the same as the configuration we just considered. At least, the potentials in the upper half planes are the same. In the lower half planes, by symmetry and with no extra work, the potential is identical to that of a wire at position $y = -y_0$ and potential $-V_0$ *below* the ground plane. Regardless, our solution for the potential due to a single line charge over a grounded plane has led us to a solution for the potential due to a biased wire over a ground plane, and the latter has now led us to the potential due to two parallel wires, one at potential V_0, the other at potential $-V_0$, separated (from center to center) by a distance $2y_0$.

To determine the capacitance per unit length C' for the two parallel wires, we need to be a little careful, since the potential difference between the two conductors is $V_0 - (-V_0) = 2V_0$. So

$$C' = \frac{\rho_\ell}{2V_0} = \frac{\pi\varepsilon_0}{\cosh^{-1}(y_0/a)}, \qquad (2.40)$$

half as large as the capacitance per unit length of the cylinder over the ground plane. We can also arrive at this same conclusion by considering the energy per unit length stored in the electric field. We leave this for the student as Prob. P2-30.

SUMMARY

A fully charged capacitor can deliver quite a jolt, if you are unfortunate enough to make contact with its terminals, displaying the might of the electrical energy stored within. In this chapter we have examined energy stored in electric fields, and applied a technique for determining the force on a body based on the variation of the energy of the system due to a small displacement. Additionally, we introduced the capacitance of a pair of conductors, a concept that we recall from circuits, but which is given new meaning here. Finally, we have introduced two powerful analytical techniques (Poisson's Equation and the Method of Images) and a graphical technique for determining the electric field and electric potential in a region of space. While this concludes our study of electrostatics, these principles will be important in the following chapters as we study electric currents, magnetic fields, and propagating waves.

Table 2.1

Collection of key formulas from Chapter 2.			
$W_e = \dfrac{1}{2}\displaystyle\int_v \rho_v(\mathbf{r})V(\mathbf{r})dv$	(2.3)	$\nabla^2 V = -\rho_v/\varepsilon$	(2.19)
$W_e = \dfrac{1}{2}\displaystyle\sum_k V_k Q_k$	(2.5)	$\nabla^2 V = 0$	(2.21)
$W_e = \dfrac{1}{2}\displaystyle\int_v \mathbf{D}\cdot\mathbf{E}\,dv$	(2.7)	$C = \dfrac{Q}{V_{12}} = \dfrac{\oint_s \mathbf{D}\cdot d\mathbf{s}}{-\int_2^1 \mathbf{E}\cdot d\ell}$	(2.27)
$w_e = \dfrac{1}{2}\mathbf{D}\cdot\mathbf{E}$	(2.8)	$C = \dfrac{2W_e}{V_{12}^2} = \dfrac{Q^2}{2W_e}$	(2.29)
$\mathbf{F}_e = \boldsymbol{\nabla}W_e\big|_{constant\ V}$	(2.14)	$C = \dfrac{2\pi\varepsilon}{\ln(b/a)}$	(2.30)
$\mathbf{F}_e = -\boldsymbol{\nabla}W_e\big|_{constant\ Q}$	(2.15)	$C = \dfrac{w\varepsilon}{d}$	(2.31)
		$C = \dfrac{\pi\varepsilon_0}{\cosh^{-1}(y_0/a)}$	(2.40)

PROBLEMS

Electrostatic Energy

P2-1 Determine the work required to assemble the three charges described in Prob. P1-4.

P2-2 Determine the work required to assemble the thick spherical shell of charge described in Probs. P1-9 and P1-45. Compare your result found using Eq. (2.3) with that found using Eq. (2.7).

P2-3 Determine the work required to assemble the spherical shell of charge described in Probs. P1-12 and P1-47. Compare your result found using Eq. (2.3) with that found using Eq. (2.7).

P2-4 Determine the energy stored in the electric field generated by a uniform sphere of charge density ρ_v and radius a. For equal total charge, compare this energy with that determined in Prob. P2-3.

P2-5 Determine the energy per unit length stored in the electric field generated by a pair of co-axial conductors. The radius of the inner conductor is a, and the inner radius of the outer conductor, which is grounded, is b. The potential of the inner conductor is V_0.

Electrostatic Forces

P2-6 In Fig. 2.26(a) we show a pair of parallel conducting plates, with a dielectric medium partially inserted into the space between the plates. The potential difference between the plates is $V_0 = 100$ V, the permittivity of the dielectric medium is $\varepsilon = 4\varepsilon_0$, and the plate spacing is $d = 0.2$ cm. The plate area is $w = 2$ cm (into the page) by 3 cm. Determine the electric force $\mathbf{F}_e$ on the dielectric. You may neglect fringe effects at the ends of the plates and at the end of the dielectric. Is the dielectric being drawn into the space between the plates, or expelled?

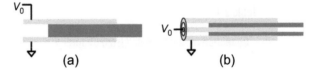

(a)   (b)

FIGURE 2.26 (a) Dielectric medium between two plane conductors. (b) Dielectric medium between two co-axial conducting cylinders. (For use in Problems P2-6 to P2-9.)

P2-7 A vertical pair of parallel conductors is inserted into a pool of water. The spacing of the plates is $d = 0.5$ mm, and a potential difference of $V_0 = 1$ kV is applied between the plates. Determine the distance the water is drawn up into the space between the plates. You may neglect fringe effects at the ends of the plates and at the end of the water. (Hint: See Table 1.1 for the dielectric constant of water. The density of water is 997 kg/m^3.)

P2-8 In Fig. 2.26(b) we show a pair of co-axial cylindrical conductors of length L, with a dielectric medium partially inserted. The potential difference between the center conductor (of radius $a = 0.2$ cm) and the inner surface of the outer conductor (of radius $b = 1.0$ cm) is $V_0 = 100$ V, the permittivity of the dielectric medium is $\varepsilon = 4\varepsilon_0$. Determine the electric force $\mathbf{F}_e$ on the dielectric. You may neglect fringe effects at the ends of the conducting cylinders and at the end of the dielectric. Is the dielectric being drawn into the space between the conducting cylinders, or expelled?

P2-9 A vertical pair of co-axial conductors is inserted into a pool of water. The radii of the conductors are $a = 1.0$ cm and $b = 1.2$ cm, and a potential difference of $V_0 = 1$ kV is applied between the conductors. Determine the distance the water is drawn up into the space between the conductors. [Hint: See Table 1.1 for the dielectric constant of water. The density of water is 997 kg/m^3.]

Poisson's Equation and Laplace's Equation

P2-10 Show that the solution for the potential $V(z)$ given in Eq. (2.25) satisfies Laplace's Equation, and that it satisfies the four conditions that we established in Example 2.8, that is: (1) $V(z) = 0$ at $z = 0$; (2) $V(z) = V_0$ at $z = (d_1 + d_2)$; (3) $V(z)$ is continuous at the interface between the dielectrics ($d_1 = d_2$); and (4) **D** is the same in the two dielectrics.

P2-11 Consider a pair of concentric spherical conductors. The inner sphere is of radius a and potential V_0, and the larger sphere is grounded and of inner radius c. The space between the conductors is filled with two non-conducting, charge-free dielectrics, of permittivity ε_1 ($a < R < b$) and ε_2 ($b < R < c$). (a) Use Poisson's Equation to determine the potential $V(R)$ in the space between the conductors. Use your result for $V(R)$ to find (b) **E** and (c) **D** in the region between the conductors, and (d) ρ_s on the surfaces of the conductors.

P2-12 Two co-axial cylindrical conductors are of length L. The inner cylinder is of radius a and potential V_0, and the larger cylinder is grounded and of inner radius c, where c is much less than L. The space between the conductors is filled with two non-conducting, charge-free dielectrics, of permittivity ε_1 ($a < \rho < b$) and ε_2 ($b < \rho < c$). (a) Use Poisson's Equation to determine the potential $V(\rho)$ in the space between the conductors. Use your result for $V(\rho)$ to find (b) **E** and (c) **D** in the region between the conductors, and (d) ρ_s on the surfaces of the conductors.

P2-13 Consider a pair of flat, rectangular, non-parallel conductors. Both conductors are 25×25 cm, and are separated by 1 cm along one side, but 1.5 cm along the opposite side. The conductor at $\phi = 0$ is grounded, and the other plate is biased to 100 V. The dielectric between the plates is air. (a) Use Poisson's Equation to determine the potential $V(\phi)$ in the space between the conductors. You may ignore fringe effects at the edges of the plates. Use your result for $V(\phi)$ to find (b) **E** and (c) **D** in the region between the conductors, and (d) ρ_s on the surfaces of the conductors.

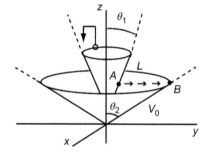

FIGURE 2.27 A pair of conducting conical surfaces. (For use in Prob. P2-14.)

P2-14 Two conical conductors are as shown in Fig. 2.27, extending to ∞. The angle θ_1 of the inner conductor, which is grounded, is 20°. The outer cone, with $\theta_2 = 50°$, is at potential 50 V. (a) Determine the potential V in the region between the conductors. (b) Determine the electric field **E** in this same region. (c) Determine the work W required to move 30 nC of charge from point A to point B along the path L shown.

A Numerical Solution of Laplace's Equation

P2-15 The conductors of Fig. 2.28 are 20 V, 50 V, 80 V, and 100 V, as labeled in the figure. Use the numerical technique described in Section 2.4 to solve for the potential at the six grid points shown. (The spacings between the grid points and the walls are uniform.)

P2-16 Repeat the numerical solution for the region between the conductors in Fig. 2.28, using a grid pattern in which the spacing is halved. (That is, the size of the grid pattern is 5×7.) Compare the changes in the potentials at the grid points.

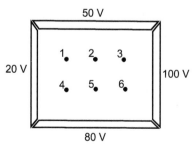

FIGURE 2.28 An assembly of conductors, with potentials of 20 V, 50 V, 80 V, and 100 V, as labeled. (For use in Prob. P2-15.)

P2-17 Repeat the solution of Example 2.10, using a grid pattern with one additional grid point to the left of point 1, and three additional grid points (one row) above points 9–11. Compare the changes in the potentials at the grid points.

Capacitance

P2-18 Use the electric and displacement fields calculated in Example 2.12 to determine the energy W_e stored in a parallel plate capacitor with two dielectrics. From W_e, determine the capacitance C. Compare your result with the capacitance for this geometry found in Example 2.12.

P2-19 Use the electric and displacement fields calculated in Example 2.13 to determine the energy W_e stored by an isolated charged spherical shell. From W_e, determine the capacitance C of the sphere. Compare your result with the capacitance for this geometry found in Example 2.13.

P2-20 Use the electric and displacement fields calculated in Example 2.14 to determine the energy W_e stored by a pair of charged co-axial conductors. From W_e, determine the capacitance C of the pair of conductors. Compare your result with the capacitance for this geometry found in the example.

P2-21 Determine the capacitance of a pair of concentric spherical conductors spaced by two dielectrics as described in Prob. P2-11.

P2-22 A pair of concentric spherical conductors, similar to that described in Prob. P2-11, are spaced by two dielectrics. In the present case, however, one dielectric (ε_1) fills the space defined by $0 < \phi < 2\pi/3$, and the second dielectric (ε_2) fills the space defined by $2\pi/3 < \phi < 2\pi$. Determine the capacitance of this structure. [Hint: **E** and **D** are still radial for this geometry, and the potential difference V_0 must, of course, be the same across one dielectric as across the second. Show that proper boundary conditions are satisfied at the interface between the two dielectrics. What do we infer about the surface charges on the conductors in the different regions?]

P2-23 A pair of co-axial cylindrical conductors of length L are spaced by two dielectrics, as described in Prob. P2-12. Determine the capacitance of this structure.

P2-24 A pair of co-axial cylindrical conductors of length L, similar to those described in Prob. P2-12, are spaced by two dielectrics. In the present case, however, one dielectric (ε_1) fills the space defined by $0 < \phi < 2\pi/3$, and the second dielectric (ε_2) fills the space defined by $2\pi/3 < \phi < 2\pi$. Determine the capacitance of this structure. [Hint: **E** and **D** are still radial for this geometry, and the potential V_0 must, of course, be the same across one dielectric as across the second. Show that proper boundary conditions are satisfied at the interface between the two dielectrics. What do we infer about the surface charges in the different regions?]

Sketching Field Lines and Equipotentials

P2-25 A long, rectangular conductor fits inside a larger, hollow rectangular conductor, as shown in Fig. 2.29(a). (This figure shows a cross-section of the conductors. The conductors extend a long distance into the page of the figure without change.) The potential of the center conductor is V_0, and the outer conductor is grounded. The dielectric between the conductors is air. (a) Sketch the equipotential lines (using dashed lines) and the electric field lines (solid lines with arrowheads to indicate the field direction) in the space between the conductors. (You only need to draw the equipotentials and field lines in one-quarter of the structure, since symmetry dictates the other quarters will be the same.) (b) Estimate the capacitance per unit length C' of these conductors. (c) Compare the surface charge density at point a

$(\rho_{s,a})$ to that at point b $(\rho_{s,b})$. (d) Compare $\rho_{s,a}$ to the surface charge density at point c $(\rho_{s,c})$. (e) Compare $\rho_{s,a}$ to the surface charge density at point d $(\rho_{s,d})$.

FIGURE 2.29 Two configurations of conductors, for use in Probs. P2-25 and P2-26.

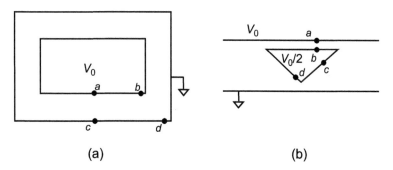

(a)   (b)

P2-26 A triangular conductor sits between two parallel conductors, as shown in Fig. 2.29(b). The potential of the top conductor is V_0, that of the triangular conductor is $V_0/2$, and the bottom conductor is grounded. (a) Sketch the equipotential lines (using dashed lines) and the electric field lines (solid arrows with arrowheads to indicate the field direction) in the space between the conductors. (b) Compare the surface charge density at point a $(\rho_{s,a})$ to that at point b $(\rho_{s,b})$. (c) Compare the surface charge density at point c $(\rho_{s,c})$ to the surface charge density at point d $(\rho_{s,d})$.

Method of Images
P2-27 A charge Q is located close to infinite, conducting planes, as shown in Fig. 2.30. Identify the charge and location of image charges induced by Q.

FIGURE 2.30 A charge Q located close to various infinite, conducting planes, for use in Prob. P2-27.

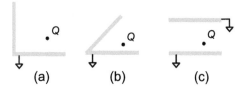

(a)   (b)   (c)

P2-28 A point charge Q is located inside a grounded, spherical conductor, as shown in Fig. 2.31. (a) Sketch four equipotential lines in the interior of the sphere. (Use dashed lines.) (b) Sketch eight electric field lines in the interior of the sphere. (Use solid lines, with arrows to indicate the direction.) Describe the surface charge induced on the spherical conductor (sign and location). Is this charge distributed or at a point? (d) The electric field and potential inside the sphere are identical to those produced by the charge Q and an image charge (but no sphere). Show the approximate location of the image charge. What is the sign of the image charge?

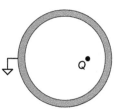

FIGURE 2.31 A charge Q located inside a grounded conducting sphere, for use in Prob. P2-28.

P2-29 A charge Q is located close to a grounded, conducting sphere centered at the origin. The sphere is of radius a, and the charge is located at $z = d$, where $d > a$. (a) Show that an image charge $Q_i = -aQ/d$ located at $z_i = a^2/d$ satisfies Laplace's Equation, satisfies the condition $V = 0$ at the surface of the sphere, and satisfies $V = 0$ at $R \to \infty$. (b) Determine the surface charge density $\rho_s(\theta)$ induced on the surface of the sphere.

P2-30 A pair of long, parallel wires of radius a are spaced by $2y_0$ (center-to-center spacing). Their potentials are $+V_0$ for one, $-V_0$ for the other. Determine the energy per unit length of this pair of conductors. [Hint: We have developed two different methods to determine this energy in Section 2.1, as given in Eqs. (2.3) and (2.7). They should, of course, give equivalent results, but Eq. (2.3) is definitely easier to use in this case!] Use this energy per unit length to determine the capacitance per unit length, C', for the pair of wires, and compare your result to Eq. (2.40).

P2-31 A long line charge of linear charge density ρ_ℓ is located close to and parallel to a long, conducting cylinder. Let the axis of the cylinder be the z-axis. The cylinder is of radius a, and the line charge is located at $x = d$, where $d > a$. The electric potential for the line charge plus an image charge $\rho_{\ell,i} = -\rho_\ell$ located at $x_i = a^2/d$ can be written as $V(x, y) = \frac{\rho_\ell}{2\pi\varepsilon_0} \ln(s_i/s_0)$, where $s_0 = \sqrt{(x-d)^2 + y^2}$ is the distance from the original line charge to a position (x, y) and $s_i = \sqrt{(x-x_i)^2 + y^2}$ is the distance from the image charge. (a) Show that the potential on the surface of the cylinder is constant. What is the potential? (b) Show that the potential $V(x, y) \to 0$ as $x, y \to \infty$. (c) Determine the surface charge density $\rho_s(\phi)$ induced on the surface of the cylinder. (d) Integrate the surface charge around the circumference of the cylinder to determine the charge per unit length induced on the cylinder.

P2-32 A long line charge of linear charge density ρ_ℓ is located close to and parallel to a long, conducting cylinder, similar to that described in Prob. P2-31. In this case, however, the conducting cylinder is electrically neutral. To determine the potential outside the cylinder, we introduce the image charge $\rho_{\ell,i} = -\rho_\ell$ located at $x_i = a^2/d$, as in Prob. P2-31, plus a second image line charge, of charge density ρ_ℓ lying on the z-axis. We can write the potential for the line charge plus the two image charges as $V(x, y) = \frac{\rho_\ell}{2\pi\varepsilon_0} \ln(s_i/s_0 s_{i,2})$, where $s_0 = \sqrt{(x-d)^2 + y^2}$ is the distance from the original line charge to a position (x, y), $s_i = \sqrt{(x-x_i)^2 + y^2}$ is the distance from the first image charge, and $s_{i,2} = \sqrt{x^2 + y^2}$ is the distance from the second image charge. (Unfortunately, this potential is not zero at an infinite distance, but we'll ignore this unfortunate behavior.) (a) Show that the potential on the surface of the cylinder is constant. What is the potential? (b) Determine the surface charge density $\rho_s(\phi)$ induced on the surface of the cylinder. (c) Integrate the surface charge around the circumference of the cylinder to determine the charge per unit length induced on the cylinder.

P2-33 In this problem, we consider one last variation of a long line charge of linear charge density ρ_ℓ located close to and parallel to a long, conducting cylinder, similar to that described in Probs. P2-31 and P2-32. In this case, however, the conducting cylinder is electrically grounded. To determine the potential outside the cylinder, we introduce the image charge $\rho_{\ell,i} = -\rho_\ell$ located at $x_i = a^2/d$, as in Prob. P2-31, plus a second image line charge, of charge density $\rho_{\ell,i,2}$ lying along the z-axis. We can write the potential for the line charge plus the two image charges as $V(x, y) = \frac{\rho_\ell}{2\pi\varepsilon_0} \ln(s_i/s_0) - \frac{\rho_{\ell,i,2}}{2\pi\varepsilon_0} \ln(s_{i,2})$, where $s_0 = \sqrt{(x-d)^2 + y^2}$ is the distance from the original line charge to a position (x, y), $s_i = \sqrt{(x-x_i)^2 + y^2}$ is the distance from the first image charge, and $s_{i,2} = \sqrt{x^2 + y^2}$ is the distance from the second image charge. (Unfortunately, this potential is not zero at an infinite distance, similar to that of Prob. P2-32, but we'll ignore this behavior.) (a) Determine the image charge density $\rho_{\ell,i,2}$ that causes the potential of the cylinder to be zero. (b) Determine the surface charge density $\rho_s(\phi)$ induced on the surface of the cylinder. (c) Integrate the surface charge around the circumference of the cylinder to determine the charge per unit length induced on the cylinder.

CHAPTER 3
Current and Resistance

KEY WORDS

current density (volume and surface); Continuity Equation; conductivity; Ohm's Law; power dissipation density.

In the previous two chapters we introduced electric charges and electric fields, using primarily the force between charged bodies as the starting point. Charges in motion, of course, constitute electrical currents, with which you have a working knowledge from your previous study in circuit analysis. In this chapter we will take a closer look at electrical currents. Specifically, we'll introduce current density, which describes the spatial distribution of a current. Then the relation between the ideas of electrostatics discussed in Chapters 1 and 2 and the laws that govern simple electric circuits, such as Kirchhoff's Current Law, Kirchhoff's Voltage Law, and Ohm's Law, will be discussed. You may have already recognized the connection between these circuits concepts and some of the ideas from electrostatics that we introduced earlier. The circuit laws are but special cases of the more general treatment that we will now discuss in detail.

In addition, we will explore means of determining the resistance of conducting elements in a general geometry, introduce the principles behind power dissipation when currents pass through conducting media, and examine methods for quantitatively determining the magnitude of the power dissipation. This discussion will set the stage for later discussions about attenuation of propagating waves in absorbing media. Finally, we will explore the boundary conditions that must be satisfied at the interface between two partially conducting dielectrics under steady-state conditions. These conditions differ from those for perfect dielectrics, since charge can accumulate at the interface for these materials.

3.1 Current and Current Density

LEARNING OBJECTIVES

After studying this section, you should be able to:
- determine the total current, given the volume current density J_v or surface current density J_s; and
- determine the current density, given the charge density and average velocity of the charges.

You are undoubtedly familiar with electrical currents, which are defined simply as electrical charges in motion. The unit of current is the ampere (A), equivalent to a coulomb per second (C/s). As you might presume from the preceding discussion of charges (see Section 1.1.2), the individuality of moving charges is seldom evident. For example, even a modest current in a wire of 1 mA corresponds to a flow of $\sim 10^{16}$ electrons per second through the wire. At this rate, the addition or elimination of a few electrons per second would be of little consequence, and we can treat the current as a continuous variable.

Similar to electric fields generated by charges, it is not simply the total current that is important in determining the fields they produce, but the *distribution* of those currents, described through the **volume current density J**$_v$ and/or the **surface current density J**$_s$. You will notice some similarities in the following discussion of the relationship between total current and current density with the preceding discussion of charge and charge density. There is an important distinction as well – that is, the current density has a *direction* associated with it – that is, the direction in which the current flows – and it must be described as a vector quantity. Charge and charge density are, of course, simply scalar quantities. Keeping that distinction in mind, let us proceed.

A current of density J$_v$ is illustrated in Fig. 3.1. This figure might represent the current within a wire, for example, or perhaps a portion of a larger current distribution flowing through some region in space, as in a vacuum tube or semiconductor. As just remarked, J$_v$ is a vector, the direction of which indicates the direction in which positive charges are moving. Its magnitude is the current flowing per unit area through the volume. In the case of negative carriers, such as electrons, the direction of the current is directly opposite their velocity. Electrically, negative charges flowing in a particular direction is completely equivalent to positive charges flowing in the opposite direction, and the current density in either case is described very nicely by the same quantity J$_v$, without further differentiation. In Fig. 3.1 we show a system of charges moving upward and to the right through a medium. If the current flowing through the small area Δs bounded by the dashed circle is ΔI, then the current density is defined as the current flow per unit area, or

$$\mathbf{J}_v = \lim_{\Delta s \to 0} \frac{\Delta I}{\Delta s}\, \mathbf{a_u},\qquad (3.1)$$

where $\mathbf{a_u}$ is a unit vector pointing in the direction of the positive charges' velocity, **u**. (For the purposes of this discussion, we will consider all the moving charges to have the same velocity. This, of course, is not usually correct, but it is sufficient for making the point without adding unnecessary complexity to the argument.) The unit of volume current density is $[\mathbf{J}_v] = \text{A/m}^2$. This unit should make sense to us when we keep the definition firmly in mind – that is, that the current density is the current per unit *area*. Some of us may initially think that, since current density represents the current flowing through three-dimensional space, the unit should be amperes per meter cubed. This is not correct, and we encourage you to reason through these units carefully. If nothing else, this will help you to firmly embed the physical picture of current density in your mind.

We introduced the flux and flux density of particles and fields earlier in Section 1.2.1 and these are also covered in Appendix C.3.2. Whether considering the flux density of a field through a surface or the particle flux density flowing through a surface, the idea is very similar. For the former, the flux density refers to the number of field lines per unit area passing through the surface. In the current discussion we are more interested in particle flow, and refer to Eq. (C.7), which shows that the flux density, or the number of particles crossing a surface s per elemental area Δs, is the product of the number density of the particles and their velocity u_z, where z is the direction normal to the surface. In general, the direction of $\mathbf{J}_v(\mathbf{r})$ will be the same as that of the particle velocity **u** (for positive charge carriers, of course; otherwise the opposite). Since the current density is just the flux density $\Delta\Phi/\Delta s$ times the charge q per particle, and the charge density ρ_v is

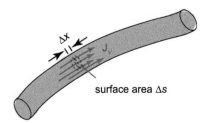

FIGURE 3.1 The current density J$_v$.

surface area Δs

just the number density n times the charge q, we can use logic similar to that leading to Eq. (C.7) to show that the current density $\mathbf{J}_v(\mathbf{r})$ is

$$\mathbf{J}_v(\mathbf{r}) = q\,\frac{\Delta\Phi}{\Delta s} = qn\mathbf{u} = \rho_v\mathbf{u}. \tag{3.2}$$

Given the current density $\mathbf{J}_v(\mathbf{r})$ in some region, we need a means of determining the total current transported across a surface s. As suggested in Fig. 3.1 and written above, the current flowing through the elemental surface area Δs is $\Delta I = \mathbf{J}_v(\mathbf{r}) \cdot \mathbf{a}_n \Delta s$, where $\mathbf{a}_n$ represents the unit vector that is normal to the surface element Δs. In order to determine the total current flowing through the cross-section s, we add the current elements ΔI_i flowing through each of the surface elements Δs_i comprising s:

$$I = \lim_{\Delta s_i \to 0} \sum_i \mathbf{J}_v(\mathbf{r}_i) \cdot \mathbf{a}_n \Delta s_i.$$

In the limit as $\Delta s_i \to 0$, the summation becomes an integral, and using $d\mathbf{s} = ds\,\mathbf{a}_n$, the current is

$$\boxed{I = \int_s \mathbf{J_v}(\mathbf{r}) \cdot d\mathbf{s}.} \tag{3.3}$$

Example 3.1 Vector Surface Integration, Current

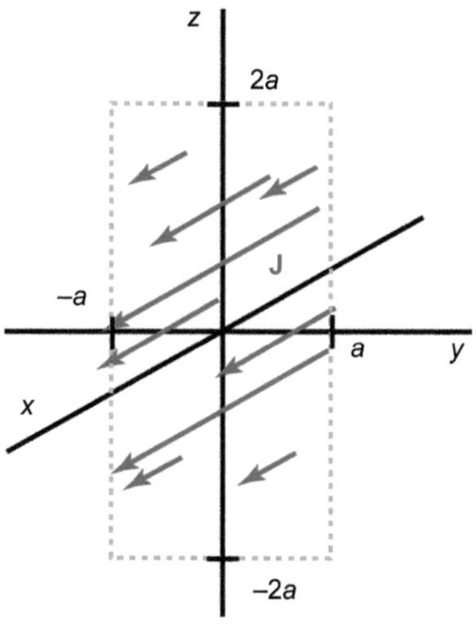

FIGURE 3.2 The current, with current density $\mathbf{J}_v$, passes through the rectangular aperture.

A current flows in the x-direction, as shown in Fig. 3.2. The density of the current, valid for $-a < y < a$ and $-2a < z < 2a$ is described by $\mathbf{J}_v(y,z) = J_{v,0} \cos(\pi y/2a) \cos(\pi z/4a)\,\mathbf{a}_x$, where $J_{v,0} = 1\ \text{mA/cm}^{-2}$ and $a = 1\ \text{cm}$. Determine the total current passing through the rectangular aperture shown in the figure.

Solution:
We will use Eq. (3.3),

$$I = \int_s \mathbf{J}_v(\mathbf{r}) \cdot d\mathbf{s},$$

in which the integration is carried out over the area of the rectangular aperture. Since the rectangular aperture lies in the y–z plane, the surface element is $d\mathbf{s} = dy\,dz\,\mathbf{a}_x$. $\mathbf{J}_v$ is given as $\mathbf{J}_v(y,z) = J_{v,0} \cos(\pi y/2a) \cos(\pi z/4a)\,\mathbf{a}_x$, so the surface integral takes the form

$$I = \int_{-2a}^{+2a} \int_{-a}^{+a} J_{v,0} \cos(\pi y/2a) \cos(\pi z/4a)\mathbf{a}_x \cdot dy\,dz\,\mathbf{a}_x.$$

$\mathbf{a}_x \cdot \mathbf{a}_x$ is 1, and carrying out the integration we obtain

$$I = J_{v,0}\,\left.\frac{\sin(\pi y/2a)}{\pi/2a}\right|_{-a}^{+a} \left.\frac{\sin(\pi z/4a)}{\pi/4a}\right|_{-2a}^{+2a} = J_{v,0}\frac{32a^2}{\pi^2},$$

which leads to $I = 3.24\ \text{mA}$.

Up to now we have considered moving charges that are distributed in three-dimensional space; they are distributed across a volume. In some cases, however, currents are confined to flow in the surface of a conductor, and we discuss this in terms of the **surface current density** $\mathbf{J}_s$. Physically, of course, these current distributions do have some thickness to them, but this thickness can be very small, and in these cases it

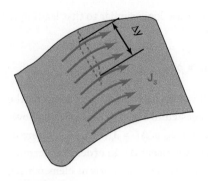

FIGURE 3.3 The surface current density $\mathbf{J}_s$.

is useful to think of the current as lying only in the surface. This is shown in Fig. 3.3. $\mathbf{J}_s$ is also a vector, but since this current is confined to the surface, we think of it as the current flowing per linear cross-sectional width, and the unit is $[\mathbf{J}_s] = A/m$. Again, take note of the units for the surface current density. This is not amperes per meter squared, and a little thought should help us feel comfortable with this:

$$\mathbf{J}_s = \lim_{\Delta y \to 0} \frac{\Delta I}{\Delta y}\, \mathbf{a_u} = \frac{dI}{dy}\, \mathbf{a_u}. \tag{3.4}$$

Inverting this, we can write the total current I in terms of the surface current density $\mathbf{J}_s$ as

$$I = \int |\mathbf{J}_s|\, dy,$$

where we have chosen the y-axis to be perpendicular to $\mathbf{J}_s$.

Traffic on an interstate is an analogue to surface current density. Automobiles are currently confined to drive on the surface of the Earth, just as a surface current is confined to a surface. If we consider a superhighway with, say, four lanes traveling in each direction, then the rate at which cars drive past a particular point in each lane of the highway is similar to the surface current density. To determine the total rate at which cars drive past us, without regard to which lane they are in, we would simply add the totals for each of the four lanes. This is analogous to integrating across the current distribution, or as in the previous equation, integrating over the dimension y.

Current can take any of three distinct forms: conduction, electrolytic, and convection. You are probably most familiar with conduction currents from your work with electrical circuits. The currents flowing through electrical wires, resistors, or even semiconductors are conduction currents. In conductors and resistors, we picture a huge number of infinitesimally small charged particles moving under the influence of an applied electric field. In most cases these particles are electrons, which we know by convention to possess a negative charge. (Since we define the direction of the current as the direction in which positive charges are moving, the current is in the direction opposite that of the average velocity of the electrons.) We will return to this picture of charged electrons in motion shortly. Electrolytic currents are characterized by the motion of positive or negative ions, as found in batteries. While batteries are, of course, widely used, their treatment in this text is confined to the discussion of Example 2.2. Finally, electric currents can be in the form of convection currents, as found in electrical discharges (fluorescent lights, neon lights, etc.) or vacuum tubes. These currents can be carried by electrons or ions, and the discharge can take place in an evacuated tube or a gas-filled tube. Unlike conduction currents, convection currents are free to move outside the confines of a conductor. Since many high-voltage functions that could previously be served only by vacuum tubes can now be performed by new types of semiconductor devices, vacuum tubes are not as common as they once were. Again, we won't be able to spend much time describing convection currents, other than to point out that these devices generally do not obey Ohm's Law. Let us now proceed to a more complete description of electrical circuits, drawing particular attention to the origin of the fundamental laws of circuit analysis and their relationships to electromagnetic fields concepts.

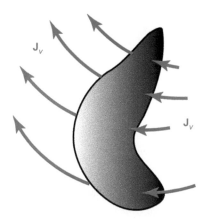

FIGURE 3.4 A charge distribution ρ_v contained within a volume v. If the current flowing out one side of v is different from the current flowing in the other, then the total charge within the volume v is changing in time. This relation is quantified through the Continuity Equation, Eq. (3.7).

3.2 Charge Conservation: The Continuity Equation

There exists an important relation, called the **Continuity Equation**, between the charge density and the current density in any region of space. This is not unexpected, since electrical currents are simply charges in motion. In this section we will derive the Continuity Equation and explore its significance.

We will use the Divergence Theorem, which is discussed in Appendix D.3, to explore this relationship quantitatively. Consider a volume v that encloses a distribution of charge described by the charge density ρ_v, as illustrated in Fig. 3.4. As we know from Eq. (1.10), the total charge within the volume is given as

$$Q = \int_v \rho_v(\mathbf{r})\, dv.$$

If there is current I flowing out of the volume v through its surface, then the charge stored within must be decreasing at a rate

$$I = -\frac{dQ}{dt}$$
$$= -\frac{d}{dt}\int_v \rho_v(\mathbf{r})\, dv.$$

Since the charge density is "well-behaved" (i.e. continuous in time and space), we can interchange the order of integration over v and differentiation with respect to t, leading to

$$I = -\int_v \frac{d\rho_v(\mathbf{r})}{dt}\, dv. \tag{3.5}$$

Using Eq. (3.3), the total current I passing through the surface s, where s is the surface that bounds the volume v, can also be written in terms of the current density $\mathbf{J}_v$:

$$I = \oint_s \mathbf{J}_v(\mathbf{r}) \cdot d\mathbf{s}.$$

Notice that s is a closed surface here, and we use the $\oint$ symbol for the integral to denote this. Applying the Divergence Theorem, the surface integral of $\mathbf{J}_v(\mathbf{r})$ is equivalent to the volume integral of $\boldsymbol{\nabla} \cdot \mathbf{J}_v$ over the volume v, so I can be written

$$I = \int_v \boldsymbol{\nabla} \cdot \mathbf{J}_v\, dv. \tag{3.6}$$

Equating Eqs. (3.5) and (3.6) results in

$$-\int_v \frac{d\rho_v(\mathbf{r})}{dt}dv = \int_v \boldsymbol{\nabla} \cdot \mathbf{J}_v\, dv.$$

Since these integrals must be equal for *any* volume v, the integrands themselves must be equal. That is,

$$\boxed{\frac{d\rho_v}{dt} + \boldsymbol{\nabla} \cdot \mathbf{J}_v = 0,} \tag{3.7}$$

a relation known as the **Continuity Equation**. $d\rho_v/dt$ represents the rate of increase of charge density, while $\boldsymbol{\nabla} \cdot \mathbf{J}_v$ is related to the rate at which charge flows across the surface bounding the region. If the current density is changing spatially, then the charge flowing into a region on one side differs from the charge flowing out the opposite side, and the total charge in that region must be changing in time. The Continuity Equation is a quantitative expression of this relation.

Example 3.2 Charge Conservation, Continuity Equation

Determine the charge density ρ_v in a region in which the current density is $\mathbf{J}_v(z, t) = \mathbf{a}_z J_0 \cos(\omega t - kz)$.

Solution:
We start by finding $\nabla \cdot \mathbf{J}_v$,

$$\nabla \cdot \mathbf{J}_v = J_0 \frac{\partial}{\partial z} \cos(\omega t - kz)$$

$$= k J_0 \sin(\omega t - kz) = -\frac{\partial \rho_v}{\partial t},$$

where the final equality follows from the Continuity Equation, Eq. (3.7). To find the charge density, we integrate $\partial \rho_v / \partial t$ to find

$$\rho_v(z, t) = -\left(\frac{k}{\omega}\right) J_0 \cos(\omega t - kz) + \rho_{v,0},$$

where $\rho_{v,0}$ is a constant value. This function for the current density and charge density is known as a traveling wave, which we will explore in various contexts much later in this text. Notice that since $\mathbf{J}_v(z, t)$ and $\rho_v(z, t)$ both vary as $\cos(\omega t - kz)$, the charge density is maximum at the same position and time as the current density.

3.3 Kirchhoff's Laws

LEARNING OBJECTIVES

After studying this section, you should be able to:
- describe the equivalence of Kirchhoff's Voltage Law in a circuit to $\oint \mathbf{E} \cdot d\boldsymbol{\ell} = 0$; and
- describe the equivalence of Kirchhoff's Current Law in a circuit to the Continuity Equation.

Key to the circuit analysis techniques that you mastered in earlier courses were Kirchhoff's Voltage (KVL) and Current (KCL) Laws. In this section we will explore the relation between these circuit laws and the equivalent, but more general, corresponding "fields" laws.

As a reminder, KVL states that for any path that forms a complete loop within an electrical circuit, the sum of the voltages across each element within the loop must be zero. This statement follows from the requirement for any system that energy must be conserved. Let's show that connection here. For each element i within the loop, the voltage V_i across the element and the electric field $\mathbf{E}_i$ within the element must be related by $V_i = -\int \mathbf{E}_i \cdot d\boldsymbol{\ell}$, where the path integral is evaluated along any path through the element. You should recall from Chapter 1 that the electric potential was originally defined in terms of the work required to move a test charge between two points A and B. In this case, the two points are on opposite sides of an electrical element within a circuit. For the example circuit shown in Fig. 3.5(a), let the path integration extend from a to b for the potential across the source, from b to c for the potential across the wire that connects the source to the resistor labeled R_1 (this potential is typically very small, of course), from c to d for the potential across resistor R_1, and so forth all the way around the left-side loop. The summation of all the voltages V_i around the loop is then

$$\sum_i V_i = \sum_i \left(-\int \mathbf{E} \cdot d\boldsymbol{\ell}\right)_i = -\int_a^b \mathbf{E} \cdot d\boldsymbol{\ell} - \int_b^c \mathbf{E} \cdot d\boldsymbol{\ell} - \int_c^d \mathbf{E} \cdot d\boldsymbol{\ell} - \cdots.$$

Since the sum extends over each of the elements within the loop, the sum of all the integrals $-\int \mathbf{E} \cdot d\boldsymbol{\ell}$ becomes the closed loop integral

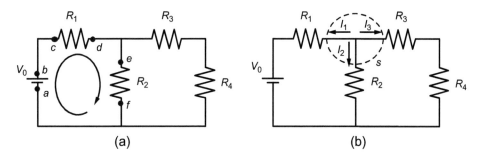

FIGURE 3.5 A simple electrical circuit in which we apply Kirchhoff's Voltage Law (KVL) and Kirchhoff's Current Law (KCL) in order to determine the voltages across or the current through the various circuit elements. In (a) we apply KVL around the loop on the left side, while in (b) we label the currents at the top center node for application of KCL.

$$\sum_i V_i = -\oint \mathbf{E} \cdot d\boldsymbol{\ell},$$

which we showed in Chapter 1 must be identically zero for static fields. Thus KVL from circuit analysis is equivalent to the requirement that $\oint \mathbf{E} \cdot d\boldsymbol{\ell} = 0$. Recall that $\oint \mathbf{E} \cdot d\boldsymbol{\ell} = 0$ follows from the Law of Conservation of Energy, in that the work required to move a charge q around a complete loop is zero since an electrostatic force is a conservative force.

Kirchhoff's Current Law in circuit analysis is equivalent to the Continuity Equation, Eq. (3.7), that we discussed in Section 3.2, as we will now show. Kirchhoff's Current Law tells us that, at any node within a circuit, the sum of the currents flowing into and out of this node must be equal. For example, consider the node at the top center of the circuit in Fig. 3.5(b). To apply this law correctly, remember the sign convention: We write KCL as $\sum_i I_i = 0$, where current flowing out of the node is positive and current flowing into the node is negative. To see the equivalence of KCL with the Continuity Equation, we construct a bounding surface around any node of the circuit. (This surface does not represent a physical structure, only a mathematical construct.) This surface is labeled s in Fig. 3.5(b). The total current flowing out through this bounding surface is $I = \oint \mathbf{J}_v \cdot d\mathbf{s}$, where the surface integral is carried out over the entire bounding surface s. In a circuit, of course, the current density is confined to the wires and the electrical elements within the circuit. The entire integral $\oint \mathbf{J}_v \cdot d\mathbf{s}$ can be separated into the sum of two parts. $\int \mathbf{J}_v \cdot d\mathbf{s}$ evaluated over the cross-section of any one of these wires is the total current within the wire; we'll call it I_i, where i is an index that labels each of the wires. $\int \mathbf{J}_v \cdot d\mathbf{s}$ evaluated over any surface element outside the wires is zero, since $\mathbf{J}_v = 0$ there. Then the total current leaving the node is $\sum_i I_i$. As we showed in the discussion of the Continuity Equation (Section 3.2), this current represents the total rate at which the charge enclosed within the surface is decreasing. In circuits, of course, where we only consider whole elements, the charge does not build up at any nodes. (You might object to this statement, citing a capacitor as a counter-example. Positive charge does indeed collect on the positive plate of a capacitor, with an equal but opposite charge collecting on the negative plate. The net charge on the capacitor, however, summing the charge on the two plates, is precisely zero, since the charges cancel one another. Since in our circuits class we only considered the *whole* capacitor, our statement that the charge does not build up at any node is correct.) Combining these results, the net current into any node is $\oint \mathbf{J}_v \cdot d\mathbf{s} = \sum_i I_i = 0$. We have therefore shown that KCL is equivalent to the Continuity Equation in steady state.

3.4 Resistance

From your prior experience with circuits you should also recall that for a resistor the current I through the resistor and the voltage V across it are proportional to one another,

$$I = V/R. \tag{3.8}$$

This is Ohm's Law, in honor of Georg Ohm. (See Biographical Note 3.1.) In Eq. (3.8) the resistance R of the resistor is a quantity that depends only on the dimensions of the resistor and the conductivity of the resistor material. For a simple geometry of a resistor of length ℓ, cross-sectional area S, and conductivity σ, the resistance is

$$R = \frac{\ell}{\sigma S}. \tag{3.9}$$

In circuits classes this law is typically presented as a reasonable, but otherwise unproven, relation. We will now examine current flow through a resistive element on a microscopic level, and derive a similar but more general relation.

Biographical Note 3.1 Georg Simon Ohm (1789–1854)

FIGURE 3.6 A portrait of Georg Ohm.

A German physicist and mathematician, Georg Ohm showed in 1827 that the current through a long wire attached to a voltage source is proportional to the potential of the source (Eq. (3.8)), a relation known as Ohm's Law. This result was controversial, as most researchers at the time accepted the notion that the current and potential were independent of one another. Ohm's extensive measurements and his use of the thermoelectric voltage source proved to be key to his conclusions. (Batteries, or voltaic piles, described in TechNote 2.2, were introduced in 1799 by Alexandro Volta, but were unsuitable for these measurements due to their internal resistance.) Ohm also showed that the current (1) decreased inversely with the length of the wire; (2) increased proportionally with the cross-sectional area of the wire; and (3) depended on the material of the wire. To carry out his measurements, Ohm used the deflection of a magnet near the wire to measure the current (see Biographical Note 4.1 on Hans Christian Oersted, who first demonstrated the deflection of a magnet by a current in 1820), and a thermocouple source (first demonstrated by Thomas Seebeck in 1821) to generate the potential. Ohm was awarded the Copley Medal of the Royal Society of London in 1841, and the unit of resistance is named in his honor. His portrait is shown in Fig. 3.6.

To understand the resistance of an element, we start with Eq. (3.2), which gives the current density $\mathbf{J}_v$ in a medium when all charged particles have the same velocity $\mathbf{u}$. The particle velocities are typically not uniform, but this is easily accommodated using

$$\boxed{\mathbf{J}_v = \rho_v \langle \mathbf{u} \rangle,} \tag{3.10}$$

where ρ_v is the charge density of the electrical carriers (typically the electrons) and $\langle \mathbf{u} \rangle$ is their average velocity. In materials in which the current is due to more than one charge carrier, this relation can be expanded as

$$\mathbf{J}_v = \sum_j \rho_{v,j} \langle \mathbf{u} \rangle_j. \tag{3.11}$$

The index j labels the different kinds of charge carriers. In the following discussion we focus our attention on electrons as the primary carrier. In the absence of an electric field $\mathbf{E}$ in the medium, the charges experience no net force, and their average velocity is zero. Upon application of an electric field, however, the electrons do experience an electric force, and they tend to drift in the direction of this force. For most conducting materials it is found that their average velocity $\langle \mathbf{u} \rangle$ is proportional to the applied field strength $\mathbf{E}$. The *mobility* of the electrons in a conducting material is defined to describe this proportionality,

$$\langle \mathbf{u} \rangle = -\mu_e \, \mathbf{E}. \tag{3.12}$$

The units of the mobility are $[\mu_e] = [\langle \mathbf{u} \rangle]/[\mathbf{E}] = (\text{m/s})/(\text{V/m}) = \text{m}^2/\text{Vs}$. Combining Eqs. (3.10) and (3.12), we arrive at the relation

$$\boxed{\mathbf{J}_v = \sigma \mathbf{E},} \tag{3.13}$$

where σ is a material parameter known as the **conductivity** of the medium, and is equal to the product of the charge density and the mobility,

$$\sigma = -\rho_v \, \mu_e. \tag{3.14}$$

The negative sign is necessary for negative charge carriers such as electrons. Equation (3.13) is known as **Ohm's Law**, not to be confused with Eq. (3.8), which is known by the same name. Upon close examination, we can note a strong similarity between these two different versions of Ohm's Law. In place of current I and voltage V, Eq. (3.13) has current density $\mathbf{J}_v$ and electric field $\mathbf{E}$, and the conductivity σ of the material appears in place of $1/R$. The primary difference between these equations is that the new "fields" version of Ohm's Law (Eq. (3.13)) is a relation between local (i.e. field) variables – that is, variables that can vary from point to point within the resistive material, whereas I and V define quantities across or through the resistive element as a whole. We will show shortly that we can apply Eq. (3.13) to determine the resistance of a wide variety of geometries of resistive elements.

The conductivities of several conducting and insulating materials are listed in Table 3.1. The units of conductivity are S/m, where a siemen (S) is the inverse of an Ohm (Ω). In Section 1.5 we discussed the "perfect" dielectric, in which the charge carriers were tightly bound to the atoms. We defined the permittivity for linear dielectric materials to characterize the polarization, or distortion of charge centers, in these materials. In this table of material conductivities we see that the conductivity of insulating materials, while many orders of magnitude smaller than that of metals, is not actually zero. There is some migration of charge carriers upon application of an electric field. In real dielectric materials, therefore, we characterize their electrical properties using both their permittivity and conductivity. For media in which the charge is carried by more than one type of carrier, the expression for the conductivity, Eq. (3.14), is generalized to read $\sigma = -\sum_j \rho_{v,j} \mu_j$, where the index j labels the different charge carrier types, and the minus sign is appropriate for negative charge carriers.

For those of you who recall the treatment in your introductory physics courses of the free acceleration of a charged body in a uniform electric field, you may feel that we have just pulled a fast one on you when we claimed that the average velocity of the charges is proportional to the field $\mathbf{E}$. So let's step back and look at this process of conduction through a resistive medium a little more closely. For a uniform electric field, the electric force on the charges is also uniform, and by Newton's Second Law we should expect the *acceleration* of the charged body to be constant. (Even the premise of a uniform field

Table 3.1

Conductivities of several conducting and insulating materials.	
Material	σ (S/m)
Metals (good conductors)	
Aluminum	3.77×10^7
Copper (annealed)	5.8×10^7
Gold	4.1×10^7
Silver	6.3×10^7
Semiconductors (moderate conductors)	
Silicon	1.7×10^{-3}
Insulators (poor conductors)	
Carbon (amorphous)	1.25 to 2×10^3
Seawater	4.80
Drinking water	5×10^{-4} to 5×10^{-2}
Glass	10^{-15} to 10^{-11}
Quartz	1.3×10^{-18}
Teflon	10^{-25} to 10^{-23}

inside the conductor is suspect, in that the local fields, which include the effects of all the nuclei and electrons in the medium, are far from uniform. It is sufficient, however, to deal with the average fields, so we can safely ignore this complication.) The velocity of a body undergoing constant acceleration, however, is increasing in time, $\mathbf{u} = \mathbf{a}t = q\mathbf{E}t/m$, which seems to be at odds with our statement in the preceding paragraph that the *velocity* is proportional to the electric field strength. The resolution of this puzzle is quite simple, once we acknowledge the numerous random collisions between the charged body and the atoms that constitute the conducting medium. Because of these multiple collisions, the motion of the charges is nothing like free acceleration. If we could view the individual electrons, we would see their motion to be almost random, moving one way for a short time, then bouncing off a massive atom within the conducting medium and rebounding in a different direction for a while, undergoing numerous collisions one after another. This is illustrated in Fig. 3.7. Between collisions, the electron will start to accelerate in the direction of the electric force, but before it can gain any appreciable velocity, bang!, it hits another atom and its velocity is randomized again. As a result, the electrons' drift velocity is proportional to $q\mathbf{E}$, but there is no net acceleration. This picture also helps us to understand the heating of resistors or wires that are conducting the current. Each time the electron collides with an atom, a large part of its kinetic energy is transferred to the atom with which it collides, which can be detected as heat.

There is an important class of materials, called superconductors, in which these collisions between the charge carriers and the atoms is inhibited, as described in TechNote 3.1.

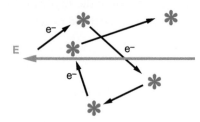

FIGURE 3.7 A simple physical model of multiple collisions of an electron (or other charge carrier) within a conductor under the influence of an applied d.c. electric field **E**.

TechNote 3.1 Superconductors

Many metals or other materials become superconducting, a process by which their resistance to electrical current becomes zero, when they are cooled to very low temperatures. Superconductivity was first observed by Heike Kamerlingh Onnes, a Dutch physicist, and colleagues in 1911 in mercury, which enters the superconducting phase when its temperature is decreased below its critical temperature of $T_c = 4.2$ K. Currently, the material with the highest critical temperature is a sulfur–hydrogen compound, for which $T_c = 203$ K ($-70\,°$C). Unfortunately, this compound forms only under extremely high pressure (1.5×10^6 atm). The search for a high-temperature superconducting material with good material parameters is an active research pursuit, as such a material could find wide application in electrical power distribution and other applications where high currents are required.

When a material is in the superconducting phase, electrical current can pass through the material without dissipation – that is, with no power loss. This becomes very important for high-field-strength electromagnets, such as those used in MRI machines, which must carry extremely high currents. (See TechNote 6.7.) For normal conductors, power losses would produce prohibitively high temperatures in the magnets. A second important property of superconductors is that magnetic fields cannot penetrate at all into a superconducting material. We will soon discuss surface currents near the surface of normal metals. In superconductors, these surface currents lie right at the surface, with a skin depth of zero. (The skin depth is the "thickness" of a surface current. We will discuss this topic in detail in Chapter 8.) This effect was discovered in 1933 by German physicists Walther Meissner and Robert Ochsenfeld, and is known as the Meissner effect.

John Bardeen, Leon Cooper, and John Schrieffer established the microscopic theory to explain superconductivity, known as the BCS theory, in 1957. At low temperature, electrons can form pairs, known as Cooper pairs, which pass through the lattice without collisions. The BCS theory works well for low-temperature superconductors, but not for high-temperature superconductors (those with $T_c > 80$ K).

In the preceding paragraphs we have described two different electron velocities. It can be instructive to estimate the magnitude of these two velocities, and we may be surprised by their respective scales. First we described the average velocity $\langle \mathbf{u} \rangle$ of a conduction electron. To be very specific, let's estimate the magnitude of this average velocity in a 12 gauge copper wire carrying 1 A of current. (Typical copper wiring in residences in the United States is 12 or 14 gauge, and 1 A is the approximate current drawn by a 100 W incandescent light bulb.) Using Eqs. (3.10) and (3.3), we can estimate the average velocity as $|\langle \mathbf{u} \rangle| = |\mathbf{J}_v|/\rho_v = I/(\Delta s\, ne)$, where $\Delta s = 3.31$ mm^2 is the cross-sectional area of the 12 gauge wire, n is the number density of the electrons, and $e = 1.602 \times 10^{-19}$ C is the charge per electron. To estimate the density n of the electrons in the copper wire, we will assume one free electron per copper atom, and use the density (8.94 g/cm^3) and atomic number ($A = 29$) of copper, and the mass of the proton or neutron (1.67×10^{-27} kg) to find $n = 1.8 \times 10^{29}$ m^{-3}. With these estimates we find the average velocity of the electron under these conditions is $|\langle \mathbf{u} \rangle| = 1.0 \times 10^{-5}$ m/s. While this velocity might seem to be much too small, remember that there are many, many electrons within the copper wire, and even such a small average velocity does in fact represent an appreciable current. Let us now estimate the magnitude of the random velocity of the individual electrons as they bounce around between collision events within the conductor. Since this motion is in all directions, we use the root-mean-square (rms, or the square root of the mean value of the square) velocity as a measure of this motion. The full explanation of this velocity is outside the topic of this course, so we will skip straight to the result, which is that under thermal equilibrium conditions, the rms velocity of an electron is $u_{\mathrm{rms}} \equiv \langle u^2 \rangle^{1/2} = \sqrt{2k_B T/m_e}$, where $k_B = 1.38 \times 10^{-23}$ J/K is the Boltzmann constant, T is the absolute temperature, which we take to be 300 K, and $m_e = 9.11 \times 10^{-31}$ kg is the electron mass. Using these terms, we find $u_{\mathrm{rms}} \sim 3 \times 10^4$ m/s. This random velocity of the electrons is many orders larger than the average velocity under these conditions. If we could imagine some means of observing the instantaneous motion of the electrons in a conductor, this random thermal velocity would totally dominate the drift velocity. Nonetheless, the *average* velocity of the electrons is reflected only in the drift velocity (since the average of the thermal velocity is zero), and it is the velocity associated with this conduction current that will be our primary focus.

We now return to the two forms of Ohm's Law, given by Eqs. (3.8) and (3.13), and combine these, along with Eq. (1.43) for the potential difference across the resistor from Chapter 1, to determine a general formulation for determining the resistance R of an element. We consider a general geometry of the form shown in Fig. 3.8. This figure, not coincidentally, reminds us of the configuration that we used to formulate the capacitance of a system, but in this case we are interested in the conductivity, rather than the dielectric properties, of the medium between the two conducting contacts. Upon application of a potential V_1 to the contact on the left and a potential V_2 to the contact on the right, the potential difference can be written in terms of the electric field using Eq. (1.43) as $\Delta V = (V_1 - V_2) = -\int_2^1 \mathbf{E} \cdot d\boldsymbol{\ell}$, where the line integral is carried out over *any* path from contact 2 to contact 1. We can also relate the total current I flowing through the medium to the current density $\mathbf{J}_v$ using $I = \int_s \mathbf{J}_v \cdot d\mathbf{s}$, where the surface integration is carried out over *any* cross-sectional area of the conductive medium between the contacts.

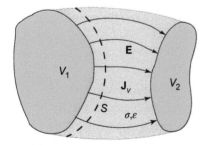

FIGURE 3.8 Current density $\mathbf{J}_v$ through a conductor of conductivity σ under the influence of an applied electric field $\mathbf{E}$.

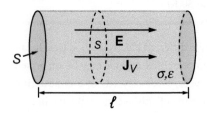

FIGURE 3.9 A simple resistor with uniform current density $\mathbf{J}_v$ and electric field $\mathbf{E}$. The length is ℓ, the cross-sectional area S, and conductivity σ.

Combining these integrals, and using Eq. (3.13), allows us to write the resistance of the medium as

$$R = \frac{\Delta V}{I} = \frac{-\int_2^1 \mathbf{E} \cdot d\ell}{\sigma \int_s \mathbf{E} \cdot d\mathbf{s}}. \tag{3.15}$$

From this expression, we see that knowledge of the electric field throughout the medium between the two contacts allows us to determine its resistance.

Let us now apply this formulation to the simple resistor geometry consisting of a medium of uniform conductivity σ positioned between two parallel conducting caps, as shown in Fig. 3.9. The distance between the conducting caps is ℓ, the cross-sectional area is S, and we will approximate the $\mathbf{E}$ field in the space between the caps as uniform. The potential difference between the end caps is $\Delta V = (V_1 - V_2) = -\int_2^1 \mathbf{E} \cdot d\ell$, which simplifies to $\Delta V = E\ell$ for uniform $\mathbf{E}$. The current integral also simplifies for uniform $\mathbf{E}$, to $I = \sigma ES$. The ratio $\Delta V/I$ is therefore $\ell/(\sigma S)$, in agreement with Eq. (3.9) for the resistance of a simple resistor.

Before we work a more interesting example, let us note that the integrals that appear in Eq. (3.15) are very similar to those that we used to determine the capacitance of a configuration consisting of a dielectric medium between two conductors. See Eq. (2.27). We should comment, however, that the similarity is strongest when the conductivity and permittivity are perfectly uniform in the entire space between the contacts. In this limit, the surface s over which we integrate to find the total charge (for the capacitance) or the total current (to find the resistance) is the same, and comparison of these two equations gives us the following relation between the resistance and capacitance of an element:

$$RC = \frac{\varepsilon}{\sigma}. \tag{3.16}$$

That is, the product of the resistance and capacitance of the element depends only on material properties ε and σ, and any dependence on field configurations precisely cancels. (In practice, it is common to have multiple media in the space between the conductors, for which this relation is not valid. Be careful that you don't apply it in these cases.)

Example 3.3 Resistance of a Radial Cylindrical Resistor

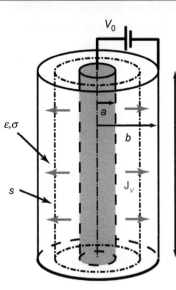

FIGURE 3.10 A co-axial resistor with current density $\mathbf{J}_v$.

Find the resistance of a conducting medium between a pair of cylindrical co-axial conductors of radii a and b and of length L, shown in Fig. 3.10. (This geometry is similar to that considered in Example 2.14, in which we determined the capacitance of such a structure.) The conductivity of the medium in the space between the conductors is σ, and its permittivity is ε. Ignore fringe field effects at the end.

Solution:
Consider application of a voltage V_0 to the center conductor, relative to the outer conductor. This causes a current to flow from the inner conductor to the outer. The current density is directed radially outward, and is cylindrically symmetric. This current density is proportional to the electric field in this region. The electric field lines originate on the positive surface charges (of density $\rho_{s,a}$) on the inner conductor and terminate on the negative surface charges (of density $\rho_{s,b}$) on the inner surface of the outer conductor. We must find the electric field $\mathbf{E}$ in the space between the cylindrical conductors, in terms of the $\rho_{s,a}$. Let's start by applying Gauss' Law to find the displacement field $\mathbf{D}$ in this region. We construct the dot-dashed Gaussian surface of length L and radius ρ, where $a < \rho < b$, centered on the axis of the cylinder. By the cylindrical symmetry of

the configuration, $\mathbf{D}$ must be directed radially, and its magnitude can depend only on ρ. Then application of Gauss' Law to this surface yields

$$\oint_s \mathbf{D} \cdot d\mathbf{s} = D_\rho 2\pi\rho L = Q_{enc} = \rho_{s,a} 2\pi a L.$$

Solving this for D_ρ yields $D_\rho = \rho_{s,a} a/\rho$. Since $\mathbf{E} = \mathbf{D}/\varepsilon$, the magnitude of the electric field is $E_\rho = \rho_{s,a} a/\varepsilon\rho$. Now that E_ρ is known, we can evaluate the two integrals in Eq. (3.15). We start with the potential difference between the conductors,

$$V_0 = -\int_b^a \mathbf{E} \cdot d\boldsymbol{\ell},$$

which we evaluate along a straight-line radial path from the outer conductor to the inner. Using $\mathbf{E} = \mathbf{a}_\rho \rho_{s,a} a/\varepsilon\rho$ and $d\boldsymbol{\ell} = \mathbf{a}_\rho \, d\rho$, this integral is

$$V_0 = -\int_b^a \left(\frac{\rho_{s,a} a}{\varepsilon\rho}\right) d\rho = -\frac{\rho_{s,a} a}{\varepsilon} \ln\left(\frac{a}{b}\right) = \frac{\rho_{s,a} a}{\varepsilon} \ln\left(\frac{b}{a}\right).$$

Next we evaluate the current integral in Eq. (3.15):

$$I = \int_s \mathbf{J}_v \cdot d\mathbf{s} = \sigma \int_s \mathbf{E} \cdot d\mathbf{s}.$$

Since we are evaluating the current flowing from the inner conductor to the outer, we choose the surface s to be a cylindrical surface between the conductors. In fact, the Gaussian surface that we used previously to find $\mathbf{D}$ serves nicely here as well. The surface element is $d\mathbf{s} = \mathbf{a}_\rho \rho \, d\phi \, dz$, with ϕ varying from 0 to 2π, and z extending the entire length of the cylinders, 0 to L. Then the current is

$$I = \sigma \int_0^L \int_0^{2\pi} \left(\frac{\rho_{s,a} a}{\varepsilon\rho}\right) \rho \, d\phi \, dz = \frac{\sigma \rho_{s,a} 2\pi a L}{\varepsilon}.$$

Notice that this current is independent of the radius ρ, as we should expect, since the current flowing outward through any surface must be the same. Otherwise, charge would accumulate in the medium, which cannot occur under steady-state conditions. This is generally true as long as the conductivity of the medium is uniform.

Having determined V_0 and I, we can now find $R = V_0/I = \ln(b/a)/\sigma 2\pi L$. Note that this result is consistent with Eq. (3.16), since we previously determined that the capacitance of this co-axial configuration is $C = 2\pi\varepsilon L/\ln(b/a)$, and $RC = \varepsilon/\sigma$, as required.

In this section we have discussed a physical picture for conduction of current through a medium, and presented a generalized form of Ohm's Law. This led us to a technique to determine the resistance of an element of arbitrary geometry, in which we must first determine the electric field throughout the medium. Then, using the electric field, the potential difference across the element, as well as the current through the element, can be determined, and the resistance is the ratio of these two quantities.

LEARNING OBJECTIVE

After studying this section, you should be able to:

- qualitatively describe current flow in non-resistive devices, such as fluorescent lights, vacuum tubes, cathode ray tubes, and ionization gauge detectors.

3.5 Non-resistive Current Flow

In the previous section we discussed resistive current flow through a medium, in which the current density obeys Ohm's Law, Eqn. (3.13). (Or, equivalently, the current obeys the macroscopic form of Ohm's Law, Eq. (3.8).) While current flow is commonly resistive, there are also several key examples of current in non-resistive elements, which we illustrate in this section. The first example is the fluorescent light tube (TechNote 3.2).

TechNote 3.2 Fluorescent Lights

In a fluorescent light, an electrical discharge in a gas-filled tube produces visible light. At either end of a fluorescent tube is a filament, as shown in Fig. 3.11, which, when heated by passing a current through it, boils off electrons in a process called thermionic emission. These electrons are accelerated toward the positive end of the tube, and through multiple collisions with the low-pressure argon gas in the tube they ionize the atoms, forming positive ions and additional electrons. In addition to the argon, the tube contains a small quantity of mercury atoms. The mercury atoms are excited to highly energetic states through collisions with the electrons and argon ions. As these excited mercury atoms relax to lower energy states, they emit ultraviolet (UV) radiation on a number of discrete lines. This UV light is converted to visible "white" light by a phosphor coating on the inside surface of the fluorescent tube.

The discharge in the gas tube does not obey Ohm's Law (i.e. the current in the discharge is not proportional to the voltage applied to the tube), and after the discharge is initiated the current in the tube could become extremely large. For this reason, an additional element known as the ballast is an integral part of the fluorescent fixture to control the current. The ballast contains an inductor (and often a transformer and power-factor improving capacitor), placed in series with the tube, to limit the current through the tube, and to restart the current in the tube after each half-cycle. Ballasts in modern compact fluorescent lights (CFLs) are quite different from the simple choke ballast, but perform the same function.

Neon lights operate in a similar manner, but these tubes are filled with neon gas. Visible light is generated directly when the neon atoms are excited in the electrical discharge, without the need for a phosphor coating on the glass tube.

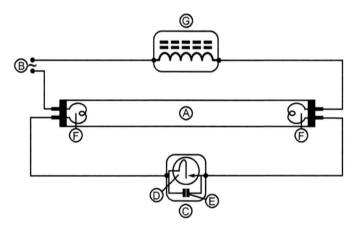

FIGURE 3.11 A fluorescent light fixture. A glass tube (A) filled with low-density argon gas and a small amount of mercury metal is fitted with filaments (F) at either end. When a.c. power (B) is applied to the filaments, they heat and eject electrons through thermionic emission. The starter (C–E) helps initiate the electrical discharge, which for long tubes requires a high voltage, as large as 1 000 V. The ballast (G) limits the current through the tube once the discharge is initiated, since the resistance of the tube becomes very small under operating conditions.

Another example of non-resistive current is found in the vacuum tube, the precursor to the semiconductor diode or transistor. These devices are described in TechNote 3.3.

TechNote 3.3 Vacuum Tubes

Vacuum tubes are electronic devices that offered the first ability to control electrical currents with voltages. The first vacuum tube was a vacuum diode, introduced by John Ambrose Fleming, a British researcher, in 1904. A diode allows current to pass through it in only one direction. See the schematic representation of a simple vacuum diode in Fig. 3.12(a). When the cathode is heated it ejects (or "boils off") electrons from its surface through the process called thermionic emission. The anode, which is biased positively with respect to the cathode, attracts the electrons, which are free to accelerate toward the anode, forming an electrical current. When the bias potential on the anode is reversed, such that it is negative with respect to the cathode, the electrons ejected by the cathode are repelled by the anode, and no current can flow. Since electrons are generated only by the heated cathode, the current can flow in only one direction through the diode.

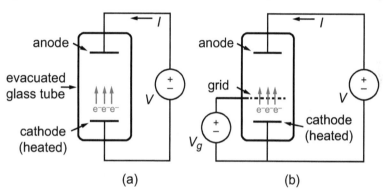

FIGURE 3.12 Schematic diagram of (a) a vacuum tube diode, and (b) a vacuum tube triode.

In 1907, Lee de Forest from New York expanded on Fleming's diode, and invented the first triode, which he used in the first electronic amplifier for telegraph and radio signals. (This early triode was called the audion, and early versions contained a partial gas, rather than a vacuum.) A triode consists of an anode and a heated cathode, similar to the diode, but also a third element positioned between the anode and cathode, called the grid. Applying a bias potential to the grid allows for control of the electrons passing from the cathode to the anode. See the triode schematic in Fig. 3.12(b). When the grid is negatively biased it repels the electrons, blocking their flow to the anode. When the grid is biased at a potential between the cathode and anode, the electrons are attracted to the grid, and due to the large openings in the grid the electrons mostly pass through and continue on to the anode. Thus, the grid potential controls the electron flow to the anode, and the anode current is a reproduction of the grid voltage. When properly configured in a circuit, the triode is the basis for an amplifier.

FIGURE 3.13 A vacuum tube amplifier.

A vacuum tube amplifier is shown in Fig. 3.13. Many styles and designs of vacuum tubes with additional grids for better control have been developed over the years. Semiconductor devices such as diodes and transistors have largely replaced vacuum tubes, but vacuum tubes still find application in some high-frequency and high-power applications. The magnetron, described in TechNote 10.3, is a form of vacuum tube for generation of 2.45 GHz microwaves.

Cathode ray tubes, or CRTs, were commonly used for early-generation television sets and oscilloscope displays. These have almost completely been replaced by LCD and LED displays, but CRTs played an important role in their time. They were based upon a beam of electrons, scanned across a phosphorescent screen.

TechNote 3.4 Cathode Ray Tube

A cathode ray tube (CRT) is a glass vacuum tube consisting of a heated cathode, steering coils or field plates, and a phosphorescent (sometimes called fluorescent) screen. (See Fig. 3.14.) The heated cathode generates electrons, which are collimated to form an electron beam, and accelerated toward the screen. The electron beam is steered using deflection coils (to produce a magnetic field) or deflecting plates (to produce an electric field). Electrons incident upon the screen excite the phosphor, producing a bright visible spot. As the electron beam is swept across the screen, its intensity is modulated to create the bright or dark points of the image to be rendered. In the late 1800s (before the discovery of the electron) an electron beam was called a cathode ray, lending its name to the CRT.

The first CRT, which used a cold cathode, was invented by Ferdinand Braun in 1897. The first hot-cathode CRT was developed by John Bertrand Johnson and Harry Weiner Weinhart at Western Electric, and the CRT became a commercial product in 1922. Early work on television CRTs was reported by Kenjiro Takayanagi and Vladimir K. Zworykin in the 1920s, and the first commercially available televisions using CRTs were produced by Telefunken in Germany in 1934.

For many decades, CRTs found wide usage in televisions, oscilloscopes, and computer monitors. CRTs are no longer available, having been made obsolete by the introduction of LED displays, plasma displays, and OLEDs (organic light emitting diodes).

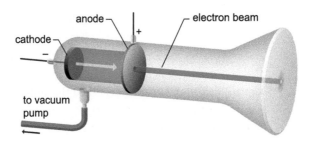

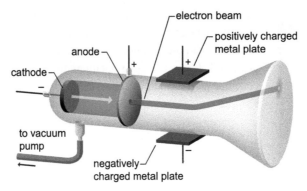

FIGURE 3.14 A cathode ray tube, commonly called a CRT. The pink lines represent the trajectories of electrons, produced by a heated cathode, accelerated to the right by the electric field between the cathode and anode to form a beam, and steered and focused onto the phosphor screen at the right. The phosphor screen emits light upon impact by the electrons.

A Bayard–Alpert ionization gauge, discussed in TechNote 3.5, is a specialized device used for reading high vacuum levels.

TechNote 3.5 Bayard–Alpert Ionization Gauge

A Bayard–Alpert ionization gauge detector is used to measure vacuum pressures in the range from 3×10^{-9} pascal to 0.1 pascal (2×10^{-11} torr to 10^{-3} torr). It operates by passing a current through the filament (the glowing U-shaped wire in Fig. 3.15), which heats it and boils off electrons, which are accelerated toward the positively biased grid (helical wire). Collisions of these 150 eV electrons with gas atoms in the detector ionize the atoms, which are collected by the thin wire in the center, called the collector. The current from the collector is measured by an electrometer (a very sensitive current meter), and from this current the pressure of the gas in the gauge is determined. The collector current is typically in the range from a few pA to a few hundred μA. The sensitivity of the Bayard–Alpert gauge detector varies with different types of background gas.

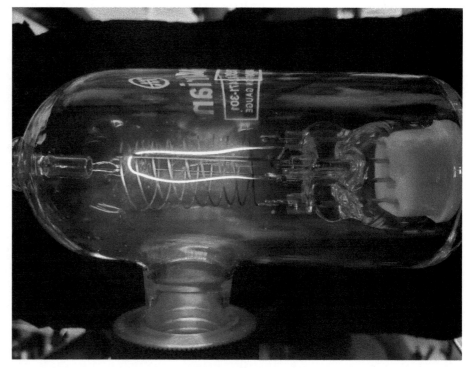

FIGURE 3.15 A Bayard–Alpert ionization gauge detector.

In this section we have described qualitatively several devices that exhibit non-resistive current flow. We leave the details of their operation to other, more advanced texts, but hope that you can appreciate their general features using the ideas we discussed earlier.

3.6 Power Dissipation in Resistors

In this section we will examine power dissipation in a current-conducting medium, which we will formulate in terms of the work performed by the electric field on the individual charge carriers within the medium. This relation is derived using the example of power dissipation in resistors, with which we are already familiar from our circuits classes, but this analysis applies more generally to the power absorbed by any system involving electric fields and charges in motion. In fact, it applies to non-resistive currents, as described in the previous section, as well as to resistive currents. These ideas will also be important later when we discuss attenuation of freely propagating waves and waves in transmission lines.

The model for power dissipation is based on the work done on individual charged particles within a conducting medium when an electric field is applied. Our primary focus in this discussion is the rate at which the field does work on the free charges within the medium. (If the charged particle is constantly colliding with atoms within the conductor, this additional kinetic energy is converted into heat, but for this part of our development we do not need to worry about the final form of the energy. Rather, we focus on the work done by the field on the charged particle.)

Consider a volume v containing a number of free charged particles, to which an electric field $\mathbf{E}(\mathbf{r})$ has been applied. The electric field $\mathbf{E}(\mathbf{r})$ applies a force to each individual charge (which we label with an index i). If the charged particle i moves a distance $\Delta\mathbf{r}_i$, the work done on that charge is $\Delta W_i = q\mathbf{E} \cdot \Delta\mathbf{r}_i$. The work ΔW_i is positive if the particle moves in the same direction as $q\mathbf{E}$, the applied electric force, consistent with increasing the kinetic energy of the particle.

We consider within this volume v one small elemental volume dv that contains a number ndv of charged particles of charge q, where n is the number density of the charged particles. The *power* absorbed from the field by charge i (i.e. the *rate* at which the field does work on the charged particle) is

$$P_i^{\mathrm{abs}} = \lim_{\Delta t \to 0} \frac{\Delta W_i}{\Delta t} = \lim_{\Delta t \to 0} \frac{q\mathbf{E} \cdot \Delta\mathbf{r}_i}{\Delta t} = q\mathbf{E} \cdot \mathbf{u}_i,$$

where $\mathbf{u}_i$ is the velocity of charged particle i. We sum this power over all the charges within the elemental volume to find the absorbed power,

$$dP^{\mathrm{abs}} = \sum_i q\,\mathbf{E} \cdot \mathbf{u}_i,$$

or, since there are ndv particles in the volume,

$$dP^{\mathrm{abs}} = q\,\mathbf{E} \cdot \langle\mathbf{u}\rangle\, ndv,$$

where $\langle\mathbf{u}\rangle$ is the *average* velocity of the charged particles. But the product qn is equivalent to the charge density ρ_v, so this power can be written as

$$dP^{\mathrm{abs}} = \mathbf{E} \cdot \langle\mathbf{u}\rangle\, \rho_v\, dv.$$

If there are multiple types of charge carriers within the volume, we sum over each type,

$$dP^{\mathrm{abs}} = \mathbf{E} \cdot \left(\sum_j \langle\mathbf{u}\rangle_j\, \rho_{v,j} \right) dv,$$

where the index j labels the different kinds of carriers. Whether we have a single type or multiple types of carriers, however, the term inside the parentheses is, by Eq. (3.11), the current density $\mathbf{J}_v$. The power absorbed within the element dv can then be written $dP^{\mathrm{abs}} = \mathbf{E} \cdot \mathbf{J}_v \, dv$, or integrating across the entire volume v,

$$P^{\mathrm{abs}} = \int_v \mathbf{E} \cdot \mathbf{J}_v \, dv. \tag{3.17}$$

The integrand $\mathbf{E} \cdot \mathbf{J}_v$ is known as the **power dissipation density**, and it signifies the rate at which energy is dissipated in the medium per unit volume. This quantity, in a more general form, will be important later as we examine attenuation of electromagnetic waves propagating through absorbing media, and we introduce the notation

$$\boxed{P_v^{\mathrm{abs}} = \mathbf{E} \cdot \mathbf{J}_v.} \tag{3.18}$$

Keeping with the theme of this chapter, we would like to confirm that this power dissipation is consistent with what we have learned previously in our circuits classes – that is, that the power absorbed by any circuit element is the product of the voltage across the element and the current flowing into the positive lead of the element. Let us start with the "fields" description of power dissipation, Eq. (3.17), and see if we can show this equivalence (at least approximately). Consider a conducting medium in which the electric and current density fields $\mathbf{E}$ and $\mathbf{J}_v$ are uniform. Let us separate the volume element dv in Eq. (3.17) into the product of $|d\boldsymbol{\ell}|$ (the elemental length of the element dv in the direction parallel to $\mathbf{E}$ or $\mathbf{J}_v$) and the surface element $d\mathbf{s}$ (the cross-sectional surface element perpendicular to $\mathbf{E}$ or $\mathbf{J}_v$). (Recall that the direction of a surface element $d\mathbf{s}$ is perpendicular to the surface.) Then we can rewrite $\mathbf{E} \cdot \mathbf{J}_v \, dv$ as $(\mathbf{E} \cdot d\boldsymbol{\ell})(\mathbf{J}_v \cdot d\mathbf{s})$, and the volume integral for the absorbed power separates into

$$P^{\mathrm{abs}} = \int_v \mathbf{E} \cdot \mathbf{J}_v \, dv = \int_L \mathbf{E} \cdot d\boldsymbol{\ell} \int_s \mathbf{J}_v \cdot d\mathbf{s}. \tag{3.19}$$

(We took some liberties with the dot products here, but all the vectors are pointing in the same direction in this simplified picture, so we should be okay with this.) We recognize the first integral as the voltage V across the element and the second as the current I through the element. Thus we have our familiar result that the power absorbed is $P^{\mathrm{abs}} = I \, V$, as we set out to show.

Example 3.4 Power Dissipation Density in a Cylindrical Resistor

Consider the cylindrical resistor described in Example 3.3. (a) Determine the power dissipation density P_v^{abs} in this medium. (b) Determine the total power dissipated in this resistor. (c) Show that this dissipated power is equivalent to the product IV_0.

Solution:
(a) In the solution to Example 3.3, we found the electric field in the conducting medium is

$$\mathbf{E} = \frac{\rho_{s,a} a}{\varepsilon \rho} \mathbf{a}_\rho,$$

so we'll use this for our solution here. Since the medium is resistive, we can also use Ohm's Law to relate $\mathbf{J}_v$ and $\mathbf{E}$:

$$\mathbf{J}_v = \sigma \mathbf{E}.$$

Then,

$$P_v^{\text{abs}} = \mathbf{E} \cdot \mathbf{J}_v = \sigma \left(\frac{\rho_{s,a} a}{\varepsilon \rho} \right)^2.$$

(b) To find the total power dissipated in this medium, we must simply integrate P_v^{abs} over the entire volume of the cylindrical resistor:

$$P^{\text{abs}} = \int_v P_v^{\text{abs}} \, dv.$$

The resistor is cylindrical in shape, so we'll use the volume element $dv = \rho d\rho \, d\phi \, dz$, where the limits of integration for ρ are from a to b (the inner and outer radii of the cylinder), for ϕ are from 0 to 2π, and for z are from 0 to L, where L is the length of the cylinder:

$$P^{\text{abs}} = \int_0^L \int_0^{2\pi} \int_a^b \sigma \left(\frac{\rho_{s,a} a}{\varepsilon \rho} \right)^2 \rho d\rho \, d\phi \, dz$$

$$= \sigma \left(\frac{\rho_{s,a} a}{\varepsilon} \right)^2 \int_a^b \frac{d\rho}{\rho} \int_0^{2\pi} d\phi \int_0^L dz$$

$$= \sigma \left(\frac{\rho_{s,a} a}{\varepsilon} \right)^2 2\pi L \ln\left(\frac{b}{a} \right).$$

(c) This result for P^{abs} is indeed identical to the product IV_0, as can be seen directly by inspection of I and V_0 derived in Example 3.3.

We will use Eq. (3.17) for power dissipation later, especially for electromagnetic waves. For example, we will develop Poynting's Theorem, which relates the power carried by electromagnetic waves, the rate at which stored energy increases, and the rate at which energy is dissipated. Also, we will consider absorption and losses when we discuss plane wave propagation and electromagnetic waves that are guided by transmission lines. For each of these, we will refer back to this expression for the rate at which electromagnetic energy is absorbed by a conducting medium.

3.7 Charge Relaxation

Throughout the text to this point, we have made several statements concerning the equilibrium location of charges and equilibrium values of fields in various media. For example, we know that free charges on a conducting body must reside on the surface of the conductor, since excess free charges in the interior would produce electric fields, which would exert forces on charges and move them around. It is therefore natural to ask, if a system is out of its equilibrium condition, what is the time scale for it to relax to its equilibrium condition. We are now in a position to shed some light on this question. We will consider a "simple" resistive medium (i.e. linear, isotropic, and homogeneous), described by a permittivity ε and conductivity σ. Thus we know that the field quantities $\mathbf{J}_v$, $\mathbf{E}$, and $\mathbf{D}$ are related by $\mathbf{J}_v = \sigma \mathbf{E}$ and $\mathbf{E} = \mathbf{D}/\varepsilon$. Under these conditions, the Continuity Equation (Eq. (3.7)), $d\rho_v/dt = -\nabla \cdot \mathbf{J}_v$, becomes

$$\frac{d\rho_v}{dt} = -\sigma \nabla \cdot \mathbf{E} = -\frac{\sigma}{\varepsilon} \nabla \cdot \mathbf{D}.$$

But the divergence of **D**, recalling Gauss' Law, is just ρ_v, giving us

$$\frac{d\rho_v}{dt} = -\frac{\sigma}{\varepsilon}\rho_v.$$

This equation is a simple first-order linear equation in one variable (ρ_v), whose solution we have encountered many times

$$\rho_v(t) = \rho_0 e^{-t/\tau}. \tag{3.20}$$

In this expression, $\tau = \varepsilon/\sigma$ is the time constant for the exponential function. This equation implies that the charge density relaxes within a time constant τ. While this model ignores propagation delays, which we will discuss later, it does give us one approximate time scale for charge relaxation. As an example of this time scale, τ for copper is $(8.854 \times 10^{-11}\ \text{F/m})/(5.97 \times 10^7\ \text{S/m}) \sim 10^{-18}$ s.

3.8 Boundary Conditions Revisited

We previously discussed the boundary conditions that the electric field **E** and the displacement field **D** must satisfy at the interface between any two media. Specifically, the normal component of **D** must be continuous across the interface, unless there is a surface charge at the boundary, and then the difference in D_n in the two media must be equal to the surface charge density ρ_s:

$$D_{2n} - D_{1n} = \rho_s. \tag{3.21}$$

Also, the tangential component of **E** must be continuous across the boundary,

$$E_{2t} = E_{1t}. \tag{3.22}$$

In our previous treatment we considered the dielectric medium to be perfect, in that it would conduct no current, and the surface charge density at the interface between dielectrics would be zero. In real dielectrics, however, currents might not be perfectly zero. We will now discuss an additional condition that must be met at the interface between conducting dielectrics in steady state.

The fields at the interface between two media are illustrated in Fig. 3.16. Previously, we applied these conditions for two very special cases: at the interface between two perfect dielectric media (i.e. the media are non-conducting, $\sigma_1 = \sigma_2 = 0$, and therefore there can be no charges at the surface, $\rho_s = 0$); and when one of the media is a perfect conductor, $\sigma_1 \to \infty$, inside which no fields can exist, $\mathbf{E}_1 = 0$. We now consider the conditions that the fields must satisfy when the two media conduct current, but with finite conductivity. We call these media "good conductors" or "lossy dielectrics," depending on the relative magnitude of the conductivity. (The distinction between lossy dielectric and good conductor is not very important for d.c. fields. Later, when we introduce oscillating fields, we'll define these two different classes of materials more precisely.) It should be expected that, as current flows through these media, surface charges may accumulate at the interface between the two media, and these charges in turn affect the electric fields in the media. We will limit our analysis to steady state conditions, in which charge densities have had sufficient time to accumulate, and have built up to such a state as they are no longer changing. Under these conditions, the current flowing into the interface from above must be equal to the current flowing away from the interface below, or

$$J_{v,1n} = J_{v,2n}. \tag{3.23}$$

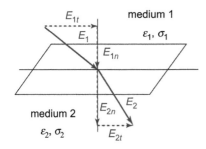

FIGURE 3.16 An interface between two media of permittivity ε_1 and ε_2 and conductivity σ_1 and σ_2, respectively.

(If this condition were not satisfied, then the charge at the surface must be changing, contrary to our steady-state assumption.) Using Ohm's Law, this becomes $\sigma_1 E_{1n} = \sigma_2 E_{2n}$. We now have three sets of conditions, Eqs. (3.21), (3.22), and (3.23), which must concurrently be satisfied. Thus the field amplitudes under these steady-state conditions can be uniquely determined, as will be shown in Example 3.5.

Example 3.5 "Leaky" Parallel Plate Capacitor

Consider the configuration shown in Fig. 3.17 consisting of two parallel conducting plates and two partially conducting media in the space between. A potential V_0 is applied between the conducting plates. The upper dielectric medium between the plates has permittivity ε_1, conductivity σ_1, and thickness d_1. For the lower dielectric medium, these same parameters are ε_2, σ_2, and d_2. Determine the steady-state values of $\mathbf{E}_1$, $\mathbf{E}_2$, $\mathbf{D}_1$, and $\mathbf{D}_2$. Ignore fringe effects at the edge of the plates.

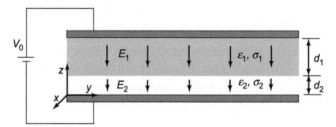

FIGURE 3.17 A "leaky" parallel plate capacitor configuration with two media of permittivity ε_1 and ε_2 and conductivity σ_1 and σ_2. Under steady-state conditions there will be surface charges at $z = 0$, $z = d_2$, and $z = d_1 + d_2$, which we label density ρ_{sb}, ρ_{si}, and ρ_{st}, respectively.

Solution:

By symmetry, the electric field points in the $-z$-direction, and is independent of x or y. Therefore, all fields are normal to the interface between the two dielectric media. The fields in the two media can be written as $\mathbf{E}_1 = -\mathbf{a}_z E_1$ in medium 1, and $\mathbf{E}_2 = -\mathbf{a}_z E_2$ in medium 2. By Eq. (3.23) the magnitudes of these fields are related as

$$\sigma_1 E_1 = \sigma_2 E_2.$$

A surface charge of magnitude

$$\rho_{si} = D_2 - D_1 = \varepsilon_2 E_2 - \varepsilon_1 E_1$$

will accumulate at the interface. Finally, the field amplitudes E_1 and E_2 are related to the voltage V_0 through

$$V_0 = -\int_0^d \mathbf{E} \cdot d\boldsymbol{\ell} = E_1 d_1 + E_2 d_2,$$

where we have used the uniform field condition to evaluate the integrals. After some algebra, these equations allow us to solve for E_1 and E_2, for which we find

$$E_1 = \frac{\sigma_2 V_0}{\sigma_2 d_1 + \sigma_1 d_2} \qquad \text{and} \qquad E_2 = \frac{\sigma_1 V_0}{\sigma_2 d_1 + \sigma_1 d_2}.$$

Since $\mathbf{J}_v = \sigma \mathbf{E}$, the current density in the upper medium is

$$J_{v,1} = \sigma_1 E_1 = \frac{\sigma_1 \sigma_2 V_0}{\sigma_2 d_1 + \sigma_1 d_2}.$$

$J_{v,2} = \sigma_2 E_2$ in the lower medium gives the same result, as expected.

The displacement field is found using $\mathbf{D} = \varepsilon\mathbf{E}$, with the results

$$\mathbf{D}_1 = -\mathbf{a}_z\,\frac{\varepsilon_1\sigma_2\,V_0}{\sigma_2 d_1 + \sigma_1 d_2} \qquad \text{and} \qquad \mathbf{D}_2 = -\mathbf{a}_z\,\frac{\varepsilon_2\sigma_1\,V_0}{\sigma_2 d_1 + \sigma_1 d_2}.$$

Finally, we can determine the surface charge densities on the inside surfaces of the two conductor plates:

$$\rho_{st} = -D_{1,z} = \frac{\varepsilon_1\sigma_2 V_0}{\sigma_2 d_1 + \sigma_1 d_2}$$

at the top plate, and

$$\rho_{sb} = D_{2,z} = \frac{-\varepsilon_2\sigma_1 V_0}{\sigma_2 d_1 + \sigma_1 d_2}$$

at the bottom plate, and the surface charge density at the interface between the two conducting dielectrics:

$$\rho_{si} = D_{1,z} - D_{2,z} = \frac{(\varepsilon_2\sigma_1 - \varepsilon_1\sigma_2)\,V_0}{\sigma_2 d_1 + \sigma_1 d_2}.$$

Let's insert some reasonable parameters to see if we can make sense of these field amplitudes and surface charges. Using parameters of $\varepsilon_{1r} = 1.5$, $\varepsilon_{2r} = 2.0$, $\sigma_1 = 4$ S/m, $\sigma_2 = 10$ S/m, $d_1 = d_2 = 0.5$ cm, and $V_0 = 10$ V, we plot $E_z(z)$ and $D_z(z)$ in Fig. 3.18. The surface charge densities are $\rho_{st} = 19.0$ nC/m^2, $\rho_{si} = -8.9$ nC/m^2, $\rho_{sb} = -10.1$ nC/m^2. We interpret these as follows. The conductivity is greater in the lower medium than in the upper (by a factor of 2.5), so in order to match the current densities in these two media the electric field E_2 is smaller than E_1 by this same factor. We see this clearly in Fig. 3.18(a). The displacement fields D_1 and D_2 follow immediately from $\mathbf{D} = \varepsilon\mathbf{E}$, and from this we see that D_1 and D_2 are quite different from one another, with D_2 only about half as large as D_1. This can only be true if there exists a surface charge at the interface between the dielectrics. This surface charge must be of magnitude $D_2 - D_1$, and must be negative to reduce the magnitude of D_2 relative to that of D_1. Effectively this surface charge partially shields the lower dielectric medium from the large positive surface charge ρ_{st}, thereby reducing the magnitude of the field in the lower region.

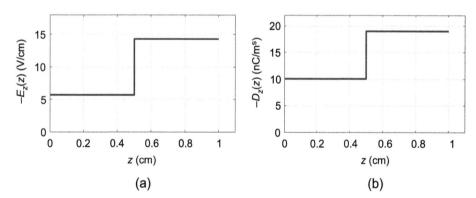

FIGURE 3.18 Plots of (a) $-E_z(z)$, and (b) $-D_z(z)$ for the "leaky" parallel plate capacitor configuration shown in Fig. 3.17. Parameters are given in the text.

In Example 3.5 we examined conduction through two media in series, and found that when these media have different conductivities a surface charge will accumulate at the interface between the two media. Now let's consider conduction through a single medium whose conductivity varies with position. As will be shown, volume charge will accumulate within the conducting medium in steady state. Once again, the effect of this charge is to balance the fields, making the current uniform.

Example 3.6 Non-uniform Conductivity

Consider a spherical, lossy, conducting medium positioned between two spherical good conductors of radius a and b, as shown in Fig. 3.19. Let the conductivity decrease with increasing R as $\sigma(R) = \sigma_0/R$. Find the resistance R_0 of this system in steady state.

Solution:

We will start by applying a potential difference V_0 between the two conductors, and asserting that, in steady state, the current through any spherical surface of radius R, where $a < R < b$, must be independent of R. Otherwise, there must be charge accumulating within the lossy medium, implying that we are not in steady state. Because of the symmetry of the configuration, we expect that the electric field **E** in the region between the conducting spheres must be directed in the radial direction, and must depend only upon R; that is, $\mathbf{E} = \mathbf{a}_R E_R(R)$. The requirement of constant current I between the good conductors allows us to determine the R dependence of the electric field E_R. I is found using the surface integral $I = \int_s \mathbf{J}_v \cdot d\mathbf{s}$, evaluated on the spherical surface s represented by the dashed line of radius R in Fig. 3.19. Using

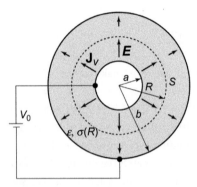

FIGURE 3.19 A lossy conducting medium positioned between two good spherical conductors. The conductivity of the medium decreases with increasing R as $\sigma(R) = \sigma_0/R$.

$$\mathbf{J}_v = \mathbf{a}_R \sigma(R) E_R(R) = \mathbf{a}_R \frac{\sigma_0}{R} E_R(R)$$

and

$$d\mathbf{s} = \mathbf{a}_R R^2 \sin\theta \, d\theta \, d\phi,$$

the current is

$$I = 4\pi\sigma_0 E_R(R) R.$$

Solving this for $E_R(R)$ yields

$$E_R(R) = \frac{I}{4\pi\sigma_0 R}.$$

Notice the R dependence of this electric field. (In a charge-free region, as would be the case with a perfect dielectric, the field should decrease as R^{-2}, rather than R^{-1}. We will return to this point in a few moments.) The electric field $E_R(R)$ can be related to V_0 using

$$V_0 = -\int_b^a \mathbf{E}(R) \cdot d\boldsymbol{\ell},$$

where we choose a radial path from the outer conductor to the inner, $d\boldsymbol{\ell} = \mathbf{a}_R \, dR$. Carrying out the integration, we find

$$V_0 = \frac{I \ln(b/a)}{4\pi\sigma_0}.$$

The steady-state resistance R_0 is therefore

$$R_0 = \frac{V_0}{I} = \frac{\ln(b/a)}{4\pi\sigma_0}.$$

We surmised at the beginning of our solution that charge will build up throughout the conducting medium in such a way as to balance the current. It is instructive, therefore, to determine this charge density, $\rho_v(R)$, using Gauss' Law, $\rho_v = \nabla \cdot \mathbf{D}$. Taking the divergence of the displacement field $\mathbf{D} = \varepsilon \mathbf{E} = \mathbf{a}_R\, \varepsilon I/(4\pi\sigma_0 R)$ (using Eq. (D.9) for the divergence in spherical coordinates), the charge density is

$$\rho_v(R) = \frac{\varepsilon I}{4\pi\sigma_0 R^2}.$$

The important point here is that charge distributed throughout the conducting region affects the displacement field $\mathbf{D}$, which in turn assures a constant value of I, independent of R.

SUMMARY

In this chapter we have discussed current and conduction. We have shown how laws that we developed earlier in this text are consistent with familiar laws, such as Kirchhoff's Voltage Law and Kirchoff's Current Law. We have also explored conduction of current by a medium in more detail, developed Ohm's Law (a new general form involving field quantities), learned how to calculate the resistance of a conducting medium, and developed the groundwork for power dissipation in a conducting medium. We will return to many of these ideas later when we consider wave propagation of various forms. Before that, however, we will introduce magnetic fields, which are produced by currents, and that apply forces on moving charges and current-carrying conductors.

Table 3.2

Collection of key formulas from Chapter 3.			
$I = \int_s \mathbf{J}_v(\mathbf{r}) \cdot d\mathbf{s}$	(3.3)	$\mathbf{J}_v = \sigma \mathbf{E}$	(3.13)
$\frac{d\rho_v}{dt} + \nabla \cdot \mathbf{J}_v = 0$	(3.7)	$R = \frac{\Delta V}{I} = \frac{-\int_2^1 \mathbf{E} \cdot d\boldsymbol{\ell}}{\sigma \int_s \mathbf{E} \cdot d\mathbf{s}}$	(3.15)
$\mathbf{J}_v = \rho_v \langle \mathbf{u} \rangle$	(3.10)	$P_v^{abs} = \mathbf{E} \cdot \mathbf{J}_v$	(3.18)

PROBLEMS

Current and Current Density

P3-1 A current of 5 A is conducted along a copper wire of diameter 2 mm and length 10 m. Assuming that the current density J_v in the wire is uniform, determine J_v.

P3-2 A stream of protons flows in the +z-direction. The particle flux density is 10^6 s^{-1} mm^{-2}, and the profile of the particle beam is Gaussian, $\exp(-\rho^2/w^2)$, where $w = 1$ mm is the beam radius. (a) Determine the current density J_v at the center of the beam. (b) Determine the total current carried by the beam. (c) Assuming a particle density of 10^9 cm^{-3}, determine the average velocity of the protons.

P3-3 A surface current lies in the x–z plane and travels in the −x-direction. The current distribution is uniform, and has a width of 5 mm. For a total current of 5 mA, determine the surface current density J_s of the current.

Charge Conservation: The Continuity Equation

P3-4 The current density in a region is given by $\mathbf{J}_v(z) = \mathbf{a}_z 5z$ A/m^3. (a) Determine the rate at which the charge density in this space changes. (b) Determine the rate of change of the total charge stored within a cube of side 1 m centered at the origin. (c) Explain the sign of dQ/dt of part (b).

P3-5 The time-varying charge density in a region is $\rho_v = \rho_0 \cos(\omega t - kz)$. Assuming that the current density $\mathbf{J}_v(z, t)$ depends only on position z and time t, determine $\mathbf{J}_v(z, t)$.

P3-6 The current density in a region is $\mathbf{J}_v(R, t) = \mathbf{a}_R J_0 \sin(\omega t - kR)/R^2$. Determine the charge density $\rho_v(R, t)$.

Resistance

P3-7 A conducting sphere has radius a and potential V_0. A larger, concentric sphere of inner radius c is grounded. The space between the conductors is filled with a conducting, charge-free medium, of permittivity ε_1 and conductivity σ_1. Determine the resistance R of the medium between the conductors.

P3-8 A pair of concentric spherical conductors is spaced by two conducting media. The inner sphere is of radius a and potential V_0, and the larger sphere is grounded and of inner radius c. The two conducting media are side by side, with one medium of conductivity σ_1 and permittivity ε_1 filling the space defined by $0 < \phi < 2\pi/3$, and the second medium (conductivity σ_2 and permittivity ε_2) filling the space defined by $2\pi/3 < \phi < 2\pi$. Determine the resistance of this structure. [Hint: E and D are still radial for this geometry, and the potential difference V_0 must, of course, be the same across one dielectric as across the second.]

P3-9 The space between two co-axial cylindrical conductors of radius a and b, where $b > a$, and length L, contains two conducting media. These media are side by side, with one medium of conductivity σ_1 and permittivity ε_1 filling the space defined by $0 < \phi < 2\pi/3$, and the second medium (conductivity σ_2 and permittivity ε_2) filling the space defined by $2\pi/3 < \phi < 2\pi$. Determine the resistance of this structure. [Hint: E and D are still radial for this geometry, and the potential difference V_0 must, of course, be the same across one dielectric as across the second.]

P3-10 Consider an inhomogeneous conducting medium between two parallel plates, as shown in Fig. 3.20. The conductivity of the medium is $\sigma(y) = \sigma_1 + \sigma_2 y$, and the area of the plates is A. The spacing between the plates d is small compared to the width of the plates, and the permittivity is uniform, $\varepsilon = 2\varepsilon_0$. The potential of the top plate is V_0 relative to the bottom plate, and the current flowing from the top to the bottom is in steady state. (a) Determine the current density $\mathbf{J}_v$ in the conducting medium in terms of the total current I. (b) Determine the electric field $E(y)$ inside the conducting medium. (c) Determine the potential difference V_0. (d) Determine the resistance of the conducting medium.

FIGURE 3.20 A conducting medium of variable conductivity between parallel plates, for use in Probs. P3-10 and P3-14.

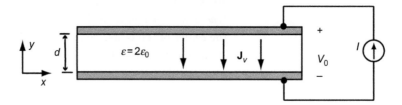

Power Dissipation in Resistors

P3-11 Consider the spherical medium described in Prob. P3-7. Determine the power dissipation density $\mathbf{J}_v \cdot \mathbf{E}$ absorbed in the conducting medium, and the total power absorbed. Compare your result for the total power absorbed to the product of $I V_0$, as expected from circuit theory.

P3-12 Consider the pair of conducting media described in Prob. P3-8. Determine the power dissipation density $\mathbf{J}_v \cdot \mathbf{E}$ absorbed in the two conducting media, and the total power absorbed. Compare your result for the total power absorbed to the product of $I V_0$, as expected from circuit theory.

P3-13 Consider the pair of conducting media described in Prob. P3-9. Determine the power dissipation density $\mathbf{J}_v \cdot \mathbf{E}$ absorbed in the two conducting media, and the total power absorbed. Compare your result for the total power absorbed to the product of $I V_0$, as expected from circuit theory.

P3-14 Consider the conductor with variable conductivity positioned between two parallel plates described in Prob. P3-10. Determine the power dissipation density $\mathbf{J}_v \cdot \mathbf{E}$ absorbed in the two conducting media, and the total power absorbed. Compare your result for the total power absorbed to the product of $I V_0$, as expected from circuit theory.

Boundary Conditions Revisited

P3-15 A spherical conductor of radius a and potential V_0 is concentric with a larger grounded spherical conductor of inner radius c. The space between the conductors is filled with two conducting media, of conductivity σ_1 and permittivity ε_1 ($a < R < b$) and the second of conductivity σ_2 and permittivity ε_2 ($b < R < c$), and the current is in steady state. (a) Determine D_1, E_1, $J_{v,1}$ in medium 1 and D_2, E_2, $J_{v,2}$ in medium 2 in terms of Q, the total charge on the inner conductor. (b) Determine the resistance of this structure. (c) Show that the surface charge density at $R = b$ equals $D_2 - D_1$, as required.

P3-16 A cylindrical conductor of length L and radius a has a potential V_0. A larger co-axial grounded cylinder of the same length has an inner radius c. The space between the conductors is filled with two conducting media, of conductivity σ_1 and permittivity ε_1 ($a < \rho < b$) and the second of conductivity σ_2 and permittivity ε_2 ($b < \rho < c$), and the current is in steady state. (a) Determine D_1, E_1, $J_{v,1}$ in medium 1 and D_2, E_2, $J_{v,2}$ in medium 2 in terms of ρ_s, the surface charge density on the inner conductor. (b) Determine the resistance of this structure. (c) Show that the surface charge density at $\rho = b$ equals $D_2 - D_1$, as required.

P3-17 In Example 3.5 we found $\mathbf{E}$, $\mathbf{D}$, and $\mathbf{J}_v$ in the two conducting media between two parallel plates. Using the conductivities, permittivities, and dimensions given at the end of that example (and a plate area of 0.3 cm^2), determine the resistance of this parallel plate configuration under steady-state conditions.

P3-18 In Example 3.6 we found the resistance of a spherical medium with a non-uniform conductivity. (a) For this resistor, find the power dissipation density. (b) Integrate this power dissipation density over the entire volume to find the total power dissipation. (c) Equate this result to $I^2 R$ to determine the resistance of this medium. Compare this result to the resistance found in the example.

CHAPTER 4
Magnetostatics I

KEY WORDS

magnetic flux density; Biot–Savart Law; Hall effect; solenoid; Ampere's Law; toroid; magnetic potential; Coulomb gauge; magnetic dipole; magnetization; effective currents; magnetic field; magnetic susceptibility; permeability; relative permeability; diamagnetic medium; paramagnetic medium; ferromagnetic medium; hysteretic behavior.

We are now ready to introduce magnetic fields, which are generated by electrical currents and which apply forces on moving charges and current-carrying wires. Historically, magnetic effects in lodestones, an iron ore that can be magnetized, have been known for a long time. The first magnetic compasses date back to about 1000 BCE, and the ancient Chinese are believed to have used such devices for navigation as early as 1100 CE. The properties of magnetic fields can be derived from a number of observations of magnetic effects that have been recorded over many years. One of the earliest such observations, by Hans Christian Oersted in 1820, was that a current-carrying wire exerts a torque on a permanent magnet (such as a compass). Current-carrying wires can also exert forces on each other, as first observed by Biot and Savart and more fully characterized by Ampère. Finally, beams of charged particles, such as electrons in a cathode ray tube (see TechNote 3.4), are deflected when in the presence of current-carrying wires. Each of these phenomena can be described quantitatively in terms of a magnetic field produced by current distributions, as we will discuss throughout this chapter.

4.1 Biot–Savart Law

The first observation that showed that electrical currents produce magnetic forces was reported by Hans Christian Oersted in 1820. His contributions are described in Biographical Note 4.1.

LEARNING OBJECTIVES

After studying this section, you should be able to:
- determine the force between two current-carrying wires;
- determine the magnetic flux density generated by a current carried by a thin wire, a surface current, or a volume current distribution, using the Biot–Savart Law; and
- determine the magnetic force on a charged body or a current-carrying wire.

Biographical Note 4.1 Hans Christian Oersted (1777–1851)

(a)

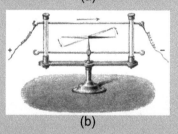

(b)

FIGURE 4.1 (a) Hans Christian Oersted, painting by Christoffer Wilhem Eckersberg (1783–1853), from the collection of Danmarks Tekniske Museum. (b) A diagram by Oersted of his apparatus showing the connection between the compass needle and the flow of electricity. When the current flowed, the compass needle deflected perpendicularly to the wire.

Hans Christian Oersted (in Danish, Ørsted), pictured in Fig. 4.1(a), was a Danish physicist who first discovered the link between electricity and magnetism. Oersted was educated in pharmacology, and became professor of physics at the University of Copenhagen in 1806.

Oersted's primary contribution to our understanding of magnetism is his observation in 1820 that a compass needle is deflected by a nearby electrical current. (The story that this discovery occurred during a lecture to the class he was teaching does not appear to be correct, as his laboratory notes indicate that he had been working on these investigations for some time.) When a current flowed through the long, straight wire, illustrated in Fig. 4.1(b), the needle of the compass lined up in a direction perpendicular to the wire. As we will discuss later, this is the direction of the magnetic flux density that circles around the wire. In September 1820, just five months after Oersted's discovery, François Arago repeated Oersted's observation in front of the French Academy. This demonstration influenced André-Marie Ampère to explore magnetic effects, carrying out a quantitative analysis of the force between two current-carrying wires. (See Biographical Note 4.4.) News of Oersted's discovery also motivated Jean Baptiste Biot and Félix Savart to investigate magnetic effects, leading to the formulation of the Biot–Savart Law. (See Biographical Notes 4.2 and 4.3.)

In unrelated studies, Oersted discovered the chemical piperine, and the first process for isolating aluminum. In 1801, shortly after the introduction of Volta's first battery, Oersted demonstrated how to determine the current flow by measuring the production rate of hydrogen gas and oxygen gas formed through electrolysis of water. Oersted was awarded the 1820 Copley Award by the British Royal Society. Oersted's Law and the unit oersted (Oe, the unit for magnetic field **H**, to be discussed later, in the cgs system of units) are named after him.

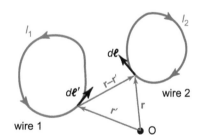

FIGURE 4.2 Two wire loops carrying currents I_1 and I_2, respectively. The position vector $\mathbf{r} - \mathbf{r}'$ points from the element $d\boldsymbol{\ell}'$ in the loop on the left to the element $d\boldsymbol{\ell}$ in the loop on the right. The force $d\mathbf{F}_2$ felt by the element $d\boldsymbol{\ell}$ due to the current I_1 in the element $d\boldsymbol{\ell}'$ is given by Eq. (4.1).

We begin our study of magnetic effects by considering two current-carrying wires, as shown in Fig. 4.2. A current I_1 flows in the wire loop on the left, while the current in the right loop is I_2. Using vector notation, the force $d\mathbf{F}_2$ felt by the element $d\boldsymbol{\ell}$ in the loop on the right due to the current I_1 in the element $d\boldsymbol{\ell}'$ of the current loop on the left can be summarized as

$$d\mathbf{F}_2 = \frac{\mu_0}{4\pi} I_1 I_2 \frac{d\boldsymbol{\ell} \times \left[d\boldsymbol{\ell}' \times (\mathbf{r} - \mathbf{r}') \right]}{|\mathbf{r} - \mathbf{r}'|^3}. \tag{4.1}$$

The position vector $\mathbf{r} - \mathbf{r}'$ follows the same convention we introduced in our study of electrostatics, and points from the element $d\boldsymbol{\ell}'$ in the loop on the left to the element $d\boldsymbol{\ell}$ in the loop on the right. We choose the direction of $d\boldsymbol{\ell}'$ and $d\boldsymbol{\ell}$ to point in the direction of the currents I_1 and I_2, respectively. μ_0 in Eq. (4.1) is a constant, known as the permeability of free space, that is necessary to properly scale the force, and is numerically equal to $\mu_0 = 4\pi \times 10^{-7}$ H/m. A current cannot exist, of course, in just one isolated element $d\boldsymbol{\ell}'$ or $d\boldsymbol{\ell}$ (at least, not without a large build-up in charge, which we will not address at this time), so we must add the contributions of all the elements $d\boldsymbol{\ell}'$ in loop 1 acting on each of the elements $d\boldsymbol{\ell}$ in loop 2. Thus we integrate over $d\boldsymbol{\ell}$ around the loop on the right to find the total force applied to loop 2 by the current I_1 in the element $d\boldsymbol{\ell}'$, and integrate over $d\boldsymbol{\ell}'$ of the left loop to find the total force $\mathbf{F}_2$ felt by loop 2 due to all the segments of loop 1.

While Eq. (4.1) might appear to be overly complicated, we can break it down into parts to understand the features of magnetic forces:

1. The magnitude of the force $d\mathbf{F}_2$ is proportional to the product of I_1 and I_2. This is reminiscent of the electrostatic force between two point charges q_1 and q_2, which is proportional to the product of charges q_1 and q_2.
2. The magnitude of the force $d\mathbf{F}_2$ decreases as the inverse of the square of the distance between the wire segments. This feature is in common with the electrostatic force between two point charges.
3. The direction of the force $d\mathbf{F}_2$ lies in the plane containing $d\boldsymbol{\ell}'$ and the position vector $\mathbf{r} - \mathbf{r}'$, and is perpendicular to the segment $d\boldsymbol{\ell}$. This direction is quite different from the electrostatic force between two point charges, which is always in the direction $\mathbf{r} - \mathbf{r}'$ or $-(\mathbf{r} - \mathbf{r}')$.
4. The force between the wires is strongest when the wires are parallel to one another. The force is repulsive when the currents flow in opposite directions, and attractive when the currents flow in the same direction. The force $d\mathbf{F}_2$ vanishes when the segment $d\boldsymbol{\ell}'$ and the position vector $\mathbf{r} - \mathbf{r}'$ are parallel to one another, or when the segments $d\boldsymbol{\ell}'$ and $d\boldsymbol{\ell}$ are perpendicular to one another.

We illustrate this force with Example 4.1, on the force between two long, straight, carrying-current wires.

Example 4.1 Force between Two Long Straight Wires

Consider the two long, straight wires shown in Fig. 4.3. A current I_1 flows to the right in wire 1, and a current I_2 to the left in wire 2. Determine the force $\mathbf{F}_2$ per unit length felt by the lower wire.

Solution:

We label the x-, y-, and z-axes as shown in Fig. 4.3, with the upper wire (wire 1) at $x = y = 0$ and the lower wire (wire 2) at $x = -d$ and $y = 0$. Then the position vectors are $\mathbf{r}' = (0, 0, z')$, which points to a segment of wire 1, and $\mathbf{r} = (-d, 0, z)$, which points to a segment of wire 2. The difference is $\mathbf{r} - \mathbf{r}' = (-d, 0, z - z')$, and the magnitude of $\mathbf{r} - \mathbf{r}'$ is $|\mathbf{r} - \mathbf{r}'| = \left[d^2 + (z - z')^2 \right]^{1/2}$. The vector elements of the wire segments are $d\boldsymbol{\ell}' = +\mathbf{a}_z dz'$ and $d\boldsymbol{\ell} = -\mathbf{a}_z dz$, where the directions of these vectors reflect the directions of the currents I_1 and I_2, respectively. Then, from Eq. (4.1), we need

$$d\boldsymbol{\ell}' \times (\mathbf{r} - \mathbf{r}') = \begin{vmatrix} \mathbf{a}_x & \mathbf{a}_y & \mathbf{a}_z \\ 0 & 0 & dz' \\ -d & 0 & z - z' \end{vmatrix} = -\mathbf{a}_y d\, dz',$$

and

$$d\boldsymbol{\ell} \times \left[d\boldsymbol{\ell}' \times (\mathbf{r} - \mathbf{r}') \right] = \begin{vmatrix} \mathbf{a}_x & \mathbf{a}_y & \mathbf{a}_z \\ 0 & 0 & -dz \\ 0 & -d\, dz' & 0 \end{vmatrix} = -\mathbf{a}_x d\, dz\, dz'.$$

Using this in Eq. (4.1), the elemental force is

$$d\mathbf{F}_2 = \frac{\mu_0}{4\pi} I_1 I_2 \frac{-\mathbf{a}_x d\, dz\, dz'}{\left[d^2 + (z - z')^2 \right]^{3/2}},$$

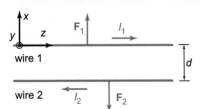

FIGURE 4.3 Two long, parallel wires carrying currents I_1 and I_2. The separation between the wires is d, and the wires feel forces $\mathbf{F}_1$ and $\mathbf{F}_2$, as shown.

which, when integrated over the entire length of wire 1 (from $z' = -\infty$ to $+\infty$) and over a length L_2 of wire 2 gives us the force per unit length of

$$\mathbf{F}_2' = \mathbf{F}_2/L_2 = -\mathbf{a}_x \frac{\mu_0 I_1 I_2}{2\pi d}.$$

This force is in the $-\mathbf{a}_x$ direction, which from Fig. 4.3 can be seen to be a repulsive force. If one of the currents I_1 or I_2 is reversed, the force between the wires becomes attractive. We did not solve for the force $\mathbf{F}_1$ applied by the lower wire on the upper, but if we had, we would have found $\mathbf{F}_1 = -\mathbf{F}_2$. That is, this force is equal but opposite the force $\mathbf{F}_2$.

You may recall from the section "SI Units" at the beginning of the book that, prior to 2019, this force formed the basis for the standard unit of current. Two long, parallel wires separated by a distance of 1 m each carrying a current of 1 A in opposite directions feel a repulsive force of 2×10^{-7} N per meter of length of the wire. The magnitude of this force decreases as $1/d$, where d is the distance between the wires.

Before we leave this example, let's put some numbers in to get a feel for the magnitude of this force. (Personally, I'm not calibrated in units of newtons, so it's hard for me to appreciate the force in that previous paragraph.) Imagine we have two long, straight wires, one above the other, as worked above, with equal but opposite currents, $I_2 = I_1$. For simplification, we use I for either current. For a touch of realism, let the wires be 22 gauge copper wires, and let the spacing between the wires be $d = 2$ mm; a little more than enough for the insulation on each of the wires. Now let's determine the current I necessary to levitate the upper wire. As we just showed, the force between the wires is repulsive; in the $-\mathbf{a}_x$ direction for the lower wire and in the $+\mathbf{a}_x$ direction for the upper wire. To levitate the upper wire, the magnetic force per unit length F_m' in the $+\mathbf{a}_z$ direction must just balance the gravitational force per unit length F_g', that is

$$F_m' = \frac{\mu_0 I^2}{2\pi d} = F_g' = m'g,$$

where m' is the mass per unit length of the copper wire, and $g = 9.8$ m/s^2 is the gravitational constant on Earth. To determine m', we need the mass density of copper and the diameter of 22 gauge wire, which we find in standard tables to be $\rho_m = 8.96$ g/cm^3 and 25.35 mil, respectively. (A mil is 0.001 inch, or 0.0254 mm.) The diameter gives us the cross-sectional area $S = 3.26 \times 10^{-7}$ m^2, and the mass per unit length is $m' = \rho_m S = 2.92 \times 10^{-3}$ kg/m. We multiply by g to get $F_g' = 0.0286$ N/m. Solving the previous equation for I, we get

$$I = \sqrt{\frac{2\pi d F_g'}{\mu_0}} = 16.9 \text{ A}.$$

This is a substantial current, especially for a 22 gauge wire, but we could probably apply a momentary pulse that the wire could withstand. The power dissipation in the wire would be rather high, however, the power per unit length amounting to

$$P' = \frac{1}{L} \int_v \mathbf{J}_v \cdot \mathbf{E} \, dv = \frac{J_v^2 S}{\sigma} = \frac{I^2}{\sigma S} = I^2 R' = 15 \text{ W/m},$$

where $\sigma = 5.8 \times 10^7$ S/m is the conductivity of copper. This wire would get very hot very quickly, so we wouldn't want to apply this current for long.

In the form expressed by Eq. (4.1), the force exerted by one wire on the other wire is an action at a distance. There is no contact between the two wires, yet one applies a force on the other. This description is perfectly adequate for the present discussion, but later, when we introduce material effects and, still later, time-varying waves, it will be much more convenient to formulate this force in terms of a vector field $\mathbf{B}$, known as the **magnetic flux density**, created by the current I_1. This magnetic flux density in turn acts on the current loop 2. The vector representation of the field $d\mathbf{B}$ at location $\mathbf{r}$ that is generated by the element $d\boldsymbol{\ell}'$ of loop 1 is

$$d\mathbf{B}(\mathbf{r}) = \frac{\mu_0}{4\pi} \frac{I_1 \, d\boldsymbol{\ell}' \times (\mathbf{r} - \mathbf{r}')}{|\mathbf{r} - \mathbf{r}'|^3}, \tag{4.2}$$

or, after adding all the contributions from individual segments $d\boldsymbol{\ell}'$,

$$\boxed{\mathbf{B}(\mathbf{r}) = \frac{\mu_0}{4\pi} \int \frac{I_1 \, d\boldsymbol{\ell}' \times (\mathbf{r} - \mathbf{r}')}{|\mathbf{r} - \mathbf{r}'|^3}.} \tag{4.3}$$

The SI unit of the magnetic flux density $\mathbf{B}$ is the tesla (T) or weber per meter squared (Wb/m^2). The latter unit emphasizes the notion of the flux of the magnetic flux density. In terms of units that we have encountered previously, 1 T is equivalent to 1 N/(Am), or equivalently to 1 Vs/m^2. Another commonly used unit of magnetic flux density is the gauss (G), equal to 10^{-4} T. This unit is often convenient to use, in that common field strengths are of this order. For example, the magnetic field of the Earth at our location is approximately 0.5 G. Equation (4.3) for calculation of the magnetic flux density $\mathbf{B}(\mathbf{r})$ is one form of the **Biot–Savart Law**, named after Jean-Baptiste Biot and Félix Savart. (See Biographical Notes 4.2 and 4.3.) The Biot–Savart Law allows us to calculate $\mathbf{B}(\mathbf{r})$ at location $\mathbf{r}$ due to a current I_1 at location $\mathbf{r}'$. The force that this field exerts on the elemental length $d\boldsymbol{\ell}$ of the current-carrying wire 2, illustrated in Fig. 4.4, is

$$d\mathbf{F}_2 = I_2 \, d\boldsymbol{\ell} \times \mathbf{B}(\mathbf{r}). \tag{4.4}$$

Replacing $\mathbf{B}(\mathbf{r})$ in this expression with $d\mathbf{B}(\mathbf{r})$ in Eq. (4.2) returns a form that is identical to Eq. (4.1).

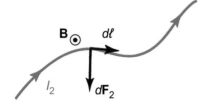

FIGURE 4.4 A segment $d\ell$ of the wire carrying a current I_2 feels a force $d\mathbf{F}_2$ applied by the magnetic flux density $\mathbf{B}$. The force $d\mathbf{F}_2$ is given in Eq. (4.4).

Biographical Note 4.2 Jean-Baptiste Biot (1774–1862)

FIGURE 4.5 Portrait of Jean-Baptiste Biot, circa 1850–1860.

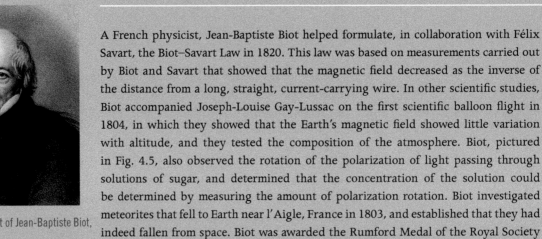

A French physicist, Jean-Baptiste Biot helped formulate, in collaboration with Félix Savart, the Biot–Savart Law in 1820. This law was based on measurements carried out by Biot and Savart that showed that the magnetic field decreased as the inverse of the distance from a long, straight, current-carrying wire. In other scientific studies, Biot accompanied Joseph-Louise Gay-Lussac on the first scientific balloon flight in 1804, in which they showed that the Earth's magnetic field showed little variation with altitude, and they tested the composition of the atmosphere. Biot, pictured in Fig. 4.5, also observed the rotation of the polarization of light passing through solutions of sugar, and determined that the concentration of the solution could be determined by measuring the amount of polarization rotation. Biot investigated meteorites that fell to Earth near l'Aigle, France in 1803, and established that they had indeed fallen from space. Biot was awarded the Rumford Medal of the Royal Society in 1840.

Biographical Note 4.3 Félix Savart (1791–1841)

Educated as a physician, Félix Savart practiced medicine for a short time before developing an interest in acoustics, and he traveled to Paris to meet Jean-Baptiste Biot to discuss these mutual interests. When news of Hans Christian Oersted's observations that a current-carrying wire deflected a magnetic compass needle reached Paris, Biot and Savart carried out a precise measurement of the magnetic force produced by a long, straight, current-carrying wire, which decreases as the inverse of the distance from the wire. These observations led to the Biot–Savart Law.

Throughout our study of magnetostatics, we will draw on parallels to our previous development of electrostatics. There are striking similarities, as well as some stark contrasts. As a first example, recall that the electric field at position $\mathbf{r}$ due to a charge dq located at $\mathbf{r}'$ is

$$d\mathbf{E}(\mathbf{r}) = \frac{1}{4\pi\varepsilon_0} \frac{dq\,(\mathbf{r} - \mathbf{r}')}{|\mathbf{r} - \mathbf{r}'|^3},$$

and compare this to the magnetic flux density given by Eq. (4.2). Features that are in common are the dependence of the field magnitude on distance (each decreases as 1 over the distance squared from the source to the test point), and each field magnitude depends on the size of the source; dq for the electric field, and $I_1\,d\boldsymbol{\ell}'$ for the magnetic field. Also, in each case, the force can be attractive or repulsive, depending on the signs of the charges (for $\mathbf{E}$) or the directions of the currents (for $\mathbf{B}$).

The directions in which these two fields point, however, are quite different. The electric field always points directly away from a positive charge or directly toward a negative charge, and the electric force is parallel or anti-parallel to this field. In contrast, the magnetic field lines always circle around the source current, and join back on themselves. There are no magnetic charges from which magnetic field lines originate or terminate. We will return to this property later when we explore some of the vector differential properties of magnetic fields. Magnetic forces are perpendicular to the magnetic flux density, in contrast to electric forces, which are parallel to the electric field.

In addition to the form of the Biot–Savart Law presented above, which is valid when calculating the field due to the current in a skinny wire, we have two additional forms, appropriate for use when the current is distributed in three dimensions, described by a current density $\mathbf{J}_v(\mathbf{r}')$:

$$\boxed{\mathbf{B}(\mathbf{r}) = \frac{\mu_0}{4\pi} \int_v \frac{\mathbf{J}_v(\mathbf{r}') \times (\mathbf{r} - \mathbf{r}')\,dv'}{|\mathbf{r} - \mathbf{r}'|^3},} \tag{4.5}$$

or when the current flows on a surface, with surface current density $\mathbf{J}_s(\mathbf{r}')$:

$$\boxed{\mathbf{B}(\mathbf{r}) = \frac{\mu_0}{4\pi} \int_s \frac{\mathbf{J}_s(\mathbf{r}') \times (\mathbf{r} - \mathbf{r}')\,ds'}{|\mathbf{r} - \mathbf{r}'|^3}.} \tag{4.6}$$

If you've forgotten the difference between volume current density $\mathbf{J}_v(\mathbf{r}')$ and surface current density $\mathbf{J}_s(\mathbf{r}')$, you should review the discussion presented in Section 3.1.

The magnetic forces that we have considered to this point have been the forces applied by one current-carrying wire on another. Magnetic forces are also felt by individual moving charged particles as they move through regions in which a $\mathbf{B}$ field is present. The form of this **Lorentz force**, as it is known, is

$$\boxed{\mathbf{F} = q\,\mathbf{u} \times \mathbf{B},} \tag{4.7}$$

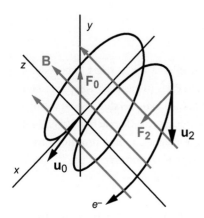

FIGURE 4.6 The trajectory of an electron as it moves through a region in space to which a uniform magnetic flux density $\mathbf{B} = |\mathbf{B}|\,\mathbf{a}_z$ has been applied.

where q and $\mathbf{u}$ are the charge and velocity of the moving charge. In Fig. 4.6, the trajectory of an electron as it travels through a region in space to which a uniform magnetic flux density $\mathbf{B}$ has been applied is shown. In this figure, $\mathbf{B}$ is directed along the $+z$-axis, and the electron initially has velocity $\mathbf{u}_0$ in the x–z plane. The force felt by the electron at this initial position is $\mathbf{a}_y\,eu_{0,x}|\mathbf{B}|$, where $u_{0,x}$ is the x-component of the initial velocity, and $-e = -1.602 \times 10^{-19}$ C is the charge of the electron. This force directed in the $+y$-direction forces the electron into a helical trajectory as shown. As the direction of the velocity $\mathbf{u}$ changes, however, the direction of the force $\mathbf{F}$ also changes, and it is always perpendicular to $\mathbf{u}$, and directed inward toward the axis of the helix. The z-component of the electron's velocity $u_{0,z}$ is unaffected by the magnetic force, and it therefore maintains a constant value. The velocity $\mathbf{u}_2$ and Lorentz force $\mathbf{F}_2$ at a later time in the trajectory are also shown. A solution for the trajectory follows from Newton's Second Law ($\mathbf{F} = m\mathbf{a} = m\,d\mathbf{u}/dt$), which when applied to the geometry of this problem gives us $\mathbf{u}(t) = u_{0,x}[\mathbf{a}_x\cos(e|\mathbf{B}|t/m_e) + \mathbf{a}_y\sin(e|\mathbf{B}|t/m_e)] + \mathbf{a}_z u_{0,z}$. Integration of the velocity yields the position of the electron,

$$\mathbf{r}(t) = \frac{m_e u_{0,x}}{e|\mathbf{B}|}\left[\mathbf{a}_x\sin(e|\mathbf{B}|t/m_e) - \mathbf{a}_y\cos(e|\mathbf{B}|t/m_e)\right] + \mathbf{a}_z u_{0,z}t, \tag{4.8}$$

consistent with the trajectory shown in Fig. 4.6.

The Lorentz force is key to several applications, such as the Hall probe, the ion velocity selector, the ion mass spectrometer, and many others. Some of these topics are discussed in TechNotes 4.1–4.7.

TechNote 4.1 Hall Effect Probe

A Hall probe provides a means of (1) determining the sign of charge carriers in a conducting medium; (2) determining the charge density of the carriers in a conducting medium; and (3) measuring the strength of magnetic flux density. The **Hall effect** was discovered by Edwin Hall at Johns Hopkins University in Baltimore, MD, in 1879 while he was a doctoral student. It is interesting to note that this was 18 years before the discovery of the electron by J.J. Thompson in 1897.

The basis for the ordinary Hall effect is as follows. Consider a current flowing through a conductor, as shown in Fig. 4.7. When a vertical magnetic flux density $\mathbf{B} = B_z\mathbf{a}_z$ is applied to the substrate, the charge carriers will feel a Lorentz force $\mathbf{F}_m = q\mathbf{u} \times \mathbf{B}$ deflecting them to the side. For negative charge carriers (electrons, for example), the carrier velocity is in the $-x$-direction (opposite the direction of current flow), q is, of course, negative, and the Lorentz force on the carrier is in the $-y$-direction, to the left. As a result, negative charges accumulate on the left side of the slab, and positive charges on the right. These charges set up an electric field $\mathbf{E} = E_y\mathbf{a}_y$ in the medium as well, which exerts an electric force $\mathbf{F}_e$ on the carriers. In steady state, the magnetic force $\mathbf{F}_m$ on the carriers is exactly balanced by the electric force $\mathbf{F}_e = q\mathbf{E}$, and the carriers feel no net force. Under these conditions, we will show that the potential difference V_H between the sides of the substrate is related to the current I through the substrate,

$$V_H = \frac{IB_z}{t\rho_v},$$

where t is the depth of the substrate and ρ_v is the charge density of the carriers, as follows. We use $V_H = wE_y$, which is, strictly speaking, valid for a pair of wide, parallel plates, but approximately correct when the electric field lines are confined to the substrate, as in high-permittivity materials; and $J_v = \rho_v u_x = I/wt$. Then,

$$V_H = wE_y = wu_xB_z = w\frac{J_v}{\rho_v}B_z = w\frac{I/wt}{\rho_v}B_z = \frac{IB_z}{t\rho_v}.$$

That is, the voltage across the substrate is proportional to the applied magnetic field, and the Hall effect provides a useful magnetic field probe.

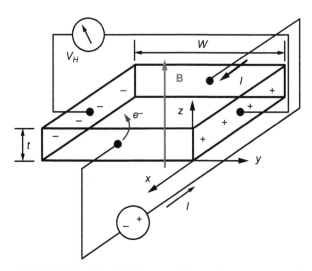

FIGURE 4.7 The Hall effect. The current I passes through the material, to which a vertical magnetic flux **B** has been applied. The carriers are deflected by the magnetic field, creating a potential across the material. The Hall voltage V_H depends on the sign of magnetic flux density, as well as the sign of the charge carriers.

When the magnetic field is known through some other means, the Hall effect provides us a means of determining the charge density of the carriers in the substrate. Furthermore, the Hall effect allows us to determine the sign of charge carriers in a material. That is, it can be used to determine whether the carriers in a semiconductor material are electrons or holes. Previously, we assumed the carriers were electrons, and we showed that the electrons were deflected to the left side of the substrate, as pictured in the figure. How does this picture change if the current is carried by positive charges, as with holes in a p-doped semiconductor, for instance? In this case, the positive charges move in the same direction as the current, the $+x$-direction. This velocity u_x is reversed from the previous case, but so is the charge, so the direction of the Lorentz force $\mathbf{F}_m = q\mathbf{u} \times \mathbf{B}$ is precisely the same as it was for the negative charge carriers, the $-y$-direction. For positive carriers, then, a net positive charge accumulates on the left side and negative on the right, and the Hall voltage is reversed from what it was for the negative carriers. Therefore, from the sign of the Hall voltage, the sign of the charge carriers can be determined.

TechNote 4.2 describes the ion velocity selector, an instrument used to selectively transmit ions of a specific velocity at an aperture, using opposing electric and magnetic forces on the ions.

TechNote 4.2 Ion Velocity Selector

In an ion velocity selector, a beam of ions experiences forces concurrently from a vertical electric field and horizontal magnetic field, as shown in Fig. 4.8. The electric force $\mathbf{F}_e = q\mathbf{E}$ and magnetic force $\mathbf{F}_m = q\mathbf{u} \times \mathbf{B}$ on the ions are in opposite directions, and those ions whose velocity equals $|u| = E/B$ experience no net force within the velocity selector. These ions therefore are able to pass through a slit, while ions of greater or smaller velocity are deflected, either upward or downward in the figure, and blocked by the slit.

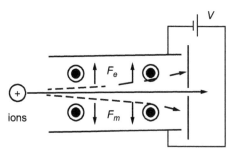

FIGURE 4.8 An ion velocity selector. Ions traveling to the right in the region between the field plates feel an upward electric force and a downward magnetic force. Those ions that feel no net force are selected by the slit.

The ion velocity selector described in TechNote 4.2 is an integral component of an ion mass spectrometer (TechNote 4.3), used to analyze the mass of the constituent particles in a beam of ions.

TechNote 4.3 Ion Mass Spectrometer

In an ion mass spectrometer, the charge-to-mass ratio of ions can be determined. Mass spectrometers find applications in such fields as chemical analysis of materials, isotope analysis, and carbon dating. Several types of mass spectrometers have been developed, such as quadrupole mass spectrometers and time-of-flight mass spectrometers. We consider here the ion mass spectrometer, shown in Fig. 4.9, in which the trajectory followed by ions in a uniform magnetic field depends upon their charge-to-mass ratio, q/m. Let ions from a source be accelerated in an electric field, after which they pass into a velocity selector. (See Tech Note 4.2 on ion velocity selectors.) In the mass spectrometer region, the velocity-selected ions feel only the magnetic force $\mathbf{F}_m$. As a result of this force, the ions follow a circular trajectory, whose radius is found by equating the Lorentz (magnetic) force $F_m = quB$, to the centripetal force $F_c = mu^2/r$ required of any object of mass m and velocity u to follow a circular orbit of radius r. The resulting trajectory radius is $r = mu/qB$, consistent with Eq. (4.8). Thus, for a given velocity u as selected by the velocity selector, and uniform magnetic flux density B, the radius of the trajectory of the ions increases with decreasing charge-to-mass ratio. Note that the entire mass spectrometer must be in vacuum so that the ions are not obstructed by collisions with background gas molecules. The mass resolution of the spectrometer (the ability to discern one species of mass m from another of slightly different mass $m + \delta m$) depends on the slit width (which determines the resolution of the velocity selector), the uniformity of the flux density B, and the spatial resolution of the detector (not shown in the figure).

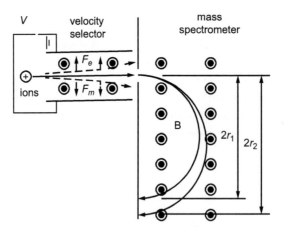

FIGURE 4.9 An ion mass spectrometer. After filtering the ions in the ion velocity selector, ions in the mass spectrometer follow a circular trajectory of radius $r = mu/qB$, where m, u, and q are the mass, velocity, and charge of the ions, and B is the magnetic flux density.

The cyclotron (TechNote 4.4), invented by Ernest Lawrence in the 1930s, promoted rapid advance in particle acceleration, primarily for studying high energy collisions between nuclei. Cyclotrons are still in widespread use today for particle beams and for nuclear medicine.

TechNote 4.4 Cyclotron

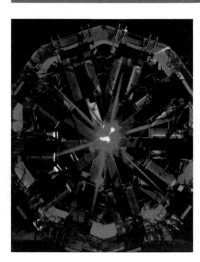

A cyclotron is a particle accelerator used to boost the kinetic energy of charged particles, such as protons. It was developed at the University of California, Berkeley, in 1929–1930 by Ernest O. Lawrence, who was awarded the Nobel Prize in physics in 1939 for this invention and for measurements of collisions he carried out with it. Cyclotrons were the most powerful particle accelerator until the 1950s, and are still in common use for particle beams and nuclear medicine.

A schematic of a cyclotron, taken from Lawrence's patent application, is shown in Fig. 4.10. The cyclotron consists of two hollow D-shaped spaced conductors, called "dees." A strong, static magnetic field is applied to the dees (directed out of the page for the figure on the left). When a large potential difference is applied between the two dees, charged particles (we'll use positively charged protons for this example) injected in the space between the dees are accelerated toward the negatively biased dee (the top dee). Due to the Lorentz force, $\mathbf{F}_m = q\mathbf{u}\times\mathbf{B}$, these particles follow a circular trajectory of radius $r = mu/qB$. (This radius is derived by recognizing that the velocity u is perpendicular to B, so $F_m = quB$, and equating the magnetic force F_m to the centripetal force mu^2/r, required of any object of mass m and velocity u to follow a circular trajectory of radius r. It is consistent with Eq. (4.8).) After completing half an orbit, which requires a time $T/2 = \pi r/u = \pi m/qB$, the proton passes from the top dee to the bottom. Notice that this time does not depend on the velocity of the ion, but only on its mass and charge, and the magnetic flux density B. By reversing the electric potential between the dees before the particle reaches the gap, the particle will again be accelerated in the gap region between the dees. Since the proton's kinetic energy, and velocity, are now greater than on the previous arc, the circular trajectory it follows in the bottom dee is larger than in the top dee. By alternating the potential difference between the dees at a frequency

$$f_c = T^{-1} = qB/2\pi m,$$

known as the **cyclotron frequency**, the proton is accelerated each and every time, gaining energy as its spiral grows. If the radius of the cyclotron is R, the maximum kinetic

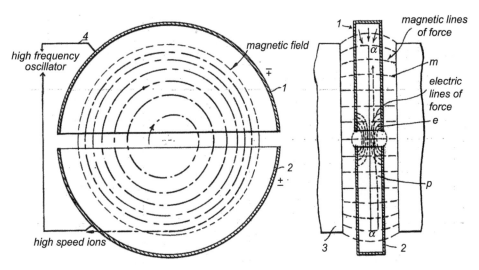

FIGURE 4.10 A sketch of the cyclotron, as it appears in the patent application, US Patent 1,948,384, by Ernest O. Lawrence.

energy of the accelerated charges is $K = mu_{max}^2/2 = m(qBR/m)^2/2 = q^2 B^2 R^2/2m$. Large cyclotrons (many meters in radius) with magnetic fields of several tesla can accelerate ions to energies of hundreds of MeV. (1 MeV $= 1.602 \times 10^{-13}$ J.)

The Lorentz force is integral in the generation of the Earth's magnetic field, which results from a magnetic dynamo effect, as described in TechNote 4.5.

TechNote 4.5 Earth's Magnetic Field

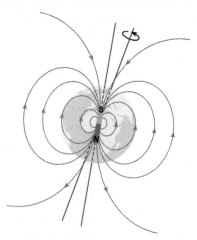

FIGURE 4.11 The shape of the magnetic field of the Earth is dipolar, pointing into the geographic North Pole, and out of the geographic South Pole.

Humans have navigated using the Earth's magnetic field since the first compasses were introduced in China in the eleventh or twelfth century, but we are only now approaching a more complete understanding of the physical origins of this field. The shape of the Earth's field is approximately dipolar, as shown in Fig. 4.11, oriented close to the rotation axis of the Earth. The magnetic North Pole differs from the geographic North Pole by about 11°, moving northward from northern Canada toward Siberia by about 10 km/year. In addition, it is known from the magnetization of rock samples near the ocean floors that the Earth's magnetic field has reversed direction 170 times in the past 76 million years. The timing of the reversals appears to be random, and the last reversal was about 770 000 years ago.

Although the Earth's magnetic field is not understood in complete detail, it is generally agreed that its primary source is a magnetic dynamo, which consists of a rotating, liquid conductor. The inner and outer cores of the Earth consist mostly of iron and nickel, and make up 31% of the Earth's mass. The temperature of the inner core is ~6 000 K, but due to the immense pressure at the center of the Earth, it is solid. This temperature is well above the Curie temperature, the maximum temperature that a material is able to maintain a permanent magnetic moment. The temperature of the outer core decreases to 3 800 K at the boundary with the mantle, but the outer core is liquid.

To explain a magnetic dynamo, we start by considering the effect of an initial "seed" magnetic field. In this description, we choose that the seed points from the North Pole toward the South Pole in the outer core. Then, due to the rotation of the Earth, the Lorentz force $\mathbf{F}_m = q\mathbf{u} \times \mathbf{B}$ on the positive and negative charges in the molten iron of the outer core causes a separation of these charges. The positive charges (iron ions) are pulled toward the inner core, while the negative charges (electrons) are expelled toward the mantle. This separation of charges in the rotating outer core represents a current, which then generates an additional magnetic field $\mathbf{B}$. Using the right-hand rule, you can see that this new additional $\mathbf{B}$ is in the same direction as the seed field, and the magnetic field grows. By this process, the resulting magnetic field can be much larger than the initial seed, and can help to maintain the total field. The complete picture is, of course, much more complex, including convection currents in the molten iron caused by the temperature gradient over the outer core, the Coriolis force causing the convection currents to swirl as they rise toward the mantle, return currents from the molten iron back toward the inner core, and energy from gravitational effects and radioactivity feeding the entire process.

Note that if the initial seed field had been pointing in the opposite direction, the magnetic dynamo would produce a field pointing toward that pole. Therefore, the dynamo effect results in a strong magnetic field roughly aligned with the Earth's axis of rotation, but without preference to its orientation. Any strong fluctuation of the field can cause its orientation to reverse, which, as we stated earlier, happens approximately every half a million years.

Manifestations of the Earth's magnetic field are the Northern and Southern Lights, as seen in Fig. 4.12, and described in TechNote 4.6.

FIGURE 4.12 A display of the Northern Lights near Bear Lake in Alaska.

TechNote 4.6 Aurora Borealis

The solar wind is a constant stream of energetic protons and electrons that have escaped the sun. The solar wind is strongest during periods of strong solar flare activity, often associated with sun spots. When these charged particles approach the Earth, most of those that approach near the equator are deflected from their paths by the Earth's magnetic field and continue into space on a new trajectory. Near the Earth's poles, however, where the magnetic field lines converge, the solar wind ions are more likely to enter the atmosphere, and we can see evidence of their presence through the aurora borealis (Northern Lights) near the North Pole, and the aurora australis (Southern Lights) near the South Pole. These nocturnal light shows are a result of collisions between the ions and molecules in the air, which ionize or excite the molecules of the air and emit fluorescent radiation as they recombine or relax to the ground state. The green glow shown in Fig. 4.12 comes primarily from fluorescence by oxygen molecules. Other hues, such as red, yellow, blue, and violet can also be seen, but are less common.

It is speculated that some birds may use the Earth's magnetic field to help navigate, as described in TechNote 4.7.

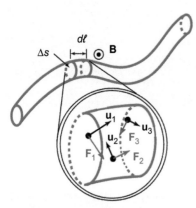

FIGURE 4.13 The current through a wire is represented by the net motion of a large number of charge carriers, typically electrons. The net force on the wire can be determined by summing the forces on the individual carriers.

We will now show the equivalence between the Lorentz force on individual moving charges and the force exerted on a current-carrying wire by a magnetic flux. A wire of cross-sectional area Δs is shown in Fig. 4.13, with a short segment magnified in the expanded view. The force applied to each individual moving charge is $q_i \mathbf{u}_i \times \mathbf{B}$, where the index i labels the individual charged carriers of charge q_i and velocity $\mathbf{u}_i$. The magnetic force felt by a segment of the wire of length $d\ell$ is found by summing the forces on the individual charges within that segment of the wire,

$$d\mathbf{F} = \sum_i \left(q_i \mathbf{u}_i \times \mathbf{B} \right).$$

When all the charges have the same value q, the force can be written $d\mathbf{F} = q \left(\sum_i \mathbf{u}_i \right) \times \mathbf{B}$, and the sum of the velocities is simply the average velocity $\langle \mathbf{u} \rangle$ times the number of carriers in the volume $\Delta v = \Delta s \, d\ell$. Then the force is

$$d\mathbf{F} = q \left(n \Delta v \langle \mathbf{u} \rangle \right) \times \mathbf{B},$$

where n is the number density of particles, and $n\Delta v$ is the total number of charged particles within volume Δv. The product nq is the charge density ρ_v, the product $\rho_v \langle \mathbf{u} \rangle$ is the current density $\mathbf{J}_v$ (Eq. (3.10)), and the product $\mathbf{J}_v \Delta s$ is the current I (Eq. (3.3)). Substituting, we get

$$d\mathbf{F} = I d\ell \times \mathbf{B}$$

for the magnetic force on the current segment, where we have used the fact that $d\ell$ is parallel to $\langle \mathbf{u} \rangle$. We saw this expression earlier as Eq. (4.4).

In Examples 4.2–4.5, the Biot–Savart Law, Eqs. (4.3), (4.5), or (4.6), will be applied to determine the magnetic flux density $\mathbf{B}$ for several simple current distributions.

Example 4.2 Biot–Savart Law, Long, Straight Wire

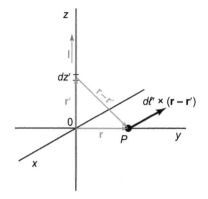

FIGURE 4.14 The current *I* along a long, straight wire generates the magnetic flux density **B** in the surrounding space.

Consider an infinitely long straight wire carrying a current I in the $+z$-direction. Use the Biot–Savart Law to determine the magnetic flux density **B** in the space outside this wire.

Solution:

Refer to the diagram shown in Fig. 4.14. We start by writing explicit expressions for each of the terms in Eq. (4.3). The element $d\boldsymbol{\ell}'$ defines a short linear segment along which the current flows. Since the current in this wire flows in the $+z$-direction, we use $d\boldsymbol{\ell}' = \mathbf{a}_z\, dz'$. The position vectors $\mathbf{r}'$ and $\mathbf{r}$ define the locations of the source element (i.e. the location of the element $d\boldsymbol{\ell}'$) and the point P at which we are determining the magnetic flux density **B**, respectively. In this example, these are $\mathbf{r}' = (0, 0, z')$ and $\mathbf{r} = (0, \rho, 0)$, where we have chosen the origin to lie along the axis of the wire, with the point P lying in the x–y plane. (The location of the origin is not especially important, since only $\mathbf{r} - \mathbf{r}'$, but not $\mathbf{r}$ or $\mathbf{r}'$ individually, appears in Eq. (4.3). Therefore this choice of the origin is convenient, but in the end does not affect the determination of **B**.) We have also chosen the point P to lie along the y-axis. (We'll return to this in a minute.) Using $\mathbf{r}'$ and $\mathbf{r}$, we find $\mathbf{r} - \mathbf{r}' = (0, \rho, -z')$, and the distance $|\mathbf{r} - \mathbf{r}'|$ is $\sqrt{\rho^2 + z'^2}$. Next we find $d\boldsymbol{\ell}' \times (\mathbf{r} - \mathbf{r}')$, which appears in the numerator of Eq. (4.3) and points in the direction of the contribution to **B** of this current segment,

$$d\boldsymbol{\ell}' \times (\mathbf{r} - \mathbf{r}') = \begin{vmatrix} \mathbf{a}_x & \mathbf{a}_y & \mathbf{a}_z \\ 0 & 0 & dz' \\ 0 & \rho & -z' \end{vmatrix} = -\mathbf{a}_x \rho\, dz'.$$

$d\boldsymbol{\ell}' \times (\mathbf{r} - \mathbf{r}')$ is shown in Fig. 4.14. This vector points in the $-x$-direction, due to our choice to place the point P along the y-axis. Using the cylindrical symmetry of the space, we can generalize this to an arbitrary point P, not necessarily on the y-axis, by making the substitution $-\mathbf{a}_x \to \mathbf{a}_\phi$.

Substituting each of these terms into Eq. (4.3), the flux density is

$$\mathbf{B}(\mathbf{r}) = \frac{\mu_0 I}{4\pi} \int_{-\infty}^{\infty} \frac{\mathbf{a}_\phi\, \rho\, dz'}{\left(\rho^2 + z'^2\right)^{3/2}}.$$

Carrying out the integration using Eq. (G.12) from Appendix G, we find

$$\boxed{\mathbf{B}(\mathbf{r}) = \frac{\mu_0 I}{2\pi\rho}\, \mathbf{a}_\phi.} \tag{4.9}$$

This result in Example 4.2 is framed, since it will be handy for future reference. Notice the $1/\rho$ dependence of the magnitude of this field and the orientation of the field. The magnetic flux density lines circle around the wire and close on themselves. Unlike electric field lines, magnetic flux density lines have no beginning or end, but rather loop around the current that produces them and close on themselves. We will see this property in the following examples as well.

Example 4.3 Biot–Savart Law, Circular Loop

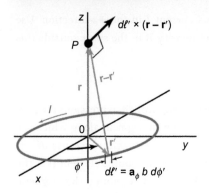

FIGURE 4.15 A circular current loop of radius b and current I lying in the x–y plane.

Consider a circular current loop of radius b and current I lying in the x–y plane. Determine the magnetic flux density $\mathbf{B}$ at a point P along the z-axis.

Solution:

The current loop is shown in Fig. 4.15. We first consider the contribution to $\mathbf{B}$ of one elemental segment $d\boldsymbol{\ell}' = \mathbf{a}_\phi\, b\, d\phi'$. Using cylindrical coordinates, the source and field position vectors are, respectively, $\mathbf{r}' = (b, 0, 0)$ and $\mathbf{r} = (0, 0, z)$, where we have chosen the origin to lie at the center of the loop. The position vector $\mathbf{r} - \mathbf{r}'$, which points from the source element to P, is $-\mathbf{a}_\rho\, b + \mathbf{a}_z\, z$, and $d\boldsymbol{\ell}' \times (\mathbf{r} - \mathbf{r}')$ is

$$d\boldsymbol{\ell}' \times (\mathbf{r} - \mathbf{r}') = \begin{vmatrix} \mathbf{a}_\rho & \mathbf{a}_\phi & \mathbf{a}_z \\ 0 & b\, d\phi' & 0 \\ -b & 0 & z \end{vmatrix} = b\, d\phi' \left(\mathbf{a}_\rho\, z + \mathbf{a}_z\, b \right).$$

This vector is shown in Fig. 4.15. (Since it may be difficult to visualize this vector correctly in a two-dimensional diagram, the following note of explanation may be helpful. For the segment $d\boldsymbol{\ell}'$ shown, the vector $d\boldsymbol{\ell}' \times (\mathbf{r} - \mathbf{r}')$ points *upward* and *out* of the page.) To complete the terms in the Biot–Savart integral, the distance $|\mathbf{r} - \mathbf{r}'| = \sqrt{b^2 + z^2}$ is needed. Note that this distance is independent of ϕ'. We are ready now to substitute each of these terms into the Biot–Savart Law and integrate, but before we do, recall that the unit vectors should first be converted to a rectangular basis. The issue is that the radial unit vector $\mathbf{a}_\rho$ points in different directions as we integrate around the loop. As discussed previously, the directions of the unit vectors $\mathbf{a}_\rho$ and $\mathbf{a}_\phi$ in cylindrical coordinates depend upon spatial location. A similar issue arises in spherical coordinates with $\mathbf{a}_R$, $\mathbf{a}_\theta$, and $\mathbf{a}_\phi$. In contrast, the unit vectors in rectangular coordinates, $\mathbf{a}_x$, $\mathbf{a}_y$, and $\mathbf{a}_z$, always point in fixed directions. Before we integrate, therefore, we will convert to rectangular components. It is convenient to continue using cylindrical *variables*, since these are the natural variables of the current loop. We convert to rectangular components using $\mathbf{a}_\rho = \mathbf{a}_x \cos\phi' + \mathbf{a}_y \sin\phi'$. With this substitution, we can write

$$\mathbf{B}(\mathbf{r}) = \frac{\mu_0 I}{4\pi} \int_0^{2\pi} \frac{\left(\mathbf{a}_x\, z \cos\phi' + \mathbf{a}_y\, z \sin\phi' + \mathbf{a}_z\, b \right) b\, d\phi'}{(b^2 + z^2)^{3/2}}.$$

Upon integration, the x- and y-components vanish since $\int \cos\phi'\, d\phi'$ and $\int \sin\phi'\, d\phi'$ are each equal to zero when integrated over one full circle, and only the z-component survives, yielding

$$\mathbf{B}(\mathbf{r}) = \mathbf{a}_z \frac{\mu_0 I b^2}{2 (b^2 + z^2)^{3/2}}. \tag{4.10}$$

This function is largest at $z = 0$, as we should expect since this point is closest to the loop and all of the contributions $d\mathbf{B}$ point in the same direction, and the peak value is $B_0 = \mu_0 I / (2b)$. B_z decreases with increasing z. We show a plot of this field amplitude in Fig. 4.16(a). The vector $\mathbf{B}$ points in the positive z-direction at any point along the z-axis, even for $z < 0$.

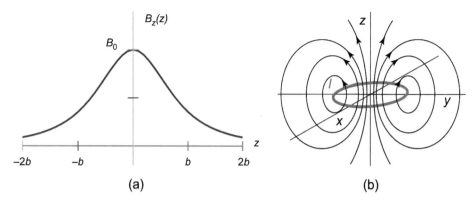

FIGURE 4.16 (a) A plot of $B_z(z)$ vs. z for the circular current loop of radius b and current I lying in the x–y plane. The maximum value of B_z is $B_0 = \mu_0 \, I/(2 \, b)$. (b) A qualitative plot of the magnetic flux density lines produced by a circular current loop of radius b and current I lying in the x–y plane.

In this solution, we calculated the flux density only at a point P that lies along the z-axis. By limiting our solution to an on-axis point, all elements $d\boldsymbol{\ell}'$ are equidistant to P, making the integrals simpler, which seems appropriate as we are learning to apply the Biot–Savart Law. It is instructive, however, to view the magnetic flux lines in a broader space, including positions away from the z-axis. A qualitative plot of these flux lines is shown in Fig. 4.16(b). The field lines form circles around the loop, and close upon themselves. In the interior of the loop, the magnetic flux density points in the positive z-direction, as indicated by the right-hand rule. (Grasp the wire with your right hand, with your thumb pointing in the direction of the current. In the middle of the loop, your fingers point upward in the positive z-direction.) Outside the loop, the magnetic flux density points in the $-z$-direction. At no point does **B** have a component pointing in the ϕ-direction, as can be seen by computing $d\boldsymbol{\ell}' \times (\mathbf{r} - \mathbf{r}')$ for an off-axis point given by the position vector $\mathbf{r} = (x, 0, z)$,

$$d\boldsymbol{\ell}' \times (\mathbf{r} - \mathbf{r}') = \begin{vmatrix} \mathbf{a}_x & \mathbf{a}_y & \mathbf{a}_z \\ -b \sin \phi' \, d\phi' & b \cos \phi' \, d\phi' & 0 \\ x - b \cos \phi' & -b \sin \phi' & z \end{vmatrix}$$

$$= \left[\left(\mathbf{a}_x \cos \phi' + \mathbf{a}_y \sin \phi' \right) bz + \mathbf{a}_z \left(b^2 - xb \cos \phi' \right) \right] d\phi'.$$

(Notice that we set up this cross product using rectangular coordinates directly, since $(\mathbf{r} - \mathbf{r}')$ is no longer expressed simply in cylindrical coordinates.) In the expression above, we recognize that $(\mathbf{a}_x \cos \phi' + \mathbf{a}_y \sin \phi')$ is simply $\mathbf{a}_\rho$, so that $d\boldsymbol{\ell}' \times (\mathbf{r} - \mathbf{r}')$ has components in the ρ- and z-directions, but not in the ϕ-direction. We conclude that even for this off-axis point P, B_ϕ is zero, as we stated above. Evaluation of the integral over $d\phi'$ that would be necessary to complete the solution for **B** at this more general point P is difficult, and since we completed our demonstration that the B_ϕ component is zero, we will conclude our example here.

In Example 4.3 we found the magnetic flux density B_z along the axis of a single loop of wire. As we showed, this flux density is strongest at the center of the loop, and falls off with increasing $|z|$ on either side of the loop. In many cases of practical importance, a uniform magnetic flux density over an extended region is needed. This can be achieved with a solenoid, which we will introduce in Example 4.5, or with a pair of loops called a Helmholtz pair. The latter configuration is considered in TechNote 4.8.

TechNote 4.8 Helmholtz Pair

A Helmholtz pair consists of two circular wire loops. Let each loop be of radius b, consist of N turns, and carry a current I in the same direction, as shown in Fig. 4.17(a). The pair is centered on the z-axis, and the loops are separated by a distance d. (One loop is placed at $z = -d/2$, the other at $z = +d/2$.) We use superposition and Eq. (4.10) for the field due to a single loop to immediately write the field at any point along the z-axis due to the pair of loops as:

$$B_z = \frac{\mu_0 N I b^2}{2} \left[\frac{1}{\left[(z + d/2)^2 + b^2\right]^{3/2}} + \frac{1}{\left[(z - d/2)^2 + b^2\right]^{3/2}} \right].$$

This flux density is the sum of two peaks (as a function of z), the first centered at $z = -d/2$, the second centered at $z = +d/2$. When the loops are well separated, $d \gg b$, $B_z(z)$ shows two distinct peaks. But as the spacing between the two coils decreases, the two peaks merge into a single peak as shown in Fig. 4.17(b). By symmetry, the slope dB/dz will always be zero at the center ($z = 0$). To find the spacing between the loops that produces the most uniform flux density over the longest possible range, we adjust d to the extent at which the second derivative d^2B/dz^2 at the center is equal to zero as well. This condition leads us to choose $d = b$. That is, for uniformity of the field near the center of the two loops, we must separate the two loops by a distance d equal to their radius b. $B_z(z)$ is plotted versus z for this optimized configuration in Fig. 4.17(b). Notice that the flux density is flat across the top of this curve for a region of width $\sim d$.

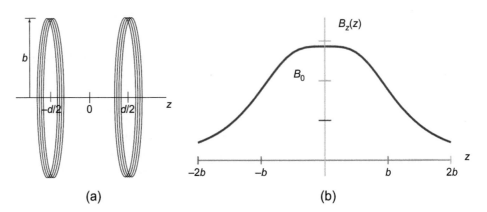

(a) (b)

FIGURE 4.17 (a) A Helmholtz pair, consisting of two circular current-carrying loops of radius b and separated by distance $d = b$. (b) The magnetic flux density $B_z(z)$ vs. z generated along the z-axis by this Helmholtz pair. This configuration produces a magnetic field that is relatively uniform over a region of width $\sim d$ near the center.

In the previous two examples we used the form of the Biot–Savart Law given by Eq. (4.3), since the current is carried by a thin wire, and we did not seek **B** inside the wire. Now let's look at applications of the Biot–Savart Law for which the *distribution* of the current is important, starting with a surface current $\mathbf{J}_s$.

Example 4.4 Biot–Savart Law, Current Sheet

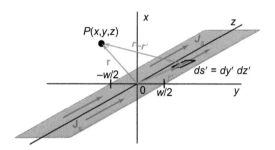

FIGURE 4.18 A sheet of current lying in the $x = 0$ plane and flowing in the $+z$-direction.

Consider an infinitely long uniform sheet of current, of width w, lying in the $x = 0$ plane, and flowing in the $+z$-direction, as shown in Fig. 4.18. The surface current density is $\mathbf{J}_s = \mathbf{a}_z J_s$. (The total current is $I = J_s w$.) Determine the magnetic flux density $\mathbf{B}$ at point $P(x, y, z)$.

Solution:

We will work our solution in rectangular coordinates. The current distribution in this problem is a sheet, so we use the form of the Biot–Savart Law given by Eq. (4.6). Since the sheet is infinitely long, our solution must be independent of z. We therefore choose the origin to be at the center of the current sheet, with the point P lying in the x–y plane (in other words, at $z = 0$). The position vector $\mathbf{r}'$ that points to an element of the source is $\mathbf{r}' = (0, y', z')$, and the point P is located at $\mathbf{r} = (x, y, 0)$. The position vector $\mathbf{r} - \mathbf{r}'$ that points from the source point to P is then $\mathbf{a}_x x + \mathbf{a}_y (y - y') - \mathbf{a}_z z'$, the distance from the source element to P is $|\mathbf{r} - \mathbf{r}'| = \sqrt{x^2 + (y - y')^2 + z'^2}$, and $\mathbf{J}_s \times (\mathbf{r} - \mathbf{r}')$ is

$$\mathbf{J}_s \times (\mathbf{r} - \mathbf{r}') = \begin{vmatrix} \mathbf{a}_x & \mathbf{a}_y & \mathbf{a}_z \\ 0 & 0 & J_s \\ x & (y - y') & -z' \end{vmatrix} = \left[-\mathbf{a}_x (y - y') + \mathbf{a}_y x \right] J_s.$$

The surface element ds' is $dy' dz'$. Substituting each of these terms into Eq. (4.6), the magnetic flux density is

$$\mathbf{B}(\mathbf{r}) = \frac{\mu_0 J_s}{4\pi} \int_{z' = -\infty}^{z' = +\infty} \int_{y' = -w/2}^{y' = +w/2} \frac{\left[-\mathbf{a}_x (y - y') + \mathbf{a}_y x \right] dy' \, dz'}{\left[x^2 + (y - y')^2 + z'^2 \right]^{3/2}}.$$

Using integral tables to evaluate the x- and y-components individually (Eqs. (G.12) and (G.11) for the x-component, and Eqs. (G.12) and (G.10) for the y-component), we can show

$$\mathbf{B}(\mathbf{r}) = \frac{\mu_0 J_s}{4\pi} \left\{ \mathbf{a}_x \ln \left(\frac{x^2 + \left(y - \frac{w}{2} \right)^2}{x^2 + \left(y + \frac{w}{2} \right)^2} \right) + \mathbf{a}_y 2 \left[\tan^{-1} \left(\frac{y + \frac{w}{2}}{x} \right) - \tan^{-1} \left(\frac{y - \frac{w}{2}}{x} \right) \right] \right\}.$$

(4.11)

This flux distribution in the x–y plane is plotted in Fig. 4.19. In this figure, the current is flowing *into* the page. Notice how the flux lines circle around the current sheet in the clockwise direction, and close on themselves.

This solution is now complete, but before we leave this example, let's look at another method of solving for $\mathbf{B}$. We will use superposition and our previous result for the magnetic flux density $\mathbf{B}$ for a long, straight wire, Eq. (4.9) of Example 4.2. For this we refer to Fig. 4.20, which shows the end-on view of the current sheet, with current flowing into

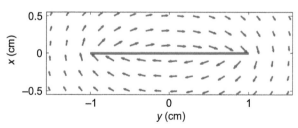

FIGURE 4.19 The magnetic flux lines surrounding a sheet of current, as given in Eq. (4.11). The current is shown in red, and flows into the page (in the $+z$-direction).

the page. The sheet of current can be treated as a series of parallel wires, each of width dy', and all lying in the $x = 0$ plane and having a current $J_s dy'$. Using Eq. (4.9), the contribution to the magnetic flux density at point $P(x, y, z)$ by the current at y' is

$$dB = a_\phi \frac{\mu_0 J_s \, dy'}{2\pi\rho},$$

where ρ is the distance between the current at y' and the point P. By the superposition principle, the total magnetic flux density **B** can be found by adding all these contributions due to each of the constituent wire segments, or rather, by integrating across the width of the current sheet. Before we do this, however, we should recognize that the direction of the unit vector a_ϕ is different for all the different current elements. In addition, the distance ρ is a function of y', so we need to write these terms in a consistent manner. Let's start with the distance ρ, which from Fig. 4.20 can be written as $\rho = \sqrt{x^2 + (y - y')^2}$. It is convenient to define the angle α, shown in Fig. 4.20 as the angle between the $x = 0$ plane and the radial arm that joins the element at y' to the point P at (x, y, z).

Since the flux density contribution dB is normal to this radial arm, dB can be reduced to its components $dB_x = -|dB| \cos \alpha$ and $dB_y = |dB| \sin \alpha$. Back to the figure, we see that $\cos \alpha$ can be written as $(y - y')/\rho$ and $\sin \alpha$ as x/ρ. Combining terms, the components of dB are

$$dB_x = -\frac{\mu_0 J_s \, dy'}{2\pi\rho} \frac{(y - y')}{\rho} = -\frac{\mu_0 J_s}{2\pi} \frac{(y - y') \, dy'}{x^2 + (y - y')^2}$$

and

$$dB_y = \frac{\mu_0 J_s \, dy'}{2\pi\rho} \frac{x}{\rho} = \frac{\mu_0 J_s x}{2\pi} \frac{dy'}{x^2 + (y - y')^2}.$$

Upon integration over y' from $-w/2$ to $+w/2$, we find

$$B_x = \frac{\mu_0 J_s}{4\pi} \left[\ln\left(x^2 + \left(y - \frac{w}{2}\right)^2\right) - \ln\left(x^2 + \left(y + \frac{w}{2}\right)^2\right) \right]$$

and

$$B_y = \frac{\mu_0 J_s}{2\pi} \left[\tan^{-1}\left(\frac{y + \frac{w}{2}}{x}\right) - \tan^{-1}\left(\frac{y - \frac{w}{2}}{x}\right) \right],$$

consistent with the results we found in our first solution, Eq. (4.11).

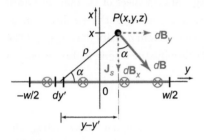

FIGURE 4.20 The end-on view of the current sheet of surface current density J_s and width w. In this view, the current is headed into the page. The flux density dB at point P is due to the element of width dy' in the current sheet. α is the angle that the radial arm makes with the $x = 0$ plane, and ρ is the distance from the element to the point P.

It is interesting to examine the flux density of this sheet of current in two limiting cases. First, for positions close to the current sheet, but not close to the ends, the limit of Eq. (4.11) as the width $w \to \infty$ shows that $B_x \to 0$ and $B_y \to \pm \mu_0 J_s/2$. (B_y is positive above the sheet, and negative below.) We will use this result later, especially when considering two parallel conductors, as in a parallel plate transmission line. The second limiting case of interest is for locations **r** far from the sheet, $|\mathbf{r}| \gg w$. In this limit, it can be shown that $\mathbf{B}(\mathbf{r}) \to a_\phi \mu_0 I/(2\pi\rho)$, which you should recognize as the flux density due to a long, straight wire carrying a current I. This should make sense, since the "shape" of the current distribution should not be important when viewed from a long distance, only its total magnitude.

Example 4.5 Biot–Savart Law, Solenoid

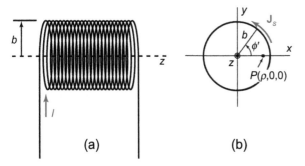

FIGURE 4.21 (a) A solenoid of finite length centered along the z-axis. The radius of the solenoid is *b*, and the number of turns per unit length, or turns density, is *n*. (b) The geometry for application of the Biot–Savart Law for determining the magnetic flux density inside (and outside) the solenoid.

Use the Biot–Savart Law to determine the magnetic flux density inside and outside an infinitely long solenoid. A **solenoid** is a long, tightly wound helical coil of wire, as shown in Fig. 4.21(a). The radius of the solenoid is *b*, the current is *I*, and the number of turns per unit length, or turns density, is *n*.

Solution:

We will approximate the current in the discrete side-by-side wires as a continuous sheet of current of current density. For a current I in the solenoid and a turns density n, the equivalent surface current density is simply nI. The direction of this current is primarily in the ϕ-direction, but there is also a small component in the z-direction, since the winding is a helix. We can write this effective surface current density as $\mathbf{J}_s = In(\mathbf{a}_\phi \cos \alpha + \mathbf{a}_z \sin \alpha)$, where $\alpha = (2\pi bn)^{-1}$ is the angle of the pitch of the helix. (This angle α comes from the ratio of n^{-1}, the spacing from one turn to the next, to $2\pi b$, the circumference of the helix.) In writing the current density in this form, we are neglecting the discrete nature of the wires, but as long as the spacing between the wires is small compared to the size of the wires themselves, this approximation is valid. Also, we will work our solution ignoring the small component of the current that flows in the z-direction, in effect setting $\alpha = 0$. In rectangular components, the current density is

$$\mathbf{J}_s = In(-\mathbf{a}_x \sin \phi' + \mathbf{a}_y \cos \phi').$$

Note that we are using cylindrical *variables* ϕ' and z', yet determining the rectangular *components* of the vectors, as will be convenient for computing vector sums and differences.

Now we are ready to find the terms needed in Eq. (4.6). Refer to the diagram in Fig. 4.21(b). We choose point P, the field point, to lie on the x-axis a distance ρ from the center of the solenoid. (We will generalize the result to other locations later.) This point is $\mathbf{r} = (\rho, 0, 0)$. The location of an element of the current on the surface of the helix at radius b and angle ϕ' is $\mathbf{r}' = (b \cos \phi', b \sin \phi', z')$. The position vector that points from a source element at $\mathbf{r}'$ to the point P is therefore

$$\mathbf{r} - \mathbf{r}' = (\rho - b \cos \phi', -b \sin \phi', -z'),$$

and the distance between these two points is

$$|\mathbf{r} - \mathbf{r}'| = \sqrt{(\rho - b \cos \phi')^2 + (-b \sin \phi')^2 + (-z')^2}$$
$$= \sqrt{\rho^2 - 2\rho b \cos \phi' + b^2 + z'^2}.$$

The vector product $\mathbf{J}_s \times (\mathbf{r} - \mathbf{r}')$ is

$$\mathbf{J}_s \times (\mathbf{r} - \mathbf{r}') = \begin{vmatrix} \mathbf{a}_x & \mathbf{a}_y & \mathbf{a}_z \\ -In \sin \phi' & In \cos \phi' & 0 \\ (\rho - b \cos \phi') & -b \sin \phi' & -z' \end{vmatrix}$$

$$= In \left[-\mathbf{a}_x z' \cos \phi' - \mathbf{a}_y z' \sin \phi' + \mathbf{a}_z (b - \rho \cos \phi') \right].$$

To integrate across the cylindrical surface of the solenoid, we use the element of surface area $ds' = b\,d\phi'\,dz'$. Substituting these terms into Eq. (4.6), the flux density is

$$\mathbf{B} = \frac{\mu_0 Inb}{4\pi} \int_{-\infty}^{\infty} \int_0^{2\pi} \frac{\left[-\mathbf{a}_x\,z'\cos\phi' - \mathbf{a}_y\,z'\sin\phi' + \mathbf{a}_z\,(b - \rho\cos\phi')\right]}{\left[\rho^2 - 2\rho b\cos\phi' + b^2 + z'^2\right]^{3/2}}\,d\phi'\,dz'.$$

As shown, this represents three separate integrals, one for each component of **B**. We can dispatch with two of these rather quickly, namely the x- and y-components, by recognizing that these integrands are odd in z', and that integration in z' over the interval $-\infty$ to ∞ will yield zero. Therefore, only the z-component of **B** survives, with

$$B_z = \frac{\mu_0 Inb}{4\pi} \int_{-\infty}^{\infty} \int_0^{2\pi} \frac{(b - \rho\cos\phi')}{\left[\rho^2 - 2\rho b\cos\phi' + b^2 + z'^2\right]^{3/2}}\,d\phi'\,dz'.$$

Carrying out the z' integration first, using Eq. (G.12) from Appendix G, we find

$$B_z = \frac{\mu_0 Inb}{4\pi} \int_0^{2\pi} \frac{2\,(b - \rho\cos\phi')\,d\phi'}{\left[\rho^2 - 2\rho b\cos\phi' + b^2\right]}.$$

This integrand can be simplified to

$$\frac{1}{b}\left[1 + \frac{b^2 - \rho^2}{\left[\rho^2 - 2\rho b\cos\phi' + b^2\right]}\right].$$

The integral of the first term is $2\pi/b$. For the second, we use the integral Eq. (G.15) in Appendix G to show that

$$\int_0^{2\pi} \frac{d\phi'}{\left[\rho^2 - 2\rho b\cos\phi' + b^2\right]} = \frac{2\pi}{|\rho^2 - b^2|}.$$

Then B_z is

$$B_z = \frac{\mu_0 In}{2}\left[1 + \frac{b^2 - \rho^2}{|\rho^2 - b^2|}\right].$$

For $\rho < b$, this gives $B_z = \mu_0 In$, and for $\rho > b$, we find $B_z = 0$.

At the beginning of the solution, we chose the point P to lie on the x-axis, but we claim this solution is valid for any location. By the cylindrical symmetry of the solenoid, it can be rotated about the z-axis without any change to the current distribution. Thus, the **B** field is unchanged as well. Also, the **B** field must be unchanged with translation along the z-axis, since the solenoid is infinitely long. So we see that **B** inside an infinitely long solenoid is directed along the z-axis, and is perfectly uniform. On the outside of the solenoid, **B** is precisely zero.

For a long but finite-length solenoid, this conclusion about the uniformity of the flux density is nearly valid as long as we are not too close to the end of the solenoid. The flux lines exiting the $+z$ end of the solenoid circle around outside the solenoid, spread out, and become very weak. They point in the $-z$-direction on the outside, and connect on the flux lines entering the $-z$ end of the solenoid. Thus the flux density is not strictly zero outside the finite-length solenoid, but as long as the radius b of the solenoid is much less than its length, $|\mathbf{B}|$ in this region is certainly very small relative to that in the interior.

In this section, we have presented the Biot–Savart Law, which gives us a means of determining the magnetic flux density $\mathbf{B}(\mathbf{r})$ due to a current I, or due to current densities

$\mathbf{J}_v(\mathbf{r})$ or $\mathbf{J}_s(\mathbf{r})$. This law is derived from observations of the magnitude and direction of the force between current-carrying wires, as well as the principle of superposition. In Section 4.2 we will use the Biot–Savart Law to derive the differential laws governing the magnetic flux density, which allows us a more general understanding of the properties of this field.

4.2 Vector Differential Properties of Magnetic Fields

While the Biot–Savart Law is very useful for explicit calculation of the magnetic flux density for known current distributions, we will also have need of a more general description of the properties of magnetic fields. We will express these in terms of the vector differentials of $\mathbf{B}$, in a manner very similar to our approach to the electric field properties.

The first differential property of $\mathbf{B}$ is that

$$\boxed{\nabla \cdot \mathbf{B} = 0.}$$

(4.12)

This can be proven by explicit differentiation of the Biot–Savart Law, Eqs. (4.3), (4.5), or (4.6), as we show in Example 4.6, and is the mathematical equivalent of the statement that there are no magnetic charges – that is, no sources or sinks of magnetic field lines. We have already observed this property in each of the examples worked in the previous section. In each case, the magnetic field lines loop around the current and close on themselves. Equation (4.12) is quite the opposite case from the divergence of the electric field, for which you should recall that $\nabla \cdot \mathbf{D} = \rho_v$, where electric field lines originate on positive charges and terminate on negative charges.

Example 4.6   Demonstration of $\nabla \cdot \mathbf{B} = 0$

Demonstrate the validity of Eq. (4.12) in rectangular coordinates.

Solution:
We demonstrate this by direct differentiation of the Biot–Savart Law, using the form given in Eq. (4.5):

$$\nabla \cdot \mathbf{B}(\mathbf{r}) = \nabla \cdot \frac{\mu_0}{4\pi} \int_v \frac{\mathbf{J}_v(\mathbf{r}') \times (\mathbf{r} - \mathbf{r}')\, dv'}{|\mathbf{r} - \mathbf{r}'|^3}.$$

As a reminder, the primed coordinates designate the source (here, the current) locations, while the unprimed coordinates designate the location of the field point. Since the derivatives represented by the divergence operator involve the unprimed coordinates only and the integration involves the primed (or source) coordinates, we can move the divergence operator inside the integral. Then,

$$\nabla \cdot \mathbf{B}(\mathbf{r}) = \frac{\mu_0}{4\pi} \int_v \nabla \cdot \frac{\mathbf{J}_v(\mathbf{r}') \times (\mathbf{r} - \mathbf{r}')\, dv'}{|\mathbf{r} - \mathbf{r}'|^3}.$$

(4.13)

Let us examine the integrand of this integral:

$$\nabla \cdot \frac{\mathbf{J}_v(\mathbf{r}') \times (\mathbf{r} - \mathbf{r}')}{|\mathbf{r} - \mathbf{r}'|^3},$$

which we write in terms of its rectangular components,

$$\frac{\partial}{\partial x}\left\{\frac{J_{v,y}(z-z')-J_{v,z}(y-y')}{D^{3/2}}\right\} + \frac{\partial}{\partial y}\left\{\frac{J_{v,z}(x-x')-J_{v,x}(z-z')}{D^{3/2}}\right\}$$
$$+\frac{\partial}{\partial z}\left\{\frac{J_{v,x}(y-y')-J_{v,y}(x-x')}{D^{3/2}}\right\}.$$

In this expression, D represents $\left[(x-x')^2+(y-y')^2+(z-z')^2\right]$, for the sake of brevity. Since each of the components of the current density $J_{v,x}$, $J_{v,y}$, and $J_{v,z}$ is a function of x', y', and z' only, each of these terms factors outside the differentials, yielding

$$\boldsymbol{\nabla}\cdot\frac{\mathbf{J}_v(\mathbf{r}')\times(\mathbf{r}-\mathbf{r}')}{|\mathbf{r}-\mathbf{r}'|^3} = -\frac{3}{2}\left\{\frac{(2)\,(x-x')\left[J_{v,y}(z-z')-J_{v,z}(y-y')\right]}{D^{5/2}}\right\}$$
$$-\frac{3}{2}\left\{\frac{(2)\,(y-y')\left[J_{v,z}\,(x-x')-J_{v,x}\,(z-z')\right]}{D^{5/2}}\right\}$$
$$-\frac{3}{2}\left\{\frac{(2)\,(z-z')\left[J_{v,x}(y-y')-J_{v,y}\,(x-x')\right]}{D^{5/2}}\right\}.$$

The terms inside the curly brackets sum to zero, telling us that the integrand of Eq. (4.13) is zero, and therefore proving Eq. (4.12).

The integral form of Eq. (4.12) is

$$\boxed{\oint_s \mathbf{B}\cdot d\mathbf{s} = 0,} \tag{4.14}$$

where the surface integral is evaluated on *any* closed surface s. This integral relation follows immediately upon taking the volume integral of $\boldsymbol{\nabla}\cdot\mathbf{B}$ for any volume v. By Eq. (4.12), this integral must be zero, since the integrand is zero. Application of the Divergence Theorem leads directly to Eq. (4.14). This integral is a remarkable result. It tells us that there are no "sources" of magnetic field lines; no magnetic charges that can be thought of like electric charges. For electric charges, electric field lines originate on positive charges and terminate on negative charges. In contrast, magnetic field lines do not originate or terminate anywhere. Instead, they must circulate around and close on themselves. The magnetic flux lines entering a volume v through one surface must exit that volume through another surface; no more and no less. We have already seen several examples of this property. For the field lines generated by the current in a long, straight wire, the field lines circulate around the wire in the ϕ direction, and close on themselves. For the current loop, we calculated the magnetic flux density $\mathbf{B}$ only along the z-axis, but we described the field lines more generally as passing through the wire loop, circling around, and closing on themselves. In each and every example, we have reached this same conclusion, and we now see through Eq. (4.14) that this must generally be fulfilled.

Our second differential property of $\mathbf{B}$ is regarding its curl:

$$\boxed{\boldsymbol{\nabla}\times\mathbf{B} = \mu_0\,\mathbf{J}_v.} \tag{4.15}$$

In words, this vector law tells us that magnetic flux lines circulate around the current that generates them. In this form, this law is valid only for static magnetic fields in vacuum; we will have to modify it later when we consider materials and time-varying effects.

Even though not yet complete, we will refer to Eq. (4.15) as **Ampère's Law**, named after André-Marie Ampère. (See Biographical Note 4.4.) But do keep in mind that we have modifications to make before it is complete.

Biographical Note 4.4 André-Marie Ampère (1775-1836)

FIGURE 4.22 André-Marie Ampère, 1825.

A French physicist and mathematician, André-Marie Ampère carried out fundamental studies of magnetic forces that helped form our understanding of magnetic fields created by electrical currents. Upon hearing the report of his friend François Arago to the French Academy, in which Arago discussed and demonstrated the observations of Hans Christian Oersted, Ampère showed that the force between two long, straight, parallel, current-carrying wires could be attractive or repulsive, depending on whether the direction of the current in the wires was the same or opposite, and started to construct the mathematical formalism to describe this. He reported the Ampère Force Law, which stated that the force between these two wires was proportional to the product of the currents, and proportional to their lengths.

Ampère, whose likeness is shown in Fig. 4.22, was elected to the French Academy in 1814, as a foreign member to the Royal Society in 1827, and to the Royal Swedish Academy of Science in 1828. The ampere, the SI unit of current, is named in his honor.

Ampère's Law can also be written in its integral form. We establish this by constructing any surface s, bounded by a contour c, as shown in Fig. 4.23. Upon taking the (vector) surface integral of both sides of Eq. (4.15) over this surface s, we find

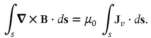

$$\int_s \boldsymbol{\nabla} \times \mathbf{B} \cdot d\mathbf{s} = \mu_0 \int_s \mathbf{J}_v \cdot d\mathbf{s}.$$

We recognize the right side as the total current, I_{enc}, passing through the surface s, and use Stokes' Theorem to write the left side as the line integral:

$$\oint_c \mathbf{B} \cdot d\boldsymbol{\ell} = \mu_0 \int_s \mathbf{J}_v \cdot d\mathbf{s}. \qquad (4.16)$$

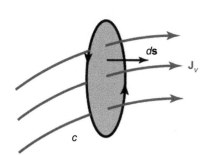

FIGURE 4.23 The contour c bounds the surface s. (For use in deriving the integral form of Ampère's Law, Eq. (4.16).)

Ampère's Law in its integral form gives us a means of determining the magnetic flux density for some very special highly symmetric cases, as we will explore more fully in the following section.

4.3 Ampère's Law in Vacuum

The integral form of Ampère's Law, Eq. (4.16), can be applied profitably to determine the magnetic flux density in many highly symmetric cases, in a manner that should remind you of application of Gauss' Law to determine electric fields. Our prescription will take the following steps:

1. Identify the symmetry of the current distribution.
2. Use the symmetry to reduce the number of components of the magnetic field, as well as the variables that the magnetic field can depend on.
3. Choose a closed path for carrying out the path integration.
4. Extract **B** from the integral $\oint_c \mathbf{B} \cdot d\boldsymbol{\ell}$, and solve for **B**.

LEARNING OBJECTIVES

After studying this section, you should be able to:

• apply Ampère's Law to determine the magnetic flux density in vacuum for geometries of high symmetry, such as infinitely long circular conductors, solenoids, toroids, and infinitely wide sheets of current.

The choice of the closed path in Step 3 is key. Using arguments based upon the symmetry of the current distribution, the magnetic flux density must be constant, or along some segments must contribute nothing to the line integral, thus making evaluation of the integral particularly simple.

We illustrate this process in Example 4.7.

Example 4.7 Ampère's Law, Long, Straight Wire

A current I flows through a long, straight round wire of radius a. The current density $\mathbf{J}_v$ is uniformly distributed within the wire. Determine the magnetic flux density $\mathbf{B}$ generated by the current at positions inside and outside the wire.

Solution:

Since the wire is cylindrically symmetric, we choose to work in cylindrical coordinates, with the z-axis along the center of the wire, and $\mathbf{J}_v$ flowing in the $+z$-direction. We immediately recognize that $\mathbf{B}$ cannot depend on the coordinates ϕ or z; only a ρ-dependence is permissible. We reach this conclusion by considering a rotation of the wire about its axis, or a translation along z. In either case, the wire and current are unchanged, and so therefore the field must be unchanged as well. The symmetry of the wire can also be used to argue that B_ρ and B_z must be zero. B_z must be zero since the current flows in the z-direction, and by the Biot–Savart Law (Eq. (4.3), (4.5), or (4.6)), $\mathbf{B}$ must be perpendicular to the direction of the current. The absence of the B_ρ is a little more subtle. Refer to the cross-sectional view of the wire in Fig. 4.24, which shows two elements within the wire and the magnetic flux density $d\mathbf{B}_1$ and $d\mathbf{B}_2$ at point P due to the current passing through each. These two elements of current are symmetrically placed, equidistant from the center of the wire, and equidistant from the field point P. Since $d\mathbf{B}$ is always perpendicular to the position vector $\mathbf{r}-\mathbf{r}'$ that joins the source to the field point P, $d\mathbf{B}$ due to each of these current elements has components in the ϕ- and ρ-directions. Since the two elements are equidistant from the point P and symmetrically placed, the ϕ components are equal in magnitude and sign, and add to one another. The ρ components are also equal in magnitude, but are opposite in sign, and these components precisely cancel one another. For each element to the left of the axis, there is a symmetrically placed element to the right side, and the net ρ component vanishes. Thus only the B_ϕ component survives. Since we already established that the magnitude of $\mathbf{B}$ can depend only on ρ, the magnetic flux density is written $\mathbf{B} = \mathbf{a}_\phi\, B_\phi(\rho)$.

We are now in a position to use Ampère's Law to determine the magnitude of $B_\phi(\rho)$. We start by finding $\mathbf{B}$ outside the wire, for which we construct a circular path c of radius $\rho > a$, concentric with the wire, as shown in Fig. 4.25(a). By choosing a circular path, $B_\phi(\rho)$ is constant everywhere on the path, and the left side of Eq. (4.16) becomes

$$\oint_c \mathbf{B} \cdot d\boldsymbol{\ell} = B_\phi\, 2\pi\rho.$$

The element $d\boldsymbol{\ell}$ is $\mathbf{a}_\phi\, \rho\, d\phi$ for a circular path. Since B_ϕ is independent of ϕ, we were able to pull this constant factor outside the integral, and then use $\int d\phi = 2\pi$ when integrating around one complete loop. The right side of Eq. (4.16) is

$$\mu_0 \int_s \mathbf{J}_v \cdot d\mathbf{s} = \mu_0 I_{enc}.$$

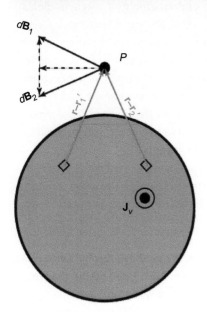

FIGURE 4.24 The cross-section of a current-carrying wire. We show two elements of current, one to the left of center and one to the right. $d\mathbf{B}_1$ and $d\mathbf{B}_2$ indicate the contribution to the total magnetic flux density at point P due to each. The components of $d\mathbf{B}_1$ and $d\mathbf{B}_2$ in the ρ direction are equal in magnitude, but opposite in sign, and perfectly cancel.

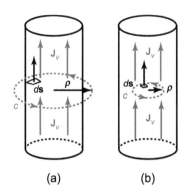

FIGURE 4.25 A long, current-carrying wire of radius a. The contours c indicate the paths around which we apply Ampère's Law. In (a), the radius of c is $\rho > a$, allowing us to find **B** at a point outside the wire, while in (b) we have chosen $\rho < a$, which we will use to find **B** inside the wire. The direction of the path contour $d\boldsymbol{\ell}$ and the surface element $d\mathbf{s}$ are related by a right-hand rule.

We have chosen the simplest surface s for this integral – that is, the flat circular surface lying in the x–y plane. Since the radius of the path is larger than that of the wire, the enclosed current I_{enc} is simply the total current carried by the wire, but let us be a bit more explicit in evaluating $\mu_0 \int_s \mathbf{J}_v \cdot d\mathbf{s}$ so that we can be sure to get the sign right. Remember that the direction of the path element $d\boldsymbol{\ell}$ and the surface element $d\mathbf{s}$ are related by a right-hand rule. The directions of each are shown in Fig. 4.25, with $d\mathbf{s} = \mathbf{a}_z\, \rho\, d\rho\, d\phi$. Since the current is uniformly distributed over the cross-section of the wire, the current density is $\mathbf{J}_v = \mathbf{a}_z\, I/\pi a^2$, where we have simply divided the total current carried by the wire by the cross-sectional area of the wire. Since $d\mathbf{s}$ and $\mathbf{J}_v$ are in the same direction, the dot product $\mathbf{J}_v \cdot d\mathbf{s}$ is positive, and we have

$$\mu_0 \int_s \mathbf{J}_v \cdot d\mathbf{s} = \mu_0 \int_{\phi=0}^{2\pi} \int_{\rho=0}^{a} \frac{I}{\pi a^2}\, \rho\, d\rho\, d\phi,$$

where the ϕ integration extends around one full circle and the ρ integration extends from 0 to a, the radius of the wire. Carrying out the integration gives us $\mu_0 \int_s \mathbf{J}_v \cdot d\mathbf{s} = +\mu_0 I$. Thus Ampère's Law becomes

$$B_\phi 2\pi\rho = \mu_0 I,$$

or after rearranging,

$$B_\phi = \frac{\mu_0 I}{2\pi\rho}, \tag{4.17}$$

valid for $\rho > a$. This result is consistent with Eq. (4.9), the magnetic flux density **B** for a long, straight wire that we found earlier using the Biot–Savart Law. (See Example 4.2.) The sign of B_ϕ is positive, consistent with the right-hand rule (grasp the wire with your right hand, with your thumb pointing in the direction of the current; **B** circulates in the direction indicated by your fingers), and its magnitude decreases as $1/\rho$.

Next we will apply Ampère's Law to determine the magnetic flux density at a point *inside* the wire. For this, we follow a similar process, evaluating Eq. (4.16) on a path c, but in this case we choose a path whose radius ρ is less than that of the wire a. This path is shown in Fig. 4.25(b). Evaluation of the integral on the left side of Eq. (4.16) is identical to what we described above, with the same result: $\oint_c \mathbf{B} \cdot d\boldsymbol{\ell} = B_\phi 2\pi\rho$. Evaluation of the current passing through the surface bounded by the contour c is a bit different, however, since the contour c is now smaller than the wire and only the current that flows through the surface inside c is included. The limits of integration when evaluating $\int_s \mathbf{J}_v \cdot d\mathbf{s}$ are 0 to 2π for ϕ and 0 to ρ for ρ, with the result

$$\mu_0 \int_s \mathbf{J}_v \cdot d\mathbf{s} = \mu_0 I \rho^2/a^2.$$

(The same result can be reached by a geometrical argument that the fraction of current passing through the circle of radius ρ is $\pi\rho^2/\pi a^2$.) Setting this equal to $B_\phi 2\pi\rho$ leads to

$$B_\phi = \frac{\mu_0 I \rho}{2\pi a^2}, \tag{4.18}$$

valid for $\rho < a$. Thus, inside the wire B_ϕ grows linearly with ρ from 0 at the center to $\mu_0 I/(2\pi a)$. Notice that B_ϕ is the same value at $\rho = a$, whether approached from the inside or from the outside. See Fig. 4.26 for a plot of B_ϕ vs. ρ.

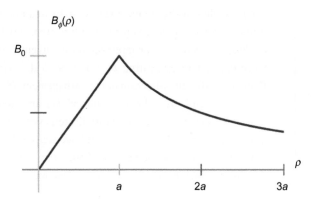

FIGURE 4.26 B_ϕ vs. ρ for a long, current-carrying wire of radius a. The value of B_ϕ at $\rho = a$ is $B_0 = \mu_0 I/2\pi a$.

Notice that in Example 4.7 we found that the magnetic flux density $\mathbf{B}(\mathbf{r})$ outside the wire is identical to that for a long, slender wire, or actually for a wire of any diameter. Let's try another application of Ampère's Law, this time for an infinite sheet of surface current (Example 4.8).

Example 4.8 Ampère's Law, Current Sheet

A sheet of current, infinite in extent in the $x = 0$ plane, is of surface current density J_s and flows in the $+z$-direction. Determine the magnetic flux density produced by this current.

Solution:

This problem is similar to, but not quite the same as, the problem that we worked previously as Example 4.4. The width of the current sheet here is infinite, while in the previous example it was w. We should expect to be able to compare these results, in certain limiting cases, at the end.

We work this problem in rectangular coordinates, and start with a symmetry argument to show that only the y-component of $\mathbf{B}$ exists, and that B_y can depend on the distance x, but not on y or z. B_z must be zero since the current flows in the z-direction, and by the Biot–Savart Law $\mathbf{B}$ must be perpendicular to the current. We show that B_x is zero using Fig. 4.27, which shows two current elements in the current sheet and $d\mathbf{B}$ at a point P produced by each. The elemental fields are labeled $d\mathbf{B}_l$ (due to the current element on the left) and $d\mathbf{B}_r$ (due to the current element on the right) in the figure. The two current elements are equidistant to the point P. Each contribution to $\mathbf{B}$ is perpendicular to the position vector $\mathbf{r} - \mathbf{r}'$ that points from the element to P. Thus both elements generate equal contributions to B_y. The contributions to B_x are also equal in magnitude, but are opposite in sign. Since the current sheet is infinite in the y dimension, the net contributions to B_x vanish. Thus $\mathbf{B}$ must be in the y-direction. To show that B_y can depend only on x, recall that the current sheet is infinite in extent in the y and z dimensions. Thus a translation in the y- or z-direction does not change the geometry of the current distribution. Therefore, B_y can depend only on x, and $\mathbf{B}$ can be written $\mathbf{a}_y B_y(x)$. Furthermore, based on the symmetry of the current sheet, $B_y(-x)$ must be equal to $-B_y(x)$ – that is, the magnetic flux density below the current sheet is directed

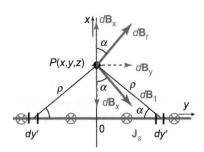

FIGURE 4.27 The magnetic flux density dB_l and dB_r generated by the current elements on the left and right sides of point P, respectively. The two current elements are equidistant from P. The y-components are the same magnitude and sign. The x-components are also the same magnitude, but are opposite in sign, and when added, the net field in the x-direction is zero.

in the opposite direction, but of equal magnitude, to the magnetic flux density above the current sheet. This is proven by rotating the current sheet about the z-axis by 180°. With this rotation, the current sheet is unchanged, so that $\mathbf{B}$ is unchanged as well. Upon the rotation, $B_y(-x)$ is transformed into $-B_y(x)$, so these quantities are identical.

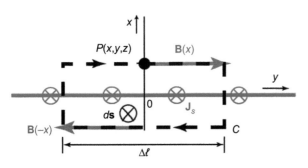

FIGURE 4.28 The infinite current sheet of density $\mathbf{J}_s$ and the path c for application of Ampère's Law.

We next construct a path c, as shown in Fig. 4.28, for application of Ampère's Law. This path is rectangular in shape, with two sides of length $\Delta\ell$ that are parallel to the y-axis, one above and the other below the current sheet, and two sides parallel to the x-axis and passing through the current sheet. The directions of the path element $d\ell$ and the surface element $d\mathbf{s}$ are related by a right-hand rule. On this path, we separate the integral $\oint \mathbf{B} \cdot d\ell$ into the sum of four integrals, one for each leg of the rectangular path. The two vertical sides, however, do not contribute to the integral since $\mathbf{B}$ is perpendicular to $d\ell = \mathbf{a}_x\, dx$ on these sides. Since $\mathbf{B} = \mathbf{a}_y\, B_y(x)$ is constant along the top of the rectangle, $\int \mathbf{B} \cdot d\ell$ for this leg is simply $B_y(x)\,\Delta\ell$. On the bottom of the rectangle, the sign of B_y is reversed, as we discussed in the previous paragraph, but the direction of $d\ell$ is reversed as well, since we are following the path c in the $-y$-direction here, and the line integral along this side is $-B_y(-x)\Delta\ell = B_y(x)\Delta\ell$. The closed-loop integral is therefore

$$\oint \mathbf{B} \cdot d\ell = 2B_y(x)\Delta\ell.$$

By Ampère's Law, this must be equal to $\mu_0 \int_s \mathbf{J}_v \cdot d\mathbf{s}$, which, since the current is a surface current, can be written as $\mu_0 \int_y |\mathbf{J}_s| dy$. Since the surface current is uniform, this integral is evaluated as $\mu_0 |\mathbf{J}_s|\Delta\ell$. We were a little cavalier regarding the sign of this term, so let's revisit that issue. The vector surface element $d\mathbf{s}$ points in the $+z$-direction (into the page), as it must by the right-hand rule for the direction of the contour c shown in Fig. 4.28. Since the current flows in the $+z$-direction as well, the dot product $\mathbf{J}_v \cdot d\mathbf{s}$ must be positive. Thus the positive sign on $\mu_0|\mathbf{J}_s|\Delta\ell$ is correct. We now equate this to $2B_y(x)\Delta\ell$, and solve for $B_y(x)$ to find

$$B_y(x) = \mu_0|\mathbf{J}_s|/2$$

for $x > 0$. It turns out that this value of B_y for an infinite sheet does not even depend on x, the distance from the current sheet. This may seem to be unphysical, but remember that the width of the current sheet itself is infinite.

Let us now return to the problem that we solved previously that can serve as a comparison for this result. Recall that in Example 4.4 we used the Biot–Savart Law to determine the magnetic flux density $\mathbf{B}$ in the space surrounding a current sheet, but that this current sheet was of finite width w. We then examined that result in two limiting cases, one of which was for positions that are close to the surface of the current sheet, but not too close to the ends. In that analysis also, we were able to see that the magnetic flux density points in the $+y$-direction above the sheet and in the $-y$-direction below the sheet, and that its magnitude is $\mu_0|J_s|/2$, in agreement with the result of this example.

In Example 4.8 we applied Ampère's Law to find the magnetic flux density $\mathbf{B}$ generated by an infinite sheet of current. Strictly speaking, this technique is not valid for a sheet of finite width, since we relied on the infinite width to be able to say that $\mathbf{B}$ only points in the y-direction, and that its magnitude depends only on x. Practically speaking, however, we see that our result is just fine for a finite sheet, as long as we are close to the sheet, and not too close to the edges.

Let's look at another application of Ampère's Law in Example 4.9.

Example 4.9 Ampère's Law, Solenoid

Consider an infinitely long solenoid, similar to that described in Example 4.5. The current through the coil is I, the radius is a, and the turns density is n. Use Ampère's Law to determine the magnetic flux density **B** produced by this configuration.

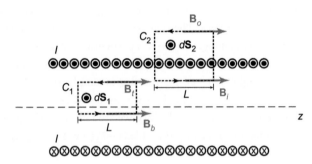

FIGURE 4.29 A side view of an infinitely long solenoid. The current is I, the turns density n, and the radius a. The two rectangular loops are paths for application of Ampère's Law.

Solution:

See Fig. 4.29 for a side view of the solenoid. The rows of circles represent the individual wires of the coil, with the current I coming out of the page at the top and going into the page at the bottom. Based on the shape of the solenoid, it is sensible to work in cylindrical coordinates, with the axis of the solenoid lying along the z-axis. To simplify the analysis, only the component of current flowing in the ϕ direction is considered. The small component of current in the z-direction is ignored. This is usually a valid approximation for a tightly wound solenoid. As with the previous examples, the first step of the solution is to determine which components of **B** can be present, and on which spatial coordinates these components of **B** can depend. Recall from Example 4.3 that **B** for a single loop lies in the ρ–z plane, circling around the wire and closing on itself. Since the B_ϕ component must be zero for each individual current loop, it must be zero for the solenoid as well. (Think of the solenoid as a stack of co-axial loops.) The B_ρ component for the solenoid must also be zero, as can be shown using the symmetry in z of the solenoid. At some point P inside the solenoid, consider the contribution to **B** due to two individual loops, one to the left of point P and another the same distance to the right. The components B_z at this point due to the two coils are equal in magnitude and sign, and these components add to one another. The components B_ρ will also be equal in magnitude, but opposite in sign, and when added together they cancel. For every loop to the left of point P, there is a symmetrically placed loop to the right, and the total contribution to B_ρ in an infinitely long solenoid will perfectly cancel. Thus **B** inside a solenoid has only a B_z component. As for the dependence of B_z on the spatial variables ρ, ϕ, and z, consider a translation of the infinitely long solenoid along the z-axis, or a rotation of the solenoid about the z-axis. In either case, the solenoid is unchanged, and therefore **B** must be unchanged as well. Therefore, B_z can depend only on the radial component ρ. The results of this paragraph are summarized by writing $\mathbf{B} = \mathbf{a}_z B_z(\rho)$.

We are now ready to apply Ampère's Law to determine the magnitude of **B**. We start by showing that **B** inside the solenoid is uniform. Consider the rectangular path C_1 of length L in Fig. 4.29. The closed path integral $\oint \mathbf{B} \cdot d\boldsymbol{\ell}$ is the sum of four integrals, one for each leg of the rectangular path. $d\boldsymbol{\ell}$ on the left or right sides of this path, however, is perpendicular to **B**, so these sides do not contribute to $\oint \mathbf{B} \cdot d\boldsymbol{\ell}$. The top and bottom do contribute, and $\oint \mathbf{B} \cdot d\boldsymbol{\ell} = B_b L - B_t L$, where B_t is the magnetic flux density along the top of the rectangular path, and B_b along the bottom. By Ampère's Law, this line integral must be equal to $\mu_0 I_{enc}$, where I_{enc} is the total current passing through the surface defined by C_1. The current in the solenoid, however, is contained entirely within the wires, and these wires do not pass through this rectangle. Since $I_{enc} = 0$, B_b and B_t must be equal. Applying this analysis to any rectangular path inside the solenoid, we conclude that **B** inside the solenoid is perfectly uniform.

Next we apply Ampère's Law on the rectangular path marked C_2 in Fig. 4.29 to determine the magnitude of this uniform flux density **B** inside the solenoid. This path is also rectangular in shape, and of length L, but for C_2 the top leg lies outside the solenoid. Following an analysis similar to that of the previous paragraph, the closed path integral around this path is $\oint \mathbf{B} \cdot d\boldsymbol{\ell} = B_i L - B_o L$, where B_i is the magnetic flux density inside the solenoid and B_o on the outside. We set this equal to $\mu_0 I_{enc}$, where now I_{enc} has a non-zero value since the current-carrying loops penetrate through the rectangle bordered by C_2. By the right-hand rule, the surface element $d\mathbf{s}_2$ points out of the page, and the current at the top of the solenoid comes out of the page as well. Therefore, I_{enc} is a positive quantity. Its magnitude is simply the current I times nL, the number of loops contained within the rectangle. (Remember that n is the turns density, or number of turns of the solenoid per unit length.) Therefore, by Ampère's Law, $B_i L - B_o L = \mu_0 I n L$, or dividing through by L,

$$B_i - B_o = \mu_0 n I.$$

Recall that we did not specify the distance from the solenoid to the top of the rectangle. So this result, which does not depend in any way on the radial distance ρ, is valid for any locations, as long as one leg of the rectangle is inside the solenoid and one leg is outside. We already know that B_i is perfectly uniform inside the solenoid, and we now conclude that B_o outside the solenoid is perfectly uniform as well. We take this one step further by insisting that B_o outside the solenoid must be zero, since at large distances away from the solenoid contributions to **B** from the top and bottom of the solenoid must perfectly cancel one another. The final result is that the magnetic flux density inside an infinitely long solenoid is

$$\mathbf{B} = \mathbf{a}_z \mu_0 n I,$$

and zero on the outside.

Note that these results for **B** inside and outside an infinitely long solenoid are in perfect agreement with the results of Example 4.5. If you aren't convinced by the argument in Example 4.9 that **B** outside the solenoid is zero, then you can revert to the more rigorous result of Example 4.5.

Example 4.10 Ampère's Law, Toroid

Consider the tightly wound **toroid** shown in Fig. 4.30. (A toroid is a wire-wound coil, similar to a solenoid, but in a circular shape. Picture wrapping the windings of wire around a doughnut.) The radius of the wire loops is a, and the radius to the center of the toroid is b. A current I passes through the windings, and there are N turns total in the toroid. Determine the magnetic flux density **B** generated by this current.

Solution:
We work this problem using cylindrical coordinates ρ, ϕ, and z, with z coming out of the page, and ϕ in the counter-clockwise direction. To simplify the analysis, the current is

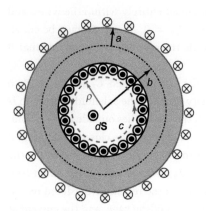

FIGURE 4.30 A tightly wound air-filled toroid treated in Example 4.10.

treated as a continuous surface current, ignoring the fact that the wires are discrete. This approximation is valid as long as the windings around the toroid are tight – that is, the spacing between wires is small. By symmetry, only the ϕ component of **B** for the toroid can exist, and the magnitude of B_ϕ can depend only on ρ. Therefore, B_ϕ is uniform on a circular path, such as path c shown in Fig. 4.30. The line integral of $\mathbf{B} = \mathbf{a}_\phi B_\phi(\rho)$ on this path, using $d\boldsymbol{\ell} = \mathbf{a}_\phi \rho d\phi$, is

$$\oint \mathbf{B} \cdot d\boldsymbol{\ell} = B_\phi 2\pi\rho.$$

B_ϕ in the interior of the toroid (the hole of the doughnut) can be determined by choosing the radius of c to be small, ρ less than $b - a$, as shown in Fig. 4.30. By Ampère's Law, $\oint \mathbf{B} \cdot d\boldsymbol{\ell}$ must be equal to $\mu_0 I_{enc}$, where I_{enc} is the current passing through the surface bounded by c. Since no current loops penetrate the surface inside this contour, the closed path integral $\oint \mathbf{B} \cdot d\boldsymbol{\ell}$ must be zero. Therefore, $B_\phi = 0$ for $\rho < b - a$.

Next, consider a contour c whose radius ρ is larger, such that the contour lies inside the toroid coil. This is defined as the region in which $b - a < \rho < b + a$. Evaluation of the closed path integral follows the same procedure, and the path integral $\oint \mathbf{B} \cdot d\boldsymbol{\ell}$ is again simply $B_\phi 2\pi\rho$. For this contour, however, the enclosed current I_{enc} is not zero, since the N windings at the inside surface of the toroid pass through the surface bounded by c. Therefore, $B_\phi 2\pi\rho = \mu_0 NI$, or solving for B_ϕ, the flux density is

$$B_\phi = \frac{\mu_0 NI}{2\pi\rho}.$$

Let's pause a moment to examine the sign of this field. We labeled the path c in the counter-clockwise direction, which is the $+\mathbf{a}_\phi$ direction. The direction of $\mathbf{a}_n$, the surface normal for the surface bounded by c, is out of the page (the $+\mathbf{a}_z$ direction). The current in the inner loop of the toroid also comes out of the page, parallel to $\mathbf{a}_n$. Therefore I_{enc} is positive, yielding $B_\phi = +\mu_0 NI/(2\pi\rho)$.

Finally, consider a contour c that is larger than the outer surface of the toroid, of radius $\rho > b + a$. The surface bounded by this contour has N windings carrying current out of the page at radius $b - a$, and N windings carrying current into the page at radius $b + a$. Summed together, the net current passing through the surface is zero, and by Ampère's Law, $B_\phi = 0$ for $\rho > b + a$.

As a sanity check on this result for **B** produced by a toroid, let us consider the similarity between a toroid and a solenoid. We could perhaps imagine taking hold of a long solenoid, of turns density n and length L, and bending it around into a circular shape so that the two ends of the solenoid join together, thus forming a toroid. As we do this, we should expect B_ϕ for the toroid to approach the same value as B_z for a solenoid. We found earlier that the latter is $\mu_0 nI$. But the turns density n is just the total number of windings N divided by the length L, which in this case is the circumference of the toroid, $L = 2\pi b$. For a large radius b, the variation in B_ϕ is small, and we substitute $\rho \sim b$, giving us the confirmation of the magnetic flux density for the toroid that we sought.

The tokamak, which is similar in some regards to a toroid, has been under development as a large-scale future energy source. This configuration is the focus of TechNote 4.9.

TechNote 4.9 Tokamak

Nuclear fusion as a future source of energy has been an intense focus of research in many countries for many decades. Nuclear fusion is the energy source of the stars, including the sun, where the internal temperature reaches levels of 15 MK (15 million kelvin). When the mass of the fused nucleus is less than the combined mass of the colliding nuclei, the excess mass is converted to energy. In order for two ions to collide, however, they must have extremely high kinetic energies, such that they can overcome the Coulomb repulsion, which at short distances becomes exceptionally large. In order to create a controlled, high-temperature plasma on Earth, confining that plasma becomes a tremendous technological problem. How does one create a container for such a high-temperature plasma? Certainly any physical bottle is out of the question at these high temperatures. This is the fundamental question of inertial confinement fusion (ICF) programs.

A Tokamak, shown in the schematic of Fig. 4.31, is a device that uses strong magnetic fields to confine a high-temperature plasma to a toroidal region. The first tokamak, developed in a program led by Natan Yavlinsky in Russia, became operational in 1958. Advances in fusion programs have been proceeding over the years, with the goal of producing more power through nuclear fusion than the input power required to operate the reactor. In the late 1980s, a multinational effort called the International Thermonuclear Experimental Reactor (ITER) was initiated. In 2020, the Joint European Tokamak (JET) reported a record power gain, producing 16 MW of output power for 24 MW of input power, not yet to the break-even point, but approaching it. (At the break-even point, the output power equals the input power.)

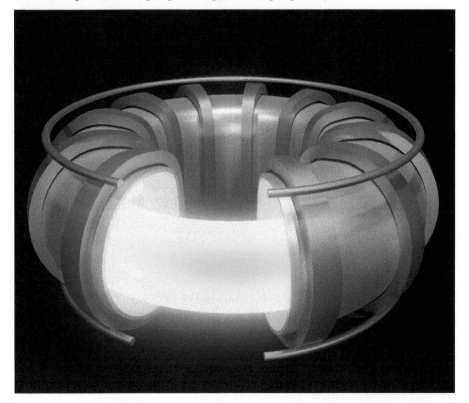

FIGURE 4.31 A schematic representation of the magnetic field coils (bronze D-shaped coils) and the confined plasma (yellow) of a tokamak, a reactor used for confinement of a high-temperature plasma to achieve large-scale nuclear fusion as a future source of energy.

A tokamak uses magnetic fields to confine the positive and negative ions that make up the high-temperature plasma. We discussed earlier how a charged particle follows a helical path when in a uniform magnetic field. We also discussed the uniformity of the magnetic field generated inside a long solenoid. If we were to inject charged particles near the center of a solenoid, these ions would in principle never reach the windings of the solenoid, but would instead spiral about an axis parallel to the axis of the solenoid. A solenoid, however, has finite length, and eventually the charged particles will reach the end and exit the solenoid. A solution to this problem is to use a toroid, as described in Example 4.10, rather than a solenoid. Geometrically, we imagine twisting the ends of a solenoid around until they meet. Since a toroid has no ends, this geometry does not suffer the end loss of the solenoid. As we showed in Example 4.10, however, the magnetic field inside a toroid is not uniform, but rather decreases as $1/\rho$, where ρ is the radial distance measured from the center axis of the toroid. Because of this field inhomogeneity, the particle trajectory inside the toroid is not stable, and particles will soon strike the walls or magnetic field coils. This issue can be addressed by adding extra magnets to the toroid, such that the magnetic field inside the toroid is not strictly in the ϕ-direction, but rather has a spiral. The ions will be guided around the toroid by the primary magnetic field, and confined to the axis of the toroid by the helical field component. This description is overly simplified, of course, and creating a tokamak that will exceed the break-even point has been a huge challenge.

In this section we have applied Ampère's Law to several important geometries for determination of the magnetic flux density. We have not yet included magnetic materials in our discussion; that will follow later. In order to apply Ampère's Law to these solutions, we rely on the symmetry of the current distribution. Ampère's Law is, of course, always valid, but without sufficient symmetry we are not able to simply evaluate $\oint \mathbf{B} \cdot d\boldsymbol{\ell}$ to determine $\mathbf{B}$ itself. Still, Ampère's Law allows us a very powerful means of determining $\mathbf{B}$ for those cases where it can be used. We can also use it to approximate field strengths in certain other cases when the symmetry conditions aren't perfect, but close enough. This will follow later, after we have introduced magnetic materials, with examples such as a toroid with an air gap.

4.4 Vector Magnetic Potential

When discussing electric fields in Chapter 1, we found it very useful to define the electric potential function $V(\mathbf{r})$. This potential gave us an alternate method for determining electric fields, as well as a means of visualizing electrical "landscapes" through a series of equipotential contours. In this section, we define a potential function useful for determination of magnetic flux densities. We will also develop further the properties of this potential, which is a vector and is denoted $\mathbf{A}(\mathbf{r})$, learn how to determine $\mathbf{A}(\mathbf{r})$ from known currents, demonstrate the calculation of magnetic flux density $\mathbf{B}(\mathbf{r})$ using $\mathbf{A}(\mathbf{r})$, and show that this form of the vector potential is consistent with the Biot–Savart Law.

Let's start by taking stock of what we know about the magnetic flux density $\mathbf{B}(\mathbf{r})$. First, as shown in Section 4.2, the magnetic flux density $\mathbf{B}(\mathbf{r})$ must satisfy the vector differential property $\nabla \cdot \mathbf{B} = 0$. (We haven't yet discussed magnetic materials or time-varying fields, but we will see that this law remains valid in these cases as well.) Second, as shown in Appendix E.2, for any vector whose divergence is zero, there exists a vector potential, which we will denote $\mathbf{A}$, and $\mathbf{B}(\mathbf{r})$ and $\mathbf{A}(\mathbf{r})$ are related through

$$\boxed{\mathbf{B}(\mathbf{r}) = \nabla \times \mathbf{A}(\mathbf{r}).}$$

(4.19)

We do not yet know how to calculate this vector potential, only that it exists. The units for $\mathbf{A}$ are $[\mathbf{A}] = \text{Tm}$ or Wb/m, as can be seen from Eq. (4.19).

The **magnetic potential** $\mathbf{A}(\mathbf{r})$ differs from the electric potential $V(\mathbf{r})$ in several significant ways. First, as we already mentioned, the magnetic potential is a vector field, whereas the electric potential is a scalar field. Additionally, the vector field $\mathbf{A}$ is not unique. This may seem like a show-stopper right there, but it turns out to be not as serious as it sounds. In Appendix E.3, it is shown that, for a general vector field, both differential operators of a vector field (its divergence *and* its curl) are needed in order to fully determine that vector field. For the vector magnetic potential, we know its curl ($\nabla \times \mathbf{A} = \mathbf{B}$), but we do not know its divergence. In fact, it will be useful later to define different functions for $\nabla \cdot \mathbf{A}$ that are useful for different types of problems. These are known as different **gauges** of the magnetic field. Since $\nabla \cdot \mathbf{A}$ is not restricted by the physics of the problems, we are free to choose $\nabla \cdot \mathbf{A}$ for our own convenience.

Another significant difference between vector magnetic potentials and electric potentials is in application to boundary value problems. With magnetic fields, we do not have an equivalent to equipotentials as we did with electric potentials. Notwithstanding these detractions, it is still extremely useful to define the vector magnetic potential function $\mathbf{A}$, to define methods for its determination, and to use $\mathbf{A}$ for determination of $\mathbf{B}$.

To this point, we know that we can define $\mathbf{A}$, but we know little else about it. In particular, we do not yet know how to determine its value. Let's see if we can make some progress in this direction. As introduced, $\mathbf{A}$ and $\mathbf{B}$ are related through $\nabla \times \mathbf{A} = \mathbf{B}$. But in Section 4.2 we introduced Ampère's Law (albeit not quite complete), which in its (present) differential form is $\nabla \times \mathbf{B} = \mu_0 \mathbf{J}_v$, where $\mathbf{J}_v$ is the current density in that region of space. We combine these two relations to form

$$\nabla \times (\nabla \times \mathbf{A}) = \mu_0 \mathbf{J}_v,$$

which becomes, using standard vector manipulation properties (see Eq. (E.29)),

$$\nabla(\nabla \cdot \mathbf{A}) - \nabla^2 \mathbf{A} = \mu_0 \mathbf{J}_v.$$

(4.20)

We have seen some of this notation before, but part of it is new. In either case, it is useful to define explicitly what each of these second-order differentials means. We will do so in rectangular coordinates, and refer you to Eqs. (E.31) and (E.33) for appropriate forms in cylindrical and spherical coordinates. The first double differential is the gradient of the divergence of $\mathbf{A}$. We have seen gradients before, and we have seen divergences before, so we simply need to combine these to find

$$\begin{aligned}
\nabla(\nabla \cdot \mathbf{A}) &= \left(\mathbf{a}_x \frac{\partial}{\partial x} + \mathbf{a}_y \frac{\partial}{\partial y} + \mathbf{a}_z \frac{\partial}{\partial z} \right) \left(\frac{\partial A_x}{\partial x} + \frac{\partial A_y}{\partial y} + \frac{\partial A_z}{\partial z} \right) \\
&= \mathbf{a}_x \left(\frac{\partial^2 A_x}{\partial x^2} + \frac{\partial}{\partial x} \frac{\partial A_y}{\partial y} + \frac{\partial}{\partial x} \frac{\partial A_z}{\partial z} \right) \\
&\quad + \mathbf{a}_y \left(\frac{\partial}{\partial y} \frac{\partial A_x}{\partial x} + \frac{\partial^2 A_y}{\partial y^2} + \frac{\partial}{\partial y} \frac{\partial A_z}{\partial z} \right) \\
&\quad + \mathbf{a}_z \left(\frac{\partial}{\partial z} \frac{\partial A_x}{\partial x} + \frac{\partial}{\partial z} \frac{\partial A_y}{\partial y} + \frac{\partial^2 A_z}{\partial z^2} \right).
\end{aligned}$$

Thus there are nine terms in all in $\nabla(\nabla \cdot \mathbf{A})$.

The second term, $\nabla^2 \mathbf{A}$, can be written compactly as

$$\nabla^2 \mathbf{A} = \mathbf{a}_x \nabla^2 A_x + \mathbf{a}_y \nabla^2 A_y + \mathbf{a}_z \nabla^2 A_z,$$

where the ∇^2 operator, known as the Laplacian, acting on a scalar function, was defined previously. (See Section E.2.) When written out explicitly, $\nabla^2 \mathbf{A}$ is

$$\nabla^2 \mathbf{A} = \mathbf{a}_x \left(\frac{\partial^2 A_x}{\partial x^2} + \frac{\partial^2 A_x}{\partial y^2} + \frac{\partial^2 A_x}{\partial z^2} \right)$$
$$+ \mathbf{a}_y \left(\frac{\partial^2 A_y}{\partial x^2} + \frac{\partial^2 A_y}{\partial y^2} + \frac{\partial^2 A_y}{\partial z^2} \right)$$
$$+ \mathbf{a}_z \left(\frac{\partial^2 A_z}{\partial x^2} + \frac{\partial^2 A_z}{\partial y^2} + \frac{\partial^2 A_z}{\partial z^2} \right).$$

Thus, $\nabla^2 \mathbf{A}$ also has nine terms. Note that several of these are common to those found in $\nabla(\nabla \cdot \mathbf{A})$, and when combined nicely cancel. The identity Eq. (E.29) can be confirmed by writing out explicitly the determinant form of $\nabla \times (\nabla \times \mathbf{A})$ and comparing to the surviving terms of $\nabla(\nabla \cdot \mathbf{A}) - \nabla^2 \mathbf{A}$. This proof is left to the reader as Prob. PE-6.

Although Eq. (4.20) can look quite intimidating, let's not throw in the towel just yet. Remember that we wrote earlier that there is no physical restriction to the quantity $\nabla \cdot \mathbf{A}$. We are therefore free to choose a value of $\nabla \cdot \mathbf{A}$ for our own convenience. One choice of gauge, as this is known, is to choose $\nabla \cdot \mathbf{A} = 0$. This is known as the **Coulomb gauge**, and is particularly useful for static magnetic fields. In the Coulomb gauge, Eq. (4.20) simplifies to

$$\nabla^2 \mathbf{A} = -\mu_0 \mathbf{J}_v, \tag{4.21}$$

or, when written in terms of each of the individual rectangular components

$$\nabla^2 A_i = -\mu_0 J_{v,i},$$

where i represents x, y, or z (when we are working in rectangular coordinates). This differential equation should remind you of Poisson's Equation (Eq. (2.19)) for the electric potential,

$$\nabla^2 V(\mathbf{r}) = -\rho_v / \varepsilon_0,$$

for which the solution is

$$V(\mathbf{r}) = \frac{1}{4\pi\varepsilon_0} \int_v \frac{\rho_v(\mathbf{r}') \, dv'}{|\mathbf{r} - \mathbf{r}'|}.$$

Without further proof, then, we assert by analogy to the scalar electric potential that the solution for each of the components of the vector magnetic potential A_i for static currents is of similar form,

$$A_i(\mathbf{r}) = \frac{\mu_0}{4\pi} \int_v \frac{J_{v,i}(\mathbf{r}') \, dv'}{|\mathbf{r} - \mathbf{r}'|},$$

or when written in vector notation:

$$\boxed{\mathbf{A}(\mathbf{r}) = \frac{\mu_0}{4\pi} \int_v \frac{\mathbf{J}_v(\mathbf{r}') \, dv'}{|\mathbf{r} - \mathbf{r}'|}.} \tag{4.22}$$

When the current flows on a surface s, the vector potential is similarly

$$\boxed{\mathbf{A}(\mathbf{r}) = \frac{\mu_0}{4\pi} \int_s \frac{\mathbf{J}_s(\mathbf{r}') \, ds'}{|\mathbf{r} - \mathbf{r}'|},} \tag{4.23}$$

or for a current I in a slender filament, such as through a wire,

$$A(\mathbf{r}) = \frac{\mu_0}{4\pi} \int_\ell \frac{I\, d\boldsymbol{\ell}'}{|\mathbf{r} - \mathbf{r}'|}.$$

(4.24)

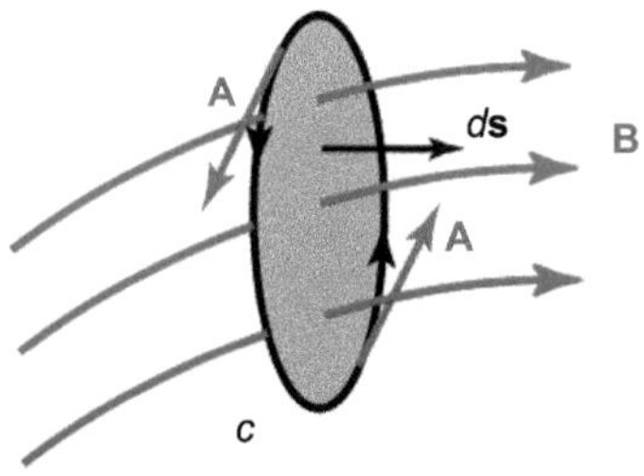

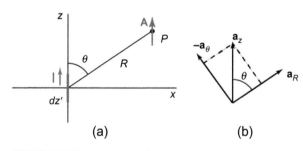

FIGURE 4.32 The magnetic flux through the surface s bounded by the contour c is $\Phi = \int_s \mathbf{B} \cdot d\mathbf{s}$.

Before working through a few examples in which we find the vector potential $A(\mathbf{r})$ using Eqs. (4.22)–(4.24), let's look at a particularly simple relationship between the vector potential $\mathbf{A}$ and the magnetic flux through some surface s. (We have already seen one relation involving the magnetic flux, as expressed in Eq. (4.14). As a reminder, that relation applied to a *closed* surface. We will also make use of the magnetic flux later when we introduce the ideas of inductance and potentials introduced by time-varying magnetic fields.) The magnetic flux through the surface s bounded by the contour c, as shown in Fig. 4.32, is defined as

$$\Phi_M = \int_s \mathbf{B} \cdot d\mathbf{s}.$$

(4.25)

But $\mathbf{B}$ can be written in terms of the vector potential $\mathbf{A}$, as in Eq. (4.19), so the flux becomes

$$\Phi_M = \int_s (\boldsymbol{\nabla} \times \mathbf{A}) \cdot d\mathbf{s}.$$

Upon application of Stokes' Theorem this surface integral becomes the path integral, and the magnetic flux through the surface s can be written as

$$\Phi_M = \oint_c \mathbf{A} \cdot d\boldsymbol{\ell},$$

(4.26)

where the path integral is performed around the path c that bounds the surface s. Thus we see that the line integral of $\mathbf{A}$ around any closed path c is equal to the magnetic flux $\int_s \mathbf{B} \cdot d\mathbf{s}$ through the surface s that is bounded by c. We'll make use of this handy relation later.

Let's look now at a few examples of $\mathbf{A}$ for specific currents, and computation of $\mathbf{B}$ from this potential function (Examples 4.11–4.13).

Example 4.11 Vector Magnetic Potential, Short Current Element

FIGURE 4.33 The geometry for finding the vector magnetic potential $\mathbf{A}$ due to a short, straight current element of length dz' and current I.

Determine the vector magnetic potential $\mathbf{A}$ at point P due to a short, straight current element of length dz' and current I at the origin. Using this function $\mathbf{A}$, determine the magnetic flux density $\mathbf{B}(\mathbf{r})$.

Solution:

A short current element is shown in Fig. 4.33(a). The current element is located at the origin, directed along the z-axis, and the distance from the source to point P is R. Application of Eq. (4.24) leads to

$$\mathbf{A} = \frac{\mu_0}{4\pi} \frac{I\, dz'}{R}\, \mathbf{a}_{z'},$$

where we have omitted the integral sign since there is only the single element of current. Notice from Eq. (4.24) that the vector potential $\mathbf{A}$ points in the same direction as the current element $d\boldsymbol{\ell}'$.

In order to determine the magnetic flux density **B**, we use Eq. (4.19). It is convenient to first convert to spherical components using

$$\mathbf{a}_z = \cos\theta\,\mathbf{a}_R - \sin\theta\,\mathbf{a}_\theta,$$

as can be verified using the illustration in Fig. 4.33(b). (We show this conversion in Example B.5, but also show it here for completeness.) In spherical coordinates, the determinant form of **B** is

$$\mathbf{B} = \nabla \times \mathbf{A} = \frac{1}{R^2\sin\theta}\begin{vmatrix} \mathbf{a}_R & \mathbf{a}_\theta R & \mathbf{a}_\phi R\sin\theta \\ \dfrac{\partial}{\partial R} & \dfrac{\partial}{\partial\theta} & \dfrac{\partial}{\partial\phi} \\ A_R & RA_\theta & (R\sin\theta)A_\phi \end{vmatrix}.$$

Using $A_R = \mu_0 I\,dz'\cos\theta/(4\pi R)$, $RA_\theta = -\mu_0 I\,dz'\sin\theta/(4\pi)$, and $(R\sin\theta)A_\phi = 0$, we find

$$\mathbf{B} = \left(\frac{\mu_0 I\,dz'}{4\pi}\right)\frac{1}{R^2\sin\theta}\left[-\mathbf{a}_\phi R\sin\theta\left(\frac{-\sin\theta}{R}\right)\right] = \frac{\mu_0 I\,dz'\sin\theta}{4\pi R^2}\,\mathbf{a}_\phi.$$

Since **B** has only a ϕ component, the magnetic flux density circles around the z-axis in the $+\phi$-direction. The strength of this flux density falls off as $1/R^2$ with increasing distance R. The form of this element of **B** is slightly different from the Biot–Savart Law of Eq. (4.3), but with just a few minutes' thought, these results can be shown to be consistent with one another.

In the next example, we return to the long, straight wire carrying a current I. Recall that in Examples 4.2 and 4.7 we found that the magnetic flux density **B** generated by this current is $\mathbf{B}(\mathbf{r}) = \mathbf{a}_\phi\mu_0 I/(2\pi\rho)$, when the length of the wire is infinite. In the following example we will determine the vector magnetic potential **A** for the current in a straight wire of finite length. We will then use this **A** to determine **B**.

Example 4.12 Vector Magnetic Potential, Long, Straight Wire

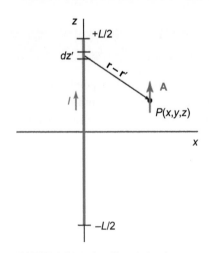

FIGURE 4.34 A wire of length L and carrying a current I lies along the z-axis.

Determine the vector magnetic potential **A** at point $P(x,y,z)$ for a straight wire of length L.

Solution:
The wire, placed along the z-axis and extending from $-L/2$ to $+L/2$, is illustrated in Fig. 4.34. The vector **r**, which points to P, is (x,y,z), and the vector $\mathbf{r}'$, which points to an element on the wire, is $(0,0,z')$. Using Eq. (4.24), the vector potential is

$$\mathbf{A} = \frac{\mu_0}{4\pi}\int_{-L/2}^{L/2}\frac{\mathbf{a}_z I\,dz'}{[\rho^2 + (z-z')^2]^{1/2}},$$

where $\rho = [x^2 + y^2]^{1/2}$. The integral is evaluated with the aid of Eq. (G.7) of Appendix G to show that

$$\mathbf{A} = \mathbf{a}_z\frac{\mu_0 I}{4\pi}\ln\left(\frac{(z+L/2)+\sqrt{(z+L/2)^2+\rho^2}}{(z-L/2)+\sqrt{(z-L/2)^2+\rho^2}}\right).$$

With this result we see that **A** points in the z-direction, as we should expect since the current flows in the z-direction as well. At the midpoint of this wire, the vector potential reduces to

$$\mathbf{A} = \mathbf{a}_z \frac{\mu_0 I}{4\pi} \ln\left(\frac{L/2 + \sqrt{(L/2)^2 + \rho^2}}{-L/2 + \sqrt{(L/2)^2 + \rho^2}}\right),$$

which can be simplified to

$$\mathbf{A} = \mathbf{a}_z \frac{\mu_0 I}{2\pi} \ln\left(\frac{L/2 + \sqrt{(L/2)^2 + \rho^2}}{\rho}\right). \tag{4.27}$$

Now let's use this potential to find $\mathbf{B}$ for an infinitely long straight wire.

Example 4.13 Finding B from A

Using the result of Example 4.12, determine the magnetic flux density in the space surrounding an infinitely long wire carrying a current I.

Solution:

Of course, we already know the result, as we have previously determined $\mathbf{B}$ for an infinitely long wire using two different methods, namely the Biot–Savart Law and Ampère's Law. In this solution we will determine $\mathbf{B} = \nabla \times \mathbf{A}$ using the vector potential at the midpoint of the finite-length wire given by Eq. (4.27). Since the length of the wire in this example is infinite, we must take the limit $L \to \infty$. It turns out that the order in which we perform these two operations is important. If we were to first find $\mathbf{A}(\mathbf{r})$ in the limit $L \to \infty$, we run into a difficulty, since the natural log function in Eq. (4.27) diverges. (It is a weak divergence, but nonetheless a divergence.) We can avoid this difficulty by first computing $\mathbf{B}$ for a finite-length wire, and then finding the limit of $\mathbf{B}$ as $L \to \infty$. In cylindrical coordinates,

$$\mathbf{B} = \nabla \times \mathbf{A} = \frac{1}{\rho} \begin{vmatrix} \mathbf{a}_\rho & \rho\mathbf{a}_\phi & \mathbf{a}_z \\ \frac{\partial}{\partial\rho} & \frac{\partial}{\partial\phi} & \frac{\partial}{\partial z} \\ A_\rho & \rho A_\phi & A_z \end{vmatrix}.$$

We get A_z from Eq. (4.27), and A_ρ and A_ϕ are zero. The result is

$$\mathbf{B} = -\mathbf{a}_\phi \frac{\mu_0 I}{2\pi}\left[\frac{1}{(L/2) + \sqrt{(L/2)^2 + \rho^2}} \frac{(1/2)\, 2\rho}{\sqrt{(L/2)^2 + \rho^2}} - \frac{1}{\rho}\right].$$

In the limit as $L \to \infty$, the first term inside the square brackets vanishes, and $\mathbf{B}$ becomes

$$\mathbf{B} = \mathbf{a}_\phi \frac{\mu_0 I}{2\pi\rho},$$

in perfect agreement with the results we derived previously through other means.

We will close this section by showing that the magnetic vector potential, as expressed in Eqs. (4.22), (4.23), or (4.24), is perfectly consistent with the Biot–Savart Law, given in Eqs. (4.5), (4.6), or (4.3), respectively. (We will show just one of these, using $\mathbf{J}_v$ as an example, but the other two follow a similar process.) We start by combining Eq. (4.19) and the integral expression Eq. (4.22) to write

$$\mathbf{B}(\mathbf{r}) = \nabla \times \frac{\mu_0}{4\pi} \int_v \frac{\mathbf{J}_v(\mathbf{r}')\, dv'}{|\mathbf{r} - \mathbf{r}'|}.$$

As a reminder, the "unprimed" Del operator includes derivatives with respect to the field point position **r** (the "unprimed" coordinates), while the integral expression involves integration over the source point coordinates **r**′ (the "primed" coordinates). Since these variables are independent of one another, we can move the Del operator inside the integral:

$$\mathbf{B}(\mathbf{r}) = \frac{\mu_0}{4\pi} \int_v \nabla \times \frac{\mathbf{J}_v(\mathbf{r}') \, dv'}{|\mathbf{r} - \mathbf{r}'|}.$$

Now we use Eq. (E.8) to write this as

$$\mathbf{B}(\mathbf{r}) = \frac{\mu_0}{4\pi} \int_v \left[\frac{1}{|\mathbf{r} - \mathbf{r}'|} \nabla \times \mathbf{J}_v(\mathbf{r}') + \nabla \frac{1}{|\mathbf{r} - \mathbf{r}'|} \times \mathbf{J}_v(\mathbf{r}') \right] dv'.$$

But $\nabla \times \mathbf{J}_v(\mathbf{r}') = 0$, since ∇ doesn't operate on the "primed" variables. Finally, we can use Eq. (D.5) to write

$$\mathbf{B}(\mathbf{r}) = \frac{\mu_0}{4\pi} \int_v \mathbf{J}_v(\mathbf{r}') \times \frac{\mathbf{r} - \mathbf{r}'}{|\mathbf{r} - \mathbf{r}'|^3} \, dv',$$

which is equivalent to the Biot–Savart Law, Eq. (4.5).

In this section we have seen how to determine the vector magnetic potential produced by a current distribution, and to use this vector potential to determine the magnetic flux density. This can be a very useful technique, in that the integrals needed to solve for $\mathbf{A}(\mathbf{r})$ can often be simpler than those for deriving $\mathbf{B}(\mathbf{r})$ directly, in much the same way that the electric potential is often simpler to find than the electric field. In the next section we will find the vector potential due to a circular current loop known as a magnetic dipole.

4.5 Magnetic Dipole

In this section we examine the vector potential **A** and magnetic flux density **B** produced by a loop of current, known as a **magnetic dipole**. We previously found **B** at a point P lying along the axis for this current loop in Example 4.3. We also showed that the ϕ component of the field must be zero everywhere, but we didn't try carrying out the complete integration of the Biot–Savart Law for an off-axis point P. Magnetic dipoles, or small loops of current, are extremely important in our development of magnetic field effects, particularly when discussing different kinds of magnetic materials, and so in this section we will develop a formalism for their properties, paying special attention to the field a large distance from the dipole. This limit is perfectly valid for atomic-scale magnetic dipoles, which are much smaller than the distances involved when considering macroscopic materials. Our recent introduction of the vector potential **A** will make this development more tractable.

Consider the circular current loop of radius b and current I lying in the $z = 0$ plane, as shown in Fig. 4.35. We start by determining the vector magnetic potential **A** at an off-axis point $P(x, y, z)$ using Eq. (4.24),

$$\mathbf{A} = \frac{\mu_0}{4\pi} \int_v \frac{I \, d\boldsymbol{\ell}'}{|\mathbf{r} - \mathbf{r}'|},$$

with the path element $d\boldsymbol{\ell}'$ written in cylindrical coordinates as

$$d\boldsymbol{\ell}' = b \, d\phi' \, \mathbf{a}_\phi = b \, d\phi' \, (-\mathbf{a}_x \sin \phi' + \mathbf{a}_y \cos \phi'). \tag{4.28}$$

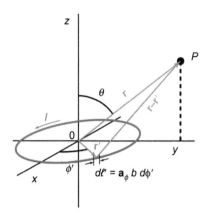

FIGURE 4.35 A current loop of radius b and current I.

The vector $\mathbf{r}$ points from the origin, which is chosen to be at the center of the current loop, to the point P, chosen as $(0, y, z)$. (Since the current loop is symmetric about the z-axis, we can set $x = 0$ for now, and generalize later.) The position vector $\mathbf{r}'$ points toward a segment of the current loop at $\mathbf{r}' = (b \cos \phi', b \sin \phi', 0)$. For variety, we determine $|\mathbf{r} - \mathbf{r}'|$ in a somewhat different way this time. The magnitude of any vector can be found as $|\mathbf{r} - \mathbf{r}'| = [(\mathbf{r} - \mathbf{r}') \cdot (\mathbf{r} - \mathbf{r}')]^{1/2}$, which can be expanded as $|\mathbf{r} - \mathbf{r}'| = [|\mathbf{r}|^2 - 2\mathbf{r} \cdot \mathbf{r}' + |\mathbf{r}'|^2]^{1/2}$. The terms in this expression are: $|\mathbf{r}| = R$, where R is the distance from the center of the current loop to the point P; $\mathbf{r} \cdot \mathbf{r}' = yb \sin \phi' = Rb \sin \theta \sin \phi'$, where θ is the polar angle measured from the z-axis, as shown in Fig. 4.35; and $|\mathbf{r}'| = b$. (We will retain the more compact notation $\mathbf{r} \cdot \mathbf{r}'$ until we need to be more explicit.) Then the inverse of $|\mathbf{r} - \mathbf{r}'|$ can be written

$$\frac{1}{|\mathbf{r} - \mathbf{r}'|} = \frac{1}{R} \frac{1}{[1 - 2\mathbf{r} \cdot \mathbf{r}'/R^2 + (b/R)^2]^{1/2}}.$$

Since our primary interest is $\mathbf{A}$ at locations far from the current loop itself, where $R \gg b$, the final term $(b/R)^2$ is negligibly small, and we can safely ignore it. Applying the Taylor expansion $(1 + \varepsilon)^{-1/2} \simeq 1 - (1/2)\varepsilon + \cdots$, where ε is a small number (see Eq. (H.5) of Appendix H), $|\mathbf{r} - \mathbf{r}'|^{-1}$ becomes

$$\frac{1}{|\mathbf{r} - \mathbf{r}'|} \simeq \frac{1}{R} \left[1 + \frac{\mathbf{r} \cdot \mathbf{r}'}{R^2} + \cdots \right].$$

Using this now in Eq. (4.24), the vector potential at $\mathbf{r}$ is

$$\mathbf{A} = \frac{\mu_0 I}{4\pi} \left[\frac{1}{R} \int d\boldsymbol{\ell}' + \frac{1}{R^3} \int d\boldsymbol{\ell}' \, \mathbf{r} \cdot \mathbf{r}' + \cdots \right].$$

The path integral $\int d\boldsymbol{\ell}'$, where the vector path element is given in Eq. (4.28), around one complete loop is zero, so the first integral vanishes. To evaluate the second integral, we substitute the explicit forms for $d\boldsymbol{\ell}'$ and $\mathbf{r} \cdot \mathbf{r}'$, using rectangular *components* while keeping spherical *variables*

$$\mathbf{A} = \frac{\mu_0 I}{4\pi} \frac{1}{R^3} \int_0^{2\pi} \left[b \, d\phi' \, (-\mathbf{a}_x \sin \phi' + \mathbf{a}_y \cos \phi') \right] \left[Rb \sin \theta \sin \phi' \right]. \tag{4.29}$$

Integrating in ϕ' gives us

$$\mathbf{A} = -\frac{\mu_0 I \left(\pi b^2 \right) \sin \theta}{4\pi R^2} \mathbf{a}_x.$$

At the beginning of this solution, we chose the point P to lie in the y–z plane. If we now relax this condition and exploit the cylindrical symmetry of the current loop, we can replace $\mathbf{a}_x \to -\mathbf{a}_\phi$, and write

$$\mathbf{A} = \frac{\mu_0 I \left(\pi b^2 \right) \sin \theta}{4\pi R^2} \mathbf{a}_\phi.$$

Additionally, we will define the magnetic dipole moment as

$$\mathbf{m} = (IS) \, \mathbf{a}_z, \tag{4.30}$$

where S is the area enclosed by the current loop (valid even for non-circular current loops) and $\mathbf{a}_z$ is the normal unit vector to the current loop, and write $\mathbf{A}$ as

$$\mathbf{A} = \frac{\mu_0 \mathbf{m} \times \mathbf{a}_R}{4\pi R^2}. \tag{4.31}$$

Now that we have found **A** for the magnetic dipole, valid for $R \gg b$, we can use this result to find the magnetic flux density **B** due to the magnetic dipole:

$$\mathbf{B} = \nabla \times \mathbf{A} = \frac{\mu_0 m}{4\pi} \frac{1}{R^2 \sin\theta} \begin{vmatrix} \mathbf{a}_R & \mathbf{a}_\theta R & \mathbf{a}_\phi R \sin\theta \\ \dfrac{\partial}{\partial R} & \dfrac{\partial}{\partial \theta} & \dfrac{\partial}{\partial \phi} \\ 0 & 0 & (R\sin\theta)\left(\dfrac{\sin\theta}{R^2}\right) \end{vmatrix}.$$

Evaluating the determinant leads to

$$\mathbf{B} = \frac{\mu_0 m}{4\pi} \frac{1}{R^2 \sin\theta} \left[\mathbf{a}_R \frac{\partial}{\partial \theta}\left(\frac{\sin^2\theta}{R}\right) - \mathbf{a}_\theta R \frac{\partial}{\partial R}\left(\frac{\sin^2\theta}{R}\right) \right],$$

which upon differentiation reduces to

$$\mathbf{B} = \frac{\mu_0 m}{4\pi R^3} \left[\mathbf{a}_R \, 2\cos\theta + \mathbf{a}_\theta \sin\theta \right]. \tag{4.32}$$

Comparison with Eq. (1.57) shows that the electric field due to an electric dipole and the magnetic flux density due to a magnetic dipole are extremely similar in the region far from the dipole itself. We show representations of these two field configurations in Fig. 4.36. Both fields, in the far-field region, have the same θ dependence, and their magnitudes decrease with increasing distance as $1/R^3$. In the next section we will make use of this description of the magnetic dipole, and the field that it generates, to develop a model of magnetic materials.

FIGURE 4.36 A graphical representation of (a) the electric field lines of an electric dipole, and (b) the magnetic flux lines of a magnetic dipole. These field patterns take on identical shapes in the far-field region ($R \gg d$ for the electric dipole, $R \gg b$ for the magnetic dipole).

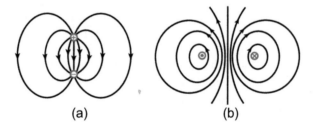

(a)   (b)

4.6 Magnetization

To this point in our discussion of magnetic effects, we have not considered any influence that materials may have on the fields. In each of the problems or examples solved, we have included a current, or a distribution of currents, and then determined the magnetic flux density **B** created by this current, with no other materials in the vicinity. With the discussion of the previous section of the magnetic potential and magnetic flux density of a single current loop, or magnetic dipole, we have laid the groundwork of a method to include magnetic materials in our considerations. Most atoms and molecules, for example, have magnetic properties due to their electronic structure. In a classical model of an atom, we picture negatively charged electrons orbiting around the positively charged nucleus much like planets orbiting the sun. (The quantum model of an atom differs significantly from this simplistic picture, but the classical picture is sufficient for motivating the idea that atoms have magnetic properties.) This motion of the electron around the nucleus constitutes a current loop, and an associated magnetic dipole moment **m**.

An assembly of dipoles is illustrated in Fig. 4.37. Let us define a vector quantity $\mathbf{M}(\mathbf{r})$, called the **magnetization** of the material, which quantifies the density, magnitude, and alignment of magnetic dipoles. Formally, the magnetization within a small volume Δv is defined as

$$\mathbf{M}(\mathbf{r}) = \lim_{\Delta v \to 0} \frac{1}{\Delta v} \sum_{k=1}^{n\,\Delta v} \mathbf{m}_k, \tag{4.33}$$

where the index k labels the individual dipoles, and the summation extends over each of the dipoles within the small elemental volume Δv. n is the number density of the dipoles, so $n\,\Delta v$ represents the total number of dipoles within the elemental volume Δv. In Fig. 4.37(a), the dipoles are randomly oriented and the vector sum $\sum_k \mathbf{m}_k$ tends to zero. In Fig. 4.37(b), however, the dipoles are aligned, and when summed lead to a net non-zero magnetization $\mathbf{M}$. Complete alignment of the dipoles would be extreme, but even partial alignment can lead to large magnetic effects.

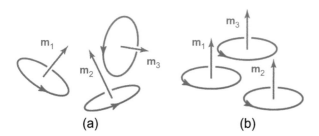

FIGURE 4.37 An assembly of microscopic magnetic dipoles. In (a), these dipoles are randomly oriented, and the net magnetization is zero. In (b), the magnetic dipoles are aligned, resulting in a net magnetization $\mathbf{M}(\mathbf{r})$.

We can quantify the effect of this magnetization field by recalling from the previous section the vector magnetic potential $\mathbf{A}$ due to a single current loop from Eq. (4.31), and extending this to include the effects of all dipoles within a small volume $\Delta v'$. This potential is called $\Delta\mathbf{A}$, since it is the potential due to only the magnetic dipoles within the volume element $\Delta v'$, and $\Delta\mathbf{A}$ can be written as

$$\Delta\mathbf{A}(\mathbf{r}) = \sum_{k=1}^{n\,\Delta v'} \frac{\mu_0 \mathbf{m}_k \times (\mathbf{r} - \mathbf{r}'_k)}{4\pi |\mathbf{r} - \mathbf{r}'_k|^3}.$$

In writing this potential, the vector $\mathbf{r}'_k$ denotes the position of each individual dipole k, and the unit vector $\mathbf{a}_R$ is written as $(\mathbf{r} - \mathbf{r}'_k)/|\mathbf{r} - \mathbf{r}'_k|$. As we let the volume $\Delta v'$ become small, however, the various position vectors $\mathbf{r}'_k$ for each dipole within that volume approach the same value, and we can approximate $\mathbf{r}'_k$ for each dipole as $\mathbf{r}'$. Then the sum $\sum_k \mathbf{m}_k \to \mathbf{M}(\mathbf{r}')dv'$ (using the definition Eq. (4.33)), and the potential $\Delta\mathbf{A}$ becomes

$$d\mathbf{A}(\mathbf{r}) = \frac{\mu_0 [\mathbf{M}(\mathbf{r})\,dv'] \times (\mathbf{r} - \mathbf{r}')}{4\pi |\mathbf{r} - \mathbf{r}'|^3}.$$

Upon integration over the entire volume of the magnetized material, the vector potential is

$$\mathbf{A}(\mathbf{r}) = \frac{\mu_0}{4\pi} \int_v \frac{\mathbf{M}(\mathbf{r}') \times (\mathbf{r} - \mathbf{r}')}{|\mathbf{r} - \mathbf{r}'|^3}\,dv'.$$

We will next reduce this integral further, and identify effective current densities, using some vector properties that are discussed in Appendices D and E. First, it is shown in Eq. (D.6) that

$$\nabla'\left(\frac{1}{|\mathbf{r} - \mathbf{r}'|}\right) = \frac{(\mathbf{r} - \mathbf{r}')}{|\mathbf{r} - \mathbf{r}'|^3}.$$

Recall that the "prime" notation on the ∇' operator indicates that the derivatives are taken with respect to the "primed" coordinates. (x', y', and z', for example, when working in rectangular coordinates.) This allows us to write $\mathbf{A}$ as

$$\mathbf{A}(\mathbf{r}) = \frac{\mu_0}{4\pi} \int_v \mathbf{M}(\mathbf{r}') \times \nabla'\left(\frac{1}{|\mathbf{r} - \mathbf{r}'|}\right) dv'. \tag{4.34}$$

Also, we can use Eq. (E.8) (this is a variant of the product rule for derivatives), in which Φ is $1/|\mathbf{r} - \mathbf{r}'|$ and $\mathbf{E}$ is $\mathbf{M}$, to write

$$\boldsymbol{\nabla}' \times \left(\frac{1}{|\mathbf{r} - \mathbf{r}'|} \mathbf{M}(\mathbf{r}') \right) = \frac{1}{|\mathbf{r} - \mathbf{r}'|} \boldsymbol{\nabla}' \times \mathbf{M}(\mathbf{r}') + \boldsymbol{\nabla}' \left(\frac{1}{|\mathbf{r} - \mathbf{r}'|} \right) \times \mathbf{M}(\mathbf{r}').$$

The second term on the right side of this relation is -1 times the integrand of Eq. (4.34). Thus the vector potential can be written

$$\mathbf{A}(\mathbf{r}) = \frac{\mu_0}{4\pi} \left[\int_v \frac{1}{|\mathbf{r} - \mathbf{r}'|} \boldsymbol{\nabla}' \times \mathbf{M}(\mathbf{r}') \, dv' - \int_v \boldsymbol{\nabla}' \times \left(\frac{1}{|\mathbf{r} - \mathbf{r}'|} \mathbf{M}(\mathbf{r}') \right) dv' \right].$$

Finally, using the following vector integral identity which we prove in SideNote 4.1, namely

$$\oint_s \mathbf{a}_n \times \mathbf{F} \, ds = \int_v \boldsymbol{\nabla} \times \mathbf{F} \, dv,$$

where $\mathbf{F}$ is any vector field, s is the surface that bounds the volume v, and $\mathbf{a}_n$ is the surface normal unit vector (pointing *out* of the magnetized medium), we have

$$\mathbf{A}(\mathbf{r}) = \frac{\mu_0}{4\pi} \left[\int_v \frac{\boldsymbol{\nabla}' \times \mathbf{M}(\mathbf{r}')}{|\mathbf{r} - \mathbf{r}'|} \, dv' + \oint_s \frac{\mathbf{M}(\mathbf{r}') \times \mathbf{a}_n'}{|\mathbf{r} - \mathbf{r}'|} \, ds' \right]. \qquad (4.35)$$

SideNote 4.1 Proof of Vector Identity

Here we present the proof of the vector integral equation presented above. That is, for any vector field $\mathbf{F}$ that is differentiable over a volume v,

$$\oint_s \mathbf{a}_n \times \mathbf{F} \, ds = \int_v \boldsymbol{\nabla} \times \mathbf{F} \, dv,$$

where s is the surface that bounds the volume v, and $\mathbf{a}_n$ is the surface normal unit vector. The geometry is illustrated in Fig. 4.38(a). Note that this proof takes us away from the primary topic of this discussion, and that our purpose of presenting it here is to take some of the mystery out of the derivation of the effective currents. If you can accept the integral relation at face value, and not be distracted by it, you could easily skip this proof and continue with the discussion of the magnetization.

Before we start, let's look at the two integrals to get a sense of what the identity is telling us. The right side is a volume integral of the curl of a vector field $\mathbf{F}$. Recall that the curl of a vector indicates how much the field is circulating, and the direction of the curl is along the axis of this rotation of the field. On the left side of the identity is the surface integral (note that this is a scalar surface element ds) of the vector $\mathbf{a}_n \times \mathbf{F}$. Recall that this cross product is sensitive only to the component of $\mathbf{F}$ that lies in the plane of the surface. (We used a similar notation when we discussed boundary conditions of the tangential component of the electric field at the interface between two media, and will do so again when we discuss the boundary conditions that magnetic fields must obey at an interface.) So if the vector field $\mathbf{F}$ is circulating around an axis, $\mathbf{a}_n \times \mathbf{F}$ on each of

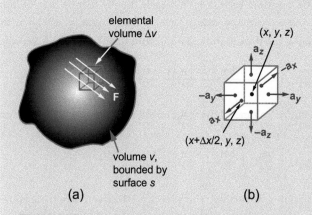

FIGURE 4.38 A volume v for proof of the integral vector relation. We illustrate the entire volume v in (a), and one elemental volume positioned at (x, y, z) in (b).

the sides is pointing in the same direction as the axis of the circulation. So with this description, the identity should at least seem plausible.

To start the proof, we break the entire volume v into small elemental volumes, Δv, as pictured in Fig. 4.38(b). Let us evaluate

$$\oint_s \mathbf{a}_n \times \mathbf{F}\, ds$$

for this elemental cube, whose dimension is $\Delta x \times \Delta y \times \Delta z$. If the vector $\mathbf{F}$ is "curling," it is slightly different on each side of the cube, so we must be careful to specify the coordinates for each side. The unit normal vector $\mathbf{a}_n$ is, of course, also different for each side of the cube, and these unit vectors are labeled in the figure. Writing out the contributions to the surface integral for each side of the cube, we have

$$\oint_s \mathbf{a}_n \times \mathbf{F}\, ds = \left[\mathbf{a}_x \times \mathbf{F}\left(x + \frac{\Delta x}{2}, y, z\right) - \mathbf{a}_x \times \mathbf{F}\left(x - \frac{\Delta x}{2}, y, z\right) \right] \Delta y \Delta z$$

$$+ \left[\mathbf{a}_y \times \mathbf{F}\left(x, y + \frac{\Delta y}{2}, z\right) - \mathbf{a}_y \times \mathbf{F}\left(x, y - \frac{\Delta y}{2}, z\right) \right] \Delta z \Delta x$$

$$+ \left[\mathbf{a}_z \times \mathbf{F}\left(x, y, z + \frac{\Delta z}{2}\right) - \mathbf{a}_z \times \mathbf{F}\left(x, y, z - \frac{\Delta z}{2}\right) \right] \Delta x \Delta y.$$

Grouping similar terms together and pulling out a common factor of $\Delta v = \Delta x\, \Delta y\, \Delta z$ leads to

$$\oint_s \mathbf{a}_n \times \mathbf{F}\, ds = \left[\mathbf{a}_x \times \frac{\mathbf{F}\left(x + \frac{\Delta x}{2}, y, z\right) - \mathbf{F}\left(x - \frac{\Delta x}{2}, y, z\right)}{\Delta x} \right.$$

$$+ \mathbf{a}_y \times \frac{\mathbf{F}\left(x, y + \frac{\Delta y}{2}, z\right) - \mathbf{F}\left(x, y - \frac{\Delta y}{2}, z\right)}{\Delta y}$$

$$\left. + \mathbf{a}_z \times \frac{\mathbf{F}\left(x, y, z + \frac{\Delta z}{2}\right) - \mathbf{F}\left(x, y, z - \frac{\Delta z}{2}\right)}{\Delta z} \right] \Delta v.$$

In the limit of small Δx, Δy, or Δz, each of these fractions is simply a partial derivative of $\mathbf{F}$, so the surface integral for the elemental cube reduces to

$$\oint_s \mathbf{a}_n \times \mathbf{F}\, ds = \left[\mathbf{a}_x \times \frac{\partial \mathbf{F}}{\partial x} + \mathbf{a}_y \times \frac{\partial \mathbf{F}}{\partial y} + \mathbf{a}_z \times \frac{\partial \mathbf{F}}{\partial z} \right] \Delta v.$$

Writing out $\mathbf{F}$ in terms of its individual components $\mathbf{F} = \mathbf{a}_x F_x + \mathbf{a}_y F_y + \mathbf{a}_z F_z$, computing the cross products of the unit vectors, and regrouping the terms, the surface integral is

$$\oint_s \mathbf{a}_n \times \mathbf{F}\, ds = \left[\mathbf{a}_x \left(\frac{\partial F_z}{\partial y} - \frac{\partial F_y}{\partial z} \right) + \mathbf{a}_y \left(\frac{\partial F_x}{\partial z} - \frac{\partial F_z}{\partial x} \right) + \mathbf{a}_z \left(\frac{\partial F_y}{\partial x} - \frac{\partial F_x}{\partial y} \right) \right] \Delta v.$$

But the terms inside the square brackets are the components of $\nabla \times \mathbf{F}$, and so

$$\oint_s \mathbf{a}_n \times \mathbf{F}\, ds = \nabla \times \mathbf{F} \Delta v$$

for the small elemental volume Δv.

Now let's look at the left and right sides of this equation as we add the contributions from two adjacent elemental cubes. That is,

$$\sum_{i=1}^{2} \oint_{s_i} \mathbf{a}_n \times \mathbf{F}\, ds = \sum_{i=1}^{2} (\nabla \times \mathbf{F})_i\, \Delta v_i.$$

On the left side, involving the surface integral, the contribution from the side that is shared between the two cubes will perfectly cancel, since $\mathbf{a}_n$ for one cube, which always points outward from the cube, points in the opposite direction for the adjacent cube. The right-hand side, involving the volume elements, simply add. As we add the contributions from more elements, extending the summation to all elements that make up the volume v, and take the limit of this summation as $\Delta v \to 0$, only the outside surfaces contribute to the surface integral on the left, proving the equality

$$\oint_s \mathbf{a}_n \times \mathbf{F}\, ds = \int_v \nabla \times \mathbf{F}\, dv.$$

Let's return to Eq. (4.35) to see what this is telling us. Using this equation, the vector magnetic potential $\mathbf{A}$ can be determined using two integrals, both involving the magnetization field $\mathbf{M}$, but one a volume integral over the volume v, and the other a surface integral over the surface s that bounds v. These may not look terribly useful, but they should remind you of integral expressions that we have seen previously. In particular, recall Eqs. (4.22) and (4.23), which allow us to determine the magnetic potential at a position $\mathbf{r}$ due to a volume current density $\mathbf{J}_v(\mathbf{r}')$ and a surface current density $\mathbf{J}_s(\mathbf{r}')$, respectively. The first (volume) integral Eq. (4.35) is precisely equivalent to Eq. (4.22), except that $\nabla' \times \mathbf{M}(\mathbf{r}')$ appears in the numerator in place of $\mathbf{J}_v(\mathbf{r}')$. Thus we interpret $\nabla' \times \mathbf{M}(\mathbf{r}')$ as an *effective* volume current density. Similarly, comparison of the surface integral in Eq. (4.35) shows that this integral is equivalent to Eq. (4.23) if $\mathbf{M}(\mathbf{r}') \times \mathbf{a}'_n$ is replaced for $\mathbf{J}_s(\mathbf{r}')$. We interpret $\mathbf{M}(\mathbf{r}') \times \mathbf{a}'_n$ as an *effective* surface current density. In summary, we have defined two effective current densities,

$$\boxed{\mathbf{J}_{v,eff}(\mathbf{r}) = \nabla \times \mathbf{M}(\mathbf{r})} \qquad (4.36)$$

and

$$\boxed{\mathbf{J}_{s,eff}(\mathbf{r}) = \mathbf{M}(\mathbf{r}) \times \mathbf{a}_n,} \qquad (4.37)$$

that are present within or on the surface of magnetized materials, and that can produce a vector potential $\mathbf{A}$ that is equivalent to that produced by real currents $\mathbf{J}_v$ and $\mathbf{J}_s$. (Note that we have dropped the "primed" notation in Eqs. (4.36) and (4.37) as it is no longer necessary.)

Our interpretation of these **effective currents** is similar to that of the effective *charge* densities that we defined when we introduced dielectric materials in our discussion of electric effects. These effective currents are a result of the microscopic magnetic dipoles within the magnetized medium. They do not represent free currents; the atomic charges are still bound to the atoms. But due to their alignment they generate magnetic fields. The magnetization $\mathbf{M}$ can be produced by applying an external field, as in an electromagnet or a transformer core, or they can be permanently aligned, as in a permanent magnet. We really haven't differentiated between these important cases to this point. Now let's look at the form of these effective currents, given by Eqs. (4.36) and (4.37), and try to explain their significance.

Consider an assembly of magnetic dipoles, as shown in Fig. 4.39. If the magnetization $\mathbf{M}$ is perfectly uniform throughout a material, then the current from any magnetic dipole within the interior of the material is perfectly canceled by the oppositely directed current of an adjacent dipole, and there is no net current inside the material. If $\mathbf{M}$ is not uniform, however, then the cancellation of these interior dipole currents is not complete, and the residual is an effective current in the interior of the material. This is reflected in the

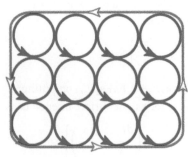

FIGURE 4.39 An array of magnetic dipoles, showing the effective currents that they represent. In this figure, all the dipoles **m** are pointing out of the page, as produced by microscopic currents in the counter-clockwise direction (represented by the red arrows). The currents of adjacent loops in the interior tend to cancel one another, but at the edges there is a net effective current, shown as the blue arrowheads, around the magnetized block in the counter-clockwise direction.

differential $\nabla \times \mathbf{M}(\mathbf{r})$. Similarly, at the surface of the magnetic material, the last dipole current loop at the surface contributes an effective surface current, wrapping around the block in the counter-clockwise direction. This effective surface current is consistent with $\mathbf{M}(\mathbf{r}) \times \mathbf{a}_n$ (Eq. (4.37)), since the magnetic dipoles point *out* of the page (use the right-hand rule), and the normal unit vectors point out from the body on all four sides. These effective currents are further explored through Example 4.14.

Example 4.14 Effective Currents of a Magnetized Cube

Consider the cube of uniform magnetization $\mathbf{M} = M_0 \mathbf{a}_x$ shown in Fig. 4.40(a). Determine the effective current densities $\mathbf{J}_{v,eff}$ and $\mathbf{J}_{s,eff}$.

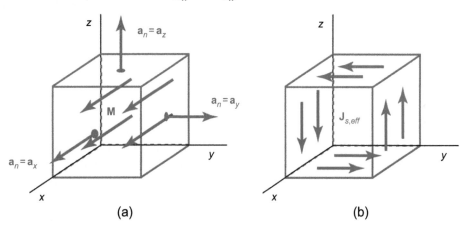

FIGURE 4.40 A magnetized cube. (a) The magnetization $\mathbf{M}$ is uniform, and points in the x-direction. In (b), we show the effective surface currents, $\mathbf{J}_{s,eff}$.

Solution:
We start with the volume effective current density $\mathbf{J}_{v,eff}$, as given by Eq. (4.36). Since $\mathbf{M}$ is uniform, its curl is zero, and $\mathbf{J}_{v,eff} = 0$. For the effective surface currents, we must consider each face individually. Using Eq. (4.37), the effective surface current on the top and bottom faces are

$$\mathbf{M} \times \mathbf{a}_n = M_0 \, \mathbf{a}_x \times (+\mathbf{a}_z) = -M_0 \, \mathbf{a}_y$$

and

$$\mathbf{M} \times \mathbf{a}_n = M_0 \, \mathbf{a}_x \times (-\mathbf{a}_z) = +M_0 \, \mathbf{a}_y,$$

respectively. For the right and left faces, we have

$$\mathbf{M} \times \mathbf{a}_n = M_0 \, \mathbf{a}_x \times (+\mathbf{a}_y) = M_0 \, \mathbf{a}_z$$

and

$$\mathbf{M} \times \mathbf{a}_n = M_0 \, \mathbf{a}_x \times (-\mathbf{a}_y) = -M_0 \, \mathbf{a}_z,$$

respectively. Finally, for the front and back faces, $\mathbf{M} \times \mathbf{a}_n = M_0 \, \mathbf{a}_x \times (\pm \mathbf{a}_x) = 0$. Thus the effective surface current goes down on the left side, to the right across the bottom, up the right side, and back to the left on top. This current pattern is shown in Fig. 4.40(b). The problem did not ask for the magnetic flux density for this current distribution, but if it had, these effective currents could have been used in the Biot–Savart Law to comply. This analysis is left to the reader in Prob. P4-23.

Example 4.15 Effective Currents of a Magnetized Cylinder

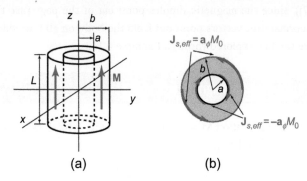

FIGURE 4.41 A magnetized cylinder. (a) The magnetization **M** is uniform, and points in the z-direction. The effective surface currents, $J_{s,eff}$, pointing in the $+a_\phi$ direction on the outer wall and in the $-a_\phi$ direction on the inner wall, are shown in (b).

Consider a cylinder of magnetized material (uniform magnetization $\mathbf{M} = M_0\,\mathbf{a}_z$) with the center drilled out, as shown in Fig. 4.41(a). The inner and outer radii of the cylinder are a and b, respectively. Determine the resulting magnetic flux density **B** along the z-axis.

Solution:

We start by determining the effective currents. Using Eq. (4.36), $\mathbf{J}_{v,eff} = \nabla \times \mathbf{M}(\mathbf{r}) = 0$, since **M** is uniform. The effective surface currents on the inner and outer surfaces are $\mathbf{M} \times \mathbf{a}_n = M_0\,\mathbf{a}_z \times (-\mathbf{a}_\rho) = -M_0\,\mathbf{a}_\phi$ and $\mathbf{M} \times \mathbf{a}_n = M_0\,\mathbf{a}_z \times (+\mathbf{a}_\rho) = M_0\,\mathbf{a}_\phi$, respectively. These currents are shown in the end-on view of the cylinder in Fig. 4.41(b).

Now that we have determined the effective surface currents on the inner and outer walls of the cylinder, we must determine the magnetic flux density along the z-axis that is produced by these currents. For this we return to the Biot–Savart Law. This will be similar to the discussion of Example 4.5, except here the cylinder is of finite length L, and the value of **B** is requested only for a point P along the z-axis. We first find the field due to just one surface current, on the surface at $\rho = b$, which is written in rectangular coordinates as $M_0(-\mathbf{a}_x \sin\phi' + \mathbf{a}_y \cos\phi')$. The position vector for an on-axis point a distance z from the center of the cylinder is $\mathbf{r} = (0, 0, z)$, and $\mathbf{r}' = (b\cos\phi', b\sin\phi', z')$ is the location of a source element. The position vector that points from the source element at $\mathbf{r}'$ to the point P is $\mathbf{r} - \mathbf{r}' = (-b\cos\phi', -b\sin\phi', z - z')$, and the distance between these two points is $|\mathbf{r} - \mathbf{r}'| = \sqrt{b^2 + (z - z')^2}$.

Next, we need the vector product $\mathbf{J}_{s,eff} \times (\mathbf{r} - \mathbf{r}')$, which is computed as

$$\mathbf{J}_{s,eff} \times (\mathbf{r} - \mathbf{r}') = \begin{vmatrix} \mathbf{a}_x & \mathbf{a}_y & \mathbf{a}_z \\ -M_0\sin\phi' & M_0\cos\phi' & 0 \\ -b\cos\phi' & -b\sin\phi' & z - z' \end{vmatrix}$$

$$= M_0\left[\mathbf{a}_x\,(z - z')\cos\phi' + \mathbf{a}_y\,(z - z')\sin\phi' + \mathbf{a}_z\,b\right].$$

Now we are ready to apply the Biot–Savart Law, as given in Eq. (4.6). The element of surface area for integration across the cylindrical surface of the solenoid is $ds' = b\,d\phi'\,dz'$. Substituting these terms into Eq. (4.6), we find

$$\mathbf{B} = \frac{\mu_0 M_0 b}{4\pi} \int_{-L/2}^{L/2} \int_0^{2\pi} \frac{\left[\mathbf{a}_x\,(z - z')\cos\phi' + \mathbf{a}_y\,(z - z')\sin\phi' + \mathbf{a}_z\,b\right]}{\left[b^2 + (z - z')^2\right]^{3/2}}\,d\phi'\,dz'.$$

Carrying out the integration in ϕ' first, the x- and y-components vanish since $\int \cos\phi'\,d\phi' = 0$ and $\int \sin\phi'\,d\phi' = 0$ when applied over one full cycle. For the z-component, $\int d\phi'$ is equal to 2π. Thus, $\mathbf{B} = \mathbf{a}_z\,B_z$, where

$$B_z = \frac{\mu_0 M_0 b^2}{2} \int_{-L/2}^{L/2} \frac{dz'}{\left[b^2 + (z - z')^2\right]^{3/2}}.$$

After integrating in z', using Eq. (G.12) from Appendix G, the magnetic field is

$$B_z = \frac{\mu_0 M_0}{2}\left[\frac{(L/2 - z)}{\sqrt{(L/2 - z)^2 + b^2}} + \frac{(L/2 + z)}{\sqrt{(L/2 + z)^2 + b^2}}\right].$$

Far from the center of the magnet in either direction (i.e. $|z| \to \infty$), this flux density drops off, as we should expect, since the two terms inside the square brackets become similar in magnitude but opposite in sign. At the center of the cylinder, B_z is at its maximum, as we should also expect, with a value of $\mu_0 M_0 (L/2)/\sqrt{(L/2)^2 + b^2}$. As a comparison with previous results, this approaches $B_z \to \mu_0 M_0$ for $L \gg b$; that is, for a long cylinder. This result is consistent with the results we discovered earlier for an infinitely long solenoid, with M_0 substituted for nI for the surface current.

We are not quite finished as yet, since we have determined the magnetic flux density due to the effective surface current on the outer wall of the cylinder, but not the inner. Fortunately, we do not need to repeat our work, since B_z due to this current can be found by multiplying our earlier result by -1 (this current circulates in the opposite direction), and by replacing the radius b by a. Summing these two flux densities, the result is then

$$B_z = \frac{\mu_0 M_0}{2}\left[\frac{(L/2 - z)}{\sqrt{(L/2 - z)^2 + b^2}} + \frac{(L/2 + z)}{\sqrt{(L/2 + z)^2 + b^2}}\right.$$
$$\left. - \frac{(L/2 - z)}{\sqrt{(L/2 - z)^2 + a^2}} - \frac{(L/2 + z)}{\sqrt{(L/2 + z)^2 + a^2}}\right].$$

A plot of this magnetic flux density is shown in Fig. 4.42. In this figure, the cylinder extends from $z/L = -1/2$ to $+1/2$, and we have used inner and outer radii of $a = L/4$ and $b = L/2$. B_z is negative at the center, since the field due to the effective current on the inner wall is greater than that of the outer wall, and a current in the $-\phi$ direction produces a magnetic flux density in the $-z$-direction. Outside the cylinder, the influence of the current on the outer wall extends to greater distances z than that of the inner wall, causing B_z to become positive.

FIGURE 4.42 A plot of B_z vs. z for the magnetized cylinder. The cylinder extends from $z = -0.5\,L$ to $0.5\,L$. We have used inner and outer radii of $a = L/4$ and $b = L/2$.

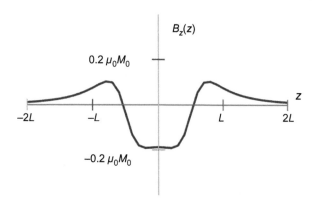

In this section we have begun to develop the methods for treating magnetic materials. We have shown that, given the magnetization $\mathbf{M}(\mathbf{r})$ of a material, we can determine the effective current densities $\mathbf{J}_{v,eff}$ and $\mathbf{J}_{s,eff}$, and use these effective currents in the Biot–Savart Law to determine the flux density $\mathbf{B}(\mathbf{r})$. In the next section we will examine how Ampère's Law must be modified to accommodate magnetic materials.

4.7 Ampère's Law Revisited

As we wrote previously, Ampère's Law in its differential and integral forms is useful to find **B** for several important problems and for exploring the properties of **B**. These solutions, however, were valid in vacuum, and did not include any discussion of materials that may be nearby. In the previous section we saw how the magnetization **M(r)** of a medium results in effective currents, and how the flux density due to these effective currents can be determined. In this section we discuss changes to Ampère's Law to incorporate the effects of magnetic materials.

We start by rewriting Ampère's Law, explicitly including as sources the free currents, as well as the effective currents that we introduced in the previous section. Recalling that **B** represents the *total* field, due to free currents and magnetized materials, the superposition principle applied to Eq. (4.15) produces

$$\frac{1}{\mu_0}(\boldsymbol{\nabla}\times\mathbf{B}) = \mathbf{J}_v + \mathbf{J}_{v,eff}.$$

Let us substitute Eq. (4.36) into the previous equation:

$$\frac{1}{\mu_0}(\boldsymbol{\nabla}\times\mathbf{B}) = \mathbf{J}_v + \boldsymbol{\nabla}\times\mathbf{M},$$

and rearrange the terms

$$\boldsymbol{\nabla}\times\left(\frac{1}{\mu_0}\mathbf{B}-\mathbf{M}\right) = \mathbf{J}_v.$$

The term in the parentheses on the left has special significance. This field is called the **magnetic field**, represented by the symbol **H**,

$$\boxed{\mathbf{H} = \frac{1}{\mu_0}\mathbf{B} - \mathbf{M}.} \tag{4.38}$$

Ampère's Law now reads

$$\boxed{\boldsymbol{\nabla}\times\mathbf{H} = \mathbf{J}_v.} \tag{4.39}$$

In an approximate sense, we can identify the magnetic field **H** as the field that is generated by free currents alone and **M** as the magnetic field that is generated by the effective currents. (This statement is not valid in a general sense, so we have to be careful not to carry it too far. It can be helpful in forming a physical picture of the various field quantities, but remember that it is only approximate.)

Ampère's Law can also be written in its integral form. To do this, consider *any* surface *s* bounded by a contour *c*, and write the surface flux integral over this surface of the left and right sides of Eq. (4.39):

$$\int_s \boldsymbol{\nabla}\times\mathbf{H}\cdot d\mathbf{s} = \int_s \mathbf{J}_v\cdot d\mathbf{s}.$$

The right side is the total current I_{enc} passing through the surface *s*, and, by Stokes' Theorem, the left side becomes the path integral of **H** along the bounding contour *c*:

$$\boxed{\oint_c \mathbf{H}\cdot d\boldsymbol{\ell} = \int_s \mathbf{J}_v\cdot d\mathbf{s} = I_{enc}.} \tag{4.40}$$

This form of Ampère's Law is valid when magnetic materials are present, and it replaces the old form of Ampère's Law, which was only valid when $\mu = \mu_0$ everywhere. You will do well to totally disregard Eqs. (4.15) and (4.16), and only use Eqs. (4.39) and (4.40) from now on.

Drawing an external magnet close to many media will cause the magnetic dipoles within that medium to align. Thus we induce a magnetization $\mathbf{M}$ in that medium. When that magnetization is proportional to the applied field $\mathbf{H}$, we say that the medium's response is linear, and we define the **magnetic susceptibility** χ_m as the proportionality between the magnetization $\mathbf{M}$ and the magnetic field $\mathbf{H}$,

$$\mathbf{M} = \chi_m \mathbf{H}. \tag{4.41}$$

If the medium is also isotropic (i.e. the susceptibility χ_m is the same for all orientations of the applied field $\mathbf{H}$), the magnetization $\mathbf{M}$ is parallel to $\mathbf{H}$. In this case of a linear, isotropic medium, $\mathbf{B}$, which by definition is equal to $\mu_0 (\mathbf{H} + \mathbf{M})$, can also be written as

$$\mathbf{B} = \mu_0 (\mathbf{H} + \chi_m \mathbf{H}).$$

Pulling out a common factor of $\mathbf{H}$, this takes the form

$$\mathbf{B} = \mu_0 (1 + \chi_m) \mathbf{H}.$$

Thus we see that for this "simple" medium, $\mathbf{B}$ and $\mathbf{H}$ are parallel to each other, and their magnitudes are proportional to one another. The ratio of their magnitudes is $\mu_0 (1 + \chi_m)$. We define this parameter as the **permeability** of the magnetic medium,

$$\mu = \mu_0 (1 + \chi_m).$$

It is often more convenient to use a parameter known as the **relative permeability**, μ_r, defined as

$$\mu_r = \frac{\mu}{\mu_0} = (1 + \chi_m).$$

We list the relative permeability and the magnetic susceptibility of several common materials in Table 4.1.

Table 4.1

Relative permeability μ_r and magnetic susceptibility χ_m of several common diamagnetic and paramagnetic materials. The permeability is $\mu = \mu_r \mu_0$, where $\mu_0 = 4\pi \times 10^{-7}$ H/m.		
Material	$\chi_m (\times 10^{-6})$	μ_r
Diamagnetic materials		
Silver	-23.1	0.999 9769
Diamond	-22	0.999 978
Copper	-9.63	0.999 990 37
Water	-9.035	0.999 990 065
Nitrogen, N_2	$-0.005\,06$	0.999 999 995
Paramagnetic materials		
Oxygen, O_2	0.373	1.000 000 373
Aluminum	22	1.000 022

Magnetic media are classified according to the magnitude of their relative permeability μ_r or magnetic susceptibility χ_m. In **diamagnetic** media, the induced magnetization opposes the applied field **H**; that is, the susceptibility is negative, and the relative permeability is less than 1. The first several entries in Table 4.1 are diamagnetic materials. In **paramagnetic** materials, μ_r is greater than, but still comparable to, 1. The alignment of dipoles in a paramagnetic material is parallel to the applied field in these media, but weak. The second set of entries in Table 4.1 are paramagnetic materials. Notice that in diamagnetic materials and paramagnetic materials, the relative permeability differs only slightly from 1. Accordingly, we can often approximate their permeability as simply μ_0, and we often refer to these media as *non-magnetic*.

In contrast, the alignment of the dipoles in **ferromagnets** is very strong. Ferromagnetic materials generally include the elements iron, cobalt, or nickel, and recent material advances have led to extremely high-strength magnets. Ferromagnetic materials are frequently nonlinear in their behavior, and exhibit **hysteretic** behavior. We show a plot of B vs. H for this type of material in Fig. 4.43. Since B and H are not proportional, the permeability of these materials is, strictly speaking, not a well-defined parameter. Still, it is often useful to define the permeability μ for these materials using the ratio B/H for some characteristic point on the hysteresis plot. The relative permeability of ferromagnetic materials $\mu_r = \mu/\mu_0$ can be as large as several thousand.

To understand this hysteresis plot, recall that (in at least an approximate picture), the magnetic field H is the result of an external current, perhaps due to current through the windings wrapped around the magnetic material, while the flux density B is the total field, resulting from these external currents as well as the magnetization of the material. The magnetic field H is therefore controllable, and as we apply a field of increasing strength (by increasing the current through the windings, for example), the increasing field H causes the magnetic dipoles in the medium to align, increasing B (the total field) even more rapidly. Eventually, all the dipoles are aligned and any further increase in H leads to only a modest increase in B, since no additional alignment of the dipoles is possible. We say that the magnetization at this point is saturated. This is the behavior that we see in the upper right quadrant of the hysteresis curve in Fig. 4.43. If we now start decreasing the external field H (by decreasing the current through the windings), the initial response of the medium is minimal, in that the aligned dipoles hold their orientation. At the point where the H field is completely turned off, there still exists a large magnetization, producing a magnetic flux density B. The material at this point is a "permanent" magnet, and the resulting magnetic flux density can be quite large, especially with some of the new materials that have been developed in recent years. The magnetic flux density under this condition is known as the residual flux density, denoted B_r in Fig. 4.43. If the external field is now reversed, the magnetic dipoles in the medium will start to reverse as well, and eventually reduce the magnetic flux density to zero. This is labeled $-H_c$ in Fig. 4.43, where H_c is known as the coercive field intensity.

A further increase in $-H$ causes the internal dipoles to line up, resulting in a large negative flux density B. Ultimately, alignment in the reverse direction is complete, or nearly so, and M saturates, so that any further increase in the field intensity can yield no more alignment. Removing the external field H, we still have a high degree of alignment of the dipoles, and large $B = -B_r$. Materials with large H_c are desirable for use as permanent magnets, since in these materials it is very hard to realign the magnetic dipoles. Alnico boasts a coercive field intensity of $\sim 10^5$ A/m. The hysteresis loop for a permanent magnet is fat. Alternatively, magnetic materials with tall hysteresis loops (i.e. large B_r) are useful in motors, generators, transformers, and magnetic memories.

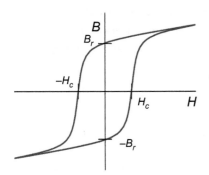

FIGURE 4.43 A hysteresis plot of B vs. H for a saturable magnetized material. The magnetic flux density B for this material depends not only on the applied field H, but also on the prior fields applied. H_c is the coercive field, the magnitude of the magnetic field that must be applied to remove most of the magnetization of the material, resulting in zero flux density. B_r is the residual magnetic flux density, the flux density due to the magnetization of the material, even in the absence of an applied magnetic field H.

These magnets produce a large B field for very little current. Magnetic memories are described in TechNote 4.10.

TechNote 4.10 Magnetic Memory

Removable discs, tapes, hard drives, and magnetic strips on credit cards are all different forms of magnetic memory. The first patent for a magnetic memory was filed by Oberlin Smith in 1878. His device recorded an analog audio signal on a thin steel wire. Valdemar Poulsen displayed a magnetic recording, also analog, at the Paris Exposition of 1900. Most data storage today is digital.

Data is recorded in magnetic domains within a thin layer of a ferromagnetic material. Iron oxide was formerly commonly used; currently a cobalt alloy is used. In either case, data is stored in sub-micron domains whose magnetization is aligned by the strong, localized magnetic field produced by a write head. The direction of the magnetization differentiates a 0 bit from a 1. The domain magnetization is read by a read head. Early forms of magnetic read heads used electromagnetic induction to read the data bits. As the bits translate past the read head, the changing magnetic field induces an electromotive field, which can be sensed electrically. More recently, magneto-resistance, in which electrical conduction of a sensor in the read head is modified by the magnetization of the memory bit, is used. While read/write heads are extremely close to the magnetic medium, as close as tens of nanometers, physical contact with the magnetic medium, which would destroy the system, is to be avoided.

It is often desirable to remove the magnetization of a ferromagnetic material that has become magnetized. This process is called degaussing, and is described in TechNote 4.11.

TechNote 4.11 Degaussing

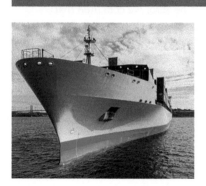

The magnetization of a material can be erased through a process known as degaussing. This process is applied to magnetic hard drives or other memory, the hulls of ships, and CRT screens that may have become magnetized through long use. The degaussing process can be understood by viewing the hysteresis curve shown in Fig. 4.43. Exposing the magnetized material to an oscillatory magnetic field $\mathbf{H}(t)$ with decreasing amplitude drives the magnetic response of the material around successive hysteresis curves, each with amplitude smaller than that of the previous cycle. This reduces the residual magnetization until a sufficiently low level is achieved.

In this section we have modified Ampère's Law to a form that is valid with magnetic materials. (We will have to modify Eq. (4.39) further to include time-varying effects, but for static fields this equation is now valid for all cases.) Equation (4.38) relates the magnetic field $\mathbf{H}$ to the magnetic flux density $\mathbf{B}$ (the total field due to external currents *and* effective, or bound, currents), through the magnetization $\mathbf{M}$, which we defined in the previous section as related to the density of magnetic dipoles and the average dipole moment. Finally, we have discussed various types of magnetic media, which are characterized through their magnetic susceptibility. In the next section we will use the differential properties of magnetic fields to determine the boundary conditions that magnetic fields must obey at the interface between two media.

4.8 Boundary Conditions for Magnetic Fields

In this section we examine the properties of magnetic fields at the interface between two magnetic media. Our purpose is to establish the relationship between fields on the two sides of the interface, known generally as the boundary conditions. We consider separately the conditions that must be satisfied by the tangential components (the components lying in the plane of the boundary) and normal components (the components perpendicular to the surface). We will use the integral relations: $\oint_s \mathbf{B} \cdot d\mathbf{s} = 0$ for any closed surface s, and $\oint_c \mathbf{H} \cdot d\boldsymbol{\ell} = I_{enc}$ for any closed path c.

The interface between two media is shown in Fig. 4.44. The two media are characterized by their permeabilities, μ_1 below the interface and μ_2 above the interface. In order to assess the conditions that the *normal* component of the magnetic flux density must satisfy, consider the pillbox-shaped structure shown with dashed lines in the figure, with one face lying in medium 1 and the other face lying in medium 2. The height of the pillbox is Δh, and the area of the two faces is Δs. Δh is very small so that the two faces of the pillbox lie very close to the interface between the two media. The arrows $\mathbf{B}_1$ and $\mathbf{B}_2$ represent the magnetic flux densities at the interface in medium 1 and 2, respectively. Also shown are the surface unit normal vectors $\mathbf{a}_{n,1}$ and $\mathbf{a}_{n,2}$, which point *out* of media 1 and 2, respectively, with $\mathbf{a}_{n,2} = -\mathbf{a}_{n,1}$. When evaluating $\oint_s \mathbf{B} \cdot d\mathbf{s}$ on this pillbox surface, recall that $d\mathbf{s}$ always points *outward* from the closed surface, in the direction $\mathbf{a}_{n,2}$ for the bottom face and in the direction $\mathbf{a}_{n,1}$ for the top face. The surface flux is then $\mathbf{B}_1 \cdot \mathbf{a}_{n,2} \Delta s + \mathbf{B}_2 \cdot \mathbf{a}_{n,1} \Delta s = (-B_{1n} + B_{2n}) \Delta s$, where the dot product between $\mathbf{B}$ and the unit normal vector picks out the normal component of $\mathbf{B}$, and B_{1n} appears with a minus sign since $\mathbf{B}_1$ is pointing into the pillbox, opposite the direction of $\mathbf{a}_{n,2}$. Since the closed surface flux must be zero, we conclude that

$$\boxed{B_{1n} = B_{2n},} \tag{4.42}$$

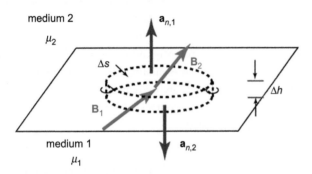

FIGURE 4.44 The interface between two magnetic media, characterized by magnetic permeability μ_1 and μ_2, as labeled. The arrows $\mathbf{B}_1$ and $\mathbf{B}_2$ represent the magnetic flux densities in medium 1 and 2, respectively, at the interface. The pillbox-shaped structure, shown with dashed lines, is not a physical boundary, but is constructed for evaluation of $\oint_s \mathbf{B} \cdot d\mathbf{s} = 0$, and determination of the boundary condition on B_n.

that is, the normal component of $\mathbf{B}$ must be continuous across any boundary. An alternative, equivalent form of this expression is

$$\mathbf{a}_n \cdot (\mathbf{B}_1 - \mathbf{B}_2) = 0, \tag{4.43}$$

where $\mathbf{a}_n$ is either of the surface normal unit vectors shown in Fig. 4.44.

We next consider the condition that must be satisfied by the *tangential* field components at the interface. The interface has been redrawn in Fig. 4.45, this time with a rectangular contour c, which has one side parallel to and immediately above the interface, and another side parallel to but immediately below the interface. Accordingly, Δh is very small; much less than Δw. The closed path integral $\oint_c \mathbf{H} \cdot d\boldsymbol{\ell}$ around this loop can be written as the sum of the line integrals for the four individual sides. The bottom side contributes $\int_c \mathbf{H} \cdot d\boldsymbol{\ell} = H_{1t} \, \Delta w$, where H_{1t} is the tangential component of $\mathbf{H}_1$ – that is, the component lying parallel to $d\boldsymbol{\ell}$, which was chosen to be parallel to the interface between the media. Along the top side, $\int_c \mathbf{H} \cdot d\boldsymbol{\ell}$ is $-H_{2t} \, \Delta w$, where the dot product gives us a minus sign since $d\boldsymbol{\ell}$ is pointing to the left, opposite to the direction of the tangential component of $\mathbf{H}_2$. We can ignore the contributions to $\oint_c \mathbf{H} \cdot d\boldsymbol{\ell}$ from the left and right sides of the rectangular path due to the smallness of Δh, and so we write

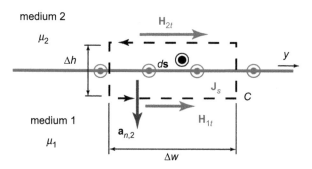

FIGURE 4.45 The interface between two magnetic media, characterized by magnetic permeability μ_1 and μ_2, as labeled. The arrows $\mathbf{H}_{1t}$ and $\mathbf{H}_{2t}$ represent the tangential component only of the magnetic fields at the interface in medium 1 and 2, respectively. The rectangular path c, shown with dashed lines, is not a physical boundary, but is constructed for our application of Ampère's Law, $\oint_c \mathbf{H} \cdot d\ell = I_{enc}$, for determination of the boundary condition on H_t.

$\oint_c \mathbf{H} \cdot d\ell = H_{1t}\,\Delta w - H_{2t}\,\Delta w$. This path integral must be equal to the total current I_{enc} passing through the surface bounded by c. Since Δh was chosen to be infinitesimally small, only surface currents can contribute to I_{enc}. The direction in which this current flows is important, and we can apply the right-hand rule to get the sign right. Remember the convention for the direction of the vector surface element $d\mathbf{s}$, shown in Fig. 4.45. Since the direction of the path c is counter-clockwise, $d\mathbf{s}$ points *out* from the page. If the current flows in the same direction as $d\mathbf{s}$, then this current is positive. The magnitude of the current I_{enc} is simply the surface current integrated across the width Δw, yielding $J_{s\perp}\,\Delta w$, where $J_{s\perp}$ is a component of the surface current parallel to $d\mathbf{s}$ (i.e. perpendicular to the path c). We can write this more compactly using the vector notation as

$$\boxed{\mathbf{a}_{n2} \times (\mathbf{H}_1 - \mathbf{H}_2) = \mathbf{J}_s.} \tag{4.44}$$

While we were not necessarily thinking about the infinite current sheet of Example 4.8 as we developed these boundary conditions, we should expect that Eq. (4.44) is consistent with that result. With the current $\mathbf{J}_s$ coming out of the page, as shown in Fig. 4.45, and no fields other than those produced by this current sheet, the tangential fields must be $\mathbf{H}_{t2} = -\mathbf{H}_{t1}$. (We use symmetry to make this statement. Recall that the magnetic field circulates around the current that produces it. So the magnetic field below the surface is directed to the right, and above the surface to the left. If we rotate the entire configuration about an axis pointing out of the page, $\mathbf{H}_{t2}$ becomes $-\mathbf{H}_{t1}$, and vice versa. But upon this rotation, the current is precisely the same as it was before the rotation, so $\mathbf{H}_{t2}$ and $-\mathbf{H}_{t1}$ must be the same.) Equation (4.44) tells us that the magnitude of each is $|\mathbf{J}_s|/2$, which is indeed consistent with that example for $\mu = \mu_0$ on both sides of the interface. In fact, this can be a handy physical picture for keeping the signs straight when we consider the discontinuity in H_t for the more general boundary condition.

Example 4.16 Magnetic Field Boundary Conditions

Consider the interface between two magnetic media, characterized by permeability μ_1 and μ_2, as shown in Fig. 4.46. The surface current at the interface is zero. The angles α_1 and α_2 (labeled only in Fig. 4.46(b)) are measured between the surface normal and the direction of $\mathbf{B}_1$ and $\mathbf{B}_2$ on either side of the interface, respectively. As shown, α_1 is greater than α_2. What does this tell us about the ratio of the permeabilities μ_1/μ_2? (Of course, we're hoping that you have failed to observe the answer that is plainly labeled on the diagram.)

Solution:
The boundary conditions $B_{1n} = B_{2n}$ and $H_{1t} = H_{2t}$ must be applied to solve this problem. Notice that $\mathbf{B}_1$ and $\mathbf{B}_2$ are illustrated for three different angles α_1 in Fig. 4.46. In (a), $\mathbf{B}_1$ is nearly normal to the surface (i.e. α_1 is small); in (c), $\mathbf{B}_1$ is nearly tangential to the surface (i.e. α_1 is nearly $\pi/2$, or 90°); in (b), α_1 is somewhere in the middle. In each case, $\mathbf{B}_1$ and $\mathbf{B}_2$ have been separated into their normal and tangential components, B_n and B_t, respectively. It is clear that B_{1n} and B_{2n} are the same in each case, while B_{2t} is only ~1/3

of B_{1t}. Since H_t must be continuous across the boundary, and since $H = B/\mu$, B_{1t}/μ_1 is equal to B_{2t}/μ_2, or upon rearranging, $B_{2t} = B_{1t}(\mu_2/\mu_1)$. From the observation that $B_{2t} \approx B_{1t}/3$, we conclude that $\mu_2/\mu_1 \approx 1/3$.

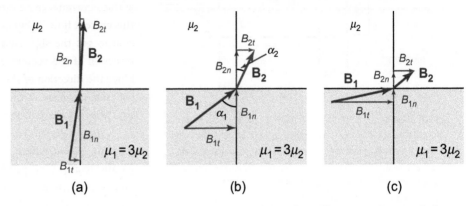

(a)　　　　　(b)　　　　　(c)

FIGURE 4.46 An interface between two magnetic materials. Vectors $\mathbf{B}_1$ and $\mathbf{B}_2$ represent the magnetic flux density in the two media at the interface. The angles α_1 and α_2, labeled only in (b), are measured between the surface normal and the vectors $\mathbf{B}_1$ and $\mathbf{B}_2$, respectively.

The relation between the angles α_1 and α_2 in terms of the permeabilities can be easily derived. From Fig. 4.46 it can be seen that $\tan\alpha_1 = B_{1t}/B_{1n}$ and $\tan\alpha_2 = B_{2t}/B_{2n}$. Dividing the first by the second gives us $\tan\alpha_1/\tan\alpha_2 = (B_{1t}/B_{1n})/(B_{2t}/B_{2n})$. But $B_{1n} = B_{2n}$, and $B_{1t}/\mu_1 = B_{2t}/\mu_2$, so the right-hand side is simply μ_1/μ_2. Thus, we have

$$\frac{\mu_1}{\mu_2} = \frac{\tan\alpha_1}{\tan\alpha_2}. \tag{4.45}$$

In Fig. 4.47(a), α_2 vs. α_1 is plotted for $\mu_1/\mu_2 = 2$ (blue plot) and $\mu_1/\mu_2 = 10$ (red plot). Notice that these plots are nonlinear, and α_2 is less than α_1 for all α_1. Thus the magnetic flux density $\mathbf{B}_2$ bends toward the surface normal, relative to the flux density $\mathbf{B}_1$ in medium 1.

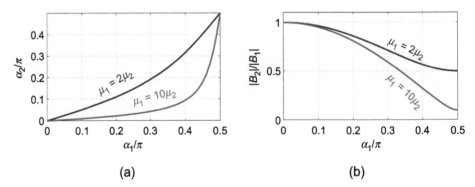

(a)　　　　　(b)

FIGURE 4.47 A plot of (a) α_2 vs. α_1 and (b) $|\mathbf{B}_2|/|\mathbf{B}_1|$ vs. α_1 for $\mu_1 = 2\mu_2$ (blue) and $\mu_1 = 10\mu_2$ (red).

We saw in Example 4.16 that when there are no surface currents at the interface, the relative orientation of the magnetic flux densities on the two sides of the interface can be determined using the boundary conditions that the fields must satisfy. We can also determine a relation between the relative magnitudes of the fields. From the example in Fig. 4.46, we see that for small angle α_1, **B** is nearly normal to the surface, and $B_{1n} = B_{2n}$ tells us that $|\mathbf{B}_1|$ and $|\mathbf{B}_2|$ are nearly the same. When **B** is nearly tangential to the surface, however, $|\mathbf{B}_2|$ is smaller than $|\mathbf{B}_1|$ by the ratio μ_2/μ_1, since H_t must be continuous.

This can be derived quantitatively using

$$|\mathbf{B}_2|^2 = B_{2n}^2 + B_{2t}^2 = B_{1n}^2 + (\mu_2/\mu_1)^2\, B_{1t}^2.$$

Using $B_{1n} = |\mathbf{B}_1| \cos\alpha_1$ and $B_{1t} = |\mathbf{B}_1| \sin\alpha_1$, this becomes

$$|\mathbf{B}_2|^2 = \left[1 + \left((\mu_2/\mu_1)^2 - 1 \right) \sin^2\alpha_1 \right] |\mathbf{B}_1|^2.$$

For $\mu_1 > \mu_2$ and $\alpha_1 \neq 0$, the term inside the square brackets must be less than 1, and so $|\mathbf{B}_2|$ is less than $|\mathbf{B}_1|$. Conversely, for $\mu_2 > \mu_1$, $|\mathbf{B}_2|$ is greater than $|\mathbf{B}_1|$. We show a plot of $|\mathbf{B}_2|/|\mathbf{B}_1|$ vs. α_1 for two values of μ_1/μ_2 in Fig. 4.47(b). Notice that for $\alpha_1 = 0$, where the fields are normal to the interface, $|\mathbf{B}_2| = |\mathbf{B}_1|$, while for $\alpha_1 = \pi/2$, where the fields are tangential to the surface, $|\mathbf{B}_2|/\mu_2 = |\mathbf{B}_1|/\mu_1$. Each of these results is consistent with the relevant boundary conditions.

The preceding discussion suggests an important use of a class of materials with extremely high permeability. That is, these materials can be used to substantially reduce magnetic fields in a region of space. These high-permeability materials can "guide" magnetic fields around a sensitive volume, reducing the flux density in the interior. This is sometimes called magnetic shielding. High-permeability materials are also useful as the core material in toroids and solenoids. In Example 4.17 we will explore the properties of a toroid in which the core is a highly permeable material with a narrow air gap. We will show that a strong magnetic flux density can be generated in the air gap region.

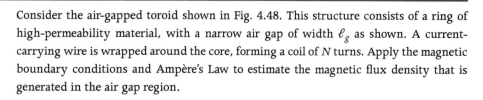

Example 4.17 Magnetic Field Boundary Conditions, Air-Gapped Toroid

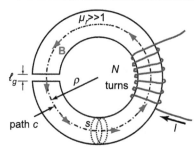

FIGURE 4.48 An air-gapped toroid. A circular magnetic material, with relative permeability μ_r, is wrapped by a winding of *N* turns which carries a current *I*. The thickness of the air-gap is ℓ_g.

Consider the air-gapped toroid shown in Fig. 4.48. This structure consists of a ring of high-permeability material, with a narrow air gap of width ℓ_g as shown. A current-carrying wire is wrapped around the core, forming a coil of *N* turns. Apply the magnetic boundary conditions and Ampère's Law to estimate the magnetic flux density that is generated in the air gap region.

Solution:

The current in the windings produces a magnetic field $\mathbf{H}_c$ in the core directed along the axis of the coil – that is, primarily in the $\mathbf{a}_\phi$ direction. (In this example, the subscript "c" denotes the magnetic core quantities and "g" the air gap.) These field lines are tangential to the walls of the high-permeability core material, so α_1 in Fig. 4.46 is $\simeq \pi/2$, where medium 1 is the magnetic core material with $\mu_r \gg 1$. We just examined the boundary conditions for $\mu_1 > \mu_2$, and saw that when $\alpha_1 \simeq \pi/2$, the flux density **B** outside the core material is reduced by a factor equal to μ_2/μ_1. Also, the magnitude of $\mathbf{B}_c$ within the core material is relatively uniform around the toroid. We conclude this by evaluating the magnetic flux $\oint \mathbf{B} \cdot d\mathbf{s}$ on the disc-shaped structure marked s in dashed lines shown in Fig. 4.48. This structure consists of three surfaces: two flat faces and the circular side. Since the flux density at the sides is very small, this contribution is negligible. Since $\oint \mathbf{B} \cdot d\mathbf{s}$ for the entire closed surface must be zero, the flux through the two faces must be equal but opposite, which indicates that $|\mathbf{B}_c|$ is unchanged from one face to the other. For $\mu_r \gg 1$, then, the magnetic flux lines are "guided" by the high-permeability material around the toroid, uniform in magnitude, and with little leakage to the outside. At the interface between the core material and the gap region, however, the angle α_1 between $\mathbf{B}$ and the surface normal is ~ 0, so $\mathbf{B}_g$ in the air gap is the same as it is in the high-permeability core. Thus we can write $B_g = B_c$, where B_c is the magnetic flux density in the high-permeability core region. The magnetic fields in these two regions are $H_g = B_g/\mu_0$ and $H_c = B_c/\mu_c = B_g/\mu_r\mu_0$.

Ampère's Law can now be applied around the dot-dashed path marked c in the figure to determine the magnitude of B_g. Since $\mathbf{B}$ is in the $\mathbf{a}_\phi$ direction, and is uniform around the loop,

$$\oint_c \mathbf{H} \cdot d\boldsymbol{\ell} \approx H_c(2\pi\rho - \ell_g) + H_g\ell_g.$$

By Ampère's Law, this quantity must be equal to I_{enc}, which, since there are N loops of the wire around the core, each carrying a current I, is NI. Substituting, we find

$$(B_g/\mu_c)(2\pi\rho - \ell_g) + (B_g/\mu_0)\ell_g = NI,$$

or, upon solving for B_g and approximating for a small air gap, $\ell_g \ll 2\pi\rho$,

$$B_g = \frac{\mu_0\mu_r NI}{2\pi\rho + \mu_r\ell_g}.$$

The two terms in the denominator can be comparable to one another, even for small ℓ_g, since μ_r can be huge. This expression shows that the air-gapped toroid can produce extremely large flux densities in the gap region.

While not circular in shape, a clampmeter guides magnetic field lines created by a current in a wire in much the same way as in the toroid just discussed. Measurement of the magnetic field then gives a means of determining the current in the wire, as described in TechNote 4.12.

In this section we have applied the integral properties of magnetic fields, $\oint_s \mathbf{B} \cdot d\mathbf{s} = 0$ and $\oint \mathbf{H} \cdot d\boldsymbol{\ell} = I_{enc}$, to determine the boundary conditions that magnetic fields must satisfy at the interface between two media. We applied these boundary conditions to understand how field lines will "bend" at the surface of magnetic materials, and have gained a qualitative understanding of the guiding of flux lines by highly permeable materials.

TechNote 4.12 Clampmeter

To measure current in a conductor without interrupting the circuit, one can use a clampmeter, as shown in Fig. 4.49. When the jaws of the clampmeter encircle a current-carrying wire, the magnetic field lines generated by the current are guided by the magnetic material in the clamps to the meter sensor. Two types of sensors are commonly found: a transformer or a Hall sensor. In the former, a second coil is wrapped around the magnetic material inside the meter, making it the secondary of a transformer. This type of clampmeter can only read alternating currents. (See Section 5.2.2.) The Hall sensor clampmeter is capable of reading d.c. currents, as well as a.c., but is generally more expensive than the transformer meter.

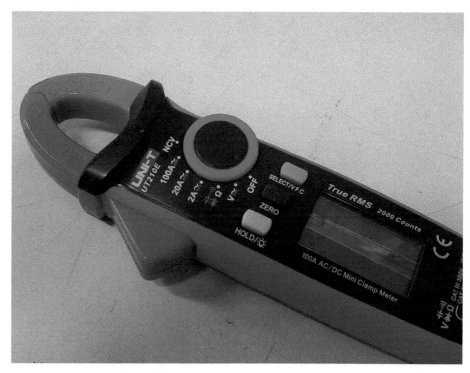

FIGURE 4.49 A digital clampmeter.

LEARNING OBJECTIVES

After studying this section, you should be able to:
- determine the scalar magnetic potential valid in a current-free region; and
- determine the magnetic flux density from the scalar magnetic potential.

4.9 Scalar Magnetic Potential

In Section 4.4 we introduced the vector magnetic potential $\mathbf{A}$, which can be used to calculate the magnetic flux density $\mathbf{B}$ using the relation $\mathbf{B} = \nabla \times \mathbf{A}$. The existence of this potential function was inferred by the property $\nabla \cdot \mathbf{B} = 0$, obeyed by magnetic fields, as we discuss in Section E.2.3.

In this section we introduce a second potential function for magnetic fields. This potential function exists only in current-free regions, and unlike $\mathbf{A}$, this potential function is a scalar field. This potential function is denoted $V_m(\mathbf{r})$, choosing notation that is similar to the electric potential $V(\mathbf{r})$ that we introduced in Chapter 1.

The existence of the scalar magnetic potential is derived from Ampère's Law, which for static fields is of the form

$$\nabla \times \mathbf{H} = \mathbf{J}_v.$$

In regions where no currents exist, $\mathbf{J}_v = 0$ and

$$\nabla \times \mathbf{H} = 0.$$

This should remind you of a similar relation for electrostatic fields, for which the electric potential $V(\mathbf{r})$ exists, due to their irrotational property. (That is, the curl of $\mathbf{E}$ is zero.) By the same reasoning, there must exist a scalar magnetic potential $V_m(\mathbf{r})$ in current-free regions, and the magnetic field can be derived from this potential using

$$\mathbf{H}(\mathbf{r}) = -\nabla V_m(\mathbf{r}).$$

The minus sign in this definition is the standard convention, in analogy to the electric potential. For a linear, isotropic, homogeneous material of permeability μ, this relation becomes

$$\mathbf{B}(\mathbf{r}) = -\mu\nabla V_m(\mathbf{r}). \tag{4.46}$$

Analogous with Eq. (1.43) for the electric potential, a potential difference between two points can be defined:

$$V_{mB} - V_{mA} = -\frac{1}{\mu}\int_A^B \mathbf{B} \cdot d\boldsymbol{\ell}.$$

Also, since $\nabla \cdot \mathbf{B} = 0$, the relation

$$\nabla \cdot \mathbf{B} = -\mu\nabla \cdot (\nabla V_m(\mathbf{r})) = -\mu\nabla^2 V_m(\mathbf{r}) = 0$$

can be derived. That is, the scalar magnetic potential satisfies Laplace's Equation,

$$\nabla^2 V_m(\mathbf{r}) = 0, \tag{4.47}$$

just as the electric potential does. Therefore, we can borrow much of what we developed for electric potentials, and apply these to the scalar magnetic potential. A word of caution, however, arises from different boundary conditions for these two potentials.

Let's determine the scalar magnetic potential due to a single magnetic dipole. We start by considering the magnetic flux density given by Eq. (4.32),

$$\mathbf{B} = \frac{\mu_0 m}{4\pi R^3}\left[\mathbf{a}_R\, 2\cos\theta + \mathbf{a}_\theta \sin\theta\right].$$

As a reminder, $\mathbf{m} = I(\pi b^2)\mathbf{a}_z$ is the magnetic dipole moment of a loop of current I of radius b. Equation (4.32) is valid for locations far from the current loop, $R \gg b$. Using arguments similar to those that we used when discussing electric dipoles (see Eqs. (1.56)–(1.57) and the accompanying text), we can show that Eq. (4.32) can be written as

$$\mathbf{B} = -\nabla\left(\frac{\mu_0}{4\pi}\frac{\mathbf{m}\cdot\mathbf{a}_R}{R^2}\right). \tag{4.48}$$

Upon comparison with Eq. (4.46), we identify the scalar magnetic potential of a single magnetic dipole as

$$V_m(\mathbf{r}) = \frac{1}{4\pi}\frac{\mathbf{m}\cdot\mathbf{a}_R}{R^2}. \tag{4.49}$$

Now that we have the scalar magnetic potential for a single magnetic dipole, we'd like to use this to determine the potential due to an assembly of dipoles, as are present in a magnetized material. This discussion will include elements of our derivation of the electric potential due to a dielectric material (Section 1.7) and the vector magnetic potential in a magnetized material (Section 4.6), and we won't repeat each step. Recalling that the magnetization $\mathbf{M}$ represents the density of magnetic dipoles, the contribution to the scalar magnetic potential due to the dipoles contained within an elemental volume dv' can be written as

$$dV_m = \frac{1}{4\pi} \frac{\mathbf{M} \cdot \mathbf{a}_R}{R^2} dv'.$$

To determine the potential due to the dipoles contained within some macroscopic volume, we must add all these contributions, which in the limit of small elemental volumes dv' becomes the volume integral

$$V_m = \frac{1}{4\pi} \int_{v'} \frac{\mathbf{M} \cdot \mathbf{a}_R}{R^2} dv'.$$

This integral is the same form as Eq. (1.59) for the electric potential of a dielectric medium. Using the same vector identities used in Section 1.7, this potential reduces to

$$V_m(\mathbf{r}) = \frac{1}{4\pi} \oint_{s'} \frac{\mathbf{M}(\mathbf{r}') \cdot d\mathbf{s}'}{|\mathbf{r} - \mathbf{r}'|} - \frac{1}{4\pi} \int_{v'} \frac{\mathbf{\nabla}' \cdot \mathbf{M}(\mathbf{r}')}{|\mathbf{r} - \mathbf{r}'|} dv'. \tag{4.50}$$

The similarity to Eq. (1.61) is striking, and by analogy with the electrostatic potential we identify two "effective" magnetic charge densities:

$$\rho_{ms} = \mathbf{M} \cdot \mathbf{a}_n \tag{4.51}$$

and

$$\rho_{mv} = -\mathbf{\nabla} \cdot \mathbf{M}. \tag{4.52}$$

The units on these quantities are $[\rho_{ms}] = $ A/m and $[\rho_{mv}] = $ A/m^2. These charges are called effective magnetic charges, recognizing that real magnetic charges do not exist. (At least, they have not been observed to this point.) Nonetheless, this formalism allows us to determine the scalar magnetic potential V_m, which allows us to determine the magnetic flux density $\mathbf{B}$. This is illustrated in Example 4.18.

Example 4.18 Scalar Magnetic Potential

Consider the cylinder of magnetized material described in Example 4.15 and illustrated in Fig. 4.41. Determine the effective magnetic charges for this cylinder. Use these charges to determine the scalar magnetic potential $V_m(z)$ along the axis of the cylinder. Use the magnetic potential to determine the magnetic flux density $B_z(z)$ along the axis.

Solution:

Since the magnetization $\mathbf{M}$ is uniform within the cylinder, its divergence is zero and the effective volume charge density ρ_{mv} given by Eq. (4.52) is also zero. Applying Eq. (4.51) to determine the effective surface charge density, we find

$$\rho_{ms} = \begin{cases} M_0 & \text{top face} \\ -M_0 & \text{bottom face} \\ 0 & \text{inner and outer side walls} \end{cases}.$$

We used the surface normal vectors $\mathbf{a}_n = +\mathbf{a}_z, -\mathbf{a}_z$, and $\pm\mathbf{a}_\rho$ for these three surfaces, respectively. These charges can be used in Eq. (4.50) to write V_m for a general field point $\mathbf{r} = (x, y, z)$:

$$V_m(\mathbf{r}) = \frac{1}{4\pi} \oint_{s'} \frac{\rho_{ms} ds'}{|\mathbf{r} - \mathbf{r}'|} \tag{4.53}$$

$$= \frac{1}{4\pi} \left\{ \int_0^{2\pi} \int_a^b \frac{M_0 \, r' dr' \, d\phi'}{\sqrt{(x - r' \cos\phi')^2 + (y - r' \sin\phi')^2 + (z - L/2)^2}} \right.$$

$$\left. - \int_0^{2\pi} \int_a^b \frac{M_0 \, r' dr' \, d\phi'}{\sqrt{(x - r' \cos\phi')^2 + (y - r' \sin\phi')^2 + (z + L/2)^2}} \right\}.$$

This integral is not pretty, but fortunately we were tasked only with finding the scalar magnetic potential at a point along the z-axis, for which x and y are zero. For this restricted solution, the integrals of Eq. (4.53) become

$$V_m(z) = \frac{M_0}{4\pi} \left\{ \int_0^{2\pi} \int_a^b \frac{r' dr' \, d\phi'}{\sqrt{r'^2 + (z - L/2)^2}} - \int_0^{2\pi} \int_a^b \frac{r' dr' \, d\phi'}{\sqrt{r'^2 + (z + L/2)^2}} \right\}.$$

Carrying out the integrals leads us to

$$V_m(z) = \frac{M_0}{2} \left\{ \sqrt{b^2 - (z - L/2)^2} - \sqrt{a^2 - (z - L/2)^2} \right.$$

$$\left. - \sqrt{b^2 - (z + L/2)^2} + \sqrt{a^2 - (z + L/2)^2} \right\}$$

for the scalar potential along the axis.

We are now in a position to determine the component B_z from this potential using Eq. (4.46). Note that since our solution for $V_m(z)$ is only valid along the z-axis, and is a function of z, but not x or y, we cannot use Eq. (4.46) to determine B_x or B_y. These components are, of course, zero along the z-axis, but we can reach this conclusion only by the symmetry of the cylinder. The magnetic flux density is

$$B_z = -\mu_0 \frac{d}{dz} V_m(z)$$

$$= \frac{\mu_0 M_0}{2} \left\{ -\frac{z - L/2}{\sqrt{b^2 - (z - L/2)^2}} + \frac{z - L/2}{\sqrt{a^2 - (z - L/2)^2}} \right.$$

$$\left. + \frac{z + L/2}{\sqrt{b^2 - (z + L/2)^2}} - \frac{z + L/2}{\sqrt{a^2 - (z + L/2)^2}} \right\}.$$

Comparison with Example 4.15 shows that we have arrived at the same result.

In this section we have presented the scalar magnetic potential as an additional tool for determining the magnetic flux density. While this potential is valid only in regions where currents are zero, it can lead to useful solutions of magnetic problems involving magnetized materials.

SUMMARY Starting with quantitative observations of the force between current-carrying wires, we have developed the properties of the magnetic flux density **B**. The Biot–Savart Law, which we presented in various forms applicable to total current, as well as volume or surface current densities, gives us a means of calculating the flux density for a given current distribution that is consistent with observations. We also discussed the differential properties of magnetic fields, which help us define more generally the properties that these magnetic fields must display. Since magnetic flux densities must have zero divergence, we were able to introduce a vector magnetic potential function

A, which gives us a new complementary tool for calculating the magnetic flux density. We introduced magnetic materials, which we modeled in terms of the potential due to a distribution of magnetic dipoles, and we explored the boundary conditions that must be satisfied at the interface between different media. Finally, we introduced a scalar magnetic potential, useful for calculating magnetic fields in current-free regions, such as with permanent magnets. We have yet to discuss the energy stored in magnetic fields, mechanical effects (forces and torques) due to magnetic fields, and inductances. This discussion is more meaningful after we have introduced Faraday's Law, and so we delay it until later.

Table 4.2

Collection of key formulas from Chapter 4.			
$\mathbf{B}(\mathbf{r}) = \dfrac{\mu_0}{4\pi} \displaystyle\int \dfrac{I_1 \, d\boldsymbol{\ell}' \times (\mathbf{r} - \mathbf{r}')}{\lvert \mathbf{r} - \mathbf{r}' \rvert^3}$	(4.3)	$\mathbf{A}(\mathbf{r}) = \dfrac{\mu_0}{4\pi} \displaystyle\int_v \dfrac{\mathbf{J}_v(\mathbf{r}') \, dV}{\lvert \mathbf{r} - \mathbf{r}' \rvert}$	(4.22)
$\mathbf{B}(\mathbf{r}) = \dfrac{\mu_0}{4\pi} \displaystyle\int_v \dfrac{\mathbf{J}_v(\mathbf{r}') \times (\mathbf{r} - \mathbf{r}') \, dV}{\lvert \mathbf{r} - \mathbf{r}' \rvert^3}$	(4.5)	$\mathbf{A}(\mathbf{r}) = \dfrac{\mu_0}{4\pi} \displaystyle\int_s \dfrac{\mathbf{J}_s(\mathbf{r}') \, ds'}{\lvert \mathbf{r} - \mathbf{r}' \rvert}$	(4.23)
$\mathbf{B}(\mathbf{r}) = \dfrac{\mu_0}{4\pi} \displaystyle\int_s \dfrac{\mathbf{J}_s(\mathbf{r}') \times (\mathbf{r} - \mathbf{r}') \, ds'}{\lvert \mathbf{r} - \mathbf{r}' \rvert^3}$	(4.6)	$\mathbf{A}(\mathbf{r}) = \dfrac{\mu_0}{4\pi} \displaystyle\int_\ell \dfrac{I \, d\boldsymbol{\ell}'}{\lvert \mathbf{r} - \mathbf{r}' \rvert}$	(4.24)
$\mathbf{F} = q\,\mathbf{u} \times \mathbf{B}$	(4.7)	$\mathbf{J}_{v,eff}(\mathbf{r}) = \nabla \times \mathbf{M}(\mathbf{r})$	(4.36)
$\mathbf{B}(\mathbf{r}) = \dfrac{\mu_0 I}{2\pi\rho}\,\mathbf{a}_\phi$	(4.9)	$\mathbf{J}_{s,eff}(\mathbf{r}) = \mathbf{M}(\mathbf{r}) \times \mathbf{a}_n$	(4.37)
$\nabla \cdot \mathbf{B} = 0$	(4.12)	$\mathbf{H} = \dfrac{1}{\mu_0}\mathbf{B} - \mathbf{M}$	(4.38)
$\oint_s \mathbf{B} \cdot d\mathbf{s} = 0$	(4.14)	$\nabla \times \mathbf{H} = \mathbf{J}_v$	(4.39)
$\nabla \times \mathbf{B} = \mu_0\,\mathbf{J}_v$	(4.15)	$\oint_c \mathbf{H} \cdot d\boldsymbol{\ell} = \int_s \mathbf{J}_v \cdot d\mathbf{s} = I_{enc}$	(4.40)
$\oint_c \mathbf{B} \cdot d\boldsymbol{\ell} = \mu_0 \int_s \mathbf{J}_v \cdot d\mathbf{s} = \mu_0 I_{enc}$	(4.16)	$B_{1n} = B_{2n}$	(4.42)
$\mathbf{B}(\mathbf{r}) = \nabla \times \mathbf{A}(\mathbf{r})$	(4.19)	$\mathbf{a}_{n2} \times (\mathbf{H}_1 - \mathbf{H}_2) = \mathbf{J}_s$	(4.44)

PROBLEMS   **Biot–Savart Law**

P4-1   (a) Use the Biot–Savart Law to determine the magnetic flux density $\mathbf{B}(\mathbf{r})$ for a finite-length current-carrying wire. The current is I, the wire lies along the z-axis from $z' = -L/2$ to $+L/2$, and $\mathbf{r}$ is any point. (b) Simplify your result for a point $\mathbf{r}$ in the $z = 0$ plane. (c) Use your result to determine $\mathbf{B}$ for an infinitely long wire.

P4-2   Show that the magnetic flux density generated by a finite-length wire carrying a current I can be written

$$\mathbf{B}(\mathbf{r}) = \mathbf{a}_\phi \frac{\mu_0 I}{4\pi\rho} \left[\sin\alpha_1 + \sin\alpha_2 \right],$$

where α_1 and α_2 are the angles shown in Fig. 4.50. The wire lies along the z-axis from $z' = -L/2$ to $+L/2$, and ρ is the distance from the z-axis to the point P.

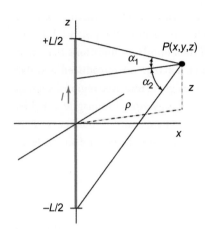

FIGURE 4.50 A wire segment, for use in Prob. P4-2.

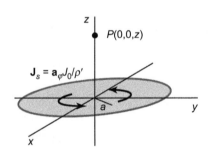

FIGURE 4.51 A disc of current, for use in Prob. P4-8.

P4-3 (a) Use superposition and your result from Prob. P4-1 to determine the magnetic flux density $\mathbf{B}(\mathbf{r})$ for a square loop of current for a point along the axis of the loop. The current in the square loop is I, and the wire loop lies in the x–y plane, centered at the origin. The length of each side is a. (b) Compare $\mathbf{B}$ along the z-axis of the square loop to $\mathbf{B}$ produced by a circular current loop of radius $b = a/2$.

P4-4 Use the Biot–Savart Law to determine the magnetic flux density $\mathbf{B}(\mathbf{r})$ for an infinite sheet of current. The surface current is $\mathbf{J}_s = \mathbf{a}_z J_s$, lying in the y–z plane. Compare your result to that of Example 4.8.

P4-5 Determine the magnetic flux density $\mathbf{B}(\mathbf{r})$ for an infinitely wide, thick sheet of current using the Biot–Savart law. The volume current density is $\mathbf{J}_v = \mathbf{a}_z J_v$ for $-d/2 < x < d/2$, and extends to infinity in the y- and z-directions. Compare your result to that of Example 4.8.

P4-6 Use superposition and the result from Example 4.4 to determine the magnetic flux density $\mathbf{B}(\mathbf{r})$ for a pair of parallel, finite-width sheets of current. The width of each current sheet is w, and the spacing between the current planes is d. The current density at $x = d/2$ is $\mathbf{J}_s = \mathbf{a}_z J_s$; at $x = -d/2$ it is $-\mathbf{J}_s$. Plot $B_y(y)$ and $B_x(y)$ at $x = -0.45\,d$, $x = 0$, and $x = 0.45\,d$, using $w = 4$ cm and $d = 1$ cm. Compare the maximum value of $B_y(y)$ in this region to J_s.

P4-7 In Example 4.5 we determined the magnetic flux density due to the current in the windings of a solenoid. In that example we ignored the small z-component of the current, which exists since the windings are helical in shape. Return to the example, and determine the magnetic flux density that results from this component of the current.

P4-8 Consider the disc of current shown in Fig. 4.51. The surface current density is $\mathbf{J}_s = \mathbf{a}_\phi J_0/\rho'$, where ρ' is the distance from the axis. The radius of the disc is a and the permeability of the surrounding medium is μ_0. Determine the magnetic flux density on the axis at position $P(0, 0, z)$.

P4-9 A disc of radius a and uniform surface charge density ρ_s, lying in the x–y plane, spins at a constant angular velocity ω. Determine $\mathbf{B}$ at a point P along the z-axis.

P4-10 A long, straight wire of radius a lies along the z-axis, and carries a current I in the $+z$-direction. The current density is uniform inside the wire. Use the Biot–Savart Law to determine the magnetic flux density $\mathbf{B}(\mathbf{r})$ for a point P (a) outside the wire, and (b) inside the wire. (c) Compare your result for part (a) to the result of Example 4.2.

Vector Differential Properties of Magnetic Fields

P4-11 Evaluate $\nabla \cdot \mathbf{B}$ to determine which of the following vector fields are allowable magnetostatic fields. Evaluate $\nabla \times \mathbf{B}$ to find the current density $\mathbf{J}_v$ associated with the field:

(a) $\mathbf{B}_1 = B_0\,(x\mathbf{a}_x + y\mathbf{a}_y + z\mathbf{a}_z)/[x^2 + y^2 + z^2]^{1/2}$
(b) $\mathbf{B}_2 = B_0\,(-y\mathbf{a}_x + x\mathbf{a}_y + z\mathbf{a}_z)/[x^2 + y^2 + z^2]^{1/2}$
(c) $\mathbf{B}_3 = B_0\,(x\mathbf{a}_x)$
(d) $\mathbf{B}_4 = B_0\,(-y\mathbf{a}_x + x\mathbf{a}_y + z\mathbf{a}_z)$.

P4-12 Evaluate $\nabla \cdot \mathbf{B}$ to determine which of the following vector fields are allowable magnetostatic fields. Evaluate $\nabla \times \mathbf{B}$ to find the current density $\mathbf{J}_v$ associated with the field:

(a) $\mathbf{B}_1 = B_0\,(\rho\mathbf{a}_\rho + z\mathbf{a}_z)/[\rho^2 + z^2]^{1/2}$
(b) $\mathbf{B}_2 = B_0\,(\rho\mathbf{a}_\phi + z\mathbf{a}_z)/[\rho^2 + z^2]^{1/2}$
(c) $\mathbf{B}_3 = B_0 \cos\phi\,\mathbf{a}_\rho$

(d) $\mathbf{B}_4 = B_0\,\mathbf{a}_\rho$.

P4-13 Evaluate $\nabla \cdot \mathbf{B}$ to determine which of the following vector fields are allowable magnetostatic fields. Evaluate $\nabla \times \mathbf{B}$ to find the current density $\mathbf{J}_v$ associated with the field:

(a) $\mathbf{B}_1 = B_0 \cos\theta\, \mathbf{a}_R / R^2$

(b) $\mathbf{B}_2 = B_0\, \mathbf{a}_R / R$

(c) $\mathbf{B}_3 = B_0\, \mathbf{a}_\phi / R^2$

(d) $\mathbf{B}_4 = B_0\,(\cos\theta\mathbf{a}_R - \sin\theta\mathbf{a}_\theta)/R^3$.

P4-14 In Example 4.2 we found the magnetic flux density due to an infinitely long current-carrying wire. Verify that the magnetic flux density $\mathbf{B}(\mathbf{r})$ given by Eq. (4.9) satisfies the requisite differential properties for magnetostatic fields in a current-free region, that is $\nabla \cdot \mathbf{B}(\mathbf{r}) = 0$ and $\nabla \times \mathbf{B}(\mathbf{r}) = 0$.

P4-15 In Example 4.4 we found the magnetic flux density due to a finite-width current-carrying sheet. Verify that the magnetic flux density $\mathbf{B}(\mathbf{r})$ given by Eq. (4.11) satisfies the requisite differential properties for magnetostatic fields in a current-free region, that is $\nabla \cdot \mathbf{B}(\mathbf{r}) = 0$ and $\nabla \times \mathbf{B}(\mathbf{r}) = 0$.

Ampère's Law in Vacuum

P4-16 Use Ampère's Law to determine the magnetic flux density $\mathbf{B}(\mathbf{r})$ for an infinitely long, thick sheet of current. The volume current density is $\mathbf{J}_v = \mathbf{a}_z J_v$ for $-d/2 < x < d/2$, and extends to infinity in the y- and z-directions. Compare your result to that of Example 4.8. For what value of $\mathbf{J}_s$ is $\mathbf{B}(\mathbf{r})$ the same in the region $|x| > d/2$?

P4-17 Consider a pair of long co-axial cylindrical conductors centered on the z-axis. A current I flows in the $+z$-direction in the inner conductor and returns in the outer conductor. The radius of the inner conductor is a, and the outer conductor has inner radius b and outer radius c. The current is uniformly distributed within each conductor. Determine the magnetic flux density $\mathbf{B}(\mathbf{r})$ for all four regions (inside the inner conductor, between the conductors, inside the outer conductor, and outside everything).

P4-18 An infinite sheet of current with current density $\mathbf{J}_s = +J_0\mathbf{a}_z$ is located at $x = +d/2$, and a second with $\mathbf{J}_s = -J_0\mathbf{a}_z$ is located at $x = -d/2$. Use Ampère's Law and superposition to determine $\mathbf{B}(\mathbf{r})$ for (a) $x > d/2$, (b) $-d/2 < x < d/2$, and (c) $x < -d/2$.

Vector Magnetic Potential

P4-19 Determine the vector magnetic potential $\mathbf{A}(\mathbf{r})$ produced by the current flowing in the square loop of wire described in Prob. P4-3. Use your result to determine $\mathbf{B}(\mathbf{r})$ for a point along the z-axis.

P4-20 (a) Determine the vector magnetic potential $\mathbf{A}(\mathbf{r})$ produced by two parallel wires carrying equal currents I in opposite directions. Let each wire be of length L and parallel to the z-axis. The wire carrying current in the $+z$-direction is at $x = d/2$, while the wire carrying current in the $-z$-direction is at $x = -d/2$. Your result should be valid for any point $(x, y, 0)$ in the $z = 0$ plane. (This plane is the middle of the wires.) (b) Evaluate your result in the limit of $L \to \infty$. (c) Use this result to find the magnetic flux density $\mathbf{B}(\mathbf{r})$ for a pair of infinitely long parallel wires.

P4-21 Determine the vector magnetic potential $\mathbf{A}(\mathbf{r})$ produced by the current flowing in a solenoid of radius a and turns density n. [Hint: Use the symmetry of the solenoid to show that $\mathbf{A}(\mathbf{r})$ can only point in the ϕ direction, and that A_ϕ can depend only on the radius ρ. Then examine $\oint_c \mathbf{A}(\mathbf{r}) \cdot d\ell$ on a circular path of radius ρ.] You should have one result valid inside the solenoid ($\rho < a$), and another valid outside the solenoid ($\rho > a$). Evaluate $\nabla \times \mathbf{A}(\mathbf{r})$ to confirm your results.

Magnetic Dipole

P4-22 Starting from Eq. (4.29), fill in the details of the steps leading to the vector potential of the magnetic dipole, Eq. (4.31).

Magnetization

P4-23 Consider the magnetized cube described in Example 4.14. Use the effective currents presented there in the Biot–Savart Law to determine the magnetic flux density $\mathbf{B}(\mathbf{r})$ at the center of the cube.

P4-24 A thin magnetized disc lies in the x–y plane and has uniform magnetization $\mathbf{M} = M_0 \mathbf{a}_z$. The radius of the disc is a and its thickness $t \ll a$. Determine the effective surface and volume currents. Use these effective currents in the Biot–Savart Law to determine the magnetic flux density $\mathbf{B}(\mathbf{r})$ along the z-axis.

Ampère's Law Revisited

P4-25 Use Ampère's Law to determine the magnetic field $\mathbf{H}(\mathbf{r})$ inside an infinitely long solenoid. The interior of the solenoid is of relative permeability $\mu_r = 1\,000$. The radius of the solenoid is $a = 3$ cm, its turns density is $n = 25$ turns/cm, and the current through the windings is $I = 2$ A. Determine the magnetic flux density $\mathbf{B}(\mathbf{r})$ inside this solenoid. Compare this result with that of an air-filled solenoid of the same geometry and current.

P4-26 Use Ampère's Law to determine the magnetic field $\mathbf{H}(\mathbf{r})$ inside a toroid at the center of the core. The interior of the toroid is of relative permeability $\mu_r = 1\,000$. The cross-section of the toroid is rectangular, with inner radius $a = 3$ cm, outer radius $b = 5$ cm, and thickness $t = 2$ cm. The winding around the core has 50 turns, and the current through the winding is $I = 2$ A. Determine the magnetic flux density $\mathbf{B}(\mathbf{r})$ inside this toroid. Compare this result with that of an air-filled toroid of the same dimension and current.

Boundary Conditions for Magnetic Fields

P4-27 Consider the interface between a simple, non-conducting medium of permeability μ_1 and vacuum, as shown in Fig. 4.52. The magnetic flux density to the left of the interface is $\mathbf{B} = (3\mathbf{a}_x + 5\mathbf{a}_y)$ T, while for $x > 0$ we have $\mathbf{B} = (3\mathbf{a}_x + 8\mathbf{a}_y)$ T. (a) Determine μ_1. (b) Determine the magnetization $\mathbf{M}$ for $x > 0$. (c) Determine the bound, or effective, surface current in the $x = 0$ plane.

P4-28 The interface between two non-conducting materials lies in the y–z plane at $x = 0$. The magnetic field $\mathbf{H}$ just inside medium 1 ($x < 0$, $\mu_1 = \mu_0$) is $\mathbf{H}_1 = (4\mathbf{a}_x - 3\mathbf{a}_y)$ A/m. (a) Determine the magnetic field $\mathbf{H}_2$ just inside medium 2 ($x > 0$, $\mu_2 = 2\mu_0$). (b) Compare $|\mathbf{H}_2|$ to $|\mathbf{H}_1|$. Which is larger, and by what factor? (c) Determine the angle α on both sides of the interface, where α is the angle between $\mathbf{H}$ and the surface normal.

P4-29 Consider an interface between two non-conducting media lying in the x–y plane. The medium for $z < 0$ has relative permeability $\mu_r = 2$, while for $z > 0$ the relative permeability is $\mu_r = 5$. The magnetic field at $z = 0^+$ is $\mathbf{H} = [\mathbf{a}_x 4 + \mathbf{a}_y 5 + \mathbf{a}_z 10]$ A/m. Determine $\mathbf{H}$ at $z = 0^-$. Determine the angle α between $\mathbf{H}$ and the surface normal on both sides of the interface. Does this result agree with Eq. (4.45)?

P4-30 The interface between two non-conducting media lies in the x–y plane. The medium for $z < 0$ has relative permeability $\mu_r = 2$, while for $z > 0$ the relative permeability is $\mu_r = 5$. The magnetic flux density at $z = 0^+$ is $\mathbf{B} =$

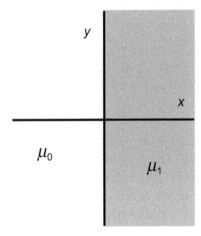

FIGURE 4.52 An interface between two magnetic materials, for use in Prob. P4-27.

$[\mathbf{a}_x 0.4 + \mathbf{a}_y 0.5 + \mathbf{a}_z 1.0]$ mT. Determine $\mathbf{B}$ at $z = 0^-$. Determine the angle α between $\mathbf{B}$ and the surface normal on both sides of the interface. Does this result agree with Eq. (4.45)?

P4-31 The interface between two media lies in the x–z plane. The medium for $y < 0$ is a conductor, while for $y > 0$ the medium is non-conducting. The permeability of the non-conductor is $5\mu_0$. In the conductor, the magnetic field is zero, but a surface current $\mathbf{J}_s = \mathbf{a}_z 10$ A/m exists. Determine the magnetic field $\mathbf{H}$ and the magnetic flux density $\mathbf{B}$ in the non-conductor.

P4-32 Consider an air-gapped toroid, similar to that shown in Fig. 4.48. The cross-section of the core material is rectangular. The inner and outer radii are $a = 5$ cm and $b = 7$ cm, the thickness is $h = 1.5$ cm, and the gap length is $\ell_g = 0.5$ cm. The permeability of the core material is $\mu = 100\mu_0$, and that of the gap region is μ_0. The current is $I = 4$ A, and the number of turns is $N = 25$. Find $\mathbf{H}_c$ in the core material in terms of N, I, ρ, ℓ_g, μ, and μ_0. Also find $\mathbf{B}_c$ in the core region, and $\mathbf{B}_g$ and $\mathbf{H}_g$ in the gap region (in terms of $\mathbf{H}_c$). Explain any approximations that you have used in your work. Using the parameter values given, find a numerical value for H_g and B_g in the gap region.

Magnetic Scalar Potential

P4-33 Consider the cylindrical magnetized material described in Examples 4.15 and 4.18. Determine the scalar magnetic potential $V_m(\mathbf{r})$ for this cylinder for large distances $R = |\mathbf{r}| = \sqrt{x^2 + y^2 + z^2}$; that is, $R \gg L, b$. Use $V_m(\mathbf{r})$ to determine the magnetic flux density $\mathbf{B}(\mathbf{r})$. Compare your result with the magnetic flux density of a single magnetic dipole of magnetic moment $\mathbf{m} = M_0 \mathbf{a}_z \pi (b^2 - a^2) L$. Explain why these results are the same.

CHAPTER 5
Faraday's Law

KEY WORDS

Faraday's Law; magnetic flux, electromotive force; Lenz' Law; Maxwell's Equations; motional EMF; self-inductance; mutual inductance; ideal transformer; coupling coefficient; generator.

LEARNING OBJECTIVES

After studying this section, you should be able to:
- describe qualitatively Faraday's Law;
- state Faraday's Law in its integral and differential forms; and
- apply Lenz' Law to determine the sign of the induced EMF.

To this point in our discussions, we have dealt solely with static fields. We started with static electric fields, in which all charges are stationary. The electric fields produced by these charges are stationary as well. With electric fields, we developed the notion of the electric potential, the energy stored by electric fields, and the capacitance of a configuration of conductors. We then moved on to introduce static magnetic fields, which are produced by stationary currents. For magnetic fields, we have also introduced potential functions, one a vector function, the other a scalar, but we have not yet discussed the energy stored by a magnetic field, or the inductance of a configuration of current-carrying wires. We will, of course, treat these important topics, but before we do so, we find it useful to take a first look at some time-varying effects. In particular, we will develop a law known as Faraday's Law, which is the basis for circuit elements such as inductors and transformers, as well as electrical generators and many other useful devices. After we have mastered Faraday's Law, we will be in a much better position to discuss the energy stored in magnetic fields and inductances, and so we will return to these topics at that time.

5.1 Faraday's Law

The relation that we now know as Faraday's Law quantifies a series of observations by Michael Faraday in the mid-1800s, as discussed in Biographical Note 5.1. This law aptly bears his name.

Biographical Note 5.1 Michael Faraday (1791–1867)

FIGURE 5.1 A photograph of Michael Faraday, circa 1861. Probably albumen carte-de-visite by John Watkins.

Despite having little formal education, Michael Faraday's (Fig. 5.1) contributions to electricity, magnetism, and electrochemistry are nothing short of foundational. At age 14, Faraday was apprenticed to a book binder, giving him the opportunity to read a wide range of topics after-hours. His interest in science took hold, especially after attending public lectures by notable scientists of the time, including Sir Humphry Davy. Faraday wrote up his lecture notes in a book, which left a strong impression on Davy, who soon hired him as an assistant. Faraday rose to the level of the Fullerian Professorship in Chemistry at the Royal Institution by 1833, a position he held until his death in 1867.

In studies relevant to electricity and magnetism, Faraday demonstrated an elementary motor, showed that the charges on a conductor reside on the outside surface of the conductor, constructed a d.c. power source based on magnetic induction (see Tech Note 5.7), studied magnetic effects in materials, and explored the effect of a magnetic field on the polarization of light. In the early 1830s, Faraday discovered what is now called **Faraday's Law**. Upon wrapping two insulated wires around a single iron ring, he found that turning the current on or off in one of the windings caused a momentary current in the other winding. The iron ring used by Faraday is still on display at the Royal Institution. In an 1846 public address to the Royal Institution, Faraday speculated that electric and magnetic lines of force might be related to light propagation, an idea later expanded upon by James Clerk Maxwell. Faraday also discovered the rotation of polarization of light in some types of glass to which a magnetic field parallel to the light propagation direction was applied. This polarization rotation is non-reciprocal; that is, the direction of rotation of the polarization of light depends on the direction in which the light is propagating. This effect is now known as Faraday rotation, and is the basis of optical and radio frequency isolators. Faraday was the first to discover diamagnetism, which we discussed in Chapter 4. Faraday's specific observation was that diamagnetic materials are repelled, not attracted, by a magnet. Faraday first proposed the ideas of electric and magnetic fields that exist in the space surrounding the sources (charges and currents), another idea that Maxwell later expanded upon.

Michael Faraday's interests and accomplishments extended to topics beyond electricity and magnetism. He is probably best known for his discovery of benzene and other hydrocarbons. He was the first to liquify chlorine and several other gases, and he developed the ammonia refrigeration cycle on which early cooling units were based. Faraday experimented with various steel alloys and optical glasses, and later used one of his glass samples to demonstrate Faraday rotation, discussed above.

Albert Einstein held Faraday in such high regard that he kept a picture of him, along with pictures of Sir Isaac Newton and James Clerk Maxwell, on his study wall. The SI unit of capacitance is named the farad in honor of Michael Faraday.

The Faraday effect can be demonstrated simply by moving a permanent magnet in the vicinity of a loop of wire, generating a momentary potential difference between the ends of the wire loop. This is illustrated in Fig. 5.2. The deflection of the voltmeter ceases if the magnet stops, and reverses if the magnet is moved in the opposite direction, or if the magnet is turned around and the measurement repeated. These observations are summarized quantitatively in the relation

$$\mathcal{E} = -\frac{d\Phi_M}{dt},$$

(5.1)

where $\mathcal{E} = \oint_c \mathbf{E} \cdot d\boldsymbol{\ell}$ is the induced potential difference across the loop, defined as the path integral of the induced electric field integrated around the complete loop c, and

$$\Phi_M = \int_s \mathbf{B} \cdot d\mathbf{s}$$

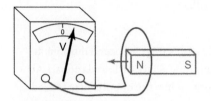

FIGURE 5.2 A permanent magnet moving in the vicinity of a loop of wire induces a potential difference between the ends of the wire.

is the **magnetic flux** integrated over the surface area that is bounded by the loop c. $\mathcal{E}$ is known as the **electromotive force**, often abbreviated EMF. (Notice that in the definition of the EMF as $\mathcal{E} = \oint_c \mathbf{E} \cdot d\boldsymbol{\ell}$, there is no minus sign, in contrast to the definition of the electric potential in Eq. (1.43)! Signs are important, and we'll come back to this soon)

This relation may seem quite strange, especially if you are in the mindset that the electric potential within a conductor is always uniform. This is only correct, however, for *static* fields, and as we are now moving into the realm of time-varying effects, this is one of the first notions that must be abandoned. We will return to this point below.

The magnetic flux Φ_M can be made to vary in several different ways. In the example described in the preceding paragraph, as the magnet moves toward or away from the wire loop, the magnetic flux density $\mathbf{B}$ in the area inside the loop varies in time, causing Φ_M to vary as well. In other cases, $\mathbf{B}$ may be constant in some region in space, but the circuit, or part of the circuit, may be moving. (If the loop moves to a region in which $\mathbf{B}$ is different, or if the area of the loop changes, the magnetic flux Φ_M passing through that loop will change.) As a third possibility, the circuit may be rotating through a constant field, as is commonly the case in an electrical generator. Rotation of the circuit is, of course, just a special form of motion of the circuit.

The minus sign in Eq. (5.1) has special significance, as it tells us the sign of the induced EMF. This sign is also expressed in terms of **Lenz' Law**, which states that the polarity of the induced EMF tends to produce a current in the circuit in such a direction as to produce a magnetic field that *opposes the change* in flux. If the magnetic flux through the loop is *increasing*, the induced EMF will try to produce a current that generates a magnetic field in the direction that opposes the flux. If the magnetic flux through the loop is *decreasing*, the induced EMF tries to produce current that generates a magnetic field that adds to the flux. If we are careful in setting up all the path integrals and surface fluxes with the proper sign conventions, Eq. (5.1) will lead us to the correct sign of the EMF. A short pause at the end of our work to verify that Lenz' Law is obeyed can be a good way to check our result. We should also remark that the EMF induced in the loop does not necessarily result in a current. For example, if the loop is open-circuited, the changing flux will cause an EMF between the ends of the loop, yet no current will flow in the loop. We can still think, however, of the direction in which the current would flow if the ends of the loop were connected through a finite resistance, and apply Lenz' Law on this basis.

We introduced Eq. (5.1) as if it were applied around the wires of an electrical circuit, but it actually is more generally applicable to any contour, not necessarily defined by any physical structures such as a loop of wire. Let us examine this equation a bit further, then, in order to derive an alternative form, known as the differential form, of Faraday's Law. We start by substituting the explicit integrals for $\mathcal{E}$ and Φ_M into Eq. (5.1):

$$\oint_c \mathbf{E} \cdot d\boldsymbol{\ell} = -\frac{d}{dt} \int_s \mathbf{B} \cdot d\mathbf{s},$$

where c is any path we choose and s is any surface that is bounded by the path c. In applying this relation to a fixed path that does not change with time, the variation in the magnetic flux Φ_M is due solely to variation in $\mathbf{B}$, and we can move the derivative with respect to time inside the integral. Also, we can apply Stokes' Theorem to the path integral of $\mathbf{E}$ to write

$$\int_s \boldsymbol{\nabla} \times \mathbf{E} \cdot d\mathbf{s} = -\int_s \frac{\partial \mathbf{B}}{\partial t} \cdot d\mathbf{s},$$

where both surface integrals are carried out over the same surface s. Note that we have written $\partial \mathbf{B}/\partial t$ using partial derivatives, since $\mathbf{B}$ may also be a function of position. Since this result is valid for any surface s, the integrals can be equal if and only if the integrands are equal. We therefore conclude that

$$\boxed{\boldsymbol{\nabla} \times \mathbf{E} = -\frac{\partial \mathbf{B}}{\partial t}.} \tag{5.2}$$

This is known as the differential form of Faraday's Law, and we immediately note a change from an important property that we discovered for static electric fields. That is, for static fields we found that $\boldsymbol{\nabla} \times \mathbf{E} = 0$. Now we see that for time-varying fields this relation must be replaced by Eq. (5.2). This relation is now complete, and is one of the four relations known collectively as **Maxwell's Equations**. Additionally, this relation, or its integral equivalent given by Eq. (5.1), implies two critical changes to the way we think of electric and magnetic fields. First, we must modify our notion of the electric potential. For static fields, recall that the closed path integral $\oint_c \mathbf{E} \cdot d\boldsymbol{\ell}$ is always equal to zero, and the work required to move a charge from point A to point B was independent of the path we chose. Now we see that this conclusion is valid only for static fields. Equivalently, our previous notion that conductors must be equipotential surfaces must be revisited. Electric potentials are still useful functions, but we do have to change how we view them and how to apply them to specific problems. We will return to this later. The second change brought about by Faraday's Law is that for static fields our discussion of electric fields and our discussion of magnetic fields were completely separate from one another. Static electric fields are generated by charges, while static magnetic fields are generated by currents. Granted, currents are just charges in motion, but even with this connection between the sources of the two different fields, there was no further apparent connection between the fields themselves. Now we have established the first connection between electric and magnetic fields in the form of Faraday's Law. As a hint of a further change, we have yet to modify Ampère's Law for time-varying fields. When Ampère's Law is complete, we will see another connection between electric and magnetic fields, but that will come later.

Despite Lenz' Law, we find that there is often a great deal of confusion regarding the sign of the induced EMF $\mathcal{E}$. We have been precise in our definitions of the EMF and magnetic fluxes, but it can be helpful to return to these definitions, emphasizing their signs as we do so. Recall that we defined the EMF as $\mathcal{E} = \oint_c \mathbf{E} \cdot d\boldsymbol{\ell}$ using the path integral of the induced electric field integrated around the complete loop, and $\Phi_M = \int_s \mathbf{B} \cdot d\mathbf{s}$, and that the directions of the path c and the surface s are tied to one another through a right-hand rule. We propose the following prescription, consistent with these conventions:

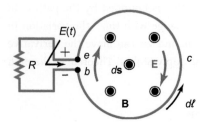

FIGURE 5.3 A single loop of wire in a changing magnetic field.

1. We first choose the direction of $d\mathbf{s}$ for determining the magnetic flux Φ_M. Often the most convenient direction of $d\mathbf{s}$ is suggested by the direction of the magnetic flux density **B** in the loop, as shown in Fig. 5.3, but this is not required.

2. Regardless of how we choose it, the direction of $d\mathbf{s}$ establishes the direction of the contour c using the right-hand rule. Since $d\mathbf{s}$ in Fig. 5.3 points out of the page, $d\boldsymbol{\ell}$ points in the counter-clockwise direction. (See Section D.5 on Stoke's Theorem if you are rusty on this right-hand rule.)

3. We define $\mathcal{E}$ around the loop as the potential of the *end* of the loop (marked with an e in Fig. 5.3) relative to that of the *beginning* (marked with a b).

To further illustrate the signs and field directions, consider a wire bent into a loop, with a time-varying magnetic field passing through the interior of the loop, as shown in Fig. 5.3. Let's be explicit as we identify the fields and their directions. We define the positive direction of the magnetic flux density **B** as shown Fig. 5.3 – that is, coming out of the page. We choose to define our surface element $d\mathbf{s}$ in the same direction. Therefore, the magnetic flux $\Phi_M = \int \mathbf{B} \cdot d\mathbf{s}$, integrated across the surface area bounded by the wire, is positive (when $B > 0$). The contour c that bounds the area s runs along the path of the wire, and is in the counter-clockwise direction, in accordance with the right-hand rule. We have labeled the EMF $\mathcal{E}$ as the potential of the end of the contour (point e) relative to the beginning (point b). Now we can attach some physical significance to the signs of the various quantities $\mathcal{E}$, **E**, etc. Let us consider the case of the increasing magnetic flux density **B**. Since **B** is pointing out of the page and increasing in magnitude, then $-\partial \mathbf{B}/\partial t$ is pointing *into* the page, and Faraday's Law in its differential form, Eq. (5.2), tells us that the induced electric field **E** is curling around in the clockwise direction. (Recall, the curl of a vector **E** points in the direction normal to the plane containing **E** consistent with the following right-hand rule. With our right thumb pointing in the direction of $\boldsymbol{\nabla} \times \mathbf{E}$, the fingers of our right hand curl around in the direction of **E**.) What do these field directions imply about the direction of the current (if any) in the loop, and the sign of induced EMF $\mathcal{E}$? Well, imagine a mobile electron in the wire, which will feel an electric force $q\mathbf{E}$ in the counter-clockwise direction, equivalent to a current in the clockwise direction. If we connect a resistor across output terminals of the wire loop, as shown in the figure, the current through the resistor will flow upward through this resistor. Current always goes into the positive lead of a resistor and comes out of the negative lead, implying that $\mathcal{E}$ as labeled in the figure is negative. Now back to the integral for $\mathcal{E}$; with the path element $d\boldsymbol{\ell}$ pointing in the counter-clockwise direction, and the electric field **E** pointing in the clockwise direction, the EMF $\mathcal{E}$, which is defined as the path integral $\oint \mathbf{E} \cdot d\boldsymbol{\ell}$, is negative, in agreement with our earlier conclusion. Finally, we examine whether we are consistent with Lenz' Law. The induced current that we just examined produces a magnetic flux density into the page in the interior of the loop, opposite the direction of the original **B** field. This opposes the *change* in the magnetic flux, just as Lenz' Law requires. If we were to return to the beginning and follow the same analysis for *decreasing* magnetic flux, we would see that the direction of **E** and the sign of the EMF and the induced current would each be reversed. So with this example, we can understand the **E** induced by the time-varying **B**, and in particular the sign of the induced EMF and current in terms of this **E**.

In the following section we will explore some specific examples of Faraday's Law. These will serve to further illustrate the law, and also help us understand some important applications of this effect.

5.2 Induced EMF through the Faraday Effect

In this section we will explore further the EMF $\mathcal{E}$ induced by a time-varying magnetic flux. We remarked earlier that there are two ways in which the magnetic flux changes: an EMF is observed when the flux density **B** varies, as well as when a circuit moves through a region to which a magnetic field is applied. We will now take a closer look at Eq. (5.1), and write it in an equivalent form that explicitly includes these two contributions,

$$\mathcal{E} = -\int_s \frac{\partial \mathbf{B}}{\partial t} \cdot d\mathbf{s} + \oint_c (\mathbf{u} \times \mathbf{B}) \cdot d\boldsymbol{\ell}. \tag{5.3}$$

The surface integral of $\partial \mathbf{B}/\partial t$ represents the EMF due to time variation in **B**, while the path integral of $\mathbf{u} \times \mathbf{B}$, evaluated around the loop specified by the path c that bounds s, represents the EMF due to motion of the circuit, or the **motional EMF**. In this expression, **u** is the velocity of each segment of the circuit. This relation can be proven very concisely using vector calculus identities, but we choose a more expository, albeit lengthier, style, presented as SideNote 5.1.

SideNote 5.1 Proof of Eq. (5.3)

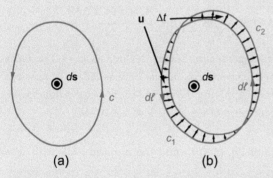

(a)   (b)

FIGURE 5.4 (a) A wire loop for derivation of Eq. (5.3). The EMF induced around this loop is due to a changing magnetic flux density **B**, as well as motion of the loop itself. In (b), the position of the wire loop at time t is shown in blue, and the position of the wire loop at time $t + \Delta t$ in red. Each segment of the loop has displaced by an incremental distance $\mathbf{u}\Delta t$ in this interval.

Consider the wire loop shown in Fig. 5.4(a). In evaluating $d\Phi_M/dt$, the flux Φ_M is evaluated over the surface bounded by this wire loop. We start by returning to the defining limit of an integral, written as

$$-\frac{d}{dt}\int_s \mathbf{B} \cdot d\mathbf{s} = -\frac{d}{dt} \lim_{\Delta s_i \to 0} \sum_i \mathbf{B}_i \cdot \mathbf{a}_{ni}\Delta s_i,$$

where we have divided the entire surface s into small elements of area Δs_i, and written the integral over the surface as the summation over each of these elements. The subscript i serves as an index to label each of the elements. $\mathbf{a}_{ni}$ is the surface normal unit vector for the element i.

We interchange the order of differentiation with respect to t and summation over the elements, and use the product rule for differentiation to write

$$-\frac{d}{dt}\int_s \mathbf{B} \cdot d\mathbf{s} = - \lim_{\Delta s_i \to 0} \left[\sum_i \frac{\partial \mathbf{B}_i}{\partial t} \cdot \mathbf{a}_{ni}\Delta s_i + \sum_i \mathbf{B}_i \cdot \frac{d}{dt}(\mathbf{a}_{ni}\Delta s_i) \right].$$

The first summation, in the limit as $\Delta s_i \to 0$, is the surface integral of $\partial \mathbf{B}/\partial t$, which appears as the first term in the right-hand side of Eq. (5.3). The second summation will take a little more work. A moving wire loop is illustrated in Fig. 5.4(b), in which the solid blue line represents the wire loop at time t, and the solid red line represents the wire loop a short time later, at time $t + \Delta t$. We call these two contours c_1 and c_2, respectively, and the surfaces bounded by these contours s_1 and s_2. Each segment of the wire in c_2 is shifted by a displacement $\mathbf{u}_i\Delta t$ from a segment in c_1, where $\mathbf{u}_i$ is the velocity of that segment of the wire. A displacement of the loop, a rotation of the loop, or a change in the shape of the loop can easily be represented in this manner. The second summation in the equation above can be written as

$$-\lim_{\Delta s_i \to 0} \sum_i \mathbf{B}_i \cdot \frac{d}{dt}\left(\mathbf{a}_{ni}\Delta s_i\right)$$

$$= -\lim_{\Delta s_i \to 0} \sum_i \lim_{\Delta t \to 0} \frac{\mathbf{B}_i \cdot \mathbf{a}_{ni}\Delta s_i(t+\Delta t) - \mathbf{B}_i \cdot \mathbf{a}_{ni}\Delta s_i(t)}{\Delta t}.$$

We have some freedom in how we define the surface s over which the flux Φ_M is evaluated, and we will exploit this freedom at this time. In general, since the magnetic flux density $\mathbf{B}$ must satisfy the condition $\nabla \cdot \mathbf{B} = 0$, the surface flux Φ_M is the same over any surface we choose, as long as the surface is bounded by the contour c. So in evaluating how the flux Φ_M changes in the time interval Δt, we will choose the surface s_2 that consists of two parts: The first part coincides precisely with the surface s_1. To this, we add the incremental surface shown in Fig. 5.4(b) as the "rim" between c_1 and c_2. Therefore, the change in the flux due to the motion of the circuit in the interval Δt is simply the surface flux through this rim region. We divide the surface of the rim into a large number of small quadrilaterals, as shown. Two sides of each quadrilateral coincide with segments of c_1 and c_2, of length $\Delta \ell_j$, while the other two sides are formed by $\mathbf{u}_j \Delta t$. Here the index j labels each of the quadrilaterals within the rim. (We changed the index labeling the element to j here, since this surface differs from s_1 or s_2.) The surface element is compactly described in magnitude and direction by $\Delta \mathbf{s}_j = (\mathbf{u}_j \Delta t) \times \Delta \boldsymbol{\ell}_j$, where $\Delta \boldsymbol{\ell}_j$ is the vector incremental length lying along the contour c. The change in the surface flux is therefore

$$-\lim_{\Delta s_j \to 0} \sum_j \mathbf{B}_j \cdot \frac{d}{dt}\left(\mathbf{a}_{nj}\Delta s_j\right) = -\lim_{\Delta \ell_j \to 0} \sum_j \lim_{\Delta t \to 0} \frac{\mathbf{B}_j \cdot (\mathbf{u}_j \Delta t \times \Delta \boldsymbol{\ell}_j)}{\Delta t}.$$

The factors Δt in the numerator and denominator cancel, the limit as $\Delta t \to 0$ becomes irrelevant, and the summation over segments $\Delta \ell_j$, in the limit as $\Delta \ell_j \to 0$ becomes the closed path integral of $\mathbf{B} \cdot (\mathbf{u} \times d\boldsymbol{\ell})$. Applying the permutation vector product rule, Eq. (A.10), equates this to the second term of Eq. (5.3), $+\oint (\mathbf{u} \times \mathbf{B}) \cdot d\boldsymbol{\ell}$.

We will now illustrate the use of Eq. (5.3) to find the induced EMF with several examples.

5.2.1 Coupled Loops

We start by considering the pair of windings shown in Fig. 5.5(a). The current $i_1(t)$ through the winding on the left, with N_1 turns, produces a flux density which, in the interior region, points to the right. These flux lines circle around on the outside of the

FIGURE 5.5 (a) A side view and (b) an end-on view of adjacent windings wound over a high-permeability rod.

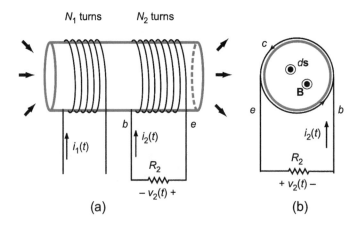

loop and close on themselves. Similarly, the current $i_2(t)$ through the winding on the right, with N_2 turns, produces a magnetic flux density that points to the right. These flux lines also circle around on the outside, and close on themselves. Since these two coils are adjacent to one another, the field lines produced by one current loop penetrates through the surface of the other. Thus we say that the two windings are magnetically coupled, and we can examine the effect of the current in one loop on the other.

Example 5.1 Faraday's Law, Coupled Windings

Consider the pair of wire windings on the highly permeable bar shown in Fig. 5.5. Determine the current $i_1(t)$ that generates a constant current $i_2(t)$, of value I_2. Let the initial value of $i_1(t)$ be $i_1(0) = 0$.

Solution:

We start by making a few observations regarding the magnetic field lines generated by the currents $i_1(t)$ and $i_2(t)$ through the loops. Calculation of the magnitude and pattern of $\mathbf{B}(\mathbf{r})$ in the bar or in the surrounding air is not a simple task, and we will not attempt to solve for this. We do expect, as we remarked previously, that the field lines in the center of the coils are pointed to the right (for $i_1(t) > 0$ and $i_2(t) > 0$), exit the rod on the right, circle around outside the rod, and re-enter the rod on the left side, closing on themselves. The few bold arrows at either end of the rod in Fig. 5.5 indicate the expected field direction at those locations. Because of the high permeability of the rod, the field lines are guided by the rod, and we can assert that the magnetic flux $\Phi_M = \int_s \mathbf{B} \cdot d\mathbf{s}$ through a single turn of coil 1 is approximately the same as Φ_M through a single turn of coil 2. (The figure shows a large open area in loop 2 between the windings and the resistor R_2, and a similar open area in loop 1 with whatever source drives this loop. We neglect any additional magnetic flux through these surfaces. Even though this flux will be small when N_1, N_2, and the permeability of the rod are large, it may have been a better technique to twist the wires leading to and from the coils about one another to further minimize this flux!)

The EMF induced across coil 1, by Faraday's Law (Eq. (5.1)), is

$$v_1(t) = -N_1 \frac{d\Phi_M(t)}{dt},$$

while that across coil 2 is

$$v_2(t) = -N_2 \frac{d\Phi_M(t)}{dt}.$$

Note the factor N_1 or N_2 in these equations, since the voltage across each coil is enhanced by the number of turns. The magnetic flux Φ_M includes contributions from the current $i_1(t)$ as well as the current $i_2(t)$, and since the flux density $\mathbf{B}(\mathbf{r})$ is proportional to the current that generates it (as long as the rod material is linear), each of these components of the flux Φ_M is linear in the currents $i_1(t)$ and $i_2(t)$, as well. This allows us to write

$$\Phi_M(t) = \frac{L_{12}}{N_2} i_1(t) + \frac{L_{22}}{N_2} i_2(t), \tag{5.4}$$

where we have introduced the **self inductance** L_{22} of coil 2, and the **mutual inductance** L_{12} between the coils. These inductances give us quantitative measures of the magnetic coupling of the loops. For simplicity of notation, upon application of d.c. currents I_1 and I_2 through coils 1 and 2, respectively, producing the flux Φ_M, the inductances L_{22} and L_{12} are defined as

$$L_{22} = \left. \frac{N_2 \Phi_M}{I_2} \right|_{I_1=0}$$

and

$$L_{12} = \left.\frac{N_2 \Phi_M}{I_1}\right|_{I_2=0}.$$

We'll spend more time discussing inductances later, but with this brief introduction we can proceed with the present example.

We are told that $i_2(t) = I_2$ is a constant, which implies that the voltage $v_2(t)$ is also a constant, of magnitude $V_2 = I_2 R_2$, since a resistor of resistance R_2 is connected across the ends of this coil. Then, since

$$\frac{d\Phi_M(t)}{dt} = -\frac{v_2(t)}{N_2} = -\frac{V_2}{N_2},$$

and the magnetic flux Φ_M must vary linearly in time t,

$$\Phi_M(t) = \frac{-V_2(t-t_0)}{N_2},$$

where t_0 is a constant of integration. Substituting this solution of $\Phi_M(t)$, together with $i_2(t) = I_2$ and $i_1(0) = 0$, into Eq. (5.4), we can solve to find

$$i_1(t) = -\frac{V_2 t}{L_{12}},$$

and $t_0 = L_{22}/R_2$.

This solution is complete, but let's pause to examine the sign of the EMF V_2 and its significance before we move on. When the current $i_1(t)$ is negative, it produces a flux density **B** pointing to the left in Fig. 5.5(a). (We have redrawn the permeable rod and the coils in Fig. 5.5(b), showing the end view from the right side. In this view, when the current $i_1(t)$ is negative, it produces **B** pointing *into* the page.) Since the magnitude of $i_1(t)$ is increasing, Lenz' Law tells us that the EMF V_2 induced across the leads of coil 2 should oppose this increase, which indicates that the component of **B** produced by I_2 points to the right in Fig. 5.5(a) (or *out* of the page in Fig. 5.5(b)). We can apply the right-hand rule again to determine that I_2 and V_2 must be positive. Thus our result $i_1(t) = -V_2 t/L_{12}$ is perfectly consistent with Lenz' Law.

5.2.2 Transformer

One common electrical element whose operation is based upon Faraday's Law is the transformer, as illustrated in Fig. 5.6. A transformer consists of two sets of current windings about a common permeable core. (You might notice that the structure described in Example 5.1 might satisfy this limited description as well. The typical transformer differs, however, in that the core material is in the shape of a ring.) A common use of the transformer is in the generation of a sinusoidally varying voltage $v_2(t)$ whose amplitude differs from that of the input $v_1(t)$. We have already discussed each of the physical properties upon which the transformer is based. First, the current $i_1(t)$ into the input side (called the "primary" winding) produces a magnetic field **H** within the permeable core. These field lines are guided by the permeable material, circling around the permeable ring and closing on themselves. To the extent that the permeability of the core is very large, the "leakage" of the field lines to the region outside the core is low, and we can approximate that the magnetic flux $\Phi_M = \int \mathbf{B} \cdot d\mathbf{s}$ through any cross-section of the core is the same anywhere around the core. In particular, the flux Φ_M penetrating the

FIGURE 5.6 (a) A transformer, consisting of two windings on a permeable core. A cross-sectional view from the top is shown in (b).

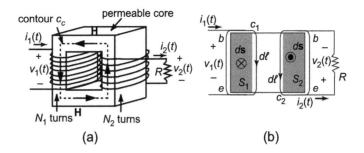

(a)   (b)

surface of the primary winding and the flux Φ_M penetrating the surface of the secondary winding (the output) are the same. A time-varying current $i_1(t)$ produces a time-varying flux $\Phi_M(t)$, and by Faraday's Law a time-varying EMF at the secondary. In Example 5.2 we will apply Faraday's Law to this device to determine the voltage $v_2(t)$ across the secondary.

Example 5.2 Transformer

The winding on the left side of the transformer shown in Fig. 5.6 consists of N_1 turns, while the winding on the right has N_2 turns. Apply Faraday's Law to the primary and secondary windings to derive a quantitative relation between the amplitudes V_1 and V_2 of the sinusoidally varying voltages $v_1(t)$ and $v_2(t)$.

Solution:

We start with application of Faraday's Law at the primary and secondary. Here we find

$$v_1(t) = +N_1 \frac{d\Phi_M}{dt} \tag{5.5}$$

and

$$v_2(t) = -N_2 \frac{d\Phi_M}{dt}, \tag{5.6}$$

where $\Phi_M(t)$ is the magnetic flux in the primary and secondary windings of the transformer. (We are assuming that the flux $\Phi_M(t)$ is the same throughout the core, as is valid when the transformer core guides the field lines without loss.)

Note the sign of these two equations, which we justify using the convention discussed in Section 5.1. Let us start with the sign in Eq. (5.5). Since we have already established that for positive $i_1(t)$ the magnetic field **H** is in the counter-clockwise direction around the permeable core, as shown in Fig. 5.6(a), it is convenient to choose the direction of $d\mathbf{s}$ to be in this same direction; that is, downward on the left side of the transformer (and upward on the right side, as we will require later when we consider the secondary winding). The EMF is defined as the potential of the end of the winding (marked with an e in the figure) relative to the potential at the beginning (marked with a b). (Recall that the direction of surface element $d\mathbf{s}$ and the path element $d\ell$ are always related by the right-hand rule. Since $d\mathbf{s}$ is pointing downward for the primary winding, $d\ell$ for the path c_1 wraps around the magnetic core in the clockwise direction, when viewed from above as in Fig. 5.6(b).) Therefore we have labeled the top lead of the primary as the beginning (marked "b"), and the bottom lead as the end (marked "e"). We see that the voltage $v_1(t)$ is defined opposite to the EMF (the $+$ and $-$ labels on $v_1(t)$ are reversed relative to the signs for $\mathcal{E}$), so $\mathcal{E} = -v_1(t)$. Thus the $+$ sign in Eq. (5.5) is appropriate. Is this consistent with Lenz' Law? An increasing magnetic flux Φ_M induces a positive voltage $v_1(t)$, which, if these leads were connected to an external resistor, would force the current $i_1(t)$ to be negative. A negative $i_1(t)$ in the primary produces a magnetic field contribution that is pointing upward in this section of the core, which does indeed oppose the change in Φ_M, as required by Lenz' Law.

Now let's apply the same ideas to the secondary to verify the sign of $v_2(t)$ in Eq. (5.6). With the surface element $d\mathbf{s}$ pointing upward in this region, the contour c_2 is counter-clockwise when viewed from above, so the EMF – that is, the potential at the end of the winding (marked e) relative to the beginning (marked b) – is consistent with the sign already labeled as $v_2(t)$. Therefore, the $-$ sign is appropriate in Eq. (5.6). Is this consistent with Lenz' Law? Indeed it is, as we see by considering the case of an increasing magnetic flux Φ_M. This produces a negative value for $v_2(t)$, which drives a negative current $i_2(t)$. A negative current $i_2(t)$ gives a magnetic field contribution in the downward direction, which opposes the change in magnetic flux Φ, consistent with Lenz' Law.

Now that we have the proper signs in Eqs. (5.5) and (5.6), let us combine them to determine the voltage amplitude ratio

$$\frac{V_2}{V_1} = -\frac{N_2}{N_1}.$$

(This expression for a transformer's voltage ratio often appears without the minus sign, but this is solely due to the direction of the windings. If we are only concerned about the ratio of amplitudes of these voltages, then, of course, we can omit the minus sign.)

So we see that the voltage ratio is equal to N_2/N_1, which is called the **turns ratio** of the transformer. This result is conditional on the magnetic flux being the same in the primary and secondary windings of the transformer. Let's next apply Ampère's Law around the path c_c shown as the dashed line in Fig. 5.6, in the counter-clockwise direction in order to determine an approximate relation between the currents $i_1(t)$ and $i_2(t)$ and the magnetic flux:

$$\oint_{c_c} \mathbf{H} \cdot d\boldsymbol{\ell} = I_{enc} = N_1 i_1(t) + N_2 i_2(t),$$

where the current enclosed by the path c_c includes both the primary and secondary currents. Note the sign of each of these terms. As we discussed earlier, we have labeled $\mathbf{H}$ in the figure in the counter-clockwise direction around the transformer core, in the same direction as $d\boldsymbol{\ell}$, so $\oint_{c_c} \mathbf{H} \cdot d\boldsymbol{\ell}$ is a positive quantity if H is positive. The surface element $d\mathbf{s}$ of the surface bounded by the contour c_c is pointing out from the page, as given by the right-hand rule. Finally, observing the direction of the primary and secondary windings, both of these currents are coming out of the page as they penetrate through the surface bounded by c_c, in the same direction as $d\mathbf{s}$, so the currents $N_1 i_1(t)$ and $N_2 i_2(t)$ appearing in the equation above as positive terms is correct.

We can derive an approximate value of the magnetic flux $\Phi_M = \int_s \mathbf{B} \cdot d\mathbf{s}$ by assuming $\mathbf{B}$ is uniform across the cross-section of the core, so $\Phi_M \simeq BS$, where S is the cross-sectional area of the core. For B and H small enough that the core material behaves linearly, then $B = \mu H$, and if the cross-sectional area of the core S is uniform over the entire path c_c, B and H will be uniform as well. This lets us approximate $\oint_{c_c} \mathbf{H} \cdot d\boldsymbol{\ell} \simeq H\ell$, where ℓ is the path length of one complete cycle around the contour c_c. Writing H in terms of Φ_M and S, we have

$$N_1 i_1(t) + N_2 i_2(t) \simeq H\ell \simeq \frac{B}{\mu}\ell \simeq \frac{\Phi_M(t)\ell}{\mu S}. \tag{5.7}$$

As we can see, the term on the right-hand side shows that the magnetic flux $\Phi_M(t)$ is multiplied by the factor $\ell/(\mu S)$. In an **ideal transformer**, the permeability is very large and $\ell/(\mu S) \to 0$, letting us approximate $N_1 i_1(t) + N_2 i_2(t) = 0$, or

$$\frac{I_2}{I_1} = -\frac{N_1}{N_2},$$

where I_1 and I_2 are the amplitudes of the sinusoidal currents $i_1(t)$ and $i_2(t)$. So we see that, as μ gets very large, the current ratio is the inverse of the voltage ratio. If the voltage amplitude is stepped up, then the current amplitude is stepped down. Thinking about conservation of energy and the power consumed or generated by an electrical device as being the product of the current through the element and the voltage across this element, this inverse relation between the voltage ratio and current ratio should seem reasonable. Let's take this one step further and ask, for a resistor of resistance R_2 connected at the secondary, what is the effective input resistance of the transformer? For this, of course, the effective input resistance is simply $R_{in} = V_1/I_1$, and Ohm's Law already tells us that $R_2 = V_2/I_2$, so

$$\frac{R_{in}}{R_2} = \frac{V_1/I_1}{V_2/I_2} = \frac{I_2/I_1}{V_2/V_1} = \frac{-N_1/N_2}{-N_2/N_1} = \left(\frac{N_1}{N_2}\right)^2.$$

Thus we see that the input resistance of the loaded transformer is the load resistance R_2 divided by the square of the turns ratio. For a reactive load, this relation can be generalized to

$$\frac{Z_{in}}{Z_2} = \left(\frac{N_1}{N_2}\right)^2.$$

We close our discussion of transformers by examining a method of characterizing how close the behavior of a real transformer approaches that of an ideal transformer. We wrote above that the magnetic flux of the transformer is related to the currents $i_1(t)$ and $i_2(t)$ through Eq. (5.7):

$$\frac{\Phi_M \ell}{\mu S} = N_1 i_1(t) + N_2 i_2(t).$$

Let's multiply each term in this expression by $N_1 \mu S/\ell$, and then take the derivative with respect to time of each term. We get

$$N_1 \frac{d\Phi_M(t)}{dt} = \left[N_1^2 \frac{di_1(t)}{dt} + N_1 N_2 \frac{di_2(t)}{dt}\right] \frac{\mu S}{\ell}.$$

By Faraday's Law, this must be the EMF across the primary winding, or $v_1(t)$. One term on the right represents the EMF proportional to $di_1(t)/dt$; the other is the EMF proportional to $di_2(t)/dt$. We can therefore identify the self-inductance of the primary L_{11} as

$$L_{11} = N_1^2 \frac{\mu S}{\ell}$$

and the mutual inductance of the secondary with the primary L_{21} as

$$L_{21} = N_1 N_2 \frac{\mu S}{\ell}.$$

Similarly, if we multiply Eq. (5.7) by $N_2 \mu S/\ell$ and take the derivative of each term with respect to time, we get

$$N_2 \frac{d\Phi_M(t)}{dt} = \left[N_1 N_2 \frac{di_1(t)}{dt} + N_2^2 \frac{di_2(t)}{dt} \right] \frac{\mu S}{\ell},$$

which is $-v_2(t)$. (Remember, the signs of $v_1(t)$ and $v_2(t)$ are consistent with how we labeled them in Fig. 5.6, as we discussed earlier.) We identify from this equation the self-inductance of coil 2 and the mutual inductance between coils 1 and 2 as

$$L_{22} = N_2^2 \frac{\mu S}{\ell}$$

and

$$L_{21} = N_1 N_2 \frac{\mu S}{\ell}.$$

Note that the mutual inductances L_{12} and L_{21} are the same. We also see from these relations $L_{12} = \sqrt{L_{11} L_{22}}$, a relation that holds true as long as the magnetic flux Φ_M is uniform around the core, and the field doesn't leak to the outside regions. If the flux is diminished around the core, the flux lines generated by one coil don't fully penetrate through the surface of the other, and the mutual inductance L_{12} is decreased. The ratio $L_{12}/\sqrt{L_{11} L_{22}}$ then gives a measure of this effect, and is known as the **coupling coefficient**

$$k = \frac{L_{12}}{\sqrt{L_{11} L_{22}}}.$$

For an ideal transformer, the coupling coefficient k is 1; for less-than-ideal behavior, the value of k is less than 1. In addition to flux leakage in transformers, other effects that can lead to non-ideal behavior include: resistance in the primary or secondary windings; hysteresis in the magnetic properties of the core material, such that the dependence of B on H is nonlinear (i.e. $B = \mu H$ is no longer valid); and the presence of eddy currents induced in the core material. The latter can produce additional magnetic fields that diminish the fields within the core, and can be mitigated by using laminated cores and core materials with low conductivity and high permeability.

In TechNotes 5.1–5.4 several common applications of the Faraday effect resulting from a time-varying magnetic flux through the surface of a fixed wire loop are described. The first are wireless charging stations and pads, coming in recent years with portable electronic devices and electric vehicles, for example.

TechNote 5.1 Wireless Charging Stations and Pads

Wireless charging stations or charging pads provide a means of recharging batteries in various portable devices, such as toothbrushes, cell phones, or electric vehicles. Wireless transfer of power is not a new concept, as Nikola Tesla demonstrated wireless resonant coupling for remote charging in the late 1800s. It is only recently, however, that broader application of his ideas has taken hold.

In concept, wireless charging systems are similar to common traditional transformers, in that a time-varying current in a primary coil of wire generates a magnetic field, whose flux lines pass through a secondary coil, inducing an EMF across the secondary. This secondary EMF is then used to recharge the battery in the portable device. Wireless chargers differ from traditional transformers, however, in that the primary and secondary in a traditional transformer are wound around a common permeable core, often in the form of a loop. For wireless charging systems, the primary is in the charging station, and the secondary is in the portable device. The portable device must be brought close to the charging station, such that the magnetic flux lines generated by the primary pass through the secondary coil. This is known as near-field coupling between the two coils. The portable device must have a non-conducting case, such as plastic or glass, so that the magnetic flux lines are not interrupted. Coupling between the coils can be enhanced by adding capacitors to the secondary loop, forming a resonant circuit whose frequency matches that of the charging station current. Resonant coupling has been shown to increase the distance over which charging can be effective. Inductive loops for electric vehicle charging, for instance, can be as large as 25 cm in diameter, and charge effectively at similar distances with efficiencies greater than 90%. Recent reports of wireless charging of the batteries in heart pacemakers offer a possible alternative to physically replacing the battery in these devices, a procedure that requires surgery.

Metal detectors for remote detection of conductors is the next application described.

TechNote 5.2 Metal Detectors

Metal detectors use magnetic induction to find conductors buried in the ground or in a person's body. They contain two multi-turn wire loops, a transmitter, and a receiver. The transmitter is driven by an oscillating or pulsed source, which produces a time-varying magnetic field. When a conductor is within the range of this field, the time-varying magnetic field produces eddy currents in the conductor. These eddy currents, in turn, produce a time-varying magnetic field, which penetrates the area of the receiver loop of the metal detector. The EMF induced across the leads of this coil through the Faraday effect indicates the presence of a conductor.

Metal detectors are commonly used in airports, courthouses, hospitals, and prisons in walk-through scanners, by hobbyists (treasure hunters), and by law enforcement and the military as mine and weapons detectors. There are three common types of metal detectors. The most common variety is the very low frequency (VLF) detector, typically operating in the 6–20 kHz range. Pulse induction (PI) detectors, which use higher-frequency pulses, can find objects buried more deeply in the ground, but aren't as good at discriminating different types of objects. Full band spectrum (FBS) detectors use multiple frequencies. The depth range over which metal detectors can find metal objects depends on multiple factors, such as the size, conductivity, and orientation of the object, the composition of the soil or body near the object, and the type of detector, but the typical range is 20–50 cm.

Gustave Trouvé, a Parisian inventor, developed a hand-held metal detector in 1874. A few years later, in 1881, Alexander Graham Bell attempted to locate an assassin's bullet in President James A. Garfield's chest. In retrospect, it is thought that the detector functioned properly, although Bell did not succeed in locating the bullet due to the metal coil spring bed that President Garfield was lying on. (Spring beds were not common at that time, and Bell was not informed of the metal springs in Garfield's bed.) President Garfield later died from infections caused by his doctors.

Autotransformers are a special class of transformer in which a single coil serves as both the primary and secondary. These are described in the TechNote 5.3.

TechNote 5.3 Variable Autotransformer

When most of us think about a transformer, the first image that comes to mind is often a pair of wire loops, the primary and secondary, wrapped around a single magnetic core. A common, useful transformer that consists of just a single coil is the variable autotransformer, commonly called a variac. (Variac is a genericized trademark from the product of a specific manufacturer.) A variable autotransformer, as its name implies, generates a variable-amplitude voltage at the output. The primary inputs are connected to two taps of the coil, which need not be at the ends of the coil. We'll let N_1 represent the number of turns on the coil between these two taps. On the output side, one of the connections to the coil is fixed, while the other is a moveable brush, which can make contact with the coil at an adjustable location. In Fig. 5.7, the brush is the roller at the upper right side of the autotransformer. We'll let N_2 represent the number of turns on the coil between the fixed tap and the brush of the secondary. Obviously, since the autotransformer consists of only a single coil, the magnetic flux passing through any single loop of the primary or secondary is the same, and the ratio of the secondary (output) voltage to the primary (input) voltage is $V_2/V_1 = N_2/N_1$. If the primary taps are not at the ends of the coil, then it is possible for N_2 to be greater than N_1, and therefore for V_2 to be greater than V_1. Note that since a single coil serves as primary and secondary in the autotransformer, the primary and secondary cannot be electrically isolated from each other, which is a disadvantage of the autotransformer in some applications.

FIGURE 5.7 An autotransformer. The brush in the rotating assembly makes contact with one winding of the coil.

The induction coil played a key role in early investigations of high-voltage effects in the mid- to late 1800s. Before the advent of the electronic ignition systems in automobiles, induction coils, commonly simply called coils, were universally used to create the high-voltage pulse that caused spark plugs to spark. The induction coil is described in TechNote 5.4.

TechNote 5.4 Ruhmkorff Induction Coil

The induction coil was first constructed in 1836 by Nicholas Callan (1799–1864), a priest and professor at St. Patrick's College in Maynooth, County Kildare, Ireland. Heinrich Daniel Ruhmkorff (1803–1877), a German-born instrument maker in Paris, made improvements to and commercialized the original induction coil of Callan, and his name is commonly associated with these high-voltage generators.

In the transformers that we've already examined, we applied a sinusoidally varying current to the input, and generated a sinusoidally varying voltage at the secondary. The operation of this induction coil differs from that of these models, in that the current through the primary (inner) coil is d.c. The operation of the induction coil is as follows. A d.c. primary current generates a magnetic field $\mathbf{B}$ in the high-permeability core, as shown in Fig. 5.8. The number of windings N_2 of the secondary (outer) coil is much larger than the number of windings N_1 of the primary. When the switch (E) opens, the primary current is suddenly interrupted, the magnetic field switches off suddenly, and a very large pulsed EMF is induced across the secondary. Induction coils were used in the late nineteenth century to study electric discharges in gases, and were widely used in automobile engines to generate the high voltage across spark plugs before the advent of automotive electronic ignition. (In the automotive field, the induction coil is commonly simply called a coil, the contacts that open to turn off the primary current are called the points, and the capacitor to avoid sparking and pitting of the points is call a condenser.) Ignition coils are often categorized according to the distance between electrodes over which they can generate a spark. These distances range from a few centimeters for small induction coils to more than a meter for a large coil. Given the dielectric strength of air ~3 kV/mm (depends on pressure, humidity, etc.), a discharge over a meter distance requires 3 MV potential!

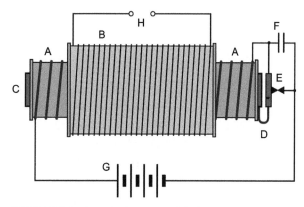

FIGURE 5.8 A schematic diagram of a Ruhmkorff induction coil. The primary coil (A) and secondary coil (B) are wound around a central iron core (C). When the primary is energized by the battery (G), the current in the primary sets up a magnetic field in the core. This field attracts the spring-mounted iron pad (D), breaking the exciter contacts (E), and turning off the current in the primary. Once the primary current is off, the spring restores the iron pad to its initial position, closing the contacts of the exciter, and restoring the primary current. This cycle repeats, turning the primary current on and off at a rate of 20–40 times per second. The capacitor helps reduce sparking between the exciter contacts. The number of turns in the secondary is much greater than the number of turns in the primary, resulting in an extremely large voltage spike build up across the secondary each time the primary turns off.

5.2.3 Generator

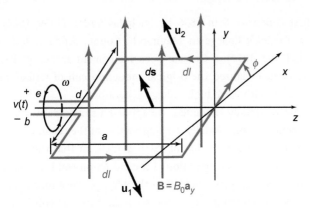

FIGURE 5.9 An alternator, consisting of a rotating winding in a uniform stationary magnetic field.

We next consider the operation of an electrical **generator** or alternator. A generator converts mechanical energy, such as the rotation of a turbine caused by passing steam, or the vanes of a windmill set in motion by the wind, to electrical energy. In this device, an EMF is generated by virtue of the motion of a loop of wire, or more accurately in most cases, a number of loops, each consisting of many turns of wire, through a region of fixed magnetic flux density. We consider the simplified geometry shown in Fig. 5.9. This figure is in many aspects similar to a simple electric motor, which we will discuss in Section 6.3. Each consists of a loop (or multiple loops) of wire rotating in a magnetic field. In a motor, however, the current through the loops produces a torque on the loop, causing it to rotate. A motor, therefore, converts electrical energy to mechanical energy, just the reverse of the operation of a generator.

In Fig. 5.9, a rectangular wire loop rotating about the z-axis in the positive ϕ direction is shown, cutting through lines of a constant uniform magnetic flux density $\mathbf{B} = B_0\mathbf{a}_y$. As we will show through the following example, Eq. (5.3) can be applied to determine the EMF $v(t)$ across the leads of the generator windings.

Example 5.3 Generator

The rectangular wire loop of the generator shown in Fig. 5.9 rotates at an angular frequency ω. Apply Eq. (5.3) to determine the EMF $v(t)$ produced across the windings.

Solution:

Since $\mathbf{B}$ is constant in time, Eq. (5.3) reduces to

$$v(t) = \oint_c (\mathbf{u} \times \mathbf{B}) \cdot d\boldsymbol{\ell}.$$

We start by determining the velocity $\mathbf{u}$ of each segment of the rectangular wire loop. Actually, we must only deal with the velocity of the two sides that are parallel to the z-axis. The two end segments, one at the plane $z = 0$ and the other at the plane $z = -a$ (note that we have placed the origin at the right side of the loop) do not contribute to the EMF, since for these segments $\mathbf{u} \times \mathbf{B}$ is perpendicular to $d\boldsymbol{\ell}$. Let us start with the side closest to us, which we call side 1. The position vector pointing from the z-axis to this side is

$$\mathbf{r}_1 = -\left(\frac{d}{2}\right)\left[\mathbf{a}_x \cos(\phi) + \mathbf{a}_y \sin(\phi)\right],$$

where d is the dimension of the loop from side to side, $\phi = \omega t$ is the angle the loop makes with the x-axis, and ω is the angular frequency of the rotation of the loop. The velocity of this side is found by taking the time derivative of $\mathbf{r}_1$, which gives us

$$\mathbf{u}_1 = \left(\frac{d\omega}{2}\right)\left[\mathbf{a}_x \sin(\omega t) - \mathbf{a}_y \cos(\omega t)\right].$$

Similarly, the position vector pointing to the opposite side is $\mathbf{r}_2 = (d/2)\left[\mathbf{a}_x \cos(\omega t) + \mathbf{a}_y \sin(\omega t)\right]$, which is precisely $-\mathbf{r}_1$, and taking the time derivative, $\mathbf{u}_2 = -\mathbf{u}_1$ as well.

With this, we are ready to find the motional electric field $\mathbf{u} \times \mathbf{B}$ for the two sides. For side 1, this is

$$\mathbf{u}_1 \times \mathbf{B} = \begin{vmatrix} \mathbf{a}_x & \mathbf{a}_y & \mathbf{a}_z \\ \dfrac{\omega d}{2}\sin(\omega t) & -\dfrac{\omega d}{2}\cos(\omega t) & 0 \\ 0 & B_0 & 0 \end{vmatrix} = \mathbf{a}_z\,\dfrac{\omega d B_0}{2}\sin(\omega t).$$

Similarly, evaluation of $\mathbf{u}_2 \times \mathbf{B}$ yields $-\mathbf{u}_1 \times \mathbf{B} = -\mathbf{a}_z(\omega d B_0/2)\sin(\omega t)$.

Now we are ready to determine $v(t)$ by integrating $\mathbf{u} \times \mathbf{B}$ around the loop. The EMF is

$$v(t) = \int_{-a}^{0} (\mathbf{u}_1 \times \mathbf{B})\cdot d\boldsymbol{\ell} + \int_{0}^{-a} (\mathbf{u}_2 \times \mathbf{B})\cdot d\boldsymbol{\ell} = \omega a d B_0 \sin(\omega t). \tag{5.8}$$

It can be instructive to picture the force felt by individual free charges in the wire loop. The Lorentz (magnetic) force $\mathbf{F}_m = q\mathbf{u} \times \mathbf{B}$ felt by an electron as the wire moves through the magnetic field tends to move the charges around the loop in the clockwise direction as we view Fig. 5.9, consistent with a current in the counter-clockwise direction and in agreement with the polarity of the EMF we just derived. We should also ask if this current is in accordance with Lenz' Law. Yes, it is, as we can see by examining the loop as the angle ϕ passes through zero. At $\phi = 0$, the magnetic flux through the loop is maximized, and as ϕ increases past this position, the magnetic flux starts to decrease. The current induced in the loop in the counter-clockwise direction produces a magnetic flux density contribution in the upward direction, which *opposes* this decrease in the magnetic flux.

Before we leave this example, let us look at an alternative means of calculating the EMF; that is, we can return to Eq. (5.1) and use this form directly. The magnetic flux is $\Phi_M = \int_s \mathbf{B}\cdot d\mathbf{s}$, integrated across the surface outlined by the wire loop. The dot product $\mathbf{B}\cdot d\mathbf{s}$ is simply $B_0\cos\phi\,ds$, and the magnetic flux density B_0 and $\cos\phi$ can be pulled outside the integral since they don't depend upon the surface variables. Thus, Φ_M is simply $B_0 ad\cos\phi$, and the EMF is

$$v(t) = -\frac{d\Phi_M}{dt} = \omega a d B_0 \sin(\omega t),$$

in agreement with Eq. (5.8).

There are many examples of generators that help fufill the need for electrical power. One example that has seen rapid growth of late are wind turbines, as described in TechNote 5.5.

TechNote 5.5 Wind Turbines

Using the wind to do work, such as pumping water or grinding grains, has been a practice for centuries, but recently large wind farms that convert wind power to electrical power have been developed as a form of renewable energy. Wind turbines (Fig. 5.10) typically have a capacity to generate 2–3 MW of power. The actual output, of course, depends on many factors such as the wind speed and the pitch of the blades. Most turbines operate in wind speeds ranging from ~8 mph to 55 mph. At higher speeds the turbines can be damaged, and so automatically turn off. Wind turbines are able to operate ~90% of the time. Since the frequency of the EMF produced by the wind turbine varies with wind velocity, this power cannot be added directly to the power grid. Instead, the wind turbine output is first rectified and then converted to 60 Hz (or 50 Hz in much of the world outside the United States), synchronized with the grid voltage. In 2019, approximately 7.3% of US electrical power was generated by wind turbines.

FIGURE 5.10 A few wind turbines, part of a vast array, in north central Indiana.

Another example of an EMF generated by a moving circuit is the dynamic microphone, as described in TechNote 5.6.

TechNote 5.6 Dynamic Microphone

Microphones convert acoustic signals into electrical signals for amplification, transmission to distant locations, or recording. As we discussed in Chapter 1, the two primary types of microphones are the dynamic and the condenser microphone. In this note, we discuss the dynamic microphone.

A dynamic microphone, illustrated in Fig. 5.11, uses the Faraday effect to convert sound waves to an electrical signal. A lightweight, multi-turn wire coil, whose cross-section is penetrated by the magnetic flux lines from an adjacent permanent magnet, is attached to the diaphragm. As the coil moves toward or away from the magnet, the total magnetic flux passing through the coil varies, generating an EMF across the coil. This EMF is amplified to produce the output signal. Dynamic microphones are capable of capturing loud sounds and are relatively inexpensive. Their response to high-frequency signals, however, is generally inferior to that of a condenser microphone.

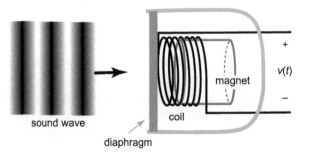

FIGURE 5.11 Diagram of a dynamic microphone.

5.2.4 Time-Varying $\mathbb{B}$ and Moving Circuit

In the examples that we have worked so far, the EMF is induced by a time-varying flux density **B** but stationary circuit, or a moving circuit in a non-varying **B**. It is possible that **B** varies in time *and* the circuit moves, and we now consider this case. In short, we still apply Eq. (5.3), but now both integrals in this expression contribute to the EMF. In certain cases we can also apply Eq. (5.1) to these solutions as well.

Example 5.4 Faraday's Law, Moving Bar

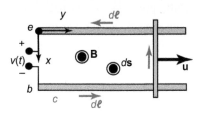

FIGURE 5.12 A pair of horizontal conducting rails joined by a moving bar. An EMF is generated around the loop by the changing magnetic flux through the circuit.

The magnetic flux density in the region between the horizontal rails in Fig. 5.12 is $\mathbf{B} = B_0 \cos(\omega t - k_y y)\, \mathbf{a}_z$, where $B_0 = 0.4$ Wb/m^2, $\omega = 10$ rad/s, and $k_y = 1$ m^{-1}. The bar on the right rests across the rails, and moves in the $\mathbf{a}_y$ direction with a velocity $u = 20$ m/s. The distance between the rails is $x_0 = 60$ cm. Determine the EMF $v(t)$ measured at the left end of the rails.

Solution:

We start by discussing orientations and signs. The voltage $v(t)$ is labeled on the diagram, with the polarity as marked. We have chosen the direction of the surface element $d\mathbf{s}$ to point out of the page, parallel to **B** in the $+z$ direction. The surface element is $d\mathbf{s} = \mathbf{a}_z\, dx\, dy$. By the right-hand rule, the path c is counter-clockwise around the circuit. The points b and e at the beginning and end of the loop are marked in the figure, and the signs of the EMF $v(t)$ are consistent with this notation.

We next evaluate $-\int_s \partial \mathbf{B}/\partial t \cdot d\mathbf{s}$, starting with the time-derivative of **B**,

$$\frac{\partial \mathbf{B}}{\partial t} = -B_0 \omega \sin(\omega t - k_y y)\, \mathbf{a}_z.$$

When integrating over the surface bounded by the rails and bar, the limits of integration for dx are 0 to x_0, and for dy are 0 to y', where y' represents the location of the sliding bar at time t. Carrying out the integration, we find

$$-\int_s \frac{\partial \mathbf{B}}{\partial t} \cdot d\mathbf{s} = \frac{B_0 \omega x_0}{k_y} \left[\cos(\omega t - k_y y') - \cos(\omega t)\right]. \tag{5.9}$$

Next we consider the path integral of $\mathbf{u} \times \mathbf{B}$. We carry out this integral around the entire loop defined by the conductors, but of course only the sliding bar is in motion, and so only the right side contributes. For $d\boldsymbol{\ell} = \mathbf{a}_x dx$, where the bottom rail is at $x = x_0$ and the top rail is at $x = 0$, the motional EMF term is

$$\oint_c \mathbf{u} \times \mathbf{B} \cdot d\boldsymbol{\ell} = \int_{x_0}^0 \left[\left(\mathbf{a}_y\, u\right) \times \mathbf{a}_z\, B_0 \cos(\omega t - k_y y')\right] \cdot (\mathbf{a}_x\, dx)$$

$$= \int_{x_0}^0 \left[u B_0 \cos(\omega t - k_y y')\, \mathbf{a}_x\right] \cdot \mathbf{a}_x\, dx.$$

The integrand does not depend on x, making the integration particularly simple, and we find

$$\oint_c \mathbf{u} \times \mathbf{B} \cdot d\boldsymbol{\ell} = -x_0 u B_0 \cos(\omega t - k_y y'). \tag{5.10}$$

Combining results of Eqs. (5.9) and (5.10), we find

$$v(t) = B_0 x_0 \left(\frac{\omega}{k_y} - u\right) \cos(\omega t - k_y y') - \frac{B_0 x_0 \omega}{k_y} \cos(\omega t). \tag{5.11}$$

Substituting in the numerical values for B_0, u, and ω, we calculate

$$v(t) = \left[-2.4 \cos(\omega t - k_y y') - 2.4 \cos(\omega t) \right] \text{ V.}$$

Making sense of the signs of this complete EMF given in Eq. (5.11) is somewhat complicated, since the magnetic flux density **B** varies with time and position. We therefore return to examine Eqs. (5.9) and (5.10) individually. Equation (5.9) represents the EMF due to just the time variation of **B**. For small values of $k_y y'$, the magnetic flux Φ_M through the circuit is increasing at $t = 0$, and we should expect, by Lenz' Law, that the EMF is directed so as to produce a clockwise current (producing a **B** field that opposes this increase). This is precisely the case, as the EMF $v(t)$ given by Eq. (5.9) is negative at time $t = 0$. The motional EMF, given by Eq. (5.10), is also negative at this instant, producing a clockwise current as well. As the bar moves to the right (again considering only small values of $k_y y'$), the area of the loop increases, which increases the magnetic flux. As the field produced by the negative current opposes the change in Φ_M, this is also in agreement with Lenz' Law, confirming our result.

For this solution, we do have an alternative method, in that we could have used Eq. (5.1) directly. Let us follow this approach and see that we arrive at an identical result. We start with the magnetic flux Φ_M,

$$\Phi_M = \int_s \mathbf{B} \cdot d\mathbf{s} = \int_0^{y'} \int_0^{x_0} B_0 \cos(\omega t - k_y y) \mathbf{a}_z \cdot \mathbf{a}_z \, dx \, dy$$

$$= \frac{B_0 x_0}{k_y} \left[\sin(\omega t) - \sin(\omega t - k_y y') \right].$$

Then we compute $d\Phi_M/dt$ as

$$\frac{d\Phi_M}{dt} = \frac{\partial \Phi}{\partial t} + \frac{\partial \Phi}{\partial y'} \frac{dy'}{dt},$$

where we use partial derivatives of Φ since the flux depends on t and y'. Computing the derivatives leads us to

$$\frac{d\Phi_M}{dt} = \frac{B_0 x_0}{k_y} \left\{ \left[\omega \cos(\omega t) - \omega \cos(\omega t - k_y y') \right] + u k_y \cos(\omega t - k_y y') \right\},$$

where we have used $dy'/dt = u$. Rearranging these terms, and multiplying by -1, leads us to the same result we found in Eq. (5.11). We do have to use and understand partial derivatives for this approach, and there are some limits to its applicability, so we will generally use the separate terms, as in Eq. (5.3).

TechNote 5.7 describes the Faraday disc generator, which produces an EMF between the rim and axis of a conducting disc as the disc rotates.

TechNote 5.7 Faraday Disc Generator

A Faraday disc generator, or homopolar generator, consists of a rotating conductor between the poles of a permanent magnet, as shown in Fig. 5.13. It was first developed by Michael Faraday in 1831, but his generator was reportedly quite inefficient, probably due to circular current distributions within the conductor. Efficient Faraday disc generators, which avoid these detrimental currents by using additional magnets distributed around the wheel, are capable of producing very large d.c. currents due to the low internal resistance of the conductor. Homopolar generators are unique as potential sources, in that they produce d.c. current directly, without the need for additional diode rectifiers or commutators.

FIGURE 5.13 A Faraday disc generator.

The operation of the Faraday generator is perhaps best understood in terms of the Lorentz force experienced by the moving electrons in a magnetic field. As introduced earlier, the Lorentz force $\mathbf{F} = q\mathbf{u} \times \mathbf{B}$ is in a direction perpendicular to the velocity $\mathbf{u}$ and the magnetic flux density $\mathbf{B}$. As the wheel rotates, the free electrons within the conductor move with velocity $\mathbf{u} = \rho\omega\mathbf{a}_\phi$, where ρ is the radius from the center of the wheel and ω is the angular velocity of the wheel. This velocity is in the ϕ, or tangential, direction. $\mathbf{B}$ is along the z, or axial, direction, and therefore the electrons are forced to move in the ρ, or radial, direction, creating a potential difference between the rim of the wheel and the axis. The sign of the potential reverses when the disc rotates in the opposite direction.

In this section we have worked several examples showing the calculation of the EMF induced by a changing magnetic flux.

SUMMARY

Faraday's Law is based upon the observation that a changing magnetic flux causes an EMF in a loop. We have presented Faraday's Law in its integral form (Eq. (5.1)) and its differential form (Eq. (5.2)). In its integral form we have expressed the EMF in a representation that separates the effect of a time-varying flux density **B** from the effect of a moving circuit, the so-called motional EMF. Each part is important, and is commonly found in various applications. The former is the basis of inductors and transformers, while the latter is the basis of alternators and generators.

We have not explored the differential form of Faraday's Law in this chapter, but we will return to this topic later. At that time, we will develop a new picture of an electric potential. (Remember, for time-varying fields the electric potential function $V(\mathbf{r})$ that we developed in Chapter 1 is no longer valid, since $\nabla \times \mathbf{E}$ is no longer equal to zero.) Also, when we introduce propagating waves in Chapters 7 and 8, Faraday's Law (in its differential form) will again play a critical role. For now, however, discussions presented here allow us to complete the study of static magnetic fields (in the next chapter), exploring the energy of magnetic systems, including magnetic forces and torques, and investigating the inductance of conducting elements in more detail.

Table 5.1

Collection of key formulas from Chapter 5.				
$\mathcal{E} = -\dfrac{d\Phi_M}{dt}$	(5.1)	$\mathcal{E} = -\displaystyle\int_s \dfrac{\partial \mathbf{B}}{\partial t} \cdot d\mathbf{s} + \oint_c (\mathbf{u} \times \mathbf{B}) \cdot d\boldsymbol{\ell}$	(5.3)
$\nabla \times \mathbf{E} = -\dfrac{\partial \mathbf{B}}{\partial t}$	(5.2)		

PROBLEMS

P5-1 A circular loop of wire rotates about the z-axis, as shown in Fig. 5.14(a). The radius of the loop is a, and the number of turns N. Show that the EMF generated between the ends of the wire loop is equivalent to Eq. (5.8), with the area of the rectangular loop ad replaced by the area of the multi-turn circular loop $N\pi a^2$.

P5-2 A uniform, sinusoidally varying magnetic flux density $\mathbf{B} = B_0 \cos(\omega t)\mathbf{a}_z$ is applied to a conducting disc centered on the z-axis, as shown in Fig. 5.14(b). The radius of the disc is a, its thickness is d, and its conductivity is σ. (a) Determine the local electric field $\mathbf{E}(\rho)$ induced by the time-varying magnetic field. (b) Determine the current density $\mathbf{J}_v(\rho)$ in the disc, the power dissipation density, and the total power absorbed by the disc. (c) Compare the power absorbed by a single disc of

FIGURE 5.14 (a) An alternator consisting of a circular wire loop of radius a and N turns, rotating in a uniform magnetic flux density **B**. (For use in Prob. P5-1.) (b) A conducting disc in a sinusoidally varying magnetic flux density $\mathbf{B}(t)$. (For use in Prob. P5-2.)

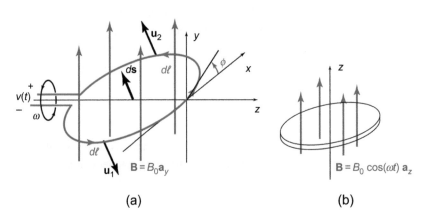

(a)

(b)

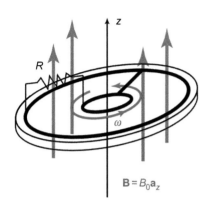

FIGURE 5.15 A non-conducting disc rotating about the z-axis. Two circular conducting traces of radius a and b $(a < b)$, and a conducting bridge between them are shown as thick black lines. (For use in Prob. P5-3.)

FIGURE 5.16 Wire loops of various shapes with N turns adjacent to a long, straight, current-carrying wire. (For use in Probs. P5-5–P5-8.)

radius a, as computed above, with the power absorbed by N discs, each of radius $a/\sqrt{N}$.

P5-3 The rotating disc shown in Fig. 5.15 is non-conducting, but has two circular conducting rings, one of radius a and the other of radius b $(b > a)$ on the top surface. In addition, a conducting bar bridges the two circular tracks. The entire disc rotates about the z-axis with angular frequency ω. A stationary resistor of resistance R makes contact with the two rings. (a) Determine the EMF V_0 induced between the two rings. (You may assume that the magnetic flux generated by the current in the "circuit" is negligible.) Which ring is at the higher potential? (b) Determine the current through the resistor. In which direction is this current? (c) What is the power dissipated in the resistor? Determine the rate of work done through the Lorentz force on the moving bar. Compare this result to the power dissipated in the resistor.

P5-4 A circular wire loop of radius a with N turns is placed inside an infinitely long solenoid. The solenoid is of radius b $(b > a)$ and turns density n. The current in the solenoid is $i(t) = I_0 \cos(\omega t)$. How must this loop be positioned/oriented within the solenoid in order to maximize the EMF induced between the ends of the loop. What is the EMF $v(t)$ induced across the ends of the loop?

P5-5 A circular wire loop of radius a with N turns is placed beside an infinitely long straight wire, as shown in Fig. 5.16(a). The distance between the center of the wire loop and the long wire is d, where $d > a$. The wire carries a current $i(t) = I_0 \cos(\omega t)$. (a) How must this loop be oriented in order to maximize the EMF induced between the ends of the loop? (b) Show that the magnetic flux through the circular loop is $\Phi_M(t) = \frac{\mu_0 I_0 \cos \omega t}{\pi d} \int_{-a}^{+a} \frac{\sqrt{a^2 + x^2}}{1 + x/d} dx$. (c) Use the Taylor expansion (Eq. (H.5)), $(1 + x/d)^{-1} = 1 - (x/d) + (x/d)^2 - (x/d)^3 + \cdots$, to evaluate this integral when $a < d$ to find an approximate EMF $v(t)$ induced between the ends of the loop.

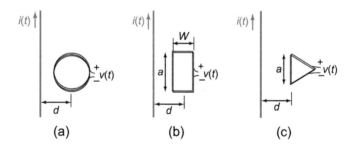

(a)   (b)   (c)

P5-6 A rectangular wire loop is placed beside an infinitely long straight wire, as shown in Fig. 5.16(b). The rectangular loop is of length a and width w, and consists of N turns. (The long wire is parallel to the side of length a.) The distance between the center of the wire loop and the long wire is d, where $d > w/2$, and the loop and the long wire lie in a common plane. The wire carries a current $i(t) = I_0 \cos(\omega t)$. What is the EMF $v(t)$ induced between the ends of the loop?

P5-7 Consider the same geometry of the rectangular loop and the long, straight wire described in Prob. P5-6, except the current in the long, straight wire is as shown in Fig. 5.17(a). Determine the EMF $v(t)$ induced between the ends of the loop.

P5-8 A triangular wire loop, each side of length a, with N turns, is placed beside an infinitely long straight wire, as shown in Fig. 5.16(c). (The long wire is parallel to the closest side of the triangular loop.) The distance between the parallel side

FIGURE 5.17 (a) Current waveform in the long, straight, current-carrying wire. (For use in Prob. P5-7.) (b) A conducting cylinder aligned with and rotating about the z-axis. (For use in Prob. P5-9.)

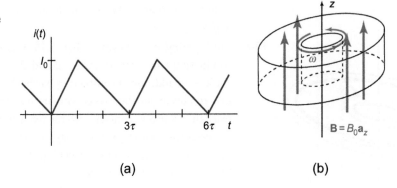

(a)   (b)

of the wire loop and the long wire is d, and the loop and the long wire lie in a common plane. The long, straight wire carries a current $i(t) = I_0 \cos(\omega t)$. What is the EMF $v(t)$ induced between the ends of the loop?

P5-9 A cylindrical conductor rotates about its axis, as shown in Fig. 5.17(b). The inner and outer radii of the cylinder are a and b, respectively, and the rotational angular frequency of the cylinder is ω. A constant, uniform magnetic flux density $\mathbf{B} = B_0 \mathbf{a}_z$ is applied parallel to the axis of the cylinder. Determine the potential difference between the inner and outer surfaces of the conductor.

P5-10 Standing in my front yard in West Lafayette, I swing a conducting bar on a string as rapidly as I can. The bar is 1 m long, and the length of the string and my arm combined is 1 m. How rapidly must I swing the bar, in revolutions per second, in order to induce a 1 mV EMF between the ends of the bar? (The Earth's magnetic field here is about 0.5 G, where 1 G is 10^{-4} T, and the plane of rotation of the bar is normal to $\mathbf{B}$.)

P5-11 A conducting bar moves back and forth along a pair of conducting rails, as shown in Fig. 5.12. The magnetic flux density $\mathbf{B}$ is uniform, constant, and perpendicular to the plane of the figure over the area between the rails. The position of the rail is given as $y(t) = y_0 + A \sin(\omega t)$. A resistor of resistance R is connected between points b and e. Determine the EMF $v(t)$ induced across the resistor R. (Assume that the magnetic flux produced by the current $i(t)$ is negligible.) Determine the current $i(t)$ through the resistor. Determine power dissipated in the resistor, and compare this to the rate at which work is done on the bar, using the Lorentz force on the bar.

P5-12 A conducting bar moves with constant velocity in the y-direction, as shown in Fig. 5.18(a). A long, straight conductor carrying a constant current I lies along the y-axis. Find the EMF induced between the ends of the metallic bar. What is its polarity?

FIGURE 5.18 (a) A moving conducting bar near a long, straight current-carrying conductor, for use in Prob. P5-12. (b) A rotating wire loop in a uniform magnetic field, for use in Prob. P5-13.

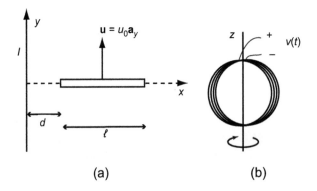

(a)   (b)

P5-13 A circular coil of radius $a = 10$ cm and $N = 50$ turns spins at a frequency $\omega = 2\pi \times 100$ rad/s about the z-axis, as shown in Fig. 5.18(b). The voltage $v(t)$ measured between the wires at the ends of the coil is of amplitude 5.0 V. Assuming that the magnetic flux density **B** is uniform, but of unknown direction, determine the minimum possible value of $|\mathbf{B}|$. (b) The measurement is repeated with the axis of rotation first along the y-axis, then along the x-axis. For the y-axis rotation measurement, the potential $v(t)$ is the same as described above. For the x-axis-rotation measurement, the potential is zero. What do these observations infer regarding the magnitude and direction of **B**?

P5-14 An infinitely long, straight wire parallel to the z-axis carries a current $i(t) = 5\cos(10^6 t)$ A. A semicircular wire loop consisting of $N = 50$ turns lies in the x–y plane. See Fig. 5.19. The radii of the inner and outer sides of the loop and $a = 1$ cm and $b = 3$ cm, respectively. The other two segments are straight and lie along the x-axis. Determine the EMF $v(t)$ induced between the ends of this loop.

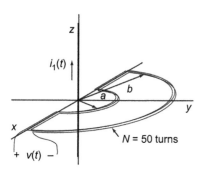

FIGURE 5.19 A semicircular loop of wire in the x–y plane, for use in Prob. P5-14.

P5-15 A rectangular loop of wire rotates in a constant magnetic field, as shown in Fig. 5.9. The number of turns of the loop is $N = 1\,000$, and the area of the loop is $a \times d = 0.1$ m². (a) Determine the magnitude B_0 of the magnetic flux density necessary to generate a root-mean-square EMF of 120 V at 60 Hz frequency. (b) Repeat for a 50 Hz frequency.

P5-16 A 10 cm long bar magnet spins perpendicular to its axis at a rate of 10 cycles/s. The axis of rotation is at the midpoint of the magnet, such that the N and S poles alternately pass a 100-turn wire coil of area $\Delta s = 1\,\mathrm{cm}^2$. The magnetic flux density B_0 at the pole faces of the magnet is 5 000 G (0.5 T). Sketch a plot of the EMF induced on the coil leads for two cycles of rotation of the magnet. Approximate the peak EMF. You may assume that the area of the pole faces of the magnet is similar to the coil area. Label the time between pulses on your plot, and mark when the N or S pole is near the coil. (Note: Don't worry about the overall sign of the voltage. You have not been told enough to determine this.)

P5-17 A 10 cm long bar magnet travels at a constant velocity of 20 cm/s through a 100-turn wire coil of area $\Delta s = 2\,\mathrm{cm}^2$. The magnet is oriented such that it enters the coil with one pole face, and exits with the other. The magnetic flux density B_0 at the pole faces of the magnet is 5 000 G (0.5 T). Sketch a plot of the EMF induced on the coil vs. time. You may assume that the area of the pole faces of the magnet is just smaller than coil area. Approximate the peak EMF. Label the time between pulses on your plot. (Note: Don't worry about the overall sign of the voltage. You have not been told enough to determine this.)

P5-18 The primary of an ideal transformer that is driven at 120 V_{rms} 60 Hz consists of 1 000 turns. The secondary produces a voltage of amplitude 15 V when connected to a 100 Ω resistor. Determine (a) the number of turns in the secondary, (b) the primary current $I_{1,rms}$, and (c) the input resistance to the primary.

P5-19 A toroidal core with rectangular cross-section is wrapped with two sets of windings. The primary winding consists of $N_1 = 100$ turns, while the secondary winding has 200 turns. The core material has permeability $100\mu_0$, inner radius $a = 5$ cm, outer radius $b = 8$ cm, and depth $d = 5$ cm. You may presume that all magnetic field lines are confined to the core material. Determine the self-inductances (a) $L_{11} = N_1\Phi_M/I_1$ when $I_2 = 0$, and (b) $L_{22} = N_2\Phi_M/I_2$ when $I_1 = 0$. Determine the mutual inductances (c) $L_{21} = N_1\Phi_M/I_2$ when $I_1 = 0$, and (d) $L_{12} = N_2\Phi_M/I_1$ when $I_2 = 0$. (e) Determine the transformer coupling coefficient k for this transformer. (Note: For this problem, the approximation that H_ϕ is constant within the core is not valid.)

P5-20 A Ruhmkorff induction coil is similar to the one shown in Fig. 5.8. The primary consists of $N_1 = 100$ turns, has a radius of 2 cm, and a length of 20 cm. The secondary consists of 10^4 turns, has a radius of 3 cm and a length of 3 cm. The secondary is centered on the primary. The core filling the interior of the primary has a permeability $\mu = 10^3 \mu_0$. The current passing through the primary windings, which is initially 100 mA, turns off in a short time $\Delta t = 1$ ms. Find an approximate value for the peak voltage developed across the secondary.

P5-21 Two co-axial solenoids are wound on a common core. The primary consists of $N_1 = 1\,000$ turns, has a radius of 2 cm, and a length of 40 cm. The secondary consists of $N_2 = 2\,500$ turns, has a radius of 3 cm, and a length of 10 cm. The core inside the primary has a permeability $\mu = 100\mu_0$. The current in the primary is $i(t) = I_0 \cos \omega t$, where $I_0 = 50$ mA and $\omega = 200$ r/s. Find the EMF generated across the secondary leads.

CHAPTER 6
Magnetostatics II

KEY WORDS

magnetostatic energy; energy density; mutual inductance; self-inductance; inductance per unit length.

In the previous chapters we saw several cases of the central role played by magnetic flux through open or closed surfaces. In this chapter we use magnetic flux and Faraday's Law to derive a very general expression for the energy stored within a magnetic field. We will also return to the inductance of current loops in Section 6.2, with several examples and illustrations, and then find the relation between the inductance of a circuit element and the energy stored within the magnetic field produced by the current in that element. Finally, we will explore magnetic forces and torques in Section 6.3.

LEARNING OBJECTIVES

After studying this section, you should be able to:
- determine the magnetostatic energy of a system, given the current distribution and magnetic potential;
- determine the magnetostatic energy density of a system, given the magnetic flux density and magnetic field; and
- determine the magnetostatic energy of a system from its energy density.

6.1 Magnetic Energy

The derivation of **magnetostatic energy** follows the same line of reasoning used in deriving the energy stored in an electric field, which was formulated by considering the work required to assemble the set of charges that generate the electric field. The energy stored in a magnetic field is equated to the work that must be done in turning on the currents that generate a magnetic field. The basic premises in this development are that: (1) a source must do work on a system to move charges across a potential difference; and (2) a changing magnetic flux is accompanied by a potential difference (i.e. Faraday's Law).

To get started, we consider a current distribution described by the current density $\mathbf{J}_v$ as shown in Fig. 6.1. The entire current distribution is divided into elemental filaments, or "tubes" of current, each of which follows a path c_k, and is of cross-sectional area $\Delta \tilde{s}_k$, where the index k labels the different tubes of current. As the current in a filament $I_k = |\mathbf{J}_{v,k}| \, \Delta \tilde{s}_k$ turns on, it generates a magnetic flux $\Phi_{M,k}$ through the surface inside that loop of current, and as we saw in Chapter 5, a changing magnetic flux induces an EMF around the loop. The work required to establish the current, then, is the work required to move the charges against this EMF. As soon as the currents are established, the magnetic flux $\Phi_{M,k}$ is constant, and the EMF drops to zero. We do not consider resistance in the loop, nor any other losses within the circuit, as these result in heating of the medium, but do not contribute to the energy stored within the magnetic field.

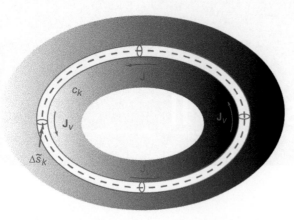

FIGURE 6.1 A current distribution of current density $\mathbf{J}_v$. From the entire distribution, we have picked out one "tube," or filament, of current, of cross-sectional area $\Delta \tilde{s}_k$ and current $I_k = |\mathbf{J}_v| \, \Delta \tilde{s}_k$.

In applying Faraday's Law, we first consider the magnetic flux through a single surface s_k defined by the interior of the loop of current indexed k,

$$\phi_{M,k}(t) = \int_{s_k} \mathbf{B} \cdot d\mathbf{s},$$

where $\mathbf{B}$ is the magnetic flux density due to *all* tubes of current. Recall that there are many surfaces that we could use here, since the only requirement is that the surface s_k is bounded by the path c_k. The simplest surface s_k in Fig. 6.1, however, lies in the plane of this page, inside the oval loop labeled c_k. (Don't be confused by the different surfaces used in this discussion. The surface s_k is different from $\Delta \tilde{s}_k$, which we introduced in the preceding paragraph as the cross-sectional area of the current tube "k.") The symbol $\phi_{M,k}(t)$ represents the magnetic flux as it turns on from zero to $\Phi_{M,k}$, where $\Phi_{M,k}$ is the final flux through the surface when the current is fully turned on. As we showed in Section 4.4, the magnetic flux can also be written in terms of the path integral of the magnetic vector potential (due to all currents) around the loop,

$$\Phi_{M,k} = \oint_{c_k} \mathbf{A} \cdot d\boldsymbol{\ell}_k.$$

This form is convenient for the present discussion.

As we saw in the previous chapter, the electromotive force (EMF) induced around this loop by the changing magnetic flux is

$$\mathcal{E}_k(t) = -\frac{d\phi_{M,k}}{dt}.$$

The work required to move a charge q through this EMF is

$$\Delta W_k = q\mathcal{E}_k,$$

and in a time Δt, the quantity of charge moved is $i_k \Delta t$, where $i_k(t)$ represents the current in tube k as it turns on from 0 to its final value of I_k. The work done in the time interval Δt as the flux increments by $\Delta \phi_{M,k}$ is

$$\Delta W_k = \mathcal{E}_k(i_k \Delta t) = \frac{\Delta \phi_{M,k}}{\Delta t}(i_k \, \Delta t) = i_k \Delta \phi_{M,k},$$

where the second step comes from Faraday's Law. We haven't been very careful with signs in this expression, but we know that the work ΔW_k is positive as the flux turns on, as the source must provide energy to move the charge against the induced EMF.

The total work W_m done in turning on the currents to their final values I_k, then, is found by summing over all k (labeling the various tubes of current), examining the limit of small $\Delta \phi_{M,k}$ (such that $\Delta W_k \to dW_k$), and integrating dW_k in the previous equation. We attach the subscript m to indicate that this work is done in establishing the *magnetic* fields. The total work is

$$W_m = \sum_k \int dW_k = \sum_k \int i_k(t) \, d(\phi_{M,k}(t)).$$

We make the simplifying assumption that all currents $i_k(t)$, and therefore all fluxes $\phi_{M,k}(t)$, turn on together, and ramp up linearly from an initial value of zero, over a time T (i.e. $i_k(t) = I_k t/T$ and $\phi_{M,k}(t) = \Phi_{M,k} t/T$). Then, $d(\phi_{M,k}(t)) = \Phi_{M,k}\, dt/T$ and

$$W_m = \sum_k \frac{I_k \Phi_{M,k}}{T^2} \int_0^T t\, dt.$$

Carrying out the integral, the magnetic energy stored in the system can then be written as

$$W_m = \frac{1}{2} \sum_k I_k \Phi_{M,k}. \tag{6.1}$$

It can be shown that this result is actually valid even when the currents turn on in some other manner, and so this result is quite generally true.

Let's review the variables here. I_k is the current within the circuit (or filament of current) "k," and $\Phi_{M,k}$ is the total magnetic flux penetrating through the surface defined by this circuit. Once again, we see a key role played by the magnetic flux Φ_M through a surface! The summation extends over all circuits k. If we write the current I_k in each filament in terms of the current density J_v (which we assume here to be flowing in the z'-direction), and take the limit as the area $\Delta \tilde{s}$ of each filament becomes small, the summation tends toward an integral of the form

$$W_m = \frac{1}{2} \int J_v(x',y') \Phi_M(x',y')\, dx'\, dy', \tag{6.2}$$

where $\Phi_{M,k}$ in the discrete filament k is replaced with $\Phi_M(x',y')$, the flux as a function of the continuous variables x' and y'.

The following example uses Eq. (6.2) to find the energy of a pair of long, straight, parallel wires. This is an important geometry, as it is related to twin-lead transmission lines, but it is also tedious. It is presented in SideNote 6.1.

SideNote 6.1 Magnetic Energy of Two Long, Parallel Wires

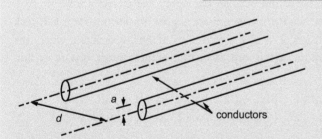

Determine the magnetic energy for a configuration of two long, parallel wires of length ℓ, as shown in Fig. 6.2. The radius of each wire a is much smaller than the distance between their centers d. The currents I in the two wires are equal in magnitude but opposite in direction.

FIGURE 6.2 A section of two long, parallel wires. The radius of each wire is a, the distance between the centers of the wires is d, and their length is ℓ. The currents I in the two wires are equal in magnitude but opposite in direction.

Solution:

We will use Eq. (6.2) to determine the magnetic energy W_m. Since the current density J_v is non-zero only within each wire, the volume of the wire defines the volume of integration. The current in a filament located at (x', y'), in the right-hand wire (centered at $(0, d/2)$), flows in the $+z$-direction, and returns in the filament at $(x', -y')$, in the left-hand wire (centered at $(0, -d/2)$). One such filament is shown as the square element inside the wires in Fig. 6.3. The first task is to evaluate the magnetic flux $\Phi_M(x', y')$ passing through the surface between these two filaments, shown as the horizontal dashed line in Fig. 6.3, and extending a distance ℓ in the z-direction. We will use the magnetic field $\mathbf{B}$ for an infinitely long wire, as given by Eqs. (4.17) and (4.18), ignoring any variations in this

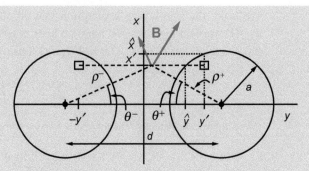

FIGURE 6.3 End-view of two long, parallel wires. The radius of each wire is a, and the distance between the centers of the wires is d.

field near the ends of the wires. Since this surface lies in the y–z plane, only the x-component of $\mathbf{B}$ is needed to determine the magnetic flux,

$$\Phi_M(x',y') = \int_s \mathbf{B} \cdot d\mathbf{s} = \int_0^\ell \int_{-y'}^{y'} B_x(x',y)\, dy\, dz = 2\ell \int_0^{y'} B_x(x',y)\, dy.$$

The final equality above is valid because of the symmetry in y of the field component B_x. The flux is the sum of three integrals,

$$\Phi_M(x',y') = 2\ell \left\{ \int_0^{y'} \frac{\mu_0 I}{2\pi\rho^-}\cos\theta^-\, dy + \int_0^{\hat{y}} \frac{\mu_0 I}{2\pi\rho^+}\cos\theta^+\, dy + \int_{\hat{y}}^{y'} \frac{\mu_0 I\rho^+}{2\pi a^2}\cos\theta^+\, dy \right\},$$

where the first integral represents the magnetic flux due to the current in the wire on the left, and the second and third integrals are the magnetic flux due to the current in the wire on the right. (Recall that $\mathbf{B}$ is different inside and outside the wire, so the flux connected with the right-hand wire leads to two separate integrals.) The distance $\rho^- = \sqrt{(y + d/2)^2 + x'^2}$ is the distance from a point (x', y) on the dashed line to the center of the wire on the left, and $\rho^+ = \sqrt{(d/2 - y)^2 + x'^2}$ is the distance from this point to the center of the wire on the right. The factors $\cos\theta^- = (y + d/2)/\rho^-$ and $\cos\theta^+ = (d/2 - y)/\rho^+$, where these angles are shown in Fig. 6.3, pick out the x-components of each. $\hat{y} = d/2 - \sqrt{a^2 - x'^2}$ is the y-coordinate where the dashed line intersects the radius of the wire. Evaluation of these integrals is direct, and the flux is

$$\Phi_M(x',y') = \frac{\mu_0 I\ell}{\pi} \left\{ \frac{1}{2}\ln\left[\frac{(y' + d/2)^2 + x'^2}{(d/2)^2 + x'^2}\right] - \frac{1}{2}\ln\left[\frac{a^2}{(d/2)^2 + x'^2}\right] \right.$$
$$\left. + \frac{1}{2a^2}\left[(a^2 - x'^2) - (d/2 - y')^2\right] \right\},$$

which simplifies to

$$\Phi_M(x',y') = \frac{\mu_0 I\ell}{\pi} \left\{ \frac{1}{2}\ln\left[(y' + d/2)^2 + x'^2\right] - \ln a + \frac{1}{2a^2}\left[(a^2 - x'^2) - (d/2 - y')^2\right] \right\}.$$

Now that we have the flux $\Phi_M(x',y')$, we use this in Eq. (6.2) to find the energy,

$$W_m = \frac{1}{2}\int J_z(x',y')\Phi_M(x',y')\, dx'\, dy',$$

where the integration must be carried out across the cross-sectional area of the wire on the right. With $J_{v,z}(x',y')$ uniform and equal to $I/\pi a^2$, this becomes

$$W_m = \frac{\mu_0 I^2 \ell}{2\pi(\pi a^2)} \int_{d/2-a}^{d/2+a} \int_{-\hat{x}}^{\hat{x}} \left\{ \frac{1}{2} \ln\left[(y' + d/2)^2 + x'^2\right] - \ln a \right.$$
$$\left. + \frac{1}{2a^2}\left[(a^2 - x'^2) - (d/2 - y')^2\right] \right\} dx' dy'.$$

The limits of integration on x are $\pm\hat{x}$, where $\hat{x} = \sqrt{a^2 - (y' - d/2)^2}$. Notice that in the first integral, the variable y' appears as $y' + d/2$, which we can rewrite as $d + (y' - d/2)$. Since we are integrating across the area of the wire, $|y' - d/2|$ has a maximum value of a, which we are told is much less than d, the spacing between the wires. The Taylor expansion (Eq. (H.8)) of this integrand is

$$\ln\left[(y' + d/2)^2 + x'^2\right] = \ln\left[d^2 + 2d(y' - d/2) + (y' - d/2)^2 + x'^2\right]$$
$$\simeq \ln d^2 + \frac{2}{d}(y' - d/2) + \cdots$$

Completing the integrations, the magnetic energy is

$$W_m = \frac{\mu_0 I^2 \ell}{2\pi}\left[\ln\left(\frac{d}{a}\right) - \frac{1}{3\pi}\frac{a}{d} + \frac{1}{4}\right].$$

Since $a \ll d$, we can safely drop the middle term, and the energy is

$$W_m = \frac{\mu_0 I^2 \ell}{2\pi}\left[\frac{1}{4} + \ln\left(\frac{d}{a}\right)\right].$$

This is the result that we set out to show.

Let's continue now to develop other forms of the magnetic energy that will be more generally useful. We continue from Eq. (6.2), substituting the path integral of the magnetic vector potential for the magnetic flux Φ_M. In this case the magnetic energy becomes

$$W_m = \frac{1}{2} \sum_k |\mathbf{J}_v|_k \, \Delta\tilde{s}_k \oint_{c_k} \mathbf{A} \cdot d\boldsymbol{\ell}_k.$$

The current density $\mathbf{J}_v$ and the path element $d\boldsymbol{\ell}_k$ are parallel to one another, since we defined the path as one of the tubes of current, allowing us to write $|\mathbf{J}_v| \, \mathbf{A} \cdot d\boldsymbol{\ell}_k$ in the equivalent form $\mathbf{J}_v \cdot \mathbf{A} \, d\boldsymbol{\ell}_k$, and in the limit as $\Delta\tilde{s}_k \to 0$, the summation over k in the previous expression becomes a surface integral over the cross-sectional area $d\tilde{s}$ of the current distribution (in the plane perpendicular to the current density $\mathbf{J}_v$). Finally, we recognize that $d\tilde{s} \, d\ell$ defines the element of volume dv, and the magnetic energy is

$$\boxed{W_m = \frac{1}{2} \int_v \mathbf{J}_v \cdot \mathbf{A} \, dv,} \tag{6.3}$$

where v is the volume over which $\mathbf{J}_v$ exists. By way of comparison, recall that for the electrostatic energy we had a similar intermediate result

$$W_e = \frac{1}{2} \int_v V(\mathbf{r})\rho_v(\mathbf{r}) \, dv.$$

These relations are similar, where we replace the electric potential with the vector magnetic potential, and the charge density with the current density.

Using a few vector relations that we have seen previously, this energy can be expressed in terms of field quantities $\mathbf{H}$ and $\mathbf{B}$. First, we use Ampère's Law (still not quite complete, but in the form that is correct to the point that we have developed it so far), $\mathbf{J}_v = \mathbf{\nabla} \times \mathbf{H}$, to write

$$W_m = \frac{1}{2} \int_v (\mathbf{\nabla} \times \mathbf{H}) \cdot \mathbf{A} \, dv.$$

Next, we make use of the vector differential relation (see Section E.2),

$$\mathbf{\nabla} \cdot (\mathbf{A} \times \mathbf{H}) = \mathbf{H} \cdot (\mathbf{\nabla} \times \mathbf{A}) - \mathbf{A} \cdot (\mathbf{\nabla} \times \mathbf{H}),$$

to write

$$W_m = \frac{1}{2} \int_v \mathbf{H} \cdot (\mathbf{\nabla} \times \mathbf{A}) \, dv - \frac{1}{2} \int_v \mathbf{\nabla} \cdot (\mathbf{A} \times \mathbf{H}) \, dv.$$

In the first integral, replace $\mathbf{B}$ for $\mathbf{\nabla} \times \mathbf{A}$, and the Divergence Theorem can be applied to the second to write

$$W_m = \frac{1}{2} \int_v \mathbf{H} \cdot \mathbf{B} \, dv - \frac{1}{2} \oint_s (\mathbf{A} \times \mathbf{H}) \cdot d\mathbf{s}.$$

The volume integral is carried out over the entire volume in which the fields $\mathbf{H}$ and $\mathbf{B}$ exist. (Note that this is generally a much more extended space than the volume over which $\mathbf{J}_v$ and $\mathbf{A}$ exist.) The surface integral is carried out over the surface s that encloses the volume v. (Apologies again, but this surface is yet another surface, unrelated to the surfaces $\Delta \tilde{s}_k$ or s_k defined earlier in this section.) As the volume of integration v become very large, the surface integral will vanish, since $|\mathbf{A}|$ decreases at least as fast as $1/R$ with increasing distance R, and $|\mathbf{H}|$ decreases as $1/R^2$ or faster. The surface area over which this integrand is evaluated grows only as R^2, so as the volume v gets large (and R gets large), the surface integral vanishes as $(1/R)(1/R^2)(R^2) \sim (1/R)$, and we can safely drop this integral. We are left with

$$\boxed{W_m = \frac{1}{2} \int_v \mathbf{H} \cdot \mathbf{B} \, dv.} \tag{6.4}$$

This integral represents the total energy stored by the magnetic field generated by the current distribution. Comparison of this with the expression for the energy contained within an electrostatic field, Eq. (2.7), shows strong similarity, with $\mathbf{H}$ replacing $\mathbf{D}$ and $\mathbf{B}$ replacing $\mathbf{E}$.

We can identify the integrand $(1/2) \, \mathbf{H} \cdot \mathbf{B}$ by reflecting upon the meaning of its integral. Since the integral of $(1/2)\mathbf{H} \cdot \mathbf{B}$ over all space yields the energy stored within that field, then the integrand itself must represent the **energy density** of the magnetic field, represented by w_m, where

$$\boxed{w_m = \frac{1}{2} \, \mathbf{H} \cdot \mathbf{B}.} \tag{6.5}$$

Example 6.1 Magnetic Energy of a Toroid

The permeability of the material in the core of a toroid, as shown in Fig. 6.4, is μ, the number of turns of wire is N, and the current in the windings is I. The cross-section of this toroidal core is rectangular, of width $b - a$ and thickness h. Determine the magnetic energy W_m of this toroid.

Solution:

We found the magnetic flux density $\mathbf{B}$ of a similar toroid previously, in Example 4.10, but we solved that example prior to our introduction of magnetic materials. Therefore we

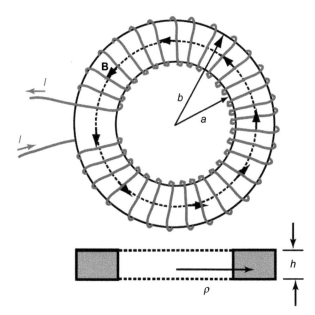

FIGURE 6.4 A toroid of rectangular cross-section. The permeability of the material in the core is μ, the thickness is h, and the number of turns of wire is N. A view showing the cross-section of the toroid is shown at the bottom.

will start from the beginning, using the appropriate form of Ampère's Law to determine $\mathbf{H}$ in the core material of the toroid, and find $\mathbf{B}$ from this. After finding $\mathbf{H}$ and $\mathbf{B}$, these will be used in Eq. (6.4) to determine the magnetic energy W_m stored in the field.

Let's start by describing the magnetic field qualitatively. The current through the wires produces a magnetic field $\mathbf{H}$ within the core region in the ϕ-direction, circling around the interior of the toroid in the counter-clockwise direction. We showed earlier that the field only exists in the interior of the windings, as long as the windings are closely spaced. This is an even better assertion when the permeability of the core is large, $\mu \gg \mu_0$. To find the magnitude of $\mathbf{H}$, we apply Ampère's Law,

$$\oint \mathbf{H} \cdot d\boldsymbol{\ell} = I_{enc},$$

on a circular path, shown as the dashed line around the toroid in Fig. 6.4. We choose $d\boldsymbol{\ell}$ in the counter-clockwise direction (in the same direction as $\mathbf{H}$), for which the element of integration $d\boldsymbol{\ell}$ for the circular path is $\mathbf{a}_\phi \rho \, d\phi$:

$$\oint \mathbf{H} \cdot d\boldsymbol{\ell} = \int_0^{2\pi} H_\phi \rho \, d\phi.$$

Using the symmetry of the toroid, the magnetic field $\mathbf{H}$ can be shown to depend only on the radial distance ρ. H_ϕ and ρ can be pulled outside the integral, and

$$\oint \mathbf{H} \cdot d\boldsymbol{\ell} = H_\phi 2\pi\rho.$$

To evaluate I_{enc}, we must recognize that the surface normal $\mathbf{a}_n$ must point out of the page, using the right-hand rule with $d\boldsymbol{\ell}$ counter-clockwise. Therefore, I_{enc} is $+NI$, where the $+$ sign results since the current comes out of the page as it passes through the surface s, in the same direction as the surface normal $\mathbf{a}_n$. Solving for H_ϕ, then,

$$H_\phi = \frac{NI}{2\pi\rho}.$$

Outside of the core ($\rho < a$ or $\rho > b$), application of Ampère's Law yields $H_\phi = 0$. The magnetic flux density in the core region is $B_\phi = \mu H_\phi = \mu NI/(2\pi\rho)$.

The energy stored in the magnetic field of the toroid, using Eq. (6.4), is

$$W_m = \frac{1}{2} \int_v \mathbf{H} \cdot \mathbf{B} \, dv = \frac{1}{2} \int_0^h \int_0^{2\pi} \int_a^b H_\phi B_\phi \rho \, d\rho \, d\phi \, dz,$$

where $\rho \, d\rho \, d\phi \, dz$ is the elemental volume dv in cylindrical coordinates. (See Appendix C.) Using H_ϕ and B_ϕ from above and carrying out the integration, we find

$$W_m = \frac{1}{2} \frac{\mu N^2 I^2 h}{2\pi} \ln(b/a).$$

We will return to work additional examples of the magnetostatic energy, but first let's examine in greater depth the inductance of a configuration of conductors, first introduced in Section 5.2.

LEARNING OBJECTIVES

After studying this section, you should be able to:

- determine the mutual inductance between two wires;
- determine the self-inductance of a toroid and solenoid; and
- determine the inductance per unit length of a pair of parallel conductors.

6.2 Inductance and Inductors

In Chapter 5 we briefly introduced the inductance between current loops as a way of describing the magnetic coupling between those loops. At the time, our interest was in determining the EMF induced between the ends of a wire loop through the Faraday effect. We return now to this topic, in order to determine the inductance of several important geometries.

We have, to this point, calculated the magnetic flux density **B** for a number of configurations, including simple loops, solenoids, toroids, and pairs of co-axial cylindrical conductors. In each case, the magnetic flux density depends linearly on the current I that generates the field. This is a general result, which can be inferred upon examining the Biot–Savart Law, given in various useful forms in Eqs. (4.3)–(4.6). In each case, we see the explicit linear dependence of **B** on I, $\mathbf{J}_v$, or $\mathbf{J}_s$. A similar conclusion can be reached by examining Ampère's Law, as in Eq. (4.39). Since the flux density **B** is linear in the current I, the magnetic flux Φ_M through any surface must also grow in proportion to the current I, and we introduced the inductance L of a configuration of wires to quantify this magnetic coupling between the wires or loops. The inductance depends only on geometric factors, such as the size of a wire loop, the distance between loops, and the number or turns density of coils, and on material properties such as the magnetic permeability of the material in or near the wire loop. The inductance does not depend upon any currents or voltages. The unit for inductance is the henry, abbreviated to H, which is equivalent to $1\ \Omega s = Vs/A$.

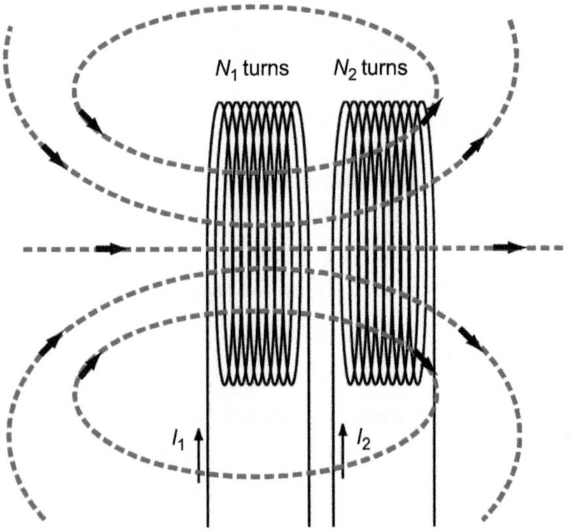

In Fig. 6.5 is a pair of wire loops adjacent to one another. The first coil consists of N_1 turns and has a current I_1 running through it, while the second consists of N_2 turns and has a current I_2 running through it. The current I_1 generates a magnetic flux density $\mathbf{B}_1(\mathbf{r})$, shown as the dashed lines, and these magnetic field lines penetrate through the cross-section of the second loop. The magnetic flux through the surface bordered by the second loop is calculated as the surface integral of the flux lines due to the current I_1 through the surface defined by loop 2, or

$$\Phi_{12} = \int_{s_2} \mathbf{B}_1 \cdot d\mathbf{s}_2.$$

This flux is, of course, proportional to the current I_1. The **mutual inductance** L_{12} quantifies the magnetic flux linkage of circuit 2 per unit current in circuit 1:

FIGURE 6.5 A pair of loops adjacent to one another. The magnetic flux lines produced by the current I_1 in the loop on the left penetrate through the surface bordered by the second loop.

$$\boxed{L_{12} = \frac{N_2 \Phi_{12}}{I_1}.} \tag{6.6}$$

Since the flux Φ_{12} is proportional to the current I_1, dividing Φ_{12} by I_1 removes the current altogether, and the inductance depends only on geometric terms (the size of the wire loops, the distance between the loops, etc.) and on the permeabilities of materials near the loops, as we wrote earlier.

The magnetic flux lines generated by the current I_1 also penetrate through the surface of loop 1 itself, and so, in addition to the mutual inductance between the loops, it is useful to define the **self-inductance** of a coil,

$$L_{11} = \frac{N_1 \Phi_{11}}{I_1}, \tag{6.7}$$

where Φ_{11} is the flux through loop 1 due to current I_1. The self-inductance of an element is often simply called its inductance.

When we introduced the mutual inductance, we set this up as the coupling to loop 2 by magnetic flux lines generated by the current I_1 in loop 1. A current I_2 in loop 2 will, of course, generate a magnetic field $\mathbf{B}_2(\mathbf{r})$ that will couple to loop 1, and a second mutual inductance is defined,

$$L_{21} = N_1 \Phi_{21} / I_2,$$

where $\Phi_{21} = \int_{s_1} \mathbf{B}_2 \cdot d\mathbf{s}_1$ is the surface flux of the field lines generated by I_2 that penetrate through the surface of wire loop 1. We will now show that the mutual inductance L_{12} is identical to the mutual inductance L_{21}. This is demonstrated by closer scrutiny of Eq. (6.6), using $\mathbf{B}_1 = \boldsymbol{\nabla} \times \mathbf{A}_1$:

$$L_{12} = \frac{N_2}{I_1} \int_{s_2} \mathbf{B}_1 \cdot d\mathbf{s}_2 = \frac{N_2}{I_1} \int_{s_2} (\boldsymbol{\nabla} \times \mathbf{A}_1) \cdot d\mathbf{s}_2.$$

Applying Stoke's Theorem to the integral, we find

$$L_{12} = \frac{N_2}{I_1} \oint_{c_2} \mathbf{A}_1 \cdot d\boldsymbol{\ell}_2,$$

where c_2 is the perimeter of loop 2. But the vector potential can be written using Eq. (4.24), so the mutual inductance is

$$L_{12} = \frac{N_2}{I_1} \oint_{c_2} \left(\frac{N_1 \mu_0}{4\pi} \oint_{c_1} \frac{I_1 d\boldsymbol{\ell}_1}{|\mathbf{r}_2 - \mathbf{r}_1|} \right) \cdot d\boldsymbol{\ell}_2,$$

where $\mathbf{r}_1$ and $\mathbf{r}_2$ are the position vectors for elemental segments within loops 1 and 2, respectively, and $d\boldsymbol{\ell}_1$ is the path element of loop 1. Pulling constants outside the integrals and simplifying, we have

$$L_{12} = N_1 N_2 \frac{\mu_0}{4\pi} \oint_{c_2} \oint_{c_1} \frac{d\boldsymbol{\ell}_1 \cdot d\boldsymbol{\ell}_2}{|\mathbf{r}_2 - \mathbf{r}_1|}.$$

Similarly, we can follow the same series of steps to determine the mutual inductance L_{21}, and show that it can be expressed as

$$L_{21} = N_1 N_2 \frac{\mu_0}{4\pi} \oint_{c_1} \oint_{c_2} \frac{d\boldsymbol{\ell}_2 \cdot d\boldsymbol{\ell}_1}{|\mathbf{r}_1 - \mathbf{r}_2|}.$$

Comparison of these integral forms for L_{12} and L_{21} shows that they are identical. (The order of $d\boldsymbol{\ell}_1$ and $d\boldsymbol{\ell}_2$ in the numerator is unimportant, the denominators $|\mathbf{r}_1 - \mathbf{r}_2|$ and $|\mathbf{r}_1 - \mathbf{r}_2|$ are equal, and the order of integration over one loop or the second is interchangeable.) That is, for any configuration of two circuits, L_{12} and L_{21} are interchangeable. This is a very useful result, in that we will sometimes encounter geometries for which one of these mutual inductances is much simpler to determine than the other, so we are free to use the simpler geometry in our calculation with complete confidence of their equivalence.

The self-inductance of a configuration is related to the magnetic energy of the energized inductor through

$$W_m = \frac{1}{2} L I^2, \tag{6.8}$$

a relation that you hopefully recall from linear circuits. (Note that the subscript on the inductance L has been dropped here.) This stored energy is, of course, precisely the magnetic energy that we discussed in the previous section. Equation (6.8) is a very direct connection between energy and inductance, and presents an alternate means of finding the inductance, or vice versa. That is,

$$L = \frac{2W_m}{I^2} = \frac{1}{I^2} \int_v \mathbf{H} \cdot \mathbf{B} \, dv, \qquad (6.9)$$

where the volume integration is carried out over the entire space in which the magnetic fields $\mathbf{H}$ and $\mathbf{B}$ exist.

Let us now determine the inductance for some specific configurations.

Example 6.2 Inductance of a Toroid

Consider the toroid shown in Fig. 6.4, for which we found the magnetic energy in Example 6.1. The permeability of the material in the core of this toroid is μ and the number of turns of wire is N. The cross-section of the toroidal core is rectangular, of width $b - a$ and thickness h, as shown. Determine the inductance L of this toroid.

Solution:

In Example 6.1 we showed that the magnetic field is confined to the core (interior) of the toroid, circling in the counter-clockwise direction, and its magnitude is

$$H_\phi = \frac{NI}{2\pi\rho}$$

inside the core, but zero outside the core ($\rho < a$ or $\rho > b$).

Let's use $B_\phi = \mu H_\phi = \mu NI/(2\pi\rho)$ to find the magnetic flux. We integrate $\mathbf{B}$ across the cross-sectional area of the core, shown as the shaded area in the view of the toroid at the bottom of Fig. 6.4. (This choice of cross-section should make sense, in that this is the area through which the magnetic field lines penetrate.) For this cross-section, the elemental surface is $d\mathbf{s} = \mathbf{a}_\phi \, d\rho \, dz$. The limits of integration on ρ are a and b, while for z they are 0 and h. Therefore the magnetic flux is

$$\Phi_M = \int_s \mathbf{B} \cdot d\mathbf{s} = \int_0^h \int_a^b B_\phi d\rho \, dz = \frac{\mu NIh}{2\pi} \ln (b/a).$$

Using this flux, the inductance is

$$L = \frac{N\Phi_M}{I} = \frac{N}{I} \left\{ \frac{\mu NIh}{2\pi} \ln (b/a) \right\} = \frac{\mu N^2 h}{2\pi} \ln (b/a). \qquad (6.10)$$

Notice that this expression for the inductance L includes only geometric factors, such as the number of turns N, the thickness h, and dimensions a and b; and material parameters, such as the permeability of the core material. The current I does not appear in Eq. (6.10), nor does any field quantity such as $\mathbf{H}$, $\mathbf{B}$, or Φ_M. Also, the dependence of L on N^2 is worthy of mention. One factor of N comes from the field amplitude, where larger N produces a proportionally larger value of $\mathbf{H}$. The second factor of N can be understood in that the flux lines penetrate through all N windings, and the EMF induced across the entirety of the toroid is N times larger than the EMF induced across each individual winding.

An alternative method of finding the inductance of a configuration is to use Eq. (6.9). In Example 6.1 we found that the energy stored in the magnetic field of the toroid is

$$W_m = \frac{1}{2} \frac{\mu N^2 I^2 h}{2\pi} \ln (b/a).$$

This gives us

$$L = \frac{2W_m}{I^2} = \frac{\mu N^2 h}{2\pi} \ln(b/a),$$

in agreement with Eq. (6.10).

In this example, we found the *self*-inductance of the toroid. Recall that the current through the windings produces a magnetic field, and these field lines penetrate through the surface area of these same windings. For any two-terminal inductance element used in circuits, when we talk about the inductance of this element, we of course are referring to its self-inductance. In Example 6.3 we will determine the mutual inductance between two wires.

Example 6.3 Mutual Inductance between a Straight Wire and a Rectangular Loop

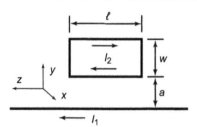

FIGURE 6.6 A rectangular loop near a long, straight wire.

A rectangular loop of wire is positioned next to an infinitely long, straight wire, as shown in Fig. 6.6. The loop is of length ℓ and its width is w, with the distance a between its near side and the long, straight wire. The long wire and the rectangular loop lie in a plane. Determine the mutual inductance between the long, straight wire and the rectangular loop.

Solution:

Let's start by recognizing that the two mutual inductances L_{12} and L_{21} are equal, and that we are free to choose which inductance we will determine. We consider a current I_1 in the long, straight wire in the $\mathbf{a}_z$-direction, and a current I_2 in the clockwise direction in the rectangular loop. As defined earlier, the magnetic flux Φ_{12} is the integral $\int_{s_2} \mathbf{B}_1 \cdot d\mathbf{s}_2$, where $\mathbf{B}_1$ is the magnetic field due to the current I_1 and s_2 is the surface area bordered by the rectangular current loop. The inductance L_{12} is then $N_2 \Phi_{12}/I_1$, where N_2 is the number of turns in the rectangular loop, which in this case is $N_2 = 1$. The alternative would be to determine the magnetic flux density $\mathbf{B}_2$ generated by the current I_2 in the rectangular loop, and determine the flux $\Phi_{21} = \int_{s_1} \mathbf{B}_2 \cdot d\mathbf{s}_1$. This sounds more difficult, for a couple of reasons. First, $\mathbf{B}_1$ due to the current in the long, straight wire is much more easily determined than $\mathbf{B}_2$; and second, the surface area s_1 over which to determine the magnetic flux Φ_{21} is extensive, and not as simple even to identify as the surface s_2. For these reasons, we will proceed by determining the inductance L_{12} rather than L_{21}.

We have previously determined the magnetic flux density $\mathbf{B}_1$ due to the current I_1 in a long, straight wire, and showed that

$$\mathbf{B}_1 = \mathbf{a}_\phi \frac{\mu_0 I_1}{2\pi\rho}.$$

In this case the radial distance ρ is y, and integrating over the surface area of the rectangular loop, we find

$$\Phi_{12} = \frac{\mu_0 I_1 \ell}{2\pi} \ln\left(\frac{a+w}{a}\right).$$

Dividing this magnetic flux by I_1, we find

$$L_{12} = \frac{N_2 \Phi_{12}}{I_1} = \frac{\mu_0 \ell}{2\pi} \ln\left(\frac{a+w}{a}\right).$$

Notice that this inductance depends only on the dimensions (ℓ, a, and w) and on the material properties (μ_0), and that the final result does not include currents or magnetic fields.

An important sensor based upon induction is in widespread use to sense traffic and to control traffic lights. Perhaps you've noticed the outline of these sensors at many urban intersections. This sensor is described in TechNote 6.1.

TechNote 6.1 Traffic Signal Sensors

Several different types of sensors are used to detect the presence of a car or other vehicle waiting for the traffic light to change at an intersection, but the most common sensor is based on a large inductor embedded in the pavement. You might have seen the large saw kerfs in the pavement, often rectangular or octagonal in shape, such as the one shown in Fig. 6.7, where these inductive sensors have been installed. Typical inductance values of the unloaded sensor are in the range 50–1 000 µH, depending on the dimensions of the loop, the gauge of the wire, and the number of turns; the resonant frequency of the L–C circuit consisting of this inductor and a capacitor is typically in the range 10–200 kHz. The presence of a car over the inductive sensor changes its inductance, and this shifts the resonant frequency of the circuit, which can be sensed by the driving circuit.

FIGURE 6.7 An octagonal inductive loop embedded in the pavement near a traffic light.

In the previous two examples we determined the self-inductance of a toroid and the mutual inductance of a rectangular wire loop near a long, straight wire. In many cases, we wish to determine a related quantity, known as the **inductance per unit length** of a configuration, typically one which has a very long length, such as a pair of long, parallel conductors (as in a transmission line, for example). Transmission lines are an important means of guiding waves, which we will discuss in detail in Chapter 9. We will see then that the inductance per unit length, for which we use the notation L', together with the capacitance per unit length C' that we introduced in Section 2.5 for a pair of infinitely long conductors, help determine the wave properties on that transmission line. We introduce this parameter by considering the example of an infinitely long solenoid. Because of its infinite length, the total energy stored would be infinite, even for a finite current I. The total inductance, then, would be infinite as well, and is therefore not a very useful parameter. The inductance per unit length, however, is the inductance of just a finite length ℓ of the solenoid, divided by that length ℓ, and this parameter can be useful indeed.

Determination of the inductance per unit length L' for a particular geometry will follow along lines very similar to determination of the total inductance L described above. For example, we can determine the magnetic flux Φ_M for the geometry, and use this in

$$L' = \frac{N\Phi_M}{\ell I}. \tag{6.11}$$

Or we can determine the magnetic energy per unit length W'_m, determined as

$$W'_m = \frac{1}{\ell}\frac{1}{2}\int_s\int_0^\ell \mathbf{H} \cdot \mathbf{B}\, dz\, ds, \tag{6.12}$$

where the axis of the solenoid is parallel to the z-axis, and then use

$$L' = \frac{2W'_m}{I^2}. \tag{6.13}$$

This concept is illustrated in Examples 6.4–6.5.

Example 6.4 Inductance per Unit Length of a Solenoid

Find the inductance per unit length L' for an infinitely long solenoid of cross-sectional area S and turns density n wrapped around a core of permeability μ.

Solution:

We found earlier (see Example 4.9), using Ampère's Law, that the magnetic flux density inside an infinitely long, air-filled solenoid is axial and of magnitude $B_z = \mu_0 nI$, where I is the current in the windings. The form of Ampère's Law that we used in that example was valid only for non-magnetic materials. With the material in the interior of the solenoid of permeability μ, Ampère's Law given by Eq. (4.40) can be applied to show that the magnetic field inside the solenoid is $H_z = nI$. The magnetic flux density is then $B_z = \mu H_z = \mu nI$, and the surface flux is

$$\Phi_M = \int_s \mathbf{B} \cdot d\mathbf{s} = B_z S = \mu nIS,$$

where we integrated over the cross-sectional area of the solenoid and used the uniformity of B_z in the interior of the solenoid to evaluate the surface integral. Finally, the inductance of a length ℓ of the solenoid, which consists of the total number of turns $N = n\ell$, is

$$L = \frac{N\Phi_M}{I} = \frac{(n\ell)(\mu nIS)}{I} = \mu n^2 S\ell.$$

Dividing by ℓ gives us

$$L' = \mu n^2 S. \tag{6.14}$$

Alternatively, the magnetic energy stored by the solenoid can be used to find L. The energy stored in a length ℓ of the solenoid is

$$W_m = \frac{1}{2}\int_0^S\int_0^\ell \mathbf{H} \cdot \mathbf{B}\, dz\, ds = \frac{1}{2}\mu n^2 I^2 S\ell.$$

The inductance of this section of the solenoid is then

$$L = \frac{2W_m}{I^2} = \mu n^2 S\ell,$$

and L' follows by dividing L by ℓ, arriving at the result given in Eq. (6.14).

It is interesting to compare this result with that of the inductance of a toroid as worked in Example 6.2. As we described previously, we can imagine constructing a toroid from a long solenoid by bending the solenoid around to form a circle and joining its ends together. Therefore, there should exist a simple relation between the inductances of these two geometries. In the toroid example, we showed that the inductance is

$$L = \frac{\mu N^2 h}{2\pi} \ln(b/a).$$

For the purpose of this comparison, we let our toroid become long and skinny, corresponding most closely with a solenoid of infinite length. In this limit, the ratio $b/a \to 1$ and we can use the approximation $\ln(b/a) \simeq b/a - 1$ (see Eq. (H.8)). The inductance per unit length for the solenoid is therefore this inductance divided by its circumference:

$$L' = \frac{L}{2\pi a} = \frac{\mu N^2 h}{(2\pi)(2\pi a)}(b/a - 1) = \frac{\mu N^2 h (b - a)}{(2\pi a)^2}.$$

(We have used $2\pi a$ for the circumference, which is just as good as $2\pi b$ for this purpose, since we have already invoked the approximation $b/a \to 1$.) Substituting the turns density n for $N/(2\pi a)$ and the cross-sectional area of the toroid S for $h(b - a)$, we find

$$L' = \mu n^2 S,$$

in agreement with Eq. (6.14).

Let's now examine a common geometry for transmission lines, a pair of co-axial cylindrical conductors. In Example 6.5 we will find the inductance per unit length for this geometry.

Example 6.5 Inductance per Unit Length of a Co-axial Cable

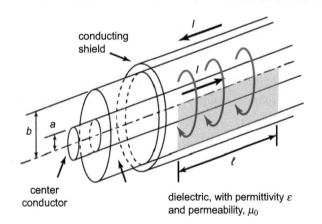

conducting shield

center conductor

dielectric, with permittivity ε and permeability, μ_0

FIGURE 6.8 A section of co-axial cable. The red arrows show the direction of the magnetic flux density field circling the axis.

A section of a co-axial cable is shown in Fig. 6.8. The center conductor is of radius a, and is surrounded by a non-conducting material of permeability μ_0. The outer conductor is of radius b. The current I is uniformly distributed in the inner conductor, traveling in the $+z$-direction, and returns on the outer conductor. The thickness of the outer conductor is very small. Determine the inductance per unit length L' of this co-axial system, (a) by determining the magnetic flux per unit length Φ'_M, and (b) by determining the magnetic energy per unit length W'_m of this configuration.

Solution:

For either solution method, we must first determine the fields $\mathbf{H}$ and $\mathbf{B}$ that are produced by the currents in the conductors. Since the current is uniformly distributed in the center conductor, the current density in this conductor can be written as $\mathbf{J}_v = \mathbf{a}_z I/(\pi a^2)$. Application of Ampère's Law is the quickest method to find $\mathbf{H}$. By the symmetry of the co-axial conductors, $\mathbf{H}$ can depend only on ρ, and it can have a component only in the $\mathbf{a}_\phi$-direction (i.e. $\mathbf{H} = \mathbf{a}_\phi H_\phi(\rho)$). We construct a circular path c of radius ρ concentric with the co-axial cable on which to apply Ampère's Law, and evaluate the path integral of $\mathbf{H}$ on c as

$$\oint_c \mathbf{H} \cdot d\boldsymbol{\ell} = 2\pi\rho H_\phi.$$

By Ampère's Law, this path integral must be equal to I_{enc}, the current passing through the surface bounded by c. This current differs for the three regions of interest. For any location inside the center conductor, defined by $\rho < a$, the total current passing through the surface bounded by c is $\int_s \mathbf{J}_v \cdot d\mathbf{s}$. Since $\mathbf{J}_v$ is uniform and parallel to the surface area element $d\mathbf{s}$, the integral reduces to

$$I_{enc} = J_v \int ds = \pi\rho^2 J_v = I\frac{\rho^2}{a^2}, \qquad \text{for } \rho < a.$$

Setting this equal to $2\pi\rho H_\phi$ and solving for H_ϕ, we find

$$H_\phi = \frac{I\rho}{2\pi a^2}, \qquad \text{for } \rho < a.$$

For the region between the conductors, where $a < \rho < b$, the evaluation of $\oint_c \mathbf{H} \cdot d\boldsymbol{\ell}$ is unchanged, but the current enclosed by the circular path is the *total* current I carried by the center conductor. For this region, H_ϕ is

$$H_\phi = \frac{I}{2\pi\rho}, \qquad \text{for } a < \rho < b.$$

Finally, to find H_ϕ outside of the outer conducting shield, $\rho > b$, we evaluate Ampère's Law using a large-radius circular contour. Since the surface enclosed by this path has the current $+I$ of the center conductor passing through it, as well as the current $-I$ of the conducting shield, the net current is zero, yielding

$$H_\phi = 0 \qquad \text{for } \rho > b.$$

In each region the permeability is μ_0, and we use $\mathbf{B} = \mu_0\mathbf{H}$.

Now we are ready to evaluate the inductance L' by the two methods requested. We start with (a), in which we must evaluate the magnetic flux Φ_M through the shaded surface shown in Fig. 6.8. This surface is in the ρ–z plane, extending from $\rho = 0$ to $\rho = b$, and is of length ℓ. We calculate this flux as

$$\Phi_M = \int_s \mathbf{B} \cdot d\mathbf{s}.$$

The surface element $d\mathbf{s}$ is of the form $\mathbf{a}_\phi\, d\rho\, dz$. (Note: This surface element is completely different from the surface element we used in our determination of the magnetic field $\mathbf{H}$ earlier in this example. In that case, we applied Ampère's Law around a circular path in the $\mathbf{a}_\phi$-direction, and $d\mathbf{s}$ (used to determine the enclosed current I_{enc}) was $\mathbf{a}_z\rho\, d\rho\, d\phi$.) Since $\mathbf{B}$ has a different form for $\rho < a$ and $a < \rho < b$, we must break this integral into two parts, written as

$$\Phi_M = \int_0^\ell \int_0^a \frac{\mu_0 I\rho}{2\pi a^2}\left(\frac{\rho^2}{a^2}\right) d\rho\, dz + \int_0^\ell \int_a^b \frac{\mu_0 I}{2\pi\rho}\, d\rho\, dz. \qquad (6.15)$$

Notice the extra factor of ρ^2/a^2 in the first integral. This factor is included because the flux linking the current in the inner conductor to the current in the outer conductor varies, depending on the location of the current within the inner conductor. We'll return to this point shortly, in SideNote 6.2. Let's pull out factors common to both parts of the flux, and carry out the integration in z,

$$\Phi_M = \frac{\mu_0 I}{2\pi}\left[\int_0^a \frac{\rho^3}{a^4}\, d\rho + \int_a^b \frac{d\rho}{\rho}\right]\ell.$$

Completing the integration, we find for the flux

$$\Phi_M = \frac{\mu_0 I}{2\pi}\left[\frac{1}{4} + \ln\left(\frac{b}{a}\right)\right]\ell. \tag{6.16}$$

From this flux, we calculate the inductance per unit length as

$$L' = \frac{\Phi_M}{I\ell} = \frac{\mu_0}{2\pi}\left[\frac{1}{4} + \ln\left(\frac{b}{a}\right)\right]. \tag{6.17}$$

Let's use parameters for RG-58 C/U co-axial cable, commonly used in laboratory settings, as an example, to see the value of L' that we compute. The diameter of the inner conductor of this cable is $2a = 0.786$ mm, and that of the outer shield is $2b = 2.95$ mm. These parameters yield $L' = 0.31$ µH/m, in good agreement with the value listed in specification charts.

We now move on to part (b), and determine the energy per unit length stored within the magnetic fields using Eq. (6.12); we then find L' using Eq. (6.13). The magnetic energy per unit length stored within the co-axial system, using Eq. (6.12), is

$$W'_m = \frac{1}{2}\frac{1}{\ell}\int_0^\ell \int_s \mathbf{H}\cdot\mathbf{B}\,ds\,dz,$$

where we must break up the integral in ρ into two parts, using the different forms of H_ϕ and B_ϕ valid in the center conductor and in the dielectric medium between the conductors:

$$W'_m = \frac{1}{2}\frac{1}{\ell}\left\{\int_0^\ell \int_0^{2\pi} \int_0^a \left(\frac{I\rho}{2\pi a^2}\right)\left(\frac{\mu_0 I\rho}{2\pi a^2}\right)\rho\,d\rho\,d\phi\,dz \right.$$
$$\left. + \int_0^\ell \int_0^{2\pi} \int_a^b \left(\frac{I}{2\pi\rho}\right)\left(\frac{\mu_0 I}{2\pi\rho}\right)\rho\,d\rho\,d\phi\,dz\right\}.$$

Carrying out the integrals, we find

$$W'_m = \frac{\mu_0 I^2}{4\pi}\left[\frac{1}{4} + \ln\left(\frac{b}{a}\right)\right].$$

Finally, we use Eq. (6.13) to find the inductance per unit length,

$$L' = \frac{2W'_m}{I^2} = \frac{\mu_0}{2\pi}\left[\frac{1}{4} + \ln\left(\frac{b}{a}\right)\right],$$

in agreement with Eq. (6.17).

In Example 6.5 we wrote in Eq. (6.15) an integral equation for the flux linking the current in the inner conductor to the current in the outer shield, and we promised to return to explain this integral more fully. Specifically, we included a factor ρ^2/a^2 in the first integral that can seem a bit arbitrary. Why is this factor inserted here? The complete answer to this question is a bit lengthy, so we'll give a brief response here and expound more fully in the optional SideNote 6.2. Recall that the magnetic flux Φ_M that we need here is the magnetic flux density $\mathbf{B}$ integrated over the surface area between the currents. But this surface area is not perfectly clear, for the following reason. Since the current is *distributed* over the inner conductor, some of the current flows right along the axis of the wire, while other parts of the current pass closer to the edge of the conductor. What surface area are we to use when computing the magnetic flux? The answer is that we use a weighted average to compute the flux, and the factor ρ^2/a^2 in the first integral is simply this weighting factor. We expand on this weighting factor in SideNote 6.2.

SideNote 6.2 Weighting Factor in the Magnetic Flux

To fully appreciate the additional weighting factor in the expression for the magnetic flux, Eq. (6.15), we return to some of the ideas that we used to determine the energy stored in the magnetic field generated by a configuration of currents in Section 6.1. There, we divided the entire current distribution into a large number of current filaments, each running parallel to the others, and each carrying a portion of the total current. To turn on the current, we have to do work against the EMF in that loop, and we showed that the energy stored in the magnetic field can be written as

$$W_m = \frac{1}{2}\sum_k I_k \Phi_{M,k},$$

which we presented as Eq. (6.1). Here, I_k is the current in each filament, and $\Phi_{M,k}$ is the magnetic flux through the surface bounded by that current path. We illustrate one of these current filaments for the co-axial conductors in Fig. 6.9(a). The location of this filament is specified by ρ' and ϕ', its surface area is $\Delta \tilde{s}'_k$, and it carries a current of $I_k = J_z \Delta \tilde{s}'_k$ in the $+z$-direction. Since the current density in the center conductor is uniform, we use $J_z = I/\pi a^2$ for the current density. The magnetic flux passing through the surface between the current filament at ρ' and the return current in the shield, shown as the dashed line in Fig. 6.9(a) and extending a length ℓ in the z-direction, is

$$\Phi_{M,k} = \int_0^\ell \int_{\rho'}^b \mathbf{B} \cdot d\mathbf{s}.$$

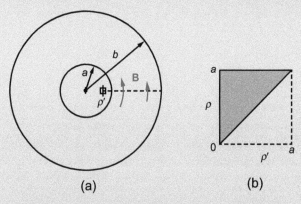

FIGURE 6.9 (a) A current filament in the center conductor (small square box), which returns in the shield of a co-axial cable. (b) An integration aid for determination of $\Phi_{M,k}$.

(Note that this surface is distinct from the surface area $\tilde{s}'_k$ of each filament.) Since $\mathbf{B}$ has a different form for $\rho < a$ and $a < \rho < b$, we divide the integral into two parts,

$$\Phi_{M,k} = \int_0^\ell \int_{\rho'}^a \frac{\mu_0 I \rho}{2\pi a^2}\, d\rho\, dz + \int_0^\ell \int_a^b \frac{\mu_0 I}{2\pi \rho}\, d\rho\, dz. \tag{6.18}$$

Carrying out the integrals, this flux is

$$\Phi_{M,k} = \frac{\mu_0 I \ell}{2\pi}\left[\frac{1}{2}\left(1 - \frac{\rho'^2}{a^2}\right) + \ln\left(\frac{b}{a}\right)\right]. \tag{6.19}$$

Remember that this is the magnetic flux for only one filament of current in the inner conductor, and we still need to extend this to include all such filaments. Before we do this, let's examine the energy W_m to make sure that we're on the right track. We insert this expression for the flux $\Phi_{M,k}$ into Eq. (6.1) to determine the magnetic energy using

$$W_m = \frac{1}{2}\sum_k I_k \Phi_{M,k} = \frac{1}{2}\sum_k J_k \Delta \tilde{s}'_k \Phi_{M,k}.$$

As we presented in Eq. (6.2), when taking the limit as $\Delta \tilde{s}'_k \to 0$, the summation over the individual filaments k becomes an integral over the cross-sectional area of the center conductor, and the energy is

$$W_m = \frac{1}{2}\frac{I}{\pi a^2}\int \Phi_M(\rho')d\tilde{s}',$$

where $d\tilde{s}' = \rho' d\rho' d\phi'$ is the surface element of the cross-sectional area of the inner conductor, and we replaced $\Phi_{M,k}$ with $\Phi_M(\rho')$. Note that we have pulled the current density outside of this integral, which is valid since the current density is uniform. We rewrite this energy as

$$W_m = \frac{1}{2}I\Phi_{M,avg},$$

where $\Phi_{M,avg}$ is the magnetic flux $\Phi_{M,k}$ linking the currents, averaged over the cross-sectional area of the center conductor

$$\Phi_{M,avg} = \frac{1}{\pi a^2}\int \Phi_M(\rho')\, ds'. \tag{6.20}$$

Using $\Phi_{M,k}$ from Eq. (6.19), this is

$$\Phi_{M,avg} = \frac{1}{\pi a^2}\frac{\mu_0 I\ell}{2\pi}\int_{\phi'=0}^{2\pi}\int_{\rho'=0}^{a}\left[\frac{1}{2}\left(1-\frac{\rho'^2}{a^2}\right)+\ln\left(\frac{b}{a}\right)\right]\rho'\, d\rho'\, d\phi',$$

which integrates to

$$\Phi_M = \frac{\mu_0 I\ell}{2\pi}\left[\frac{1}{4}+\ln\left(\frac{b}{a}\right)\right].$$

This result is in perfect agreement with the flux given in Eq. (6.16). It does not, however, satisfy our goal of explaining why we inserted that extra factor of ρ^2/a^2 in the first integral of Eq. (6.15). For this, let's examine the average flux in a slightly different way; that is, we insert the integral form of $\Phi_{M,k}$, given by Eq. (6.18), into Eq. (6.20). This gives us

$$\Phi_{M,avg} = \frac{1}{\pi a^2}\frac{\mu_0 I}{2\pi}\int\left[\int_0^\ell\int_{\rho'}^a\frac{\rho}{a^2}\, d\rho\, dz + \int_0^\ell\int_a^b\frac{1}{\rho}\, d\rho\, dz\right]d\tilde{s}', \tag{6.21}$$

where the common factor of $\mu_0 I/2\pi$ has been pulled outside the integrals. The surface element $d\tilde{s}'$ is, as before, $d\tilde{s}' = \rho'\, d\rho'\, d\phi'$, with limits of integration over the area of the center conductor; for ρ' from 0 to a, and for ϕ' from 0 to 2π. If we can carry out the integration in ρ' and ϕ', we should expect to reproduce Eq. (6.15). Let's see if we can accomplish this. The integration in ϕ' just gives a factor of 2π, of course. So let's focus on the two-dimensional integral in ρ and ρ'. We would like to reverse the order of integration; that is, to first carry out the integration over the $d\rho'$, leaving the integral in $d\rho$. For the second integral inside the square brackets, this is no issue, and we simply reverse the order of integration. For the first integral, however, we have to be a little more careful, since the lower limit of integration for the ρ integral is ρ'. We isolate just this integral as

$$\int_{\rho'=0}^a\left[\int_{\rho=\rho'}^a\frac{\rho d\rho}{a^2}\right]\rho' d\rho',$$

and illustrate the "surface" of this two-dimensional integral as the shaded area in Fig. 6.9(b). Since the lower limit of integration on ρ is ρ', the surface of integration only includes the region above the diagonal, for which $\rho > \rho'$. When we integrate first in ρ', we must integrate from the left side ($\rho = 0$) to the diagonal, so the limits of integration are from 0 to ρ, and the integral becomes

$$\int_{\rho=0}^a\frac{\rho}{a^2}\left[\int_{\rho'=0}^\rho\rho' d\rho'\right]d\rho.$$

We can now carry out the integration in ρ',

$$\int_{\rho=0}^{a} \frac{\rho}{a^2} \left[\frac{\rho'^2}{2} \right]_0^{\rho} d\rho = \int_{\rho=0}^{a} \frac{\rho}{a^2} \left(\frac{\rho^2}{2} \right) d\rho.$$

Inserting this integral into Eq. (6.21) leads us precisely to Eq. (6.15), including the extra factor of ρ^2/a^2. So we see that this extra factor is necessary to account for the distribution of the current over the cross-sectional area of the center conductor, since the magnetic flux linking current at the outer portions of the center conductor does not include the field within the inner regions of the wire.

We move on now to another transmission line geometry, consisting of two long, parallel, flat conducting plates, as shown in Fig. 6.10. For a width w and separation d, with $d \ll w$, and permeability μ_0 in the region between the plates, the inductance per unit length of this geometry is

$$\boxed{L' = \frac{\mu_0 w}{d}}. \tag{6.22}$$

Example 6.6 Inductance per Unit Length of Parallel Plate Conductors

Derive Eq. (6.22).

Solution:

To find the inductance per unit length for this parallel plate geometry, we will determine the magnetic flux density $\mathbf{B}$ in the region between the plates resulting from a current I in the upper plate and the return current of the same magnitude in the lower plate. We then integrate $\mathbf{B}$ across the rectangular $\ell \times d$ shaded surface shown in Fig. 6.10 to determine the flux density Φ_M. The inductance per unit length L' is then

$$L' = \frac{1}{\ell} \frac{\Phi_M}{I}.$$

We must first find the flux density $\mathbf{B}$, which, for $d \ll w$, is approximately uniform throughout the region between the plates, with $\mathbf{B} \approx -\mathbf{a}_y \mu_0 I/w$, and zero elsewhere.

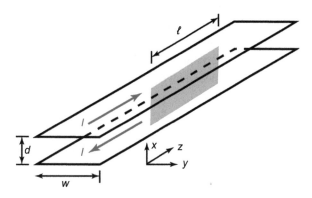

FIGURE 6.10 A pair of plane parallel conducting strips.

We will show this using two different methods. In either case, we depend on results found earlier, so most of the work is already done.

In the first method, we use the result for **B** that we found in Example 4.8; that is, that the flux density generated by a single infinite sheet of current lying in the y–z plane, with the current flowing in the $+z$-direction, is $\mathbf{B} = \mathbf{a}_y \mu_0 J_s / 2$ above the sheet and $\mathbf{B} = -\mathbf{a}_y \mu_0 J_s / 2$ below the sheet. The magnetic flux density generated by the current in the bottom conductor is the same magnitude, but opposite in direction, since the current in the bottom conductor flows in the $-z$-direction. We expect these results to be valid in the present case as long as we are not too close to the edge of the conducting plates. So in the region between the plates, the flux density is $\mathbf{B} = -\mathbf{a}_y \mu_0 J_s / 2$ from the upper plate, and the same from the bottom plate. By superposition, the total field is the sum of these two contributions, $\mathbf{B} = -\mathbf{a}_y \mu_0 J_s$. With a total current in either plate of I, and width w, the current density is $J_s = I/w$, where we assume the current is uniformly distributed across the plate, and the magnetic flux density is $\mathbf{B} = -\mathbf{a}_y \mu_0 I/w$ in the region between the plates. Above or below the two plates, the contributions of the two currents are opposite one another, and the net field is zero.

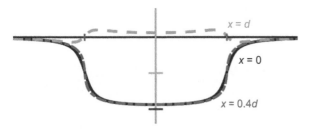

FIGURE 6.11 Plots of B_y vs. y generated by the current in a pair of plane parallel conducting strips. The three lines represent B_y midway between the plates ($x = 0$), between the plates but close to the top plate ($x = 0.4d$), and outside the plates ($x = d$).

Alternatively, we can use the result for **B** that we found earlier in Example 4.4 for a sheet of current of finite width w. We gave this result in Eq. (4.11), and plotted this distribution in Fig. 4.19. Again, in this example we found **B** due to just a single sheet of current, but we can adapt that result to our present example by writing the field due to the top plate, substituting $x \to (x - d/2)$, and the field due to the bottom plate by substituting $x \to (x + d/2)$. We multiply the latter by -1 since the current is in the $-z$-direction, and add the results. The resulting B_y is plotted in Fig. 6.11 at three different elevations: $x = 0$ is exactly in the center between the plates, $x = 0.4d$ is close to the top plate but still in the space between the plates, and $x = d$ is outside the pair of plates. From this plot, the primary field can be seen to be located in the region between the plates, pointing to the left (the $-\mathbf{a}_y$-direction), and is relatively uniform in this region. The small positive B_y component at $x = d$ is consistent with the idea that the field lines for $x > d/2$ circle up and around the upper plate and close on themselves. Field lines for $x < -d/2$ are symmetrical, circling down and around. Outside the plates, therefore, the magnitude of the field is greatly reduced, and by comparison we can ignore the field in these outside regions. We can evaluate the leading terms of this field in the region between the plates by considering $|x| < d/2$ and $|y| < w/2$. In this region, the x-component is very small, and we ignore it, and the y-component in this region is $B_y \to -\mu_0 J_s = -\mu_0 I/w$.

Now that we know **B** generated by the current in the two parallel conductors, we must determine the magnetic flux Φ_M through the rectangular cross-section of length l, and the inductance per unit length L'. The flux is

$$\Phi_M = \int \mathbf{B} \cdot d\mathbf{s} = \int_0^\ell \int_{-d/2}^{d/2} \left(\frac{\mu_0 I}{w} \right) dx \, dz = \frac{\mu_0 I d \, \ell}{w},$$

and the inductance per unit length is

$$L' = \frac{1}{\ell} \frac{\Phi_M}{I} = \frac{\mu_0 d}{w},$$

as we set out to show.

The final geometry that we will analyze is the twin-lead transmission line; that is, a pair of long, parallel wires separated by a uniform spacing d. The wires are each of circular cross-section, of radius a. We already treated this geometry in Example 6.1, in which we solved for the energy W_m. You will find an illustration of a short segment of this parallel wire twin-lead transmission line in Fig. 6.2. The inductance per unit length for this system is

$$L' = \frac{\mu_0}{\pi} \left[\frac{1}{4} + \ln\left(\frac{d}{a}\right) \right], \tag{6.23}$$

as derived in Example 6.7.

Example 6.7 Inductance per Unit Length of a Twin-Lead Transmission Line

Derive Eq. (6.23), the inductance per unit length of a twin-lead transmission line.

Solution:

We will determine the inductance per unit length L' for this parallel twin-lead transmission line using the energy per unit length W'_m and Eq. (6.13). For a current I in the two wires, equal in magnitude but opposite in direction, we found in Example 6.1 that the energy W_m for a finite length is

$$W_m = \frac{\mu_0 I^2 \ell}{2\pi} \left[\frac{1}{4} + \ln\left(\frac{d}{a}\right) \right].$$

Dividing W_m by the length of the wires ℓ yields the energy per unit length, W'_m. Inserting this into Eq. (6.13), we arrive at

$$L' = \frac{2W'_m}{I^2} = \frac{\mu_0}{\pi} \left[\frac{1}{4} + \ln\left(\frac{d}{a}\right) \right]$$

for the inductance per unit length. This is the result that we set out to show.

So we see that we can determine the inductance of a system by either of two methods. We can use either the magnetic flux in the region between the conductors, or the total energy stored in the magnetic field. The former approach is complicated in some cases in which the current is distributed over a region. In these cases, it may be easier to determine the magnetic energy, and from this solve for the inductance. We have also introduced the inductance per unit length L' as a parameter that we will need later in our discussion of transmission lines. Now that we have introduced magnetic energy and inductance, we move on to examine techniques that will allow us to determine the magnitude and direction of forces and torques that can be applied to magnetic systems.

6.3 Magnetic Forces and Torques

LEARNING OBJECTIVE

After studying this section, you should be able to:
- determine the magnetostatic force or torque on a body from the variation of the energy of a system under constant flux or constant current conditions.

We started our entire conversation regarding magnetic fields by considering the force that current-carrying wires exert on each other, and it was through quantifiable observations of the direction and magnitude of these forces that the formalism of magnetic fields was developed. In this section, we return to magnetic forces, exploring them in much more detail.

For the case of the force felt by a current-carrying wire in a magnetic field, the primary means of describing this force will be through use of Eq. (4.4). An alternative form for determining the magnetic force on a body is based upon the energy of a system in various configurations. This is a much more general approach, as it will allow us to determine the magnetic force exerted on a much broader range of objects, not just current-carrying wires. As introduced in Section 2.2, the Principle of Virtual Displacement allows us to calculate the force exerted on a body by determining how a small displacement of that body changes the energy (in this case the magnetic energy) of the system. We will not repeat the derivation presented previously, but rather present the results as they apply to the case of magnetic systems:

$$\boxed{\mathbf{F} = -\boldsymbol{\nabla} W_m \big|_{constant\ \Phi_M}} \tag{6.24}$$

determined under the condition of fixed magnetic flux of the system, or

$$\boxed{\mathbf{F} = \boldsymbol{\nabla} W_m \big|_{constant\ I}} \tag{6.25}$$

when the currents that generate the fields remain constant. To describe these, let's consider a specific system: one consisting of a solenoid with an iron rod of permeability μ and cross-sectional area S, as shown in Fig. 6.12.

Example 6.8 Magnetic Force on a Bar Inside a Solenoid

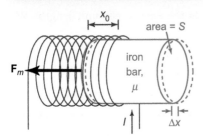

FIGURE 6.12 A solenoid with a movable iron bar.

Approximate the force $\mathbf{F}$ experienced by the iron bar in Fig. 6.12. The iron rod is of permeability μ and cross-sectional area S, and the solenoid is of length ℓ and consists of N turns carrying a current I.

Solution:

The current through the windings generates a magnetic field, the exact pattern of which is difficult to determine with the tools we have developed so far. But we should agree that the magnetic energy stored in the system consisting of the windings and the iron bar depends upon the placement of the bar. We will start by trying to understand the significance of the two conditions attached to Eqs. (6.24) and (6.25). In each expression, we must consider the change in the magnetic energy of the configuration upon an incremental displacement of one of the objects. In this case, we are interested in the force on the iron bar, so it is this object that we allow to move a little. In Eq. (6.24) we compute the change in the energy when the bar moves, holding the magnetic flux Φ_M constant. Alternatively, we can compute the force by examining the change in energy of the system upon a small displacement of the bar while holding the current I constant, using Eq. (6.25). You may recall that to determine the electrostatic forces we also used the gradient of the system energy, but in that case we determined the gradient while holding either the system charges Q constant or holding the system potentials V constant. Those cases corresponded to whether or not the source was connected to the system or not connected. In the present case, the source must remain connected to the solenoid; otherwise it is hard to imagine how the current can continue passing through the windings of the solenoid. But even with the source connected, the critical factor is whether the source *does any work* on the system. For example, if the source provides a current I through the windings, but the magnetic flux does not change, Faraday's Law tells us that no EMF is induced across the windings. Therefore, the power delivered by the source to the system, which is equal to the product of the current I and the EMF, is zero. If the power delivered to the system is zero, then, of course, the work done by the source over that time interval is also zero. Since the source does no work, the work done on the bar by the magnetic force is equal to the change in energy stored in the magnetic field. By the way, if the magnetic flux is fixed while the configuration changes, the current must necessarily change as well. This may seem strange, but keep in mind that the inductance of the system also changes.

Conversely, if the current through the system is constant, it follows that the average value of the magnetic field $\mathbf{H}$ is constant. (Recall that Ampère's Law relates the path integral of $\mathbf{H}$ to the enclosed current I!) In this case, as the position of the iron bar changes, the magnetic flux of the system changes, inducing an EMF across the windings of the system, and the source does work on the system. Equation (6.24) was derived with the work done by the source accounted for, under the condition of constant current.

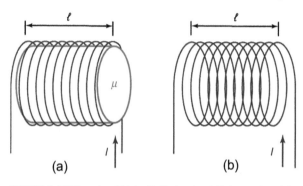

FIGURE 6.13 The solenoid (a) with the iron bar fully inserted, and (b) with the iron bar completely removed.

We stated above that the magnetic force on the iron bar will draw it into the center of the solenoid. We can motivate this through a qualitative comparison of the energy of the system for two configurations of the system. In one configuration, the iron bar is positioned completely inside the solenoid, while in the second the iron bar is completely removed from the solenoid. We call these two configurations, which are illustrated in Fig. 6.13, "bar in" and "bar out," respectively. Now let's compare the relative energy of these two cases. Recall that the energy can be computed as Eq. (6.4),

$$W_m = \frac{1}{2}\int_v \mathbf{H}\cdot\mathbf{B}\,dv.$$

For the purpose of this estimate, we approximate the magnetic field $|\mathbf{H}|$ to be equal to nI inside the solenoid, and zero at any point outside the solenoid. (n is the turns density, equal to N/ℓ.) When the core of the solenoid is air, the permeability in the interior region is μ_0, $\mathbf{B}=\mu_0\mathbf{H}$, and the magnetic flux through the cross-sectional area of the solenoid is $\Phi_M = \int_s \mathbf{B}\cdot d\mathbf{s} \simeq |\mathbf{B}|\,S$. Then the magnetic energy can be approximated by

$$W_m \simeq \frac{1}{2}\frac{1}{\mu_0}|\mathbf{B}|^2 S\ell \simeq \frac{\Phi_M^2\ell}{2\mu_0 S},$$

where ℓ is the length of the solenoid. Now we insert the iron bar into the solenoid, while maintaining the fixed magnetic flux Φ_M. As we discussed previously, the source provides no energy to the system in this case. We ask whether the energy of the system increases or decreases upon insertion of the iron bar. Since Φ_M is unchanged, the magnetic field $|\mathbf{B}|$ is unchanged. On the other hand, $\mathbf{H}$, which is equal to $\mathbf{B}/\mu$, where μ is the permeability of the iron bar, must change. Specifically, it must decrease significantly. The magnetic energy of the system after insertion of the iron bar is

$$W_m \simeq \frac{1}{2}\frac{1}{\mu}|\mathbf{B}|^2 S\ell \simeq \frac{\Phi_M^2\ell}{2\mu S}.$$

Since μ for the iron bar is much greater than μ_0 of the air core, we conclude that, for a fixed magnetic flux Φ_M, the energy of the solenoid with the iron bar inserted is much less than that of the solenoid with the iron bar removed. Systems always tend toward configurations of lower potential energy, from which we conclude that the magnetic force on the iron bar draws it into the center of the solenoid, where its energy is minimized. Note that in order for the magnetic flux to remain constant, $\Phi_M = \mu_0 nI_{out}S$, where I_{out} is the current through the windings of the solenoid when the iron bar is absent, must be equal to $\Phi_M = \mu n I_{in} S$, where I_{in} is the current through the windings of the solenoid when the iron bar is inside the solenoid. Therefore, in order for the magnetic flux to remain constant, the current will decrease by an enormous factor μ_r, the relative permeability or the iron, upon insertion of the iron core.

Now let's estimate the magnitude of the magnetic force on the bar. In order to apply Eq. (6.24) or (6.25), we must estimate the change in the magnetic energy of the system resulting from a small displacement Δx of the iron bar. We can work this under constant flux Φ_M or constant current I conditions; either way is equally valid. In this example, we will work this holding the current from the source fixed. We must also be clear about the sign of Δx. We define a positive Δx to mean that the iron bar moves a little bit *into* the solenoid.

As we said above, the field pattern with the iron bar partially inserted is very complex, and we will not attempt a precise calculation. Still, we can arrive at a reasonable approximation using the following estimates. In the regions of the interior of the solenoid that are far from the ends of the windings, and far from the end of the iron bar, we can apply Ampère's Law to show that the magnitude of **H** is approximately nI, where n is the turns density and I the current in the windings. (This approximation is only valid in the regions specified, since we needed the symmetry of an infinitely long solenoid to argue that the magnetic field is uniform and directed parallel to the axis. Clearly this symmetry breaks down close to the ends.) Near the center of the left half of the solenoid in Fig. 6.12, $|\mathbf{H}| \sim nI$ and $|\mathbf{B}| \sim \mu_0 nI$, and near the center of the right half of the solenoid, $|\mathbf{H}| \sim nI$ is the same, but $|\mathbf{B}| \sim \mu nI$ is much larger. Also, we expect that along the axis of the solenoid, where **B** is normal to the end of the iron bar, the boundary condition that B_n must be continuous, Eq. (4.42), tells us that $|\mathbf{B}|$ must decrease smoothly and monotonically from $\sim \mu nI$ on the right side to $\sim \mu_0 nI$ on the left.

We haven't tried to work out precisely how $|\mathbf{B}|$ transitions in this region, but for the purpose of this estimate of the change in magnetic energy of the system, we really don't need to know this. The important points are: (1) we can reasonably estimate $|\mathbf{H}|$ and $|\mathbf{B}|$ in the regions in the middle of the air-filled side and the iron-filled side of the solenoid; and (2) after the bar moves an incremental distance Δx to the left, the volume of the air-filled side is decreased by $S\Delta x$ and the volume of the iron-filled side is increased by the same amount. Since the energy density in either region is $w_m = \frac{1}{2}\mathbf{H} \cdot \mathbf{B}$, the magnetic energy stored within the bar increases by $w_m S \Delta x = \frac{1}{2}\mu(nI)^2 S \Delta x$, and the magnetic energy stored within the air-filled side decreases by $w_m S \Delta x = \frac{1}{2}\mu_0(nI)^2 S \Delta x$. Therefore, the change in the magnetic energy of the system is approximately

$$\Delta W_m = \frac{1}{2}(\mu - \mu_0)(nI)^2 S \Delta x.$$

Note that the energy stored in the field of the solenoid *increases* when Δx is positive – that is, the iron bar moves a bit further into the solenoid. (Remember, the source does work on the solenoid under this constant current condition.) Finally, we use Eq. (6.25) to determine the force on the iron bar:

$$\mathbf{F}_m = \nabla W_m \big|_{constant\, I} = \frac{\Delta W_m}{\Delta x}\, \mathbf{a}_x = \frac{1}{2}(\mu - \mu_0)(nI)^2 S\, \mathbf{a}_x.$$

This force is in the $+x$-direction, drawing it into the center of the solenoid, consistent with our conclusion above.

In this example, we used the magnetic energy of the system, or rather the variation of the magnetic energy with position of the iron bar, to approximate the force drawing the bar into the center of the solenoid. It would be good to form a more intuitive picture of this magnetic force as well. For this, let's consider the force between two circular, co-axial wire loops, carrying currents I_1 and I_2, respectively. If the currents in these two loops are both in the same direction, then the force on these loops is attractive, while if the currents are in opposite directions, then the loops repel one another. (We can reach this conclusion by considering the magnetic force, given by Eq. (4.1), between two wire segments, which is attractive when the currents are parallel to one another. In a wire loop, the direction of the current varies around the loop, but the largest contribution to the force comes from the closest segments.) How can we use this force between current loops to understand the force on the bar? The current in the solenoid in Fig. 6.12 (viewed from the right side) is in the counter-clockwise direction. This current produces a magnetic field **H** in the interior of the solenoid that points to the right. The magnetic field applied to the bar causes the magnetic dipoles in the iron bar to line up in the same direction, creating a magnetization **M**, also pointing to the right. The effect of this magnetization can be understood in terms of the effective current induced at the surface of the iron bar. In this case, this effective current is in the counter-clockwise direction. So the real currents in the solenoid, and the effective current in the surface of the iron bar, are circulating with the same sense, and are therefore attractive. If the bar is only partially inserted into the interior of the solenoid, there are more current loops on the left side pulling the bar into the center than there are current loops on the right, and the net force on the bar pulls it into the center. In this way, we can understand at least the direction of the magnetic force on the iron bar. (With a great deal of effort, we could probably also use this model to estimate the magnitude of this force, but we choose to move on at this point to our next example.)

Let's now consider the force that a simple U-shaped electromagnet can exert on a bar, as shown in Fig. 6.14. In Example 6.9 the magnetic core and the bar that it lifts serve to guide the magnetic flux lines around the loop, and we can use this feature to make a reasonable estimation of the field strength within the permeable materials.

Example 6.9 Electromagnet

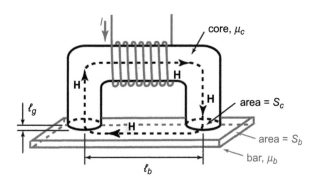

FIGURE 6.14 A U-shaped electromagnet picking up a bar made of permeable material.

Estimate the magnetic force exerted by the U-shaped electromagnet on the iron bar shown in Fig. 6.14. The U-shaped magnet is of high permeability μ_c, where the subscript c designates the core, or central part of the electromagnet. A current-carrying wire of N windings and current I is wrapped around the central region of the core. The cross-sectional area of the core is S_c from end to end. The bar has permeability μ_b (the subscript b designates the bar), and the cross-sectional area of the bar is S_b. The spacing ℓ_g between the bar and the electromagnet is small.

Solution:

Our goal is to determine the lifting capacity of this simple structure, and the approach we use will be similar to that of Example 6.8. We will use principles that we discussed earlier to approximate the magnitudes of **H** and **B** in different regions of the system, and then use these field strengths to estimate the magnetic energy of the system. The current through the windings will generate a magnetic field $\mathbf{H}_c$ in the core region. With the current I in the direction shown in Fig. 6.14, $\mathbf{H}_c$ is directed to the right in this region. As discussed in Example 6.2, these field lines are "guided" by the permeable core material down toward the flat surface of the core. When the air gap between the U-shaped electromagnet and the bar is small, the field lines are guided by the bar over to the opposite face, where they re-enter the U-shaped magnet, and close on themselves. In order to estimate the field strengths, we start with the integral condition $\oint \mathbf{B} \cdot d\mathbf{s} = 0$, valid on any closed surface, to argue that $|\mathbf{B}_c|$, the magnitude of the magnetic flux density in the core region, is uniform throughout the core. (The cross-sectional area of the core is uniform.) Let's use the simpler notation B_c for $|\mathbf{B}_c|$. Similarly, B_b, the magnitude of the magnetic flux density in the bar, is uniform throughout the bar, at least in the regions between the two pole faces of the core. Furthermore, we use the condition $\oint \mathbf{B} \cdot d\mathbf{s} = 0$ to argue that $B_c S_c \sim B_b S_b$. (As long as the gap length ℓ_g is small, the flux lines go from the core to the bar with little "leakage," or loss of flux.) Since the normal component of **B** must be continuous at any interface, we can also state that the magnetic flux density in the gap region between the core and the bar, B_g, must be the same as B_c. Next we relate the strengths of H in each of these regions, since in each region $H = B/\mu$. To summarize, then, we have

$$B_c = \mu_c H_c$$
$$B_g = \mu_0 H_g = B_c \tag{6.26}$$
$$B_b = \mu_b H_b = B_c S_c / S_b.$$

Now we can apply Ampère's Law around the path shown by the dotted line in Fig. 6.14:

$$\oint_c \mathbf{H} \cdot d\boldsymbol{\ell} = NI \simeq H_c \ell_c + H_g \ell_g + H_b \ell_b + H_g \ell_g,$$

where ℓ_c is the length of the core of the electromagnet from one pole face to the other, ℓ_b is the effective length of the field within the bar, which we approximate as the distance from the center of one pole face of the core to the other, and ℓ_g is the gap spacing, which we take to be equal on the two sides. Using the relations of Eq. (6.26), this can be written as

$$NI = B_c S_c \left(\frac{\ell_c}{\mu_c S_c} + \frac{2\ell_g}{\mu_0 S_c} + \frac{\ell_b}{\mu_b S_b} \right).$$

We can turn this around and solve for the flux in the core region $\Phi_c \simeq B_c S_c$, giving us

$$B_c S_c = \frac{NI}{\left(\ell_c / \mu_c S_c + 2\ell_g / \mu_0 S_c + \ell_b / \mu_b S_b \right)}. \tag{6.27}$$

Next we estimate the energy contained in the magnetic field of this system:

$$W_m = \frac{1}{2} \int_V \mathbf{H} \cdot \mathbf{B} \, dv,$$

where the volume integral is carried out over all space. Fortunately, the fields are non-zero only in the core, the air gaps, and the bar, and we can estimate this total energy as

$$W_m \simeq \frac{1}{2} \left[B_c H_c S_c \ell_c + 2 B_g H_g S_g \ell_g + B_b H_b S_b \ell_b \right].$$

Using the uniformity of the fields, and using $H = B/\mu$, this magnetic energy is approximately

$$W_m \simeq \frac{1}{2} \left[\frac{B_c^2}{\mu_c} S_c \ell_c + \frac{2 B_g^2}{\mu_0} S_g \ell_g + \frac{B_b^2}{\mu_b} S_b \ell_b \right],$$

and using the relations of Eq. (6.26) and $S_g \simeq S_c$, this energy is

$$W_m \simeq \frac{B_c^2 S_c^2}{2} \left[\frac{\ell_c}{\mu_c S_c} + \frac{2 \ell_g}{\mu_0 S_c} + \frac{\ell_b}{\mu_b S_b} \right].$$

We use Eq. (6.27) to write this in terms of the current I:

$$W_m \simeq \frac{N^2 I^2}{2 \left[\ell_c/\mu_c S_c + 2 \ell_g/\mu_0 S_c + \ell_b/\mu_b S_b \right]}.$$

This expression for the energy of the electromagnet when current I passes through the windings allows us to determine the magnetic force. Let's use z as the vertical position of the bar, and determine the change in energy of the system upon an incremental upward displacement of the bar. We apply Eq. (6.25), which for a displacement in the z-direction reads

$$\mathbf{F} = \boldsymbol{\nabla} W_m = \mathbf{a}_z \frac{\partial}{\partial z} W_m.$$

An increment in z, raising the bar, decreases the gap size by the same amount, so we substitute $\partial/\partial z \to -\partial/\partial \ell_g$, and write

$$\mathbf{F} = -\mathbf{a}_z \frac{\partial}{\partial \ell_g} \left[\frac{N^2 I^2}{2 \left[\ell_c/\mu_c S_c + 2 \ell_g/\mu_0 S_c + \ell_b/\mu_b S_b \right]} \right],$$

and carrying out the differentiation, we find

$$\mathbf{F} = \mathbf{a}_z \frac{N^2 I^2 \left(2/\mu_0 S_c \right)}{2 \left[\ell_c/\mu_c S_c + 2 \ell_g/\mu_0 S_c + \ell_b/\mu_b S_b \right]^2}.$$

This force is greatest, logically, when the core is directly in contact with the bar (i.e. $\ell_g = 0$), and the strongest electromagnets have large cross-sectional areas S_c and high permeability μ_c. Under these conditions, the first two terms in the denominator become small, and the force is approximately

$$\mathbf{F} \sim \mathbf{a}_z \frac{N^2 I^2 \mu_b^2 S_b^2}{\ell_b^2 \mu_0 S_c}.$$

This is an upward force on the bar, indicating that bar is indeed attracted to the electromagnet, consistent with expectation.

Magnetic forces on large objects have important applications, such as in electromagnets (described quantitatively in Example 6.9), magnetic levitation, and switchable magnetic bases. These applications are described in TechNotes 6.2–6.4.

TechNote 6.2 Electromagnets

Electromagnets are the working element behind scrap metal separators, cranes, relays, doorbells, many forms of ringers and buzzers, acoustic speakers, and steering and collimating magnets for beams of charged particles. A common form of electromagnet consists of a simple helical winding of wire with a central soft-iron plunger. The strongest, continuous magnetic field generated on Earth by a resistive (non-superconducting) magnet was 37.5 T, produced at the center of a Bitter magnet at the Radboud University High Field Magnet Laboratory in Nijmegen, the Netherlands. A Bitter magnet, named after its inventor Francis Bitter, is a special design in which the windings are replaced by copper discs fitted with water-cooling holes. This design gives the magnet superior mechanical strength to prevent it from expanding, and provides water-cooling conduits to carry away the massive amounts of heat generated as the current passes through the conductor. A hybrid system consisting of a Bitter magnet inside a superconducting magnet at the National Magnetics Laboratory in Tallahassee, FL can produce a field strength of 45 T. The superconducting beam-steering magnets at the Large Hadron Collider (LHC) at CERN in Switzerland produce a field strength of 8.3 T over an extended region. For comparison, the strongest field produced by a permanent magnet is 0.8 T, or 8 000 G, produced by an alnico (a compound of aluminum, nickel, and cobalt) magnet, and a strong refrigerator magnet produces a flux density of 0.01 T, or ~100 G.

Magnetic forces can be used to cause large objects, such as permanent magnets or high-speed trains, to float, without physical contact with the base or railroad tracks, thus minimizing frictional losses. Magnetic levitation is described in TechNote 6.3.

TechNote 6.3 Magnetic Levitation

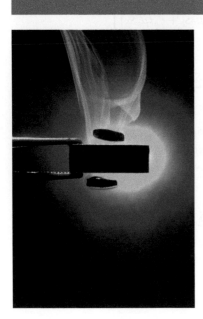

Magnetic levitation (maglev) is attractive in many applications due to the absence of frictional losses. Maglev transportation systems, such as the train at Pudong International Airport shown in Fig. 6.15, are currently operating in China, Japan, and South Korea. Magnetic bearings provide frictionless operation for high-speed rotation, such as in high-speed vacuum pumps.

FIGURE 6.15 A magnetically levitated train near Pudong International Airport, Shanghai, China.

In any application of magnetic levitation, it is necessary to (1) provide enough lift to overcome gravity, and (2) stabilize the system mechanically to prevent it from flipping over. A variety of technologies are used in various systems, some based on high-field permanent magnets, others based on strong electromagnets, including superconducting magnets.

A simple demonstration of one form of magnetic levitation is shown in Fig. 6.16(a), which shows a magnet floating over a cuprate type $YBa_2Cu_3O_7$ superconductor cooled to 77 K ($-196\,°C$). As the magnet first approaches the superconductor, its flux lines attempt to penetrate into the superconductor. In response, the superconductor develops surface eddy currents that prevent the flux lines from the magnet from passing to the interior of the superconductor. The magnetic flux density generated by these surface currents exerts a force on the magnet, which is sufficient to balance the force of gravity on the magnet, and the magnet is levitated over the superconductor. The example shown in Fig. 6.16(b) may be helpful in understanding this levitation. Consider a magnet above the superconductor (SC), with its north pole pointed down, such that the magnetic flux lines are as shown. The effective surface currents $J_{s,eff}$ on this magnet are shown. (Looking from above, these currents are in the clockwise direction.) As the magnet approaches the superconductor, the flux lines induce the surface current J_s, which by Lenz' Law are in a direction such that they oppose the change in the flux through the surface of the super-conductor. This induced current is in the counter-clockwise direction when viewed from above. Currents flowing in opposite directions to one another create a repulsive force, which pushes the magnet upward. This magnetic force opposes the gravitational force on the magnet, and the magnet floats above the superconductor. Since the superconductor has no resistance, the induced current is persistent, and the magnet will continue to float indefinitely.

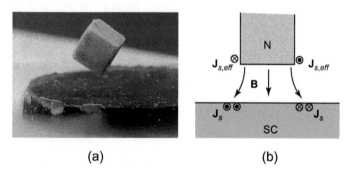

(a) (b)

FIGURE 6.16 (a) Magnetic levitation of a rare earth permanent magnet over a superconductor cooled to 77 K (-196°C). (Photograph used with permission.) (b) Sketch of currents and magnetic flux density of a magnet levitated over a superconductor (SC). As the magnet approaches the superconductor, it induces surface currents in the latter, marked J_s. These induced currents apply an upward magnetic force on the magnet, holding it in levitation.

Magnetic bases, described in TechNote 6.4, are used in machine shops and research laboratories for rigid mounting and rapid relocation of components. The magnetic attraction of these bases to an iron surface can be turned on or off with the flip of a mechanical switch.

TechNote 6.4 Magnetic Base

A magnetic base, or a magnetic switchable device, allows for firm and rapid mounting or dismounting of components in a workshop or laboratory setting. These magnets can be easily turned on or off to attach or release the base from a surface of iron or other magnetizable material. The base consists of a magnetized cylinder (the yellow circle in Fig. 6.17) between two iron pieces (dark gray). The filler (light gray) between the iron pieces is of non-ferrous metal, such as brass or aluminum. When the switch is in the "Off" position, as in Fig. 6.17(a), the north and south pole faces of the magnetized cylinder are vertical, and the magnetic field lines emerge from the north pole of the magnetized cylinder, pass through the non-ferrous filler, circle down through the iron sides, and return to the south pole through the non-ferrous filler. Few magnetic field lines extend into the iron platform (not shown) on which the base sits. When the switch is rotated to the "On" position, however, the field lines that emerge from the north pole face circle down through the iron on the left side, cross over through the iron platform beneath the base (again, not shown), and return to the south pole through the iron on the right side. In this configuration, the magnetic field is very strong, and the magnetic energy stored in the fields must increase significantly if the base is to be separated from the iron platform, resulting in an extremely large force holding the base firmly to the platform.

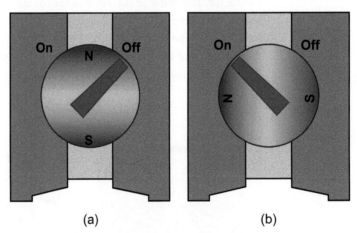

(a)   (b)

FIGURE 6.17 Magnetic bases in the (a) 'Off' and (b) 'On' positions.

The magnetic forces in Examples 6.8 and 6.9 would likely result in a translational motion of a body. For the iron bar in the solenoid, the bar is drawn into the interior of the solenoid. With the U-shaped electromagnet, the iron bar is attracted to the electromagnet. Magnetic forces can also be used to apply a torque to a system, causing it to rotate, as in a current loop in a constant magnetic field, shown in Fig. 6.18. This, of course, is the basic model of an electric motor, and by analyzing the forces exerted on each segment of this current loop, motor operation can be understood. In Example 6.10 we examine the torque on a current loop.

Example 6.10 Torque

A current I passes through a rectangular loop of wire, which is of dimensions a by b, and is free to rotate about the z-axis. The total number of turns in the current loop is N.

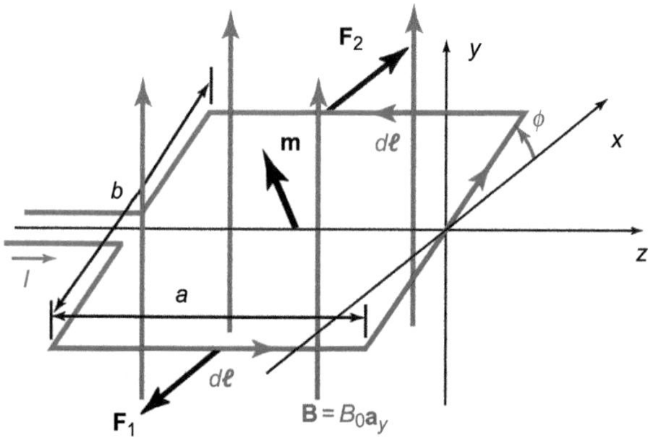

FIGURE 6.18 A current loop in a constant magnetic field. (For use in Example 6.10.)

(For clarity, only a single loop is shown in Fig. 6.18.) Determine the torque felt by this wire loop.

Solution:

As discussed previously, near Eq. (4.4), the magnetic force exerted on a segment $d\boldsymbol{\ell}$ of wire carrying a current I is

$$d\mathbf{F} = I \, d\boldsymbol{\ell} \times \mathbf{B}(\mathbf{r}).$$

Applied to the front segment of the loop (called segment 1), where $d\boldsymbol{\ell}$ is $\mathbf{a}_z \, dz$, and using $\mathbf{B} = B_0 \mathbf{a}_y$, the force is

$$\mathbf{F}_1 = NI \int_{-a}^{0} (\mathbf{a}_z \, dz) \times \left(B_0 \mathbf{a}_y \right) = -NIaB_0 \mathbf{a}_x.$$

The factor N is included since there are N turns in the wire loop. On the rear segment (called segment 2) the analysis is similar, but now we integrate from 0 to $-a$, since the current in this branch flows in the $-z$-direction. The force on this segment is

$$\mathbf{F}_2 = NI \int_{0}^{-a} (\mathbf{a}_z \, dz) \times \left(B_0 \mathbf{a}_y \right) = +NIaB_0 \mathbf{a}_x,$$

equal in magnitude but opposite in direction to $\mathbf{F}_1$. The two segments of the loop in the x–y planes (the ends) feel a force in the $\pm \mathbf{a}_z$-direction, but no torque, and the two forces are equal in magnitude but opposite in direction. We will consider them no further.

Since $\mathbf{F}_1$ and $\mathbf{F}_2$ are equal in magnitude but opposite in direction, there is no net force on this current loop. There is, however, a **torque** on the loop, causing it to spin about the z-axis. To find this torque, we use $\mathbf{T} = \mathbf{r}_1 \times \mathbf{F}_1 + \mathbf{r}_2 \times \mathbf{F}_2$, where $\mathbf{r}_1$ and $\mathbf{r}_2$ are the position vectors pointing from the axis of rotation to the wire segments. These vectors can be written as

$$\mathbf{r}_1 = -\mathbf{a}_x(b/2)\cos\phi - \mathbf{a}_y(b/2)\sin\phi$$

and

$$\mathbf{r}_2 = \mathbf{a}_x(b/2)\cos\phi + \mathbf{a}_y(b/2)\sin\phi = -\mathbf{r}_1,$$

where b is the width of the wire loop. The contributions to the torque from the two sides, $\mathbf{r}_1 \times \mathbf{F}_1$ and $\mathbf{r}_2 \times \mathbf{F}_2$, are equal, and the total torque on the wire loop about the z-axis is

$$\mathbf{T} = 2\mathbf{r}_1 \times \mathbf{F}_1 = 2 \begin{vmatrix} \mathbf{a}_x & \mathbf{a}_y & \mathbf{a}_z \\ -\dfrac{b}{2}\cos\phi & -\dfrac{b}{2}\sin\phi & 0 \\ -NIaB_0 & 0 & 0 \end{vmatrix} = -\mathbf{a}_z NIabB_0 \sin\phi.$$

This torque in the $-\mathbf{a}_z$-direction causes the current-carrying loop to accelerate in the $-\phi$-direction when $\sin\phi$ is positive. The wire loop will accelerate about this axis until $\sin\phi$ becomes negative (i.e. $\phi < 0$), at which point the torque reverses and points in the $+\mathbf{a}_z$-direction, slowing down the rotational motion. This behavior is not terribly useful for real d.c. motors, so motors typically employ a means of reversing the direction of the current through the wire loop to keep the loop spinning in the same direction. (See TechNote 6.5.) This current-carrying loop is the basis for conversion of electrical to mechanical energy as described.

From the previous equation, we see that the torque **T** depends on several factors, including several that characterize the loop: the dimensions of the loop, the current through the loop, and the orientation of the loop relative to the external magnetic flux density. This should remind us of our previous discussion of the magnetic moment **m** of a current loop, which we defined (see Section 4.5) as $\mathbf{m} = (NIS)\mathbf{a}_n$, where S is the area of the loop and $\mathbf{a}_n$ is the surface unit normal vector. Remember, the surface normal is always defined relative to the direction of the current in the loop using the right-hand rule. In the present case, with the current in the direction shown, $\mathbf{a}_n = -\mathbf{a}_x \sin \phi + \mathbf{a}_y \cos \phi$. You should convince yourself that this normal vector is correct by studying Fig. 6.18, on which the magnetic moment vector **m** is labeled. With this definition of **m**, an alternative definition of the torque on the wire loop as $\mathbf{T} = \mathbf{m} \times \mathbf{B}$ gives the same result as we found previously.

We state without proof that the potential energy of a magnetic dipole in a magnetic field **B** is

$$\boxed{W_m = -\mathbf{m} \cdot \mathbf{B}.} \tag{6.28}$$

Let's examine this energy to see what we can learn from it, specifically what it tells us of the behavior of the dipole in a magnetic field **B**. First, the energy of the magnetic dipole is minimized when **m** points in the same direction as **B**, corresponding to $\phi = 0$ in Fig. 6.18. In this orientation, the wire loop is perfectly stable, just as a marble sitting at the bottom of a bowl is stable. A small angular displacement of the wire loop in one direction or the other increases W_m a bit. Since systems always seek to minimize their potential energy, we should expect this displacement will result in a torque that tends to restore the orientation of the dipole to $\phi = 0$. This is consistent with the torque $T = -\mathbf{a}_z NabIB_0 \sin \phi$ that we found above. If there are no mechanical losses, of course, the dipole will oscillate back and forth *ad infinitum*. But add a little friction, and the wire loop will settle back to steady state with **m** aligned with **B**. At the other extreme, when **m** is anti-parallel to **B**, the torque is zero at this precise point, but any slight disturbance to one side or the other introduces a torque that further accelerates the rotation of the current loop. We say this is an unstable configuration.

The potential energy $W_m = -\mathbf{m} \cdot \mathbf{B}$ also presents us a means of calculating the torque on the wire loop. Earlier we used the energy of a system in order to determine the magnetic *force* it will feel, expressed in terms of the gradient of the magnetic potential energy. A similar expression for the *torque* the system feels is

$$\boxed{T_z = \frac{\partial W_m}{\partial \phi},} \tag{6.29}$$

when determined under the condition of constant current I. With $W_m = -\mathbf{m} \cdot \mathbf{B} = -mB_0 \cos \phi$, we find $T_z = -mB_0 \sin \phi$, in agreement with our result in the previous paragraph.

Electric motors can be designed to operate on direct or alternating current, as described in TechNotes 6.5 and 6.8.

TechNote 6.5 d.c. Motors

We have just finished showing how a current-carrying loop of wire in a permanent magnetic field experiences a torque, tending to rotate the wire loop to align its magnetic moment with the external magnetic field. Our discussion to this point, however, falls short of telling us how to construct a working electric motor, which we expect should be able to provide a continuous rotational motion about an axis, hopefully at a somewhat constant angular velocity. How can we achieve this goal? In this TechNote, we'll examine a d.c. motor.

Figure 6.19 shows a d.c. motor with many of the same working parts that we discussed above. In place of a single loop, we see multiple windings (in blue), with a common current passing through each. This loop is called the rotor, since it rotates about the axis of the motor. The magnets, labeled N and S, produce a magnetic flux density pointing from left to right. With the current passing through the windings in the counter-clockwise direction, as viewed from the right side, the rotor accelerates in the clockwise direction, as indicated by the round blue arrow. In terms of the magnetic moment of the rotor, $\mathbf{m}$, which points up and to the right along the axis of the rotor coils, the torque tries to align $\mathbf{m}$ with $\mathbf{B}$. Let's label the angle between $\mathbf{m}$ and $\mathbf{B}$ as ϕ. When ϕ reaches zero, the rotor continues turning, but it starts slowing down since the torque changes direction after the rotor passes through $\phi = 0$. If we could reverse the direction of the current through the rotor coils at this moment, however, then the torque would be maintained in the same direction, and the rotor would continue to accelerate in the clockwise direction. This is exactly the function of the commutator, shown as the pair of half-rings around the rotation axis of the rotor. The contacts abutting the commutator are called brushes, through which the d.c. potential is applied to the windings. As the rotor passes through $\phi = 0$, the commutator and brushes reverse the direction of the current in the rotor, and the rotor continues to accelerate in the clockwise direction. The speed of the motor is controlled through the voltage of the source providing the current to the rotor. A d.c. motor can rotate in either direction, often depending on whether ϕ is positive or negative at the instant the source is applied, and also on the polarity of the d.c. source. If the operation of the d.c. motor were reversed, using it to generate power, the EMF across the commutator would appear as a rectified sinusoidal voltage, whose amplitude increases with increasing rotational velocity.

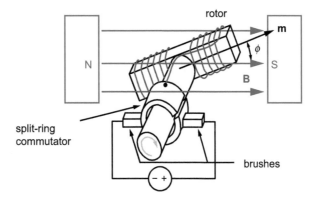

FIGURE 6.19 Schematics of a d.c. motor.

Galvanometers are described in TechNote 6.6. These devices produce an angular deflection proportional to a current through a winding, and were commonly used in analog ammeters.

TechNote 6.6 Galvanometer (Including Analog Voltmeter and Ammeter)

A galvanometer is an instrument that produces an angular deflection whose magnitude and direction are proportional to the magnitude and sign of the current passing through the coil windings. Analog voltmeters, analog ammeters, and rotary control devices are based on galvanometers. The first galvanometer was built by Johann Schweigger, a German mathematician and physicist, just months after Hans Christian Oersted discovered that a current-carrying wire caused a deflection of a nearby magnetic compass. (See Biographical Note 4.1.) Georg Ohm used a simple home-built galvanometer to measure the current through long wires produced by a thermoelectric voltage source, leading to what we today call Ohm's Law, which he reported in 1827. (See Biographical Note 3.1.) The French physicist Jacques Arsene D'Arsonval developed the first moving-coil galvanometer, shown in Fig. 6.20. The galvanometer is named in honor of Luigi Galvani.

When current passes through the coil in the galvanometer, it produces a torque on the central core. (Recall that a magnetic moment always tries to line up with the external magnetic field, which in this case is created by the north and south pole faces of a permanent magnet.) The central core is spring-loaded to oppose the deflection, and reaches a steady-state deflection when the magnetic torque is equal but opposite to the torque of the spring. A pointer that rotates with the central core allows one to read the deflection on a scale. The scale often includes a mirror to aid in reading the deflection. Voltmeters and ammeters were long based upon the galvanometer, but these have largely been replaced by electronic instruments. Galvanometers for controlling the orientation of objects mounted on their axes are still in common use.

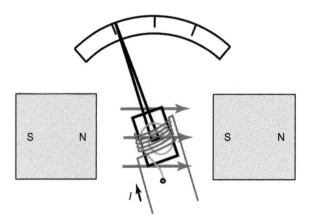

FIGURE 6.20 A schematic diagram of a galvanometer. Current through the windings generates a torque on the rotor.

Magnetic resonance imaging (MRI) is widely used in medical diagnostics. Although based on quantum effects (nuclear spins), some aspects of MRI can be understood in terms of the classical energy of a magnetic dipole in a magnetic field. MRI is described in TechNote 6.7.

TechNote 6.7 Magnetic Resonance Imaging

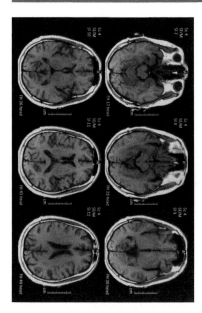

Magnetic resonance imaging is a technique for creating an image of the internal organs, joints, and tissue of the human body using measurements based on a technique known as nuclear magnetic resonance (NMR). Nuclear magnetic resonance is based on the precession of a proton, or any nucleus having non-zero "spin" in a static magnetic field. Since water is prevalent in the human body, and every water molecule contains two hydrogen atoms, MRI instruments use the precession of the proton to generate the image. A particle's spin is a form of angular momentum, which can be thought of in terms of its classical analog, in which the proton is a sphere and the spin represents the rotation of this sphere about its own axis. Due to the distribution of charge in a proton, a spinning proton represents an electrical current, and the spin angular momentum is related to a magnetic moment $\mathbf{m}_p$ of the proton. As discussed above, the energy of a magnetic moment in a static magnetic field $\mathbf{B}$ is $W_m = -\mathbf{m}_p \cdot \mathbf{B}$. In a static magnetic field, quantum physics tells us that there are two spin states of the proton: the low-energy state in which $\mathbf{m}_p$ is aligned with $\mathbf{B}$, and the high-energy state with $\mathbf{m}_p$ opposed to $\mathbf{B}$. In steady-state conditions, the proton magnetic moment $\mathbf{m}_p$ tends to the low-energy state, with its spin parallel to the magnetic field. By applying a radio frequency wave whose frequency is resonant with the transition between the two spin states, the proton can be excited to the higher-energy state. This frequency is 21.3 MHz in a 1 T field, and the amplitude of $\mathbf{B}$ applied in MRI is typically in the range of 0.5–1.5 T. (The magnetic moment of a proton is $|\mathbf{m}_p| = 14.1 \times 10^{-27}$ J/T.) Upon application of an inhomogeneous magnetic field to the body, the protons in different regions of the body absorb at different frequencies. The NMR frequency also depends on the local environment of the proton, so different tissues and organs can be distinguished.

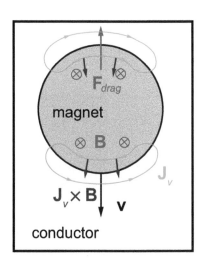

FIGURE 6.21 The current $\mathbf{J}_v$ induced in a conductor by a moving magnet (blue traces), and the resulting force felt by the conductor (proportional to $\mathbf{J}_v \times \mathbf{B}$). The magnetic drag force $\mathbf{F}_{drag}$ felt by the magnet is equal but opposite to the force on the conductor, as required by Newton's Third Law.

In the examples of the TechNotes 6.5–6.7, the mechanical torque on the system is due to the magnetic moment of a current loop and an external magnetic field. In each of TechNotes 6.8–6.10, the force or torque results from an induced current produced by a time-varying magnetic field. To understand these applications, it is instructive to first introduce the notion of a drag force between a moving magnet and a conductor. Consider, for example, the magnet moving down the conductor shown in Fig. 6.21. The magnetic flux density $\mathbf{B}$ is strongest across the face of the magnet, and in this figure is oriented into the conductor. As the magnet moves downward, the magnetic flux density at a point in the conductor increases as the magnet approaches that point, producing a circulating electric field $\mathbf{E}$ (by Faraday's Law), and an induced current density $\mathbf{J}_v = \sigma\mathbf{E}$. This current density is shown in blue in the figure. The local force density on the conductor is given by $\mathbf{J}_v \times \mathbf{B}$. By Newton's Third Law, the force on the magnet is equal in magnitude but opposite in direction to the force on the conductor. An equal drag force results near the trailing edge of the magnet. Because of this drag force, a strong magnet allowed to slide down an inclined conducting sheet is seen to move very slowly, even though there is no magnetic attraction between a stationary magnet and the conductor. TechNotes 6.8–6.10 highlight applications of forces or torques involving induced currents.

TechNote 6.8 a.c. Induction Motors

An a.c. motor is quite different from a d.c. motor. There are many different types, of course, so we will restrict our attention to the induction motor, which is the most common a.c. motor. In this a.c. motor, the source current is applied to windings around the rotor, which generate the magnetic field. In Fig. 6.22, these are shown by the magenta, blue, and yellow segments of the circle near the outside, each representing a different phase of a three-phase supply. These windings are called the stator, since they are non-moving, or static. When the current in the blue segments is at its maximum, the magnetic field generated by these windings is shown by the black, arrowed loops. As the currents vary sinusoidally, the orientation of the magnetic field rotates at a fraction (one-half for the motor shown in the figure) of line frequency. This is accomplished by winding the stator with a series of poles, and controlling the sequence of current passing through the windings for each of the poles. As the magnetic field orientation rotates, it induces currents in the rotor (the rotatable central element of the motor) through the Faraday effect, and the force between these induced eddy currents and the magnetic field generated by the stator forces rotation of the rotor. Note that the rotor of the induction motor is not an electromagnet nor a permanent magnet, but simply a conductor. Induction motors are asynchronous, meaning that the rate of rotation of the rotor is less than the rate of rotation of the magnetic field. The difference in these rotation rates, as a fraction of the rotation rate of the field, is known as slip. Some slip is required in the induction motor, as without slip the magnetic flux density in the rotor would be constant, and no current would be induced in the rotor. The greater the load on the motor, the greater the slip.

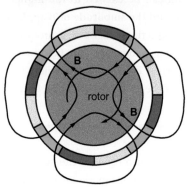

FIGURE 6.22 An a.c. induction motor.

Induced currents are also pivotal in the operation of the eddy-current speedometer, as described in Technote 6.9.

TechNote 6.9 Speedometer

Patented by Otto Schultze in 1902, the eddy-current speedometer was the primary type of speedometer used in cars until recently, when electronic speedometers, which generate a voltage pulse upon each rotation of the drive train, started becoming common. In the eddy-current speedometer, shown in Fig. 6.23, a cable runs from the transmission or wheels of the automobile to the instrument panel. As this cable spins, it causes a permanent magnet within the speedometer to spin as well. The spinning magnet is enclosed within a "speed cup," a spring-loaded conducting (commonly aluminum) assembly that is attached to a dial indicator. As the magnet rotates, its changing magnetic field induces an eddy current in the speed cup. The drag force created by the magnetic field of the magnet and the current induced in the speed cup causes the speed cup to rotate in the same direction as the magnet rotation. The speed cup rotates to the extent

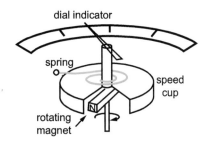

FIGURE 6.23 A schematic diagram of an eddy-current speedometer.

that the torque caused by the eddy currents is balanced by the opposing torque caused by the spring. The faster the rotation of the magnet, the greater the torque applied to the speed cup, and the greater the deflection of the speed indicator. The driver reads the speed of the vehicle from the dial indicator. Note that there is no direct mechanical connection between the spinning magnet and the speed cup. The only connection is through the magnetic field of the magnet and the eddy current induced in the speed cup.

In recent years, digital electronic speedometers have become more commonly used than eddy-current speedometers. Voltage pulses generated upon each rotation of the drive train indicate the distance traveled, and the speed is determined by counting these pulses during a specified time interval, and dividing the distance traveled by the elapsed time. Eddy-current and electronic speedometers are each subject to calibration errors, which may result from the size of the tires used on the vehicle. If the tires on the vehicle are changed for tires of a different radius, or if the tires are under- or over-inflated, the accuracy of the speedometer is affected.

A vehicle's speed can also be based upon GPS determinations of its location; these speed measurements are not subject to calibration errors resulting from the vehicle's tires.

Finally, regenerative brakes slow a vehicle, converting mechanical energy to electrical energy, and storing the charge for future use.

TechNote 6.10 Regenerative Brakes

In recent years, regenerative braking systems can be found on many hybrid and electric vehicles. In traditional braking systems, a car's velocity is decreased by bringing the brake pads into contact with the wheel. Through friction between the brake pads and a rotating disc or drum, the kinetic energy of the moving vehicle is converted into heat, which is then dissipated to the environment. Regenerative brakes provide a means of recovering and storing some of the kinetic energy of a moving vehicle, increasing the fuel efficiency of hybrid vehicles or the range of electric vehicles.

We have already discussed electric generators and electric motors, which physically are very similar to one another. In fact, an electric motor can be made to function as a generator, and a generator can be made to function as a motor. As a reminder, both motors and generators consist of a rotating coil, or armature, and a set of permanent magnets. When a current is supplied to the armature of a motor, the motor generates a torque about its axis, and it converts electrical energy to mechanical energy. Conversely, when the windings of a generator rotate, an EMF is induced across the armature through the Faraday effect, and mechanical energy is converted to electrical energy. The principle behind regenerative braking is based upon this dual functionality of a motor/generator. When we want the vehicle to accelerate, a current is supplied to the windings of the motor, and the energy stored in the battery is used to drive the motor. Conversely, to brake the vehicle, the drive train rotates the windings of the motor, which now functions as a generator. The motor thus generates an EMF, which produces a current to recharge the battery or a supercapacitor. Since the kinetic energy of the vehicle is stored in the battery or capacitor, to be reused later, regenerative braking can improve the efficiency of the vehicle by 5–20%, depending on driving conditions and other factors. Since regenerative brakes are not effective at low velocities, and since they often are not adequate for quick stops, regenerative brakes typically augment, but don't replace, conventional brakes. In addition to the benefit of saving energy, the pads of conventional brakes in vehicles with regenerative brakes don't wear as rapidly, since they are used much less than in cars with only conventional brakes.

While regenerative brakes have received a great deal of attention lately, the idea is not new. The Krieger electric landaulet, introduced in 1894 and shown in Fig. 6.24, used regenerative brakes. In this application, the motor had a pair of windings, one for use as the motor and the other for use as the generator.

FIGURE 6.24 Photograph taken in Washington, DC, circa 1906, of Senator George P. Wetmore, Rhode Island. This photo is from the Harris and Ewing Collection at the Library of Congress.

SUMMARY In this chapter we discussed the energy stored in the magnetic field, and explored several methods for calculating this energy. We explored the self- and mutual inductance of conductors, and we determined the mechanical force or torque on bodies within magnetic systems. With this, we have now completed our discussion of magnetostatic fields. As you think back over our development of magnetics, you can hopefully see the parallels with, as well as the differences from, electrostatic effects. The entire development is based on the observation that current-carrying wires exert forces on one another, and a quantitative description, known as the Biot–Savart Law, of the magnitude and direction of that force. For most of us, magnetic fields are much less intuitive than electric fields. We are constantly applying a right-hand rule in one form or another. We also introduced the potential function for magnetic fields, which is much more a mathematical construct than the potential for electric fields. Still, we have used this potential to our advantage, and will continue to do so later in our studies of time-varying fields.

Table 6.1

Collection of key formulas from Chapter 6.			
$W_m = \dfrac{1}{2} \displaystyle\int_v \mathbf{J}_v \cdot \mathbf{A}\, dv$	(6.3)	$L' = \dfrac{\mu_0 w}{d}$	(6.22)
$W_m = \dfrac{1}{2} \displaystyle\int_v \mathbf{H} \cdot \mathbf{B}\, dv$	(6.5)	$L' = \dfrac{\mu_0}{\pi}\left[\dfrac{1}{4} + \ln\left(\dfrac{d}{a}\right)\right]$	(6.23)
$w_m = \dfrac{1}{2}\,\mathbf{H} \cdot \mathbf{B}$	(6.5)	$\mathbf{F} = -\nabla W_m \mid_{constant\,\Phi_M}$	(6.24)
$L_{12} = \dfrac{N_2 \Phi_{12}}{I_1}$	(6.6)	$\mathbf{F} = \nabla W_m \mid_{constant\,I}$	(6.25)
$L_{11} = \dfrac{N_1 \Phi_{11}}{I_1}$	(6.7)	$W_m = -\mathbf{m} \cdot \mathbf{B}$	(6.28)
$L = \dfrac{2W_m}{I^2} = \dfrac{1}{I^2}\displaystyle\int_v \mathbf{H} \cdot \mathbf{B}\, dv$	(6.9)	$T_z = \dfrac{\partial W_m}{\partial \phi}$	(6.29)
$L' = \dfrac{\mu_0}{2\pi}\left[\dfrac{1}{4} + \ln\left(\dfrac{b}{a}\right)\right]$	(6.17)		

PROBLEMS

Magnetic Energy

P6-1 A uniform electric field of magnitude $E_0 = 100$ V/m is applied to a simple material of relative permittivity $\varepsilon_r = 4$ and permeability $\mu_r = 2$. Determine the amplitude of the magnetic field H_0 that must be applied such that the energy densities of the electric and magnetic fields are equivalent.

P6-2 A pair of long, plane parallel conductors with a width $w = 5$ cm are separated by a distance $d = 0.5$ cm. The current in the top plate is $I_0 = 2$ A; this current returns in the bottom plate. The permeability of the medium between the plates is μ_0. Assuming that the current is a surface current uniformly distributed on the inside surfaces of the conductors, and that the fields are uniform in the region between the conductors and zero elsewhere, determine (a) the energy density of the magnetic field, and (b) the energy per unit length of the configuration.

P6-3 A pair of long, plane parallel conductors with a width $w = 5$ cm are separated by a distance $d = 0.5$ cm. The current in the top plate is $I_0 = 2$ A; this current returns in the bottom plate. The permeability of the medium between the plates is μ_0 in the upper half and $3\mu_0$ in the lower. Assuming that the current is a surface current uniformly distributed on the inside surfaces of the conductors, and that the fields are uniform in the region between the conductors and zero elsewhere, determine (a) the energy density of the magnetic field in the two media, and (b) the energy per unit length of the configuration.

P6-4 Consider a pair of long, co-axial, cylindrical conductors. The radius of the inner conductor is $a = 2$ mm, and the inner radius of the outer conductor is $c = 2.2$ cm. The current in the center conductor is $I_0 = 2$ A; this current returns in the outer conductor. The permeability of the medium between the conductors is μ_0 for $\rho < b$, where $b = 1.2$ cm, and $3\mu_0$ for $b < \rho < c$. Assuming that the current is uniformly distributed in the center conductor, determine (a) the energy density of the magnetic field in the three regions, and (b) the energy per unit length of the configuration.

P6-5   A toroid consists of a winding of N turns carrying a current I wrapped on a rectangular cross-section of two materials. The permeability of the inner medium ($a < \rho < b$) is μ_1, while that of the outer medium ($b < \rho < c$) is μ_2. The thickness of the toroid (in the z-direction) is h. (a) Determine the energy density in the two media. (b) Use your results from part (a) to determine the total magnetic energy stored by the toroid.

Inductance and Inductors

P6-6   A pair of long, plane parallel conductors of width $w = 5$ cm are separated by $d = 0.5$ cm. The current in the top plate is $I_0 = 2$ A; this current returns in the bottom plate. The permeability of the medium between the plates is μ_0 in the upper half and $3\mu_0$ in the lower. (a) Assuming that the current is a surface current uniformly distributed on the inside surfaces of the conductors, and that the fields are uniform in the region between the conductors and zero elsewhere, determine the magnetic flux through a rectangular surface of height d (between the top and bottom conductors) and length ℓ (in the direction of the current). (A numerical answer is not required.) (b) Use the result of part (a) to determine the inductance per unit length of the configuration in terms of μ_0, d, and w. (c) What is the numerical value of L'?

P6-7   For the parallel conductors described in Prob. P6-6, determine (a) the magnetic energy density in the two media (a numerical answer is not required), (b) the magnetic energy per unit length of the configuration, (c) the inductance per unit length L', and (d) the numerical value of L'.

P6-8   Determine the inductance per unit length of a pair of long, co-axial conductors. The inner conductor is hollow and its current is uniformly distributed on the surface, at radius a, while on the outer conductor the current is on the inner surface at radius b. The permeability everywhere is μ_0.

P6-9   Determine the inductance per unit length of a pair of long, parallel hollow wires. The wires are of radius a, and the distance between centers is d. Let the current on each wire be uniformly distributed on the surface of the wires. The permeability everywhere is μ_0.

P6-10   Consider the pair of long, co-axial conductors described in Prob. P6-4. The radius of the inner conductor is $a = 2$ mm, and the inner radius of the outer conductor is $c = 2.2$ cm. The current in the center conductor is $I_0 = 2$ A; this current returns in the outer conductor. The permeability of the medium between the conductors is μ_0 for $a < \rho < b$, for $b = 1.2$ cm, and $3\mu_0$ for $b < \rho < c$. (a) Assuming that the current is uniformly distributed in the center conductor, determine the magnetic flux through a rectangular surface of height c (between the axis and the outer conductor) and of length ℓ (in the direction of the current). (A numerical answer is not required.) (b) Use the result of part (a) to determine the inductance per unit length of the configuration in terms of μ_0, I, a, b, and c. (c) What is the numerical value of L'?

P6-11   For the co-axial conductors described in Probs. P6-4 and P6-10, use the energy per unit length stored in the magnetic field to determine the inductance per unit length. Express your answer in terms of μ_0, a, b, and c, and calculate a numerical value.

P6-12   Determine the inductance per unit length of an infinitely long, air-core solenoid. The turns density of the solenoid is $n = 10$ turns/cm, the current through the winding is $I = 0.5$ A, and the solenoid is aligned with the z-axis. The cross-sectional area of the solenoid is 2 cm^2.

Magnetic Forces and Torques

P6-13 Derive an expression for the torque felt by a permanent magnetic needle in a static, uniform magnetic flux density **B**. The needle pivots on its center axis, has a permanent uniform magnetization **M** aligned with its long axis, cross-sectional area S, and is length L.

P6-14 Consider an air-gapped toroid like that described in Example 4.17. In this problem, we will find the energy stored in the magnetic field of the toroid, comparing the total energy under two conditions: (a) with the air gap empty, and (b) with the gap filled with a permeable disc whose diameter, thickness, and permeability match those of the core. Compare the energies of these two different configurations (disc removed vs. disc inserted) under conditions of (i) unchanged magnetic flux Φ, and (ii) unchanged current I in the toroid windings. What does this tell us about the direction of the force on the disc pulling it into the air gap?

P6-15 A rectangular wire loop of dimension 2 by 3 cm consists of $N = 20$ turns, and has a current of $I_0 = 5$ A. It is immersed in a uniform magnetic field of flux density $\mathbf{B} = B_0 \mathbf{a}_z$, where $B_0 = 10$ mT. Determine the magnetic energy W_m of this loop when the surface normal vector of the loop is at an angle θ with respect to $\mathbf{a}_z$. Use W_m to determine the torque $\boldsymbol{\tau}$ (a vector) on this loop. Describe (in words) the motion of the loop, assuming it is initially at rest at an angle θ.

P6-16 Consider a current-carrying loop in a non-uniform magnetic flux density $\mathbf{B} = \mathbf{a}_z B_0 y$, as shown in Fig. 6.25. The current loop is centered at the origin and lies in the x–y plane. Its dimensions are ℓ by w, in the x- and y-directions, respectively, and the current is I. We have labeled the sides of this loop 1–4. (a) Determine the force $\mathbf{F}_1$ exerted by the field on side 1. (b) Determine the force $\mathbf{F}_3$ exerted by the field on side 3. (c) Determine the total force $\mathbf{F}$ experienced by the entire loop. (d) Determine the torque experienced by side 1. (e) Determine the total torque experienced by the loop.

FIGURE 6.25 A rectangular current loop in a constant, non-uniform, magnetic field, as described in Prob. P6-16.

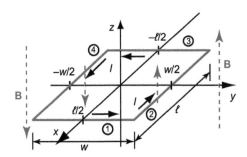

CHAPTER 7
Maxwell's Equations

As we have been discussing electric and magnetic effects throughout this text, we have been developing a set of equations, known collectively as **Maxwell's Equations**, that describe the properties of these fields in a very general sense. These equations are named after James Clerk Maxwell, whose contributions are discussed in Biographical Note 7.1. The development of Maxwell's Equations has been critical to our understanding and application of electromagnetic effects, as they govern such diverse effects as are present in capacitors, transformers, and electric generators, which we have already examined, and free-wave propagation, transmission lines, waveguides, and antennas, which we have not yet discussed. Before we can undertake our study of these new topics, we must first complete the development of Maxwell's Equations, which are not quite finished. As we will show shortly, there is an inconsistency in these equations as they stand to this point, an inconsistency that can be rectified by introducing a new term, known as the **displacement current**, to Ampère's Law. This additional term is the final piece of the puzzle, and with its inclusion Maxwell's Equations can be used to describe wave propagation, allowing us to understand (at an overview level, at least) the principles that govern our wireless routers, microwave ovens, and cable and satellite TV systems. In this chapter, we will introduce the displacement current, redefine the potential functions for time-varying fields, and re-examine the boundary conditions that must be satisfied at the interface between two different materials.

Biographical Note 7.1 James Clerk Maxwell (1831–1879)

By the mid-1800s, the foundational observations by Coulomb (and Cavendish, who didn't publish his studies and so went unrecognized) of electrostatic forces, and by Oersted, Biot, Savart, and Ampère of magnetostatic forces, were well established. In addition, Faraday's observation that a time-varying magnetic field produced an electromotive force, or an electric potential, was the first hint that electric and magnetic effects were in some way coupled to one another. Separate from these advances were discussions of the nature of light, primarily as to whether it was a particle or a wave, and if it were a wave, what was the nature of the wave. In his unified theory of electromagnetic effects and propagating waves in 1861, Scottish physicist James Clerk Maxwell suggested for the first time that light was a form of an oscillating electromagnetic wave. He introduced the displacement current density (to be discussed in the following section), derived a wave equation for electromagnetic waves (discussed in the following chapter), and showed theoretically that the velocity of these waves in vacuum was in close agreement with the best measurements of the velocity of light. Maxwell also suggested that electromagnetic waves at various other frequencies should be possible.

Maxwell, pictured in Fig. 7.1, described electromagnetic effects in terms of their potential functions, rather than the vector fields. In fact, his formalism used 20 variables and included 20 differential equations. In 1873, Oliver Heaviside reduced the complexity of Maxwell's theory to the present-day form that we call Maxwell's Equations, which number just four equations with four field quantities as variables.

In other areas of science and engineering, Maxwell developed a statistical model of the velocity distribution of molecules in a gas, known as the Maxwell–Boltzmann distribution; analyzed the stability of the rings of Saturn to show that they could not be solid or fluid, but rather a system of independent bodies; mathematically analyzed the governor on steam engines, which controlled their speed; and created color images using filters of the primary colors red, green, and blue.

In the 1870s, Maxwell acquired the notes of Henry Cavendish, which he organized, edited, and published. (See Biographical Note 1.2 on Cavendish.) Many, perhaps most, of Cavendish's expansive studies did not come to light until this undertaking by Maxwell.

Maxwell died of abdominal cancer in 1879, at the age of only 48. He is recognized by many as the greatest scientific mind of the nineteenth century.

James Clerk Maxwell.

FIGURE 7.1 Photograph of James Clerk Maxwell.

7.1 Displacement Current Density

While the set of laws in the form that we have developed them to this point describes the behavior of electric and magnetic fields with remarkable accuracy, in one particular aspect they lead to a critical inconsistency. The resolution of this inconsistency is the introduction of one additional term, known as the displacement current, to Ampère's Law. The introduction of the displacement current will be essential to the description of wave propagation, discussed in following chapters. In this chapter, we examine the displacement current and study its implications.

To this point in our development, Maxwell's Equations consist of the following set of equations:

Gauss' Law: $\qquad \boldsymbol{\nabla} \cdot \mathbf{D} = \rho_v \qquad \oint_s \mathbf{D} \cdot d\mathbf{s} = Q$

No magnetic charges: $\quad \boldsymbol{\nabla} \cdot \mathbf{B} = 0 \qquad \oint_s \mathbf{B} \cdot d\mathbf{s} = 0$

Faraday's Law: $\qquad \boldsymbol{\nabla} \times \mathbf{E} = -\dfrac{\partial \mathbf{B}}{\partial t} \qquad \oint_c \mathbf{E} \cdot d\boldsymbol{\ell} = -\dfrac{d}{dt} \int_s \mathbf{B} \cdot d\mathbf{s}$

Ampère's Law: $\qquad \boldsymbol{\nabla} \times \mathbf{H} = \mathbf{J}_v \qquad \oint_c \mathbf{H} \cdot d\boldsymbol{\ell} = I \quad \text{(but not yet complete)}$

In each case, these laws are shown in their differential (center column) and integral (right column) forms, and, as we remarked above, Ampère's Law is not quite finished.

In addition to these differential and integral laws, the constitutive relations

$$\mathbf{D} = \varepsilon_0 \mathbf{E} + \mathbf{P} = \varepsilon \mathbf{E}$$

and

$$\mathbf{H} = (1/\mu_0)\mathbf{B} + \mathbf{M} = (1/\mu)\mathbf{B}$$

relate **D** and **E**, and **H** and **B**, respectively. The right-hand equality in each line applies for linear media. Finally, the Continuity Equation,

$$\boldsymbol{\nabla} \cdot \mathbf{J}_v + \frac{\partial \rho_v}{\partial t} = 0,$$

is an expression of charge conservation.

Now let us apply Ampère's Law to a simple section of an electrical circuit, and show that we have a problem. Consider the section of a circuit shown in Fig. 7.2, a simple capacitor with the wire leads conducting the current I to and from the capacitor plates. We apply Ampère's Law to the contour c drawn around the wire lead on the left side, and write

$$\oint_c \mathbf{H} \cdot d\boldsymbol{\ell} = I,$$

where I is the current in the wire, passing through the surface labeled s_1 in the figure,

$$I = \oint_{s_1} \mathbf{J}_v \cdot d\mathbf{s}.$$

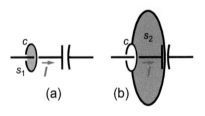

(a)　　(b)

FIGURE 7.2 A simple capacitor in a circuit with a current I. We apply Ampère's Law to the contour c, which bounds the surface s_1 in (a) and the surface s_2 in (b).

$\mathbf{J}_v$ is the current density, which is confined to the wire in this circuit. Everything seems just fine so far. But now consider the surface s_2, which is also bounded by the same contour c, but which passes down through the middle of the capacitor. (Remember, there are infinitely many surfaces s bounded by the contour c.) As such, Ampère's Law should also hold true for this surface. Through this surface, however, no free current passes. As we know, charge builds up on one side of the capacitor and is depleted from the other side, but no current passes from plate to plate. So it appears that Ampère's Law is violated here. The path integral $\oint_c \mathbf{H} \cdot d\boldsymbol{\ell}$ that we determined above, and which we agreed must be equal to the current I in the wire, now appears to be equal to zero. How do we resolve this dilemma?

The answer can be found through application of the Continuity Equation,

$$\boldsymbol{\nabla} \cdot \mathbf{J}_v + \frac{\partial \rho_v}{\partial t} = 0,$$

which we introduced in Eq. (3.7). Since the surfaces s_1 and s_2 are both bounded by the same contour c, we can combine these two surfaces to make one closed surface. The volume integral of the Continuity Equation computed over the volume v bounded by this closed surface is

$$\int_v \left(\boldsymbol{\nabla} \cdot \mathbf{J}_v + \frac{\partial \rho_v}{\partial t} \right) dv = 0.$$

By Gauss' Law, the charge density ρ_v is equal to $\boldsymbol{\nabla} \cdot \mathbf{D}$, which we can substitute into the second term in the integrand,

$$\int_v \left(\boldsymbol{\nabla} \cdot \mathbf{J}_v + \frac{\partial}{\partial t} \boldsymbol{\nabla} \cdot \mathbf{D} \right) dv = 0,$$

and since the displacement field $\mathbf{D}$ is generally well-behaved (i.e. continuous in space and time), the order of differentiation can be interchanged to write

$$\int_v \boldsymbol{\nabla} \cdot \left(\mathbf{J}_v + \frac{\partial \mathbf{D}}{\partial t} \right) dv = 0.$$

Upon application of Stokes' Theorem, this volume integral becomes a surface integral, evaluated over the surface $s_1 + s_2$ that bounds the volume v:

$$\oint_{s_1+s_2} \left(\mathbf{J}_v + \frac{\partial \mathbf{D}}{\partial t} \right) \cdot d\mathbf{s} = 0.$$

In this problem, as we set it up, a free current $\mathbf{J}_v$ passes through the surface s_1 (in the wire leading to the capacitor), but the displacement field, which is confined to the space between the capacitor plates, is absent there. In contrast, no free current passes through the surface s_2, but there is indeed a displacement field in this region. So the closed surface integral of the previous equation can be separated into two parts, of the form

$$\int_{s_1} \mathbf{J}_v \cdot d\mathbf{s} = \int_{s_2} \frac{\partial \mathbf{D}}{\partial t} \cdot d\mathbf{s}. \tag{7.1}$$

The signs of the two terms in this last equation are correct due to the conventions that must be applied. For the closed surface $s_1 + s_2$, the surface element $d\mathbf{s}$ points outward from the interior – that is, to the left for surface s_1 and to the right for surface s_2. For the individual surfaces s_1 and s_2, however, the right-hand rule applied to the contour c is used to determine the correct direction of the vector surface element $d\mathbf{s}$. Thus, $d\mathbf{s}$ must be reversed in direction for the surface s_1, and this minus sign has been absorbed in Eq. (7.1).

From this application of the Continuity Equation to the capacitor geometry, then, we conclude that Ampère's Law must be generalized to include what is called the **displacement current density**, $\partial \mathbf{D}/\partial t$. Ampère's Law now reads

$$\boxed{\nabla \times \mathbf{H} = \mathbf{J}_v + \frac{\partial \mathbf{D}}{\partial t},}$$ (7.2)

and is finally complete. The displacement current does not represent the flow of any charges, and in this regard is not what we customarily think of as a current at all. It does, however, serve as the source of a circulating magnetic field $\mathbf{H}$, just as the free current density $\mathbf{J}_v$ does.

Ampère's Law given in Eq. (7.2) is in the form known as the differential form. It can, of course, be expressed in its integral form, which is completely equivalent. This is derived by taking the surface integral of Eq. (7.2) over *any* open surface s,

$$\int_s \nabla \times \mathbf{H} \cdot d\mathbf{s} = \int_s \left(\mathbf{J}_v + \frac{\partial \mathbf{D}}{\partial t} \right) \cdot d\mathbf{s}.$$

Applying Stokes' Law, this becomes

$$\boxed{\oint_c \mathbf{H} \cdot d\boldsymbol{\ell} = I + \int_s \frac{\partial \mathbf{D}}{\partial t} \cdot d\mathbf{s},}$$ (7.3)

where the contour c forms the boundary of the surface s, and the current I is the total free current passing through the surface s. Maxwell's Equations, as listed in Table 7.1, are now complete.

Before we move on, let us return to the capacitor problem that we considered earlier in this section, and show that inclusion of the displacement current term in Ampère's Law does indeed remove the dilemma. Let the capacitor be a parallel plate capacitor of plate area A and separation d, with d small enough that edge effects can be ignored, and let the permittivity of the material between the plates be ε. Let us confirm for this capacitor the result we obtained above in Eq. (7.1), that is,

$$\int_{s_1} \mathbf{J}_v \cdot d\mathbf{s} = \int_{s_2} \frac{\partial \mathbf{D}}{\partial t} \cdot d\mathbf{s}.$$

Table 7.1

Maxwell's Equations in their complete form.		
Law	**Differential form**	**Integral form**
Gauss' Law	$\nabla \cdot \mathbf{D} = \rho_v$	$\oint_s \mathbf{D} \cdot d\mathbf{s} = Q$
No magnetic charges	$\nabla \cdot \mathbf{B} = 0$	$\oint_s \mathbf{B} \cdot d\mathbf{s} = 0$
Faraday's Law	$\nabla \times \mathbf{E} = -\dfrac{\partial \mathbf{B}}{\partial t}$	$\oint_c \mathbf{E} \cdot d\boldsymbol{\ell} = -\dfrac{d}{dt} \int_s \mathbf{B} \cdot d\mathbf{s}$
Ampère's Law	$\nabla \times \mathbf{H} = \mathbf{J}_v + \dfrac{\partial \mathbf{D}}{\partial t}$	$\oint_c \mathbf{H} \cdot d\boldsymbol{\ell} = I + \int_s \dfrac{\partial \mathbf{D}}{\partial t} \cdot d\mathbf{s}$

Clearly the surface integral over s_1 on the left represents the current I passing along the wire. The integral over surface s_2 on the right can be written as

$$\int_{s_2} \frac{\partial \mathbf{D}}{\partial t} \cdot d\mathbf{s} = \frac{\partial |\mathbf{D}|}{\partial t} A,$$

where we have used the fact that the displacement field $\mathbf{D}$ is uniform in the region between the capacitor plates, and that $\mathbf{D}$ points in the same direction as the vector surface element $d\mathbf{s}$. For a linear medium, $\mathbf{D} = \varepsilon \mathbf{E}$, and since $\mathbf{E}$ is uniform, the potential difference across the capacitor plates is $V = -\int \mathbf{E} \cdot d\boldsymbol{\ell} = Ed$. With these substitutions, we have

$$\int_{s_2} \frac{\partial \mathbf{D}}{\partial t} \cdot d\mathbf{s} = \frac{\varepsilon}{d}\left(\frac{dV}{dt}\right) A.$$

But for a parallel plate capacitor, $\varepsilon A/d$ is the capacitance C, yielding

$$I = \int_{s_2} \frac{\partial \mathbf{D}}{\partial t} \cdot d\mathbf{s} = C\frac{dV}{dt}.$$

This relation is the familiar equation relating the current flowing onto a capacitor with the rate at which the voltage across the capacitor grows. Thus we see that the introduction of the displacement current density $\partial \mathbf{D}/\partial t$ is perfectly consistent with our previous knowledge of the behavior of capacitors.

Now that each of Maxwell's Equations are complete, we must revisit some of our earlier discussions, in which Maxwell's Equations were applied to determine potential functions and boundary conditions. Those discussions were based on the incomplete set of Maxwell's Equations, so we'll reconsider these conditions in the following sections.

7.2 Potential Functions for Time-Varying Fields

For static electric fields, we defined potential functions that were very useful in cases for determining the fields and their properties. The electric potential, V, is a scalar function, and its existence is made possible through the irrotational nature of $\mathbf{E}$ – that is, for any static field $\mathbf{E}$, $\nabla \times \mathbf{E} = 0$. The electric field is determined from its potential through the relation

$$\mathbf{E} = -\nabla V.$$

The existence of the magnetic potential function $\mathbf{A}$ was assured through the absence of magnetic charges, $\nabla \cdot \mathbf{B} = 0$, and the magnetic flux density is related to $\mathbf{A}$ through

$$\boxed{\mathbf{B} = \nabla \times \mathbf{A}.} \tag{7.4}$$

Now that we have modified Maxwell's Equations to include time-varying effects, we must revisit these relations.

For time-varying fields, the relation $\nabla \cdot \mathbf{B} = 0$ is still valid, and so the vector magnetic potential $\mathbf{A}$ exists, and the magnetic flux density $\mathbf{B}$ is still related in the same way, $\mathbf{B} = \nabla \times \mathbf{A}$. (The explicit form of this potential may change, so we'll have to revisit that later.) For the electric potential V, however, we must change our approach much more radically. Time-varying electric fields are not irrotational, since $\nabla \times \mathbf{E}$, as given in Faraday's Law, is

$$\nabla \times \mathbf{E} = -\frac{\partial \mathbf{B}}{\partial t}.$$

There is, however, an alternative, which we can find by deeper application of Faraday's Law. Since the vector potential for **B** still exists, let's use that potential, and substitute $\boldsymbol{\nabla} \times \mathbf{A}$ for **B** in Faraday's Law,

$$\boldsymbol{\nabla} \times \mathbf{E} = -\frac{\partial}{\partial t}\left(\boldsymbol{\nabla} \times \mathbf{A}\right).$$

The vector potential **A** is a continuous function of space and time, and so the order of differentiation is interchangeable. Combining terms, we have

$$\boldsymbol{\nabla} \times \left(\mathbf{E} + \frac{\partial \mathbf{A}}{\partial t}\right) = 0.$$

Now this is an interesting development. While **E** is no longer irrotational, the field given by $\mathbf{E} + \partial \mathbf{A}/\partial t$ is, and we can define a potential function V for this vector field, such that

$$\mathbf{E} + \frac{\partial \mathbf{A}}{\partial t} = -\boldsymbol{\nabla} V.$$

Rearranging these terms gives us

$$\boxed{\mathbf{E} = -\boldsymbol{\nabla} V - \frac{\partial \mathbf{A}}{\partial t}.} \tag{7.5}$$

Thus, if we can find the potential functions V and **A**, we can determine the electric field **E** using Eq. (7.5), and **B** using Eq. (7.4). As with **A**, the scalar potential V will likely be of a different form from what we derived previously for static fields, but we do see that, with proper modifications, we can still make use of potential functions for time-varying fields.

So now let's take a look at these potential functions. We limit our discussion to linear, isotropic, homogeneous media, for simplicity. Our goal is to develop a set of differential equations for these potential functions $V(t, \mathbf{r})$ and $\mathbf{A}(t, \mathbf{r})$ that will allow us to solve for the potentials. Based on our previous experience, it is expected that these functions will depend on the distribution of charges $\rho_v(t, \mathbf{r}')$ and currents $\mathbf{J}_v(t, \mathbf{r}')$. We start with Gauss' Law, which is still valid for time-varying fields without modification:

$$\boldsymbol{\nabla} \cdot \mathbf{E} = \rho_v/\epsilon.$$

Taking the divergence of each term in Eq. (7.5), this becomes

$$-\nabla^2 V - \frac{\partial}{\partial t}\boldsymbol{\nabla} \cdot \mathbf{A} = \rho_v/\epsilon. \tag{7.6}$$

This equation is similar to Poisson's Equation, except for the additional term involving the vector potential **A**. Let's put this equation aside for a few moments, and derive a second differential equation for potential functions starting with Ampère's Law. For a linear, homogeneous, isotropic medium, Ampère's Law becomes

$$\boldsymbol{\nabla} \times \mathbf{H} = \mathbf{J}_v + \varepsilon\frac{\partial \mathbf{E}}{\partial t}.$$

Using Eq. (7.5) for **E**, this becomes

$$\boldsymbol{\nabla} \times \mathbf{H} = \mathbf{J}_v + \varepsilon\frac{\partial}{\partial t}\left(-\boldsymbol{\nabla} V - \frac{\partial \mathbf{A}}{\partial t}\right).$$

We can interchange the order of differentiation of the electric potential V, and since $\mathbf{H} = \mathbf{B}/\mu$ and $\mathbf{B} = \boldsymbol{\nabla} \times \mathbf{A}$, this becomes

$$\boldsymbol{\nabla} \times (\boldsymbol{\nabla} \times \mathbf{A}) = \mu\mathbf{J}_v - \varepsilon\mu\boldsymbol{\nabla}\frac{\partial V}{\partial t} - \varepsilon\mu\frac{\partial^2 \mathbf{A}}{\partial t^2}.$$

Using Eq. (E.29) for $\nabla \times (\nabla \times \mathbf{A})$ and rearranging terms leads to

$$\nabla (\nabla \cdot \mathbf{A}) - \nabla^2 \mathbf{A} + \varepsilon\mu\nabla\frac{\partial V}{\partial t} + \varepsilon\mu\frac{\partial^2 \mathbf{A}}{\partial t^2} = \mu \mathbf{J}_v. \tag{7.7}$$

So Eqs. (7.6) and (7.7) each contain second-order differentials of the potentials on the left sides, and sources (charge density ρ_v or current density $\mathbf{J}_v$) on the right side. And, we have to admit, they look pretty formidable in these forms. Our goal is to arrive at some simple expression for these potential functions that we can use for specific sources. But let's not throw in the towel just yet. Recall that we have some flexibility in how we define the vector potential $\mathbf{A}$. The curl of this potential gives us the magnetic flux density $\mathbf{B}$, but its divergence is unconstrained by any physical quantities, and we are free to choose it at our convenience. For static fields, we found it convenient to choose $\nabla \cdot \mathbf{A} = 0$, a condition that we called the Coulomb gauge. With this choice, we were led to the integral expressions of Eq. (4.22) for the vector potential. By examining Eqs. (7.6) and (7.7), we see that these equations can be simplified significantly by a different choice of gauge – that is

$$\nabla \cdot \mathbf{A} = -\varepsilon\mu\frac{\partial V}{\partial t}. \tag{7.8}$$

This choice is known as the **Lorenz gauge**, and with this choice Eqs. (7.6) and (7.7) take on the much simpler forms

$$\boxed{\nabla^2 V - \varepsilon\mu\frac{\partial^2 V}{\partial t^2} = -\rho_v/\varepsilon} \tag{7.9}$$

and

$$\boxed{\nabla^2 \mathbf{A} - \varepsilon\mu\frac{\partial^2 \mathbf{A}}{\partial t^2} = -\mu \mathbf{J}_v.} \tag{7.10}$$

Remarkably, these expressions are not so different from those for the time-independent potential functions. For example, Eq. (7.9) differs from Poisson's Equation only in the term with the time derivative. Otherwise, these expressions are identical. Similarly, Eq. (7.10) should remind us of Eq. (4.22), except again we have an additional term involving the time derivative of the magnetic potential.

The solutions to these equations for the potential functions will have similar form to the integral expressions we derived for static fields. Without further proof, we present these solutions as

$$\boxed{V(t, \mathbf{r}) = \frac{1}{4\pi\varepsilon} \int_v \frac{\rho_v(t', \mathbf{r}')}{|\mathbf{r} - \mathbf{r}'|}\, dv'} \tag{7.11}$$

and

$$\boxed{\mathbf{A}(t, \mathbf{r}) = \frac{\mu}{4\pi} \int_v \frac{\mathbf{J}_v(t', \mathbf{r}')\, dv'}{|\mathbf{r} - \mathbf{r}'|},} \tag{7.12}$$

where $t' = t - \sqrt{\varepsilon\mu}\,|\mathbf{r} - \mathbf{r}'|$. We see from these expressions that the potentials at location $\mathbf{r}$ and time t depend on the sources, as expected, but not at the same time t; rather, they depend on the sources at an *earlier* time t'. The difference between t and t' is simply the delay time required for information about the source distribution to propagate out to the distant field point $\mathbf{r}$. (We're being a bit vague about how this "information" propagates. We'll be in a better position to be specific later, so please hang on until then.) Because of the time delay, these potentials are called the **retarded potentials**.

Let's review what we have just discovered. First, we showed that, since for time-varying field $\nabla \times \mathbf{E}$ is no longer equal to zero, an electric potential cannot be defined in the same manner as for static fields. We were able to remedy this, however, by showing that the curl of $\mathbf{E} + \partial \mathbf{A}/\partial t$ is equal to zero. This led to Eq. (7.5), which allows the electric field to be determined from the pair of potential functions V and $\mathbf{A}$. The relation $\mathbf{B} = \mathbf{\nabla} \times \mathbf{A}$ is still valid for time-varying fields. Then we set out to determine the properties of the potential functions, which upon invoking the Lorenz gauge are formulated through the differential equations Eqs. (7.9) and (7.10). The solutions for these differential equations, known as the retarded potentials, are presented in Eqs. (7.11) and (7.12).

7.3 Boundary Conditions for Time-Varying Fields

LEARNING OBJECTIVE

After studying this section, you should be able to:
- state and apply the boundary conditions that time-varying electric and magnetic fields must satisfy at the interface between two media.

When we introduced static electric and magnetic fields earlier in this text, we derived a set of boundary conditions – that is, conditions that fields must satisfy at the interface between two media. We discussed these boundary conditions for static electric fields in Section 1.8, and for static magnetic fields in Section 4.8. As you will recall, we based the derivation of these boundary conditions on Maxwell's Equations, but as we have since discovered, those equations were valid only for static fields. Now that we have added time-varying terms to Faraday's Law and Ampère's Law, we must return to reconsider the boundary conditions anew. Our approach will follow lines similar to those that we employed previously, in that we will apply Maxwell's Equations (now in their completed form) to carefully chosen closed contours or closed surfaces that straddle an interface between two media. As we will see in this section, the boundary conditions that we derived for static fields are valid without modification for time-varying fields.

7.3.1 Tangential $\mathbf{E}$

We first consider the tangential field component of the electric field. As before, we construct a rectangular contour as shown in Fig. 7.3(a), and evaluate $\oint_c \mathbf{E} \cdot d\boldsymbol{\ell}$ on this path. For time-varying fields, this path integral must be equal to $-\frac{d}{dt} \int_s \mathbf{B} \cdot d\mathbf{s}$, where the surface s is bounded by the rectangular contour c. Since the two long sides of the rectangular path lie very close to the surface, $\Delta w \gg \Delta h$, and the path integral is

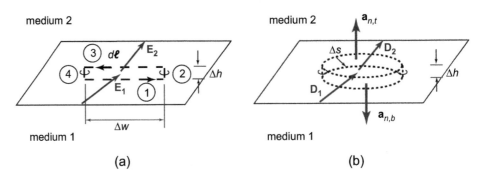

FIGURE 7.3 An interface between two different media for determination of the boundary conditions that must be satisfied by tangential and normal time-varying electric fields.

$$\oint_c \mathbf{E} \cdot d\boldsymbol{\ell} = E_{t,1}\Delta w + E_{n,2}\Delta h - E_{t,3}\Delta w - E_{n,4}\Delta h.$$

Since Δh is so small, the second and fourth terms from this expression can be dropped. The small height of the rectangle also implies that the magnetic flux $\int_s \mathbf{B} \cdot d\mathbf{s}$ through this surface is exceedingly small. Therefore, at the interface between two media, the tangential components of $\mathbf{E}$ must be continuous, or

$$E_{1,t} = E_{2,t}. \tag{7.13}$$

Thus the first of the boundary conditions for time-varying fields is unchanged from the static case.

7.3.2 Normal D

For static fields, we constructed the pillbox-shaped volume at the interface, shown in Fig. 7.3(b), and applied Gauss' Law,

$$\oint_s \mathbf{D} \cdot d\mathbf{s} = Q,$$

to this volume. By this means, we were able to show that the difference in the normal components of the displacement field $\mathbf{D}$ across the boundary is equal to the surface charge density ρ_s at the interface. Gauss' Law remains unchanged when we move to time-varying fields, and as a consequence this boundary conditions is also unchanged. For completeness of the present discussion, a brief proof of this result is presented anew. We construct a shape as shown in Fig. 7.3(b). This shape has one face just above the interface and one face just below the interface. Since these faces are close to the surface, the surface area of the sides of the shape is very small. Applying Gauss' Law to this closed surface leads to

$$(\mathbf{D}_2 - \mathbf{D}_1) \cdot \mathbf{a}_{n,1}\Delta s = \rho_s\,\Delta s,$$

where only the surface charge on the interface contributes to Q, since the height Δh of the pillbox is so small. Dividing by Δs, which is common to each term, and recognizing that $\mathbf{D} \cdot \mathbf{a}_n$ picks out the normal component only, we have

$$D_{2,n} - D_{1,n} = \rho_s. \tag{7.14}$$

7.3.3 Normal B

We showed for static magnetic fields that the normal component of $\mathbf{B}$ must be continuous across the boundary between two media. This was a result of the law $\oint_s \mathbf{B} \cdot d\mathbf{s} = 0$, which tells us that magnetic charges do not exist. This law is unchanged for time-varying fields, and so we expect that our boundary condition on the normal component of $\mathbf{B}$ is unchanged as well.

We apply the condition $\oint_s \mathbf{B} \cdot d\mathbf{s} = 0$ to the surface of the pillbox-shaped structure shown in Fig. 7.4(a). The height of the pillbox is Δh, and the area of the two faces is Δs. The two faces of the pillbox lie very close to the interface between the two media, so that Δh is very small. The closed surface flux is $(-B_{1n} + B_{2n})\Delta s$, where the dot product between $\mathbf{B}$ and the unit normal vector picks out the normal component of $\mathbf{B}$. Since the closed surface flux must be zero, we conclude that

$$B_{1n} = B_{2n}, \tag{7.15}$$

that is, the normal component of $\mathbf{B}$ must be continuous across any boundary.

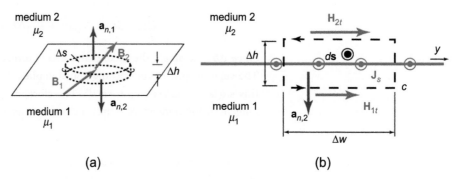

FIGURE 7.4 An illustration of the interface between two magnetic media, characterized by magnetic permeability μ_1 and μ_2, as labeled. (a) The arrows $\mathbf{B}_1$ and $\mathbf{B}_2$ represent the magnetic flux densities at the interface in medium 1 and 2, respectively. The closed "pillbox" surface is for determination of the boundary condition on B_n. (b) The same interface, showing the rectangular path for application of Ampère's Law, leads to the boundary condition on the tangential component of $\mathbf{H}$.

7.3.4 Tangential H

Finally, we consider the condition that must be satisfied by the tangential magnetic field components at the interface. Refer to the interface shown in Fig. 7.4(b), and the rectangular contour c, constructed with one side immediately above the interface and another side immediately below the interface. The closed path integral $\oint_c \mathbf{H} \cdot d\ell$ around this loop can be written as $H_{1t} \Delta w - H_{2t} \Delta w$, where we omit small terms contributed by the sides, which are negligible since $\Delta h \ll \Delta w$. This path integral must be equal to the total current $\int_s \mathbf{J} \cdot d\mathbf{s}$ passing through the surface s bounded by c, plus a new displacement current term $\int_s \partial \mathbf{D}/\partial t \cdot d\mathbf{s}$. Since the path c circles around in the counter-clockwise direction, $d\mathbf{s}$ points *out* from the page. Since Δh is small, only the surface current contributes to $\int_s \mathbf{J} \cdot d\mathbf{s}$, yielding $J_{s\perp} \Delta w$, where $J_{s\perp}$ is the component of the surface current that is parallel to $d\mathbf{s}$ (i.e. perpendicular to the path c). The displacement current term is proportional to Δh, and therefore negligibly small. (The displacement current represents, in a manner of speaking, a *volume* distribution, analogous to the volume current density $\mathbf{J}_v$. There is no equivalent of the displacement current analogous to the surface current density $\mathbf{J}_s$. Therefore, the displacement current term does not contribute to the boundary condition.) Dropping this term, and dividing the remaining terms by Δw, we have (using a compact vector notation style to remove ambiguity about directions and signs),

$$\mathbf{a}_{n2} \times (\mathbf{H}_1 - \mathbf{H}_2) = \mathbf{J}_s. \tag{7.16}$$

These four boundary conditions for electric and magnetic fields, whether static or time-varying, are summarized in Table 7.2. In this table, boundary conditions are listed for three cases. In the first column we list the general boundary conditions – that is, the conditions that must always be satisfied across the interface between any pair of media. In the middle column are the boundary conditions that must be satisfied when both media are perfect lossless dielectrics. In these media, there are no free charges, and so all charge and current densities are identically zero. In the right-hand column are the conditions that must be satisfied at the boundary between a lossless dielectric (medium 1) and a perfect conductor (medium 2). In the perfect conductor, of course, all field amplitudes must be zero, since the free charges that are present in a perfect conductor will immediately redistribute themselves in response to any applied field, such that fields are canceled.

Table 7.2

Summary of boundary conditions at the interface between two media.		
General	**Two lossless dielectrics** $(\rho_s = 0, \mathbf{J}_s = 0)$	**Lossless dielectric (1) + perfect conductor (2)**
$E_{1t} = E_{2t}$	$E_{1t} = E_{2t}$	$E_{1t} = E_{2t} = 0$
$\mathbf{a}_{n2} \times (\mathbf{H}_1 - \mathbf{H}_2) = \mathbf{J}_s$	$H_{1t} = H_{2t}$	$\mathbf{a}_{n2} \times \mathbf{H}_1 = \mathbf{J}_s, H_{2t} = 0$
$(\mathbf{D}_2 - \mathbf{D}_1) \cdot \mathbf{a}_{n,1} = \rho_s$	$D_{1n} = D_{2n}$	$\mathbf{a}_{n2} \cdot \mathbf{D}_1 = \rho_s, D_{2n} = 0$
$B_{1n} = B_{2n}$	$B_{1n} = B_{2n}$	$B_{1n} = B_{2n} = 0$

SUMMARY

In this chapter we have introduced the displacement current, finally completing the development of Maxwell's Equations. The displacement current, associated with a time-varying electric field, generates a magnetic field, just as a real current does, even without the flow of free charges. With this term added to Ampère's Law, we will be able to describe propagation of electromagnetic waves, which will be the topic of the remainder of this text.

We have also, in this chapter, returned to our discussions of potential functions and boundary conditions, which we treated earlier for static fields. We find that we can still define and use potential functions for time-varying fields, although we have to modify them somewhat. We introduced the Lorenz gauge for this purpose, and defined the retarded potentials. We will not need the retarded potential until we talk about antennas in the final chapter, but it does play an important role at that time. We also showed that the boundary conditions that fields must satisfy at the interface between two media are the same for time-varying fields as they were for static fields. With these topics in hand, we are now ready to talk about wave propagation, which will be introduced in the following chapter.

Table 7.3

Collection of key formulas from Chapter 7.			
$\nabla \times \mathbf{H} = \mathbf{J}_v + \dfrac{\partial \mathbf{D}}{\partial t}$	(7.2)	$\nabla^2 V - \varepsilon \mu \dfrac{\partial^2 V}{\partial t^2} = -\rho_v/\varepsilon$	(7.9)
$\oint_c \mathbf{H} \cdot d\ell = I + \int_s \dfrac{\partial \mathbf{D}}{\partial t} \cdot d\mathbf{s}$	(7.3)	$\nabla^2 \mathbf{A} - \varepsilon \mu \dfrac{\partial^2 \mathbf{A}}{\partial t^2} = -\mu \mathbf{J}_v$	(7.10)
See Table 7.1 for Maxwell's Equations in their complete form.		$V(t, \mathbf{r}) = \dfrac{1}{4\pi\varepsilon} \int_v \dfrac{\rho_v(t', \mathbf{r}')}{|\mathbf{r} - \mathbf{r}'|} dv'$	(7.11)
$\mathbf{B} = \nabla \times \mathbf{A}$	(7.4)	$\mathbf{A}(t, \mathbf{r}) = \dfrac{\mu}{4\pi} \int_v \dfrac{\mathbf{J}_v(t', \mathbf{r}') \, dv'}{|\mathbf{r} - \mathbf{r}'|}$	(7.12)
$\mathbf{E} = -\nabla V - \dfrac{\partial \mathbf{A}}{\partial t}$	(7.5)	See Table 7.2 for a summary of boundary conditions that fields must satisfy at the interface between two media.	

PROBLEMS

Displacement Current Density

P7-1 Consider a capacitor consisting of two co-axial, cylindrical conductors of radii a and b and length L. The space between the conductors is filled with a non-conducting dielectric of permittivity ε. (a) Presuming a charge $Q(t)$ on the inner conductor, find $\mathbf{D}$ in the region between a and b, (b) $\mathbf{E}$ in the same region, (c) the displacement current density, (d) the total displacement current I_d. (e) Find $v(t)$, the voltage difference between the conductors. (f) Find the capacitance C of the conductors. (g) Show that $I_d = C\,dv/dt$.

P7-2 Consider a capacitor consisting of two concentric spherical conductors of radii a and b. The space between the conductors is filled with a non-conducting dielectric of permittivity ε. (a) Presuming a charge $Q(t)$ on the inner conductor, find $\mathbf{D}$ in the region between a and b, (b) $\mathbf{E}$ in the same region, (c) the displacement current density, (d) the total displacement current I_d. (e) Find $v(t)$, the voltage difference between the conductors. (f) Find the capacitance C of the conductors. (g) Show that $I_d = C\,dv/dt$.

Potential Functions for Time-Varying Fields

P7-3 For potential functions $\mathbf{A}(\mathbf{r}, t) = \mathbf{A}_0 e^{j(\omega t - kz)}$ and $V(\mathbf{r}, t) = V_0 e^{j(\omega t - kz)}$, where $k = \omega\sqrt{\varepsilon\mu}$, find the relation between $A_{0,z}$ and V_0 such that the Lorenz gauge condition (Eq. (7.8)) is satisfied.

Relative Current Densities

P7-4 An electromagnetic wave whose electric field amplitude is $E_0 = 100$ V/m and frequency is $\omega = 1 \times 10^8$ r/s propagates in air. Find the amplitude of the displacement current density.

P7-5 An electric field in a non-magnetic medium of conductivity $\sigma = 10$ μS/m and permittivity $\varepsilon = 4\varepsilon_0$ is $\mathbf{E}(t, z) = \mathbf{a}_y 10 \cos(\omega t - \beta z)e^{-\alpha z}$ V/m, where α and β are known parameters. Determine the frequency ω at which the magnitude of the free current density $\mathbf{J}_v$ equals that of the displacement current density.

P7-6 A typical conductivity of seawater is ~ 4 S/m, and its permittivity is $80\,\varepsilon_0$. Determine the frequency below which the conduction current density is greater than the displacement current density.

P7-7 A wave of amplitude 100 V/m and frequency 1 GHz passes through a dielectric material. Determine the free current density J_v and displacement current density when the dielectric is (a) quartz, (b) silicon, (c) Teflon, and (d) drinking water. (Use data from Tables 1.1 and 3.1.) For each case, determine the frequency at which the real current density and displacement current density are equal.

CHAPTER 8
Propagation of Waves

KEY WORDS

wave equation; wave number; propagation constant; wavelength; phase velocity; wavefront; time-harmonic field; phasor representation; wave vector; transverse; intrinsic impedance; Doppler shift; polarization; liquid crystals; loss tangent; attenuation constant; skin depth; Poynting's Theorem; Poynting vector; power density; reflection coefficient; transmission coefficient; surface impedance; surface resistance; transverse electric (TE); plane of incidence; transverse magnetic (TM); angles of incidence, refraction, and reflection; Law of Reflection; Snell's Law of refraction; critical angle; total internal reflection; Brewster's angle; standing wave; standing wave ratio.

Having examined many useful and interesting properties of first static electric fields, then static magnetic fields, and most recently the combination of electric and magnetic fields through the introduction of time-varying effects, we have reached a turning point in our studies. Specifically, we will introduce what is perhaps the most revolutionary concept in electromagnetism: propagation of electromagnetic waves. Electromagnetic waves can carry information and energy, and their properties are described in full using Maxwell's Equations. We will explore these properties in detail in this and the following chapters.

We will describe electromagnetic waves in terms of their associated electric and magnetic fields. These field quantities depend on spatial coordinates and time in a very specific way, in accordance with Maxwell's Equations. In the preceding chapters we saw how a time-varying magnetic field can generate a curling electric field. This is described through Faraday's Law:

$$\nabla \times \mathbf{E} = -\frac{\partial \mathbf{B}}{\partial t}.$$

We also introduced the displacement current, which, along with free current density $\mathbf{J}_v$, serves to produce a curling magnetic field:

$$\boldsymbol{\nabla} \times \mathbf{H} = \mathbf{J}_v + \frac{\partial \mathbf{D}}{\partial t}.$$

This relation, of course, is Ampère's Law. When these two laws are put together, we have a powerful synergy, in that the time-varying magnetic field can generate the electric field, and the time-varying electric field can generate the magnetic field. These two fields support one another to form a propagating wave. Our goal within this chapter is to discover the properties of these propagating waves. As we will see, the electric field and the magnetic field of a propagating wave:

1. are perpendicular to one another, and each is perpendicular to the direction in which the wave is traveling (we call this a transverse wave);
2. have amplitudes that are of a unique ratio determined by the properties of the medium through which the wave is propagating; and
3. are precisely in phase with one another (for propagation through a lossless medium).

In this chapter we will explore these properties of propagating waves. We will not worry yet about how to generate these waves. This will come later, when we consider the generation of electromagnetic waves through oscillating currents within linear antennas.

8.1 Propagation in Lossless Media

Let us start by considering a source-free, linear, isotropic, homogeneous, non-conducting medium. We have introduced each of these terms earlier in this text, but to put it briefly, these conditions define a simple, lossless medium. We can relax any or all of these criteria later. Be assured that Maxwell's Equations still serve to describe wave propagation in these more complicated media, but that is better left until after we have a good working knowledge of wave propagation in simple media. For source-free media, there are no free charges, or

$$\rho_v = 0$$

and no free currents,

$$\mathbf{J}_v = 0.$$

For linear media, the displacement field **D** and the electric field **E** are proportional to one another,

$$\mathbf{D} = \varepsilon \mathbf{E},$$

as are the magnetic field **H** and the magnetic flux density **B**,

$$\mathbf{H} = \frac{1}{\mu}\mathbf{B},$$

where ε and μ are the electric permittivity and the magnetic permeability, respectively. Also, these material parameters are the same for all directions of the field quantities (since the medium is isotropic), and are the same for all locations (since the medium is homogeneous). Finally, we explicitly state that the conductivity σ is zero, consistent with $\mathbf{J}_v = 0$. Under these conditions, Faraday's Law can be written as

$$\boldsymbol{\nabla} \times \mathbf{E} = -\mu \frac{\partial \mathbf{H}}{\partial t}, \tag{8.1}$$

and Ampère's Law as

$$\mathbf{\nabla} \times \mathbf{H} = \varepsilon \frac{\partial \mathbf{E}}{\partial t}. \tag{8.2}$$

In addition, Gauss' Law $\mathbf{\nabla} \cdot \mathbf{D} = \rho_v$ becomes

$$\mathbf{\nabla} \cdot \mathbf{E} = 0, \tag{8.3}$$

and $\mathbf{\nabla} \cdot \mathbf{B} = 0$ becomes

$$\mathbf{\nabla} \cdot \mathbf{H} = 0. \tag{8.4}$$

We will use Eqs. (8.1)–(8.4) to find solutions for the electric and magnetic fields associated with propagating waves.

In their present form, a solution of this set of differential equations is somewhat hindered in that they contain two variables, $\mathbf{E}$ and $\mathbf{H}$. These equations would be much simpler to solve if we could in some way combine them to produce a single equation that involves only one variable, $\mathbf{E}$ or $\mathbf{H}$. That is precisely the approach that we will adopt in the following discussion. As we go through this derivation, you may at several points along the way find yourself wondering why we want to take this particular step or that step. In many steps, the reason may not be obvious at all. Clearly some pretty smart people went before us and forged the path, and found very useful and interesting results. We encourage you, as we follow their footsteps, to not get lost in questions of the motivation for certain steps, other than that it leads us to the result, and focus instead on the validity of each step. If we can complete our derivation with you agreeing that each step is correct, then we will take that as a positive, and go on to explore solutions that describe wave propagation.

We start with Ampère's Law, in the form of Eq. (8.2), taking the curl of both sides to write

$$\mathbf{\nabla} \times (\mathbf{\nabla} \times \mathbf{H}) = \mathbf{\nabla} \times \left(\varepsilon \frac{\partial \mathbf{E}}{\partial t} \right).$$

The curl of the curl of any vector field involves second-order derivatives with respect to spatial coordinates, as are introduced in Appendix E. This term can be written using vector identity Eq. (E.29) as

$$\mathbf{\nabla} \times (\mathbf{\nabla} \times \mathbf{H}) = \mathbf{\nabla}(\mathbf{\nabla} \cdot \mathbf{H}) - \nabla^2 \mathbf{H}.$$

On the right-hand side of the previous equation, the permittivity ε can be brought outside the derivative, since ε is a constant, and the order of the differentiation interchanged, since $\mathbf{E}$ is continuous in time and position:

$$\mathbf{\nabla}(\mathbf{\nabla} \cdot \mathbf{H}) - \nabla^2 \mathbf{H} = \varepsilon \frac{\partial}{\partial t}(\mathbf{\nabla} \times \mathbf{E}).$$

But $\mathbf{\nabla} \cdot \mathbf{H} = 0$ (by Eq. (8.4)), and $\mathbf{\nabla} \times \mathbf{E} = -\mu \partial \mathbf{H}/\partial t$ (by Eq. (8.1)), which, upon insertion, gives us

$$-\nabla^2 \mathbf{H} = -\varepsilon \mu \frac{\partial^2 \mathbf{H}}{\partial t^2}.$$

Placing both terms on the same side results in

$$\boxed{\nabla^2 \mathbf{H} - \varepsilon \mu \frac{\partial^2 \mathbf{H}}{\partial t^2} = 0.} \tag{8.5}$$

This equation is known as the **wave equation**, and we will spend a great deal of time exploring various forms of its solutions. Of the two terms in this equation, one involves only a derivative with respect to time, and the other a derivative with respect to the spatial coordinates. Take note of what we have just accomplished. Starting with the various Maxwell's Equations, which are each first-order differential equations, but which describe a coupling between **E** and **H**, we have derived a single equation that includes just a single variable, in this case **H**. If we had started with Faraday's Law, and used Ampère's Law to eliminate **H**, then the result would be an identical wave equation for **E**:

$$\boxed{\nabla^2 \mathbf{E} - \varepsilon\mu \frac{\partial^2 \mathbf{E}}{\partial t^2} = 0.}$$

(8.6)

So now we have a differential equation in a single variable, **E** or **H**. This is not without cost, of course, as the single equation is of second order, whereas we started with a set of first-order equations. Still, we have seen second-order differential equations of similar form on many previous occasions, and the solution of this equation should be familiar to you. If not, it might be a good idea to revisit your linear circuit analysis notes and review the solutions for the current and voltage in simple LC or RLC networks. These differ somewhat, of course, in that (1) currents and voltages are described as scalar functions, whereas **E** and **H** are vectors, and (2) currents and voltages in circuits are functions of time, but not position, whereas **E** and **H** depend on position and time. Still, our previous treatment of second-order linear differential equations in circuits will prove useful as we explore solutions of the wave equation in the present context!

8.1.1 Example Solution

A simple example of a solution of the wave equation in $\mathbf{E}(t, z)$ is

$$\mathbf{E}(t, z) = \mathbf{E}_0 \sin\left(\omega t - kz\right).$$

(8.7)

This expression describes a function that oscillates sinusoidally as a function of time at a frequency ω, and oscillates sinusoidally as a function of position z as well. In addition, it travels, or propagates, in the $+z$ direction. The constant k is known as the **wave number** or **propagation constant**. The amplitude $\mathbf{E}_0$ is a vector, in that it points in a particular direction (which we must still explore). Certainly Eq. (8.7) is only one example of a solution to the wave equation, but in the following we will demonstrate that it does indeed satisfy the wave equation as advertised, and then we'll use this example to start uncovering a few of the properties of propagating waves. Let us start by verifying that it satisfies Eq. (8.6). To do this, we need the Laplacian, $\nabla^2 \mathbf{E}$, which in rectangular coordinates represents

$$\nabla^2 \mathbf{E} = \left(\frac{\partial^2}{\partial x^2} + \frac{\partial^2}{\partial y^2} + \frac{\partial^2}{\partial z^2} \right) \mathbf{E}$$

(see Eqs. (E.30) and (E.19)). Since the sample solution $\mathbf{E}(t, z)$ given in Eq. (8.7) is independent of x and y, the derivatives with respect to these variables are zero. Taking the derivative twice with respect to z, we have

$$\nabla^2 \mathbf{E} = -k^2 \mathbf{E}.$$

The second derivative with respect to t, which becomes $-\omega^2 \mathbf{E}$, is also required. Inserting these into the wave equation leads to

$$\nabla^2 \mathbf{E} - \varepsilon\mu \frac{\partial^2 \mathbf{E}}{\partial t^2} = \left[-k^2 + \varepsilon\mu\omega^2 \right] \mathbf{E} = 0.$$

The only way that $\left[-k^2 + \varepsilon\mu\omega^2\right]\mathbf{E}$ can be zero in a general sense is if the term inside the square brackets is itself zero. (There are a couple of other ways that this equation is satisfied, but these are not worth pursuing. For example, the amplitude of the wave $\mathbf{E}_0$ could be zero, but this is a trivial and not very useful solution. Also, we could find some specific values of z and t for which the equation is satisfied, but we are looking for solutions in which the equation is correct for any values of z and t.) Therefore, $\mathbf{E}(t,z)$ given by Eq. (8.7) satisfies the wave equation, provided the condition

$$k = \omega\sqrt{\varepsilon\mu} \tag{8.8}$$

holds. Thus, for any frequency of oscillation ω, there is a specific value of the wave number k for that wave.

We have now verified that the wave given by the electric field $\mathbf{E}(t,z)$ in Eq. (8.7) is a valid solution of the wave equation, as long as Eq. (8.8) is satisfied. Now let us look a little more closely at this solution in order to learn more about it. A plot of $E(t,z)$, the scalar magnitude of $\mathbf{E}(t,z)$, as a function of t is shown in Fig. 8.1(a). Since the electric field $\mathbf{E}(t,z)$ in Eq. (8.7) is also a function of position z, we must fix the value of z before we plot the wave as a function of t. For the plot in Fig. 8.1(a), the position is $z = 0$. This plot represents the electric field strength at just one spot ($z = 0$) as a function of time. It is, of course, the familiar-looking sine wave. If we had chosen to plot the electric field strength as a function of t at a slightly different location, say at $z = \tilde{z}$, then the plot would have looked almost exactly the same, only shifted in phase, as shown in Fig. 8.1(b). We can see this explicitly by examining Eq. (8.7) at these two locations:

$$\mathbf{E}(t, z = 0) = \mathbf{E}_0 \sin(\omega t)$$

and

$$\mathbf{E}(t, z = \tilde{z}) = \mathbf{E}_0 \sin(\omega t - k\tilde{z}).$$

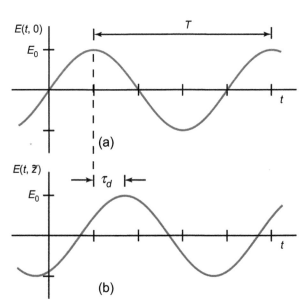

FIGURE 8.1 A plot of Eq. (8.7) as a function of time t, at location (a) $z = 0$ and (b) $z = \tilde{z}$. This function is one solution of the wave equation, given in Eq. (8.6).

Due to the presence of the additional phase term $k\tilde{z}$, the electric field $\mathbf{E}(t, z = \tilde{z})$ at $\tilde{z}$ is delayed by a time $\tau_d = k\tilde{z}/\omega$ relative to the electric field $\mathbf{E}(t, z = 0)$ at $z = 0$. This delay time is most easily seen by writing

$$\mathbf{E}(t, z = \tilde{z}) = \mathbf{E}_0 \sin(\omega(t - \tau_d)),$$

and equating the arguments of the sine function within both forms of $E(t, z = \tilde{z})$. The delay time τ_d is shown in Fig. 8.1(b).

Let us now define the other parameters contained in Eq. (8.7). Certainly the frequency ω is a term that is known to most of us. The electric field $\mathbf{E}$ oscillates sinusoidally, and the frequency tells us how rapidly in time the wave is oscillating. For any periodic function, the period T of the function is defined as the elapsed time after which the wave repeats itself. The period T of the field given by Eq. (8.7) is indicated in Fig. 8.1(a). Of course, the starting point for the period is arbitrary. We could have shown the period T as the elapsed time from one minimum of the wave to the next, for example. For a function oscillating sinusoidally at frequency ω, the period is given by

$$T = \frac{2\pi}{\omega}, \tag{8.9}$$

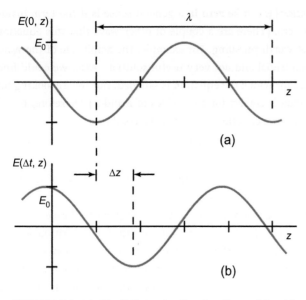

FIGURE 8.2 A plot of Eq. (8.7) as a function of position z, at time (a) $t = 0$ and (b) $t = \Delta t$.

which can be demonstrated by setting $E_0 \sin(\omega t - kz)$ equal to $E_0 \sin(\omega(t + T) - kz)$. Setting $\omega T = 2\pi$ leads to Eq. (8.9).

It is natural to ask if we can think about the wave number k in a manner that is analogous to the frequency ω. Of course, the answer is yes. A "snapshot" of the wave is shown in Fig. 8.2(a). By snapshot, we mean that the electric field $|E(z, t)|$ is plotted as a function of z, holding time t at a fixed value. Let's set $t = 0$. The **wavelength** of a wave λ is a measure of the distance over which the wave repeats itself. In this regard, the wavelength λ is analogous to the period T for the time-dependence of the wave. Thus the wavelength λ and the wave number k are related by $k\lambda = 2\pi$, or

$$\lambda = \frac{2\pi}{k}. \tag{8.10}$$

The wave number k is a measure of how rapidly the wave oscillates spatially, and in this regard it plays a role in the spatial variation of the wave that is similar to the role played by the frequency ω in describing the time variation of the wave.

Example 8.1 Plane wave: ω, T, k, and λ

Consider an electromagnetic wave of the form given by Eq. (8.7) of frequency $f = 100$ MHz propagating through vacuum. Determine (a) the angular frequency ω, (b) the period T, (c) the wave number k, and (d) the wavelength λ of this wave.

Solution:

We start with a note about notation. In this text, the symbol f is used to denote the cyclical frequency, or the number of cycles per second. The unit of cyclical frequency is the hertz, abbreviated Hz. Radial frequency, on the other hand, is the number of radians per second. Since there are 2π radians in one cycle, radial and cyclical frequency are related using $\omega = 2\pi f$. Also note that radians and cycles are not units in the same sense as meters or seconds. They are simply placeholder units that are used to denote an angle or a phase. To illustrate this, recall that the length s of a circular arc made by sweeping through an angle α at a radius ρ is simply $s = \rho\alpha$. Since s and ρ are both distances, they each have a unit of length, the m. The angle α, whose unit is the radian, is equivalent to the unit $[s/\rho] = $ m/m $= 1$. Thus the units of the angle in radians is equivalent to no unit at all. Accordingly, the unit of angular frequency is often written as rad/s $= $ s^{-1}. Another note regarding notation is that we often use, for example, $\omega = 2\pi \times 100$ MHz. Of course, Hz is not a valid unit for radial frequency, so strictly speaking this is a bit of a mixture of conventions. Still, it a convenient means of stating the radial frequency, using the cyclical frequency f.

(a) We are given $f = 100$ MHz, so the radial frequency is $\omega = 2\pi f = 2\pi \times 10^8$ s^{-1}.
(b) The period T of the oscillation is $f^{-1} = 10^{-8}$ s $= 10$ ns.
(c) The wave number is $k = \omega\sqrt{\varepsilon\mu}$. Since we are told that the wave is propagating through vacuum, we know that $\varepsilon = \varepsilon_0 = 8.854 \times 10^{-12}$ F/m and $\mu = \mu_0 = 4\pi \times 10^{-7}$ H/m. These values give us $k = 2.1$ m^{-1}.
(d) Finally, the wavelength λ is found using $\lambda = 2\pi/k$, which is 3 m.

We stated earlier that the wave given in Eq. (8.7) is traveling in the $+z$-direction. Let's take a look at the velocity at which this wave propagates. We start by plotting the wave $E(t, z) = E_0 \sin(\omega t - kz)$ as a function of z at two different times. Since $E(0, z)$ is already plotted in Fig. 8.2(a), we need only add to this, in Fig. 8.2(b), a snapshot of E a short time Δt later – that is $E(\Delta t, z)$. Note that this plot looks identical to the plot shown for $t = 0$, except that the wave has shifted to the right. In other words, the wave has moved. The velocity of the wave can be determined by calculating the ratio of the distance Δz that the wave has traveled by the time Δt. So let's park ourselves at a peak (or any other fixed point) of the wave, and go "surfing." By this, we mean that we will move with the wave such that we stay at the peak of the wave. Thus the argument of the sine function in Eq. (8.7),

$$\omega t - kz,$$

stays fixed. So if at time t_1 the peak is at position z_1, and at time t_2 the same peak is at position z_2, then

$$\omega t_1 - kz_1 = \omega t_2 - kz_2,$$

and upon rearranging terms,

$$\omega(t_2 - t_1) = k(z_2 - z_1).$$

Substituting $\Delta t = t_2 - t_1$ and $\Delta z = z_2 - z_1$, the velocity of the wave is

$$u_p = \frac{\Delta z}{\Delta t} = \frac{\omega}{k}.$$

This velocity is often called the **phase velocity** of the wave, motivating the subscript p in the symbol u_p. Finally, we can use $k = \omega\sqrt{\varepsilon\mu}$ (Eq. (8.8)) to write the phase velocity as

$$u_p = \frac{1}{\sqrt{\varepsilon\mu}}. \tag{8.11}$$

This is interesting in that the phase velocity of the wave depends only on material properties: the permittivity ε and the permeability μ. For a wave propagating through vacuum, the phase velocity is

$$c = \frac{1}{\sqrt{\varepsilon_0\mu_0}}. \tag{8.12}$$

Using the known constants for ε_0 and μ_0, we compute $c = 2.998 \times 10^8$ m/s. This value of the phase velocity has special significance, and is usually represented by the symbol c, as used here.

Combining Eqs. (8.11) and (8.12), the phase velocity through a medium can be written as

$$u_p = \frac{c}{\sqrt{\varepsilon_r\mu_r}}.$$

In other words, the phase velocity of a wave through a medium is typically reduced from the speed of a wave in vacuum by $\sqrt{\varepsilon_r\mu_r}$. In the field of optics, this reduction factor is called the refractive index, which for non-magnetic ($\mu_r = 1$) materials is

$$n = \sqrt{\varepsilon_r}. \tag{8.13}$$

(Do not confuse refractive index with number density, both commonly denoted n. These are completely distinct, unrelated quantities.)

The demonstration of radio wave generation and propagation and exposition of some of the wave properties by Heinrich Hertz in the 1880s showed that radio waves and visible light are related, differing only in frequency, and verified the theories of James Clerk Maxwell. Hertz' contributions are reviewed in Biographical Note 8.1.

Biographical Note 8.1 Heinrich Rudolf Hertz (1857–1894)

(a)

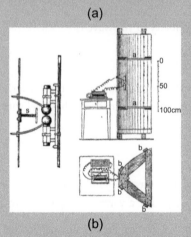

(b)

FIGURE 8.3 (a) Photograph of Heinrich Rudolf Hertz. (b) Sketch of Hertz' spark gap radio transmitter and parabolic reflector used in his 1888 demonstration of radio waves, from his 1893 book. A Hertzian dipole antenna is suspended along the focal line of the parabolic reflector.

By the late 1800s, Maxwell's mathematical description of electric and magnetic fields, and his suggestion that light was an electromagnetic wave, were well known but unproven. This proof was produced by Heinrich Hertz, pictured in Fig. 8.3(a), who designed and carried out a series of experiments at Karlsruhe in the period 1886–1889. In these works, Hertz showed convincingly that radio waves displayed the same properties as light, and established the validity of Maxwell's theory.

Hertz earned his PhD at the University of Berlin in 1880, where he studied under Hermann von Helmholtz. He continued at Berlin for a few years after completion of his PhD studies, until he accepted a faculty position in mathematical physics at Kiel. At Kiel, he studied Maxwell's works in depth, and in 1885, when he moved to Karlsruhe as a full professor, he began his experimental studies of radio waves. Hertz constructed what we would today call a radio frequency transmitter, consisting of a Ruhmkorff induction coil, a spark gap, and a cylindrical parabolic reflector. See diagrams of Hertz' apparatus from his notebook in Fig. 8.3(b). He used a second spark gap as a detector, and when radio waves from the transmitter were incident upon the receiver, Hertz observed an electrical spark generated in the gap. By varying the gap distance between the conductors, and determining the distance over which a spark was generated, he could measure the relative radio wave amplitude. Hertz carried out many key measurements of the properties of these waves, including the decreasing amplitude of the wave with increasing distance, and he showed that the waves could be reflected by a conductor. Using a reflected wave interfering with the incident wave, he was able to generate a standing wave pattern of the wave (to be discussed in the next chapter), and by measuring the distance between minima in this pattern he was able to determine the 5.1 m wavelength of the wave. Hertz calculated inductance and capacitance of the elements in his transmitter, and from these determined that the frequency of the radio wave was 62 MHz. Hertz then determined the velocity of the radio wave from the product of the wavelength and the frequency, yielding a value of $\sim 3.15 \times 10^8$ m/s, within a few percent of the speed of light, supporting the idea that radio waves and light were the same phenomena, differing only in frequency.

Hertz also demonstrated the field polarization properties of the radio waves, using a type of polarizer that consisted of a series of parallel wires strung across a wooden frame. When inserted between the transmitter and the receiver, this structure transmitted the waves when the wires were perpendicular to the axis of the spark gap, but absorbed them when the wires were parallel to this axis.

It is notable that Hertz did not foresee the widespread applications of radio waves that we now take for granted in everyday life. He famously wrote in 1890, "I do not think that the wireless waves I have discovered will have any practical application."

Hertz is also credited with discovery that a charged sphere will rapidly discharge when illuminated by ultraviolet light. We now recognize this as the photoelectric effect, in which electrons are ejected from a conductor surface when ultraviolet light is incident upon it. Hertz' discovery was in 1887, before the discovery of the electron by J.J. Thompson in 1897, so interpretation of the effect had to wait for the contributions of others, including Albert Einstein, who was awarded the Nobel Prize in physics in 1921 for his explanation of the photoelectric effect.

In 1890, Hertz moved to the University of Bonn. Unfortunately, he developed an infection a short time later, which proved to be fatal, and he died before reaching his 37th birthday. The unit of frequency, the hertz, is named in his honor.

8.1.2 Examples of Wave Propagation

Before we dive any more deeply into wave propagation theory, let's take a look at a series of examples in which these ideas are found in practice. In this section, we look at a few of these, including free-wave propagation through the atmosphere, global positioning systems, and satellite TV.

TechNote 8.1 Wave Propagation in the Atmosphere

Electromagnetic wave propagation through the atmosphere is typically categorized as one of four different modes: line-of-sight propagation; surface modes; ionospheric modes; and satellite communications. Here, we will discuss these different modes of wave propagation.

Line-of-sight (LoS) propagation, as its name implies, describes the direct propagation of a wave from a transmitting antenna to a receiving antenna. Cell phones, wireless networks, FM radio and television broadcasting, and radar all use LoS propagation. The maximum range for LoS transmission on the surface of the Earth depends on a number of factors, including the height of the antenna, the power of the transmitter, the sensitivity of the receiver, and the absorption or scattering properties of the atmosphere between the transmitter and receiver. Under the best conditions, the power density of an unguided wave is expected to decrease as $1/R^2$, where R is the distance from the transmitter. (We'll discuss power density later in this chapter, and the $1/R^2$ dependence in Sections 11.1 and 11.2.) So the received signal decreases with increasing distance. Rain or other forms of precipitation can further reduce the power of the transmitted wave. Another factor to consider is the curvature of the Earth. In Fig. 8.4(a) a transmitting antenna of height h_t and a receiving antenna of height h_r are shown. To receive a direct signal from the transmitter, the antenna heights must be great enough that the Earth's surface does not block the signal path. The distance d_t from the transmitting antenna to the horizon satisfies the condition $R_e^2 + d_t^2 = (R_e + h_t)^2$, where $R_e = 6\,378$ km is the radius of the Earth. Similarly, the distance d_r from the horizon to the receiving antenna must satisfy $R_e^2 + d_r^2 = (R_e + h_r)^2$. The former reduces to $d_t = \sqrt{2h_t R_e + h_t^2} \approx \sqrt{2h_t R_e}$, while the latter reduces to $d_r = \sqrt{2h_r R_e + h_r^2} \approx \sqrt{2h_r R_e}$. If the separation between the antennas is greater than $d_t + d_r \approx \sqrt{2h_t R_e} + \sqrt{2h_r R_e}$, then the horizon obscures the direct LoS of one antenna from the other. As an example, a transmitting antenna of height 100 m has a range of 36 km (~22 miles) for a receiver at ground level, presuming the surface of the Earth is perfectly spherical. Note that "LoS" propagation is perhaps not a completely accurate title for these waves, since wireless network signals can be received a room or two away from the transmitter. This is because materials that block optical waves, such

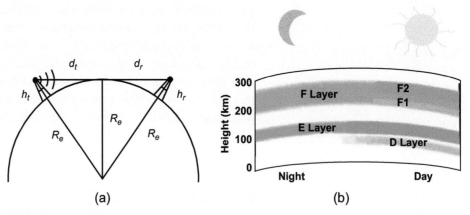

FIGURE 8.4 (a) Two tall antennas on the surface of the Earth, for determination of the range of the signal. R_e is the radius of the Earth, h_t and h_r the heights of the transmitting and receiving antenna towers, respectively, and d_t and d_r the distance from the respective antennas to the horizon. (b) The structure of the atmosphere at night and during the day. The ionosphere plays a critical role in propagation of radio waves over long distances, especially at night, when the absorbing D layer is reduced.

as walls, can still transmit radio-frequency waves. But still, in this mode of propagation the idea is that the waves propagate directly from the transmitting antenna to the receiving antenna, without substantial bending or reflection by barriers.

At low frequencies, because of diffraction effects (which we will not treat in this text), waves can travel greater distances, beyond the horizon, by following the curvature of the surface of the Earth. These are often called **surface modes**, or ground waves. The idea is that large wavelength waves can bend around obstructions, and reach regions that are out of direct LoS. Also, low-frequency waves can penetrate into the earth, with two consequences: (a) very low frequencies (VLF, 3–30 kHz) to extremely low frequencies (ELF, 3–30 Hz) are used for communication in mines and with submerged submarines; and (2) surface currents induced in the earth can lead to losses in low-frequency waves propagating through the atmosphere.

We've probably all heard radio signals from broadcast stations or shortwave radio transmitters that are located halfway across the continent. How do we square that observation with the previous discussion of LoS propagation? And what is the basis for transatlantic communications, starting with the telegraph messages in 1902 between Poldhu, Cornwall, UK, and Glace Bay, Nova Scotia, Canada by Guglielmo Marconi. (Marconi was awarded the Nobel Prize in physics in 1909 for this feat, shared with Karl Ferdinand Braun.)

Clearly LoS propagation over these great distances ($\sim$4 300 km, or 2 600 miles) is not possible. Receiving these signals would be impossible were it not for the presence of the Earth's ionosphere. The ionosphere is a layer of the atmosphere, between 60 and 500 km above the surface of the Earth, in which the air is highly ionized, forming a plasma of positively charged molecular ions and an equal density of negatively charged electrons. See Fig. 8.4(b) for a simplified model of the ionosphere. These ions are formed by interactions with UV radiation from the sun. Under the right conditions, electromagnetic waves can reflect from the upper regions of the ionosphere (the F layer) and from the surface of the Earth, and after multiple bounces can be received by distant receiving antennas. These waves are called **ionospheric modes**, or skywave or skip propagation modes. Waves incident upon a plasma whose frequency is below the plasma frequency of the medium are strongly reflected. The plasma frequency is determined as

$$\omega_p = \sqrt{\frac{n_e e^2}{m_e \varepsilon_0}},$$

where n_e is the electron number density and m_e the mass of the electron. (See Section 8.6.1 and Example 8.10 for more on the plasma frequency.) For plasmas in the ionosphere, the plasma frequency is typically in the range 1–10 MHz. (Metals can also be considered to be a plasma, due to the "sea" of free electrons. The plasma frequency of most metals is in the UV range, giving metals their shiny appearance at optical frequencies.) The D layer of the ionosphere is very absorptive for medium- to high-frequency waves, limiting the range of AM radio stations in the daytime. The nighttime ionosphere, however, has a quite different structure, and the D layer is greatly reduced, allowing reception from distant sources.

Satellite communication can be considered an LoS mode, in which a ground station transmits a signal to an Earth-orbiting satellite, and the satellite retransmits the signal to another ground station. These signals must be at high frequencies to improve their transmission through the ionosphere.

Radio waves that propagate from the ground station to a set of geostationary satellites in orbit around the Earth, and which are retransmitted from the satellites to your dish antenna, deliver television signals for home entertainment, as described in TechNote 8.2.

TechNote 8.2 Satellite TV

TV signals are transmitted from a constellation of geostationary satellites orbiting high above the surface of the Earth at the equator. These satellites appear to be stationary in the sky, since they orbit the Earth at a rate of one revolution per day. The radius of the geostationary orbit (the distance from the center of the Earth) is 42.24×10^3 km, equivalent to 6.6 Earth radii, so these satellites are really out a long distance. Signals transmitted from a ground station on Earth toward the satellite are received, shifted in frequency, and transmitted back toward Earth. At the user location, the signal is collected, typically with an 18-inch (45 cm) diameter dish that reflects the microwave signal onto a feed horn receiver. The time delay for the signal to propagate from the ground station to the satellite and back to the user is ~0.25 s. The frequency range used for the TV signal is typically in the C band (4–8 GHz), Ku band (12–18 GHz), and Ka band (26.5–40 GHz).

The Global Positioning System (GPS) depends on precise timing of clock signals received from four or more satellite-based transmitters. The difference in the arrival times of these signals is used to determine the receiver's location. The GPS system is described in more detail in TechNote 8.3.

TechNote 8.3 Global Positioning System (GPS)

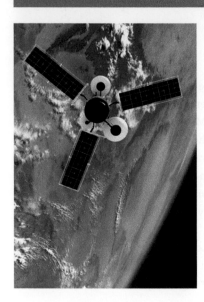

GPS is a system of 24 or more (currently 31) satellites in orbit around Earth at an altitude of 20 200 km that emit one-way transmissions for purposes of location, navigation, and timing. See the network of satellites in Fig. 8.5(a), and one such satellite in Fig. 8.5(b). With this coverage, a person at any location on Earth, such as the young motorist in Fig. 8.6(a), is in the field of view of at least four satellites at any time. Each satellite is equipped with an atomic clock for accurate time keeping, and continuously broadcasts its location and the time at which its signal is sent. This allows a GPS receiver on Earth to determine its distance from the satellite through trilateration, provided its own internal clock is up to date. For this reason, the internal clocks on GPS receivers are constantly updated to keep them accurate. The atomic clocks (see TechNote 8.4) on GPS satellites generate a highly stable signal at 10.23 MHz. The GPS satellites transmit at two frequencies in the L-band microwave range: L1 (1 575.42 MHz) and L2 (1 227.60 MHz). These frequencies are the 154th and 120th harmonic of the stable clock frequency.

It is interesting to understand that the time-keeping clocks on the satellites must be corrected for relativistic effects to maintain the high precision needed for the applications for which they are intended. In addition to the satellites launched and maintained by the US government, a number of other countries and unions, including Russia, the European Union, China, Japan, and India, have launched their own satellites and/or systems.

While GPS was the result of many talented scientists and engineers, we will use this opportunity to highlight the contributions of Dr. Gladys West, shown in Fig. 8.6(b). Dr. West was one of the first female African American mathematicians hired by the Defense Department to carry out detailed calculations for flight simulations in the 1950s in the era before the advent of electronic computers. She took on increasing responsibility, and in the 1970s and 1980s led a project that used satellite measurements and detailed mathematical techniques to derive an accurate model of the shape of Earth. This model is key to the precision of modern GPS. African American women hired as "computers" by the National Advisory Committee for Aeronautics (NACA), a precursor to NASA, were the subject of *Hidden Figures*, the book by author Margot Lee Shetterly and subsequent movie.

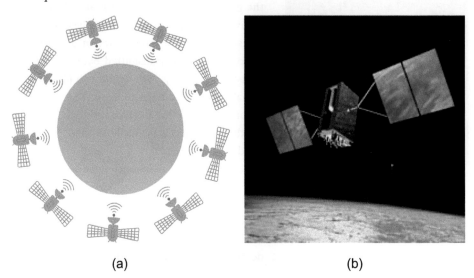

(a)   (b)

FIGURE 8.5 (a) The global constellation of satellites. (b) GPS III satellite.

(a)   (b)

FIGURE 8.6 (a) A young motorist checking her destination on the GPS-based navigator. (b) Dr. Gladys West, a human computer and NASA mathematician whose precise models of the shape of the Earth were critical to the modern-day GPS system.

The precision of the GPS system depends upon very precise atomic clocks, the most accurate time pieces available. Atomic clocks are described in TechNote 8.4.

TechNote 8.4 Atomic Clocks

The ability to keep accurate time is essential for communications, navigation, and commerce. Historically, measurements of time have always been based upon observations of some periodic event, such as the cycles of day and night, the swing of a pendulum, or electrical oscillations stabilized to a quartz crystal. Since 1967, the unit of time, the second, has been defined as the time required for 9 192 631 770 oscillations of the radio frequency (rf) field resonant with the ground state hyperfine transition in atomic cesium. Cesium is the heaviest of the stable alkali metal atoms, and its ground state consists of two components separated by 9.192 631 770 GHz. An rf field at this frequency can drive a transition between the two ground state components. In an atomic clock, if the rf frequency of a source shifts slightly away from the resonance, this shift is quickly detected as a decrease in the transition strength. The frequency of the source can then be adjusted to bring it back into resonance, and the oscillations of the wave are counted electronically. The NIST F1 atomic clock in Boulder, CO is so stable that the error in time is as little as 0.03 ns per day, or 3 parts in 10^{15}. Research toward more advanced atomic clocks has been able to limit this uncertainty even further. Precise atomic clocks require very narrow bandwidth atomic resonances, as well as extreme reduction of factors that are known to shift the atomic transition frequency, such as electric fields, magnetic fields, gas collisions, and Doppler shifts.

These examples serve to illustrate the impact of propagating electromagnetic waves on our modern world. In the following section, we will return to our development of the wave propagation with a more general solution to the wave equation.

8.1.3 General Solution

While the solution of the wave equation given in Eq. (8.7) is valid, it is by no means the only solution. More generally, any function $f(t')$, which is a function of t' where

$$t' = t - \hat{\mathbf{k}} \cdot \mathbf{r}\sqrt{\varepsilon\mu}, \tag{8.14}$$

satisfies the wave equation. $f(t')$ is a function of t and $\mathbf{r}$, but in a very specific combination. In the expression for t', $\hat{\mathbf{k}}$ is a vector which we will show (1) has unit magnitude and (2) points in the direction in which the wave is traveling. The function $f(t')$ can actually take many different forms, such as a sinusoidal wave or even a non-periodic pulse, as long as it depends on t' as defined in Eq. (8.14). To show that $f(t')$ satisfies the wave equation, we must determine each of its second-order derivatives. We will work in rectangular coordinates for this demonstration, in which

$$\nabla^2 f(t') = \left(\frac{\partial^2}{\partial x^2} + \frac{\partial^2}{\partial y^2} + \frac{\partial^2}{\partial z^2}\right) f(t')$$

Expanding t' from the compact scalar product notation of Eq. (8.14), t' is

$$t' = t - \left(\hat{k}_x x + \hat{k}_y y + \hat{k}_z z\right)\sqrt{\varepsilon\mu}.$$

Computing the first derivative of $f(t')$ with respect to x yields

$$\frac{\partial f(t')}{\partial x} = \frac{df(t')}{dt'}\frac{\partial t'}{\partial x} = \left(-\hat{k}_x\sqrt{\varepsilon\mu}\right)f'(t'),$$

where the chain rule was used in the first step, and the prime symbol on the function $f'(t')$ represents $df(t')/dt'$ – that is, the derivative of $f(t')$ with respect to its full argument. Similarly, the second-order derivative is

$$\frac{\partial^2 f(t')}{\partial x^2} = \frac{\partial}{\partial x}\left[\left(-\hat{k}_x\sqrt{\varepsilon\mu}\right)f'(t')\right]$$

$$= \left(-\hat{k}_x\sqrt{\varepsilon\mu}\right)\frac{df'(t')}{dt'}\frac{\partial t'}{\partial x} = \left(-\hat{k}_x\sqrt{\varepsilon\mu}\right)^2 f''(t'),$$

where we have used the chain rule again, and $f''(t')$ represents $d^2 f(t')/dt'^2$. Expressions for $\partial^2 f(t')/\partial y^2$ and $\partial^2 f(t')/\partial z^2$ are similar. Next, the second-order derivative with respect to time t is

$$\frac{\partial^2 f(t')}{\partial t^2} = f''(t').$$

Substituting these derivatives into Eq. (8.6), the wave equation becomes

$$\left[\left(-\hat{k}_x\sqrt{\varepsilon\mu}\right)^2 + \left(-\hat{k}_y\sqrt{\varepsilon\mu}\right)^2 + \left(-\hat{k}_z\sqrt{\varepsilon\mu}\right)^2 - \varepsilon\mu\right]f''(t') = 0. \tag{8.15}$$

The first three terms contract to give

$$\left(-\hat{k}_x\sqrt{\varepsilon\mu}\right)^2 + \left(-\hat{k}_y\sqrt{\varepsilon\mu}\right)^2 + \left(-\hat{k}_z\sqrt{\varepsilon\mu}\right)^2 = \varepsilon\mu|\hat{\mathbf{k}}|^2,$$

and Eq. (8.15) becomes

$$\left[\varepsilon\mu|\hat{\mathbf{k}}|^2 - \varepsilon\mu\right]f''(t') = 0.$$

So when $|\hat{\mathbf{k}}| = 1$, the terms inside the square brackets in Eq. (8.15) sum to zero, and the function $f(t')$ satisfies the wave equation, Eq. (8.6). With this, we have shown that $\hat{\mathbf{k}}$ is a unit vector.

Now let's see what we can learn about the *direction* of the vector $\hat{\mathbf{k}}$. To start, let's consider the function $f(t')$ at two different locations, $\mathbf{r}_1$ and $\mathbf{r}_2$, but at the same time t.

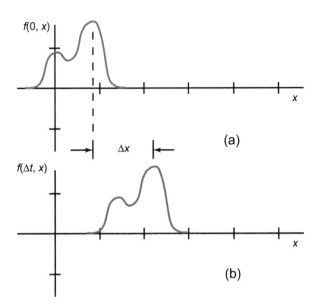

FIGURE 8.7 A plot of an arbitrary pulse $f(t')$ as a function of position at time (a) $t = 0$ and (b) $t = \Delta t$.

How does $f(t')$ compare at these two points? Well, this depends on t' at these two points. Let's represent these values of t' with t'_1 and t'_2. If $\hat{\mathbf{k}} \cdot \mathbf{r}_1$ and $\hat{\mathbf{k}} \cdot \mathbf{r}_2$ are the same, then t'_1 and t'_2 are also the same, and $f(t')$ is unchanged as well. So the set of all points $\mathbf{r}_1$ and $\mathbf{r}_2$ such that $\hat{\mathbf{k}} \cdot (\mathbf{r}_2 - \mathbf{r}_1) = 0$ defines a surface over which the value of $f(t')$ is constant. This surface is simply the plane perpendicular to $\hat{\mathbf{k}}$. This set of points is often called a **wavefront**. Conversely, the variation in $f(t')$ between points $\mathbf{r}_1$ and $\mathbf{r}_2$ is maximal when $\hat{\mathbf{k}} \cdot (\mathbf{r}_2 - \mathbf{r}_1)$ is greatest, and this is the case when $\mathbf{r}_2 - \mathbf{r}_1$ is parallel to $\hat{\mathbf{k}}$. So the vector $\hat{\mathbf{k}}$ plays a defining role in this waveform. It points in the direction in which the function varies most rapidly in space, and the planes normal to $\hat{\mathbf{k}}$ are the wavefronts, over which $f(t')$ is perfectly uniform.

Furthermore, if we were to now let the clock run and watch the wave move, its motion would be in the direction perpendicular to its wavefronts – that is, in the direction of the unit vector $\hat{\mathbf{k}}$. Let's now ask how fast this wavefront is propagating. In Fig. 8.7, we have plotted a pulse $f(t')$ of rather arbitrary shape. (For the purpose of this plot only, we have used x to designate the position, but don't intend this to necessarily coincide with the x-axis. Rather, using the logic from the previous paragraph, we recognize that this should be the distance traveled in the direction parallel to $\hat{\mathbf{k}}$, and perpendicular to the wavefront.) So $\hat{\mathbf{k}}$ defines for us the direction of propagation of the traveling wave. The plot in Fig. 8.7(a) represents a snapshot of the pulse at $t = 0$, while that in (b) is the pulse at some later time, $t = \Delta t$. The pulse has moved a distance Δx in the time Δt, so its velocity is $\Delta x / \Delta t$. Following the motion of the peak, or any fixed point on this pulse, the value of $t' = t - \hat{\mathbf{k}} \cdot \mathbf{r}\sqrt{\varepsilon\mu}$ is unchanged, and we can write

$$\Delta t' = 0 = \Delta t - \hat{\mathbf{k}} \cdot \Delta \mathbf{r}\sqrt{\varepsilon\mu}.$$

As we already discussed, the direction of propagation is parallel to $\hat{\mathbf{k}}$, so $\hat{\mathbf{k}} \cdot \Delta \mathbf{r} = \tilde{k}\Delta x$, and

$$0 = \Delta t - \tilde{k}\Delta x \sqrt{\varepsilon\mu}.$$

Thus we have the velocity of this wave as $u = \Delta x / \Delta t = 1/\sqrt{\varepsilon\mu}$, as it was in the first sample solution given by the sinusoidal field in Eq. (8.7). (If $f(t')$ is a pulse, as shown in the figure, this is strictly true only when ε and μ are independent of the frequency, but we ignore that detail for now.)

We have shown in this section that there are many solutions to the wave equation, and that the primary condition that must be fulfilled is that the wave or pulse depends on time and position in a very specific, coupled manner; we interpret this as a traveling wave or pulse. We will return to sinusoidally varying waves in the following sections, but we should keep in mind that, while sinusoidal waves are important in many aspects, many other waveforms are possible. We have tried in this section to gain a little insight into this more general class of waveforms.

8.1.4 Time-Harmonic Fields

We used the sinusoidally varying wave of Eq. (8.7) as our first example of a propagating wave. This is one example of a **time-harmonic** field, which we define as a function that varies sinusoidally as a function of time. In fact, time-harmonic functions play an important role in our treatment of electromagnetic waves. Certainly many generators or oscillators naturally produce sinusoidally varying fields. It is also the case that filters, elements that block some waves and transmit others, are well described by their responses to single-frequency waves. Finally, we know through our study of Fourier analysis that any arbitrary function can be expressed as a sum of a series of harmonic waves, and this opens the door to applying harmonic waves to a broad range of signals. Therefore we redirect our attention back to these special cases.

A time-harmonic function can be written as

$$E(t) = E_0 \cos(\omega t + \phi),$$

where the amplitude E_0 may depend on position, or may be constant. While we have written this as a scalar function, everything that follows applies to vector quantities as well. An equivalent form for the function $E(t)$ is

$$E(t) = A \cos(\omega t) + B \sin(\omega t).$$

In the first equation, a single amplitude E_0 and a phase ϕ are used to specify the sinusoidal function, while in the second we use two amplitudes A and B. These two expressions are completely equivalent, as one can be related to the other through simple trigonometric identities. The former is often referred to as the amplitude and phase form, and the latter as the quadrature amplitude form. If we know E_0 and the phase ϕ, we can determine the amplitudes A and B through

$$A = E_0 \cos \phi \qquad \text{and} \qquad B = -E_0 \sin \phi.$$

Conversely, if we know the amplitudes A and B, we can use the inverse relations

$$E_0 = \sqrt{A^2 + B^2} \qquad \text{and} \qquad \phi = -\tan^{-1}(B/A)$$

to determine E_0 and ϕ.

We can further simplify our analysis by using complex notation. See Appendix I for a brief summary of complex numbers and operations. In adopting the complex exponential notation for time-dependent functions, using $\mathfrak{R}(e^{j(\omega t + \varphi)})$ in place of $\cos(\omega t + \varphi)$, we must always keep the insignificance of the imaginary part in mind. We understand that field quantities, as abstract as they may seem at times, must always be real quantities, just as any physically observable quantity must be real. The imaginary part of $E(t)$ or $H(t)$, for example, when its time-dependence is expressed as $e^{j(\omega t + \varphi)}$, has no physical meaning. Yet by adding in this imaginary part ($j \sin(\omega t + \varphi)$) to our harmonic wave function, we can simplify our analysis of the real fields substantially. This seemingly paradoxical statement is explained as follows. When working with $\cos(\omega t + \varphi)$ or $\sin(\omega t + \varphi)$, derivatives of the cosine give us the sine, and derivatives of the sine give us the cosine. We quickly end up with a trigonometric forest, and all the complexities that go along with that. With the complex exponential function $e^{j(\omega t + \varphi)}$, however, all these complications go away. Derivatives and integrals of $e^{j(\omega t + \varphi)}$ simply bring out a factor of $j\omega$ or $1/j\omega$, respectively, but leave the time-dependence as $e^{j(\omega t + \varphi)}$. All of the terms in Maxwell's Equations have this identical time-dependence, and this term can be safely factored away. We are then left with algebraic equations, which are in general much simpler to solve. The only cost for this simplification is that we must accept in our minds the complex notation and the complex algebra that goes along with it. As you work with this notation, you will soon appreciate the beauty of complex algebra, and the simplification that it brings to us. You can find a brief review of complex numbers in Appendix I.

The next step in our progression for time-harmonic waves is to adopt what is known as the **phasor representation** for the time-harmonic field quantities. In fact, we hinted at this in the previous paragraph. When we write any of the relations between complex, time-harmonic field quantities and sources for linear media, such as Maxwell's Equations, as we will do shortly, each of the field or source quantities includes precisely the same time-dependent factor, $e^{j\omega t}$. Since this factor is quickly removed from each term, it can simply be dropped altogether from our notation. That is, each field or source quantity is expressed as a complex amplitude, with no explicit time-dependence shown. This complex quantity is called the **phasor** representation. It is important to remember that these phasor functions represent time-dependent (sinusoidally varying) functions, even though we don't write the time-dependence explicitly. To convert from a phasor quantity to its time-dependent equivalent, we multiply the complex phasor quantity by $e^{j\omega t}$ and take the real part. For example, the phasor $\tilde{\mathbf{E}} = |\mathbf{E}|e^{j\varphi}$ represents the time-dependent function

$$\Re\left[|\mathbf{E}|e^{j\varphi}\, e^{j\omega t}\right] = |\mathbf{E}|\,\Re\left[e^{j(\omega t+\varphi)}\right] = |\mathbf{E}|\cos(\omega t + \varphi) = \mathbf{E}(t).$$

(We will use the tilde to indicate a phasor quantity.) Conversely, the time-dependent magnetic field $\mathbf{H}(t) = \mathbf{a}_x[H_0'\cos(\omega t) - H_0''\sin(\omega t)]$ can be expressed as a phasor by

$$\mathbf{H}(t) = \mathbf{a}_x[H_0'\Re(e^{j\omega t}) + H_0''\Re(je^{j\omega t})] = \mathbf{a}_x\Re[(H_0' + jH_0'')e^{j\omega t}]$$
$$\rightarrow \tilde{\mathbf{H}} = \mathbf{a}_x(H_0' + jH_0'').$$

In this prescription, we have converted the cosines and sines to complex exponential terms, and in the final step dropped the $e^{j\omega t}$ factor and $\Re$ part notation. The phasor is expressed in terms of its real (H_0') and imaginary (H_0'') parts here. The equivalent magnitude and phase form of the phasor representing the magnetic field is $\tilde{\mathbf{H}} = \mathbf{a}_x H_0 \exp(j\varphi)$, where $H_0 = [H_0'^2 + H_0''^2]^{1/2}$ and $\varphi = \tan^{-1}(H_0''/H_0')$.

Let's return to Maxwell's Equations, and tailor them to a specialized form that is valid for time-harmonic functions. As we just showed, when time-harmonic field quantities are written as complex exponential functions, the time derivatives of the exponentials in Maxwell's Equations can be replaced with $j\omega$ times the exponential. That is,

Gauss' Law:	$\nabla \cdot \tilde{\mathbf{D}} = \tilde{\rho}_v$	(8.16)
No magnetic charges:	$\nabla \cdot \tilde{\mathbf{B}} = 0$	(8.17)
Faraday's Law:	$\nabla \times \tilde{\mathbf{E}} = -j\omega\tilde{\mathbf{B}}$	(8.18)
Ampère's Law:	$\nabla \times \tilde{\mathbf{H}} = \tilde{\mathbf{J}}_v + j\omega\tilde{\mathbf{D}}.$	(8.19)

Now that Maxwell's Equations have been written in forms that are valid for time-harmonic fields; we would like to use these to develop a wave equation that governs the propagation of time-harmonic electromagnetic waves. We start by treating the case of wave propagation through a simple (linear, isotropic, homogeneous), source-free ($\tilde{\mathbf{J}}_v = 0$, $\tilde{\rho}_v = 0$) medium. Following the same procedure that we followed previously to find the wave equation for the general case, we start with Faraday's Law, take the curl of each term, and use Ampère's Law to eliminate $\tilde{\mathbf{H}}$, and arrive at

$$\boxed{\nabla^2\tilde{\mathbf{E}} + \varepsilon\mu\omega^2\tilde{\mathbf{E}} = 0.} \tag{8.20}$$

On the other hand, if we start with Ampère's Law, take the curl of each term, and use Faraday's Law to eliminate $\tilde{\mathbf{E}}$, we can show

$$\boxed{\nabla^2\tilde{\mathbf{H}} + \varepsilon\mu\omega^2\tilde{\mathbf{H}} = 0.} \tag{8.21}$$

These wave equations are similar to those presented in Eqs. (8.6) and (8.5), respectively. In both cases, $\partial^2/\partial t^2$ in the latter is replaced by $(j\omega)^2$. Notice also that the wave equation satisfied by the harmonic field $\tilde{E}$ is identical to that for $\tilde{H}$, just as was the case for the general fields.

Before we move on to special time-harmonic solutions of the wave equation, known as uniform plane waves, we will address one source of confusion that can often confound us. In this section, we have introduced quantities such as the electric field $\tilde{E}$ or the magnetic field $\tilde{H}$ that are both vectors and phasors (i.e. complex quantities). As a vector, the quantity has an associated direction, and can be expressed using components in three orthogonal directions. The magnitude of a vector, as we have used throughout this text, is $|\mathbf{E}| = \sqrt{\mathbf{E} \cdot \mathbf{E}} = \sqrt{E_x^2 + E_y^2 + E_z^2}$, where the last form is valid in rectangular coordinates. Complex numbers, of course, also have magnitudes (see Appendix I), $|z| = \sqrt{zz^*} = \sqrt{(\Re(z))^2 + (\Im(z))^2}$. For phasor vectors, then, we confess that there can be two uses of the term "magnitude" of this quantity. In some cases we will require the magnitude of the phasor, retaining its vector properties. In other cases we will retain the phasor properties, but combine the vector components to find the magnitude of the vector. We will try to be clear about which meaning we have in mind as we develop these ideas, but hope that, with this word of caution as to the various uses of the term "magnitude," we can avoid confusion in the future as we discuss these quantities.

8.2 Uniform Plane Waves

Within the broad class of time-harmonic waves, the **uniform plane wave**, described by the relation

$$\mathbf{E}(\mathbf{r}, t) = \mathbf{E}_0 e^{j(\omega t - \mathbf{k} \cdot \mathbf{r})}, \tag{8.22}$$

is a widely treated special case. The amplitude $\mathbf{E}_0$ of this wave is a constant, independent of position $\mathbf{r}$. The uniform plane wave is a simple model of propagating radiation, and it lets us determine and discuss the general properties of electromagnetic wave propagation. This wave propagates in a single direction, indicated by the direction of the vector $\mathbf{k}$, known as the **wave vector** or **propagation vector**, and has perfectly flat wavefront (surfaces over which the phase of the wave is constant), which are perpendicular to $\mathbf{k}$. Strictly speaking, $\mathbf{k}$ represents the direction in which the phase $(-\mathbf{k} \cdot \mathbf{r})$ of the wave changes most rapidly. As seen in Eq. (8.22), the wave amplitude and phase are perfectly uniform across the plane perpendicular to $\mathbf{k}$, which also makes the wave somewhat unrealistic from a physical viewpoint. For example, when we later consider the energy carried by propagating waves, we will discover that the infinite extent of a uniform plane wave implies that the energy of this wave is infinite as well. Still, the uniform plane wave is useful for many of its features, and its treatment at this time is instructive.

In our treatment of the uniform plane wave in lossless, isotropic media, we will discover several features that apply to electromagnetic waves in general. We will return to each of these features throughout this section, but list them here as an introduction:

1. The electric field $\mathbf{E}$, the associated magnetic field $\mathbf{H}$, and the wave vector $\mathbf{k}$ are orthogonal to one another, with $\mathbf{E} \times \mathbf{H}$ pointing in the direction $\mathbf{k}$. We say that electromagnetic waves are **transverse** waves, in that the "disturbance" is perpendicular to the direction of propagation.

2. The ratio of the phasor amplitudes $\tilde{E}_0$ and $\tilde{H}_0$ of the electric and magnetic fields, $\tilde{E}_0/\tilde{H}_0$, is constrained to a value that depends only on the permittivity and permeability of the medium through which the wave is propagating. We call this ratio the **intrinsic impedance** of the medium η, and will show that $\eta = \sqrt{\mu/\varepsilon}$.

3. The sinusoidal oscillation of the electric field **E** is in phase with the sinusoidal oscillation of the magnetic field **H**.

4. The phase velocity of the wave is $u_p = 1/\sqrt{\varepsilon\mu}$. We showed this earlier for our first example of a propagating wave, and will show here that it is true generally.

5. Electromagnetic waves carry energy. We will characterize this feature in terms of the **Poynting vector**, whose direction indicates the direction of the flow of energy and whose magnitude is the density of power flow.

Let us start by considering a special case of Eq. (8.22) in the phasor form (i.e. no explicit time-dependence shown),

$$\tilde{\mathbf{E}}(z) = \mathbf{a}_x E_0 e^{-jkz}. \tag{8.23}$$

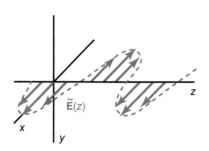

FIGURE 8.8 A snapshot of the uniform plane wave representation at $t = 0$, as described by Eq. (8.23). The electric field vector points in the $\pm x$-direction, varying sinusoidally as a function of z.

This wave, illustrated in Fig. 8.8, propagates in the $+z$-direction, with the electric field pointing in the x-direction. We first show that this wave does indeed satisfy the wave equation, which for a time-harmonic wave is given by Eq. (8.20):

$$\nabla^2\tilde{\mathbf{E}}(z) + \varepsilon\mu\omega^2\tilde{\mathbf{E}}(z) = 0.$$

Since $\tilde{\mathbf{E}}(z)$ does not depend on x or y, the derivatives with respect to these variables are zero, and $\nabla^2\tilde{\mathbf{E}}(z)$ reduces to $\partial^2\tilde{\mathbf{E}}(z)/\partial z^2$. Each partial differentiation with respect to z brings out a factor $-jk$, and the wave equation becomes

$$(-jk)^2\tilde{\mathbf{E}}(z) + \varepsilon\mu\omega^2\tilde{\mathbf{E}}(z) = 0.$$

Simplifying, we have

$$\left[-k^2 + \varepsilon\mu\omega^2\right]\tilde{\mathbf{E}}(z) = 0,$$

and the wave equation is satisfied when $k = \omega\sqrt{\varepsilon\mu}$, consistent with our previous conclusion based upon the wave of Eq. (8.7), whose time-dependence was expressed in terms of a real (i.e. not complex) sine function.

The velocity of this wave can be determined in a manner similar to our earlier treatment. Consider the wave given by Eq. (8.23), which must first be converted back to its time-dependent form. (Recall that this conversion is made by multiplying by $e^{j\omega t}$, and taking the real part.) We consider this wave at two different times, t_1 and t_2, and move with the peak from z_1 to z_2. Since (t_1, z_1) and (t_2, z_2) are both at the peak of the wave, $\omega t - kz$ of the cosine (or the complex exponential) function must be the same for each case, so

$$\omega t_2 - kz_2 = \omega t_1 - kz_1.$$

Rearranging terms, the velocity is

$$u_p = \frac{\Delta z}{\Delta t} = \frac{z_2 - z_1}{t_2 - t_1} = \frac{\omega}{k} = \frac{1}{\sqrt{\varepsilon\mu}}.$$

We have written this uniform plane wave in terms of its electric field **E**, but of course there is a magnetic field **H** associated with this wave as well. Let us now use Faraday's Law to find this associated magnetic field. For time-harmonic fields, we use Eq. (8.18),

$$\nabla \times \tilde{\mathbf{E}} = -j\omega\tilde{\mathbf{B}},$$

which for simple media (linear, isotropic) becomes

$$\boldsymbol{\nabla} \times \tilde{\mathbf{E}} = -j\omega\mu\tilde{\mathbf{H}}.$$

Solving for $\tilde{\mathbf{H}}$,

$$\tilde{\mathbf{H}} = \frac{j}{\omega\mu}\boldsymbol{\nabla} \times \tilde{\mathbf{E}} = \frac{jE_0}{\omega\mu} \begin{vmatrix} \mathbf{a}_x & \mathbf{a}_y & \mathbf{a}_z \\ \dfrac{\partial}{\partial x} & \dfrac{\partial}{\partial y} & \dfrac{\partial}{\partial z} \\ e^{-jkz} & 0 & 0 \end{vmatrix} = \mathbf{a}_y \frac{jE_0}{\omega\mu}(-jk)e^{-jkz}.$$

Simplifying, the phasor magnetic field is

$$\tilde{\mathbf{H}} = +\mathbf{a}_y \frac{k}{\omega\mu}E_0 \exp(-jkz).$$

But we showed earlier that the wave number k must be equal to $\omega\sqrt{\varepsilon\mu}$, which upon substitution gives us

$$\tilde{\mathbf{H}}(z) = +\mathbf{a}_y E_0 \sqrt{\frac{\varepsilon}{\mu}} \exp(-jkz). \tag{8.24}$$

This wave propagates in the $+z$-direction, the same as $\tilde{\mathbf{E}}(z)$, of course, with the magnetic field pointing in the y-direction, as shown in Fig. 8.9. We make three observations from this expression:

1. The magnetic field $\tilde{\mathbf{H}}$ is orthogonal to $\tilde{\mathbf{E}}$ and $\mathbf{k}$. More specifically, the cross product $\tilde{\mathbf{E}} \times \tilde{\mathbf{H}}$ points in the direction $\mathbf{k}$, as we will later show must be the case for propagation through an isotropic medium.

2. The amplitude of the magnetic field H_0 is proportional to that of the electric field,

$$H_0 = E_0\sqrt{\varepsilon/\mu}.$$

The ratio of these amplitudes is called the **intrinsic impedance** of the medium, denoted as

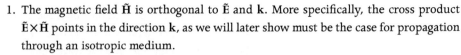

$$\boxed{\eta = \frac{E_0}{H_0} = \sqrt{\frac{\mu}{\varepsilon}}.} \tag{8.25}$$

In other words, the magnetic field amplitude H_0 associated with a wave must be E_0/η, where η depends only on the medium through which the wave propagates. The unit of the intrinsic impedance is $[\eta] = \sqrt{[\mu]/[\varepsilon]} = \sqrt{(\mathrm{H/m})/(\mathrm{F/m})} = \sqrt{(\Omega\,\mathrm{s})/(\mathrm{s}/\Omega)} = \Omega$, as suggested by the name "impedance." (Also, the term impedance is reminiscent of a circuit element, for which the impedance is the ratio of the phasor voltage V across that element to the phasor current I through the element. We see a correspondence with the present case when we associate the electric field with V and magnetic field with I.) For propagation through vacuum, the intrinsic impedance is

$$\eta_0 = \sqrt{\frac{\mu_0}{\varepsilon_0}} = \sqrt{\frac{4\pi \times 10^{-7}\ \mathrm{H/m}}{8.854 \times 10^{-12}\ \mathrm{F/m}}} = 120\pi\ \Omega \approx 377\ \Omega.$$

3. The sinusoidal oscillation of the magnetic field $\mathbf{H}(z, t)$ is in phase with that of the electric field $\mathbf{E}(z, t)$. This is true since the impedance η of a lossless medium is real. Note that $\tilde{\mathbf{E}}$ in Fig. 8.8 and $\tilde{\mathbf{H}}$ in Fig. 8.9 each reach a maximum at $z = 0$.

Let's now return to the general uniform plane wave, described by Eq. (8.22),

$$\mathbf{E}(\mathbf{r}, t) = \mathbf{E}_0 e^{j(\omega t - \mathbf{k} \cdot \mathbf{r})},$$

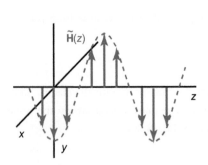

FIGURE 8.9 A snapshot at $t = 0$ of the magnetic field associated with $\tilde{\mathbf{E}}$ shown in Fig. 8.8. The magnetic field vector points in the $\pm y$-direction, varying sinusoidally as a function of z.

and show that these wave properties are generally true, not only for the specific case of Eq. (8.23). The phasor form of this wave is

$$\tilde{\mathbf{E}}(\mathbf{r}) = \tilde{\mathbf{E}}_0 e^{-j\mathbf{k}\cdot\mathbf{r}}. \tag{8.26}$$

We start by showing that the wave vector $\mathbf{k}$ tells us the direction in which the spatial rate of change of the phase of the wave is maximized. We show this by recalling that the gradient operation, discussed in Appendix D, when operating on any scalar function (such as the phase term $\mathbf{k}\cdot\mathbf{r}$ that describes the uniform plane wave), gives us the spatial rate of change of that scalar, and points in the direction in which the rate of change is maximized. In rectangular coordinates, the gradient $\nabla(\mathbf{k}\cdot\mathbf{r})$ is given by Eq. (D.1):

$$\nabla(\mathbf{k}\cdot\mathbf{r}) = \left(\mathbf{a}_x\frac{\partial}{\partial x} + \mathbf{a}_y\frac{\partial}{\partial y} + \mathbf{a}_z\frac{\partial}{\partial z}\right)\left(k_x x + k_y y + k_z z\right).$$

Carrying out the differentiation, we find

$$\nabla(\mathbf{k}\cdot\mathbf{r}) = \left(\mathbf{a}_x k_x + \mathbf{a}_y k_y + \mathbf{a}_z k_z\right).$$

But this vector is simply $\mathbf{k}$, written in terms of its x-, y-, and z-components. This result confirms our statement that the direction of $\mathbf{k}$ tells us the direction in which the phase of the wave varies most rapidly, from which we infer that this is the direction of propagation of the wave, and its magnitude tells us how rapidly, as a function of $\mathbf{r}$, the phase varies.

We next would like to show that the general uniform plane wave, given by Eq. (8.26), satisfies the wave equation. We will start by introducing a shortcut for the spatial derivatives that is valid for the uniform plane wave. We will be able to apply this to the wave equation, as well as to the Maxwell's Equations in which curl and divergence operations abound. Since the only spatial dependence of the plane wave is included in the $\exp(-j\mathbf{k}\cdot\mathbf{r})$ term, taking the derivative of this function with respect to x is just like multiplying by $-jk_x$, that is,

$$\frac{\partial}{\partial x}e^{-j\mathbf{k}\cdot\mathbf{r}} = \frac{\partial}{\partial x}e^{-j\left(k_x x + k_y y + k_z z\right)} = -jk_x e^{-j\mathbf{k}\cdot\mathbf{r}}.$$

Similar substitutions apply for $\partial/\partial y \to -jk_y$ and $\partial/\partial z \to -jk_z$, and the wave equation for time-harmonic waves, Eq. (8.20), becomes

$$\nabla^2\tilde{\mathbf{E}} + \varepsilon\mu\omega^2\tilde{\mathbf{E}} = \left(\frac{\partial^2}{\partial x^2} + \frac{\partial^2}{\partial y^2} + \frac{\partial^2}{\partial z^2} + \varepsilon\mu\omega^2\right)\tilde{\mathbf{E}} = 0$$

$$= \left((-jk_x)^2 + (-jk_y)^2 + (-jk_z)^2 + \varepsilon\mu\omega^2\right)\tilde{\mathbf{E}}.$$

This simplifies to

$$\left(-|\mathbf{k}|^2 + \varepsilon\mu\omega^2\right)\tilde{\mathbf{E}} = 0,$$

which leads to the condition

$$|\mathbf{k}| = \omega\sqrt{\varepsilon\mu}.$$

This condition is identical to the previous result, but now we see that it is equally valid for a wave propagating in an arbitrary general direction $\mathbf{k}$.

The above substitutions can also be used to rewrite Maxwell's Equations in forms that are valid for uniform plane waves. We first examine the divergence of $\tilde{\mathbf{E}}$, when $\tilde{\mathbf{E}}$ is given by Eq. (8.26):

$$\nabla \cdot \tilde{\mathbf{E}} = \left[\mathbf{a}_x \frac{\partial}{\partial x} + \mathbf{a}_y \frac{\partial}{\partial y} + \mathbf{a}_z \frac{\partial}{\partial z} \right] \cdot \tilde{\mathbf{E}}$$

$$= \left[(-jk_x)\tilde{E}_x + (-jk_y)\tilde{E}_y + (-jk_z)\tilde{E}_z \right]$$

$$= -j\mathbf{k} \cdot \tilde{\mathbf{E}}.$$

By a similar process, $\nabla \times \tilde{\mathbf{E}}$ can be shown to equal $-j\mathbf{k} \times \tilde{\mathbf{E}}$ when applied to a plane wave. Similar results apply to the vector magnetic field $\tilde{\mathbf{H}}$. So when operating on a uniform plane wave, $-j\mathbf{k}\cdot$ can be substituted directly for the differential operator $\nabla \cdot$ (i.e. the divergence), and $-j\mathbf{k}\times$ for $\nabla \times$ (i.e. the curl). Applying these to Maxwell's Equations for propagation through a source-free, lossless, homogeneous, isotropic medium gives us

Gauss' Law:	$\mathbf{k} \cdot \tilde{\mathbf{E}} = 0$	(8.27)
No magnetic charges:	$\mathbf{k} \cdot \tilde{\mathbf{H}} = 0$	(8.28)
Faraday's Law:	$\mathbf{k} \times \tilde{\mathbf{E}} = \omega\mu\tilde{\mathbf{H}}$	(8.29)
Ampère's Law:	$\mathbf{k} \times \tilde{\mathbf{H}} = -\omega\varepsilon\tilde{\mathbf{E}}.$	(8.30)

Turning Faraday's Law around to find $\tilde{\mathbf{H}}$ leads to

$$\tilde{\mathbf{H}} = \frac{1}{\omega\mu} \mathbf{k} \times \tilde{\mathbf{E}},$$

and since $\mathbf{k}$ has magnitude $\omega\sqrt{\varepsilon\mu}$, $\tilde{\mathbf{H}}$ is

$$\tilde{\mathbf{H}} = \sqrt{\frac{\varepsilon}{\mu}} \, \hat{\mathbf{k}} \times \tilde{\mathbf{E}},$$

where $\hat{\mathbf{k}}$ is the unit vector pointing in the same direction as $\mathbf{k}$. Since the intrinsic impedance of the medium is $\eta = \sqrt{\mu/\varepsilon}$, this becomes

$$\tilde{\mathbf{H}} = \frac{1}{\eta} \, \hat{\mathbf{k}} \times \tilde{\mathbf{E}}.$$

This relation implies that $\tilde{\mathbf{H}}$ is perpendicular to $\mathbf{k}$, and that $\tilde{\mathbf{H}}$ is perpendicular to $\tilde{\mathbf{E}}$, and that the ratio of the magnitudes of $\tilde{\mathbf{E}}$ to $\tilde{\mathbf{H}}$ is equal to the intrinsic impedance η, as we discovered for the previous example.

We can also invert Ampère's Law, as expressed in Eq. (8.30), to write

$$\tilde{\mathbf{E}} = -\frac{1}{\omega\varepsilon} \, \mathbf{k} \times \tilde{\mathbf{H}}.$$

Using the same relations that we used to simplify Faraday's Law, this gives us

$$\tilde{\mathbf{E}} = -\sqrt{\frac{\mu}{\varepsilon}} \, \hat{\mathbf{k}} \times \tilde{\mathbf{H}} = -\eta \, \hat{\mathbf{k}} \times \tilde{\mathbf{H}}.$$

This relation leads to the same ratio of $\tilde{\mathbf{E}}$ to $\tilde{\mathbf{H}}$, and tells us that $\tilde{\mathbf{E}}$ is perpendicular to $\mathbf{k}$ and that $\tilde{\mathbf{E}}$ is perpendicular to $\tilde{\mathbf{H}}$. Combining these results with those of the previous paragraph, $\tilde{\mathbf{E}}$, $\tilde{\mathbf{H}}$, and $\mathbf{k}$ must be orthogonal to each other, with $\tilde{\mathbf{E}} \times \tilde{\mathbf{H}}$ pointing in the direction $\mathbf{k}$. We often use illustrations such as those shown in Fig. 8.10 to represent uniform plane waves. In Fig. 8.10(a), the straight parallel lines represent the wavefronts of the wave, which are perpendicular to the wave vector $\mathbf{k}$. The electric and magnetic field vectors lie within the planes defined by the wavefronts, and the field strength anywhere in these planes is the same. In Fig. 8.10(b), three orthogonal vectors are shown, representing the electric field $\tilde{\mathbf{E}}$, the magnetic field $\tilde{\mathbf{H}}$, and the wave vector $\mathbf{k}$. $\tilde{\mathbf{E}}$ and $\tilde{\mathbf{H}}$,

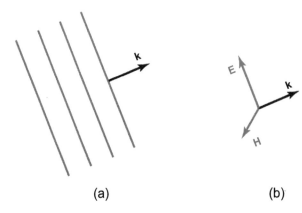

FIGURE 8.10 Illustration of a uniform plane wave.

of course, are phasors representing the time-dependent field quantities, while **k** is a constant. The cross product $\tilde{\mathbf{E}} \times \tilde{\mathbf{H}}$ is parallel to **k**. These conclusions are all perfectly consistent with the conclusions that we drew for the first example, but now we see that they hold in the general case of the uniform plane wave expressed by Eq. (8.26).

In this section, we have introduced propagation of electromagnetic waves, as governed by Maxwell's Equations. We have also introduced the uniform plane wave as a simple model for wave propagation. This model has allowed us to explore many of the properties of wave phenomena. We have seen that electromagnetic waves are transverse, that the electric and magnetic fields must be perpendicular to one another, and that the ratio of their magnitudes must equal the impedance of the medium.

LEARNING OBJECTIVES

After studying this section, you should be able to:
- determine the Doppler shift due to the motion of an object; and
- describe applications of the Doppler shift, including traffic radar, Doppler weather radar, and Doppler astronomy.

8.3 Doppler Effect

Let's take a break from all that heavy-duty vector calculus, and examine a wave phenomenon with wide-ranging applications: the Doppler shift. The **Doppler shift** is a change in the apparent frequency of a wave that results from the motion of an object. This effect applies to wave phenomena in general, not just electromagnetic waves. The Doppler shift of electromagnetic waves allows us to measure atmospheric conditions (Doppler weather radar), makes air travel safer through wind shear detection, provides a means of traffic speed control, and allows for measurements of the velocity of distant stars and galaxies.

Most of us are familiar with the Doppler effect as manifested in acoustic waves. Imagine a train whistle that might be heard when standing by the side of railroad tracks (at a safe distance, of course) with the train approaching. The pitch (a.k.a. the frequency) of that whistle changes from high to low as the train passes by. The wave, which is acoustic in this example, is a series of compressions and rarefactions of the air (the medium that carries the wave). If you were sitting in the cab of the train, you would hear the whistle at its frequency f_0. We call this the rest frame frequency, since this is the frequency of the wave emitted by the whistle, or the emitter, as observed by an observer who is at rest relative to the emitter. But if you were standing by the side of the tracks, with the train approaching, the compressions and rarefactions of the acoustic wave reach you at a higher frequency. In the time between one compression and the next, the whistle has moved closer, and so the next compression has less distance to travel to reach you than does an earlier one, and therefore takes less time. After the train passes by, just the opposite effect takes place. The acoustic wave is shifted to a lower frequency since the time from one compression to the next is increased, and therefore the wave appears to be at a lower frequency.

The Doppler shift can be quantified in the following way. The time T observed between crests of the wave, or rather the period of that wave, as the emitter moves toward an observer is

$$T = T_0 - \frac{\Delta x}{u_p},$$

where T_0 is the period of the wave in its rest frame, Δx is the distance the emitter has moved in the time T_0, and u_p is the velocity of the wave. The time $\Delta x/u_p$ represents the decrease in the time between crests of the wave due to the motion of the emitter. But the distance Δx depends on the period as $\Delta x = u T_0$, where u is the velocity of the emitter. So the period measured by a stationary observer is

$$T = T_0 \left(1 - \frac{u}{u_p} \right).$$

Since the frequency of the wave is the inverse of its period

$$f = f_0 \left(1 + \frac{u}{u_p} \right),$$

where we have used the Taylor expansion, $(1 - \varepsilon)^{-1} \approx 1 + \varepsilon$, Eq. (H.5), valid for small numbers ε. For an emitter moving *away* from an observer, the time between crests is increased, and the sign of the Doppler shift is reversed. Any motion in a transverse direction (i.e. perpendicular to the line between the source and the observer) doesn't cause a Doppler shift, at least not to first order. These results can be summarized as

$$f = f_0 \left(1 + \frac{u \cos \theta}{u_p} \right),$$

where θ is the angle between the emitter's velocity $\mathbf{u}$ and the line pointing from the emitter to the observer. $\theta = 0$ indicates the emitter is moving toward the observer, and $\theta = \pi$ indicates the emitter is moving away from the observer. This Doppler-shifted frequency can also be written very compactly as

$$\omega = \omega_0 + \mathbf{k} \cdot \mathbf{u},$$

where $\omega = 2\pi f$ is the angular frequency and $\mathbf{k} = (\omega/u_p)\hat{\mathbf{k}}$ is the wave vector, as defined in the previous section. Applications in many arenas abound, as discussed in TechNotes 8.5–8.7.

TechNote 8.5 Doppler Traffic Radar

Most radar detectors used today by the police for monitoring traffic operate in the Ka band. (See Table 8.1 for definitions and applications of various frequency bands.) The FCC authorized this band for traffic radars in the United States in 1983, and expanded the band for this purpose to 33.4–36.0 GHz in 1992. Most models have 100 MHz or 200 MHz bandwidth. Some radar models include a feature for frequency hopping as a means of frustrating radar detectors, but these are used only infrequently today. Other frequency bands that are less commonly used for traffic speed detection include the K band (24.150 GHz), the Ku band (13.45 GHz), and the X band (10.525 GHz). The Ka band radar detector has a narrower, more collimated beam due to its higher frequency than the other bands, but its effective range can be decreased by poor weather conditions.

Example 8.2 Doppler Traffic Radar

A car is traveling over the posted speed limit at a speed of 75 mph. What is the frequency shift of the radio wave reflected head-on by this speeding auto? Assume that the radar detector is in the K band ($\sim$24.150 GHz). Determine the frequency resolution required of the detector to record the velocity with a ± 1 mph uncertainty. What is the minimum time required to make this measurement?

Solution:

We will have to apply the Doppler shift twice to solve this problem. The frequency f of the radio wave as would be observed in the rest frame of the car is $f_0(1 + u/c)$, where f_0 is the frequency emitted by the radar gun, and u is the velocity of the car. Note that the apparent frequency is shifted up since the car is moving toward the detector. This radio frequency wave reflects off the car back toward the detector, which is colocated with the emitter. In the rest frame of the speeding car, the frequency of the reflected wave is the same as the frequency f, and the detector will observe this reflected wave with a Doppler shift as well, since the car is moving toward the detector. The frequency of the reflected wave as seen at the detector is $f(1 + u/c) = f_0(1 + u/c)^2$. Since u/c is small, we can approximate this as $f_0(1 + 2u/c)$. The Doppler shift is then

$$\Delta f = f - f_0 = 2f_0 u/c$$
$$= 2(24.150\,\text{GHz})(75\,\text{mph})(1\,600\,\text{m/mile})(\text{hr}/3\,600\,\text{sec})/(3 \times 10^8\,\text{m/s})$$
$$= 5.3\,\text{kHz}.$$

The minimum uncertainty in this measurement if we require an accuracy of $\delta u = 1$ mph for the velocity measurement is $\delta f = 2f_0 \delta u/c = 70$ Hz.

The minimum time necessary to achieve this frequency resolution comes from the Fourier transform relations, which tell us that in order to achieve a frequency resolution δf, the duration of a measurement must be at least δt, where $\delta t \delta f = 1/2\pi$. From this we determine $\delta t = 2$ ms.

The Doppler shift is also exploited to help monitor weather conditions, as described in TechNote 8.6.

TechNote 8.6 Doppler Weather Radar

We've all seen the Doppler radar images on the local television weather reports. To construct these images, a sequence of repetitive (1 kHz repetition rate), high-power (450 kW), short-duration (1.57 μs) radar pulses is emitted by a transmitter, and a receiver "listens" for any waves reflected by bodies in the path of the outgoing wave. Radar signals at different elevations must also be collected in order to detect weather patterns at different altitudes. A full volume scan is collected in 4–6 minutes. The time delay of the reflected signal yields the distance to the weather system, while the size and/or density of precipitation shows up as the intensity of the reflected wave. Additionally, the Doppler frequency shift determines the velocity of the scatterers. Rotating weather patterns, as indicative of precursors to tornados, are indicated in the Doppler signal as well. Different Doppler weather radar systems operate in different frequency bands, L (1–2 GHz), S (2–4 GHz), C (4–8 GHz), X (8–12 GHz), or K (18–27 GHz). These and other frequency bands are listed in Table 8.1.

Commercial aircraft are equipped with related technology for detection of a phenomenon known as wind shear. Wind shear occurs from microbursts, producing changing wind patterns that affect the lift of the wings of the aircraft. Wind shear can be especially hazardous during take-off and landing.

The Doppler shift of light emitted by distant galaxies is an important tool used by astronomers to determine their velocity, as is explored in TechNote 8.7 and Example 8.3.

TechNote 8.7 Doppler Astronomy

By measuring the Doppler shift of spectral lines emitted by distant stars and galaxies, their radial velocity relative to our position can be determined. The light from sources moving toward us is blue-shifted (the apparent frequency is greater than that of light emitted from a body at rest), while light from sources moving away from us is red-shifted (the apparent frequency is less than that of light emitted from a body at rest). The Doppler shift can also reveal the existence of binary star systems, in which two stars orbit one another. In this case, the Doppler shift of one star of the system would be to the red and the other to the blue. As these stars orbit one another, the Doppler shift varies, and the period of the oscillation reveals the period of the orbit of the stars.

Note that we cannot determine the Doppler shift by observing just a single spectral line, since we must be able to identify the element that is emitting the light. When we observe several lines from the same element, each shifted by a Doppler shift consistent with a single velocity, this is sufficient to determine the velocity of the body. Note also that the Doppler shift is sensitive primarily to the radial velocity (i.e. the velocity in the direction toward us or away from us) of the star.

Example 8.3 Doppler Astronomy

Light from a distant galaxy contains a characteristic spectral line of atomic oxygen. The wavelength of this line from a source on Earth is $\lambda_0 = 513$ nm, but it is displaced to $\lambda = 525$ nm in the spectrum from the distant galaxy. Find the velocity **u** of this galaxy.

Solution:

Measurement of the Doppler shift can only be used to determine the component of the velocity **u** that is directly toward or away from us, labeled $u_\parallel$. Since the observed spectral line is at a longer wavelength than it would be in the laboratory (i.e. it is red-shifted), this is a lower frequency, and we conclude that the galaxy is moving away from us, consistent with an expanding universe. The ratio of wavelengths λ_0/λ is

$$\frac{\lambda_0}{\lambda} = \frac{c/f_0}{c/f} = \frac{f}{f_0} = 1 + u_\parallel/c.$$

Solving this for the velocity, we find $u_\parallel = -6.8 \times 10^6$ m/s, where the minus sign also indicates that the galaxy is moving away from us. This velocity can also be expressed as $-0.023c$, a non-negligible velocity when compared to the velocity of light in vacuum.

LEARNING OBJECTIVE

After studying this section, you should be able to:
- describe qualitatively various applications of electromagnetic waves.

8.4 The Electromagnetic Frequency Spectrum

It is safe to say that, of all the frequency ranges of electromagnetic waves, we are primarily impacted by the optical range, which spans in wavelength from only 400 nm (violet) to 700 nm (red) (frequency range $4.3 - 7.5 \times 10^{14}$ Hz). Our eyes are sensitive to these wavelengths, and vision is extremely important to our everyday functioning. While electromagnetic waves from d.c. (0 Hz) to γ-rays (10^{25} Hz) (and beyond?) fill the space

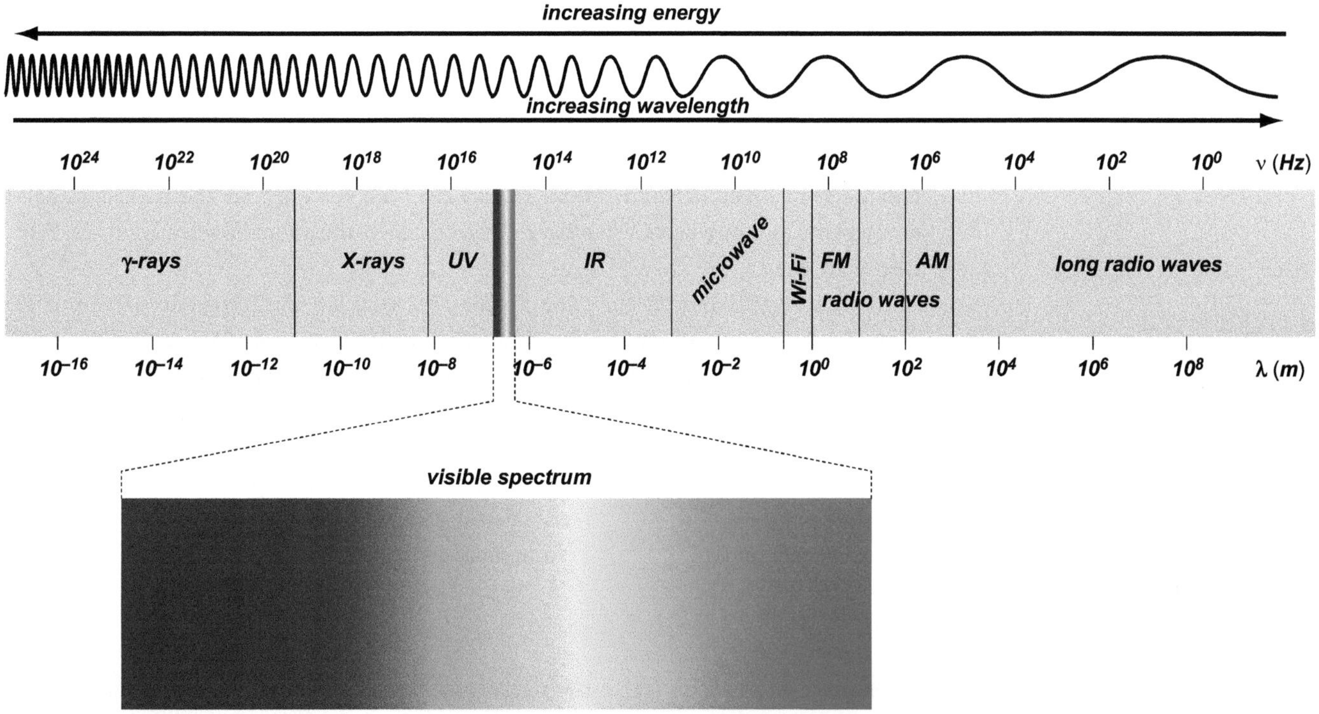

FIGURE 8.11 The electromagnetic spectrum, ranging from long radio waves (up to 100 kHz) to γ-rays (10^{20}–10^{24} Hz).

around us, it is probably safe to say that we are mostly unaware of their presence. But these waves are virtually always present, and many methods for using them for a multitude of purposes have been developed. We will review a few of these in this section.

Applications of electromagnetic waves depend upon their frequency, and methods of generating and detecting waves in different frequency bands are required. These important technical differences notwithstanding, electromagnetic waves are all fundamentally the same, as described by Maxwell's Equations, or equivalently by the wave equation. In this section we take a closer look at the frequency spectrum of electromagnetic waves, and specifically at some of their applications, and how we can learn about the universe surrounding us from them.

The visible light spectrum that we can see makes up only a very small part of the electromagnetic spectrum, as shown in Fig. 8.11. Longer-wavelength radiation includes the infrared (IR), microwave, and radio frequencies. Shorter wavelengths include ultraviolet (UV), X-rays, and gamma (γ) rays. Each of these regions of the spectrum occur naturally in the physical world, and can be generated, detected, and used for different applications using a multitude of different sources, detectors, and techniques.

The IR spectrum abuts the visible spectrum on the long wavelength side, ranging in wavelength from 0.7 μm to 1 mm (in frequency from 4.3×10^{14} Hz to 3×10^{11} Hz, or 300 GHz). The all-important optical communications band around 1.55 μm, where losses in optical fibers are minimal, is in the IR region, as are the majority of molecular vibrational resonances of molecules. The peak radiance of the blackbody spectrum emitted by a room-temperature object falls in the IR region, prompting some to equate heat with IR light. This is, of course, a false equality, in that heat is the quantity of thermal energy of a body, while the IR spectrum is the range of its emitted radiation. Besides warm objects, whose emission spectrum is very broad, IR light-emitting diodes (LEDs) and several types of lasers can be used to generate light in the IR range. Detection of IR light can be achieved using photodiodes (devices that convert incident light into an output current, usually on a one-photon-in to one-electron-out basis), fabricated from materials such as indium gallium arsenide (InGaAs) or germanium (Ge).

The long wavelength range of the IR spectrum, say from wavelengths of 0.01–1 mm (frequencies from 300 GHz to 30 THz), has been receiving a great deal of interest lately, because sources and detectors in this range have been missing from the arsenal of tools of scientists and engineers. This band is often called the submillimeter band, or THz frequency band.

At wavelengths shorter than the visible spectrum are the ultraviolet (UV) range, X-rays, and γ-rays (in order of increasing frequency). Our means of generating UV light is limited to specific wavelengths, typically by using visible wavelength lasers, followed by nonlinear optical conversion techniques to shift the frequency into the UV range. A strong effort to generate shorter wavelengths has been ongoing, largely motivated by photolithography for semiconductor fabrication, which requires short-wavelength light to produce finer feature sizes. Optical data storage on CDs, DVDs, and Blu-ray devices benefit for the same reason from shorter wavelengths. The primary uses of X-rays

Table 8.1

A summary of some of the IEEE designation of radar bands, with frequency and wavelength ranges, and common uses for each entry.			
Letter designation	Frequency range (GHz)	Wavelength (cm)	Common uses
L band	1–2	15–30	GPS, radio, telecommunications, radar, weather radar, and aircraft surveillance radar
S band	2–4	7.5–15	Wi-Fi, Bluetooth, air traffic control, weather radar, communications satellites
C band	4–8	3.75–7.5	Satellite communications, Wi-Fi, cordless telephones, weather radar, aircraft surveillance radar
X band	8–12	2.5–3.75	Weather radar, air traffic control, maritime tracking, police traffic speed radar
Ku band	12–18	1.67–2.5	Downlink for satellite television broadcast, other satellite communications
K band	18–26.5	1.13–1.67	Satellite communications, cellular telephones, terrestrial microwave communications, police traffic speed radar, weather radar
Ka band	26.5–40	0.75–1.13	Communication satellite uplinks, military radar, police traffic speed radar
Q band	33–50	0.60–0.90	Satellite communications, terrestrial microwave communications, radio astronomy
V band	50–75	0.40–0.60	Open use in many countries
W band	75–110	0.27–0.40	Satellite communications, millimeter-wave radar, military radar

are in medical imaging, airport screening, and material inspection. γ-rays emitted from neutron stars, pulsars, and supernova explosions are collected and studied to unlock the mysteries of the universe, and to learn of its origins.

At longer wavelengths, from 1 mm to 1 m (300 MHz to 300 GHz) and >1 m (<300 MHz) are the microwave and radio frequency bands, respectively. These bands are heavily used for communications purposes, in many different forms, as we will cover in some of the following TechNotes. Generation and detection of waves in these regions uses currents in antennas, in forms which are fundamentally different from generation and detection of waves in the IR, visible, and UV spectra. In the following notes we explore many different applications of electromagnetic waves in some of these frequency bands.

In Table 8.1, some of the microwave frequency bands, as defined by the IEEE, are listed. In order to avoid spectral congestion, specific bands are regulated for specific purposes, some of which are listed in this table. We will cover a few of these applications in more detail in the following notes. The first application is Wi-Fi, common in homes and offices.

TechNote 8.8 Wi-Fi

Wi-Fi networks are commonly used to connect computers, laptops, smart phones, and smart TVs in the home, restaurant, or business. A wireless router for home use is shown in Fig. 8.12. Wi-Fi networks usually operate in one of two frequency bands centered at either 2.4 GHz or 5.8 GHz, and the transmitted power is relatively low, typically ~100 mW. The range for the Wi-Fi signal is typically up to 20 m, but this does vary, depending on the local environment and type of antenna. Walls and other obstructions can also weaken the signal. Data speeds can be as large as 1 Gbit/s, but these also vary considerably.

FIGURE 8.12 A wireless router.

Electromagnetic waves, either in the IR or radio-frequency spectrum, generated by remote controls are used to control many devices, as described in TechNote 8.9.

TechNote 8.9 Remote Controls

Remote controllers for your television, garage door opener, car door locks, or toys, as shown in Fig. 8.13, are based on either LEDs that emit electromagnetic waves in the IR region of the spectrum, or radio-frequency (rf) electronics. LED controllers are commonly used to control televisions, where short-range, same-room operation is adequate. The IR radiation produced by the LED is somewhat directional, so the controller must be pointed in a direction close to the receiver, where the signal is detected by an IR-sensitive photodiode. rf waves produced by rf controllers can have longer range, can pass through walls (which are opaque to IR radiation, but not to rf waves), and are not as directional as the IR controller. For either type of controller, the wave is modulated to encode the device address and command in the signal.

FIGURE 8.13 A young control expert negotiates a tight curve with his rf remote-controlled vehicle.

In TechNote 8.10, rf identification systems are described.

TechNote 8.10 Radio Frequency Identification

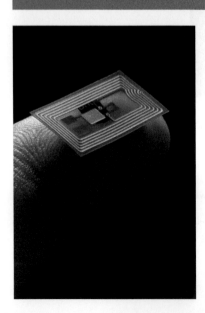

Radio frequency identification, often referred to simply as rfid, finds applications for identifying and tracking in manufacturing, supply chain management, self-check-out and shoplifting detection in retail stores, controlling access to facilities or rooms, identification and tracking in the railway system, electronic passports, mass-transit fare collection, E-ZPass toll collection, bike lockers and Zipcars, book identification in libraries, and many more. An rfid tag can be attached to, embedded in, or implanted in an item, animal, or person to track, identify, or provide information to the reader. The rfid tag consists of an electronic chip, a radio transponder (receiver and transmitter), and a high-frequency antenna, typically on a substrate. The tag communicates with a reader, which is tuned to collect the emitted waves at the frequency of the tag.

Passive tags, which have no battery or other power source of their own, are powered by the radiation from the reader itself. The waves from the reader induce an EMF in the transponder coil, which is rectified and powers the chip. This EMF also powers the transmission of the waves back to the reader. A passive tag requires a higher-power reader than an active tag, and the range of the signal from a passive tag is often quite limited, as small as 10 cm. This limited range is actually an advantage for applications where privacy or security are required.

Active tags, which contain a battery, are larger, more expensive, and periodically require replacement of the battery. To their advantage, they have a longer range, sometimes 100 m or more. Direct line of sight is not required to read an rfid tag. Some rfid tags contain memory and logic, which allows the chip to store information about the tagged unit. The frequency of rfid varies, spanning from LF (120–150 kHz) used for animal identification and inventory tracking, with a range of 10 cm, to microwave (2.45–5.80 GHz).

The spectrum of electromagnetic waves emitted by an object can be used to determine the temperature of the object. This offers the capability of contactless thermometers, which are described in TechNote 8.11.

TechNote 8.11 IR Thermometer

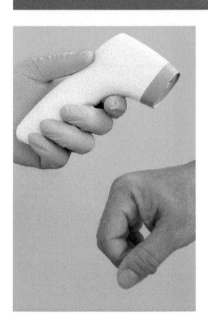

An IR thermometer, such as the one demonstrated in Fig. 8.14(a), collects and measures the blackbody radiation emitted at a surface to determine the temperature of that surface. Any body at temperature T emits radiation in a characteristic spectrum known as the blackbody radiation (BBR) pattern. Wien's Displacement Law gives the radiance per unit area of the emitter per unit solid angle

$$P(\lambda, T) = \frac{2hc^2}{\lambda^5} \frac{1}{e^{hc/\lambda k_B T} - 1},$$

where $h = 6.626 \times 10^{-34}$ Js is the Planck constant and $k_B = 1.381 \times 10^{-23}$ J/K is the Boltzmann constant. (The appearance of the Planck constant in this expression is an indication that its derivation is based on quantum theory. In fact, the development of this expression by Max Planck in 1901, and its dependence on quantization of the energy of the electromagnetic modes, is generally recognized as the beginning of the quantum era. Planck was awarded the Nobel Prize in physics in 1918 for this discovery.) A semi-log plot of the BBR spectrum at several temperatures is shown in Fig. 8.14(b). Notice that as the temperature increases, the spectrum increases in amplitude, and the peak moves closer to the visible region of the spectrum. Thus, molten metal seems to glow, and we say it is "red hot," or even "white hot." For objects at room temperature, the peak of the spectrum is in the IR region. An IR thermometer collects the BBR radiation emitted from the surface, and measures the power of this radiation using a thermopile. (A thermopile is a set of thermocouples, usually connected in series, which produces a voltage dependent upon the temperature of the thermocouples. A thermocouple is a junction between two dissimilar metals, which generates a voltage across the junction that varies with temperature.)

The amplitude of BBR emitted by most objects is actually less than that of the ideal spectra shown in the figure. This is characterized by a factor known as the emissivity of the surface, a coefficient that scales from 0 to 1. For example, a highly reflective surface (such as a mirror) simply reflects the radiation incident upon it, without regard to the temperature of the reflector. Its emissivity is close to zero. On the other hand, skin, water, or paper emit BBR that approaches the ideal spectrum, and the emissivity of these materials is close to 1. Regardless, the emissivity of the surface whose temperature is being measured must be included, and an IR source and a second thermopile are used in an IR thermometer to correct the measurement for the surface emissivity.

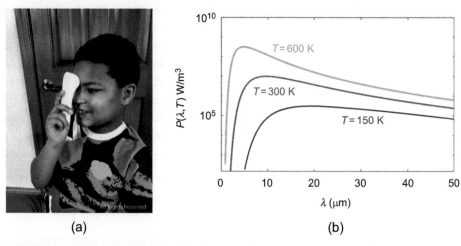

(a)

(b)

FIGURE 8.14 (a) A young public health official demonstrating the proper use of an IR thermometer. (b) A calculated semi-log plot of the blackbody radiation spectrum, as given by the Wien's Displacement Law equation, at three temperatures. Notice how the radiant power grows with increasing temperature, and how the peak shifts to shorter wavelengths.

X-rays, a form of electromagnetic waves of extremely short wavelength (high frequency), helped usher in the era of modern physics, and are routinely used in medical diagnostics, security screening, and material testing. X-ray imaging is the topic of TechNote 8.12.

TechNote 8.12 X-Ray Imaging

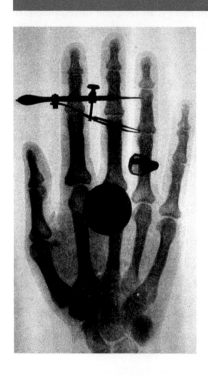

X-ray imaging is used extensively for medical diagnostics, airport screening, and microscopic examination of materials. X-rays are a form of electromagnetic waves, similar to light and radio signals, but at a much shorter wavelength and higher frequency. The wavelength range of X-rays is 0.01–10 nm, corresponding to a frequency of 3×10^{19}–3×10^{16} Hz.

X-rays were first observed by Wilhelm Röntgen (1845–1923) in 1895, when he was studying electrical discharges in a low-pressure gas at the University of Würzburg. Similar discharges, and the properties of cathode rays, as these currents were called at the time, had been studied previously by several other investigators, but Röntgen, pictured in Fig. 8.15(a), noted that paper coated with barium platinocyanide sitting on an adjacent bench fluoresced (glowed) when placed in the path of the rays, even though the gas discharge was completely enclosed within a thick black carton. He further noted that objects of different thicknesses affected the brightness of the radiation as recorded on a photographic plate, and recorded an image of the hand of Anna Bertha Röntgen (his wife), showing the shadow of the bones in her hand and a ring on one finger, when exposed to the radiation. This image, reproduced in Fig. 8.15(b), is recognized as the first medical X-ray image. In recognition of his discovery of X-rays and their use, he was awarded the first Nobel Prize in physics, in 1901.

(a)

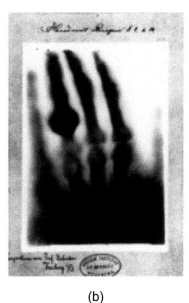

(b)

X-ray tube

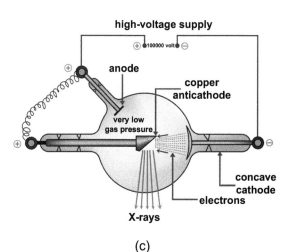

(c)

FIGURE 8.15 (a) Wilhelm Röntgen. (b) The first medical X-ray image, showing the hand of Anna Bertha Röntgen. The shadow of the bones in her hand and a ring on one finger are visible in this image. (c) Illustration of a Crookes X-ray tube.

Modern X-ray machines operate using a similar process in an X-ray tube, illustrated in Fig. 8.15(c). A heated filament (cathode) generates free electrons, which are accelerated toward the anode in the tube. The anode is metal, commonly tungsten, which emits X-rays when high-energy electrons strike it.

Computed tomography (CT) is a technique for constructing three-dimensional images of the body using multiple X-ray images. The 1979 Nobel Prize in physiology or medicine was awarded to Allan MacLeod Cormack for his theoretical studies leading to, and Godfrey Hounsfield for his demonstration of, the first CT scanner.

The spectrum of microwaves and their distribution over the sky holds important information about the beginning of the universe, as described in TechNote 8.13.

TechNote 8.13 Cosmic Microwave Background

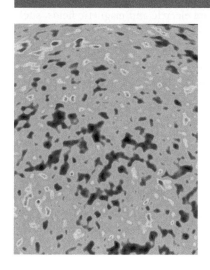

When we look up to the night sky, we see the vast array of stars and galaxies, if we're lucky enough to be outside the range of nightglow from nearby towns or cities. The visible light that can be seen from the stars is only part of the broad spectrum of electromagnetic waves incident upon the Earth. In particular, there exists a broad background of microwaves, first observed by Andrew McKellar in his studies of the cyano (CN) radical in interstellar space in 1941, and rediscovered by Arno Penzias and Robert Wilson in 1964. Penzias and Wilson first thought that the noisy radio signal they observed was due to dirt on the Holmdel Horn Antenna used for their measurement, but after a thorough cleaning the signal persisted, and it was soon recognized that this microwave signal was a remnant of the early universe. This radiation spectrum, known as the cosmic microwave background (CMB), peaks at ∼160 GHz, corresponding to an average temperature at the edges of the universe of 2.7 K. In 1978, Penzias and Wilson were awarded a share of the Nobel Prize in physics for their work.

More recent measurements have helped physicists explore the beginnings of the universe through the CMB. Figure 8.16 shows results from the space-based measurements of the Wilkinson Microwave Anisotropy Probe (WMAP), a temperature map of the universe. This map was generated from measurements of the spectrum of the microwave background at different azimuths and elevations of the sky. The color map shows temperature deviations from the average temperature of 2.7 K, with the signal from our galaxy (the Milky Way) subtracted. The magnitude of the fluctuations is small, only $\sim 200\,\mu K$. Analysis of this temperature map shows evidence of the oscillations of the hot gas of the early universe, the curvature of the universe, and the predominance of dark matter in the universe.

FIGURE 8.16 The detailed, all-sky picture of the infant universe created from nine years of WMAP data. The image reveals 13.77 billion-year-old temperature fluctuations (shown as color differences) that correspond to the seeds that grew to become the galaxies. The signal from our galaxy was subtracted using the multi-frequency data. This image shows a temperature range of $\pm 200\,\mu K$.

Out constant immersion in a sea of electromagnetic waves has led to concerns over the effects of these waves on our health. This is the topic of TechNote 8.14.

TechNote 8.14 Health Effects of Electric and Magnetic Fields

Concerns over potential health risks to humans from electric and magnetic fields, such as those produced by high-tension transmission lines, microwave ovens, or cell phones, have sometimes been raised over the past several decades. (Absorption of waves by your body, of course, can lead to local heating effects, which can be detrimental if extreme. We consider here, however, the direct effect of electric and magnetic fields on tissues of your body.) The website of the National Institute of Environmental Health Sciences (NIEHS) reports that there is little evidence to support these concerns from low-frequency sources. Low-frequency fields, from d.c. to visible light, are labeled non-ionizing radiation fields. These include the radiation produced by power lines, radio and television broadcast antennas, microwave ovens, wireless networks, and MRIs. High-tension power lines can carry currents as large as 1 000 A, but at a distance of 30 m the magnetic flux density of this field has dropped to $\sim$ 70 mG. As a point of reference, the magnetic field of the Earth is about 500 mG. Cell phones might be a source of concern, since we typically hold them directly to our ear when used for conversations. While no evidence of health risks for cell phones has been found yet, research on this issue continues.

The risk associated with ionizing radiation is greater than that of non-ionizing radiation. Ionizing radiation includes fields in UV, X-ray, and γ-rays, originating from sources such as sunlight, tanning beds, and X-ray machines. Cellular and DNA damage in some circumstances under prolonged exposure have been linked to ionizing radiation. This is the reason the X-ray technician or dental hygienist may cover parts of your body that are not to be imaged with a lead apron, and operates the X-ray machine remotely when taking X-ray images of your body or teeth.

8.5 Polarization

In Section 8.2, we showed that the electric field of an electromagnetic wave propagating through an isotropic medium must lie in the plane perpendicular to the direction of propagation, which is defined by the wave vector **k**. We call a wave with this property a transverse wave. We also showed that the associated magnetic field **H** must be perpendicular to **E** and **k**. But even within this limited space, there is still a great deal of freedom, for **E** can point in *any* direction within the plane perpendicular to **k**. There are an infinite number of directions that qualify. In addition, **E** need not point in a single direction at all points, but can spiral around in a helical pattern. We call the direction and helicity of the electric field **E** the **polarization** of the wave. We can control and make use of the polarization state of the wave for a variety of applications, and so in this section we will investigate this important property.

We start with a word of notational caution here. The term *polarization* was used previously, in Chapter 1. At that time, the polarization of a dielectric medium, denoted with the symbol **P**, was defined as the product of the density of electric dipoles and the average dipole moment within that medium. Now the same term polarization is used to describe the direction in which the electric field of an electromagnetic wave points. *These two terms have nothing to do with one another*, and we apologize for the ambiguity that this repeated usage brings. This is, however, the common nomenclature. It will be clear, we hope, which definition is intended through the way in which it is used.

Many sources generate electromagnetic waves that are naturally polarized in one direction or another. We will discuss linear antennas later in this text, where it will be seen that the polarization of the radiation that is produced by a vertical antenna and that propagates in a horizontal plane is vertical. We will explore this more later, but for now we should appreciate the polarization of the wave, which for this example is vertical.

Polarization effects were instrumental to Heinrich Hertz in his classic experiments of the 1880s, in which he confirmed that radio waves are electromagnetic waves, similar to light. (See Biographical Note 8.1.) In one experiment, he constructed a pair of polarizers for his rf electromagnetic waves, each consisting of a series of parallel wires. A **polarizer** is an element that allows transmission of waves polarized in one direction, but absorbs or reflects the waves polarized in the orthogonal direction. With his two polarizers oriented parallel to one another, he was able to detect waves transmitted through the polarizer pair from a source to a detector. But by rotating one of the polarizers so that its transmission axis was perpendicular to the transmission axis of the other, the amplitude of the transmitted wave was attenuated, and no power was detected. This effect can be understood by considering the action of each polarizer in turn. If the initial electromagnetic wave is unpolarized, we can think of this wave as consisting of two polarization components, one parallel to the transmission axis of the first polarizer, the other perpendicular, in equal proportions. The first polarizer polarizes the wave – that is, it allows only one polarization component to be transmitted. (The polarization component that is parallel to the wires can induce currents in the wire, doing work on the charges. This polarization component is therefore strongly absorbed. The orthogonal polarization component of the field, however, cannot drive any current, since the electrons are confined to the wire, and so this component is transmitted through the polarizer without attenuation. For reasons that we won't try to explain, this argument is valid when the spacing between the wire elements is small compared to the wavelength of the wave. Thus the radio waves in Hertz' demonstration were polarized by the parallel-wire polarizer, while visible light, whose wavelength is much smaller, passes through largely unaffected.) Then the second polarizer allows this wave to pass through only if the second polarizer is oriented along this same direction. If the second polarizer is crossed with the first, then it absorbs the wave, and no wave is transmitted.

Polarizing sheets such as found in some sunglasses operate in a similar fashion, but on a molecular scale. These are described in TechNote 8.15.

TechNote 8.15 Absorptive Polarizers

Polarizing sunglasses can greatly improve your vision on bright, sunny days by reducing the glare from shiny surfaces in your field of view. Polarizing sunglasses are effective because the light reflected from horizontal surfaces, such as a wet road surface or oncoming cars, is partially horizontally polarized. Direct sunlight itself, of course, is not polarized. Polarized sunglasses, which strongly absorb the horizontally polarized component of light, then reduce glare.

The concept behind polarizing sunglasses for optical waves is very similar to that of the original polarizer that Heinrich Hertz constructed for application with radio waves. Originally, the composition of polaroid films developed by Edwin Land in 1929 was of microscopic herapathite crystals, which are long, needle-shaped molecules that are aligned with one another. Modern polaroid films consist of a polyvinyl alcohol plastic doped with iodine. When heated, and subsequently stretched while cooling, the long molecules align themselves with one another. The electrons in the material are free to oscillate along the length of the molecules, but not in the transverse direction. Thus, the polarizer strongly absorbs the polarization component parallel to the molecules, but transmits the light polarized in the perpendicular direction.

You may have noticed that, with some LCD displays in the instrument panel of your car, you cannot see the display when wearing polarizing sunglasses. LCD displays, of course, use polarization control to function, and if the polarization of the light reflected from the LCD is perpendicular to the transmission axis of your polarizing sunglasses, the display is not visible. See TechNote 8.17 for more on LCDs.

Recently, a new form of absorptive polarizer based on silver nanoparticles embedded in glass has been developed. Similar to the polarizers described above, the silver nanoparticles are elongated, and are therefore strongly absorbing to light whose polarization direction is aligned with the nanoparticles. These polarizers can be of very high optical quality, and the ratio of the absorption for the two polarization components of light can be as large as 10^5.

Let's explore a quantitative description of polarized waves. For a wave propagating in the z-direction, two states of **linear polarization** can be written as

$$\tilde{\mathbf{E}}_x(z) = \mathbf{a}_x E_{0,x}\, e^{-jkz} \tag{8.31}$$

and

$$\tilde{\mathbf{E}}_y(z) = \mathbf{a}_y E_{0,y}\, e^{-jkz}, \tag{8.32}$$

where $E_{0,x}$ and $E_{0,y}$ are the amplitudes of the two components. These, of course, are written in the phasor form, and are illustrated as a function of z in Fig. 8.17. In time-dependent notation, these phasors represent

$$\mathbf{E}_x(t, z) = \mathbf{a}_x\, |E_{0,x}|\cos\left(\omega t - kz + \phi_x\right)$$

and

$$\mathbf{E}_y(t, z) = \mathbf{a}_y\, |E_{0,y}|\cos\left(\omega t - kz + \phi_y\right),$$

where ϕ_x and ϕ_y are the phases of the complex amplitudes $E_{0,x}$ and $E_{0,y}$, respectively. (The phases ϕ_x and ϕ_y are often unimportant, in which case they can be dropped. But there are situations when the phases must be retained, so we have introduced them here.) These waveforms travel in the $+z$-direction, as we discussed earlier. The electric field vector always points in the $\pm\mathbf{a}_x$-direction for the former, and in the $\pm\mathbf{a}_y$-direction for the latter. You may see some texts refer to these states of polarization as plane polarization, since the electric field lines all lie in a common plane for each individual case.

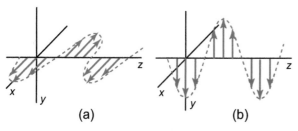

FIGURE 8.17 (a) Horizontally and (b) vertically polarized waves, each varying sinusoidally as a function of z. The arrows represent the electric field vector at varying locations.

The polarization states given by Eqs. (8.31) and (8.32) can be used to describe quantitatively any other polarization state as well. For example, a state of linear polarization that is oriented at 45° to the x-axis can be described as

$$\tilde{\mathbf{E}}_{45} = \frac{1}{\sqrt{2}}\left(\mathbf{a}_x + \mathbf{a}_y\right) E_0 e^{-jkz}.$$

More generally, a state of linear polarization at an angle θ relative to the x-axis can be written

$$\tilde{\mathbf{E}} = \left(\mathbf{a}_x \cos\theta + \mathbf{a}_y \sin\theta\right) E_0 e^{-jkz}.$$

Two polarization states that are perpendicular to one another and that can be used to describe any other polarization state are called an **orthogonal basis set**. The vectors $\mathbf{E}_x$ and $\mathbf{E}_y$ form one such orthogonal basis set. This set is not unique, but it is a convenient, commonly used basis set.

Example 8.4 Orientation of Linear Polarization

The electric field of a wave is given by $\tilde{\mathbf{E}} = \left(\mathbf{a}_x\, 10\text{ V/m} + \mathbf{a}_y\, 30\text{ V/m}\right) e^{-jkz}$. Determine the angle θ between the electric field $\tilde{\mathbf{E}}$ and the x-axis.

Solution:
Since $E_{0,x} = 10$ V/m and $E_{0,y} = 30$ V/m, the angle θ is

$$\theta = \tan^{-1}(E_{0,y}/E_{0,x}) = \tan^{-1}(3) = 71.6°.$$

Now we'd like to explore in a little more detail the action of the polarizer, introduced earlier in the discussion of the experiments performed by Hertz. As described there, a polarizer transmits one polarization component but blocks the orthogonal polarization component. So if a polarizer is lined up with its transmission axis parallel to a_x (i.e. its wires parallel to the y-axis), and an unpolarized traveling wave is incident upon it, the polarizer transmits the component of the wave whose electric field is parallel to a_x, but blocks the a_y-component. So the amplitude of the transmitted wave is reduced. (We haven't discussed power density yet, but if you are familiar with this term, you will agree that half of the power is transmitted and half is blocked.) Furthermore, we should recognize that the transmitted beam is linearly polarized along the x-direction. If this wave is now incident upon a second polarizer, what can we expect to be transmitted here? Well, this depends upon the orientation of this second polarizer, relative to the x-axis. If the transmission axis of the second polarizer is parallel to a_x, then the wave is completely transmitted. If the transmission axis is perpendicular to a_x, however, the wave is completely blocked. This much should be clear from our previous discussions. What now happens when the polarizer is at some other orientation, say its transmission axis is at an angle θ from a_x? To understand this, we must first project the polarization state of the incident wave onto the axes of the polarizer. Let's call these x' and y'. Transformation from the x–y coordinates to x'–y' can be executed by a simple rotation, by an angle θ, about the z-axis. Simple geometry gives us $a'_x = a_x \cos\theta + a_y \sin\theta$ and $a'_y = -a_x \sin\theta + a_y \cos\theta$, as illustrated in Fig. 8.18(a). In the primed coordinates, we can express E_x as $E'_x a'_x + E'_y a'_y$, as shown in Fig. 8.18(b), where $E'_x = E_x \cos\theta$ and $E'_y = -E_x \sin\theta$. Since the polarizer only transmits the $E'_x a'_x$ component, the amplitude of the transmitted wave is reduced by the factor $\cos\theta$. Furthermore, the polarization state of the transmitted wave has been altered. The polarization is linear, as before, but it is now aligned with the x' axis.

FIGURE 8.18 The x–y coordinate system rotated into the x'–y' coordinate system. In (a) the projection of unit vectors is shown; in (b) the transformation of field components.

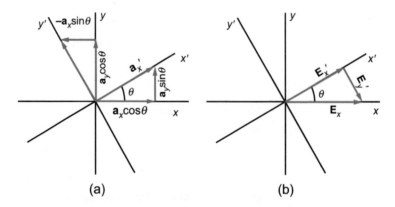

(a)   (b)

Example 8.5 Frame Rotation

An x-polarized wave is incident upon a polarizer whose transmission axis has been rotated by 20° from the x-axis. The amplitude of the incident electric field is 10 V/m. Determine the amplitude of the electric field along the x' and y' axes of the polarizer.

Solution:
The amplitude in the x' direction is

$$E'_x = E_x \cos\theta = 9.4 \text{ V/m},$$

while that in the y' direction is

$$E'_y = -E_x \sin\theta = -3.4 \text{ V/m}.$$

Polarization properties of waves can lead to some seemingly bizarre effects – if we don't think about them properly, that is. For example, if you just think of a polarizer as an element that can fully or partially block the transmission of a wave, then the following experiment might seem very strange. Consider a pair of polarizers oriented orthogonally to each other. Let's say the first transmits a wave polarized along the *x*-axis, while the second transmits the polarization component along the *y*-axis. As we have already described, any wave incident upon these polarizers is completely blocked. What happens, then, if a third polarizer is inserted in the space between the first two polarizers? If the transmission axis of this polarizer is oriented at an angle $\theta = 45°$ from the *x*-axis, partial transmission of the wave is restored. This can be understood by recalling that the polarizer doesn't just block part of the wave, but it also alters the polarization state of the transmitted wave. Without the middle polarizer, the polarization of the wave transmitted by the first polarizer is along the *x*-axis, and the last polarizer rejects this wave. With the middle polarizer inserted, however, the amplitude of the wave is reduced, of course, but also its polarization is altered, aligned now at 45°. This wave is then partially transmitted by the final polarizer, since it is no longer orthogonal to the transmission axis. Thus the amplitude of the transmitted wave is partially restored upon insertion of the middle polarizer.

Let's now look at another polarization state of electromagnetic waves; that is, circular polarization. There are two orthogonal states, called right- and left-circular polarization. Two unit vectors are defined as

$$\mathbf{a}_R = \frac{1}{\sqrt{2}}\left(\mathbf{a}_x - j\mathbf{a}_y\right)$$

and

$$\mathbf{a}_L = \frac{1}{\sqrt{2}}\left(\mathbf{a}_x + j\mathbf{a}_y\right),$$

where the subscripts R and L represent "right" and "left" for the sense of the rotation of the vectors. A right-circularly polarized wave is represented by

$$\tilde{\mathbf{E}}_R = \mathbf{a}_R E_0 e^{-jkz},$$

while a left-circularly polarized wave is given by

$$\tilde{\mathbf{E}}_L = \mathbf{a}_L E_0 e^{-jkz}.$$

These waves are represented in Fig. 8.19.

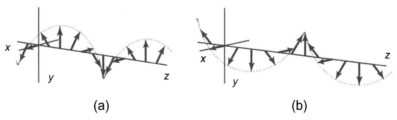

(a)   (b)

FIGURE 8.19 (a) Right- and (b) left-circularly polarized waves. The arrows represent the electric field vector at varying locations. When separated into *x*- and *y*-components, each can be seen to vary sinusoidally as a function of *z*, with a $\pm\pi/2$ phase difference between these components.

In time-dependent notation, $\mathbf{E}_R(t, z)$ can be written as

$$\mathbf{E}_R(t, z) = \Re\left[\mathbf{a}_R E_0 e^{-jkz} e^{j\omega t}\right]$$

$$= \Re\left[\frac{1}{\sqrt{2}}\left(\mathbf{a}_x - j\mathbf{a}_y\right) E_0 e^{j(\omega t - kz)}\right]$$

$$= \frac{E_0}{\sqrt{2}}\left[\mathbf{a}_x \cos(\omega t - kz) + \mathbf{a}_y \sin(\omega t - kz)\right],$$

when the amplitude E_0 is real. In a similar fashion, the time-dependent form of $\mathbf{E}_L$ is

$$\mathbf{E}_L(t, z) = \Re\left[\mathbf{a}_L E_0 e^{-jkz} e^{j\omega t}\right]$$

$$= \Re\left[\frac{1}{\sqrt{2}}\left(\mathbf{a}_x + j\mathbf{a}_y\right) E_0 e^{j(\omega t - kz)}\right]$$

$$= \frac{E_0}{\sqrt{2}}\left[\mathbf{a}_x \cos(\omega t - kz) - \mathbf{a}_y \sin(\omega t - kz)\right].$$

The electric field *direction* for these two polarization states varies as a function of t or z, circling around the z-axis, as illustrated in Fig. 8.19(a) for right- and Fig. 8.19(b) for left-circular polarization states. In each case, the amplitude of the electric field is a constant value of $E_0/\sqrt{2}$, as seen by computing $|\mathbf{E}_R(t, z)|$,

$$|\mathbf{E}_R(t, z)| = \frac{E_0}{\sqrt{2}}\sqrt{\cos^2(\omega t - kz) + \sin^2(\omega t - kz)} = \frac{E_0}{\sqrt{2}},$$

valid for any time t or position z. Similarly, $|\mathbf{E}_L(t, z)| = E_0/\sqrt{2}$. Only the direction in which the electric field points varies. The tip of the arrow representing $\mathbf{E}_R(t, z)$ sweeps through a circular arc in the counter-clockwise direction as a function of time t, while for $\mathbf{E}_L(t, z)$ it sweeps in the clockwise direction, when viewed toward the source.

Example 8.6 Orthogonality of $\mathbf{E}_R$ and $\mathbf{E}_L$

Show that $\mathbf{E}_R$ and $\mathbf{E}_L$ are orthogonal to one another.

Solution:
We can show orthogonality of these two polarization states by computing the time-averaged value of $\mathbf{E}_R(t, z) \cdot \mathbf{E}_L(t, z)$:

$$\overline{\mathbf{E}_R(t, z) \cdot \mathbf{E}_L(t, z)} = \overline{\left[\mathbf{a}_x E_0 \cos(\omega t - kz) + \mathbf{a}_y E_0 \sin(\omega t - kz)\right]}$$
$$\overline{\cdot \left[\mathbf{a}_x E_0 \cos(\omega t - kz) - \mathbf{a}_y E_0 \sin(\omega t - kz)\right]}.$$

Using $\mathbf{a}_x \cdot \mathbf{a}_x = 1$, $\mathbf{a}_y \cdot \mathbf{a}_y = 1$, and $\mathbf{a}_x \cdot \mathbf{a}_y = 0$, this reduces to

$$\overline{\mathbf{E}_R(t, z) \cdot \mathbf{E}_L(t, z)} = \overline{E_0^2\left(\cos^2(\omega t - kz) - \sin^2(\omega t - kz)\right)},$$

and since $\cos^2(\omega t - kz) - \sin^2(\omega t - kz) = \cos[2(\omega t - kz)]$, whose average value is equal to zero, we have

$$\overline{\mathbf{E}_R(t, z) \cdot \mathbf{E}_L(t, z)} = 0.$$

Thus the circular polarization states $\mathbf{E}_R$ and $\mathbf{E}_L$ are orthogonal to one another.

We can reach this same conclusion, perhaps in a simpler fashion, using the phasor state representations for $\mathbf{E}_R$ and $\mathbf{E}_L$. To show orthogonality, we must show $\tilde{\mathbf{E}}_R \cdot (\tilde{\mathbf{E}}_L)^* = 0$, where the superscript * represents the complex conjugate.

$$\tilde{\mathbf{E}}_R \cdot (\tilde{\mathbf{E}}_L)^* = \mathbf{a}_R E_0 e^{-jkz} \cdot \left(\mathbf{a}_L E_0 e^{-jkz}\right)^*$$

$$= \frac{1}{\sqrt{2}}\left(\mathbf{a}_x - j\mathbf{a}_y\right) \cdot \frac{1}{\sqrt{2}}\left(\mathbf{a}_x + j\mathbf{a}_y\right)^* E_0^2$$

$$= \frac{1}{2}\left(\mathbf{a}_x \cdot \mathbf{a}_x - j\mathbf{a}_y \cdot \mathbf{a}_x - j\mathbf{a}_x \cdot \mathbf{a}_y + j^2\mathbf{a}_y \cdot \mathbf{a}_y\right) E_0^2 = 0.$$

Since the states $\mathbf{E}_R$ and $\mathbf{E}_L$ are orthogonal to one another, they can also be used as a basis set, as an alternative to, but in a similar way as, the vectors $\mathbf{E}_x$ and $\mathbf{E}_y$. For example, we can combine $\mathbf{E}_R$ and $\mathbf{E}_L$ to represent linear polarization. In the following example, we construct three different states of linear polarization, using only the unit vectors $\mathbf{a}_R$ and $\mathbf{a}_L$.

Example 8.7 $\mathbf{a}_R$ and $\mathbf{a}_L$ as Basis States

Find combinations of $\mathbf{a}_R$ and $\mathbf{a}_L$ that are equivalent to (a) horizontal linear polarization, (b) vertical linear polarization, and (c) linear polarization at $-45°$.

Solution:
(a) We start by combining $\mathbf{a}_R$ and $\mathbf{a}_L$ in phase with one another:

$$\mathbf{a}_R + \mathbf{a}_L = \frac{1}{\sqrt{2}}\left(\mathbf{a}_x - j\mathbf{a}_y\right) + \frac{1}{\sqrt{2}}\left(\mathbf{a}_x + j\mathbf{a}_y\right) = \sqrt{2}\mathbf{a}_x.$$

This state points in the x-direction, as required, and has an amplitude and phase of $\sqrt{2}$ and zero, respectively.

(b) With a simple π phase shift of the $\mathbf{a}_L$ component, we can recover vertical polarization $\mathbf{a}_y$.

$$\mathbf{a}_R - \mathbf{a}_L = \frac{1}{\sqrt{2}}\left(\mathbf{a}_x - j\mathbf{a}_y\right) - \frac{1}{\sqrt{2}}\left(\mathbf{a}_x + j\mathbf{a}_y\right) = -j\sqrt{2}\mathbf{a}_y.$$

This state points in the y-direction, as required, and has an amplitude and phase of $\sqrt{2}$ and $-\pi/2$, respectively.

(c) For $-45°$,

$$\mathbf{a}_R + j\mathbf{a}_L = \frac{1}{\sqrt{2}}\left(\mathbf{a}_x - j\mathbf{a}_y\right) + \frac{j}{\sqrt{2}}\left(\mathbf{a}_x + j\mathbf{a}_y\right) = (1+j)\frac{\mathbf{a}_x - \mathbf{a}_y}{\sqrt{2}}.$$

This state is linear, pointing in the direction $\theta = -45°$, as required, and has an amplitude and phase of $\sqrt{2}$ and $\pi/4$, respectively.

Notice that in each case, we are combining the left- and right-circular polarization unit vectors $\mathbf{a}_R$ and $\mathbf{a}_L$. The only difference between these is the phase difference between the basis states. In fact, the possibilities are endless, and we can create any linear polarization state just by choosing the right phase shift between the right- and left-circular states.

Finally, we mention in closing a set of polarization states known as elliptical polarization. These states can be decomposed into two linear components that are 90° out of phase with one another, just as circular polarization, but differ in that the amplitudes of these two components are unequal. If one follows the evolution of the instantaneous electric field of this polarization state, one would observe it to sweep around in a right- or left-handed direction, tracing out an elliptical pattern (thus the name), rather than a circular pattern as is characteristic of a state of circular polarization.

Applications based on the polarization of waves are common. We discuss several of these in TechNotes 8.16 and 8.17, starting with CD, DVD, and Blu-ray disc readers.

TechNote 8.16 CDs, DVDs, and Blu-ray Discs

Compact discs (CDs), digital video discs (DVDs), and Blu-ray discs store vast amounts of data digitally, to be read out optically using a laser diode and photodiode sensor. The disc itself consists of polycarbonate plastic, of diameter 12 cm and thickness 1.2 mm. A thin layer of highly reflective aluminum is located close to the labeled side (top) of the CD, as shown in Fig. 8.20. A laser beam, which is used to read the data, is incident from the non-labeled side (bottom). Data is encoded on the CD through the sequence of "pits" and "lands." A pit is a well of depth ~120 nm (on a CD) formed when part of the aluminum layer has been removed. The dimension of a pit on a CD is ~0.5 μm wide by 0.83 μm or greater long. At a land, the aluminum surface is flat. These pits and lands are arranged in a spiral geometry, starting near the center and expanding to the outer rim. The spacing between adjacent spirals is 1.6 μm.

DVDs are able to store seven times more than a CD, while Blu-ray discs store 40 times more than a CD. The higher density of data on DVDs and Blu-rays is possible since these devices use shorter wavelength lasers for readout. In each case, the laser used is a diode laser, but the operating wavelength for CD, DVD, and Blu-ray discs is 780 nm, 650 nm, and 405 nm, respectively. Shorter-wavelength beams can be focused to a tighter focal spot, due to diffraction effects. (We will not discuss diffraction effects in this text. Briefly stated, the smallest size a beam of any wave can be focused is approximately equal to its wavelength.)

Light from the laser is focused onto the disc, and the light reflected by the disc is collected and directed toward a photodiode, which registers the relative optical power. When the read laser input is incident upon a pit, the reflection from the disc is small, while a large reflection results when the laser input is incident upon a land. How can we understand this? First, we must recognize that the focused spot from the laser is larger than the pit. About half the light's power falls within the pit, and the remainder is incident on the aluminum surface surrounding the pit. The weak reflection from a pit is at least partially due to the destructive interference between the light reflected from these two regions. These waves interfere destructively since the well depth is 120 nm, the permittivity of the polycarbonate $\varepsilon_r \sim 2.5$, and the wavelength of the light in the disc material is $\lambda_0/\sqrt{\varepsilon_r} \sim 490$ nm. Therefore, the wave reflected from the pit must travel about half a wavelength farther than light reflected from the surrounding surface, and these waves interfere destructively and partially cancel. This explanation is somewhat simplified, in that it ignores further diffraction effects, but it is sufficient to get a general idea of the reduced reflection.

The function of the polarizing prism is to separate the light generated by the diode laser from the light reflected from the disc. The light from the diode laser is linearly polarized, oriented such that its transmission through the polarizer is high. Between the prism and the disc is a quarter-wave retardation plate, which induces a $\pi/2$ phase delay between the linear components of the light, changing its polarization state to circular. The polarization state of the reflected beam is the same as that of the incident wave,

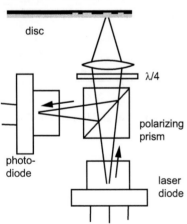

FIGURE 8.20 The optical configuration for reading a CD, DVD, or Blu-ray disc. The output of a diode laser is focused onto the disc, where the power of the reflected beam is weak or strong, depending on whether the beam is incident upon a pit or land, respectively. Polarization optics are used to separate the reflected beam from the incident beam.

but upon passing through the quarter-wave retarder a second time, the polarization is changed to linear, in a direction orthogonal to that of the incident beam. Therefore, when this reflected wave reaches the polarizer, it is reflected toward the photodiode. In this way, polarization effects help keep the optical setup very compact and are effective at steering the reflected beam in the required direction.

Liquid crystal displays also depend on polarization effects of light, as described in TechNote 8.17.

TechNote 8.17 Liquid Crystal Displays

Liquid crystal displays (LCDs) are used in high-resolution color television screens and computer monitors, and digital displays, such as wristwatches, kitchen appliances, and car instrument panels. These display elements function through their ability to alter the polarization of light passing through them. **Liquid crystals** were first discovered by Friedrich Reinitzer (1857–1927), an Austrian chemist and botanist. Many researchers contributed to the study of liquid crystals over the years, but we should certainly recognize the work of Pierre-Gilles de Gennes, who applied his theoretical expertise with superconductivity to study this class of materials, earning the Nobel Prize in physics in 1991 "for discovering that methods developed for studying order phenomena in simple systems can be generalized to more complex forms of matter, in particular to liquid crystals and polymers." The first working LCD was introduced by J. Fergason in 1970.

The name liquid crystal sounds like a paradox in itself, since we usually think of a crystal as a solid material with long-range order (repeating periodic structure), and a liquid as a material that has no structure, that is not compressible (similar to a solid), and that lacks a distinct shape (not similar to a solid). This worldview, in which there are but three phases of matter – solid, liquid, and gas – is an oversimplification of reality. Phases of different materials can be defined by different physical properties, such as conductivity, crystalline structure, or physical appearance. Liquid crystals are characterized by properties that are normally associated with liquids and solids. In a phase known as the nematic phase, long, needle-shaped molecules are free to move around, as in a liquid, but they retain their alignment. In another phase, known as the smectic phase, the molecules arrange themselves in layers, and the layers are free to move individually. Most LCDs are based on nematic liquid crystals.

A single pixel of an LCD is shown in Fig. 8.21. In this figure, a pair of polarizers whose transmission axes are crossed relative to one another are shown. Without the liquid crystal in the space between the polarizers, the transmission through the crossed polarizers would be near zero, and the pixel would be dark. With the liquid crystal, however, the transmission through the polarizers can be controlled by applying a potential difference between the two electrodes. Note the orientation of the liquid crystal molecules in Fig. 8.21(a), in which the switch is open and no electric field is applied to the liquid crystals. (The molecules are represented by the long red lines.) In this case, the molecules line up with the etched grating on the glass substrate at the top, and also with the etched grating on the glass substrate at the bottom. These etchings are perpendicular to one another, and the alignment of the liquid crystals in the space between changes smoothly from top to bottom as shown. The polarization of the light passing from bottom to top will also rotate, such that the polarization of the light is parallel to the transmission axis of the top polarizer, and the light is transmitted by this polarizer. That is, the pixel appears bright. This is the "off" state for the pixel, in that the voltage across the pixel is zero. In contrast, Fig. 8.21(b) shows the "on" state of the pixel, in which the potential is applied across the liquid crystal. The electric field in this space causes the liquid crystal molecules to align themselves vertically, as shown, with the result that the polarization of the light passing through the liquid crystal is unchanged. The light in this state is not transmitted by the second polarizer, and the pixel appears dark.

Displays consist of an array of pixels, with the pixel size determining the resolution of the image. For color displays, each pixel consists of individual sub-pixels, each with a red, green, or blue light filter, corresponding to the three primary colors.

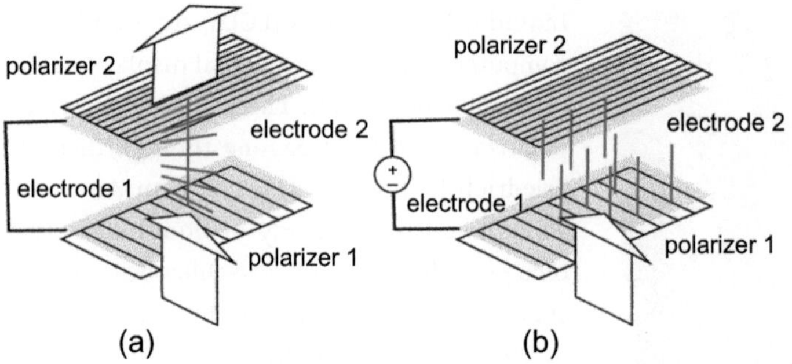

FIGURE 8.21 A diagram showing the principle of operation of an LCD. The red lines represent the liquid crystal molecules, the yellow sheets are the electrodes, and the black parallel lines represent the polarizers.

In this section, we have introduced the polarization state of electromagnetic waves, including linear polarization, circular polarization, and, briefly, elliptical polarization. We have also examined a few applications of wave polarization, such as polarizing sunglasses, CD, DVD, and Blu-ray discs, and LCDs.

LEARNING OBJECTIVES

After studying this section, you should be able to:

- determine the power dissipation density for a time-harmonic wave in an absorbing medium;
- determine if a medium is a good conductor; and
- determine the complex wave number, the attenuation constant, the phase constant, the impedance, and the phase velocity for a time-harmonic wave in
  - a general absorbing medium;
  - a general non-conducting, absorbing medium;
  - a non-conducting, weakly absorbing medium; and
  - a good conductor.

8.6 Plane Waves in Lossy Media

Our treatment of plane wave propagation to this point is valid only for media that do not absorb or scatter the wave. We should expect, however, that the medium through which the wave propagates might (and usually does) absorb energy from the wave, and that correspondingly the amplitude of the wave will be reduced upon propagation through the material. The rate of this attenuation depends on the material, of course, as well as on the frequency of the wave. In this section we will investigate losses, or attenuation of the wave, and relate this to the material parameters of the medium through which the wave propagates. We will also re-examine the wave number k, the intrinsic impedance of the medium η, and the phase velocity of the wave u_p, all of which are important parameters that characterize propagating waves, and must be modified somewhat for wave propagation through absorbing media.

We will consider two types of media in this discussion of losses of a propagating wave. These are the absorbing, or lossy, dielectric medium, and the good (but not perfect) conductor. In fact, we have already laid the groundwork for this discussion in Section 3.6, when we discussed power dissipation in conducting media in d.c. circuits. In the present discussion, of course, we are considering absorption from sinusoidally varying fields. In a non-conducting dielectric material, we will show that when the electric field $\mathbf{E}(\mathbf{r})$ and the displacement current density $\partial \mathbf{D}(\mathbf{r})/\partial t$ (both of which oscillate sinusoidally at the same frequency ω) are precisely $\pi/2$ out of phase with one another, the wave propagates through the medium without loss. But when the current density has a component that is in phase with $\mathbf{E}(\mathbf{r})$, then this is the signature that the medium will absorb energy from the field, and the field amplitude is consequently attenuated as it propagates through the medium. In a conductor, $\mathbf{E}(\mathbf{r})$ and the current density $\mathbf{J}_v(\mathbf{r})$ are inherently in phase with one another, and this leads to strong absorption of the wave.

8.6.1 Power Dissipation Density

Let's first look at wave propagation through a dielectric. As we introduced earlier when we discussed dielectric media and d.c. electric fields in Section 1.5, the charges in a perfect dielectric medium are not free, but are bound rather tightly to the atoms within the medium. When we apply an oscillating electric field to a dielectric medium, the individual charges are displaced in response to the field, but they remain bound to their respective atoms. The primary response within the material is that of the negatively charged electrons, which are displaced in the direction opposite to that of the field, thus inducing an oscillating electric dipole. We previously defined, through Eq. (1.58), the polarization field $\mathbf{P}$ of the medium as the density of these electric dipole moments within the dielectric material,

$$\mathbf{P} = \lim_{\Delta v \to 0} \frac{1}{\Delta v} \sum_{k=1}^{n\Delta v} \mathbf{p}_k,$$

where n is the number density of the dipoles in the dielectric medium, $n\Delta v$ is the number of dipoles in the volume Δv, and $\mathbf{p}_k$ is the vector dipole moment of kth dipole. The polarization $\mathbf{P}$ can be written in terms of the average molecular dipole moment $\langle \mathbf{p} \rangle$ as

$$\mathbf{P} = n\langle \mathbf{p} \rangle = nq\langle \mathbf{r} \rangle, \tag{8.33}$$

where $\langle \mathbf{r} \rangle$ indicates the average displacement of the charges. (This average is computed over the dipoles within Δv.) When the electric field varies with time, the polarization $\mathbf{P}$ also varies as a function of time. For a medium that is linear (i.e. one in which the polarization field is proportional to the electric field strength) and isotropic (i.e. one in which $\mathbf{P}$ is parallel to $\mathbf{E}$), and when the electric field is a harmonic field, the polarization field is

$$\mathbf{P}(t) = \varepsilon_0 \chi_e \mathbf{E}_0 \exp(j\omega t). \tag{8.34}$$

In words, when the electric field varies sinusoidally with time t at frequency ω, the polarization of the medium varies sinusoidally as well, at the same frequency ω. (Remember, in writing Eq. (8.34), we are using complex notation, and only the real part of this equation has any significance.) If the oscillation of the electric dipoles is perfectly in phase with the oscillation of the electric field $\mathbf{E}$, then the susceptibility χ_e is a real quantity. This corresponds to the perfectly non-absorbing case, as the field does no net work on the system under this condition. We motivate this statement as follows. (Note that in some respects the following discussion parallels that of Chapter 3 in which we showed that $\mathbf{E} \cdot \mathbf{J}_v$, known as the power dissipation density, is the rate at which the d.c. electric field $\mathbf{E}$ does work on free charges in the medium. Recall that the current density $\mathbf{J}_v$ is equal to the product of the free charge density ρ_v and the average charge velocity $\langle \mathbf{u} \rangle$. The charges are not free in the dielectric, but the formulation of the work done by the field $\mathbf{E}$ on the *bound* charges of the dielectric follows in a similar fashion.)

In general, when a force $\mathbf{F}$ is applied to an object whose velocity is $\mathbf{u}$, the *rate* at which the energy of that body increases is given by $\mathbf{F} \cdot \mathbf{u}$. Note that by using the scalar product (dot product) notation of $\mathbf{F}$ and $\mathbf{u}$, all possible relative orientations between these vectors are automatically accommodated. First, only the component of $\mathbf{F}$ in the direction of $\mathbf{u}$ changes the energy of the object, and this is properly reflected in the scalar product. Also, the sign of the scalar product properly accounts for whether the object's energy is increasing or decreasing. If the force $\mathbf{F}$ is in the same direction as $\mathbf{u}$ (i.e. $\mathbf{F} \cdot \mathbf{u} > 0$), then we push it in the direction it is already moving, its velocity *increases*, as does its kinetic energy, and we've done positive work on the object. Conversely, when the direction of $\mathbf{F}$ opposes the direction of $\mathbf{u}$ (i.e. $\mathbf{F} \cdot \mathbf{u} < 0$), we slow the body down, and its kinetic energy decreases. Using the same logic as in Section 3.6, one can show that the rate at which work is done on the system, per unit volume Δv, is

$$P_v^{\text{abs}}(t) = \lim_{\Delta v \to 0} \frac{1}{\Delta v} \sum_{k=1}^{n\Delta v} \mathbf{F}(t) \cdot \mathbf{u}_k(t), \tag{8.35}$$

where $\mathbf{u}_k(t)$ is the velocity of the kth bound charge, and the summation extends over all such charges within the volume Δv. We are often more interested in the time-average of the power density absorbed, rather than the instantaneous power density given in Eq. (8.35), which we write as

$$\overline{P_v^{\text{abs}}} = \lim_{\Delta v \to 0} \frac{1}{\Delta v} \sum_{k=1}^{n\Delta v} \overline{\mathbf{F} \cdot \mathbf{u}_k}. \tag{8.36}$$

The overbar represents the time-averaged quantity. Our interest in the present discussion is, of course, an electric force, for which $\mathbf{F} = q\mathbf{E}$, where q is the electric charge of the object, and where $\mathbf{E}$ is a time-harmonic electric field. Since $\mathbf{F}$ is the same for all the charges within Δv, the force can be pulled outside the summation, changing Eq. (8.36) to

$$\overline{P_v^{\text{abs}}} = n\overline{\mathbf{F} \cdot \langle \mathbf{u} \rangle}, \tag{8.37}$$

where

$$\langle \mathbf{u} \rangle = \lim_{\Delta v \to 0} \frac{1}{n\Delta v} \sum_{k=1}^{n\Delta v} \mathbf{u}_k$$

is the average velocity of the charges within the volume Δv. Furthermore, since the force is harmonic, the only velocity $\langle \mathbf{u} \rangle$ that contributes to the time-average in Eq. (8.37) is the harmonic component at the same frequency as the field. All other components vanish when averaged. This component of the average velocity $\langle \mathbf{u} \rangle$ is related to the polarization $\mathbf{P}$ induced in the dielectric medium by the electric field as

$$\langle \mathbf{u} \rangle = \frac{d\langle \mathbf{r} \rangle}{dt} = \frac{1}{nq} \frac{d\mathbf{P}}{dt},$$

where we used Eq. (8.33). Substituting for the force $\mathbf{F}$ and velocity $\langle \mathbf{u} \rangle$ in Eq. (8.37), we get

$$\overline{P_v^{\text{abs}}} = \overline{\mathbf{E} \cdot \frac{d\mathbf{P}}{dt}}.$$

When the polarization $\mathbf{P}$ is precisely in phase with the electric field $\mathbf{E}$, as it is when the susceptibility χ_e in Eq. (8.34) is real, the average power density absorbed by the medium is

$$\overline{P_v^{\text{abs}}} = \mathbf{E}_0 \cos(\omega t) \cdot \frac{d}{dt}\left(\varepsilon_0 \chi_e \mathbf{E}_0 \cos(\omega t)\right)$$

$$= -\varepsilon_0 \omega \chi_e \left|\mathbf{E}_0\right|^2 \overline{\cos(\omega t)\sin(\omega t)}.$$

But the average value of $\cos(\omega t)\sin(\omega t)$ is zero, as we can compute formally using

$$\overline{\cos(\omega t)\sin(\omega t)} = \frac{1}{T}\int_0^T \cos(\omega t)\sin(\omega t)dt$$

$$= \frac{1}{2T}\int_0^T \sin(2\omega t)dt = \left.\frac{-\cos(2\omega t)}{4\omega T}\right|_0^T = 0,$$

where the period T is $2\pi/\omega$. The average power density is illustrated with the aid of Fig. 8.22(a). Here, $\mathbf{E}(t)$ and $\mathbf{P}(t)$ are plotted for a harmonic field in a lossless dielectric. With the susceptibility χ_e real, $\mathbf{E}(t)$ and $\mathbf{P}(t)$ are precisely in phase with one another, and the velocity of the charges $\mathbf{u}(t)$ leads $\mathbf{E}(t)$ by $\pi/2$, or 90°. Now let's look at the work done by the electric field on the charges over the course of one complete cycle of oscillation of the field. We must examine this quarter by quarter. In the first quarter-cycle, the electric field $\mathbf{E}(t)$ is positive, while the velocity is negative. Thus the product $\mathbf{E} \cdot \mathbf{u}$ is negative during this quarter. During the next quarter-cycle, both $\mathbf{E}$ and $\mathbf{u}$ are negative, and so $\mathbf{E} \cdot \mathbf{u} > 0$. In the third quarter, $\mathbf{E} \cdot \mathbf{u}$ is again negative, while in the fourth it is positive. In each quarter-cycle, the average values of the magnitude of the rate of work $|\mathbf{E} \cdot \mathbf{u}|$ is the same; only the sign toggles, and so when averaged over one complete cycle the field has done no net work on the charges.

How does this picture change when the polarization field lags behind the driving electric field $\mathbf{E}(t)$? This situation is shown in Fig. 8.22(b), in which $\mathbf{E}(t)$ is unchanged from (a), but the polarization field $\mathbf{P}(t)$, as well as the charge velocity $\mathbf{u}(t)$, are retarded. As we will show, this phase shift of $\mathbf{P}(t)$ causes $\mathbf{E} \cdot \mathbf{u}$, when averaged over one complete cycle, to become positive, telling us that the electric field is doing work on the medium. As we hinted previously, the phase shift between the harmonic electric field $\mathbf{E}(t)$ and the polarization field $\mathbf{P}(t)$ can be expressed by allowing the susceptibility χ_e of the medium to become complex,

$$\chi_e = \chi_e' - j\chi_e''.$$

Let's show this connection here. When we write χ_e in this form, then the time-dependent polarization field becomes

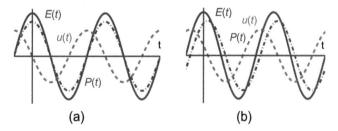

(a)   (b)

FIGURE 8.22 Plots of harmonic electric field $\mathbf{E}(t)$, polarization field $\mathbf{P}(t)$, and charge velocity $\mathbf{u}(t)$. In (a), $\mathbf{E}(t)$ and $\mathbf{P}(t)$ are in phase with one another, so the field does no net work on the charges. In (b), $\mathbf{P}(t)$ is phase-shifted relative to $\mathbf{E}(t)$, and on average work is done by the field on the moving charges, resulting in attenuation of the wave.

$$\mathbf{P}(t) = \Re\left[\varepsilon_0\left(\chi'_e - j\chi''_e\right)\mathbf{E}_0\exp(j\omega t)\right]$$

$$= \varepsilon_0\mathbf{E}_0\left[\chi'_e\cos(\omega t) + \chi''_e\sin(\omega t)\right].$$

This polarization field is still sinusoidal, as shown in Fig. 8.22(b), but delayed by the phase $\varphi = \tan^{-1}(\chi''_e/\chi'_e)$. Over one complete cycle of this field, the average rate of work done on the bound charges per unit volume is

$$\overline{P_v^{\mathrm{abs}}} = \overline{\mathbf{E}\cdot\frac{\partial\mathbf{P}}{\partial t}} \tag{8.38}$$

$$= \overline{\mathbf{E}_0\cos(\omega t)\cdot\frac{\partial}{\partial t}\left(\varepsilon_0\mathbf{E}_0\left[\chi'_e\cos(\omega t) + \chi''_e\sin(\omega t)\right]\right)}$$

$$= \varepsilon_0\omega\left|\mathbf{E}_0\right|^2\left[-\chi'_e\overline{\cos(\omega t)\sin(\omega t)} + \chi''_e\overline{\cos^2(\omega t)}\right].$$

We showed previously that the average value of $\cos(\omega t)\sin(\omega t)$ vanishes. The average value of $\cos^2(\omega t)$ taken over one complete cycle, however, is $1/2$, and the average rate at which the field does work on the dielectric medium, per unit volume, is

$$\overline{P_v^{\mathrm{abs}}} = \frac{1}{2}\varepsilon_0\omega\chi''_e\left|\mathbf{E}_0\right|^2.$$

This relation shows that the imaginary part of the susceptibility χ''_e is related to power absorption by the dielectric.

Referring to Fig. 8.22(b), it can be seen that the balance of $E(t)u(t)$ from one quarter-cycle to the next that we saw in Fig. 8.22(a) is upset by the phase lag of $\mathbf{P}(t)$. While the work done in one quarter-cycle in Fig. 8.22(a) is always given back in the subsequent quarter-cycle when χ_e is real, this is no longer the case when $\mathbf{P}(t)$ is delayed. As a result, the net work done by the field on the charges is positive.

To this point, we have formulated this work done by the field on the dielectric in terms of the polarization field $\mathbf{P}(t)$ of the medium. For later inclusion in Maxwell's Equations, remember that this affects the displacement field $\mathbf{D}$, which for a linear, absorbing, dielectric medium, is

$$\mathbf{D} = \varepsilon_0\mathbf{E} + \mathbf{P}.$$

When written using the complex time-dependent notation, or the phasor representation, we make use of the complex permittivity

$$\mathbf{D} = \varepsilon_0\mathbf{E} + \varepsilon_0\left(\chi'_e - j\chi''_e\right)\mathbf{E} = \left(\varepsilon' - j\varepsilon''\right)\mathbf{E}, \tag{8.39}$$

where $\varepsilon' = \varepsilon_0(1 + \chi'_e)$ and $\varepsilon'' = \varepsilon_0\chi''_e$. The imaginary part of the permittivity leads to a displacement field $\mathbf{D}(t)$ that is phase-shifted with respect to the electric field $\mathbf{E}(t)$. As we will soon see, this leads to attenuation of the amplitude of a wave as it propagates through this medium.

Before we explore the relation between wave attenuation and the complex permittivity, let's express the power dissipation density $P_v^{\mathrm{abs}}(t)$ in terms of the displacement current. In order to make the notation more concise, the displacement current density is denoted $\mathbf{J}_d(t)$ in place of $\partial\mathbf{D}(t)/\partial t$. We return to Eq. (8.38), and use Eq. (8.39) to write the time-average power dissipation density as

$$\overline{P_v^{\mathrm{abs}}} = \overline{\mathbf{E}(t)\cdot\frac{\partial\mathbf{P}(t)}{\partial t}} = \overline{\mathbf{E}(t)\cdot\frac{\partial}{\partial t}\left[\mathbf{D}(t) - \varepsilon_0\mathbf{E}(t)\right]}.$$

For harmonic fields, $E(t)$ and $\partial E(t)/\partial t$ are always $\pi/2$ out of phase with one another, so the average value of the product is zero, leaving us with

$$\overline{P_v^{\text{abs}}} = \overline{E(t) \cdot \frac{\partial D(t)}{\partial t}} = \overline{E(t) \cdot J_d(t)}.$$

When $E(t)$ and $J_d(t)$ are represented as phasors, this power absorption density is written as

$$\overline{P_v^{\text{abs}}} = \frac{1}{2}\Re\left[\tilde{E} \cdot \tilde{J}_d^*\right].$$

As is customary, the asterisk is used to represent the complex conjugate, and the factor $1/2$ comes from the time average of $\cos^2 \omega t$. Since in phasor form for linear media the displacement current density $\tilde{J}_d$ is $j\omega\varepsilon\tilde{E}$, and for absorbing media $\varepsilon = \varepsilon' - j\varepsilon''$, the time-average power dissipation density for an absorbing dielectric medium can be written as

$$\overline{P_v^{\text{abs}}} = \frac{1}{2}\omega\varepsilon''|\tilde{E}|^2. \tag{8.40}$$

Example 8.8 Dielectric Absorption

A 10 GHz wave of amplitude $E_0 = 10$ V/m propagates through a dielectric medium. The complex permittivity of the medium is $\varepsilon = 4\varepsilon_0 - j\varepsilon''$. The wave deposits an average power of 100 mW/m^3 in the medium. Determine ε'' of this medium.

Solution:
We use Eq. (8.40), which we solve for ε'' as

$$\varepsilon'' = \frac{2\overline{P_v^{\text{abs}}}}{\omega|\tilde{E}|^2} = \frac{2(100 \text{ mW/m}^3)}{(2\pi \times 10^{10} \text{ s}^{-1})(10 \text{ V/m})^2} = 0.32 \times 10^{-13} \text{ F/m}.$$

The relative value ε_r'' is

$$\varepsilon_r'' = \frac{\varepsilon''}{\varepsilon_0} = 0.036.$$

To this point in the discussion of power dissipation, we have focused on dielectric media. Let's now examine wave propagation and power dissipation in conducting media. As we discussed before, charge carriers are free and unconstrained within the conductor, and the current density J_v is proportional to the applied field E,

$$J_v = \sigma E,$$

where σ is the conductivity of the conductor. A simple model of free electrons in a conductor known as the Drude model can give some insight into the basis for the conductivity. Recall that in Chapter 3 we derived that current density (Eq. (3.2)) as

$$J_v(\mathbf{r}) = \rho_v \mathbf{u} = qn\mathbf{u} = -en_e\mathbf{u}.$$

The final equality is specific to electrons, and includes the electron charge $q = -e$ and the electron number density n_e. In a conductor, or in a plasma (a highly ionized unstructured medium consisting of equal densities of positive ions and electrons), the electrons, as free carriers, are free to move under the influence of an external electric field E. (The ions of a plasma are mobile as well, of course, but since they are so massive we can ignore their effect.) Using Newton's Second Law, the equation of motion of an electron is

$$m_e \frac{d\mathbf{u}}{dt} = -e\mathbf{E} - m_e \gamma \mathbf{u}.$$

The left side of this equation is the mass of the electron times its acceleration, and the right side contains the electric force and the damping force, where γ is the damping rate of the electron motion due to collisions with the atomic lattice (metal) or background gas (plasma). When the electric field oscillates sinusoidally at frequency ω, $\mathbf{E}(t) = \mathbf{E}_0 e^{j\omega t}$, the average electron velocity is oscillatory as well, with $\mathbf{u}(t) = \mathbf{u}_0 e^{j\omega t}$. Inserting this into Newton's Second Law, and solving for the amplitude of the average velocity, $\mathbf{u}_0$ is

$$\mathbf{u}_0 = \frac{-e\mathbf{E}_0}{m_e \left(j\omega + \gamma\right)}.$$

The current density under these conditions is

$$\mathbf{J}_v = -en_e\mathbf{u} = \frac{n_e e^2 \left(\gamma - j\omega\right)}{m_e \left(\gamma^2 + \omega^2\right)}\mathbf{E}_0.$$

The conductivity of the conductor is identified from the real part of this expression,

$$\sigma = \frac{n_e e^2 \gamma}{m_e \left(\gamma^2 + \omega^2\right)}.$$

Now we apply Ampère's law for a time-harmonic wave in this medium:

$$\nabla \times \tilde{\mathbf{H}} = \tilde{\mathbf{J}}_v + j\omega\tilde{\mathbf{D}}$$

$$= \frac{n_e e^2 \left(\gamma - j\omega\right)}{m_e \left(\gamma^2 + \omega^2\right)}\tilde{\mathbf{E}}_0 + j\omega\varepsilon\tilde{\mathbf{E}}_0$$

$$= j\omega\left\{\left(\varepsilon_b' - \frac{n_e e^2 \left(\gamma - j\omega\right)}{m_e \left(\gamma^2 + \omega^2\right)}\right) - j\left(\varepsilon_b'' + \frac{n_e e^2 \gamma}{m_e \omega \left(\gamma^2 + \omega^2\right)}\right)\right\}\tilde{\mathbf{E}}_0.$$

ε_b here is the permittivity of the material due to the bound charges and positive ions. So the real part of the effective permittivity ε is

$$\varepsilon' = \varepsilon_b' - \frac{n_e e^2}{m_e \left(\gamma^2 + \omega^2\right)},$$

and the imaginary part is

$$\varepsilon'' = \varepsilon_b'' + \frac{n_e e^2 \gamma}{m_e \omega \left(\gamma^2 + \omega^2\right)}.$$

From the definition of the conductivity above, the second term of the imaginary part can also be written as σ/ω.

For materials with modest conductivity and high collision rates, such as semiconductors, the effective permittivity is

$$\varepsilon = \varepsilon_b - j\frac{\sigma}{\omega},$$

while for low-density materials (small collision rates) with $\varepsilon_b = \varepsilon_0$, the real part of the permittivity is

$$\varepsilon' \approx \varepsilon_0 - \frac{n_e e^2}{m_e \omega^2},$$

valid for frequencies $\omega > \gamma$. Using the plasma frequency

$$\omega_p = \sqrt{\frac{n_e e^2}{m_e \varepsilon_0}}$$

defined in TechNote 8.1, the permittivity is

$$\varepsilon' \approx \varepsilon_0 \left(1 - \frac{\omega_p^2}{\omega^2} \right).$$

For frequencies ω less than the plasma frequency, the real part of the permittivity is negative, and free propagation of electromagnetic waves is inhibited.

Example 8.9 Plasma Frequency in Silver

Determine the plasma frequency of metallic silver.

Solution:

To estimate the number density of electrons, we will assume that there is one free electron in the conduction band for each silver nucleus, and use the mass density, $\rho_m = 10.49$ gm/cm^3, and the atomic weight $A = 107.9$ of silver to find

$$n_e = \frac{\rho_m N_A}{A} = \frac{\left(10.49 \text{ gm/cm}^3\right)\left(6.022 \times 10^{23} \text{ atoms/mole}\right)}{107.9 \text{ gm/mole}} \times (1 \text{ electron/atom})$$

$$= 0.585 \times 10^{23} \text{ electrons/cm}^3.$$

Then the plasma frequency is

$$\omega_p = \sqrt{\frac{n_e e^2}{m_e \varepsilon_0}} = \sqrt{\frac{(0.585 \times 10^{29} \text{ electrons/m}^3)(1.602 \times 10^{-19} \text{ C})^2}{(9.11 \times 10^{-31} \text{ kg})(8.854 \text{ pF/m})}} = 1.36 \times 10^{16} \text{ s}^{-1}.$$

The equivalent cyclical frequency is 2.2×10^{15} Hz, which lies in the UV region of the spectrum, at a wavelength of 140 nm. Thus, silver does not transmit any waves whose wavelength is longer than 140 nm.

Example 8.10 Plasma Frequency of the Ionosphere

Determine the plasma frequency of the ionosphere.

Solution:

The ionosphere is very complex, of course, but simply stated it is created by UV radiation from the sun which ionizes molecules in the upper regions of the atmosphere. The density of free electrons in the ionosphere depends on many factors, but is typically in the range 10^4–10^6 cm^{-3}. Then the plasma frequency is

$$\omega_p = \sqrt{\frac{n_e e^2}{m_e \varepsilon_0}} = \sqrt{\frac{(10^{10} - 10^{12} \text{ electrons/m}^3)(1.602 \times 10^{-19} \text{ C})^2}{(9.11 \times 10^{-31} \text{ kg})(8.854 \text{ pF/m})}} = 6 \times 10^6 - 6 \times 10^7 \text{ s}^{-1}.$$

This is a cyclical frequency of 1–10 MHz. Thus, radio wave propagation is affected by the ionosphere. In particular, the ionosphere can reflect radio waves, allowing for propagation over large distances as the radio waves bounce back and forth between the ionosphere and the surface of the Earth. Lightning and solar flare activity can have a strong effect on the ionosphere, and therefore on EM wave propagation. The ionosphere changes dramatically from night to day, also affecting radio wave propagation.

A model of power dissipation similar to that introduced above for dielectrics can be applied here as well. For frequencies well below the plasma frequency ω_p, the conductivity is real, the average velocity **u** of the charges is perfectly in phase with the electric field $E(t)$, and the electric field does positive work on the charges during each of the four quarters of each cycle. Once again, we apply Eq. (3.17) for the power dissipation density, which we computed from the rate at which the electric field does work on the free charges within a volume Δv. For time-harmonic waves of the form $\mathbf{E}(t) = \mathbf{E}_0 \exp^{j\omega t}$, the time-averaged rate of work done on the charges is

$$\overline{P_v^{\text{abs}}} = \overline{\mathbf{E}(t) \cdot \mathbf{J}_v(t)} = \overline{(\mathbf{E}_0 \cos(\omega t)) \cdot (\sigma \mathbf{E}_0 \cos(\omega t))}.$$

The electric field amplitude is constant, and the average value of $\cos^2(\omega t)$ taken over one cycle is 1/2, leading to

$$\overline{P_v^{\text{abs}}} = \frac{1}{2}\sigma \left|\mathbf{E}_0\right|^2. \tag{8.41}$$

This can also be represented, using phasor amplitudes, as

$$\overline{P_v^{\text{abs}}} = \frac{1}{2}\tilde{\mathbf{E}} \cdot \tilde{\mathbf{J}}_v^* = \frac{1}{2}\sigma|\tilde{\mathbf{E}}|^2. \tag{8.42}$$

Regardless of which form we use to represent this work, the result is that the energy of the wave is diminished – that is, the wave amplitude decreases.

In the general case, in which absorption is due to finite conductivity *and* dielectric absorption, we can combine Eqs. (8.40) and (8.42) to write

$$\boxed{\overline{P_v^{\text{abs}}} = \frac{1}{2}\mathfrak{R}\left[\tilde{\mathbf{E}} \cdot \left(\tilde{\mathbf{J}}_v + \tilde{\mathbf{J}}_d\right)^*\right] = \frac{1}{2}\left(\sigma + \omega\varepsilon''\right)|\tilde{\mathbf{E}}|^2.} \tag{8.43}$$

This expression gives us the rate at which the electric field does work on the free and bound charges in the medium, per unit volume, and we can apply it to any medium, conducting or dielectric (or both). In practice, it can be difficult to separate the contributions of real conduction currents and bound dielectric currents from losses.

Example 8.11 Absorption due to Conduction Current

A 10 GHz wave of amplitude $E_0 = 10$ V/m propagates through a conducting medium. The conductivity of the medium is 250 S/m. Determine the time-averaged power dissipation density due to conduction currents in this medium.

Solution:
We use Eq. (8.43), with ε'' set to zero:

$$\overline{P_v^{\text{abs}}} = \frac{1}{2}\sigma|\tilde{\mathbf{E}}|^2 = \frac{1}{2}(250 \text{ S/m})(10 \text{ V/m})^2 = 12.5 \text{ kW/m}^3.$$

This value of $\overline{P_v^{\text{abs}}}$ is valid only in the small volume for which the amplitude of the wave is still $E_0 = 10$ V/m.

Losses are often characterized in terms of a parameter known as the **loss tangent**. For a non-conducting dielectric the loss tangent is defined as

$$\tan\delta = \frac{\varepsilon''}{\varepsilon'}, \tag{8.44}$$

Table 8.2

Relative dielectric constant and loss tangent of several dielectric materials.			
Material	ε'_r	tan δ	Frequency
Alumina: 96–99.5%	10.0	0.0002	1 GHz
	9.6	0.0002	100 MHz
		0.0003	10 GHz
Beryllium oxide	6.7	0.006	10 GHz
Fused quartz	3.8	0.0002	100 MHz
		0.000 06	3 GHz
Gallium arsenide (GaAs)	13.1	0.0016	10 GHz
Glass (Corning 7059)	5.75	0.0036	10 GHz
Polyethylene	2.26	0.0031	3 GHz
Silicon	11.7–12.9	0.005	1 GHz
		0.015	10 GHz
Teflon (PTFE)	2.0–2.1	0.000 28	3 GHz
Water (distilled)	76.7–78.2	0.005	100 MHz
		0.157	3 GHz

while for a conductor the loss tangent is

$$\tan \delta = \frac{\sigma}{\omega \varepsilon}. \tag{8.45}$$

The permittivities and loss tangents of several common materials are listed in Table 8.2.

Example 8.12 Loss Tangent

Determine ε'' for fused quartz at $f = 100$ MHz.

Solution:
From Table 8.2, we see that $\varepsilon' = 3.8\varepsilon_0$, and tan $\delta = 0.0002$. We solve Eq. (8.44) for ε'':

$$\varepsilon'' = (\tan \delta)\varepsilon' = (0.0002)(3.8\varepsilon_0) = 0.000\,76\varepsilon_0.$$

Although losses in dielectric materials and losses in conductors are physically distinct from one another, it is not uncommon in some disciplines to combine the parameters for these two process, and group both types of losses in a single term ε''. This practice has the virtue of simplifying some of the expressions that we will encounter shortly for the propagation constant of a wave or the intrinsic impedance of a medium. We will continue in this book to write these terms separately in order to highlight the physical effects and to facilitate later discussions, but you should be aware of both approaches, and learn to recognize which is in use in a particular discussion or application.

In summary, we have determined $\overline{P_v^{\text{abs}}}$, the time-averaged power absorbed per unit volume by the medium from a propagating wave in this section. This absorption can result from the displacement current in a dielectric and/or from free currents in a conductor. We might expect that power absorption by the medium leads to a decrease in the amplitude of the wave, the topic addressed in the next section.

8.6.2 Wave Propagation in Absorbing Media

Now that we have developed a qualitative picture that allows us to visualize absorption of energy from the wave, let's return to Maxwell's Equations to see what they can tell us about the waves themselves. We still consider a time-harmonic wave with electric field

$$\mathbf{E}(\mathbf{r}, t) = \mathbf{E}(\mathbf{r}) \, e^{j\omega t},$$

and magnetic field

$$\mathbf{H}(\mathbf{r}, t) = \mathbf{H}(\mathbf{r}) \, e^{j\omega t},$$

where (as always) only the real part is physically significant. Consistent with our earlier treatment of wave propagation through a lossless medium, we consider the medium to be linear, isotropic, and charge-neutral ($\rho_v = 0$), and we represent the field quantities as phasors. As we have just seen, losses are associated with a conduction current $\mathbf{J}_v = \sigma\mathbf{E}$ and a phase-shifted displacement current $\partial\mathbf{D}/\partial t$, so we include these currents in our derivation. (It might seem odd to have current density $\mathbf{J}_v \neq 0$ in a neutral medium, $\rho_v = 0$, but these are not contradictory. Imagine, for example, a metal, which we described earlier as charge-neutral in the interior. There are equal densities of free electrons and immobile positively charged nuclei, resulting in no net charge in the interior. But the electrons are free to move under the influence of an electric field, resulting in a current density $\mathbf{J}_v$.) Maxwell's Equations for these time-harmonic waves in this medium become

Gauss' Law:	$\boldsymbol{\nabla} \cdot \tilde{\mathbf{E}}(\mathbf{r}) = 0$	(8.46)
No magnetic charges:	$\boldsymbol{\nabla} \cdot \tilde{\mathbf{H}}(\mathbf{r}) = 0$	(8.47)
Faraday's Law:	$\boldsymbol{\nabla} \times \tilde{\mathbf{E}}(\mathbf{r}) = -j\omega\mu\tilde{\mathbf{H}}(\mathbf{r})$	(8.48)
Ampère's Law:	$\boldsymbol{\nabla} \times \tilde{\mathbf{H}}(\mathbf{r}) = \sigma\tilde{\mathbf{E}}(\mathbf{r}) + j\omega\varepsilon\tilde{\mathbf{E}}(\mathbf{r}).$	(8.49)

Development of a wave equation for $\tilde{\mathbf{E}}(\mathbf{r})$ in a lossy medium parallels that followed earlier in this chapter for lossless wave propagation. We turn Faraday's Law, Eq. (8.48), around to write

$$\tilde{\mathbf{H}}(\mathbf{r}) = \frac{j}{\omega\mu}\boldsymbol{\nabla} \times \tilde{\mathbf{E}}(\mathbf{r}).$$

Then we take the curl of both sides,

$$\boldsymbol{\nabla} \times \tilde{\mathbf{H}}(\mathbf{r}) = \boldsymbol{\nabla} \times \left[\frac{j}{\omega\mu}\boldsymbol{\nabla} \times \tilde{\mathbf{E}}(\mathbf{r})\right].$$

On the left side we have $\boldsymbol{\nabla} \times \tilde{\mathbf{H}}(\mathbf{r})$, which we can replace using Ampère's Law, Eq. (8.49). On the right side we use the vector identity Eq. (E.29), leading to

$$(\sigma + j\omega\varepsilon)\,\tilde{\mathbf{E}}(\mathbf{r}) = \frac{j}{\omega\mu}\left[\boldsymbol{\nabla}(\boldsymbol{\nabla} \cdot \tilde{\mathbf{E}}(\mathbf{r})) - \nabla^2\tilde{\mathbf{E}}(\mathbf{r})\right].$$

By Gauss' Law, $\boldsymbol{\nabla} \cdot \tilde{\mathbf{E}}(\mathbf{r}) = 0$, and rearranging terms gives us

$$\left[\nabla^2 + \omega^2\varepsilon\mu\left(1 - \frac{j\sigma}{\omega\varepsilon}\right)\right]\tilde{\mathbf{E}}(\mathbf{r}) = 0. \tag{8.50}$$

This is the wave equation that governs the propagation of a wave through a lossy medium. An identical wave equation can be determined for the magnetic field of this wave:

$$\left[\nabla^2 + \omega^2 \varepsilon \mu \left(1 - \frac{j\sigma}{\omega\varepsilon}\right)\right] \tilde{\mathbf{H}}(\mathbf{r}) = 0. \tag{8.51}$$

The plane-wave solutions to these equations are

$$\tilde{\mathbf{E}}(\mathbf{r}) = \mathbf{E}_0 e^{-j\boldsymbol{\kappa}\cdot\mathbf{r}} \tag{8.52}$$

and

$$\tilde{\mathbf{H}}(\mathbf{r}) = \mathbf{H}_0 e^{-j\boldsymbol{\kappa}\cdot\mathbf{r}}. \tag{8.53}$$

Although these solutions appear to be identical to our solution for waves propagating through lossless media, they differ in the wave vector $\boldsymbol{\kappa}$. Let's take a closer look.

In order for Eqs. (8.52) and (8.53) to be valid solutions of the wave equation, the expression inside the square brackets of Eqs. (8.50) and (8.51) must be zero, which requires that

$$\kappa^2 = \omega^2 \varepsilon \mu \left(1 - \frac{j\sigma}{\omega\varepsilon}\right) \tag{8.54}$$

or

$$\boxed{\kappa = \omega\sqrt{\varepsilon\mu}\sqrt{1 - j\sigma/\omega\varepsilon}.} \tag{8.55}$$

From this expression, it is clear that the wave number κ becomes complex for either of the lossy media we discussed earlier. For an absorbing dielectric, the permittivity $\varepsilon = \varepsilon' - j\varepsilon''$ is complex, while for a conductor the conductivity σ is non-zero. In either case, the magnitude of the wave vector $\boldsymbol{\kappa}$ in Eq. (8.55) is a complex quantity. We have replaced $\mathbf{k}$ with $\boldsymbol{\kappa}$ to designate this distinction, and denote the real and imaginary parts of κ as

$$\boxed{\kappa = \beta - j\alpha,} \tag{8.56}$$

where

$$\beta = \mathfrak{R}\left[\kappa\right] = \mathfrak{R}\left[\omega\sqrt{\varepsilon\mu}\sqrt{1 - j\sigma/\omega\varepsilon}\right] \tag{8.57}$$

and

$$\alpha = -\mathfrak{I}\left[\kappa\right] = -\mathfrak{I}\left[\omega\sqrt{\varepsilon\mu}\sqrt{1 - j\sigma/\omega\varepsilon}\right]. \tag{8.58}$$

Without repeating the proof that we presented for propagation through a lossless medium, we simply state that the direction of $\boldsymbol{\kappa}$ is the direction of propagation of the wave.

In order to interpret the significance of the constants β and α, let us suppose that the wave is propagating in the $+z$-direction. A wave of the form

$$E(z) = E_0 e^{-j\kappa z} = E_0 e^{-\alpha z} e^{-j\beta z} \tag{8.59}$$

is illustrated in Fig. 8.23. We have seen the factor $\exp(-j\beta z)$ previously (where of course only the real part of the complex exponential is significant). The sinusoidal oscillations in Fig. 8.23 result from this factor. If we examine the factor β in Eq. (8.57) carefully, with real permittivity ($\varepsilon'' = 0$) and zero conductivity ($\sigma = 0$), we recover the phase term $k = \omega\sqrt{\varepsilon\mu}$ of old. β is called the **phase constant**, describing the sinusoidal oscillations of the wave. As before, the wavelength of these oscillations is defined as

$$\lambda = 2\pi/\beta.$$

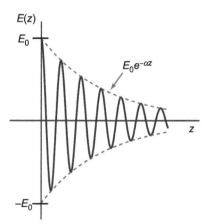

FIGURE 8.23 An attenuated oscillating wave $\tilde{E}(z)$, as given in Eq. (8.59).

The new feature here, however, is the decreasing amplitude of the wave, as represented by the term $E_0 \exp(-\alpha z)$. As the wave progresses in the z-direction, its amplitude decreases exponentially. The factor α, defined in Eq. (8.58), indicates the (spatial) rate at which the wave amplitude decreases. In a distance $\Delta z = \alpha^{-1}$, the amplitude of the wave decreases to $1/e$ of its original value. α is called the **attenuation constant**. As shown by Eq. (8.58), the attenuation constant is zero when the permittivity is real ($\varepsilon'' = 0$) and $\sigma = 0$.

For convenience, the **propagation constant γ** for the wave is defined as

$$\gamma = \alpha + j\beta = j\kappa. \tag{8.60}$$

The last equality in this line shows that the propagation constant γ and the complex wave number κ differ only by the factor j. Both terms are in common usage, and we will choose to use whichever is more convenient on a case-by-case basis.

The field orientations are unchanged from those for a wave propagating through lossless media. Faraday's Law, Eq. (8.48), tells us that $\tilde{\mathbf{H}}$ is perpendicular to $\tilde{\mathbf{E}}$ and $\boldsymbol{\kappa}$, while Ampère's Law, Eq. (8.49), tells us that $\tilde{\mathbf{E}}$ is perpendicular to $\tilde{\mathbf{H}}$ and $\boldsymbol{\kappa}$. Put together, $\tilde{\mathbf{E}}$, $\tilde{\mathbf{H}}$, and $\boldsymbol{\kappa}$ are mutually orthogonal to one another.

Now let us examine the intrinsic impedance of an absorbing medium. As a reminder, the intrinsic impedance is

$$\eta = \frac{\tilde{E}}{\tilde{H}},$$

where $\tilde{E}$ and $\tilde{H}$ are the phasor representations of $\mathbf{E}(t)$ and $\mathbf{H}(t)$, as introduced in Eqs. (8.52) and (8.53), respectively. (As a word of caution here, these phasor amplitudes $\tilde{E}$ and $\tilde{H}$ are scalar complex numbers, not vectors. $\mathbf{E}(t)$ and $\mathbf{H}(t)$, of course, are vectors, as they have direction, but the intrinsic impedance η does not include any information about direction, nor does it vary in time. Specifically, if we were to write η as the ratio of $\mathbf{E}(t)$ over $\mathbf{H}(t)$, as we sometimes see students try to do, this would be wrong in two regards. First, vectors in the denominator of a fraction have no meaning! And second, the ratio of the time-dependent field quantities is itself time-dependent, and thus meaningless.) Faraday's Law, Eq. (8.48), tells us that

$$\kappa \tilde{E} = \omega \mu \tilde{H},$$

or

$$\eta = \frac{\tilde{E}}{\tilde{H}} = \frac{\omega \mu}{\kappa} = \frac{\omega \mu}{\beta - j\alpha} = \sqrt{\frac{\mu}{\varepsilon} \left[\frac{1}{1 - j\sigma/\omega\varepsilon} \right]}. \tag{8.61}$$

η reduces to the familiar $\sqrt{\mu/\varepsilon}$ in the limit when the permittivity is real ($\varepsilon'' = 0$) and the conductivity σ is zero, as we expect for propagation through lossless media. This result for the intrinsic impedance of the medium is clearly a complex quantity, and so we should pause here to reflect on what this means. For this, let's return to the initial definition of the intrinsic impedance,

$$\eta = |\eta|\, e^{j\theta_\eta} = \frac{\tilde{E}}{\tilde{H}},$$

where the first equality is simply the impedance written in terms of its *magnitude* and *phase*. The magnitude $|\eta|$ therefore represents the ratio of magnitudes $|\tilde{E}|/|\tilde{H}|$,

$$|\eta| = \frac{\sqrt{\mu}}{[\varepsilon'^2 + (\varepsilon'' + \sigma/\omega)^2]^{1/4}}, \tag{8.62}$$

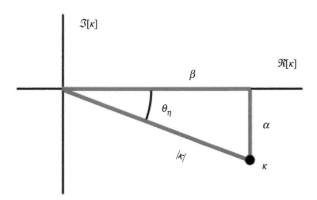

FIGURE 8.24 The wave number $|\kappa|$, phase constant β, attenuation constant α, and phase angle $\theta_\eta = -\tan^{-1}(\alpha/\beta)$ in the complex κ plane for a non-magnetic material. For a material with real permeability μ, the phase angle of complex κ is $-\theta_\eta$, since $\eta = \omega\mu/\kappa$.

and its phase θ_η represents the phase difference between the electric field $\tilde{E}$ and the magnetic field $\tilde{H}$. Notice that as long as the magnetic permeability μ is real (as it almost always is), the phase θ_η is precisely the same as minus the phase of the complex wave number κ, or $\theta_\eta = -\tan^{-1}(\alpha/\beta)$. The relation between κ, β, α, and θ_η is illustrated in Fig. 8.24. We will return to this point when we discuss the specific cases of the lossy dielectric medium and the good conductor.

To determine the phase velocity for a wave propagating through an absorbing medium, we follow the same prescription as was used for the lossless case. That is, we follow a wave propagating through the medium by traveling along with the crest of the wave (or any other point of constant phase). We must first restore Eq. (8.59) to its time-dependent form (multiply by $\exp(j\omega t)$, and take the real part), which gives us

$$E(z, t) = \Re\left[E_0 e^{-\alpha z} e^{-j\beta z} e^{j\omega t}\right] = E_0 e^{-\alpha z} \cos(\omega t - \beta z).$$

(For this discussion, the phase of E_0 is not important, so we have let the amplitude E_0 be real.) A point of constant phase, then, is defined by

$$\omega t - \beta z = \text{constant}.$$

So if at time t_1 the peak is at z_1, and at time t_2 the peak is at z_2, then

$$\omega t_1 - \beta z_1 = \omega t_2 - \beta z_2.$$

In a time interval $\Delta t = t_2 - t_1$, to remain at this point on the wave, we must have moved by $\Delta z = z_2 - z_1$, and our velocity is

$$u_p = \frac{\Delta z}{\Delta t} = \frac{\omega}{\beta}. \tag{8.63}$$

In the general case, we will use Eq. (8.57) to determine β, and then use Eq. (8.63) to determine the phase velocity.

Example 8.13 Wave Number, Impedance, and Phase Velocity in a General Absorbing Medium

A 10 GHz wave of amplitude $E_0 = 10$ V/m propagates through a non-magnetic medium of permittivity $\varepsilon = (4-j)\varepsilon_0$ and conductivity $\sigma = 0.5$ S/m. Determine (a) the complex wave number, (b) the propagation constant γ, (c) the wavelength, (d) the distance in which the wave amplitude is reduced to 2.5 V, (e) the impedance of the medium, (f) the phase lag of the magnetic field relative to the electric field, and (g) the phase velocity for this wave.

Solution:

(a) Using Eq. (8.55), the wave number is

$$\kappa = \omega\sqrt{\varepsilon\mu}\sqrt{1 - j\sigma/\omega\varepsilon} = \omega\sqrt{\mu}\sqrt{\varepsilon - j\sigma/\omega}.$$

Each of these quantities is given, so κ is

$$\kappa = \left(2\pi \times 10^{10}\ \text{s}^{-1}\right)\sqrt{4\pi \times 10^{-7}\ \text{H/m}}\sqrt{(4-j) \times 8.854 \times 10^{-12}\ \text{F/m} - j\frac{0.5\ \text{S/m}}{2\pi \times 10^{10}\ \text{s}^{-1}}}$$

$$= 441\,e^{-j13.3°}\ \text{m}^{-1} = 429 - j101\ \text{m}^{-1}.$$

The real part of κ is the phase constant of the wave, $\beta = 429$ m^{-1}, and the imaginary part is the attenuation constant, $\alpha = 101$ m^{-1}. For comparison, the phase constant of a non-absorbing medium with $\varepsilon = 4\varepsilon_0$ is

$$\beta = \omega\sqrt{\varepsilon\mu} = \frac{2\omega}{c} = \frac{2\left(2\pi \times 10^{10} \text{ s}^{-1}\right)}{3 \times 10^8 \text{ m/s}} = 419 \text{ m}^{-1}.$$

(b) $\gamma = j\kappa = 101 + j429$ m^{-1}.

(c) $\lambda = 2\pi/\beta = 0.0146$ m.

(d)

$$10 \text{ V}e^{-\alpha z} = 2.5 \text{ V} \quad \Rightarrow \quad z = \frac{\ln 4}{101 \text{ m}^{-1}} = 0.0137 \text{ m}.$$

(e)

$$\eta = \frac{\omega\mu}{\kappa} = \frac{(2\pi \times 10^{10} \text{ s}^{-1})(4\pi \times 10^{-7} \text{ H/m})}{441 \, e^{-j13.3°} \text{ m}^{-1}} = 174 \, \Omega \, e^{j13.3°}.$$

(f) The phase lag of H is the same as the phase of the impedance, 13.3°.

(g) $u_p = \omega/\beta = 1.46 \times 10^8$ m/s.

We have now found, for the general case, expressions for the complex wave number κ, the closely related propagation constant γ, the intrinsic impedance of the medium η, and the phase velocity of the wave u_p. We next look at these important parameters for the special cases that we identified earlier, that is, the lossy dielectric and the good conductor.

8.6.3 Non-conducting, Absorbing Dielectric Media

We showed earlier that we can characterize a lossy (or absorbing) dielectric material by zero conductivity and complex permittivity: $\varepsilon = \varepsilon' - j\varepsilon''$. The wave number κ, which is defined in Eq. (8.55) for the general case, becomes

$$\kappa = \omega\sqrt{\left(\varepsilon' - j\varepsilon''\right)\mu}. \tag{8.64}$$

Explicit expressions for α and β are useful, so let's work with this expression for κ to determine these. The constants α and β are most easily found by first computing κ^2, which gives us

$$\kappa^2 = \omega^2\left(\varepsilon' - j\varepsilon''\right)\mu = (-j\alpha + \beta)^2 = -\alpha^2 - 2j\alpha\beta + \beta^2,$$

where we used Eq. (8.56) to expand κ^2 on the right side. Equating the real and imaginary parts gives us two equations:

$$\omega^2\varepsilon'\mu = -\alpha^2 + \beta^2$$

and

$$\omega^2\varepsilon''\mu = 2\alpha\beta.$$

After some algebra, one can show that β and α for a lossy dielectric are

$$\beta = \omega\sqrt{\frac{(\varepsilon' + |\varepsilon|)\mu}{2}}, \tag{8.65}$$

where $|\varepsilon| = \sqrt{(\varepsilon')^2 + (\varepsilon'')^2}$ is the magnitude of the complex permittivity, and

$$\alpha = \omega\varepsilon''\sqrt{\frac{\mu}{2\left(\varepsilon' + |\varepsilon|\right)}}. \tag{8.66}$$

Note that the attenuation constant α increases with increasing ε''.

In weakly absorbing media, $\varepsilon'' \ll \varepsilon'$, so $|\varepsilon|$ is only slightly larger than ε', and with the aid of a few applications of the Taylor expansion, Eq. (H.6), it can be shown that

$$\beta \approx \omega\sqrt{\mu\varepsilon'}\left[1 + \frac{1}{8}\left(\frac{\varepsilon''}{\varepsilon'}\right)^2\right] \tag{8.67}$$

and

$$\alpha \approx \frac{\omega\varepsilon''}{2}\sqrt{\frac{\mu}{\varepsilon'}}. \tag{8.68}$$

As we might have expected, the phase constant β is only slightly affected by the imaginary part of the permittivity.

We next evaluate the intrinsic impedance for a lossy dielectric. To do this, we start with the general expression for the intrinsic impedance, Eq. (8.61), set $\sigma = 0$, and use κ for the lossy dielectric as given in Eq. (8.64). With this, we can write

$$\eta = |\eta|\,e^{j\theta_\eta} = \frac{\omega\mu}{\kappa} = \frac{\omega\mu}{\omega\sqrt{(\varepsilon' - j\varepsilon'')\,\mu}} = \sqrt{\frac{\mu}{(\varepsilon' - j\varepsilon'')}}. \tag{8.69}$$

Recall the significance of the magnitude and phase of this complex impedance. Its magnitude gives us the ratio of the magnitudes of the phasor amplitudes for the electric and magnetic fields,

$$|\eta| = \frac{|E_0|}{|H_0|} = \sqrt{\frac{\mu}{|\varepsilon|}},$$

where

$$\tilde{E}(\mathbf{r}) = E_0 e^{-j\boldsymbol{\kappa}\cdot\mathbf{r}} = |E_0|e^{j\varphi_E}e^{-j\boldsymbol{\kappa}\cdot\mathbf{r}}$$

and

$$\tilde{H}(\mathbf{r}) = H_0 e^{-j\boldsymbol{\kappa}\cdot\mathbf{r}} = |H_0|e^{j\varphi_H}e^{-j\boldsymbol{\kappa}\cdot\mathbf{r}}.$$

The phase of η is equivalent to the phase lag of the magnetic field H_0 relative to the electric field E_0:

$$\theta_\eta = \varphi_E - \varphi_H = \frac{1}{2}\tan^{-1}\left(\frac{\varepsilon''}{\varepsilon'}\right).$$

For a medium that is weakly absorbing, $\varepsilon'' \ll \varepsilon'$, and the impedance is approximately

$$\eta \approx \sqrt{\frac{\mu}{\varepsilon'}}\left[1 + \frac{j\varepsilon''}{2\varepsilon'}\right]. \tag{8.70}$$

The phase velocity of an electromagnetic wave in a lossy dielectric is, from Eq. (8.63),

$$u_p = \frac{\omega}{\beta} = \left\{\mu\left[\frac{\varepsilon' + |\varepsilon|}{2}\right]\right\}^{-1/2},$$

where we have used β from Eq. (8.65). For weakly absorbing media, the Taylor expansion, Eq. (H.7), of this expression gives

$$u_p \approx \frac{1}{\sqrt{\varepsilon'\mu}}\left[1 - \frac{1}{8}\left(\frac{\varepsilon''}{\varepsilon'}\right)^2\right], \tag{8.71}$$

valid for $\varepsilon'' \ll \varepsilon'$, which is only slightly reduced from the phase velocity in a lossless medium.

Example 8.14 Wave Number, Impedance, and Phase Velocity in a Weakly Absorbing Dielectric

A 10 GHz wave of amplitude $E_0 = 10$ V/m propagates through a non-magnetic, weakly absorbing dielectric of permittivity $\varepsilon = (4 - j0.16)\varepsilon_0$. Determine (a) the wave number, (b) the impedance, and (c) the phase velocity for this wave.

Solution:

Since $\varepsilon'' \ll \varepsilon'$, this dielectric is weakly absorbing, and the results derived in this section can be used.

(a) To find the wave number $\kappa = \beta - j\alpha$, refer to Eqs. (8.67) and (8.68). For β, we find

$$\beta \approx \omega\sqrt{\mu\varepsilon'}\left[1 + \frac{1}{8}\left(\frac{\varepsilon''}{\varepsilon'}\right)^2\right]$$

$$= (2\pi \times 10^{10}\text{ s}^{-1})\sqrt{\mu_0(4\varepsilon_0)}\left[1 + \frac{1}{8}\left(\frac{0.16\varepsilon_0}{4\varepsilon_0}\right)^2\right]$$

$$= (2\pi \times 10^{10}\text{ s}^{-1})\left(\frac{2}{c}\right)[1.0002] = 419\text{ m}^{-1}. \tag{8.72}$$

This result is barely distinguishable from that of a non-absorbing dielectric. The attenuation constant is

$$\alpha \approx \frac{\omega\varepsilon''}{2}\sqrt{\frac{\mu}{\varepsilon'}} = \frac{(2\pi \times 10^{10}\text{ s}^{-1})(0.16\varepsilon_0)}{2}\sqrt{\frac{\mu_0}{4\varepsilon_0}}$$

$$= \frac{(2\pi \times 10^{10}\text{ s}^{-1})(0.16)}{2}\left(\frac{1}{2c}\right) = 8.43\text{ m}^{-1}.$$

(b) The intrinsic impedance is, using Eq. (8.70),

$$\eta \approx \sqrt{\frac{\mu}{\varepsilon'}}\left[1 + \frac{j\varepsilon''}{2\varepsilon'}\right] = \sqrt{\frac{\mu_0}{4\varepsilon_0}}\left[1 + \frac{j0.16\varepsilon_0}{8\varepsilon_0}\right] = (60\pi\ \Omega)\left[1 + 0.02j\right].$$

This impedance tells us that the magnetic field of this wave lags the electric field by a phase angle of $\tan^{-1}(0.02) \approx 0.02$ radians.

(c) The phase velocity of this wave is

$$u_p \approx \frac{1}{\sqrt{\varepsilon'\mu}}\left[1 - \frac{1}{8}\left(\frac{\varepsilon''}{\varepsilon'}\right)^2\right] = \frac{1}{\sqrt{4\varepsilon_0\mu_0}}\left[1 - \frac{1}{8}\left(\frac{0.16\varepsilon_0}{4\varepsilon_0}\right)^2\right]$$

$$= \frac{c}{2}[0.9998] = 1.499 \times 10^8\text{ m/s},$$

as comes from Eq. (8.71). This result is very close to $c/2$.

We next turn our attention to wave propagation in a medium that is a good conductor.

8.6.4 Wave Propagation in a Good Conductor

In a conducting medium, the electromagnetic wave is attenuated by the free currents within the medium. The electric field of the wave applies a force to the free charges and generates currents, thereby doing work on these charges. As the field does work, the wave amplitude is attenuated. We examine κ, α, β, η, and u_p for a wave propagating through a conducting medium of conductivity σ in this section. A good conductor is defined as a material in which the free current density $J_v = \sigma E$ is much larger than the displacement current density $j\omega\varepsilon E$, or more compactly, $\sigma/\omega\varepsilon \gg 1$.

We start with κ, which from Eq. (8.55), is

$$\kappa = \omega\sqrt{\varepsilon\mu}\sqrt{1 - j\sigma/\omega\varepsilon} = \beta - j\alpha.$$

Since $\sigma/\omega\varepsilon \gg 1$, the 1 under the square root sign can be ignored, and upon squaring both sides of this equation we have

$$\omega^2\varepsilon\mu\left(-j\sigma/\omega\varepsilon\right) = \left(\beta - j\alpha\right)^2,$$

or

$$-j\omega\mu\sigma = \beta^2 - 2j\alpha\beta - \alpha^2.$$

Equating the real parts of this expression leads to

$$0 = \beta^2 - \alpha^2,$$

or

$$\alpha = \beta.$$

Equating the imaginary parts, we find

$$\omega\mu\sigma = 2\alpha\beta.$$

Thus, for a good conductor,

$$\beta = \alpha = \sqrt{\frac{\omega\mu\sigma}{2}} = \sqrt{\pi f\mu\sigma}, \tag{8.73}$$

where $f = \omega/2\pi$ is the cyclical frequency of the wave.

With the phase constant β equal to the attenuation constant α, the wave is strongly attenuated, and it shows only a few oscillations before it has nearly completely diminished. This is illustrated in Fig. 8.25.

When we started this discussion of wave propagation in conductors, some of you may have had the thought that this concept was nonsensical, that waves cannot possibly

FIGURE 8.25 An attenuated oscillating wave $E(z)$ for a good conductor, with $\beta = \alpha$.

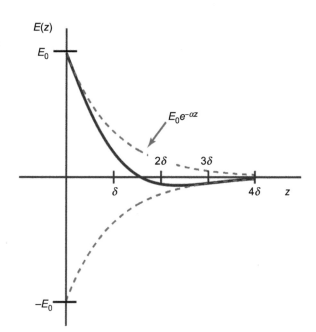

propagate through a conductor. After all, we did make the argument in Chapter 1 that upon application of an electric field to a conductor, free charges within the conductor would immediately adjust themselves in response to the field, in such a way as to remove the field. That argument, of course, is still valid for d.c. fields. We now see that for sinusoidally varying fields, the wave can propagate within the conductor, but that the wave is rapidly attenuated as it propagates. The characteristic distance over which the amplitude decreases is called the **skin depth**, denoted by δ, and defined as

$$\delta = 1/\alpha = \frac{1}{\sqrt{\pi f \mu \sigma}}. \tag{8.74}$$

In a distance of one skin depth, the wave amplitude decreases to a value $1/e \approx 0.37$ of its initial amplitude.

So for an electromagnetic wave incident on the surface of a conductor, the wave penetrates into the conductor a short distance and induces a volume current density $\mathbf{J}_v = \sigma \mathbf{E}$, which decreases exponentially as the wave penetrates inside the conductor. In high-frequency circuits the skin depth plays an important role, since the current in conductors is confined to the surfaces of the conductors, with the current density decreasing exponentially toward the interior of the conductor. In the limit of a perfect conductor, $\sigma \to \infty$, we see from Eq. (8.74) that the skin depth decreases, and the current becomes more and more a true surface current.

Example 8.15 Wave Propagation in a Good Conductor

Determine the phase constant, the attenuation constant, and the skin depth for a 100 MHz wave in copper, a good conductor.

Solution:
We should perhaps start by verifying that copper is indeed a good conductor at this frequency. We'll evaluate $\sigma/(\omega\varepsilon)$. The conductivity of copper is $\sigma = 5.8 \times 10^7$ S/m, and we'll use $\varepsilon = \varepsilon_0$. Then,

$$\frac{\sigma}{\omega\varepsilon} = \frac{5.8 \times 10^7 \text{ S/m}}{(2\pi \times 10^8 \text{ s}^{-1})(8.854 \text{ pF/m})} = 1.0 \times 10^{10}.$$

Since this quantity is much larger than 1, copper is a good conductor at this frequency.

Now let's find β. Copper is non-magnetic, so we use $\mu = \mu_0$ for the permeability in Eq. (8.73) to find

$$\beta = \alpha = \sqrt{\pi f \mu \sigma} = \sqrt{\pi(10^8 \text{ s}^{-1})(4\pi \times 10^{-7} \text{ H/m})(5.8 \times 10^7 \text{ S/m})}$$
$$= 1.51 \times 10^5 \text{ m}^{-1}.$$

The skin depth is

$$\delta = 1/\alpha = 6.6 \text{ µm}.$$

The intrinsic impedance of a good conductor is found using the general expression, Eq. (8.61),

$$\eta = \frac{\omega\mu}{\beta - j\alpha} = \frac{\omega\mu}{\sqrt{\omega\mu\sigma/2}\,(1-j)},$$

where β and α come from Eq. (8.73). This simplifies to

$$\eta = \sqrt{\frac{\omega\mu}{2\sigma}}\,(1+j) = \sqrt{\frac{\pi f \mu}{\sigma}}\,(1+j). \tag{8.75}$$

Since the conductivity σ appears in the denominator of this expression, the impedance of a good conductor is typically very small, just as the real electrical resistance is small. This should not be surprising, since in a conductor under static conditions, electric fields inside the conductor vanish completely. The complex factor $(1+j)$ tells us that $\mathbf{H}$ lags behind $\mathbf{E}$ by 45° as the wave propagates through the conductor.

Example 8.16 Intrinsic Impedance of a Good Conductor

Determine the intrinsic impedance for copper for a 100 MHz wave.

Solution:
The conductivity of copper is $\sigma = 5.8 \times 10^7$ S/m and the permeability is $\mu = \mu_0$. We use Eq. (8.75) to find

$$\eta = \sqrt{\frac{\pi f \mu}{\sigma}}\,(1+j) = \sqrt{\frac{\pi (10^8\ \text{s}^{-1})(4\pi \times 10^{-7}\ \text{H/m})}{(5.8 \times 10^7\ \text{S/m})}}\,(1+j)$$

$$= 2.6\,(1+j)\ \text{m}\Omega.$$

The magnitude of this impedance is $|\eta| = 3.7$ mΩ.

The phase velocity of the wave in a good conductor is determined using Eq. (8.63)

$$u_p = \frac{\omega}{\beta} = \frac{\omega}{\sqrt{\omega\mu\sigma/2}} = \sqrt{\frac{2\omega}{\mu\sigma}}. \tag{8.76}$$

Owing to the large conductivity σ in good conductors, the phase velocity is $u_p \ll c$.

Example 8.17 Phase Velocity in a Good Conductor

Determine the phase velocity of a 100 MHz wave in copper.

Solution:
Use the conductivity of copper $\sigma = 5.8 \times 10^7$ S/m and the permeability is $\mu = \mu_0$ in Eq. (8.76) to find

$$u_p = \sqrt{\frac{2\omega}{\mu\sigma}} = \sqrt{\frac{2(2\pi \times 10^8\ \text{s}^{-1})}{(4\pi \times 10^{-7}\ \text{H/m})(5.8 \times 10^7\ \text{S/m})}}$$

$$= 4.15 \times 10^3\ \text{m/s}.$$

When compared to $c = 3 \times 10^8$ m/s, this phase velocity is indeed very small.

For quick reference, the expressions for β, α, η, and u_p in different types of absorbing media are summarized in Table 8.3.

A prime example of electromagnetic wave propagation through a highly absorbing medium is used for location of underground utilities lines, as described in TechNote 8.18.

Table 8.3

Summary of expressions for β, α, η, and u_p for wave propagation through different media.

$$\kappa = \omega\sqrt{\varepsilon\mu}\sqrt{1 - j\sigma/\omega\varepsilon} = \beta - j\alpha, \ |\varepsilon| = \sqrt{(\varepsilon')^2 + (\varepsilon'')^2}$$

	General form	Non-conducting dielectric		Good conductor
		General	Weakly absorbing $(\varepsilon'' \ll \varepsilon')$	
β	$\Re[\kappa]$	$\omega\sqrt{\dfrac{(\varepsilon' + |\varepsilon|)\mu}{2}}$	$\omega\sqrt{\varepsilon'\mu}\left[1 + \dfrac{1}{8}\left(\dfrac{\varepsilon''}{\varepsilon'}\right)^2\right]$	$\sqrt{\pi f\mu\sigma}$
α	$-\Im[\kappa]$	$\omega\varepsilon''\sqrt{\dfrac{\mu}{2(\varepsilon' + |\varepsilon|)}}$	$\dfrac{\omega\varepsilon''}{2}\sqrt{\dfrac{\mu}{\varepsilon'}}$	$\sqrt{\pi f\mu\sigma}$
η	$\dfrac{\omega\mu}{\kappa}$	$\sqrt{\dfrac{\mu}{(\varepsilon' - j\varepsilon'')}}$	$\sqrt{\dfrac{\mu}{\varepsilon'}}\left[1 + \dfrac{j\varepsilon''}{2\varepsilon'}\right]$	$\sqrt{\dfrac{\pi f\mu}{\sigma}}(1 + j)$
u_p	$\dfrac{\omega}{\beta}$	$\sqrt{\dfrac{2}{(\varepsilon' + |\varepsilon|)\mu}}$	$\dfrac{1}{\sqrt{\varepsilon'\mu}}\left[1 - \dfrac{1}{8}\left(\dfrac{\varepsilon''}{\varepsilon'}\right)^2\right]$	$\sqrt{\dfrac{2\omega}{\mu\sigma}}$

TechNote 8.18 Sensing Instruments for Underground Utility Location

You have probably seen the warnings from the utility companies (gas, water, electric, telephone) that you should "call before you dig." Utility lines are routinely buried beneath the surface of the ground, and accidental contact between your hand-held shovel or more powerful excavation equipment can result in disruption of service, the need for expensive repairs, or even personal injury or death. Therefore, it is a good idea, and in many cases your legal responsibility, to call the utility companies, who will dispatch a technician to locate and mark the underground utility lines around your dig site. There are several different methods used for locating these underground utility lines.

- A common technique is electromagnetic induction, in which a transmitter induces an rf current in a conducting utility line at one point, and a receiver picks up this signal from further along the line. This can be used for any conductor, such as electrical, gas lines, telephone lines, or water lines. For non-conducting pipes, a metal tracer wire is often strung along with the pipe for locating purposes.
- Ground penetrating radar (GPR) is also used to locate underground structures such as utility lines. In GPR systems, a transmitter projects an rf wave, typically in the frequency range of 10 MHz to 3 GHz, into the ground. When the wave encounters an interface between the soil and another material whose electrical permittivity and/or conductivity differ from that of the surrounding soil, a reflected wave is generated. (Reflections are discussed in Sections 8.8 and 8.9.) This reflected wave is detected by a receiving antenna, which uses the strength and delay time of the reflected wave to create an image of the underground domain. Ground penetrating radar detects non-conductors, such as PVC, as well as conductors. It has also been used to locate

gravesites and archeological sites. The useful range of GPR systems is limited by absorption of the wave between the transmitter and the receiver.

• rfid tags can be affixed to underground utilities to aid in their location. We discussed rfid in TechNote 8.10.

8.7 Power Flow

LEARNING OBJECTIVES

After studying this section, you should be able to:

• identify the physical significance of the various terms in Poynting's Theorem;
• determine the time-dependent Poynting vector, or power density, of a wave;
• determine the time-average Poynting vector, or power density, of a harmonic wave; and
• determine the total power incident upon a specified cross-section, given the Poynting vector.

In the discussion in the previous section, we talked about the electric fields doing work on the bound charges within a dielectric medium or on the free charges within a conducting medium. We also showed that the field amplitude is attenuated as the wave propagates through an absorbing medium. As the fields do work on these charges, we should expect that the energy stored in the field decreases, but we haven't yet made that connection. We did define, in our treatment of static fields, the energy stored in the fields, with the energy density of an electric field given by

$$w_e = \frac{1}{2}\mathbf{D} \cdot \mathbf{E}$$

and the energy density of a magnetic field by

$$w_m = \frac{1}{2}\mathbf{H} \cdot \mathbf{B}.$$

So it is natural to ask, as a wave is traveling, does it transport energy? And if a wave traveling through a medium becomes attenuated, doesn't the decreased amplitude of the wave indicate that the energy it stores is decreasing? If so, then where does that energy go? Finally, we are all familiar with circuit elements getting hot when a large current flows through them. We imagine that, somehow, energy that is stored in the source is transmitted to the load. How can we understand this flow of energy from the source to the load? These are all questions that we can address using Maxwell's Equations, as we will now explore.

Before we start, this will be another one of those derivations for which the *motivation* for each step may not be transparent, and we can easily get bogged down if we worry about that. If you find yourself getting lost because of these kinds of questions, we encourage you to focus instead on the *validity* of each step, which should be easier to grasp. The motivation for each step will be in the final results, and our interpretation of the physical significance that we can offer at the end. Again, we have the benefit of following some very smart people who had the foresight to develop this formalism, and we will happily stand on their shoulders and learn from their prior work.

We start by considering a volume v in space, such as that shown in Fig. 8.26, in which an electric field $\mathbf{E}(t, \mathbf{r})$ and a magnetic field $\mathbf{H}(t, \mathbf{r})$ exist. Note that $\mathbf{E}(t, \mathbf{r})$ and $\mathbf{H}(t, \mathbf{r})$ are functions of time t and position $\mathbf{r}$; we are not restricting ourselves to time-harmonic waves or uniform plane waves, but rather keeping the discussion general. We let the medium filling this space be simple (homogeneous and isotropic), but we do allow for currents in this space. The permittivity, permeability, and conductivity of the medium are given by ε, μ, and σ, respectively. Let's start by examining the quantity

$$\nabla \cdot (\mathbf{E} \times \mathbf{H})$$

within this volume. We first apply a vector differential identity, Eq. (E.7), introduced in Appendix E, that is similar to the product rule for differentiation for scalar functions. That is,

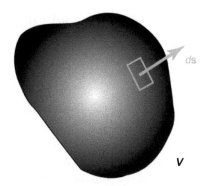

FIGURE 8.26 A volume v bounded by the surface s. We evaluate $\nabla \cdot (\mathbf{E} \times \mathbf{H})$ over this volume.

$$\nabla \cdot (\mathbf{E} \times \mathbf{H}) = \mathbf{H} \cdot (\nabla \times \mathbf{E}) - \mathbf{E} \cdot (\nabla \times \mathbf{H}).$$

But we know $\nabla \times \mathbf{E}$ from Faraday's Law (see Table 7.1),

$$\nabla \times \mathbf{E} = -\frac{\partial \mathbf{B}}{\partial t},$$

and $\nabla \times \mathbf{H}$ from Ampère's Law

$$\nabla \times \mathbf{H} = \mathbf{J}_v + \frac{\partial \mathbf{D}}{\partial t}.$$

Substituting these terms leads to

$$\nabla \cdot (\mathbf{E} \times \mathbf{H}) = \mathbf{H} \cdot \left(-\frac{\partial \mathbf{B}}{\partial t}\right) - \mathbf{E} \cdot \left(\mathbf{J}_v + \frac{\partial \mathbf{D}}{\partial t}\right).$$

In a simple medium, the two terms involving the time derivatives can be written as

$$\mathbf{H} \cdot \frac{\partial \mathbf{B}}{\partial t} = \mu \mathbf{H} \cdot \frac{\partial \mathbf{H}}{\partial t} = \frac{\mu}{2} \frac{\partial}{\partial t} (\mathbf{H} \cdot \mathbf{H})$$

and

$$\mathbf{E} \cdot \frac{\partial \mathbf{D}}{\partial t} = \varepsilon \mathbf{E} \cdot \frac{\partial \mathbf{E}}{\partial t} = \frac{\varepsilon}{2} \frac{\partial}{\partial t} (\mathbf{E} \cdot \mathbf{E}).$$

Substituting these terms and integrating the result over the entire volume v, we find

$$\int_v \nabla \cdot (\mathbf{E} \times \mathbf{H}) \, dv = -\int_v \left\{\frac{\mu}{2} \frac{\partial}{\partial t} (\mathbf{H} \cdot \mathbf{H}) + \mathbf{E} \cdot \mathbf{J}_v + \frac{\varepsilon}{2} \frac{\partial}{\partial t} (\mathbf{E} \cdot \mathbf{E})\right\} dv.$$

We apply the Divergence Theorem to the integral on the left side of this equation, rearrange terms, and reverse the order of integration and differentiation to find

$$\oint_s \mathbf{E} \times \mathbf{H} \cdot d\mathbf{s} = -\frac{d}{dt} \int_v \left\{\frac{\varepsilon}{2} (\mathbf{E} \cdot \mathbf{E}) + \frac{\mu}{2} (\mathbf{H} \cdot \mathbf{H})\right\} dv - \int_v \mathbf{E} \cdot \mathbf{J}_v \, dv. \quad (8.77)$$

This result is known as **Poynting's Theorem**, and as we will show next, it is a statement of the conservation of energy for electromagnetic fields.

Several of the terms contained in Poynting's Theorem should be familiar to us. Let's start with these. First, we have the time-rate-of-change of $\varepsilon \mathbf{E} \cdot \mathbf{E}/2$ and $\mu \mathbf{H} \cdot \mathbf{H}/2$. These quantities represent the density of energy stored in the electric and magnetic fields, respectively, as we reviewed at the beginning of this section. When these energy densities are integrated over the volume v, we find the total energy stored in the fields. Poynting's Theorem has the time derivative of this stored energy, which is the rate at which the energy is changing. (With the minus sign in front of this term, it more specifically represents that rate at which the stored energy decreases.)

Second, we see the volume integral of $\mathbf{E} \cdot \mathbf{J}_v$. This product is the power dissipation density, discussed in Sections 3.6 and 8.6, and when integrated over the volume v we recognize this term as the total rate at which power is dissipated in the medium. This is the rate at which the fields do work on the charges inside the volume.

We have now identified the two integrals that appear on the right-hand side of Poynting's Theorem. Can we use these two known integrals to identify the surface integral on the left side? One known term represents the rate at which the energy stored in the fields is decreasing. How can the field energy decrease? One way would be if the fields are doing work on the charges within the volume. We have already identified this term, $\mathbf{J}_v \cdot \mathbf{E}$. The other mechanism that can lead to a decrease in the energy stored in the fields is energy being transported out of the volume via the electromagnetic waves, and this is precisely the correct interpretation of the surface integral on the left. The quantity $\mathbf{E} \times \mathbf{H}$ is known as the **Poynting vector**,

$$\mathcal{P}(t, \mathbf{r}) = \mathbf{E}(t, \mathbf{r}) \times \mathbf{H}(t, \mathbf{r}), \qquad (8.78)$$

represented with the script $\mathcal{P}$. The surface integral of $\mathcal{P}(t, \mathbf{r})$ gives the total rate of energy leaving the volume, so $\mathcal{P}$ is interpreted as the power flow *per unit area* through the surface. The Poynting vector is often called the **power density** of the wave. In optics this is often called the **intensity** of the wave. The surface integral uses the dot product between $\mathcal{P}$ and $d\mathbf{s}$ to pick out the component of $\mathcal{P}$ that is perpendicular to the surface. Only the normal component of $\mathcal{P}$ results in net transfer of energy out of the volume. The Poynting vector $\mathcal{P}$ is perpendicular to $\mathbf{E}$ and to $\mathbf{H}$, and so in simple media is parallel to the wave vector $\mathbf{k}$. Its magnitude is equal to $EH = E^2/\eta$. The units of the Poynting vector are $[\mathcal{P}] = [E][H] = (V/m)(A/m) = W\ m^{-2}$, consistent with our interpretation of $\mathcal{P}$ as the power density.

Example 8.18 Poynting Vector

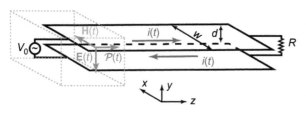

FIGURE 8.27 A voltage source connected to a resistor R through a pair of flat, parallel conducting plates.

A pair of long, parallel conducting plates of width w and separation d is shown in Fig. 8.27. A voltage source producing a sinusoidal voltage of amplitude V_0 and frequency ω is connected to the plates at one end, and a resistor R at the other. Evaluate the Poynting vector in the region between the plates, and relate this to the power absorbed by the resistor. Let $d \ll w$, and ignore the resistance of the conducting plates.

Solution:
The potential difference between the plates is $v(t) = V_0 \cos \omega t$, and the current $i(t) = I_0 \cos \omega t$, where $I_0 = V_0/R$, flows to the resistor in the upper plate and returns in the bottom plate. In the region between the plates, the electric field is $\mathbf{E}(t) = \mathbf{a}_y E_y(t)$, where $E_y(t) = -(V_0/d) \cos \omega t$, while the magnetic field is $\mathbf{H}(t) = \mathbf{a}_x H_x(t)$, where $H_x(t) = (I_0/w) \cos \omega t$. (We solved for this magnetic field in Example 6.6.) Outside this region, the electric and magnetic fields are small, and we ignore them. The Poynting vector in the region between the plates is

$$\mathcal{P}(t) = \mathbf{E}(t) \times \mathbf{H}(t) = \begin{vmatrix} \mathbf{a}_x & \mathbf{a}_y & \mathbf{a}_z \\ 0 & E_y(t) & 0 \\ H_x(t) & 0 & 0 \end{vmatrix} = -\mathbf{a}_z E_y(t) H_x(t)$$

$$= +\mathbf{a}_z \frac{V_0 I_0 \cos^2 \omega t}{wd}.$$

This represents the density of the power transported from the source to the resistive load. The total power generated by the source and delivered to the load can be found by constructing a surface surrounding the source, as shown in Fig. 8.27 using dashed lines, and integrating the power density $\mathcal{P}(t)$ over this surface. Since the electric and magnetic fields are confined to the region between the conducting plates, the total surface integral reduces to the surface integral carried out over the rectangular region between the conductors of dimension w by d. For this surface, the surface element is $d\mathbf{s} = \mathbf{a}_z\, dx\, dy$, and the transmitted power is

$$P(t) = \oint_s \mathcal{P}(t) \cdot d\mathbf{s} = \int_0^d \int_0^w \left[\mathbf{a}_z \frac{V_0 I_0 \cos^2 \omega t}{wd} \right] \cdot \mathbf{a}_z \, dx \, dy = V_0 I_0 \cos^2 \omega t.$$

This power should be familiar from your circuit analysis class, where the instantaneous power generated by the source, or absorbed by the load, is the product $v(t)i(t)$.

We are now ready to answer one of the questions that we posed at the beginning of this section. That is, how is the energy that is generated by the source delivered to the load in an electric circuit? The answer is that the fields carry the energy. In the previous example we used the known electric and magnetic fields for the parallel plate conductor configuration to compute the power density of the electromagnetic wave, and showed that the total power transmitted by this wave is perfectly consistent with the familiar energy absorbed by a load in terms of the current through the load and the voltage across the load. We chose this example for simplicity, since the fields are uniform in the region between the plates and negligible elsewhere, but the result is perfectly general.

While we're on the topic of power transferred by electromagnetic waves, let's examine another facet of this power density. In the previous example we treated the parallel plate conductors as if they were perfect conductors. Under normal conditions, of course, metals are very good conductors, but are certainly not perfect, and are characterized by the conductivity of the material. As the conductor has a small resistance, the two parallel conducting plates of the previous example will absorb some of the power generated by the source. We can understand this absorption in terms of the Poynting vector of the wave as well, as we examine in Example 8.19.

Example 8.19 Poynting Vector, Resistive Losses

Consider the source, the pair of long, flat, parallel conducting plates, and the resistive load of Example 8.18, as shown in Fig. 8.27. Treating the conducting plates that carry the current to the load as good, but not perfect, conductors, of conductivity σ, evaluate the Poynting vector that represents power flow into the conductors, and the total power absorbed. Let the length of the conducting plates be ℓ, and the thickness of the conducting plates be h.

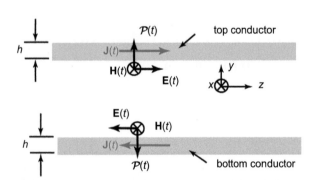

FIGURE 8.28 An expanded view of the top and bottom plates from Fig. 8.27. The electric field vector shown in this figure is the small z-component due to the finite conductivity of the conducting plates. The primary component **E**, as shown in Fig. 8.27, is not shown here.

Solution:
In Example 8.18 we found the primary field components E_y and H_x that exist in the region between the conducting plates. In addition, there must also exist a small electric field component in the z-direction, E_z. After all, the conducting plates carry the current, $i(t)$, and since the conductivity σ of the conductors is finite, a small electric field in the conductor is necessary to drive this current. E_z is illustrated in the cross-sectional view of the conductors shown in Fig. 8.28. (Only the E_z component is shown in this figure. Remember that the primary field component is in the y-direction.) We estimate the magnitude of this electric field component E_z as follows. At low frequencies, the current density in the conductor is uniform, so the current density in the upper plate is $J_{v,z}(t) = i(t)/wh$, where h is the thickness of the conductors and w is their width, as defined in Example 8.18.

The current density in the bottom plate is $-J_{v,z}(t)$. The electric field in the conducting plates required to drive this current is $E_z = J_{v,z}/\sigma$ in the upper plate, $-J_{v,z}/\sigma$ in the lower. This field component is very small, owing to the large conductivity σ of the material of which the parallel plate conductors consist, but it is finite. Note that this electric field component is tangential to the surface of the conductor. Tangential electric fields must be continuous at the boundary, so E_z of this magnitude is also present just below the bottom surface of the upper conductor. A small component E_z of the opposite sign exists just above the bottom plate as well. The component E_z, taken together with the magnetic field component H_x, gives a component of the Poynting vector in the y-direction:

$$\mathcal{P}_y(t) = (\mathbf{E}(t) \times \mathbf{H}(t))_y = E_z(t)H_x(t) - E_x(t)H_z(t)$$

$$= +\left[\frac{i(t)}{\sigma wh}\right]\left[\frac{i(t)}{w}\right] = \frac{i^2(t)}{\sigma w^2 h}.$$

A Poynting vector of the opposite sign can be found directed downward into the bottom plate. These Poynting vectors represent the density of power flow into the conducting strips, and are in addition to the large power density representing power flow to the resistive load that we calculated in the last example. The total power flow into the top conducting strip is $P(t) = \int_s \mathcal{P}_y(t)ds$, where the surface integral is carried out over the entire bottom surface of the top conductor, and $ds = dx\,dz$,

$$P(t) = \int_s \mathcal{P}_y(t)ds = \int_0^w \int_0^\ell \left[\frac{i^2(t)}{\sigma w^2 h}\right] dx\,dz = i^2(t)\left(\frac{\ell}{\sigma wh}\right).$$

$\ell/\sigma wh$ is the resistance R_{plate} of the conducting plate of cross-sectional area $w \times h$ and length ℓ, and $i^2(t)R_{plate}$ is the power dissipated in a resistor R_{plate} when a current $i(t)$ passes through it. The power absorbed by the lower plate is the same. Thus this analysis of the small y-component of the Poynting vector leads to an expression for the resistive power losses in the conductors that is consistent with our expectations.

We see from these examples that we can characterize the rate at which energy is transmitted (per unit area) by a wave through the power density of that wave, given by the Poynting vector. The power density increases as the square of the amplitude of the field, and is an important feature of wave propagation.

8.7.1 Time-Averaged Power Density

To this point, we have discussed the time-dependent Poynting vector, often called the instantaneous Poynting vector. In many cases, however, we are interested only in the time-averaged value of the Poynting vector. This is especially true for high-frequency fields, in the terahertz or higher frequency range, for which standard instrumentation is not capable of responding to the instantaneous fields. Even at lower frequencies, however, the time-averaged power density is often a useful parameter for characterizing the electromagnetic wave.

The average power density can be determined, of course, by averaging the instantaneous Poynting vector over one cycle of oscillation,

$$\overline{\mathcal{P}(\mathbf{r})} = \frac{1}{T}\int_0^T \mathcal{P}(t,\mathbf{r})\,dt = \frac{1}{T}\int_0^T \mathbf{E}(t,\mathbf{r}) \times \mathbf{H}(t,\mathbf{r})\,dt,$$

where $T = 2\pi/\omega$ is the period of the field. In order to simplify the notation, we consider a wave traveling in the z-direction. Then $\mathbf{E}(t,\mathbf{r})$ can be written as $\mathbf{E}_0 e^{-\alpha z}\cos(\omega t - \beta z)$ and

$H(t, \mathbf{r})$ as $H_0 e^{-\alpha z} \cos(\omega t - \beta z - \theta_\eta)$, where θ_η is the phase difference between the electric and magnetic fields. Remember, in the time-dependent notation that we are using now, the field amplitudes E_0 and H_0 are *real* vector amplitudes. Also, in writing $H(t, \mathbf{r})$ in this form, we have used the fact that the phase difference between the electric and magnetic fields is θ_η, the phase of the intrinsic impedance η of the medium, as we discussed earlier. The average power density of the wave then becomes

$$\overline{\mathcal{P}(z)} = E_0 \times H_0 \, e^{-2\alpha z} \frac{1}{T} \int_0^T \cos(\omega t - \beta z) \cos(\omega t - \beta z - \theta_\eta) \, dt.$$

In order to evaluate the time average of $\cos(\omega t - \beta z) \cos(\omega t - \beta z - \theta_\eta)$, the second cosine term $\cos(\omega t - \beta z - \theta_\eta)$ must be expanded as $\cos(\omega t - \beta z) \cos \theta_\eta + \sin(\omega t - \beta z) \sin \theta_\eta$, and the integral in the previous equation becomes

$$\frac{1}{T} \int_0^T \left[\cos^2(\omega t - \beta z) \cos \theta_\eta + \cos(\omega t - \beta z) \sin(\omega t - \beta z) \sin \theta_\eta \right] \, dt.$$

The first term is $(1/2) \cos \theta_\eta$, the second is zero, and the time-averaged Poynting vector is

$$\overline{\mathcal{P}(z)} = \frac{1}{2} E_0 \times H_0 \, e^{-2\alpha z} \cos \theta_\eta.$$

Several features of this time-averaged power density are of interest. Note the factor $1/2$ out front of the expression. This factor is a direct result of taking the time average of $\cos^2(\omega t - \beta z)$. The factor $\exp(-2\alpha z)$ describes the attenuation of the power density as the wave propagates in the z-direction. As the wave moves in the z-direction, the power density decreases. The rate of attenuation of the power density is -2α, where α is the attenuation constant for the field amplitude. We can understand this extra factor of 2 in front of α in terms of the dependence of the power density on E^2. Since the electric field amplitude decreases as $\exp(-\alpha z)$, the power density decreases as $\exp(-2\alpha z)$. Finally, the factor $\cos \theta_\eta$ shows the role of the phase difference between E and H. Only the component of H that is in phase with E represents average power carried by the wave.

Phasor representations of the electric and magnetic fields offer an alternative method of expressing the average power density, in the form

$$\boxed{\overline{\mathcal{P}} = \frac{1}{2} \mathfrak{R} \left[\tilde{E} \times \tilde{H}^* \right].} \tag{8.79}$$

To show that these forms are equivalent, refer to Eqs. (8.52) and (8.53), and use Eq. (8.56) to write

$$\tilde{E}(\mathbf{r}) \times \tilde{H}^*(\mathbf{r}) = E_0 e^{(-\alpha - j\beta)\hat{k} \cdot \mathbf{r}} \times \left(H_0 e^{(-\alpha - j\beta)\hat{k} \cdot \mathbf{r}} \right)^*.$$

We introduced $\hat{k}$ earlier as the unit vector pointing in the direction of propagation of the wave. Recall that now, since we are using phasor notation for the fields, the field amplitudes E_0 and H_0 are complex terms, and the phase difference between these amplitudes is θ_η. For a wave propagating in the z-direction, then, the time-averaged power Poynting vector is

$$\overline{\mathcal{P}(z)} = \frac{1}{2} \mathfrak{R} \left[E_0 \times H_0^* \right] e^{-2\alpha z},$$

which can also be written as

$$\overline{\mathcal{P}(z)} = \frac{1}{2} |E_0 \times H_0^*| e^{-2\alpha z} \cos \theta_\eta,$$

in agreement with the result we found above. As is commonly done, we will in the future omit the explicit inclusion of the real part symbol in this expression, and simply write

$$\overline{\mathcal{P}(z)} = \frac{1}{2}\mathbf{E}(\mathbf{r}) \times \mathbf{H}^*(\mathbf{r}).$$

Note that, since the explicit time-dependence has been pulled out of the phasor format, we cannot use the phasor representation to produce the instantaneous power density (unless we first convert the field amplitudes to their time-dependent forms).

The magnitude of the power density can be written in a compact form as

$$|\overline{\mathcal{P}}| = \frac{1}{2}\left|\mathbf{E} \times \mathbf{H}^*\right| = \frac{|\mathbf{E}|^2}{2|\eta|}\cos\theta_\eta.$$

For a lossless medium, the intrinsic impedance η is real, and this becomes

$$\boxed{|\overline{\mathcal{P}}| = \frac{E_0^2}{2\eta}.} \tag{8.80}$$

Example 8.20 Field Amplitude from Power Density

Determine the amplitude of the electric field associated with sunlight at the surface of the Earth. The power density of sunlight reaching us at sea level on a clear day is $|\overline{\mathcal{P}}|$ = 1.3 kW/m². (Clearly, sunlight is not a single-frequency harmonic wave, but for the purpose of this example you may treat it as such.)

Solution:
We invert Eq. (8.80) to write

$$E_0 = \sqrt{2\eta\,|\overline{\mathcal{P}}|} = \sqrt{2(120\,\pi\,\Omega)(1.3\,\text{kW/m}^2)} = 990\,\text{V/m}.$$

In this calculation, we used $\eta_0 = 120\pi\;\Omega$ for the intrinsic impedance, presuming that the sunlight is traveling through air or vacuum.

We have now discussed several aspects of power and absorption of waves. For example, we showed (in Section 8.6.1) that an electric field does work on the free and bound charges within a medium. Then we showed (in Section 8.6.2) that the amplitude of the wave decreases as the wave propagates through the absorbing medium. And in this section, we showed that the density of power transported by a wave, given by the Poynting vector, is related to the field amplitude, such that as the amplitude decreases, the power density carried by the wave also decreases. Since energy must be conserved, the work done by the field on the medium should be precisely the same as the loss of energy transported by the field. Let's see if this works out.

Example 8.21 Equivalence of Power Absorbed by the Medium to Power Lost by the Wave

Show that for a time-harmonic, uniform plane wave traveling through an absorbing medium, the power absorbed by the medium is equivalent to the power lost by the wave.

Solution:
Let's start with the latter — that is, the loss of power carried by the uniform plane wave. Consider a uniform plane wave propagating through an absorbing medium of thickness d. The medium could be a dielectric or a conductor; we won't restrict it to one or the other. Either way, the electric field of the propagating wave can be written as

$$\tilde{\mathbf{E}}(z) = \mathbf{a}_x E_0 \, e^{-\alpha z} \, e^{-j\beta z},$$

where the phasor representation is used for the fields. We chose this field, which propagates in the z-direction and points in the x-direction, to simplify the notation, but the result is general, and it is not necessary to limit the wave in this way.

When the wave enters the medium through the face at $z = 0$, its power density is the magnitude of the Poynting vector, given by

$$\overline{\mathcal{P}} = \frac{|E_0|^2}{2|\eta|} \cos \theta_\eta \, \mathbf{a}_z,$$

where the impedance is

$$\eta = \frac{\omega\mu}{\kappa} = \frac{\omega\mu}{\beta - j\alpha}.$$

After propagating through the medium, the amplitude of the wave is diminished, and the Poynting vector representing the power density at the opposite face is

$$\overline{\mathcal{P}} = \frac{|E_0|^2}{2|\eta|} \, e^{-2\alpha d} \cos \theta_\eta \, \mathbf{a}_z.$$

The decrease in the power density is just the difference between these two,

$$\Delta|\overline{\mathcal{P}}| = \frac{|E_0|^2}{2|\eta|} \left(1 - e^{-2\alpha d}\right) \cos \theta_\eta. \tag{8.81}$$

Can we relate this decrease in the power density to the work done by the field on the medium?

To do this, let's return to our previous discussion of this work, in which we showed that the rate of that work per unit volume is, as given in Eq. (8.42),

$$P_v^{\mathrm{abs}}(z) = \frac{1}{2}\mathfrak{R}\left[\tilde{\mathbf{E}}(\mathbf{r}) \cdot (\tilde{\mathbf{J}}_v(\mathbf{r}) + \tilde{\mathbf{J}}_d(\mathbf{r}))^*\right].$$

As we discussed earlier, the free current density is due to conduction, $\tilde{\mathbf{J}}_v(\mathbf{r}) = \sigma\tilde{\mathbf{E}}(\mathbf{r})$, and the displacement current density is $\tilde{\mathbf{J}}_d(\mathbf{r}) = j\omega\varepsilon\tilde{\mathbf{E}}(\mathbf{r})$. So the power dissipation density is

$$P_v^{\mathrm{abs}}(z) = \frac{1}{2}\left(\sigma + \omega\varepsilon''\right)|E_0|^2 e^{-2\alpha z}.$$

Since P_v^{abs} is the rate of work done per unit *volume*, we can find the work done per unit *area* by integrating $P_v^{\mathrm{abs}}(z)$ in z through the thickness of the absorbing medium,

$$\int_0^d P_v^{\mathrm{abs}}(z)dz = \int_0^d \frac{1}{2}\left(\sigma + \omega\varepsilon''\right)|E_0|^2 e^{-2\alpha z}dz$$

$$= \frac{1}{2}\frac{\sigma + \omega\varepsilon''}{2\alpha}|E_0|^2 \left(1 - e^{-2\alpha d}\right). \tag{8.82}$$

But $\sigma + \omega\varepsilon''$ is just $-\mathfrak{I}(\kappa^2)/\omega\mu$ (see Eq. (8.54)), and $\mathfrak{I}(\kappa^2) = -2\alpha\beta$ (square both sides of Eq. (8.56)), so

$$\frac{\sigma + \omega\varepsilon''}{2\alpha} = \frac{\beta}{\omega\mu}.$$

Since $\beta = |\kappa| \cos \theta_\eta$ (see Fig. 8.24)) and $\kappa/\omega\mu = 1/\eta$ (see Eq. (8.61)),

$$\frac{\sigma + \omega\varepsilon''}{2\alpha} = \frac{|\kappa| \cos \theta_\eta}{\omega\mu} = \frac{1}{|\eta|} \cos \theta_\eta,$$

and Eqs. (8.81) and (8.82) are the same. Therefore the power density lost by the wave is indeed equal to the power absorbed by the medium, as we expected it should be.

Before we conclude our discussion of the power density of a propagating wave, we have one last task to complete. Earlier we derived Poynting's Theorem for time-dependent fields, and interpreted this important result as a statement that energy is conserved in these systems of electromagnetic fields in media. We have also treated the special case of time-harmonic waves, and introduced the phasor representation of these fields. It can be useful, then, to derive Poynting's Theorem in a form that is valid for time-harmonic fields for linear, homogeneous, absorbing media. We address this in the following.

The derivation will follow steps that parallel the derivation for Poynting's Theorem for the general case, Eq. (8.77). We start by considering the divergence of the Poynting vector,

$$\boldsymbol{\nabla} \cdot [\tilde{\mathbf{E}}(\mathbf{r}) \times \tilde{\mathbf{H}}^*(\mathbf{r})] = \tilde{\mathbf{H}}^*(\mathbf{r}) \cdot [\boldsymbol{\nabla} \times \tilde{\mathbf{E}}(\mathbf{r})] - \tilde{\mathbf{E}}(\mathbf{r}) \cdot [\boldsymbol{\nabla} \times \tilde{\mathbf{H}}^*(\mathbf{r})],$$

where we have used Eq. (E.7) to expand the divergence of the Poynting vector. By Faraday's Law, Eq. (8.48), $\boldsymbol{\nabla} \times \tilde{\mathbf{E}}(\mathbf{r})$ is $-j\omega\mu\tilde{\mathbf{H}}(\mathbf{r})$, and by Ampère's Law, Eq. (8.49), $\boldsymbol{\nabla} \times \tilde{\mathbf{H}}^*(\mathbf{r})$ is $\tilde{\mathbf{J}}_v^*(\mathbf{r}) - j\omega\varepsilon^*\tilde{\mathbf{E}}^*(\mathbf{r})$, so the divergence of the Poynting vector becomes

$$\boldsymbol{\nabla} \cdot [\tilde{\mathbf{E}}(\mathbf{r}) \times \tilde{\mathbf{H}}^*(\mathbf{r})] = -j\omega\mu|\tilde{\mathbf{H}}(\mathbf{r})|^2 - \tilde{\mathbf{E}}(\mathbf{r}) \cdot \tilde{\mathbf{J}}_v^*(\mathbf{r}) + j\omega\varepsilon^*|\tilde{\mathbf{E}}(\mathbf{r})|^2.$$

Over any volume where these derivatives are finite, we take the volume integral of both sides and divide by two to find

$$\frac{1}{2}\int_v \boldsymbol{\nabla} \cdot [\tilde{\mathbf{E}}(\mathbf{r}) \times \tilde{\mathbf{H}}^*(\mathbf{r})]\,dv = \frac{j\omega}{2}\int_v \left[\varepsilon^*|\tilde{\mathbf{E}}(\mathbf{r})|^2 - \mu|\tilde{\mathbf{H}}(\mathbf{r})|^2\right]dv - \frac{1}{2}\int_v \tilde{\mathbf{E}}(\mathbf{r}) \cdot \tilde{\mathbf{J}}_v^*(\mathbf{r})dv.$$

Finally, we apply the Divergence Theorem to the volume integral on the left side, resulting in

$$\oint_s \overline{\mathcal{P}} \cdot d\mathbf{s} = \frac{j\omega}{2}\int_v \left[\varepsilon^*|\tilde{\mathbf{E}}(\mathbf{r})|^2 - \mu|\tilde{\mathbf{H}}(\mathbf{r})|^2\right]dv - \frac{1}{2}\int_v \tilde{\mathbf{E}}(\mathbf{r}) \cdot \tilde{\mathbf{J}}_v^*(\mathbf{r})dv, \qquad (8.83)$$

where $\overline{\mathcal{P}}$ is the Poynting vector defined in Eq. (8.79). As is typical with complex quantities, we find physical significance in the real parts of each of these terms, and the interpretation of each of these terms is similar to that of Poynting's Theorem for instantaneous fields. The surface integral of $\overline{\mathcal{P}}$ is the (time-averaged) rate at which the waves carry energy out of the volume v. The volume integral of $\frac{1}{2}\tilde{\mathbf{E}}(\mathbf{r}) \cdot \tilde{\mathbf{J}}_v^*(\mathbf{r})$ is the average rate at which the fields do work on the free charges in the volume. The first integral on the left side, at first glance, might appear to be purely imaginary, but remember that the permittivity ε is complex, $\varepsilon = \varepsilon' - j\varepsilon''$. This term simplifies to the volume integral of $\omega\varepsilon''|\tilde{\mathbf{E}}(\mathbf{r})|^2$, and we recognize this as dielectric absorption. This is precisely the power dissipation term that we discussed previously. In a similar way, the second term on the right side can be simplified to $\omega\mu''|\tilde{\mathbf{H}}(\mathbf{r})|^2$, where $-\mu''$ is the imaginary part of the magnetic permeability. We interpret this as work done by the magnetic field on the medium. This is less common, and we will not consider it any further here.

In this section, we have discussed the flow of power, as represented through the Poynting vector. This power transfer is real, as you will experience when sitting outside on a sunny day, or when heating your dinner in the microwave oven. We showed that the power density of the wave is proportional to the square of the field amplitude, a result that we will apply in later discussions.

8.8 Reflection of Uniform Plane Waves, Normal Incidence

To this point, we have discussed plane waves propagating through various media, but we haven't worried about what happens when the wave arrives at the end of the medium. As a practical matter, a wave can only propagate so far before it encounters an interface with another medium. It is natural, therefore, to ask what happens to the wave when it is incident upon such an interface. Experience tells us that part of the wave is reflected and part is transmitted. For example, when you stand inside your house and look through a window, you can see the lawn and trees in the front yard, examples of waves transmitted by the glass window. Some of the sunlight that is scattered off these objects travels toward you and passes through the window, and you can sense this light with your eyes. If it is nighttime, however, and therefore dark outside your house, and you stand in your well-lit room and look outside, you probably won't see any objects in the yard, but rather you will see the reflected waves that originate from objects inside the house. Light from those objects that is incident upon the windows is partially reflected, and you are able to see these reflected waves. Those reflections, of course, are present in the daytime as well, but you often don't notice them since they are relatively weak in comparison with the bright light from the outside when the sun is shining. The example we have just given was centered on optical waves (i.e. light), to which our eyes are sensitive. Reflection and transmission of waves in other frequency bands behave precisely the same, however, and our treatment will be general. In this section, we will study the transmission and reflection of waves at an interface between two media.

In this section, we treat the case of *normal* incidence. By normal incidence, we mean that the direction of propagation of the wave is parallel to the surface normal, and that the electric and magnetic fields of the incident wave both lie in the plane of the surface, as shown in Fig. 8.29. A uniform plane wave propagates to the right through medium 1, and is incident on the interface of this medium with medium 2. The electric field $\tilde{\mathbf{E}}_i$, the magnetic field $\tilde{\mathbf{H}}_i$, and the wave vector $\mathbf{k}_i$, for this incident wave, where the subscript "i" indicates the incident field, are shown in this figure. In phasor notation, the electric field $\tilde{\mathbf{E}}_i(z)$ of the incident wave is

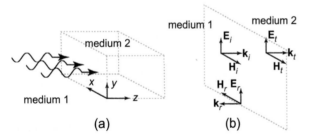

FIGURE 8.29 (a) A wave propagates in the z-direction through medium 1, and is normally incident on the interface with medium 2. (b) **E**, **H**, and **k** for the incident, transmitted, and reflected waves.

$$\tilde{\mathbf{E}}_i(z) = \mathbf{a}_y E_i e^{-j\beta_1 z}, \qquad (8.84)$$

and the magnetic field $\tilde{\mathbf{H}}_i(z)$ is

$$\tilde{\mathbf{H}}_i(z) = -\mathbf{a}_x \frac{E_i}{\eta_1} e^{-j\beta_1 z}. \qquad (8.85)$$

The direction of propagation of the incident wave is labeled as the $+z$-direction, and, for simplification, medium 1 is lossless, such that $\mathbf{k}_i = \mathbf{a}_z \beta_1$. Note that $\tilde{\mathbf{E}}_i \times \tilde{\mathbf{H}}_i$ points in the direction of $\mathbf{k}_i$, as we know it must. The intrinsic impedance of medium 1 is η_1. When this wave strikes the surface (located at $z = 0$), part of the wave is transmitted and part is reflected. The electric field $\tilde{\mathbf{E}}_t$, the magnetic field $\tilde{\mathbf{H}}_t$, and the wave vector $\mathbf{k}_t$ for the transmitted wave are shown in Fig. 8.29. The subscript "t" indicates the transmitted field. The electric field $\tilde{\mathbf{E}}_t(z)$ of the transmitted incident wave is

$$\tilde{\mathbf{E}}_t(z) = \mathbf{a}_y E_t e^{-jk_2 z}, \qquad (8.86)$$

and the magnetic field $\tilde{\mathbf{H}}_t(z)$ is

$$\tilde{\mathbf{H}}_t(z) = -\mathbf{a}_x \frac{E_t}{\eta_2} e^{-jk_2 z}. \tag{8.87}$$

η_2 is the intrinsic impedance of medium 2, and $k_2 = -j\alpha_2 + \beta_2$ is the complex wave number of the wave in medium 2.

Finally, the electric and magnetic fields of the reflected wave are

$$\tilde{\mathbf{E}}_r(z) = \mathbf{a}_y E_r e^{j\beta_1 z} \tag{8.88}$$

and

$$\tilde{\mathbf{H}}_r(z) = \mathbf{a}_x \frac{E_r}{\eta_1} e^{j\beta_1 z}. \tag{8.89}$$

Note several important features of this wave. First, it propagates in the $-z$-direction. Thus the complex exponential has an argument $+j\beta_1 z$. Second, this wave propagates in medium 1, so we use the parameters β_1 and η_1; the same constants as for the incident wave. And finally, since $\tilde{\mathbf{E}}_r \times \tilde{\mathbf{H}}_r$ must point in the direction of $\mathbf{k}_r$, and $\mathbf{k}_r$ is opposite $\mathbf{k}_i$, either $\tilde{\mathbf{E}}_r$ or $\tilde{\mathbf{H}}_r$ must be reversed as well, relative to the incident wave. We chose to define $\tilde{\mathbf{E}}_r$ parallel to $\tilde{\mathbf{E}}_i$ (i.e. in the $+\mathbf{a}_y$-direction) and $\tilde{\mathbf{H}}_r$ opposite to $\tilde{\mathbf{H}}_i$ (i.e. in the $+\mathbf{a}_x$-direction). Be aware that this is a choice we make, and that other texts may choose the opposite notation. If you are, at some point later in your career, having trouble with the signs, be sure to check on which sign convention is being used.

Now that the incident wave, the transmitted wave, and the reflected wave are defined, the next order of business is to determine the amplitudes of the transmitted and reflected waves. For this we will apply the boundary conditions that must be satisfied at the interface between medium 1 and medium 2. When we consider normal incidence, both the electric field and the magnetic field are tangential fields. They lie in the plane of the surface. Recall that tangential electric fields must be continuous across the boundary, and that, unless there are surface currents, tangential magnetic fields must also be continuous. As we discussed in the previous section, true surface currents exist only for perfect conductors. In real conductors, the fields and current densities decay exponentially, with a characteristic depth equal to the skin depth δ. The skin depth is small, of course, but finite, and we will treat the magnetic field $\tilde{\mathbf{H}}$ field as continuous as well.

What do we mean when we say that the field must be continuous across the boundary? Well, on the left side, in medium 1, the fields consist of the incident wave and the reflected wave. Put together, these give us the total field amplitude on the incident side of the boundary. On the right side, there is only the transmitted wave. Thus the condition that the electric field must satisfy at the boundary is

$$E_i + E_r = E_t, \tag{8.90}$$

where we have used Eqs. (8.84), (8.86), and (8.88), evaluated at the interface between the two media, $z = 0$. Similarly, the condition that the magnetic field must satisfy at the boundary is

$$\frac{E_i}{\eta_1} - \frac{E_r}{\eta_1} = \frac{E_t}{\eta_2}. \tag{8.91}$$

Notice the minus sign for the term for the reflected wave here, since $\tilde{\mathbf{H}}_r$ is flipped relative to the incident and transmitted magnetic fields.

The **reflection coefficient** Γ at the surface is defined as

$$\Gamma = \frac{E_r}{E_i}, \tag{8.92}$$

and the **transmission coefficient** τ as

$$\tau = \frac{E_t}{E_i}. \tag{8.93}$$

Γ is the ratio of the reflected wave amplitude to the incident wave amplitude, while τ is the ratio of the transmitted wave amplitude to the incident wave amplitude. Then Eq. (8.90) becomes

$$\boxed{1 + \Gamma = \tau,} \tag{8.94}$$

and Eq. (8.91) becomes

$$1 - \Gamma = \tau \frac{\eta_1}{\eta_2}. \tag{8.95}$$

These two linear equations can be solved for Γ to find

$$\boxed{\Gamma = \frac{\eta_2 - \eta_1}{\eta_2 + \eta_1}} \tag{8.96}$$

and for the transmission coefficient τ,

$$\tau = \frac{2\eta_2}{\eta_2 + \eta_1}. \tag{8.97}$$

From these results we see that the fraction of the wave amplitude that is reflected or transmitted depends only on the intrinsic impedances of the two media, and that as these two impedances approach one another the amplitude of the reflected wave decreases. Remember that we allowed for medium 2 to be a lossy medium, so these expressions are quite generally valid. For non-conducting dielectrics, the impedance is

$$\eta = \sqrt{\frac{\mu}{(\varepsilon' - j\varepsilon'')}} \approx \sqrt{\frac{\mu}{\varepsilon'}} \left[1 + \frac{j\varepsilon''}{2\varepsilon'} \right],$$

which comes from Eq. (8.69). The second approximate equality is valid for a weakly absorbing medium. In a good conductor, the intrinsic impedance is

$$\eta = \sqrt{\frac{\pi f \mu}{\sigma}} \, (1 + j),$$

which comes from Eq. (8.75). These expressions for the intrinsic impedance of dielectric or conducting media will be useful as we calculate the reflection coefficients at interfaces between media.

Example 8.22 Reflection Coefficient, Dielectric

Light is normally incident from air upon a glass window of permittivity $\varepsilon = 2.25\varepsilon_0$. Determine the reflection coefficient at the first and second surfaces as the light passes through the glass.

Solution:
At the first surface, the light is incident from air onto glass, as shown in Fig. 8.30(a). So medium 1 is the air, of intrinsic impedance

$$\eta_1 = \eta_0 = \sqrt{\mu_0/\varepsilon_0} = 120\pi \ \Omega \approx 377 \ \Omega,$$

and medium 2 is glass, of intrinsic impedance

$$\eta_2 = \sqrt{\mu_0/\varepsilon_2} = \sqrt{\mu_0/2.25\varepsilon_0} = \eta_0/\sqrt{2.25} = 2\eta_0/3.$$

FIGURE 8.30 A wave is incident on a glass window. The wave is partially reflected at the first surface, as well as at the second surface. We treat the reflection at the first surface in (a), and the reflection at the second surface in (b).

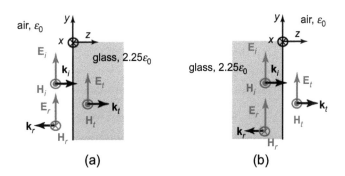

Using Eq. (8.96), the reflection coefficient at the first surface is therefore

$$\Gamma_1 = \frac{\eta_2 - \eta_1}{\eta_2 + \eta_1} = \frac{2\eta_0/3 - \eta_0}{2\eta_0/3 + \eta_0} = \frac{-1/3}{5/3} = -0.2.$$

So the amplitude of the reflected wave is only 1/5 that of the incident wave. But what is the significance of the minus sign? To answer this, let's examine the amplitude of the reflected field,

$$E_r = \Gamma E_i = -0.2 E_i,$$

or

$$\tilde{E}_r(z) = -\mathbf{a}_y\, 0.2 E_i e^{j\beta_1 z},$$

where we have used the coordinate system defined in Fig. 8.29. A negative reflection coefficient Γ indicates that the phasor representing the electric field of the reflected wave is actually opposite that shown in Fig. 8.30. This reversal is equivalent to a π phase shift upon reflection, since displacing a sine wave by half a cycle is equivalent to inverting the wave. The magnetic field is also inverted, relative to that shown for the reflected wave in Fig. 8.30,

$$\tilde{H}_r(z) = -\mathbf{a}_x \frac{E_r}{\eta_1} e^{j\beta_1 z} = -\mathbf{a}_x \frac{0.2 E_i}{\eta_1} e^{j\beta_1 z},$$

so the Poynting vector of the reflected wave,

$$\overline{\mathcal{P}}_r = \frac{1}{2}\tilde{E}_r(z) \times \tilde{H}_r^*(z) = \frac{1}{2}\left(-\mathbf{a}_y\, 0.2 E_i e^{j\beta_1 z}\right) \times \left(-\mathbf{a}_x \frac{0.2 E_i}{\eta_1} e^{j\beta_1 z}\right)^*$$

$$= -\mathbf{a}_z \frac{|0.2 E_i|^2}{2\eta_1},$$

points in the $-z$-direction, as expected. The power density of the reflected wave is $(0.2)^2 = 4\%$ of the power density of the incident wave. You may have heard reference to a 4% reflection of the light at the first surface; this is where that comes from. The transmission coefficient at this surface is $\tau_1 = 1 + \Gamma_1 = 0.8$.

The portion of the wave that is transmitted at the first surface is incident upon the second surface, and will again be partially transmitted and partially reflected. To determine the reflection coefficient at this second surface, we again apply Eq. (8.96), but now the identities of η_1 and η_2 are reversed. After all, the wave is propagating through glass (medium 1), and incident upon the interface to the air (medium 2). Therefore, we have $\eta_1 = 2\eta_0/3$, $\eta_2 = \eta_0$, and

$$\Gamma_2 = \frac{\eta_2 - \eta_1}{\eta_2 + \eta_1} = \frac{\eta_0 - 2\eta_0/3}{\eta_0 + 2\eta_0/3} = \frac{+1/3}{5/3} = +0.2.$$

The reflection coefficient at this second surface therefore has the same magnitude as Γ at the first surface, but the sign is now positive; 4% of the wave's power density is reflected, but there is no π phase shift of this reflected wave. The transmission coefficient at this surface is $\tau_2 = 1 + \Gamma_2 = 1.2$.

Notice that the transmitted wave amplitude can actually be greater than the amplitude of the incident wave. By Eq. (8.94), when Γ is positive, the transmission coefficient $\tau = 1 + \Gamma$ is greater than 1, and the transmitted field amplitude $E_t = \tau E_i$ is greater than E_i. This may seem incredible, but it is correct. No conservation laws or other valid physical laws of nature are violated. We should, of course, expect that energy is conserved, as this is one of the fundamental tenets of the physical world. Let's look at the power density flowing to and from any surface. The incident wave is the only wave that travels toward the surface, and its power density is of magnitude

$$|\overline{\mathcal{P}_i}| = \frac{|E_i|^2}{2\eta_1}.$$

The power density flowing away from the surface is contained in the transmitted wave and the reflected wave, and is of magnitude

$$|\overline{\mathcal{P}_t}| + |\overline{\mathcal{P}_r}| = \frac{|E_t|^2}{2\eta_2} + \frac{|E_r|^2}{2\eta_1}.$$

But $E_t = \tau E_i$ and $E_r = \Gamma E_i$, where τ and Γ are given by Eqs. (8.97) and (8.96), respectively, and the net power density leaving the surface becomes

$$|\overline{\mathcal{P}_t}| + |\overline{\mathcal{P}_r}| = \frac{|E_i|^2}{2}\left[\left(\frac{2\eta_2}{\eta_2+\eta_1}\right)^2\frac{1}{\eta_2} + \left(\frac{\eta_2-\eta_1}{\eta_2+\eta_1}\right)^2\frac{1}{\eta_1}\right].$$

A little algebra shows that the term inside the square brackets reduces to $1/\eta_1$, and therefore the combined power density of the waves traveling away from the surface is indeed equal to the power density of the incident wave, $|\overline{\mathcal{P}_i}|$. Note that this is valid even when η_1 or η_2 are complex, as they are for absorbing media.

Example 8.23 Reflection Coefficient, Good Conductor

Determine the reflection coefficient Γ for a uniform plane wave traveling through air and normally incident upon the plane surface of a good conductor. Find a numerical result for Γ for aluminum at a wavelength of the wave $\lambda = 0.5\ \mu\text{m}$.

Solution:

The incident, transmitted, and reflected waves are shown in Fig. 8.31. Let the conductivity of the conductor be σ, and $\mu = \mu_0$ (as is valid for most metals). The frequency of the wave is f. Medium 1 is air, for which $\eta_1 = \eta_0$, and medium 2 is aluminum, with

$$\eta_2 = \sqrt{\frac{\pi f \mu_0}{\sigma}}\,(1+j).$$

Then, Γ is

$$\Gamma = \frac{\eta_2 - \eta_1}{\eta_2 + \eta_1} = \frac{\sqrt{\pi f \mu_0/\sigma}\,(1+j) - \eta_0}{\sqrt{\pi f \mu_0/\sigma}\,(1+j) + \eta_0}.$$

After dividing all terms by $\eta_0 = \sqrt{\mu_0/\varepsilon_0}$, this becomes

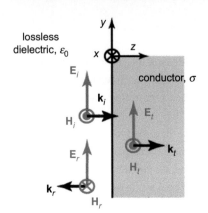

FIGURE 8.31 A wave propagating through air is incident upon a good conductor of conductivity σ.

$$\Gamma = \frac{\sqrt{\pi f \varepsilon_0 / \sigma}\,(1+j) - 1}{\sqrt{\pi f \varepsilon_0 / \sigma}\,(1+j) + 1}.$$

Recall that, for a good conductor, $\sigma \gg \omega\varepsilon$, so in the preceding equation $\pi f \varepsilon_0 / \sigma \ll 1$, and the Taylor expansion Eq. (H.5) can be used to derive $(1 + \epsilon)/(1 - \epsilon) = 1 + 2\epsilon + \cdots$, valid for any small number ϵ. The reflection coefficient is then

$$\Gamma = -1 + 2\sqrt{\frac{\pi f \varepsilon_0}{\sigma}}\,(1+j).$$

Now we are ready to analyze this reflection coefficient for the specific case of aluminum at the specified wavelength of $\lambda = 0.5\ \mu\text{m}$. (By the way, this wavelength is in the visible region of the spectrum – that is, the range of frequencies to which our eyes are sensitive – and is a pretty green color.) The frequency f for this wavelength is $f = c/\lambda = (3 \times 10^8$ m/s)/(0.5 $\times 10^{-6}$ m) = 6×10^{14} Hz. The conductivity of aluminum, $\sigma_{\text{Al}} = 3.77 \times 10^7$ S/m, is listed in Table 3.1. The quantity $2\sqrt{\pi f \varepsilon_0 / \sigma_{\text{Al}}}$ is

$$2\sqrt{\frac{\pi f \varepsilon_0}{\sigma_{\text{Al}}}} = 2\sqrt{\frac{\pi (6 \times 10^{14}\ \text{Hz})(8.854 \times 10^{-12}\ \text{F/m})}{(3.77 \times 10^7\ \text{S/m})}} = 0.04.$$

(Let's be sure to check the units here. This quantity should be unitless, since the reflectivity is a pure number. The conversion is

$$\frac{\text{Hz(F/m)}}{\text{S/m}} = \frac{\text{s}^{-1}\,\text{F}}{\Omega^{-1}} = \frac{\text{s}^{-1}\,(\text{C/V})}{(\text{V/A})^{-1}},$$

and since an ampere is 1 coulomb per second, all units cancel, and the fraction is indeed unitless.)

The final result then for the reflectivity Γ is

$$\Gamma = -1 + 0.04(1 + j) = -0.96 + j0.04.$$

The fraction of the power density that is reflected at this surface is

$$|\Gamma|^2 = |-0.96 + j0.04|^2 = 0.92,$$

and the phase shift of the reflected wave is

$$\tan^{-1}\frac{0.04}{-0.96} = 0.99\pi.$$

Most of the power is therefore reflected, and the phase shift of the reflected wave is close to (but not quite equal to) π. With 92% of the wave power reflected, there must also be partial transmission at the surface. We were not asked to find the transmission coefficient, but let's find it anyway. That is,

$$\tau = 1 + \Gamma = 2\sqrt{\frac{\pi f \varepsilon_0}{\sigma}}\,(1+j) = 0.04(1+j).$$

The transmitted power is absorbed by the aluminum, and this wave attenuates within a short distance, characterized by the skin depth δ.

Through the previous example, we found a reflection coefficient whose magnitude is just a little less than 1, and whose phase is slightly less than π. As the conductivity gets larger, or the frequency of the wave gets smaller, the reflection coefficient $\Gamma \to -1$. This should make sense, since for a perfect conductor the tangential electric field at the boundary must be zero. (Remember, electric fields inside a perfect conductor must be zero, and tangential electric fields must be continuous across the interface.) How can

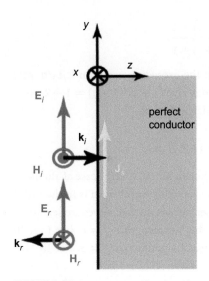

FIGURE 8.32 A wave propagating through air is incident upon a perfect conductor.

the tangential electric field outside the conductor be zero, you might ask, when a wave is incident upon the surface? The answer rests with the reflected field, as is illustrated in Fig. 8.32. Remember that the total field on the incident side of the interface is the superposition of the incident and the reflected field. When these two fields are equal in magnitude, but π out of phase with one another, then they perfectly cancel one another at the interface, and the net field is zero. And if the reflected wave is the same magnitude as the incident wave, with a π phase difference, well, that is exactly what a reflection coefficient of $\Gamma = -1$ tells us. Furthermore, in this limit of a perfect conductor, the transmitted field amplitude is reduced to zero.

We haven't talked much about the magnetic fields, but these are important as well. For a wave incident upon a perfect conductor, for which $\Gamma = -1$, the amplitude of the magnetic field of the reflected wave is $-\Gamma = +1$ times the magnetic field of the incident wave,

$$\tilde{\mathbf{H}}_r(z) = (-\Gamma)\left(-\mathbf{a}_x \frac{E_i}{\eta_1}e^{j\beta_1 z}\right) = -\mathbf{a}_x \frac{E_i}{\eta_1}e^{j\beta_1 z}.$$

At the interface, $z = 0$, the magnetic field for the reflected wave is in the same direction as $\tilde{\mathbf{H}}_i(0)$, and these two fields add constructively at the surface. The total magnetic field on the incident side is therefore $2\tilde{\mathbf{H}}_i(0)$. This might, at first glance, appear to violate the necessary boundary condition, since

$$\tilde{\mathbf{H}}_t(z) = 0$$

inside the perfect conductor. Therefore, $\tilde{\mathbf{H}}$ is not continuous across the boundary. Remember, however, that for a perfect conductor the skin depth $\delta \to 0$, and the current in the conductor becomes a true surface current. So now we apply the boundary condition on the magnetic field in this limit, and we find a surface current

$$\tilde{\mathbf{J}}_s = |\tilde{\mathbf{H}}_i + \tilde{\mathbf{H}}_r|\,\mathbf{a}_y = \frac{2E_i}{\eta_1}\mathbf{a}_y. \tag{8.98}$$

With this current in the surface of the perfect conductor, the boundary condition for the tangential magnetic field is indeed satisfied.

Before we conclude this section, let's take a closer look at the notion of a surface current, and in particular, the transition between the description of the current $\tilde{\mathbf{J}}_v(z)$ distributed within the skin depth δ of the surface for a good conductor, and the surface current $\tilde{\mathbf{J}}_s$ for the perfect conductor. We should expect that $\tilde{\mathbf{J}}_s$ is a limiting case of $\tilde{\mathbf{J}}_v(z)$ as $\sigma \to \infty$, so let's see if this is the case. (As an added benefit, perhaps we can gain a better appreciation of the whole concept of volume current and surface current densities.) For the good conductor, with the electric field of the incident plane wave of the form given in Eq. (8.84),

$$\tilde{\mathbf{E}}_i(z) = \mathbf{a}_y E_i e^{-j\beta_1 z},$$

we solved for the transmission coefficient $\tau = 2\eta_2/(\eta_1 + \eta_2)$ at the interface, where η_1 is the impedance of the lossless dielectric on the incident side of the interface, and $\eta_2 = \sqrt{\pi f\mu/\sigma}(1+j)$ is the intrinsic impedance of the conductor. The transmitted wave propagates into the conductor, but as we discovered in Section 8.6.4, this wave is rapidly attenuated. Written in a form similar to Eq. (8.86), the electric field of the transmitted wave is

$$\tilde{\mathbf{E}}_t(z) = \mathbf{a}_y E_t e^{-\alpha_2 z}e^{-j\beta_2 z},$$

where $\alpha_2 = \beta_2 = \sqrt{\pi f\mu\sigma}$. The current density that is driven by the electric field $\tilde{\mathbf{E}}_t(z)$ in this conductor must be

$$\tilde{\mathbf{J}}_v(z) = \sigma\tilde{\mathbf{E}}_t(z) = \mathbf{a}_y \sigma E_t e^{-\alpha_2 z}e^{-j\beta_2 z}.$$

The transmitted electric field is attenuated within a distance that is characterized by the skin depth δ, and so the bulk of the current lies within δ of the surface as well. Let's integrate this current density $\tilde{J}_v(z)$ over z to find the total net current that lies near the surface. We'll call this $\tilde{J}_{int}$. We'd like to see if we find a result similar to the surface current $\tilde{J}_s$ that we found above for a perfect conductor:

$$\tilde{J}_{int} = \int_0^\infty \tilde{J}_v(z)\, dz = \int_0^\infty \sigma \tilde{E}_t(z)\, dz = \mathbf{a}_y\, \sigma E_t \int_0^\infty e^{-\alpha_2 z} e^{-j\beta_2 z}\, dz = \mathbf{a}_y\, \frac{\sigma E_t}{\alpha_2 + j\beta_2}.$$

But $\alpha_2 = \beta_2 = \sqrt{\omega\mu\sigma/2}$ (Eq. 8.73), and $\eta_2 = \sqrt{\omega\mu/(2\sigma)}(1 + j)$ (Eq. 8.75), so this integrated current becomes simply

$$\tilde{J}_{int} = \int_0^\infty \tilde{J}_v(z)\, dz = \mathbf{a}_y \frac{E_t}{\eta_2}. \tag{8.99}$$

Now this is an interesting result indeed. First, this effective surface current is in the direction $\mathbf{a}_y$. No issue here, since this is the direction of the incident electric field, and it is also consistent via the right-hand rule with the direction of $\tilde{H}_i + \tilde{H}_r$ just outside the conductor. So this direction sounds good. How about its magnitude, E_t/η_2? Notice that as the conductivity σ of the conductor gets larger, the amplitude E_t gets smaller (the transmission coefficient becomes smaller), but the intrinsic impedance η_2 also gets smaller. What about the limiting value of the integrated surface current density, E_t/η_2? It turns out that this approaches a finite value. Let's investigate. We use $E_t = \tau E_i$, with $\tau = 2\eta_2/(\eta_1 + \eta_2)$ (Eq. (8.97)) to write

$$\tilde{J}_{int} = \mathbf{a}_y \frac{2E_i}{\eta_1 + \eta_2}.$$

But $\eta_2 \ll \eta_1$, even for a good conductor (i.e. σ large but not infinite), and we have

$$\tilde{J}_{int} \to \mathbf{a}_y \frac{2E_i}{\eta_1},$$

in agreement with the surface current $\tilde{J}_s$ that we derived in Eq. (8.98) using boundary conditions for the magnetic field $\tilde{H}$ at the surface of a perfect conductor. Thus we find a perfectly consistent result for the current near the surface of the conductor when a plane wave is incident upon it. As a result, we will refer to this current simply as a surface current $\tilde{J}_s$, keeping in mind, of course, that a truer characterization, as described above, is more complicated.

This integrated current density is much more than just an idle curiosity, as we will now show. It is an important cog in our understanding of losses within conductors, and it plays an important role in the attenuation of waves in transmission lines and waveguides, and resistive losses in radiating antenna systems. To see this, let's consider a uniform plane wave incident upon a plane conducting surface, and examine the time-average power density of the wave transmitted at the surface,

$$\overline{\mathcal{P}}_t = \frac{1}{2} \Re\left[\tilde{E}_t \times \tilde{H}_t^*\right]. \tag{8.100}$$

As the Poynting vector, this represents the power flow, per unit area, just inside the conductor. As we already discussed, this transmitted wave is rapidly attenuated, within a characteristic skin depth of $\delta = 1/\alpha = 1/\sqrt{\pi f \mu \sigma}$. This skin depth depends on the frequency of the wave and the conductivity of the metal, but it is often on the scale of tens of microns or smaller. Assuming that the thickness of the conductor is greater than a few times δ, the wave is completely attenuated, and the power transmitted at the surface, Eq. (8.100), is completely absorbed by the conductor. This is worth repeating:

Eq. (8.100) gives us the absorbed power per unit area at the surface of the conductor. But we can write $|\tilde{H}_t|$ in this equation as $|\tilde{E}_t|/\eta_2$, which, by Eq. (8.99), is just J_s. Combining these results, the power absorbed at the conductor surface, per unit area, is

$$\overline{P_s^{\text{abs}}} = \frac{1}{2}\Re\{E_t J_s^*\}. \tag{8.101}$$

Since E_t and J_s are proportional to one another, this can be written as

$$\overline{P_s^{\text{abs}}} = \frac{1}{2}\Re\{|J_s|^2 Z_s\}, \tag{8.102}$$

where the **surface impedance** is defined as

$$Z_s \equiv \frac{E_t}{J_s} = R_s + jX_s. \tag{8.103}$$

By comparison with Eq. (8.99), and the subsequent identification of this as the surface current, the surface impedance Z_s is, in fact, identical to the impedance of the conductor η_2. While it may seem silly to define the surface impedance Z_s as a new variable, when the intrinsic impedance η of the conductor would do very nicely, this is standard notation. These two impedances do have very different physical significance, so they are in fact distinct parameters. E_t, of course, is the electric field amplitude just inside the conductor. Using Eq. (8.103) in Eq. (8.101), the absorbed power per unit area can be written

$$\overline{P_s^{\text{abs}}} = \frac{1}{2}|J_s|^2 R_s. \tag{8.104}$$

R_s is known as the **surface resistance**, and is equal to

$$R_s = \sqrt{\frac{\pi f \mu}{\sigma}} = \sqrt{\frac{\omega \mu}{2\sigma}}. \tag{8.105}$$

The power loss is written in terms of the surface current J_s. This result will be useful in future discussions, and we will return to it at that time.

Example 8.24 Power Absorbed, Good Conductor

Consider the aluminum surface of Example 8.23, on which a $\lambda = 0.5$ µm wave is incident. Let the amplitude of this wave be $E_0 = 10$ V/m. Determine the power absorbed by this surface.

Solution:
We start by finding the surface current density induced in the aluminum surface by this wave. From Eq. (8.98), this current density is

$$\tilde{J}_s = \frac{2E_i}{\eta_1} = \frac{2(10 \text{ V/m})}{120\pi \ \Omega} = 0.053 \text{ A/m}.$$

The surface resistance of aluminum at this frequency is

$$R_s = \sqrt{\frac{\pi f \mu}{\sigma}} = \sqrt{\frac{\pi(6 \times 10^{14} \text{ s}^{-1})(4\pi \times 10^{-7} \text{ H/m})}{(3.5 \times 10^7 \text{ S/m})}} = 8.2 \ \Omega.$$

The time-averaged power absorbed by this surface from the incident wave, per unit surface area, is

$$\overline{P_s^{\text{abs}}} = \frac{1}{2}R_s|J_s|^2 = \frac{1}{2}(8.2 \ \Omega)|0.053 \text{ A/m}|^2 = 12 \text{ mW/m}^2.$$

The reflectivity of a surface can be altered by applying a thin transparent film to the surface; either a single-layer or multi-layer thin film. The interference between the waves reflected from the different surfaces can lead to a reduction or enhancement of the total reflection, depending on the thickness of the layers, the wavelength of the wave, and the relative refractive indices of the film and substrate materials. These are described in TechNotes 8.19 and 8.20.

TechNote 8.19 Thin Film Coatings

A thin layer of a transparent dielectric material can be applied to a reflecting surface to alter its reflection coefficient. An incident wave is reflected from each surface, and the net amplitude of the reflected wave depends on the interference between these two waves. As an example, consider a wave propagating through air and incident upon the surface of a dielectric medium of impedance η_2. If a thin layer of impedance $\eta_{tf} \approx \sqrt{\eta_0 \eta_2}$ and thickness $\lambda_{tf}/4$, where λ_{tf} is the wavelength of the wave in the thin film, is deposited on the dielectric medium, the reflection coefficients at the first interface (air/thin film) and the second interface (thin film/dielectric medium) are approximately the same, and of the same sign. The two reflected waves will therefore interfere destructively, since the reflection from the second surface must travel half a wavelength farther than the reflection from the second surface, resulting in a π phase shift between them. This can be an effective means of reducing the reflection at a specified wavelength.

As useful as single-layer thin films are, multi-layer thin films are even more impressive. These are described in TechNote 8.20.

TechNote 8.20 Multi-layer Thin Film Coatings

In an extension of this concept, surfaces can be made highly reflective, non-reflective, or partially reflective by depositing on the surface many layers of thin transparent films. For example, consider multiple layers of thin dielectric coatings, with each layer alternately of one material or another. The thickness of each layer is $d_i = \lambda_i/4$, where λ_i is the wavelength of the wave in medium i, and the material impedance alternates between η_1 and η_2. The alternation of η causes the reflection coefficients from one interface to the next to alternate in sign. The partial waves reflected from adjacent surfaces add in phase with one another, since the π phase shift difference upon reflection and the $2\pi(2d_i/\lambda_i) = \pi$ phase shift difference due to propagating an extra distance of $2d_i$ add to give a 2π phase shift. We considered only two partial waves, but by extension all these waves add in phase with one another, with the net result that reflection coefficients of multi-layer thin films can be exceptionally large, as large as 99.99%. Multi-layer thin films can also be designed to reduce the reflection. This is appropriately called an anti-reflection coating. Note that the reflectivity of these films depends on the wavelength of the incident wave, and specified only within a certain range of wavelengths, or frequencies (the bandwidth), and at a specified angle of incidence. (We have only treated normal incidence in this section, but incidence at oblique angles, discussed in the next section, is also important, and found in a wide range of applications.)

In this section we have used the boundary conditions for tangential electric and magnetic fields to determine the reflection coefficient and the transmission coefficient for waves normally incident upon an interface between two media. We have examined this for waves incident on dielectrics and conductors, and for the latter introduced the idea of surface resistance, which is key in determining the power absorbed by a conductor.

8.9 Reflection of Uniform Plane Waves, Oblique Incidence

When an electromagnetic wave is incident at an oblique angle upon a plane surface of a second medium, the propagation directions of the transmitted and reflected waves are altered, and the reflection and transmission coefficients are modified, relative to the normal incidence case. This influences the amplitude of the waves (of course) and the polarization of the waves, and can lead to a process called total internal reflection. In this section we will explore these properties of the transmitted and reflected waves.

Similar to the analysis in the previous section of the reflection and transmission coefficients for a normally incident wave, we will apply the boundary conditions on transverse electric and magnetic fields that must be obeyed at the interface between the two media. The incident, transmitted, and reflected waves are illustrated in Fig. 8.33. As suggested by this figure, we must treat separately two cases, depending on the orientation of the polarization of the wave. In Fig. 8.33(a), the electric field $\mathbf{E}_i$ is completely tangential to the surface, while the magnetic field $\mathbf{H}_i$ has a tangential component and a component in the z-direction. Since $\mathbf{E}_i$ is tangential to the surface, this is often called a **transverse electric**, abbreviated TE, or **s-polarized** wave. (The notation s-polarized comes from the German word *senkrecht*, which means perpendicular. The field polarization of an s-polarized wave is perpendicular to the **plane of incidence**, which is defined as the plane containing the surface normal and the propagation vector $\mathbf{k}_i$ of the incident wave.) The second geometry is illustrated in Fig. 8.33(b). Note that in this figure, the magnetic field $\mathbf{H}_i$ of the incident wave is completely tangential to the surface, while the electric field $\mathbf{E}_i$ has both a tangential and a z-component. Since $\mathbf{H}_i$ is tangential at the surface this is often called a **transverse magnetic** (abbreviated TM) or **p-polarized** wave. (The term p-polarized comes from *parallel*, which in German is the same word. This field polarization lies parallel to the plane of incidence.) While the following initial treatment is general, we will primarily be interested in non-absorbing, non-magnetic media.

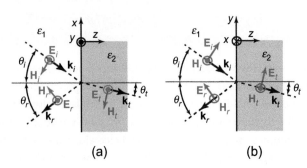

(a)   (b)

FIGURE 8.33 A wave incident from medium 1 on the interface with medium 2 at an oblique angle. A TE wave is illustrated in (a) and a TM wave in (b).

8.9.1 TE Waves

We start with TE waves. Referring to Fig. 8.33(a), the electric field of the incident wave is

$$\tilde{\mathbf{E}}_i = \mathbf{a}_y E_i e^{-j\mathbf{k}_i \cdot \mathbf{r}}, \tag{8.106}$$

the magnetic field $\tilde{\mathbf{H}}_i$ is

$$\tilde{\mathbf{H}}_i = -\frac{E_i}{\eta_1} \left(\mathbf{a}_x \cos\theta_i + \mathbf{a}_z \sin\theta_i \right) e^{-j\mathbf{k}_i \cdot \mathbf{r}}, \tag{8.107}$$

and the propagation vector is

$$\mathbf{k}_i = k_i \left(\mathbf{a}_z \cos \theta_i - \mathbf{a}_x \sin \theta_i \right).$$

The magnitude of $\mathbf{k}_i$ is $k_i = \omega \sqrt{\varepsilon_1 \mu_1}$. The angle θ_i, called the **angle of incidence**, is the angle between $\mathbf{k}_i$ and the surface normal, which is parallel to the z-axis in the figure. Normal incidence corresponds to $\theta_i = 0$. Note that $\tilde{\mathbf{E}}_i \times \tilde{\mathbf{H}}_i$ is parallel to $\mathbf{k}_i$.

Expressions for $\tilde{\mathbf{E}}_t$, $\tilde{\mathbf{H}}_t$, and $\mathbf{k}_t$ for the transmitted wave are

$$\tilde{\mathbf{E}}_t = \mathbf{a}_y E_t e^{-j\mathbf{k}_t \cdot \mathbf{r}}, \tag{8.108}$$

$$\tilde{\mathbf{H}}_t = -\frac{E_t}{\eta_2} \left(\mathbf{a}_x \cos \theta_t + \mathbf{a}_z \sin \theta_t \right) e^{-j\mathbf{k}_t \cdot \mathbf{r}}, \tag{8.109}$$

and

$$\mathbf{k}_t = k_t \left(\mathbf{a}_z \cos \theta_t - \mathbf{a}_x \sin \theta_t \right).$$

The magnitude of $\mathbf{k}_t$ is $k_t = \omega \sqrt{\varepsilon_2 \mu_2}$, and the angle θ_t, often called the **angle of refraction**, is the angle between $\mathbf{k}_t$ and the surface normal. When propagating from a medium of smaller ε to greater ε, the direction of propagation bends toward the surface normal. When propagating from a medium of larger ε to a medium of smaller ε, the wave propagation direction bends away from the normal.

Finally, the electric and magnetic fields of the reflected wave are

$$\tilde{\mathbf{E}}_r = \mathbf{a}_y E_r e^{-j\mathbf{k}_r \cdot \mathbf{r}}, \tag{8.110}$$

and

$$\tilde{\mathbf{H}}_r = \frac{E_t}{\eta_1} \left(\mathbf{a}_x \cos \theta_r - \mathbf{a}_z \sin \theta_r \right) e^{-j\mathbf{k}_r \cdot \mathbf{r}}, \tag{8.111}$$

and the propagation vector is

$$\mathbf{k}_r = -k_r \left(\mathbf{a}_z \cos \theta_i + \mathbf{a}_x \sin \theta_i \right).$$

Since the reflected wave travels through medium 1, the magnitude of its propagation vector is $|\mathbf{k}_r| = k_r = \omega \sqrt{\varepsilon_1 \mu_1} = k_i$. The angle θ_r of the reflected wave is called the **angle of reflection**.

Now that the electric and magnetic fields are defined, we are in a position to apply the boundary conditions at the interface. Specifically, at $z = 0$, the tangential electric field must be the same in medium 1 and medium 2, and the tangential magnetic field must also be continuous. Let's start with the former, and see what we can learn about the angles of the reflected wave θ_r and of the transmitted wave θ_t. The tangential electric field boundary condition is

$$E_i e^{jk_i \sin \theta_i x} + E_r e^{jk_r \sin \theta_r x} = E_t e^{jk_t \sin \theta_t x}, \tag{8.112}$$

which comes directly from Eqs. (8.106), (8.108), and (8.110), after setting z equal to zero. The boundary condition must be satisfied at all locations x, of course, which requires that

$$k_i \sin \theta_i = k_r \sin \theta_r = k_t \sin \theta_t. \tag{8.113}$$

The first equality in this expression leads to the **Law of Reflection**,

$$\boxed{\theta_r = \theta_i,} \tag{8.114}$$

since $k_r = k_i$. That is, the angle of reflection must be equal to the angle of incidence. The wave number for the transmitted wave k_t differs from $k_i = k_r$, of course, so the final equality in Eq. (8.113) leads to

$$\boxed{\sqrt{\varepsilon_1\mu_1}\,\sin\theta_i = \sqrt{\varepsilon_2\mu_2}\,\sin\theta_t.} \tag{8.115}$$

This relation is equivalent to **Snell's Law of refraction** (which is commonly written using the refractive indices of the media).

We next apply the boundary conditions to determine the reflection coefficient Γ and the transmission coefficient τ. Using Eq. (8.113) in Eq. (8.112), the exponential factors cancel out, leaving

$$E_i + E_r = E_t.$$

Dividing through by E_i, and using $\Gamma = E_r/E_i$ and $\tau = E_t/E_i$, this reduces to

$$1 + \Gamma = \tau. \tag{8.116}$$

The tangential magnetic field must also be continuous at the surface, or

$$H_{i,x} + H_{r,x} = H_{t,x}.$$

Note that only the x-components of H appear in this equation, since only the x-components are tangential to the surface. Using Eqs. (8.107), (8.109), and (8.111), this equation leads to

$$-\frac{E_i}{\eta_1}\cos\theta_i + \frac{E_r}{\eta_1}\cos\theta_r = -\frac{E_t}{\eta_2\cos\theta_t},$$

and dividing through by E_i/η_1, to

$$1 - \Gamma = \tau\frac{\eta_1/\cos\theta_i}{\eta_2/\cos\theta_t}. \tag{8.117}$$

Solving Eqs. (8.116) and (8.117) for Γ produces

$$\boxed{\Gamma = \frac{\eta_2\sec\theta_t - \eta_1\sec\theta_i}{\eta_2\sec\theta_t + \eta_1\sec\theta_i}.} \tag{8.118}$$

The transmission coefficient τ follows directly using Eq. (8.116). See Fig. 8.34 for a plot of Γ as a function of θ_i, where the solid lines apply for TE waves. Both media are non-absorbing and non-magnetic for this plot. In Fig. 8.34(a), the permittivity ε_2 is greater than ε_1, while in Fig. 8.34(b), $\varepsilon_2 < \varepsilon_1$. Notice that Γ is negative for small angles θ_i when $\varepsilon_2 > \varepsilon_1$, and positive when $\varepsilon_2 < \varepsilon_1$, consistent with Γ for normal incidence.

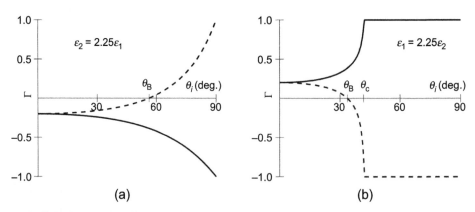

FIGURE 8.34 Reflection coefficient Γ vs. θ_i for (a) $\varepsilon_2 = 2.25\varepsilon_1$ and (b) $\varepsilon_1 = 2.25\varepsilon_2$. In each plot, the solid line is for a TE wave (s-polarization) and the dashed line for a TM wave (p-polarization).

Eq. (8.118) can be simplified a bit by introducing a wave parameter known as the wave impedance. For a TE wave, this is defined as

$$Z_{TE,i} = \frac{E_i}{H_{i,x}} = \eta_1 \sec\theta_i$$

for the medium on the incident side, and

$$Z_{TE,t} = \frac{E_t}{H_{t,x}} = \eta_2 \sec\theta_t$$

on the transmission side. Note that these impedances are based only on the tangential field components, which for the magnetic field is only the x-component. Then, Eq. (8.118) for the reflection coefficient Γ takes the form

$$\Gamma = \frac{Z_{TE,t} - Z_{TE,i}}{Z_{TE,t} + Z_{TE,i}}. \tag{8.119}$$

Since the secant functions of Eq. (8.118) have been absorbed by the wave impedances, this equation for the reflection coefficient closely resembles Eq. (8.96) for the reflection coefficient at normal incidence, in which impedances η have been replaced by Z_{TE}.

Example 8.25 TE Wave Reflection and Refraction

A 100 GHz TE wave propagating through quartz (SiO_2) is incident upon an interface with methanol. The angle of incidence of the wave is 60°, and its amplitude is 10 V/m. The permittivities of quartz and methanol are listed in Table 1.1. Determine (a) the angle of refraction; (b) the reflection coefficient Γ; (c) the transmission coefficient; and (d) the wave impedance of the wave in the quartz and in the methanol. For a TE wave propagating through methanol and incident upon an interface with quartz at an angle of incidence 18.2°, determine (e) the angle of refraction, (f) the reflection coefficient, and (g) the transmission coefficient.

Solution:

(a) From Table 1.1, $\varepsilon_{r,1} = 3.9$ and $\varepsilon_{r,2} = 30$. Then

$$\theta_t = \sin^{-1}\left(\sqrt{\varepsilon_1/\varepsilon_2}\,\sin\theta_i\right) = \sin^{-1}\left(\sqrt{3.9/30}\,\sin 60°\right) = 18.2°.$$

(b)

$$\Gamma = \frac{\eta_2 \sec\theta_t - \eta_1 \sec\theta_i}{\eta_2 \sec\theta_t + \eta_1 \sec\theta_i},$$

where $\eta_1 = \sqrt{\mu_1/\varepsilon_1} = 120\pi\,\Omega/\sqrt{\varepsilon_{r,1}} = 190.9\,\Omega$ and $\eta_2 = \sqrt{\mu_2/\varepsilon_2} = 120\pi\,\Omega/\sqrt{\varepsilon_{r,2}} = 68.8\,\Omega$. Then,

$$\Gamma = \frac{(68.8\,\Omega)(\sec 18.2°) - (190.9\,\Omega)(\sec 60°)}{(68.8\,\Omega)(\sec 18.2°) + (190.9\,\Omega)(\sec 60°)} = \frac{72.45 - 381.8}{72.45 + 381.8} = -0.68.$$

(c)

$$\tau = 1 + \Gamma = 0.32.$$

(d)

$$Z_{TE,i} = \eta_1 \sec\theta_i = (190.9\,\Omega)(\sec 60°) = 381.8\,\Omega,$$

$$Z_{TE,t} = \eta_2 \sec\theta_t = (68.8\,\Omega)(\sec 18.2°) = 72.4\,\Omega.$$

(e) The values of η_1 and η_2 are reversed from those used for parts (a)–(d).

$$\theta_t = \sin^{-1}\left(\sqrt{\varepsilon_1/\varepsilon_2}\,\sin\theta_i\right) = \sin^{-1}\left(\sqrt{30/3.9}\,\sin 18.2°\right) = 60°.$$

(f)

$$\Gamma = \frac{(190.9\ \Omega)(\sec 60°) - (68.8\ \Omega)(\sec 18.2°)}{(190.9\ \Omega)(\sec 60°) + (68.8\ \Omega)(\sec 18.2°)} = \frac{381.8 - 72.4}{381.8 + 72.4} = +0.68.$$

(g)

$$\tau = 1 + \Gamma = 1.68.$$

You should note in Fig. 8.34 that the magnitude $|\Gamma|$ of the reflection coefficient for the TE wave increases monotonically with increasing θ_i, reaching an ultimate value of 1. When $\varepsilon_2 > \varepsilon_1$, $|\Gamma|$ reaches 1 at $\theta_i = 90°$, or $\pi/2$, but for $\varepsilon_2 < \varepsilon_1$, $|\Gamma| = 1$ for any angle greater than the **critical angle** θ_c, whose value is

$$\boxed{\theta_c = \sin^{-1}\sqrt{\frac{\varepsilon_2}{\varepsilon_1}}.} \tag{8.120}$$

This angle is easily derived from Eq. (8.115) (with μ_1 and μ_2 set equal to μ_0 for nonmagnetic media). When $\varepsilon_2 < \varepsilon_1$, the direction of the transmitted beam bends away from the surface normal, so the refraction angle θ_2 is always greater than θ_i. When θ_i reaches the critical angle θ_c, the angle of the transmitted (refracted) beam is 90°, or $\pi/2$, and the transmitted beam propagates in the plane of the interface. If θ_i exceeds θ_c, there is no angle θ_t that satisfies Eq. (8.115), as this would require $\sin\theta_t$ to exceed 1. Thus, when $\theta_i \geq \theta_c$, the transmitted wave vanishes and the incident beam is completely reflected. This phenomenon is called **total internal reflection**.

This is not to say that there is no field in medium 2 under conditions of total internal reflection. There is no *propagating* wave; that much is clear. But there is a tangential electric field in medium 1, and therefore there must also be a tangential electric field in medium 2. Let's use Eq. (8.113) to see if we can learn more about E_t. First, since there is no propagating wave in medium 2, let's rename $k_t \sin\theta_t$ in Eq. (8.113), and call it simply $k_{t,x}$. But when $\theta_i \geq \theta_c$, the value of $k_{t,x}$ exceeds $\omega\sqrt{\varepsilon_2\mu_2}$. But a valid solution to the wave equation requires that

$$k_t^2 = k_{t,x}^2 + k_{t,z}^2 = \omega^2\varepsilon_2\mu_2.$$

Solving this for $k_{t,z}$ gives us

$$k_{t,z} = \sqrt{\omega^2\varepsilon_2\mu_2 - k_{t,x}^2}.$$

Since the second term under the radical is larger than the first, a factor of -1 can be pulled out front, so

$$k_{t,z} = \pm j\sqrt{k_{t,x}^2 - \omega^2\varepsilon_2\mu_2}. \tag{8.121}$$

Therefore, $k_{t,z}$ is seen to be purely imaginary.

With this value of $k_{t,z}$, the electric field in medium 2 (Eq. (8.108)) takes the form

$$\tilde{\mathbf{E}}_t = \mathbf{a}_y E_t e^{-j\mathbf{k}_t \cdot \mathbf{r}} = \mathbf{a}_y E_t e^{-jk_{t,x}x - jk_{t,z}z} = \mathbf{a}_y E_t e^{-jk_{t,x}x}e^{-\alpha z},$$

where

$$\alpha = \sqrt{k_{t,x}^2 - \omega^2\varepsilon_2\mu_2}.$$

This solution for $\tilde{E}_t$ shows sinusoidal modulation in x and exponential decay in z. (The positive imaginary solution for $k_{t,z}$ in Eq. (8.121) leads to an exponentially *increasing* function for $\tilde{E}_t$, which is rejected as physically unreasonable.) The wave in medium 2, called an **evanescent wave**, is not a propagating wave, and it carries no net power in the z-direction.

The decay constant α is simplified using Eq. (8.113),

$$k_{t,x} = k_{i,x} = k_i \sin\theta_i = \omega\sqrt{\varepsilon_1\mu_1}\,\sin\theta_i,$$

which casts α in the form

$$\alpha = \omega\sqrt{\varepsilon_1\mu_1 \sin^2\theta_i - \varepsilon_2\mu_2}.$$

The minimum value of α is zero, which occurs for $\theta_i = \theta_c$. The maximum value of α is $\omega\sqrt{\varepsilon_1\mu_1 - \varepsilon_2\mu_2}$, which occurs as θ_i approaches $\pi/2$, or grazing incidence. The distance the evanescent wave penetrates into medium 2 is characterized by α^{-1}.

Example 8.26 Total Internal Reflection

A 100 GHz TE wave propagating through Pyrex glass is incident upon an interface with air. The angle of incidence of the wave is 60°, and its amplitude is 10 V/m. The permittivity of Pyrex glass is listed in Table 1.1. (a) Determine the critical angle θ_c. (b) Determine the decay constant α. (c) Determine the amplitude of the electric field a distance $z = 1$ mm into the air. (d) Determine the distance z at which the wave amplitude is 1 V/m.

Solution:

(a) The critical angle is $\theta_c = \sin^{-1}\sqrt{\varepsilon_2/\varepsilon_1} = \sin^{-1}\sqrt{1/4.7} = 27.5°$.
(b) The decay constant is

$$\alpha = \omega\sqrt{\varepsilon_1\mu_1 \sin^2\theta_i - \varepsilon_2\mu_2} = \frac{\omega}{c}\sqrt{\varepsilon_{r,1}\mu_{r,1}\sin^2\theta_i - \varepsilon_{r,2}\mu_{r,2}}$$

$$= \frac{2\pi \times 10^{11}\ \text{s}^{-1}}{3\times 10^8\ \text{m/s}}\sqrt{4.7\sin^2 60° - 1} = 3.3\ \text{mm}^{-1}.$$

(c) $E_0 e^{-\alpha z} = (10\ \text{V/m})e^{-(3.3\ \text{mm}^{-1})(1\ \text{mm})} = 0.37\ \text{V/m}.$
(d) $E_0 e^{-\alpha z} = 1\ \text{V/m} \quad \Rightarrow \quad z = \alpha^{-1}\ln\left(\frac{10\ \text{V/m}}{1\ \text{V/m}}\right) = 0.70\ \text{mm}.$

8.9.2 TM Waves

The second oblique incidence case is with TM waves, or p-polarization. As discussed earlier and illustrated in Fig. 8.33(b), the magnetic field $\tilde{H}_i$ of the incident wave is purely tangential, while the electric field consists of a tangential component and a z-component. This electric field is expressed as

$$\tilde{E}_i = E_i\left(\mathbf{a}_y\cos\theta_i + \mathbf{a}_z\sin\theta_i\right)e^{-j\mathbf{k}_i\cdot\mathbf{r}}, \tag{8.122}$$

and the magnetic field $\tilde{H}_i$ is

$$\tilde{H}_i = -\mathbf{a}_x\frac{E_i}{\eta_1}e^{-j\mathbf{k}_i\cdot\mathbf{r}}. \tag{8.123}$$

The propagation vector of the incident wave is

$$\mathbf{k}_i = k_i \left(-\mathbf{a}_y \sin \theta_i + \mathbf{a}_z \cos \theta_i \right),$$

with magnitude $k_i = \omega \sqrt{\varepsilon_1 \mu_1}$, the same as for a TE wave.

The electric field $\tilde{\mathbf{E}}_t$ of the wave transmitted at the interface between the two media can be written as

$$\tilde{\mathbf{E}}_t = E_t \left(\mathbf{a}_y \cos \theta_t + \mathbf{a}_z \sin \theta_t \right) e^{-j\mathbf{k}_t \cdot \mathbf{r}}, \tag{8.124}$$

and the magnetic field $\tilde{\mathbf{H}}_t$ is

$$\tilde{\mathbf{H}}_t = -\mathbf{a}_x \frac{E_t}{\eta_2} e^{-j\mathbf{k}_t \cdot \mathbf{r}}. \tag{8.125}$$

The propagation vector of the transmitted wave is

$$\mathbf{k}_t = k_t \left(-\mathbf{a}_y \sin \theta_t + \mathbf{a}_z \cos \theta_t \right),$$

with magnitude $k_t = \omega \sqrt{\varepsilon_2 \mu_2}$.

Finally, the electric and magnetic fields of the reflected wave are

$$\tilde{\mathbf{E}}_r = E_r \left(\mathbf{a}_y \cos \theta_r - \mathbf{a}_z \sin \theta_r \right) e^{-j\mathbf{k}_r \cdot \mathbf{r}} \tag{8.126}$$

and

$$\tilde{\mathbf{H}}_r = \mathbf{a}_x \frac{E_r}{\eta_1} e^{-j\mathbf{k}_r \cdot \mathbf{r}}, \tag{8.127}$$

with propagation vector

$$\mathbf{k}_r = -k_r \left(\mathbf{a}_y \sin \theta_i + \mathbf{a}_z \cos \theta_i \right).$$

The magnitude of $\mathbf{k}_r$ is $k_r = \omega \sqrt{\varepsilon_1 \mu_1} = k_i$, just as for TE waves.

TM waves are governed by the same Laws of Reflection and Refraction as TE waves. For TM waves, the requirement that the tangential magnetic field is continuous across the interface at $z = 0$ leads to

$$\frac{E_i}{\eta_1} e^{jk_i \sin \theta_i x} + \frac{E_r}{\eta_1} e^{jk_r \sin \theta_r x} = \frac{E_t}{\eta_2} e^{jk_t \sin \theta_t x}, \tag{8.128}$$

which comes directly from Eqs. (8.123), (8.125), and (8.127). This condition must be satisfied at all locations x, of course, which requires that

$$k_i \sin \theta_i = k_r \sin \theta_r = k_t \sin \theta_t. \tag{8.129}$$

The first equality in this expression leads to the Law of Reflection, Eq. (8.114); the finale to Snell's Law, Eq. (8.115).

To determine the reflection coefficient Γ for TM waves, the requirement that the tangential electric field is continuous at the interface is

$$E_{i,y} + E_{r,y} = E_{t,y},$$

where only the y-components of the electric field are tangential to the surface. Dividing through by $E_{i,y}$, this reduces to

$$1 + \Gamma = \tau \frac{\cos \theta_t}{\cos \theta_i}. \tag{8.130}$$

The tangential magnetic field must also be continuous at the surface, or

$$H_i + H_r = H_t.$$

This equation leads to

$$-\frac{E_i}{\eta_1} + \frac{E_r}{\eta_1} = -\frac{E_t}{\eta_2},$$

and dividing through by E_i/η_1, to

$$1 - \Gamma = \tau \frac{\eta_1}{\eta_2}. \tag{8.131}$$

Solving Eqs. (8.130) and (8.131) for Γ produces

$$\boxed{\Gamma = \frac{\eta_2 \cos\theta_t - \eta_1 \cos\theta_i}{\eta_2 \cos\theta_t + \eta_1 \cos\theta_i}.} \tag{8.132}$$

The transmission coefficient τ follows directly using Eq. (8.130) or (8.131). The reflection coefficient Γ as a function of θ_i is plotted for TM waves as dashed lines in Fig. 8.34. Notice that at $\theta_i = 0$, Γ for the TM wave matches that for the normal incidence case, for $\varepsilon_2 > \varepsilon_1$ and for $\varepsilon_2 < \varepsilon_1$.

For internal reflection, $\varepsilon_2 < \varepsilon_1$, large angles of incidence result in total reflection, just as is observed for TE waves. The derivations and results for the critical angle θ_c and the decay constant α are the same as for TE waves.

Equation (8.132) can be written in a simpler form by defining the wave impedance for TM waves,

$$Z_{TM,i} = \frac{E_{i,y}}{H_i} = \eta_1 \cos\theta_i$$

for the medium on the incident side, and

$$Z_{TM,t} = \frac{E_{t,y}}{H_t} = \eta_2 \cos\theta_t$$

on the transmission side. As with Z_{TE}, these impedances are based only on the tangential field components, which for the electric field is only the y-component. Then Eq. (8.132) for the reflection coefficient Γ takes the form

$$\Gamma = \frac{Z_{TM,t} - Z_{TM,i}}{Z_{TM,t} + Z_{TM,i}}. \tag{8.133}$$

The plots for the reflection coefficient Γ for TM waves do differ in an important regard from those for TE waves. That is, Γ passes through zero and changes sign at an angle called **Brewster's angle**, denoted θ_B. Brewster's angle is determined using Eq. (8.132), which when set to zero and combined with Snell's Law, Eq. (8.115), yields

$$\theta_B = \tan^{-1}\sqrt{\frac{\varepsilon_2}{\varepsilon_1}}, \tag{8.134}$$

valid for non-magnetic (i.e. $\mu = \mu_0$) media.

Example 8.27 TM Wave Reflection and Refraction

A 100 GHz TM wave propagating through quartz (SiO$_2$) is incident upon an interface with methanol. The angle of incidence of the wave is 60°, and its amplitude is 10 V/m. The permittivities of quartz and methanol are listed in Table 1.1. Determine (a) the angle of refraction; (b) the reflection coefficient Γ; (c) Brewster's angle; (d) the transmission coefficient; and (e) the wave impedance of the wave in the quartz and in the methanol. For a TM wave propagating through methanol and incident upon an interface with quartz at an angle of incidence 18.2°, determine (f) the angle of refraction, (g) the reflection coefficient, (h) Brewster's angle, and (i) the transmission coefficient. (j) What is the sum of the two Brewster's angles from parts (c) and (j)?

Solution:

(a) From Table 1.1, $\varepsilon_{r,1} = 3.9$ and $\varepsilon_{r,2} = 30$. Then,

$$\theta_t = \sin^{-1}\left(\sqrt{\varepsilon_1/\varepsilon_2}\sin\theta_i\right) = \sin^{-1}\left(\sqrt{3.9/30}\sin 60°\right) = 18.2°.$$

This result is the same as that for the TE wave of Example 8.25.

(b)

$$\Gamma = \frac{\eta_2\cos\theta_t - \eta_1\cos\theta_i}{\eta_2\cos\theta_t + \eta_1\cos\theta_i},$$

where $\eta_1 = \sqrt{\mu_1/\varepsilon_1} = 120\pi\,\Omega/\sqrt{\varepsilon_{r,1}} = 190.9\,\Omega$ and $\eta_2 = \sqrt{\mu_2/\varepsilon_2} = 120\pi\,\Omega/\sqrt{\varepsilon_{r,2}} = 68.8\,\Omega$. Then,

$$\Gamma = \frac{(68.8\,\Omega)(\cos 18.2°) - (190.9\,\Omega)(\cos 60°)}{(68.8\,\Omega)(\cos 18.2°) + (190.9\,\Omega)(\cos 60°)} = \frac{65.4 - 95.5}{65.4 + 95.5} = -0.19.$$

(c)

$$\theta_B = \tan^{-1}\sqrt{\frac{\varepsilon_2}{\varepsilon_1}} = \tan^{-1}\sqrt{\frac{30}{3.9}} = 70.2°.$$

(d)

$$\tau = (1-\Gamma)\frac{\eta_2}{\eta_1} = 0.42.$$

(e)

$$Z_{TM,i} = \eta_1\cos\theta_i = (190.9\,\Omega)(\cos 60°) = 95.5\,\Omega,$$

$$Z_{TM,t} = \eta_2\cos\theta_t = (68.8\,\Omega)(\cos 18.2°) = 65.4\,\Omega.$$

(f) The values of ε_1 and ε_2 (as well as η_1 and η_2) are reversed from those used for parts (a)–(d).

$$\theta_t = \sin^{-1}\left(\sqrt{\varepsilon_1/\varepsilon_2}\sin\theta_i\right) = \sin^{-1}\left(\sqrt{30/3.9}\sin 18.2°\right) = 60°.$$

(g)

$$\Gamma = \frac{(190.9\,\Omega)(\cos 60°) - (68.8\,\Omega)(\cos 18.2°)}{(190.9\,\Omega)(\cos 60°) + (68.8\,\Omega)(\cos 18.2°)} = \frac{95.5 - 65.4}{95.5 + 65.4} = +0.19.$$

(h)

$$\theta_B = \tan^{-1}\sqrt{\frac{\varepsilon_2}{\varepsilon_1}} = \tan^{-1}\sqrt{\frac{3.9}{30}} = 19.8°.$$

(i)

$$\tau = (1-\Gamma)\frac{\eta_2}{\eta_1} = 2.44.$$

(j)

$$70.2° + 19.8° = 90.0°.$$

In most media, wave propagation follows the rules that we have discussed in this section, refracting or reflecting at surfaces, and obeying Snell's Law and the Law of Reflection. There has been a significant body of work in recent years to develop materials that have different properties, leading to quite strange behavior. These materials are known as negative refractive index materials, or left-handed materials, due to their negative permittivity and permeability. This phenomenon is described in TechNote 8.21.

TechNote 8.21 Negative Refractive Index Materials

In a negative index material, both the permittivity ε and permeability μ are negative. These materials were first considered theoretically by Victor Veselago, who hypothesized many counter-intuitive effects of electromagnetic waves propagating in such materials. Experimental demonstrations of these effects have been more recent, and have fueled the interest in these novel materials.

At first thought, it may seem strange for a material to have a negative permittivity or negative permeability. We have, however, already discussed the effective negative permittivity of metals at low frequencies. (See Section 8.6.) In addition, the magnetic permeability of ferromagnetic materials can become negative near resonances of effective current loops. As these only occur near resonances, however, their frequency range is limited to specific regions of the spectrum. Also, the frequencies at which the permeability of any natural materials is negative differ from the frequency at which the material permittivity is negative. Thus, there are no known examples of naturally occurring negative index materials.

Design and fabrication of negative index materials is a non-trivial challenge for researchers. In the microwave frequency range, the structure shown in Fig. 8.35, consisting of an array of linear (straight) conductors and C-shaped conductors known as split-ring resonators has been explored. An oscillating field sets up currents in the conductors, turning them into microscopic LC resonators. At the resonant frequency, it can be shown that the permittivity and permeability of this medium are both negative. Efforts to create negative index materials in the visible range are ongoing.

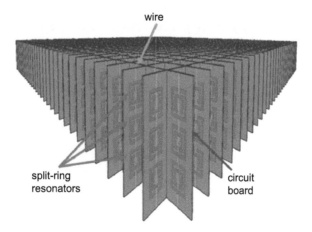

FIGURE 8.35 Synthetic negative index material, consisting of arrays of linear conductors and split ring resonators.

How can we understand the effect of these structures on the propagation of electromagnetic waves through the medium? To answer this question, let's return to the wave equation, as given by Eqs. (8.1) or (8.2). We discovered that an electromagnetic wave of the form

$$\mathbf{E}(t, z) = \mathbf{E}_0 \sin{(\omega t - kz)}$$

satisfies the wave equation when

$$k = \omega\sqrt{\varepsilon\mu}.$$

Without giving it much thought at the time, we took this wave number as the positive root of $k^2 = \omega^2\varepsilon\mu$, which is the correct choice when $\varepsilon > 0$ and $\mu > 0$. When $\varepsilon < 0$ and $\mu < 0$, the product $\varepsilon\mu$ is still positive, so we might dismiss this as an uninteresting avenue to pursue. But remember that the propagating wave is built on Ampère's and Faraday's Laws, that a time-varying magnetic field circulates around an electric field, and a time-varying electric field circulates around the magnetic field, and together these fields sustain each other. For a material with $\varepsilon < 0$, Ampère's Law tells us that $\mathbf{H}$ circulates around the time-varying electric field in the opposite sense from that of a normal material. Similarly, Faraday's Law tells us that $\mathbf{E}$ circulates around a time-varying magnetic field, and that the sense of this rotation is opposite for a material with negative μ. Reversing the direction in which $\mathbf{E}$ circulates around $\mathbf{H}$, and vice versa, has the effect that the phase of the wave varies opposite in a negative index material to that in a normal material. We can justify the choice of the negative root for a negative index material by writing ε as $\varepsilon = |\varepsilon|e^{j\pi}$ and μ as $\mu = |\mu|e^{j\pi}$. Then,

$$k = \omega\sqrt{\varepsilon\mu} = \omega\sqrt{|\varepsilon|e^{j\pi}|\mu|e^{j\pi}} = \omega\sqrt{|\varepsilon||\mu|}\sqrt{e^{j2\pi}} = \omega\sqrt{|\varepsilon||\mu|}e^{j\pi} = -\omega\sqrt{|\varepsilon||\mu|}.$$

The net result is that the phase kz of the wave decreases with increasing z. This can lead to some very counter-intuitive phenomena. That is, when a wave is incident upon a surface of a normal material at an oblique angle, we know that wave will be bent, or refracted, at the surface. An example of normal refraction is shown in Fig. 8.36(a), where a ray incident upon a normal surface at an angle θ_i is refracted at an angle θ_r. The ray crosses over the surface normal, and is bent toward the normal if the refractive index on the right side of the surface is greater than the refractive index of the material on the incident side. The angle of refraction obeys Snell's Law. For a negative index material, however, the refraction angle is negative, as shown in Fig. 8.36(b). This strange behavior can be understood on the basis of boundary conditions at the surface. In the case of a negative index material, Eq. (8.113) must be modified as

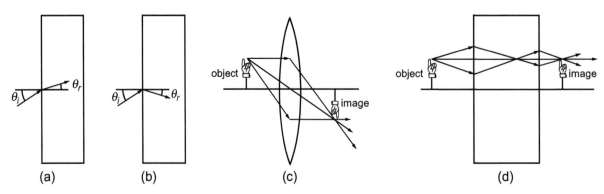

(a) (b) (c) (d)

FIGURE 8.36 (a) Normal refraction at a plane interface. (b) Refraction at the surface of a negative index material. (c) Imaging by a normal convex lens. (d) Imaging by a negative index material.

$$k_i \sin \theta_i = k_r \sin \theta_r = -\omega \sqrt{|\varepsilon||\mu|} \sin \theta_t,$$

since $k = -\omega \sqrt{|\varepsilon||\mu|}$. A very important application of normal refraction is a lens, as shown in Fig. 8.36(c), which is fabricated from a transparent material such as glass, and forms an image of an object by refracting the rays emerging from the object, and bringing them to cross at the image location. Figure 8.36(d) shows a scheme using a negative index material which can also form an image. In this case, however, the image can be of much greater resolution than can be attained with normal materials. Another intriguing idea is the notion of cloaking, or hiding an object in plain sight, by using the extraordinary powers of negative refraction. Ultimately, problems with the limited band of frequencies of negative materials will have to be dealt with.

In this section we have explored reflection and transmission of a plane wave at an interface between two media, focusing primarily on two non-absorbing, non-magnetic dielectric media. The reflection coefficients for TE and TM waves are given by Eqs. (8.118) and (8.132). (These are also expressed in terms of wave impedances using Eqs. (8.119) and (8.133).) Total internal reflection results when the angle of incidence exceeds the critical angle θ_c, and the reflection coefficient of a TM wave vanishes when the angle of incidence equals Brewster's angle, resulting in a reflected wave that is purely TE polarized.

8.10 Standing Waves

When a wave is incident upon a surface, the wave is partially or completely reflected from the surface, as we discussed in the previous two sections. The total field on the incident side of the interface (in medium 1), then, is the sum of these two waves, the incident wave and the reflected wave. In this section we will explore the properties of the combination of these two traveling waves, called a **standing wave**, when the incident wave is normally incident upon the surface.

To determine the time- and space-dependence of the total field on the incident side of the interface, let's start by writing the time-dependent traveling waves, that is, the incident wave

$$\mathbf{E}_i(z, t) = \mathbf{a}_y \, \Re \left[E_i e^{-j\beta_1 z} e^{j\omega t} \right]$$

and the reflected wave

$$\mathbf{E}_r(z, t) = \mathbf{a}_y \, \Re \left[E_r e^{+j\beta_1 z} e^{j\omega t} \right],$$

where we have used the time-dependent complex exponential notation. The total field in medium 1 (to the left of the reflecting surface) is the sum of these two waves, and using $E_r = \Gamma E_i$, the net field $\mathbf{E}_1(z, t)$ is

$$\mathbf{E}_1(z, t) = \mathbf{E}_i(z, t) + \mathbf{E}_r(z, t) = \mathbf{a}_y \, \Re \left[E_i e^{-j\beta_1 z} e^{j\omega t} + \Gamma E_i e^{+j\beta_1 z} e^{j\omega t} \right].$$

Pulling out common factors, this is

$$\mathbf{E}_1(z, t) = \mathbf{a}_y \, \Re \left[E_i \left(e^{-j\beta_1 z} + \Gamma e^{+j\beta_1 z} \right) e^{j\omega t} \right]. \tag{8.135}$$

What does this wave look like? We can start to answer this question by examining the plots in Fig. 8.37. In these plots, "snapshots" of the electric field as described by Eq. (8.135) are shown, plotted as a function of z for $\Gamma = 0.5$ at five different instants in time: $t = 0$, $t = T/8$, $t = T/4$, $t = 3T/8$, and $t = T/2$, where $T = 2\pi/\omega$ is the period of the

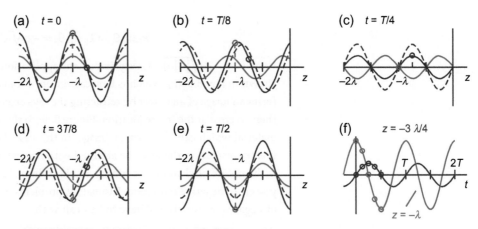

FIGURE 8.37 Plots of Eq. (8.135) for $\Gamma = 0.5$ at five instances: (a) $t = 0$; (b) $t = T/8$; (c) $t = T/4$; (d) $t = 3T/8$; and (e) $t = T/2$. The dot-dash violet line represents the field strength of the incident wave, the red line is the field strength of the reflected wave, and the blue line is the total field. In these plots, the total field is highlighted at two locations, $z = -3\lambda/4$ and $z = -\lambda$ with circles; blue and red, respectively. In plot (f), the time-dependent electric field at these two distinct locations is shown; $z = -3\lambda/4$ in blue, and $z = -\lambda$ in red. The blue curve is the minimum in the standing wave pattern, while the red is the maximum. These two time-dependent traces are $\pi/2$ out of phase with one another.

wave. The dot-dash violet line represents the field strength of the incident wave, and the red line is the field strength of the reflected wave. The blue line is the total field, found by adding the incident and reflected beams. Notice that the incident (dot-dash violet) wave moves to the right from one plot to the next, while the reflected (red) wave moves to the left. Also, notice that the amplitude of the reflected wave is a factor of Γ smaller than that of the incident wave. Now let's focus on the behavior of the total electric field $E_1(z, t)$ as a function of time t at a particular location. On each of these plots, the total field strength at two locations is marked: $z = -3\lambda/4$ with the blue circles and $z = -\lambda$ with the red circles. The field at either of these locations, or any other location, oscillates sinusoidally (always with frequency ω of course), but with amplitude that depends on location. At $z = -3\lambda/4$, marked with the blue circles, the amplitude of the oscillation is relatively small (actually, its a minimum for this example), while at $z = -\lambda$, marked with the red circles, the field amplitude is large (actually, a maximum). The last plot in Fig. 8.37 shows the time-dependent electric field at these two locations: $z = -3\lambda/4$ in blue and $z = -\lambda$ in red. The points on each z-dependent curve represented by red or blue circles are similarly represented on these time-dependent plots.

Let's return to Eq. (8.135) to see if we can understand some of these observations. In this expression, the time-dependence of the field is separate from the z-dependence. The standing wave field is the product of a z-dependent amplitude, $E_1(z)$ and a factor $\exp(j\omega t)$. The latter is the sinusoidal time-dependence. The amplitude $E_1(z)$ of the field at any location z is given by

$$E_1(z) = E_i \left(e^{-j\beta_1 z} + \Gamma e^{+j\beta_1 z} \right). \tag{8.136}$$

This function is periodic in z, and displays a series of maxima and minima, as shown in Fig. 8.38. The maxima are located at positions where the incident wave and the reflected wave add in phase with one another. In Eq. (8.136), the first term is of amplitude E_i, the second of amplitude $|\Gamma|E_i$. When these two terms are in phase with one another, they combine to give an amplitude $E_i(1 + |\Gamma|)$. (As we have seen, the reflection coefficient Γ can be complex. In the present context we are not interested in its phase, only

FIGURE 8.38 Plots of the magnitude $|E_1(z)|$ vs. z for different values of the reflection coefficient Γ. This function represents the amplitude of the sinusoidally varying field at each location z. The reflection coefficient is (a) $\Gamma = 0.1$, (b) $\Gamma = 0.5$, (c) $\Gamma = 1.0$, (d) $\Gamma = -0.1$, (e) $\Gamma = -0.5$, and (f) $\Gamma = -1.0$. Notice that for positive Γ, the maximum field amplitude is located at $z = 0$, while for negative Γ, $z = 0$ is the location of the field minima.

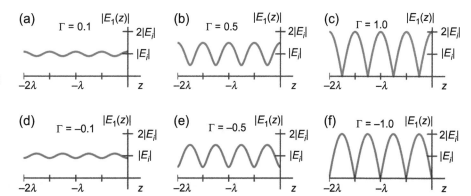

its amplitude $|\Gamma|$.) Similarly, the incident and reflected waves will combine to form a minimum when these two terms are π out of phase with one another. In this event, the amplitudes combine to give an amplitude of $E_i(1 - |\Gamma|)$.

The **standing wave ratio**, abbreviated SWR and denoted S, is used to characterize the magnitude of the "ripples" on the plots of the field amplitude. The SWR is defined as the ratio of the maximum amplitude of $|E_1(z)|$ to its minimum. As was discussed in the previous paragraph, the maximum and minimum values of $|E_1(z)|$ depend on the reflection coefficient Γ, and the SWR is

$$S = \frac{|E_1(z)|_{max}}{|E_1(z)|_{min}} = \frac{E_i(1 + |\Gamma|)}{E_i(1 - |\Gamma|)} = \frac{1 + |\Gamma|}{1 - |\Gamma|}. \tag{8.137}$$

Note that the SWR does not depend on the wave properties at all, only on the magnitude of the reflection coefficient Γ. The plots in Fig. 8.38 show the standing wave pattern for a variety of values of Γ. The magnitude of $|\Gamma|$ increases from left to right in these plots. For small $|\Gamma|$, the maxima and minima of the standing wave pattern differ only modestly, and S is only slightly greater than 1. For $|\Gamma| = 0.1$ as shown, S is $1.1/0.9 = 1.22$. As the reflected wave increases in magnitude, the variations in the maxima and minima of the standing wave pattern increase as well. In the center plots, where $|\Gamma| = 0.5$, the standing wave ratio is $S = 3$, and as $|\Gamma| \to 1$, the standing wave ratio $S \to \infty$.

Example 8.28 SWR for Glass–Air Interface

In Example 8.22 we found that the reflection coefficients Γ at either of the glass–air interfaces of a window are ± 0.2, when the relative permittivity of the glass is 2.25. Determine the SWR of the standing wave formed by this reflection.

Solution:
We solve this using Eq. (8.137):

$$S = \frac{1 + |\Gamma|}{1 - |\Gamma|} = \frac{1.2}{0.8} = 1.5.$$

This ratio S tells us that the maximum electric field amplitude is 1.5 times greater than the minimum electric field amplitude. Note that this result is valid for reflection from either interface, since $|\Gamma| = 0.2$ in either case.

We draw your attention now to the spacing between the maxima in the plots of Fig. 8.38. This spacing, which is $\lambda/2$, where $\lambda = 2\pi/\beta$, can be understood as follows. The maxima are located at positions z such that the incident wave and the reflected wave amplitudes, as given in Eq. (8.136), add in phase with one another. But these two waves are periodic in z, and if we move a distance $\Delta z = \lambda/2$ from one maximum, the phase of the incident wave changes by $-\beta_1\Delta z = -(2\pi/\lambda)(\lambda/2) = -\pi$, while the phase of the reflected wave changes by $\beta_1\,\Delta z = (2\pi/\lambda)(\lambda/2) = \pi$. The phase *difference* between the incident and reflected waves then is $\pi - (-\pi) = 2\pi$, and this is the location of the next maximum of the pattern. By a similar argument, minima must always lie halfway between two maxima, and the spacing from a maximum and the adjacent minimum is $\lambda/4$.

The reflection coefficients Γ for each of the plots in Fig. 8.38 are real, and are positive in the top row and negative in the bottom row. Look carefully at these curves and you will see that the only effect of changing $\Gamma \to -\Gamma$ is a shift of the entire pattern by $\Delta z = \lambda/4$. When Γ is real and positive, the incident and reflected waves add together in phase at the reflecting surface, and the total field amplitude at this location is a maximum. When Γ is real and negative, however, the incident and reflected waves at the surface add out of phase, and a minimum in the field amplitude occurs. Thus, these standing wave patterns are shifted by $\lambda/4$ relative to one another.

Example 8.29 Location of Maxima in Standing Wave Pattern

In Example 8.28 we found that the SWR for the wave pattern formed by reflection from an interface between glass ($\varepsilon_r = 2.25$) and air is 1.5. Determine the location of the maximum in the field pattern that lies closest to the surface.

Solution:

There are two surfaces to this window, and each must be considered separately. We'll start with the first surface, for which the wave travels through air before being incident upon the glass. We found, in Example 8.22, that the reflection coefficient at this surface is $\Gamma_1 = -0.2$. The negative value of Γ_1 tells us that the reflected wave is π out of phase with the incident wave at the surface. These two waves combine out of phase with one another, producing a minimum at the surface. The first maximum in the field pattern would therefore be a distance $\lambda/4$ to the left of the surface.

For the second interface, the wave is traveling through the glass, and is incident upon the interface with the air. For this surface, we found that the reflection coefficient is $\Gamma_2 = +0.2$. Since $\Gamma_2 > 0$, the reflected wave adds in phase with the incident wave at the interface, giving a maximum to the field pattern right at the interface.

This discussion of the standing wave pattern to this point has been focused on the electric field. The magnetic field $\mathbf{H}$ shows similar affects. The following formulation for the total magnetic field is very similar to the development of the standing wave electric field pattern. The total time-dependent magnetic field is $\mathbf{H}_1(z, t) = \mathbf{H}_i(z, t) + \mathbf{H}_r(z, t)$, where

$$\mathbf{H}_i(z, t) = -\mathbf{a}_x\,\Re\left[\frac{E_i}{\eta_1}e^{-j\beta_1 z}e^{j\omega t}\right]$$

and

$$\mathbf{H}_r(z, t) = \mathbf{a}_x\,\Re\left[\Gamma\frac{E_i}{\eta_1}e^{+j\beta_1 z}e^{j\omega t}\right].$$

The total magnetic field in medium 1 is therefore

$$\mathbf{H}_1(z, t) = -\mathbf{a}_x \, \Re\left[\frac{E_i}{\eta_1}\left(e^{-j\beta_1 z} - \Gamma e^{+j\beta_1 z}\right) e^{j\omega t}\right].$$

(8.138)

This differs from Eq. (8.135) for $\mathbf{E}_1(z, t)$ in three ways:

1. The magnetic field is oriented in the x-direction, perpendicular to the direction of $\mathbf{E}$.
2. The magnitude of each traveling wave amplitude is E/η_1.
3. The incident and reflected wave are *subtracted* in the present case, whereas they were added in Eq. (8.135).

This last difference results since, where Γ gives the ratio E_r/E_i for the electric field amplitudes, it gives us $-H_r/H_i$ for the magnetic field amplitudes. This sign difference shifts the magnetic field standing wave pattern, such that the minima of the $H_1(z)$ pattern coincide with the maxima of the $E_1(z)$ pattern, and the maxima of the $H_1(z)$ pattern coincide with the minima of the $E_1(z)$ pattern.

A special case that is worth examining closer is the case of reflection from a perfect conductor. As we already discovered, the reflection coefficient for a perfect conductor is $\Gamma = -1$. Using this in Eqs. (8.135) and (8.138) leads to

$$\mathbf{E}_1(z, t) = \mathbf{a}_y \, 2E_i \sin(\beta_1 z) \sin(\omega t)$$

and

$$\mathbf{H}_1(z, t) = -\mathbf{a}_x \, 2\frac{E_i}{\eta_1} \cos(\beta_1 z) \cos(\omega t).$$

The form of these fields is relatively simpler, compared to the case of an arbitrary reflection coefficient Γ, so the essential features are perhaps a bit easier to analyze, but this is truly just a special case of the earlier results given by Eqs. (8.135) and (8.138). At the risk of repetition, notice the following characteristics:

1. The electric field has a node at the reflecting surface. (Remember, this is at the surface of a perfect conductor, so the total electric field must be zero at $z = 0$. Otherwise, the current density $\tilde{\mathbf{J}}_y$ would become infinite, which is nonsense.)
2. The magnetic field has an anti-node at the reflecting surface. Applying the boundary conditions that must prevail at any interface, the surface current density in the surface of the conductor is $\mathbf{J}_s = \mathbf{H}_1(0, t) \times \mathbf{a}_n = \mathbf{a}_y \, 2(E_i/\eta_1) \cos(\omega t)$.
3. The maxima of the $\mathbf{E}_1(z, t)$ standing wave pattern coincide with the minima of the $\mathbf{H}_1(z, t)$ pattern, and vice versa.
4. The standing wave ratio S for this pattern is infinite.

Finally, let us consider the power density of such standing waves. As we showed earlier, the power density of an electromagnetic field is given by the Poynting vector. We will limit our discussion to the average power density, which we can find using the fields in phasor notation in Eqs. (8.136) and (8.138) to write

$$\overline{\mathcal{P}} = \frac{1}{2}\Re\left[\tilde{\mathbf{E}}_1 \times \tilde{\mathbf{H}}_1^*\right]$$

$$= \frac{1}{2}\Re\left[\mathbf{a}_y \left\{E_i\left(e^{-j\beta_1 z} + \Gamma e^{+j\beta_1 z}\right)\right\} \times \left\{-\mathbf{a}_x \frac{E_i}{\eta_1}\left(e^{-j\beta_1 z} - \Gamma e^{+j\beta_1 z}\right)\right\}^*\right].$$

Multiplying through, the cross terms cancel one another (after retaining only the real part), and several common terms factor out, giving us

$$\overline{\mathcal{P}} = \mathbf{a}_z \frac{|E_i|^2}{2\eta_1}\left(1 - |\Gamma|^2\right).$$

The Poynting vector for the time-averaged power density reduces to this simple sum of two terms. We have seen these two terms previously. The first term is simply the Poynting vector for the incident wave of amplitude E_i, while the second term is the Poynting vector for the reflected wave of amplitude $E_r = \Gamma E_i$. The net Poynting vector for the

Table 8.4

Collection of key formulas from Chapter 8.			
$\nabla^2 \mathbf{H} - \varepsilon\mu \dfrac{\partial^2 \mathbf{H}}{\partial t^2} = 0$	(8.5)	$\nabla^2 \tilde{\mathbf{E}} + \varepsilon\mu\omega^2 \tilde{\mathbf{E}} = 0$	(8.20)
$\nabla^2 \mathbf{E} - \varepsilon\mu \dfrac{\partial^2 \mathbf{E}}{\partial t^2} = 0$	(8.6)	$\nabla^2 \tilde{\mathbf{H}} + \varepsilon\mu\omega^2 \tilde{\mathbf{H}} = 0$	(8.21)
$k = \omega\sqrt{\varepsilon\mu}$	(8.8)	$\eta = \dfrac{E_0}{H_0} = \sqrt{\dfrac{\mu}{\varepsilon}}$	(8.25)
$T = \dfrac{2\pi}{\omega}$	(8.9)	$\overline{P_v^{abs}} = \dfrac{1}{2}(\sigma + \omega\varepsilon'')\,|\tilde{\mathbf{E}}|^2$	(8.43)
$\lambda = \dfrac{2\pi}{k}$	(8.10)	$\kappa = \omega\sqrt{\varepsilon\mu}\sqrt{1 - j\sigma/\omega\varepsilon}$	(8.55)
$u_p = \dfrac{1}{\sqrt{\varepsilon\mu}}$	(8.11)	$\kappa = \beta - j\alpha$	(8.56)
$c = \dfrac{1}{\sqrt{\varepsilon_0\mu_0}}$	(8.12)		

See Table 8.3 for a summary of expressions for β, α, η, and u_p for wave propagation through different media.

$\oint_s \mathbf{E} \times \mathbf{H} \cdot d\mathbf{s} = -\dfrac{d}{dt}\int_v \left\{\dfrac{\varepsilon}{2}(\mathbf{E}\cdot\mathbf{E}) + \dfrac{\mu}{2}(\mathbf{H}\cdot\mathbf{H})\right\}dv - \int_v \mathbf{E}\cdot\mathbf{J}_v\,dv$	(8.77)
$\overline{\mathcal{P}} = \dfrac{1}{2}\mathfrak{R}\,[\tilde{\mathbf{E}} \times \tilde{\mathbf{H}}^*]$	(8.79)
$|\overline{\mathcal{P}}| = \dfrac{E_0^2}{2\eta}$	(8.80)
$\oint_s \overline{\mathcal{P}} \cdot d\mathbf{s} = \dfrac{j\omega}{2}\int_v \left[\varepsilon^*|\tilde{\mathbf{E}}(\mathbf{r})|^2 - \mu|\tilde{\mathbf{H}}(\mathbf{r})|^2\right]dv - \dfrac{1}{2}\int_v \tilde{\mathbf{E}}(\mathbf{r})\cdot\tilde{\mathbf{J}}_v^*(\mathbf{r})dv$	(8.83)

$1 + \Gamma = \tau$	(8.94)	$\sqrt{\varepsilon_1\mu_1}\sin\theta_i = \sqrt{\varepsilon_2\mu_2}\sin\theta_t$	(8.115)
$\Gamma = \dfrac{\eta_2 - \eta_1}{\eta_2 + \eta_1}$	(8.96)	$\Gamma = \dfrac{\eta_2\sec\theta_t - \eta_1\sec\theta_i}{\eta_2\sec\theta_t + \eta_1\sec\theta_i}$	(8.118)
$\overline{P_s^{abs}} = \dfrac{1}{2}|J_s|^2 R_s$	(8.104)	$\theta_c = \sin^{-1}\sqrt{\dfrac{\varepsilon_2}{\varepsilon_1}}$	(8.120)
$R_s = \sqrt{\dfrac{\pi f\mu}{\sigma}} = \sqrt{\dfrac{\omega\mu}{2\sigma}}$	(8.105)	$\Gamma = \dfrac{\eta_2\cos\theta_t - \eta_1\cos\theta_i}{\eta_2\cos\theta_t + \eta_1\cos\theta_i}$	(8.132)
$\theta_r = \theta_i$	(8.114)	$S = \dfrac{|E_1(z)|_{max}}{|E_1(z)|_{min}} = \dfrac{1 + |\Gamma|}{1 - |\Gamma|}$	(8.137)

total field in medium 1 is the *difference* between these two individual Poynting vectors since these waves travel in opposite directions; the incident wave in the $+z$-direction, the reflected wave in the $-z$-direction. In the limiting case of complete reflection at the interface between the media, $|\Gamma| = 1$ and the Poynting vector vanishes. The incident wave and the reflected wave carry identical power, and the net power density of the standing wave pattern is zero.

SUMMARY
We have shown in this chapter how Maxwell's Equations can be used to describe propagating electromagnetic waves. We have seen that these waves must be transverse waves ($\mathbf{E}(t, \mathbf{r})$ and $\mathbf{H}(t, \mathbf{r})$ are perpendicular to the direction of propagation); the amplitudes of these fields are related through the intrinsic impedance η of the medium through which they propagate, and their phase velocity is governed by material properties as well. These waves carry energy, and we discussed the power density of the wave. When waves propagate through an absorbing medium, the absorption of energy is related to the attenuation of the wave. And finally, we discussed reflection of a wave upon normal and oblique incidence with an interface between two media. In each of the topics discussed in this chapter, the electromagnetic wave is a freely propagating wave, unguided by any conductors. This is relevant, of course, for discussions of many applications in our everyday lives: garage door opener remote control, the remote control for televisions, the wireless internet connection, and reception of signals from broadcast networks. The receiver or transmitter in each of these is untethered by any wires. In the next chapter, we will discuss wave propagation in which the waves are guided by conductors, termed transmission lines. Each of the topics discussed in the present chapter for free-wave propagation has prepared us for similar treatments for transmission lines.

PROBLEMS

Propagation in Lossless Media

P8-1 In Eq. (8.5), we completed the derivation of the wave equation that governs the behavior of the magnetic field in non-absorbing media. We presented a similar result, without proof, of the identical wave equation for the electric field in Eq. (8.6). Use Maxwell's Equations to prove Eq. (8.6) for a simple, source-free medium (Eqs. (8.1)–(8.4)).

P8-2 In Eq. (8.7), one example of an electric field that satisfies the wave equation for lossless media, Eq. (8.6), was presented. Determine the magnetic field that accompanies this electric field.

P8-3 The carrier frequency of a local FM public radio station, WBAA, is 101.3 MHz. (a) Determine the angular frequency, the wavelength, the wave number, and the period of this wave as it propagates through air. (b) Repeat for a wave of the same frequency as it propagates through a glass window, with $\varepsilon = 2.25\varepsilon_0$.

P8-4 The carrier frequency of a local AM public radio station, WBAA, is 920 kHz. (a) Determine the angular frequency, the wavelength, the wave number, and the period of this wave as it propagates through air. (b) Repeat for this same wave as it propagates through a glass window, with permittivity $\varepsilon = 2.25\varepsilon_0$.

P8-5 Show that a pulse of the form $\mathbf{E}(t, z) = \mathbf{a}_y E_0 e^{-|t/\tau - kz|}$ satisfies the wave equation. Determine k in terms of τ. Use Faraday's Law to determine the associated magnetic field $\mathbf{H}(t, z)$.

Uniform Plane Waves

P8-6 The magnetic field of an electromagnetic wave propagating through air is $\mathbf{H}(t, z) = \mathbf{a}_y H_0 \cos(\omega t - kz)$, where $H_0 = 50$ mA/m, and $\omega = 10^9$ r/s. (a) In which direction is this wave propagating? Explain. (b) Determine k, λ, and f. (c) Use Ampère's Law to find the associated electric field $\mathbf{E}(t, z)$.

P8-7 Repeat Prob. P8-6, using the magnetic field $\mathbf{H}(t, z) = \mathbf{a}_y H_0 \cos(\omega t + kz)$.

P8-8 A uniform plane propagating through a lossless medium is described by $\mathbf{E}(t, y) = \mathbf{a}_x E_0 \cos(\omega t - ky)$ V/m, where $E_0 = 4$ V/m, $\omega = 3 \times 10^9$ r/s, and $k = 20$ m^{-1}. The permeability of the medium is μ_0. (a) What is the direction in which this wave is propagating? (b) What is the cyclical frequency of this wave (in MHz)? (c) What is the wavelength of this wave? (d) What is the relative permittivity ε_r of this medium?

P8-9 The magnetic field of a uniform plane wave is $\mathbf{H}(t, x) = \mathbf{a}_z H_0 \cos(\omega t + kx)$, where $H_0 = 10$ mA/m and $k = 5$ m^{-1}. The medium is non-magnetic, with relative permittivity $\varepsilon_r = 4$. Determine ω, f, and λ for this wave. Use Ampère's Law to find $\mathbf{E}(t, \mathbf{r})$. What is the amplitude E_0 and direction $\mathbf{a}_E$ of $\mathbf{E}(t, \mathbf{r})$?

P8-10 A plane wave whose electric field is $\mathbf{E}(t, \mathbf{r}) = \mathbf{a}_z E_0 \sin(\omega t - \mathbf{k} \cdot \mathbf{r})$, with $E_0 = 20$ V/m and $\mathbf{k} = (4\mathbf{a}_x + 3\mathbf{a}_y)$ m^{-1}, propagates through a lossless, non-magnetic medium with relative permittivity $\varepsilon_r = 4$. Determine ω, f, and λ for this wave. Use Faraday's Law to find $\mathbf{H}(t, \mathbf{r})$. What is the amplitude H_0 and direction $\mathbf{a}_H$ of $\mathbf{H}(t, \mathbf{r})$?

P8-11 A uniform plane wave propagates through a lossless, non-magnetic medium. The magnetic field of this wave is given by $\mathbf{H}(t, \mathbf{r}) = \mathbf{a}_z H_0 \cos(\omega t - (k_x x + k_y y))$, where $H_0 = 5$ mA/m and $\mathbf{k} = \mathbf{a}_x k_x + \mathbf{a}_y k_y + \mathbf{a}_z k_z = (3\mathbf{a}_x - 4\mathbf{a}_y)$ m^{-1}. The permittivity of the medium is ε_0. (a) Determine the wavelength λ of this wave. (b) Determine the frequency ω of this wave. (c) The electric field associated with the wave can be written $\mathbf{E}(t, \mathbf{r}) = \mathbf{a}_n E_0 \cos(\omega t - (k_x x + k_y y))$, where E_0 is the amplitude and $\mathbf{a}_n$ is a unit vector. Determine E_0 and $\mathbf{a}_n$.

P8-12 A uniform plane wave propagates in the z-direction through air. The electric field of this wave, in phasor form, is $\tilde{\mathbf{E}}(z) = \mathbf{a}_x E_0 e^{-jkz}$, where $E_0 = 10$ V/m and $k = 200\pi$ m^{-1}. (a) Determine the associated magnetic field $\tilde{\mathbf{H}}(z)$ for this wave, the wavelength λ, the radial frequency ω, and the cyclical frequency f. (b) Repeat for this wave traveling through glass, with $\varepsilon = 2.25\varepsilon_0$.

P8-13 A uniform plane wave propagates in the $-y$-direction through air. The magnetic field of this wave, in phasor form, is $\tilde{\mathbf{H}}(y) = \mathbf{a}_x H_0 e^{jky}$, where $H_0 = 0.5$ A/m and $k = 4\,000\pi$ m^{-1}. (a) Determine the associated electric field $\tilde{\mathbf{E}}(y)$ for this wave, the wavelength λ, the radial frequency ω, and the cyclical frequency f. (b) Repeat for a wave of the same wave number k traveling through glass, with $\varepsilon = 2.25\varepsilon_0$.

P8-14 Convert $\mathbf{H}(t, z)$ of Prob. P8-6 to its phasor form. Use Ampère's Law for time-harmonic fields to find the associated electric field $\tilde{\mathbf{E}}(z)$. Compare this result with $\mathbf{E}(t, z)$ found in Prob. P8-6.

P8-15 Convert $\mathbf{E}(t, z)$ of Prob. P8-10 to its phasor form. Use Faraday's Law for time-harmonic fields to find the associated magnetic field $\tilde{\mathbf{H}}(\mathbf{r})$. Compare this result with $\mathbf{H}(t, z)$ found in Prob. P8-10.

Polarization

P8-16 A uniform plane wave of amplitude E_0 and vertical linear polarization (in the z-direction) is incident upon a stack of N linear polarizers. The transmission axis of the first polarizer is at an angle $\pi/2N$ from vertical, the second is at an angle

$2\pi/2N$, the third at $3\pi/2N$, and so forth, until the final polarizer, which is at $\pi/2$ (i.e. horizontal). Determine the polarization and the amplitude of the transmitted wave for $N = 2, 3, 4$, and 10. What is the limiting value of the transmitted amplitude as N gets arbitrarily large?

P8-17 Determine the amplitude of two circularly polarized waves, one right- and the other left-circular, which, when combined, produce a vertically polarized wave of amplitude $E_0 = 10$ V/m. What is the phase difference between these two circularly polarized waves? (b) Describe how these same two circularly polarized waves can be modified to form a horizontally spolarized wave.

P8-18 A right-circularly polarized wave of amplitude E_0 is incident upon a linear polarizer. The polarizer is rotated about an axis parallel to the propagation direction of the wave, with θ being the angle between the transmission axis of the polarizer and the vertical. Describe the amplitude and polarization of the transmitted wave as a function of θ.

P8-19 A vertically polarized wave of amplitude E_0 is incident upon a linear polarizer. The polarizer is rotated about an axis parallel to the propagation direction of the wave. We define θ as the angle between the transmission axis of the polarizer and the vertical. Describe the amplitude and polarization of the transmitted wave as a function of θ.

Power Dissipation Density

P8-20 Calculate the time-averaged power absorbed per unit volume from a wave of amplitude $E = 10$ V/m by a medium of conductivity $\sigma = 0.05$ S/m and permittivity $\varepsilon = (9.0 - j0.1)\,\varepsilon_0$ at frequencies (a) $\omega = 10^{10}$ s^{-1}, (b) $\omega = 10^{11}$ s^{-1}, and (c) $\omega = 10^{12}$ s^{-1}.

P8-21 A medium absorbs 10 mW/cm^{-3} from a wave of amplitude 5 V/m, independent of the frequency of the wave. Determine the conductivity of the material.

Wave Propagation in Absorbing Media

P8-22 A time-harmonic wave propagates through a non-magnetic medium, whose conductivity is $\sigma = 10$ mS/m and whose permittivity is $\varepsilon = (2.25 - j0.1)\varepsilon_0$. The frequency of the wave is 1 GHz. Determine the phase constant β and the attenuation constant α, the intrinsic impedance η, the phase velocity u_p, and the wavelength λ of this wave. Determine the distance over which the amplitude of the wave decreases by a factor 1/2.

P8-23 Repeat Prob. P8-22 for a wave whose frequency is 10 GHz.

P8-24 A time-harmonic wave propagates through a non-magnetic medium, whose conductivity is $\sigma = 10$ mS/m and whose permittivity is $\varepsilon = (2.25 - j0.1)\varepsilon_0$. The amplitude of the wave is 10 V/m. Determine and plot the conduction current density $\tilde{J}_v$ and the displacement current density $\tilde{J}_d$ versus frequency. At what frequency are these two current densities of equal magnitude?

Non-conducting Absorbing Dielectric Media

P8-25 A time-harmonic wave propagates through a non-conducting absorbing dielectric medium whose permittivity is $\varepsilon = (4.0 - j0.1)\varepsilon_0$. The frequency of the wave is 100 MHz. Determine the phase constant β and the attenuation constant α, the intrinsic impedance η, the phase velocity u_p, and the wavelength λ of this wave. Determine the distance over which the amplitude of the wave decreases to 1/2 its initial value. Determine the loss tangent for this medium.

P8-26 A time-harmonic wave propagates through a non-conducting absorbing dielectric medium. The frequency of the wave is 100 MHz, the wavelength is 2.0 m, and the amplitude of the wave decreases to 1/10 its initial value after propagating a distance of 20 m. Determine the complex permittivity ε. Determine the loss tangent for this medium.

P8-27 The electric field of a time-harmonic wave propagating through a non-conducting absorbing dielectric medium is shown in Fig. 8.39. The frequency of the wave is $f = 1$ GHz. Determine (a) the phase constant β and the attenuation constant α for this wave, (b) the wavelength λ, (c) the phase velocity, (d) the complex permittivity ε of the medium, and (e) the loss tangent for this medium.

FIGURE 8.39 The electric field $E(z)$ vs. z for a wave propagating through a non-conducting absorbing dielectric medium in Prob. P8-27.

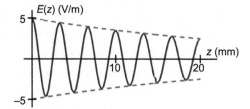

Wave Propagation in a Good Conductor

P8-28 A 100 MHz wave is incident upon a copper surface. The electric field amplitude just inside the copper is $E_0 = 1$ V/m. (a) Determine the skin depth for this wave. (b) At what depth is the wave amplitude decreased to 1% of its initial value? (c) What is the phase velocity of this wave? (d) What is the amplitude of the associated magnetic field H_0? (e) What is the volume current density J_v at the initial point?

P8-29 (a) At what frequency is the skin depth of a wave propagating through aluminum equal to 1 μm? (b) Given an amplitude of the magnetic field $H_0 = 10$ mA/m, determine the amplitude of the associated electric field. (c) At what depth is the wave amplitude decreased to 1% of its initial value? (d) What is the phase velocity of this wave? (e) What is the volume current density J_v at the initial point?

Power Flow

P8-30 A d.c. source produces a voltage V_0 that drives a current I_0 through a load resistor R_L. The source is connected to the load resistor through a co-axial pair of conductors. The radius of the inner conductor is a, the inner radius of the outer conductor is b, and the permittivity of the dielectric medium between the two conductors is ε. The current density in the inner conductor is uniform. Evaluate the Poynting vector in the inner conductor and in the region between the conductors, and relate this to the power absorbed by the load resistor. Ignore the resistance of the conductors.

P8-31 Consider the source, co-axial conductor, and load resistor described in Prob. P8-30. Here we will consider the resistive losses of the conductors. Let the conductivity of these conductors be σ, which is large but finite. Evaluate the Poynting vector that represents power flow into the conductors, and the total power absorbed by the conductors. Let the length of the co-axial conductors be ℓ, and the thickness of the outer conductor be h.

P8-32 Determine the electric field amplitude E_0 of a 10 mW laser beam propagating through air. The diameter of the beam is 1 mm, and the power density is uniform. Find the associated magnetic field amplitude H_0.

P8-33 A uniform plane wave propagates through a dielectric medium of relative permittivity $\varepsilon_r = 2.25$. The power density of the wave is 100 kW/m^2. Determine the amplitudes of the electric and magnetic fields.

P8-34 A uniform plane wave travels through a lossy non-magnetic dielectric medium in the $+z$-direction. Its frequency and wavelength are $\omega = 2\pi \times 10^9$ s^{-1} and $\lambda = 0.1$ m. The power density of this wave in the $z = 0$ plane is 10 W/m^2, but is reduced to 9 W/m^2 at $z = 100$ m. (a) Determine the attenuation constant α for this wave. (b) Determine the relative permittivity ε_r for this medium.

Reflection of Uniform Plane Waves, Normal Incidence

P8-35 A uniform plane wave is normally incident from air on a dielectric medium of permittivity $\varepsilon = 4.0\,\varepsilon_0$. The power density of the incident wave is 100 W/m^2. (a) Determine the reflection coefficient Γ and the transmission coefficient τ. (b) Determine the electric field amplitudes of the reflected and transmitted waves. (c) Determine the power density of the reflected and transmitted waves. Is power conserved?

P8-36 A uniform plane wave is normally incident upon a plane interface between two dielectric media. The reflection coefficient Γ for the surface is -0.3. Determine the ratio of permittivities of the two media.

P8-37 A uniform plane wave of frequency $\omega = 6\pi \times 10^{15}$ r/s is incident upon a conductor surface from glass, whose permittivity is $\varepsilon = 2.25\varepsilon_0$. The magnitude of the reflection coefficient is $|\Gamma| = 0.9$. Determine the conductivity of the conductor, and the phase of the reflection coefficient.

P8-38 A uniform plane wave of frequency $\omega = 2\pi \times 10^{14}$ r/s is incident upon a conductor surface from air. The conductivity of the conductor is 4×10^6 S/m, and electric field amplitude of the incident wave is 10 V/m. (a) Determine the reflection coefficient Γ (magnitude and phase, or real and imaginary parts) for the surface. (b) Determine the phasor for the transmitted electric field amplitude. (c) Write the current density $J_v(z)$ vs z. (d) Determine the skin depth for the wave in the conductor. (e) Integrate the current density $J_v(z)$ in z to find an effective surface current density J_s. (f) Determine the surface impedance $E(0)/J_s$. (g) Determine the power per unit area absorbed by the conductor.

Reflection of Uniform Plane Waves, Oblique Incidence

P8-39 A TE wave is incident from medium 1 ($\varepsilon_{r,1} = 2.0$) onto the surface of medium 2 ($\varepsilon_{r,2} = 4.0$) at an angle of incidence $\theta_i = 30°$. Both media are non-magnetic. Determine (a) the angle of reflection, (b) the angle of refraction, (c) the reflection coefficient, and (d) the transmission coefficient. Now let the wave be incident from medium 2 onto medium 1. The angle of incidence is 20.7°. Determine (e) Γ and (f) τ. (g) Determine the critical angle at this interface.

P8-40 A TE wave is incident from a high-permittivity medium ($\varepsilon_{r,1} = 9.0$) onto the surface with air at an angle of incidence of $\theta_i = 75°$. The frequency of the wave is 10 GHz. Determine (a) the critical angle θ_c for total internal reflection, (b) the decay constant α, and (c) the distance into the air at which the evanescent wave decreases to one-half its value at the surface. (d) Compare the transverse component of the propagation number k_x inside medium 1 with the propagation number k_2 for a propagating wave at this frequency in medium 2.

P8-41 A TM wave is incident from medium 1 ($\varepsilon_{r,1} = 1.5$) onto the surface of medium 2 ($\varepsilon_{r,2} = 6.0$) at an angle of incidence $\theta_i = 30°$. Both media are non-magnetic. Determine (a) the angle of reflection, (b) the angle of refraction, (c) the reflection coefficient, and (d) the transmission coefficient. Now let the wave be incident from medium 2 onto medium 1. The angle of incidence is 14.5°. Determine (e) Γ and (f) τ. Determine (g) the critical angle at this interface, and (h) Brewster's angle.

Standing Waves

P8-42 The reflection coefficient Γ for a surface is -0.6. Determine the standing wave ratio S for this surface. Plot the total electric field amplitude $E(z)$ vs. z, given the incident wave amplitude is $E_i = 4$ V/m. In terms of the wavelength λ, determine the locations of the first minimum and the first maximum of the standing wave pattern. Repeat for $\Gamma = +0.6$.

P8-43 The reflection coefficient Γ for a surface is $-0.6 + j0.4$. Determine the standing wave ratio S for this surface. Plot the total electric field amplitude $E(z)$ vs. z, given the incident wave amplitude is $E_i = 4$ V/m. In terms of the wavelength λ, determine the locations of the first minimum and the first maximum of the standing wave pattern.

P8-44 The reflection coefficient Γ for a surface is -0.95. Determine the standing wave ratio S for this surface. Plot the total electric field amplitude $E(z)$ vs. z, given the incident wave amplitude is $E_i = 4$ V/m. In terms of the wavelength λ, determine the locations of the first minimum and the first maximum of the standing wave pattern.

P8-45 A uniform plane wave propagating in the $+y$-direction through vacuum is normally incident upon a partially reflecting interface at $y = 0$. The electric field of the incident wave is $\mathbf{E}_i(t, y) = \mathbf{a}_z E_i \cos(\omega t - ky)$, where $E_i = 10$ V/m and $\omega = 2\pi \times 10^8$ s^{-1}. The amplitude of the wave reflected at the interface is $E_r = -5$ V/m. (a) Determine the standing wave ratio of the field formed in the region $y < 0$. (b) Is the total electric field amplitude at $y = 0$ a maximum or a minimum? Explain. (c) What is the electric field amplitude of the wave that is transmitted at this interface? (d) Assuming that the medium in the region $y > 0$ is a non-magnetic perfect dielectric, determine the relative permittivity ε_r of this medium.

CHAPTER 9
Transmission Lines

In the previous chapter we discussed freely propagating waves: waves that propagate through a medium without any guidance from supporting conducting or dielectric surfaces. We can, however, effectively guide waves from a source to a load using a pair of parallel, uniform conductors known as a **transmission line.** Transmission lines consist of two or more conductors extending a long distance along one axis, and maintaining a uniform cross-sectional geometry in the plane transverse to this axis. In this chapter we will consider several different types of transmission line systems. Common examples that each of you has probably encountered in everyday experience are co-axial conductors, often used for delivering a signal from the cable provider to your TV; microstrip, often seen in printed circuit board applications; and twin-lead transmission lines, often used with portable "rabbit-ears" antennas for TV receivers. In this chapter we will explore the properties of waves on transmission line systems, including such important aspects as traveling waves, standing waves, characteristic impedances, impedance matching, losses, and much more.

KEY WORDS

transmission line (co-axial, twin-lead, microstrip, parallel plate); characteristic impedance; resistance per unit length; conductance per unit length; traveling wave; standing wave; impedance matching; short-circuit load; open-circuit load; quarter-wave transformer; stub lines; Smith chart; normalized impedance; bounce diagram.

LEARNING OBJECTIVE

After studying this section, you should be able to:
- identify the common transmission line types illustrated in Fig. 9.1.

9.1 Examples of Transmission Lines

There are several common transmission line geometries. We will not attempt to be exhaustive in this review, but instead will focus on just a few representative styles. We have, in fact, already introduced most of these examples earlier in our studies as we calculated the capacitance or inductance of certain configurations of conductors.

The geometries of several common co-axial line geometries are shown in Fig. 9.1.

FIGURE 9.1 Examples of common transmission lines: (a) co-axial transmission line; (b) twin-lead transmission line; (c) microstrip transmission line; and (d) parallel plate transmission line.

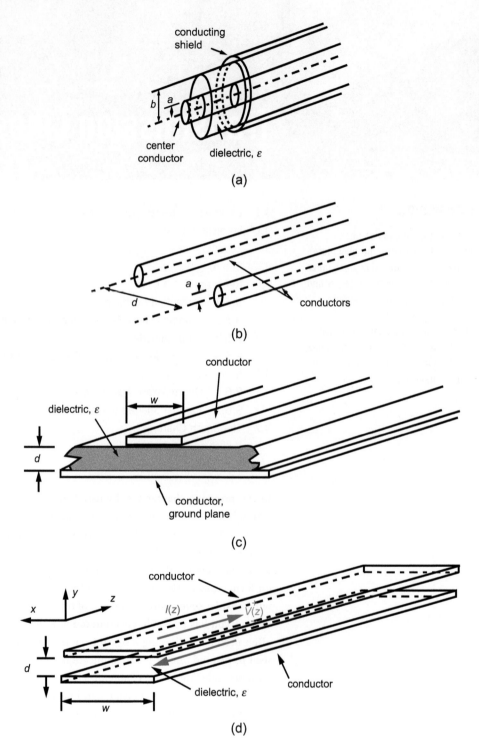

(a)

(b)

(c)

(d)

This figure includes examples of:

1. a co-axial transmission line;
2. a twin-lead transmission line;
3. a microstrip transmission line; and
4. a parallel plate transmission line.

In each case, the structure extends a long distance in the z-direction.

The co-axial line consists of two cylindrical conductors: one of radius a in the center and the other, often called the shield, of radius $b > a$. The space between the

Table 9.1

Transmission line parameters for a few common types of transmission lines. R_s in this table is the surface resistance of the conductor, defined in Eq. (8.105) as $\sqrt{\pi f \mu / \sigma_c}$, where σ_c is the conductivity of the conductor. σ_d in the expressions for the conductance per unit length G' is the conductivity of the dielectric material. The expressions for the inductance per unit length L' and resistance per unit length R' listed in this table are valid when the thickness of the conductors is greater than the skin depth δ, where δ is defined in Eq. (8.74).

| | Type of line | | | |
	Co-axial line	Twin-lead line	Parallel plate line	Units
C'	$\dfrac{2\pi\varepsilon'}{\ln(b/a)}$	$\dfrac{\pi\varepsilon'}{\cosh^{-1}(d/2a)}$	$\dfrac{\varepsilon' w}{d}$	F/m
L'	$\dfrac{\mu}{2\pi}\ln\left(\dfrac{b}{a}\right)$	$\dfrac{\mu}{\pi}\cosh^{-1}\left(\dfrac{d}{2a}\right)$	$\dfrac{\mu d}{w}$	H/m
R'	$\dfrac{R_s}{2\pi}\left(\dfrac{1}{a}+\dfrac{1}{b}\right)$	$2\left(\dfrac{R_s}{2\pi a}\right)$	$\dfrac{2R_s}{w}$	Ω/m
G'	$\dfrac{2\pi(\sigma_d+\omega\varepsilon'')}{\ln(b/a)}$	$\dfrac{\pi(\sigma_d+\omega\varepsilon'')}{\cosh^{-1}(d/2a)}$	$(\sigma_d+\omega\varepsilon'')\dfrac{w}{d}$	S/m

conductors contains a dielectric material of permittivity ε. We treated this conducting system previously in the discussions of electrostatics and magnetostatics. Recall that, if the conductors are perfect conductors, then the electric field is confined to the region between the conductors ($a < \rho < b$), with the electric field directed in the radial ($\mathbf{a}_\rho$) direction. The magnetic field circulates azimuthally about the axis (along the direction $\mathbf{a}_\phi$). We also found the capacitance per unit length C' and the inductance per unit length L', in terms of a, b, ε, and μ, which are repeated in Table 9.1.

The twin-lead transmission line, shown in Fig. 9.1(b), also consists of a pair of conductors extending a long distance in the z-direction. The spacing between the conductors (each of radius a) is uniform and equal to d. As can be seen, this is an open geometry, and the electric and magnetic field lines for this geometry extend over larger distances from the conductors than is the case for the co-axial system. We might also expect that this line is more readily influenced by neighboring structures that are not necessarily intended to be part of the transmission line system, or susceptible to outside interference from other sources.

In Fig. 9.1(c), a configuration known as microstrip, commonly used on printed circuit boards, is shown. The inductance per unit length L' and the capacitance per unit length C' depend upon the width w of the top conductor, the permittivity of the printed circuit board, ε, and the thickness of the dielectric material d, but they are not easily cast in simple analytic formulas, and we do not include these parameters in Table 9.1. Still, these lines are in common use, and tables and software packages are easily found to aid in the design of microstrip transmission line systems.

Finally, Fig. 9.1(d) shows a parallel plate transmission line. This transmission line geometry is not commonly used, but it is perhaps the easiest to visualize, and it is convenient to use this geometry to develop transmission line properties. As we have come to appreciate, the electric field lines for this geometry are vertical, while the magnetic field lines are horizontal, and each is essentially uniform in the region between the conductors.

So if a transmission line is simply a pair of conductors extending a long distance in the z-direction, as we have already stated, you may be wondering what it is that requires special treatment. After all, we connect a source to each of the conductors at one end of the transmission line, and a load at the other end, and current passes along one conductor to the load and returns via the other conductor. When the source voltage is d.c., or oscillates at a low frequency, this picture is just fine, and we can treat the transmission line as nothing more than a pair of conductors. But as the frequency of the sinusoidal voltage produced by the source increases, we must consider the time that it takes for a signal to propagate down the line, and the d.c. picture becomes inadequate. The potential difference between the conductors is no longer uniform, but rather varies along the transmission line. Likewise, the current in the conductors varies with z. Our treatment of these conductors as a transmission line system properly allows us to include these effects. How do we decide if the frequency is high enough that we need to treat the system as a transmission line rather than just a pair of wires? As a general rule of thumb, transmission line effects start to become important when the length of the line becomes comparable to or greater than about one-quarter of the wavelength of the radiation. At a frequency of 1 GHz, for example, transmission line effects become important when the circuit dimensions exceed a few centimeters.

Transmission lines, regardless of type, share several noteworthy features:

- The electric and magnetic fields are primarily in the region between the conductors, and are orthogonal to one another.
- The electric field lines are normal to the conductor surfaces, and are directed from one conductor to the other.
- The magnetic field lines are tangential to the conductor surfaces.
- The magnitudes of the fields generally share the same spatial variation. For example, in the co-axial system, **E** and **H** each decrease as $1/\rho$.
- The currents and charges are distributed in or on the surfaces of the conductors.
- The electric field lines originate at the positive charges in one of the conductors, and terminate on the negative charges in the other.
- Similarly, currents in the conductors are associated with the tangential **H** field at the surface.

In this chapter we will study the properties of the electromagnetic waves guided by such transmission line systems. In particular, we will examine such properties as:

- the wave impedance, $|\tilde{\mathbf{E}}|/|\tilde{\mathbf{H}}|$;
- the distributions of currents and charges in the conductors;
- the characteristic impedance of the transmission line;
- the propagation velocity of the waves; and
- the rate of attenuation of the waves – that is, losses.

We shall use, as a representative model for all transmission lines, the parallel plate transmission line. Other types of lines will differ in some of the details, but the parallel plate line is relatively simple, in that the fields are localized to the region directly between the plates, and the fields are uniform in magnitude and direction in this region. Thus this system allows us to develop the important attributes of wave propagation on transmission lines without introducing unnecessary complications.

The outline of the discussion is as follows. We will first develop the important properties of propagation of waves that are guided by the conductors. In this discussion, we will consider a wave traveling only in one direction, the +z-direction. These waves must obey all of the features that we have previously discussed for free-wave propagation, and, in addition, they must satisfy the known boundary conditions at the surfaces of the conductors. We will start the development using the electric and magnetic fields that transmission lines support, but then make a transition to a description based upon the potential difference between the conductors and the currents in the conductors, both of which display "wave" properties. We will then discuss losses of the waves on transmission lines, and learn how to incorporate losses in the description of wave propagation. Next, we examine the properties of standing waves on transmission lines, which result when a wave traveling in the +z-direction reaches the end of the line and is partially or completely reflected. This leads to various important applications and compensation techniques. Finally, we close the discussion of transmission lines by introducing pulse propagation, which becomes especially relevant in this era of high-speed digital systems.

9.2 Lossless Transmission Line

We start the study of wave propagation on transmission lines by using the parallel plate transmission line as an example. This structure, with $w \gg d$, is shown in Fig. 9.1(d). When the conductors are perfect conductors and the dielectric medium between the conductors is non-absorbing, an electromagnetic wave can propagate along this structure without losses. (We will consider the more general case of wave propagation with losses later.)

An oscillatory potential difference of amplitude V_0 and frequency ω is applied between the two conductors at the beginning of the transmission line, designated as $z = 0$. This source generates a wave traveling along the transmission line in the +z-direction. What are the properties of this wave? Well, we expect that any electromagnetic wave guided by this structure must satisfy two sets of conditions, each of which we have studied in detail previously. First, the time-varying electric and magnetic fields must obey the wave equation, as we discussed in Chapter 8; and second, the fields must satisfy the specific boundary conditions at the surfaces of the conductors. The conductors support currents and charges, which in the lossless case reside on the inside surfaces of the two conductors, and are quantified through the surface current density $\mathbf{J}_s$ and the surface charge density ρ_s. For the parallel plate configuration, with the conductors extending in the z-direction and currents in the conductors pointing in the ±z-direction, the boundary conditions that must be satisfied at both inside surfaces of the conductors, located at $y = 0$ and $y = d$, are

$$E_t = 0 \rightarrow E_x = E_z = 0$$
$$D_n = \rho_s \rightarrow E_y = -\rho_s/\varepsilon$$
$$H_t = J_s \rightarrow H_x = J_{s,z} \text{ and } H_z = 0$$
$$B_n = 0 \rightarrow H_y = 0,$$

where ρ_s and $J_{s,z}$ are the surface charge density and surface current density on the *upper* conductor. Since $\tilde{\mathbf{E}}$ and $\tilde{\mathbf{H}}$ must satisfy the wave equation, we expect that the z-dependence of these quantities should be of the form $\exp(-\gamma z)$, where $\gamma = \alpha + j\beta$ is the complex propagation constant introduced in Eq. (8.60). Initially, as we consider lossless propagation, the attenuation constant α is zero, and $\exp(-\gamma z)$ simplifies to $\exp(-j\beta z)$.

One solution that satisfies each of these conditions, written in phasor notation, is

$$\tilde{\mathbf{E}}(z) = -\mathbf{a}_y E_0 e^{-j\beta z} \tag{9.1}$$

and

$$\tilde{\mathbf{H}}(z) = \mathbf{a}_x \frac{E_0}{\eta} e^{-j\beta z}, \tag{9.2}$$

where $\eta = \sqrt{\mu/\varepsilon}$ is the intrinsic impedance of the dielectric medium between the conductors. This, as we surely recognize, is simply one of our plane wave solutions, propagating in the z-direction in the space between the two conductors. The electric and magnetic field vectors are normal to each other, the direction of $\tilde{\mathbf{E}} \times \tilde{\mathbf{H}}$ is along the direction of propagation of the wave (the z-direction), and the amplitude of the electric field is η times the amplitude of the magnetic field. The electric field lines originate at positive charges on the surface of one conductor and terminate at negative charges on the other. The surface charge density on the inside surface of the upper conductor can be written as

$$\tilde{\rho}_{s,u}(z) = -\varepsilon \tilde{E}_y(z) = \varepsilon E_0 e^{-j\beta z}, \tag{9.3}$$

and on the inside surface of the bottom conductor as

$$\tilde{\rho}_{s,b}(z) = +\varepsilon \tilde{E}_y(z) = -\varepsilon E_0 e^{-j\beta z}.$$

Similarly, the surface current densities on the two conductor plates can be derived using the boundary condition for the magnetic field $\tilde{\mathbf{H}}$, given by Eq. (7.16). In applying this equation, we take medium 1 as the dielectric medium between the conductors, medium 2 as the conductor, and $\mathbf{a}_{n2}$ is the unit normal vector pointing out of medium 2. For the top conductor, $\mathbf{a}_{n2} = -\mathbf{a}_y$, and the current density is

$$\tilde{\mathbf{J}}_{s,u}(z) = -\mathbf{a}_y \times \tilde{\mathbf{H}} = -\mathbf{a}_y \times \left(\mathbf{a}_x \frac{E_0}{\eta} e^{-j\beta z} \right) = \mathbf{a}_z \frac{E_0}{\eta} e^{-j\beta z} \tag{9.4}$$

on the inside surface of the top conductor. Similarly, for the inside surface of the bottom conductor, $\mathbf{a}_{n2} = \mathbf{a}_y$, and

$$\tilde{\mathbf{J}}_{s,b}(z) = +\mathbf{a}_y \times \tilde{\mathbf{H}} = +\mathbf{a}_y \times \left(\mathbf{a}_x \frac{E_0}{\eta} e^{-j\beta z} \right) = -\mathbf{a}_z \frac{E_0}{\eta} e^{-j\beta z}.$$

Each of these quantities is illustrated in Fig. 9.2, which shows a side view of the parallel plate transmission line, including the electric field $\tilde{\mathbf{E}}(z)$ and the magnetic field $\tilde{\mathbf{H}}(z)$ in the region between the plates, and the surface charge densities $\tilde{\rho}_{s,u}$ and $\tilde{\rho}_{s,b}$ and surface current densities $\tilde{\mathbf{J}}_{s,u}$ and $\tilde{\mathbf{J}}_{s,b}$. Notice that each of these quantities varies sinusoidally as a function of z, as well as a function of time t. Of course, when written in their phasor forms, we don't see the time-dependence explicitly, but we should never lose sight of the fact that these are harmonic waves, all varying in time together.

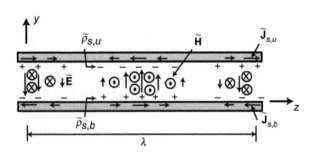

FIGURE 9.2 A side view of the parallel plate transmission line, showing the electric field $\tilde{\mathbf{E}}(z)$ and the magnetic field $\tilde{\mathbf{H}}(z)$ in the region between the plates, and the surface charge densities $\tilde{\rho}_{s,u}$ and $\tilde{\rho}_{s,b}$ and surface current densities $\tilde{\mathbf{J}}_{s,u}$ and $\tilde{\mathbf{J}}_{s,b}$ on the conductors.

Example 9.1 Relationship between $J_{s,z}(z,t)$ and $\rho_s(z,t)$ in the Conductors

Since the current represents charges in motion, the surface current density $J_{s,z}(z, t)$ and the surface charge density $\rho_s(z, t)$ in the conductors are related through an equation similar to the Continuity Equation. (We introduced the Continuity Equation in Section 3.2.) Derive this relation, and then show that the charge density given by Eq. (9.3) and the current density given by Eq. (9.4) satisfy this relation.

Solution:

We are dealing here with *surface* currents and *surface* charges, whereas the Continuity Equation dealt with volume distributions, so we won't blindly accept the Continuity Equation without question. In fact, as we will show, the surface current density $J_{s,u,z}(z,t)$ and the surface charge density $\rho_{s,u}(z,t)$ in the upper conductor must obey the relation

$$\frac{\partial J_{s,u,z}(z,t)}{\partial z} + \frac{\partial \rho_{s,u}(z,t)}{\partial t} = 0, \tag{9.5}$$

and a similar relation for the bottom plate. We consider a section of the upper conductor with surface current density $J_{s,u,z}(z,t)$ and the surface charge density $\rho_{s,u}(z,t)$, as shown in Fig. 9.3. The total charge Q contained within the shaded area of width Δx and length Δz is $Q = \rho_{s,u} \Delta x \, \Delta z$, and if $\rho_{s,u}$ varies in time, then Q also varies in time, with

$$\frac{\Delta Q}{\Delta t} = \frac{\Delta \rho_{s,u}}{\Delta t} \Delta x \, \Delta z.$$

Why isn't this charge Q just a constant, you might ask. Well, there is current flowing in on the left side of the shaded area (of density $J_{s,u,z}(z,t)$), and current flowing out on the right (of density $J_{s,u,z}(z+\Delta z,t)$). If these two current densities are unequal, then the total charge must be changing. The current flowing into the area increases Q, and the current flowing out decreases Q, so we write

$$\frac{\Delta Q}{\Delta t} = J_{s,u,z}(z,t)\Delta x - J_{s,u,z}(z+\Delta z,t)\Delta x = -\Delta J_{s,u,z}\Delta x,$$

where $\Delta J_{s,u,z}$ is written for $J_{s,u,z}(z+\Delta z,t) - J_{s,u,z}(z,t)$. Equating the two forms of $\Delta Q/\Delta t$, we have

$$\frac{\Delta \rho_{s,u}}{\Delta t} = -\frac{\Delta J_{s,u,z}}{\Delta z},$$

where the common factors Δx cancel one another. Taking the limit as $\Delta t \to 0$ and $\Delta z \to 0$, and putting both terms on the left side, we have

$$\frac{\partial J_{s,u,z}(z,t)}{\partial z} + \frac{\partial \rho_{s,u}(z,t)}{\partial t} = 0,$$

as we set out to show.

Now let's show that the time-dependent forms of the phasors given in Eqs. (9.3) and (9.4) do indeed satisfy Eq. (9.5):

$$\frac{\partial J_{s,u,z}(z,t)}{\partial z} + \frac{\partial \rho_{s,u}(z,t)}{\partial t} = \frac{\partial}{\partial z}\left(\frac{E_0}{\eta} e^{j\omega t} e^{-j\beta z}\right) + \frac{\partial}{\partial t}\left(\varepsilon E_0 e^{j\omega t} e^{-j\beta z}\right)$$

$$= E_0 e^{j\omega t} e^{-j\beta z}\left[\frac{1}{\eta}(-j\beta) + \varepsilon(j\omega)\right].$$

But the terms inside the square brackets reduce to zero,

$$-j\frac{\beta}{\eta} + j\omega\varepsilon = -j\frac{\omega\sqrt{\varepsilon\mu}}{\sqrt{\mu/\varepsilon}} + j\omega\varepsilon = 0,$$

where we have used the definitions of β and η from our study of plane wave properties. Therefore we see that Eqs. (9.3) and (9.4) do satisfy Eq. (9.5). The currents and charges on the bottom conductor are precisely the same in magnitude to, but opposite in sign from, those of the top plate, and so these charges and currents will likewise satisfy Eq. (9.5).

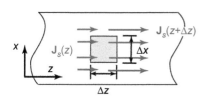

FIGURE 9.3 A view of an elemental section of the conductor of area $\Delta x \times \Delta z$.

Recall from Chapter 8 that the electric field $\tilde{\mathbf{E}}$ given in Eq. (9.1) and the magnetic field $\tilde{\mathbf{H}}$ given in Eq. (9.2) each satisfy the wave equation. As we will show next, the potential difference $v(z, t)$ between the upper and lower conductors, and the current $i(z, t)$ carried by the conductors, will also satisfy a similar wave equation. We start, of course, with Maxwell's Equations, using the phasor form appropriate for time-harmonic waves in simple media. Faraday's Law, $\nabla \times \tilde{\mathbf{E}} = -j\omega\mu\tilde{\mathbf{H}}$, applied to the parallel plate transmission line, gives us

$$\frac{\partial \tilde{E}_y}{\partial z} = j\omega\mu\tilde{H}_x, \tag{9.6}$$

where we have used

$$\nabla \times \tilde{\mathbf{E}} = \begin{vmatrix} \mathbf{a}_x & \mathbf{a}_y & \mathbf{a}_z \\ \dfrac{\partial}{\partial x} & \dfrac{\partial}{\partial y} & \dfrac{\partial}{\partial z} \\ 0 & \tilde{E}_y(z) & 0 \end{vmatrix} = -\mathbf{a}_x \frac{\partial \tilde{E}_y(z)}{\partial z}.$$

This form of Faraday's Law can be used to find a relation between the potential difference $\tilde{V}(z)$ between the plates, and the current $\tilde{I}(z)$ in the plates. (Both of these quantities are expressed in phasor form.) We start with the familiar relation between potential and the path integral of E,

$$\tilde{V}(z) = -\int_0^d \tilde{E}_y(z)\, dy = -\tilde{E}_y(z)d,$$

where the last equality follows since the field $\tilde{E}_y(z)$ is uniform along the path from the bottom plate to the top plate. We have already reviewed the boundary conditions that relate $\tilde{H}_x(z)$ to the surface current density $\tilde{J}_{s,u,z}(z)$ in the upper plate, and the total current in the upper plate follows by integrating across the width of the plate,

$$\tilde{I}(z) = \int_0^w \tilde{J}_{s,u,z}(z)\, dx = \tilde{H}_x(z)w.$$

Substituting $\tilde{E}_y(z) = -\tilde{V}(z)/d$ and $\tilde{H}_x(z) = \tilde{I}(z)/w$ into Eq. (9.6) leads to

$$\frac{d\tilde{V}(z)}{dz} = -j\omega \left(\frac{\mu d}{w}\right)\tilde{I}(z).$$

But the term inside the parentheses is the inductance per unit length L' for the parallel plate transmission line, and we have

$$\frac{d\tilde{V}(z)}{dz} = -j\omega L'\tilde{I}(z). \tag{9.7}$$

In a similar manner, we can start with Ampère's Law to find a complementary differential relation between $\tilde{V}(z)$ and $\tilde{I}(z)$. Ampère's Law, $\nabla \times \tilde{\mathbf{H}} = j\omega\varepsilon\tilde{\mathbf{E}}$, applied to the parallel plate transmission line, gives us

$$\frac{\partial \tilde{H}_x}{\partial z} = j\omega\varepsilon\tilde{E}_y, \tag{9.8}$$

where we have used

$$\nabla \times \tilde{\mathbf{H}} = \begin{vmatrix} \mathbf{a}_x & \mathbf{a}_y & \mathbf{a}_z \\ \dfrac{\partial}{\partial x} & \dfrac{\partial}{\partial y} & \dfrac{\partial}{\partial z} \\ \tilde{H}_x(z) & 0 & 0 \end{vmatrix} = \mathbf{a}_y \frac{\partial \tilde{H}_x(z)}{\partial z}.$$

As before, we substitute $\tilde{E}_y(z) = -\tilde{V}(z)/d$ and $\tilde{H}_x(z) = \tilde{I}(z)/w$ into Eq. (9.8) to find

$$\frac{d\tilde{I}(z)}{dz} = -j\omega \left(\frac{\varepsilon w}{d}\right) \tilde{V}(z).$$

The term inside the parentheses in this case is the capacitance per unit length C' for the parallel plate transmission line, and we have

$$\frac{d\tilde{I}(z)}{dz} = -j\omega C' \tilde{V}(z). \tag{9.9}$$

To this point in the discussion we have dealt solely with the parallel plate transmission line. A similar analysis for *any* type of transmission line, however, leads to the same differential relations given by Eqs. (9.7) and (9.9), where the only difference is that we must, of course, choose the proper form of L' and C' appropriate to that form of transmission line. You should now see the relevance of all the previous derivations of inductance and capacitance toward our understanding of the properties of wave propagation on transmission lines.

Our next goal will be to combine Eqs. (9.7) and (9.9) to form a wave equation that governs the current and voltage on a transmission line. The process should seem familiar. We start by taking the derivative of Eq. (9.7) with respect to z:

$$\frac{d}{dz}\left(\frac{d\tilde{V}(z)}{dz}\right) = \frac{d}{dz}\left(-j\omega L' \tilde{I}(z)\right).$$

Substituting for $d\tilde{I}(z)/dz$ using Eq. (9.9) gives us

$$\frac{d^2 \tilde{V}(z)}{dz^2} = -j\omega L'\left(-j\omega C' \tilde{V}(z)\right).$$

Simplifying, we have

$$\frac{d^2 \tilde{V}(z)}{dz^2} + \omega^2 L' C' \tilde{V}(z) = 0. \tag{9.10}$$

In a similar way, we can take the derivative of Eq. (9.9) and substitute using Eq. (9.7) to find the same wave equation for $\tilde{I}(z)$:

$$\frac{d^2 \tilde{I}(z)}{dz^2} + \omega^2 L' C' \tilde{I}(z) = 0. \tag{9.11}$$

You should be familiar with solutions to these wave equations, as we have seen them before in the study of plane wave propagation:

$$\tilde{V}(z) = V_0 e^{-j\beta z} \tag{9.12}$$

and

$$\tilde{I}(z) = I_0 e^{-j\beta z}. \tag{9.13}$$

These solutions can be verified by direct substitution into the appropriate wave equations,

$$\left[(-j\beta)^2 + \omega^2 L' C'\right] \tilde{V}(z) = 0,$$

which is satisfied as long as

$$\boxed{\beta = \omega\sqrt{L'C'}.} \tag{9.14}$$

Substituting for L' and C' using the values for the parallel plate transmission line, the phase constant is

$$\beta = \omega\sqrt{\left(\frac{\mu d}{w}\right)\left(\frac{\varepsilon w}{d}\right)} = \omega\sqrt{\varepsilon\mu}, \tag{9.15}$$

and we recover the form of the phase constant β that we recall for plane waves. As before, this β is related to the wavelength λ of the wave through $\beta = 2\pi/\lambda$.

We will now introduce a new parameter, known as the **characteristic impedance** of the line, that is extremely important for transmission lines:

$$Z_0 = \frac{V_0}{I_0}. \tag{9.16}$$

As can be seen from its definition, the characteristic impedance is simply the ratio of the amplitudes of the voltage wave to the current wave. Just as with the electric and magnetic fields for propagating waves, these amplitudes are inextricably connected to one another. The characteristic impedance depends, as we will see next, only on properties of the transmission line; namely the dimensions of the conductors and the permittivity ε and permeability μ of the dielectric material. We start by taking the derivative of Eq. (9.12):

$$\frac{d}{dz}\tilde{V}(z) = -j\beta V_0 e^{-j\beta z}.$$

But Eq. (9.7) also tells us that

$$\frac{d\tilde{V}(z)}{dz} = -j\omega L' \tilde{I}(z) = -j\omega L' I_0 e^{-j\beta z}.$$

Equating these two, and canceling common factors, we have

$$\beta V_0 = \omega L' I_0.$$

Rearranging for $V_0/I_0 = Z_0$, we have

$$Z_0 = \frac{V_0}{I_0} = \frac{\omega L'}{\beta} = \frac{\omega L'}{\omega\sqrt{L'C'}},$$

where we have used Eq. (9.14) in the last equality. Eliminating common factors, we are left with

$$\boxed{Z_0 = \sqrt{\frac{L'}{C'}}.} \tag{9.17}$$

For the parallel plate transmission line, this characteristic impedance is

$$Z_0 = \sqrt{\frac{\mu d/w}{\varepsilon w/d}} = \sqrt{\frac{\mu}{\varepsilon}}\frac{d}{w} = \eta\frac{d}{w}.$$

Note that the characteristic impedance Z_0 has nothing to do with losses; rather, as its definition tells us, it is simply the ratio of the voltage and current amplitudes V_0/I_0 for a wave propagating along the transmission line in the $+z$-direction. Also, notice that for a lossless transmission line, Z_0 is a real quantity. As will soon be seen, this expression for Z_0 must be modified when losses in the transmission line are included.

Finally, we address the phase velocity of a wave on this transmission line. To do this, we must convert the voltage waveform, which is written in Eq. (9.12) in phasor form, back into its time-dependent form,

$$v(z,t) = \Re\left[\tilde{V}(z)e^{j\omega t}\right] = \Re\left[V_0 e^{-j\beta z}e^{j\omega t}\right] = V_0\cos(\omega t - \beta z).$$

Using arguments similar to those for the phase velocity of plane waves in Chapter 8, the velocity of this voltage waveform can be shown to be

$$u_p = \frac{\omega}{\beta}.$$

We showed earlier (Eq. (9.14)) that β is just $\omega\sqrt{L'C'}$, which leads us to

$$\boxed{u_p = \frac{1}{\sqrt{L'C'}}.}$$

(9.18)

A quick inspection of L' and C' for each of the different transmission line geometries in Table 9.1 shows this form of the phase velocity is perfectly equivalent to

$$u_p = \frac{1}{\sqrt{\varepsilon\mu}},$$

(9.19)

a result that you should recall from plane wave propagation.

Example 9.2 Transmission Line Parameters

A co-axial transmission line has a characteristic impedance of $Z_0 = 50\ \Omega$. The dielectric material between the conductors has permittivity $\varepsilon = 2.25\varepsilon_0$, and is non-magnetic ($\mu = \mu_0$). A 100 MHz harmonic voltage of amplitude $V_0 = 2$ V is applied to one end, launching a wave in the $+z$-direction. Determine (a) u_p, (b) β, (c) λ, (d) L', (e) C', (f) the ratio of radii of the conductors b/a, and (g) the amplitude of the current waveform on the line.

Solution:
We take these in order.

(a) $u_p = 1/\sqrt{\varepsilon\mu}$, from Eq. (9.19). Using μ and ε given, we have $u_p = 1/\sqrt{2.25\ \varepsilon_0\mu_0} = c/1.5 = 2 \times 10^8$ m/s.

(b) Use Eq. (9.15) to write $\beta = \omega\sqrt{\varepsilon\mu} = \omega/u_p$. The angular frequency is $\omega = 2\pi f$, where f is given as 100 MHz, and we found u_p in (a), which gives us $\beta = 3.14$ m^{-1}.

(c) Use $\lambda = 2\pi/\beta$ to find $\lambda = 2$ m.

(d) Divide Eqs. (9.17) and (9.18) to find $L' = Z_0/u_p = 0.25\ \mu$H/m.

(e) Multiply Eqs. (9.17) and (9.18) to find $C' = 1/Z_0 u_p = 0.10$ nF/m.

(f) Use $C' = 2\pi\varepsilon/\ln(b/a)$ to show $b/a = \exp(2\pi\varepsilon/C') = e^{1.25} = 3.5$.

(g) The amplitude I_0 of the sinusoidal current is $V_0/Z_0 = 2$ V/50 Ω = 40 mA.

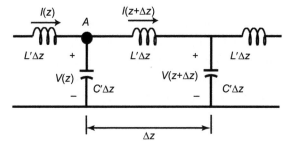

FIGURE 9.4 A model of a lossless transmission line as an infinitely long "ladder" of series inductors of inductance $L'\Delta z$ and parallel capacitors of capacitance $C'\Delta z$, where Δz is the length of one segment of the transmission line.

An alternative derivation of the voltage $\tilde{V}(z)$ and current $\tilde{I}(z)$ on the transmission line is based on the circuit model of a lossless transmission line, illustrated in Fig. 9.4. In this figure the transmission line is shown as an infinitely long "ladder" constructed of inductor elements in line with one of the conductors of the transmission line, and capacitor elements between the conductors. For each segment of the transmission line of length Δz, the inductance of each inductor is $L'\Delta z$ and the capacitance of each capacitor is $C'\Delta z$. We can then understand the charging and discharging of the inductors and capacitors using elementary linear circuit analysis techniques. Applying Kirchhoff's Voltage Law to the central loop of this circuit leads to

$$-\tilde{V}(z) + j\omega L' \, \Delta z \, \tilde{I}(z + \Delta z) + \tilde{V}(z + \Delta z) = 0.$$

Grouping the voltage terms together and dividing by Δz, we have

$$\frac{\tilde{V}(z + \Delta z) - \tilde{V}(z)}{\Delta z} = -j\omega L' \tilde{I}(z + \Delta z),$$

and taking the limit as $\Delta z \to 0$,

$$\frac{d\tilde{V}(z)}{dz} = -j\omega L' \tilde{I}(z).$$

This differential relation between $\tilde{V}(z)$ and $\tilde{I}(z)$ is in agreement with Eq. (9.7), which was derived using the plane wave solution for the fields in the region between the conductors.

Similarly, Kirchhoff's Current Law can be applied at the Node labeled "A" in Fig. 9.4. Summing the currents entering this node,

$$\tilde{I}(z) - j\omega C' \, \Delta z \, \tilde{V}(z) - \tilde{I}(z + \Delta z) = 0.$$

Grouping the current terms together and dividing by Δz, we have

$$\frac{\tilde{I}(z + \Delta z) - \tilde{I}(z)}{\Delta z} = -j\omega C' \tilde{V}(z),$$

and taking the limit as $\Delta z \to 0$,

$$\frac{d\tilde{I}(z)}{dz} = -j\omega C' \tilde{V}(z).$$

This differential relation between $\tilde{V}(z)$ and $\tilde{I}(z)$ is in agreement with Eq. (9.9).

A qualitative picture that accompanies this model is as follows. A potential $\tilde{V}(z)$ between the conductors at location z and $\tilde{V}(z + \Delta z)$ an incremental distance Δz removed implies that a potential difference exists across the elemental inductor. The resulting increasing current through the inductor charges the next elemental capacitor along the line, which increases the potential $\tilde{V}(z + \Delta z)$. A current $\tilde{I}(z)$ leads to a variation in the potential, and the potential $\tilde{V}(z)$ leads to a variation of the current, and the result is a propagating voltage and current waveform governed by the wave equation of Eqs. (9.10) or (9.11). The utility of the ladder model is that it gives an alternative picture leading to the wave equations for the voltage $\tilde{V}(z)$ and current $\tilde{I}(z)$ on a transmission line, and it can help us form an intuitive physical model for wave propagation along transmission lines. We will also see, in the following section, that losses on transmission lines can be included with a simple modification to this model.

So now we have determined by two different means the behavior of the voltage difference between the conductors, and the current in the conductors, for a lossless transmission line. Each are sinusoidally varying functions, and form traveling waves. Recall that the solution derived from the electric and magnetic fields in the space between the conductors is simply the transverse plane wave solution. An important feature of electromagnetic waves that we discovered in the previous chapter is that they carry energy. The time-averaged power density is given by the Poynting vector $\overline{\mathcal{P}} = \frac{1}{2}\tilde{\mathbf{E}} \times \tilde{\mathbf{H}}^*$, and the total power transmitted by this wave is found by integrating the power density over the cross-sectional area of the transmission line. For the parallel plate transmission line this is

$$P = \int_s \overline{\mathcal{P}} \cdot d\mathbf{s},$$

where the surface element is $d\mathbf{s} = \mathbf{a}_z \, dx \, dy$. Using the uniform fields $\tilde{\mathbf{E}}(z) = \mathbf{a}_y (V_0/d) \exp(-j\beta z)$ and $\tilde{\mathbf{H}}(z) = -\mathbf{a}_x (I_0/w) \exp(-j\beta z)$, this becomes

$$P = \int_{y=0}^{w} \int_{x=0}^{d} \frac{1}{2} \left\{ \mathbf{a}_y \frac{V_0}{d} e^{-j\beta z} \times \left(-\mathbf{a}_x \frac{I_0}{w} e^{-j\beta z} \right)^* \right\} \cdot \mathbf{a}_z \, dx \, dy,$$

which simplifies to

$$P = \int_{y=0}^{w} \int_{x=0}^{d} \frac{1}{2}\left(\frac{V_0}{d}\right)\left(\frac{I_0^*}{w}\right) dx\, dy.$$

After integration, the power transmitted along the transmission line is

$$P = \frac{1}{2} V_0 I_0^*. \tag{9.20}$$

While we worked this result for the parallel plate transmission line, it is actually quite general, and we can find the power transmitted along any transmission line using Eq. (9.20).

Example 9.3 Power Transmitted by Transmission Line

For the co-axial transmission line described in Example 9.2, determine the power transmitted by the wave.

Solution:
We are given that the voltage amplitude on this line is $V_0 = 2$ V, and that the characteristic impedance is $Z_0 = 50\ \Omega$. From this we determine that the amplitude of the current wave is $I_0 = V_0/Z_0 = 40$ mA, and the transmitted power is 40 mW.

In this section we have established a basic model for propagation of a wave that is "guided" by a transmission line. We started out by using results that were derived for plane wave propagation from Chapter 8, and chose results that can also satisfy the boundary conditions that must be obeyed at the conductor surfaces. We then reformulated the model of the transmission lines to discuss them in terms of voltage and current waves that propagate along the line. In the coming sections, we will explore losses in the transmission lines, and reflections that are formed at the end of the transmission line.

LEARNING OBJECTIVES

After studying this section, you should be able to:
- determine the resistance and conductance per unit length for a transmission line;
- determine the power loss per unit length for a wave propagating on a transmission line; and
- determine the complex propagation constant, the attenuation constant, the phase constant, the characteristic impedance, and the phase velocity for a wave on a lossy transmission line.

9.3 Losses

In the previous section we introduced the propagation of waves along a transmission line by considering the special case of perfectly conducting conductors ($\sigma_c \to \infty$, where we use the subscript "c" for the conductor), and a non-absorbing dielectric material ($\varepsilon'' = 0$ and $\sigma_d = 0$, where the subscript "d" indicates the dielectric) between the conductors. The wave supported by this ideal structure propagates without loss. When the conductivity of the conductor is finite, however, or when the dielectric medium is conductive or absorbing, these material properties lead to losses of the propagating wave. How can we understand these losses quantitatively?

Let's first address the losses that result from the finite conductivity of the conductors. Recall Example 8.19, in which we considered a pair of flat parallel conductors delivering current from a source to a load and back, and showed that a small component of the Poynting vector in the transverse direction represents power flow into the conductor. This power is absorbed by the conductors, and represents a loss of power carried by the wave. We revisited the idea of power losses in conductors in Section 8.8, when we discussed surface currents and resistive losses due to these currents. While the context for that discussion was a uniform plane wave normally incident upon a conducting surface, the argument and results are equally valid for the present context as well. We won't repeat the entire discussion here, but there is value in reviewing a few of the important features. Clearly, the primary field components present for the transmission line structure are the transverse fields in the dielectric medium that we described in the last section. For the parallel plate transmission line, these are the electric field in the y-direction (pointing from one conductor to the other) and the magnetic field in the x-direction. In addition, however, there is also a small component of the electric field pointing in the z-direction. We know that this component must be present since the current in the conductors is flowing in the z-direction, and the conduction current density $\tilde{J}_v$ in a good conductor requires a small field $\tilde{E}$ to drive it, $\tilde{J}_v = \sigma\tilde{E}$. This component is small, much smaller than the primary $\tilde{E}_y$ component, but it is present at the surface, decreasing exponentially into the conductor. The effective thickness of this current distribution is the skin depth, $\delta = 1/\sqrt{\pi f \mu \sigma_c}$. We showed in Section 8.8 that the power density absorbed in the surface of the conductor is

$$\overline{P_s^{\text{abs}}} = \frac{1}{2}|\tilde{J}_s|^2 R_s.$$

This appeared as Eq. (8.104), and it presumes that the thickness of the conductor is much greater than the skin depth δ. $\overline{P_s^{\text{abs}}}$ is the time-averaged power absorbed per unit area of the conductor. The surface resistance R_s is equal to

$$R_s = \sqrt{\frac{\pi f \mu}{\sigma_c}} = \sqrt{\frac{\omega\mu}{2\sigma_c}},$$

and $\tilde{J}_s$ is the effective surface current density, equal to I/w for the parallel plate transmission line.

To determine the power loss per unit length of the transmission line, we integrate $\overline{P_s^{\text{abs}}}$ over the width w of the transmission line conductor, and multiply by 2 to include the power loss in the two conductors. The fields and currents are uniform over this width, and substituting $\tilde{J}_s = \tilde{I}/w$, we find a power loss per unit length,

$$P' = 2\int_0^w \overline{P_s^{\text{abs}}}dx = 2w\frac{1}{2}\left|\frac{\tilde{I}}{w}\right|^2 R_s = \frac{1}{2}|\tilde{I}|^2\left(\frac{2R_s}{w}\right).$$

This should look familiar, since the time-averaged power dissipated in a resistor of resistance R when a sinusoidal current of amplitude I passes through it is $P = (1/2)I^2 R$. We therefore identify the term in parentheses as the **resistance per unit length**,

$$R' = 2R_s/w, \tag{9.21}$$

for the parallel plate transmission line system due to conductor losses. This resistance per unit length can be understood from a purely geometrical approach. For frequencies large enough that the skin depth δ is much less than the thickness of the conductor, the current lies primarily close to the surface, and the effective thickness of the conductor is δ. In this case, the resistance per unit length R' becomes

$$R' = \frac{2}{\sigma_c w \delta} = \frac{2}{w}\sqrt{\frac{\pi f \mu}{\sigma_c}},$$

where $w\delta$ is the effective cross-sectional area of the conductor, and the factor 2 is included to account for losses in the two conductors. Using the definition of the surface resistance given above, we see that this expression for the resistance per unit length R' agrees with Eq. (9.21).

The resistance per unit length R' for other transmission line geometries can be calculated using similar logic. Simple formulas for this parameter for the co-axial, twin-lead, and parallel plate geometries are included in Table 9.1.

Example 9.4 Resistance per Unit Length, R'

A co-axial transmission line carries a wave of voltage amplitude $V_0 = 10$ V and frequency 500 MHz. The conductors are of copper, of inner radius $a = 0.419$ mm and shield radius $b = 1.473$ mm. (These dimensions correspond to those of RG-58 transmission line.) Determine the resistance per unit length R' for this transmission line.

Solution:
The resistance per unit length for a co-axial transmission line, as listed in Table 9.1, is

$$R' = \frac{R_s}{2\pi}\left(\frac{1}{a}+\frac{1}{b}\right).$$

We are given the dimensions a and b, but must solve for

$$R_s = \sqrt{\frac{\pi f \mu}{\sigma_c}}.$$

Copper is non-magnetic, so $\mu = \mu_0$, and the conductivity of copper is $\sigma_c = 5.8 \times 10^7$ S/m, as listed in Table 3.1. Using these values, we find

$$R_s = \sqrt{\frac{\pi(500 \times 10^6 \text{ s}^{-1})(4\pi \times 10^{-7} \text{ H/m})}{5.8 \times 10^7 \text{ S/m}}} = 5.8 \text{ m}\Omega.$$

Then R' for this transmission line is

$$R' = \frac{R_s}{2\pi}\left(\frac{1}{a}+\frac{1}{b}\right) = \frac{5.8 \text{ m}\Omega}{2\pi}\left(\frac{1}{0.419 \text{ mm}}+\frac{1}{1.473 \text{ mm}}\right) = 2.83 \text{ }\Omega/\text{m}.$$

The second important loss mechanism in transmission lines is due to absorption by the dielectric material between the conductors. For lossless systems, the dielectric is perfectly insulating ($\sigma_d = 0$) and the permittivity is purely real ($\varepsilon'' = 0$). (The subscript "d" on the conductivity of the dielectric is used to distinguish it from the conductivity σ_c of the conductors in the transmission line.) If either of these conditions is relaxed, losses in the wave occur, as can be seen by referring back to the discussion of Section 8.6. As we showed in Eq. (8.43), the rate at which the field does work, per unit volume, on the free and bound charges in a medium is given as

$$\overline{P_v^{\text{abs}}} = \frac{1}{2}(\sigma_d + \varepsilon''\omega)|\tilde{\mathbf{E}}|^2.$$

Since the field amplitude is uniform in the region between the conducting plates, the power lost per unit length Δz is

$$P' = \int_0^w \int_0^d \overline{P_v^{\text{abs}}}\, dx\, dy = \frac{1}{2}(\sigma_d + \varepsilon''\omega)\, wd\, |\tilde{\mathbf{E}}|^2,$$

and since $\tilde{E} = \tilde{V}/d$,

$$P' = \frac{1}{2}\left[\frac{(\sigma_d + \varepsilon''\omega)\, w}{d}\right]|\tilde{V}|^2.$$

The term inside the square brackets is identified as the **conductance per unit length** of the parallel plate transmission line,

$$G' = \frac{(\sigma_d + \varepsilon'' \omega)\, w}{d}.$$

In a similar manner, the conductance per unit length can be determined for other transmission line geometries, and is included in Table 9.1.

Example 9.5 Conductance per unit length G'

We return to the co-axial transmission line described in Example 9.4. The dielectric between the conductors is polyethylene. (See Table 8.2 for the permittivity and loss tangent of polyethylene. These values are valid at a frequency of 3 GHz, but we'll use them for this example, despite the difference in frequency.) Determine the conductance per unit length G' for this transmission line.

Solution:

The conductance per unit length for a co-axial transmission line, as listed in Table 9.1, is

$$G' = \frac{2\pi\,(\sigma_d + \omega\varepsilon'')}{\ln\,(b/a)}.$$

Polyethylene is a non-conducting dielectric, so we'll ignore σ_d relative to $\omega\varepsilon''$. Using ε_r' and $\tan\delta$ from Table 8.2, we find

$$\varepsilon_r'' = \varepsilon_r' \tan\delta = (2.26)(0.0031) = 0.0070.$$

Then $\omega\varepsilon''$ is

$$\omega\varepsilon'' = (2\pi \times 500 \times 10^6 \text{ s}^{-1})(0.0070 \times 8.854 \text{ pF/m}) = 195 \text{ μS/m},$$

and the conductance per unit length is

$$G' = \frac{2\pi\,(\sigma_d + \omega\varepsilon'')}{\ln\,(b/a)} = \frac{2\pi(195 \text{ μS/m})}{\ln\,(1.473/0.419)} = 0.973 \text{ mS/m}.$$

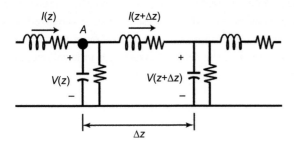

FIGURE 9.5 A circuit model of the lossy transmission line. This model consists of an infinitely long "ladder" of series inductors ($L'\Delta z$) and resistors ($R'\Delta z$), and shunt capacitors ($C'\Delta z$) and conductors ($G'\Delta z$), where Δz is the length of one element of the transmission line.

Let's return to the circuit model of a transmission line, a ladder structure consisting of series inductors and shunt capacitors, described in the previous section. We successfully applied Kirchhoff's Voltage and Current Laws to determine simple differential equations relating the voltage and current. We will now, with only two small modifications shown in Fig. 9.5, include losses in this transmission line model. Comparison with the lossless transmission line model shown in Fig. 9.4 reveals a resistor of value $R'\Delta z$ in series with each inductor, and a shunt resistor of conductance $G'\Delta z$ in parallel with each capacitor. The combined impedance of the inductor and resistor is $(R' + j\omega L')\Delta z$, and the conductance of the parallel combination of the capacitor and conductor is $(G' + j\omega C')\Delta z$. Applying Kirchhoff's Voltage Law around the center loop, we have

$$-\tilde{V}(z) + \left[R' + j\omega L'\right]\Delta z\,\tilde{I}(z + \Delta z) + \tilde{V}(z + \Delta z) = 0,$$

which, in the limit as $\Delta z \to 0$, leads us to

$$\frac{d\tilde{V}(z)}{dz} = -\left[R' + j\omega L'\right]\tilde{I}(z). \tag{9.22}$$

Application of Kirchhoff's Current Law at the node marked "A" in the figure gives us

$$\tilde{I}(z) - \left[G' + j\omega C'\right]\Delta z \, \tilde{V}(z) - \tilde{I}(z + \Delta z) = 0,$$

which leads to

$$\frac{d\tilde{I}(z)}{dz} = -\left[G' + j\omega C'\right]\tilde{V}(z). \tag{9.23}$$

We can develop a wave equation from these two linear first-order equations by taking the derivative of each term in Eq. (9.22) with respect to z,

$$\frac{d^2\tilde{V}(z)}{dz^2} = -\left[R' + j\omega L'\right]\frac{d\tilde{I}(z)}{dz},$$

and substituting for $d\tilde{I}/dz$,

$$\frac{d^2\tilde{V}(z)}{dz^2} = -\left[R' + j\omega L'\right]\left\{-\left[G' + j\omega C'\right]\tilde{V}(z)\right\}.$$

This equation can be rewritten as

$$\frac{d^2\tilde{V}(z)}{dz^2} - \gamma^2\tilde{V}(z) = 0, \tag{9.24}$$

where

$$\boxed{\gamma = \sqrt{\left[R' + j\omega L'\right]\left[G' + j\omega C'\right]}} \tag{9.25}$$

The solution of Eq. (9.24) is

$$\tilde{V}(z) = V_0\,e^{-\gamma z} = V_0\,e^{-\alpha z}\,e^{-j\beta z}, \tag{9.26}$$

where V_0 is the amplitude of the voltage at $z = 0$. In the second equality, γ is separated into its real and imaginary parts

$$\gamma = \alpha + j\beta, \tag{9.27}$$

where

$$\alpha = \Re\sqrt{\left[R' + j\omega L'\right]\left[G' + j\omega C'\right]}.$$

and

$$\beta = \Im\sqrt{\left[R' + j\omega L'\right]\left[G' + j\omega C'\right]}.$$

γ, α, and β are the propagation constant, the attenuation constant, and the phase constant, respectively, which we introduced in the discussion of uniform plane waves.

In a similar manner, an identical wave equation for the current $\tilde{I}(z)$ can be developed,

$$\frac{d^2\tilde{I}(z)}{dz^2} - \gamma^2\tilde{I}(z) = 0,$$

the solution of which is

$$\tilde{I}(z) = I_0 e^{-\gamma z}. \tag{9.28}$$

Using these solutions for the voltage $\tilde{V}(z)$ and current $\tilde{I}(z)$ on the lossy transmission line, we next find the characteristic impedance Z_0 of the transmission line. We start by computing the derivative with respect to z of Eq. (9.26),

$$\frac{d\tilde{V}(z)}{dz} = -\gamma V_0 e^{-\gamma z}.$$

Comparing this to Eq. (9.22) gives us

$$\gamma V_0 e^{-\gamma z} = \left[R' + j\omega L'\right] I_0 e^{-\gamma z}.$$

This relation yields the characteristic impedance $Z_0 = V_0/I_0$,

$$Z_0 = \frac{R' + j\omega L'}{\gamma},$$

and using the definition of γ from Eq. (9.25),

$$Z_0 = \sqrt{\frac{R' + j\omega L'}{G' + j\omega C'}}. \tag{9.29}$$

Notice that when either R' or G' is non-zero, then the characteristic impedance Z_0 is complex. Since Z_0 denotes the ratio V_0/I_0 (where $V_0 = |V_0|\exp(j\phi_v)$ and $I_0 = |I_0|\exp(j\phi_i)$ are the complex voltage and current amplitudes, respectively, written in terms of their magnitudes and phases), the magnitude of Z_0 is $|V_0|/|I_0|$ and its phase is $\exp[j(\phi_v - \phi_i)]$. So the phase of the complex characteristic impedance Z_0 is a direct measure of the phase *difference* between the voltage and current waves.

Example 9.6 α, β, and Z_0 for a Co-axial Transmission Line

Determine α, β, and Z_0 for the 500 MHz wave on the co-axial transmission line described in Examples 9.4 and 9.5.

Solution:
We'll use Eq. (9.25) for the propagation constant γ. $R' = 2.83\,\Omega/\text{m}$ and $G' = 0.973\,\text{mS/m}$ are already known from Examples 9.4 and 9.5, so we only need L' and C'. Refer to Table 9.1 for these formulas:

$$L' = \frac{\mu}{2\pi}\ln\left(\frac{b}{a}\right) = \frac{4\pi\times10^{-7}\,\text{H/m}}{2\pi}\ln\left(\frac{1.473}{0.419}\right) = 0.251\,\mu\text{H/m}$$

and

$$C' = \frac{2\pi\varepsilon'}{\ln(b/a)} = \frac{2\pi(2.26\times8.854\,\text{pF/m})}{\ln(1.473/0.419)} = 100\,\text{pF/m}.$$

We use these parameters in Eq. (9.25). Let's first find

$$R' + j\omega L' = 2.83\,\Omega/\text{m} + j(2\pi\times500\times10^6\,\text{s}^{-1})(0.251\,\mu\text{H/m}) = (2.83 + j789)\,\Omega/\text{m}.$$

Next we find

$$G' + j\omega C' = 0.973\,\text{mS/m} + j(2\pi\times500\times10^6\,\text{s}^{-1})(100\,\text{pF/m}) = (0.973 + j314)\,\text{mS/m}.$$

These two terms, converted to their magnitude and phase form, are

$$R' + j\omega L' = 789\,e^{j89.79°}\,\Omega/\text{m}$$

and

$$G' + j\omega C' = 314\,e^{j89.82°}\,\text{mS/m}.$$

With this, we are ready to find γ:

$$\gamma = \sqrt{\left[R' + j\omega L'\right]\left[G' + j\omega C'\right]} = \sqrt{[789\,e^{j89.79°}\,\Omega/\text{m}][314\,e^{j89.82°}\,\text{mS/m}]} = 15.74\,e^{j89.80°}\,\text{m}^{-1}.$$

The attenuation constant α is found from the real part of γ,

$$\alpha = 15.74 \cos(89.80°)\ \text{m}^{-1} = 0.0549\ \text{m}^{-1},$$

and the phase constant β from the imaginary part of γ,

$$\beta = 15.74 \sin(89.80°)\ \text{m}^{-1} = 15.74\ \text{m}^{-1}.$$

All that remains is to find the characteristic impedance Z_0, for which we use Eq. (9.29):

$$Z_0 = \sqrt{\frac{R' + j\omega L'}{G' + j\omega C'}} = \sqrt{\frac{789\ e^{j89.79°}\ \Omega/\text{m}}{314\ e^{j89.82°}\ \text{mS/m}}} = 50.1\ e^{-j0.015°}\ \Omega.$$

The phase velocity u_p of the wave on the transmission line comes directly from β, much as we derived earlier in other contexts. The voltage wave, with losses, as given by Eq. (9.26), must first be written in its time-dependent form,

$$v(z, t) = \Re\left\{\tilde{V}(z)\, e^{j\omega t}\right\}.$$

Inserting the phasor voltage in terms of its magnitude and phase, this becomes

$$v(z, t) = \Re\left\{|V_0|\, e^{j\theta_v}\, e^{-\alpha z}\, e^{-j\beta z}\, e^{j\omega t}\right\},$$

and taking the real part

$$v(z, t) = |V_0|\, e^{-\alpha z} \cos(\omega t - \beta z + \theta_v).$$

Focus on the phase term $\omega t - \beta z + \theta_v$. We select a position on this sinusoidal waveform at time t_1, let's say at the peak, which we designate as position z_1. If we move along the transmission line with the wave, keeping at the peak of the wave, then at a short time later $t_2 = t_1 + \Delta t$, the peak is at position $z_2 = z_1 + \Delta z$, such that the phase has not changed, and $\omega t_2 - \beta z_2 + \theta_v = \omega t_1 - \beta z_1 + \theta_v$. Rearranging, we have $\omega(t_2 - t_1) = \beta(z_2 - z_1)$. The velocity of the wave is the distance moved Δz divided by the time elapsed Δt, and we have

$$u_p = \frac{\Delta z}{\Delta t} = \frac{\omega}{\beta}. \tag{9.30}$$

The wavelength λ of the wave at frequency ω is related to the propagation term β by $\lambda = 2\pi/\beta$, leading us to $u_p = \omega/\beta = \omega\,\lambda/2\pi = f\lambda$, where $f = \omega/2\pi$ is the cyclical frequency of the wave, measured in hertz, or cycles per second.

Example 9.7 Current Amplitude, Wavelength, Attenuation Length, and Phase Velocity

For the sinusoidal wave on the co-axial transmission line described in Examples 9.4, 9.5, and 9.6, determine (a) the amplitude I_0 and phase of the current, (b) the wavelength λ, (c) the distance $\Delta z_{1/2}$ over which the wave amplitude decreases to one-half its initial amplitude, and (d) the phase velocity u_p of the wave.

Solution:
We take these in order.

(a) The current amplitude is

$$I_0 = \frac{V_0}{Z_0} = \frac{10\text{V}}{50.1\ e^{-j0.015°}\ \Omega} = 200\ e^{+j0.015°}\ \text{mA}.$$

The current leads the voltage by just 0.015°.

(b)

$$\lambda = \frac{2\pi}{\beta} = \frac{2\pi}{15.74 \text{ m}^{-1}} = 0.399 \text{ m}.$$

(c) Since the wave amplitude decreases as $V_0 e^{-\alpha z}$, the distance $\Delta z_{1/2}$ is found by solving

$$e^{-\alpha \Delta z_{1/2}} = 1/2 \quad \Rightarrow \quad \Delta z_{1/2} = \frac{1}{\alpha} \ln(2) = \frac{\ln(2)}{0.0549 \text{ m}^{-1}} = 12.6 \text{ m}.$$

(d)

$$u_p = \frac{\omega}{\beta} = \frac{2\pi \times (500 \times 10^6 \text{ s}^{-1})}{15.74 \text{ m}^{-1}} = 1.99 \times 10^8 \text{ m/s}.$$

Before we move on, let's examine the power transferred by the electromagnetic wave guided by the lossy transmission line. We showed earlier that the transmitted power, based upon the Poynting vector $\overline{\mathcal{P}} = (1/2)\tilde{\mathbf{E}} \times \tilde{\mathbf{H}}^*$, can quite generally be written in terms of the voltage $\tilde{V}(z)$ and current $\tilde{I}(z)$ waveforms as

$$P(z) = \frac{1}{2}\mathfrak{R}\left[\tilde{V}(z)\tilde{I}^*(z)\right].$$

Using the solutions for $\tilde{V}(z)$ in Eq. (9.26) and the current waveform in Eq. (9.28), the transmitted power is

$$P(z) = \frac{1}{2}\mathfrak{R}\left[V_0 e^{-\alpha z} e^{-j\beta z}(I_0 e^{-\alpha z} e^{-j\beta z})^*\right] = \frac{1}{2}V_0 I_0^* e^{-2\alpha z}.$$

Taking the derivative with respect to z of the total transmitted power, we find

$$\frac{dP(z)}{dz} = (-2\alpha)\frac{1}{2}V_0 I_0^* e^{-2\alpha z} = (-2\alpha)P(z).$$

Therefore, the fractional rate at which power is lost from the traveling wave, 2α, is simply twice that of the electric field, α. Upon a brief moment of reflection, this should make sense, in that the transmitted power is proportional to the square of the electric field amplitude. Since the electric field amplitude decreases as $e^{-\alpha z}$, the power must decrease as $e^{-2\alpha z}$; that is, at twice the rate as the amplitude.

In this section we have included losses in the model of wave propagation on transmission lines. These losses include those due to the finite conductivity σ_c of the conductors, as well as the absorption by the dielectric medium between the conductors. We have examined how these losses affect the propagation constant for the wave, which we write as the complex parameter $\gamma = \alpha + j\beta$.

9.4 Standing Waves

In the previous section we derived a wave equation for the voltage waveform,

$$\frac{d^2 \tilde{V}(z)}{dz^2} - \gamma^2 \tilde{V}(z) = 0,$$

as well as for the current waveform,

$$\frac{d^2 \tilde{I}(z)}{dz^2} - \gamma^2 \tilde{I}(z) = 0.$$

For a line that is infinitely long in the $+z$-direction (so that we don't have to worry about a reflection at the end of the line!) and that is driven by a source located at $z = 0$, the solutions for the voltage and current waveforms are given by Eqs. (9.26) and (9.28), representing a wave traveling in the $+z$-direction. Another perfectly valid solution of the wave equation, however, is a wave that propagates in the $-z$-direction, which can be represented as the voltage wave

$$\tilde{V}(z) = V_0 e^{+\gamma z}$$

and the current wave

$$\tilde{I}(z) = I_0 e^{+\gamma z}.$$

The general solutions of the wave equations, then, must include waves traveling in the $+z$ *and* $-z$ directions,

$$\tilde{V}(z) = V_0^+ e^{-\gamma z} + V_0^- e^{+\gamma z}, \qquad (9.31)$$

for the potential difference between the conductors at location z, and

$$\tilde{I}(z) = I_0^+ e^{-\gamma z} + I_0^- e^{+\gamma z} \qquad (9.32)$$

for the current in each of the conductors at location z. Note the superscripts "+" or "−" on the amplitudes to distinguish between the amplitude of the wave component traveling in the $+z$-direction or the $-z$-direction, respectively. These two wave components are individually known as **traveling waves**, since they represent a wave that is clearly traveling in one direction or the other. If attenuation of the wave is negligible, the amplitude of the traveling wave is constant along the line, V_0^+ for the voltage wave traveling in the $+z$-direction, and V_0^- for the voltage wave traveling in the $-z$-direction.

An example of a plot of $V_0^+ e^{-\gamma z}$, $V_0^- e^{\gamma z}$, and $\tilde{V}(z)$ is shown in Fig. 9.6(a). In this figure the attenuation constant α is set to zero, and we show a "snapshot" at $t = 0$ of the traveling waves, $V_0^+ e^{-\gamma z}$ as the red dashed line, and $V_0^- e^{+\gamma z}$ as the purple dot-dashed line. The total voltage – that is, the sum of these two traveling waves as given by Eq. (9.31) – gives $\tilde{V}(z)$, and is shown as the solid blue line.

The total z-dependent current $\tilde{I}(z)$ can be plotted in a similar fashion. But we need to think a little more carefully about the sign of the $-z$-traveling current waveform. As we discussed in the last section, the amplitudes of the voltage and the current waveforms for the $+z$-traveling wave must be related by

$$\frac{V_0^+}{I_0^+} = Z_0,$$

FIGURE 9.6 A plot of (a) $V_0^+ e^{-\gamma z}$, $V_0^- e^{\gamma z}$, and $\tilde{V}(z)$, and (b) $I_0^+ e^{-\gamma z}$, $I_0^- e^{\gamma z}$, and $\tilde{I}(z)$ along a transmission line. The attenuation constant α is zero for these plots, so $\gamma = j\beta$. In (a) and (b), the red dashed line is the wave traveling to the right, the purple dot-dashed line is the wave traveling to the left, and the sum of these two traveling waves yields $\tilde{V}(z)$ and $\tilde{I}(z)$, as given in Eqs. (9.31) and (9.32).

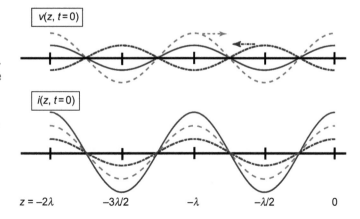

where Z_0 is the characteristic impedance of the transmission line. The ratio of the amplitudes of the $-z$-traveling wave is similar, but of opposite sign

$$\frac{V_0^-}{I_0^-} = -Z_0. \tag{9.33}$$

The minus sign here can be understood by remembering that the Poynting vector $\overline{\mathcal{P}} = (1/2)\tilde{\mathbf{E}} \times \tilde{\mathbf{H}}^*$ points in the $+z$-direction for the wave traveling in the $+z$-direction, and in the $-z$-direction for the wave traveling in the $-z$-direction. This implies that either $\tilde{\mathbf{E}}$ or $\tilde{\mathbf{H}}$, but not both, must be inverted for the $-z$-traveling wave when compared to the fields of the $+z$-traveling wave. By extension, either V_0 or I_0 (but not both) must be inverted as well, and this leads to the minus sign in Eq. (9.33).

With this result, we can write Eq. (9.32) for the current $\tilde{I}(z)$ in the useful alternate form of

$$\tilde{I}(z) = \frac{1}{Z_0} \left[V_0^+ e^{-\gamma z} - V_0^- e^{+\gamma z} \right]. \tag{9.34}$$

The current waveforms corresponding to $\tilde{I}(z)$ are shown in Fig. 9.6(b). Note that the current waveform traveling in the $-z$-direction is inverted with respect to the $-z$-traveling voltage wave. The consequence of this inversion is that at locations where the $V_0^+ e^{-\gamma z}$ and $V_0^- e^{+\gamma z}$ add destructively, giving a reduced value of $\tilde{V}(z)$, the traveling current waves $I_0^+ e^{-\gamma z}$ and $I_0^- e^{+\gamma z}$ add constructively, giving an increased value of $\tilde{I}(z)$, and vice versa.

In these plots we have examined the total voltage and the total current, but only at a single time $t = 0$. Let's examine these voltages and currents as a function of time. We will discover in the following that the voltage and current form a **standing wave** pattern, characterized by a periodic sequence of maxima and minima. This pattern results from the superposition of the two traveling waves. The phasor voltage $\tilde{V}(z)$ or the current $\tilde{I}(z)$ at any position z represents a sinusoidally oscillating function (of course), but, in contrast to a traveling wave, the standing wave displays strong modulation of the amplitude over the spacing of the wavelength λ of the wave. This is seen graphically by determining $v(z, t)$ from Eq. (9.31) and plotting this voltage as a function of z at several different times t, as shown in Fig. 9.7. In this example, $V_0^- = 0.5V_0^+$, and for simplicity, we have again set the attenuation constant α equal to zero. The red dashed line is the wave traveling in the $+z$-direction, $V_0^+ e^{-\gamma z}$. Notice that with increasing time t, this wave moves to the right. The purple dot-dashed line is the wave traveling to the left, $V_0^- e^{+\gamma z}$. The sum of these two traveling waves, $\tilde{V}(z)$, as given in Eq. (9.31), is shown as the solid blue line. At $t = 0$, the waveforms for $V_0^+ e^{-\gamma z}$ and $V_0^- e^{+\gamma z}$ are of opposite sign at any position z, so the total voltage is reduced. But as the red dashed line moves to the right and the purple dot-dashed line moves to the left, the two waves add to one another in some regions, and the constructive interference leads to an increase in the total voltage. After a quarter-cycle, labeled $\omega t = \pi/2$, the two waveforms add constructively everywhere, and the total voltage is quite large. Over the next quarter-cycle, the voltage waveform evolves back to one of destructive interference between the traveling waves. In the following half-cycle, not shown in the figure, the pattern continues, but with the voltage inverted; negative voltage goes to positive, and positive becomes negative. At any location z, the voltage varies sinusoidally as a function of time, with a minimum amplitude of oscillation located at $z = m\lambda/2$, where m is an integer (in this example), and the maximum amplitude is located at $z = m\lambda/2 + \lambda/4$. These plots show the evolution of the pattern over just half of a cycle.

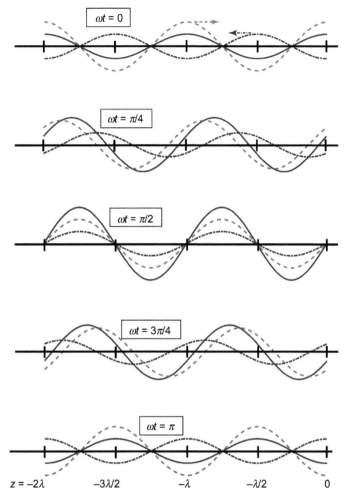

FIGURE 9.7 Plots of $V_0^+ \cos(\omega t - \beta z)$, $V_0^- \cos(\omega t + \beta z)$, and $\tilde{V}(z)$ versus z (along a transmission line) for different times. In this example, $V_0^- = 0.5 V_0^+$ and $\alpha = 0$. The red dashed line is the wave traveling to the right, $V_0^+ e^{-\gamma z}$. Notice that with increasing time t, this wave moves to the right. The purple dot-dashed line is the wave traveling to the left, $V_0^- e^{+\gamma z}$. The sum of these two traveling waves yields $\tilde{V}(z)$, as given in Eq. (9.31). Over the course of a half-cycle, the two traveling waves move a distance of $\lambda/2$, and the voltage waveform evolves from one of destructive interference between the traveling waves to one of constructive interference between the traveling waves, and back to one of destructive interference. At any location z, the voltage varies sinusoidally as a function of time, with a minimum amplitude of oscillation located at $z = m\lambda/2$, where m is an integer (in this example), and the maximum amplitude is located at $z = m\lambda/2 + \lambda/4$. These plots show the evolution of the pattern over just one half of a cycle.

Now let's derive an analytic expression that describes what we have just seen graphically. To simplify the analysis, the amplitudes V_0^+ and V_0^- are purely real, and the transmission line is lossless – that is, the attenuation constant $\alpha = 0$. We start by converting the phasor voltage $\tilde{V}(z)$ to its time-dependent form. As usual, this requires multiplication by $\exp(j\omega t)$, and retention of only the real part of the product:

$$v(z, t) = \Re\left\{\tilde{V}(z)\, e^{j\omega t}\right\} = \Re\left\{\left[V_0^+ e^{-j\beta z} + V_0^- e^{+j\beta z}\right] e^{j\omega t}\right\}.$$

Since the amplitudes are real, $\Re\{e^{j(\omega t \pm \beta z)}\} = \cos(\omega t \pm \beta z)$ can be used to write

$$v(z, t) = V_0^+ \cos(\omega t - \beta z) + V_0^- \cos(\omega t + \beta z),$$

and the trigonometric identity $\cos(a \pm b) = \cos a \cos b \mp \sin a \sin b$, with $a = \omega t$ and $b = \beta z$, to write

$$v(z, t) = \left(V_0^+ + V_0^-\right)\cos(\beta z)\cos(\omega t) + \left(V_0^+ - V_0^-\right)\sin(\beta z)\sin(\omega t). \tag{9.35}$$

Notice that the form of this wave pattern is fundamentally different from that of the traveling wave. In this standing wave expression, the dependence on time t and the dependence on z (or more generally on position) are separate functions. Recall that for the traveling wave, the waveform was always a function of the two variables together, as $\omega t - \beta z$, for example.

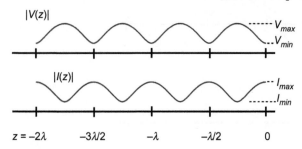

FIGURE 9.8 Plots of the voltage and current standing wave amplitude. The amplitude varies with z due to the interference between the two traveling waves. The voltage amplitude minima are located, in this example, at $z = m\lambda/2$, where m is an integer, and the maxima are located at $z = m\lambda/2 + \lambda/4$. The minima in the current pattern coincide with the maxima in the voltage pattern, and vice versa.

Let's look at what this standing wave function looks like in a little more detail. At any location z, the voltage $v(z, t)$ oscillates sinusoidally as a function of time t, but the *amplitude* of this oscillating voltage depends strongly on z, as shown in Fig. 9.8. The voltage amplitude maxima occur at locations where the two traveling waves add in phase with one another, leading to a total amplitude of $V_{max} = |V_0^+| + |V_0^-|$. Conversely, minima in the voltage amplitude occur at locations where the two traveling waves add out of phase with one another, yielding an amplitude minimum of $V_{min} = ||V_0^+| - |V_0^-||$. Adjacent maxima are separated by a distance $\lambda/2$, and minima occur at locations halfway between maxima. For example, if V_0^+ and V_0^- are both real and positive, and $V_0^+ > V_0^-$, Eq. (9.35) tells us that the maximum amplitude occurs where $\cos(\beta z) = \pm 1$, or $\beta z = 0, \pm\pi, \pm 2\pi, \ldots$. Since $\beta = 2\pi/\lambda$, the distance between adjacent maxima is $\Delta z = \pi/\beta = \lambda/2$. Similarly, voltage minima occur at locations at which $\sin(\beta z) = \pm 1$, or $\beta z = \pm\pi/2, \pm 3\pi/2, \ldots$.

Example 9.8 Standing Waves

The amplitude of a standing wave pattern on an air-spaced transmission line varies from a maximum of $V_{max} = 10$ V to a minimum of $V_{min} = 2$ V. The spacing between adjacent maxima is 50 cm. Determine (a) the amplitudes of the two counter-propagating voltage waves, and (b) the frequency of the waves.

Solution:

(a) We solve $V_{max} = |V_0^+| + |V_0^-|$ and $V_{min} = ||V_0^+| - |V_0^-||$ to show $|V_0^+| = \frac{1}{2}(V_{max} + V_{min})$, and $|V_0^-| = \frac{1}{2}(V_{max} - V_{min})$. (Actually, this is only correct if $|V_0^+| > |V_0^-|$. If this is not the case, then the expressions for $|V_0^+|$ and $|V_0^-|$ must be reversed. With the information given in the problem statement, we don't know which case is correct.) Using the values of V_{max} and V_{min} given, we find $|V_0^+| = 6$ V and $|V_0^-| = 4$ V.

(b) We are told that the distance between adjacent maxima, which we know must be $\lambda/2$, is 50 cm. Therefore the wavelength of the wave is 1.0 m. We are also told that the dielectric between the conductors is air, which tells us that the phase velocity of the wave is c. Therefore the frequency is $f = c/\lambda = (3.0 \times 10^8 \text{ m/s})/(1 \text{ m}) = 300$ MHz.

Using a slotted line, described in TechNote 9.1, one can measure the standing wave pattern generated at a load and determine the quality of the impedance matching of the load.

TechNote 9.1 Slotted Line

A slotted line, such as the one shown schematically in Fig. 9.9, is a device used to measure the standing wave pattern on a transmission line. The probe wire on top of the slotted line registers the voltage amplitude at the location of the probe. By translating the probe along the line, and reading the voltage amplitude as a function of position, a pattern such as that shown in Fig. 9.8 can be generated. The SWR $S = V_{max}/V_{min}$ and λ (twice the distance between maxima) can be read from this plot.

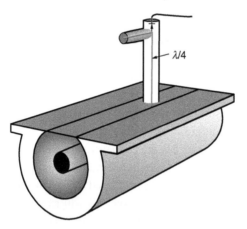

FIGURE 9.9 A slotted line for measurement of the standing wave pattern on a transmission line. The voltage sensor can be translated along the line, mapping out the voltage waveform.

In addition to the voltage standing wave pattern, which we have just discussed, the current also forms a standing wave pattern, and this pattern can be calculated using a similar process. The current can be written in its time-dependent form as

$$i(z, t) = \Re\left\{ \tilde{I}(z)\, e^{j\omega t} \right\} = \Re\left\{ \frac{1}{Z_0}\left[V_0^+ e^{-j\beta z} - V_0^- e^{+j\beta z} \right] e^{j\omega t} \right\},$$

which, for real amplitudes V_0^+ and V_0^- and real impedance Z_0 becomes

$$i(z, t) = \frac{1}{Z_0}\left\{ V_0^+ \cos(\omega t - \beta z) - V_0^- \cos(\omega t + \beta z) \right\}.$$

Using trigonometric identities, this standing pattern becomes

$$i(z, t) = \left(\frac{V_0^+ - V_0^-}{Z_0} \right) \cos(\beta z)\cos(\omega t) + \left(\frac{V_0^+ + V_0^-}{Z_0} \right) \sin(\beta z)\sin(\omega t).$$

This current standing wave pattern has some characteristics that are similar to those of the voltage, but there are important differences as well. This amplitude is plotted in Fig. 9.8. First, note that the maxima in the current standing wave pattern (again taking the traveling wave amplitudes V_0^+ and V_0^- to both be positive, with $V_0^+ > V_0^-$) occur at locations for which $\sin\beta z = \pm 1$. These are precisely the locations at which the voltage standing wave pattern goes through a minimum. Similarly, the minima in the current pattern coincide with the maxima in the voltage standing wave pattern. This reversal of the maxima and minima is a direct consequence of the reversal of the sign of the I_0^- amplitude.

In this section we have introduced voltage and current standing waves on transmission lines. We treated these standing waves in a manner very similar to our treatment of standing waves formed by uniform plane waves in Chapter 8. We found that the amplitude of a standing wave varies significantly over the range of a wavelength. This is in contrast to a traveling wave, whose amplitude is constant over this distance (ignoring attenuation effects, of course). Standing waves are formed by the constructive and destructive superposition of two traveling waves, one propagating in the $+z$-direction and the other in the $-z$-direction. Standing waves, and their differences for potential and current waveforms, will have a strong impact on the input impedance of a transmission line, which we will discuss in Section 9.5.3.

9.5 Reflections at a Load

Now that we have examined the voltage and current standing wave patterns formed by the superposition of two counter-propagating traveling waves, let's consider the transmission line of length ℓ and characteristic impedance Z_0, as shown in Fig. 9.10. In this configuration, the source launches a wave that travels in the $+z$-direction toward the load. The load is an impedance element, of impedance Z_L. The wave incident upon the load can be partially reflected to form a wave traveling in the $-z$-direction, propagating toward the source. This reflected wave can then be reflected at the source end of the transmission line, and so on. We will examine this transmission line geometry in the steady state – that is, after the waves have settled to fixed amplitudes. There are four important features that we would like to explore with this configuration:

1. What is the amplitude of the reflected wave?
2. What is nature of the standing wave along the line?
3. What is the input impedance of the loaded transmission line?
4. How can we determine the amplitudes of the $+z$- and $-z$-traveling waves?

We consider these features in this section.

9.5.1 Reflection Coefficient

The amplitude of the reflected wave V_0^- is proportional to the amplitude of the incident wave V_0^+. A convenient parameter to describe the reflection, then, is the reflection coefficient Γ_L, which is defined as the ratio of the amplitudes of these two traveling waves:

$$\Gamma_L = \frac{V_0^-}{V_0^+}. \tag{9.36}$$

We have encountered a similar reflection coefficient previously, in the discussion of freely propagating waves incident upon an interface between two media in Chapter 8. The idea is similar here.

How can we determine the amplitudes of the waves V_0^+ and V_0^- in this configuration? The source, which generates a voltage of amplitude V_g and possesses an internal series impedance Z_g, drives the line at the input (the left end in Fig. 9.10). The position of this end of the line is labeled as $z = -\ell$. (This notation is a change from the previous notation that we used in Section 9.2, in which this position was $z = 0$. Either way is fine. We choose this notation here simply for convenience.) At the opposite end, labeled $z = 0$, a load impedance element of impedance Z_L is connected between the two

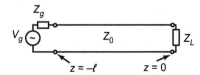

FIGURE 9.10 A transmission line of characteristic impedance Z_0 is driven by a source at the left end and terminated in a load of impedance Z_L at the right end. The source generates a voltage of amplitude V_g and has an internal series impedance Z_g. The transmission line is depicted as the two long, straight parallel lines between the small open circles.

conductors of the transmission line. This is commonly referred to as the **termination** of the transmission line.

Let's start by representing the total voltage at the load as $\tilde{V}_L$, where

$$\tilde{V}_L \equiv \tilde{V}(0) = V_0^+ + V_0^-,$$

which is found simply by evaluating the waveform of Eq. (9.31) at $z = 0$. Similarly, from Eq. (9.34), the total current at the load is I_L,

$$I_L \equiv I(0) = \frac{V_0^+}{Z_0} - \frac{V_0^-}{Z_0}.$$

Now remember that the current I_L and voltage V_L are the transmission line current and voltage at the load, but these must also be the current through and voltage across the load, and so they must be related by the load impedance itself. That is, V_L must be equal to $I_L Z_L$. Rearranging, the two conditions that must be met by V_0^+ and V_0^- are

$$V_0^+ + V_0^- = I_L Z_L$$

and

$$V_0^+ - V_0^- = I_L Z_0.$$

The solutions to these two conditions are

$$V_0^+ = \frac{I_L}{2}(Z_L + Z_0)$$

and

$$V_0^- = \frac{I_L}{2}(Z_L - Z_0).$$

Even without yet knowing the load current I_L, this brings us to our first important result – that is, that the reflection coefficient of the load, which we defined in Eq. (9.36), is

$$\boxed{\Gamma_L = \frac{Z_L - Z_0}{Z_L + Z_0}.}$$

(9.37)

That is, the amplitude of the wave reflected by the load depends on the difference between the load impedance and the characteristic impedance of the transmission line. As you might expect, it is often desirable to eliminate, or at least reduce the amplitude of, the reflected wave. From this equation, this can be achieved by choosing the termination impedance to be equal to the characteristic impedance of the line, $Z_L = Z_0$. This condition is called **impedance matching**.

Example 9.9 Reflection

A sinusoidal wave of amplitude 5 V is launched along a lossless 50 Ω transmission line, and incident upon a 45 Ω load at the end of the line. Determine the amplitude of the reflected wave.

Solution:
The reflection coefficient Γ_L is given by Eq. (9.37). Using $Z_0 = 50\ \Omega$ and $Z_L = 45\ \Omega$, we find $\Gamma_L = (45\ \Omega - 50\ \Omega)/(45\ \Omega + 50\ \Omega) = -5/95 \approx -0.053$. Then we find the amplitude of the reflected wave using $V_0^- = \Gamma V_0^+ = -0.053 \times 5\text{V} = -0.26$ V.

Notice in Example 9.9 that the reflection coefficient is negative. This implies that the reflected wave is inverted relative to the incident wave. In general, the magnitude of the reflection coefficient must always be less than or equal to unity. A couple of important special cases are useful to keep in mind: the short-circuit load and the open-circuit load.

If we **short circuit** the transmission line (i.e. $Z_L = 0$), then Γ_L will be -1. This implies that the reflected wave has the same amplitude as the incident wave, but that it is inverted. How can we understand this? Well, remember that the total wave at the load is $V_L = V_0^+ + V_0^-$. With $V_0^- = -V_0^+$, the voltage at the load is $V_L = 0$. This should make perfect sense, since, after all, the two conductors are shorted at the end of the transmission line. What can be said about the current at the load, I_L? Since $I_L = V_0^+/Z_0 - V_0^-/Z_0$, the load current is $I_L = 2V_0^+/Z_0$. This hefty current should not bother us, since the load has zero resistance, so any current at all should be fine. The total load current consists of the current of the incident wave V_0^+/Z_0 plus the current of the reflected wave, which is equal in magnitude and sign to that of the incident wave.

In the opposite limit of an **open-circuit load**, $Z_L = \infty$, the reflection coefficient becomes $+1$. The reflected wave has a magnitude equal to that of the incident wave, but in this case the wave amplitude is of the same sign, $V_0^- = V_0^+$. The total voltage amplitude at the load is $V_L = 2V_0^+$, and the load current is $I_L = 0$. The zero load current is the only possible result for an open-circuit load, through which no current may pass.

While for the general case the phase of Γ_L can be virtually any value, if Z_0 is real (as it will be for a lossless transmission line) and the load impedance Z_L is real, then the reflection coefficient Γ_L will also be real.

9.5.2 Standing Wave Patterns due to Impedance Mismatch

In Section 9.4 we saw that waves traveling in the $+z$- and $-z$-directions produce a standing wave on the line, and in Section 9.5.1 we saw that an impedance mismatch at the end of a transmission line produces a wave traveling in the $-z$-direction by partially reflecting the wave that is incident upon the load. In this section, then, we'd like to combine these two topics, and examine the standing wave pattern produced by an impedance mismatch at the end of a transmission line. We will see that, through measurements of the standing wave pattern produced by the reflection, the load impedance can be fully characterized. We will follow this with a discussion of the input impedance of the transmission line, which is strongly impacted by reflections.

The wave generated by the source at the input to the transmission line shown in Fig. 9.10 travels down the line of length ℓ and is incident upon the load of impedance Z_L. As we showed, a reflected wave of amplitude $V_0^- = \Gamma_L V_0^+$ is generated at the load, where Γ_L is the reflection coefficient, and V_0^+ is the amplitude of the incident wave at the load. It is useful to write the voltage $\tilde{V}(z)$ and current $\tilde{I}(z)$ waveforms in a format that explicitly includes these relations. Let us write the voltage $\tilde{V}(z)$ using Eq. (9.31),

$$\tilde{V}(z) = V_0^+ e^{-\gamma z} + V_0^- e^{+\gamma z},$$

and using $V_0^- = \Gamma_L V_0^+$ and factoring out V_0^+,

$$\tilde{V}(z) = V_0^+ \left[e^{-\gamma z} + \Gamma_L e^{+\gamma z} \right]. \tag{9.38}$$

The standing wave ratio is defined as $S = V_{max}/V_{min}$, where V_{max} and V_{min} are the maximum value and minimum value, respectively, of the voltage $\tilde{V}(z)$. For a lossy transmission line, for which $\alpha \neq 0$, the amplitudes of the traveling waves vary exponentially with distance z, and V_{max} and V_{min} vary along the length of the transmission line. Let's start with the easier case of the lossless line, that is, $\alpha = 0$. In this case, we can write

$$V_{max} = |V_0^+|(1 + |\Gamma_L|),$$

since the $e^{\pm \gamma z}$ terms become simply $e^{\pm j\beta z}$, and the voltage maxima occur when these two phase factors add in phase with one another. Similarly, the voltage minimum is the result of the two traveling waves subtracting from one another, partially canceling each other, and

$$V_{min} = |V_0^+| - |V_0^-| = |V_0^+|(1 - |\Gamma_L|).$$

Using these forms of V_{max} and V_{min}, we can write the standing wave ratio as

$$S = \frac{(1 + |\Gamma_L|)}{(1 - |\Gamma_L|)}. \tag{9.39}$$

This equation is a direct relation between the standing wave ratio S and the magnitude of the reflection coefficient $|\Gamma_L|$.

In a similar manner, we can examine the standing wave pattern of the current $\tilde{I}(z)$, using

$$\tilde{I}(z) = \frac{1}{Z_0}\left[V_0^+ e^{-\gamma z} - V_0^- e^{+\gamma z}\right),$$

which, when the $-z$-traveling wave is the reflection of the $+z$-traveling wave, can be written as

$$\tilde{I}(z) = \frac{V_0^+}{Z_0}\left[e^{-\gamma z} - \Gamma_L e^{+\gamma z}\right]. \tag{9.40}$$

For the lossless line case, the current maxima and minima have values

$$I_{max} = \frac{|V_0^+|}{Z_0}(1 + |\Gamma_L|)$$

and

$$I_{min} = \frac{|V_0^+|}{Z_0}(1 - |\Gamma_L|).$$

Remember that the maxima of the current standing wave pattern occur at the minima of the voltage standing wave pattern, and vice versa. That notwithstanding, the standing wave ratio S found from the current maxima and minima is, of course, identical to that of the voltage standing wave pattern, given above by Eq. (9.39).

To this point, we have discussed the magnitudes of the voltage (or current) maxima and minima, but we haven't worried about their locations. The locations can tell us the *phase* of the load impedance. Let's see how this works.

We know that the reflection coefficient at the load is given by Eq. (9.37). For simplicity, we will limit our discussion to a lossless transmission line, for which $Z_0 = R_0$ is real. If the load impedance is complex, then the reflection coefficient Γ_L will be complex as well. We can write this as

$$\Gamma_L = |\Gamma_L|e^{j\theta_\Gamma},$$

where the reflection coefficient is represented in terms of its magnitude $|\Gamma_L|$ and phase θ_Γ. Now we write the voltage $\tilde{V}(z)$ at location z using Eq. (9.31), with $\gamma = j\beta$ since this is a lossless line, and substituting $\Gamma_L V^+$ for V^-, as

$$\tilde{V}(z) = V_0^+\left(e^{-j\beta z} + \Gamma_L e^{j\beta z}\right).$$

Where does this voltage attain its maximum value? This occurs at any location where the two terms (representing the $+z$- and $-z$-traveling waves) add in phase with one another. That is, whenever the *difference* in phase between the two terms is zero or an integer multiple of 2π,

$$-\beta z_{max} - (\theta_\Gamma + \beta z_{max}) = 2m\pi,$$

where m is an integer. We solve this for the phase of the reflection coefficient,

$$\theta_\Gamma = -2\beta z_{max} - 2m\pi.$$

If z_{max} is the location of the first maximum (closest to the load), then we use $m = 0$ to find

$$\theta_\Gamma = -2\beta z_{max}.$$

(Remember that we labeled the location of the source as $z = -\ell$ and the location of the load as $z = 0$, so z_{max} will be negative. If the first maximum is located more than one-quarter wavelength from the load, you are free to subtract 2π from this result for θ_Γ to keep it within a convenient range.) Conversely, if we are told the location of the first *minimum* of the standing wave pattern, we can show that

$$\theta_\Gamma = -2\beta z_{min} \pm \pi,$$

through a similar argument. (We will choose the plus or minus sign to find the minimum in the range $-\lambda/2 < z_{min} \leq 0$.)

Example 9.10 Standing Wave Ratio

Consider a lossless transmission line of characteristic impedance $R_0 = 50\,\Omega$, terminated with a load impedance of Z_L. The standing wave ratio for this load is measured to be $S = 1.5$, and the first voltage maximum is located a distance $\lambda/16$ from the load. Find Z_L.

Solution:
We will first find the magnitude of the reflection coefficient of the load Γ_L using Eq. (9.39), which we invert to find

$$|\Gamma_L| = \frac{S-1}{S+1}.$$

This yields $|\Gamma_L| = 0.2$ for $S = 1.5$. Then we use the location of the first maximum of the voltage standing wave pattern to find the phase of the reflection coefficient $\theta_\Gamma = -2\beta z_{max} = -2(2\pi/\lambda)(-\lambda/16) = \pi/4$. So now we have the magnitude and phase of the reflection coefficient, and we can use these to find the load impedance using Eq. (9.37). We invert this equation to solve for the load impedance

$$Z_L = R_0 \frac{1 + \Gamma_L}{1 - \Gamma_L}. \tag{9.41}$$

All that remains is an exercise in manipulation of the complex numbers, with the result that $Z_L = (63 + j18)\,\Omega$. The impedance of an inductive load, such as this load, has a positive phase, which gives the reflection coefficient Γ_L a positive phase, which shifts the location of the first maximum a little to the left of the load.

9.5.3 Input Impedance

We next examine the input impedance Z_{in} of the transmission line of characteristic impedance Z_0 and length ℓ, terminated in a load impedance Z_L. The input impedance of the line is defined as the ratio of the input voltage to the input current of the transmission line, or

$$Z_{in} = \frac{V_{in}}{I_{in}} = \frac{V(-\ell)}{I(-\ell)}. \tag{9.42}$$

(Recall that we defined the beginning of the transmission line, where the source is located, as $z = -\ell$.) This ratio of $V(-\ell)/I(-\ell)$ then depends on the length ℓ of the transmission line. This may seem a strange result, but remember that the voltage amplitude and the current amplitude each vary strongly in magnitude and phase with ℓ (related to the standing wave pattern of each), and since their dependence on ℓ is not the same, the ratio $V(-\ell)/I(-\ell)$ shows strong variation as well, in magnitude and phase.

Starting with Eq. (9.38) for the input voltage $V_{in} \equiv V(-\ell)$, and Eq. (9.40) for the input current $I_{in} \equiv I(-\ell)$, the input impedance Z_{in} is

$$Z_{in} = \frac{V_0^+ \left[e^{\gamma\ell} + \Gamma_L e^{-\gamma\ell} \right]}{\left(V_0^+/Z_0 \right) \left[e^{\gamma\ell} - \Gamma_L e^{-\gamma\ell} \right]}.$$

Using Eq. (9.37) for the reflection coefficient leads to

$$Z_{in} = Z_0 \frac{(Z_L + Z_0) e^{\gamma\ell} + (Z_L - Z_0) e^{-\gamma\ell}}{(Z_L + Z_0) e^{\gamma\ell} - (Z_L - Z_0) e^{-\gamma\ell}}.$$

Rearranging terms makes this

$$Z_{in} = Z_0 \frac{Z_L \left(e^{\gamma\ell} + e^{-\gamma\ell} \right) + Z_0 \left(e^{\gamma\ell} - e^{-\gamma\ell} \right)}{Z_L \left(e^{\gamma\ell} - e^{-\gamma\ell} \right) + Z_0 \left(e^{\gamma\ell} + e^{-\gamma\ell} \right)}.$$

This can be written more compactly using the hyperbolic sine and hyperbolic cosine functions,

$$Z_{in} = \frac{V(-\ell)}{I(-\ell)} = Z_0 \frac{Z_L \cosh(\gamma\ell) + Z_0 \sinh(\gamma\ell)}{Z_L \sinh(\gamma\ell) + Z_0 \cosh(\gamma\ell)}.$$

(See Appendix J for a review and summary of properties of hyperbolic trigonometric functions.)

Finally, we divide through by $\cosh(\gamma\ell)$ to write

$$\boxed{Z_{in} = Z_0 \frac{Z_L + Z_0 \tanh(\gamma\ell)}{Z_0 + Z_L \tanh(\gamma\ell)}.} \tag{9.43}$$

The input impedance for a lossless line can be written as

$$\boxed{Z_{in} = R_0 \frac{Z_L + jR_0 \tan(\beta\ell)}{R_0 + jZ_L \tan(\beta\ell)}.} \tag{9.44}$$

Note that, even for the lossless transmission line, the input impedance Z_{in} is generally a complex quantity, due to the phase variation of $\tilde{V}(z)$ and $\tilde{I}(z)$ along the line.

There are a few cases of Eqs. (9.43) and (9.44) that merit special mention. First, when the transmission line is terminated with a load impedance Z_L equal to Z_0, the reflection coefficient Γ_L is zero, and the voltage and current standing wave patterns along the transmission line are flat – that is, there is no variation in $|\tilde{V}(z)|$ or $|\tilde{I}(z)|$. As we introduced in Section 9.5.1, this condition is called impedance matching. Inserting $Z_L = Z_0$ in Eq. (9.43), we find $Z_{in} = Z_0$, regardless of the length ℓ of the transmission

FIGURE 9.11 The input impedance for a lossless transmission line of length ℓ, when the load is (a) $Z_L = 0$ (short) and (b) $Z_L = \infty$ (open).

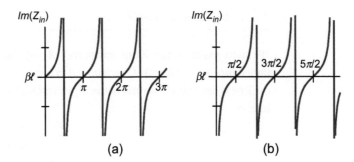

(a)   (b)

line. It is tempting to conclude that Z_{in} is independent of frequency as well, and as long as we limit the bandwidth over which we make such a claim this might be valid. Recall, however, that the load impedance Z_L will in general be a function of frequency as well, and so it is in general difficult to match Z_L to the characteristic impedance Z_0 for all frequencies, so we do need to be careful in this regard. Nonetheless, impedance matching is often desirable, as it will usually lead to broadband performance of the transmission line system.

When we choose a **short-circuit** load, $Z_L = 0$, Eq. (9.43) yields an input impedance of $Z_{in} = Z_0 \tanh(\gamma\ell)$. For the lossless case, this becomes

$$Z_{in} = jR_0 \tan(\beta\ell), \tag{9.45}$$

which is plotted in Fig. 9.11(a). For this lossless case, the impedance is purely imaginary, with its magnitude varying from zero for $\beta\ell = n\pi$, where n is any integer, to infinite for $\beta\ell = (n + 1/2)\pi$. The reason for this variation is again related to the standing wave pattern of the voltage and current. When the length of the transmission line places a node in the voltage standing wave pattern at the input to the line, and an anti-node of the current pattern, the impedance is zero. (A node is a position at which the amplitude is zero; the amplitude is maximal at an anti-node. Since the voltage is zero, but the current is not, the impedance is zero; that is, a short circuit.) Conversely, the input impedance is infinite when the current standing wave pattern has a node at the input to the line. With no current going in, even when an input voltage is applied, the input impedance is infinite. Notice that the reactance of this shorted line alternates between positive (inductive, lagging) and negative (capacitive, leading). Just as with a circuit element, a capacitive impedance implies that the current leads the voltage at the input, while an inductive impedance tells us that the current lags the voltage. Referring back to the discussion of standing wave patterns, you should not be surprised that either case, depending on the length of the line, is possible.

Let's take a little closer look at the shorted ($Z_L = 0$), lossless ($\alpha = 0$, $\gamma = j\beta$) transmission line in the case that the line is short ($\ell \ll \beta^{-1}$), and see if we can't gain some insight into this geometry. Starting with $Z_{in} = jR_0 \tan(\beta\ell)$, applying the approximation $\tan(\beta\ell) \simeq (\beta\ell)$, valid for $\beta\ell \ll 1$ (see Eq. (H.4)), and using Eqs. (9.17) and (9.14), the input impedance of a short, shorted line is

$$Z_{in} \simeq jR_0(\beta\ell) = j\sqrt{\frac{L'}{C'}}\left(\omega\sqrt{L'C'}\ell\right) = j\omega L'\ell.$$

A short, shorted line, of course, is little more than the two parallel conductors of the transmission line, connected to one another at the end, which effectively makes the line a wire loop, whose inductance is $L'\ell$ – that is, the product of the inductance per unit length and the length of the line.

The final special case for Z_{in} is for an **open-circuit** load, $Z_L = \infty$. Examining Eq. (9.43) in the limit as Z_L gets very large, the terms proportional to Z_0 become negligible in comparison, and the input impedance is $Z_{in} = Z_0/\tanh(\gamma\ell)$. For the lossless case, this becomes

$$Z_{in} = \frac{-jR_0}{\tan(\beta\ell)}. \qquad (9.46)$$

This impedance, which is plotted versus $\beta\ell$ in Fig. 9.11(b), is also purely imaginary, as we found for the $Z_L = 0$ case, but in this case the input impedance is infinite for $\beta\ell = m\pi$, and 0 for $\beta\ell = (m + 1/2)\pi$. This is reversed from the $Z_L = 0$ case. An interpretation of this behavior in terms of the placement of the zeros of the voltage and current standing wave pattern is identical.

How does the open line behave when the line is short? For short lines, $\tan(\beta\ell) \simeq \beta\ell$ (see Eq. (H.4)), and

$$Z_{in} \simeq -jR_0\frac{1}{(\beta\ell)} = -j\sqrt{\frac{L'}{C'}}\left(\frac{1}{\omega\sqrt{L'C'}\ell}\right) = \frac{-j}{\omega C'\ell}.$$

The interpretation of this result follows a line similar to that of the shorted line. The short, open-circuited line consists simply of the two parallel conductors of the transmission line. Since the capacitance per unit length is C', the total capacitance is $C'\ell$, and Z_{in} is the impedance of this "capacitor."

Example 9.11 Reflection, Input Impedance

Consider a lossless transmission line of characteristic impedance $Z_0 = 50\,\Omega$ and length ℓ, terminated with a load impedance of $Z_L = 100\,\Omega$. Let the voltage at the load be of amplitude $V_L = 2$ V. Find (a) the reflection coefficient Γ_L, (b) V_{max}, (c) V_{min}, (d) S, (e) I_{max}, and (f) I_{min}. (g) Find Z_{in} for $\ell = \lambda/4$, $3\lambda/8$, and $\lambda/2$.

Solution:

(a) The reflection coefficient Γ_L is, using Eq. (9.37),

$$\Gamma_L = \frac{Z_L - Z_0}{Z_L + Z_0} = \frac{100\,\Omega - 50\,\Omega}{100\,\Omega + 50\,\Omega} = \frac{1}{3}.$$

So the amplitude V_0^- of the reflected voltage wave is one-third that of the incident voltage wave, V_0^+.

(b) Since Γ_L is real and positive, the voltage at the load is a voltage maximum. This follows from $V_L = V_0^+ + V_0^- = V_0^+(1 + \Gamma_L)$. Therefore $V_{max} = V_L = 2$ V.

(c) The voltage minima are located at positions at which the two traveling waves add out of phase with one another. Since $\Gamma_L = 1/3$ is real and positive, the minimum voltage V_{min} is equal to $V_0^+ - V_0^- = V_0^+(1 - \Gamma_L)$. Therefore, we need V_0^+. At the load, $V_L = V_{max} = V_0^+(1 + \Gamma_L) = 2$ V. This can be solved to find $V_0^+ = 1.5$ V. (Also, $V_0^- = \Gamma_L V_0^+ = 0.5$ V.) At the minima, the voltage amplitudes subtract from one another and the voltage amplitude is $V_{min} = |V_0^+| - |V_0^-| = V_0^+(1 - \Gamma_L) = 1.5\ \text{V}\ (2/3) = 1.0$ V.

(d) To find the standing wave ratio $S = V_{max}/V_{min}$, use V_{max} and V_{min} found in (b) and (c) to compute $S = 2.0$.

(e) I_{max} is found in a similar manner to V_{max}. We have already found $V_0^+ = 1.5$ V and $V_0^- = 0.5$ V, so these give us directly $I_0^+ = V_0^+/Z_0 = 30$ mA and $I_0^- = -V_0^-/Z_0 = -10$ mA. The anti-node in the current standing wave pattern occurs when these two current waves combine constructively, giving $I_{max} = |I_0^+| + |I_0^-| = 40$ mA.

(f) The node in the current standing wave pattern occurs when the two traveling wave current amplitudes subtract from one another, giving $I_{min} = |I_0^+| - |I_0^-| = 20$ mA. Notice that if we had wanted to calculate the standing wave ratio S using I_{max} and I_{min}, this would have been equally valid, and we would have obtained the same result as we found in part (d).

(g) This is a lossless line, so Eq. (9.44) can be used to find Z_{in}. We need $\beta\ell$ for each of the three lengths given. Since each length is given in terms of the wavelength λ, we will use $\beta = 2\pi/\lambda$, giving us $\beta\ell = (2\pi/\lambda)(\lambda/4) = \pi/2$, $\beta\ell = (2\pi/\lambda)(3\lambda/8) = 3\pi/4$, and $\beta\ell = (2\pi/\lambda)(\lambda/2) = \pi$. For each of these three cases, then, $\tan(\beta\ell)$ is equal to ∞, -1, or 0, respectively. Using $R_0 = 50\ \Omega$ and $R_L = 100\ \Omega$ in Eq. (9.44), we have

$$Z_{in} = 50\ \Omega \frac{100\ \Omega + j(50\ \Omega)\tan(\beta\ell)}{50\ \Omega + j(100\ \Omega)\tan(\beta\ell)}.$$

As $\tan(\beta\ell) \to \infty$, the other terms in this expression become negligible, and

$$Z_{in} \to 50\ \Omega \frac{j(50\ \Omega)\tan(\beta\ell)}{j(100\ \Omega)\tan(\beta\ell)} = 50\ \Omega \left(\frac{1}{2}\right) = 25\ \Omega.$$

For $\tan(\beta\ell) = -1$, the input impedance is

$$Z_{in} = 50\ \Omega \frac{100\ \Omega + j(50\ \Omega)(-1)}{50\ \Omega + j(100\ \Omega)(-1)} = 50\ \Omega \left(\frac{2-j}{1-2j}\right) = (40 + j30)\ \Omega.$$

At this length, the transmission line is inductive (the imaginary part of the input impedance is positive). Finally, for $\tan(\beta\ell) = 0$, the input impedance is

$$Z_{in} = 50\ \Omega \frac{100\ \Omega + j(50\ \Omega)(0)}{50\ \Omega + j(100\ \Omega)(0)} = 50\ \Omega\ (2) = 100\ \Omega,$$

or more simply, R_L. This result is expected for a lossless transmission line whose length is an integer multiple of a half wavelength.

9.5.4 Amplitudes of Traveling Waves V_0^+ and V_0^-

Many of the results of this section have been expressed in terms of the amplitudes V_0^+ and V_0^- of the individual traveling waves on the line, but we have not yet discussed how to determine these amplitudes for a specific transmission line circuit. In this section we will address this issue, using what we have learned about the input impedance of a loaded transmission line. Our goal is to determine the amplitudes of the individual traveling waves on the line, V_0^+ and V_0^-, in terms of the source voltage and impedance, the length and characteristic impedance of the line, and the load impedance.

Let's consider the loaded transmission line shown in Fig. 9.12(a), and its equivalent network shown in Fig. 9.12(b). The voltage and current at the input to the transmission line are V_{in} and I_{in}, respectively. Starting at the input end of the transmission line, and using simple voltage division to find V_{in} yields

$$V_{in} = V_g \frac{Z_{in}}{Z_g + Z_{in}}. \tag{9.47}$$

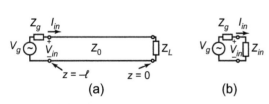

(a)   (b)

FIGURE 9.12 (a) The transmission line of length ℓ, and (b) its equivalent network.

It might seem overly simplistic to use voltage division here. After all, we have gone to lengths to show that the transmission line is quite different from a circuit element. If we think just a little about the meaning of the input impedance Z_{in} of the transmission line, however, we quickly conclude that this is the correct approach. The input current I_{in} passes through the source impedance Z_g, so the voltage across this element is $I_{in}Z_g$. Since

$$V_{in} = V_g - I_{in}Z_g$$

(by Kirchhoff's Voltage Law), and

$$V_{in} = I_{in}Z_{in},$$

then

$$I_{in} = \frac{V_g}{Z_g + Z_{in}}$$

and

$$V_{in} = \frac{V_g Z_{in}}{Z_g + Z_{in}},$$

as we asserted above.

Let's remind ourselves what this voltage V_{in} represents. This is the total voltage at the input end of the transmission line. But we showed earlier that at any position z, the total wave can be written as the sum of the two traveling waves,

$$V(z) = V_0^+ e^{-\gamma z} + V_0^- e^{\gamma z},$$

where V_0^+ and V_0^- are the amplitudes of the waves traveling in the $+z$- and $-z$-directions, respectively. Applied to the input of the transmission line at $z = -\ell$, this gives

$$V_{in} = V(-\ell) = V_0^+ e^{\gamma\ell} + V_0^- e^{-\gamma\ell} = V_0^+ e^{\gamma\ell}\left(1 + \Gamma_L e^{-2\gamma\ell}\right), \tag{9.48}$$

where the last equality results from $V_0^- = \Gamma_L V_0^+$. That is, the wave traveling in the $-z$-direction is the reflection of the $+z$-traveling wave. Equating V_{in} in Eqs. (9.47) and (9.48), it can be shown that the amplitude of the $+z$-traveling wave on this loaded line is

$$V_0^+ = V_g \left(\frac{Z_{in}}{Z_g + Z_{in}}\right)\left(\frac{e^{-\gamma\ell}}{1 + \Gamma_L e^{-2\gamma\ell}}\right). \tag{9.49}$$

The amplitude of the $-z$-traveling wave follows from this, simply multiplying by the reflection coefficient Γ_L. We illustrate this in the following example, in which we also examine the power delivered to the load.

Example 9.12 Wave Amplitudes, Power

Consider a lossless transmission line, such as shown in Fig. 9.12(a), of characteristic impedance $Z_0 = R_0 = 50\ \Omega$ and length $\ell = 3\lambda/4$, terminated with a load impedance of $Z_L = 100\ \Omega$. This line is driven with a sinusoidal source consisting of an ideal source of amplitude 16 V and an internal series resistor $R_g = 15\ \Omega$. Determine the amplitudes of the traveling waves V_0^+ and V_0^-, and the average power delivered to the load.

Solution:
We'll start by finding the voltage V_{in} at the input end of the line. To find V_{in}, we need the input impedance Z_{in} for this line. For $\ell = 3\lambda/4$,

$$\beta\ell = \left(\frac{2\pi}{\lambda}\right)\left(\frac{3\lambda}{4}\right) = \frac{3\pi}{2},$$

and $\tan(\beta\ell)$ becomes infinite. The input impedance is

$$Z_{in} = R_0\frac{Z_L + jR_0\tan(\beta\ell)}{R_0 + jZ_L\tan(\beta\ell)} = \frac{R_0^2}{Z_L} = 25\ \Omega.$$

Z_{in} can be used to find V_{in} using simple voltage division,

$$V_{in} = V_g\frac{Z_{in}}{Z_g + Z_{in}} = 16\ \text{V}\frac{25\ \Omega}{15\ \Omega + 25\ \Omega} = 10\ \text{V}.$$

The input voltage V_{in} (at $z = -\ell$) depends on the amplitudes V_0^+ and V_0^- through

$$V_{in} = 10\ \text{V} = V(-\ell) = V_0^+e^{j\beta\ell} + V_0^-e^{-j\beta\ell} = -jV_0^+ + jV_0^-,$$

where we used $e^{\pm j\beta\ell} = e^{\pm j3\pi/2} = \mp j$. This expression for V_{in} is one condition that must be satisfied by V_0^+ and V_0^-. The other condition that V_0^+ and V_0^- must satisfy comes from the reflection coefficient at the load, Γ_L. Since R_0 and Z_L in this example are identical to those used in Example 9.11, we will borrow $\Gamma_L = 1/3$ from that example. Solving simultaneous equations, we find

$$V_0^+ = j15\ \text{V}\qquad\text{and}\qquad V_0^- = j5\ \text{V}.$$

(To check our arithmetic, we can verify that these amplitudes do indeed give us the correct input voltage V_{in} and the correct reflection coefficient Γ_L.) Alternatively, we could have applied Eq. (9.49) directly to find

$$V_0^+ = 16\ \text{V}\left(\frac{25\ \Omega}{15\ \Omega + 25\ \Omega}\right)\left(\frac{+j}{1 + (1/3)(j)^2}\right) = j15\ \text{V},$$

and

$$V_0^- = V_0^+\Gamma_L = j5\ \text{V}.$$

We've now done all the hard work. From here, we can find nearly any other relevant quantity for this loaded transmission line. For example, the load voltage is

$$V_L = V_0^+ + V_0^- = j20\ \text{V},$$

and the load current is

$$I_L = \frac{1}{Z_0}\left(V_0^+ - V_0^-\right) = j0.2\ \text{A}.$$

Of course, the ratio V_L/I_L must be Z_L, which was given as $100\ \Omega$, so this is an important consistency check. The input current (at $z = -\ell$) is

$$I_{in} = I(-\ell) = \frac{1}{Z_0}\left(V_0^+e^{j\beta\ell} - V_0^-e^{-j\beta\ell}\right)$$

$$= \frac{1}{50\Omega}\left[j15\ \text{V}(-j) - j5\ \text{V}(+j)\right] = \frac{15\ \text{V} + 5\ \text{V}}{50\ \Omega} = 0.4\ \text{A}.$$

This current I_{in} can be used for another check on the consistency of our solution, in that V_{in}/I_{in} is $25\ \Omega$, the same result as the input impedance for our transmission line.

Finally, we were asked to find the power delivered to the load. We have already determined everything we need to find this, and need only compute

$$P_L = \frac{1}{2}|I_L|^2R_L = \frac{1}{2}|j0.2A|^2(100\ \Omega) = 2\ \text{W}.$$

We have determined the amplitudes of the $+z$- and $-z$-traveling waves, which in general depend on quite a few parameters, including the length and characteristic impedance of the line, the load impedance, and the amplitude and impedance of the source.

9.6 Impedance Matching

As we have already pointed out, it is often desirable to match the impedance of a load to a transmission line so as to minimize reflections. By matching the impedance, of course, the power delivered to the load is maximized, and the dependence of the signal on frequency or on the length ℓ of the transmission line is reduced. But we don't always have the option of adjusting the load impedance itself. In this section, we will review a few techniques that have been developed to reduce reflections at a load.

9.6.1 Quarter-Wave Transformer

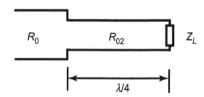

FIGURE 9.13 Impedance matching a load to a lossless transmission line using a quarter-wave transformer.

A short segment of transmission line known as a **quarter-wave transformer** can be inserted in a line to minimize reflections from a load. For example, imagine that a load of impedance Z_L is to be driven via a transmission line of characteristic impedance Z_0. For simplicity, we will treat the line as lossless. If we were to connect this load directly to the transmission line, then the amplitude of the wave reflected by the load is Γ_L times the incident wave, where Γ_L is given by Eq. (9.37). Instead, if an additional transmission line of characteristic impedance R_{02} (taken as lossless here) and length $\lambda/4$ is inserted between the transmission line and the load, then the reflection can be significantly reduced. This configuration is illustrated in Fig. 9.13.

We start by determining the input impedance of the loaded quarter-wave line, using Eq. (9.44). For $\ell = \lambda/4$, $\beta\ell$ is $(2\pi/\lambda)(\lambda/4) = \pi/2$. Since $\tan(\beta\ell)$ goes to infinity as $\beta\ell \to \pi/2$, let's just let $\tan(\beta\ell)$ be large, and take the limit of Z_{in} as $\beta\ell \to \pi/2$. In this limit, clearly the tangent terms in Eq. (9.44) dominate, and the input impedance is

$$Z_{in} = \lim_{\tan(\beta\ell)\to\infty} R_{02}\frac{Z_L + jR_{02}\tan(\beta\ell)}{R_{02} + jZ_L\tan(\beta\ell)} = \frac{R_{02}^2}{Z_L}.$$

Now consider this quarter-wave line as the load on the long transmission line of characteristic impedance R_0. To eliminate the reflected wave from this load, R_{02} should be chosen to make $Z_{in} = R_0$. (R_{02} can be adjusted, for example, with microstrip by adjusting the width of the trace on a printed circuit board.) This value of R_{02} is

$$R_{02} = \sqrt{R_0 Z_L}. \tag{9.50}$$

How can we understand this effect? The wave incident from the source is partially transmitted at the junction between the two transmission lines, and partially reflected. When the transmitted wave is incident on the load Z_L, that wave is partially reflected as well. So the wave traveling back toward the source consists of primarily two components, the wave that was reflected by the junction between the transmission lines and the wave that was reflected by the load. By choosing R_{02} to be the geometric mean of R_0 and Z_L, the reflection coefficients from these two points are the same. This is seen by computing Γ_2, the reflection coefficient for the wave incident from the left on the junction between the two transmission lines:

$$\Gamma_2 = \frac{R_{02} - R_0}{R_{02} + R_0} = \frac{\sqrt{R_0 Z_L} - R_0}{\sqrt{R_0 Z_L} + R_0} = \frac{\sqrt{Z_L} - \sqrt{R_0}}{\sqrt{Z_L} + \sqrt{R_0}}.$$

Similarly, the reflection coefficient at the load is

$$\Gamma_L = \frac{Z_L - R_{02}}{Z_L + R_{02}} = \frac{Z_L - \sqrt{R_0 Z_L}}{Z_L + \sqrt{R_0 Z_L}} = \frac{\sqrt{Z_L} - \sqrt{R_0}}{\sqrt{Z_L} + \sqrt{R_0}},$$

which is identical to Γ_2 in magnitude and sign. So the two primary reflected waves are the same in amplitude, and since the length of the inserted line is $\lambda/4$, these two waves are π out of phase. (The wave reflected from the load picks up an additional phase of $2\beta\ell = \pi$ relative to the phase of the wave reflected from the junction between the transmission lines.) Therefore these two waves add out of phase with each other, and they cancel one another. In this way, the total amplitude of the reflected wave is canceled. (This explanation is not complete, in that it ignores multiple reflections between the surfaces. Still, it gives us a good physical picture for the quarter-wave stub.)

Example 9.13 Quarter-Wave Transformer

Find the (a) length and (b) characteristic impedance of a quarter-wave transformer designed to impedance-match a 100 Ω load to a lossless 50 Ω transmission line at a frequency of 100 MHz. The dielectric material of the transmission line is non-magnetic and has a relative permittivity of 2.25. (c) Find the reflection coefficient of this transformer for a wave at a frequency of 200 MHz. Compare this to the reflection coefficient of the load without the quarter-wave transformer.

Solution:

(a) We start by finding the wavelength of the wave at the frequency of $f = 100$ MHz. We'll use $\lambda = u_p/f$, but we must first find the phase velocity u_p:

$$u_p = \frac{1}{\sqrt{\varepsilon\mu}} = \frac{1}{\sqrt{2.25\varepsilon_0\mu_0}} = \frac{c}{\sqrt{2.25}} = 2.0 \times 10^8 \text{ m/s}.$$

Then the wavelength is

$$\lambda = \frac{u_p}{f} = \frac{2.0 \times 10^8 \text{ m/s}}{1.0 \times 10^8 \text{ s}^{-1}} = 2.0 \text{ m},$$

and the length of the quarter-wave transformer should be

$$\ell = \frac{\lambda}{4} = 0.5 \text{ m}.$$

(b) To find the characteristic impedance of the transformer, we use Eq. (9.50),

$$R_{02} = \sqrt{R_0 Z_L} = \sqrt{(50 \ \Omega)(100 \ \Omega)} = 70.7 \ \Omega.$$

(c) At $f = 200$ MHz, the wavelength will be half as much, or $\lambda_2 = 1.0$ m. Thus, $\beta\ell$ is

$$\beta\ell = \left(\frac{2\pi}{1 \text{ m}}\right)(0.5 \text{ m}) = \pi,$$

and the input impedance to the transformer is

$$Z_{in} = R_{02} \frac{Z_L + jR_{02}\tan(\beta\ell)}{R_{02} + jZ_L\tan(\beta\ell)} = Z_L.$$

The reflection coefficient is therefore

$$\Gamma = \frac{Z_{in} - R_0}{Z_{in} + R_0} = \frac{50 \ \Omega}{150 \ \Omega} = \frac{1}{3}.$$

This reflection coefficient is exactly the same as the reflection coefficient of the load without the transformer.

This example serves to illustrate that the performance of the transformer does depend upon the frequency of the wave. At the frequency of 200 MHz, the reflection from the

load and the reflection from the junction between the lines of characteristic impedance R_0 and R_{02} add in phase with one another, so no reduction in the net reflection is achieved.

9.6.2 Parallel Shorted or Open Stub Lines

Another common impedance-matching technique is based on parallel **stub lines**, either shorted or open at the end. These are illustrated in Fig. 9.14. Here is how a stub line works. When a wave is incident from the left, it reaches the junction between the loaded transmission line, terminated with the impedance element Z_L, and the stub, which is represented in the figure with the line that is directed down and to the right. The stub in Fig. 9.14(a) is shorted at the end, while in (b) it is open. The characteristic impedance of each of the lines is labeled as the same value, Z_0, in this figure, but these could in fact be different. We recognize these two lines as being a parallel connection to the main transmission line, in that they share the same voltage at the input, and the current incident from the left is divided between the two lines. When two elements are added in parallel, the admittance $(Y_{in} = Z_{in}^{-1})$ of the combination is the sum of the admittances of the individuals. Also, the input impedance, and therefore the input admittance, of a shorted or open stub is purely imaginary. Refer back to Eqs. (9.45) or (9.46). So the scheme for impedance matching with a stub is to adjust the length of the loaded line ℓ_L and the length of ℓ_{St} to make the input impedance Z_{in} of the parallel combination of these two lines equal to Z_0, the characteristic impedance of the line. As we will see in the following, we will adjust R_{in} using the length ℓ_L, and then cancel the reactive part X_{in} using the length of the stub. When Z_{in} is equal to Z_0, the amplitude of the reflected wave is zero. This technique is demonstrated in Example 9.14.

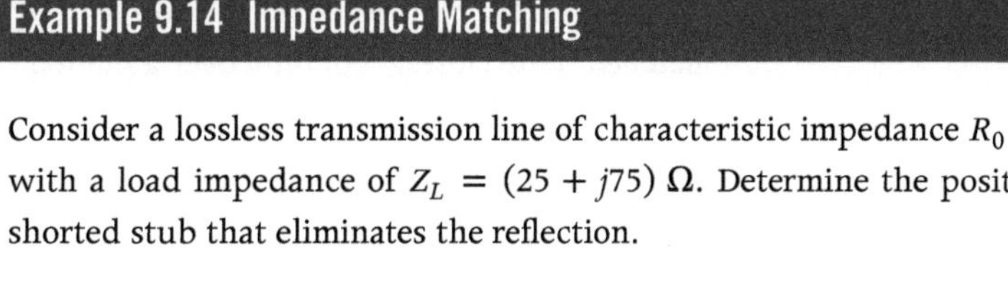

FIGURE 9.14 Impedance matching a load on a lossless transmission line by adding a parallel (a) shorted or (b) open stub.

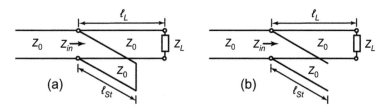

Example 9.14 Impedance Matching

Consider a lossless transmission line of characteristic impedance $R_0 = 50\,\Omega$, terminated with a load impedance of $Z_L = (25 + j75)\,\Omega$. Determine the position and length of a shorted stub that eliminates the reflection.

Solution:
See Fig. 9.14(a) for an illustration of this configuration. The shorted stub is positioned a distance ℓ_L from the load, and the length of the stub is ℓ_{St}. We start by inverting Eq. (9.44) to write the input admittance as

$$Y_{in} = Z_{in}^{-1} = Y_0 \frac{R_0 + jZ_L \tan(\beta\ell_L)}{Z_L + jR_0 \tan(\beta\ell_L)} + Y_{St},$$

where $Y_0 = Z_0^{-1}$. The first term here is the admittance of the loaded line, and the second is the admittance of the shorted stub,

$$Y_{St} = Z_{St}^{-1} = -jY_0 \cot(\beta\ell_{St}).$$

For the characteristic impedance and load impedance given, this gives

$$Y_{in} = Y_0 \frac{50\,\Omega + j(25 + j75)\,\Omega\tan(\beta\ell_L)}{(25 + j75)\,\Omega + j50\,\Omega\tan(\beta\ell_L)} + Y_{St}.$$

This is simplified a bit by factoring out 25 Ω,

$$Y_{in} = Y_0 \frac{2 + j(1 + j3)\tan(\beta\ell_L)}{(1 + j3) + j2\tan(\beta\ell_L)} + Y_{St},$$

and rationalizing

$$Y_{in} = Y_0 \frac{(1 - j3) + 3j\tan(\beta\ell_L) + (1 + j3)\tan^2(\beta\ell_L)}{5 + 6\tan(\beta\ell_L) + 2\tan^2(\beta\ell_L)} + Y_{St}.$$

To achieve our goal of matching the impedance, then, we must make the real part of the right-hand side of this expression be equal to Y_0, and the imaginary part equal to zero. The first condition is satisfied when

$$Y_0 = Y_0 \frac{1 + \tan^2(\beta\ell_L)}{5 + 6\tan(\beta\ell_L) + 2\tan^2(\beta\ell_L)}.$$

(Remember that Y_0 is purely real for a lossless line, and Y_{St} is purely imaginary.) The value of $\tan(\beta\ell_L)$ that satisfies this equation is −5.24, which we found using the quadratic formula. Using $\beta = 2\pi/\lambda$, the value of $\beta\ell_L$ is 0.560π. The length of this line is $\ell_L = 0.280\lambda$.

Now we examine $\Im(Y_{in})$, which must be zero to achieve impedance matching. This leads to

$$0 = Y_0 \frac{-3 + 3\tan(\beta\ell_L) + 3\tan^2(\beta\ell_L)}{5 + 6\tan(\beta\ell_L) + 2\tan^2(\beta\ell_L)} - Y_0\cot(\beta\ell_{St}).$$

Since we have already solved for $\tan(\beta\ell_L)$, we can use this value to determine $\cot(\beta\ell_{St}) = 2.24$, resulting in $\beta\ell_{St} = 0.131\pi$, or $\ell_{St} = 0.066\lambda$.

In this section we have introduced two techniques for matching the impedance of a load to the characteristic impedance of the transmission line. Impedance matching is often desirable, since this condition minimizes reflections, maximizes the power delivered to the load, and reduces the dependence of the signal on the frequency of the wave or the length ℓ of the transmission line. In each of the techniques discussed in this section (the quarter-wave transformer and the parallel stub), the impedance can be matched quite closely, but only over a limited range of frequencies.

9.7 The Smith Chart

In this section we will describe a graphical technique for determining the reflection coefficient of a load, the input impedance or admittance of transmission line, and the proper location and length of a shorted or open stub to impedance-match a load. This graphical technique is based upon the **Smith Chart**, introduced independently by Mizuhashi Tosaku (in 1937) and Phillip H. Smith (in 1939), and shown in Fig. 9.15. The Smith Chart lets us plot the "translated" reflection coefficient $\Gamma(z)$, as seen at different locations z along the transmission line, in a complex plane, and consists of a series of non-concentric circles.

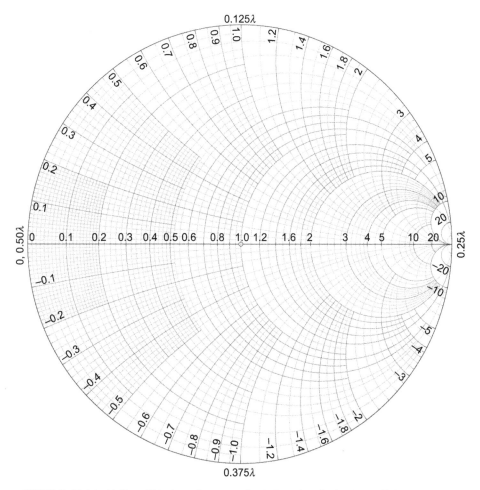

FIGURE 9.15 A Smith Chart. This chart allows us to plot the "translated" reflection coefficient on a complex plane, and consists of series of circles, each of which is a contour of constant resistance or reactance.

To understand the Smith Chart and how to use it, we first present a simple (but perhaps tedious) derivation of the different series of circles shown in Fig. 9.15. These circles are contours of constant real and imaginary parts of the transmission line impedance $Z(z)$. This impedance is a generalization of the input impedance that we discussed in Section 9.5.3. The input impedance of a lossless line of characteristic impedance R_0, length ℓ, and load impedance Z_L is given by Eq. (9.44),

$$Z_{in} = R_0 \frac{Z_L + jR_0 \tan(\beta\ell)}{R_0 + jZ_L \tan(\beta\ell)}.$$

Let's generalize this to define the impedance at any position z along the transmission line in terms of the voltage $\tilde{V}(z)$ and current $\tilde{I}(z)$,

$$Z(z) = \frac{\tilde{V}(z)}{\tilde{I}(z)} = R_0 \frac{Z_L + jR_0 \tan(-\beta z)}{R_0 + jZ_L \tan(-\beta z)}.$$

Using

$$\tan(-\beta z) = \frac{\sin(-\beta z)}{\cos(-\beta z)} = -j\frac{e^{-j\beta z} - e^{j\beta z}}{e^{-j\beta z} + e^{j\beta z}}$$

and the reflection coefficient from Eq. (9.37) with Z_0 replaced by R_0 (since we are dealing with a lossless line),

$$\Gamma_L = \frac{Z_L - R_0}{Z_L + R_0},$$

the impedance can be written

$$Z(z) = R_0 \frac{1 + \Gamma_L e^{2j\beta z}}{1 - \Gamma_L e^{2j\beta z}}. \tag{9.51}$$

Defining the **normalized input impedance** $z(z)$ as $Z(z)/R_0$ leads to

$$z(z) = \frac{Z(z)}{R_0} = \frac{1 + \Gamma_L e^{2j\beta z}}{1 - \Gamma_L e^{2j\beta z}}.$$

(Note an unfortunate double usage of z here. z represents the coordinate along the axis of the transmission line, as it has throughout this chapter. Do not confuse the coordinate z with the normalized impedance, also denoted by z. The proper meaning should be evident from the context, but be aware of this possible ambiguity in advance.) The term $\Gamma_L e^{2j\beta z}$ in the numerator and denominator of this expression is simply the reflection coefficient of the load translated a distance z from the load. At the input, $z = -\ell$, this becomes

$$z_{in} = \frac{1 + \Gamma_L e^{-2j\beta\ell}}{1 - \Gamma_L e^{-2j\beta\ell}}. \tag{9.52}$$

To streamline the notation, let's rename the translated reflection coefficient w, such that

$$w(z) = \Gamma_L e^{2j\beta z}, \tag{9.53}$$

and assign $u(z)$ and $v(z)$ as the real and imaginary parts of $w(z)$,

$$w(z) = u(z) + jv(z).$$

So the normalized line impedance is

$$z(z) = \frac{1 + w}{1 - w} = \frac{(1 + u) + jv}{(1 - u) - jv}.$$

After rationalizing this complex fraction and setting the real and imaginary parts of $z(z)$ equal to the input resistance $r(z)$ (the real part of $z(z)$) and the input reactance $x(z)$ (the imaginary part of $z(z)$), we arrive at

$$r(z) = \frac{1 - \left(u^2 + v^2\right)}{(1 - u)^2 + v^2} \tag{9.54}$$

and

$$x(z) = \frac{2v}{(1 - u)^2 + v^2}. \tag{9.55}$$

The Smith Chart, as shown in Fig. 9.15, is a plot in the complex (u, v) plane of contours of constant $r(z)$ and contours of constant $x(z)$. The u-axis (real part of w) is horizontal and the v-axis (imaginary part of w) is vertical, with the origin at the center of the chart. These axes are not shown explicitly on the plot. Two series of circles are evident on the Smith Chart. The circles that are centered on the u-axis ($v = 0$) represent lines of constant $r(z)$, as can be shown by solving Eq. (9.54). This solution yields a series of circles of radius $(1 + r(z))^{-1}$ centered at $(u, v) = (r(z)/(1 + r(z)), 0)$. A second series of circles on the Smith Chart is centered on a vertical line $u = 1$ at the right side. These contours are lines of constant $x(z)$, which can be shown by solving Eq. (9.55). These circles are of radius

$|x(z)|^{-1}$ centered at $(u, v) = (1, x(z)^{-1})$. Given these contours, any load impedance Z_L can be located on the Smith Chart by normalizing the load impedance (i.e. $z_L = Z_L/R_0$), finding the circular contour for the real part r_L, the contour for the imaginary part x_L, and following these contours to their intersection. This point on the Smith Chart is the reflection coefficient at the load $w_L = u_L + jv_L$. Then translation along the transmission line is suggested by inspection of Eq. (9.53), the reflection coefficient translated to z. In the complex plane, $w(z)$ is a circle centered on the origin $(u = 0, v = 0)$. Translation from the load (at $z = 0$) to the source (at $z = -\ell$) is in the $-z$-direction, which on the Smith Chart is in the clockwise direction. We follow this circular path to the load by rotation about the origin (the center of the Smith Chart, $(u, v) = (0, 0)$, or $(r, x) = (1, 0)$). Note that, since the argument in the exponential of Eq. (9.53) is $2j\beta z$, one rotation about the Smith Chart is a translation of $\lambda/2$. After rotating to $z = -\ell$, we read r_{in} and x_{in} from the contours.

Points of special significance on the Smith Chart are:

1. For an impedance-matched load, $Z_L = R_0$, and $\Gamma_L = 0$. This point is at the origin, $u = 0$ and $v = 0$ of the Smith Chart. Rotation from this point about the origin causes no change, as expected, since $Z_{in} = R_0$ for an impedance-matched transmission line, independent of the transmission line length.
2. For a shorted transmission line, $Z_L = 0$ and $\Gamma_L = -1$. This point on the Smith Chart is $u = -1$ and $v = 0$, on the left side.
3. For an open transmission line, $Z_L = \infty$ and $\Gamma_L = +1$. This point on the Smith Chart is $u = +1$ and $v = 0$, on the right side.
4. For an inductive load, $Z_L = j\omega L$ and $|\Gamma_L| = +1$. This point on the Smith Chart is $|w| = 1$ and $v > 0$. This is the largest radius circle, on the top side of the Smith Chart.
5. For a capacitive load, $Z_L = 1/j\omega C$ and $|\Gamma_L| = +1$. This point on the Smith Chart is $|w| = 1$ and $v < 0$. This is the largest radius circle on the bottom side of the Smith Chart.

Now let's apply the Smith Chart to solutions of a few transmission line problems. In our first example, we use the Smith Chart to find the reflection coefficient of a load graphically.

Example 9.15 Reflection Coefficient

A lossless transmission line of characteristic impedance $R_0 = 50\,\Omega$ is terminated with a load impedance $Z_L = (25 + j75)\,\Omega$. Use the Smith Chart to determine the magnitude and phase of the reflection coefficient.

Solution:
We find the normalized load impedance as

$$z_L = \frac{Z_L}{R_0} = 0.5 + j1.5.$$

On the Smith Chart of Fig. 9.16, this normalized impedance is located at the intersection of the $r = 0.5$ circle (red dashed) and the $x = 1.5$ circle (blue dashed). To read the magnitude and phase of the reflection coefficient, draw a straight line from the origin through the impedance point $(r, x) = (0.5, 1.5)$ (the solid black line). The length of this line, normalized to the radius of the Smith Chart, which is $|\Gamma| = 1$, gives us $|\Gamma_L| = 0.74$. The phase angle of the reflection coefficient, $\theta_\Gamma = 63.5°$, is measured as the angle between the real axis and this line, as indicated in Fig. 9.16.

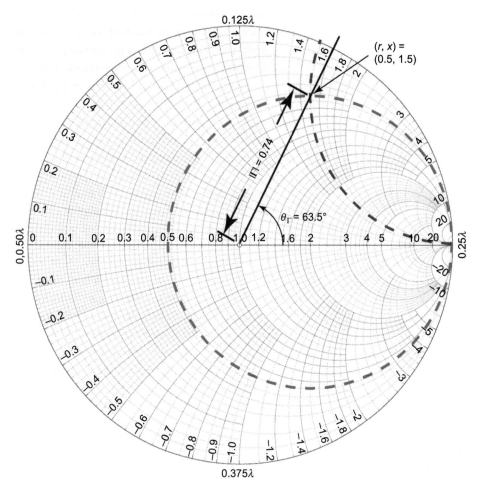

FIGURE 9.16 A Smith Chart showing the reflection coefficient for the load of Example 9.15. The red dashed line is $r = 0.5$, while the blue dashed line is $x = 1.5$.

To compare the precision of the result and the amount of effort required, we can compute this reflection coefficient using Eq. (9.37),

$$\Gamma_L = \frac{Z_L - R_0}{Z_L + R_0} = \frac{1 + 2j}{3}. \tag{9.56}$$

The magnitude and phase of Γ_L are $|\Gamma_L| = \sqrt{5}/3 = 0.745$ and $\theta_\Gamma = \tan^{-1}\left(\frac{2}{1}\right) = 63.4°$, in good agreement with the graphical results.

In Example 9.16 the Smith Chart is used to determine the input impedance of a transmission line.

Example 9.16 Input Impedance

Let the length of the loaded, lossless transmission line of Example 9.15 be 2 m, and the wave velocity on this line be $c = 3.0 \times 10^8$ m/s. The frequency of the source driving the line is $f = 200$ MHz. Use the Smith Chart to determine the impedance Z_{in} at the input of this line.

Solution:

In Example 9.15 we found the point on the Smith Chart corresponding to $(r_L, x_L) = (0.5, 1.5)$. We use that result, plotted in Fig. 9.17, as the starting point for this example. We also need the length of the transmission line in terms of the wavelength λ. The wavelength is

$$\lambda = \frac{c}{f} = \frac{3.0 \times 10^8 \ \text{m/s}}{2.0 \times 10^8 \ \text{s}^{-1}} = 1.5 \ \text{m},$$

and therefore the length of the line is

$$\ell = \frac{\ell}{\lambda} \times \lambda = \frac{2 \ \text{m}}{1.5 \ \text{m}} \times \lambda = \frac{4\lambda}{3}.$$

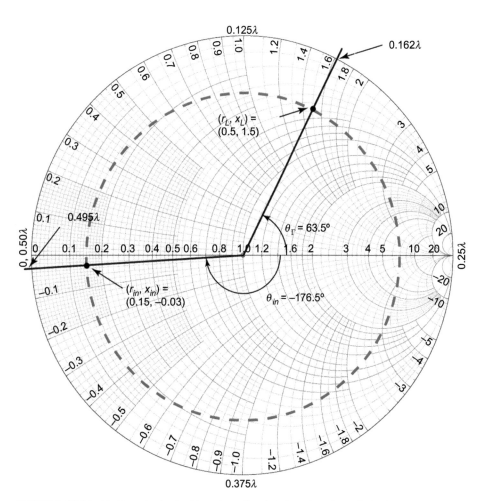

FIGURE 9.17 A Smith Chart showing load impedance Z_L and the input impedance Z_{in} for the transmission line of Example 9.16. The red dashed line is a circle centered on the origin that passes through the input impedance point $(r, x) = (0.5, 1.5)$.

Translation along the transmission line from the load to the source is affected by rotating clockwise on the red dashed curve, which is a circle centered on the origin and passing through the initial point (r_L, x_L). In Example 9.15 we found the phase $\theta_\Gamma = 63.5°$. Since one rotation about the origin of the Smith Chart corresponds to a $\lambda/2$ translation along the transmission line, a $4\lambda/3$ translation is affected by a rotation of 8/3 cycles. An integer number of rotations has no effect, of course, and 2/3 of a cycle is 240°. Starting at $\theta_\Gamma = 63.5°$ and rotating 240° in the clockwise direction leads to $\theta_{in} = -176.5°$, shown in Fig. 9.17. The intersection between the red dashed line and the straight line at θ_{in} tells us $(r_{in}, x_{in}) = (0.15, -0.03)$. We recover the input impedance from the normalized input impedance using

$$Z_{in} = R_0 z_{in} = 50\ \Omega\ (0.15 - j0.03) = (7.5 - j1.5)\ \Omega.$$

To contrast this graphical technique with the direct calculation, we start with the reflection coefficient at the load, calculated in Eq. (9.56). Then we find the translated reflection coefficient

$$\Gamma_L e^{-2j\beta\ell} = \left(\frac{1+2j}{3}\right) e^{-2j[2\pi(4/3)]}.$$

After a few lines of complex arithmetic, this reduces to

$$\Gamma_L e^{-2j\beta\ell} = \frac{\sqrt{5}}{3} e^{-j0.98\pi}.$$

Then we insert this into Eq. (9.52) to find

$$z_{in} = \frac{1 + \Gamma_L e^{-2j\beta\ell}}{1 - \Gamma_L e^{-2j\beta\ell}} = \frac{1 + \frac{\sqrt{5}}{3} e^{-j0.98\pi}}{1 - \frac{\sqrt{5}}{3} e^{-j0.98\pi}},$$

which, after a few more lines of complex arithmetic, becomes

$$z_{in} = 0.15 - j0.03.$$

Multiplication by $R_0 = 50\ \Omega$ gives us $Z_{in} = (7.5 - j1.5)\ \Omega$, which agrees with the result using the Smith Chart.

We can also use the Smith Chart to choose the length of a transmission line to match specific requirements, as illustrated in Example 9.17.

Example 9.17 Input Impedance

Use the Smith Chart to find the length ℓ of the loaded, lossless transmission line of Example 9.15 that results in an input impedance that is real and greater than 50 Ω. Determine this input impedance.

Solution:
We borrow the result of Example 9.15 to locate the load impedance $(r_L, x_L) = (0.5, 1.5)$ on the Smith Chart. The red dashed circle centered on the origin of Fig. 9.18 includes all points on the complex w plane accessible by translation along the transmission line. We must find the intersection of this circle with the positive u-axis. (On this axis, the impedance is real since $v = 0$, and $r > 1$ corresponding to $R_{in} > R_0$.) By inspection we see that this intersection point is $(r_{in}, x_{in}) = (7.0, 0.0)$. The phase θ_{in} is zero at this point, so the phase difference $\theta_\Gamma - \theta_{in} = 63.5°$, and the length of the line must be

$$\ell = 63.5° \times \frac{\lambda/2}{360°} = 0.088\lambda.$$

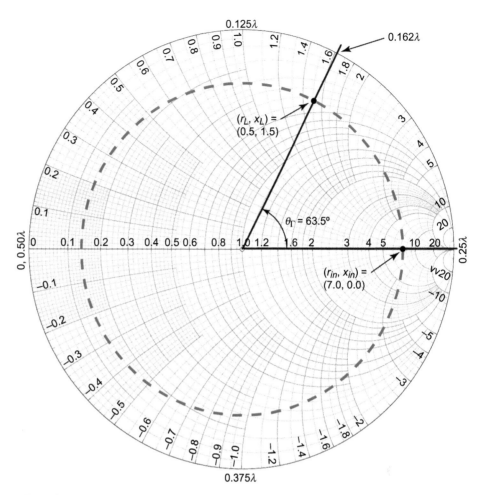

FIGURE 9.18 A Smith Chart showing load impedance Z_L and the input impedance Z_{in} for the transmission line of Example 9.17. The red dashed line is a circle centered on the origin that passes through the input impedance point $(r_{in}, x_{in}) = (0.5, 1.5)$.

This is the shortest cable length that would achieve the specified requirement. Other lengths that would achieve the same goal are $0.088\lambda + 0.5\lambda$, $0.088\lambda + 1.0\lambda$, $0.088\lambda + 1.5\lambda$, etc., although these longer transmission lines would be more sensitive to changes in the frequency of the source. The input impedance of this transmission line is $Z_{in} = R_0 z_{in} = 350\ \Omega$.

To compute this result analytically, examine Eq. (9.52) for the normalized input impedance,

$$z_{in} = \frac{1 + \Gamma_L e^{-2j\beta\ell}}{1 - \Gamma_L e^{-2j\beta\ell}}.$$

This impedance will be real only when the phase $-2\beta\ell$ is equal in magnitude but opposite in sign to the phase of Γ_L, $63.5° = 0.353\pi$ radians. The phase and line length are related by

$$2\beta\ell = 2\left(\frac{2\pi}{\lambda}\right)\ell = 0.353\pi \quad \Rightarrow \quad \ell = \frac{0.353}{4}\lambda = 0.088\lambda.$$

At this location the translated reflection coefficient $\Gamma_L e^{-2j\beta\ell}$ is real, and

$$z_{in} = \frac{1 + |\Gamma_L|}{1 - |\Gamma_L|}.$$

We have previously determined $|\Gamma_L| = \sqrt{5}/3$, so

$$z_{in} = \frac{1 + \sqrt{5}/3}{1 - \sqrt{5}/3} = 6.89,$$

and $Z_{in} = R_0 z_{in} = 345 \ \Omega$, in good agreement with the graphical result.

A Smith Chart can also be used with the admittance of transmission lines. We demonstrate this in Example 9.18.

Example 9.18 Impedance Matching with a Stub

Use the Smith Chart to find the length ℓ_{St} and position ℓ_L of a shorted stub that will impedance-match the loaded, lossless transmission line of Example 9.15. See the transmission line geometry of Fig. 9.14(a). Compare the results with those of Example 9.14.

Solution:
Since the shorted stub and the loaded line segment (i.e., the portion of the transmission line between the junction with the shorted stub and the load) are connected in parallel, it is convenient to work with admittances of the stub and line, rather than impedances. The admittance is, of course, simply the inverse of the impedance,

$$Y_0 = Z_0^{-1}$$

and

$$Y_{in} = Z_{in}^{-1}.$$

The input admittance of the parallel combination Y_{in} is the sum of the individual admittances

$$Y_{in} = Y_{St} + Y_{line},$$

where Y_{St} and Y_{line} are the input admittances of the stub and the loaded line segment, respectively. The former is

$$Y_{St} = Z_{St}^{-1} = -jY_0 \cot(\beta \ell_{St}),$$

where we used Eq. (9.45) for the input impedance of the shorted stub. The normalized input admittance of the stub,

$$y_{St} = \frac{Y_{St}}{Y_0} = -j \cot(\beta \ell_{St}),$$

is purely negative imaginary for $\beta \ell_{St} < \pi/2$. The normalized input admittance of the loaded line segment is, in general, complex, and can be written in terms of the conductance g_{line} and susceptance b_{line}:

$$y_{line} = g_{line} + jb_{line}.$$

We'll find y_{line} graphically using the Smith Chart. We start with the load admittance:

$$y_L = \frac{R_0}{Z_L} = \frac{50 \ \Omega}{(25 + j75) \ \Omega} = \frac{2}{1 + j3} = 0.2 - j0.6 = g_L + jb_L.$$

This admittance is shown on the Smith Chart of Fig. 9.19. The angle $\theta_L = -116.64°$, measured from the real axis, is also shown. Starting from this point, we translate toward the source (clockwise rotation following the red dashed circle centered at the origin of the Smith Chart), until we intersect with the circle for $g = 1.0$. At this point, the conductance is the correct value needed for impedance matching,

$$G_{in} = G_{line} = g_{line}Y_0 = Y_0.$$

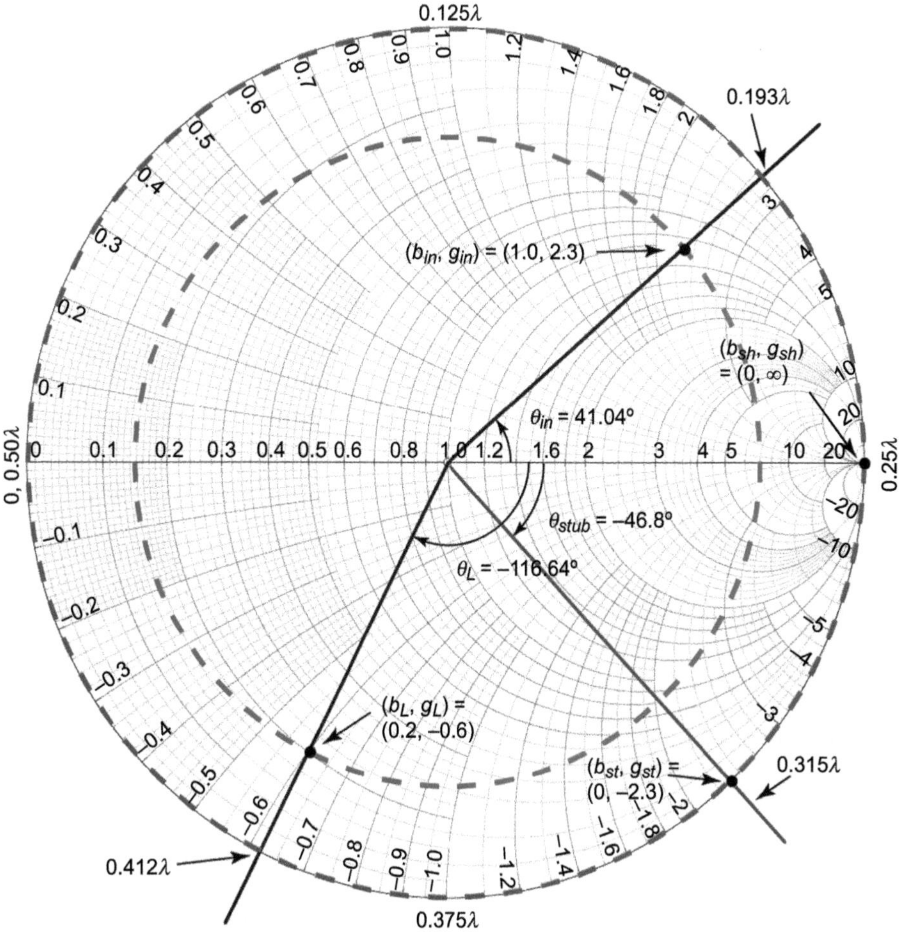

FIGURE 9.19 A Smith Chart used for impedance matching a transmission line with a shorted stub in Example 9.18. The red dashed line is a circle centered on the origin that passes through the input admittance point $(g_{in}, b_{in}) = (0.2, -0.6)$. The blue dashed line represents the admittance of the shorted stub.

The angle $\theta_{in} = 41.04°$ to this line is marked on Fig. 9.19. The length of the loaded line can be determined from the difference in angles $\theta_L - \theta_{in} + 360° = 202.32°$. The additional $360°$ term appears since we started at angle θ_L, and rotated clockwise until we reached θ_{in}. The physical length of the transmission line is

$$\ell = 202.32° \times \frac{\lambda/2}{360°} = 0.281\lambda.$$

We must choose the length of the shorted stub such that its purely imaginary admittance exactly balances the susceptance b_{line} of the loaded line:

$$b_{St} = -b_{line} = -2.3.$$

For a stub of arbitrary length, the admittance of the stub lies somewhere on the outer rim of the Smith Chart, $|\Gamma| = 1$. Its location on the outer rim depends upon its length. For a stub of zero length, the admittance is

$$y_{St} = \lim_{\ell_{St} \to 0} -j\cot(\beta\ell_{St}) = 0 + j\infty.$$

This point is shown on the Smith Chart at $(u, v) = (1, 0)$. We rotate clockwise on the blue dashed circle until we arrive at $b_{St} = -2.3$. This point is also shown on the Smith Chart. The length of the line is determined from the angle $\theta_{stub} = -46.8°$, measured from the real axis of the Smith Chart:

$$\ell_{St} = (0° - \theta_{stub}) \times \frac{\lambda/2}{360°} = 0.065\lambda.$$

The values determined for ℓ_L and ℓ_{St} using the Smith Chart are in very good agreement with the values we calculated in Example 9.14: $\ell_L = 0.280\lambda$ and $\ell_{St} = 0.066\lambda$.

In this section we have introduced graphical techniques to find important transmission line values using the Smith Chart, including the reflection coefficient of a load, the input impedance and admittance of a loaded transmission line, and the lengths of lines to achieve various conditions. The Smith Chart can help you form an intuitive picture for solutions to problems, and with practice can save you time in arriving at results. It can also give us a means of checking our computed results.

9.8 Pulses on Transmission Lines

LEARNING OBJECTIVES

After studying this section, you should be able to:
- plot the voltage $v(t)$ and current $i(t)$ at a specified position along a transmission line on which a pulse has been launched; and
- interpret a plot of the voltage $v(t)$ or current $i(t)$ at a specified position along a transmission line to determine the reflection coefficient of the load and the length of the line.

We conclude this chapter with a discussion of pulse propagation on transmission lines. This topic is, of course, of increasing importance in this digital age, and with the current high speeds of computers and communication systems, even chip-to-chip pulse propagation on circuit boards is subject to transmission line effects. As a demonstration of this, recall that at the outset of this chapter we made the statement that transmission line effects become important whenever the physical size of the circuit is a quarter of the wavelength of the wave, or larger. For 1 GHz clock speeds, keeping just the lowest few Fourier components that comprise the square pulse, a physical size (such as the spacing between chips) of even a few centimeters is enough to exceed this limit. Propagation effects, such as those we introduce in this section, can lead to severe distortion of the signals, and dropped bits.

The basis for this discussion of pulse propagation is similar to that of wave propagation. That is, when a pulse is launched on a transmission line, it takes a finite time for the pulse to travel down that line to the load. Furthermore, when that pulse reaches the load, a portion of the incident pulse is reflected at the load back toward the source. For this discussion we will consider only lossless transmission lines and real (resistive) loads. Also, we will ignore variations in the velocity of the different Fourier components of the pulse. In real systems, any or all of these dispersion effects can be significant, but we will defer those discussions to higher-level treatments.

Consider a pulse generated by the source shown in Fig. 9.20. The transmission line is of length ℓ, the velocity of the pulse is u, and initially we consider the pulse duration τ_p to be short compared to ℓ/u, the time it takes for the pulse to travel the length of the line. The source end of the transmission line is labeled as $z = 0$, and the load end as $z = \ell$.

It is useful to pause here and think about what this pulse really looks like as it propagates along the line. The pulse generator produces a voltage of pulse height

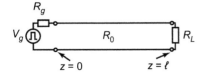

FIGURE 9.20 A transmission line driven by a source that produces a short square pulse of voltage V_g. The internal resistance of the source is R_g, the load resistance is R_L, and the characteristic impedance of the transmission line, taken to be lossless, is R_0.

V_g, which starts propagating along the transmission line. Remember, at points where the potential difference between the conductors is non-zero, there is an electric field pointing from the positive charges on one conductor to the negative charges on the other.

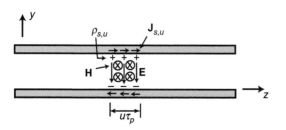

FIGURE 9.21 A snapshot of a voltage pulse on a transmission line. This pulse propagates to the right, and the pulse length is $u\tau_p$, where u is the velocity and τ_p is the pulse duration.

As the pulse propagates along the line, this "charge pulse" must move as well, and charges in motion constitute a current. A snapshot of the transmission line a short time after the pulse is launched, as if we could view the voltage and current along the line, is shown in Fig. 9.21. The potential difference between the two conductors of the transmission line is V_1^+ within the pulse, and zero everywhere else. Similarly, the current in the conductors is I_1^+ in one conductor (reverse this in the other conductor) within the pulse, and zero in either conductor everywhere else. (A word on notation: In the event that one end or the other of the transmission line is not perfectly impedance matched, reflections at the ends will produce multiple pulses, so the notation used here helps us keep track of the various pulses. A superscript "+" is used for a pulse traveling in the $+z$-direction, and a superscript "−" for a pulse traveling in the $−z$-direction. The numerical subscript indicates the number of trips the pulse has made.) In Fig. 9.21, the electric field **E**, the surface charge $\rho_{s,u}$ in the top conductor and $−\rho_{s,u}$ in the bottom conductor, the magnetic field **H**, and the surface current $\mathbf{J}_{s,u}$ in the top conductor and $−\mathbf{J}_{s,u}$ in the bottom conductor are shown. All of these quantities are zero at any other location. A short time later, the pulse has moved a bit closer to the load, and the entire voltage and current distribution has moved as a package as well. The pulse height of the voltage pulse V_1^+ and that of the current pulse I_1^+ are, of course, related to one another, since their ratio must be R_0, the characteristic impedance of the transmission line.

What is it that determines the pulse height of this initial pulse? A little thought leads us to the conclusion that we can think of a simple resistive divider of values R_g and R_0, so that the first pulse amplitude must be

$$V_1^+ = V_g \frac{R_0}{R_g + R_0},\qquad (9.57)$$

where V_g is the pulse height produced by the ideal voltage source. Here is how we support this. The current through the source resistor R_g must be the same as the current pulse launched on the transmission line, I_1^+. As this current passes through the source resistor, the voltage drop across this resistor must be $I_1^+ R_g$ by Ohm's Law, and the potential V_1^+ applied across the input to the transmission line is $V_g − I_1^+ R_g$. But this pulse height must also be $I_1^+ R_0$, as we explained in the previous paragraph. So equating these two forms of V_1^+ yields

$$I_1^+ = \frac{V_g}{R_g + R_0},$$

and the product of this current with R_0 yields Eq. (9.57).

As this pulse travels down the transmission line it comes upon the load of resistance R_L. The voltage V_L across the load and the current I_L through the load must be in the ratio $V_L/I_L = R_L$, which in general is not the same as the characteristic impedance of the transmission line R_0. A mismatch between R_L and R_0 necessarily results in a reflected pulse of amplitude V_1^-, with an associated current pulse of amplitude I_1^-. The ratio V_1^-/I_1^- must still be of magnitude R_0, but the sign is now negative, for reasons identical to those described earlier for single-frequency wave propagation. Either the voltage or the current, but not both, must be inverted upon reflection. So the ratio V_1^-/I_1^- is

$$\frac{V_1^-}{I_1^-} = -R_0.$$

Since the total voltage V_L at the load is the incident pulse amplitude V_1^+ plus the reflected pulse amplitude V_1^-, and the total current amplitude is a similar sum $I_1^+ + I_1^-$, and since the ratio of these must equal R_L, we can write

$$R_L = \frac{V_L}{I_L} = \frac{V_1^+ + V_1^-}{I_1^+ + I_1^-}.$$

The reflection coefficient at the load is defined similarly to the reflection coefficient for single-frequency waves,

$$\Gamma_L = \frac{V_1^-}{V_1^+} = \frac{-I_1^-}{I_1^+}.$$

Using this definition, R_L can be written as

$$R_L = \frac{V_1^+ + \Gamma_L V_1^+}{I_1^+ - \Gamma_L I_1^+} = \frac{V_1^+}{I_1^+}\frac{1 + \Gamma_L}{1 - \Gamma_L}.$$

But the ratio V_1^+/I_1^+ is just R_0. Finally, turning this expression around and solving for Γ_L gives us

$$\Gamma_L = \frac{R_L - R_0}{R_L + R_0}. \tag{9.58}$$

With a pulse of amplitude V_1^+ incident on the load, the reflected pulse is of amplitude $V_1^- = \Gamma_L V_1^+$. This result is very similar to that for the reflection coefficient for sinusoidal waves, and it shows the importance of terminating the transmission line with a load resistor whose resistance matches R_0, as this eliminates the reflection of the pulse from the load. With proper matching of the load to the line, the voltage and current of the incident pulse are of just the right amplitude as required by the load resistor, such that no reflection results.

If the pulse is partially reflected by the load, then this reflected pulse travels in the $-z$-direction, reaching the source end of the transmission line at time $t = 2\ell/u$. Here, we must expect that the pulse is once again partially reflected, with the amplitude of the reflection depending on the mismatch between the source resistance R_g and the characteristic resistance of the transmission line R_0. We will not repeat the derivation of this reflection coefficient, since it follows precisely along the same lines as the reflection coefficient of the load. The result is

$$\Gamma_g = \frac{R_g - R_0}{R_g + R_0}. \tag{9.59}$$

The amplitude of the pulse reflected from the source end of the transmission line, denoted V_2^+, is equal to $\Gamma_g V_1^-$, and in terms of the initial pulse amplitude, $\Gamma_L \Gamma_g V_1^+$. This pulse propagates back toward the load, where it is once again reflected, with pulse amplitude $V_2^- = \Gamma_L^2 \Gamma_g V_1^+$, and so forth and so on. Each time the pulse is reflected from one end or the other, its pulse height is decreased by a factor Γ_L for reflection at the load end, or by a factor Γ_g for reflection at the source end. If the load and source nearly match the line impedance, the pulse amplitude decreases to insignificance after only a few trips back and forth. But if the load and source are not closely matched to the line, then the reflections can last for many transits.

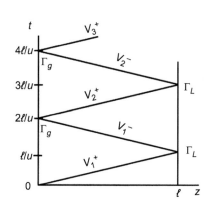

FIGURE 9.22 A bounce diagram, useful for tracking multiply reflected pulses on a transmission line.

We have described this process as if we could watch the pulse bouncing back and forth from source to load to source and on and on. This, of course, is not practical. More commonly, we would observe these effects on an oscilloscope, which is able to view the potential difference on the transmission line at one location, and give us a view of the voltage at this one location as a function of time. The question that we would next like to address, then, is how does this voltage appear when observed at a single location? The answer, of course, depends on the location at which we sample the voltage. We are aided in this analysis by creating a **bounce diagram**, as shown in Fig. 9.22. In this plot, the position z along the transmission line is the abscissa, and the time t the ordinate. The zig-zag "bounce" line represents the position of the leading edge of the pulse as a function of time. If we want to know the arrival time of the various initial or reflected pulses at a position z_0, we simply draw a vertical line at this position, and read off the arrival times along the vertical axis at the intersections with the bounce line. Furthermore, this diagram is useful for keeping track of the amplitudes of the successive pulses as well. These are marked on the diagram for each transit.

With the aid of this diagram, we have constructed a plot of the voltage $v(t)$ at three different locations, (1) the source end, (2) the middle, and (3) the load end, which are shown in Fig. 9.23. For the purposes of this plot, we chose $\Gamma_L = -0.8$ and $\Gamma_g = +0.8$. The voltage at the source is shown in Fig. 9.23(a). Notice the red vertical dashed line in the bounce diagram at $z = 0$. Reading off the time axis, this red line intersects the bounce line at $t = 0$ for the initial pulse, twice at $t = 2\ell/u$, for the pulse which is reflected by the load, requiring one round-trip time to arrive back at the source, and the reflection from the source end, which is at the same time, and then again at $t = 4\ell/u$, equal to two round-trip times. The amplitude of the first pulse is simply V_1^+, since this pulse is alone. The second pulse, at time $t = 2\ell/u$, however, is a composite of V_1^- and V_2^+, for a total amplitude of $V_1^- + V_2^+ = (1 + \Gamma_g)\Gamma_L V_1^+$. The third pulse has amplitude $V_2^- + V_3^+ = (1 + \Gamma_g)\Gamma_g\Gamma_L^2 V_1^+$, and so forth.

FIGURE 9.23 The voltage $v(t)$ versus t seen on the transmission line when measured (a) at the source end $z = 0$, (b) in the middle $z = \ell/2$, and (c) at the load $z = \ell$. We have used reflection coefficients $\Gamma_L = -0.8$ and $\Gamma_g = +0.8$ for this illustration.

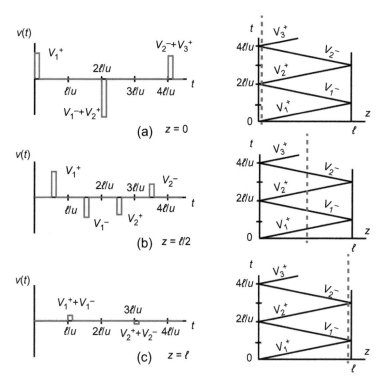

The plot of $v(t)$ at the center of the transmission line versus t, which we show in Fig. 9.23(b), appears to have little in common with the voltage at the beginning of the line. But indeed, they are derived from the same bounce diagram. At this position, the arrival time of each pulse is read from the intersection of the red, dashed vertical line at $z = \ell/2$ with the zig-zag bounce line. Here, the initial pulse does not arrive until a later time, $t = \ell/2u$, since this much time is required for the initial pulse to arrive at this position. The amplitude of this pulse is V_1^+, the same amplitude as we observed at $z = 0$. The second pulse arrives at time $t = 3\ell/2u$, representing the first reflection from the load. This pulse arrives alone, unencumbered by other pulses, and its amplitude is V_1^-. Successive pulses have amplitudes V_2^+, V_2^-, V_3^+, and so forth, arriving at regular intervals of $\Delta t = \ell/u$.

Finally, the voltage $v(t)$ at $z = \ell$ is shown in Fig. 9.23(c), with the bounce diagram to its right. Pulses are once again merged, since at this point the first pulse launched by the source is overlapped by the first reflection, giving us a pulse height of $V_1^+ + V_1^- = (1 + \Gamma_L)V_1^+$, the second has height $V_2^+ + V_2^- = (1 + \Gamma_L)\Gamma_L\Gamma_g V_1^+$, and so forth. In this way, the plot of voltage versus time is constructed at any location. Notice the small amplitudes of the pulses at this location. This is a direct result of the choice of $\Gamma_L = -0.8$, such that the overlapping incident and reflected pulses nearly cancel one another at this location.

Example 9.19 Short Pulse Propagation on a Transmission Line

A single short pulse is launched on a long transmission line at $t = 0$. The velocity of the pulse is $u = 2 \times 10^8$ m/s, and the real characteristic impedance of the transmission line is $R_0 = 50\ \Omega$. When observed at the center on the transmission line, the voltage shows a series of pulses, whose arrival time (in ns) and amplitude (in V) are:

1st	50	10
2nd	150	8
3rd	250	−6
4th	350	−4.8
5th	450	3.6

Determine (a) the load resistance R_L, (b) the source resistance R_g, (c) the source voltage V_g, and (d) the length of the transmission line ℓ.

Solution:

We start by identifying the various pulses. The first pulse must be the initial pulse launched by the source, traveling toward the load. This pulse amplitude is V_1^+, and it arrives at the middle of the line at $t = 50$ ns, which must be equal to $0.5\ell/u$. This pulse continues on to the load, is reflected, and arrives back at the middle of the line ℓ/u later, or at $t = 1.5\ell/u$. This pulse amplitude is V_1^-. A time ℓ/u after this, at $t = 2.5\ell/u$, the reflection from the source end arrives, and so forth. So we identify the amplitudes of the third, fourth, and fifth pulses as V_2^+, V_2^-, and V_3^+, respectively. Now we are ready to determine the quantities requested.

(a) To find the load resistor, R_L, we start by finding the reflection coefficient Γ_L. We obtain this from $\Gamma_L = V_1^-/V_1^+$ or V_2^-/V_2^+, either of which gives us $\Gamma_L = 0.8$. Solving Eq. (9.58) for R_L gives us

$$R_L = R_0 \frac{1 + \Gamma_L}{1 - \Gamma_L} = R_0 \frac{1.8}{0.2} = 9R_0 = 450\ \Omega.$$

(b) We follow a similar procedure to find the source resistor R_g. We need the reflection coefficient at the source, which we find using $\Gamma_g = V_2^+/V_1^-$ or V_3^+/V_2^-, either of which gives us $\Gamma_g = -0.75$. We solve Eq. (9.59) for R_g to obtain

$$R_g = R_0 \frac{1 + \Gamma_g}{1 - \Gamma_g} = R_0 \frac{0.25}{1.75} = \frac{R_0}{7} = 7.14\ \Omega.$$

(c) We can find the source voltage using the voltage division equation Eq. (9.57), which we solve for V_g to show

$$V_g = V_1^+ \frac{R_g + R_0}{R_0} = (10\ \mathrm{V}) \frac{\frac{8}{7}R_0}{R_0} = 11.4\ \mathrm{V}.$$

(d) We use the arrival time of the first pulse, $t_1 = 50$ ns, to find the length of the transmission line

$$\ell = \frac{u t_1}{2} = \frac{(2 \times 10^8\ \mathrm{m/s})(50\ \mathrm{ns})}{2} = 20\ \mathrm{m}.$$

To this point we have discussed short pulses, defined as pulses whose duration is short in comparison with the time required to travel the length of the transmission line. Under this condition the pulses are non-overlapping (except at the ends of the line) and the amplitudes of the pulses observed are simple. What happens when the pulse duration is not this short? In fact, we have already addressed something similar to this when we evaluated the overlapping short pulses at the ends of the lines. The observed voltages are found through superposition; they are simply the sum of the amplitudes of the individual pulses. For example, for a source that turns on at $t = 0$ and stays on for all $t > 0$, the resulting waveform is of the form shown in Fig. 9.24. Similar to short pulses, the voltage $v(t)$ varies depending on where it is observed, and we are aided by

FIGURE 9.24 The voltage $v(t)$ versus t seen on the transmission line when measured (a) at the source end ($z = 0$), (b) in the middle ($z = \ell/2$), and (c) at the load ($z = \ell$) for a step function pulse which turns on at $t = 0$ and stays on. We have used reflection coefficients $\Gamma_L = -0.8$ and $\Gamma_g = +0.8$ for this illustration.

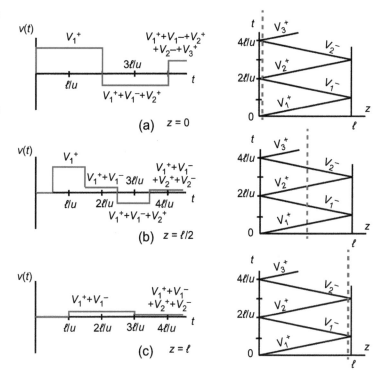

use of a bounce diagram. The voltage $v(t)$ at the source end of the transmission line is shown in Fig. 9.24(a). At $t = 0$, the voltage increases to V_1^+ and stays at this value until one complete round-trip time $2\ell/u$ later. At that time, the first reflection from the load arrives, as well as the reflection of the reflection at the source, so the total voltage becomes $V_1^+ + V_1^- + V_2^+$. The voltage stays at this value until the next pair of reflections arrives at $t = 4\ell/u$. The amplitude then is $V_1^+ + V_1^- + V_2^+ + V_2^- + V_3^+$.

What do we see if we measure the voltage $v(t)$ at the middle of the line? The story goes much the same, except the first pulse does not arrive until $t = \ell/2u$, and the voltage changes at intervals of $t = \ell/u$. This is shown in Fig. 9.24(b). The voltage first turns on to a value V_1^+, the next step increments the voltage by V_1^- (which, in this example, is negative), then by V_2^+ (also negative), then V_2^-, and so forth. We can follow the voltage at any point along the transmission line in a similar manner. The resulting voltage $v(t)$ at the load, $z = \ell$, is shown in Fig. 9.24(c) without further description.

At any position on this line, the voltage after we wait a long time will settle to a steady-state value. In steady state, all transients have long since died away, nothing changes anymore, and the transmission line behaves simply as a pair of wires. In this limit, the voltage across the load is found by simple circuit analysis. For example, voltage division tells us that the voltage across the load should be $V_L = V_g R_L/(R_g + R_L)$. In this example we didn't tell you the values of the load resistor or the source resistor, but we did tell you the reflection coefficient at the load was -0.8, and the reflection coefficient at the source was $+0.8$. If you work this out, you should be able to show that $R_L = R_0/9$, and $R_g = 9R_0$. This tells us that the steady-state value of V_L is $V_g/82$. Note that since the reflection coefficients in this example were so large, many bounces are required before the voltage settles into this value.

An alternative means of viewing the steady-state voltage reached on the transmission line is to write this as

$$V_L = V_1^+ + V_1^- + V_2^+ + V_2^- + V_3^+ + V_3^- + \cdots$$
$$= V_1^+ + \Gamma_L V_1^+ + \Gamma_g \Gamma_L V_1^+ + \Gamma_g \Gamma_L^2 V_1^+ + \Gamma_g^2 \Gamma_L^2 V_1^+ + \Gamma_g^2 \Gamma_L^3 V_1^+ + \cdots.$$

Grouping these terms two by two, and factoring out a common factor of $V_1^+(1+\Gamma_L)$, this becomes

$$V_L = V_1^+(1 + \Gamma_L)\left(1 + \Gamma_L \Gamma_g + (\Gamma_L \Gamma_g)^2 + \cdots\right).$$

Finally, the last infinite series in this expression can be contracted to $(1 - \Gamma_L \Gamma_g)^{-1}$ using the Taylor expansion, Eq. (H.5), so the load voltage is

$$V_L = V_1^+ \frac{1 + \Gamma_L}{1 - \Gamma_L \Gamma_g}.$$

Using Eqs. (9.58) and (9.59) to replace the reflection coefficients yields

$$V_L = V_g \frac{R_L}{R_g + R_L},$$

as we found earlier.

Example 9.20 Open-Circuit Load, an Oscillator

Consider a transmission line of length ℓ with an open-circuit load. The source driving this transmission line turns on to a value of V_g at $t = 0$, and remains on. The internal resistance of the source is $R_g = 0\ \Omega$. Plot the voltage $v_L(t)$ at the load.

Solution:

Using Eq. (9.58), the reflection coefficient at the load end is found as $\Gamma_L = 1.0$, and from Eq. (9.59), $\Gamma_g = -1.0$. Using a bounce diagram such as the one in Fig. 9.24(c), it can be shown that $v_L(t) = 0$ from $t = 0$ until the time that the first pulse arrives at the load, $t = \ell/u$, where u is the velocity of the pulse on the transmission line. As soon as this pulse arrives, a reflected wave of magnitude V_g is generated at the load, propagating back toward the source. The value of $v_L(\ell/u)$ is therefore $V_1^+ + V_1^- = V_g + \Gamma_L V_g = 2V_g$. This value of v_L persists until one round-trip time later, $t = 3\ell/u$, at which time the next reflection arrives at the load. The amplitude of the voltage is then

$$v_L(3\ell/u) = V_1^+ + V_1^- + V_2^+ + V_2^- = V_g + \Gamma_L V_g + \Gamma_g \Gamma_L V_g + \Gamma_g \Gamma_L^2 V_g = 0.$$

By similar reasoning, the load voltage at time $t = 5\ell/u$ turns back on to $2V_g$, and at time $t = 7\ell/u$ turns off once again. With no losses in the system, the voltage v_L is a periodic rectangular waveform, oscillating between values of $2V_g$ and zero with a period of $4\ell/u$. A plot of this voltage is shown in Fig. 9.25.

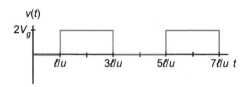

FIGURE 9.25 The voltage $v_L(t)$ versus t for Example 9.20. The load is an open circuit, and the voltage is measured at the load ($z = \ell$). The source generates a step function pulse, which turns on at $t = 0$ and stays on. The reflection coefficients at the load and source are $\Gamma_L = 1.0$ and $\Gamma_g = -1.0$.

In this section we have described pulse propagation on transmission lines. When a pulse is launched by the source on the transmission line, the pulse (voltage and current) propagates along the line until it reaches the load, at which point the pulse is reflected (unless, of course, the load resistance matches the characteristic resistance of the transmission line) back toward the source. Upon reaching the source end, this pulse is reflected back toward the load, and the story goes on. The time-dependent voltage $v(t)$ is interpreted at any point along the transmission line in terms of the sum of all the pulses that still persist at that point and time. This analysis is aided by use of a bounce diagram, such as those shown in Figs. 9.23 and 9.24.

SUMMARY

Transmission lines consist of two or more conductors with uniform spacing that guide electromagnetic waves from a source to a load, allowing much more focused distribution of the signal and power than freely propagating waves. Transmission line effects account for the delay time needed for the changing voltage or current to propagate from the source. As a rule of thumb, it is necessary to consider transmission line effects for any conductors that are longer than $\sim \lambda/4$, where λ is the wavelength of the propagating wave.

In this chapter we have developed a model for these transmission lines by starting with freely propagating waves, initially introduced in Chapter 8, and discovering wave modes that also satisfy the boundary conditions at the surfaces of the conductors of the transmission line. Although we started this discussion in terms of the electric and magnetic fields of the propagating mode, we quickly transformed to a description based on the potential difference between the conductors and the current in the conductors. This voltage/current description can be much more intuitive. The transmission line analysis has allowed us to quantify losses on the transmission line, measure reflections at the load, determine the input impedance, and follow transient effects in pulse propagation.

Table 9.2

Collection of key formulas from Chapter 9.			
See Table 9.1 for transmission line parameters for co-axial, twin-lead, and parallel-plate transmission lines.			
$\beta = \omega\sqrt{L'C'}$	(9.14)	$\tilde{I}(z) = I_0^+ e^{-\gamma z} + I_0^- e^{+\gamma z}$	(9.32)
$Z_0 = \sqrt{\dfrac{L'}{C'}}$	(9.17)	$\Gamma_L = \dfrac{Z_L - Z_0}{Z_L + Z_0}$	(9.37)
$u_p = \dfrac{1}{\sqrt{L'C'}}$	(9.18)	$S = \dfrac{(1 + |\Gamma_L|)}{(1 - |\Gamma_L|)}$	(9.39)
$\gamma = \sqrt{[R' + j\omega L'][G' + j\omega C']}$	(9.25)	$Z_{in} = Z_0 \dfrac{Z_L + Z_0\tanh(\gamma\ell)}{Z_0 + Z_L\tanh(\gamma\ell)}$	(9.43)
$Z_0 = \sqrt{\dfrac{R' + j\omega L'}{G' + j\omega C'}}$	(9.29)	$Z_{in} = R_0 \dfrac{Z_L + jR_0\tan(\beta\ell)}{R_0 + jZ_L\tan(\beta\ell)}$	(9.44)
$\tilde{V}(z) = V_0^+ e^{-\gamma z} + V_0^- e^{+\gamma z}$	(9.31)		

PROBLEMS   **Lossless Transmission Line**

P9-1 A parallel plate transmission line is constructed of perfectly conducting plates separated by a non-absorbing, non-magnetic dielectric of permittivity $\varepsilon = 4\varepsilon_0$. The plate width is $w = 3$ mm, and the plate separation is $d = 0.6$ mm. The amplitude of the wave traveling in the $+z$-direction is $V_0^+ = 10$ V, and its frequency is 200 MHz. Determine (a) the capacitance per unit length C', (b) the inductance per unit length L', (c) the characteristic impedance Z_0, (d) the propagation constant β, (e) the phase velocity u_p, (f) the amplitude of the current wave I_0^+, and (g) the amplitude and direction of the electric field and the magnetic field in the region between the conductors.

P9-2 A co-axial transmission line is constructed of perfectly conducting cylindrical conductors separated by a non-absorbing, non-magnetic dielectric of permittivity $\varepsilon = 3\varepsilon_0$. The radius of the inner conductor $a = 1.5$ mm, and the inner radius of the outer conductor is $b = 4.5$ mm. The amplitude of the wave traveling in the $+z$-direction is $V_0^+ = 10$ V, and its frequency is 200 MHz. Determine (a) the capacitance per unit length C', (b) the inductance per unit length L', (c) the characteristic impedance Z_0, (d) the propagation constant β, (e) the phase velocity u_p, (f) the amplitude of the current wave I_0^+, and (g) the amplitude and direction of the electric field and the magnetic field outside the inner conductor at $\rho = a$.

P9-3 A twin-lead transmission line is constructed of two perfectly conducting wires separated by a non-absorbing, non-magnetic dielectric of permittivity $\varepsilon = 2\varepsilon_0$. The radius of the conductors is $a = 1.5$ mm, and the center-to-center spacing of the outer conductors is $d = 12$ mm. The amplitude of the wave traveling in the $+z$-direction is $V_0^+ = 5$ V, and its frequency is 200 MHz. Determine (a) the capacitance per unit length C', (b) the inductance per unit length L', (c) the characteristic impedance Z_0, (d) the propagation constant β, (e) the phase velocity u_p, and (f) the amplitude of the current wave I_0^+.

Losses

P9-4 A parallel plate transmission line consists of a pair of copper planar conductors of width 4 mm and separation 1 mm. The non-conducting material between the copper conductors has relative permittivity $\varepsilon_r = 2 - j0.04$, and permeability μ_0. The amplitude of the wave traveling in the $+z$-direction is $V_0^+ = 2$ V, and its frequency is 500 MHz. You may ignore edge effects. (a) Find C', L', R', and G' for this transmission line. (b) Find the characteristic impedance Z_0 and the complex propagation constant γ at a frequency of 500 MHz. (c) Find the wavelength of the wave, and the phase velocity at this same frequency. (d) Find the amplitude and phase of the current wave I_0^+. (e) How far must a wave travel on this line before the field amplitude is halved?

P9-5 The center conductor of a co-axial transmission line has radius $a = 2$ mm, while the outer conductor has inner radius $b = 5$ mm. Both conductors are copper. The non-conducting material between the copper conductors has relative permittivity $\varepsilon_r = 4 - j0.04$, and permeability μ_0. The amplitude of the wave traveling in the $+z$-direction is $V_0^+ = 5$ V, and its frequency is 400 MHz. (a) Find C', L', R', and G' for this transmission line. (b) Find the characteristic impedance Z_0 and the complex propagation constant γ at a frequency of 400 MHz. (c) Find the wavelength of the wave and the phase velocity at this same frequency. (d) Find the amplitude and phase of the current wave I_0^+. (e) How far must a wave travel on this line before the field amplitude is halved?

P9-6 A twin-lead transmission line is constructed of two copper wires separated by a non-absorbing, non-magnetic dielectric of permittivity $\varepsilon = (2 - j0.05)\varepsilon_0$. The radius of the conductors is $a = 1.5$ mm, and the center-to-center spacing of the outer conductors is $d = 12$ mm. The amplitude of the wave traveling in the $+z$-direction is $V_0^+ = 10$ V, and its frequency is 200 MHz. (a) Find C', L', R', and G' for this transmission line. (b) Find the characteristic impedance Z_0 and the complex propagation constant γ at a frequency of 200 MHz. (c) Find the wavelength of the wave, and the phase velocity at this same frequency. (d) Find the amplitude and phase of the current wave I_0^+. (e) How far must a wave travel on this line before the field amplitude is halved?

P9-7 A transmission line has characteristic impedance $Z_0 = (50 + j)\,\Omega$, and the propagation constant is $\gamma = (0.1 + j0.5\pi)$ m^{-1} for a wave of frequency $\omega = 10^8$ rad/s. Determine R', L', G', and C' for this transmission line.

Standing Waves

P9-8 The voltage amplitude $|V(z)|$ measured along a 50 Ω transmission line is shown in Fig. 9.26. Determine the standing wave ratio S, the amplitudes of the traveling waves V_0^+ and V_0^-, and the wavelength of the wave. Draw the current standing wave pattern for this same line.

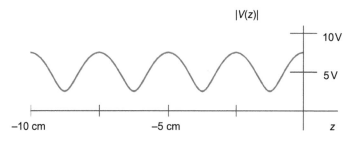

FIGURE 9.26 The voltage amplitude $V(z)$ measured on a transmission line, for use in Prob. P9-8.

P9-9 Plot the standing wave voltage $|V(z)|$ that results from the interference of two counter-propagating waves of amplitude $V_0^+ = 5$ V and $V_0^- = 4.5$ V on a 50 Ω transmission line. Use $\beta = 2$ m^{-1}, let the line be lossless, and place the first voltage minimum at $z = 0$. The domain of your plot should be $-2\lambda \leq z \leq 0$. Determine the standing wave ratio S.

Reflections at a Load

P9-10 Find the input impedance Z_{in} for the transmission line whose characteristic impedance is $R_{01} = 75$ Ω, length is $\ell_1 = 2$ m, and the load impedance is $Z_L = (75 - 15j)$ Ω. The wavelength of the wave is 0.6 m.

P9-11 Repeat Prob. P9-10 for a lossy transmission line whose characteristic impedance is $Z_0 = (50 + 2.5j)$ Ω, and with $\gamma\ell = 0.1 + j0.5\pi$.

P9-12 Find the input impedance Z_{in} for the transmission line configuration shown in Fig. 9.27(a). The characteristic impedances of the two lossless sections are $R_{01} = 75$ Ω and $R_{02} = 50$ Ω, the lengths of the segments are $\ell_1 = 3\lambda/4$ and $\ell_2 = \lambda/4$, and the load impedance is $Z_L = (75 - j15)$ Ω.

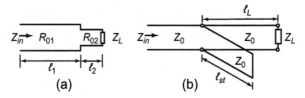

FIGURE 9.27 Transmission line circuits for input impedance problems. (a) Two segments of transmission line with different characteristic impedances, for Prob. P9-12, and (b) loaded line with parallel stub for Prob. P9-13.

P9-13 Find the input impedance Z_{in} for the transmission line configuration shown in Fig. 9.27(b). The characteristic impedance of each section of line is 50 Ω, the length of the shorted stub is 0.3λ, and this stub is connected to the primary line a distance 0.4λ from the load. The impedance of the load is 100 Ω, and the total length of the primary line, from input to load, is 2λ.

P9-14 Consider the lossless transmission line in Fig. 9.28. Its length is ℓ and its characteristic impedance is R_0. (a) What load impedance Z_L minimizes the reflection of the waveform at the load end of the line? (b) For a different Z_L, the voltage $V(z)$ is measured at several positions z along the transmission line. A voltage minimum is located at the load, and the standing wave ratio S is 5. Determine Z_L. (c) For a load $Z_L = 9R_0$, determine the input impedance Z_{in} for the following two cases: (1) $\ell = 1.25\lambda$, and (2) $\ell = 1.5\lambda$, where λ is the wavelength of the wave on the transmission line.

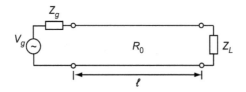

FIGURE 9.28 Transmission line circuits for input impedance problems, for Probs. P9-14 and P9-21.

P9-15 A voltage wave of amplitude 5 V is incident on a load whose impedance is real, and of value R_L. The line is lossless, and has characteristic resistance $R_0 = 50\ \Omega$. The reflected wave has amplitude 4 V. Determine the two possible values for R_L.

P9-16 A voltage wave of amplitude 5 V is incident upon a load, producing a reflected wave of amplitude 2.5 V. The transmission line is lossless, with characteristic resistance $R_0 = 50\ \Omega$. Determine the load impedance for the following cases: (a) V_0^- is in phase with V_0^+. (b) V_0^- is π out of phase with V_0^+. (c) V_0^- leads V_0^+ by $\pi/2$. (d) V_0^- lags V_0^+ by $\pi/4$.

P9-17 A voltage wave of amplitude 5 V propagates on a lossless 75 Ω line. The load impedance is $Z_L = 60e^{j\theta_z}\ \Omega$, where the phase θ_z is between $-\pi/2$ and $+\pi/2$. Determine the amplitude V_0^- of the reflected wave. Your answer will be a function of θ_z. Evaluate V_0^- for $\theta_z = \pi/2$, $\pi/4$, 0, $-\pi/4$, and $-\pi/2$.

P9-18 The standing wave ratio measured on a lossless transmission line of characteristic impedance R_0 is $S = 1.8$. Assuming that the reflection coefficient Γ_L at the load is real, find Γ_L. There are two results. Find both, identify the location of the first voltage maximum for each, and determine the ratio R_L/R_0 that results in this value of Γ_L.

P9-19 A load impedance $Z_L = 75\ \Omega$ terminates a transmission line whose characteristic impedance is nominally 50 Ω, but slightly lossy. The attenuation constant α on this line is 0.02 m^{-1}. (a) Determine S when measured close to the load. (b) Determine S when measured 10 m from the load.

P9-20 The standing wave ratio measured on a transmission line is $S = 2$ when measured close to the load, but 1.5 when measured 2 m from the load. Determine the attenuation constant α for this line.

P9-21 A 200 MHz voltage source is connected to a lossless 50 Ω co-axial transmission line, as shown in Fig. 9.28. The length of the transmission line is $\ell = 1$ m, and the amplitude of the oscillating voltage along the transmission line is plotted in Fig. 9.29. (The source is at $z = -\ell$ and the load is at $z = 0$.) (a) Determine the standing wave ratio. (b) Determine the amplitudes of the traveling waves V_0^+ and V_0^-. (c) Determine the reflection coefficient at the load, Γ_L. (d) Determine the load impedance Z_L. (e) What is the wavelength of this wave? (f) Determine the relative permittivity ε_r of the dielectric material in the space between the conductors.

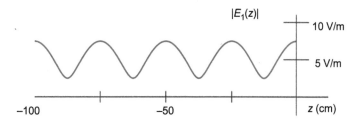

FIGURE 9.29 Standing wave pattern for Prob. P9-21.

P9-22 A 15 V voltage source with internal 35 Ω resistor drives a lossless 50 Ω transmission line. The length of the line is 15.2λ, and the line is terminated by a $Z_L = (50 - j15)$ Ω impedance element. Determine the amplitudes V_0^+ and V_0^- of the traveling waves, the load voltage V_L and load current I_L, and the average power absorbed by the load.

P9-23 A 10 V voltage source with internal 65 Ω resistor drives a lossless 50 Ω transmission line. The length of the line is 0.75λ, and the line is terminated by a $Z_L = (50 - j15)$ Ω impedance element. Determine the amplitudes V_0^+ and V_0^- of the traveling waves, the load voltage V_L and load current I_L, and the average power absorbed by the load.

Impedance Matching

P9-24 A configuration of three sections of lossless transmission lines is shown in Fig. 9.30. The characteristic impedances are $R_{01} = 75$ Ω and $R_{03} = 50$ Ω, the length ℓ_3 is $\lambda/4$, and the load impedance is $Z_L = 25$ Ω. Determine the impedance R_{02} of the center section that eliminates the reflection from this line.

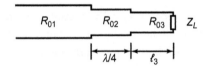

FIGURE 9.30 Transmission line configuration for Prob. P9-24.

P9-25 A wave of amplitude 10 V is incident from the left on the configuration shown in Fig. 9.13. The characteristic impedance R_0 is 50 Ω, the load impedance is $(100 + j0)$ Ω, and R_{02} is chosen to minimize reflections. Determine the amplitudes V_{02}^+ and V_{02}^- of the traveling waves in the quarter-wave segment, and the power absorbed by the load resistor.

P9-26 A lossless 50 Ω transmission line is terminated with a 75 Ω load resistor. A 5 V amplitude wave is incident from a source. A shorted stub, as illustrated in Fig. 9.14(a), is to be connected to eliminate the reflected wave. Determine the length of the loaded line ℓ_L and the length of the shorted stub in terms of the wavelength λ.

P9-27 A lossless 50 Ω is terminated with a $(40 - j40)$ Ω load resistor. A 5 V amplitude wave is incident from a source. A shorted stub, as illustrated in Fig. 9.14(a), is to be connected to eliminate the reflected wave. Determine the length of the loaded line ℓ_L and the length of the shorted stub, in terms of the wavelength λ.

P9-28 A lossless 50 Ω transmission line is terminated with a 75 Ω load resistor. A 5 V amplitude wave is incident from a source. An open stub, as illustrated in Fig. 9.14(b), is to be connected to eliminate the reflected wave. Determine the length of the loaded line ℓ_L and the length of the open stub ℓ_{St}, in terms of the wavelength λ.

P9-29 A lossless 50 Ω transmission line is terminated with a $(40 - j40)$ Ω load resistor. A 5 V amplitude wave is incident from a source. An open stub, as illustrated in Fig. 9.14(b), is to be connected to eliminate the reflected wave. Determine the length of the loaded line ℓ_L and the length of the open stub, in terms of the wavelength λ.

Smith Chart

P9-30 Repeat Prob. P9-24 using the Smith Chart for your analysis. Compare your result to the result of Prob. P9-24.

P9-31 Repeat Prob. P9-26 using the Smith Chart. Find the two shortest line lengths ℓ_L, and the shortest stub length ℓ_{St} for each. Compare your result to the result of Prob. P9-26.

P9-32 Repeat Prob. P9-27 using the Smith Chart. Find the two shortest line lengths ℓ_L, and the shortest stub length ℓ_{St} for each. Compare your result to the result of Prob. P9-27.

P9-33 Repeat Prob. P9-28 using the Smith Chart. Find the two shortest line lengths ℓ_L, and the shortest stub length ℓ_{St} for each. Compare your result to the result of Prob. P9-28.

P9-34 Repeat Prob. P9-29 using the Smith Chart. Find the two shortest line lengths ℓ_L, and the shortest stub length ℓ_{St} for each. Compare your result to the result of Prob. P9-29.

Pulses

P9-35 In Fig. 9.31(a), the switch connecting the $V_g = 100$ V source to the 50 Ω transmission line closes at $t = 0$. The internal resistance of the source is $R_g = 16.7$ Ω. The lossless line is of length 10 m, and the propagation velocity of the pulse is 2×10^8 m/s. The transmission line is open-circuited at the load end. Plot the voltages $v_{in}(t)$ and $v_L(t)$ versus time t.

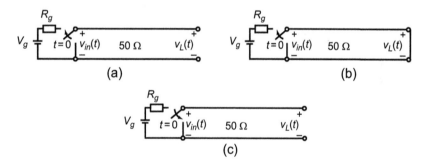

FIGURE 9.31 Transmission line configurations for Probs. (a) P9-35, (b) P9-36, and (c) P9-37.

P9-36 Repeat Prob. P9-35 for the transmission line configuration shown in Fig. 9.31(b), in which the load is shorted.

P9-37 In the circuit shown in Fig. 9.31(c), the switch at the source end is initially closed, and the line fully charged to 100 V. At $t = 0$, the switch switches to ground. Using the parameters given in Prob. P9-35, determine and plot the voltages $v_{in}(t)$ and $v_L(t)$ versus time t.

P9-38 A pulsed voltage source generates a 12 V, 10 ns duration pulse, which is launched on a lossless 50 Ω transmission line through a 40 Ω internal resistor. The line is of length 4 m, the propagation velocity of the pulse is 2×10^8 m/s, and the line is loaded with a 60 Ω resistor. Plot $v_{in}(t)$ at the input end of the line, $v_{mid}(t)$ at the middle of the line, $v_L(t)$ at the load end of the line versus time t.

P9-39 Repeat Prob. P9-38, replacing the 10 ns pulse with a 40 ns pulse.

P9-40 Consider the lossless transmission line system shown in Fig. 9.20. The characteristic impedance of the line is $R_0 = 50$ Ω, and the source and load resistances are R_g and R_L, respectively. The voltage pulse generated (10 V amplitude, 50 ns duration) propagates along the transmission line. The voltage $v_m(t)$ recorded at the precise middle of the transmission line is shown in Fig. 9.32. (a) Assuming that the velocity of the pulse on the transmission line is $2c/3$, determine the overall length ℓ of the transmission line. (b) Determine the load resistance R_L. (c) Determine the source resistance R_g.

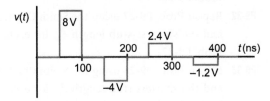

FIGURE 9.32 The voltage measured at the middle of the transmission line as a function of time, Prob. P9-40.

CHAPTER 10
Waveguides

In the previous chapter we introduced transmission lines, structures consisting of a pair of conductors that guide a high-frequency electromagnetic wave in the space between the conductors from a source to a load. We considered a number of different geometries (parallel plate, twin-lead, co-axial), each with a uniform cross-section, that guide the wave in the $\pm z$-direction, and we discussed the properties of signals carried by the transmission lines, largely in terms of the potential difference $\tilde{V}(z)$ between the conductors and the current $\tilde{I}(z)$ flowing in one conductor and returning in the other. For many high-power, high-frequency applications, **waveguides** are often used to guide electromagnetic waves from source to load. Unlike a transmission line, a waveguide typically consists only of a single conductor, which is hollow, with the wave existing in the interior of the waveguide. Waveguides are often used for frequencies ranging from 10 to 100 GHz, and suffer lower losses in this range than do transmission lines. The typical transverse dimension of a waveguide is a few centimeters, on the order of the wavelength of the wave inside. A few examples of waveguides are illustrated in Fig. 10.1. The parallel plate waveguide shown in Fig. 10.1(a) is not in common usage, but serves as a simple geometry for developing the properties of general waveguides. (Parallel plates differ from most other waveguides, in that they consist of two conductors and can function as transmission lines or waveguides.) Also shown in this figure are (b) rectangular and

FIGURE 10.1 Examples of waveguides: (a) parallel plate, (b) rectangular, and (c) circular waveguides. Each of the surfaces shown are conductors. The rectangular and circular waveguides are hollow on the inside.

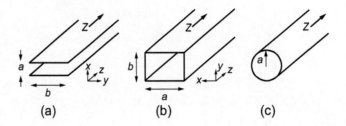

(a) (b) (c)

(c) circular waveguides. While the electromagnetic principles for the circular waveguide are similar to those of the rectangular waveguide, the mathematical functions needed for this geometry are not familiar to most undergraduate students, so we'll need to introduce these functions. Also, while we devote our attention here to *conducting* waveguides, be aware that *dielectric* waveguides, such as optical fibers and photonic crystals can be treated in a similar manner. With the exception of TechNotes 10.1 and 10.2, we will not consider dielectric waveguides in this text. Optical fibers are common conduits for optical signals, and find application in medical instruments and communications systems.

TechNote 10.1 Optical Fibers

Optical fibers are a form of waveguide for waves in the optical and near-infrared regions of the spectrum. Optical fibers differ from the conducting waveguides that we will discuss in detail in this chapter, in that they are fabricated from dielectric materials, such as silica or germania. They differ in their dimensions as well. The cross-sectional diameter of an optical fiber is quite small, suitable for carrying optical waves, whose wavelength is a few microns, rather than the centimeter-scale wavelengths of waves carried by conducting waveguides. The primary communications wavelength band is around 1.55 µm, where high-quality fibers with low absorption and scattering losses and low dispersion effects have been developed. Fibers typically consist of a central core, whose refractive index is high to confine the optical mode, surrounded by cladding, a glass material of slightly lower refractive index. (See Eq. (8.13) for the definition of refractive index n.)

For long-distance communications systems, single-mode fibers, whose core diameter is as small as 8 µm, are used. Short pulses of light can propagate long distances without dispersing (or spreading out) in single-mode fibers. Multi-mode fibers, whose core diameter can be 60 µm or larger, can be used for short-range communications (up to ~1 km). For applications for which the polarization of the wave is important, polarization-maintaining fibers, whose cross-sections are slightly elliptical in shape, are available. In addition to communications, optical fibers have found important applications in medical fields, such as endoscopy, in which high-quality images from inside the body can be formed, or for surgery, in which the fibers deliver high optical power into the body for minimally invasive procedures.

Wave propagation in photonic crystal materials is modified by adding periodic structure to the medium. The description of electromagnetic waves in these materials is similar to that of electronic waves in semiconductors. These structures are introduced in TechNote 10.2.

TechNote 10.2 Photonic Crystals

Photonic crystals are materials in which the electric permittivity, or equivalently the refractive index, is spatially modulated, leading to a band structure for the waves propagating through them. Electromagnetic wave propagation in these materials can be described in a manner that is very similar to the description of electron propagation through semiconductor materials, which are characterized by a highly ordered arrangement of atoms in a lattice structure. Just as in semiconductors, the band structure for wave propagation in photonic crystals leads to reflection of waves at some frequencies, while other frequencies are transmitted. Photonic structures can exhibit periodic behavior in one, two, or three dimensions. Three-dimensional structures, are, of course, more difficult to fabricate. An example of a one-dimensional photonic structure is a multilayer thin film, described in TechNote 8.20. The optical fiber shown in Fig. 10.2(a) is an example of an engineered photonic structure, which guides and confines the optical mode to the central region of the fiber.

Photonic structures occur in nature, as for example in the Brazilian weevil *Entimus imperialis* and butterfly wings. Such a structure can give the surface a luminous appearance that varies depending on the angle at which it is viewed.

FIGURE 10.2 A photonic crystal optical fiber, which guides and confines optical waves to the central region of the fiber.

Waveguide properties become important when the cross-sectional dimensions of the waveguide, *a* and *b* in Fig. 10.1(b), for example, become comparable to or greater than the wavelength of a freely propagating wave at the same frequency. For frequencies in the range 10–100 GHz, the corresponding wavelength range is 3 to 0.3 cm.

As we discuss specific waveguides and electromagnetic modes carried by these waveguides in this chapter, we will examine the following properties of wave propagation:

- the spatial distribution and orientation of the electric and magnetic fields;
- the propagation constant γ, and its dependence on the frequency ω;
- the minimum frequency, known as the cutoff frequency, necessary for wave propagation;
- the phase variation of the wave;
- the attenuation of the wave;
- the wave velocity; and
- the waveguide impedance, defined as the ratio of the transverse electric and magnetic field components.

In this chapter we will develop the general formalism to describe wave propagation on waveguides, and then apply this formalism to three specific types of waveguides: the parallel plate, rectangular, and circular waveguides, each pictured in Fig. 10.1. We will also use what we have learned about waveguide modes to introduce a related structure known as a **resonant cavity**, a structure which can store electromagnetic energy when excited at its resonant frequency.

10.1 General Approach

In Sections 10.2–10.4 we will explore the properties of specific waveguide systems, including parallel plate, rectangular, and circular waveguides. Each of these studies is based upon a common formalism, which is established in detail here. In this section we will develop the methods, based upon Maxwell's Equations, to determine the electromagnetic modes supported by a general waveguide, the propagation constant, phase and group velocities, wavelength, and wave impedance of the mode, and the power transmitted and lost by the wave. Application of this formalism to the specific waveguide geometries will then follow in a straightforward manner in the subsequent sections.

As we seek solutions for the spatial profile (the shape) and other properties of an electromagnetic wave guided by a waveguide, we must, of course, base the analysis on Maxwell's Equations, Faraday's Law and Ampère's Law in particular. We will take, as our starting point, for the electric field of the wave,

$$\mathbf{E}(x, y, z, t) = \tilde{\mathbf{E}}(x, y)e^{j\omega t - \gamma z}, \tag{10.1}$$

and similarly for the magnetic field,

$$\mathbf{H}(x, y, z, t) = \tilde{\mathbf{H}}(x, y)e^{j\omega t - \gamma z}. \tag{10.2}$$

In writing $\mathbf{E}(x, y, z, t)$ and $\mathbf{H}(x, y, z, t)$ in this form, we have made the following reasonable presumptions:

1. the wave is time-harmonic;
2. the wave propagates in the $+z$-direction, perhaps with attenuation; and
3. the field is separable – that is, it can be written as the product of a transverse part $\tilde{\mathbf{E}}(x, y)$ or $\tilde{\mathbf{H}}(x, y)$, and the time- and z-dependent part, $e^{j\omega t - \gamma z}$.

Recall that the notation $\tilde{\mathbf{E}}$ and $\tilde{\mathbf{H}}$ represent the phasor amplitudes of the electric and magnetic fields, in which the sinusoidal time-dependence has been extracted from the complex field amplitude. In this case, we have taken this one step further and extracted the z-dependence as well.

In treating the parallel plate or rectangular waveguide geometry, rectangular coordinates are, of course, the appropriate system for the analysis, while for circular waveguides we'll use cylindrical coordinates. $\tilde{\mathbf{E}}(x, y)$ and $\tilde{\mathbf{H}}(x, y)$ can, in general, have *components* in the x-, y-, and z-directions, and they contain information about the *transverse* pattern of the fields. That is, they describe the spatial profile of the fields in the x–y plane.

As before, γ is a complex quantity known as the propagation constant, representing the phase and attenuation of the wave,

$$\gamma = \alpha + j\beta.$$

The attenuation constant $\alpha = \mathfrak{R}(\gamma)$ represents the decreasing amplitude of the wave along the axis (in the z-direction), while the phase constant $\beta = \mathfrak{I}(\gamma)$ represents the sinusoidal variation of the wave along the axis, with wavelength $\Lambda = 2\pi/\beta$. This mode wavelength Λ is generally different from the wavelength λ of a freely propagating uniform plane wave at the same frequency, a relationship that we'll explore later.

The wave equations for time-harmonic waves, Eqs. (8.20) and (8.21), are

$$\nabla^2 \mathbf{E}(x,y,z,t) + k^2 \mathbf{E}(x,y,z,t) = 0 \tag{10.3}$$

and

$$\nabla^2 \mathbf{H}(x,y,z,t) + k^2 \mathbf{H}(x,y,z,t) = 0, \tag{10.4}$$

where k is

$$k = \omega\sqrt{\varepsilon\mu}. \tag{10.5}$$

μ and ε are the permeability and permittivity of the medium inside the waveguide. As a reminder, ∇^2 in rectangular coordinates represents

$$\nabla^2 = \frac{\partial^2}{\partial x^2} + \frac{\partial^2}{\partial y^2} + \frac{\partial^2}{\partial z^2},$$

while in cylindrical coordinates it is

$$\nabla^2 = \frac{1}{\rho}\frac{\partial}{\partial \rho}\left(\rho\frac{\partial}{\partial \rho}\right) + \frac{1}{\rho^2}\frac{\partial^2}{\partial \phi^2} + \frac{\partial^2}{\partial z^2}.$$

In either case, the transverse derivatives can be separated from the z-derivatives to write

$$\nabla^2 = \nabla_t^2 + \frac{\partial^2}{\partial z^2},$$

where ∇_t^2 represents the derivatives in the transverse directions,

$$\nabla_t^2 = \frac{\partial^2}{\partial x^2} + \frac{\partial^2}{\partial y^2}$$

in rectangular coordinates, and

$$\nabla_t^2 = \frac{1}{\rho}\frac{\partial}{\partial \rho}\left(\rho\frac{\partial}{\partial \rho}\right) + \frac{1}{\rho^2}\frac{\partial^2}{\partial \phi^2}$$

in cylindrical coordinates.

Since the z-dependence of $\mathbf{E}(x,y,z,t)$ is contained entirely in the factor $\exp(-\gamma z)$ in Eq. (10.1), the second-order z-derivative in the wave equation is easily determined as

$$\frac{\partial^2 \mathbf{E}(x,y,z,t)}{\partial z^2} = \gamma^2 \mathbf{E}(x,y,z,t). \tag{10.6}$$

Combining Eq. (10.3) with (10.6) leads to

$$\nabla_t^2 \mathbf{E} = \nabla^2 \mathbf{E} - \frac{\partial^2}{\partial z^2}\mathbf{E} = -k^2\mathbf{E} - \gamma^2\mathbf{E} = -k_c^2\mathbf{E}, \tag{10.7}$$

where k_c^2 is defined as

$$k_c^2 = \gamma^2 + k^2 = \gamma^2 + \omega^2\varepsilon\mu. \tag{10.8}$$

Remember that γ is the propagation constant of the wave in the waveguide, which we don't yet know, and $k = \omega\sqrt{\varepsilon\mu}$ is the propagation constant of a freely propagating wave at the same frequency ω. Also, while $\mathbf{E}$ in Eq. (10.7) was the full (x,y,z,t)-dependent field quantity, ∇_t involves only the transverse derivatives, and the z- and t-dependence is contained entirely in the factor $e^{j\omega t - \gamma z}$ (see Eqs. (10.1) and (10.2)), so the relation

$\nabla_t^2 \mathbf{E} = -k_c^2 \mathbf{E}$ is equally valid for the phasor amplitude $\tilde{\mathbf{E}}(x, y)$. Equation (10.7) written for each component of $\tilde{\mathbf{E}}$ individually is

$$\nabla_t^2 \tilde{E}_x = -k_c^2 \tilde{E}_x, \qquad \nabla_t^2 \tilde{E}_y = -k_c^2 \tilde{E}_y, \qquad \text{and} \qquad \nabla_t^2 \tilde{E}_z = -k_c^2 \tilde{E}_z. \tag{10.9}$$

Similarly, it can be shown that the magnetic field $\mathbf{H}(x, y, z, t)$ obeys the same transverse wave equation,

$$\nabla_t^2 \mathbf{H} = \nabla^2 \mathbf{H} - \frac{\partial^2}{\partial z^2} \mathbf{H} = -k_c^2 \mathbf{H}, \tag{10.10}$$

which, when applied to the phasor amplitude $\tilde{\mathbf{H}}(x, y)$ and expanded for individual components, reads

$$\nabla_t^2 \tilde{H}_x = -k_c^2 \tilde{H}_x, \qquad \nabla_t^2 \tilde{H}_y = -k_c^2 \tilde{H}_y, \qquad \text{and} \qquad \nabla_t^2 \tilde{H}_z = -k_c^2 \tilde{H}_z. \tag{10.11}$$

The first step in finding the complete field patterns for waveguide modes will be to use the z-component of Eq. (10.9),

$$\nabla_t^2 \tilde{E}_z(x, y) = -k_c^2 \tilde{E}_z(x, y), \tag{10.12}$$

to solve for $\tilde{E}_z(x, y)$, or the z-component of Eq. (10.11),

$$\nabla_t^2 \tilde{H}_z(x, y) = -k_c^2 \tilde{H}_z(x, y), \tag{10.13}$$

to solve for $\tilde{H}_z(x, y)$. These are relatively simple second-order differential equations, which we can solve using the following boundary conditions applied at each of the conducting walls of the waveguide:

- The normal component B_n of **B** must be continuous across the conductor surface. In the limit of the conductor being a perfect conductor (conductivity $\sigma \to \infty$), in which no fields exist, B_n of the magnetic flux density just outside the conductor (i.e. in the interior of the waveguide) must be zero as well.
- The tangential component E_{tan} of **E** must be continuous across the conductor surface. Again, in the limit of the conductor being a perfect conductor (conductivity $\sigma \to \infty$), in which no fields exist, E_{tan} of the electric field just outside the conductor (i.e. in the interior of the waveguide) must be zero.

Once the longitudinal field components $\tilde{E}_z(x, y)$ and $\tilde{H}_z(x, y)$ have been found, these components can be used to find solutions for the transverse components $\tilde{E}_x(x, y)$, $\tilde{E}_y(x, y)$, $\tilde{H}_x(x, y)$, and $\tilde{H}_y(x, y)$. To help in this, we will first find a set of relations between these transverse field components and the *transverse derivatives* of the z-components $\partial \tilde{E}_z(x, y)/\partial x$, $\partial \tilde{E}_z(x, y)/\partial y$, $\partial \tilde{H}_z(x, y)/\partial x$, and $\partial \tilde{H}_z(x, y)/\partial y$. These are found using Faraday's Law and Ampère's Law for fields in a linear, homogeneous, isotropic medium inside a waveguide, using the fields written in the specific forms of Eqs. (10.1) and (10.2) for a wave traveling in the $+z$-direction:

$$\nabla \times \left(\tilde{\mathbf{E}}(x, y)e^{-\gamma z} \right) = -j\omega\mu\tilde{\mathbf{H}}(x, y)e^{-\gamma z} \qquad \text{Faraday's Law}$$

and

$$\nabla \times \left(\tilde{\mathbf{H}}(x, y)e^{-\gamma z} \right) = j\omega\varepsilon\tilde{\mathbf{E}}(x, y)e^{-\gamma z} \qquad \text{Ampère's Law.}$$

Writing out these equations explicitly component-by-component gives us

$$x: \quad \frac{\partial \tilde{E}_z}{\partial y} + \gamma \tilde{E}_y = -j\omega\mu\tilde{H}_x, \quad (10.14a) \qquad \frac{\partial \tilde{H}_z}{\partial y} + \gamma \tilde{H}_y = j\omega\varepsilon\tilde{E}_x, \quad (10.15a)$$

$$y: -\gamma\tilde{E}_x - \frac{\partial \tilde{E}_z}{\partial x} = -j\omega\mu\tilde{H}_y, \quad (10.14b) \qquad -\gamma\tilde{H}_x - \frac{\partial \tilde{H}_z}{\partial x} = j\omega\varepsilon\tilde{E}_y, \quad (10.15b)$$

$$z: \quad \frac{\partial \tilde{E}_y}{\partial x} - \frac{\partial \tilde{E}_x}{\partial y} = -j\omega\mu\tilde{H}_z, \quad (10.14c) \qquad \frac{\partial \tilde{H}_y}{\partial x} - \frac{\partial \tilde{H}_x}{\partial y} = j\omega\varepsilon\tilde{E}_z. \quad (10.15c)$$

(The (x, y)-dependence notation of each of the field components is suppressed in Eqs. (10.14a)–(10.15c) for compactness.)

Now regroup these to express the x- and y-field components in terms of derivatives of $\tilde{E}_z(x, y)$ and $\tilde{H}_z(x, y)$. For example, the y-component of Faraday's Law (Eq. (10.14b)) can be used to replace $\tilde{H}_y(x, y)$ in the x-component of Ampère's Law (Eq. (10.15a)) to write

$$j\omega\varepsilon\tilde{E}_x(x, y) = \frac{\partial \tilde{H}_z(x, y)}{\partial y} + \gamma\tilde{H}_y(x, y)$$

$$= \frac{\partial \tilde{H}_z(x, y)}{\partial y} + \frac{\gamma}{-j\omega\mu}\left(-\gamma\tilde{E}_x(x, y) - \frac{\partial \tilde{E}_z(x, y)}{\partial x}\right).$$

Rearranging terms, multiplying through by a factor $-j\omega\mu$, and using Eq. (10.5), leads to

$$\tilde{E}_x(x, y) = -\frac{1}{k^2 + \gamma^2}\left(\gamma\frac{\partial \tilde{E}_z(x, y)}{\partial x} + j\omega\mu\frac{\partial \tilde{H}_z(x, y)}{\partial y}\right). \quad (10.16a)$$

Here, $\tilde{E}_x(x, y)$ is expressed as a particular combination of transverse derivatives ($\partial/\partial x$ and $\partial/\partial y$) of $\tilde{E}_z(x, y)$ and $\tilde{H}_z(x, y)$. Similarly, other components of Faraday's Law and Ampère's Law can be combined to derive the following additional relations:

$$\tilde{E}_y(x, y) = \frac{1}{k^2 + \gamma^2}\left(-\gamma\frac{\partial \tilde{E}_z(x, y)}{\partial y} + j\omega\mu\frac{\partial \tilde{H}_z(x, y)}{\partial x}\right), \quad (10.16b)$$

$$\tilde{H}_x(x, y) = \frac{1}{k^2 + \gamma^2}\left(j\omega\varepsilon\frac{\partial \tilde{E}_z(x, y)}{\partial y} - \gamma\frac{\partial \tilde{H}_z(x, y)}{\partial x}\right), \quad (10.16c)$$

and

$$\tilde{H}_y(x, y) = -\frac{1}{k^2 + \gamma^2}\left(j\omega\varepsilon\frac{\partial \tilde{E}_z(x, y)}{\partial x} + \gamma\frac{\partial \tilde{H}_z(x, y)}{\partial y}\right). \quad (10.16d)$$

Each of these four equations then gives us a field component $\tilde{E}_x(x, y)$, $\tilde{E}_y(x, y)$, $\tilde{H}_x(x, y)$, or $\tilde{H}_y(x, y)$ in terms of transverse derivatives ($\partial/\partial x$ and $\partial/\partial y$) of the longitudinal field components $\tilde{E}_z(x, y)$ and $\tilde{H}_z(x, y)$.

For each of the different waveguide geometries that we will consider in this chapter, our procedure will be to:

1. Solve for the electric field component $\tilde{E}_z(x, y)$ using Eq. (10.12) or the magnetic field component $\tilde{H}_z(x, y)$ using Eq. (10.13).
2. Use Eqs. (10.16a)–(10.16d) to determine the transverse field components $\tilde{E}_x(x, y)$, $\tilde{E}_y(x, y)$, $\tilde{H}_x(x, y)$, and $\tilde{H}_y(x, y)$.
3. Use these solutions for the electric and magnetic fields inside a waveguide to determine the wave propagation properties, such as the propagation constant γ, the waveguide wavelength Λ, the velocity of the wave in the waveguide (actually two velocities: the phase velocity and the group velocity), the impedance of the waveguide, and the attenuation due to losses.

As we study waveguide modes, or spatial patterns of the fields inside the waveguide, we will classify them as one of three types:

- **Transverse electric magnetic**, or **TEM**, modes. As the name implies, the electric and magnetic fields are each precisely in the transverse direction. That is, $\tilde{E}_z(x, y) = 0$ and $\tilde{H}_z(x, y) = 0$. In our discussion of transmission lines in Chapter 9 all of the modes were TEM modes, although we didn't make a point of this at the time. We will show later that TEM modes, while obviously permitted for the parallel plate waveguide system, are not allowed in single-conductor waveguides, such as rectangular or circular waveguides. TEM modes are the only modes allowed for which the free-space wavelength λ is much greater than the dimensions of the waveguide, or equivalently, are allowed at lower frequencies.
- **Transverse magnetic**, or **TM**, modes. In this case, the magnetic field is perpendicular to the z-axis, $\tilde{H}_z(x, y) = 0$, while $\tilde{E}(x, y)$ has a component in the z-direction.
- **Transverse electric**, or **TE**, modes. In this case, the electric field is perpendicular to the z-axis, $\tilde{E}_z(x, y) = 0$, while $\tilde{H}(x, y)$ has a component in the z-direction.

For each of the waveguide modes that we explore, we are interested in its propagation properties. For example, we will determine the wavelength of the waveguide mode as

$$\Lambda = \frac{2\pi}{\beta}$$

and the phase velocity as

$$u_p = \frac{\omega}{\beta}.$$

For any wave whose z- and t-dependence is $e^{j(\omega t - \beta z)}$, the derivations of Λ and u_p follow precisely the same approach as we described in Chapters 8 and 9, so they won't be repeated here.

In addition to the phase velocity, there exists another important wave velocity known as the **group velocity**, denoted u_g. In every case that we have considered to this point, we haven't worried about the distinction between these two parameters. But in many cases, the group velocity and the phase velocity differ from one another, and the group velocity can be just as important as the phase velocity. We address this distinction as follows. The phase velocity, as we defined previously, is the speed of the pattern of a single-frequency wave. That is, for a pure sinusoid. A wave that carries any information, however, must be modulated, either in phase/frequency or in amplitude. Either way, this modulation of the wave adds new frequency components to the spectrum of the wave, such that it consists not of a single frequency, but of multiple frequencies. Let's consider a wave consisting of just two sinusoids, for simplicity, whose frequencies ω_1 and ω_2 differ slightly. We'll denote the average frequency of these two waves as $\overline{\omega}$, and the difference in frequency as $\Delta\omega$. Similarly, the phase constants β_1 and β_2 differ slightly, and we'll use $\overline{\beta}$ to denote the average, and $\Delta\beta$ as their difference. Letting the amplitudes A of these two waves be equal, the superposition of the waves is written as

$$A_1(t, z) + A_2(t, z) = A\{\sin(\omega_1 t - \beta_1 z) + \sin(\omega_2 t - \beta_2 z)\}$$

$$= A\left\{\sin\left[\left(\overline{\omega} - \frac{\Delta\omega}{2}\right)t - \left(\overline{\beta} - \frac{\Delta\beta}{2}\right)z\right] + \sin\left[\left(\overline{\omega} + \frac{\Delta\omega}{2}\right)t - \left(\overline{\beta} + \frac{\Delta\beta}{2}\right)z\right]\right\}.$$

With a little trigonometry, this superposition wave can be written as

$$2A\sin(\overline{\omega}t - \overline{\beta}z)\cos(\Delta\omega t/2 - \Delta\beta z/2). \tag{10.17}$$

FIGURE 10.3 Plots of (a) two sine functions of slightly different frequency $\Delta\omega$, and (b) the superposition of these two sine functions, showing a distinct beat at the frequency difference.

(a)
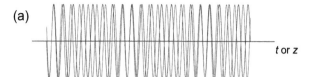
t or z

(b)
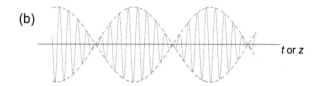
t or z

This is an interesting waveform, as shown in Fig. 10.3. The two individual sinusoids are plotted in Fig. 10.3(a), and their superposition in Fig. 10.3(b). The superposition wave oscillates at frequency $\bar{\omega}$, with a varying amplitude, or envelope, as shown by the red dashed lines. This envelope shows strong modulation at frequency $\Delta\omega/2$. Notice in Fig. 10.3(a) that the phase of the two original sinusoids is constantly changing, since their frequencies differ slightly. Where the individual sinusoids are approximately in phase with one another, the amplitude of the superposition in (b) is at a maximum, but where the two sinusoids are out of phase, the amplitude of the superposition wave is quite small. This plot can be viewed as a function of time t or position z. Its behavior is the same either way. Thus the abscissa of this plot is labeled with "t or z."

Example 10.1 Beat Signal

For two sinusoids, one of frequency 100 Hz and the other of frequency 99 Hz, determine the period of the slow envelope.

Solution:
From Eq. (10.17), the frequency of the envelope, sometimes called the beat signal, is

$$\frac{\Delta\omega}{2} = \frac{2\pi\Delta\nu}{2} = \pi \ \text{r/s},$$

where $\Delta\nu = 1$ Hz is the difference in frequency between the two signals. Then the period is

$$T = \frac{2\pi}{\Delta\omega/2} = 2 \ \text{s}.$$

Now we can ask how fast this wave moves. There are two features of interest. First, the rapidly oscillating structure beneath the broader envelope will move. The velocity of this feature of the wave is found by setting

$$\bar{\omega}t - \bar{\beta}z = \text{a constant},$$

so that if a particular peak is at z_1 at time t_1, and has moved to z_2 at time t_2, then

$$\bar{\omega}(t_2 - t_1) - \bar{\beta}(z_2 - z_1) = 0.$$

Defining $\Delta t = t_2 - t_1$ and $\Delta z = z_2 - z_1$, this relation is written as

$$\bar{\omega}\Delta t - \bar{\beta}\Delta z = 0.$$

So this part of the wave moves with velocity

$$\frac{\Delta z}{\Delta t} = \frac{\overline{\omega}}{\overline{\beta}}.$$

This is the average phase velocity u_p, discussed at length previously.

Next, we ask how fast the broad envelope moves. This envelope is described by the function $\cos(\Delta\omega t/2 - \Delta\beta z/2)$ in Eq. (10.17). The location of a peak (or any position, really) of the envelope moves such that

$$\Delta\omega t/2 - \Delta\beta z/2 = \text{a constant,}$$

or

$$\Delta\omega\Delta t/2 - \Delta\beta\Delta z/2 = 0.$$

Here the velocity of the envelope is seen to be

$$\frac{\Delta z}{\Delta t} = \frac{\Delta\omega}{\Delta\beta}.$$

This is the velocity called the **group velocity** u_g. In the limit of small differences $\Delta\omega$ and $\Delta\beta$, this becomes

$$u_g = \frac{d\omega}{d\beta}.$$

It can be shown that the group velocity is equal to the time-averaged power (energy flow per unit time) carried by the waveguide mode divided by the time-averaged energy density per unit length of the field. In other words, the group velocity represents the rate at which energy is carried by the waveguide mode. We'll determine the phase and group velocities for each of the waveguides discussed later.

Example 10.2 Phase Velocity and Group Velocity

Determine the phase velocity and group velocity of a freely propagating wave passing through a non-magnetic material whose permittivity is $\varepsilon(\omega) = \varepsilon_0(1 - a\omega)^{-2}$, where a is a constant.

Solution:
We start by finding the phase constant β as a function of frequency ω in this material,

$$\beta = \omega\sqrt{\varepsilon\mu},$$

where we use $\mu = \mu_0$ since the material is non-magnetic, and the permittivity as given,

$$\beta = \omega\sqrt{\varepsilon_0\mu_0}(1 - a\omega)^{-1} = \frac{\omega}{c}(1 - a\omega)^{-1}.$$

Then the phase velocity is

$$u_p = \frac{\omega}{\beta} = c(1 - a\omega).$$

To find the group velocity $u_g = d\omega/d\beta$, we first find the derivative,

$$\frac{d\beta}{d\omega} = \frac{1}{c}(1 - a\omega)^{-1} - \frac{\omega}{c}(-a)(1 - a\omega)^{-2} = \frac{1}{c}(1 - a\omega)^{-2}.$$

Then we use

$$u_g = \frac{d\omega}{d\beta} = \left(\frac{d\beta}{d\omega}\right)^{-1} = c(1 - a\omega)^2.$$

Another important characteristic of waveguide modes is known as the **wave impedance**, defined as the ratio of the transverse components of the electric and magnetic fields. Wave impedance was first introduced in Section 8.9. We will discover in the following examples that the transverse electric and magnetic field magnitudes are proportional to one another (as was the case for freely propagating waves and waves carried by transmission lines). For each of the different waveguide geometries, we will examine the waveguide impedance, denoted Z_{TEM}, Z_{TM}, and Z_{TE}, for the different modes supported by the waveguide.

The **total power** P carried by the waveguide mode is computed using the power density of the wave,

$$P = \int \frac{1}{2} \tilde{\mathbf{E}}(x,y) \times \tilde{\mathbf{H}}^*(x,y) \cdot \mathbf{a}_z \, dx \, dy, \tag{10.18}$$

where $(1/2)\tilde{\mathbf{E}}(x,y) \times \tilde{\mathbf{H}}^*(x,y)$ is the time-averaged Poynting vector, whose magnitude is the power density. The scalar product of the Poynting vector with $\mathbf{a}_z$ isolates the power flowing in the z-direction.

To this point, we have ignored any losses on the waveguide. But real dielectric media can absorb power from an electromagnetic wave, and the currents in the conductors can lead to losses as well. We introduced power losses of waves traveling through lossy dielectric media near conductors with finite conductivity in Sections 8.6.1 and 8.8, and applied them to losses in transmission lines in Section 9.3. Our treatment of losses in waveguides is very similar to that for transmission lines. The power lost per unit volume in a lossy dielectric medium is given by Eq. (8.43),

$$\overline{P_v^{\text{abs}}} = \frac{1}{2} \left(\sigma_d + \omega \varepsilon'' \right) \left| \tilde{\mathbf{E}}(x,y) \right|^2,$$

where σ_d is the conductivity of the dielectric and ε'' is the imaginary part of the permittivity. In an ideal dielectric, σ_d and ε'' are both zero. For a wave propagating in the z-direction in a waveguide, the power lost per unit length due to dielectric absorption P_d' can be easily determined by integrating $\overline{P_v^{\text{abs}}}$ over the cross-sectional area of the waveguide

$$P_d' = \int \frac{1}{2} \left(\sigma_d + \omega \varepsilon'' \right) \left| \tilde{\mathbf{E}}(x,y) \right|^2 \, dx \, dy. \tag{10.19}$$

Then we use

$$\alpha_d = \frac{P_d'(z)}{2P(z)}, \tag{10.20}$$

which follows directly from

$$P(z) = P_0 e^{-2\alpha z} \quad \text{and} \quad P'(z) = -\frac{dP(z)}{dz},$$

two relations that were introduced in Chapter 8.

We can also find the power lost per unit length P_c' due to the finite conductivity of the conductors. Recall the power loss per unit area at the surface of a conductor presented in Eq. (8.104),

$$\overline{P_s^{\text{abs}}} = \frac{1}{2} |\tilde{J}_s|^2 R_s,$$

where $R_s = 1/\sigma\delta = \sqrt{\omega\mu/2\sigma}$ is the surface resistance defined in Eq. (8.105), and $\tilde{J}_s$ is the surface current density in the conductor. We showed earlier that $\tilde{J}_s$ is equal in magnitude to the tangential magnetic field H_{tan}. (The *direction* of the surface current is perpendicular to that of the magnetic field, obeying a right-hand rule.) These losses must be integrated across dx or dy. For a conductor lying in the x–z plane, the total power lost per unit length is

$$P'_c = \int_0^a \frac{1}{2}|\tilde{J}_s|^2 R_s \, dx, \qquad (10.21a)$$

while for a conductor lying in the y–z plane,

$$P'_c = \int_0^b \frac{1}{2}|\tilde{J}_s|^2 R_s \, dy. \qquad (10.21b)$$

Finally, the attenuation constant α_c due to the conductor losses can be computed as

$$\alpha_c = \frac{P'_c}{2P}. \qquad (10.22)$$

We will use this approach in the following sections.

With this development of the general waveguide properties, we are now ready to consider waveguides of specific geometries. We will start with very wide parallel conducting plates, due to their simplicity. We follow this with treatments of rectangular and circular waveguide structures.

LEARNING OBJECTIVES

After studying this section, you should be able to:
- describe a parallel plate waveguide;
- determine the cutoff frequency for TM and TE modes on a parallel plate waveguide;
- for frequencies greater than the cutoff frequency, calculate the phase constant, wavelength, phase velocity, and group velocity of a TEM, TM, and TE mode on a parallel plate waveguide;
- for frequencies less than the cutoff frequency, determine the attenuation coefficient;
- calculate the wave impedance for a TEM, TM, and TE mode on a parallel plate waveguide;
- calculate the power density and total transmitted power for a TEM, TM, and TE mode on a parallel plate waveguide; and
- calculate the power loss and attenuation coefficient for a TEM, TM, and TE mode on a parallel plate waveguide.

10.2 Parallel Plate Waveguide

The first waveguide structure that we consider is the parallel plate waveguide, shown in Fig. 10.1(a). We will show that this structure supports TEM, TM, and TE modes. The width of the two parallel conductors is b, and the spacing a. The conductors extend to infinity in the z-direction. The discussion is limited to very wide conductors, $b \gg a$. In this limit, variations of the fields in the y-dimension can be ignored. (See the coordinate axes in Fig. 10.1.) We study TEM, TM, and TE modes in succession.

10.2.1 TEM Modes of a Parallel Plate Waveguide

For a TEM mode, the electric and magnetic fields are, by definition, transverse to the propagation direction. That is, $\tilde{E}_z(x,y) = 0$ and $\tilde{H}_z(x,y) = 0$. It turns out that Eqs. (10.16a)–(10.16d) are not valid for TEM waves, since each of these equations contains a factor $(k^2 + \gamma^2)^{-1}$, and, as we show below, $k^2 + \gamma^2 = 0$ for TEM modes. Therefore, we must return to Faraday's Law and Ampère's Law, Eqs. (10.14a)–(10.15c), which we repeat here with $\tilde{E}_z(x,y)$ and $\tilde{H}_z(x,y)$ set equal to zero:

$$x: \quad \gamma\tilde{E}_y = -j\omega\mu\tilde{H}_x, \quad (10.23a) \qquad\qquad \gamma\tilde{H}_y = j\omega\varepsilon\tilde{E}_x, \quad (10.24a)$$

$$y: \quad -\gamma\tilde{E}_x = -j\omega\mu\tilde{H}_y, \quad (10.23b) \qquad\qquad -\gamma\tilde{H}_x = j\omega\varepsilon\tilde{E}_y, \quad (10.24b)$$

$$z: \quad \frac{\partial\tilde{E}_y}{\partial x} - \frac{\partial\tilde{E}_x}{\partial y} = 0, \quad (10.23c) \qquad\qquad \frac{\partial\tilde{H}_y}{\partial x} - \frac{\partial\tilde{H}_x}{\partial y} = 0. \quad (10.24c)$$

(Once again, the (x,y)-dependence notation of each of these field quantities has been suppressed for compactness.)

Eqs. (10.23a) and (10.24b) can be rearranged as

$$\tilde{H}_x(x,y) = \frac{\gamma}{-j\omega\mu}\tilde{E}_y(x,y) = \frac{j\omega\varepsilon}{-\gamma}\tilde{E}_y(x,y),$$

while Eqs. (10.23*b*) and (10.24*a*) give us

$$\tilde{H}_y(x,y) = \frac{\gamma}{j\omega\mu}\tilde{E}_x(x,y) = \frac{j\omega\varepsilon}{\gamma}\tilde{E}_x(x,y).$$

From either of these equations we see that

$$\gamma^2 = -\omega^2\varepsilon\mu = -k^2,$$

as was stated above. Taking the square root of both sides gives us

$$\gamma = \pm j\omega\sqrt{\varepsilon\mu},$$

consistent with the propagation constant for unguided waves discussed in Chapter 8, as well as for wave propagation on transmission lines in Chapter 9. The "+" sign is used for waves propagating in the +*z*-direction, and the "−" sign for waves propagating in the −*z*-direction. For non-absorbing media, for which ε and μ are real, γ is purely imaginary for all frequencies ω, and can be written $\gamma = j\beta$, where the phase constant β is $\omega\sqrt{\mu\varepsilon}$. This feature tells us that TEM modes *of any frequency* can propagate on parallel plate waveguides. For TM and TE modes, we will show later that only waves whose frequency is greater than a certain cutoff frequency ω_c can propagate. Below this frequency, the wave is sharply attenuated in the *z*-direction. We will discuss this cutoff frequency for those modes later, but for now simply state that the cutoff frequency ω_c for TEM modes is zero.

Now let's examine the *z*-component equations: Eqs. (10.23*c*) and (10.24*c*). These two conditions, along with the boundary conditions that the fields must satisfy at the conductor surface, tell us that the electric field has only an *x*-component in the region between the conductors, and the magnetic field only a *y*-component, as we now show. We start by showing that $\tilde{\mathbf{E}}(x,y)$ can have only an *x*-component. We already discussed that for wide conductor plates, $b \gg a$, the fields are uniform across the width of the waveguide. That is,

$$\frac{\partial \tilde{E}_x(x,y)}{\partial y} = 0 \qquad \text{and} \qquad \frac{\partial \tilde{H}_x(x,y)}{\partial y} = 0.$$

Therefore, by Eqs. (10.23*c*) and (10.24*c*), we conclude that

$$\frac{\partial \tilde{E}_y(x,y)}{\partial x} \qquad \text{and} \qquad \frac{\partial \tilde{H}_y(x,y)}{\partial x},$$

respectively, must also be equal to zero. That is, the field components $\tilde{E}_y(x,y)$ and $\tilde{H}_y(x,y)$ are uniform in the *x*-dimension. You will recall that we asserted this in Chapter 9, but here we see that this is required by Faraday's and Ampère's Laws. Also, since $\partial\tilde{E}_y(x,y)/\partial x = 0$ everywhere inside the waveguide, and $\tilde{E}_y(x,y) = 0$ at either of the conductors, then $\tilde{E}_y(x,y)$ must be zero everywhere. To summarize, the electric field **E** of the TEM mode inside the parallel plate waveguide has only an *x*-component, $\tilde{E}_x(x,y)$, and this component must be uniform, not varying with *x* or *y*. The electric field points directly from one conductor to the other, as expected.

We have to work a little harder to show that **H** can point only in the *y*-direction. In this case, we'll use the Maxwell Equation that tells us there are no magnetic charges, $\nabla \cdot \mathbf{B} = 0$, which for a linear, isotropic, homogeneous material we can write as

$$\nabla \cdot \mathbf{H} = \frac{\partial H_x}{\partial x} + \frac{\partial H_y}{\partial y} + \frac{\partial H_z}{\partial z} = 0.$$

Since $\tilde{H}_z(x,y) = 0$ for a TEM wave, and fields are independent of y in the wide parallel plate waveguide (i.e. $\partial \tilde{H}_y(x,y)/\partial y = 0$), $\nabla \cdot \tilde{\mathbf{H}} = 0$ reduces to

$$\frac{\partial \tilde{H}_x(x,y)}{\partial x} = 0.$$

And since $\tilde{H}_x(x,y)$ must be zero at the conductor surfaces (H_x is the component normal to the conductor surface), $\tilde{H}_x(x,y)$ must also be zero everywhere between the conductors, as we stated earlier. In summary, $\tilde{\mathbf{E}}(x,y)$ and $\tilde{\mathbf{H}}(x,y)$ for the TEM mode must be uniform in the region between the conductors, and they must be perpendicular to one another, consistent with the property of freely propagating waves and waves on transmission lines. We write these conditions compactly as

$$\tilde{\mathbf{E}}(x,y) = E_x \mathbf{a}_x \tag{10.25a}$$

and

$$\tilde{\mathbf{H}}(x,y) = H_y \mathbf{a}_y, \tag{10.25b}$$

where E_x and H_y are constants.

Using $\beta = \omega\sqrt{\varepsilon\mu}$, the wavelength Λ of the TEM wave on the waveguide is determined as

$$\Lambda = \frac{2\pi}{\beta} = \frac{2\pi}{\omega\sqrt{\varepsilon\mu}},$$

the phase velocity is

$$u_p = \frac{\omega}{\beta} = \frac{1}{\sqrt{\varepsilon\mu}},$$

and the group velocity is

$$u_g = \frac{d\omega}{d\beta} = \frac{1}{\sqrt{\varepsilon\mu}}.$$

The final equality is valid when ε and μ are independent of frequency. For the TEM mode, the phase velocity u_p and group velocity u_g are equal to one another.

As noted earlier, the field components $\tilde{E}_x(x,y)$ and $\tilde{H}_y(x,y)$ are each uniform in the space between the conductors. The wave impedance of the TEM mode, defined as the ratio of these field strengths, is

$$Z_{TEM} = \left| \frac{\tilde{E}_x}{\tilde{H}_y} \right| = \left| \frac{E_x}{(\gamma/(j\omega\mu))E_x} \right| = \left| \frac{j\omega\mu}{\gamma} \right| = \left| \frac{j\omega\mu}{j\omega\sqrt{\varepsilon\mu}} \right| = \sqrt{\frac{\mu}{\varepsilon}}.$$

From this it is seen that the wave impedance for this waveguide mode,

$$Z_{TEM} = \sqrt{\frac{\mu}{\varepsilon}} = \eta, \tag{10.26}$$

is the same as the intrinsic impedance η of the dielectric medium for a freely propagating wave in this medium.

The total power carried by the TEM waveguide mode is, from Eq. (10.18),

$$P = \int_0^b \int_0^a \frac{1}{2} \tilde{\mathbf{E}}(x,y) \times \tilde{\mathbf{H}}^*(x,y) \cdot \mathbf{a}_z \, dx \, dy = \frac{1}{2} E_x H_y^* \, ab = \frac{|E_x|^2}{2Z_{TEM}} \, ab.$$

Finally, we examine the attenuation constants α_d due to absorption by the dielectric and α_c due to the finite conductivity of the conductors. For this, we need the power loss per unit length of the waveguide. For dielectric absorption, we use Eq. (10.19) to find

$$P_d' = \frac{1}{2}(\sigma_d + \omega\varepsilon'')|\tilde{E}_x|^2 \, ab.$$

Table 10.1

Summary of results for TEM modes on a very wide ($b \gg a$) parallel plate waveguide.		
$\tilde{E}_z(x, y) = 0$		$\tilde{H}_z(x, y) = 0$
$\tilde{E}_x(x, y) = E_x$		$\tilde{H}_x(x, y) = 0$
$\tilde{E}_y(x, y) = 0$		$\tilde{H}_y(x, y) = \dfrac{E_x}{Z_{TEM}}$
$\omega_c = 0$	$u_p = u_g = \dfrac{1}{\sqrt{\mu\varepsilon}}$	$Z_{TEM} = \eta$
$P = \dfrac{\|E_x\|^2}{2Z_{TEM}} ab$	$\alpha_d = \dfrac{\eta(\sigma_d + \omega\varepsilon'')}{2}$	$\alpha_c = \dfrac{R_s}{a\eta}$

Then, using Eq. (10.20), we find

$$\alpha_d = \frac{P_d'}{2P} = \frac{\frac{1}{2}(\sigma_d + \omega\varepsilon'')|E_x|^2 \, ab}{2\left(|E_x|^2/2Z_{TEM}\right) ab}$$

$$= \frac{\eta(\sigma_d + \omega\varepsilon'')}{2}. \tag{10.27}$$

From Eq. (10.21b), the power loss per unit length due to the finite conductivity of the conductor is

$$P_c' = \int_0^b \frac{1}{2}|\tilde{J}_s|^2 R_s \, dy \times 2,$$

where we integrate across the width b of the conductor in the y-direction, and multiply by 2 since there are two conductors. The surface current $\tilde{J}_s$ is equal in magnitude to the tangential $\tilde{H}_y$ field, so

$$P_c' = |H_y|^2 R_s b.$$

Using P_c' and P in Eq. (10.22), the attenuation coefficient is

$$\alpha_c = \frac{P_c'}{2P} = \frac{|H_y|^2 R_s b}{E_x H_y^* ab} = \frac{R_s H_y}{E_x a} = \frac{R_s}{a\eta}.$$

In summary, the TEM mode in a wide parallel plate waveguide is characterized by uniform fields $\tilde{\mathbf{E}}(x, y) = E_x \mathbf{a}_x$ and $\tilde{\mathbf{H}}(x, y) = H_y \mathbf{a}_y$, with $H_y = E_x/Z_{TEM}$, where Z_{TEM} is equal to the impedance of the dielectric medium η. The propagation constant β is simply $\omega\sqrt{\varepsilon\mu}$, and the phase velocity is independent of frequency (to the extent that ε and μ for the dielectric medium inside the waveguide do not depend on the frequency ω of the fields). Using Faraday's and Ampère's Laws, we have verified the mode pattern that we simply asserted in Chapter 9. The key results for TEM modes on a wide parallel plate waveguide are collected in Table 10.1.

In the following sections we will examine the TM and TE modes in these same structures. These modes have very different behaviors to that of the TEM mode developed here.

10.2.2 TM Modes on a Parallel Plate Waveguide

We turn our attention now to transverse magnetic, or TM, modes on a parallel plate waveguide. For a TM mode, $\tilde{H}_z(x, y)$ is zero (by definition), but $\tilde{E}_z(x, y)$ is not. As outlined in Section 10.1, we will solve Eq. (10.12) to determine $\tilde{E}_z(x, y)$, use this solution to find the transverse field components $\tilde{E}_x(x,y)$, $\tilde{E}_y(x,y)$, $\tilde{H}_x(x,y)$, and $\tilde{H}_y(x,y)$, and determine the parameters that describe the propagation of this wave.

We start with Eq. (10.12),

$$\nabla_t^2 \tilde{E}_z(x,y) = -k_c^2 \tilde{E}_z(x,y),$$

where $k_c^2 = \gamma^2 + k^2$. We again use the condition that the fields cannot depend on the variable y in the wide parallel plate waveguide, so

$$\frac{\partial^2 \tilde{E}_z(x,y)}{\partial y^2} = 0,$$

and Eq. (10.12) then simplifies to

$$\frac{\partial^2 \tilde{E}_z(x)}{\partial x^2} = -k_c^2 \tilde{E}_z(x).$$

Note that $\tilde{E}_z(x,y)$ is now written $\tilde{E}_z(x)$ since the phasor amplitude depends only on x, but not on y. We have seen differential equations of this form many times previously. As long as k_c^2 is positive, its solution is

$$\tilde{E}_z(x) = A\sin(k_c x) + B\cos(k_c x), \tag{10.28}$$

where A and B are amplitudes that must be determined using boundary conditions applied at the top and bottom conductor surfaces.

Since

$$k_c^2 = k^2 + \gamma^2$$

must be positive for this solution to be valid, and γ^2 must be negative for wave propagation in general (i.e. γ is purely imaginary), then k_c for the waveguide mode must be less than k for the freely propagating wave (i.e. $k > k_c$). Therefore, only high-frequency waves can propagate. We'll return to this point shortly. As discussed earlier, the relevant boundary condition at the surface of a conductor is that E must be perpendicular to the conductor, which tells us that $\tilde{E}_z(x)$ must be zero at $x = 0$ and at $x = a$. The first condition (that $\tilde{E}_z(x)$ must be zero at $x = 0$) implies that the amplitude B in Eq. (10.28) must be zero. Then, at $x = a$, the second boundary condition requires that

$$\tilde{E}_z(a) = A\sin(k_c a) = 0,$$

which is satisfied only when

$$k_c = \frac{m\pi}{a}, \tag{10.29}$$

where m is a positive integer. m is used to label the various modes, or patterns, of the field inside the waveguide. $m = 1$ labels the lowest TM mode, for which half of a cycle of $\tilde{E}_z(x)$ fills the space from one conductor to the other. This is denoted as the TM_1 mode, using $m = 1$ as a subscript on TM. For $m = 2$, one full cycle of the TM_2 mode fits into this space; for $m = 3$, one and a half cycles, and so forth. The z-component of the electric field for a TM wave must therefore be of the form

$$\tilde{E}_z(x) = A\sin\left(\frac{m\pi x}{a}\right). \tag{10.30a}$$

Notice that $m = 0$ is not allowed for a TM mode, since $\tilde{E}_z(x)$ would be zero everywhere.

Now $\tilde{E}_z(x)$ from Eq. (10.30a), along with $\tilde{H}_z(x) = 0$, can be used in Eq. (10.16a) to determine

$$\tilde{E}_x(x) = -\frac{1}{k^2 + \gamma^2}\left(\gamma\frac{\partial \tilde{E}_z(x,y)}{\partial x} + j\omega\mu\frac{\partial \tilde{H}_z(x,y)}{\partial y}\right) = -j\beta\left(\frac{a}{m\pi}\right)\cos\left(\frac{m\pi x}{a}\right). \tag{10.30b}$$

In completing Eq. (10.30b), we have used $k^2 + \gamma^2 = k_c^2 = (m\pi/a)^2$ and $\gamma = j\beta$, valid for high frequencies as long as losses are negligible. Similarly, Eq. (10.16d) can be used to find

$$\tilde{H}_y(x) = -\frac{1}{k^2 + \gamma^2}\left(j\omega\varepsilon\frac{\partial \tilde{E}_z(x,y)}{\partial x} + \gamma\frac{\partial \tilde{H}_z(x,y)}{\partial y}\right) = -j\omega\varepsilon\left(\frac{a}{m\pi}\right)A\cos\left(\frac{m\pi x}{a}\right).$$

(10.30c)

Equations (10.16b) and (10.16c) tell us that $\tilde{E}_y(x)$ and $\tilde{H}_x(x)$ are each zero between the conductors.

Now that we have found the phasors $\tilde{E}_z(x)$, $\tilde{E}_x(x)$, and $\tilde{H}_y(x)$ for a TM mode on the parallel plate waveguide, we'd like to create a visual representation of this mode. What is the spatial pattern of the electric and magnetic fields? This visualization is hampered a bit by the factor of j in the expressions for the phasors $\tilde{E}_x(x)$ and $\tilde{H}_y(x)$. This factor represents a $\pi/2$ phase shift of these field components relative to the $\tilde{E}_z(x)$. Let's proceed by first determining the full x-, z-, and t-dependent form of these field components. As a reminder, the field amplitudes $\tilde{E}_x(x)$ and $\tilde{H}_y(x)$ only give us the x-dependence of the fields. For the full x-, z-, and t-dependence, refer to Eqs. (10.1) and (10.2). Briefly, we must multiply $\tilde{E}_z(x)$, $\tilde{E}_x(x)$, and $\tilde{H}_y(x)$ by $\exp\{j(\omega t - \beta z)\}$ and take the real part. Following this process, the field components can be written

$$E_z(x, z, t) = A\sin\left(\frac{m\pi x}{a}\right)\cos\left(\omega t - \beta z\right),$$

$$E_x(x, z, t) = \beta\left(\frac{a}{m\pi}\right)A\cos\left(\frac{m\pi x}{a}\right)\sin\left(\omega t - \beta z\right),$$

and

$$H_y(x, z, t) = \omega\varepsilon\left(\frac{a}{m\pi}\right)A\cos\left(\frac{m\pi x}{a}\right)\sin\left(\omega t - \beta z\right).$$

Note that the impact of the factor j in the expressions for $\tilde{E}_x(x)$ and $\tilde{H}_y(x)$ is that $E_x(x, z, t)$ and $H_y(x, z, t)$ vary as $\sin\left(\omega t - \beta z\right)$, whereas $E_z(x, z, t)$ varies as $\cos\left(\omega t - \beta z\right)$. An x–z plot of $\mathbf{E}(x, z, t = 0)$ (a "snapshot" at $t = 0$) for the TM$_1$ mode is shown in Fig. 10.4. Notice how the electric field lines meet the top and bottom conductors at a right angle. The wavelength Λ of the repeating waveform is marked on the figure. $E_z(x, z, t)$ and $E_x(x, z, t)$ are $\pi/2$ out of phase with one another, with $E_z(x, z, t)$ at its maximum at the

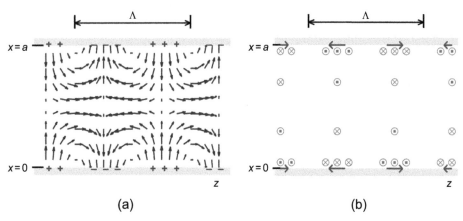

(a)   (b)

FIGURE 10.4 (a) The electric field pattern and (b) the magnetic field pattern of the TM$_1$ mode of a parallel plate waveguide. The yellow bars represent the conductors, spaced by a. The magenta + and − signs in (a) represent the surface charge on the inside surface of the conductors. The magenta arrows in (b) represent the surface current on the inside surface of the conductors. The wave propagates in the z-direction. Only a short segment of the waveguide is shown.

beginning of the cycle shown, and $E_x(x, z, t)$ reaching its maximum magnitude a quarter-cycle to the right. If we were to examine the wave at a slightly later time, we would observe that this entire field pattern has moved as a unit to the right.

What can we say about the wavelength Λ shown in Fig. 10.4? We start by examining the complex propagation term γ,

$$\gamma = \sqrt{k_c^2 - k^2}.$$

Inserting Eqs. (10.29) and (10.5) for k_c and k, respectively, this becomes

$$\gamma = \sqrt{\left(\frac{m\pi}{a}\right)^2 - \omega^2 \varepsilon \mu}.$$

For a propagating wave, γ is must be purely imaginary (ignoring losses), implying that

$$\left(\frac{m\pi}{a}\right)^2 - \omega^2 \varepsilon \mu < 0,$$

or

$$\omega > \frac{m\pi}{a\sqrt{\varepsilon \mu}}.$$

The right-hand side of this equation is called the **cutoff frequency** of the waveguide mode,

$$\omega_c = \frac{k_c}{\sqrt{\varepsilon \mu}} = \frac{m\pi}{a\sqrt{\varepsilon \mu}}, \tag{10.31}$$

and it represents the lowest frequency at which the mode can propagate on the waveguide. In cyclical frequency units, the cutoff frequency is

$$f_c = \frac{\omega_c}{2\pi} = \frac{m}{2a\sqrt{\varepsilon \mu}}. \tag{10.32}$$

For large frequencies $\omega > \omega_c$, $\beta = -j\gamma$ is

$$\beta = \omega\sqrt{\varepsilon \mu}\sqrt{1 - \left(\frac{\omega_c}{\omega}\right)^2}. \tag{10.33}$$

A plot of β vs. frequency f is shown in Fig. 10.5(a). For large frequencies, $\omega \gg \omega_c$, the phase constant β of the waveguide mode approaches the phase constant $k = \omega\sqrt{\varepsilon \mu}$ of the unguided wave, represented by the straight, red, dashed line in the figure. As the frequency of the wave decreases, however, the propagation constant β decreases, dropping to zero at frequency $\omega = \omega_c$.

FIGURE 10.5 A plot of (a) $\beta/2\pi$ and (b) α vs. frequency f for the TM$_1$ mode on a parallel plate waveguide. The waveguide is very wide, $b \gg a$, air-filled, and the plate spacing is $a = 1$ cm. The cutoff frequency of this waveguide is $f_c = 15$ GHz.

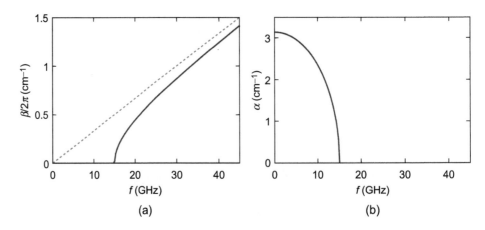

We can use β to determine the wavelength Λ of the mode pattern of a TM_m wave, using

$$\Lambda = \frac{2\pi}{\beta} = \frac{\lambda}{\sqrt{1 - (\omega_c/\omega)^2}}, \tag{10.34}$$

where $\lambda = 2\pi/\omega\sqrt{\mu\varepsilon}$ is the wavelength of an unguided wave at this frequency. For $\omega \gg \omega_c$, the wavelength of the wave in the waveguide approaches that of the unguided wave. As the frequency ω approaches ω_c, however, the wavelength gets extremely large, approaching ∞ at $\omega = \omega_c$.

We can gain some insight into these propagation effects by picturing a uniform plane wave (UPW) bouncing back and forth at an angle θ between the two conductors of the waveguide, as illustrated in Fig. 10.6(a). The plane wave travels in the x–z plane, and can be defined by its propagation vector $\mathbf{k} = k_x\mathbf{a}_x + k_z\mathbf{a}_z$. k_x can be determined by following the wave through a round-trip in the x-direction from one conductor to the other and back, after which the wave adds constructively with the original wave. That is, $k_x a = m\pi$, where m is an integer. (m is simply the mode index.) Similarly, we can find k_z by following the wave along the waveguide a distance Λ in the z-direction, corresponding to a 2π phase shift. So $k_z\Lambda = 2\pi$. Using these values of k_x and k_z, the magnitude of $|\mathbf{k}|$ is

$$|\mathbf{k}| = k = \sqrt{k_x^2 + k_z^2} = \sqrt{\left(\frac{m\pi}{a}\right)^2 + \left(\frac{2\pi}{\Lambda}\right)^2} = 2\pi/\lambda.$$

Since $\lambda = 2\pi/\omega\sqrt{\varepsilon\mu}$ is the wavelength of the plane wave, and $a = m\pi/\omega_c\sqrt{\varepsilon\mu}$, we can show with a little algebra that $\Lambda = \lambda/\sqrt{1 - (\omega_c/\omega)^2}$, in agreement with Eq. (10.34). At high frequencies (small wavelengths), the UPW propagates nearly straight along the waveguide axis in the z-direction, with the angle θ nearly zero. In this case, the wavelength Λ approaches the wavelength λ of an unguided wave. As the frequency decreases, however, the UPW propagates closer to a perpendicular direction, $\theta \sim \pi/2$, bouncing from one conductor to the other and making very little headway in the z-direction. The phase fronts of the waves shown in Fig. 10.6 extend far down the waveguide before they cross the axis in this case, so the distance between wavefronts, measured along the z-axis, is quite long. When the wave frequency equals the cutoff frequency $\omega = \omega_c$, the spacing between the conductors a is equal to $\lambda/2$ for the $m = 1$ mode, and half a wavelength of the UPW just fits between the parallel plates, with no propagation along the z-axis at all.

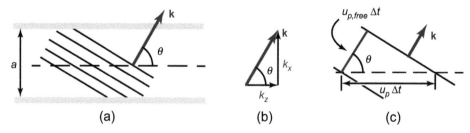

(a)   (b)   (c)

FIGURE 10.6 (a) A model of the parallel plate waveguide mode as a uniform plane wave bouncing back and forth between the two conductors. (b) The x- and z-components of the propagation vector $\mathbf{k}$ of the uniform plane wave. (c) Supporting diagram for the calculation of the phase velocity u_p.

Example 10.3 Cutoff Frequencies, Parallel Plate

Consider an air-filled parallel plate waveguide of width $b = 10$ cm and plate separation $a = 0.5$ cm. (a) Determine the cutoff frequencies for the TM_1, TM_2, and TM_3 modes. (b) Find the corresponding free-space wavelengths. How do these wavelengths relate to the waveguide dimensions?

Solution:

(a) We'll use Eq. (10.32) for the cutoff frequencies. Since the material inside the waveguide is air, $\varepsilon = \varepsilon_0$ and $\mu = \mu_0$ (to a good approximation), so $\sqrt{\varepsilon\mu} = c^{-1}$, and

$$\frac{1}{2a\sqrt{\varepsilon\mu}} = \frac{c}{2a} = \frac{3 \times 10^{10} \text{ cm/s}}{2(0.5 \text{ cm})} = 30 \text{ GHz.}$$

Then the cutoff frequencies of the three modes requested are

$$f_c = m\frac{c}{2a} = \begin{cases} 30 \text{ GHz,} & \text{for } m = 1 \\ 60 \text{ GHz,} & \text{for } m = 2. \\ 90 \text{ GHz,} & \text{for } m = 3 \end{cases}$$

(b) The free-space wavelengths at these frequencies are

$$\lambda_c = \frac{c}{f_c} = \begin{cases} 1 \text{ cm,} & \text{for } m = 1 \\ 0.5 \text{ cm,} & \text{for } m = 2. \\ 0.33 \text{ cm,} & \text{for } m = 3 \end{cases}$$

Each of these wavelengths is an integer fraction of $2a$. That is, $\lambda = 2a/m$.

Let's also examine the phase constant and mode wavelength on this waveguide.

Example 10.4 Phase Constant and Wavelength, Parallel Plate, TM

Consider the air-filled parallel plate waveguide described in Example 10.3. Determine the phase constant β and the mode wavelength Λ for a TM_1 mode of frequency 45 GHz.

Solution:
To find the phase constant, we use Eq. (10.33)

$$\beta = \omega\sqrt{\mu\varepsilon}\sqrt{1 - \left(\frac{\omega_c}{\omega}\right)^2} = \frac{2\pi(45 \times 10^9 \text{ s}^{-1})}{2.998 \times 10^{10} \text{ cm/s}}\sqrt{1 - \left(\frac{30 \text{ GHz}}{45 \text{ GHz}}\right)^2} = 7.03 \text{ rad/cm.}$$

Notice that this value is less than the unguided wave phase constant of 9.43 rad/cm.
We use this value of β to determine the mode wavelength,

$$\Lambda = \frac{2\pi}{\beta} = 0.894 \text{ cm.}$$

This mode wavelength is the distance along the z-axis after which the field pattern repeats itself.

We still need to examine the velocity and impedance of this wave, but before we do that, let's look at the TM mode at low frequencies, $\omega < \omega_c$. The solution Eq. (10.28) for $\tilde{E}_z(x)$ is valid only for high frequencies, such that $k > k_c$. This condition is the same as requiring that the frequency ω is greater than the cutoff frequency ω_c, for which the propagation constant γ is purely imaginary. For low frequencies, however, γ is purely real, as is seen from

$$\gamma = \sqrt{k_c^2 - k^2} = \sqrt{\left(\frac{m\pi}{a}\right)^2 - \omega^2 \mu \varepsilon}.$$

In this case, the field, rather than forming a propagating wave, decays exponentially in the z-direction. In terms of the cutoff frequency ω_c, the attenuation constant $\alpha = Re(\gamma)$ can be shown to be

$$\alpha = \left(\frac{m\pi}{a}\right)\sqrt{1 - \left(\frac{\omega}{\omega_c}\right)^2}. \tag{10.35}$$

A plot of α vs. frequency $f = \omega/2\pi$ is shown in Fig. 10.5(b). In our physical picture of this mode, the wavelength λ is "too big to fit" into the space between the conductors. So for low frequencies, $\omega < \omega_c$, the field amplitudes are attenuated with increasing z, and the wave cannot propagate. This attenuation is independent of absorption by the conductors or the dielectric material between the conductors, but rather is simply a matter of the wave not fitting in the space between the conductors.

Example 10.5 Attenuation Coefficient, Parallel Plate, TM

(a) For the air-filled parallel plate waveguide described in Example 10.3, determine the attenuation coefficient α for a TM_1 mode of frequency 15 GHz. (b) Determine the distance Δz over which the field amplitude decreases to 1% of its initial value.

Solution:

(a) For the attenuation coefficient, we use Eq. (10.35):

$$\alpha = \left(\frac{m\pi}{a}\right)\sqrt{1 - \left(\frac{\omega}{\omega_c}\right)^2} = \frac{\pi}{0.5 \text{ cm}}\sqrt{1 - \left(\frac{15 \text{ GHz}}{30 \text{ GHz}}\right)^2} = 5.44 \text{ cm}^{-1}.$$

(b) The distance Δz that we seek satisfies the condition

$$e^{-\alpha \Delta z} = 0.01.$$

Taking the natural logarithm of both sides and dividing by α, we get

$$\Delta z = \frac{-\ln(0.01)}{\alpha} = \frac{\ln(100)}{5.44 \text{ cm}^{-1}} = 0.846 \text{ cm}.$$

We now return to the propagating mode with frequency ω greater than the cutoff frequency ω_c. As stated earlier, the waveform shown in Fig. 10.4 moves in the $+z$-direction as time t advances. What is the velocity of this waveform? For unguided waves, waves

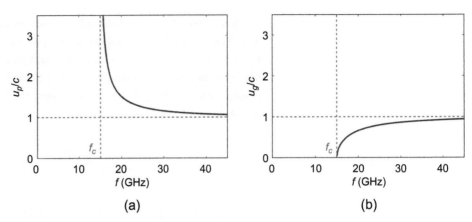

FIGURE 10.7 A plot of (a) the phase velocity u_p and (b) the group velocity u_g vs. frequency f for a TM_1 mode on a parallel plate waveguide.

on transmission lines, and even for the TEM mode on the parallel plate waveguide, the phase velocity is $u_p = \omega/\beta$. Using Eq. (10.33), we find

$$u_p = \frac{\omega}{\beta} = \frac{1}{\sqrt{\varepsilon\mu}} \frac{1}{\sqrt{1 - (\omega_c/\omega)^2}}. \tag{10.36}$$

Recall that $1/\sqrt{\varepsilon\mu}$ is the phase velocity of a free wave, $u_{p,free}$. The phase velocity vs. frequency f of the TM mode is plotted in Fig. 10.7. The phase velocity in the waveguide approaches the free value for large frequencies $f \gg f_c$, but becomes extremely large as $f \to f_c$.

This seems strange. How do we understand a phase velocity u_p that is greater than c? For the answer to this question, we return to the uniform plane wave model of the waveguide mode illustrated in Fig. 10.6. In Fig. 10.6(c), two wavefronts of a UPW that is moving at the angle θ from the z-axis of the waveguide are shown. In time Δt the second wavefront has traveled a distance $u_{p,free}\Delta t$ from the first. The phase velocity of the mode is a measure of how quickly the point at which the wavefront crosses the z-axis moves. From Fig. 10.6(c), it can be seen that $u_p = u_{p,free}/\cos\theta$, where $\theta = \tan^{-1}(k_x/k_z)$. (See Fig. 10.6(b).) Using $k_x = m\pi/a$, $k_z = 2\pi/\Lambda$, Λ from Eq. (10.34), and a little algebra, we can show that $u_p = u_{p,free}\left(1 - (\omega_c/\omega)^2\right)^{-1/2}$, in agreement with Eq. (10.36). So while the phase front of this plane wave moves with velocity $u_{p,free}$, since the plane wave moves at the angle θ, the intersection point between the wavefront and the z-axis of the waveguide moves at a faster rate, and the latter is given by the phase velocity u_p.

In contrast, the group velocity of the TM_m mode on the parallel plate waveguide is

$$u_g = \frac{d\omega}{d\beta} = \left(\frac{d\beta}{d\omega}\right)^{-1}$$
$$= \frac{1}{\sqrt{\varepsilon\mu}}\sqrt{1 - \left(\frac{\omega_c}{\omega}\right)^2} = u_{p,free}\sqrt{1 - \left(\frac{\omega_c}{\omega}\right)^2}. \tag{10.37}$$

A plot of the group velocity vs. frequency f is shown in Fig. 10.7(b). Notice how this differs from the plot of the phase velocity. For frequencies just greater than ω_c, the group velocity becomes very small. In terms of the UPW model, this results from the plane wave bouncing back and forth between the conductors, but making very little headway in the z-direction. This can be computed as $u_g = (k_z/k)u_{p,free} = (k_z/\sqrt{k_x^2 + k_z^2})u_{p,free}$. Using $k_x = m\pi/a$, $k_z = 2\pi/\Lambda$, and Λ from Eq. (10.34), one can easily confirm the group velocity of Eq. (10.37).

Example 10.6 Phase and Group Velocity, Parallel Plate, TM

Determine the phase and group velocity for the TM_1 mode at a frequency of 45 GHz on the waveguide described in Example 10.3.

Solution:

The phase velocity is given in Eq. (10.36) as

$$u_p = \frac{u_{p,free}}{\sqrt{1 - (\omega_c/\omega)^2}} = \frac{2.998 \times 10^{10} \text{ cm/s}}{\sqrt{1 - (30 \text{ GHz}/45 \text{ GHz})^2}} = 4.02 \times 10^{10} \text{ cm/s}.$$

Using Eq. (10.37), the group velocity is

$$u_g = u_{p,free}\sqrt{1 - (\omega_c/\omega)^2}$$
$$= 2.998 \times 10^{10} \text{ cm/s}\sqrt{1 - (30 \text{ GHz}/45 \text{ GHz})^2} = 2.23 \times 10^{10} \text{ cm/s}.$$

Next, we examine the wave impedance Z_{TM} of the waveguide for a TM wave. This is computed as the ratio of the transverse components $\tilde{E}_x$ and $\tilde{H}_y$. Using the forms of these field quantities given by Eqs. (10.30b) and (10.30c), this impedance is

$$Z_{TM} = \frac{\tilde{E}_x(x)}{\tilde{H}_y(x)} = \frac{\beta}{\omega\varepsilon} = \eta\sqrt{1 - \left(\frac{\omega_c}{\omega}\right)^2}, \qquad (10.38)$$

where we used Eq. (10.33) for β, and $\eta = \sqrt{\mu/\varepsilon}$ is the impedance of the dielectric medium interior to the waveguide. This impedance is plotted in Fig. 10.8(a).

This impedance approaches that of an unguided wave for large frequencies $\omega \gg \omega_c$, and becomes zero as ω approaches ω_c. Note that this impedance:

- is independent of spatial position (x, y, z), since $\tilde{E}_x(x)$ has the same profile as $\tilde{H}_y(x)$;
- is purely real for high frequencies $\omega > \omega_c$, since $\tilde{E}_x(x)$ and $\tilde{H}_y(x)$ are in phase with one another; and
- is purely imaginary for low frequencies $\omega < \omega_c$, the frequency space in which the waveguide does not support a propagating mode, indicating that $\tilde{E}_x(x)$ and $\tilde{H}_y(x)$ are

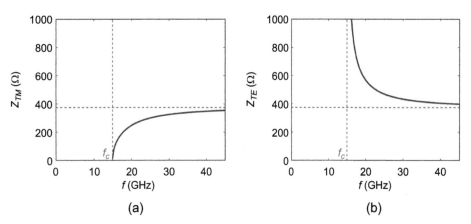

FIGURE 10.8 A plot of the mode impedance (a) Z_{TM} for the TM_1 mode and (b) Z_{TE} for the TE_1 mode on a wide air-filled parallel plate waveguide of spacing $a = 1$ cm.

$\pi/2$ out of phase with one another. A determination of the time-averaged Poynting vector $Re\left(\tilde{\mathbf{E}}(x) \times \tilde{\mathbf{H}}(x)\right)$ yields zero in this case, as there is no energy carried into the waveguide.

Example 10.7 Wave Impedance, Parallel Plate, TM

Determine the wave impedance Z_{TM} for the TM$_1$ mode at a frequency of 45 GHz on the waveguide described in Example 10.3.

Solution:
The wave impedance is given by Eq. (10.38):

$$Z_{TM} = \eta \sqrt{1 - \left(\frac{\omega_c}{\omega}\right)^2} = (120\pi \ \Omega) \sqrt{1 - (30 \text{ GHz}/45 \text{ GHz})^2} = 281 \ \Omega.$$

The total power carried by the TM waveguide mode is, from Eq. (10.18),

$$P = \int_0^b \int_0^a \frac{1}{2} \tilde{\mathbf{E}}(x) \times \tilde{\mathbf{H}}^*(x) \cdot \mathbf{a}_z \, dx \, dy = \frac{b}{2} \int_0^a \tilde{E}_x(x) \tilde{H}_y^*(x) \, dx.$$

Using $E_x(x)$ and $H_y(x)$ from Eqs. (10.30b) and (10.30c), the power for a TM waveguide mode on a parallel plate waveguide is

$$P = \frac{1}{4} \left(\frac{\beta a}{m\pi}\right) \left(\frac{\omega \varepsilon a}{m\pi}\right) A^2 ab = \frac{1}{4} \beta \omega \varepsilon \left(\frac{a}{m\pi}\right)^2 A^2 ab. \tag{10.39a}$$

Using β from Eq. (10.33) and k_c from Eq. (10.29), this total transmitted power can also be written as

$$P = \frac{1}{4} \left(\frac{\omega}{\omega_c}\right)^2 \sqrt{1 - \left(\frac{\omega_c}{\omega}\right)^2} \frac{A^2 ab}{\eta}. \tag{10.39b}$$

Example 10.8 Group Velocity as the Rate of Energy Flow

We stated earlier that the group velocity u_g of a waveguide mode can be determined as the ratio of the transmitted power to the energy stored in the waveguide mode per unit length. Demonstrate this for the TM$_m$ mode on a parallel plate waveguide.

Solution:
We already found the total power P transmitted by the waveguide mode, given in Eq. (10.39a) or (10.39b). The time-averaged energy density of the mode, consisting of the electric energy and the magnetic energy, is

$$w(x) = w_e(x) + w_m(x) = \frac{\varepsilon}{4} |\tilde{\mathbf{E}}(x)|^2 + \frac{\mu}{4} |\tilde{\mathbf{H}}(x)|^2.$$

(The energy density of a *static* field is given by $w_e = (1/2)\mathbf{D} \cdot \mathbf{E}$ and $w_m = (1/2)\mathbf{B} \cdot \mathbf{H}$. For time-harmonic fields, time averaging introduces another factor of $1/2$.) Using $\tilde{E}_z(x)$, $\tilde{E}_x(x)$, and $\tilde{H}_y(x)$ from Eqs. (10.30a)–(10.30c), the energy density is

$$w(x) = \frac{\varepsilon}{4} \left[\sin^2\left(\frac{m\pi x}{a}\right) + \beta^2 \left(\frac{a}{m\pi}\right)^2 \cos^2\left(\frac{m\pi x}{a}\right) \right] A^2$$

$$+ \frac{\mu}{4} (\omega \varepsilon)^2 \left(\frac{a}{m\pi}\right)^2 A^2 \cos^2\left(\frac{m\pi x}{a}\right).$$

The energy density $w(x)$ is the energy stored in the fields per unit volume. W', the energy stored per unit length, is found by integrating $w(x)$ over the cross-sectional area of the waveguide,

$$W' = \int_0^b \int_0^a w(x)\, dx\, dy = \frac{ab\varepsilon}{8}\left[1 + \beta^2\left(\frac{a}{m\pi}\right)^2\right]A^2 + \frac{ab\mu(\omega\varepsilon)^2}{8}\left(\frac{a}{m\pi}\right)^2 A^2.$$

Using β from Eq. (10.33) and k_c from Eq. (10.29), W' becomes

$$W' = \frac{ab\varepsilon}{4}\left(\frac{\omega}{\omega_c}\right)^2 A^2.$$

Now we are ready to evaluate the ratio P/W' as

$$\frac{P}{W'} = \frac{(1/4)(\omega/\omega_c)^2\sqrt{1 - (\omega_c/\omega)^2}\,A^2 ab/\eta}{(ab\varepsilon/4)(\omega/\omega_c)^2 A^2} = \frac{1}{\sqrt{\varepsilon\mu}}\sqrt{1 - \left(\frac{\omega_c}{\omega}\right)^2}.$$

The final equality follows when we use $\eta = \sqrt{\mu/\varepsilon}$. This result matches the group velocity given by Eq. (10.37).

As our final step in this section, we examine the losses of a TM_m mode in waveguides due to absorption by the dielectric material and the losses due to the finite conductivity of the conductors. To find the power loss due to dielectric absorption, we use Eq. (10.19), and find

$$P'_d = \int_0^b \int_0^a \frac{1}{2}(\sigma_d + \omega\varepsilon'')\left[|\tilde{E}_x(x)|^2 + |\tilde{E}_z(x)|^2\right]dx\, dy.$$

Keep in mind that, although not shown explicitly, P'_d is a weak function of z, since the field amplitudes decrease with increasing z. Using Eqs. (10.30b) and (10.30a) for $\tilde{E}_x(x)$ and $\tilde{E}_z(x)$, the power loss per unit length is

$$P'_d = \frac{1}{4}(\sigma_d + \omega\varepsilon'')\left(\frac{\omega}{\omega_c}\right)^2 A^2 ab.$$

(Equation (10.33) for β and Eq. (10.29) for $(m\pi/a)$ were used to reach this result.) Then, Eq. (10.20) becomes

$$\alpha_d = \frac{P'_d}{2P} = \frac{\eta\,(\sigma_d + \omega\varepsilon'')}{2\sqrt{1 - (\omega_c/\omega)^2}}.$$

We remarked earlier that P'_d decreases with increasing z. The exponential decrease of the transmitted power P with increasing z is precisely the same, so that the attenuation coefficient α_d is independent of z, as we should expect it to be.

The power loss due to the conductivity of the conductors is found following a similar process. From Eq. (10.21b), the power loss per unit length is

$$P'_c = \int_0^b \frac{1}{2}|\tilde{J}_s|^2 R_s dy \times 2. \tag{10.40}$$

As with the TEM mode, we include the factor ×2 to account for the two conductors. The surface current $\tilde{J}_s$ is equal in magnitude to the tangential $\tilde{H}_y$ field, which by Eq. (10.30c) is

$$\tilde{H}_y(x = 0) = -\tilde{H}_y(x = a) = -j\omega\varepsilon\frac{a}{m\pi}A.$$

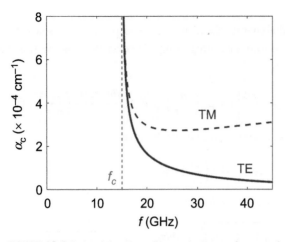

FIGURE 10.9 A plot of the attenuation constant α_c vs. frequency for the TM$_1$ mode (blue dashed line) and (b) for the TE$_1$ mode (purple solid line) on a wide air-filled parallel plate waveguide of spacing $a = 1$ cm. The conductors are copper, with conductivity $\sigma = 5.8 \times 10^7$ S/m.

$\tilde{J}_s$ is uniform in y, so the integral in Eq. (10.40) gives us

$$P'_c = \left| -j\omega\varepsilon \left(\frac{a}{m\pi}\right) A \right|^2 R_s b = (\omega\varepsilon)^2 \left(\frac{a}{m\pi}\right)^2 A^2 R_s b. \quad (10.41)$$

Using P'_c and P in Eq. (10.22), we find the attenuation coefficient

$$\alpha_c = \frac{P'_c}{2P} = \frac{(\omega\varepsilon)^2(a/m\pi)^2 A^2 R_s b}{2\left(\frac{1}{4}\beta\omega\varepsilon(a/m\pi)^2 A^2\right)ab} = \frac{2\omega\varepsilon R_s}{a\beta} = \frac{2R_s}{\eta a\sqrt{1-(\omega_c/\omega)^2}}. \quad (10.42)$$

A plot of α_c vs. frequency $f = \omega/2\pi$ for a TM$_1$ mode is shown as the blue dashed line in Fig. 10.9. Remember that the surface resistance $R_s = \sqrt{\omega\mu/2\sigma}$ is also a function of frequency, causing the positive slope in the attenuation constant α_c for high frequencies.

Example 10.9 Attenuation Coefficient, Parallel Plate, TM

For the parallel plate waveguide as described in Example 10.3 constructed of copper conducting plates, carrying a TM$_1$ mode of amplitude $A = 10$ V/m and frequency 45 GHz, determine (a) the total transmitted power, (b) the power loss per unit length due to the copper conductors, and (c) the attenuation coefficient α_c due to conductor losses.

Solution:

(a) We find the total power transmitted by the waveguide mode using Eq. (10.39a):

$$P = \frac{1}{4}\beta\omega\varepsilon\left(\frac{a}{m\pi}\right)^2 A^2 ab$$

$$= \frac{1}{4}(703 \text{ m}^{-1})(2\pi \times 45 \times 10^9 \text{ s}^{-1})(8.854 \times 10^{-12} \text{ F/m})\left(\frac{0.005 \text{ m}}{\pi}\right)^2$$

$$\times (10 \text{ V/m})^2(0.005 \text{ m})(0.1\text{m}) = 5.57 \times 10^{-5} \text{ W}.$$

We have used $\beta = 703$ m^{-1}, which we derived in Example 10.4.

(b) To find the losses due to the conductors, we start by determining the surface resistance of copper at this frequency:

$$R_s = \sqrt{\frac{\omega\mu}{2\sigma}}.$$

Using $\sigma = 5.8 \times 10^7$ S/m, as listed in Table 3.1 and $\mu = \mu_0$, we find $R_s = 0.0553$ Ω. We use this in Eq. (10.41) to find

$$P'_c = R_s(\omega\varepsilon)^2\left(\frac{a}{m\pi}\right)^2 A^2 b$$

$$= (0.0553 \text{ }\Omega)(2\pi \times 45 \times 10^9 \text{ s}^{-1})^2(8.854 \times 10^{-12} \text{ F/m})^2$$

$$\times \left(\frac{0.005 \text{ m}}{\pi}\right)^2 (10 \text{ V/m})^2(0.1 \text{ m}) = 8.78 \times 10^{-6} \text{ W/m}.$$

(c) Finally, we calculate the attenuation coefficient. We have two means of finding this parameter. Since we have already found P'_c and P, the simpler form is to use

$$\alpha_c = \frac{P'_c}{2P} = \frac{8.78 \times 10^{-6} \text{ W/m}}{2 \times 5.57 \times 10^{-5} \text{ W}} = 0.079 \text{ m}^{-1}.$$

Alternatively, we could use Eq. (10.42) to find

$$\alpha_c = \frac{2R_s}{\eta a \sqrt{1 - (\omega_c/\omega)^2}} = \frac{2 \times (0.0553 \ \Omega)}{(377 \ \Omega)(0.005 \ \text{m}) \sqrt{1 - (30 \ \text{GHz}/45 \ \text{GHz})^2}}$$

$$= 0.079 \ \text{m}^{-1}.$$

The results for each of the parameters of the TM mode on a parallel plate waveguide are collected in Table 10.2.

Table 10.2

Summary of results for TM modes on a very wide ($w \gg a$) parallel plate waveguide. See Table 10.3 for mode impedance, phase constant, attenuation constant, phase velocity, and group velocity.	
$\tilde{E}_z(x) = A \sin\left(\dfrac{m\pi x}{a}\right)$	$\tilde{H}_z(x) = 0$
$\tilde{E}_x(x) = -j\beta\left(\dfrac{a}{m\pi}\right) A \cos\left(\dfrac{m\pi x}{a}\right)$	$\tilde{H}_x(x) = 0$
$\tilde{E}_y(x) = 0$	$\tilde{H}_y(x) = \dfrac{\tilde{E}_x(x)}{Z_{TM}} = -j\omega\varepsilon\left(\dfrac{a}{m\pi}\right) A \cos\left(\dfrac{m\pi x}{a}\right)$
$k_c = \dfrac{m\pi}{a}$	$\omega_c = \dfrac{k_c}{\sqrt{\varepsilon\mu}} = \dfrac{m\pi}{a\sqrt{\varepsilon\mu}} \qquad m = 1, 2, 3, \ldots$
$P = \dfrac{1}{4}\beta\omega\varepsilon\left(\dfrac{a}{m\pi}\right)^2 A^2 ab = \dfrac{1}{4}\left(\dfrac{\omega}{\omega_c}\right)^2 \sqrt{1 - \left(\dfrac{\omega_c}{\omega}\right)^2}\ \dfrac{A^2 ab}{\eta}$	
$\alpha_d = \dfrac{\eta\left(\sigma_d + \omega\varepsilon''\right)}{2\sqrt{1 - (\omega_c/\omega)^2}} \qquad \alpha_c = \dfrac{2R_s}{\eta a\sqrt{1 - (\omega_c/\omega)^2}}$	

As we will see in the following sections, several parameters of waveguide waves are the same for the different geometries of waveguides. These include the phase constant β, the attenuation constant α, the phase and group velocities u_p and u_g, and the mode impedances Z_{TM} and Z_{TE}. We collect and display these parameters in Table 10.3.

10.2.3 TE Modes on a Parallel Plate Waveguide

In the previous sections we derived the properties of TEM and TM modes on parallel plate waveguides. We now address **transverse electric**, or TE, modes on these structures. As the name implies, the electric field lies entirely in the transverse plane; that is, the z-component of the electric field is by definition zero. The magnetic field, however, possesses an $\tilde{H}_z(x, y)$ component. The derivation of TE modes parallels that of the TM modes, so we will abbreviate our discussion and focus on the aspects of these modes that differ from one another.

We return to Eq. (10.13) to determine the $\tilde{H}_z(x, y)$ component:

$$\nabla_t^2 \tilde{H}_z(x, y) = -k_c^2 \tilde{H}_z(x, y).$$

Table 10.3

A compilation of the parameters of wave impedance, phase constant, attenuation constant, phase velocity, and group velocity for TE and TM modes, which are common to all the various types of waveguides.		
Parameter	**TM**	**TE**
Impedance	$Z_{TM} = \eta\sqrt{1 - \left(\dfrac{\omega_c}{\omega}\right)^2}$	$Z_{TE} = \dfrac{\eta}{\sqrt{1 - (\omega_c/\omega)^2}}$
	TM and TE	
Phase constant	$\beta = \omega\sqrt{\varepsilon\mu}\sqrt{1 - \left(\dfrac{\omega_c}{\omega}\right)^2}$, for $\omega > \omega_c$	
Attenuation constant	$\alpha = k_c\sqrt{1 - \left(\dfrac{\omega}{\omega_c}\right)^2}$, for $\omega < \omega_c$	
Phase velocity	$u_p = \dfrac{1}{\sqrt{\varepsilon\mu}}\dfrac{1}{\sqrt{1 - (\omega_c/\omega)^2}}$	
Group velocity	$u_g = \dfrac{1}{\sqrt{\varepsilon\mu}}\sqrt{1 - \left(\dfrac{\omega_c}{\omega}\right)^2}$	

For the wide parallel plate structure, the fields do not vary as a function of y, so the derivatives $\partial/\partial y$ must vanish, and Eq. (10.13) becomes

$$\frac{\partial^2}{\partial x^2}\tilde{H}_z(x) = -k_c^2\tilde{H}_z(x).$$

The general solution of this differential equation, when $k_c^2 > 0$, is

$$\tilde{H}_z(x) = A\sin(k_c x) + B\cos(k_c x),$$

where A and B are amplitudes that must be determined from boundary conditions. The $\tilde{H}_z(x)$ magnetic field component is parallel to the surfaces of the conductors, so we are not yet able to apply any boundary conditions to simplify $\tilde{H}_z(x)$. (Recall that the boundary condition for a magnetic field at the surface of a perfect conductor is that the normal component of **H** must be zero. Tangential magnetic fields are allowed, as long as there is a surface current in the conductor perpendicular to the magnetic field.) Therefore, we must first determine $\tilde{E}_y(x)$ using Eq. (10.16b), which, since $\tilde{E}_z$ is zero, simplifies to

$$\tilde{E}_y(x) = \frac{1}{k^2 + \gamma^2}\left(-\gamma\frac{\partial\tilde{E}_z(x)}{\partial y} + j\omega\mu\frac{\partial\tilde{H}_z(x)}{\partial x}\right) = \frac{j\omega\mu}{k_c}\left[A\cos(k_c x) - B\sin(k_c x)\right].$$

Since $\tilde{E}_y(x)$ is tangential to the conductor surface, this field component must be zero at $x = 0$ and $x = a$. The first of these boundary conditions implies that the amplitude A must be zero, while the second determines that k_c is

$$k_c = \frac{m\pi}{a},$$

where m is an integer equal to or greater than 1. $\tilde{H}_z(x)$ therefore simplifies to

$$\tilde{H}_z(x) = B\cos\left(\frac{m\pi x}{a}\right), \tag{10.43a}$$

and $\tilde{E}_y(x)$ to

$$\tilde{E}_y(x) = -j\omega\mu\left(\frac{a}{m\pi}\right)B\sin\left(\frac{m\pi x}{a}\right).\qquad(10.43b)$$

For $\tilde{H}_x(x)$, we use Eq. (10.16c) to find

$$\tilde{H}_x(x) = \frac{1}{k^2 + \gamma^2}\left(j\omega\varepsilon\frac{\partial\tilde{E}_z(x,y)}{\partial y} - \gamma\frac{\partial\tilde{H}_z(x,y)}{\partial x}\right) = j\beta\left(\frac{a}{m\pi}\right)B\sin\left(\frac{m\pi x}{a}\right).\qquad(10.43c)$$

$\tilde{H}_y(x)$ and $\tilde{E}_x(x)$ are zero everywhere (see Eqs. (10.16d) and (10.16a). Notice that $\tilde{H}_x(x)$ is zero at the top and bottom conductor, as it must be, since this component is normal to these surfaces.

To find the full x-, z-, and t-dependent field components, we multiply $\tilde{H}_z(x)$, $\tilde{E}_y(x)$, and $\tilde{H}_x(x)$ by $e^{j(\omega t - \beta z)}$ and take the real part, as shown in Eqs. (10.1) and (10.2). The results for $H_z(x, z, t)$, $E_y(x, z, t)$, and $H_x(x, z, t)$ are

$$H_z(x, z, t) = B\cos\left(\frac{m\pi x}{a}\right)\cos(\omega t - \beta z),$$

$$E_y(x, z, t) = \omega\mu\left(\frac{a}{m\pi}\right)B\sin\left(\frac{m\pi x}{a}\right)\sin(\omega t - \beta z),$$

and

$$H_x(x, z, t) = -\beta\left(\frac{a}{m\pi}\right)B\sin\left(\frac{m\pi x}{a}\right)\sin(\omega t - \beta z).$$

x–z plots of $\mathbf{E}(x, z, t = 0)$ and $\mathbf{H}(x, z, t = 0)$ for the TE$_1$ mode at $t = 0$ are shown in Fig. 10.10(a) and (b), respectively. Since the tangential magnetic field of this mode at the conductor surfaces is in the z-direction, the surface currents in the conductor are in the $\pm y$-direction. There are no surface charges since the normal electric field at the conductor surfaces is zero.

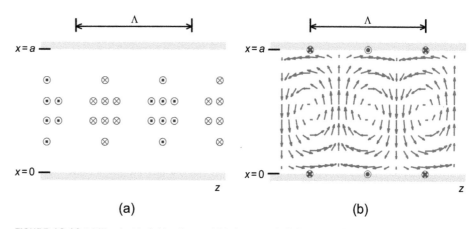

FIGURE 10.10 (a) The electric field pattern and (b) the magnetic field pattern of the TE$_1$ mode of a parallel plate waveguide. The yellow bars represent the conductors, spaced by a. The magenta vectors in the inside surfaces of the conductors of (b) represent the surface current. The wave propagates in the z-direction. Only a short segment of the waveguide is shown.

We can find γ using

$$\gamma = \sqrt{k_c^2 - k^2} = \begin{cases} \alpha = \dfrac{m\pi}{a}\sqrt{1 - \left(\dfrac{\omega}{\omega_c}\right)^2} & \text{when } \omega < \omega_c \\[4mm] j\beta = j\omega\sqrt{\mu\varepsilon}\sqrt{1 - \left(\dfrac{\omega_c}{\omega}\right)^2} & \text{when } \omega > \omega_c \end{cases} \qquad (10.44)$$

Note that the phase constant β, the attenuation constant α, and the cutoff frequency ω_c are the same for TE and TM modes.

The phase velocity u_p, group velocity u_g, and mode wavelength Λ are also each the same as the corresponding quantities for the TM modes, and you may refer to those derivations in the previous section. The waveguide impedance for the TE mode, defined as the ratio of the transverse field components, is

$$Z_{TE} = -\frac{\tilde{E}_y(x)}{\tilde{H}_x(x)} = \frac{\omega\mu}{\beta} = \eta\frac{1}{\sqrt{1 - (\omega_c/\omega)^2}}. \qquad (10.45)$$

For TE modes, the impedance Z_{TE} is greater than η for $\omega > \omega_c$. A plot of Z_{TE} vs. frequency f is shown in Fig. 10.8(b).

Example 10.10 Wave Impedance, Parallel Plate, TE

Determine the wave impedance Z_{TE} for the TE$_1$ mode at a frequency of 45 GHz on the waveguide described in Example 10.3.

Solution:
The wave impedance is given by Eq. (10.45). We found the cutoff frequency for this mode in Example 10.3.

$$Z_{TE} = \frac{\eta}{\sqrt{1 - (\omega_c/\omega)^2}} = \frac{377\ \Omega}{\sqrt{1 - (30\ \text{GHz}/45\ \text{GHz})^2}} = 506\ \Omega.$$

To determine the total power carried by the TE waveguide mode, we use Eq. (10.18),

$$P = \int_0^b \int_0^a \frac{1}{2}\tilde{\mathbf{E}} \times \tilde{\mathbf{H}}^* \cdot \mathbf{a}_z\, dx\, dy = \int_0^b \int_0^a \frac{1}{2}(-\tilde{E}_y(x)\tilde{H}_x^*(x))\, dx\, dy.$$

Using $\tilde{E}_y(x)$ and $\tilde{H}_x(x)$ from Eqs. (10.43b) and (10.43c), and carrying out the integration, we find

$$P = \frac{1}{4}(\omega\mu)\beta\left(\frac{a}{m\pi}\right)^2 B^2 ab. \qquad (10.46a)$$

Using β from Eq. (10.33) and k_c from Eq. (10.29), the total power transmitted in a TE mode can also be written as

$$P = \frac{1}{4}\left(\frac{\omega}{\omega_c}\right)^2\sqrt{1 - \left(\frac{\omega_c}{\omega}\right)^2}\, \eta B^2 ab. \qquad (10.46b)$$

Our final step for the TE mode of the parallel plate waveguide is to determine the attenuation constant due to dielectric absorption and due to the finite conductivity of the conductors. For the dielectric losses, we use Eq. (10.19) to compute

$$P'_d = \int_0^b \int_0^a \frac{1}{2} (\sigma_d + \omega\varepsilon'') \left| \tilde{E}_y(x) \right|^2 \, dx \, dy.$$

Using $\tilde{E}_y(x)$ from Eq. (10.43b), the power loss is

$$P'_d = \frac{1}{4} (\sigma_d + \omega\varepsilon'')(\omega\mu)^2 \left(\frac{a}{m\pi} \right)^2 B^2 ab.$$

Using P and P'_d, we determine the attenuation coefficient using Eq. (10.22) as

$$\alpha_d = \frac{P'_d}{2P} = \frac{(1/4)(\sigma_d + \omega\varepsilon'')(\omega\mu)^2 (a/m\pi)^2 B^2 ab}{2 \, (1/4)(\omega\mu)\beta \, (a/m\pi)^2 B^2 ab}.$$

Many terms in this expression cancel. Using β from Eq. (10.33), this becomes

$$\alpha_d = \frac{\eta \, (\sigma_d + \omega\varepsilon'')}{2\sqrt{1 - (\omega_c/\omega)^2}}.$$

Next we look at losses due to the conductivity of the parallel plates. Using Eq. (10.21b), the power lost per unit length is

$$P'_c = \int_0^b \frac{1}{2} |\tilde{J}_s|^2 R_s dy \times 2 = B^2 R_s b, \tag{10.47}$$

where we have evaluated $\tilde{H}_z(x)$ at $x = 0$ and $x = a$ to determine the magnitude of the surface current $|\tilde{J}_s|$. ($\tilde{H}_z(x)$ is the only component of $\tilde{\mathbf{H}}$ that is tangential to the conductor surface since $\tilde{H}_y(x) = 0$ everywhere for this mode, and $\tilde{H}_x(x)$ is a normal field, not tangential.)

Finally, we use Eq. (10.22) to compute the attenuation constant:

$$\alpha_c = \frac{P'_c}{2P} = \frac{2R_s}{a\omega\mu\beta(a/m\pi)^2}.$$

Using β from Eq. (10.44) and $\omega_c = m\pi/a\sqrt{\varepsilon\mu}$, the attenuation coefficient simplifies to

$$\alpha_c = \frac{2R_s \, (\omega_c/\omega)^2}{\eta a\sqrt{1 - (\omega_c/\omega)^2}}. \tag{10.48}$$

A plot of the attenuation coefficient α_c vs. frequency for the TE_1 mode is shown as the solid purple line in Fig. 10.9. Notice that the conductor losses for the TE mode are significantly less than those of the TM mode at all frequencies. We also see that α_c tends to zero as $\omega \to \infty$, since the surface current density $\tilde{J}_s$, which equals $\tilde{H}_z$, also becomes small as the waveguide mode propagates nearly parallel to the waveguide axis for high frequencies.

Example 10.11 Attenuation Coefficient, Parallel Plate, TE

For the parallel plate waveguide as described in Example 10.3 constructed of copper conducting plates, carrying a TE_1 mode of amplitude $B = 1.0$ A/m and frequency 45 GHz, determine (a) the total transmitted power, (b) the power loss per unit length due to the conductors, and (c) the attenuation coefficient α_c due to conductor losses.

Solution:

(a) The total power transmitted by the waveguide mode is, using Eq. (10.46a),

$$P = \frac{1}{4}(\omega\mu)\beta\left(\frac{a}{m\pi}\right)^2 B^2 ab$$

$$= \frac{1}{4}(2\pi \times 45 \times 10^9 \text{ s}^{-1})(4\pi \times 10^{-7} \text{ H/m})(703 \text{ m}^{-1})\left(\frac{0.005 \text{ m}}{\pi}\right)^2$$

$$\times (1.0 \text{ A/m})^2(0.005 \text{ m})(0.1 \text{ m}) = 0.0791 \text{ W}.$$

We have used $\beta = 703 \text{ m}^{-1}$, which was derived in Example 10.4.

(b) We found the surface resistance $R_s = 0.0553 \ \Omega$ in Example 10.9, and use this in Eq. (10.47) to find

$$P'_c = B^2 R_s b$$

$$= (1.0 \text{ A/m})^2(0.0553 \ \Omega)(0.1 \text{ m}) = 0.00553 \text{ W/m}.$$

(c) Finally, we calculate the attenuation coefficient. Using P'_c and P in Eq. (10.22), we find

$$\alpha_c = \frac{P'_c}{2P} = \frac{0.00553 \text{ W/m}}{2 \times 0.0791 \text{ W}} = 0.0350 \text{ m}^{-1}.$$

Alternatively, we could use Eq. (10.48) to find

$$\alpha_c = \frac{2R_s (\omega_c/\omega)^2}{\eta a \sqrt{1 - (\omega_c/\omega)^2}} = \frac{2(0.0553 \ \Omega)(30 \text{ GHz}/45 \text{ GHz})^2}{(377 \ \Omega)(0.005 \text{ m})\sqrt{1 - (30 \text{ GHz}/45 \text{ GHz})^2}}$$

$$= 0.0350 \text{ m}^{-1}.$$

Notice that the attenuation coefficient α_c for the TE_1 mode is less than that of the TM_1 mode by a factor of 0.44.

The results for the TE mode on a parallel plate waveguide are collected in Table 10.4.

Table 10.4

Summary of results for TE modes on a very wide ($b \gg a$) parallel plate waveguide. See Table 10.3 for wave impedance, phase constant, attenuation constant, phase velocity, and group velocity.	
$\tilde{E}_z(x) = 0$	$\tilde{H}_z(x) = B\cos\left(\frac{m\pi x}{a}\right)$
$\tilde{E}_x(x) = 0$	$\tilde{H}_x(x) = -\frac{\tilde{E}_y(x)}{Z_{TE}} = j\beta\left(\frac{a}{m\pi}\right)B\sin\left(\frac{m\pi x}{a}\right)$
$\tilde{E}_y(x) = -j\omega\mu\left(\frac{a}{m\pi}\right)B\sin\left(\frac{m\pi x}{a}\right)$	$\tilde{H}_y(x) = 0$
$k_c = \frac{m\pi}{a} \qquad \omega_c = \frac{k_c}{\sqrt{\varepsilon\mu}} = \frac{m\pi}{a\sqrt{\varepsilon\mu}} \qquad m = 1, 2, 3, \ldots$	
$P = \frac{1}{4}(\omega\mu)\beta\left(\frac{a}{m\pi}\right)^2 B^2 ab = \frac{1}{4}\left(\frac{\omega}{\omega_c}\right)^2\sqrt{1 - \left(\frac{\omega_c}{\omega}\right)^2}\,\eta B^2 ab$	
$\alpha_d = \frac{\eta\,(\sigma_d + \omega\varepsilon'')}{2\sqrt{1 - (\omega_c/\omega)^2}}$	$\alpha_c = \frac{2R_s (\omega_c/\omega)^2}{\eta a\sqrt{1 - (\omega_c/\omega)^2}}$

10.3 Rectangular Waveguides

Now that we have mastered the basics of waveguides using the simple parallel plate waveguide as an introductory example, we turn our attention to rectangular waveguides. As their name implies, the cross-section of these waveguides is rectangular in shape; this geometry is the most commonly employed in microwave systems, by far. A rectangular waveguide is shown in Fig. 10.1(b). We will follow the same general development for rectangular waveguides as we introduced for parallel plate waveguides. Features that are common to both types of waveguides are that only frequencies greater than the cutoff frequencies of TM and TE modes can propagate, and that the velocity of the wave depends strongly on the frequency of excitation. TEM waves are not supported on rectangular waveguides (or any single-conductor waveguide).

10.3.1 No TEM Modes on Rectangular Waveguides

Our first order of business is to show that TEM modes are not allowed on single-conductor (such as rectangular or circular) waveguides. We return to Faraday's and Ampère's Laws for waveguide modes for $\tilde{E}_z(x,y) = 0$ and $\tilde{H}_z(x,y) = 0$, as presented in Section 10.2.1. In particular, we examine Eqs. (10.23c) and (10.24c). Since

$$\frac{\partial \tilde{E}_y(x,y)}{\partial x} - \frac{\partial \tilde{E}_x(x,y)}{\partial y} = 0$$

and

$$\frac{\partial \tilde{H}_y(x,y)}{\partial x} - \frac{\partial \tilde{H}_x(x,y)}{\partial y} = 0,$$

and since $\tilde{E}(x,y)$ and $\tilde{H}(x,y)$ depend only on x and y (but not on z), the curl of these fields is zero,

$$\nabla \times \tilde{E}(x,y) = 0,$$

and

$$\nabla \times \tilde{H}(x,y) = 0.$$

(Remember that $\tilde{E}(x,y)$ and $\tilde{H}(x,y)$ are not the complete electric and magnetic fields, respectively, but rather only the phasor amplitudes as defined in Eqs. (10.1) and (10.2).) We also use Gauss' Law ($\nabla \cdot D = \rho_v$) and the non-existence of magnetic charges ($\nabla \cdot B = 0$), which are valid for time-varying fields. For homogeneous, charge-free media such as the dielectric medium inside the waveguide, Gauss' Law reduces to $\nabla \cdot \tilde{E}(x,y) = 0$, and $\nabla \cdot B(x,y) = 0$ simplifies to $\nabla \cdot \tilde{H}(x,y) = 0$. So $\tilde{E}(x,y)$ and $\tilde{H}(x,y)$ have no curl (they are irrotational), and they have no divergence (they are solenoidal). Any vector field that is irrotational and solenoidal can be at most a uniform field. (This is analogous to a one-dimensional function $f(x)$ whose derivative $df(x)/dx$ is zero for all x. This function can only be a constant, $f(x) = f_0$.) But the tangential electric field E_{tan} and normal magnetic field H_n at the conductor surfaces must be zero. So for a TEM mode in a hollow waveguide, the transverse fields $\tilde{E}(x,y)$ and $\tilde{H}(x,y)$ must also be zero for all x and y, and there is no TEM mode allowed. For these waveguides, then, we only need to examine TM and TE modes, as we do in the following sections.

10.3.2 TM Modes of a Rectangular Waveguide

We now proceed to modes that *are* supported by hollow waveguides, that is, TM and TE modes. In this section we explore the properties of TM modes for rectangular waveguides, and in the following section we explore TE modes. We have already laid much of the groundwork for this in Sections 10.1 and 10.2.2, so we will frequently rely on those results. Modes on the rectangular waveguide are similar to parallel plate waveguide modes in that waves can propagate only when the wave frequency is greater than the cutoff frequency ω_c for the mode. We'll develop a new expression to evaluate the cutoff frequency. Rectangular waveguide modes differ from parallel plate modes in at least one important aspect. That is, the electric and magnetic components of the parallel plate waveguide were uniform in the dimension y, whereas for the rectangular waveguide we will show that the components of the fields vary sinusoidally in the y-direction.

We start with Eq. (10.12), written in rectangular coordinates,

$$\nabla_t^2 \tilde{E}_z(x,y) = \frac{\partial^2 \tilde{E}_z(x,y)}{\partial^2 x} + \frac{\partial^2 \tilde{E}_z(x,y)}{\partial^2 y} = -k_c^2 \tilde{E}_z(x,y), \qquad (10.49)$$

which must be satisfied by the longitudinal $\tilde{E}_z(x,y)$ component. (Remember, $\tilde{H}_z(x,y) = 0$ for a TM mode by definition.) $\tilde{E}_z(x,y)$ will in general be a function of x and y. Using the separation-of-variables technique, we find the solution as the product of an x-dependent term and a y-dependent term. The general solution of Eq. (10.49) is

$$\tilde{E}_z(x,y) = \{A' \sin(k_x x) + B' \cos(k_x x)\} \times \{C' \sin(k_y y) + D' \cos(k_y y)\}, \qquad (10.50)$$

where A', B', C', and D' are amplitudes that must be determined from boundary conditions or total power considerations, and k_x and k_y are parameters that tell us how rapidly (as a function of position x and y) the field varies. We can confirm that Eq. (10.50) does indeed satisfy Eq. (10.49) by simply carrying out the second-order derivatives in this expression, and this solution is valid as long as k_x and k_y satisfy the condition

$$k_x^2 + k_y^2 = k_c^2,$$

where, from Eq. (10.8),

$$k_c^2 = \gamma^2 + k^2 = \gamma^2 + \omega^2 \varepsilon \mu.$$

We'll use these relations in a few moments.

First, let's see what we can learn about the amplitudes A', B', C', and D' and the spatial frequencies k_x and k_y by applying the boundary conditions that must be satisfied by the tangential electric field at the conductor walls,

$$\tilde{E}_z(x,y) = 0 \qquad \text{at } x = 0 \text{ and } x = a$$

and

$$\tilde{E}_z(x,y) = 0 \qquad \text{at } y = 0 \text{ and } y = b.$$

Applying $\tilde{E}_z(x,y) = 0$ at $x = 0$ to Eq. (10.50) tells us that B' must be zero, and applying $\tilde{E}_z(x,y) = 0$ at $y = 0$ tells us that D' must be zero. This lets us simplify Eq. (10.50) to write

$$\tilde{E}_z(x,y) = A \sin(k_x x) \sin(k_y y), \qquad (10.51)$$

where the product of A' and C' have been merged into a single amplitude A.

Next we use the boundary condition that $\tilde{E}_z(x, y) = 0$ at $x = a$ to determine that

$$k_x a = m\pi, \quad \text{or } k_x = \frac{m\pi}{a}, \tag{10.52}$$

where $m = 1, 2, 3, \ldots$ is a positive integer. (Note that $m = 0$ is *not* permitted. Otherwise, E_z would be zero everywhere.) Similarly, the boundary condition that $E_z = 0$ at $y = b$ tells us that

$$k_y b = n\pi, \quad \text{or } k_y = \frac{n\pi}{b}, \tag{10.53}$$

where $n = 1, 2, 3, \ldots$ is a positive integer. (Again, $n = 0$ is not permitted.) We will use the integers m and n to label the various TM modes using the notation TM_{mn}. By convention, the larger dimension of an $a \times b$ rectangular waveguide is chosen to be a, and the smaller is b.

Now we are ready to examine the cutoff frequency, the minimum frequency for mode propagation, for TM modes. For high frequencies, Eq. (10.8) can be used to write the mode propagation constant as

$$\gamma = \sqrt{k_c^2 - k^2} = j\sqrt{k^2 - k_c^2} = j\sqrt{\omega^2 \varepsilon \mu - \left(\frac{m\pi}{a}\right)^2 - \left(\frac{n\pi}{b}\right)^2},$$

where we have used

$$k_c^2 = k_x^2 + k_y^2 = \left(\frac{m\pi}{a}\right)^2 + \left(\frac{n\pi}{b}\right)^2. \tag{10.54}$$

The propagation constant γ of the mode is imaginary only when $k^2 > k_c^2$, or equivalently when $\omega > \omega_c$, where

$$\omega_c = \frac{k_c}{\sqrt{\varepsilon \mu}} = \frac{1}{\sqrt{\varepsilon \mu}}\sqrt{\left(\frac{m\pi}{a}\right)^2 + \left(\frac{n\pi}{b}\right)^2} \tag{10.55}$$

is the cutoff frequency for the TM_{mn} mode.

Example 10.12 Cutoff Frequencies, Rectangular, TM

Consider an air-filled rectangular waveguide of cross-sectional dimension $a = 3$ cm by $b = 2$ cm. (a) Determine the cutoff frequencies for the TM_{11}, TM_{21}, TM_{12}, TM_{22}, and TM_{13} modes. (b) Find the corresponding free-space wavelengths at these frequencies. Is there a simple relationship between these wavelengths and the waveguide dimensions?

Solution:

(a) We'll use Eq. (10.55) for the cutoff frequencies, and find cyclical frequencies using $f_c = \omega_c/2\pi$. Since the material inside the waveguide is air, we'll approximate $\sqrt{\varepsilon \mu} = c^{-1}$. Then the cutoff frequencies of the five modes requested are

$$f_c = \frac{\omega_c}{2\pi} = \frac{c}{2}\sqrt{\left(\frac{m}{a}\right)^2 + \left(\frac{n}{b}\right)^2} = \begin{cases} 9.01 \text{ GHz}, & \text{for } TM_{11} \\ 12.5 \text{ GHz}, & \text{for } TM_{21} \\ 15.8 \text{ GHz}, & \text{for } TM_{12} \\ 18.0 \text{ GHz}, & \text{for } TM_{22} \\ 23.0 \text{ GHz}, & \text{for } TM_{13} \end{cases}$$

(b) The free-space wavelengths are

$$\lambda_c = \frac{c}{f_c} = \begin{cases} 3.33 \text{ cm,} & \text{for } \text{TM}_{11} \\ 2.40 \text{ cm,} & \text{for } \text{TM}_{21} \\ 1.90 \text{ cm,} & \text{for } \text{TM}_{12} \\ 1.67 \text{ cm,} & \text{for } \text{TM}_{22} \\ 1.30 \text{ cm,} & \text{for } \text{TM}_{13} \end{cases}$$

The wavelengths are, of course, related to the dimensions a and b, but the relation is not simple.

In terms of this cutoff frequency, the mode propagation constant is

$$\gamma = j\sqrt{\varepsilon\mu}\sqrt{\omega^2 - \omega_c^2},$$

when the mode frequency ω is greater than the cutoff frequency ω_c. Keep in mind that the cutoff frequency is different for the different modes of the waveguide. Pulling out a factor of ω, the phase constant β can be written compactly as

$$\beta = \omega\sqrt{\varepsilon\mu}\sqrt{1 - \left(\frac{\omega_c}{\omega}\right)^2} = k\sqrt{1 - \left(\frac{\omega_c}{\omega}\right)^2}. \tag{10.56}$$

For low frequencies, $\omega < \omega_c$, the propagation constant γ is purely real, signifying attenuation of the mode, rather than propagation, and we write the attenuation constant α as

$$\alpha = k_c\sqrt{1 - \left(\frac{\omega}{\omega_c}\right)^2}.$$

Notice that α and β for the rectangular waveguide have the same frequency dependence as we saw for the parallel plate waveguide, aside from the different form for the cutoff frequency ω_c.

Now that we have established γ for the TM_{mn} mode (β for $\omega > \omega_c$ and α for $\omega < \omega_c$) and the cutoff frequencies, let's move on with our analysis and determine the transverse field components $\tilde{E}_x(x,y)$, $\tilde{E}_y(x,y)$, $\tilde{H}_x(x,y)$, and $\tilde{H}_y(x,y)$ for $\omega > \omega_c$. For this, we use Eqs. (10.16a)–(10.16d), with $\tilde{E}_z(x,y)$ for the TM_{mn} mode given by Eq. (10.51). This is straightforward, and we simply present the results:

$$\tilde{E}_x(x,y) = \frac{-j\beta k_x}{k_c^2}A\cos(k_x x)\sin(k_y y), \tag{10.57a}$$

$$\tilde{E}_y(x,y) = \frac{-j\beta k_y}{k_c^2}A\sin(k_x x)\cos(k_y y), \tag{10.57b}$$

$$\tilde{H}_x(x,y) = \frac{j\omega\varepsilon k_y}{k_c^2}A\sin(k_x x)\cos(k_y y), \tag{10.57c}$$

and

$$\tilde{H}_y(x,y) = \frac{-j\omega\varepsilon k_x}{k_c^2}A\cos(k_x x)\sin(k_y y). \tag{10.57d}$$

The z and t dependence in these field components can be included by multiplying by $e^{j(\omega t - \beta z)}$ and taking the real part, as prescribed in Eqs. (10.1) and (10.2). Plots of these field patterns for the TM_{11} mode are shown in Fig. 10.11.

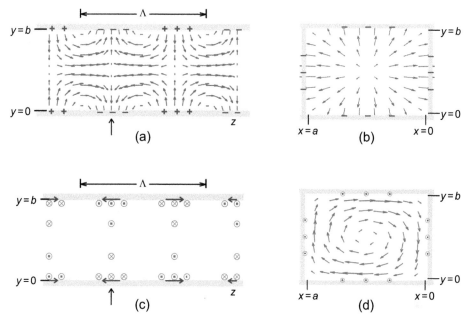

FIGURE 10.11 Plots of the field patterns for a TM$_{11}$ mode on a rectangular waveguide. (a) y–z plot of the electric field at x = a/2. The magenta + and − signs represent the surface charge on the inside surface of the conductors; (b) x–y plot of the electric field at z = λ/4; (c) y–z plot of the magnetic field at x = a/2. The magenta arrows represent the surface current on the inside surface of the conductors. (d) x–y plot of the magnetic field at z = λ/4. The arrows beneath the waveguide in (a) and (c) indicate the z-location for plots (b) and (d).

Next, we examine the waveguide impedance Z_{TM}. Notice that $\tilde{E}_x(x,y)$ and $\tilde{H}_y(x,y)$ have the same dependence on x and y, as do $\tilde{E}_y(x,y)$ and $\tilde{H}_x(x,y)$. The waveguide impedance is the ratio of the transverse electric and magnetic fields,

$$Z_{TM} = \frac{\tilde{E}_x(x,y)}{\tilde{H}_y(x,y)} = \frac{-j\beta k_x}{k_c^2} \frac{k_c^2}{-j\omega\varepsilon k_x} = \frac{\beta}{\omega\varepsilon} = \eta\sqrt{1 - \left(\frac{\omega_c}{\omega}\right)^2}. \tag{10.58}$$

Alternatively, we can use $\tilde{E}_y(x,y)$ and $\tilde{H}_x(x,y)$ to compute the impedance,

$$Z_{TM} = -\frac{\tilde{E}_y(x,y)}{\tilde{H}_x(x,y)} = \frac{j\beta k_y}{k_c^2} \frac{k_c^2}{j\omega\varepsilon k_y} = \frac{\beta}{\omega\varepsilon} = \eta\sqrt{1 - \left(\frac{\omega_c}{\omega}\right)^2},$$

consistent with Eq. (10.58).

Example 10.13 Wave Impedance, Rectangular, TM

Determine the wave impedance Z_{TM} for the TM$_{11}$ mode at a frequency of 12 GHz on the waveguide described in Example 10.12.

Solution:
The wave impedance is given by Eq. (10.58):

$$Z_{TM} = \eta\sqrt{1 - \left(\frac{\omega_c}{\omega}\right)^2} = (120\pi\ \Omega)\sqrt{1 - (9.01\ \text{GHz}/12\ \text{GHz})^2} = 249\ \Omega.$$

The total power carried by the waveguide mode is

$$P = \int_0^b \int_0^a \frac{1}{2} \tilde{\mathbf{E}} \times \tilde{\mathbf{H}}^* \cdot \mathbf{a}_z \, dx \, dy = \frac{1}{2} \int_0^b \int_0^a \left(\tilde{E}_x \tilde{H}_y^* - \tilde{E}_y \tilde{H}_x^* \right) dx \, dy. \qquad (10.59)$$

Using $\tilde{E}_x(x,y)$, $\tilde{E}_y(x,y)$, $\tilde{H}_x(x,y)$, and $\tilde{H}_y(x,y)$ from Eqs. (10.57a)–(10.57d) and evaluating the integrals, we find

$$P = \frac{1}{8} \frac{\beta \omega \varepsilon}{k_c^2} A^2 ab. \qquad (10.60a)$$

Using β from Eq. (10.56) and $k_c = \omega_c \sqrt{\varepsilon \mu}$, this total transmitted power can also be written as

$$P = \frac{1}{8} \left(\frac{\omega}{\omega_c} \right)^2 \sqrt{1 - \left(\frac{\omega_c}{\omega} \right)^2} \frac{A^2 ab}{\eta}. \qquad (10.60b)$$

We close this discussion of TM modes on the rectangular waveguide with a calculation of the attenuation constants due to dielectric absorption and due to conductor losses. For this we need the power loss per unit length P'. For dielectric absorption, we use Eq. (10.19) and find

$$P'_d = \int_0^b \int_0^a \frac{1}{2} (\sigma_d + \omega \varepsilon'') \left[|\tilde{E}_z(x,y)|^2 + |\tilde{E}_x(x,y)|^2 + |\tilde{E}_y(x,y)|^2 \right] dx \, dy.$$

Using $\tilde{E}_z(x,y)$, $\tilde{E}_x(x,y)$, and $\tilde{E}_y(x,y)$ from Eqs. (10.51), (10.57a), and (10.57b), the power loss is

$$P'_d = \frac{1}{8} (\sigma_d + \omega \varepsilon'') \left[1 + \left(\frac{\beta}{k_c} \right)^2 \right] A^2 ab.$$

Using β from Eq. (10.56), this simplifies to

$$P'_d = \frac{1}{8} (\sigma_d + \omega \varepsilon'') \left(\frac{\omega}{\omega_c} \right)^2 A^2 ab.$$

Then we use Eq. (10.20) to find

$$\alpha_d = \frac{P'_d}{2P} = \frac{(1/8)(\sigma_d + \omega \varepsilon'')(\omega/\omega_c)^2 A^2 ab}{2 (1/8)(\beta \omega \varepsilon / k_c^2) A^2 ab}.$$

Using k_c from Eq. (10.8) with $\gamma = j\beta$ and β from Eq. (10.56), this simplifies to

$$\alpha_d = \frac{\eta (\sigma_d + \omega \varepsilon'')}{2 \sqrt{1 - (\omega_c/\omega)^2}}.$$

Notice that α_d for TM modes of a rectangular waveguide is precisely the same as α_d for TM or TE modes of a parallel plate waveguide.

To find the power loss due to conductor losses P'_c, we use Eqs. (10.21a) and (10.21b). The magnitude of the surface current density $\tilde{J}_s$ is equal to that of the tangential magnetic field $\tilde{H}_{tan}$ at the conductor walls. Evaluating Eq. (10.57c) at $y = 0$ and $y = b$ gives us the magnitude of the surface current in the left and right sides, and evaluating Eq. (10.57d) at $x = 0$ and $x = a$ gives us the magnitude of the surface current in the top and bottom. Then the power loss per unit length is

$$P'_c = \frac{1}{2} \left[\left| \frac{j\omega \varepsilon k_y}{k_c^2} \right|^2 A^2 \int_0^a \sin^2(k_x x) \, dx + \left| \frac{-j\omega \varepsilon k_x}{k_c^2} \right|^2 A^2 \int_0^b \sin^2(k_y y) \, dy \right] R_s \times 2,$$

Table 10.5

Summary of results for TM modes on a rectangular waveguide. See Table 10.3 for mode impedance, phase constant, attenuation constant, phase velocity, and group velocity.	
$\tilde{E}_z(x, y) = A \sin(k_x x) \sin(k_y y)$	$\tilde{H}_z(x, y) = 0$
$\tilde{E}_x(x, y) = \dfrac{-j\beta k_x}{k_c^2} A \cos(k_x x) \sin(k_y y)$	$\tilde{H}_x(x, y) = -\tilde{E}_y(x, y)/Z_{TM}$
$\tilde{E}_y(x, y) = \dfrac{-j\beta k_y}{k_c^2} A \sin(k_x x) \cos(k_y y)$	$\tilde{H}_y(x, y) = \tilde{E}_x(x, y)/Z_{TM}$
where $k_x = \dfrac{m\pi}{a}$ $\quad k_y = \dfrac{n\pi}{b} \quad m, n = 1, 2, 3, \ldots$	
$k_c = \sqrt{k_x^2 + k_y^2} \qquad \omega_c = \dfrac{k_c}{\sqrt{\varepsilon\mu}} = \dfrac{1}{\sqrt{\varepsilon\mu}} \sqrt{\left(\dfrac{m\pi}{a}\right)^2 + \left(\dfrac{n\pi}{b}\right)^2}$	
$P = \dfrac{1}{8} \dfrac{\beta\omega\varepsilon}{k_c^2} A^2 ab = \dfrac{1}{8} \left(\dfrac{\omega}{\omega_c}\right)^2 \sqrt{1 - \left(\dfrac{\omega_c}{\omega}\right)^2} \dfrac{A^2 ab}{\eta}$	
$\alpha_d = \dfrac{\eta (\sigma_d + \omega\varepsilon'')}{2\sqrt{1 - (\omega_c/\omega)^2}} \qquad \alpha_c = \dfrac{2R_s}{\eta\sqrt{1 - (\omega_c/\omega)^2}} \left[\dfrac{m^2 b^2/a + n^2 a^2/b}{m^2 b^2 + n^2 a^2}\right]$	

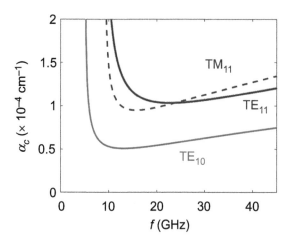

FIGURE 10.12 Plots of the attenuation coefficient α_c for the TM_{11} (Eq. (10.62), blue dashed), TE_{11} (Eq. (10.70), purple solid), and TE_{10} (Eq. (10.73), red solid) mode on a rectangular waveguide. The waveguide dimensions are 3×2 cm, and the walls are copper.

where the final factor of $\times 2$ accounts for the two sides, and the two surfaces at the top and bottom. Carrying out the integrals gives us

$$P'_c = \frac{1}{2} R_s \left(\frac{\omega\varepsilon}{k_c^2}\right)^2 A^2 \left(k_y^2 a + k_x^2 b\right). \qquad (10.61)$$

Finally, we determine the attenuation coefficient using Eq. (10.22). Using β from Eq. (10.56) and k_x, k_y, and k_c from Eqs. (10.52), (10.53), and (10.54), and a fair bit of manipulation, one can show that

$$\alpha_c = \frac{2R_s}{\eta\sqrt{1 - (\omega_c/\omega)^2}} \left[\frac{m^2 b^2/a + n^2 a^2/b}{m^2 b^2 + n^2 a^2}\right]. \qquad (10.62)$$

α_c for the TM_{11} mode on a copper waveguide of dimensions 3×2 cm is plotted versus frequency in Fig. 10.12.

The results for the TM mode on a rectangular waveguide are collected in Table 10.5.

Example 10.14 Attenuation Coefficient, Rectangular, TM

For the rectangular waveguide as described in Example 10.12, constructed of copper conducting plates, carrying a TM_{11} mode of amplitude $A = 10$ V/m and frequency 12 GHz, determine (a) the total transmitted power, (b) the power lost due to conductor losses per unit length, and (c) the attenuation coefficient α_c.

Solution:

(a) The total power transmitted by the waveguide mode is given in Eq. (10.60a). This power is

$$P = \frac{1}{8}\frac{\beta\omega\varepsilon}{k_c^2}A^2ab$$

$$= \frac{1}{8}\frac{(164\ \text{m}^{-1})(2\pi \times 12 \times 10^9\ \text{s}^{-1})(8.854 \times 10^{-12}\ \text{F/m})}{(189\ \text{m}^{-1})^2}(10\ \text{V/m})^2(0.03\ \text{m})(0.02\ \text{m})$$

$$= 2.30 \times 10^{-5}\ \text{W},$$

where we have used

$$\beta = \omega\sqrt{\varepsilon\mu}\sqrt{1-\left(\frac{f_c}{f}\right)^2} = \frac{2\pi \times 12 \times 10^9\ \text{s}^{-1}}{2.998 \times 10^8\ \text{m/s}}\sqrt{1-\left(\frac{9.08\ \text{GHz}}{12\ \text{GHz}}\right)^2} = 164\ \text{m}^{-1}$$

and

$$k_c = \pi\sqrt{a^{-2}+b^{-2}} = 189\ \text{m}^{-1}.$$

(b) We must next find the power lost per unit length P_c'. Using the conductivity of copper $\sigma = 5.8 \times 10^7$ S/m from Table 3.1 and $\mu = \mu_0$, the surface resistance of copper at the frequency of 12 GHz is

$$R_s = \sqrt{\frac{\omega\mu}{2\sigma}} = \sqrt{\frac{(2\pi \times 12 \times 10^9\ \text{s}^{-1})(4\pi \times 10^{-7}\ \text{H/m})}{2 \times 5.8 \times 10^7\ \text{S/m}}} = 0.0286\ \Omega.$$

We use this in Eq. (10.61) to find

$$P_c' = \frac{1}{2}R_s\left(\frac{\omega\varepsilon}{k_c^2}\right)^2 A^2\left[k_y^2 a + k_x^2 b\right]$$

$$= \frac{1}{2}(0.0286\ \Omega)\left(\frac{(2\pi \times 12 \times 10^9\ \text{s}^{-1})(8.854 \times 10^{-12}\ \text{F/m})}{(189\ \text{m}^{-1})^2}\right)^2 (10\ \text{V/m})^2$$

$$\times\left[(157\ \text{m}^{-1})^2(0.03\ \text{m}) + (105\ \text{m}^{-1})^2(0.02\ \text{m})\right]$$

$$= 0.479 \times 10^{-6}\ \text{W/m}.$$

We have used $k_x = m\pi/a = 1.05\ \text{cm}^{-1}$, $k_y = n\pi/b = 1.57\ \text{cm}^{-1}$, and $k_c = \sqrt{k_x^2 + k_y^2} = 1.89\ \text{cm}^{-1}$ in this calculation.

(c) Finally, we calculate the attenuation coefficient using two methods. Since we have already found P_c' and P, the simpler form is to use

$$\alpha_c = \frac{P_c'}{2P} = \frac{0.479 \times 10^{-6}\ \text{W/m}}{2 \times 2.30 \times 10^{-5}\ \text{W}} = 0.0104\ \text{m}^{-1}.$$

Alternatively, we could use Eq. (10.62) to find

$$\alpha_c = \frac{2R_s}{\eta\sqrt{1-(f_c/f)^2}}\left[\frac{m^2b^2/a + n^2a^2/b}{m^2b^2 + n^2a^2}\right]$$

$$= \frac{2(0.0286\ \Omega)}{(377\ \Omega)\sqrt{1-(9.08\ \text{GHz}/12\ \text{GHz})^2}}\left[\frac{(0.02\ \text{m})^2/(0.03\ \text{m}) + (0.03\ \text{m})^2/(0.02\ \text{m})}{(0.02\ \text{m})^2 + (0.03\ \text{m})^2}\right]$$

$$= 0.0104\ \text{m}^{-1}.$$

10.3.3 TE Modes of a Rectangular Waveguide

In this final discussion of rectangular waveguides we examine the properties of TE modes, for which $\tilde{E}_z(x,y)$ is, by definition, equal to zero, but for which the longitudinal $\tilde{H}_z(x,y)$ component exists. We start this discussion with the governing differential equation for this field component, Eq. (10.13),

$$\nabla_t^2 \tilde{H}_z(x,y) = \frac{\partial^2 \tilde{H}_z(x,y)}{\partial^2 x} + \frac{\partial^2 \tilde{H}_z(x,y)}{\partial^2 y} = -k_c^2 \tilde{H}_z(x,y), \tag{10.63}$$

when written using rectangular coordinates. Using the separation of variables technique to solve for $\tilde{H}_z(x,y)$, the result is

$$\tilde{H}_z(x,y) = \{A'' \sin(k_x x) + B'' \cos(k_x x)\}\{C'' \sin(k_y y) + D'' \cos(k_y y)\}, \tag{10.64}$$

where A'', B'', C'', and D'' are amplitudes yet to be determined, and k_x and k_y are spatial frequencies. We'll use boundary conditions to simplify this expression, but since this $\tilde{H}_z(x,y)$ component is tangential to the conductor surface, and we don't yet know the surface currents in the conductors, we'll first solve for $\tilde{E}_x(x,y)$ and $\tilde{E}_y(x,y)$, and then apply the boundary condition that E_{tan} must be zero at the surface of the conductor. Before we do this, let's insert Eq. (10.64) into Eq. (10.63) to verify that $\tilde{H}_z(x,y)$ is indeed a good solution. This exercise leads to the condition

$$k_x^2 + k_y^2 = k_c^2.$$

This relation must be satisfied by k_x and k_y jointly, and is identical to the condition that we discovered for k_x and k_y of TM modes of rectangular waveguides.

Using $\tilde{H}_z(x,y)$ from Eq. (10.64) and $\tilde{E}_z(x,y) = 0$ in Eq. (10.16a), $\tilde{E}_x(x,y)$ is

$$\tilde{E}_x(x,y) = \frac{-j\omega\mu}{k_c^2} \frac{\partial \tilde{H}_z(x,y)}{\partial y} \tag{10.65}$$

$$= \frac{-j\omega\mu}{k_c^2} k_y \{A'' \sin(k_x x) + B'' \cos(k_x x)\}\{C'' \cos(k_y y) - D'' \sin(k_y y)\}.$$

Similarly, using Eq. (10.16b), we find

$$\tilde{E}_y(x,y) = \frac{j\omega\mu}{k_c^2} \frac{\partial \tilde{H}_z(x,y)}{\partial x} \tag{10.66}$$

$$= \frac{j\omega\mu}{k_c^2} k_x \{A'' \cos(k_x x) - B'' \sin(k_x x)\}\{C'' \sin(k_y y) + D'' \cos(k_y y)\}.$$

Now we are in a position to apply the boundary condition for $\tilde{E}_{tan}$ at the conductor walls. These conditions require that $\tilde{E}_{tan} = \tilde{E}_y(x,y)$ must be zero at (1) $x = 0$ and at (2) $x = a$, and that $\tilde{E}_{tan} = \tilde{E}_x(x,y)$ must be zero at (3) $y = 0$ and at (4) $y = b$. We'll start with the first of these, that $\tilde{E}_y(x,y)$ must be zero at $x = 0$. Evaluating Eq. (10.66) at $x = 0$ and setting it equal to zero, we see that the amplitude A'' must be zero. Similarly, using the third boundary condition, we evaluate $\tilde{E}_x(x,y)$ from Eq. (10.65) at $y = 0$, set it equal to zero, and see that the amplitude C'' must be zero. These two results allow $\tilde{H}_z(x,y)$ to be written in the simplified form

$$\tilde{H}_z(x,y) = B \cos(k_x x) \cos(k_y y), \tag{10.67a}$$

where the amplitude B replaces the product of B'' and D''.

Next we examine the second and fourth boundary conditions, and find that k_x must equal $m\pi/a$ and k_y must equal $n\pi/b$, where m and n are positive integers. For the TE mode, one of these integers m or n can be zero, but not both. Otherwise the field $\tilde{H}_z(x,y)$

would be uniform inside the waveguide, and the electric field components $\tilde{E}_x(x,y)$ and $\tilde{E}_y(x,y)$ would be zero. Recall that for TM modes in a rectangular waveguide neither m nor n could be zero, so this is one important distinction between TE and TM modes.

Now that we know $\tilde{H}_z(x,y)$, k_x, and k_y, we can return to the transverse electric field components in a simplified form, with

$$\tilde{E}_x(x,y) = \frac{j\omega\mu}{k_c^2} k_y B \cos(k_x x) \sin(k_y y) \tag{10.67b}$$

and

$$\tilde{E}_y(x,y) = \frac{-j\omega\mu}{k_c^2} k_x B \sin(k_x x) \cos(k_y y). \tag{10.67c}$$

The final components that must be found are $\tilde{H}_x(x,y)$ and $\tilde{H}_y(x,y)$. Using Eqs. (10.16c) and (10.16d), these terms are

$$\tilde{H}_x(x,y) = \frac{-\gamma}{k_c^2} \frac{\partial \tilde{H}_z(x,y)}{\partial x} = \frac{j\beta}{k_c^2} k_x B \sin(k_x x) \cos(k_y y) \tag{10.67d}$$

and

$$\tilde{H}_y(x,y) = \frac{-\gamma}{k_c^2} \frac{\partial \tilde{H}_z(x,y)}{\partial y} = \frac{j\beta}{k_c^2} k_y B \cos(k_x x) \sin(k_y y). \tag{10.67e}$$

To plot the TE mode pattern on the waveguide, the field amplitudes must be converted to their full x-, y-, z-, and t-dependent form using Eqs. (10.1) and (10.2) (i.e. multiply by $e^{j(\omega t - \beta z)}$ and take the real part). For example, $H_z(x,y,z,t)$ for the TE_{11} mode becomes

$$H_z(x,y,z,t) = B\cos(k_x x)\cos(k_y y)\cos(\omega t - \beta z),$$

and $E_x(x,y,z,t)$ becomes

$$E_x(x,y,z,t) = -\frac{\omega\mu}{k_c^2} k_y B \cos(k_x x)\sin(k_y y)\sin(\omega t - \beta z).$$

The full results for this mode are plotted in Fig. 10.13. Notice that $\mathbf{E}(x,y,z,t)$ at the conductor surfaces is always normal to these surfaces, and $\mathbf{H}(x,y,z,t)$ is always tangential, as they must be.

To find the propagation constant γ, we return to

$$\gamma^2 = k_c^2 - k^2.$$

γ^2 is less than zero, as required for propagating waves, only when $k^2 > k_c^2$. Since $k^2 = \omega^2\sqrt{\varepsilon\mu}$ and $k_c^2 = (m\pi/a)^2 + (n\pi/b)^2$, this defines the cutoff frequency as

$$\omega_c = \frac{k_c}{\sqrt{\varepsilon\mu}} = \frac{1}{\sqrt{\varepsilon\mu}} \sqrt{\left(\frac{m\pi}{a}\right)^2 + \left(\frac{n\pi}{b}\right)^2}.$$

When $\omega > \omega_c$, γ is imaginary, the waveguide solution is a propagating wave, and $\beta = -j\gamma$ leads to

$$\beta = \omega\sqrt{\varepsilon\mu}\sqrt{1 - \left(\frac{\omega_c}{\omega}\right)^2}.$$

When $\omega < \omega_c$, γ is purely real, the field decays exponentially, and $\alpha = \gamma$ gives us

$$\alpha = k_c\sqrt{1 - \left(\frac{\omega}{\omega_c}\right)^2}.$$

In fact, these results are identical to those for TM modes on a rectangular waveguide, and to TM and TE modes on a parallel plate waveguide. Therefore, the derivation of Λ, u_p, or u_g are not repeated here. You can simply refer back to these previous discussions.

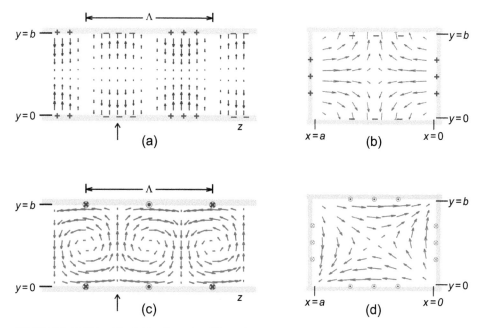

FIGURE 10.13 Plots of the field patterns for a TE$_{11}$ mode on a rectangular waveguide. (a) y–z plot of the electric field at $x = a/2$; (b) x–y plot of the electric field at $z = \lambda/4$; (c) y–z plot of the magnetic field at $x = a/2$; and (d) x–y plot of the magnetic field at $z = \lambda/4$. The magenta + and − signs in (a) and (b) represent the surface charge on the inside surface of the conductors. The magenta arrows in (c) and (d) represent the surface current on the inside surface of the conductors. The arrows beneath the waveguide in (a) and (c) indicate the z-location for plots (b) and (d).

Example 10.15 Cutoff Frequencies, Rectangular

For an air-filled rectangular waveguide with $a = 2$ cm and $b = 1$ cm, determine the cutoff frequencies of the five lowest-frequency modes.

Solution:

For TM or TE modes, the cutoff frequencies are computed as

$$\omega_c = \frac{1}{\sqrt{\varepsilon\mu}}\left[\left(\frac{m\pi}{a}\right)^2 + \left(\frac{n\pi}{b}\right)^2\right]^{1/2},$$

and $f_c = \omega_c/2\pi$. Remember that, for TM$_{mn}$ modes, m and n must each be 1 or larger, but for TE$_{mn}$ modes, either m or n, but not both, can be 0. The cutoff frequencies for this waveguide are listed in Table 10.6. Notice that four of the five lowest-frequency modes are TE modes, and the other is TM.

Table 10.6

Cutoff frequencies for the lowest TE$_{mn}$ and TM$_{mn}$ modes of the rectangular waveguide of Example 10.15.			
m	n	f_c (GHz)	modes
1	0	7.5	TE only
0	1	15.0	TE only
2	0	15.0	TE only
1	1	16.8	TE and TM

The wave impedance for the TE mode can be found from

$$Z_{TE} = \frac{\tilde{E}_x(x,y)}{\tilde{H}_y(x,y)} = \frac{j\omega\mu k_y/k_c^2}{j\beta k_y/k_c^2} = \frac{\omega\mu}{\beta} = \frac{\eta}{\sqrt{1-(\omega_c/\omega)^2}}.$$

Alternatively, the wave impedance can be found using

$$Z_{TE} = -\frac{\tilde{E}_y(x,y)}{\tilde{H}_x(x,y)} = \frac{j\omega\mu k_x/k_c^2}{j\beta k_x/k_c^2} = \frac{\omega\mu}{\beta} = \frac{\eta}{\sqrt{1-(\omega_c/\omega)^2}},$$

producing the same result. This expression for the wave impedance Z_{TE} is the same as it was for the parallel plate waveguide.

The total power transmitted by the waveguide mode is

$$P = \int_0^b \int_0^a \frac{1}{2}(\tilde{E}_x(x,y)\tilde{H}_y^*(x,y) - \tilde{E}_y(x,y)\tilde{H}_x^*(x,y))\,dx\,dy.$$

Using $\tilde{E}_x(x,y)$, $\tilde{E}_y(x,y)$, $\tilde{H}_x(x,y)$, and $\tilde{H}_y(x,y)$ from Eqs. (10.67b)–(10.67e), we find

$$P = \frac{1}{8}\frac{\omega\mu\beta}{k_c^2}B^2 ab. \tag{10.68a}$$

(This result, and several of the following results, are valid only when $m \neq 0$ and $n \neq 0$. We'll return to consider the case when m or n is zero shortly.) We can use β from Eq. (10.56) and $k_c = \omega_c\sqrt{\varepsilon\mu}$ to write an equivalent form of the total power carried by the waveguide as

$$P = \frac{1}{8}\left(\frac{\omega}{\omega_c}\right)^2\sqrt{1-\left(\frac{\omega_c}{\omega}\right)^2}\,\eta B^2 ab. \tag{10.68b}$$

The attenuation due to absorption by the dielectric is found using Eq. (10.19):

$$P_d' = \int_0^b \int_0^a \frac{1}{2}(\sigma_d + \omega\varepsilon'')\left[\left|\tilde{E}_x(x,y)\right|^2 + \left|\tilde{E}_y(x,y)\right|^2\right]dx\,dy.$$

Using $\tilde{E}_z(x,y)$ and $\tilde{E}_y(x,y)$ from Eqs. (10.67b) and (10.67c), the power loss is

$$P_d' = \int_0^b \int_0^a \frac{1}{2}(\sigma_d + \omega\varepsilon'')\left[\left|\frac{j\omega\mu}{k_c^2}k_y B\cos(k_x x)\sin(k_y y)\right|^2 + \left|\frac{-j\omega\mu}{k_c^2}k_x B\sin(k_x x)\cos(k_y y)\right|^2\right]dx\,dy.$$

Carrying out the integrals, and using $k_x^2 + k_y^2 = k_c^2$, the power loss becomes

$$P_d' = \frac{1}{8}(\sigma_d + \omega\varepsilon'')\left(\frac{\omega\mu}{k_c}\right)^2 B^2 ab.$$

We then find the attenuation constant due to dielectric absorption using Eq. (10.20):

$$\alpha_d = \frac{P_d'}{2P} = \frac{(1/8)(\sigma_d + \omega\varepsilon'')(\omega\mu/k_c)^2 B^2 ab}{2(1/8)(\omega\mu\beta/k_c^2)B^2 ab}.$$

Canceling similar terms, and using β from Eq. (10.56), this simplifies to

$$\alpha_d = \frac{\eta(\sigma_d + \omega\varepsilon'')}{2\sqrt{1-(\omega_c/\omega)^2}}.$$

Notice that this attenuation coefficient is the same as we found before for the TM mode on the rectangular waveguide and the TM and TE mode on the parallel plate waveguide.

To find the attenuation coefficient due to the finite conductivity of the waveguide walls, we use Eq. (10.22). The power loss per unit length is

$$P'_c = \int_0^a \frac{1}{2}|\tilde{J}_s|^2 R_s \, dx \times 2 + \int_0^b \frac{1}{2}|\tilde{J}_s|^2 R_s \, dy \times 2. \tag{10.69}$$

For $|\tilde{J}_s|^2$ in the left and right walls, we use $|\tilde{H}_z(x, y = 0)|^2 + |\tilde{H}_x(x, y = 0)|^2$, while for $|\tilde{J}_s|^2$ in the top and bottom conductors we use $|\tilde{H}_z(x = 0, y)|^2 + |\tilde{H}_y(x = 0, y)|^2$. Upon carrying out the integration, we find

$$P'_c = \frac{1}{2}\left[a + b + \frac{\beta^2}{k_c^4}\left(k_x^2 a + k_y^2 b\right)\right]B^2 R_s.$$

Finally, we use P'_c and P in Eq. (10.22) to find α_c. The result is

$$\alpha_c = \frac{2R_s}{\eta\sqrt{1 - (\omega_c/\omega)^2}}\left\{\frac{\left(m^2 b^2/a + n^2 a^2/b\right)(\omega_c/\omega)^2 + \left(m^2 b + n^2 a\right)}{m^2 b^2 + n^2 a^2}\right\}. \tag{10.70}$$

α_c for the TE_{11} mode on a copper waveguide of dimensions 3×2 cm is plotted versus frequency in Fig. 10.12. The results for the TE mode on a rectangular waveguide are collected in Table 10.7.

Table 10.7

Summary of results for TE modes on a rectangular waveguide. See Table 10.3 for mode impedance, phase constant, attenuation constant, phase velocity, and group velocity.	
$\tilde{E}_z(x, y) = 0$	$\tilde{H}_z(x, y) = B\cos(k_x x)\cos(k_y y)$
$\tilde{E}_x(x, y) = \dfrac{j\omega\mu}{k_c^2}k_y B\cos(k_x x)\sin(k_y y)$	$\tilde{H}_x(x, y) = -\tilde{E}_y(x, y)/Z_{TE}$
$\tilde{E}_y(x, y) = \dfrac{-j\omega\mu}{k_c^2}k_x B\sin(k_x x)\cos(k_y y)$	$\tilde{H}_y(x, y) = \tilde{E}_x(x, y)/Z_{TE}$
where $k_x = \dfrac{m\pi}{a}$ $\qquad$ $k_y = \dfrac{n\pi}{b}$ $\qquad$ $m, n = 0, 1, 2, 3, \ldots$ but m and n cannot both be zero.	
$k_c = \sqrt{k_x^2 + k_y^2}$ $\qquad$ $\omega_c = \dfrac{k_c}{\sqrt{\varepsilon\mu}} = \dfrac{1}{\sqrt{\varepsilon\mu}}\sqrt{\left(\dfrac{m\pi}{a}\right)^2 + \left(\dfrac{n\pi}{b}\right)^2}$	
$P = \begin{cases} \dfrac{1}{8}\dfrac{\omega\mu\beta}{k_c^2}B^2 ab = \dfrac{1}{8}\left(\dfrac{\omega}{\omega_c}\right)^2\sqrt{1 - \left(\dfrac{\omega_c}{\omega}\right)^2}\,\eta B^2 ab, & \text{for } m, n \neq 0 \\[3mm] \dfrac{1}{4}\dfrac{\omega\mu\beta}{k_x^2}B^2 ab = \dfrac{1}{4}\left(\dfrac{\omega}{\omega_c}\right)^2\sqrt{1 - \left(\dfrac{\omega_c}{\omega}\right)^2}\,\eta B^2 ab, & \text{for } n = 0 \end{cases}$ $\alpha_d = \dfrac{\eta(\sigma_d + \omega\varepsilon'')}{2\sqrt{1 - (\omega_c/\omega)^2}}$	
$\alpha_c = \begin{cases} \dfrac{2R_s}{\eta\sqrt{1 - (\omega_c/\omega)^2}}\left\{\dfrac{\left(m^2 b^2/a + n^2 a^2/b\right)(\omega_c/\omega)^2 + \left(m^2 b + n^2 a\right)}{m^2 b^2 + n^2 a^2}\right\}, & \text{for } m, n \neq 0 \\[3mm] \dfrac{2R_s}{\eta\sqrt{1 - (\omega_c/\omega)^2}}\left[\left(\dfrac{\omega_c}{\omega}\right)^2\dfrac{1}{a} + \dfrac{1}{2b}\right], & \text{for } n = 0 \end{cases}$	

10.3.4 TE$_{10}$ and TE$_{01}$ Modes of a Rectangular Waveguide

The TE$_{10}$ and TE$_{01}$ modes are special for a couple of reasons, and deserve to be treated separately. The first reason for the special treatment of these modes is that the cutoff frequency of the TE$_{10}$ mode is lower than that of all other TE and TM modes (when $a > b$). (Notice the cutoff frequencies in Example 10.15.) This low cutoff frequency is important, and leads to wide usage of this mode. In applications where single-mode excitation is required, it is advantageous to choose a waveguide whose cutoff frequencies ω_c for all modes except the TE$_{10}$ mode are greater than the excitation frequency. In this way, only the TE$_{10}$ mode can propagate, and single-mode operation of the waveguide is assured.

The second reason for treating the TE$_{10}$ and TE$_{01}$ modes separately is that some of the parameters (specifically, the transmitted power P and the attenuation constant α_c) describing this mode differ from the general TE$_{mn}$ mode result. When n is zero (i.e. for the TE$_{10}$ mode), the field amplitudes $\tilde{E}_x(x,y)$ and $\tilde{H}_y(x,y)$ vanish, and the amplitudes $\tilde{E}_y(x,y)$, $\tilde{H}_x(x,y)$, and $\tilde{H}_z(x,y)$ depend only on x. Similarly, when m is zero, as it is for the TE$_{01}$ mode, the field amplitudes $\tilde{E}_y(x,y)$ and $\tilde{H}_x(x,y)$ vanish, and the amplitudes $\tilde{E}_x(x,y)$, $\tilde{H}_y(x,y)$, and $\tilde{H}_z(x,y)$ depend only on y.

We'll focus on the TE$_{10}$ mode in the following. Starting with the field components, which follow from Eqs. (10.67a)–(10.67e) by simply setting $n = 0$, the TE$_{m0}$ amplitudes are:

$$\tilde{E}_z(x) = 0, \qquad (10.71a)$$

$$\tilde{E}_x(x) = 0, \qquad (10.71b)$$

$$\tilde{E}_y(x) = \frac{-j\omega\mu}{k_x} B \sin(k_x x), \qquad (10.71c)$$

$$\tilde{H}_z(x) = B\cos(k_x x), \qquad (10.72a)$$

$$\tilde{H}_x(x) = -\tilde{E}_y(x)/Z_{TE}, \qquad (10.72b)$$

$$\tilde{H}_y(x) = 0. \qquad (10.72c)$$

The field patterns for the TE$_{10}$ mode are plotted in Fig. 10.14. These are found by converting the field amplitudes, Eqs. (10.71a)–(10.72c) to their full y-, z-, and t-dependent form using Eqs. (10.1) and (10.2) (i.e. multiply by $e^{j(\omega t-\beta z)}$ and take the real part).

The propagation constant γ, cutoff frequency, wave velocities, and mode impedance are each as discussed above for the general modes, and those discussions or derivations won't be repeated here.

The total transmitted power for this mode is found to be

$$P = \frac{1}{4}\frac{\omega\mu\beta}{k_x^2}B^2 ab.$$

We can use β from Eq. (10.56) and $k_c = \omega_c\sqrt{\mu\varepsilon}$ to write an equivalent form of the total power carried by the waveguide as

$$P = \frac{1}{4}\left(\frac{\omega}{\omega_c}\right)^2\sqrt{1-\left(\frac{\omega_c}{\omega}\right)^2}\,\eta B^2 ab.$$

This power for the TE$_{10}$ mode differs from Eqs. (10.68a) and (10.68b) for the general TE$_{mn}$ mode by a factor of two.

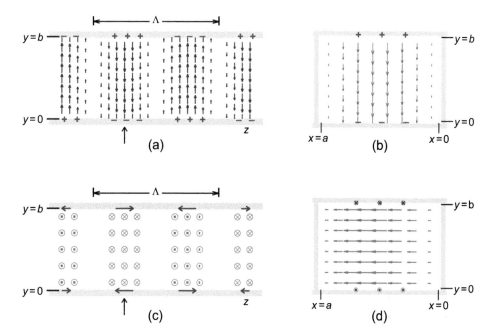

FIGURE 10.14 Plots of the field patterns for a TE_{10} mode on a rectangular waveguide. (a) y–z plot of the electric field at $x = a/2$; (b) x–y plot of the electric field at $z = \lambda/4$; (c) y–z plot of the magnetic field at $x = a/2$; and (d) x–y plot of the magnetic field at $z = \lambda/4$. The magenta $+$ and $-$ signs in (a) and (b) represent the surface charge on the inside surface of the conductors. The magenta arrows in (c) and (d) represent the surface current on the inside surface of the conductors. The arrows beneath the waveguide in (a) and (c) indicate the z-location for plots (b) and (d).

The loss per unit length due to dielectric absorption is, using Eq. (10.19),

$$P'_d = \int_0^b \int_0^a \frac{1}{2}(\sigma_d + \omega\varepsilon'') |\tilde{E}_y(x)|^2 \, dx \, dy$$

$$= \int_0^b \int_0^a \frac{1}{2}(\sigma_d + \omega\varepsilon'') \left| \frac{-j\omega\mu}{k_x} B \sin(k_x x) \right|^2 \, dx \, dy.$$

Upon integration, this is

$$P'_d = \frac{1}{4}(\sigma_d + \omega\varepsilon'') \left(\frac{\omega\mu}{k_x} \right)^2 B^2 ab.$$

(This is twice the power loss P'_d that we found for a TE_{mn} mode.)

The attenuation coefficient due to dielectric absorption is, using Eq. (10.20),

$$\alpha_d = \frac{P'_d}{2P} = \frac{(1/4)(\sigma_d + \omega\varepsilon'')(\omega\mu/k_x)^2 B^2 ab}{2(1/4)(\omega\mu\beta/k_x^2) B^2 ab}.$$

Canceling similar terms, and using β from Eq. (10.56), this simplifies to

$$\alpha_d = \frac{\eta(\sigma_d + \omega\varepsilon'')}{2\sqrt{1 - (\omega_c/\omega)^2}}.$$

This attenuation constant for the TE_{10} mode is the same as α_d that we found for the general TE_{mn} mode.

The loss per unit length due to the finite conductivity of the conductors for the TE_{10} mode is

$$P'_c = \left[\frac{a}{2} \left(\frac{\omega}{\omega_c} \right)^2 + b \right] B^2 R_s.$$

This is found using Eq. (10.69), but with the simplified expressions of field amplitudes for the TE_{10} mode. Then the attenuation coefficient is

$$\alpha_c = \frac{P'_c}{2P} = \frac{2R_s}{\eta\sqrt{1-(\omega_c/\omega)^2}}\left[\left(\frac{\omega_c}{\omega}\right)^2\frac{1}{a}+\frac{1}{2b}\right].\tag{10.73}$$

α_c for the TE_{10} mode on a copper waveguide of dimensions 3×2 cm is plotted versus frequency in Fig. 10.12. Note that the losses for the TE_{10} mode are significantly lower than those for the TE_{11} or TM_{11} modes for all frequencies.

In the preceding discussion we identified several compelling reasons for exciting the TE_{10} mode of a waveguide. Perhaps foremost is that the cutoff frequency of this mode is lower than that of all other TE or TM modes on the waveguide, so exciting the waveguide at a frequency greater than the cutoff frequency for the TE_{10} mode, but below that of the next mode, is the best way to achieve single-mode operation of the waveguide. In addition, we have seen that the attenuation coefficient α_c is much smaller for the TE_{10} mode than all other modes. The practical operating frequency range for a waveguide is $\sim 25\%$ above the cutoff frequency for the TE_{10} mode to a little below the cutoff frequency for the next higher mode. The minimum value of this range is due to large attenuation losses as $\omega\to\omega_c$, as seen in Eq. (10.73) and Fig. 10.12, and strong dispersion effects due to u_p or u_g changing rapidly with frequency ω. (See Fig. 10.7.) The high-frequency limit of the operating range is set by the cutoff frequency of the next waveguide mode. By choosing the dimensions $b=a/2$, the cutoff frequency of the TE_{01} mode is two times that of the TE_{10} mode, and equal to that of the TE_{20} mode, maximizing the frequency range for single-mode operation of the waveguide.

This concludes our discussion of TM and TE modes on rectangular waveguides. In this section we have derived expressions for the various modes, labeled by the indices m and n, for the electric and magnetic field patterns, and the propagation parameters, such as the cutoff frequencies, impedances Z_{TM}, Z_{TE}, phase constant β, and attenuations constants α, etc., and listed these in Tables 10.5 and 10.7. In the following section we introduce circular waveguides.

10.4 Circular Waveguides

The final waveguide geometry that we will treat is the circular waveguide, as pictured in Fig. 10.1(c). Similar to the rectangular waveguide, the circular waveguide is composed of a single conductor, and is hollow on the inside. These waveguides are suitable for transmitting circularly polarized radiation fields. This analysis will follow a similar approach to that used for the parallel plate and rectangular waveguides, but for circular waveguides cylindrical coordinates should be used. As a result, we need to introduce a new set of functions, known as the Bessel functions, to describe the waveguide modes.

Our approach to determine the modes (TM and TE) supported by a circular waveguide is the same as the approach we applied in the preceding sections for parallel plate and rectangular waveguides. We start with Eq. (10.12),

$$\nabla_t^2\tilde{E}_z(\rho,\phi)=-k_c^2\tilde{E}_z(\rho,\phi),$$

to find $\tilde{E}_z(\rho,\phi)$, and Eq. (10.13),

$$\nabla_t^2\tilde{H}_z(\rho,\phi)=-k_c^2\tilde{H}_z(\rho,\phi),$$

to find $\tilde{H}_z(\rho,\phi)$. As before, k_c, k, and γ are related through

$$k_c^2 = \gamma^2 + k^2.$$

When working in cylindrical coordinates, the proper form of ∇_t^2 is

$$\nabla_t^2 = \frac{1}{\rho}\frac{\partial}{\partial\rho}\left(\rho\frac{\partial}{\partial\rho}\right) + \frac{1}{\rho^2}\frac{\partial^2}{\partial^2\phi}. \tag{10.74}$$

These differential equations include derivatives that are second-order with respect to the two transverse coordinates, ρ and ϕ. Compare this to Eqs. (10.49) and (10.63), the same equations written in rectangular coordinates, which included second-order derivatives with respect to x and y.

After $\tilde{E}_z(\rho,\phi)$ and $\tilde{H}_z(\rho,\phi)$ have been found, these longitudinal components will be used to determine the transverse field components $\tilde{E}_\rho(\rho,\phi)$, $\tilde{E}_\phi(\rho,\phi)$, $\tilde{H}_\rho(\rho,\phi)$, and $\tilde{H}_\phi(\rho,\phi)$. This is accomplished using Faraday's Law and Ampère's Law, which, when written out as the individual components in cylindrical coordinates for a wave traveling in the $+z$-direction, are

$$\rho:\quad \frac{1}{\rho}\frac{\partial\tilde{E}_z}{\partial\phi} + \gamma\tilde{E}_\phi = -j\omega\mu\tilde{H}_\rho, \tag{10.75a}$$

$$\phi:\quad -\gamma\tilde{E}_\rho - \frac{\partial\tilde{E}_z}{\partial\rho} = -j\omega\mu\tilde{H}_\phi, \tag{10.75b}$$

$$z:\quad \frac{1}{\rho}\frac{\partial(\rho\tilde{E}_\phi)}{\partial\rho} - \frac{1}{\rho}\frac{\partial\tilde{E}_\rho}{\partial\phi} = -j\omega\mu\tilde{H}_z, \tag{10.75c}$$

$$\frac{1}{\rho}\frac{\partial\tilde{H}_z}{\partial\phi} + \gamma\tilde{H}_\phi = j\omega\varepsilon\tilde{E}_\rho, \tag{10.76a}$$

$$-\gamma\tilde{H}_\rho - \frac{\partial\tilde{H}_z}{\partial\rho} = j\omega\varepsilon\tilde{E}_\phi, \tag{10.76b}$$

$$\frac{1}{\rho}\frac{\partial(\rho\tilde{H}_\phi)}{\partial\rho} - \frac{1}{\rho}\frac{\partial\tilde{H}_\rho}{\partial\phi} = j\omega\varepsilon\tilde{E}_z. \tag{10.76c}$$

(The (ρ,ϕ) dependence notation of each of these field quantities is suppressed for compactness.)

Similar to our treatment in rectangular coordinates, these equations must be combined in such a way as to find expressions for each of the transverse field components $\tilde{E}_\rho(\rho,\phi)$, $\tilde{E}_\phi(\rho,\phi)$, $\tilde{H}_\rho(\rho,\phi)$, and $\tilde{H}_\phi(\rho,\phi)$ in terms of the transverse derivatives ($\partial/\partial\rho$ and $\partial/\partial\phi$) of $\tilde{E}_z(\rho,\phi)$ and $\tilde{H}_z(\rho,\phi)$. For example, Eq. (10.76a) can be used to eliminate $\tilde{H}_\phi(\rho,\phi)$ in Eq. (10.75b),

$$\gamma\tilde{E}_\rho(\rho,\phi) = j\omega\mu\tilde{H}_\phi(\rho,\phi) - \frac{\partial\tilde{E}_z(\rho,\phi)}{\partial\rho}$$
$$= \frac{j\omega\mu}{\gamma}\left(j\omega\varepsilon\tilde{E}_\rho(\rho,\phi) - \frac{1}{\rho}\frac{\partial\tilde{H}_z(\rho,\phi)}{\partial\phi}\right) - \frac{\partial\tilde{E}_z(\rho,\phi)}{\partial\rho}.$$

Then combining the terms that contain $\tilde{E}_\rho(\rho,\phi)$ and dividing both sides by $k^2 + \gamma^2$, we find

$$\tilde{E}_\rho(\rho,\phi) = -\frac{1}{k^2+\gamma^2}\left(\gamma\frac{\partial\tilde{E}_z(\rho,\phi)}{\partial\rho} + \frac{j\omega\mu}{\rho}\frac{\partial\tilde{H}_z(\rho,\phi)}{\partial\phi}\right). \tag{10.77a}$$

Similar expressions for $\tilde{E}_\phi(\rho,\phi)$, $\tilde{H}_\rho(\rho,\phi)$, and $\tilde{H}_\phi(\rho,\phi)$ can be found:

$$\tilde{E}_\phi(\rho,\phi) = -\frac{1}{k^2+\gamma^2}\left(\frac{\gamma}{\rho}\frac{\partial\tilde{E}_z(\rho,\phi)}{\partial\phi} - j\omega\mu\frac{\partial\tilde{H}_z(\rho,\phi)}{\partial\rho}\right), \tag{10.77b}$$

$$\tilde{H}_\rho(\rho,\phi) = \frac{1}{k^2+\gamma^2}\left(\frac{j\omega\varepsilon}{\rho}\frac{\partial\tilde{E}_z(\rho,\phi)}{\partial\phi} - \gamma\frac{\partial\tilde{H}_z(\rho,\phi)}{\partial\rho}\right), \tag{10.77c}$$

and

$$\tilde{H}_\phi(\rho,\phi) = -\frac{1}{k^2+\gamma^2}\left(j\omega\varepsilon\frac{\partial\tilde{E}_z(\rho,\phi)}{\partial\rho} + \frac{\gamma}{\rho}\frac{\partial\tilde{H}_z(\rho,\phi)}{\partial\phi}\right). \tag{10.77d}$$

Similar to the rectangular waveguide case, TEM modes are not allowed for the circular waveguide. In the following sections we will solve for the electric and magnetic field components of the TM and TE modes.

10.4.1 TM Modes of a Circular Waveguide

For TM waves, the magnetic field has only transverse components; that is, $\tilde{H}_z(\rho, \phi) = 0$. The electric field component in the z-direction is governed by Eq. (10.12), with ∇^2 in cylindrical coordinates given by Eq. (10.74). We will again use the separation-of-variables technique, as we did with the rectangular coordinate solution, to solve this two-dimensional equation, and write the solution in a general form,

$$\tilde{E}_z(\rho, \phi) = \{A' J_n(k_c\rho) + B' N_n(k_c\rho)\} \times \{C' \cos(n\phi) + D' \sin(n\phi)\},$$

where $J_n(k_c\rho)$ and $N_n(k_c\rho)$ are special functions known as the **Bessel functions**. $J_n(k_c\rho)$ is called the Bessel function of the first kind, and $N_n(k_c\rho)$ is the Bessel function of the second kind. The subscript n is the order of the Bessel function, and must be an integer $n = 0, 1, 2, \ldots$. These functions are probably new to most readers of this text, although they are well-known functions in engineering and physics, and you may encounter them in a number of different problems and applications. We will not attempt a complete development of their properties here, but do show plots of several members of this family of functions at a scaled radius $u = k_c\rho$ in Fig. 10.15. Note that, in general, these functions oscillate as a function of ρ, with decreasing amplitude. Note also that the magnitudes of the functions $N_n(k_c\rho)$ become extremely large as $\rho \to 0$; that is, they diverge at the z-axis. Since electric fields must always be finite in magnitude, the "boundary condition" at $\rho = 0$ can only be satisfied if the amplitudes B' are zero for all n. Also, since the waveguide is cylindrically symmetric, we are free to choose its orientation such that only the $\cos(n\phi)$ terms are needed. Therefore, $\tilde{E}_z(\rho, \phi)$ is simplified as

$$\tilde{E}_z(\rho, \phi) = A J_n(k_c\rho) \cos(n\phi), \quad \text{for } n = 0, 1, 2, \ldots \tag{10.78a}$$

Next, we use $\tilde{E}_z(\rho, \phi)$ of Eq. (10.78a), along with $\tilde{H}_z(\rho, \phi) = 0$ (by definition for a TM mode), in Eqs. (10.77a)–(10.77d) to determine $\tilde{E}_\rho(\rho, \phi)$, $\tilde{E}_\phi(\rho, \phi)$, $\tilde{H}_\rho(\rho, \phi)$, and $\tilde{H}_\phi(\rho, \phi)$. For example, $\tilde{E}_\rho(\rho, \phi)$ can be found from Eq. (10.77a) as

$$E_\rho(\rho, \phi) = -\frac{1}{k_c^2}\left[\gamma \frac{\partial E_z}{\partial \rho}\right] = -\frac{\gamma}{k_c} A J_n'(k_c\rho) \cos(n\phi). \tag{10.78b}$$

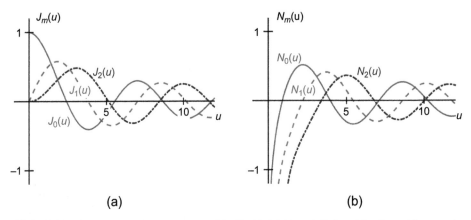

(a)   (b)

FIGURE 10.15 Plots of the (a) Bessel functions of the first kind $J_n(u)$, and (b) Bessel functions of the second kind $N_n(u)$, for $n = 0$, 1, and 2.

The "prime" notation is used on $J'_n(k_c\rho)$ to indicate the derivative of the Bessel function with respect to its argument,

$$J'_n(u) = \frac{d}{du}J_n(u).$$

Similarly, we use Eq. (10.77b) to find $\tilde{E}_\phi(\rho, \phi)$,

$$\tilde{E}_\phi(\rho, \phi) = -\frac{1}{k_c^2}\left[\frac{\gamma}{\rho}\frac{\partial \tilde{E}_z(\rho, \phi)}{\partial \phi}\right] = \frac{\gamma n}{k_c^2 \rho}AJ_n(k_c\rho)\sin(n\phi); \qquad (10.78c)$$

Eq. (10.77c) to find $\tilde{H}_\rho(\rho, \phi)$,

$$\tilde{H}_\rho(\rho, \phi) = \frac{1}{k_c^2}\left[\frac{j\omega\varepsilon}{\rho}\frac{\partial \tilde{E}_z(\rho, \phi)}{\partial \phi}\right] = -\frac{j\omega\varepsilon n}{k_c^2 \rho}AJ_n(k_c\rho)\sin(n\phi); \qquad (10.78d)$$

and Eq. (10.77d) to find $\tilde{H}_\phi(\rho, \phi)$,

$$\tilde{H}_\phi(\rho, \phi) = -\frac{1}{k_c^2}\left[j\omega\varepsilon\frac{\partial \tilde{E}_z(\rho, \phi)}{\partial \rho}\right] = \frac{-j\omega\varepsilon}{k_c}AJ'_n(k_c\rho)\cos(n\phi). \qquad (10.78e)$$

Notice that $\tilde{H}_\phi(\rho, \phi)$ and $\tilde{E}_\rho(\rho, \phi)$ depend on ρ and ϕ in precisely the same way, and we can write

$$\tilde{H}_\phi(\rho, \phi) = \frac{\tilde{E}_\rho(\rho, \phi)}{Z_{TM}},$$

where Z_{TM} is the mode impedance defined as

$$Z_{TM} = \frac{\beta}{\omega\varepsilon} = \eta\sqrt{1 - \left(\frac{\omega_c}{\omega}\right)^2}.$$

This is precisely the same Z_{TM} as we found for rectangular waveguides modes. Similarly, the dependence of $\tilde{H}_\rho(\rho, \phi)$ on ρ and ϕ is the same as that of $\tilde{E}_\phi(\rho, \phi)$, and can be written as

$$\tilde{H}_\rho(\rho, \phi) = -\frac{\tilde{E}_\phi(\rho, \phi)}{Z_{TM}}.$$

So we have found the transverse components $\tilde{E}_\rho(\rho, \phi)$, $\tilde{E}_\phi(\rho, \phi)$, $\tilde{H}_\rho(\rho, \phi)$, and $\tilde{H}_\phi(\rho, \phi)$ in terms of the Bessel functions and their derivatives, but we don't yet know k_c. For this, we apply the boundary condition that $\tilde{E}_z(\rho, \phi)$ and $\tilde{E}_\phi(\rho, \phi)$, which are the components of $\tilde{E}(\rho, \phi)$ that are tangential to the conductor at that surface, must both be zero at $\rho = a$, where a is the radius of the waveguide conductor. So the allowed values of k_c are those values that satisfy

$$k_c a = p_{n\ell},$$

where $p_{n\ell}$ are the zeros of the Bessel functions. (The zeros of the Bessel functions are those values of u for which $J_n(u) = 0$.) The index n is the order of the radial mode, represented by the Bessel function $J_n(k_c\rho)$, and the index ℓ tells us which zero of $J_n(k_c\rho)$ coincides with the conducting wall. The first three zeros of $J_n(u)$ for $n = 0$, 1, and 2 are listed in Table 10.8. Each of these zeros can be seen in Fig. 10.15(a) as the points at which the different curves cross the u-axis.

A plot of the field patterns for the TM_{01} mode of a cylindrical waveguide is shown in Fig. 10.16. Notice that the electric field lines are normal to the conducting surface, and the magnetic field lines are tangential.

How can we use the values $p_{n\ell}$ of these zeros to determine the cutoff frequencies for the various modes of the circular waveguide? The following discussion is similar to that

Table 10.8

ℓ	$p_{0\ell}$	$p_{1\ell}$	$p_{2\ell}$	ℓ	$p'_{0\ell}$	$p'_{1\ell}$	$p'_{2\ell}$
The first three zeros $p_{n\ell}$ of the Bessel functions $J_n(u)$ and of the derivatives of the Bessel functions $J'_n(u)$ for $n = 0$, 1, and 2.							
1	2.4048	3.8317	5.1356	1	3.8317	1.8412	3.0542
2	5.5201	7.0156	8.4172	2	7.0156	5.3314	6.7061
3	8.6537	10.1735	11.6198	3	10.1735	8.5363	9.9695

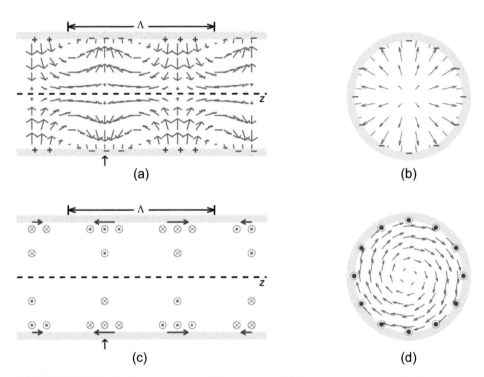

FIGURE 10.16 Plots of the field patterns for a TM$_{01}$ mode on a circular waveguide. (a) ρ–z plot of the electric field; (b) ρ–ϕ plot of the electric field at $z = \lambda/4$; (c) ρ–z plot of the magnetic field; and (d) ρ–ϕ plot of the magnetic field at $z = \lambda/4$. The magenta + and − signs in (a) and (b) represent the surface charge on the inside surface of the conductors. The magenta arrows in (c) and (d) represent the surface current on the inside surface of the conductors. The arrows beneath the waveguide in (a) and (c) indicate the z-location for plots (b) and (d).

of cutoff frequencies of TM and TE modes of parallel plate and rectangular waveguides. The propagation constant is

$$\gamma = \sqrt{k_c^2 - k^2},$$

where $k = \omega\sqrt{\varepsilon\mu}$ and $k_c = \omega_c\sqrt{\varepsilon\mu} = p_{n\ell}/a$. Therefore the cutoff frequency is

$$\omega_c = \frac{k_c}{\sqrt{\varepsilon\mu}} = \frac{p_{n\ell}}{a\sqrt{\varepsilon\mu}}. \tag{10.79}$$

Table 10.9

Summary of results for TM modes on a circular waveguide. See Table 10.3 for mode impedance, phase constant, attenuation constant, phase velocity, and group velocity.	
$\tilde{E}_z(\rho,\phi) = AJ_n(k_c\rho)\cos(n\phi)$	$\tilde{H}_z(\rho,\phi) = 0$
$\tilde{E}_\rho(\rho,\phi) = -\dfrac{j\beta}{k_c}AJ_n'(k_c\rho)\cos(n\phi)$	$\tilde{H}_\rho(\rho,\phi) = -\tilde{E}_\phi(\rho,\phi)/Z_{TM}$
$\tilde{E}_\phi(\rho,\phi) = \dfrac{j\beta n}{k_c^2\rho}AJ_n(k_c\rho)\sin(n\phi)$	$\tilde{H}_\phi(\rho,\phi) = \tilde{E}_\rho(\rho,\phi)/Z_{TM}$
where $k_c = \dfrac{p_{n\ell}}{a},\ n = 0, 1, 2, \dots$ $\qquad \omega_c = \dfrac{1}{\sqrt{\varepsilon\mu}}\dfrac{p_{n\ell}}{a}$	

For high frequencies, $\omega > \omega_c$,

$$\gamma = j\omega\sqrt{\varepsilon\mu}\sqrt{1 - \left(\frac{\omega_c}{\omega}\right)^2} = j\beta \tag{10.80}$$

is imaginary, and gives us the phase constant of the field amplitudes. For low frequencies, $\omega < \omega_c$,

$$\gamma = \omega_c\sqrt{\varepsilon\mu}\sqrt{1 - \left(\frac{\omega}{\omega_c}\right)^2} = \alpha \tag{10.81}$$

is real, and gives us the attenuation coefficient of the field amplitudes. These expressions differ from those for the rectangular waveguide only in the determination of the cutoff frequencies. The results for the TM mode on a circular waveguide are collected in Table 10.9.

Example 10.16 TM Modes on a Circular Waveguide

Consider an air-filled circular waveguide of diameter 4 cm. (a) Determine the cutoff frequency of the TM_{01}, TM_{11}, and TM_{21} modes. (b) Determine the phase constant β of the TM_{01} mode when the waveguide is excited at a frequency of 8 GHz. (c) Determine the attenuation constant α of the TM_{11} mode when the waveguide is excited at a frequency of 8 GHz.

Solution:

(a) From Table 10.8, we read that the zeros $p_{n\ell}$ of the Bessel functions for these modes are $p_{01} = 2.4048$, $p_{11} = 3.8317$, and $p_{21} = 5.1356$. Using Eq. (10.79), the three cutoff frequencies requested are computed as

$$f_c = \frac{\omega_c}{2\pi} = \frac{p_{n\ell}}{2\pi a\sqrt{\varepsilon\mu}} = \frac{c p_{n\ell}}{2\pi a} = \begin{cases} 5.74\,\text{GHz} & \text{for the } TM_{01} \text{ mode} \\ 9.14\,\text{GHz} & \text{for the } TM_{11} \text{ mode} \\ 12.25\,\text{GHz} & \text{for the } TM_{21} \text{ mode} \end{cases},$$

where we have used $c = 2.998 \times 10^8$ m/s and $a = 0.02$ m.

(b) To find the phase constant β of the TM_{01} mode, we use Eq. (10.80) and the cutoff frequency we found in part (a):

$$\beta = \omega\sqrt{\varepsilon\mu}\sqrt{1 - \left(\frac{\omega_c}{\omega}\right)^2} = \frac{\omega}{c}\sqrt{1 - \left(\frac{\omega_c}{\omega}\right)^2} = \frac{2\pi \times 8 \times 10^9 \text{ s}^{-1}}{2.998 \times 10^8 \text{ m/s}}\sqrt{1 - \left(\frac{5.74 \text{ GHz}}{8 \text{ GHz}}\right)^2} = 117 \text{ m}^{-1}.$$

Note that β is smaller than the phase constant of a freely propagating wave at this same frequency, which is 168 m^{-1}.

(c) We compute the attenuation constant α of TM_{11} mode using Eq. (10.81):

$$\alpha = \omega_c\sqrt{\varepsilon\mu}\sqrt{1 - \left(\frac{\omega}{\omega_c}\right)^2} = \frac{\omega_c}{c}\sqrt{1 - \left(\frac{\omega}{\omega_c}\right)^2}$$

$$\alpha = \frac{2\pi \times 9.14 \times 10^9 \text{ s}^{-1}}{2.998 \times 10^8 \text{ m/s}}\sqrt{1 - \left(\frac{8 \text{ GHz}}{9.14 \text{ GHz}}\right)^2} = 92.6 \text{ m}^{-1}.$$

In a distance of just 10 cm, this mode amplitude would be decreased by a factor

$$e^{-\alpha z} = e^{(92.6 \text{ m}^{-1})(0.1 \text{ m})} = 9.5 \times 10^{-5}.$$

10.4.2 TE Modes of a Circular Waveguide

For the TE modes, as the name implies, the electric field has only transverse components, so $\tilde{E}_z(\rho, \phi) = 0$. Therefore, there must exist a longitudinal component $\tilde{H}_z(\rho, \phi)$ of the magnetic field, which satisfies Eq. (10.13):

$$\nabla_t^2 \tilde{H}_z(\rho, \phi) = -k_c^2 \tilde{H}_z(\rho, \phi).$$

The general solution to this differential equation in cylindrical coordinates is of the form

$$\tilde{H}_z(\rho, \phi) = \{A' J_n(k_c\rho) + B' N_n(k_c\rho)\} \times \{C' \cos(n\phi) + D' \sin(n\phi)\}.$$

Since $N_n(k_c\rho)$ becomes infinite as $\rho \to 0$, the center of the waveguide, B' must be zero. In addition, the waveguide is cylindrically symmetric, so we can choose the orientation of our coordinate system to allow only the $\cos(n\phi)$ terms, and set D' equal to zero. Then we write $H_z(\rho, \phi)$ in the simplified form:

$$\tilde{H}_z(\rho, \phi) = B J_n(k_c\rho) \cos(n\phi), \qquad \text{for } n = 0, 1, 2, \ldots \qquad (10.82a)$$

Using $\tilde{H}_z(\rho, \phi)$ from Eq. (10.82a) and $\tilde{E}_z(\rho, \phi) = 0$ in Eqs. (10.77a)–(10.77d), $\tilde{E}_\rho(\rho, \phi)$, $\tilde{E}_\phi(\rho, \phi)$, $\tilde{H}_\rho(\rho, \phi)$, and $\tilde{H}_\phi(\rho, \phi)$ are found as

$$\tilde{E}_\rho(\rho, \phi) = \frac{j\omega\mu n}{k_c^2}\frac{B}{\rho} J_n(k_c\rho) \sin(n\phi), \qquad (10.82b)$$

$$\tilde{E}_\phi(\rho, \phi) = \frac{j\omega\mu}{k_c} B J_n'(k_c\rho) \cos(n\phi), \qquad (10.82c)$$

$$\tilde{H}_\rho(\rho, \phi) = -\frac{E_\phi(\rho, \phi)}{Z_{TE}}, \qquad (10.82d)$$

and

$$\tilde{H}_\phi(\rho, \phi) = \frac{E_\rho(\rho, \phi)}{Z_{TE}}, \qquad (10.82e)$$

where

$$Z_{TE} = \frac{\omega\mu}{\beta} = \frac{\eta}{\sqrt{1 - (\omega_c/\omega)^2}}.$$

Now we are ready to apply the boundary condition $E_{tan} = 0$ at the conducting surface. The components of $\tilde{\mathbf{E}}(\rho, \phi)$ that are tangential to the conductor surface are $\tilde{E}_z(\rho, \phi)$ and $\tilde{E}_\phi(\rho, \phi)$. $\tilde{E}_z(\rho, \phi)$, of course, is zero for a TE mode, so $\tilde{E}_{tan}$ consists only of $\tilde{E}_\phi(\rho, \phi)$. From Eq. (10.82c), therefore, we see that the condition that $\tilde{E}_\phi(\rho, \phi)$ must be zero at $\rho = a$ tells us that

$$J_n'(k_c a) = 0,$$

or

$$k_c = \frac{p_{n\ell}'}{a},$$

where $p_{n\ell}'$ are the zeros of $J_n'(u)$, or the values of u such that $J_n'(u) = 0$. The first three zeros of $J_n'(u)$ for $n = 0, 1$, and 2 are tabulated in Table 10.8(b). The index ℓ on $p_{n\ell}'$ indicates which zero of the function $J_n'(u)$ describes the mode. Plots of the field patterns for the TE$_{11}$ mode are shown in Fig. 10.17.

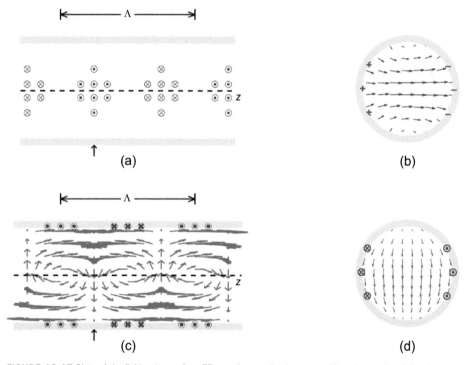

FIGURE 10.17 Plots of the field patterns for a TE$_{11}$ mode on a circular waveguide. (a) ρ–z plot of the electric field; (b) ρ–ϕ plot of the electric field at $z = \lambda/4$; (c) ρ–z plot of the magnetic field; and (d) ρ–ϕ plot of the magnetic field at $z = \lambda/4$. The magenta $+$ and $-$ signs in (b) represent the surface charge on the inside surface of the conductor. The magenta $\times$ signs and dots in (c) and (d) represent the surface current on the inside surface of the conductor. The arrows beneath the waveguide in (a) and (c) indicate the z-location for plots (b) and (d).

The cutoff frequencies are related to k_c using

$$\omega_c = \frac{k_c}{\sqrt{\varepsilon\mu}} = \frac{p'_{n\ell}}{a\sqrt{\varepsilon\mu}}. \tag{10.83}$$

The mode with the lowest cutoff frequency on a circular waveguide is the TE_{11} mode. This may at first seem counter-intuitive, but a quick examination of Table 10.8 and/or Fig. 10.15 shows that p'_{11} is less than all $p_{n\ell}$ and all other $p'_{n\ell}$. The cutoff frequency for the TE_{11} mode is

$$\omega_c = \frac{1.8412}{a\sqrt{\varepsilon\mu}},$$

while for the TM_{01} mode the cutoff frequency is

$$\omega_c = \frac{2.4048}{a\sqrt{\varepsilon\mu}}.$$

Example 10.17 TE Modes on a Circular Waveguide

Consider the air-filled circular waveguide of diameter 4 cm that we described in Example 10.16. (a) Determine the cutoff frequency of the TE_{01}, TE_{11}, and TE_{21} modes. (b) Determine the phase constant β of the TE_{11} mode when the waveguide is excited at a frequency of 8 GHz. (c) Determine the attenuation constant α of the TE_{01} mode when the waveguide is excited at a frequency of 8 GHz.

Solution:

(a) From Table 10.8, we read that the zeros $p'_{n\ell}$ of the derivatives of the Bessel functions for these modes are $p'_{01} = 3.8317$, $p'_{11} = 1.8412$, and $p'_{21} = 3.0542$. Then using Eq. (10.83), the three cutoff frequencies requested are

$$f_c = \frac{\omega_c}{2\pi} = \frac{p'_{n\ell}}{2\pi a\sqrt{\varepsilon\mu}} = \frac{cp'_{n\ell}}{2\pi a} = \begin{cases} 9.14 \text{ GHz} & \text{for the } TE_{01} \text{ mode} \\ 4.39 \text{ GHz} & \text{for the } TE_{11} \text{ mode} \\ 7.28 \text{ GHz} & \text{for the } TE_{21} \text{ mode} \end{cases},$$

where we have used $c = 2.998 \times 10^8$ m/s and $a = 0.02$ m. Notice that the cutoff frequency of the TE_{11} mode is the lowest of all TE and TM modes. Note also that the cutoff frequency of the TE_{01} mode matches that of the TM_{11} mode, since $p'_{01} = p_{11}$.

(b) To find the phase constant β of the TE_{11} mode, we use Eq. (10.80), which is valid for TE modes as well as TM modes, and the cutoff frequency we found in part (a):

$$\beta = \omega\sqrt{\varepsilon\mu}\sqrt{1 - \left(\frac{\omega_c}{\omega}\right)^2} = \frac{\omega}{c}\sqrt{1 - \left(\frac{\omega_c}{\omega}\right)^2} = \frac{2\pi \times 8 \times 10^9 \text{ s}^{-1}}{2.998 \times 10^8 \text{ m/s}}\sqrt{1 - \left(\frac{4.39 \text{ GHz}}{8 \text{ GHz}}\right)^2} = 140 \text{ m}^{-1}.$$

(c) The attenuation constant α of TE_{01} mode is computed using Eq. (10.81), which is also valid for TE modes:

$$\alpha = \omega_c\sqrt{\varepsilon\mu}\sqrt{1 - \left(\frac{\omega}{\omega_c}\right)^2} = \frac{\omega_c}{c}\sqrt{1 - \left(\frac{\omega}{\omega_c}\right)^2} = \frac{2\pi \times 9.14 \times 10^9 \text{ s}^{-1}}{2.998 \times 10^8 \text{ m/s}}\sqrt{1 - \left(\frac{8 \text{ GHz}}{9.14 \text{ GHz}}\right)^2}$$

$$= 92.6 \text{ m}^{-1}.$$

Table 10.10

Summary of results for TE modes on a circular waveguide. See Table 10.3 for mode impedance, phase constant, attenuation constant, phase velocity, and group velocity.	
$\tilde{E}_z(\rho, \phi) = 0$	$\tilde{H}_z(\rho, \phi) = BJ_n(k_c\rho)\cos(n\phi)$
$\tilde{E}_\rho(\rho, \phi) = \dfrac{j\omega\mu n}{k_c^2}\dfrac{B}{\rho}J_n(k_c\rho)\sin(n\phi)$	$\tilde{H}_\rho(\rho, \phi) = -\tilde{E}_\phi(\rho, \phi)/Z_{TE}$
$\tilde{E}_\phi(\rho, \phi) = \dfrac{j\omega\mu}{k_c}BJ_n'(k_c\rho)\cos(n\phi)$	$\tilde{H}_\phi(\rho, \phi) = \tilde{E}_\rho(\rho, \phi)/Z_{TE}$
where $k_c = \dfrac{p_{n\ell}'}{a}$, $n = 0, 1, 2, \dots$ $\omega_c = \dfrac{1}{\sqrt{\varepsilon\mu}}\dfrac{p_{n\ell}'}{a}$	

We will not determine the total power transmitted by the mode or the losses of the mode on circular waveguides, since those discussions require a deeper knowledge of Bessel functions. That treatment is deferred to more advanced texts. The results for the TE mode on a circular waveguide are collected in Table 10.10.

10.5 Resonant Cavities

The final discussion of this chapter will be an introduction to **resonant cavities**. A cavity is a completely enclosed conducting structure, such as the rectangular cavity shown in Fig. 10.18. Cavities can act as a tuned element (or filter) for radio-frequency waves, as the field amplitude inside the cavity depends strongly on the frequency of the electromagnetic field. When driven at their **resonant** frequency, the field strengths in the interior of the cavity can reach extremely high values. The electric and magnetic fields form standing waves within the cavity, and understanding these patterns will help us to decipher the cavity properties. Cavities are often called resonators, and we will use these two names interchangeably. Only rectangular cavities are treated here, although cavities of other geometries are also possible. A common example of a microwave cavity is found in many kitchens, as described in TechNote 10.3.

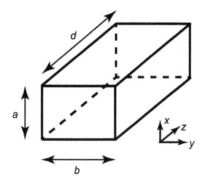

FIGURE 10.18 A rectangular resonant cavity. Each of the surfaces shown are conductors, with vacuum or a dielectric material filling the inside.

TechNote 10.3 Microwave Oven

A common example of a resonant cavity is found in many household kitchens. A microwave oven, as shown in Fig. 10.19(a), is useful for rapidly heating our leftover pizza. The primary components of a microwave oven are a magnetron that generates a powerful electromagnetic wave at a frequency of 2.45 GHz, a waveguide that transmits this radiation to the oven, and the microwave cavity where one places the dinner to be heated.

The magnetron, invented by J.T. Randall and H.A.H. Boot at the University of Birmingham in the UK in 1940 for use in radar systems, is a vacuum tube device that produces a high-power, high-frequency oscillating signal. See Fig. 10.19(b) for a diagram of a magnetron. The cathode of this device is a negatively biased filament, which when heated emits electrons in a process called thermionic emission. These electrons are accelerated outward toward the anode (the entire outer assembly), which is positively biased relative to the cathode. The electrons follow a curved arc (the red trajectory) due to a magnetic field (directed into the page) created by a pair of permanent magnets. As the electrons fly past one of the copper fins inside the tube, they induce positive charges in that fin. Those charges are drawn from the adjacent fins, which are then left with a negative charge. This induced charge imbalance initiates an oscillating current in each "loop," or resonator, of the anode. (This resonator consists of a capacitor formed by the opposing faces of the conductors in adjacent fins, and an inductor formed by the circular surface highlighted by the circular red arrow. The capacitance C of the capacitor is determined by the dimensions of the gaps between adjacent fins, while the inductance L of the inductor is determined by the dimension of the round cavity formed between the fins.) The current in this loop oscillates at the resonant frequency of the loop, $1/\sqrt{LC}$. The rf signal is coupled out of the magnetron through the brown hook-shaped antenna, shown on the right side of the figure.

The microwave wave generated by the magnetron is coupled into a waveguide, which carries the wave to the cavity, a metal rectangular enclosure that confines the electromagnetic waves. The wavelength of a 2.45 GHz wave in air (or vacuum) is 12.2 cm, so the cavity dimensions $a \times b \times d$ must be chosen to be resonant with this frequency/wavelength. To be effective at heating our dinner, we depend on the

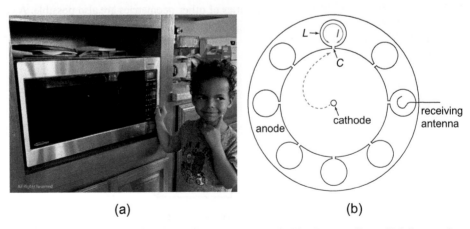

(a)   (b)

FIGURE 10.19 (a) A young chef ponders the proper timing to set for his microwave dinner. (b) A diagram of a magnetron, a vacuum tube device that generates a high-power, electromagnetic wave at a microwave frequency for use in radar and microwave ovens.

absorption by water at this frequency. To demonstrate this, let's examine the complex permittivity of water at this frequency, which is $\varepsilon = \varepsilon' - \varepsilon'' \approx \left(80 - j10\right)\varepsilon_0$. (See Table 8.2.) Using Eq. (8.68), this relates to an absorption coefficient of $\alpha = 0.29$ cm^{-1}, so the waves are completely absorbed by the water in the food within a few centimeters, and the electromagnetic energy generated by the magnetron is converted to heat.

10.5.1 Electromagnetic Modes of a Rectangular Cavity

To derive the electromagnetic modes of the rectangular resonant cavity, we are not required to go back to first principles to determine the field patterns supported by the rectangular cavity. Rather, we will borrow from the results that we developed for the rectangular waveguide structure in Section 10.3. There we found that for a TE$_{mn}$ mode, the electric and magnetic fields are described by

$$\tilde{E}_z(x,y) = 0, \qquad\qquad\qquad \tilde{H}_z(x,y) = B\cos(k_x x)\cos(k_y y),$$

$$\tilde{E}_x(x,y) = \frac{j\omega\mu}{k_c^2}k_y B\cos(k_x x)\sin(k_y y), \qquad \tilde{H}_x(x,y) = -\tilde{E}_y(x,y)/Z_{TE},$$

$$\tilde{E}_y(x,y) = \frac{-j\omega\mu}{k_c^2}k_x B\sin(k_x x)\cos(k_y y), \qquad \tilde{H}_y(x,y) = \tilde{E}_x(x,y)/Z_{TE}.$$

where $k_x = m\pi/a$, $k_y = n\pi/b$, and $Z_{TE} = \eta/\sqrt{1-(\omega_c/\omega)^2}$. In the rectangular resonant cavity the geometry is similar to that of the waveguide, in that both have conducting walls at $x = 0$ and a, and at $y = 0$ and b. The resonant cavity, however, also has conducting walls at $z = 0$ and d, which clearly are absent in the waveguide structure. The effects of these "extra" conducting walls are two-fold:

1. First, waves traveling in both the $+z$- and $-z$-directions must be considered. For the waveguide structures, the discussion was limited to waves traveling in the $+z$-direction only. Of course, waves traveling in the $-z$-direction are perfectly allowed, but when considering a waveguide with a source at one end and a load at the other, it makes sense to focus on the wave traveling from the source to the load (i.e. in the $+z$-direction). For a resonant cavity, however, a wave traveling in the $+z$-direction, upon reflection from the conductor at $z = d$, becomes a wave traveling in the $-z$-direction, and a wave traveling in the $-z$-direction, upon reflection from the conductor at $z = 0$, becomes a wave traveling in the $+z$-direction.

2. The second effect of these walls is that we now have additional boundary conditions that the fields must satisfy: that is, $E_{tan} = 0$ and $H_n = 0$ at $z = 0$ and $z = d$. For these walls, the normal direction is the z-direction, so $H_n = H_z$ must be zero at these walls.

We account for the first of these modifications (i.e. waves traveling in the $+z$- and $-z$-directions) by replacing $e^{-j\beta z}$ in the expressions for waveguide fields by

$$H_+ e^{-j\beta z} + H_- e^{+j\beta z}.$$

(We are considering lossless propagation here, allowing us to substitute $j\beta$ for γ.) $H_+ e^{-j\beta z}$ represents a wave traveling in the $+z$-direction, and $H_- e^{+j\beta z}$ represents a wave traveling in the $-z$-direction. The sum of these two traveling waves is a standing wave, similar to those that we discussed for UPWs in Chapter 8 and for transmission lines in Chapter 9.

With this z-dependence of the wave, $\tilde{H}_z(x, y)$ from Eq. (10.67a) can be modified to write

$$\tilde{H}_z(x, y, z) = B\cos(k_x x)\cos(k_y y)\left(H_+ e^{-j\beta z} + H_- e^{+j\beta z}\right).$$

Now we apply the boundary conditions that must be satisfied by $\tilde{H}_z(x, y, z)$ at $z = 0$ and $z = d$. Using the former, we conclude that

$$H_+ + H_- = 0. \tag{10.84}$$

This condition lets us write

$$H_+ e^{-j\beta z} + H_- e^{+j\beta z} = H_+\left(e^{-j\beta z} - e^{+j\beta z}\right) = -2jH_+ \sin(\beta z).$$

Next we apply the condition that $\tilde{H}_z(x, y, z)$ must be zero at $z = d$ to show that

$$\beta = \frac{p\pi}{d},$$

where p is an integer 1, 2, 3, …. . (Note that $p = 0$ is not allowed, as this would force the field amplitude to be zero everywhere.) So far, we have only considered a lossless cavity, but note that for this case the phase constant β can take only discrete values, similar to $k_x = m\pi/a$ and $k_y = n\pi/b$. Finally, substituting C for $-2jBH_+$, the $\tilde{H}_z(x, y, z)$ component is written as

$$\tilde{H}_z(x, y, z) = C\cos\left(\frac{m\pi x}{a}\right)\cos\left(\frac{n\pi y}{b}\right)\sin\left(\frac{p\pi z}{d}\right). \tag{10.85}$$

We must still determine the other field components of this mode, but before doing so let's explore the frequency ω of the electromagnetic wave inside the resonator. For this, recall the following relations, which we derived in the discussion of waveguides,

$$k_x^2 + k_y^2 = k_c^2 = k^2 + \gamma^2 = k^2 - \beta^2.$$

Recall that k is the propagation constant for a freely propagating wave in the dielectric medium of the interior of the waveguide,

$$k = \omega\sqrt{\varepsilon\mu}.$$

Rearranging, the allowed values of k are those that satisfy

$$k^2 = k_x^2 + k_y^2 + \beta^2.$$

Substituting the known values of k_x, k_y, and β,

$$k^2 = \left(\frac{m\pi}{a}\right)^2 + \left(\frac{n\pi}{b}\right)^2 + \left(\frac{p\pi}{d}\right)^2. \tag{10.86}$$

Using $k = \omega\sqrt{\varepsilon\mu}$, the resonant frequency ω of the TE$_{mnp}$ cavity mode, where the subscripts m, n, p are used to label the mode, is uniquely defined as

$$\boxed{\omega = \frac{1}{\sqrt{\varepsilon\mu}}\left[\left(\frac{m\pi}{a}\right)^2 + \left(\frac{n\pi}{b}\right)^2 + \left(\frac{p\pi}{d}\right)^2\right]^{1/2}.} \tag{10.87}$$

The corresponding wavelength is

$$\lambda = \frac{2\pi}{k} = 2\left[\left(\frac{m}{a}\right)^2 + \left(\frac{n}{b}\right)^2 + \left(\frac{p}{d}\right)^2\right]^{-1/2}.$$

The electromagnetic field inside a resonant cavity has a set of very specific, or discrete, frequencies, which depend on the dimensions a, b, and d, as given by Eq. (10.87). This is in contrast to waveguide modes, for which any frequency greater than the cutoff frequency ω_c is permitted. Likewise, the wavelength of this mode is unique. Recall the plane wave model that we introduced in Section 10.2. It can be shown that resonance occurs when the uniform plane wave reflects off each of the cavity walls, and after one round-trip around the cavity the wave finds itself to be perfectly phased to the original plane wave.

Example 10.18 Resonant Frequencies of TE modes in a Rectangular Cavity

Consider an air-filled rectangular cavity of dimensions 6 × 8 × 10 cm. Determine the resonant frequencies of the five lowest-frequency TE modes.

Solution:
We use Eq. (10.87) for the resonant frequencies. Since the cavity is air-filled, we'll replace $1/\sqrt{\varepsilon\mu}$ by c. There is some guesswork needed to determine which mode indices lead to the lowest frequencies, so we'll calculate an extra to be sure we found the five lowest frequencies. Recall also that m or n can be zero, but not both, and p cannot be zero.

m	n	p	$f = \dfrac{\omega}{2\pi}$ (GHz)
0	1	1	2.40
1	0	1	2.91
1	1	1	3.46
0	1	2	3.54
1	0	2	3.90
1	1	2	4.33

With this, we have found the lowest five TE mode resonant frequencies for this cavity, identified as the TE_{011}, TE_{101}, TE_{111}, TE_{012}, and TE_{102} modes.

Let's take stock of where we are in this development of the fields for a TE mode of a rectangular resonant cavity. To this point, we have found the form of the $\tilde{H}_z(x,y,z)$ component, and the resonant frequency of the cavity. Our next order of business is to find the transverse field components $\tilde{E}_x(x,y,z)$, $\tilde{E}_y(x,y,z)$, $\tilde{H}_x(x,y,z)$, and $\tilde{H}_y(x,y,z)$. Our approach will be the same as we employed to find $\tilde{H}_z(x,y,z)$. That is, we start with the relevant component from the rectangular waveguide solution, and replace $e^{-j\beta z}$ with $H_+ e^{-j\beta z} + H_- e^{+j\beta z} = -2jH_+ \sin(\beta z)$. Substituting C for $-2jBH_+$, the result is

$$\tilde{E}_x(x,y,z) = \frac{j\omega\mu}{k_c^2}\left(\frac{n\pi}{b}\right) C \cos\left(\frac{m\pi x}{a}\right)\sin\left(\frac{n\pi y}{b}\right)\sin\left(\frac{p\pi z}{d}\right) \tag{10.88a}$$

and

$$\tilde{E}_y(x,y,z) = \frac{-j\omega\mu}{k_c^2}\left(\frac{m\pi}{a}\right) C \sin\left(\frac{m\pi x}{a}\right)\cos\left(\frac{n\pi y}{b}\right)\sin\left(\frac{p\pi z}{d}\right). \tag{10.88b}$$

For $\tilde{H}_x(x,y,z)$ and $\tilde{H}_y(x,y,z)$, there is an additional twist that must be included. That is, $e^{-j\beta z}$ in the waveguide solutions must be replaced with $H_+ e^{-j\beta z} - H_- e^{+j\beta z}$, where the sign of the H_- term has been changed. This change in sign comes from the condition that $\partial/\partial z \to +j\beta$ for the wave traveling in the $-z$-direction. Recall that in deriving Eqs. (10.14a), (10.14b), (10.15a), and (10.15b), we used $\partial/\partial z \to -\gamma$, which of course is $-j\beta$ for lossless wave propagation. This derivative is valid for a wave traveling in the $+z$-direction. For a wave traveling in the $-z$-direction, however, we must reverse the sign of $\gamma = j\beta$, which reverses the sign of γ in each of its instances in Eqs. (10.16a)–(10.16d). The net result is a change in sign of the H_- terms in $\tilde{H}_x(x, y, z)$ and $\tilde{H}_y(x, y, z)$, as stated above. Then, for $H_- = -H_+$, as we established in Eq. (10.84),

$$H_+ e^{-j\beta z} - H_- e^{+j\beta z} = 2H_+ \cos(\beta z)$$

and

$$\tilde{H}_x(x, y, z) = -\frac{1}{k_c^2}\left(\frac{m\pi}{a}\right)\left(\frac{p\pi}{d}\right) C \sin\left(\frac{m\pi x}{a}\right)\cos\left(\frac{n\pi y}{b}\right)\cos\left(\frac{p\pi z}{d}\right) \qquad (10.88c)$$

and

$$\tilde{H}_y(x, y, z) = -\frac{1}{k_c^2}\left(\frac{n\pi}{b}\right)\left(\frac{p\pi}{d}\right) C \cos\left(\frac{m\pi x}{a}\right)\sin\left(\frac{n\pi y}{b}\right)\cos\left(\frac{p\pi z}{d}\right). \qquad (10.88d)$$

Notice that each of the magnetic field phasor amplitudes (Eqs. (10.85), (10.88c), and (10.88d)) is real, while the phasor amplitudes of each of the electric field components (Eqs. (10.88a) and (10.88b)) are imaginary. When we determine the full time-dependent fields (multiply the phasor amplitude by $e^{j\omega t}$ and take the real part), we will see that the time-dependent magnetic and electric fields each oscillate sinusoidally at the frequency ω, but that these fields are 90° out of phase with each other. We will return to this point later when we consider the energy stored in the fields.

To this point, we have discussed only the TE modes of the rectangular cavity. We can treat TM modes with a similar approach, starting with the TM modes of the rectangular waveguide, and creating a superposition of $+z$- and $-z$-traveling waves to form cavity modes. We will not present a step-by-step analysis of the TM modes, but simply present the fields, which we leave to the reader to derive in Prob. P10-32. The fields of the TM modes are

$$\tilde{E}_z(x, y, z) = A \sin\left(\frac{m\pi x}{a}\right)\sin\left(\frac{n\pi y}{b}\right)\cos\left(\frac{p\pi z}{d}\right), \qquad (10.89a)$$

$$\tilde{E}_x(x, y, z) = \frac{-k_x}{k_c^2}\left(\frac{p\pi}{d}\right) A \cos\left(\frac{m\pi x}{a}\right)\sin\left(\frac{n\pi y}{b}\right)\sin\left(\frac{p\pi z}{d}\right), \qquad (10.89b)$$

$$\tilde{E}_y(x, y, z) = \frac{-k_y}{k_c^2}\left(\frac{p\pi}{d}\right) A \sin\left(\frac{m\pi x}{a}\right)\cos\left(\frac{n\pi y}{b}\right)\sin\left(\frac{p\pi z}{d}\right), \qquad (10.89c)$$

$$\tilde{H}_z(x, y, z) = 0, \qquad (10.90a)$$

$$\tilde{H}_x(x, y, z) = \frac{j\omega\varepsilon k_y}{k_c^2} A \sin\left(\frac{m\pi x}{a}\right)\cos\left(\frac{n\pi y}{b}\right)\cos\left(\frac{p\pi z}{d}\right), \qquad (10.90b)$$

$$\tilde{H}_y(x, y, z) = \frac{-j\omega\varepsilon k_x}{k_c^2} A \cos\left(\frac{m\pi x}{a}\right)\sin\left(\frac{n\pi y}{b}\right)\cos\left(\frac{p\pi z}{d}\right). \qquad (10.90c)$$

Remember that for TM modes, m and n must be integers 1, 2, 3, ... That is, neither m nor n can be zero. The index p, however, can be zero, as well as any positive integer. Note that some low-order TE modes and TM modes are related, in that with the proper choice of rotation of coordinates, a TE mode transforms into a TM mode, and vice versa.

10.5.2 TE$_{101}$ Mode of a Rectangular Cavity

With this, we have found each of the electric and magnetic field components for the general TE$_{mnp}$ and TM$_{mnp}$ modes of the rectangular waveguide. Let's now look at the specific case of the TE$_{101}$ mode. (The subscripts indicate $m = 1$, $n = 0$, and $p = 1$.) The field components for this mode are

$$\tilde{E}_z(x, z) = 0, \tag{10.91a}$$

$$\tilde{E}_x(x, z) = 0, \tag{10.91b}$$

$$\tilde{E}_y(x, z) = \frac{-j2a\eta}{\lambda} C \sin\left(\frac{\pi x}{a}\right) \sin\left(\frac{\pi z}{d}\right), \tag{10.91c}$$

$$\tilde{H}_z(x, z) = C \cos\left(\frac{\pi x}{a}\right) \sin\left(\frac{\pi z}{d}\right), \tag{10.92a}$$

$$\tilde{H}_x(x, z) = \frac{-a}{d} C \sin\left(\frac{\pi x}{a}\right) \cos\left(\frac{\pi z}{d}\right), \tag{10.92b}$$

$$\tilde{H}_y(x, z) = 0. \tag{10.92c}$$

These field expressions follow directly from Eqs. (10.85) and (10.88a)–(10.88d), with the substitutions $m = 1$, $n = 0$, and $p = 1$, and using the simple relations

$$k_c = \sqrt{(m\pi/a)^2 + (n\pi/b)^2} = \pi/a,$$

$$\omega = 2\pi u_p/\lambda,$$

and

$$\mu u_p = \frac{\mu}{\sqrt{\varepsilon \mu}} = \sqrt{\frac{\mu}{\varepsilon}} = \eta.$$

The field patterns for the TE_{101} mode are shown in Fig. 10.20. The electric field is strictly in the y-direction, with maximum value at the center of the cavity, diminishing to zero at $x = 0$ and a, and at $z = 0$ and d. The magnetic field is in the x–z plane circling about the center of the cavity, with maximum value near the walls of the cavity. The electric and magnetic fields are 90° out of phase with each other, as evidenced by the factor j in $\tilde{E}_y(x, z)$, absent from $\tilde{H}_x(x, z)$ and $\tilde{H}_z(x, z)$. Thus when the electric field is at its maximum value, the magnetic field is zero, and vice versa.

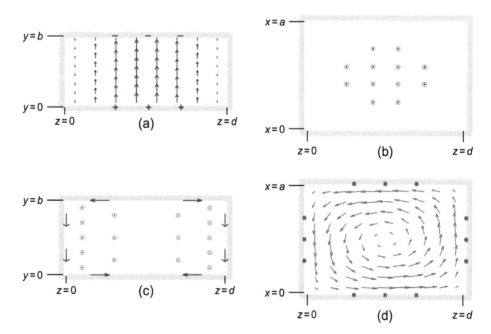

FIGURE 10.20 Plots of the field patterns for a TE_{101} mode on a rectangular resonator. (a) y–z plot of the electric field at $x = a/2$; (b) x–z plot of the electric field; (c) y–z plot of the magnetic field at $x = a/2$; and (d) x–z plot of the magnetic field.

The resonant frequency of the TE_{101} mode is

$$\omega = \frac{1}{\sqrt{\varepsilon\mu}}\left[\left(\frac{m\pi}{a}\right)^2 + \left(\frac{n\pi}{b}\right)^2 + \left(\frac{p\pi}{d}\right)^2\right]^{1/2} = \frac{\pi}{\sqrt{\varepsilon\mu}}\left[\left(\frac{1}{a}\right)^2 + \left(\frac{1}{d}\right)^2\right]^{1/2},$$

and the wavelength of the unguided wave at this frequency is

$$\lambda = \frac{2\pi}{\omega\sqrt{\varepsilon\mu}} = 2\left[\left(\frac{1}{a}\right)^2 + \left(\frac{1}{d}\right)^2\right]^{-1/2}. \tag{10.93}$$

10.5.3 Energy Stored in Cavity Modes

We next examine the electric and magnetic *energy* stored in the TE_{101} cavity mode. We showed earlier that the electric and magnetic fields are out of phase with each other, such that when the electric field is maximal, the magnetic field is zero, and vice versa. We start then by examining the energy stored in the electric field at this maximum point in time. We showed in Chapter 1 that the energy density (i.e. the energy per unit volume) of a static electric field is

$$w_e = \frac{1}{2}\mathbf{D} \cdot \mathbf{E} = \frac{\varepsilon}{2}|\mathbf{E}|^2,$$

where the second form is valid for simple, linear, isotropic media. For sinusoidally varying fields, as we have in the present case, the energy density stored in the electric field at the maximum of the cycle can be shown to be

$$w_{e,max} = \frac{\varepsilon}{2}|\tilde{E}(x,y,z)|^2.$$

(Don't confuse this maximum value with the time-averaged energy density, which is smaller by a factor of one-half.)

The total *maximum* stored energy is found by integrating the electric energy density $w_{e,max}$ over the volume inside the cavity,

$$W_{e,max} = \int_v w_{e,max}\, dv.$$

Applied to the TE_{101} cavity mode at its maximum amplitude, the stored energy is

$$W_{e,max} = \frac{\varepsilon}{2}\int_0^d \int_0^b \int_0^a |\tilde{E}_y(x,z)|^2\, dx\, dy\, dz$$

$$= \frac{\varepsilon}{2}\int_0^d \int_0^b \int_0^a \left|\frac{-2ja\eta}{\lambda}\right|^2 C^2 \sin^2\left(\frac{\pi x}{a}\right)\sin^2\left(\frac{\pi z}{d}\right) dx\, dy\, dz.$$

After carrying out the integrations, the energy is

$$W_{e,max} = \frac{\varepsilon}{2}\left(\frac{2a}{\lambda}\right)^2\left(\frac{\mu}{\varepsilon}\right)C^2\left(\frac{abd}{4}\right) = \frac{\mu}{8}\left(\frac{2a}{\lambda}\right)^2 C^2\, abd. \tag{10.94}$$

A change of variable can make this energy seem more familiar. Upon inspection of Eq. (10.91c), the amplitude of the $\tilde{E}_y(x,z)$ component can be identified as $E_0 = 2a\eta C/\lambda$. Substituting this into $W_{e,max}$ leads to

$$\boxed{W_{e,max} = \frac{\varepsilon E_0^2}{8}abd.} \tag{10.95}$$

Since E_0 is the maximum electric field amplitude (which occurs at the center of the resonator), $\frac{1}{2}\varepsilon E_0^2$ is the maximum energy density. The spatially averaged energy density is one-quarter of the maximum energy density, and the total electric energy is just the product of the spatially averaged energy density and the volume, abd.

Since $E_y(x, z, t)$ varies as $\sin(\omega t)$, as can be seen by multiplying $\tilde{E}_y(x, z)$ in Eq. (10.91c) by $e^{j\omega t}$ and taking the real part, the energy stored in the electric field in the cavity mode is

$$W_e(t) = \frac{\varepsilon E_0^2}{8} abd \ \sin^2(\omega t).$$

The electric energy stored in the cavity mode oscillates sinusoidally in time, reaching its maximum value twice in every cycle of the field, when $E_y(x, z, t)$ is at its maximum positive value and at its maximum negative value. Where does the energy go when the electric energy is less than its maximum value? To answer this, we look at the energy stored in the magnetic field of the cavity mode.

We showed in Chapter 6 that the energy density of a magnetic field can be written as

$$w_m = \frac{1}{2}\mathbf{H} \cdot \mathbf{B} = \frac{\mu}{2}|\mathbf{H}|^2.$$

The second form is valid for simple, linear, isotropic media. For sinusoidally varying fields, this maximum energy density becomes

$$w_{m,max} = \frac{\mu}{2}|\tilde{\mathbf{H}}(x, y, z)|^2.$$

Integrating the magnetic energy density $w_{m,max}$ over the entire volume inside the cavity, the total maximum stored energy is

$$W_{m,max} = \int_v w_{m,max} \ dv.$$

For the cavity fields for the TE_{101} mode, we find the maximum magnetic energy

$$W_{m,max} = \frac{\mu}{2} \int_0^d \int_0^b \int_0^a (|\tilde{H}_z(x, z)|^2 + |\tilde{H}_x(x, z)|^2) \ dx \ dy \ dz$$

$$= \frac{\mu}{2} \int_0^d \int_0^b \int_0^a \left[C^2 \cos^2\left(\frac{\pi x}{a}\right) \sin^2\left(\frac{\pi z}{d}\right) + \left(\frac{-a}{d}\right)^2 C^2 \sin^2\left(\frac{\pi x}{a}\right) \cos^2\left(\frac{\pi z}{d}\right) \right] dx \ dy \ dz.$$

We carry out the integrations to find

$$W_{m,max} = \frac{\mu}{2} C^2 \left(\frac{abd}{4}\right)\left[1 + \left(\frac{a}{d}\right)^2\right].$$

But from Eq. (10.93) we can substitute

$$\left[1 + \left(\frac{a}{d}\right)^2\right] = \left(\frac{2a}{\lambda}\right)^2,$$

which leads us to

$$W_{m,max} = \frac{\mu}{8}\left(\frac{2a}{\lambda}\right)^2 C^2 \ abd.$$

This maximum energy $W_{m,max}$ stored in the magnetic field of the cavity mode is precisely the same as $W_{e,max}$, as presented in Eq. (10.94). What is the time-dependence of this magnetic energy? For this, we need the time-dependence of the magnetic fields $H_x(x, z, t)$ and $H_z(x, z, t)$, which we find by multiplying $\tilde{H}_x(x, z)$ and $\tilde{H}_z(x, z)$ in Eqs. (10.92b) and (10.92a) by $e^{j\omega t}$ and taking the real part of the result. We find that the magnetic field varies as $\cos(\omega t)$ (in contrast to the $\sin(\omega t)$ time-dependence of the electric field). Since the energy scales as the square of the field, the energy stored in the magnetic field of the cavity mode is

$$W_m(t) = \frac{\mu}{8}\left(\frac{2a}{\lambda}\right)^2 C^2 \ abd \cos^2(\omega t).$$

That is, the magnetic energy oscillates sinusoidally in time, just as the electric energy does, but the magnetic energy and electric energy are out of phase with one another due to the 90° phase difference between the electric and magnetic fields. When the electric and magnetic energies are added together, the total energy stored in the electromagnetic field is constant,

$$W(t) = W_e(t) + W_m(t) = W_{e,max} \sin^2(\omega t) + W_{m,max} \cos^2(\omega t)$$

$$= W_{e,max} \left[\sin^2(\omega t) + \cos^2(\omega t) \right] = \frac{\mu}{8} \left(\frac{2a}{\lambda} \right)^2 C^2 \, abd. \quad (10.96)$$

During one cycle of the field, the energy stored in the cavity converts in form, from magnetic to electric, back to magnetic, back to electric, and finally to magnetic, but the total energy remains constant.

Example 10.19 Energy Stored in a Cavity

The energy stored in the TE_{101} mode of an air-filled resonant cavity is 1 nJ. The dimensions of the cavity are $20 \times 25 \times 30$ cm. Determine the peak electric field amplitude for this mode.

Solution:

We'll use Eq. (10.96) to solve this question. We are given the cavity dimensions, and since the cavity is air-filled, we can use $\mu = \mu_0$. To find the free-space wavelength, we'll use Eq. (10.93),

$$\lambda = 2 \left[\left(\frac{1}{a} \right)^2 + \left(\frac{1}{d} \right)^2 \right]^{-1/2} = 33.3 \text{ cm.}$$

Then, solving Eq. (10.93) for C, the amplitude of $\tilde{H}_z(x, z)$, is

$$C = \sqrt{\frac{8W}{\mu \, abd}} \left(\frac{\lambda}{2a} \right)$$

$$= \sqrt{\frac{8(1 \text{ nJ})}{(4\pi \times 10^{-7} \text{ H/m})(0.20 \times 0.25 \times 0.30) \text{ m}^3}} \left(\frac{0.333 \text{ m}}{2(0.20 \text{ m})} \right) = 0.542 \text{ A/m.}$$

Using Eq. (10.91c), the maximum amplitude of the electric field, which occurs at the center of the cavity, is

$$\frac{2a\eta}{\lambda} C = \frac{2(0.20 \text{ m})(120\pi \text{ } \Omega)}{0.333 \text{ m}} \left(0.542 \text{ A/m} \right) = 245 \text{ V/m.}$$

10.5.4 Losses

We based the discussion of the field energy on the field amplitudes given by Eqs. (10.91a)–(10.92c), which we derived without considering losses within the cavity. There is, however, energy lost within each cycle, as the dielectric can absorb energy from the mode, and currents in the conductor walls can dissipate energy. Let's examine the rate of these energy losses.

The time-averaged power lost per unit volume in the dielectric $\overline{P_v^{abs}}$ is given by Eq. (8.43). As a reminder, we integrated this term across the cross-sectional area of a waveguide to determine the power loss per unit length P_d' for those elements. In the present analysis we are interested in the rate at which the energy W stored in the cavity mode changes. This is found as

$$-\left.\frac{dW}{dt}\right|_d = \int_v \overline{P_v^{abs}}\, dv = \frac{1}{2}\left(\sigma_d + \omega\varepsilon''\right)\int_v |\tilde{\mathbf{E}}|^2\, dv.$$

(The qualifier d on $-dW/dt$ indicates that this is the power lost due to dielectric absorption.) Notice that dW/dt is negative, since losses result in a decrease of W, the energy stored in the cavity mode. Notice also that dW/dt depends on the volume integral of $|\tilde{\mathbf{E}}|^2$, just as the total energy W does,

$$W = \frac{\varepsilon}{2}\int_v |\tilde{\mathbf{E}}|^2\, dv.$$

Power losses of the cavity, relative to the total energy stored in the cavity, are characterized with a dimensionless parameter known as the **cavity quality factor**, also known as the **Q factor** or, more simply, the **cavity Q**. The quality factor is defined as

$$Q = \frac{2\pi}{T}\frac{W}{-dW/dt} = \omega_0 \frac{W}{-dW/dt}, \tag{10.97}$$

where W is the energy stored in the cavity fields, $-dW/dt$ is the power loss, and T is the period of oscillation at the resonant frequency ω_0,

$$T = \frac{2\pi}{\omega_0}. \tag{10.98}$$

The cavity Q_d due to dielectric absorption is

$$Q_d = \omega_0 \frac{(\varepsilon'/2)\int_v |\tilde{\mathbf{E}}|^2\, dv}{(1/2)(\sigma_d + \omega_0\varepsilon'')\int_v |\tilde{\mathbf{E}}|^2\, dv} = \frac{\omega_0\varepsilon'}{(\sigma_d + \omega_0\varepsilon'')}.$$

This result is valid for any cavity mode. Notice that Q_d does not explicitly depend on the dimensions of the cavity, other than the dependence of the resonant frequency on dimensions.

We next consider the energy losses due to conduction currents in the walls of the cavity. Figure 10.21 shows these surface currents for the TE_{101} mode of a rectangular cavity. Remember that the surface currents are perpendicular to the tangential magnetic fields just inside the cavity, with their direction given by the right-hand rule. The magnetic field circles around in the x–z plane, as shown by the red dashed circles. The surface current $\tilde{\mathbf{J}}_s$ diverges from the center of the right side, flows to the left in the top, bottom, front, and back surfaces, and converges to the center of the left side. We must consider power loss in each of the six sides when computing the total power loss.

On the top and bottom walls, the power loss is

$$-\left.\frac{dW}{dt}\right|_{c,top} = -\left.\frac{dW}{dt}\right|_{c,bot} = \frac{1}{2}\int_0^d \int_0^b |\tilde{J}_{s,y}|_{x=0}^2 R_s\, dy\, dz$$

$$= \frac{1}{2}\int_0^d \int_0^b |\tilde{H}_z(x=0,z)|^2 R_s\, dy\, dz,$$

where

$$\tilde{H}_z(x=0,z) = C\sin\left(\frac{\pi z}{d}\right).$$

Carrying out the integration, the power loss for the top and bottom walls is

$$-\left.\frac{dW}{dt}\right|_{c,top} = -\left.\frac{dW}{dt}\right|_{c,bot} = \frac{R_s b d}{4}C^2.$$

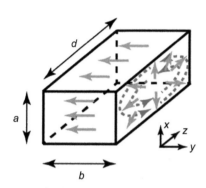

FIGURE 10.21 Plot of the surface current (green arrows) on the conducting walls for a TE_{101} mode on a rectangular resonator. The red dashed lines show the magnetic field $\tilde{\mathbf{H}}(x,z)$ inside the cavity. The surface current $\tilde{\mathbf{J}}_s$ diverges from the center of the right side, flows to the left on the top, bottom, front, and back surfaces, and converges to the center of the right side. As a function of time, the current oscillates sinusoidally at the frequency of the cavity electric and magnetic fields. Not shown are the electric field lines of the mode or the electric charges that accumulate on the left and right surfaces, which terminate the electric field lines.

The power loss in the front and back sides is

$$-\frac{dW}{dt}\bigg|_{c,fr} = -\frac{dW}{dt}\bigg|_{c,back} = \frac{1}{2}\int_0^b\int_0^a |\tilde{J}_{s,y}|^2_{z=0}R_s\,dx\,dy$$

$$= \frac{1}{2}\int_0^b\int_0^a |\tilde{H}_{x,z=0}|^2 R_s\,dx\,dy,$$

where

$$H_x(x,z=0) = \frac{-a}{d}C\sin\left(\frac{\pi x}{a}\right).$$

Carrying out the integration, the power loss for the front and back surfaces is

$$-\frac{dW}{dt}\bigg|_{c,fr} = -\frac{dW}{dt}\bigg|_{c,back} = \frac{R_s ab}{4}\left(\frac{a}{d}\right)^2 C^2.$$

Finally, for the left and right sides, the power loss is

$$-\frac{dW}{dt}\bigg|_{c,left} = -\frac{dW}{dt}\bigg|_{c,right} = \frac{1}{2}\int_0^d\int_0^a \left[|\tilde{J}_{s,x}|^2_{y=0} + |\tilde{J}_{s,z}|^2_{y=0}\right]R_s\,dx\,dz$$

$$= \frac{1}{2}\int_0^d\int_0^a \left[|\tilde{H}_z(x,y=0,z)|^2 + |\tilde{H}_x(x,y=0,z)|^2\right]R_s\,dx\,dz,$$

where

$$\tilde{H}_x(x,y=0,z) = \frac{-a}{d}C\sin\left(\frac{\pi x}{a}\right)\cos\left(\frac{\pi z}{d}\right)$$

and

$$\tilde{H}_z(x,y=0,z) = C\cos\left(\frac{\pi x}{a}\right)\sin\left(\frac{\pi z}{d}\right).$$

Substituting these fields, and carrying out the integration, the power loss for the left and right sides is

$$-\frac{dW}{dt}\bigg|_{c,left} = -\frac{dW}{dt}\bigg|_{c,right} = \frac{R_s ad}{8}\left[1+\left(\frac{a}{d}\right)^2\right]C^2.$$

Adding the power losses for the six individual surfaces gives the total power loss due to conduction currents as

$$-\frac{dW}{dt}\bigg|_c = 2R_s\left\{\frac{bd}{4} + \frac{ab}{4}\left(\frac{a}{d}\right)^2 + \frac{ad}{8}\left[1+\left(\frac{a}{d}\right)^2\right]\right\}C^2$$

$$= \frac{R_s}{4}\left\{2bd + 2\frac{a^3b}{d^2} + ad + \frac{a^3}{d}\right\}C^2. \tag{10.99}$$

The cavity Q_c due to these conduction losses for the resonator operating on the TE_{101} mode is found using Eq. (10.97),

$$Q_c = \omega_0\frac{W}{-dW/dt|_c}$$

$$= \frac{\omega_0\,(\mu\,abd/8)(2a/\lambda)^2 C^2}{(R_s/4)\{2bd + 2a^3b/d^2 + ad + a^3/d\}\,C^2},$$

where we have used Eq. (10.96) for W and Eq. (10.99) for the power loss $-dW/dt|_c$. Canceling common factors and substituting

$$\frac{2a}{\lambda} = \sqrt{1+\left(\frac{a}{d}\right)^2}$$

gives us

$$Q_c = \frac{\omega_0\mu\,ab\left(a^2+d^2\right)}{2R_s\{2bd^2 + 2a^3b/d + ad^2 + a^3\}}. \tag{10.100}$$

Finally, we evaluate the product of ω_0 and μ as

$$\omega_0 \mu = \left(\omega_0 \sqrt{\varepsilon \mu}\right) \sqrt{\mu/\varepsilon}.$$

But $\omega_0 \sqrt{\varepsilon \mu}$ is the propagation constant k for an unguided wave, and $\sqrt{\mu/\varepsilon}$ is the impedance of the medium, so $\omega_0 \mu$ is equivalent to

$$\omega_0 \mu = k\eta = \eta \sqrt{\left(\frac{m\pi}{a}\right)^2 + \left(\frac{n\pi}{b}\right)^2 + \left(\frac{p\pi}{d}\right)^2} = \pi\eta \sqrt{\frac{1}{a^2} + \frac{1}{d^2}} = \frac{\pi\eta}{ad}\sqrt{a^2 + d^2}.$$

(We have used Eq. (10.86) for the discrete value of k of the TE_{mnp} mode in a cavity, and then evaluated this term for the specific case of the TE_{101} mode.) Upon substitution into Eq. (10.100), we find

$$Q_c = \frac{\pi\eta}{2R_s} \frac{b\left(a^2 + d^2\right)^{3/2}}{\{2bd^3 + 2a^3b + ad^3 + a^3d\}}. \tag{10.101}$$

Example 10.20 Cavity Quality Factor

Determine the cavity Q of the TE_{101} mode for an empty cubical cavity fabricated from copper. The frequency of the cavity mode is 10 GHz.

Solution:

We start by evaluating the dimensional factor in Eq. (10.101). Since we're told that the cavity is cubical, we can set $a = b = d$:

$$\frac{b\left(a^2 + d^2\right)^{3/2}}{\{2bd^3 + 2a^3b + ad^3 + a^3d\}} = \frac{2^{3/2}}{6} = \frac{\sqrt{2}}{3} = 0.471.$$

For a copper cavity, resonating at 10 GHz, the surface resistance is

$$R_s = \sqrt{\frac{\omega\mu}{2\sigma}} = \sqrt{\frac{(2\pi \times 10^{10}\ s^{-1})(4\pi \times 10^{-7}\ H/m)}{2(5.8 \times 10^7\ S/m)}} = 0.0260\ \Omega.$$

(We have used $\sigma = 5.8 \times 10^7$ S/m from Table 3.1.) Since the cavity is empty, we use $\eta = 377\ \Omega$, and find a cavity Q_c of

$$Q_c = \frac{\pi(377\ \Omega)}{2 \times 0.0260\ \Omega} \times 0.471 \approx 10^4.$$

Higher cavity Q factors can be achieved by cooling the cavity, which increases the conductivity of the walls and decreases the surface resistance.

Cavity losses often come in different forms, and so we need to be able to combine Q-factors. For instance, when we have losses due to dielectric absorption and conduction losses, then

$$-\frac{dW}{dt} = -\frac{dW}{dt}\bigg|_d - \frac{dW}{dt}\bigg|_c,$$

and the cavity Q is

$$Q = \omega_0 \frac{W}{-dW/dt} = \frac{1}{-(1/\omega_0 W)dW/dt|_d - (1/\omega_0 W)dW/dt|_c} = \frac{1}{1/Q_d + 1/Q_c}.$$

In other words, the total cavity Q is the inverse of the sum of the inverses of the individual Qs.

Let's now explore the following scenario. Picture that we have a resonant cavity, whose Q we know, and that this cavity is energized such that the initial energy stored in the fields is W_0. Upon turning off the source that drives the fields in the cavity, how does the energy stored in the cavity change? We can answer this by rearranging Eq. (10.97) to write

$$\frac{dW(t)}{dt} = -\frac{\omega_0}{Q} W(t),$$

a simple differential equation whose solution we immediately recognize as

$$W(t) = W_0 e^{-\omega_0 t/Q} = W_0 e^{-t/\tau},$$

where τ is the decay constant, equal to Q/ω_0. Since the stored energy scales as the square of the field amplitude, the field amplitude also decays exponentially,

$$E(t) = E_0 e^{-\omega_0 t/2Q} e^{j\omega_0 t}. \tag{10.102}$$

The time-dependence of this field amplitude has implications for its spectral composition. Specifically, when driven by an input (a radiating antenna protruding slightly through one of the walls, for example) whose frequency is close to, but not quite equal to, the resonant frequency ω_0, we should still expect excitation of the resonator mode, although at a reduced level. We can see this through the Fourier transform relations

$$E(t) = \frac{1}{\sqrt{2\pi}} \int_{-\infty}^{\infty} E(\omega) e^{j\omega t} \, d\omega$$

and

$$E(\omega) = \frac{1}{\sqrt{2\pi}} \int_{0}^{\infty} E(t) e^{-j\omega t} \, dt.$$

For the time-dependent field of Eq. (10.102), the Fourier transform is

$$E(\omega) = \frac{1}{\sqrt{2\pi}} \int_{0}^{\infty} E_0 e^{-\omega_0 t/2Q} e^{j(\omega_0 - \omega)t} dt$$

$$= \frac{1}{\sqrt{2\pi}} E_0 \frac{e^{-\omega_0 t/2Q} e^{j(\omega_0 - \omega)t} \big|_0^{\infty}}{[-\omega_0/2Q + j(\omega_0 - \omega)]} = \frac{E_0}{\sqrt{2\pi}} \frac{1}{[\omega_0/2Q - j(\omega_0 - \omega)]}.$$

The energy stored in the field scales as the magnitude squared of the field amplitude,

$$W(\omega) \propto \frac{1}{(\omega_0 - \omega)^2 + (\omega_0/2Q)^2}. \tag{10.103}$$

This lineshape function is known as a Lorentzian, whose value is maximum at the resonant frequency $\omega = \omega_0$, and whose linewidth is

$$\boxed{\Gamma = \frac{\omega_0}{Q}.} \tag{10.104}$$

There are a variety of ways to define the linewidth of a resonant function. A common definition is the full-width-at-half-maximum, abbreviated FWHM. When the frequency of the field ω differs from the resonant frequency by half of the FWHM, that is,

$$|\omega_0 - \omega| = \frac{\Gamma}{2} = \frac{\omega_0}{2Q},$$

the stored energy W is only half as large as the maximum value at resonance, when $\omega = \omega_0$. A plot of the Lorentzian lineshape function of Eq. (10.103) is shown in Fig. 10.22.

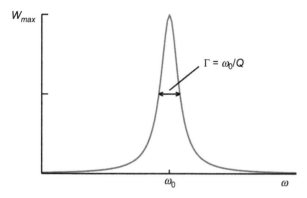

FIGURE 10.22 Plot of Lorentzian function. The resonant frequency is ω_0 and the full-width-at-half-maximum (FWHM) is $\Gamma = \omega_0/Q$.

In this section we have explored the properties of a rectangular resonator. We determined the electromagnetic modes of the cavity, and found that each mode has a specific frequency, which is related to the dimensions of the cavity. We solved for the spatial profile of the electric and magnetic fields, and found the total energy stored in the fields. Finally, we introduced the cavity Q as a measure of losses within a cavity, with high Q corresponding to low losses. When a cavity is excited by an input signal at a frequency that is equal or close to the resonant frequency of one of the cavity modes, the electromagnetic field inside the cavity can build up, and the cavity can store large amounts of energy in the fields. The cavity Q also defines the linewidth of the frequency response of the resonator, with large Q corresponding to a narrow linewidth.

SUMMARY

In this chapter we have explored wave propagation on conducting waveguides. Waveguides are commonly used for guiding high-frequency signals from a source to a load. Hollow single-conductor waveguides, such as the rectangular or circular waveguides, can support only TM and TE modes, never TEM modes, and only allow wave propagation for frequencies above a specific cutoff frequency. This cutoff frequency is governed by

Table 10.11

Collection of key formulas from Chapter 10.	
Table	**Summary of results for:**
10.1	TEM modes on a very wide ($b \gg a$) parallel plate waveguide
10.2	TM modes on a very wide ($w \gg a$) parallel plate waveguide
10.3	The parameters of wave impedance, phase constant, attenuation constant, phase velocity, and group velocity for TE and TM modes
10.4	TE modes on a very wide ($b \gg a$) parallel plate waveguide
10.5	TM modes on a rectangular waveguide
10.7	TE modes on a rectangular waveguide
10.9	TM modes on a circular waveguide
10.10	TE modes on a circular waveguide
$$\omega = \frac{1}{\sqrt{\varepsilon\mu}}\left[\left(\frac{m\pi}{a}\right)^2 + \left(\frac{n\pi}{b}\right)^2 + \left(\frac{p\pi}{d}\right)^2\right]^{1/2} \qquad (10.87)$$	
$$W_{e,max} = \frac{\varepsilon E_0^2}{8}abd \qquad (10.95)$$	
$$Q = \omega_0 \frac{W}{-dW/dt} \qquad (10.97)$$	
$$\Gamma = \frac{\omega_0}{Q} \qquad (10.104)$$	

the transverse dimensions of the waveguide, and limits the excitation to modes whose unguided wavelength is less than twice the waveguide size. Lower-frequency waves are quickly attenuated. When exciting a waveguide mode, as with an antenna feed, all modes whose cutoff frequencies are lower than the excitation frequency can be excited. For this reason, when single-mode operation is required, it is common to use the mode with the lowest cutoff frequency, such as the TE_{10} mode for a rectangular waveguide.

For each mode, important waveguide properties include the waveguide impedance, Z_{TM} and Z_{TE}, the mode attenuation constant α, the phase constant β, the wavelength Λ, the phase velocity u_p, and the group velocity u_g. We explored each of these parameters for the parallel plate, rectangular, and circular waveguides. We also explored the total power carried by the fields, and losses due to dielectric absorption and conduction currents for the parallel plate and rectangular waveguides.

Finally, we examined the resonant cavity, a structure related to waveguides, which can store energy in the fields, and, due to its sharp frequency dependence, acts as a filter for waves. We examined losses in the resonator, which we quantified in terms of the cavity Q.

PROBLEMS

General Approach

P10-1 Determine the phase velocity and group velocity of a wave passing through a non-magnetic material whose permittivity is $\varepsilon(\omega) = \varepsilon_0 (1 + a\omega)^2$, where a is a constant.

P10-2 Show that the ratio of the transmitted power over the stored electromagnetic energy per unit length for a TE_m mode on a parallel plate waveguide yields the group velocity u_g.

Parallel Plate Waveguide

P10-3 Determine the cutoff frequencies of the three lowest TM or TE modes of an air-filled parallel plate waveguide of plate spacing $a = 0.5$ cm and plate width $b = 5$ cm.

P10-4 (a) For the parallel plate waveguide of Prob. P10-3, determine the phase constant β and the mode wavelength Λ of the TM_1 mode at a frequency of 10 GHz. (b) Repeat for the TM_3 mode.

P10-5 For the parallel plate waveguide of Prob. P10-3, determine the phase velocity, group velocity, and appropriate wave impedance (Z_{TM} or Z_{TE}) for a TM_1 and TE_1 wave at a frequency of 10 GHz.

P10-6 For the parallel plate waveguide of Prob. P10-3, determine the phase velocity, group velocity, and appropriate wave impedance (Z_{TM} or Z_{TE}) for a TM_3 and TE_3 wave at a frequency of 10 GHz.

P10-7 For the parallel plate waveguide of Prob. P10-3, determine (a) the total transmitted power for a TM_1 wave at a frequency of 10 GHz and field amplitude $A = 100$ V/m, (b) the power lost per unit length P'_c when the conductors are made of copper, and (c) the attenuation coefficient.

P10-8 For the parallel plate waveguide of Prob. P10-3, determine (a) the total transmitted power for a TE_1 wave at a frequency of 10 GHz and field amplitude $B = 250$ mA/m, (b) the power lost per unit length P'_c when the conductors are made of copper, and (c) the attenuation coefficient.

P10-9 Sketch the electric and magnetic fields for the TM_2 mode of a parallel plate waveguide.

P10-10 Sketch the electric and magnetic fields for the TE_2 mode of a parallel plate waveguide.

P10-11 (a) Determine the conductor spacing of an air-spaced parallel plate waveguide whose TM_1 mode cutoff frequency is 5 GHz. (b) Repeat for a parallel plate waveguide with a dielectric material of permittivity $\varepsilon = 3\varepsilon_0$ between the conductors.

Rectangular Waveguide

P10-12 Determine the cutoff frequencies of the five lowest-frequency modes of an air-filled rectangular waveguide of dimensions $a = 3.0$ cm and $b = 2.0$ cm. (Be sure to identify TE or TM modes.)

P10-13 (a) For the rectangular waveguide of Prob. P10-12, determine the phase constant β and the mode wavelength Λ of a TE_{10} at a frequency of 10 GHz. (b) Repeat for a TE_{01} mode.

P10-14 Use $\tilde{E}_z(x, y)$ of Eq. (10.51) and Eqs. (10.16a)–(10.16d) to verify $\tilde{E}_x(x, y)$, $\tilde{E}_y(x, y)$, $\tilde{H}_x(x, y)$, and $\tilde{H}_y(x, y)$ for the TM modes on a rectangular waveguide.

P10-15 Starting with Eqs. (10.51)–(10.57d) for the field components of the TM mode on a rectangular waveguide, find the full z- and t-dependent forms $E_z(x, y, z, t)$, $E_x(x, y, z, t)$, $E_y(x, y, z, t)$, $H_x(x, y, z, t)$, and $H_y(x, y, z, t)$.

P10-16 For the rectangular waveguide of Prob. P10-12, determine the phase velocity, group velocity, and appropriate wave impedances (Z_{TM} or Z_{TE}) for a TM_{11} and TE_{11} wave at a frequency of 10 GHz.

P10-17 For the rectangular waveguide of Prob. P10-12, determine the phase velocity, group velocity, and appropriate wave impedances (Z_{TM} or Z_{TE}) for a TM_{21} and TE_{12} wave at a frequency of 20 GHz.

P10-18 (a) Starting from Eq. (10.59) for the power transmitted by a mode on a rectangular waveguide, verify the result shown in Eq. (10.60a). (b) Use this result to prove Eq. (10.60b).

P10-19 For the TM_{11} mode of the rectangular waveguide of Prob. P10-12, determine (a) the total transmitted power for a wave at a frequency of 10 GHz and field amplitude $A = 100$ V/m, (b) the power lost per unit length P'_c due to conductor losses when the conductors are made of copper, and (c) the attenuation coefficient α_c.

P10-20 (a) Starting from Eq. (10.59) for the power transmitted by a mode on a rectangular waveguide, verify the result shown in Eq. (10.68a). (b) Use this result to prove Eq. (10.68b).

P10-21 For the TE_{11} mode of the rectangular waveguide of Prob. P10-12, determine (a) the total transmitted power for a wave at a frequency of 10 GHz and field amplitude $B = 250$ mA/m, (b) the power lost per unit length P'_c due to conductor losses when the conductors are made of copper, and (c) the attenuation coefficient α_c.

P10-22 Sketch the electric and magnetic fields for the TM_{12} mode of a rectangular waveguide.

P10-23 Sketch the electric and magnetic fields in the $x = a/2$ plane for the TE_{20} mode of a rectangular waveguide.

P10-24 (a) Determine the dimensions of an empty rectangular waveguide whose cutoff frequency for the lowest-frequency mode is 5 GHz. (The answer is not unique.) (b) Repeat part (a) with the additional requirement that the second lowest mode frequency should be 9 GHz. (c) Repeat part (a) for a rectangular waveguide filled with a dielectric material of permittivity $\varepsilon = 3\varepsilon_0$ between the conductors.

Circular Waveguide

P10-25 Determine the cutoff frequencies of the five lowest modes of an air-filled circular waveguide of radius $a = 2.0$ cm.

P10-26 For the circular waveguide of Prob. P10-25, determine the phase velocity, group velocity, and appropriate wave impedance (Z_{TM} or Z_{TE}) for a TE_{11} and TM_{01} mode wave at a frequency of 10 GHz.

P10-27 Sketch the electric and magnetic fields for the TE_{21} mode of a circular waveguide.

Resonant Cavity

P10-28 Determine the five lowest TE mode frequencies for a rectangular cavity of dimension $a = 2.0$ cm, $b = 3.0$ cm, and $d = 4.0$ cm.

P10-29 Determine the five lowest TM mode frequencies for a rectangular cavity of dimension $a = 2.0$ cm, $b = 3.0$ cm, and $d = 4.0$ cm.

P10-30 Determine the cavity Q for the TE_{101} mode of the cavity described in Prob. P10-28, made of copper.

P10-31 (a) Sketch the electric and magnetic field patterns for the TE_{201} mode of a rectangular resonator. (For the y–z plots, use $x = a/2$.) (b) Sketch the surface currents in the conductor walls for this mode.

P10-32 Confirm the electric and magnetic fields for TM modes of a rectangular resonant cavity, as presented in Eqs. (10.89a)–(10.90c).

CHAPTER 11
Antennas

KEY WORDS

elemental dipole antenna; near-field region; far-field region; radiation resistance; efficiency; directionality; directivity; radiation intensity; long linear antenna; half-wave dipole antenna; antenna array.

LEARNING OBJECTIVES

After studying this section, you should be able to:

- describe qualitatively the radiation pattern generated by a short dipole antenna;
- determine the electric and magnetic field strength generated by a short dipole antenna;
- determine the time-averaged power density and total power radiated by a short dipole antenna;
- determine the radiation resistance, real resistance, and efficiency of a short dipole antenna; and
- determine the directivity and the radiation intensity of a short dipole antenna.

In the last few chapters we have examined the propagation of electromagnetic waves; freely propagating waves in Chapter 8, waves guided along transmission lines in Chapter 9, and waves guided within waveguides in Chapter 10. But we paid no attention in these discussions to the *generation* of these waves. In this chapter our goal is to remedy this shortcoming. As we will show, an oscillating current in an open-ended wire can produce an electromagnetic wave. We will examine the distribution of the radiated power, the total radiated power, the efficiency of the power generation, the polarization of the wave, and the input impedance of a few simple radiating systems. We will start by examining a short, or **elemental**, dipole antenna, and then expand this to longer, more efficient, antennas. We will also look at the field distribution and power density produced by an array of antennas, and show how the distribution varies with the relative phase of the radiators.

11.1 Elemental Dipole Antenna

We start this discussion by considering a short, linear dipole antenna driven by a sinusoidal current at frequency f. We should pause a moment to explain what is meant by *short*, *linear*, and *dipole*. A **short** antenna is one whose total length $\Delta\ell$ is much less than the wavelength $\lambda = c/f$ of a propagating wave at that same frequency f. The current along the length of a short antenna is taken to be uniform. A **linear** antenna is one which lies along a straight line. Finally, the term **dipole** antenna derives from its characteristics, such as the distribution of charge and field patterns, reminiscent of the electric dipole that we introduced in Chapter 1, and in particular Eq. (1.55).

The current $i(t)$ applied to the elemental antenna varies sinusoidally as a function of time. As already stated, the current is uniform along the length of the antenna. For those students who are still thinking in terms of d.c. circuits, this may again seem like a strange concept. In a d.c. circuit, of course, the current always follows a closed path, and if a path is not closed, no current flows. Yet in this antenna, once the current reaches the end of the antenna, there is no place for the current to go. What happens to this charge? The answer, of course, is that the charge accumulates at the ends of this short wire.

Writing the antenna current as $i(t) = I_0 \cos(\omega t)$, the charge at either end of the wire must be $\pm q(t)$, where $q(t) = (I_0/\omega) \sin(\omega t)$. The current $i(t)$ is, of course, just $dq(t)/dt$, the rate at which the charge at the end of the antenna changes. So the charge $q(t)$

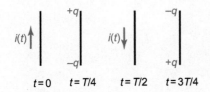

FIGURE 11.1 A progression of images of a short, linear, dipole antenna, showing the oscillating current $i(t)$ in the antenna, and the charge $q(t)$ that accumulates at the ends of the antenna.

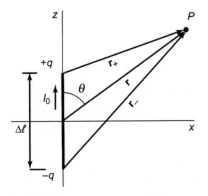

FIGURE 11.2 The elemental dipole antenna positioned at the origin, lying along the z-axis.

accumulates at one end of the dipole antenna, and an equal but opposite charge $-q(t)$ at the other (see Fig. 11.1). As the current oscillates sinusoidally, the charge oscillates sinusoidally as well, 90° out of phase with the current. Since charges and currents are the sources of electric and magnetic fields, these sinusoidally varying charges and currents produce sinusoidally varying electric and magnetic fields in the space surrounding the antenna.

With this physical picture in mind, let's see if we can be more quantitative. Our goal here is to determine the electric field $\mathbf{E}(t, \mathbf{r})$ and the magnetic field $\mathbf{H}(t, \mathbf{r})$ that are generated by the current and charge distributions in the elemental dipole antenna. We have already developed the framework for this derivation, so we'll rely on that. That is, we showed in Section 7.2 that, for time-varying fields, the potential functions that we introduced for static fields must be redefined. The good news in that section was that the potential functions $V(t, \mathbf{r})$ and $\mathbf{A}(t, \mathbf{r})$ still exist, but there are some differences that we have to acknowledge. First, the potentials are retarded by the time required for a wave to propagate from the source (the charge or current) to the field point. These are given in Eqs. (7.11) and (7.12). And second, the electric field and magnetic field are now coupled, and we calculate them from the potential functions using Eqs. (7.4) and (7.5).

We'll start with the magnetic field, using the parameters as shown in Fig. 11.2. The retarded vector magnetic potential $\tilde{\mathbf{A}}(\mathbf{r})$, using Eq. (7.12), takes the particularly simple form for harmonic fields,

$$\tilde{\mathbf{A}}(\mathbf{r}) = \frac{\mu_0}{4\pi} \int_v \frac{\tilde{\mathbf{J}}_v(\mathbf{r}')e^{-j\beta|\mathbf{r}-\mathbf{r}'|}\,dv'}{|\mathbf{r} - \mathbf{r}'|}. \tag{11.1}$$

This equation is, of course, identical to Eq. (4.22), except for the factor $e^{j\beta|\mathbf{r}-\mathbf{r}'|}$ in the numerator of the integrand. This factor represents the retardation of the field, as described in the previous paragraph, due to the time required for the field pattern to propagate a distance $|\mathbf{r} - \mathbf{r}'|$ from an element of the antenna (source) at $\mathbf{r}'$ to the distant field point $\mathbf{r}$. For the elemental dipole antenna aligned with the z-axis, located at the origin, and carrying a current $\tilde{I}_0$ ($\tilde{I}_0$ is the phasor amplitude representing the time-harmonic current), the term in the numerator of Eq. (11.1), $\tilde{\mathbf{J}}_v(\mathbf{r}')dv'$, becomes $\tilde{I}_0\mathbf{a}_z\Delta\ell$, and the vector potential simplifies to

$$\tilde{\mathbf{A}}(R) = \tilde{A}_z(R)\mathbf{a}_z, \tag{11.2}$$

where

$$\tilde{A}_z(R) = \frac{\mu_0}{4\pi} \frac{\tilde{I}_0\Delta\ell e^{-j\beta R}}{R}. \tag{11.3}$$

Since this potential depends on the distance R from the origin to the field point P, it makes sense to work in spherical coordinates to determine the magnetic field. The spherical components of the potential are

$$\tilde{A}_R(R, \theta) = \tilde{A}_z(R)\cos\theta = \frac{\mu_0}{4\pi} \frac{\tilde{I}_0\Delta\ell e^{-j\beta R}}{R}\cos\theta$$

and

$$\tilde{A}_\theta(R, \theta) = -\tilde{A}_z(R)\sin\theta = -\frac{\mu_0}{4\pi} \frac{\tilde{I}_0\Delta\ell e^{-j\beta R}}{R}\sin\theta.$$

(This conversion from rectangular to spherical components is most easily seen by writing the vector $\tilde{\mathbf{A}}$ in both coordinate systems,

$$\tilde{\mathbf{A}} = \tilde{A}_z\mathbf{a}_z = \tilde{A}_R\mathbf{a}_R + \tilde{A}_\theta\mathbf{a}_\theta + \tilde{A}_\phi\mathbf{a}_\phi.$$

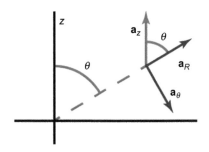

FIGURE 11.3 A diagram showing the unit vectors $\mathbf{a}_z$, $\mathbf{a}_R$, and $\mathbf{a}_\theta$.

Taking the scalar product of both sides with $\mathbf{a}_R$, $\mathbf{a}_\theta$, and $\mathbf{a}_\phi$, with the aid of Fig. 11.3, gives us $\tilde{A}_R$, $\tilde{A}_\theta$, and $\tilde{A}_\phi$, respectively.) The final result for the magnetic potential expressed in spherical components is

$$\tilde{\mathbf{A}}(R,\theta) = \frac{\mu_0}{4\pi} \frac{\tilde{I}_0 \Delta\ell \, e^{-j\beta R}}{R} \left[\mathbf{a}_R \cos\theta - \mathbf{a}_\theta \sin\theta \right]. \tag{11.4}$$

The curl of the magnetic potential in spherical coordinates, using Eq. (D.15), is

$$\tilde{\mathbf{H}}(\mathbf{r}) = \frac{1}{\mu_0} \boldsymbol{\nabla} \times \tilde{\mathbf{A}} = \frac{1}{\mu_0} \frac{1}{R^2 \sin\theta} \begin{vmatrix} \mathbf{a}_R & R\mathbf{a}_\theta & R\sin\theta\,\mathbf{a}_\phi \\ \dfrac{\partial}{\partial R} & \dfrac{\partial}{\partial \theta} & \dfrac{\partial}{\partial \phi} \\ \tilde{A}_R & R\tilde{A}_\theta & R\sin\theta\,\tilde{A}_\phi \end{vmatrix}.$$

Since $\tilde{A}_\phi = 0$, and $\tilde{A}_R(R,\theta)$ and $\tilde{A}_\theta(R,\theta)$ do not depend on the coordinate ϕ, only the H_ϕ component survives,

$$\tilde{\mathbf{H}}(\mathbf{r}) = \frac{1}{\mu_0} \frac{R\sin\theta}{R^2 \sin\theta} \left[\frac{\partial}{\partial R}(R\tilde{A}_\theta) - \frac{\partial}{\partial \theta}\tilde{A}_R \right] \mathbf{a}_\phi = \tilde{H}_\phi \mathbf{a}_\phi.$$

Carrying out the derivatives, we find

$$\tilde{H}_\phi = -\frac{\tilde{I}_0 \Delta\ell \beta^2}{4\pi} \left[\frac{1}{j\beta R} + \frac{1}{(j\beta R)^2} \right] e^{-j\beta R} \sin\theta. \tag{11.5}$$

Now that we have determined $\tilde{\mathbf{H}}(\mathbf{r})$ generated by an elemental antenna, let's see what we can learn about the electric field $\tilde{\mathbf{E}}(\mathbf{r})$. One approach for this is to use Ampère's Law to determine $\tilde{\mathbf{E}}(\mathbf{r})$ from $\tilde{\mathbf{H}}(\mathbf{r})$. This approach is left to the end-of-chapter problems (Prob. P11-4). Instead, we'll take this opportunity to further illustrate the retarded potentials. Recall Eq. (7.5), which for time-harmonic fields is

$$\tilde{\mathbf{E}} = -\boldsymbol{\nabla}\tilde{V} - j\omega\tilde{\mathbf{A}}.$$

We have already found the potential $\tilde{\mathbf{A}}$, but not the retarded scalar electric potential $\tilde{V}$. This potential is generated by the charges $\pm q(t)$ at the ends of the elemental antenna as well. This charge $q(t) = (I_0/\omega)\sin(\omega t)$ can be written in its complex exponential form as $(I_0/\omega)(-j)e^{j\omega t}$, which leads to its phasor representation as $\tilde{q} = -j\tilde{I}_0/\omega$. Then, using Eq. (7.11), with one charge $\tilde{q}$ at $\mathbf{r}' = (\Delta\ell/2)\mathbf{a}_z$ and another charge $-\tilde{q}$ at $\mathbf{r}' = -(\Delta\ell/2)\mathbf{a}_z$, the potential is

$$\tilde{V}(\mathbf{r}) = \frac{\tilde{q}}{4\pi\varepsilon_0} \frac{e^{-j\beta r_+}}{r_+} - \frac{\tilde{q}}{4\pi\varepsilon_0} \frac{e^{-j\beta r_-}}{r_-}. \tag{11.6}$$

The distances r_+ and r_- from the charges to the field point P, as shown in Fig. 11.4, are

$$r_+ = |\mathbf{r}_+| = \left| \mathbf{r} - \frac{\Delta\ell}{2}\mathbf{a}_z \right|$$

and

$$r_- = |\mathbf{r}_-| = \left| \mathbf{r} + \frac{\Delta\ell}{2}\mathbf{a}_z \right|.$$

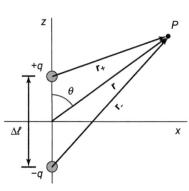

FIGURE 11.4 The charges $+q(t)$ and $-q(t)$ at the ends of the elemental dipole antenna.

We have studied distances such as these before, but for completeness the details are repeated here. The length of the vector $\mathbf{r}_+$ is

$$|\mathbf{r}_+| = \left[\left(\mathbf{r} - \frac{\Delta\ell}{2}\mathbf{a}_z \right) \cdot \left(\mathbf{r} - \frac{\Delta\ell}{2}\mathbf{a}_z \right) \right]^{1/2},$$

which when expanded is

$$r_+ = \left[|\mathbf{r}|^2 - \mathbf{r} \cdot \Delta\ell\mathbf{a}_z + \frac{\Delta\ell^2}{4} \right]^{1/2}.$$

In terms of the spherical coordinates of the point P (R, θ, and ϕ), $|\mathbf{r}| = R$, and $\mathbf{r} \cdot \mathbf{a}_z = R \cos \theta$. Furthermore, the final term $\Delta \ell^2 / 4$ is negligible in comparison to the other terms for $R \gg \Delta \ell$, and it can safely be omitted. The distance r_+ is then

$$r_+ \simeq \left[R^2 - R\Delta \ell \cos \theta \right]^{1/2} .$$

Finally, pulling out a factor of R^2 from both terms and using the Taylor expansion of Eq. (H.6), $\sqrt{1 + \epsilon} \simeq 1 + \epsilon/2$, valid when $\epsilon \ll 1$, r_+ is

$$r_+ = R \left[1 - \frac{\Delta \ell}{R} \cos \theta \right]^{1/2} \simeq R - \frac{\Delta \ell}{2} \cos \theta .$$

Using similar steps, the distance r_- can be approximated as

$$r_- \simeq R + \frac{\Delta \ell}{2} \cos \theta .$$

Now we are ready to use r_+ and r_- in Eq. (11.6). Factoring common terms, this equation becomes

$$\tilde{V}(\mathbf{r}) = \frac{\tilde{q}}{4\pi\epsilon_0} \left(\frac{e^{-j\beta(R - \Delta \ell \cos \theta / 2)}}{R - \frac{\Delta \ell}{2} \cos \theta} - \frac{e^{-j\beta(R + \Delta \ell \cos \theta / 2)}}{R + \frac{\Delta \ell}{2} \cos \theta} \right) .$$

Substituting $\tilde{q} \to -j\tilde{I}_0 / \omega$, and pulling out the common factor of $e^{-j\beta R}$, this potential is

$$\tilde{V}(\mathbf{r}) = \frac{-j\tilde{I}_0}{4\pi\epsilon_0 \omega} e^{-j\beta R} \left(\frac{e^{j\beta \Delta \ell \cos \theta / 2}}{R - \frac{\Delta \ell}{2} \cos \theta} - \frac{e^{-j\beta \Delta \ell \cos \theta / 2}}{R + \frac{\Delta \ell}{2} \cos \theta} \right) . \tag{11.7}$$

In many previous derivations that may appear to be similar, we argued that the small terms $\pm \Delta \ell \cos \theta / 2$ in the denominators were inconsequential, and we dropped them. For this retarded potential, however, we have to be a bit more careful. These terms are certainly small, but as we'll soon show, not negligible, so let's proceed by using another Taylor expansion, Eq. (H.5), to write

$$\frac{1}{R \mp \frac{\Delta \ell}{2} \cos \theta} = \frac{1}{R} \frac{1}{(1 \mp \frac{\Delta \ell}{2R} \cos \theta)} \simeq \frac{1}{R} \left[1 \pm \frac{\Delta \ell}{2R} \cos \theta \right] .$$

Using this expansion, the retarded potential becomes

$$\tilde{V}(\mathbf{r}) = \frac{-j\tilde{I}_0}{4\pi\epsilon_0 \omega} \frac{e^{-j\beta R}}{R} \left(e^{j\beta \Delta \ell \cos \theta / 2} \left[1 + \frac{\Delta \ell}{2R} \cos \theta \right] \right.$$
$$\left. - e^{-j\beta \Delta \ell \cos \theta / 2} \left[1 - \frac{\Delta \ell}{2R} \cos \theta \right] \right) ,$$

and applying Euler's Identities, this simplifies to

$$\tilde{V}(\mathbf{r}) = \frac{-j\tilde{I}_0}{4\pi\epsilon_0 \omega} \frac{e^{-j\beta R}}{R} \left[2j \sin \left(\frac{\beta \Delta \ell}{2} \cos \theta \right) + \left(\frac{\Delta \ell}{2R} \cos \theta \right) 2 \cos \left(\frac{\beta \Delta \ell}{2} \cos \theta \right) \right] .$$

Finally, since the length $\Delta \ell$ of the elemental antenna is much less than the wavelength λ of the wave, and $\beta = 2\pi/\lambda$, the factor $\frac{1}{2}\beta \Delta \ell \cos \theta$ is much less than 1, and we can approximate $\sin \left(\frac{1}{2}\beta \Delta \ell \cos \theta \right) \approx \frac{1}{2}\beta \Delta \ell \cos \theta$, and $\cos \frac{1}{2} (\beta \Delta \ell \cos \theta) \approx 1$. In this limit, the retarded potential is

$$\tilde{V}(\mathbf{r}) = \frac{\tilde{I}_0 \Delta \ell}{4\pi} \eta_0 \cos \theta \frac{e^{-j\beta R}}{R} \left[1 + \frac{1}{j\beta R} \right] . \tag{11.8}$$

In arriving at this expression we used the conversion

$$\frac{\beta}{\varepsilon_0 \omega} = \frac{\omega\sqrt{\varepsilon_0 \mu_0}}{\varepsilon_0 \omega} = \sqrt{\frac{\mu_0}{\varepsilon_0}} = \eta_0,$$

where $\eta_0 = \sqrt{\mu_0/\varepsilon_0} = 120\pi\ \Omega$ is the intrinsic impedance of vacuum.

From Eq. (11.8) we can start to see the importance of the small difference between the denominators in Eq. (11.7). This difference led to the second term inside the square brackets of Eq. (11.8). As we saw in Eq. (11.5), we have kept terms of various orders in $1/(j\beta R)$, and as we move on to calculate $\mathbf{E}$ from V and $\mathbf{A}$, this second term makes important non-negligible contributions.

So now we have the retarded scalar potential $\tilde{V}(\mathbf{r})$ for the elemental dipole antenna, and we want to use this, along with the magnetic potential given by Eq. (11.4), to determine the electric field generated by this antenna. We use Eq. (D.3) for the gradient in spherical coordinates. Then the radial component of $\tilde{\mathbf{E}}$ is

$$\tilde{E}_R = -\frac{\partial \tilde{V}}{\partial R} - j\omega \tilde{A}_z \cos\theta.$$

Carrying out the derivative, we find

$$\boxed{\tilde{E}_R = -\frac{2\tilde{I}_0 \Delta\ell}{4\pi}\eta_0 \beta^2 \left[\frac{1}{(j\beta R)^2} + \frac{1}{(j\beta R)^3}\right]e^{-j\beta R}\cos\theta.} \tag{11.9}$$

Similarly, the θ component of $\tilde{\mathbf{E}}$ is

$$\tilde{E}_\theta = -\frac{1}{R}\frac{\partial \tilde{V}}{\partial\theta} + j\omega \tilde{A}_z \sin\theta,$$

leading to the result

$$\boxed{\tilde{E}_\theta = -\frac{\tilde{I}_0 \Delta\ell}{4\pi}\eta_0 \beta^2 \left[\frac{1}{j\beta R} + \frac{1}{(j\beta R)^2} + \frac{1}{(j\beta R)^3}\right]e^{-j\beta R}\sin\theta.} \tag{11.10}$$

Finally, the component $\tilde{E}_\phi$ is zero, since $\partial\tilde{V}/\partial\phi = 0$ and $\tilde{A}_\phi = 0$.

It is interesting to examine Eqs. (11.5), (11.9), and (11.10) for the electric and magnetic fields in the region close to the antenna, $R \ll \lambda$, and in the region far from the antenna, $R \gg \lambda$. These are often called the **near-field region** and **far-field region**, respectively. In the former, $\beta R = 2\pi R/\lambda$ is very small, and the $(j\beta R)^{-3}$ terms in the equations above dominate. Remember, however, that in deriving the retarded scalar potential $\tilde{V}$, we made expansions that were valid only for $R \gg \Delta\ell$. Therefore, the near-field expression is valid only for a limited range of distances R. We won't explore this limit any further, as our primary interest is in the radiation of waves at long distances, that is, in the far-field region.

In the far-field region, the product βR is very large, and the dominant terms in the electric and magnetic fields are those that vary as $(j\beta R)^{-1}$. We will focus on these terms, since they in fact represent the outgoing wave that is generated by the antenna. The magnetic field in the far-field region is

$$\boxed{\tilde{H}_\phi = \frac{j\tilde{I}_0 \Delta\ell\beta}{4\pi R}e^{-j\beta R}\sin\theta,} \tag{11.11}$$

while the leading term of the electric field is

$$\boxed{\tilde{E}_\theta = \frac{j\tilde{I}_0 \Delta\ell\beta}{4\pi R}\eta_0 e^{-j\beta R}\sin\theta.} \tag{11.12}$$

There are several notable features in these relations. First, observe that $\tilde{\mathbf{E}}$ and $\tilde{\mathbf{H}}$ in the far-field region are in phase with one another, they are orthogonal to one another, and the ratio of their magnitudes is constant and equal to η_0. They each decrease with increasing distance as $1/R$, and they are independent of the azimuthal angle ϕ. The $\sin\theta$ dependence of each indicates that the field strengths vanish at the poles ($\theta = 0$ and $\theta = \pi$), and they attain their maximum value at the equator, $\theta = \pi/2$. A polar plot of $\tilde{E}_\theta(\theta)$ versus θ is shown in Fig. 11.5(a). Notice also the orientation of these field components. The primary magnetic field component is the ϕ component, circling around the current in the antenna. This is perfectly consistent with our observations of static magnetic fields generated by the current in a straight wire. Conversely, the primary electric field component in the far-field region is the θ component, as might be expected for field lines originating on the positive charge and terminating on the negative charge at the ends of the antenna.

FIGURE 11.5 Polar plots of (a) the far-field electric field amplitude $|\tilde{E}_\theta|$ and (b) the power density distribution generated by a short linear antenna lying on the z-axis. The antenna (not to scale) is depicted by the red line.

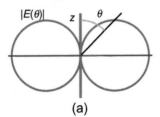
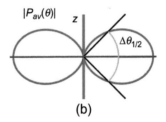

(a)　　　　　　　(b)

Now that we have expressions for the far-field $\tilde{\mathbf{E}}$ and $\tilde{\mathbf{H}}$, we are ready to address the question of the power density of the radiated wave. As shown in Chapter 8, the power density (energy transferred per unit time per unit area) is given by the Poynting vector, Eq. (8.79):

$$\overline{\mathcal{P}} = \frac{1}{2}\Re\left[\tilde{\mathbf{E}} \times \tilde{\mathbf{H}}^*\right].$$

Using the far-field expressions for $\tilde{\mathbf{H}}$ (Eq. (11.11)) and $\tilde{\mathbf{E}}$ (Eq. (11.12)), the time-averaged power density is

$$\overline{\mathcal{P}} = \frac{1}{2}\Re\left[\mathbf{a}_\theta \frac{j\tilde{I}_0\Delta\ell\beta}{4\pi R}\eta_0 e^{-j\beta R}\sin\theta \times \left(\mathbf{a}_\phi \frac{j\tilde{I}_0\Delta\ell\beta}{4\pi R}e^{-j\beta R}\sin\theta\right)^*\right].$$

Using $\mathbf{a}_\theta \times \mathbf{a}_\phi = \mathbf{a}_R$, and combining other terms, this Poynting vector is

$$\boxed{\overline{\mathcal{P}} = \mathbf{a}_R \frac{1}{2}\left(\frac{I_0\Delta\ell\beta}{4\pi R}\right)^2 \eta_0 \sin^2\theta.}$$
(11.13)

The power flow is therefore in the radial direction, as indicated by the unit vector $\mathbf{a}_R$, directly outward from the antenna. The maximum power density is radiated in the direction $\theta = \pi/2$, and no power is radiated in the direction of the poles, $\theta = 0$ or π. A three-dimensional polar plot of this power density distribution is shown in Fig. 11.5(b), and a surface plot in Fig. 11.6.

The *total power* radiated by the elemental antenna can be easily determined from the power density of Eq. (11.13). To determine the total power radiated P_r, we construct a spherical surface centered on the antenna, whose radius R is large compared to β^{-1}, and integrate the power density over this surface:

$$P_r = \oint_s \overline{\mathcal{P}} \cdot d\mathbf{s}.$$

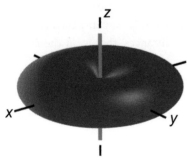

FIGURE 11.6 The power density distribution generated by a short linear antenna lying on the z-axis. The antenna is depicted by the red line, while the surface gives the power density as a function of direction.

The surface element $d\mathbf{s}$ is of magnitude $ds = R^2 \sin\theta \, d\theta \, d\phi$. The direction of $d\mathbf{s}$ is normal to the surface of the sphere, which is given by $\mathbf{a}_R$. The radiated power then becomes

$$P_r = \oint_s \mathbf{a}_R \frac{1}{2} \left(\frac{I_0 \Delta\ell\beta}{4\pi R} \right)^2 \eta_0 \sin^2\theta \cdot \mathbf{a}_R \, R^2 \sin\theta \, d\theta \, d\phi.$$

Notice that the factor $1/R^2$ in the power density perfectly cancels the factor of R^2 in $d\mathbf{s}$, leaving a result that is independent of the radius R of the sphere of integration. We should breathe a collective sigh of relief with this result, since this means that the total power transmitted through one surface, say of radius R_1, is identical to the power transmitted through a second surface of a different radius, say R_2, consistent with our expectation that energy is conserved. (The total power transmitted through these two surfaces would differ, of course, if the medium through which our wave propagates is absorbing. But we are only treating the problem of propagation through a vacuum, which does not absorb power from the wave.) Carrying out the integration over θ from 0 to π and the integration over ϕ from 0 to 2π, these integrals yield 4/3 and 2π, respectively, and the total radiated power is

$$\boxed{P_r = \frac{(I_0 \Delta\ell\beta)^2}{12\pi} \eta_0.} \tag{11.14}$$

This expression for P_r is the total power radiated by the short antenna. Let's look at this a bit closer. First, note that the radiated power grows as the current squared, I_0^2. This should not be surprising, since the electric field generated by an antenna is proportional to the current, as is the magnetic field. Since the power density $\overline{\mathcal{P}}$ depends on the product of $\tilde{\mathbf{E}}$ and $\tilde{\mathbf{H}}^*$, we expect the I_0^2 dependence quite directly. Note also that the radiated power includes the factor $(\beta\Delta\ell)^2$. This factor indicates that the power radiated by the short antenna must, by definition, be quite small. Remember that the short antenna is defined as one whose length $\Delta\ell$ is small compared to the wavelength λ of the radiation. Since the wave number β is $2\pi/\lambda$, the dimensionless product $\beta\Delta\ell$ is $2\pi\Delta\ell/\lambda$, which will always be much less than 1.

Example 11.1 Short Dipole Antenna

Consider a vertical antenna driven by a sinusoidal current of amplitude $I_0 = 10$ A at a frequency of 100 MHz. The antenna is of length 10 cm, and is surrounded by air. Determine (a) the electric field amplitude E_θ, and (b) the power density of this wave at a distance of 50 m and $\theta = \pi/2$, and (c) the total radiated power.

Solution:
Let's start by finding the wavelength of the radiated wave, $\lambda = c/f = 3$ m. Note that $\Delta\ell$ is indeed much less than λ, so this antenna satisfies the "shortness" condition.
(a) To find the electric field amplitude, we use Eq. (11.12):

$$|\tilde{E}_\theta(\theta = \pi/2)| = \frac{I_0\Delta\ell\beta}{4\pi R}\eta_0 = \frac{I_0\Delta\ell}{2\lambda R}\eta_0 = \frac{(10 \text{ A})(0.1 \text{ m})}{2(3 \text{ m})(50 \text{ m})}(120\pi \text{ }\Omega) = 1.26 \text{ V/m}.$$

(b) To find the power density, we have two options. Since we have already found $|\tilde{E}_\theta(\theta = \pi/2)|$, perhaps the simpler approach is to use

$$\overline{\mathcal{P}}(\theta = \pi/2) = \frac{|\tilde{E}_\theta(\theta = \pi/2)|^2}{2\eta_0} = \frac{(1.26 \text{ V/m})^2}{2(120\pi \text{ }\Omega)} = 0.0021 \text{ W/m}^2.$$

Alternatively, we can use Eq. (11.13),

$$\overline{\mathcal{P}}(\theta = \pi/2) = \frac{1}{2}\left(\frac{I_0\Delta\ell\beta}{4\pi R}\right)^2\eta_0 = \frac{1}{2}\left(\frac{I_0\Delta\ell}{2\lambda R}\right)^2\eta_0$$
$$= \frac{1}{2}\left(\frac{(10\text{ A})(0.1\text{ m})}{2(3\text{ m})(50\text{ m})}\right)^2(120\pi\ \Omega) = 0.0021\text{ W/m}^2.$$

Either approach gives the same result, of course.

(c) To find the total power radiated by this antenna, we use Eq. (11.14),

$$P_r = \frac{(I_0\Delta\ell\beta)^2}{12\pi}\eta_0 = \frac{(I_0\Delta\ell)^2}{3\lambda^2}\pi\eta_0 = \frac{((10\text{ A})(0.1\text{ m}))^2}{3(3\text{ m})^2}\pi(120\pi\ \Omega) = 43.9\text{ W}.$$

Let's return to the dependence of the radiated power on I_0^2. The harder the antenna is driven, the more power is radiated. This much makes perfect sense. But it would be useful to define a parameter that allows us to describe the power radiated by a system, one which includes all the factors other than I_0^2, which, as we will see later, is always a factor in the radiated power for any antenna. Just such a parameter is found in the **radiation resistance**, R_r, which is defined by comparing the power radiated by the antenna to the power absorbed by any circuit element when a sinusoidal current of amplitude I_0 passes through it. You should recall that when a sinusoidally varying current of amplitude I_0 passes through a circuit impedance element, the average power absorbed by that element is $P^{abs} = \frac{1}{2}I_0^2R$, where R is the resistive part of its impedance. While the final form of the energy here is not quite the same (the circuit element heats up, whereas the current in the antenna generates the outgoing electromagnetic wave), the analogy is still quite good. In either case, electrical power is supplied by some source, and the power depends on the current in the same way (P^{abs} proportional to I_0^2). Therefore, the radiation resistance of the antenna is defined through the radiated power,

$$P_r = \frac{1}{2}I_0^2R_r.$$

This can be turned around to write the radiation resistance in terms of the radiated power,

$$R_r = \frac{2P_r}{I_0^2}. \tag{11.15}$$

For the short dipole antenna, this radiation resistance becomes

$$R_r = \frac{2P_r}{I_0^2} = \frac{2}{I_0^2}\frac{(I_0\Delta\ell\beta)^2}{12\pi}\eta_0 = \frac{\eta_0}{6\pi}(\Delta\ell\beta)^2, \tag{11.16}$$

or, in terms of the wavelength,

$$\boxed{R_r = \frac{2\pi}{3}\eta_0\left(\frac{\Delta\ell}{\lambda}\right)^2.} \tag{11.17}$$

It is clear from this last form that the radiated power, expressed in this case through the radiation resistance, is quite small, since $\Delta\ell/\lambda$ is small for an elemental dipole antenna.

Of course, not all the power that is input to the antenna is radiated in the wave. We have to expect that there are resistive losses in the antenna as well. The antenna is, after all, simply a conducting wire with a current passing through it, and as a current-carrying conductor with finite conductivity, it will experience resistive losses. At high frequencies, the current is located primarily close to the surface of the conductor, falling off exponentially with increasing depth, as we have discussed in Chapter 8. We defined the skin depth,

$$\delta = \frac{1}{\sqrt{\pi f \mu \sigma}},$$

earlier (Eq. (8.74)) as the effective thickness at the surface of the conductor through which the current passes, and the surface resistance,

$$R_s = \sqrt{\frac{\pi f \mu_0}{\sigma}},$$

as originally defined in Eq. (8.105). As long as the radius a of the antenna wire is large compared to the skin depth δ, the effective resistance of the conductor is

$$R_L = R_s \frac{\Delta \ell}{2\pi a},$$

where $\Delta \ell$ is the length of the conductor, and $2\pi a$ is its circumference.

With increasing frequency f, the skin depth of the conductor becomes smaller (the current is confined to narrower regions near the surface of the conductor), the surface resistance R_s increases, and the resistive losses of the antenna also increase. These losses are related to the resistance R_L as

$$P_L = \frac{1}{2}I_0^2 R_L. \tag{11.18}$$

The **efficiency** η_r of the radiating antenna is defined as the ratio of the radiated power to the total power input to the antenna. That is,

$$\eta_r = \frac{P_r}{P_{in}} = \frac{P_r}{P_L + P_r}. \tag{11.19}$$

In terms of the resistances, this efficiency can be written as

$$\eta_r = \frac{R_r}{R_L + R_r}.$$

We would, of course, like this efficiency to be as close to 1 as possible. In Example 11.2 we will determine the efficiency of a short dipole antenna.

Example 11.2 Short Dipole Antenna

Consider an antenna surrounded by air consisting of a 1 mm diameter copper wire that is 10 cm in length and is driven at a frequency of 100 MHz. Determine the radiation resistance R_r, the loss resistance R_L, and the efficiency η_r of this antenna.

Solution:
The length of this antenna and the drive frequency are the same as those of Example 11.1, so we'll use the wavelength of the radiated wave, $\lambda = 3$ m, that we found in that example. Also, we confirmed in that example that this antenna is short, so we'll use Eq. (11.17) for the radiation resistance,

$$R_r = \frac{2\pi}{3}\eta_0 \left(\frac{\Delta \ell}{\lambda}\right)^2 = \frac{2\pi}{3}(120\pi \ \Omega)\left(\frac{0.1 \ \text{m}}{3 \ \text{m}}\right)^2 = 0.88 \ \Omega.$$

To find the loss resistance, we start with the surface resistance $R_s = \sqrt{\pi f \mu_0 / \sigma}$. The conductivity of copper is $\sigma = (5.8 \times 10^7$ S/m$)$, leading to $R_s = 2.6$ mΩ. Therefore the loss resistance is

$$R_L = R_s \frac{\Delta \ell}{2\pi a} = 2.6 \text{ m}\Omega \frac{0.1 \text{ m}}{2\pi (0.0005 \text{ m})} = 83 \text{ m}\Omega.$$

Using R_r and R_L in Eq. (11.19), the efficiency of this radiator is about 91%.

Before we finish with the short dipole antenna, let's go back to the power density given by Eq. (11.13). Specifically, we'd like to develop some tools for describing the **angular distribution** of the radiated power, or the **directionality** of the radiation. Imagine a highly focused beam of radiation that propagates from a source to some target. In terms of getting as much as possible of the total power produced by the source to the target, this would be highly effective. Conversely, if our goal is to broadcast a signal broadly to reach the maximum number of receivers that are widely distributed around the antenna, then this highly directional beam may not be a good choice. Either way, we'd like to be able to quantify the directionality of the radiated power. The power density, or Poynting vector, gives us the starting point for achieving this. Several different parameters are available for this purpose, and while they differ in detail, they all tell us pretty much the same thing. First, we can define the angular spread $\Delta \theta_{1/2}$ over which most of the power is directed. This is illustrated in Fig. 11.5(b), which shows the radiated power density as a function of the polar angle θ. The two solid lines originating at the origin intersect the power density curve at the half-power points. That is, the power density at that intersection is half the maximum power density, which, as we pointed out previously, occurs in the direction given by $\theta = \pi/2$. Since the power density given in Eq. (11.13) depends on $\sin^2 \theta$, which has a maximum value of 1, we must find the angles θ at which $\sin^2 \theta = 1/2$. The $\sin^{-1}(1/\sqrt{2})$ is $\theta = \pi/4$ and $\theta = 3\pi/4$. These angles define the two solid lines in Fig. 11.5(b), and their difference gives us $\Delta \theta_{1/2} = \pi/2$, or 90°.

The directionality of the radiated power can also be described through the **directivity** D, which is defined as the ratio of the power density $\overline{\mathcal{P}}$ (the scalar magnitude of the Poynting vector given in Eq. (11.13)), to the mean value of the power density $\langle \overline{\mathcal{P}} \rangle$. The angle brackets here denote the average value when averaged over all directions (θ, ϕ):

$$\langle \overline{\mathcal{P}} \rangle = \frac{\oint \overline{\mathcal{P}} \sin \theta \, d\theta \, d\phi}{\oint \sin \theta \, d\theta \, d\phi} = \frac{1}{4\pi} \oint \overline{\mathcal{P}} \sin \theta \, d\theta \, d\phi.$$

But the total radiated power P_r is also related to this mean value through

$$P_r = \oint \overline{\mathcal{P}} R^2 \sin \theta \, d\theta \, d\phi,$$

so $P_r = 4\pi R^2 \langle \overline{\mathcal{P}} \rangle$. Therefore, the directivity can be written as

$$D = \frac{\overline{\mathcal{P}}(\theta, \phi)}{\langle \overline{\mathcal{P}} \rangle} = \frac{4\pi R^2 \overline{\mathcal{P}}(\theta, \phi)}{P_r}.$$

If the direction (θ, ϕ) is not specified, the direction of the maximum power density is implied. The directivity really contains the same information about the angular distribution as is contained in the power density directly, but it is simply scaled differently. If the power density is integrated over all angles (θ, ϕ), the result is the total radiated power. If instead the directivity is integrated over all angles, the result is 1. This can be simpler to interpret if all we really want is information about the relative power density in various directions. For the short dipole antenna, Eqs. (11.13) and (11.14) are used to determine

$$D = \frac{4\pi R^2 \left[\frac{1}{2} \left(\frac{I\Delta\ell\beta}{4\pi R} \right)^2 \eta_0 \sin^2\theta \right]}{\frac{(I\Delta\ell\beta)^2}{12\pi}\eta_0}.$$

After canceling common factors, the directivity simplifies to

$$\boxed{D = \frac{3}{2}\sin^2\theta.}$$
(11.20)

The maximum directivity for the elemental dipole antenna is $D_0 = 1.5$.

Finally, the **radiation intensity** is defined as the power radiated *per unit solid angle*, given in W/steradian. The utility of this quantity can be seen by examining Eq. (11.13), which shows that the power density decreases with increasing distance as $1/R^2$. The radiation intensity, defined as

$$U(\theta, \phi) = R^2 \,\overline{\mathcal{P}}(\theta, \phi),$$

factors out this R dependence, which can make it a useful quantity when all that is really required is the angular distribution. For the short dipole antenna the radiation intensity is

$$\boxed{U(\theta, \phi) = \frac{1}{2} \left(\frac{I\Delta\ell\beta}{4\pi} \right)^2 \eta_0 \sin^2\theta.}$$
(11.21)

The maximum value of this radiation intensity, at $\theta = \pi/2$, is

$$U_{max} = \frac{1}{2} \left(\frac{I\Delta\ell\beta}{4\pi} \right)^2 \eta_0.$$

The free-electron laser (FEL) is described in TechNote 11.1. While an FEL physically is quite different from an antenna, it does share the common property that oscillating charges radiate an electromagnetic wave.

TechNote 11.1 Free Electron Laser

A free electron laser (FEL; Fig. 11.7) is a device in which a high-energy beam of electrons is forced into oscillatory motion, which then radiates an electromagnetic wave. Generation of an electromagnetic wave by an FEL is similar in some regards to wave generation by a physical antenna, as described in this section. The most common form of FEL employs a long array of permanent magnets whose orientation alternates, called an "undulator" or "wiggler." As the electrons navigate the alternating magnetic field of the wiggler, they begin to alternate back and forth in response. This oscillatory motion of the electrons generates an electromagnetic wave, emitted in the same direction as the velocity of the electrons. FEL operation was originally confined to the microwave region of the spectrum, but FELs today are capable of generating waves through the infrared, visible, ultraviolet, and X-ray regions. FELs find application in biology, surgery, X-ray crystallography, and defense.

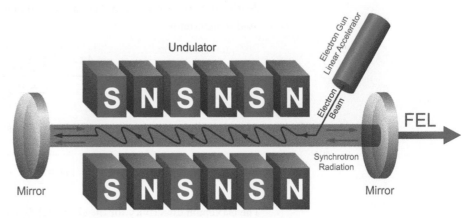

FIGURE 11.7 A schematic of a free electron laser (FEL). An electron beam passing through an undulator, or wiggler, picks up an oscillatory motion in the transverse direction. Just like the oscillatory current in a physical antenna, these electrons generate an electromagnetic wave at the frequency of oscillation.

Now that we have examined the elemental linear dipole antenna, we turn our attention in the next section to long antennas. The elemental antenna is a quick and easy radiating source, but not very effective. We will see that longer antennas, most notably the half-wave dipole antenna, can generate an outgoing wave of much higher power density. Still, this discussion of the short antenna has served several important purposes. First, it has laid the groundwork for the general principles on which a radiating dipole system is based. The physical picture that we have developed for the short antenna applies nicely to the longer antenna as well. Also, we will use the elemental dipole antenna as a building block in the analysis of this longer antenna.

11.2 Long Antennas

Now that we have laid the groundwork for generation of waves with the short dipole antenna, let's see if we can apply these results to understand the generation of waves by longer antennas; that is, antennas whose length is comparable to the wavelength λ of the generated wave. We will keep our discussion somewhat general, in that it will apply to a range of lengths, but we will also focus particular attention on the half-wave dipole antenna, which is of special interest due to its efficiency in generating waves, and its common usage. As we will see, the directivity of the half-wave dipole is $D = 1.64$, only slightly larger than that of the short dipole antenna, but its radiation resistance, which characterizes the total power radiated for a given input current, is $R_r \simeq 73 \ \Omega$.

Radio and microwave antennas, as needed for broadcasting and long-distance communications systems, are everywhere. A few well-known applications are described in TechNotes 11.2–11.4.

TechNote 11.2 Radio and Television Broadcast Antennas

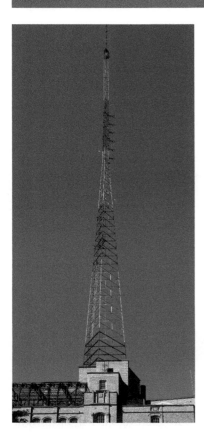

AM (amplitude modulated) radio (550 kHz to 1 720 MHz) transmission antennas are typically constructed vertically, as shown in Fig. 11.8(a), and therefore produce a vertically polarized wave. Vertically polarized waves are less susceptible to fading effects, caused by reflections from the Earth's surface, which can add destructively to the direct wave at the receiver antenna and cause weak signals. For best reception of a vertically polarized wave, the receiving antenna should also be vertical.

FM (frequency modulated) radio waves (87.5 MHz to 108.0 MHz in the United States) are usually of mixed polarization, consisting of components of both vertical and horizontal polarization. (This is often erroneously stated as circular polarization.) This polarization allows best penetration into buildings and other structures.

Analog television signals are often horizontally polarized, particularly in urban areas, in order to reduce interference from reflections from tall buildings. A common receiving antenna for a horizontally polarized wave is the Yagi-Uda antenna, shown in Fig. 11.8(b). The active element in this antenna is the second from the right. The other elements are passive, and help direct the wave to the active element.

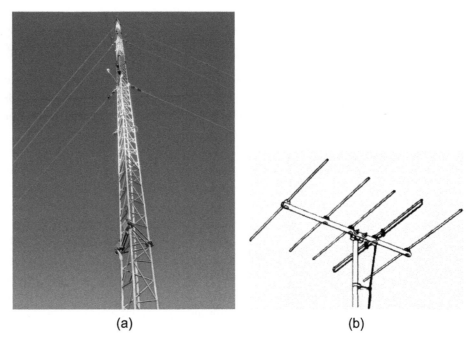

(a)   (b)

FIGURE 11.8 (a) An AM radio broadcast antenna for station KBRC, Mount Vernon, Washington (state). This antenna, with a length of ~60 m, is a quarter-wavelength long. (b) A Yagi-Uda antenna, from a 1954 advertisement by the Radiart Corp., Cleveland, Ohio.

Generation of very low frequency waves requires extremely large antenna systems, which are discussed in TechNote 11.3.

TechNote 11.3 Extremely Low Frequency Communications

Due to the high conductivity of seawater, the oceans strongly absorb electromagnetic waves in the usual communication frequency bands, making communication with submerged submarines impossible. The US Navy has developed a communication system in the 40–80 Hz range, part of the extremely low frequency (ELF) band. Dipole antennas are most efficient when the length of the antenna is half the wavelength of the radiation, so very long antennas are required, both for the transmitter and the receiver. For this purpose, a 222 km dipole antenna in Wisconsin is used. One feature of this system is that ELF and very low frequency (VLF) waves are reflected by the ionosphere and by Earth, so signals can be carried around the globe. Data rates on low-frequency communication bands are, of course, very limited.

Microwaves have long played an important role in communication systems. Microwave relay systems are described in TechNote 11.4.

TechNote 11.4 Microwave Relay Stations

The first demonstration of a microwave communication over a significant distance was across the English Channel between the UK and France in 1931. Development of radar systems during the Second World War led to improvements in microwave technology, which made large networks more practical, and in the years following the war, a network across Europe and another across the United States were initiated for long-distance transmission of telephone and television signals. The US network was constructed by AT&T, and the first telephone call was carried by this network in 1951. Thus, signals were carried by free-space line-of-sight microwave transmission, rather than conducting wires. The distance between adjacent towers in this network was typically around 50 km, or 30 miles. Microwave transmissions are very directional, since the wavelength of microwaves is in the range 0.03–30 cm, which is smaller than the dimensions of the reflective elements on the antennas used to form beams of microwaves. The high frequency of microwaves also allows larger bandwidth for carrying information. The AT&T towers were used until 1999, when newer technologies, such as fiber optic networks and satellite communication systems (starting in the 1960s, and also using microwaves), were used to replace the free-space microwaves.

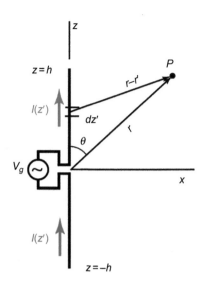

FIGURE 11.9 A long, linear, center-fed antenna. The antenna lies along the z-axis, extending from $z = -h$ to $z = +h$. The sinusoidally varying (in time) current, represented by the phasor $\tilde{I}(z')$, varies along the length of the antenna.

A model of a long dipole antenna is shown in Fig. 11.9. The antenna lies along the z-axis, with an overall length of $2h$. The source provides the current at the center of the antenna, which is labeled $z' = 0$. The current $\tilde{I}(z')$ is symmetric about $z' = 0$, that is $\tilde{I}(-z') = \tilde{I}(z')$, since the current into the source in one half of the antenna is identical to the current out of the source into the other half of the antenna.

To solve for the radiation pattern correctly, we need to know the distribution of current in the dipole antenna. Recall that for the short dipole we asserted that the current, which was sinusoidally oscillating as a function of time t, was uniform along the length of the antenna. That was reasonable, in that the antenna was short, $\Delta\ell \ll \lambda$, and little variation over the length is expected. For long antennas, however, we must expect that the current varies over the length of the antenna, and a proper solution of the antenna radiation pattern must satisfy the known boundary conditions between the fields just outside the antenna and the currents and charges within the antenna. This problem is well beyond the scope of this textbook, and we will settle for an approximate solution (that is actually relatively close to the more rigorous solution). That is, we will presume that the current $\tilde{I}(z')$:

1. varies sinusoidally (with wavelength λ) along the length of the antenna;
2. equals zero at the ends of the antenna; and
3. is symmetric in z', as already noted above.

The functional form of this current is

$$\tilde{I}(z') = I_0 \sin\left(\beta\left[h - z'\right]\right)$$

for $z' > 0$, and

$$\tilde{I}(z') = I_0 \sin\left(\beta\left[h + z'\right]\right)$$

for $z' < 0$. See Fig. 11.10 for a plot of this current, for which the total length is $2h = 0.75\lambda$. Note that this current goes to 0 at $z' = \pm h$, and that it is indeed symmetric in z', since if we replace $z' \to -z'$, the current is unchanged. This current can be written more compactly as

$$\tilde{I}(z') = I_0 \sin\left(\beta\left[h - |z'|\right]\right). \tag{11.22}$$

Now let's apply the result that we derived for the short dipole antenna to find the field generated by this long dipole at point P specified by the position vector **r**. Treating only the far-field region, and using Eq. (11.12), the contribution $d\tilde{E}_\theta$ to the total electric field by an elemental segment of the antenna at $z = z'$ with current $\tilde{I}(z')$ and length dz' is

FIGURE 11.10 The current $\tilde{I}(z')$ on a long, linear, center-fed antenna, as given in Eq. (11.22). In this plot, the total length of the antenna is $2h = 0.75\lambda$.

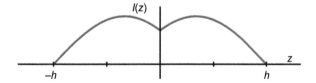

$$dÉ_\theta = \frac{j\tilde{I}(z')dz'\beta}{4\pi|\mathbf{r}-\mathbf{r}'|}\eta_0 e^{-j\beta|\mathbf{r}-\mathbf{r}'|}\sin\theta. \tag{11.23}$$

Consistent with the notation used throughout this text, $\mathbf{r}$ points to the field point (the position at which we are determining the field), $\mathbf{r}' = \mathbf{a}_z z'$ is the source location, and $\mathbf{r} - \mathbf{r}'$ points from the source to the field point. The distance $|\mathbf{r} - \mathbf{r}'|$ is

$$|\mathbf{r}-\mathbf{r}'| = \sqrt{R^2 - 2Rz'\cos\theta + z'^2},$$

which in the far-field region $(R \gg z')$ is approximately $R - z'\cos\theta$. Similarly,

$$d\tilde{H}_\phi = \frac{1}{\eta_0}dÉ_\theta \tag{11.24}$$

is the elemental contribution to the magnetic field due to this same segment of the antenna. To find the total fields $É_\theta$ and $\tilde{H}_\phi$ at $\mathbf{r}$, we must simply integrate Eqs. (11.23) or (11.24) over the length of the antenna:

$$É_\theta = \frac{jI_0}{4\pi}\beta\eta_0 e^{-j\beta R}\sin\theta \int_{-h}^{h} \frac{\sin(\beta[h-|z'|])}{R-z'\cos\theta} e^{j\beta z'\cos\theta}\,dz'.$$

In the integrand, the z' dependence of the denominator is relatively unimportant. Remember, we are calculating the field in the region far from the antenna itself, so $R \gg z'$. So we approximate this integrand by replacing $R - z'\cos\theta \to R$ in the denominator. It is *not* valid, however, to make this same substitution in the phase term of the complex exponential function. Here, a variation in z' of one wavelength λ makes a huge difference. With this simplification in the denominator, the integral becomes tractable, and it can be shown that

$$\boxed{É_\theta = \frac{j\eta_0 I_0}{2\pi R}e^{-j\beta R}F(\theta),} \tag{11.25}$$

where

$$F(\theta) = \frac{\cos(\beta h\cos\theta) - \cos(\beta h)}{\sin\theta}. \tag{11.26}$$

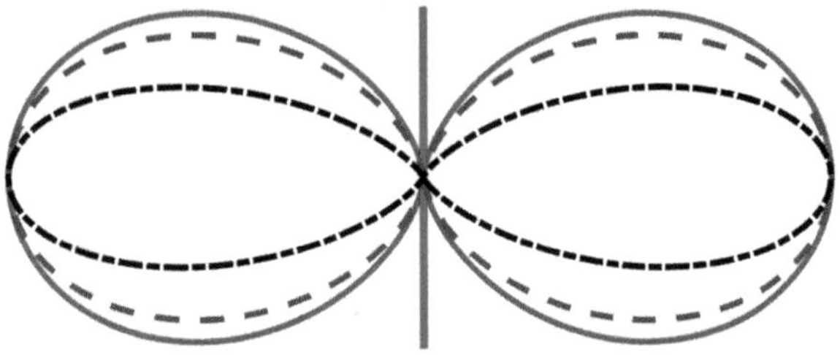

FIGURE 11.11 Comparison of the angular distributions of the power density for the short dipole antenna (blue solid), the half-wave dipole (magenta dashed), and the full-wave dipole (black dot-dashed) antenna. Each distribution is normalized to its value at $\theta = \pi/2$.

In the far-field region, $\tilde{H}_\phi = É_\theta/\eta_0$. This radiation pattern is independent of the azimuthal angle ϕ, which we expected since the antenna lies along the z-axis, and the antenna geometry is perfectly symmetric about this axis. The dependence of E_θ on R is similar to that (in the far-field region) of the short antenna. That is, the magnitude drops off as $1/R$, and the phase term is $e^{-j\beta R}$. No surprises here. Now let's look at the dependence of this field on the polar angle θ. This dependence is contained entirely within the function $F(\theta)$, which admittedly is not very transparent. We really need to plot this out for specific antenna lengths $2h$ in order to visualize it. Polar plots of $F^2(\theta)$ for three examples are shown in Fig. 11.11.

The magenta dashed plot in Fig. 11.11 shows $F^2(\theta)$ for the **half-wave dipole** antenna. The total length $2h$ of this antenna is equal to $\lambda/2$; hence its name. Since $\beta = 2\pi/\lambda$, βh for this antenna is $\pi/2$, and the function $F(\theta)$ simplifies to the form

$$F(\theta) = \frac{\cos\left(\frac{\pi}{2}\cos\theta\right)}{\sin\theta}. \tag{11.27}$$

This function is zero in the direction of the poles ($\theta = 0$ or π) and has a value of 1 at the equator ($\theta = \pi/2$). In fact, this distribution is similar to that of the short dipole, only a little narrower. We'll calculate the power density before we compare these distributions in detail.

The Poynting vector for the half-wave dipole is

$$\overline{\mathcal{P}} = \frac{1}{2}\Re\left[\tilde{\mathbf{E}}\times\tilde{\mathbf{H}}^*\right] = \frac{|\tilde{E}_\theta|^2}{2\eta_0}\mathbf{a}_R. \tag{11.28}$$

Using $\tilde{E}_\theta$ from Eq. (11.25) and $F(\theta)$ from Eq. (11.27), this is

$$\overline{\mathcal{P}} = \frac{\eta_0}{2}\left(\frac{I_0}{2\pi R}\right)^2\frac{\cos^2\left(\frac{\pi}{2}\cos\theta\right)}{\sin^2\theta}\mathbf{a}_R. \tag{11.29}$$

For comparison, the angular distribution for the short dipole antenna is also shown, as the blue solid curve, in Fig. 11.11. Notice that the power density for the half-wave dipole antenna is narrower than that of the short antenna, but only slightly. We will examine the directivity and the angular spread $\Delta\theta_{1/2}$ shortly, and see that these parameters and functions lead us to the same conclusion.

The final plot shown in Fig. 11.11 is the black dot-dashed line, which represents $F^2(\theta)$ for the full-wave dipole antenna. The full length of this antenna is λ. Therefore, βh is π, and the function $F(\theta)$ is

$$F(\theta) = \frac{\cos(\pi\cos\theta) - 1}{\sin\theta}. \tag{11.30}$$

Notice how narrow, relative to the other examples in this figure, the angular distribution of power is for the full-wave antenna.

While the angular distribution of the power densities for the half-wave and short dipole antennas differ only slightly, these power densities differ strongly in another regard; that is, in the *magnitude* of the radiated power densities. To quantify this, let us examine the total power radiated by the half-wave antenna, and compare this result to the total power radiated by the short antenna, given in Eq. (11.14). The total radiated power can be found by integrating the power density across the surface of a sphere whose radius R is large, much greater than the wavelength,

$$P_r = \oint_s \overline{\mathcal{P}}\cdot d\mathbf{s},$$

in the style precisely the same as we followed for the short dipole. The surface element $d\mathbf{s}$ is $\mathbf{a}_R R^2\sin\theta\, d\theta\, d\phi$, and the radiated power then becomes

$$P_r = \oint_s \frac{\eta_0}{2}\left(\frac{I_0}{2\pi R}\right)^2\frac{\cos^2\left(\frac{\pi}{2}\cos\theta\right)}{\sin^2\theta}\mathbf{a}_R\cdot\mathbf{a}_R\, R^2\sin\theta\, d\theta\, d\phi.$$

Canceling the R^2 factors, and carrying out the integration in ϕ, this is

$$P_r = \frac{\eta_0 I_0^2}{4\pi}\int_0^\pi\frac{\cos^2\left(\frac{\pi}{2}\cos\theta\right)}{\sin\theta}\, d\theta.$$

The remaining integral must be evaluated numerically, and has a value of 1.218. Thus the total power radiated is

$$P_r = 1.218 \left(\frac{\eta_0}{4\pi}\right) I_0^2 = (36.5 \ \Omega) I_0^2. \tag{11.31}$$

Recall that we defined the radiation resistance of an antenna system in the discussion of the short dipole antenna. That is, the power radiated in terms of the current in the antenna is

$$P_r = \frac{1}{2} I_{in}^2 R_r. \tag{11.32}$$

Note that we're being a little more careful in the definition of radiation resistance here. For the short antenna, we presumed the current to be uniform along the length of the antenna. For the long antenna, however, the current varies along the length. In thinking of the antenna as an electrical element, the manner in which we define the "input" resistance should be in terms of the input current, that is I_{in}. This is reflected in the way in which the radiation resistance is defined in Eq. (11.32). The input current is just $I(z = 0)$, where we defined the antenna current in Eq. (11.22):

$$I_{in} = I(z = 0) = I_0 \sin\left(\beta \left[h - |z|\right]\right)_{z=0} = I_0 \sin\left(\beta h\right).$$

For the half-wave dipole, $\beta h = \pi/2$, and $I_{in} = I_0$, so

$$R_r = \frac{2P_r}{I_{in}^2} \approx 73 \ \Omega. \tag{11.33}$$

It can be shown, with more advanced treatment than we pursue here, that the input impedance of a half-wave dipole antenna has a relatively large reactive component to it as well as the resistive component that we derived here. This reactive component is related to energy that is stored in the electric and magnetic fields in the near-field region, but which do not lead to radiated power in the far-field region. These more precise analyses of long antennas show that, for a slightly shorter antenna of length $\sim 0.485\lambda$, this reactive component vanishes, leaving only the real radiation resistance, whose magnitude is still approximately the same value, $R_r = 73 \ \Omega$. In practice, then, half-wave antennas are typically a bit shorter than half a wavelength long.

In our discussion of the short antenna we defined several useful functions and parameters that help us quantify the angular distribution of the radiated power. These include the radiation intensity $U(\theta, \phi)$, the directivity D, and the angular width $\Delta\theta_{1/2}$. Let us examine these terms for the case of the half-wave antenna.

The radiation intensity $U(\theta, \phi)$ was defined earlier as the power radiated into a solid angle of one steradian. At a distance R from the emitting antenna the radiation intensity is related to the power density through $U(\theta, \phi) = R^2|\overline{\mathcal{P}}|$. For the half-wave antenna, the radiation intensity is

$$U(\theta, \phi) = \frac{\eta_0 I_0^2}{8\pi^2} \frac{\cos^2\left(\frac{\pi}{2} \cos\theta\right)}{\sin^2\theta}. \tag{11.34}$$

The maximum value of this emission is at an angle $\theta = \pi/2$, in which direction

$$U_{max} = \frac{\eta_0 I_0^2}{8\pi^2}.$$

Comparing this to the radiation intensity for the short antenna, Eq. (11.21), shows that the radiation intensity for the half-wave antenna is larger by a factor of $(2/\beta\Delta\ell)^2$, primarily due to the much larger total power radiated by the former.

The directivity for the half-wave dipole antenna is

$$D = \frac{U(\theta,\phi)}{U_{av}} = \frac{U(\theta,\phi)}{P_r/4\pi} = \frac{\left(\eta_0 I_0^2/8\pi^2\right)\left(\cos^2\left(\frac{\pi}{2}\cos\theta\right)/\sin^2\theta\right)}{1.218\,\eta_0 I_0^2/(4\pi)^2}.$$

Simplifying, this becomes

$$D = 1.64\frac{\cos^2\left(\frac{\pi}{2}\cos\theta\right)}{\sin^2\theta}. \tag{11.35}$$

The maximum directivity is $D_0 = 1.64$. Compare D and D_0 of the half-wave dipole to the same quantities for the short antenna, which are $D = 1.5\sin^2\theta$ and $D_0 = 1.5$. These parameters indicate that the half-wave antenna is more directional, but only slightly, than the short dipole antenna.

Finally, the angular width $\Delta\theta_{1/2}$ is defined as the full width in θ between the directions at which the power density is reduced to $1/2$ its maximum value. These angles θ are found by setting

$$\frac{\cos^2\left(\frac{\pi}{2}\cos\theta\right)}{\sin^2\theta} = \frac{1}{2},$$

and solving for θ. This must be done numerically, as it is not solvable analytically, yielding $\Delta\theta_{1/2} \approx 78°$. Recall that for the short antenna, $\Delta\theta_{1/2}$ is $\pi/2$ or $90°$. Thus we have one more measure that the radiation pattern emitted by the half-wave dipole antenna is slightly more directional than that of the short antenna.

Before completing this discussion of the half-wave dipole antenna, we mention a very similar system. That is, the $\lambda/4$ monopole antenna, illustrated in Fig. 11.12. As its name implies, the length of this antenna is one-quarter of the wavelength of the emitted radiation. The current drives it from one end of the antenna, and it is positioned perpendicular to and just above a grounded plane, such as the surface of the Earth. Its operation can be understood by recalling our earlier discussion of image charges. At that time, of course, we were not concerned about currents, only charges, so we need to extend that conversation just a bit. Recall that a charge suspended above a grounded plane surface will induce charges in the grounded surface, and the potential and electric field above the grounded plane are equivalent to that generated by the much simpler distribution of the charge and its image charge. The image charge is of opposite sign to that of the original charge, and it is located below the ground plane a distance equal to that of the original charge above the ground plane. Now considering that a current is simply charges in motion, a *positive* current in the quarter-wave antenna above the ground plane induces a *positive* image current below the ground plane. (Remember that the image current is a negative charge moving downward, which is equivalent to a positive charge moving upward.) So with this image current, the radiated fields for the quarter-wave monopole antenna are completely equivalent to those of the half-wave dipole antenna; at least they are for any location above the ground plane. The fields are essentially zero for any position below the ground plane. Invoking this equivalence,

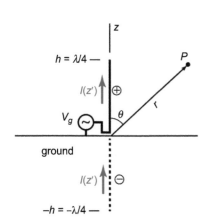

FIGURE 11.12 A quarter-wave monopole antenna.

there is no need to recalculate field patterns, power densities, or any of the other parameters that we discussed above. Instead, we can immediately state that the power density for the quarter-wave monopole antenna in the far-field region for any point above the ground plane ($0 \leq \theta < \pi/2$) is

$$\overline{\mathcal{P}} = \frac{\eta_0}{2} \left(\frac{I_0}{2\pi R} \right)^2 \frac{\cos^2 \left(\frac{\pi}{2} \cos \theta \right)}{\sin^2 \theta} \mathbf{a}_R.$$

We can also state that the average power radiated by this antenna is precisely half the power radiated by the half-wave dipole, and that the radiation resistance is similarly one-half of the radiation resistance of the half-wave dipole antenna. These latter results follow since the power radiated by the quarter-wave monopole is only radiated into the upper half-plane, and so the radiated power is only half as much.

We close this section with a brief description of antennas that are even longer than the half-wave dipole antenna. We already mentioned a full-wave dipole antenna whose total length is $2h = \lambda$, for which the power density is plotted as the black dot-dashed line in Fig. 11.11. As the length of the antenna increases even further, the central lobe of the power density becomes more and more directional, and the pattern develops side lobes. This is shown in Fig. 11.13, in which $F^2(\theta)$ for six different lengths of dipole antennas are compared. The qualitative features of each of these radiation patterns can be understood in terms of the interference between the field contributions from different segments of the current $I(z')$ in the antenna. For example, let's examine the power density at the equator ($\theta = \pi/2$) for three of these antennas, with lengths $2h = \lambda$, $2h = 2\lambda$, and $2h = 3\lambda$. The current $\tilde{I}(z')$ is plotted for each of these antennas in Fig. 11.14. In the $\theta = \pi/2$ plane, the distance between the observation point and any segment of the wire is essentially uniform. (Remember, this distance shows up in the phase term, which is $e^{-j\beta z' \cos \theta}$, which in this direction is $e^{-j0} = 1$.) For the full-wave antenna, the current $\tilde{I}(z')$ is positive for $z' > 0$ and positive for $z' < 0$, and these positive currents generate field contributions in the $\theta = \pi/2$ which add constructively. This leads to a maximum in the power density in this direction. For the antenna whose length is 2λ, however, the current is positive over half the antenna and negative over the other half. The contributions to the electric field from the positive current and those from the negative current perfectly cancel one another in the $\theta = \pi/2$ direction, and the power density in this direction is zero. For the 3λ antenna, there is again more positive current than negative in the antenna, and power is emitted in the $\theta = \pi/2$ direction.

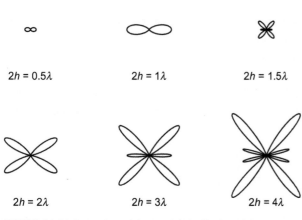

FIGURE 11.13 Comparison of the angular distributions of the power density for long dipole antennas of various lengths.

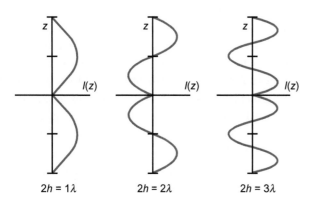

FIGURE 11.14 Comparison of the current distributions $\tilde{I}(z')$ for center-fed antennas of varying lengths: $2h = \lambda$, $2h = 2\lambda$, and $2h = 3\lambda$.

In other directions, $\theta \neq \pi/2$, a similar analysis can help us predict the positions of the lobes in the distributions shown in Fig. 11.13, but the details are considerably more difficult.

Cell phones contain several antennas to facilitate their connections with cell towers, Wi-Fi systems, and near-field signals. In TechNote 11.5 we describe some of the features of these systems.

TechNote 11.5 Smart Phone Antennas

Smart phones have changed our lives in many ways. They keep us in touch with friends, family, and business associates no matter where we are, they help us navigate in strange places, and they give us virtually unlimited internet access. To carry out all these functions, cell phones have a number of different antennas for transmitting and receiving a variety of signals. Cell phones connect with the service provider network through cell towers, such as the one shown in Fig. 11.15(a). Service regions, or cells, within the network are typically hexagonal in shape, so the cell towers typically have three sides to them to give the best coverage. As the user moves around, the call will switch to a different cell tower when that new signal is stronger. The signal from your phone can be picked up from as far away as 70 km (45 miles). The spacing between cell towers must be much less than this in towns and cities, however, as the capacity of a single tower would not be adequate for the number of calls needed at a time. Also, cell phone towers carry only a single service provider, leading to an increase in the number of towers you might see in an area.

There can be as many as five different antennas within a smart phone, each performing a different function. Since the primary function of a telephone might be regarded to be the ability to make and receive telephone calls, and in recent times, to send texts, we'll start with the *primary cellular antenna*. This antenna is the primary antenna for transmitting signals to and receiving signals from the cell tower. Often this antenna must

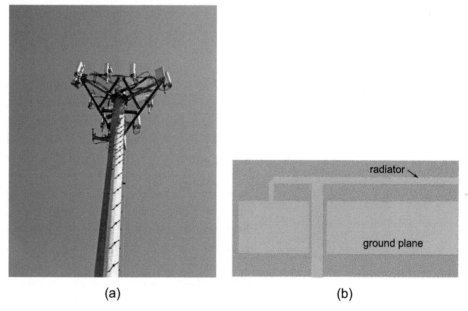

(a)   (b)

FIGURE 11.15 (a) A cell phone tower. (b) A printed inverted F antenna (PIFA), a common cellular antenna inside a typical cell phone.

be capable of operating in multiple frequency bands, as there are two primary bands over which cell phone systems operate: the 1900 band, covering 1 850–1 990 MHz; and the PCS band at 824–894 MHz. The frequency of the transmitted signal differs from that of the received signal, allowing both parties to talk at the same time. This is known as a duplex system, which differs from walkie-talkies or CB radio, examples of singlet systems, which allow the user to either talk or listen, but not both, at a given time. The primary cellular antenna in your cell phone is often a printed inverted F antenna (PIFA), examples of which are shown in Fig. 11.15(b). The signal is input from the right on the horizontal trace just below the center. The rectangular patches are grounded planes. The oscillating potential between the vertical trace and the ground plane produces an electric field pointing from one to the other, which propagates outward and forms the radiating wave. The name "inverted F antenna" is derived from the shape of the trace in Fig. 11.15(b), ignoring the ground plane, which resembles the letter F. When this antenna is mounted horizontally in the phone, the radiation pattern generated by this antenna is a doughnut-shaped mode, symmetric about a vertical axis. This pattern is well suited for transmission and reception in all directions close to ground level, including the direction of the nearest cell tower. The long dimension of the IFA antenna is $\sim \lambda/4$ in length, which at 800 MHz is 9.4 cm. This primary antenna is usually located at the bottom region of the cell phone, away from the head of the user, to reduce the absorption of the signal. Many cell phones can transmit signals at one of two strengths: either 0.6 W or 3 W. Transmitting low power is normally sufficient in urban settings, where cell phone towers are more closely spaced. Low-power transmission conserves battery energy, of course, and decreases the possibility of interference with signals transmitted in nearby cells.

A second antenna in a cell phone is the *diversity cellular antenna*. This antenna is for reception only, and helps monitor the signal power on other frequency bands to optimize the performance of the primary cellular signal.

A third antenna is dedicated to the *GPS* system. The frequency of the signal from the GPS satellites is 1.575 GHz. This antenna is receive-only, as no transmission back to the GPS satellites is needed.

The *Wi-Fi antenna* must be capable of transmitting and receiving information. There are primarily two frequency bands used for Wi-Fi: 2 400–2 484 MHz, and 5 150–5 850 MHz. Bluetooth connections operate in the low-frequency band.

Finally, most cell phones include a Near Field Communications (NFC) antenna. This system can transmit and receive signals. The frequency of the NFC is $f = 13.56$ MHz, which is very low compared to the other antennas we have discussed. In fact, this frequency is so low that you may see some describe this as not an antenna at all, but rather an inductor that can couple magnetically to a second external inductor. What is the distinction? At the frequency of 13.56 MHz, the wavelength of the wave is $\lambda = 22$ m. Clearly it is not practical to include a half-wave dipole antenna, or any other efficient antenna system in a cell phone whose dimensions are of order 10 cm. As a long-range radiator, these antennas are very inefficient. We discussed this in connection with the short linear antenna. We didn't discuss much about the near-field pattern for a short antenna, or for a loop antenna, but the field amplitudes formed by these simple antennas drop off rapidly as a function of distance. In the near field, however, for distances $R \ll \lambda$, they can couple to a nearby inductor without forming a long-range outgoing wave, and this is the basis for the NFC antenna.

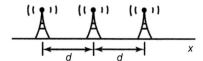

FIGURE 11.16 A linear array of $N = 3$ dipole antennas. Each antenna is driven by a current of amplitude I_0, with the phase difference between neighboring antenna equal to ξ. The spacing between antenna is given by d.

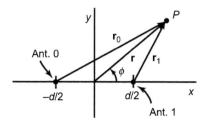

FIGURE 11.17 A top view of a linear array of two dipole antennas. The position vectors $\mathbf{r}$, $\mathbf{r}_0 = \mathbf{r} + \mathbf{a}_x d/2$, and $\mathbf{r}_1 = \mathbf{r} - \mathbf{a}_x d/2$ point from the origin, from antenna 0, and from antenna 1, respectively, to the field point P. The angle ϕ measures the azimuthal angle of vector $\mathbf{r}$ from the x-axis.

11.3 Antenna Arrays

In this section we examine the field patterns generated by an **antenna array**. We will see here that we can make a much more directional output wave, and that we can control the direction of the outgoing wave through careful choice of several parameters that influence the interference between the outgoing waves.

An array of identical antennas is shown in Fig. 11.16. In this figure, only three antennas are shown, but we can imagine extending this to any integer N. The current in each antenna flows in the z-direction, consistent with our treatment of the single antennas, and we choose to position the individual antennas along the x-axis, side by side, with uniform spacing d between adjacent antennas. We will let the magnitude I_0 of the sinusoidally varying current in each antenna be the same, but allow for variation of the relative phase of the currents. Furthermore, we will consider only the far-field solution, in which the distance to the field point (the point at which we observe the field) is large compared to the wavelength λ of the wave as well as the spacing d between the individual antennas.

We start by considering an array of just two antennas. This will allow us to introduce the essential ideas of the pattern generated by a larger array. A top view of this geometry, with antenna "0" at $x = -d/2$, and antenna "1" at $x = +d/2$, is shown in Fig. 11.17. The field generated by an array is simply the superposition of the fields radiated by the individual antennas in the array.

Each individual antenna generates a field pattern described by the formalism that we introduced in the last section, given in Eq. (11.25); that is,

$$\tilde{E}_0 = \frac{j\eta_0 I_0}{2\pi R_0} e^{-j\beta R_0} F(\theta)$$

and

$$\tilde{E}_1 = \frac{j\eta_0 I_0 e^{j\xi}}{2\pi R_1} e^{-j\beta R_1} F(\theta).$$

Note the phase term $e^{j\xi}$ in the expression for $\tilde{E}_1$, where ξ represents the phase of the current in antenna 1 relative to that in antenna 0. In these expressions, R_0 is the distance from antenna 0 to the field point P, and R_1 is the distance from antenna 1 to the field point. These distances R_0 and R_1 are most easily written in terms of the vector $\mathbf{r}$, which points from the origin to the point P, and the vector $\pm d\mathbf{a}_x/2$, which points from the origin to antennas 1 or 0,

$$R_0 = \left| \mathbf{r} + \frac{d}{2}\mathbf{a}_x \right|$$

and

$$R_1 = \left| \mathbf{r} - \frac{d}{2}\mathbf{a}_x \right|.$$

The distance R_0 is

$$R_0 = \left[\left(\mathbf{r} + \frac{d}{2}\mathbf{a}_x \right) \cdot \left(\mathbf{r} + \frac{d}{2}\mathbf{a}_x \right) \right]^{1/2},$$

which expands to

$$R_0 = \left[|\mathbf{r}|^2 + \mathbf{r} \cdot d\mathbf{a}_x + \frac{d^2}{4} \right]^{1/2}.$$

In terms of the spherical coordinates of the point P (R, θ, and ϕ), $|\mathbf{r}|$ is equal to R, and $\mathbf{r} \cdot \mathbf{a}_x$ is equal to $R \sin\theta \cos\phi$. Furthermore, the final term $d^2/4$ is negligible in comparison to the other terms for $R \gg d$, and it can be safely dropped. The distance R_0 is then

$$R_0 \simeq \left[R^2 + Rd \sin\theta \cos\phi \right]^{1/2}.$$

Finally, we pull out a factor of R^2 and use the Taylor expansion of Eq. (H.6), $\sqrt{1+\epsilon} \simeq 1 + \epsilon/2$, valid when $\epsilon \ll 1$, to write

$$R_0 = R \left[1 + \frac{d}{R} \sin\theta \cos\phi \right]^{1/2} \simeq R + \frac{d}{2} \sin\theta \cos\phi.$$

Using similar arguments, the distance R_1 can be written as

$$R_1 \simeq R - \frac{d}{2} \sin\theta \cos\phi.$$

Now we are ready to return to the superposition field,

$$\tilde{E}_T = \tilde{E}_0 + \tilde{E}_1 = \frac{j\eta_0 I_0}{2\pi} \left[\frac{e^{-j\beta R_0}}{R_0} + \frac{e^{j\xi} e^{-j\beta R_1}}{R_1} \right] F(\theta).$$

The distances R_0 and R_1 appear in this equation in two places: in the complex exponentials and in the denominators. Since the complex exponential functions vary sinusoidally with wavelength $\lambda = 2\pi/\beta$, and the distance d between the two antennas is comparable to the wavelength λ, the distinction between R_0 and R_1 here is important, and we must retain these distinct distances. In the denominators, however, in which the field strength at the distant location P is decreasing as $1/R$, the distinction between $1/R_0$ and $/R_1$ is insignificant, and R_0 and R_1 in the denominators can be replaced with simply R, the average distance. Pulling out the common factors in $\tilde{E}_T$, the superposition field is

$$\tilde{E}_T = \frac{j\eta_0 I_0}{2\pi R} e^{-j\beta R} \left[e^{-j\beta d \sin\theta \cos\phi/2} + e^{j\xi} e^{j\beta d \sin\theta \cos\phi/2} \right] F(\theta).$$

To simplify the notation, we define a phase ψ, which is the phase difference between the two field components at the distant $(R \gg d)$ position P,

$$\psi = \beta d \sin\theta \cos\phi + \xi. \tag{11.36}$$

As a reminder, the angles θ and ϕ are the polar and azimuthal angles of the position P. In terms of this phase difference ψ, the total field becomes

$$\tilde{E}_T = \frac{j\eta_0 I_0}{2\pi R} e^{-j\beta R} e^{j\xi/2} \left[e^{-j\psi/2} + e^{j\psi/2} \right] F(\theta).$$

Using Euler's identities, the term in the square brackets is just $2\cos(\psi/2)$, and the magnitude of the far field (omitting the phase) is

$$|\tilde{E}_T| = \frac{\eta_0 I_0}{\pi R} \cos(\psi/2) F(\theta).$$

$|\tilde{E}_T|^2$ vs. θ and ϕ are plotted for three examples of two-element antenna arrays in Fig. 11.18(a–c). For clarity, these are shown as two-dimensional plots in the $\theta = \pi/2$ plane in Fig. 11.18(d–f). Notice that, as expected, the power density is no longer symmetric upon rotation about the z-axis. (We expected this, or at least we should have, since there is a pair of antennas located at different positions along the x-axis, so the geometry of the antennas themselves is no longer symmetric about the z-axis.)

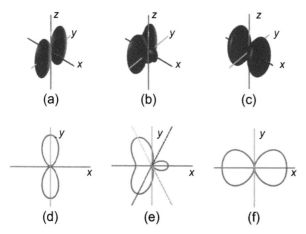

FIGURE 11.18 Comparison of the angular distributions of the power density for a two-element array of half-wave dipole antennas. The spacing d between the elements is $\lambda/2$, and the phase difference of the two currents is (a) $\xi = 0$, (b) $\xi = \pi/2$, and (c) $\xi = \pi$. The two-dimensional plots in (d–f) show these same power density distributions, restricted to the x–y plane for clarity.

Let's look at these distributions to see if we can't understand their features qualitatively. For all three cases, the spacing between the elements is $d = \lambda/2$. In Fig. 11.18(a) and (d), the currents in the two antenna elements are in phase with one another. So for any position along the $\pm x$-axis, the field contributions $\tilde{E}_0$ and $\tilde{E}_1$ are out of phase with one another due to their spacing. Along the $+x$-axis the wave from antenna 0 must travel half a wavelength more than the wave from antenna 1, resulting in a phase difference of $\beta d = \pi$. Similarly, along the $-x$-axis the wave from antenna 0 travels half a wavelength less, resulting in a phase difference of $-\pi$. Since the waves add out of phase in these two directions, and the amplitudes are equal, the total field is zero. For any point along the y-axis, however, the distances from the two antennas to that point are the same, and the fields add in phase with one another. Therefore, a maximum is observed in the power density in this direction.

On the other hand, when the phase difference between the currents is $\xi = \pi$, as in Fig. 11.18(c) and (f), we reach just the opposite conclusion. That is, along the $\pm x$-axis, the waves $\tilde{E}_0$ and $\tilde{E}_1$ are now in phase with one another and they add constructively. In these directions, there is a $\pm\pi$ phase shift due to the different distance traveled from one antenna or the other, as we described in the previous paragraph. In addition, however, there is an additional $\xi = \pi$ phase shift introduced by the currents, so the total phase shift in these directions is zero along the $+x$-axis and 2π along the $-x$-axis. Either way, the fields $\tilde{E}_0$ and $\tilde{E}_1$ add in phase with one another, and a maximum is observed in the power density in these directions. For a point along the y-axis, however, the distances from one antenna or the other to point P are the same, so the net phase difference between the field contributions is π, due only to the current phase ξ. Therefore, the waves from the two antenna add destructively in these directions, and a minimum is observed in the power density.

For the intermediate phase $\xi = \pi/2$, the shape of the power density is much more interesting. Note that the power density lobes are shifted away from the axes, as seen in Fig. 11.18(b) and (e). Let's see if we can explain these maxima and minima in a coherent manner. Minima in the power density plots are expected to occur when the two waves $\tilde{E}_0$ and $\tilde{E}_1$ add out of phase with one another. Let's not worry about the three-dimensional distribution, but rather focus on the two-dimensional plot in Fig. 11.18(e). Using Eq. (11.36) with $\theta = \pi/2$ (x–y plane), $\xi = \pi/2$, and $\beta d = (2\pi/\lambda)(\lambda/2) = \pi$, the condition for destructive interference, producing a null in the power density, is

$$\psi = \pi \cos\phi + \pi/2 = \pm\pi,$$

which leads to $\cos\phi = 1/2$. Thus, minima should be observed at $\phi = \pm\pi/3$, or $\pm60°$. These directions are marked with the dot-dashed lines in Fig. 11.18(e). Conversely, maxima are expected at positions for which

$$\psi = \pi \cos\phi + \pi/2 = 0 \text{ or } \pm 2\pi.$$

This is satisfied by $\cos\phi = -1/2$, or $\phi = \pm2\pi/3$ ($\pm120°$). These directions are marked with the dotted straight lines in the figure.

Example 11.3 Two-Element Antenna Array

For two-element arrays of linear dipole antennas, with (a) in-phase currents ($\xi = 0$) and separation $d = \lambda$ and (b) out-of-phase currents ($\xi = \pi$) and separation $d = \lambda$, determine the angular dependence of the total field amplitude. Identify the minima and maxima of these distributions in the $\theta = \pi/2$ plane.

Solution:
For case (a) or (b), the separation between the antennas is $d = \lambda$, so $\beta d = (2\pi/\lambda)(\lambda) = 2\pi$. With $\xi = 0$, as in (a), the total phase difference between $\tilde{E}_0$ and $\tilde{E}_1$ is $\psi = 2\pi \cos\phi$. We should find minima at

$$\psi = \pm\pi,$$

which corresponds to $\cos\phi = \pm 1/2$. Therefore, we expect minima at $\phi = \pm 60°$ and $\pm 120°$.

Conversely, we expect maxima when

$$\psi = 0 \text{ or } \pm 2\pi.$$

$\psi = 0$ gives us $\phi = \pm 90°$, and $\psi = \pm 2\pi$ corresponds to maxima at $\cos\phi = \pm 1$, or $\phi = 0$ or $180°$. The plot of $|\tilde{E}_T(\phi)|^2$ for this array, shown in Fig. 11.19(a), displays minima and maxima at the predicted angles.

For case (b), the spacing is the same, $\beta d = 2\pi$, but now there is a $\xi = \pi$ phase difference between the currents. Therefore, the total phase difference between $\tilde{E}_0$ and $\tilde{E}_1$ is $\psi = 2\pi \cos\phi + \pi$. Minima will occur at

$$\psi = 2\pi \cos\phi + \pi = \pm\pi,$$

which corresponds to $\phi = \pm 90°$, $\phi = 0$ and $180°$, while maxima occur at

$$\psi = 2\pi \cos\phi + \pi = 0 \text{ or } \pm 2\pi,$$

which is satisfied at $\phi = \pm 60°$ and $\pm 120°$. Plots of these power density distributions are shown in Fig. 11.19(b).

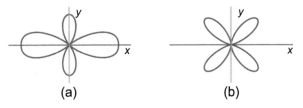

(a)　　　　　(b)

FIGURE 11.19 The angular distributions of the power density for two-element arrays of linear dipole antennas in the $\theta = \pi/2$ plane of Example 11.3. The spacing d between the elements is λ, and the phase difference between the two currents is $\xi = $ (a) 0 and (b) π.

Now that we have introduced antenna arrays using a simple two-element array, let's extend this treatment to N elements, where N is any integer greater than (or equal to) 2. As with the two-element array, we will consider all of the individual antennas to be identical, equally spaced by a distance d along the x-axis, and consider all of the currents to be the same amplitude. The phase difference between the currents in adjacent antennas is ξ. Applying superposition, with N terms in the sum, one for each antenna, the total field amplitude is

$$\tilde{E}_T = \tilde{E}_0 + \tilde{E}_1 + \cdots + \tilde{E}_{N-1}.$$

Since the amplitude generated by each antenna is the same, and the phase difference between the field contributions $\tilde{E}_n$ and $\tilde{E}_{n-1}$ of any two adjacent antennas is ψ, where the phase ψ is given in Eq. (11.36), the total field can be written as

$$\tilde{E}_T = \tilde{E}_0 \left[1 + e^{j\psi} + e^{2j\psi} + \cdots + e^{(N-1)j\psi} \right].$$

Now let's evaluate this finite sum. To do this, we use the infinite series

$$1 + x + x^2 + x^3 + \cdots,$$

which can be written using the Taylor expansion (Eq. (H.5)) as

$$\frac{1}{1-x}.$$

This expansion is valid for any x whose magnitude is less than or equal to unity. Of course, we're trying to evaluate a finite sum, and this Taylor expansion has an infinite number of terms. But this can be remedied by subtracting the infinite series

$$x^N + x^{N+1} + x^{N+2} + \cdots.$$

Factoring out x^N from each term, this series is

$$x^N \left(1 + x + x^2 + \cdots \right),$$

and since the series inside the parentheses here is $1/(1-x)$, it is written simply as

$$\frac{x^N}{1-x}.$$

Therefore, the finite sum of N terms is

$$1 + x + x^2 + x^3 + \cdots + x^{N-1} = \frac{1}{1-x} - \frac{x^N}{1-x} = \frac{1-x^N}{1-x}.$$

Applying this to our problem, with $x = e^{j\psi}$, and keeping only the magnitude of the field, we have

$$|\tilde{E}_T| = E_0 \left| \frac{1 - e^{jN\psi}}{1 - e^{j\psi}} \right|.$$

Factoring out $e^{jN\psi/2}$ from the numerator and $e^{j\psi/2}$ from the denominator, the total field magnitude is

$$|\tilde{E}_T| = E_0 \left| \frac{e^{-jN\psi/2} - e^{jN\psi/2}}{e^{-j\psi/2} - e^{j\psi/2}} \right|,$$

and applying Euler's Identity,

$$|\tilde{E}_T| = E_0 \left| \frac{\sin(N\psi/2)}{\sin(\psi/2)} \right|.$$

The power density radiated by the antenna array is proportional to the square of the field amplitude:

$$\mathcal{P} = \frac{|E_0|^2}{2\eta} \left| \frac{\sin(N\psi/2)}{\sin(\psi/2)} \right|^2. \tag{11.37}$$

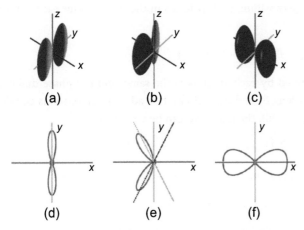

FIGURE 11.20 The angular distributions of the power density for four-element antenna arrays. The spacing between adjacent elements here is $d = \lambda/2$, and ξ is (a) 0, (b) $\pi/2$, and (c) π. Two-dimensional plots in the x–y plane for each are shown in (d), (e), and (f).

This power density is much more directional than that of the two-element array. Distributions for three examples of a four-element array of antennas are plotted in Fig. 11.20(a–c). The spacing between adjacent elements here is $d = \lambda/2$, and ξ is (a) 0, (b) $\pi/2$, and (c) π. For clarity, these same distributions, in the x–y plane (i.e. $\theta = \pi/2$), are plotted in Fig. 11.20(d–f). Notice how slender the lobes of these distributions are, relative to those of the two-element array shown in Fig. 11.18. Finding the maxima and minima for this array is a bit more complex than it was for the two-element array, but the idea is similar. $\psi = 0$ or $\pm\pi$ identifies the primary lobes of the power density distribution. Although the numerator of Eq. (11.37) is zero for this case, the denominator also approaches zero, and the field amplitude approaches N in the limit as $\psi \to 0$ or $\pm\pi$. For the three distributions shown in Fig. 11.20, the phase ξ is 0, $\pi/2$, or π. Setting $\psi = 0, \pm\pi, \ldots$, the maximum in the power density occurs at

$$\beta d \cos\phi + \xi = \pi \cos\phi + \xi = 0, \ \pm\pi, \ \ldots,$$

where we have used $\beta d = \pi$, since $d = \lambda/2$.

- For (d) $\xi = 0$, and $\cos\phi = 0$ is satisfied at $\phi = \pm\pi/2$ or $\pm90°$.
- For (e) $\xi = \pi/2$, and $\cos\phi = -1/2$ is satisfied at $\phi = \pm2\pi/3$ or $\pm120°$.
- For (f) $\xi = \pi$. Here $\psi/2$ can be 0 or $+1$. For the former, $\cos\phi = -1$ is satisfied at $\phi = \pi$ or $180°$, while for the latter $\cos\phi = +1$ is satisfied at $\phi = 0$ radians or $0°$.

Inspection of the plots in Fig. 11.20 shows the primary maxima at each of these predicted directions. The zeros in the power density occur where $N\psi/2$ is equal to $n\pi$, where n is any positive or negative integer.

In this section we have treated arrays of antennas by applying superposition. The total field emitted by the array is simply the sum of the fields emitted by the individual antenna elements. We limited our discussion to a linear array of identical antennas, varying only in the relative phase of their currents, but the approach can be applied to other, more complex geometries as well. The net result that we have seen is that arrays of antennas tend to produce outputs that are much more directed than that of single antennas, and that the direction of the outgoing wave can be controlled by controlling the relative phase difference between the currents.

SUMMARY In this chapter we have examined the generation of waves by simple linear currents. We first considered short "elemental" antennas, defined as antennas whose length $\Delta\ell$ is much shorter than the wavelength of the wave at the frequency of the oscillating current. We found simple expressions for the radiated fields, the power density, the total radiated power, and the radiation resistance. We applied what we learned about the elemental dipole antenna to the long dipole antenna, which is a much more efficient radiator than the short antenna. The half-wave dipole is a common radiating antenna. Finally, we used superposition to understand the radiation pattern generated by a linear array of identical antennas. We found explicit formulas for the radiated field pattern, and interpreted the locations of minima and maxima in terms of the phase difference between radiated fields.

We should also point out that, while we only considered the generation of an outgoing electromagnetic wave by currents in the antenna, antennas are also used as receivers of incoming waves. That is, an electromagnetic wave incident upon an antenna will produce a current in the antenna, which can then be filtered, amplified, and processed. This is the topic of another textbook.

Table 11.1

Collection of key formulas from Chapter 11.		
$$\tilde{H}_\phi = -\frac{\tilde{I}_0 \Delta \ell \beta^2}{4\pi}\left[\frac{1}{j\beta R} + \frac{1}{(j\beta R)^2}\right]e^{-j\beta R}\sin\theta$$		(11.5)
$$\tilde{E}_R = -\frac{2\tilde{I}_0 \Delta \ell}{4\pi}\eta_0\beta^2\left[\frac{1}{(j\beta R)^2} + \frac{1}{(j\beta R)^3}\right]e^{-j\beta R}\cos\theta$$		(11.9)
$$\tilde{E}_\theta = -\frac{\tilde{I}_0 \Delta \ell}{4\pi}\eta_0\beta^2\left[\frac{1}{j\beta R} + \frac{1}{(j\beta R)^2} + \frac{1}{(j\beta R)^3}\right]e^{-j\beta R}\sin\theta$$		(11.10)

	Elemental dipole		Half-wave dipole	
$\tilde{H}_\phi$	$\dfrac{j\tilde{I}_0\Delta\ell\beta}{4\pi R}e^{-j\beta R}\sin\theta$	(11.11)		
$\tilde{E}_\theta$	$\dfrac{j\tilde{I}_0\Delta\ell\beta}{4\pi R}\eta_0 e^{-j\beta R}\sin\theta$	(11.12)	$\dfrac{j\eta_0 I_0}{2\pi R}e^{-j\beta R}F(\theta)$	(11.25)
$\overline{\mathcal{P}}$	$\mathbf{a}_R\dfrac{1}{2}\left(\dfrac{I_0\Delta\ell\beta}{4\pi R}\right)^2\eta_0\sin^2\theta$	(11.13)	$\dfrac{\eta_0}{2}\left(\dfrac{I_0}{2\pi R}\right)^2\dfrac{\cos^2\left(\frac{\pi}{2}\cos\theta\right)}{\sin^2\theta}\mathbf{a}_R$	(11.29)
P_r	$\dfrac{(I_0\Delta\ell\beta)^2}{12\pi}\eta_0$	(11.14)	$(36.5\ \Omega)I_0^2$	(11.31)
R_r	$\dfrac{2\pi}{3}\eta_0\left(\dfrac{\Delta\ell}{\lambda}\right)^2$	(11.17)	$73\ \Omega$	(11.33)
$D(\theta,\phi)$	$\dfrac{3}{2}\sin^2\theta$	(11.20)	$1.64\dfrac{\cos^2\left(\frac{\pi}{2}\cos\theta\right)}{\sin^2\theta}$	(11.35)
$U(\theta,\phi)$	$\dfrac{1}{2}\left(\dfrac{I\Delta\ell\beta}{4\pi}\right)^2\eta_0\sin^2\theta$	(11.21)	$\dfrac{\eta_0 I_0^2}{8\pi^2}\dfrac{\cos^2\left(\frac{\pi}{2}\cos\theta\right)}{\sin^2\theta}$	(11.34)
$$\mathcal{P} = \frac{|E_0|^2}{2\eta}\left|\frac{\sin(N\psi/2)}{\sin(\psi/2)}\right|^2$$		(11.37)

PROBLEMS

Elemental Dipole Antenna

P11-1 A short, linear antenna of length 2.5 cm in air radiates 10 W of total power at a frequency of 100 MHz. Determine the current I_0 in the antenna.

P11-2 A short, linear dipole antenna in air lies along the z-axis, is 0.3 cm long, and is driven by a current of amplitude 5 A at a frequency of $f = 1$ GHz. The antenna consists of a copper wire of radius 0.2 mm. (a) How far from the antenna must we be before we are in the far-field region? (b) Determine the maximum value of the electric field amplitude at a distance $R = 20$ m. (c) Determine the maximum value of the time-average power density at this same distance. (d) Determine the total power radiated by the antenna, the radiation resistance, and the resistive power loss.

P11-3 Consider two short, linear dipole antennas in air. They are identical in length and have the same current at the same frequency. One antenna lies along the x-axis, the other along the z-axis. Answer the following questions in the far-field region. (a) In what direction is the power density radiated by this pair of antennas the maximum? Explain your response. (b) Find the ratio $\overline{\mathcal{P}}_x/\overline{\mathcal{P}}_z$, where $\overline{\mathcal{P}}_x$ is the time-averaged power density radiated in the x-direction, and $\overline{\mathcal{P}}_z$ is the time-averaged power density radiated in the z-direction. (c) Describe the polarization of the wave along the x-axis, along the y-axis, and along the z-axis, when the two currents are in phase with one another. (d) Repeat part (c) when the two antenna currents are $\pi/2$ out of phase with one another.

P11-4 Use Ampère's Law to determine $\tilde{E}_R$, $\tilde{E}_\theta$, and $\tilde{E}_\phi$ from $\tilde{H}_\phi$ (Eq. (11.11)) for the elemental dipole antenna.

Long Dipole Antenna

P11-5 (a) Determine the current amplitude I_0 in a linear half-wave antenna that radiates 10 kW of power at a frequency 101.3 MHz in air. (b) Determine the length of this antenna. (c) Determine the maximum power density radiated by this antenna a distance 10 km away.

P11-6 Consider a Hertzian dipole lying along the z-axis, as shown in Fig. 11.4. The length of the dipole is $\Delta\ell = 0.01\lambda$. (a) Find the radiation resistance of the dipole. (b) Assuming the dipole is fed with a current $i(t) = 1\,\text{A}\cos(\omega t)$, find the total power radiated by it. (c) Suppose the Hertzian dipole is replaced by a half-wave dipole fed at its center with the same current I. Find the ratio $E_{\lambda/2}/E_{elem}$, where $E_{\lambda/2}$ is the field radiated by the half-wave dipole at a distance R, and E_{elem} is the field radiated by the Hertzian dipole at that same distance. $R \gg \beta^{-1}$ for this problem.

Antenna Arrays

P11-7 Two half-wave electric dipole antenna elements, each with current of amplitude I_0 and lying along the z-axis, are located at $(x,y) = (\pm d/2, 0)$. (a) For antenna currents that are π out of phase with one another and $d = \lambda/4$, determine and plot the power density in the $\theta = \pi/2$ plane (i.e. the x–y plane). (b) Repeat for $d = \lambda/2$.

P11-8 An array of four half-wave electric dipole antenna elements, each with current of amplitude I_0 directed along the z-axis, is located at $(x,y) = (\pm 3d/2, 0)$ and $(\pm d/2, 0)$. For $d = \lambda/4$, determine and plot the power density in the $\theta = \pi/2$ plane (i.e. the x–y plane) when the phase difference between adjacent antennas is (a) 0, (b) $\pi/8$, (c) $\pi/4$, (d) $3\pi/8$, and (e) $\pi/2$.

P11-9 Two parallel linear dipole antennas (aligned with the z-axis) each have a current of amplitude I_0 and frequency $f = 20$ MHz. They are separated by a distance 22.5 m. Determine all angles ϕ at which the far-field radiated power density is a maximum when the two antenna currents are 180° out of phase with one another.

CHAPTER 12
Wrap Up

We have now reached the end of our journey exploring the fundamentals and simple applications of electromagnetics. We are surrounded by applications of these concepts in our daily lives. A partial list includes electric motors and generators, microwave ovens, remote controls for our television or garage door opener, magnetic resonance imaging, broadcast, satellite, or cable television, high-speed chip-to-chip communications on printed circuits, and many, many more. While we have not dealt much here with the specific engineering principles of many of these devices, we have tried to lay the fundamental concepts on which they are based.

It is remarkable to think back on the basis for all the ideas that we have developed and explored, and discover that everything we have discussed has been the product of just a few, rather simple, empirical observations. For example, the Coulomb force between two charged bodies, which is always directed along the line between the charged bodies and falls off as the square of the inverse of the distance between the charges, plus the principle of superposition, led us to the full development of electrostatics. We discovered the differential properties of electrostatic fields (Gauss' Law, $\nabla \cdot \mathbf{D} = \rho_v$; and the irrotational property of electrostatic fields, $\nabla \times \mathbf{E} = 0$). We applied Gauss' Law to find quick solutions for electric fields, and the property $\nabla \times \mathbf{E} = 0$ allowed us to define the electric potential. We'll stop detailing this now, but each of the following developments of electrostatic fields followed from these simple properties.

Similarly, the development of magnetostatics followed fully from the quantitative observations of magnetic forces. Observations of the electric potential (i.e. the electromotive force) that results from a changing magnetic flux led to Faraday's Law, which describes electric field lines that circle around time-varying magnetic fields. Electric and magnetic fields are also coupled through Ampère's Law, which in its complete form shows that magnetic field lines circle around time-varying electric fields (displacement current), as well as real currents.

All of these rather fundamental properties led us to a description of electromagnetic wave propagation. We explored free-wave propagation, and later, waves guided by transmission lines and waveguides. And finally, in Chapter 11, we briefly introduced the notion of wave generation through currents in the conductor of an antenna. There are many relevant topics, of course, that we touched on only briefly, or were not able to cover at all. Our goal has been to develop the key concepts and tools to allow you to understand and apply these notions to the topics and problems with which modern electrical engineers must be well acquainted, and to base these studies on the proper relevant physical principles. We have tried to strike a balance in our discussion between developing a valid physical mental image, on one hand, and carrying out detailed calculations of the fields, currents, etc., that are necessary to develop and apply these ideas, on the other. We hope that we have succeeded, at least partially, in that endeavor, and invite your constructive comments if you feel there are sections that we could improve. With that, we close.

APPENDIX A
Scalars and Vectors

LEARNING OBJECTIVE

After studying this appendix, you should be able to:

- add and subtract two vectors, and determine the scalar (dot) product and vector (cross) product of vectors expressed in rectangular components.

Many of the quantities that we study and use through our studies are fully characterized by a magnitude only, while other quantities have a magnitude and direction. The former are called **scalar** quantities, while the latter are **vector** quantities. Examples of scalars are the temperature of a pot of water, the air pressure in the tires of your car, the height of a building, and the weight of my dog. Each of these quantities is fully described by a magnitude alone; there is no associated direction. These quantities may change with time, or with any of a variety of other variables, but when we make a measurement and determine their values, a single value of that quantity is all that is needed. On the other hand, there are many quantities that, in addition to a magnitude, have an associated direction. We use vectors to describe these. Examples of quantities that must be expressed as a vector are the velocity of cars on the interstate and the force of gravity holding me in my seat. The direction associated with the vector velocity of a car is the direction in which the car is moving. Driving at a velocity of 65 mph to the north is different from driving at 65 mph to the east, and the vectors that represent these two velocities are unequal. The direction of the gravitational force holding me in place is downward, toward the center of the Earth (conditional on the simplification that the Earth is a perfectly spherical distribution of mass, of course).

Vector notation, and vector equations, can simplify our work, in that we can represent the three orthogonal **components** in a single symbol or equation, rather than writing each separately. Without vectors we would need to write separate equations for each component of the vector separately. For example, Newton's Second Law relating the force acting on a body of mass m and the acceleration $\mathbf{a}$ that the body experiences is $\mathbf{F} = m\mathbf{a}$, where we use the bold notation to represent a vector. Without vector notation, we would need to write separately three equations, one relating the x-component of the force to the x-component of the acceleration, another for the y-components, and a third for the z-components. Our vector notation allows us to be very much more compact in our notation. Furthermore, and perhaps much more importantly, we will develop our studies without necessarily specifying a particular coordinate system. Whether we are working in rectangular, cylindrical, or spherical coordinates, Newton's Second Law is $\mathbf{F} = m\mathbf{a}$. If we were to write this in rectangular coordinates we would specify three equations, one each for the x-, y-, and z-components. In cylindrical coordinates, however, we would specify relations for the ρ-, ϕ-, and z-components of the vectors, so this law would appear different. Our vector notation will allow us to develop the laws of electric and magnetic fields without restricting ourselves to a particular set of coordinates. We will need to be more specific when we apply the laws that govern electric and magnetic fields to the solution to a particular problem, of course, but our analysis can be quite general until that point.

The rules that govern the operations of vector quantities are rather simple, and can each be determined by considering the operations among the individual components. A vector can be expressed in terms of three orthogonal components, which depend on the coordinate system in which we have chosen to work. For example, if we are considering the vector representing the position of an object, and are working in a rectangular coordinate system, then we will express the position in terms of its three components along the directions defined by the x-, y-, and z-axes of the coordinate system. If the measure of its position relative to the origin (i.e. the point whose x-, y-, and z-coordinates are each 0) is given by $x = x_1, y = y_1,$ and $z = z_1$, then we would write the vector representing this position as $\mathbf{r}_1 = x_1\mathbf{a}_x + y_1\mathbf{a}_y + z_1\mathbf{a}_z$, or more compactly as (x_1, y_1, z_1). The vector $\mathbf{a}_x$ is the **unit vector** (i.e. a vector whose magnitude is precisely 1) that points in the x-direction; similar definitions apply for $\mathbf{a}_y$ and $\mathbf{a}_z$. These unit vectors are perpendicular (i.e. orthogonal) to each other, and form a **basis set** in the rectangular coordinate system. The components of any vector written in terms of a basis set are independent of one another, and using this basis set we are able to express any vector quantity. We illustrate these unit vectors in Fig. A.1. We wrote earlier that a vector represents a quantity that has a magnitude and direction, and we will show a bit later that writing it in terms of these three components is an equivalent, and sometimes more convenient, form.

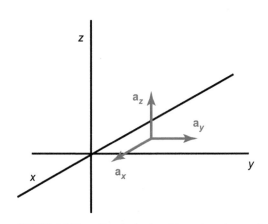

FIGURE A.1 The unit vector basis set for rectangular coordinates, $\mathbf{a}_x$, $\mathbf{a}_y$, and $\mathbf{a}_z$.

A.1 Vector Addition and Subtraction

Let's review some rules for simple arithmetic involving vector quantities. Two vectors can be added or subtracted by adding or subtracting the individual components.

Example A.1 Vector Addition and Subtraction

For $\mathbf{A} = 5\mathbf{a}_x + 3\mathbf{a}_y - 2\mathbf{a}_z$, and $\mathbf{B} = 3\mathbf{a}_x - 4\mathbf{a}_y + 2\mathbf{a}_z$, determine $\mathbf{A} + \mathbf{B}$ and $\mathbf{A} - \mathbf{B}$.

Solution:
We have $\mathbf{A} + \mathbf{B} = (5 + 3)\mathbf{a}_x + (3 - 4)\mathbf{a}_y + (-2 + 2)\mathbf{a}_z = 8\mathbf{a}_x - \mathbf{a}_y$, and $\mathbf{A} - \mathbf{B} = (5 - 3)\mathbf{a}_x + (3 + 4)\mathbf{a}_y + (-2 - 2)\mathbf{a}_z = 2\mathbf{a}_x + 7\mathbf{a}_y - 4\mathbf{a}_z$.

Vector addition is commutative (i.e. $\mathbf{A} + \mathbf{B} = \mathbf{B} + \mathbf{A}$) and associative (i.e. $\mathbf{A} + [\mathbf{B} + \mathbf{C}] = [\mathbf{A} + \mathbf{B}] + \mathbf{C}$).

We can multiply a scalar times a vector by multiplying the scalar times each of the vector components. Thus, for a scalar b, and a vector $\mathbf{A}$, $b\mathbf{A} = b(A_x\mathbf{a}_x + A_y\mathbf{a}_y + A_z\mathbf{a}_z) = bA_x\mathbf{a}_x + bA_y\mathbf{a}_y + bA_z\mathbf{a}_z$.

Example A.2 Vector Multiplication by a Scalar

For **A** from the previous example, determine 2**A**.

Solution:

$$2\mathbf{A} = 2\left(5\mathbf{a}_x + 3\mathbf{a}_y - 2\mathbf{a}_z\right) = 10\mathbf{a}_x + 6\mathbf{a}_y - 4\mathbf{a}_z.$$

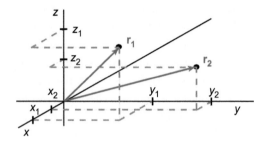

FIGURE A.2 Two position vectors $\mathbf{r}_1$ and $\mathbf{r}_2$, represented graphically.

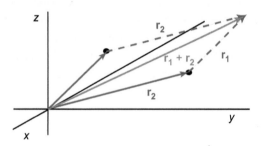

FIGURE A.3 The sum of two vectors $\mathbf{r}_1 + \mathbf{r}_2$. We place the tail of $\mathbf{r}_2$ to the head of $\mathbf{r}_1$, or vice versa, to find the resultant sum.

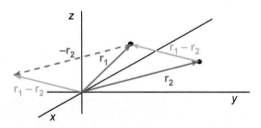

FIGURE A.4 The difference of two vectors $\mathbf{r}_1 - \mathbf{r}_2$. We invert $\mathbf{r}_2$ to find to $-\mathbf{r}_2$, and then add $\mathbf{r}_1 + (-\mathbf{r}_2)$ tail-to-head, as in vector sums.

Vector addition and scalar multiplication obey the distributive law: $b(\mathbf{A} + \mathbf{B}) = b\mathbf{A} + b\mathbf{B}$.

A graphical representation of these vectors can help us to visualize a physical model. We return to our example of the vector $\mathbf{r}_1$ that represents the position of an object, and define a second position vector $\mathbf{r}_2 = x_2\mathbf{a}_x + y_2\mathbf{a}_y + z_2\mathbf{a}_z$. We show these two vectors, with their individual components, in Fig. A.2. We wrote previously that to find the vector sum, we add the individual components. We can show that this is equivalent to placing the vectors tail-to-head, as in Fig. A.3. For instance, if $\mathbf{r}_1$ represents the vector pointing from a dog to a tree, and $\mathbf{r}_2$ represents the vector pointing from the tree to a fire hydrant, then $\mathbf{r}_1 + \mathbf{r}_2$ gives us the vector pointing from the dog to the fire hydrant.

We can also represent vector differences graphically. With our example of $\mathbf{r}_1$ and $\mathbf{r}_2$, we can find the difference graphically by inverting $\mathbf{r}_2$ to find $-\mathbf{r}_2$, and then adding $\mathbf{r}_1 + (-\mathbf{r}_2) = \mathbf{r}_1 - \mathbf{r}_2$. We illustrate this in Fig. A.4. As an example, if we let $\mathbf{r}_1$ be the position of Indianapolis relative to our location in West Lafayette, and $\mathbf{r}_2$ represents the position of Chicago, again referenced from our location, then the vector difference $\mathbf{r}_1 - \mathbf{r}_2$ represents the position of Indianapolis relative to Chicago. Using our laws of vector subtraction, we can write $\mathbf{r}_2 - \mathbf{r}_1 = (x_2 - x_1)\mathbf{a}_x + (y_2 - y_1)\mathbf{a}_y + (z_2 - z_1)\mathbf{a}_z$.

A.2 Multiplication of Vectors

We have two forms of vector multiplication that are useful in our calculations. We will denote these two operations as scalar multiplication, represented by a form called the *dot* product, and vector multiplication, called the *cross* product.

A.2.1 Scalar Product or Dot Product

For two vectors $\mathbf{A} = A_x\mathbf{a}_x + A_y\mathbf{a}_y + A_z\mathbf{a}_z$ and $\mathbf{B} = B_x\mathbf{a}_x + B_y\mathbf{a}_y + B_z\mathbf{a}_z$, the dot, or scalar, product, represented by $\mathbf{A} \cdot \mathbf{B}$, is defined as

$$\mathbf{A} \cdot \mathbf{B} = A_xB_x + A_yB_y + A_zB_z. \tag{A.1}$$

The dot product is always a scalar – that is, it has only a single component, with no direction associated to it.

Example A.3 Scalar Product or Dot Product

For vectors $\mathbf{A} = 5\mathbf{a}_x - 4\mathbf{a}_y - \mathbf{a}_z$ and $\mathbf{B} = 2\mathbf{a}_x + 3\mathbf{a}_y - 5\mathbf{a}_z$, determine $\mathbf{A} \cdot \mathbf{B}$.

Solution:
$\mathbf{A} \cdot \mathbf{B} = (5)(2) + (-4)(3) + (-1)(-5) = 10 - 12 + 5 = 3.$

Computing the dot product of a vector with itself gives us the square of the magnitude of the vector, or

$$|\mathbf{A}| = \sqrt{A_x^2 + A_y^2 + A_z^2}$$
$$= \sqrt{\mathbf{A} \cdot \mathbf{A}}, \tag{A.2}$$

where the first equality follows from the Pythagorean theorem (applied twice), and the second equality from the definition of the scalar product, Eq. (A.1).

Example A.4 Vector Magnitude

For vectors $\mathbf{A}$ and $\mathbf{B}$ defined in Example A.3, determine $|\mathbf{A}|$ and $|\mathbf{B}|$.

Solution:
$|\mathbf{A}| = \sqrt{5^2 + (-4)^2 + (-1)^2} = \sqrt{42}$ and $|\mathbf{B}| = \sqrt{2^2 + 3^2 + (-5)^2} = \sqrt{38}.$

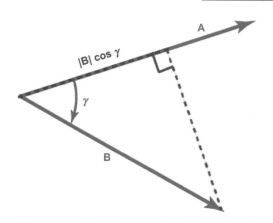

FIGURE A.5 A graphical representation of the scalar, or dot, product of two vectors, **A · B**. We separate the vector **B** into two components, one of which is parallel to the vector **A**, and of length |**B**|cosγ. (Note the right triangle in the figure, with the vector **B** as the hypotenuse.) This component is often called the *projection* of **B** onto **A**. The scalar product of two vectors **A** and **B** is the product of the magnitudes of **A** and the projection of **B** onto **A**.

Applying Eq. (A.1) to the unit bases vectors gives us

$$\mathbf{a}_x \cdot \mathbf{a}_x = \mathbf{a}_y \cdot \mathbf{a}_y = \mathbf{a}_z \cdot \mathbf{a}_z = 1, \tag{A.3}$$

and

$$\mathbf{a}_x \cdot \mathbf{a}_y = \mathbf{a}_y \cdot \mathbf{a}_z = \mathbf{a}_z \cdot \mathbf{a}_x = 0. \tag{A.4}$$

Another useful form of the dot product is

$$\mathbf{A} \cdot \mathbf{B} = |\mathbf{A}||\mathbf{B}|\cos\gamma, \tag{A.5}$$

where γ is the angle between **A** and **B**. We represent this form graphically in Fig. A.5.

Example A.5 Angle between Vectors

For vectors **A** and **B** defined in Example A.3, determine the angle γ between **A** and **B**.

Solution:
We already determined $\mathbf{A} \cdot \mathbf{B}$ in Example A.3 and $|\mathbf{A}|$ and $|\mathbf{B}|$ in Example A.4, so we'll use these results in

$$\cos \gamma = \frac{\mathbf{A} \cdot \mathbf{B}}{|\mathbf{A}||\mathbf{B}|} = \frac{3}{\sqrt{42}\sqrt{38}} = 0.07509.$$

Then the angle γ is

$$\gamma = \cos^{-1}(0.07509) = 85.69°.$$

This result tells us that **A** and **B** are nearly perpendicular to one another, which we could have anticipated, since **A** is mostly in the x–y plane and **B** largely in the z direction.

Example A.6 Proof of Eq. (A.5)

Prove Eq. (A.5).

Solution:
Choose a rectangular coordinate system (x', y', z') in which vector **A** lies along the z'-axis, such that $\mathbf{A} = A_{z'}\mathbf{a}_{z'}$. In this coordinate system, the vector **B** will have components $B_{x'}$, $B_{y'}$, and $B_{z'}$, such that $\mathbf{B} = B_{x'}\mathbf{a}_{x'} + B_{y'}\mathbf{a}_{y'} + B_{z'}\mathbf{a}_{z'}$. Then $\mathbf{A} \cdot \mathbf{B} = 0 \cdot B_{x'} + 0 \cdot B_{y'} + A_{z'}B_{z'} = A_{z'}B_{z'}$. In other words, only the component of **B** parallel to **A**, whose magnitude is $|\mathbf{B}| \cos \gamma$, survives, and we arrive at Eq. (A.5). (Notice that our proof depends on the notion that the dot product $\mathbf{A} \cdot \mathbf{B}$ is the same, no matter which coordinate system we choose to describe the vectors **A** and **B**. This is correct, but we state this here without proof.)

The dot product gives us the projection of one vector onto another, and from either definition Eq. (A.1) or (A.5), we see that the order of multiplication for the scalar product is not important, as $\mathbf{A} \cdot \mathbf{B} = \mathbf{B} \cdot \mathbf{A}$.

Example A.7 Unit Vectors, Perpendicular Vectors

Find a unit vector perpendicular to the vector $\mathbf{C} = 5\mathbf{a}_x - 2\mathbf{a}_y + \mathbf{a}_z$.

Solution:
Let us label the vector we seek as $\mathbf{D} = D_x\mathbf{a}_x + D_y\mathbf{a}_y + D_z\mathbf{a}_z$. Since **D** must be perpendicular to **C**, we know that $\gamma = \pi/2$ (where γ is the angle between **C** and **D**), and therefore $\mathbf{C} \cdot \mathbf{D} = 5D_x - 2D_y + D_z = 0$. Since **D** must be a unit vector, we also know that $D_x^2 + D_y^2 + D_z^2 = 1$. There are actually infinitely many vectors **D** that will satisfy these two conditions, all lying in a plane. To find one example, let us choose $D_x = 0$. Then our two conditions reduce to $-2D_y + D_z = 0$ and $D_y^2 + D_z^2 = 1$, which lead us to $(D_y, D_z) = (1/\sqrt{5}, 2/\sqrt{5})$ or $(-1/\sqrt{5}, -2/\sqrt{5})$. Therefore, two possible solutions are $\mathbf{D} = 1/\sqrt{5}\,\mathbf{a}_y + 2/\sqrt{5}\,\mathbf{a}_z$ and

$\mathbf{D} = -1/\sqrt{5}\,\mathbf{a}_y - 2/\sqrt{5}\,\mathbf{a}_z$. We can confirm these solutions by computing $\mathbf{C} \cdot \mathbf{D}$ and showing that this dot product is zero, and showing that $\mathbf{D} \cdot \mathbf{D}$ yields 1.

We wrote earlier of two equivalent forms in which to express vector quantities, either magnitude and direction, or using the individual components along three orthogonal directions. Let us explore the equivalence of these two forms a bit more. Consider the vector $\mathbf{r}_1 = x_1\mathbf{a}_x + y_1\mathbf{a}_y + z_1\mathbf{a}_z$, perhaps representing the position of an object. The magnitude of this vector is $|\mathbf{r}_1| = \sqrt{x_1^2 + y_1^2 + z_1^2}$, and we find the unit vector pointing in the same direction dividing $\mathbf{r}_1$ by its magnitude,

$$\mathbf{a}_{r_1} = \frac{\mathbf{r}_1}{|\mathbf{r}_1|}. \tag{A.6}$$

Turning this around, we can write the vector $\mathbf{r}_1$ as $\mathbf{r}_1 = |\mathbf{r}_1|\mathbf{a}_{r_1}$. We will routinely express vectors in either of these forms, depending on which is more convenient at the time.

In summary, the dot product of two vectors is a quantity that depends on the component of one vector that is parallel to the other. When two vectors are parallel to one another, the magnitude of the dot product is simply the product of their magnitudes, but when the vectors are perpendicular to one another, their dot product is zero. The dot product of two vectors is a scalar quantity, which has no direction.

A.2.2 Vector Product or Cross Product

The cross product of two vectors is complementary to the dot product. We represent this product as $\mathbf{A} \times \mathbf{B}$, and define it (in rectangular coordinates) as

$$\mathbf{A} \times \mathbf{B} = \begin{vmatrix} \mathbf{a}_x & \mathbf{a}_y & \mathbf{a}_z \\ A_x & A_y & A_z \\ B_x & B_y & B_z \end{vmatrix} \tag{A.7}$$

$$= \mathbf{a}_x\,(A_yB_z - A_zB_y) + \mathbf{a}_y\,(A_zB_x - A_xB_z) + \mathbf{a}_z\,(A_xB_y - A_yB_x).$$

In the first line we have written this vector product using the determinant form, which we then write out using its individual components. In contrast to the dot product, which gives a scalar, the cross product is a vector, whose direction is normal (perpendicular) to the plane containing the vectors $\mathbf{A}$ and $\mathbf{B}$. We illustrate this in Fig. A.6. There are, of course, two directions that are normal to any plane, one directly opposite the other, so we use a right-hand rule to help us determine the proper direction. We point the fingers of our right hand in the direction of vector $\mathbf{A}$, and sweep them through the *small* angle to point along vector $\mathbf{B}$. The thumb of our right hand then points in the direction of the vector $\mathbf{A} \times \mathbf{B}$. As you can tell from this rule, the order in which $\mathbf{A}$ and $\mathbf{B}$ appear in the cross product does matter, since the right-hand rule applied to $\mathbf{B} \times \mathbf{A}$ gives a vector in the direction precisely opposite to $\mathbf{A} \times \mathbf{B}$. Then $\mathbf{B} \times \mathbf{A} = -\mathbf{A} \times \mathbf{B}$. The unit bases vectors $\mathbf{a}_x$, $\mathbf{a}_y$, and $\mathbf{a}_z$ shown in Fig. A.1 satisfy the relations

$$\mathbf{a}_x \times \mathbf{a}_y = \mathbf{a}_z, \qquad \mathbf{a}_y \times \mathbf{a}_z = \mathbf{a}_x, \qquad \mathbf{a}_z \times \mathbf{a}_x = \mathbf{a}_y, \tag{A.8}$$

as must be true for a right-handed orthogonal coordinate system. Notice that the order of the indices is x, y, z, or one of its cyclic permutations y, z, x or z, x, y. (A cyclic permutation of the indices is a reordering that can be achieved by repeating the original

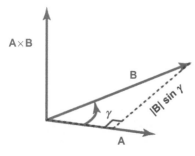

FIGURE A.6 The vector, or cross, product of two vectors **A** and **B** is perpendicular to both vectors **A** and **B** (direction given by a right-hand rule), and its magnitude is $|\mathbf{A}||\mathbf{B}|\sin\gamma$, where γ is the angle between the vectors.

sequence $xyz \rightarrow \ldots xyzxyx\ldots$ and extracting all unique sequences of the same number of elements; three in the present example.)

The magnitude of $\mathbf{A} \times \mathbf{B}$ is the area of a parallelogram of sides $|\mathbf{A}|$ and $|\mathbf{B}|$,

$$|\mathbf{A} \times \mathbf{B}| = |\mathbf{A}||\mathbf{B}|\sin\gamma, \tag{A.9}$$

where γ is the angle between $\mathbf{A}$ and $\mathbf{B}$. In contrast to the dot product, which gives a measure of the component of $\mathbf{B}$ that is parallel to $\mathbf{A}$, the cross product indicates the magnitude of the component of $\mathbf{B}$ that is perpendicular to $\mathbf{A}$.

Example A.8 Vector Product or Cross Product

(a) For vectors $\mathbf{A}$ and $\mathbf{B}$ defined in Example A.3, determine the cross product $\mathbf{A} \times \mathbf{B}$. (b) Use this result to determine the angle γ between $\mathbf{A}$ and $\mathbf{B}$.

Solution:

(a) Using Eq. (A.7), we compute the cross product as

$$\mathbf{A} \times \mathbf{B} = \begin{vmatrix} \mathbf{a}_x & \mathbf{a}_y & \mathbf{a}_z \\ 5 & -4 & -1 \\ 2 & 3 & -5 \end{vmatrix}$$

$$= \mathbf{a}_x\left[(-4)(-5) - (-1)(3)\right] + \mathbf{a}_y\left[(-1)(2) - (5)(-5)\right]$$
$$+ \mathbf{a}_z\left[(5)(3) - (-4)(2)\right]$$
$$= \mathbf{a}_x 23 + \mathbf{a}_y 23 + \mathbf{a}_z 23.$$

(b) We use Eq. (A.9) to write

$$\sin\gamma = \frac{|\mathbf{A} \times \mathbf{B}|}{|\mathbf{A}||\mathbf{B}|} = \frac{23\sqrt{3}}{\sqrt{42}\sqrt{38}} = 0.99718.$$

Then the angle γ is

$$\gamma = \sin^{-1}(0.99718) = 85.69°.$$

This result agrees precisely with that of Example A.5.

Example A.9 Demonstration that the Cross Product $\mathrm{A} \times \mathrm{B}$ is Perpendicular to Vectors A and B.

Show that $\mathbf{A} \times \mathbf{B}$ determined in Example A.8 is perpendicular to both vector $\mathbf{A}$ and vector $\mathbf{B}$.

Solution:
The easiest way to show that two vectors are perpendicular to one another is to show that their scalar product is zero. For $\mathbf{A} = 5\mathbf{a}_x - 4\mathbf{a}_y - \mathbf{a}_z$ and $\mathbf{B} = 2\mathbf{a}_x + 3\mathbf{a}_y - 5\mathbf{a}_z$, we found that $\mathbf{A} \times \mathbf{B} = \mathbf{a}_x 23 + \mathbf{a}_y 23 + \mathbf{a}_z 23$. Now we compute

$$\mathbf{A} \cdot (\mathbf{A} \times \mathbf{B}) = (5)(23) + (-4)(23) + (-1)(23) = 0$$

and

$$\mathbf{B} \cdot (\mathbf{A} \times \mathbf{B}) = (2)(23) + (3)(23) + (-5)(23) = 0,$$

from which we conclude that $\mathbf{A} \times \mathbf{B}$ is indeed perpendicular to both vector $\mathbf{A}$ and vector $\mathbf{B}$.

Example A.10 Proof of Eq. (A.9)

Prove Eq. (A.9).

Solution:
This is proven by choosing the rectangular coordinate system (x', y', z') introduced in Example A.6, in which vector $\mathbf{A}$ lies along the z'-axis, such that $\mathbf{A} = A_{z'}\mathbf{a}_{z'}$. As we wrote before, the vector $\mathbf{B}$ in this coordinate system will have components $B_{x'}$, $B_{y'}$, and $B_{z'}$, such that $\mathbf{B} = B_{x'}\mathbf{a}_{x'} + B_{y'}\mathbf{a}_{y'} + B_{z'}\mathbf{a}_{z'}$. Then,

$$\mathbf{A} \times \mathbf{B} = (0 \cdot B_{z'} - A_{z'}B_{y'})\,\mathbf{a}_{x'} + (A_{z'}B_{x'} - 0 \cdot B_{z'})\,\mathbf{a}_{y'} + (0 \cdot B_{y'} - 0 \cdot B_{x'})\mathbf{a}_{z'}$$
$$= A_{z'}\left(-B_{y'}\mathbf{a}_{x'} + B_{x'}\mathbf{a}_{y'}\right).$$

Since the magnitude of $(-B_{y'}\mathbf{a}_{x'} + B_{x'}\mathbf{a}_{y'})$ is $\sqrt{B_{y'}^2 + B_{x'}^2}$, and the projection of $\mathbf{B}$ onto the x'–y' plane, which is $|\mathbf{B}|\sin\gamma$, gives the same result, then Eq. (A.9) follows.

From Eq. (A.9) we see that the magnitude of the cross product of two parallel vectors is zero, and the magnitude of the cross product of two perpendicular vectors is the maximum possible value of $|\mathbf{A}||\mathbf{B}|$.

In summary, then, the cross product of two vectors is a product that reflects the component of one vector that is perpendicular to the other. In this way, it is complementary to the dot product. When two vectors are orthogonal to one another, the magnitude of the cross product is the product of their magnitudes. The cross product of two vectors is itself a vector, and its direction is perpendicular to both, and given by a right-hand rule. Together, the dot product and cross product of two vector products are sufficient to describe any products that we will need. We will complete our discussion of vector products in the following with a pair of identities regarding triple products of vectors.

A.2.3 Triple Products

Through our use of vectors in later chapters, we will find useful two relations involving triple products of vectors (or products of three vector quantities, say $\mathbf{A}$, $\mathbf{B}$, and $\mathbf{C}$). These come in two flavors: $\mathbf{A} \cdot (\mathbf{B} \times \mathbf{C})$ and $\mathbf{A} \times (\mathbf{B} \times \mathbf{C})$. The first of these forms a scalar, while the second forms a vector. With a little not-very-deep thought, you can convince yourself that these are the only triple products that make sense. For the scalar triple product, one can show that

$$\mathbf{A} \cdot (\mathbf{B} \times \mathbf{C}) = \mathbf{B} \cdot (\mathbf{C} \times \mathbf{A}) = \mathbf{C} \cdot (\mathbf{A} \times \mathbf{B}). \qquad (A.10)$$

We see that these three triple products are related to one another through *cyclic permutations* of the vectors $\mathbf{A}$, $\mathbf{B}$, and $\mathbf{C}$. (We defined a cyclic permutation earlier, but repeat it here. With this operation, we rearrange the vectors, exchanging $\mathbf{A} \to \mathbf{B}$, $\mathbf{B} \to \mathbf{C}$, and $C \rightarrow A$. If we exchange just two of the vectors with one another, and leave the third alone, this is not a cyclic permutation, and we would have to change the sign of the product as well.) Eq. (A.10) can be proven in rectangular coordinates by writing the products out in terms of the x-, y-, and z-coordinates for vectors A, B, and C, and comparing the results. We leave this exercise for Prob. PA-11 at the end of this Appendix.

Example A.11 Scalar Triple Product

For vectors $A = 1a_x + 2a_y - 3a_z$, $B = 4a_x + 5a_y + 6a_z$, and $C = 7a_x + 8a_y + 9a_z$, verify the first equality in Eq. (A.10).

Solution:

We start by computing the vector product

$$B \times C = \begin{vmatrix} a_x & a_y & a_z \\ 4 & 5 & 6 \\ 7 & 8 & 9 \end{vmatrix}$$

$$= a_x [(5)(9) - (6)(8)] + a_y [(6)(7) - (4)(9)] + a_z [(4)(8) - (5)(7)]$$
$$= a_x(-3) + a_y 6 + a_z(-3).$$

Then,

$$A \cdot (B \times C) = (1)(-3) + (2)(6) + (-3)(-3) = 18.$$

Next we find the vector product

$$C \times A = \begin{vmatrix} a_x & a_y & a_z \\ 7 & 8 & 9 \\ 1 & 2 & -3 \end{vmatrix}$$

$$= a_x [(8)(-3) - (9)(2)] + a_y [(9)(1) - (7)(-3)] + a_z [(7)(2) - (8)(1)]$$
$$= a_x(-42) + a_y 30 + a_z 6.$$

Then we take the scalar product of B with $C \times A$ to find

$$B \cdot (C \times A) = (4)(-42) + (5)(30) + (6)(6) = 18.$$

With this result, we have showed that $A \cdot (B \times C)$ is equal to $B \cdot (C \times A)$.

The relation involving the vector triple product is

$$A \times (B \times C) = B(A \cdot C) - C(A \cdot B). \tag{A.11}$$

This rule is called the BAC CAB rule, a mnemonic for the rule itself. The result of this triple product is a vector that lies in the plane defined by vectors B and C. (This should make sense to us, since $B \times C$ must be perpendicular to vectors B and C, and the triple product must be perpendicular to $B \times C$ (as well as A, of course), which places it right back in the plane containing B and C.) Similar to the scalar triple product, this rule can be proven in rectangular coordinates by writing the products out in terms of the x-, y-, and z-coordinates for vectors A, B, and C. Upon regrouping the terms, the scalar products

$\mathbf{A} \cdot \mathbf{C}$ and $\mathbf{A} \cdot \mathbf{B}$ emerge. (You must also add and subtract a term to complete each dot product, as you will see when you carry out Prob. PA-13.)

Example A.12 BAC CAB Rule

Using the vectors $\mathbf{A}$, $\mathbf{B}$, and $\mathbf{C}$ defined in Example A.11, verify the BAC CAB rule, Eq. (A.11).

Solution:

We have already found $\mathbf{B} \times \mathbf{C}$ in the previous example, so we will borrow that result, and solve for

$$\mathbf{A} \times (\mathbf{B} \times \mathbf{C}) = \begin{vmatrix} \mathbf{a}_x & \mathbf{a}_y & \mathbf{a}_z \\ 1 & 2 & -3 \\ -3 & 6 & -3 \end{vmatrix}$$

$$= \mathbf{a}_x \left[(2)(-3) - (-3)(6) \right] + \mathbf{a}_y \left[(-3)(-3) - (1)(-3) \right]$$
$$+ \mathbf{a}_z \left[(1)(6) - (2)(-3) \right]$$
$$= \mathbf{a}_x 12 + \mathbf{a}_y 12 + \mathbf{a}_z 12.$$

Next we must find

$$\mathbf{B} (\mathbf{A} \cdot \mathbf{C}) = \left(4\mathbf{a}_x + 5\mathbf{a}_y + 6\mathbf{a}_z \right) \left[(1)(7) + (2)(8) + (-3)(9) \right]$$
$$= -4 \left(4\mathbf{a}_x + 5\mathbf{a}_y + 6\mathbf{a}_z \right)$$

and

$$\mathbf{C} (\mathbf{A} \cdot \mathbf{B}) = \left(7\mathbf{a}_x + 8\mathbf{a}_y + 9\mathbf{a}_z \right) \left[(1)(4) + (2)(5) + (-3)(6) \right]$$
$$= -4 \left(7\mathbf{a}_x + 8\mathbf{a}_y + 9\mathbf{a}_z \right).$$

The difference $\mathbf{B} (\mathbf{A} \cdot \mathbf{C}) - \mathbf{C} (\mathbf{A} \cdot \mathbf{B})$ is easily found to be $\mathbf{a}_x 12 + \mathbf{a}_y 12 + \mathbf{a}_z 12$, which matches the result we found for $\mathbf{A} \times (\mathbf{B} \times \mathbf{C})$.

PROBLEMS

Vector Addition and Subtraction

PA-1 Using the vectors $\mathbf{A}$, $\mathbf{B}$, and $\mathbf{C}$ listed in Table A.1, determine:

(a) $\mathbf{A} - \mathbf{B}$

(b) $\mathbf{A} - 2\mathbf{C}$

(c) $\mathbf{A} + \mathbf{B}$

(d) $2\mathbf{A} - \mathbf{C}$

(e) $\mathbf{C} - 2\mathbf{A}$.

Table A.1

Vectors for use with Problems PA-1–PA-12.
$\mathbf{A} = 2\mathbf{a}_x - 3\mathbf{a}_y + 2\mathbf{a}_z$
$\mathbf{B} = 5\mathbf{a}_x - 5\mathbf{a}_y - 2\mathbf{a}_z$
$\mathbf{C} = -2\mathbf{a}_x + 3\mathbf{a}_y + 4\mathbf{a}_z$

Scalar Product or Dot Product

PA-2 Using the vectors A, B, and C listed in Table A.1, determine:

(a) $\mathbf{A} \cdot \mathbf{B}$

(b) $\mathbf{A} \cdot (2\mathbf{B})$

(c) $\mathbf{B} \cdot (\mathbf{A} - \mathbf{B})$

(d) $(\mathbf{A} + \mathbf{B}) \cdot \mathbf{C}$

(e) $(2\mathbf{A}) \cdot (\frac{1}{2}\mathbf{B})$.

PA-3 Using the vectors A, B, and C listed in Table A.1, determine the magnitude of the following vectors:

(a) $\mathbf{A}$

(b) $\mathbf{A} - (2\mathbf{C})$

(c) $\mathbf{B} - 2\mathbf{C}$

(d) $\mathbf{A} + 2\mathbf{B}$

(e) $2\mathbf{A}$.

PA-4 Using the vectors A, B, and C listed in Table A.1, determine the angle between the directions of the following pairs of vectors:

(a) $\mathbf{A}$ and $\mathbf{B}$

(b) $\mathbf{A}$ and $-\mathbf{B}$

(c) $\mathbf{B}$ and $\mathbf{C}$

(d) $\mathbf{A}$ and $\mathbf{C}$

(e) $2\mathbf{A}$ and $3\mathbf{B}$.

PA-5 (a) Determine the angle γ between the vector $\mathbf{r}_{Madrid}$, which points from the center of Earth to Madrid, and the vector $\mathbf{r}_{NYC}$, which points from the center of Earth to New York City. (b) Find the shortest distance (in kilometers, and constrained to the surface of Earth, of course) from Madrid to New York City. (c) Find the distance you must travel if you follow a path that is directly to the west the entire way. [Hint: You will need the longitude and latitude of Madrid and New York City, and the radius of Earth. You may assume that the Earth is perfectly spherical, and for part (c) you may ignore the small difference in latitude of the two cities.]

PA-6 A vector D is perpendicular to vector A listed in Table A.1, has a magnitude of 5 units, and has a y-component of -3. Determine the vector D. (There is more than one solution.)

PA-7 Using the vectors A, B, and C listed in Table A.1, determine the unit vectors parallel to:

(a) $\mathbf{A}$

(b) $\mathbf{A} - (2\mathbf{C})$

(c) $\mathbf{B} - 2\mathbf{C}$

(d) $\mathbf{A} - \mathbf{B}$

(e) $2\mathbf{A}$.

Vector Product or Cross Product

PA-8 Using the vectors A, B, and C listed in Table A.1, determine the following cross products:

(a) $\mathbf{A} \times \mathbf{B}$

(b) $\mathbf{A} \times (2\mathbf{C})$

(c) $\mathbf{B} \times 2\mathbf{C}$

(d) $\mathbf{C} \times 2\mathbf{B}$

(e) $2\mathbf{A} \times (\frac{1}{2}\mathbf{B})$.

PA-9 Using vector C listed in Table A.1, find the vector D for which $\mathbf{C} \cdot \mathbf{D} = 5.5$ and $\mathbf{C} \times \mathbf{D} = (-1/2, -3, 2)$.

Triple Products of Vectors

PA-10 Show explicitly that $A \cdot (B \times C) = B \cdot (C \times A) = C \cdot (A \times B)$ for the vectors A, B, and C listed in Table A.1.

PA-11 Show that $A \cdot (B \times C) = B \cdot (C \times A) = C \cdot (A \times B)$ for general vectors A, B, and C by writing the products out in terms of the x-, y-, and z-coordinates for vectors A, B, and C, and comparing the results.

PA-12 Show explicitly that $A \times (B \times C) = B(A \cdot C) - C(A \cdot B)$ for the vectors A, B, and C listed in Table A.1.

PA-13 Show that $A \times (B \times C) = B(A \cdot C) - C(A \cdot B)$ for general vectors A, B, and C, by writing the products out in terms of the x-, y-, and z-coordinates for vectors A, B, and C. Upon regrouping the terms, the scalar products $A \cdot C$ and $A \cdot B$ emerge. [Hint: You must also add and subtract a term to complete each dot product.]

APPENDIX B
Coordinate Systems

B.1 Rectangular, Cylindrical, and Spherical Coordinate Systems

Many different coordinate systems have been devised for description of physical systems. Three are of particular relevance to geometries that we will consider in this study of elementary electromagnetic devices. These are the rectangular, cylindrical, and spherical coordinate systems. In this section we will review the three orthogonal coordinates for each, and the transformations from one set of coordinates to another. We will also define a basis set of unit vectors in each of these coordinate systems. Later in this appendix we will review integral and differential properties in each of these systems, as we will rely heavily on these operations in our development and applications of electric and magnetic fields.

In **rectangular coordinates**, a point in space relative to the origin is defined in terms of its coordinates measured along the x-, y-, and z-axes, as shown in Fig. B.1(a). We represent the coordinates of a point P_1 in this figure as (x_1, y_1, z_1). The range of values of x, y, and z to define any point in space is from $-\infty$ to $+\infty$. The distance d between point P_1 and a second point P_2, located at (x_2, y_2, z_2) is given by

$$d = [(x_2 - x_1)^2 + (y_2 - y_1)^2 + (z_2 - z_1)^2]^{1/2}. \tag{B.1}$$

We can relate this distance to the vectors that we reviewed in the last section. If $\mathbf{r}_1$ is the vector pointing from the origin to point P_1, and $\mathbf{r}_2$ is the vector pointing from the origin to point P_2, then the vector $\mathbf{r}_1 - \mathbf{r}_2$ points from P_2 to P_1, and distance d is the magnitude of $\mathbf{r}_1 - \mathbf{r}_2$. We represent this as $|\mathbf{r}_1 - \mathbf{r}_2| = [(x_2 - x_1)^2 + (y_2 - y_1)^2 + (z_2 - z_1)^2]^{1/2}$. As a reminder, we defined the magnitude of a vector in Eq. (A.2), where here we use $\mathbf{A} = \mathbf{r}_1 - \mathbf{r}_2$. We will find it convenient to work in rectangular coordinates when the geometry of objects within our system conforms to rectangular planes or spaces.

FIGURE B.1 (a) Two points P_1 and P_2 in a rectangular coordinate system, and a single point P in (b) cylindrical and (c) spherical coordinate systems.

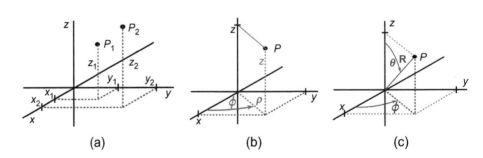

(a)　　　　(b)　　　　(c)

In **cylindrical coordinates**, a point in space is defined in terms of the coordinates ρ, ϕ, and z, as shown in Fig. B.1(b). ρ represents the distance of the point from the z-axis, called the radial distance. The angle ϕ is the azimuthal angle of this radius line measured from the x-axis. The coordinate z is the distance of the point P from the x–y plane, identical to the coordinate of that same label for rectangular coordinates. The range of ρ is $0 \rightarrow \infty$, ϕ ranges from 0 to 2π, and z from $-\infty$ to $+\infty$. Transformation from rectangular to cylindrical coordinates is affected by the following relations:

- $\rho = \sqrt{x^2 + y^2}$
- $\phi = \tan^{-1}(y/x)$
- $z = z$.

Remember that the $\tan^{-1}(y/x)$ function has two values, so we need to examine the signs of x and y to choose the proper value of ϕ. When x is negative, this places our point in the second or third quadrants of the coordinates system, and the angle ϕ must be in the range $\pi/2 < \phi < 3\pi/2$. For example, if $y/x = -1$, and you calculate $\tan^{-1}(-1)$ with your calculator, it does not know if $x > 0$ and $y < 0$, or vice versa. It will return an answer $\tan^{-1}(-1) = -\pi/4$, or -45°. This result is fine if $x > 0$, and you are done. But if $x < 0$ and $y > 0$, then of course you must add π to this result to get a second quadrant answer. So as always, examine your result to make sure that it makes sense.

Conversely, we can transform from cylindrical to rectangular coordinates using:

- $x = \rho \cos \phi$
- $y = \rho \sin \phi$
- $z = z$.

Physical systems that display cylindrical symmetry will in general be more easily analyzed using cylindrical coordinates than they will using other coordinate systems.

If we need to determine the distance between two points given in cylindrical coordinates, we must first transform to rectangular coordinates, and then apply Eq. (B.1).

Example B.1 Distance between Points Specified in Cylindrical Coordinates

Find the distance d between two points P_1 at $\mathbf{r}_1 = (\rho_1, \phi_1, z_1) = (2, \pi, 2)$ and P_2 at $\mathbf{r}_2 = (\rho_2, \phi_2, z_2) = (2\sqrt{2}, \pi/4, 0)$.

Solution:

We illustrate these two points in Fig. B.2(a). We must first convert these vectors $\mathbf{r}_1$ and $\mathbf{r}_2$ to rectangular components, which we do using $x = \rho \cos \phi$, $y = \rho \sin \phi$, and $z = z$. This gives us $\mathbf{r}_1 = (-2, 0, 2)$ and $\mathbf{r}_2 = (2, 2, 0)$. Then,

$$d = |\mathbf{r}_2 - \mathbf{r}_1| = |(4, 2, -2)| = \sqrt{(4)^2 + (2)^2 + (-2)^2} = \sqrt{24} \approx 4.90.$$

Notice that without converting the vectors $\mathbf{r}_1$ and $\mathbf{r}_2$ to rectangular components first, we do not know how to compute the difference $\mathbf{r}_2 - \mathbf{r}_1$.

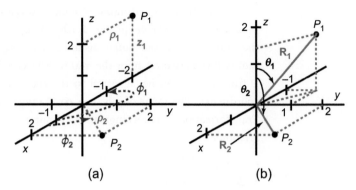

FIGURE B.2 Points P_1 and P_2 in (a) cylindrical and (b) spherical coordinate systems. (For use in Examples B.1 and B.2.)

Our final coordinate system is the **spherical coordinate** system, in which a point in space is defined using the three coordinates R, θ, and ϕ. R is the distance from the origin O to the point P, θ is the polar angle measured from the positive z-axis, and ϕ is the azimuthal angle about the z-axis measured from the x-axis. R, θ, and ϕ range $0 \rightarrow +\infty$, $0 \rightarrow \pi$, and $0 \rightarrow 2\pi$, respectively. Transformation from rectangular to spherical coordinates follows from the following relations:

- $R = \sqrt{x^2 + y^2 + z^2}$
- $\theta = \cos^{-1}(z/R)$
- $\phi = \tan^{-1}(y/x)$.

Note that ϕ denotes the same angle in spherical and cylindrical coordinates. We repeat the caution when evaluating $\tan^{-1}(y/x)$ that we described above. That is, $\tan^{-1}(y/x)$ is two-valued, so we need to examine the sign of x and y to choose the proper value of ϕ.

To transform from spherical to rectangular coordinates, we use the following relations:

- $x = R \sin\theta \cos\phi$
- $y = R \sin\theta \sin\phi$
- $z = R \cos\theta$.

We will choose to work with spherical coordinates when the symmetry of the problem is best described this way.

As with points given in cylindrical coordinates, to find the distance between two points given in spherical coordinates, we must first convert to rectangular coordinates and apply Eq. (B.1).

Example B.2 Distance between Points Specified in Spherical Coordinates

Find the distance d between two points P_1 at $\mathbf{r}_1 = (R_1, \theta_1, \phi_1) = (2, \pi/4, 3\pi/4)$ and P_2 at $\mathbf{r}_2 = (R_2, \theta_2, \phi_2) = (2\sqrt{2}, \pi/2, \pi/4)$.

Solution:
We illustrate these two points in Fig. B.2(b). Just as in our last example, we must first convert the vectors $\mathbf{r}_1$ and $\mathbf{r}_2$ to rectangular components, which we do using $x = R \sin\theta \cos\phi$, $y = R \sin\theta \sin\phi$, and $z = R \cos\theta$. This gives us $\mathbf{r}_1 = (-1, 1, \sqrt{2})$ and $\mathbf{r}_2 = (2, 2, 0)$. Then,

$$d = |\mathbf{r}_2 - \mathbf{r}_1| = |(3,\ 1,\ -\sqrt{2})| = \sqrt{3^2 + 1^2 + (-\sqrt{2})^2} = \sqrt{12} \approx 3.46.$$

Once again, without converting the vectors $\mathbf{r}_1$ and $\mathbf{r}_2$ to rectangular components first, we do not know how to compute the difference $\mathbf{r}_2 - \mathbf{r}_1$.

B.2 Unit Vectors in Rectangular, Cylindrical, and Spherical Coordinate Systems

In the discussion in the previous section in which we reviewed vectors, we used vectors written in rectangular components, and we defined the unit vectors $\mathbf{a}_x$, $\mathbf{a}_y$, and $\mathbf{a}_z$, which each have unit magnitude (see Eq. (A.3)), and which point in orthogonal directions (see Eq. (A.4)). We said that these vectors form a basis set, since we can write any vector in terms of its x-, y-, and z-components. We also noted that $\mathbf{a}_x$, $\mathbf{a}_y$, and $\mathbf{a}_z$ form a right-handed system, as expressed through the cross products of Eqs. (A.8).

FIGURE B.3 The unit vector basis sets for (a) cylindrical coordinates, $\mathbf{a}_\rho$, $\mathbf{a}_\phi$, and $\mathbf{a}_z$; and (b) spherical coordinates, $\mathbf{a}_R$, $\mathbf{a}_\theta$, and $\mathbf{a}_\phi$.

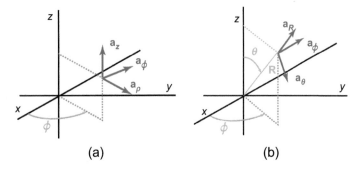

(a)　　　　　　　(b)

We can also describe vector quantities using their cylindrical or spherical components, and we show the unit basis vectors for these systems in Fig. B.3(a) and (b), respectively. In cylindrical coordinates, $\mathbf{a}_\rho$, $\mathbf{a}_\phi$, and $\mathbf{a}_z$ point in the direction of increasing ρ, ϕ, and z, respectively. Similar to the unit vectors in a rectangular system, these unit vectors satisfy the conditions

$$\mathbf{a}_\rho \cdot \mathbf{a}_\rho = \mathbf{a}_\phi \cdot \mathbf{a}_\phi = \mathbf{a}_z \cdot \mathbf{a}_z = 1, \tag{B.2}$$

$$\mathbf{a}_\rho \cdot \mathbf{a}_\phi = \mathbf{a}_\phi \cdot \mathbf{a}_z = \mathbf{a}_z \cdot \mathbf{a}_\rho = 0, \tag{B.3}$$

and

$$\mathbf{a}_\rho \times \mathbf{a}_\phi = \mathbf{a}_z, \qquad \mathbf{a}_\phi \times \mathbf{a}_z = \mathbf{a}_\rho, \qquad \mathbf{a}_z \times \mathbf{a}_\rho = \mathbf{a}_\phi. \tag{B.4}$$

Eq. (B.2) tells us that $\mathbf{a}_\rho$, $\mathbf{a}_\phi$, and $\mathbf{a}_z$ are of unit magnitude, Eq. (B.3) tells us that they are orthogonal to one another, and Eqs. (B.4) tell us that they form a right-handed coordinate system, just as $\mathbf{a}_x$, $\mathbf{a}_y$, and $\mathbf{a}_z$ define a right-handed coordinate system.

To transform from unit vectors in a rectangular system to unit vectors in a cylindrical system, we can use

$$\begin{aligned} \mathbf{a}_\rho &= \cos\phi\,\mathbf{a}_x + \sin\phi\,\mathbf{a}_y \\ \mathbf{a}_\phi &= -\sin\phi\,\mathbf{a}_x + \cos\phi\,\mathbf{a}_y \\ \mathbf{a}_z &= \mathbf{a}_z, \end{aligned} \tag{B.5}$$

and to transform back,

$$\mathbf{a}_x = \cos\phi\,\mathbf{a}_\rho - \sin\phi\,\mathbf{a}_\phi$$
$$\mathbf{a}_y = \sin\phi\,\mathbf{a}_\rho + \cos\phi\,\mathbf{a}_\phi \tag{B.6}$$
$$\mathbf{a}_z = \mathbf{a}_z.$$

Example B.3 Converting Cylindrical Unit Vectors to Rectangular Unit Vectors

The velocity of a particle circling around the z-axis is given as $\mathbf{u} = u_0\mathbf{a}_\phi$, where u_0 is constant. Write this particle velocity in terms of $\mathbf{a}_x$ and $\mathbf{a}_y$ when the particle position is (a) $\mathbf{r} = (1, 0, 0)$, and (b) $\mathbf{r} = (-1, 1, 0)$.

Solution:

(a) At $\mathbf{r} = (1, 0, 0)$, the azimuthal angle ϕ is 0, and Eq. (B.5) tells us that $\mathbf{a}_\phi = \mathbf{a}_y$ at this angle. Therefore, the particle velocity is $\mathbf{u} = u_0\mathbf{a}_y$.

(b) Similarly, at point $\mathbf{r} = (-1, 1, 0)$, the angle ϕ is $3\pi/4$, and evaluation of Eq. (B.5) at this angle shows that $\mathbf{a}_\phi = -\mathbf{a}_x/\sqrt{2} - \mathbf{a}_y/\sqrt{2}$. The particle velocity is therefore $\mathbf{u} = -u_0(\mathbf{a}_x + \mathbf{a}_y)/\sqrt{2}$.

Note that in either case the magnitude of the particle velocity is simply u_0, whether computed using the velocity $\mathbf{u}$ expressed in rectangular or cylindrical components.

In spherical coordinates, we use the basis vectors $\mathbf{a}_R$, $\mathbf{a}_\theta$, and $\mathbf{a}_\phi$, shown in Fig. B.3(b), which point in the direction of increasing R, θ, and ϕ, respectively. These unit vectors are of unit magnitude

$$\mathbf{a}_R \cdot \mathbf{a}_R = \mathbf{a}_\theta \cdot \mathbf{a}_\theta = \mathbf{a}_\phi \cdot \mathbf{a}_\phi = 1; \tag{B.7}$$

they are orthogonal to one another,

$$\mathbf{a}_R \cdot \mathbf{a}_\theta = \mathbf{a}_\theta \cdot \mathbf{a}_\phi = \mathbf{a}_\phi \cdot \mathbf{a}_R = 0; \tag{B.8}$$

and they form a right-handed coordinate system,

$$\mathbf{a}_R \times \mathbf{a}_\theta = \mathbf{a}_\phi,\ \mathbf{a}_\theta \times \mathbf{a}_\phi = \mathbf{a}_R,\ \mathbf{a}_\phi \times \mathbf{a}_R = \mathbf{a}_\theta. \tag{B.9}$$

To transform from rectangular unit basis vectors to spherical unit basis vectors, we can use

$$\mathbf{a}_R = \sin\theta\cos\phi\,\mathbf{a}_x + \sin\theta\sin\phi\,\mathbf{a}_y + \cos\theta\,\mathbf{a}_z$$
$$\mathbf{a}_\theta = \cos\theta\cos\phi\,\mathbf{a}_x + \cos\theta\sin\phi\,\mathbf{a}_y - \sin\theta\,\mathbf{a}_z \tag{B.10}$$
$$\mathbf{a}_\phi = -\sin\phi\,\mathbf{a}_x + \cos\phi\,\mathbf{a}_y,$$

and to transform back,

$$\mathbf{a}_x = \sin\theta\cos\phi\,\mathbf{a}_R + \cos\theta\cos\phi\,\mathbf{a}_\theta - \sin\phi\,\mathbf{a}_\phi$$
$$\mathbf{a}_y = \sin\theta\sin\phi\,\mathbf{a}_R + \cos\theta\sin\phi\,\mathbf{a}_\theta + \cos\phi\,\mathbf{a}_\phi \tag{B.11}$$
$$\mathbf{a}_z = \cos\theta\,\mathbf{a}_R - \sin\theta\,\mathbf{a}_\theta.$$

These conversions of the unit vectors from one coordinate system to another are very handy. One use of these is to transform an arbitrary vector $\mathbf{A}$ between coordinate systems, which we will have to do frequently. To transform a vector from rectangular to cylindrical components, we use the following relations:

$$A_\rho = \mathbf{a}_\rho \cdot \mathbf{A} = \mathbf{a}_\rho \cdot \left(\mathbf{a}_x A_x + \mathbf{a}_y A_y + \mathbf{a}_z A_z \right)$$
$$= \cos\phi A_x + \sin\phi A_y$$
$$A_\phi = \mathbf{a}_\phi \cdot \mathbf{A} = \mathbf{a}_\phi \cdot \left(\mathbf{a}_x A_x + \mathbf{a}_y A_y + \mathbf{a}_z A_z \right)$$
$$= -\sin\phi A_x + \cos\phi A_y \tag{B.12}$$
$$A_z = \mathbf{a}_z \cdot \mathbf{A} = \mathbf{a}_z \cdot \left(\mathbf{a}_x A_x + \mathbf{a}_y A_y + \mathbf{a}_z A_z \right)$$
$$= A_z.$$

We used the unit vector relations of Eqs. (B.5) to evaluate the dot products such as $\mathbf{a}_\rho \cdot \mathbf{a}_x$, $\mathbf{a}_\rho \cdot \mathbf{a}_y$, etc., above.

To transform in the opposite direction, from cylindrical to rectangular coordinates, we use

$$A_x = \mathbf{a}_x \cdot \mathbf{A} = \cos\phi A_\rho - \sin\phi A_\phi$$
$$A_y = \mathbf{a}_y \cdot \mathbf{A} = \sin\phi A_\rho + \cos\phi A_\phi \tag{B.13}$$
$$A_z = \mathbf{a}_z \cdot \mathbf{A} = A_z.$$

Here we used Eqs. (B.6) to evaluate the dot products $\mathbf{a}_x \cdot \mathbf{a}_\rho$, $\mathbf{a}_x \cdot \mathbf{a}_\phi$, etc.

Example B.4 Vector Conversion: Rectangular to Cylindrical Coordinates

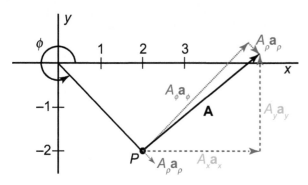

FIGURE B.4 The vector $\mathbf{A} = 5\,\mathbf{a}_x + 4\,\mathbf{a}_y - 3\,\mathbf{a}_z$ at point $P = (2, -2, 3)$, used in Example B.4.

The vector $\mathbf{A}$ at point $P = (2, -2, 3)$ is expressed in rectangular components as $\mathbf{A} = 5\mathbf{a}_x + 4\mathbf{a}_y - 3\mathbf{a}_z$. See an x–y projection of this vector in Fig. B.4. The blue dashed lines in this figure show the rectangular components of $\mathbf{A}$. Express $\mathbf{A}$ using cylindrical components.

Solution:

We use Eqs. (B.12) to convert this vector from rectangular to cylindrical components. The angle ϕ at point P is $7\pi/4$, so $\cos\phi = 1/\sqrt{2}$ and $\sin\phi = -1/\sqrt{2}$. Then,

$$A_\rho = 5/\sqrt{2} - 4/\sqrt{2} = 1/\sqrt{2},$$
$$A_\phi = 5/\sqrt{2} + 4/\sqrt{2} = 9/\sqrt{2},$$

and

$$A_z = A_z = -3.$$

We show the ρ and ϕ components of $\mathbf{A}$ in Fig. B.4 in red dotted lines. Note that adding these components head-to-tail gives the same result as we saw with the rectangular components.

Similarly, we transform from rectangular to spherical components using

$$A_R = \mathbf{a}_R \cdot \mathbf{A} = \sin\theta\cos\phi A_x + \sin\theta\sin\phi A_y + \cos\theta A_z$$
$$A_\theta = \mathbf{a}_\theta \cdot \mathbf{A} = \cos\theta\cos\phi A_x + \cos\theta\sin\phi A_y - \sin\theta A_z \tag{B.14}$$
$$A_\phi = \mathbf{a}_\phi \cdot \mathbf{A} = -\sin\phi A_x + \cos\phi A_y,$$

or from spherical to rectangular components using

$$A_x = \mathbf{a}_x \cdot \mathbf{A} = \sin\theta\cos\phi A_R + \cos\theta\cos\phi A_\theta - \sin\phi A_\phi$$
$$A_y = \mathbf{a}_y \cdot \mathbf{A} = \sin\theta\sin\phi A_R + \cos\theta\sin\phi A_\theta + \cos\phi A_\phi \qquad \text{(B.15)}$$
$$A_z = \mathbf{a}_z \cdot \mathbf{A} = \cos\theta A_R - \sin\theta A_\theta.$$

Example B.5 Vector Conversion

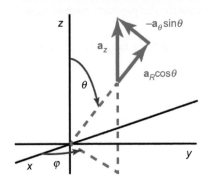

FIGURE B.5 A graphical illustration of the equivalence of $\mathbf{a}_z$ and $\mathbf{a}_R\cos\theta - \mathbf{a}_\theta\sin\theta$.

Consider the vector $\mathbf{A}$, which in rectangular components is of the form $\mathbf{A} = 5\mathbf{a}_z$. Write this same vector in terms of its (a) spherical components, and (b) cylindrical components.

Solution:

(a) In spherical components, we start by illustrating the unit vector $\mathbf{a}_z$ in Fig. B.5. We will show that $\mathbf{a}_z$ is equivalent to $\mathbf{a}_R\cos\theta - \mathbf{a}_\theta\sin\theta$. Our first step is to write the unit vector as $\mathbf{a}_z = c_R\mathbf{a}_R + c_\theta\mathbf{a}_\theta + c_\phi\mathbf{a}_\phi$, where c_R, c_θ, and c_ϕ are magnitudes to be determined. To isolate c_R, we take the scalar product of $\mathbf{a}_z$ and $\mathbf{a}_R$; that is

$$\mathbf{a}_R \cdot \mathbf{a}_z = \mathbf{a}_R \cdot \left(c_R\mathbf{a}_R + c_\theta\mathbf{a}_\theta + c_\phi\mathbf{a}_\phi\right) = c_R,$$

where we used Eqs. (B.7) and (B.8) to evaluate the dot products of the unit vectors. But $\mathbf{a}_R \cdot \mathbf{a}_z$, from the first line of Eq. (B.10), is simply $\cos\theta$. By a similar process, we use $c_\theta = \mathbf{a}_\theta \cdot \mathbf{a}_z$ to find $c_\theta = -\sin\theta$, and $c_\phi = \mathbf{a}_\phi \cdot \mathbf{a}_z$ to find $c_\phi = 0$. Therefore, our unit vector $\mathbf{a}_z$ written in its spherical components is $\mathbf{a} = \mathbf{a}_R\cos\theta - \mathbf{a}_\theta\sin\theta$, and the vector $\mathbf{A} = 5\mathbf{a}_z$ is $\mathbf{a}_R 5\cos\theta - \mathbf{a}_\theta 5\sin\theta$. Notice that while A_z is just a constant, independent of location, A_R and A_θ depend on our location (through the polar angle θ).

(b) To convert the vector $\mathbf{A}$ to cylindrical coordinates, we follow a similar procedure. Now we want to find the components A_ρ, A_ϕ, and A_z, where $\mathbf{A} = A_\rho\mathbf{a}_\rho + A_\phi\mathbf{a}_\phi + A_z\mathbf{a}_z$. Using Eqs. (B.2) and (B.3), we have $A_\rho = \mathbf{a}_\rho \cdot \mathbf{A}$, $A_\phi = \mathbf{a}_\phi \cdot \mathbf{A}$, and $A_z = \mathbf{a}_z \cdot \mathbf{A}$. Then using Eqs. (B.5), we find $A_\rho = 0$, $A_\phi = 0$, and $A_z = 5$. So $\mathbf{A}$ written in cylindrical components is $\mathbf{A} = 5\mathbf{a}_z$. Notice that this is identical to its form in rectangular components. In retrospect, we could have predicted this, since the z-components in cylindrical and rectangular bases are the same.

In Appendix A we discussed simple vector arithmetic operations, such as vector addition, subtraction, and multiplication, and worked several examples. At that time, we only used vectors represented in their rectangular components. As examples, consider the explicit forms for scalar (Eq. (A.1)) and vector (Eq. (A.7)) multiplication of vectors in rectangular components. These same operations are performed equally well in cylindrical or spherical components. In cylindrical components, the dot (scalar) product is

$$\mathbf{A} \cdot \mathbf{B} = A_\rho B_\rho + A_\phi B_\phi + A_z B_z, \qquad \text{(B.16)}$$

and the cross (vector) product is

$$\mathbf{A} \times \mathbf{B} = \begin{vmatrix} \mathbf{a}_\rho & \mathbf{a}_\phi & \mathbf{a}_z \\ A_\rho & A_\phi & A_z \\ B_\rho & B_\phi & B_z \end{vmatrix} \qquad \text{(B.17)}$$

$$= \mathbf{a}_\rho\left(A_\phi B_z - A_z B_\phi\right) + \mathbf{a}_\phi\left(A_z B_\rho - A_\rho B_z\right) + \mathbf{a}_z\left(A_\rho B_\phi - A_\phi B_\rho\right).$$

For vectors in spherical coordinates, the dot product is

$$\mathbf{A} \cdot \mathbf{B} = A_R B_R + A_\theta B_\theta + A_\phi B_\phi, \tag{B.18}$$

and the cross product is

$$\mathbf{A} \times \mathbf{B} = \begin{vmatrix} \mathbf{a}_R & \mathbf{a}_\theta & \mathbf{a}_\phi \\ A_R & A_\theta & A_\phi \\ B_R & B_\theta & B_\phi \end{vmatrix} \tag{B.19}$$

$$= \mathbf{a}_R \left(A_\theta B_\phi - A_\phi B_\theta \right) + \mathbf{a}_\theta \left(A_\phi B_R - A_R B_\phi \right) + \mathbf{a}_\phi \left(A_R B_\theta - A_\theta B_R \right). \tag{B.20}$$

Notice the similarity of these equations to those for rectangular coordinates. Notice also the ordering of the coordinates. That is, (ρ, ϕ, z) for cylindrical coordinates and (R, θ, ϕ) for spherical coordinates. This ordering helps us maintain our right-handed coordinate systems.

We will revisit vectors and the various coordinate systems later as we develop some higher-level vector operations and review setting up integrals of charge densities and electric and magnetic fluxes, but for now we have provided the basic principles that will get us started with our studies.

PROBLEMS

Conversion between Coordinate Systems

PB-1 Consider a point P given in rectangular coordinates as $(4, 3, -5)$. Determine the (a) cylindrical (ρ, ϕ, z) and (b) spherical (R, θ, ϕ) coordinates of this point.

PB-2 Consider the point Q given in cylindrical coordinates as $(3, 3\pi/4, -4)$. Determine the (a) rectangular (x, y, z) and (b) spherical (R, θ, ϕ) coordinates of this point.

PB-3 Consider the point R given in spherical coordinates as $(4, 3\pi/4, 5\pi/4)$. Determine the (a) rectangular (x, y, z) and (b) cylindrical (ρ, ϕ, z) coordinates of this point.

Conversion of Sets of Unit Vectors

PB-4 At the point P defined in Prob. PB-1, write the unit vectors $\mathbf{a}_x$, $\mathbf{a}_y$, and $\mathbf{a}_z$ in terms of (a) $\mathbf{a}_\rho$, $\mathbf{a}_\phi$, and $\mathbf{a}_z$, and (b) in terms of $\mathbf{a}_R$, $\mathbf{a}_\theta$, and $\mathbf{a}_\phi$. Show that your results have unit length.

PB-5 At the point Q defined in Prob. PB-2, write the unit vectors $\mathbf{a}_\rho$, $\mathbf{a}_\phi$, and $\mathbf{a}_z$ in terms of (a) $\mathbf{a}_x$, $\mathbf{a}_y$, and $\mathbf{a}_z$, and (b) in terms of $\mathbf{a}_R$, $\mathbf{a}_\theta$, and $\mathbf{a}_\phi$. Show that your results have unit length.

PB-6 At the point R defined in Prob. PB-3, write the unit vectors $\mathbf{a}_R$, $\mathbf{a}_\theta$, and $\mathbf{a}_\phi$ in terms of (a) $\mathbf{a}_x$, $\mathbf{a}_y$, and $\mathbf{a}_z$, and (b) $\mathbf{a}_\rho$, $\mathbf{a}_\phi$, and $\mathbf{a}_z$. Show that your results have unit length.

APPENDIX C
Scalar and Vector Integration

LEARNING OBJECTIVES

After studying this appendix, you should be able to:
- set up and evaluate scalar and vector path integrals using rectangular, cylindrical, and spherical coordinates;
- set up and evaluate scalar and vector surface integrals using rectangular, cylindrical, and spherical coordinates; and
- set up and evaluate scalar volume integrals using rectangular, cylindrical, and spherical coordinates.

C.1 Elements of Integration

We will frequently need to set up and carry out integrations of various quantities, such as forces, charges, currents, or electric or magnetic fluxes, to be defined later, over lines, surfaces or volumes. We review here the tools that we will need to set up these integral equations. We must define line, surface, and volume integrals, recognizing differences when the integrand is a scalar or vector quantity, and define line, surface, and volume elements in each of the three coordinate systems that we reviewed in the previous section, necessary to properly set up these integrals. We will introduce these according to the different kinds of integrals that we need to evaluate.

We will define, and give examples of, each of these different types of integrals shortly, but first, let's examine the different *elements* of length, surface area, and volume that we will use in each of the three coordinate systems. An element of length is a smooth, somewhat linear segment, such as Δx, chosen to be short enough that any quantity we seek to integrate is essentially constant over the segment. As you might anticipate from your previous experience with integral calculus, integration over a finite length will involve a summation over these elements, and in the limit that each segment becomes infinitesimally small as the number of elements becomes infinitely large, this summation becomes an integral. Similarly, a surface element is one small quasi-rectangular "patch" of area Δs, while a volume element is a small, nearly cubic volume Δv, which we define to facilitate integration over a surface and volume, respectively.

We start with the rectangular coordinate system. For example, if we know the linear mass density ρ_ℓ (we use a subscript ℓ to indicate a "linear" density, such as, in this case, the mass per unit length) of a length of rope that lies along the x-axis, then we could determine the total mass of the rope by dividing it into short segments of length Δx_i, finding the mass of each segment $\rho_{\ell,i}\Delta x_i$, and adding each of these individual masses. (We are using the index i here to label the different segments of the rope.) As we take the limit of this sum when the individual segments become shorter and shorter, the total mass is just the integral of the density over the length of the rope,

$$m = \int \rho_\ell(x)dx.$$

This integral is easily set up since we were told that the rope lies along the x-axis, and the element of integration is dx. Similarly, we could have evaluated this mass with ease if the path over which we integrate is parallel to the y- or z-axis of our coordinate system, and our element of integration is simply dy or dz, respectively. We can also use these three

FIGURE C.1 Surface and volume elements in (a) rectangular, (b) cylindrical, and (c) spherical coordinate systems.

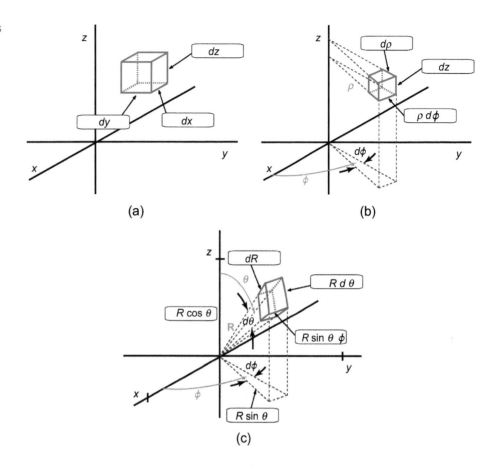

(a)

(b)

(c)

elements, dx, dy, and dz, for evaluation of a surface integral. For example, if we have a quantity that is distributed within the x–y plane, and we are working in rectangular coordinates, we effectively divide the entire surface into infinitesimal rectangles of side Δx in the x-direction by Δy in the y-direction, and add the contributions of all these patches that comprise the total surface area. In the limit as the area of each patch becomes infinitesimal, and the number of patches increases accordingly, our surface element of integration will be $ds = \lim_{\Delta x, \Delta y \to 0} \Delta x \Delta y = dx\, dy$. By similar argument, we will use $ds = dy\, dz$ when we integrate a quantity in the y–z plane, and $ds = dz\, dx$ for surface integration in the z–x plane. Finally, we will often need to integrate some quantity over a volume in x–y–z space. For this, we divide the volume into three-dimensional elements, effectively cubes, of dimension Δx by Δy by Δz. Our volume element of integration will therefore be $dv = dx\, dy\, dz$. The elemental cube in Fig. C.1(a) illustrates these elements, where the line element $\Delta \ell$ is represented by one of the lines of length Δx, Δy, Δz, the surface element is the area of one of the three faces of the cube, and the volume element is the volume of the cube.

When the geometry of a problem is cylindrical, we will find it convenient to use cylindrical coordinates for our solution. We illustrate line, surface, and volume elements used for the cylindrical coordinate system in Fig. C.1(b). Here, the element is nearly a cube, a pseudo-cube, if you will, of dimension $\Delta \rho$ by $\rho \Delta \phi$ by Δz. $\Delta \rho$ comes simply from extending a bit in the ρ direction, while Δz signifies an increase in the vertical direction, parallel to the axis of a cylinder. The second of these dimensions, $\rho \, \Delta \phi$, may require a bit of clarification. We can see that this side is formed by sweeping the radial arm of length ρ through an incremental angle $\Delta \phi$ about the z-axis. Recall that the length of an arc formed in this way is equal to the product of the radius ρ times the angular width $\Delta \phi$.

When we are required to evaluate an integral along a line in the ρ-, ϕ-, or z-direction, we will choose $d\ell = d\rho$, $d\ell = \rho \, d\phi$, or $d\ell = dz$, respectively. As with the elemental cube in rectangular coordinates, the three surface elements that can arise in various calculations are represented by the area of one of the three faces. For integration over a surface that lies on the outside surface of the cylinder, we use $ds = \rho \, d\phi \, dz$; that is, a patch on this surface is of height dz and of base $\rho d\phi$, so the element of area is $ds = (dz)(\rho d\phi)$. For integration on the plane that contains a radial axis and the z-axis, the surface element is $d\rho \, dz$; and for integration over the surface parallel to the x–y plane, we use $ds = \rho \, d\rho \, d\phi$. And finally, for volume integration in the cylindrical system, the appropriate form for the elemental volume is the product of the elemental lengths of the three sides of the pseudo-cube, $dv = \rho \, d\rho \, d\phi \, dz$.

In a spherical system, we use the elemental pseudo-cube shown in Fig. C.1(c). The sides of this element are of length ΔR by $R \, \Delta\theta$ by $R \sin\theta \, \Delta\phi$. The first side is formed by incrementing the radial distance R by a distance ΔR. The other two sides are formed by sweeping through arcs, so they may require further explanation. Recall that the polar angle θ is measured down from the z-axis (or pole), pivoting about the origin on the surface of a sphere of radius R. The length of the arc corresponding to an incremental angle $\Delta\theta$ is therefore $R \, \Delta\theta$. We show this in Fig. C.1(c) as the "height" of the cube. We form the curved bottom of this cube by sweeping in the ϕ-direction. Recall that ϕ is the azimuthal angle, measured from the x-axis. The effective radius of this arc is $R \sin\theta$, as is most easily seen by projecting this side onto the x–y plane. (Alternatively, we can understand this effective radius by recognizing that the pivot point when sweeping out this arc is a point on the z-axis, rather than the origin.) Therefore, the length of this arc is the radius $R \sin\theta$ times the incremental angle $\Delta\phi$, or $R \sin\theta \, \Delta\phi$. We will use the three sides of this cube to visualize line, surface, and volume elements in much the same way as we did in rectangular and cylindrical coordinates. For surface integration over a surface that lies on the outside of the sphere, our surface element is $ds = \lim_{\Delta\phi, \Delta\theta \to 0} (R \sin\theta \, \Delta\phi)(R \, \Delta\theta) = R^2 \sin\theta \, d\theta \, d\phi$. For integration over a side plane of the cube, we use $ds = R \, dR \, d\theta$, and for integration over the surface of a cone, we use $ds = R \sin\theta \, dR \, d\phi$. For volume integration, the volume element is the product of the three sides of the cube, $dv = R^2 \sin\theta \, dR \, d\theta \, d\phi$.

C.2 Integration along a Path

C.2.1 Scalar Path Integrals

In the last section, as we defined the elements of integration for rectangular, cylindrical, and spherical coordinate systems, we motivated the discussion with the example of the mass of a rope extended along the x-axis, and showed that the total mass is $m = \int \rho_\ell(x) dx$, where $\rho_\ell(x)$ is the linear mass density, given in kilograms per meter. In case the rope lies along some path other than the x-axis, we can write this more generally as $m = \int \rho_\ell(\ell) d\ell$, where ℓ gives some measure of the distance along the rope. We call this a scalar path integral, since the density of the rope is a scalar quantity (i.e. there is no direction associated with the density), and upon integrating the density along the length of the rope our result is the total mass of the rope, which is also a scalar quantity. We will choose the integration element to match the geometry of the problem. For example, if the rope forms a circle of radius a, then we would choose ϕ as our element of integration, with $d\ell = a \, d\phi$.

Example C.1 Scalar Linear Integration

Consider a rope of non-uniform density that forms a circle of radius a. Let the density be $\rho_\ell(\phi) = \rho_0 + \rho_1 \cos \phi$. Find the total mass of this rope, and the coordinates of the center of mass of the rope.

Solution:

To find the mass from the density, we simply have to integrate the mass density of the rope over the length of the rope. Since the rope is formed in a circle, we will use the angle ϕ as our variable of integration, with the path element $d\ell = a\, d\phi$. Then

$$m = \int \rho_\ell \, d\ell = \int_0^{2\pi} [\rho_0 + \rho_1 \cos \phi] \, a \, d\phi = 2\pi a \rho_0,$$

since the integral of the $\cos \phi$ term is zero. This answer should make sense if we recognize that the average density is ρ_0, since the mass is just the circumference of the circle $(2\pi a)$ times this average density. A word of caution here: Generally we are not allowed to simply compute the total mass as the density times the length, since the density of the rope varies along its length. This example is a special case, but you should not presume that result. Unless the density is uniform, we must evaluate the integral expression.

For the center-of-mass coordinates, we use

$$x_{c.m.} = \frac{1}{m} \int x\, \rho_\ell \, d\ell = \frac{1}{m} \int_0^{2\pi} a \cos \phi \, [\rho_0 + \rho_1 \cos \phi] \, a \, d\phi = \frac{a\rho_1}{2\rho_0},$$

where we had to convert $x = a \cos \phi$ to circular coordinates before integrating, and

$$y_{c.m.} = \frac{1}{m} \int y\, \rho_\ell \, d\ell = \frac{1}{m} \int_0^{2\pi} a \sin \phi \, [\rho_0 + \rho_1 \cos \phi] \, a \, d\phi = 0,$$

where we used $y = a \sin \phi$. With a little forethought, perhaps we should have anticipated that $y_{c.m.} = 0$, since the density is symmetric about the x-axis.

Example C.2 Scalar Linear Integration

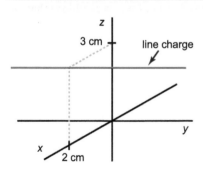

FIGURE C.2 A line of charge of density $\rho_\ell(\mathbf{r})$ as given in Example C.2.

Charge is distributed along the line parallel to the y-axis shown in Fig. C.2, defined by $x = 2$ cm and $z = 3$ cm, with $\rho_\ell(y) = \rho_{\ell,0}[1 + (y/b)^2]^{-1}$, where $\rho_{\ell,0} = 1$ mC/cm is the charge density at $y = 0$, and $b = 1$ cm defines the width of the charge distribution. Determine the total charge on this line within the range $y = -2$ cm to $y = +2$ cm.

Solution:

We will use Eq. (1.8), which differs from the integral in the previous example only through the charge density rather than the mass density. Since the charge is distributed along a line parallel to the y-axis, the variable of integration is $d\ell = dy$. To simplify the notation of the integral, we will define $y_0 = 2$ cm.

$$Q = \int_\ell \rho_\ell(\mathbf{r}) \, d\ell$$

$$= \rho_{\ell,0} \int_{-y_0}^{+y_0} \frac{1}{1 + (y/b)^2} \, dy.$$

We use Eq. (G.10) from Appendix G to find $Q = 2.21$ mC. (Be sure to use radians, not degrees, as you evaluate $\tan^{-1}(y_0/b)$.) Notice that the position of the line, given as $x = 2$ cm and $z = 3$ cm, does not enter into our calculation.

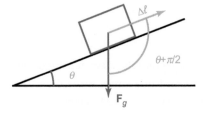

FIGURE C.3 A block on an incline. The force of gravity acting on this block is $\mathbf{F}_g$, acting in the downward direction. We will ignore friction, and push on the block just enough to maintain a constant velocity up the plane. How much work must we do on the block to move a distance $\Delta\boldsymbol{\ell}$?

C.2.2 Vector Path Integration

Until now, the integrals that we have considered have all been integrals of simple scalar quantities. There are important cases in which we must introduce new vector integration techniques and notation. One of these is called vector path integration, which you may have encountered previously to determine the work done when moving an object against a force. For example, consider a block that we would like to move up the side of a hill, as shown in Fig. C.3. We will let gravity be the only force acting on the object, of magnitude $|\mathbf{F}_g| = mg$ where m is the mass of the block and g the acceleration due to gravity, and ignore other forces such as friction, etc. As we push on the object, we will push just hard enough to overcome the gravitational force, but not enough to give it any appreciable kinetic energy. In moving the object up the slope a distance $\Delta\boldsymbol{\ell}$, we will have to do work on the object, thereby increasing its potential energy. We might guess that the work we have done is equal to $|\mathbf{F}_g|\Delta\boldsymbol{\ell}$. But let's be careful; we didn't say that we were lifting the object directly upward. Rather, we are pushing it up the inclined slope of the hill. It should make sense that the direction in which we move the object should somehow be included in our computation. At this point, we will skip directly to the result that correctly accounts for this – that is, the work done when moving an object a short distance $\Delta\boldsymbol{\ell}$ against a force $\mathbf{F}_g$ is $\Delta W = -\mathbf{F}_g \cdot \Delta\boldsymbol{\ell}$. We write $\mathbf{F}_g$ and $\Delta\boldsymbol{\ell}$ as vectors, since their relative directions, as well as their magnitudes, are important. In this example the direction of $\mathbf{F}_g$ is vertically down (remember, we are doing work against the gravitational force), while the direction of the displacement $\Delta\boldsymbol{\ell}$ is the direction in which the object moves. Use of the dot (scalar) product of $\mathbf{F}$ and $\Delta\boldsymbol{\ell}$ in this determination is just what we need, since only the component of $\Delta\boldsymbol{\ell}$ in the direction of $\mathbf{F}$ (or the opposite direction) represents any work done. For example, in our example of pushing the object on the hillside, if we move the object to the side, going neither up nor down the hill, we do not change its potential energy, and we do no work. In vector notation, we would say that for the path we chose, $\Delta\boldsymbol{\ell}$ is perpendicular to $\mathbf{F}$, and therefore $\Delta W = -\mathbf{F} \cdot \Delta\boldsymbol{\ell} = 0$. We should also explain the minus sign in our definition of the work done. Here, $\mathbf{F}$ is the force we are working against (in this example, gravity). If we move the object in a direction opposite the gravitational force, we increase its energy, and so the work we do is positive. The minus sign is consistent with this convention. For the block on the hillside in the figure, with the slope of the incline given by the angle θ, $\Delta W = -\mathbf{F}_g \cdot \Delta\boldsymbol{\ell} = -|\mathbf{F}_g||\boldsymbol{\ell}|\cos(\theta + \pi/2)$, where $\theta + \pi/2$ is the angle between $\mathbf{F}_g$ and $\Delta\boldsymbol{\ell}$, as shown in the figure. This simplifies to $\Delta W = |\mathbf{F}_g||\boldsymbol{\ell}|\sin\theta$, which is $mg\Delta h$, where Δh is the change in elevation of the block. This should be a familiar result.

Finally, the total work that we do when we move the object over some long path is found by dividing the long path into many short intervals $\Delta\boldsymbol{\ell}_i$ and summing the work done over each short interval:

$$W = -\sum_i \mathbf{F}_i \cdot \Delta\boldsymbol{\ell}_i.$$

In the limit of $\Delta\ell_i \to 0$, the summation becomes the integral

$$W = -\int \mathbf{F} \cdot d\boldsymbol{\ell}. \tag{C.1}$$

Now let's talk about how to evaluate such a path integral of a vector $\mathbf{F}$ over a path joining two points A and B. In rectangular coordinates, we write $d\boldsymbol{\ell}$ as

$$d\boldsymbol{\ell} = \mathbf{a}_x\, dx + \mathbf{a}_y\, dy + \mathbf{a}_z\, dz. \tag{C.2}$$

In words, this simply says that the component of $d\boldsymbol{\ell}$ in the x-direction is dx, the component of $d\boldsymbol{\ell}$ in the y-direction is dy, and the component of $d\boldsymbol{\ell}$ in the z-direction is dz. Then $\mathbf{F} \cdot d\boldsymbol{\ell}$ is $F_x\, dx + F_y\, dy + F_z\, dz$, which indicates that the integration that we must carry out is actually the sum of three integrals, one in dx, one in dy, and one in dz. In general, we should expect that F_x, F_y, and F_z will each be functions of x, y, and z, and we must use the equation of the path to eliminate the dependent variables. We show this by example.

Example C.3 Vector Path Integration

For the vector $\mathbf{F} = \mathbf{a}_x(x - y) + \mathbf{a}_y(xy)$, determine $\int_A^B \mathbf{F} \cdot d\boldsymbol{\ell}$ for the two paths shown in Fig. C.4, where $A = (0, 0)$ and $B = (1, 2)$.

Solution:
Since the two paths are along straight line segments, we will determine the path integrals using rectangular coordinates. $d\boldsymbol{\ell}$ is given in Eq. (C.2), where we can drop the z-component since our paths are restricted to the x–y plane. We have

$$\mathbf{F} \cdot d\boldsymbol{\ell} = F_x\, dx + F_y\, dy = (x - y)\, dx + (xy)\, dy.$$

For path 1, we break our integral into two parts, one for each segment

$$\int_A^B \mathbf{F} \cdot d\boldsymbol{\ell} = \int_A^{(1,0)} \mathbf{F} \cdot d\boldsymbol{\ell} + \int_{(1,0)}^B \mathbf{F} \cdot d\boldsymbol{\ell}.$$

The first term on the right side is

$$\int_A^{(1,0)} \mathbf{F} \cdot d\boldsymbol{\ell} = \int_A^{(1,0)} \left(F_x\, dx + F_y\, dy \right) = \int_0^1 F_x\, dx + \int_0^0 F_y\, dy,$$

where the limits of integration on the two integrals reflect the starting and ending coordinates for this segment: $(x, y) = (0, 0)$ to $(1, 0)$. Since the starting and ending limits of integration on the y integral are identical, this integral must be zero, regardless of the value of F_y, and we need consider it no further. Therefore,

$$\int_A^{(1,0)} \mathbf{F} \cdot d\boldsymbol{\ell} = \int_0^1 F_x\, dx = \int_0^1 (x - y)\Big|_{y=0}\, dx = \int_0^1 x\, dx = \frac{x^2}{2}\Big|_0^1 = \frac{1}{2}.$$

Notice that y appeared in the integrand, but that the variable of integration was x. Therefore, it was necessary to evaluate y on the path, which in this case was quite simple, since $y = 0$ anywhere along this segment.

Next we consider the integral along the vertical segment of the path from $(x, y) = (1, 0)$ to point B at $(1, 2)$. Here we write

$$\int_{(1,0)}^B \mathbf{F} \cdot d\boldsymbol{\ell} = \int_1^1 F_x\, dx + \int_0^2 F_y\, dy,$$

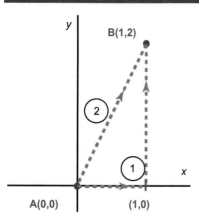

FIGURE C.4 Two paths leading from point $A = (0, 0)$ to point $B = (1, 2)$.

where the first term on the right is 0 for the same reason mentioned above — that is, the limits of integration are the same, so the area under the curve is zero. Now we evaluate the y integral:

$$\int_{(1,0)}^{B} \mathbf{F} \cdot d\boldsymbol{\ell} = \int_{0}^{2} F_y \, dy = \int_{0}^{2} (xy) \Big|_{x=1} \, dy = \int_{0}^{2} y \, dy = \frac{y^2}{2} \Big|_{0}^{2} = 2.$$

Notice here that x appears in the integrand, while y is the variable of integration. Since $x = 1$ for any point along this segment, the integrand is simply y. Combining the results for the two segments, we get $\int_{A}^{B} \mathbf{F} \cdot d\boldsymbol{\ell} = 5/2$ for path 1.

For path 2, the process is similar, but there is only a single segment, and we must use the equation of the straight line joining A and B, $y = 2x$, to reduce the variables:

$$\int_{A}^{B} \mathbf{F} \cdot d\boldsymbol{\ell} = \int_{0}^{1} F_x \, dx + \int_{0}^{2} F_y \, dy$$

$$= \int_{0}^{1} (x - y) \, dx + \int_{0}^{2} (xy) \, dy.$$

Here we see that the x integral includes a term in y, and since y and x are related through the equation of the path, $y = 2x$, we must use this to replace y in the integrand. Similarly, the y integral includes a term in x, which must be written in terms of y. The path integral becomes

$$\int_{A}^{B} \mathbf{F} \cdot d\boldsymbol{\ell} = \int_{0}^{1} [x - (2x)] \, dx + \int_{0}^{2} \left(\frac{1}{2}y\right) y \, dy = -\frac{x^2}{2} \Big|_{0}^{1} + \frac{y^3}{6} \Big|_{0}^{2} = \frac{5}{6}.$$

Since the path integral $\int_{A}^{B} \mathbf{F} \cdot d\boldsymbol{\ell}$ is different for the two paths that we evaluated, we would say that this integral is path-dependent. We will show later that the path integral of a static electric field will always be **path-independent**. This is an extremely important property, since that allows us to define the electric potential. We will return to this idea later, when we introduce the electric field properly.

We can, of course, also determine path integrals in cylindrical or spherical coordinates, and it will be most convenient to do so when the path is easier to express in these variables. For example, if our path were along a spiral-shaped curve, rising in the z-direction as it circles around the axis, then we should choose to work in cylindrical coordinates. The primary difference will be in how we define the path element $d\boldsymbol{\ell}$. In cylindrical coordinates we use

$$d\boldsymbol{\ell} = \mathbf{a}_\rho \, d\rho + \mathbf{a}_\phi \rho \, d\phi + \mathbf{a}_z \, dz. \tag{C.3}$$

Our interpretation of this $d\boldsymbol{\ell}$ is exactly parallel to that of Eq. (C.2). The ρ-component is the elemental length $d\rho$, the step size in the ϕ-direction is $\rho \, d\phi$, and in the z-direction it is dz, just as it was in rectangular coordinates. In spherical coordinates, we use

$$d\boldsymbol{\ell} = \mathbf{a}_R \, dR + \mathbf{a}_\theta R \, d\theta + \mathbf{a}_\phi R \sin\theta \, d\phi. \tag{C.4}$$

Each of the angular components in these expressions of $d\boldsymbol{\ell}$ can be seen to be the elemental length of an arc segment when sweeping through the incremental angle $d\theta$ or $d\phi$.

C.3 Integration over a Surface

C.3.1 Scalar Surface Integration

We routinely use surface densities (i.e. the measure of some quantity per unit area) in many different disciplines. Examples include the number of residents of a city or nation per square kilometer (or mile), the number of bushels of corn produced per acre of farmland, and, of direct relevance to the present study, the surface charge density, as often encountered at the surface of a conductor. We define the last of these as the quantity of charge per unit area,

$$\rho_s = \lim_{\Delta s \to 0} \frac{\Delta q}{\Delta s} = \frac{dq}{ds}, \tag{C.5}$$

where Δq is the charge contained within an infinitesimal surface element of area Δs. We illustrate this in Fig. C.5. The unit for the surface charge density ρ_s is $[\rho_s] = C/m^2$. Whereas we can automatically interpret the term "population density" as a surface density (people are nearly always located on or near the surface of the Earth), with charge densities we will be more careful to differentiate between the various flavors (volume, surface, or linear) of density.

If the surface charge density $\rho_s(\mathbf{r})$ is known over some region of surface s, then the total charge on the surface s is found by summing the charge $\rho_s(\mathbf{r}_i)\Delta s_i$ of each of the surface elements over the entire surface s. In the limit as the area of the surface elements become small, the summation

$$Q = \lim_{\Delta s_i \to 0} \sum_i \rho_s(\mathbf{r}_i)\Delta s_i$$

becomes the integral

$$Q = \int_s \rho_s(\mathbf{r})ds. \tag{C.6}$$

Here the subscript s on the integral sign indicates that we are to carry out the integration over the entire surface s. We choose the form of the surface element ds according to the geometry of the problem, as we will show later through examples.

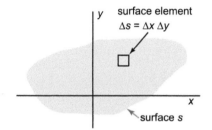

FIGURE C.5 Surface charge of density $\rho_s(\mathbf{r})$. The charge inside the surface element Δs is $\rho_s(\mathbf{r})\Delta s$.

Example C.4 Scalar Surface Integration

The surface charge density on the $x - y$ plane within the square of side $b = 1$ cm, with one corner at the origin and the opposite corner at $(b, b, 0)$, is given by $\rho_s(\mathbf{r}) = \rho_{s,0} \exp(-x/b) \exp(y/b)$, where $\rho_{s,0} = 1$ nC/m^2 is the surface charge density at the origin. Determine the total charge Q inside the square.

Solution:
We will use Eq. (C.6):

$$Q = \int_s \rho_s(\mathbf{r})\, ds$$

$$= \rho_{s,0} \int_0^b \int_0^b \exp(-x/b) \exp(y/b)\, dx\, dy.$$

Carrying out the integration, we find

$$Q = \rho_{s,0} \left.\frac{\exp(-x/b)}{-1/b}\right|_0^b \left.\frac{\exp(y/b)}{1/b}\right|_0^b = \rho_{s,0} b^2 \left(1 - e^{-1}\right)(e-1),$$

which leads to $Q = 1.09 \times 10^{-13}$ C.

C.3.2 Vector Surface Integration

A second useful surface integral is of the form

$$\int_s \mathbf{F} \cdot d\mathbf{s}.$$

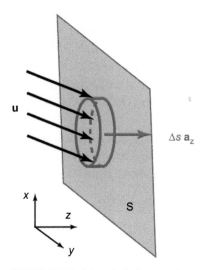

FIGURE C.6 Particles of velocity **u** crossing through a boundary from left to right. The vector d**s** represents an element of surface area, whose direction is normal to the surface.

Notice that the surface element $d\mathbf{s}$ in this case is written as a vector. We'll expand on this idea soon. In contrast to a surface charge density, which is a scalar quantity, the integrand of this surface integral is a vector quantity. The result of this integral, however, is a scalar. As an example, consider the flow of particles in a stream, for which we wish to measure the total rate of flow across a boundary s, as shown in Fig. C.6. This rate at which the particles flow across the entire surface is known as the **flux**, represented by the symbol Φ. It should be clear from Fig. C.6 that the direction in which the particles are flowing relative to the surface is important.

We start by defining the surface s as the x–y plane, and we show a small circle of area Δs, lying in the surface. We construct a volume to the left of the surface (in the shape of a disc in the figure), whose face is of area Δs and of thickness Δz. The volume of this element is $\Delta v = \Delta s \Delta z$. The number of particles within Δv is $\Delta N = n \Delta v = n \Delta z \Delta s$, where n is the number density, or the number of particles per unit volume, inside the disc. (The disc is small enough that we can treat the density as being uniform.)

Now let's consider the **flux density**, or the rate at which the particles cross the boundary s per unit area. We represent this quantity as $\Delta \Phi / \Delta s$. For simplicity, we will initially let the velocity **u** of all the particles within the disc be uniform. Then, within a time interval $\Delta t = \Delta z / u_z$, all of the particles within the disc will pass through the surface s, and we can write the flux density as

$$\frac{\Delta \Phi}{\Delta s} = \frac{\Delta N}{\Delta s \, \Delta t} = \frac{\Delta N}{\Delta s} \frac{u_z}{\Delta z}.$$

But as we wrote above, ΔN is $n\Delta z \Delta s$, so

$$\frac{\Delta \Phi}{\Delta s} = n u_z. \tag{C.7}$$

Notice that this flux depends only on the component of **u** in the z-direction, which should make sense, since only this component represents the flow of particles across the boundary. To find the total flux, we consider all surface elements of area Δs_i of the surface s, and sum over all elements i to find

$$\Phi = \sum_i (n u_z)_i \, \Delta s_i.$$

In the limit $\Delta s_i \to 0$, the summation becomes an integral, and the flux is

$$\Phi = \int_s n u_z \, ds.$$

At the beginning of our discussion we chose the x–y plane parallel to the surface s, and, as a result, we found that only the z-component of the velocity **u** mattered. If the surface were in a different orientation we should choose the component of **u** normal to that surface. A short-hand notation that accommodates any orientation of the surface is

$$\Phi = \int_s n\mathbf{u} \cdot \mathbf{a}_n \, ds,$$

where $\mathbf{a}_n$ is the unit vector normal to surface. A more compact notation of writing this is

$$\Phi = \int_s n\mathbf{u} \cdot d\mathbf{s}, \tag{C.8}$$

in which we define the *vector* surface element $d\mathbf{s} = \mathbf{a}_n \, ds$, whose magnitude is ds, the same as the *scalar* surface element ds introduced earlier, and whose direction is normal to the surface, given by $\mathbf{a}_n$. The scalar product $\mathbf{u} \cdot d\mathbf{s}$ picks out the component of **u** that is normal to the surface, as we require for this example.

Another assumption that we made earlier in our discussion was that the velocity of all of the particles is uniform. We can relax that requirement now. Since the flux Φ is linear in the velocity of the particles, it can be argued that Eq. (C.8) is valid for a distribution of velocities, if we interpret **u** as the average velocity of the particles.

Before we conclude this discussion, we should acknowledge an unfortunate dual usage of the word "flux." In some fields, such as fluid dynamics, the term flux is defined as the flow rate *per unit area* across a boundary. Since we are at present studying electromagnetics, we will adhere to the common usage within this field, as introduced earlier in this section. That is, the *flux density* is the rate of flow per unit area, and the *flux* is the total rate of flow through the entire surface. We use this terminology when we discuss electric currents, electric fields, magnetic fields, and power flow.

Example C.5 Vector Surface Integration

Determine $\int_s \mathbf{F} \cdot d\mathbf{s}$ for the vector $\mathbf{F} = \mathbf{a}_\rho \sin^2 \phi + \mathbf{a}_z z \cos^2 \phi$ over a cylindrical surface of radius a and height h co-axial with the z-axis.

Solution:
For the wall of the cylinder, we use $d\mathbf{s} = \mathbf{a}_\rho a \, d\phi \, dz$. Note that the magnitude of $d\mathbf{s}$ is the same as the scalar surface element for the outside surface of a cylinder, and the direction $\mathbf{a}_\rho$ is radially outward, normal to surface. Then $\mathbf{F} \cdot d\mathbf{s}$ is $(\mathbf{a}_\rho \sin^2 \phi + \mathbf{a}_z z \cos^2 \phi) \cdot \mathbf{a}_\rho a \, d\phi \, dz = a \sin^2 \phi \, d\phi \, dz$. We integrate this as

$$\int_s \mathbf{F} \cdot d\mathbf{s} = \int_{\phi=0}^{2\pi} \int_{z=0}^{h} a \sin^2 \phi \, dz \, d\phi = \pi a h.$$

Notice that the z-component of **F** does not enter into this calculation since this component lies within the surface, perpendicular to $d\mathbf{s}$, and therefore does not contribute to the flow through the surface.

Example C.6 Vector Surface Integration, Water Current

The water flow in a pipe of radius $b = 1$ cm is $\mathbf{J}_v(\rho) = J_0(1 - \rho^2/b^2)\mathbf{a}_z$, where $J_0 = 1$ l/cm² is the current density at the center, and the center of the pipe lies on the z-axis. Determine the total flow rate through the pipe.

Solution:

This problem deals with fluid flow, but the same ideas apply to water flow and charge flow. Since $\mathbf{J}_v(\rho)$ gives the current per unit area of water in the pipe, we need only integrate this quantity across the cross-section of the pipe to get the total flow rate. Since the pipe is round, cylindrical coordinates are most convenient, and the vector surface element is $d\mathbf{s} = \rho d\rho\, d\phi\, \mathbf{a}_z$. Then,

$$
\begin{aligned}
I &= \int \mathbf{J}_v(\rho) \cdot d\mathbf{s} \\
&= \int_0^{2\pi} \int_0^b J_0\left(1 - \frac{\rho^2}{b^2}\right)\mathbf{a}_z \cdot \mathbf{a}_z\, \rho d\rho\, d\phi \\
&= J_0 \int_0^b \left(\rho - \frac{\rho^3}{b^2}\right) d\rho \int_0^{2\pi} d\phi = J_0 b^2 \pi / 2 = 1.57 \;\; \text{l/s}.
\end{aligned}
$$

C.4 Integration over a Volume

Finally, we will often need to carry out integration of various scalar and vector quantities over a volume in space. You are no doubt already familiar with scalar volume integration. For example, given a density $\rho_v(\mathbf{r})$ (this could be a charge density, for example), then the total quantity (charge) is

$$
q = \int_v \rho_v(\mathbf{r}) dv.
$$

(Recall that $\mathbf{r}$ used here is short-hand notation for spatial coordinates, and is interpreted as x, y, and z when working in rectangular coordinates, ρ, ϕ, and z when in cylindrical coordinates, and R, θ, and ϕ when in rectangular coordinates.)

Where does this integral relation come from? Picture an infinitesimal region in space of volume Δv that contains a charge Δq. The charge density, that is, the charge per unit volume, is defined as

$$
\rho_v = \lim_{\Delta v \to 0} \frac{\Delta q}{\Delta v} = \frac{dq}{dv}. \tag{C.9}
$$

The second equality follows immediately from the formal definition of the derivative of a function.

Then, if the charge density $\rho_v(x, y, z)$ is known over some region of space v, the total charge within the entire volume v is found by adding all of the charges within each of the elements of the volume v. Consider the volume illustrated in Fig. C.7. The charge within one element of volume Δv_i is $\rho_v(\mathbf{r}_i)\Delta v_i$, where $\rho_v(\mathbf{r}_i)$ is the charge density within that element. The subscript i is an index used to label each of the different elements. Summing these bits of charge then gives us

$$
q = \sum_i \rho_v(\mathbf{r_i})\Delta v_i,
$$

where the summation extends over all of the elements in the volume v. In the limit as $\Delta v_i \to 0$, this summation becomes a volume integral of ρ_v, and

$$
q = \int_v \rho_v(\mathbf{r}) dv, \tag{C.10}
$$

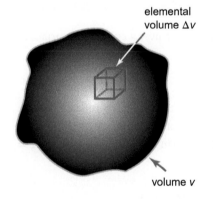

elemental volume Δv

volume v

FIGURE C.7 A distribution of charge within a volume v. The charge within the elemental volume Δv located at position $\mathbf{r}$ is $\rho_v(\mathbf{r})\Delta v$.

where the subscript v on the integral sign indicates that the integral is to be carried out over the entire volume v. We reviewed the explicit forms of the element of integration dv in rectangular, cylindrical, and spherical coordinate systems in Section C.1.

Example C.7 Scalar Volume Integration

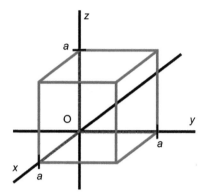

FIGURE C.8 A cube of charge. The charge density inside the cube is $\rho_v(\mathbf{r}) = 1$ nC/m$^3 \times \sin(\pi x/a)$.

The charge density in a cube of side $a = 1$ cm, with one corner at the origin and the opposite corner at (a, a, a) as shown in Fig. C.8, is given by $\rho_v(\mathbf{r}) = \rho_{v,0}\sin(\pi x/a)$, where $\rho_{v,0} = 1$ nC/m^3. Determine the total charge Q contained inside the cube.

Solution:
We will use Eq. (C.10) to find the total charge:

$$Q = \int_v \rho_v(\mathbf{r})dv$$

$$= \rho_{v,0} \int_0^a \int_0^a \int_0^a \sin(\pi x/a)\, dx\, dy\, dz.$$

Note that we have used rectangular coordinates, since this is the most convenient system for this geometry, and therefore $dv = dx\, dy\, dz$. Carrying out the integration, we find

$$Q = \rho_{v,0} \left. \frac{-\cos(\pi x/a)}{\pi/a} \right|_0^a a^2,$$

which leads to $Q = 2/\pi \times 10^{-15}$ C.

When taking the volume integral of a vector quantity, such as

$$\mathbf{A}(\mathbf{r}) = \frac{\mu}{4\pi} \int_v \frac{\mathbf{J}_v(\mathbf{r}')}{|\mathbf{r} - \mathbf{r}'|}dv',$$

we interpret this as three separate equations, one for the x-component,

$$A_x = \frac{\mu}{4\pi} \int_v \frac{J_{v,x}(\mathbf{r}')}{|\mathbf{r} - \mathbf{r}'|}dv',$$

one for the y-component,

$$A_y = \frac{\mu}{4\pi} \int_v \frac{J_{v,y}(\mathbf{r}')}{|\mathbf{r} - \mathbf{r}'|}dv',$$

and a third for the z-component,

$$A_z = \frac{\mu}{4\pi} \int_v \frac{J_{v,z}(\mathbf{r}')}{|\mathbf{r} - \mathbf{r}'|}dv'.$$

(Don't worry about the difference between $\mathbf{r}$ and $\mathbf{r}'$, or the significance of the prime on dv'. We get into this in Sections 1.1 with Coulomb's Law and 4.1 with the Biot–Savart Law. For now, we are just interested in the meaning of the integration of the vector quantity, which in this case is $\mathbf{J}_v/|\mathbf{r} - \mathbf{r}'|$.) In other words, this is short-hand notation for three equations involving integrals of scalar quantities ($J_x/|\mathbf{r} - \mathbf{r}'|$, $J_y/|\mathbf{r} - \mathbf{r}'|$, or $J_z/|\mathbf{r} - \mathbf{r}'|$), and we can apply the integration techniques with which we are already familiar.

PROBLEMS   **Integration Along a Path**

PC-1   Consider a vector field $\mathbf{F} = \mathbf{a}_x xy + \mathbf{a}_y yz + \mathbf{a}_z zx$. For the two paths shown in Fig. C.9(a), determine the path integral $\int_A^B \mathbf{F} \cdot d\boldsymbol{\ell}$, where $A = (0, 0, 0)$ and $B = (1, 1, 1)$. Path 1 consists of three straight segments: from A to $(1, 0, 0)$ along the x-axis; from $(1, 0, 0)$ to $(1, 1, 0)$; and from $(1, 1, 0)$ to B. Path 2 is a straight direct line from A to B, along which $x = y = z$.

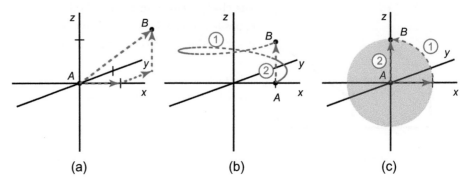

FIGURE C.9 Paths leading from point A to point B for path integration in Probs. PC-1 to PC-3.

PC-2   Consider a vector field $\mathbf{B} = \mathbf{a}_\rho \rho^2 \cos\phi - \mathbf{a}_\phi z^2 \rho + \mathbf{a}_z z \rho \phi^2$. For the two paths shown in Fig. C.9(b), determine the path integral $\int_A^B \mathbf{B} \cdot d\boldsymbol{\ell}$, where $A = (1, 0, 0)$ and $B = (1, 0, 1)$. Path 1 consists of a smooth one-turn spiral, described by $z = \phi/2\pi$, while Path 2 is a straight direct line from A to B, along which $\phi = 0$.

PC-3   Consider a vector field $\mathbf{D} = \mathbf{a}_R R + \mathbf{a}_\theta R \cos\theta + \mathbf{a}_\phi R \sin\theta \cos\phi$. For the two paths shown in Fig. C.9(c), determine the path integral $\int_A^B \mathbf{D} \cdot d\boldsymbol{\ell}$, where $A = (0, 0, 0)$ and $B = (1, 0, 0)$ (i.e. the "North Pole"). Path 1 consists of two segments: from A to $(1, \pi/2, 0)$ along the x-axis; and "due North" from $(1, \pi/2, 0)$ to B. Path 2 is a straight direct line from A to B, along which $\theta = 0$.

Integration Over a Surface

PC-4   Determine the surface area on the cylindrical wall of a cylinder of radius $a_0 = 2$ m, bounded by $0 < \phi < 2\pi$ and $0 < z < a_0 \phi/2\pi$. Sketch this surface.

PC-5   Determine the surface area of a segment of the surface of a sphere of radius 10 m bounded by $0 < \theta < \pi/4$ and $0 < \phi < \pi/2$. Sketch this surface.

PC-6   Determine the total charge on a cylinder of radius a and length L when the surface charge density is (a) $\rho_s(\phi, z) = \rho_0 \sin(2\phi)e^{-2z/L}$, and (b) $\rho_s(\phi, z) = \rho_0 \sin(\phi/2)e^{-2z/L}$.

PC-7   During a rainstorm, the density of raindrops in the sky is 200 drops per cubic meter, and the average raindrop volume is 0.01 ml. The wind is strong, causing the drops to fall at an angle 60° from vertical. The average velocity of these drops is 10 cm/s. Determine the rate of increase of the depth of water on the surface in cm/hr, assuming no drainage of the rain water into or out of this area.

PC-8   (a) Determine the surface flux $\int_{s_a} \mathbf{E} \cdot d\mathbf{s}$ for the vector field $\mathbf{E} = E_0 \mathbf{a}_R/R^2$ for the surface defined by a planar circle of radius a_0 centered on the z-axis a distance z_0 from the origin, as shown in Fig. C.10(a). Write your answer in terms of the angle $\theta_0 = \tan^{-1}(a_0/z_0)$. (b) Repeat for the surface s_b shown in Fig. C.10(b). The surface s_b is a portion of the surface of a sphere of radius $r_0 = \sqrt{a_0^2 + z_0^2}$ centered at the origin. Note that the surfaces s_a and s_b have the same edge. Compare your results for the two surface fluxes.

FIGURE C.10 Surfaces for evaluation of the surface flux $\int_s \mathbf{E} \cdot d\mathbf{s}$ in Prob. PC-8.

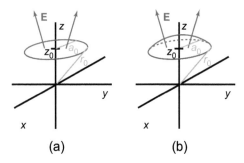

(a)　　　　　(b)

Integration Over a Volume

PC-9　A cone of height h and base radius a has a density given by $\rho_0 \cos^2(\phi)\, e^{-z/h}$. Its axis is parallel to the z-axis, and its base is in the x–y plane, as shown in Fig. C.11. Determine the mass of this cone.

FIGURE C.11 A cone-shaped object for Prob. PC-9.

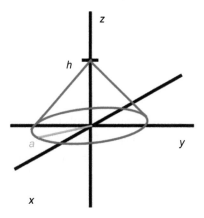

PC-10　A sphere of radius a is centered at the origin, and has a density given by $\rho_0 e^{-R/a}$. Determine the mass of this sphere.

APPENDIX D
Gradient, Divergence, and Curl

When we began our review of vectors in Appendix A, we wrote these vectors as constants, such as the vector $\mathbf{A} = 5\mathbf{a}_x + 3\mathbf{a}_y - 2\mathbf{a}_z$. The components $A_x = 5$, $A_y = 3$, and $A_z = -2$ did not vary with position x, y, and z. These worked well for us as we reviewed addition and subtraction of vectors, or multiplication of vectors using the dot or cross product. Real physical quantities, however, whether scalars (such as temperature, pressure, or charge densities) or vectors (such as electric fields, magnetic fields, ocean currents, or electrical current densities), typically vary from place to place. We call these **scalar fields** or **vector fields**, respectively. In order to describe the spatial variation of a scalar field, a differential operator known as the **gradient** is used. (A differential operator, in mathematics, is an operator that calls for taking derivatives of a function.) Similarly, it is useful to define two specific forms of differential operators of vector fields: These are the **divergence** and **curl** operators. We introduce these operators in this Appendix.

D.1 The Gradient of a Scalar Field

You will recall from your introductory calculus class that the slope of a function $f(x)$ that depends on only a single variable x can be found by taking the derivative of the function f with respect to x,

$$f'(x) \equiv \frac{df(x)}{dx} = \lim_{\Delta x \to 0} \frac{f(x + \Delta x) - f(x)}{\Delta x}.$$

The slope is a measure of the change in $f(x)$ when x changes by a small increment Δx (i.e. $x \to x + \Delta x$), divided by that distance Δx. In the case of a scalar field, $\psi(\mathbf{r})$, which is a function of variables x, y, and z (or ρ, ϕ, and z, or R, θ, and ϕ), this is a bit more complicated, but the idea is the same. When working in rectangular coordinates, we would like a measure of how much $\psi(\mathbf{r})$ changes for a displacement Δx in the x-direction, as well as for a displacement Δy in the y-direction, and for a displacement Δz in the z-direction. The gradient of a scalar field, represented as $\nabla\psi(\mathbf{r})$ or grad $\psi(\mathbf{r})$, is a vector quantity that represents each of these increments. Its magnitude is the maximum rate of change of the scalar, and its direction points in the direction of this maximum rate of change. For example, if I am standing on a hillside midway between the base and the peak of the hill, and the peak is southwest from my position, then the steepest ascent of the hillside from my position is to the southwest. If I were to walk due south, I would likely be going uphill, but not as rapidly as if I were to walk directly toward the peak to the southwest. Likewise, if I were to walk to the southeast or northwest, I would probably remain at the same elevation, going neither uphill or downhill. So the direction in which I walk is important in determining the rate at which I change my elevation. The gradient of a scalar function allows us to describe the change in the function in different directions.

It has a different form in each of the three coordinate systems that we introduced in Appendix B. It is

$$\boldsymbol{\nabla}\psi = \mathbf{a}_x \frac{\partial\psi}{\partial x} + \mathbf{a}_y \frac{\partial\psi}{\partial y} + \mathbf{a}_z \frac{\partial\psi}{\partial z},$$ (D.1)

as expressed in rectangular coordinates. In cylindrical coordinates, the gradient is

$$\boldsymbol{\nabla}\psi = \mathbf{a}_\rho \frac{\partial\psi}{\partial\rho} + \mathbf{a}_\phi \frac{1}{\rho}\frac{\partial\psi}{\partial\phi} + \mathbf{a}_z \frac{\partial\psi}{\partial z},$$ (D.2)

and in spherical coordinates

$$\boldsymbol{\nabla}\psi = \mathbf{a}_R \frac{\partial\psi}{\partial R} + \mathbf{a}_\theta \frac{1}{R}\frac{\partial\psi}{\partial\theta} + \mathbf{a}_\phi \frac{1}{R\sin\theta}\frac{\partial\psi}{\partial\phi}.$$ (D.3)

Example D.1 Gradient

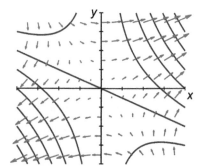

FIGURE D.1 A contour plot of the function $g(x, y) = x^2 + 2xy$. The ranges of x and y in this plot are from -2 to $+2$. The arrows show the gradient of $g(x, y)$, $\boldsymbol{\nabla}g(x, y)$.

Consider the scalar field defined by $g(x,y) = x^2 + 2yx$. Determine the gradient of $g(x, y)$.

Solution:
Since $g(x, y) = x^2 + 2yx$ is given to us using rectangular coordinates, we will determine $\boldsymbol{\nabla}g(x, y)$ using Eq. (D.1). We must use $\partial g/\partial x = 2x + 2y$ and $\partial g/\partial y = 2x$, giving us $\boldsymbol{\nabla}g(x, y) = \mathbf{a}_x(2x + 2y) + \mathbf{a}_y 2x$. We illustrate a contour plot of $g(x, y)$ and $\boldsymbol{\nabla}g(x, y)$ (shown as the arrows) in Fig. D.1.

The gradient of a function is always perpendicular to the contours of that function. By contours, we mean lines along which that function is constant. Think in terms of a contour map, for those of you who have done some hiking. The contour lines on this map denote lines of constant elevation. So when we walk directly along a contour line, we are going neither up nor down the slope. From this direction, if we were to turn 90° to our left or right, and proceed, then in this direction our rate of ascent is greatest (in a negative sense, of course, if we turned downhill). We illustrate this in Fig. D.1, in which we have plotted contours for the function $g(x, y) = x^2 + 2xy$ that we used in Example D.1. The ranges of x and y in this plot are from -2 to $+2$, and the solid lines show the values of $g(x, y)$. For example, along the straight, blue contour that passes through the origin, the function $g(x, y)$ is zero. The light blue lines show where $g(x, y)$ is $+2$, while along the green lines, $g(x, y)$ is $+4$. We also show $\boldsymbol{\nabla}g(x, y)$ on this plot, marked by the arrows. The length of the arrows gives the relative magnitude of $\boldsymbol{\nabla}g(x, y)$, while the direction in which the arrows point shows the direction of $\boldsymbol{\nabla}g(x, y)$. In the first quadrant, for example, the arrows point up and to the right. Notice that the arrows always cross the contours at right angles. (There may be a few arrows in this plot that don't look quite perpendicular to the contours. This is actually just a matter of the spatial resolution of the plot. Each arrow shows the gradient $\boldsymbol{\nabla}g(x, y)$ at the *base* of the arrow. If the arrow crosses the contour closer to the arrowhead, then the gradient at that position has already changed, and that is the reason that some arrows may appear to be not quite perpendicular to the contour.)

Throughout our studies, we will often have need to determine the unit vector that is normal to a surface, often denoted $\mathbf{a}_n$. The gradient operator gives us a simple means to determine this. We usually define a surface using an equation of the form $\psi(x, y, z) = C$, where C is a constant. Simple examples of equations that define a surface include: $y = 0$, $y = 5$, $x + 2y - 2z = 2$, or $x^2 + y^2 + z^2 = 3^2$. For instance, $y = 0$ defines the plane in which $y = 0$, but x and z can be anything. The plane defined by $y = 5$ is parallel to this plane, shifted by five units in the $+y$-direction. The unit vector normal to either of these planes is $\mathbf{a}_y$. We can find this formally by using

$$\mathbf{a}_n = \pm \frac{\boldsymbol{\nabla}\psi}{|\boldsymbol{\nabla}\psi|}. \qquad (D.4)$$

We motivate this expression as follows. Since we define a surface by an expression of the form $\psi(\mathbf{r}) = C_1$, and surfaces parallel to this would be $\psi(\mathbf{r}) = C_2$, $\psi(\mathbf{r}) = C_3$, etc., where C_2 and C_3 are constants different from C_1, we can think of these surfaces in precisely the same way as we think about contours of a function ψ, as we discussed above. The gradient of $\psi(\mathbf{r})$ tells us the direction to move most rapidly from one surface to the next, which must be the normal to the surface. We divide the gradient function by its magnitude in Eq. (D.4) to give it unit magnitude. There are, of course, two unit normal vectors for any surface, pointing in directions opposite one another, leading to the $\pm$ sign in Eq. (D.4). We will discuss the convention guiding our choice of sign for a "closed" surface later.

Example D.2 Unit Normal Vector

Consider the surfaces defined by (a) $x + 2y - 2z = 2$ and (b) $x^2 + y^2 + z^2 = 3^2$. Find the unit normal vector for these two surfaces.

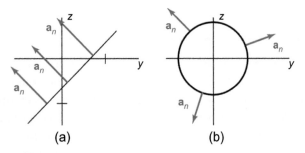

FIGURE D.2 Two surfaces, for calculation of the surface normal as in Example D.2.

Solution:

(a) $x + 2y - 2z = 2$ defines a plane as illustrated in Fig. D.2(a). (For clarity, we only show the projection onto the y–z plane in this figure.) We compute $\boldsymbol{\nabla}(x + 2y - 2z) = \mathbf{a}_x + 2\mathbf{a}_y - 2\mathbf{a}_z$ using Eq. (D.1). The magnitude of this vector is $|\mathbf{a}_x + 2\mathbf{a}_y - 2\mathbf{a}_z| = \sqrt{1^2 + (2)^2 + (-2)^2} = 3$, so our unit normal is $\mathbf{a}_n = \pm(\mathbf{a}_x + 2\mathbf{a}_y - 2\mathbf{a}_z)/3$. We show the result with the $-$ sign in the figure. Notice that the normal vector is the same anywhere on this surface. This is exactly as we should expect for a planar surface.

(b) $x^2 + y^2 + z^2 = 3^2$ defines a sphere of radius 3, centered at the origin. We illustrate this in Fig. D.2(b). To determine the unit normal vector, we find $\boldsymbol{\nabla}(x^2 + y^2 + z^2) = 2x\mathbf{a}_x + 2y\mathbf{a}_y + 2z\mathbf{a}_z$ and $|2x\mathbf{a}_x + 2y\mathbf{a}_y + 2z\mathbf{a}_z| = \sqrt{(2x)^2 + (2y)^2 + (2z)^2}$. Thus, we find $\mathbf{a}_n = (x\mathbf{a}_x + y\,\mathbf{a}_y + z\mathbf{a}_z)/\sqrt{x^2 + y^2 + z^2} = \mathbf{a}_R$. This is the unit vector pointing radially outward from the origin. We show several examples of the outwardly directed surface normal vector.

We could also have easily solved this example in spherical coordinates. When we convert the equation of the surface $x^2 + y^2 + z^2 = 3^2$ to spherical coordinates, we get $R^2 = 3^2$, or $R = 3$. Using Eq. (D.3) for the gradient of a function in spherical coordinates, and noting that $\partial\psi/\partial\theta$ and $\partial\psi/\partial\phi$ vanish, we are left with only the radial term. Since $\partial\psi/\partial R = 3$, we have $\boldsymbol{\nabla}\psi = 3\mathbf{a}_R$, and Eq. (D.4) gives us $\mathbf{a}_n = \mathbf{a}_R$, in agreement with our result found using rectangular coordinates.

Example D.3 Gradient

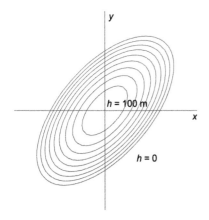

y

$h = 100$ m

x

$h = 0$

FIGURE D.3 Contour plot of an elliptically shaped island, for use in Example D.3.

Consider the elliptically shaped island illustrated in Fig. D.3. The elevation above sea level, $h(x, y)$, is given by the function

$$h(x, y) = h_0 \left(1 - \frac{(x - y)^2}{2a^2} - \frac{(x + y)^2}{2b^2} \right),$$

where $h_0 = 100$ m is the peak elevation, and $a = 2$ km and $b = 4$ km define the dimensions of the island. Determine the maximum slope for any pathway on this island.

Solution:

We need to determine the gradient of $h(x, y)$ to solve this problem. We could, of course, determine the gradient directly in the x–y coordinates, but it seems simpler if we define $x' = (x - y)/\sqrt{2}$ and $y' = (x + y)/\sqrt{2}$, such that

$$h(x', y') = h_0 \left(1 - \frac{x'^2}{a^2} - \frac{y'^2}{b^2} \right).$$

Then,

$$\boldsymbol{\nabla}' h(x', y') = \frac{\partial h}{\partial x'} \mathbf{a}_{x'} + \frac{\partial h}{\partial y'} \mathbf{a}_{y'} = h_0 \left(-\frac{2x'}{a^2} \mathbf{a}_{x'} - \frac{2y'}{b^2} \mathbf{a}_{y'} \right).$$

This slope is maximized at the coastline (where $h = 0$ and $x'^2/a^2 + y'^2/b^2 = 1$), and since $a < b$, the maximum value is found at $y' = 0$, $x' = \pm a$. The slope at this location is $2h_0/a$.

We will later use the gradient operator when we discuss static electric fields and electric potentials. Specifically, we will show that the electric field $\mathbf{E}$ is equal to $-\boldsymbol{\nabla} V(\mathbf{r})$, where $V(\mathbf{r})$ is the electric potential. This is discussed in more detail in Chapter 1.

Before we leave this discussion of the gradient operator, there is a relation involving the gradient that will come in handy later, and so we will derive it now and refer back to this relation later when needed. The relation we will need is

$$\boldsymbol{\nabla} \left(\frac{1}{|\mathbf{r} - \mathbf{r}'|} \right) = \frac{-(\mathbf{r} - \mathbf{r}')}{|\mathbf{r} - \mathbf{r}'|^3}. \tag{D.5}$$

In this equation, notice that are taking the gradient of a function, but that function includes two kinds of position vectors, $\mathbf{r}$ and $\mathbf{r}'$. This may seem strange, so let's motivate this before we evaluate the gradient. The vector $\mathbf{r}'$ might represent the location of a charge, for example, and the vector $\mathbf{r}$ might be the point at which we want to determine a field. Then the vector $\mathbf{r} - \mathbf{r}'$ simply points from that charge to the field point, and $|\mathbf{r} - \mathbf{r}'|$ is the distance between these two points. The gradient $\boldsymbol{\nabla}$ (with no primes) tells us to differentiate $|\mathbf{r} - \mathbf{r}'|^{-1}$ with respect to the unprimed coordinates.

Let's now prove Eq. (D.5) by direct calculation in rectangular coordinates:

$$\boldsymbol{\nabla} \left(\frac{1}{|\mathbf{r} - \mathbf{r}'|} \right) = \left(\mathbf{a}_x \frac{\partial}{\partial x} + \mathbf{a}_y \frac{\partial}{\partial y} + \mathbf{a}_z \frac{\partial}{\partial z} \right) \frac{1}{D},$$

where we have abbreviated $D = |\mathbf{r} - \mathbf{r}'| = \left[(x - x')^2 + (y - y')^2 + (z - z')^2 \right]^{1/2}$. Then,

$$\nabla\left(\frac{1}{|\mathbf{r}-\mathbf{r}'|}\right) = \left(-\frac{1}{2}\right)\frac{2\mathbf{a}_x(x-x') + 2\mathbf{a}_y(y-y') + 2\mathbf{a}_z(z-z')}{D^3}$$

$$= \frac{-\mathbf{a}_x(x-x') - \mathbf{a}_y(y-y') - \mathbf{a}_z(z-z')}{D^3}$$

$$= \frac{-(\mathbf{r}-\mathbf{r}')}{|\mathbf{r}-\mathbf{r}'|^3},$$

which is the result that we set out to show. How do we interpret this result? The function $1/|\mathbf{r}-\mathbf{r}'|$ is a function that diverges at $\mathbf{r} = \mathbf{r}'$, and decreases monotonically in all directions from this peak. We should expect the gradient of this function to point inward toward the peak, and its magnitude should be the slope. With a little thought, we see that Eq. (D.5) satisfies this expectation.

Note that, upon interchanging $\mathbf{r}$ and $\mathbf{r}'$, we can rewrite Eq. (D.5) as

$$\nabla'\left(\frac{1}{|\mathbf{r}-\mathbf{r}'|}\right) = \frac{(\mathbf{r}-\mathbf{r}')}{|\mathbf{r}-\mathbf{r}'|^3}, \tag{D.6}$$

where ∇' is the gradient operator involving the derivatives with respect to the three components of $\mathbf{r}'$. We will refer back to this result later in our development of static electric fields, and again for static magnetic fields.

D.2 The Divergence of a Vector Field

The first of the differential operators for vector fields that we will discuss is the divergence. This operator, in broad terms, gives us a measure of the growth or diminution of the field in the direction in which it points. We will present the divergence of a vector field directly using its explicit forms in rectangular, cylindrical, and spherical coordinates:

$$\nabla \cdot \mathbf{A} = \frac{\partial A_x}{\partial x} + \frac{\partial A_y}{\partial y} + \frac{\partial A_z}{\partial z}, \tag{D.7}$$

$$\nabla \cdot \mathbf{A} = \frac{1}{\rho}\frac{\partial(\rho A_\rho)}{\partial \rho} + \frac{1}{\rho}\frac{\partial A_\phi}{\partial \phi} + \frac{\partial A_z}{\partial z}, \tag{D.8}$$

and

$$\nabla \cdot \mathbf{A} = \frac{1}{R^2}\frac{\partial(R^2 A_R)}{\partial R} + \frac{1}{R\sin\theta}\frac{\partial(A_\theta \sin\theta)}{\partial \theta} + \frac{1}{R\sin\theta}\frac{\partial A_\phi}{\partial \phi}, \tag{D.9}$$

respectively. While there are more mathematically rigorous and general definitions of the divergence, these forms are sufficient for our purposes, and they have the advantage that we can apply them directly to vector fields to determine their divergence. A vector field that is perfectly uniform has a zero divergence. Note that in each of the expressions for divergence, the derivative that we require for each term matches the particular component of the vector field. For instance, in rectangular coordinates, the divergence of $\mathbf{A}$ includes a term $\partial A_x/\partial x$, but not $\partial A_x/\partial y$ or $\partial A_x/\partial z$. (Remember these last two terms, however. They will show up in the next section when we discuss the curl of a vector field.)

Example D.4 Divergence

Determine $\nabla \cdot \mathbf{A}$ for $\mathbf{A} = xy\mathbf{a}_x + 3zx\mathbf{a}_y - 5z^2\mathbf{a}_z$.

Solution:

Since $\mathbf{A}$ is given in rectangular coordinates, we will use Eq. (D.7) to find its divergence. We need $\partial A_x/\partial x = \partial(xy)/\partial x = y$, $\partial A_y/\partial y = \partial(3zx)/\partial y = 0$, and $\partial A_z/\partial z = \partial(-5z^2)/\partial z = -10\,z$. Thus we have $\nabla \cdot \mathbf{A} = y - 10\,z$.

Notice that the divergence operation always yields a scalar field. The result $\nabla \cdot \mathbf{A} = y - 10z$ depends on position, but it has no direction associated to it.

Example D.5 will be relevant to the electric field produced by a point charge.

Example D.5 Divergence of Radial Inverse Square Field

For the vector $\mathbf{K} = C\left(x\mathbf{a}_x + y\mathbf{a}_y + z\mathbf{a}_z\right)/\left(x^2 + y^2 + z^2\right)^{3/2}$, where C is a constant, determine $\nabla \cdot \mathbf{K}$.

Solution:

We will work this two ways. First, since $\mathbf{K}$ is given to us in rectangular coordinates, we will use Eq. (D.7), as we did in the previous example. We will start by finding $\partial K_x/\partial x$, where $K_x = Cx\left(x^2 + y^2 + z^2\right)^{-3/2}$. Using the product rule for derivatives, we find

$$\frac{\partial K_x}{\partial x} = C\left[\left(x^2 + y^2 + z^2\right)^{-3/2} + x\,(-3/2)(2x)\left(x^2 + y^2 + z^2\right)^{-5/2}\right].$$

We factor out a common denominator, finding

$$\frac{\partial K_x}{\partial x} = C\frac{\left[\left(x^2 + y^2 + z^2\right) - 3x^2\right]}{\left(x^2 + y^2 + z^2\right)^{5/2}}.$$

Since K_y and K_z are of similar form, we do not need to repeat our derivative, and can immediately write

$$\frac{\partial K_y}{\partial y} = C\frac{\left[\left(x^2 + y^2 + z^2\right) - 3y^2\right]}{\left(x^2 + y^2 + z^2\right)^{5/2}}$$

and

$$\frac{\partial K_z}{\partial z} = C\frac{\left[\left(x^2 + y^2 + z^2\right) - 3z^2\right]}{\left(x^2 + y^2 + z^2\right)^{5/2}}.$$

Summing these three terms gives us $\nabla \cdot \mathbf{K} = 0$. Note that this result is not valid at the origin, since $|\mathbf{K}| \to \infty$ at this point and the derivatives are undefined.

As we promised, we will also work $\nabla \cdot \mathbf{K}$ a second way – that is, using spherical coordinates. To do this, notice that the numerator of $\mathbf{K}$ is, when converted to spherical coordinates, $\left(x\mathbf{a}_x + y\mathbf{a}_y + z\mathbf{a}_z\right) = R\,\mathbf{a}_R$, and the denominator is $\left(x^2 + y^2 + z^2\right)^{3/2} = R^3$. Then we can write $\mathbf{K} = CR^{-2}\mathbf{a}_R$, which has only a radial component. Using Eq. (D.9) for the divergence of a vector field in spherical coordinates, we need

$$\frac{1}{R^2}\frac{\partial(R^2 K_R)}{\partial R} = \frac{1}{R^2}\frac{\partial(R^2 \times CR^{-2})}{\partial R},$$

which is zero. Since K_θ and K_ϕ are zero, clearly the other two derivatives in Eq. (D.9) are zero as well. Therefore, we again find $\nabla \cdot K = 0$, a result that is valid for $R > 0$, but not at the origin.

We previously made the statement that the divergence operator gives us a means to quantify the growth of a vector field, such that a perfectly uniform field has a zero divergence. Yet in the previous example we examined a vector field that was pointing radially outward, with its radial component decreasing with increasing distance. This field is certainly not uniform, so we have to ask: Does this result contradict our picture of what the divergence gives us? The answer is no, not at all. A uniform field is only one example of a field that is not growing. We need to wait a bit to expand on this more fully, but the explanation rests with the surface flux of this vector field through the surface of a volume that encloses the origin. For example, we will show that the flux of K through a surface of radius R that is centered at the origin does not depend on the radius R, and this is what allows us to understand the zero divergence of this important field.

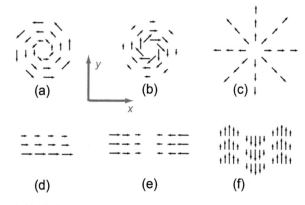

Finally, we show in Fig. D.4 diagrams of several sample vector fields. We can examine these fields to evaluate their divergence. In this figure and throughout this text, we use the notation that an arrow shows the direction of the vector field at the location indicated by its base, and the length of the arrow represents the relative magnitude of the vector. In this figure, fields (a) and (b) have zero divergence, since any variation in the components is in a direction orthogonal to the vector direction. Notice how the y-component varies in the x-direction, and vice versa. Field (c) appears to point radially outward from the center, and this field definitely has non-zero divergence. If the radial component were shown to decrease with increasing R, then we might not be certain, since a $1/R^2$ dependence of the radial component, as shown in the prior example, does have zero divergence. Of the remaining fields in this figure, only field (e) has a non-zero divergence.

FIGURE D.4 Samples of vector fields in two dimensions. Identify those whose divergence is zero.

D.3 The Divergence Theorem

Now that we have mastered the concept of the divergence of a vector field, and previously integration, we are ready to introduce an important theorem known as the Divergence Theorem. This applies to many problems that we will encounter, starting in Section 3.2 with the Continuity Equation, a statement in mathematical terms that electric charge must be conserved. The Divergence Theorem is extremely critical to our development of fields concepts, and we will work through a simple proof, perhaps not so much to convince ourselves of its truth, but more to become familiar with its meaning and significance, and to emphasize definitions of surface integrals, volume integrals, and the divergence of a vector field. Our proof will be stated using rectangular coordinates, but of course it is much more general than that and it is equally valid in any coordinate system. The Divergence Theorem states that, for any vector field $\mathbf{E}(\mathbf{r})$ that is continuous and differentiable over some region in space v, the volume integral of $\boldsymbol{\nabla} \cdot \mathbf{E}$ is equal to the surface integral of $\mathbf{E}$ over the surface s that forms the boundary of v, that is:

$$\int_v \boldsymbol{\nabla} \cdot \mathbf{E}\, dv = \oint_s \mathbf{E} \cdot d\mathbf{s}. \tag{D.10}$$

We use the surface integral symbol, consisting of the usual $\int$-sign modified with the open circle, to signify that the surface over which $\mathbf{E} \cdot d\mathbf{s}$ is integrated is a *closed* surface – that is, a surface that forms the entire boundary surrounding the volume v. For example, if we consider a volume that is the shape of a watermelon, the closed surface bounding that watermelon is the entire rind. An example of a surface which is *not* closed would be the top surface of my desk, for there is no volume that is enclosed by this surface. We must also comment on the vector surface element $d\mathbf{s}$ that appears in Eq. (D.10). Recall that the direction of $d\mathbf{s}$ is always normal to the surface. But, as we pointed out earlier, there are two normal vectors $\mathbf{a}_n$ for a surface. For a closed surface, we follow the convention that $d\mathbf{s}$ always points *outward* from the interior of the volume. This becomes important in the proof shown in SideNote D.1, as well as in the final results given in Eq. (D.10), as we will introduce a sign error if we don't use the outward normal vector.

SideNote D.1 Proof of Divergence Theorem

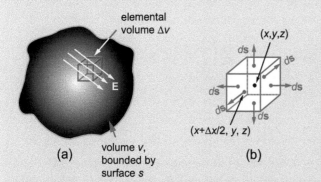

(a)

elemental volume Δv

$\mathbf{E}$

volume v, bounded by surface s

(b)

(x,y,z)

$d\mathbf{s}$

$(x+\Delta x/2, y, z)$

FIGURE D.5 Volume v bounded by surface s. The vector field $\mathbf{E}$ is shown in (a). In (b), we show one element of this larger volume, with $d\mathbf{s}$ indicated for each of the six faces.

To prove the Divergence Theorem, we consider a volume v in space in which a vector field $\mathbf{E}$ is defined and continuous, as shown in Fig. D.5. We divide the entire volume into small volume elements, as shown, of dimension Δx by Δy by Δz. We will start by computing the flux $\mathbf{E} \cdot d\mathbf{s}$ on all sides of one of these small volume elements, as shown in Fig. D.5(b), whose center is located at the point (x, y, z). Consider first the front face of the cube, lying in the y–z plane at $x + \Delta x/2$. The surface element for this face is $\mathbf{a}_n \Delta s = \mathbf{a}_x \Delta y \Delta z$. Therefore the surface flux is $\mathbf{E} \cdot \mathbf{a}_n \Delta s = E_x(x + \Delta x/2, y, z)\, \Delta y\, \Delta z$. On the opposite face, the surface element is $\mathbf{a}_n\, \Delta s = -\mathbf{a}_x\, \Delta y\, \Delta z$. (Remember, the surface normal $\mathbf{a}_n$ of a closed surface always points out from the interior of the volume, leading to the minus sign here.) The flux at this face is therefore $\mathbf{E} \cdot \mathbf{a}_n \Delta s = -E_x(x - \Delta x/2, y, z)\, \Delta y\, \Delta z$. Notice also that the magnitudes of $\mathbf{E} \cdot \mathbf{a}_n \Delta s$ on two opposite sides of this elemental cube are slightly different from one another if the vector field is growing or diminishing. This gives us a hint in advance of the ultimate entry of $\boldsymbol{\nabla} \cdot \mathbf{E}$ in our proof. Writing similar terms for all of the faces and adding them together to find the total flux through the surfaces of the entire elemental cube, we have

$$\sum_{k=1}^{6} \mathbf{E} \cdot (\mathbf{a}_n \Delta s)_k = \left[E_x(x + \Delta x/2, y, z) - E_x(x - \Delta x/2, y, z) \right] \Delta y \, \Delta z$$

$$+ \left[E_y(x, y + \Delta y/2, z) - E_y(x, y - \Delta y/2, z) \right] \Delta x \, \Delta z$$

$$+ \left[E_z(x, y, z + \Delta z/2) - E_z(x, y, z - \Delta z/2) \right] \Delta x \, \Delta y.$$

The index k in this sum runs over the six sides of the elemental cube. We now group terms together by pulling out a common factor of $\Delta x \, \Delta y \, \Delta z$, yielding

$$\sum_{k=1}^{6} \mathbf{E} \cdot (\mathbf{a}_n \Delta s)_k = \left[\frac{E_x(x + \Delta x/2, y, z) - E_x(x - \Delta x/2, y, z)}{\Delta x} \right.$$

$$+ \frac{E_y(x, y + \Delta y/2, z) - E_y(x, y - \Delta y/2, z)}{\Delta y}$$

$$\left. + \frac{E_z(x, y, z + \Delta z/2) - E_z(x, y, z - \Delta z/2)}{\Delta z} \right] \Delta x \, \Delta y \, \Delta z,$$

which we can write as

$$\sum_{k=1}^{6} \mathbf{E} \cdot (\mathbf{a}_n \Delta s)_k = \left[\frac{\Delta E_x}{\Delta x} + \frac{\Delta E_y}{\Delta y} + \frac{\Delta E_z}{\Delta z} \right] \Delta x \, \Delta y \, \Delta z, \tag{D.11}$$

where $\Delta E_x = E_x(x + \Delta x/2, y, z) - E_x(x - \Delta x/2, y, z)$, with similar expressions for ΔE_y and ΔE_z.

Next we sum these surface fluxes for all the elements Δv_i that comprise the total volume v, where the index i labels each of the cubes. Note that when we add the surface flux for two adjacent elements, the result is the flux of the entire surface of the combined volume, since the fluxes for the shared surfaces of the two elements are equal but opposite, and cancel when added. Thus, when we add the surface fluxes for all the volume elements that comprise the volume v, only the flux through the exterior surface elements survive, and in the limit as the volume of the elements become very small, and the number of volume elements increases, we have that

$$\oint_s \mathbf{E} \cdot d\mathbf{s} = \lim_{\Delta x, \Delta y, \Delta z \to 0} \sum_i \left(\sum_{k=1}^{6} \mathbf{E} \cdot (\mathbf{a}_n \Delta s)_k \right)_i.$$

Using Eq. (D.11), this becomes

$$\oint_s \mathbf{E} \cdot d\mathbf{s} = \lim_{\Delta x, \Delta y, \Delta z \to 0} \sum_i \left(\left[\frac{\Delta E_x}{\Delta x} + \frac{\Delta E_y}{\Delta y} + \frac{\Delta E_z}{\Delta z} \right] \Delta x \, \Delta y \, \Delta z \right)_i.$$

But in the limit $\Delta x, \Delta y, \Delta z \to 0$, the sum inside the square brackets becomes $\nabla \cdot \mathbf{E}$ and the summation over all the volume elements becomes the volume integral, leaving us with the expression of the Divergence Theorem as given in Eq. (D.10),

$$\oint_s \mathbf{E} \cdot d\mathbf{s} = \int_v \nabla \cdot \mathbf{E} \, dv.$$

With this, we have completed our proof of the Divergence Theorem.

We will apply this theorem often in our treatment of electric and magnetic fields, such as in Section 3.2, where we derive a relation between charge densities and current densities known as the Continuity Equation. But first, let us work an example that demonstrates the validity of the Divergence Theorem to a simple vector field.

Example D.6 Divergence Theorem

Let $\mathbf{D}(\mathbf{r}) = \mathbf{a}_R R \sin\theta + \mathbf{a}_\phi c \cos\theta \sin\phi$, where c is a constant. (a) Determine $\oint_s \mathbf{D}(\mathbf{r}) \cdot d\mathbf{s}$ for the spherical surface of radius $b = 3$ centered at the origin. (b) Determine $\int_v \boldsymbol{\nabla} \cdot \mathbf{D}\, dv$ for the volume inside this surface. (c) Is the Divergence Theorem satisfied? (d) Repeat (a), (b), and (c) for the same vector $\mathbf{D}$ for the region between two concentric spherical surfaces, one of radius b and the other of radius $a = 1$, each centered at the origin.

Solution:

(a) We start by determining the surface flux of $\mathbf{D}(\mathbf{r})$ through the spherical surface of radius $b = 3$. We use $d\mathbf{s} = b^2 \sin\theta\, d\theta\, d\phi\, \mathbf{a}_R$. (See Fig. C.1(c) for the surface element $d\mathbf{s}$ for a spherical surface. The base of this elemental surface is $b \sin\theta d\phi$, and the height is $bd\theta$, so the magnitude of the surface element is $ds = (b \sin\theta d\phi)\,(bd\theta) = b^2 \sin\theta d\theta d\phi$. For the direction of $d\mathbf{s}$, remember that this surface element always points outward from a closed volume. That is, radially outward for this case.) Then,

$$\int_{R=b} \mathbf{D}(\mathbf{r}) \cdot d\mathbf{s} = \int \left(\mathbf{a}_R b \sin\theta + \mathbf{a}_\phi c \cos\theta \sin\phi\right) \cdot \left(\mathbf{a}_R b^2 \sin\theta\, d\theta\, d\phi\right),$$

where we have substituted $R = b$ in $\mathbf{D}(\mathbf{r})$, since here we are only interested in $\mathbf{D}(\mathbf{r})$ on the surface of the sphere. Since $\mathbf{a}_R \cdot \mathbf{a}_R = 1$ and $\mathbf{a}_R \cdot \mathbf{a}_\phi = 0$, the integral simplifies to

$$\int_{R=b} \mathbf{D}(\mathbf{r}) \cdot d\mathbf{s} = b^3 \int_0^\pi \sin^2\theta\, d\theta \int_0^{2\pi} d\phi.$$

Completing the integration, we find

$$\int_{R=b} \mathbf{D}(\mathbf{r}) \cdot d\mathbf{s} = b^3 \left(\frac{\pi}{2}\right)(2\pi) = b^3 \pi^2,$$

which for $b = 3$ is $27\pi^2$.

(b) Now we need to find $\boldsymbol{\nabla} \cdot \mathbf{D}(\mathbf{r})$, and then integrate this over the volume inside the sphere. Since we are working in spherical coordinates, we use Eq. (D.9) to determine the divergence,

$$\boldsymbol{\nabla} \cdot \mathbf{D} = \frac{1}{R^2}\frac{\partial(R^2 D_R)}{\partial R} + \frac{1}{R \sin\theta}\frac{\partial(D_\theta \sin\theta)}{\partial\theta} + \frac{1}{R \sin\theta}\frac{\partial D_\phi}{\partial\phi},$$

where $D_R = R \sin\theta$, $D_\theta = 0$, and $D_\phi = c \cos\theta \sin\phi$. Taking the derivatives, we find $\boldsymbol{\nabla} \cdot \mathbf{D} = 3 \sin\theta + c \cot\theta \cos\phi/R$. The volume integral of $\boldsymbol{\nabla} \cdot \mathbf{D}$ is

$$\int_v \boldsymbol{\nabla} \cdot \mathbf{D}\, dv = \int_{\phi=0}^{2\pi} \int_{\theta=0}^\pi \int_{R=0}^b (3 \sin\theta + c \cot\theta \cos\phi/R)\, R^2\, dR \sin\theta\, d\theta\, d\phi.$$

Notice that we have used the elemental volume of a quasi-cube in spherical coordinates, $dv = R^2\, dR \sin\theta\, d\theta\, d\phi$. This quasi-cube, of base $b \sin\theta d\phi$, of height $bd\theta$, and of depth dR, can be seen in Fig. C.1(c). Integrating over ϕ simplifies this nicely, since $\int d\phi = 2\pi$ and $\int \cos\phi\, d\phi = 0$. Completing the integration results in $\int_v \boldsymbol{\nabla} \cdot \mathbf{D}\, dv = 27\pi^2$.

(c) Since evaluation of $\oint_s \mathbf{D}(\mathbf{r}) \cdot d\mathbf{s}$ and $\int_v \boldsymbol{\nabla} \cdot \mathbf{D}\, dv$ yields the same result, we have shown that this vector field $\mathbf{D}$ does indeed satisfy the Divergence Theorem.

(d) We can borrow much from our result from the first volume. The surface bounding this volume has two parts – that is, the surface at $b = 3$ and the surface at $a = 1$. We already evaluated the surface flux through the outer surface in part (a), so we only need to evaluate the flux through the inner surface, and add the two fluxes. On the inside surface, we use $d\mathbf{s} = -a^2 \sin\theta \, d\theta \, d\phi \, \mathbf{a}_R$. The minus sign here comes from the fact that the surface normal always points *outward* from the interior of the volume, which in this case is radially *inward*. Then,

$$\int_{R=a} \mathbf{D(r)} \cdot d\mathbf{s} = \int \left(\mathbf{a}_R a \sin\theta + \mathbf{a}_\phi c \cos\theta \sin\phi \right) \cdot \left(-a^2 \sin\theta \, d\theta \, d\phi \, \mathbf{a}_R \right)$$

$$= -a^3 \int_0^\pi \sin^2\theta \, d\theta \int_0^{2\pi} d\phi$$

$$= -a^3 \left(\frac{\pi}{2} \right) (2\pi)$$

$$= -a^3 \pi^2.$$

The total surface flux is $-a^3\pi^2 + b^3\pi^2 = 26\pi^2$. We evaluate the volume integral in a manner similar to that shown in part (b), except the lower limit on the R integral is now $R = 1$. Our result becomes $\int_v \nabla \cdot \mathbf{D} \, dv = 26\pi^2$. Since evaluation of $\oint_s \mathbf{D(r)} \cdot d\mathbf{s}$ and $\int_v \nabla \cdot \mathbf{D} \, dv$ over this modified volume are again equal to one another, we have shown that the Divergence Theorem is again satisfied.

We conclude our discussion of the divergence by using the Divergence Theorem to determine a general definition of the divergence of a vector field. You will recall that we defined it simply, but not particularly generally, by presenting specific forms to calculate the divergence, using Eqs. (D.7)–(D.9), in rectangular, cylindrical, and spherical coordinates. Let us apply the Divergence Theorem (Eq. (D.10)) to an infinitesimal volume Δv, which is small enough that $\nabla \cdot \mathbf{E}$ is relatively uniform. Then we have

$$\int_v \nabla \cdot \mathbf{E} \, dv = \oint_s \mathbf{E} \cdot d\mathbf{s} \simeq \nabla \cdot \mathbf{E} \, \Delta v.$$

We can divide by Δv and take the limit as $\Delta v \to 0$ to determine

$$\nabla \cdot \mathbf{E} = \lim_{\Delta v \to 0} \frac{1}{\Delta v} \oint_s \mathbf{E} \cdot d\mathbf{s}, \tag{D.12}$$

which is the formal definition of the divergence of the vector field $\mathbf{E}$. Besides being general, and therefore applicable to any coordinate system, this definition illustrates the connection between the divergence of a vector field and the growth of that field, which causes the flux of this field through the surface of a closed volume to be non-zero.

Vector fields whose divergence is zero are known as **solenoidal** fields. As we show in Chapter 1, static electric fields have a zero divergence in charge-free regions, and in Chapter 4 we show that magnetic fields always have zero divergence.

D.4 The Curl of a Vector Field

We turn our attention now to the second of the differential vector field operators for a vector field $\mathbf{A}$, known as the curl and denoted as $\nabla \times \mathbf{A}$. We adopt the following explicit forms of this operation in rectangular, cylindrical, and spherical coordinates as its definition. In rectangular coordinates, the curl is given as

$$\boldsymbol{\nabla} \times \mathbf{A} = \begin{vmatrix} \mathbf{a}_x & \mathbf{a}_y & \mathbf{a}_z \\ \dfrac{\partial}{\partial x} & \dfrac{\partial}{\partial y} & \dfrac{\partial}{\partial z} \\ A_x & A_y & A_z \end{vmatrix} \qquad (D.13)$$

$$= \mathbf{a}_x \left(\frac{\partial A_z}{\partial y} - \frac{\partial A_y}{\partial z} \right) + \mathbf{a}_y \left(\frac{\partial A_x}{\partial z} - \frac{\partial A_z}{\partial x} \right) + \mathbf{a}_z \left(\frac{\partial A_y}{\partial x} - \frac{\partial A_x}{\partial y} \right).$$

In cylindrical coordinates, the curl is

$$\boldsymbol{\nabla} \times \mathbf{A} = \frac{1}{\rho} \begin{vmatrix} \mathbf{a}_\rho & \rho\mathbf{a}_\phi & \mathbf{a}_z \\ \dfrac{\partial}{\partial \rho} & \dfrac{\partial}{\partial \phi} & \dfrac{\partial}{\partial z} \\ A_\rho & \rho A_\phi & A_z \end{vmatrix} \qquad (D.14)$$

$$= \mathbf{a}_\rho \left(\frac{1}{\rho}\frac{\partial A_z}{\partial \phi} - \frac{\partial A_\phi}{\partial z} \right) + \mathbf{a}_\phi \left(\frac{\partial A_\rho}{\partial z} - \frac{\partial A_z}{\partial \rho} \right) + \frac{\mathbf{a}_z}{\rho} \left(\frac{\partial(\rho A_\phi)}{\partial \rho} - \frac{\partial A_\rho}{\partial \phi} \right).$$

In spherical coordinates, the curl is

$$\boldsymbol{\nabla} \times \mathbf{A} = \frac{1}{R^2 \sin\theta} \begin{vmatrix} \mathbf{a}_R & R\mathbf{a}_\theta & R\sin\theta\,\mathbf{a}_\phi \\ \dfrac{\partial}{\partial R} & \dfrac{\partial}{\partial \theta} & \dfrac{\partial}{\partial \phi} \\ A_R & RA_\theta & R\sin\theta A_\phi \end{vmatrix} \qquad (D.15)$$

$$= \frac{\mathbf{a}_R}{R\sin\theta} \left(\frac{\partial(\sin\theta A_\phi)}{\partial \theta} - \frac{\partial A_\theta}{\partial \phi} \right) + \frac{\mathbf{a}_\theta}{R} \left(\frac{1}{\sin\theta}\frac{\partial A_R}{\partial \phi} - \frac{\partial(RA_\phi)}{\partial R} \right) + \frac{\mathbf{a}_\phi}{R} \left(\frac{\partial(RA_\theta)}{\partial R} - \frac{\partial A_R}{\partial \theta} \right).$$

A less precise, but perhaps more descriptive, definition of the curl operator is that the curl gives us a quantitative means of describing the degree to which a vector field rotates, or circulates, about an axis. The direction of the curl of a vector field points along the axis of this rotation. Let's illustrate this with a couple of examples.

Example D.7 Curl in Rectangular Coordinates

Consider a vector field given by $\mathbf{A}(\mathbf{r}) = -y\mathbf{a}_x + x\mathbf{a}_y$. Determine $\boldsymbol{\nabla} \times \mathbf{A}(\mathbf{r})$.

Solution:
We show an illustration of this field in Fig. D.4(a). Clearly this vector field circulates around the z-axis in a counter-clockwise direction, and we should expect this field to have an appreciable curl. In addition, we can use a right-hand rule (curl the fingers of your right hand in the direction of the circulating field $\mathbf{A}(\mathbf{r})$; your thumb points in the direction of $\boldsymbol{\nabla} \times \mathbf{A}(\mathbf{r})$) to predict that $\boldsymbol{\nabla} \times \mathbf{A}(\mathbf{r})$ points in the $+z$-direction. To compute $\boldsymbol{\nabla} \times \mathbf{A}(\mathbf{r})$ explicitly, we use Eq. (D.13), since we are given $\mathbf{A}(\mathbf{r})$ in rectangular coordinates. Using $A_x = -y$ and $A_y = x$, the only non-zero derivatives in Eq. (D.13) are $\partial A_x / \partial y = -1$ and $\partial A_y / \partial x = +1$, and we obtain $\boldsymbol{\nabla} \times \mathbf{A}(\mathbf{r}) = 2\mathbf{a}_z$. As expected, $\boldsymbol{\nabla} \times \mathbf{A}(\mathbf{r})$ points in the $+z$-direction.

Note that we might also have worked this example in cylindrical or spherical coordinates by first expressing $\mathbf{A}(\mathbf{r})$ in one of these sets of variables, $\mathbf{A}(\mathbf{r}) = \rho\mathbf{a}_\phi$ (cylindrical, using Eqs. (B.12)) or $\mathbf{A}(\mathbf{r}) = R\sin\theta\mathbf{a}_\phi$ (spherical, using Eqs. (B.14)). The result, of course, is the same. You might try these for practice in taking the curl of a simple vector in these alternative coordinates, as well as for practice in converting vectors from one coordinate system to another.

Example D.8 Curl in Cylindrical or Spherical Coordinates

Consider a vector field given by $\mathbf{B}(\mathbf{r}) = \mathbf{a}_\phi$. Determine $\nabla \times \mathbf{B}(\mathbf{r})$.

Solution:

This field is expressed in cylindrical or spherical coordinates; we can't really tell for this field. No matter, we will work it either way, and find the same result. We will start with Eq. (D.14) for the curl in cylindrical coordinates. Before we start, note that this field is similar to, but not quite the same as, the field $\mathbf{A}(\mathbf{r})$ that we considered in Example D.7. If we were to convert $\mathbf{A}(\mathbf{r})$ to cylindrical coordinates, we would find $\mathbf{A}(\mathbf{r}) = \rho \mathbf{a}_\phi$, which has an additional factor ρ not present in $\mathbf{B}(\mathbf{r})$. From Eq. (D.14), we see that we require the following six derivatives: $\partial B_z/\partial \phi$, $\partial B_\phi/\partial z$, $\partial B_\rho/\partial z$, $\partial B_z/\partial \rho$, $\partial(\rho B_\phi)/\partial \rho$, and $\partial B_\rho/\partial \phi$, where $B_\phi = 1$ and B_ρ and B_z are zero. The only derivative from this list that is not zero is $\partial(\rho B_\phi)/\partial \rho$, which is 1. Therefore, we find $\nabla \times \mathbf{B}(\mathbf{r}) = \mathbf{a}_z/\rho$. Similar to the previous example, this curl points in the positive z-direction, since the field $\mathbf{B}(\mathbf{r})$ circulates about the z-axis in the counter-clockwise sense.

Now let's rework this solution using spherical coordinates. For this, we refer to Eq. (D.15) and see that we must evaluate six derivatives for this solution. But B_R and B_θ are zero in this example, so there are really only two derivatives that we must determine: $\partial(\sin\theta B_\phi)/\partial\theta$ and $\partial(R B_\phi)/\partial R$. Since B_ϕ is simply 1, we find

$$\nabla \times \mathbf{B}(\mathbf{r}) = \mathbf{a}_R \frac{1}{R\sin\theta}\cos\theta - \mathbf{a}_\theta \frac{1}{R} = \frac{1}{R\sin\theta}\left(\mathbf{a}_R\cos\theta - \mathbf{a}_\theta\cos\theta\right).$$

On first glance this appears to be a different result, but let's take a closer look. Recall Example B.5 (including Fig. B.5), in which we showed the equivalence of $\mathbf{a}_z$ in rectangular or cylindrical coordinates and $\mathbf{a}_R\cos\theta - \mathbf{a}_\theta\cos\theta$ (in spherical coordinates). You will also recall that ρ is equivalent to $R\sin\theta$. With these two conversions, we see that our result for $\nabla \times \mathbf{B}(\mathbf{r})$ is $\mathbf{a}_z/\rho$, identical to the result that we computed using cylindrical coordinates.

Example D.9 Curl in Spherical Coordinates

Consider a vector field given by $\mathbf{C}(\mathbf{r}) = \mathbf{a}_R/R^2$. Determine $\nabla \times \mathbf{C}(\mathbf{r})$.

Solution:

This field is expressed in spherical coordinates, so we will use Eq. (D.15) to find its curl. You might recall that we have encountered this field previously, as we have already calculated its divergence. (See Example D.5.) From Eq. (D.15), we see that we need to evaluate six different derivatives, but since only the R-component of $\mathbf{C}(\mathbf{r})$ is non-zero, our task is somewhat simplified, and we need only examine $\partial C_R/\partial\phi$ and $\partial C_R/\partial\theta$. Since $C_R = R^{-2}$, these derivatives are each zero, and so the $\nabla \times \mathbf{C}(\mathbf{r}) = 0$. We might have anticipated this result, since this vector field points radially outward from the origin, and shows no sign of circulation about any axis.

Through the preceding three examples we have examined the curl of three of the fields illustrated in Fig. D.4, which we originally presented in order to draw a physical picture of the divergence of fields. Fields (a) and (b) have a non-zero curl, while the curl of field (c) is zero. Through examination of fields (d), (e), and (f), you should be able to demonstrate that fields (d) and (f) are circulating, while field (e) has zero curl.

D.5 Stokes' Theorem

We now examine **Stokes' Theorem**, which is in some sense parallel to the Divergence Theorem, and applies to the curl differential operator. We use this widely in our discussions of fields. Consider the surface s illustrated in Fig. D.6(a). The path around the edge of this surface is labeled c, and some vector field $\mathbf{E}(\mathbf{r})$ is defined and well behaved in this region of space. (By "well behaved," we mean that it is continuous and differentiable.) Stokes' Theorem states that

$$\oint_c \mathbf{E} \cdot d\boldsymbol{\ell} = \int_s \boldsymbol{\nabla} \times \mathbf{E} \cdot d\mathbf{s}. \tag{D.16}$$

Notice the open circle on the integral sign for the path integral on the left side of Eq. (D.16). This indicates that the integral is to be taken around a *closed* path – that is, a path that starts and ends at the same point. (Recall that we previously used an integral sign with an open circle when we discussed integration over a *closed* surface. This meaning is similar, but now applied to a closed path rather than a surface.) As always, signs are important and we have to be careful to define the *direction* of the path, for which we resort to yet another right-hand rule. Specifically, we must be consistent between the direction of the line element $d\boldsymbol{\ell}$ and the direction of the surface normal $\mathbf{a}_n$, where $d\mathbf{s} = ds\, \mathbf{a}_n$. We grasp the surface contour c with our right hand, with our thumb pointing in the direction of the line element $d\boldsymbol{\ell}$. Then the correct surface normal $\mathbf{a}_n$ is identified by the direction in which the fingers of our right hand point through the interior of the loop. In this example, our fingers point upward through this surface, consistent with $d\mathbf{s}$ as labeled.

SideNote D.2 Proof of Stokes' Theorem

To prove Eq. (D.16), let us break up the entire surface s into a mesh consisting of a large number of small elemental loops. We have drawn one of these elements in Fig. D.6(b). We will consider the line integral around each of the individual loops,

$$\left(\oint \mathbf{E} \cdot d\boldsymbol{\ell} \right)_k,$$

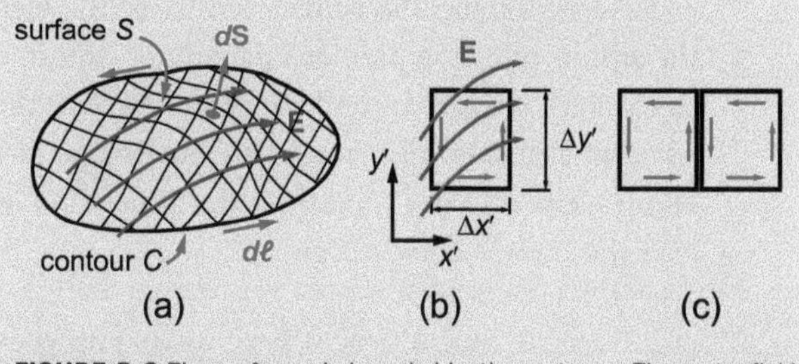

FIGURE D.6 The surface s is bounded by the contour c. The vector field $\mathbf{E}$ is shown. In (b) we show one element of this larger surface, with $d\boldsymbol{\ell}$ indicated for each of the four sides. (c) When combining elements, only the external line segments survive.

where the index k labels the particular elemental loop that we are considering. We define the plane in which this element lies as the x'–y' plane, and the dimension of this element is $\Delta x'$ by $\Delta y'$. (We should remark here that the surface s is not necessarily planar. We take each element k, however, to be small enough such that its surface is very close to being a plane.) The path integral that we seek is the sum of four terms, one for each side of the elemental rectangle. Recall that in rectangular coordinates we can express $\Delta \boldsymbol{\ell}$ as $\mathbf{a}_{x'}\Delta x' + \mathbf{a}_{y'}\Delta y' + \mathbf{a}_{z'}\Delta z'$, but we can ignore the last term since our path lies in the x'–y' plane. Then the closed path integral around element k is

$$\left(\oint \mathbf{E}\cdot d\boldsymbol{\ell}\right)_k = E_{x'}(x',y'-\Delta y'/2,z')\,\Delta x' \; + \; E_{y'}(x'+\Delta x'/2,y',z')\,\Delta y'$$
$$-E_{x'}(x',y'+\Delta y'/2,z')\,\Delta x' \; - \; E_{y'}(x'-\Delta x'/2,y',z')\,\Delta y'.$$

$E_{x'}(x',y'-\Delta y'/2,z')$ is the x' component of the field $\mathbf{E}$ evaluated along the bottom of the rectangle, with similar definitions for the other three sides. Grouping similar terms together and pulling out a factor of $\Delta x'\Delta y'$, this becomes

$$\left(\oint \mathbf{E}\cdot d\boldsymbol{\ell}\right)_k = \left[-\frac{E_{x'}(x',y'+\Delta y'/2,z')-E_{x'}(x',y'-\Delta y'/2,z')}{\Delta y'}\right.$$
$$\left. +\frac{E_{y'}(x'+\Delta x'/2,y',z')-E_{y'}(x'-\Delta x'/2,y',z')}{\Delta x'}\right]\Delta x'\Delta y'.$$

We recognize that the terms inside the square brackets on the right side become the differentials $-\partial E_{x'}/\partial y'$ and $\partial E_{y'}/\partial x'$, respectively, and that in combination these are the z'-component of $\boldsymbol{\nabla}\times\mathbf{E}$,

$$\left(\oint \mathbf{E}\cdot d\boldsymbol{\ell}\right)_k = \left((\boldsymbol{\nabla}\times\mathbf{E})_{z'}\,\Delta x'\Delta y'\right)_k = \left((\boldsymbol{\nabla}\times\mathbf{E})_{z'}\,\Delta s'\right)_k, \tag{D.17}$$

where z' is the direction perpendicular to the x'–y' plane, and $\Delta s'$ is the elemental surface area $\Delta x'\Delta y'$.

Now we consider this integral over two adjacent elements. Picture two elements, such as shown in Fig. D.6(c), sitting side by side. On the left side of Eq. (D.17) we add the path integrals for the two elements. Each element has four sides, so in combination there are eight sides. But two of these sides are shared, and $\int \mathbf{E}\cdot d\boldsymbol{\ell}$ along this shared side for one element is equal but opposite to that for the adjacent element, since the direction of $d\boldsymbol{\ell}$ is reversed. Therefore, $\oint_c \mathbf{E}\cdot d\boldsymbol{\ell}$ for the combined elements (i.e. evaluated around the perimeter of the combination) is just the sum of the path integrals for the two individual elements. On the right side of Eq. (D.17) we get the sum of the two terms $(\boldsymbol{\nabla}\times\mathbf{E})_{z'}\Delta s'$.

Now we extend the summation to *all* the elements k covering the surface s. By the argument of the previous paragraph, the summation of the path integrals gives us the path integral around the contour c. (All the interior paths are shared between adjacent elements, and these contributions perfectly cancel one another.) The sum of all the elements $(\boldsymbol{\nabla}\times\mathbf{E})_{z'}\Delta s'$, in the limit of $\Delta s'$ getting small, is the surface integral over the surface s, and we get

$$\oint_c \mathbf{E}\cdot d\boldsymbol{\ell} = \int_s (\boldsymbol{\nabla}\times\mathbf{E})_{z'}\,ds_{z'}.$$

Finally, we can use the vector surface element notation $d\mathbf{s}$ in order to write this compactly,

$$\oint_c \mathbf{E} \cdot d\boldsymbol{\ell} = \int_s \boldsymbol{\nabla} \times \mathbf{E} \cdot d\mathbf{s},$$

as we wrote in Eq. (D.16), completing our proof of Stokes' Theorem.

Now let's work an example to illustrate Stokes' Theorem.

Example D.10 Stokes' Theorem

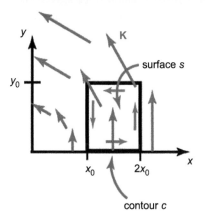

FIGURE D.7 In Example D.10 we evaluate $\int_s \boldsymbol{\nabla} \times \mathbf{K} \cdot d\mathbf{s}$ over the surface lying in the $z = 0$ plane bounded by the rectangular contour shown, as well as $\oint_c \mathbf{K} \cdot d\boldsymbol{\ell}$, as a demonstration of Stokes' Theorem. The blue arrows show the vector $\mathbf{K}$, while the red arrows show the elements $d\boldsymbol{\ell}$. By the right-hand rule, the surface element $d\mathbf{s}$ points out of the page.

Consider the vector field $\mathbf{K}(\mathbf{r}) = \mathbf{a}_\phi \rho$. (a) Evaluate $\int_s \boldsymbol{\nabla} \times \mathbf{K} \cdot d\mathbf{s}$ over the surface lying in the $z = 0$ plane bounded by the rectangular contour shown in Fig. D.7. (b) Evaluate $\oint_c \mathbf{K} \cdot d\boldsymbol{\ell}$ for this same vector $\mathbf{K}$ and rectangular path. (c) Does this obey Stokes' Theorem?

Solution:

(a) We find $\boldsymbol{\nabla} \times \mathbf{K}$ using Eq. (D.14), $\boldsymbol{\nabla} \times \mathbf{K} = 2\mathbf{a}_z$. Since the surface is bounded by a rectangular contour aligned with the x and y axes, it is most convenient to work in rectangular coordinates. Our surface element is $d\mathbf{s} = \mathbf{a}_z \, dx \, dy$. Note that this direction of $d\mathbf{s}$ (i.e. the $+z$-direction) is consistent with the counter-clockwise contour c shown in Fig. D.7, by the right-hand rule. Then the surface integral is

$$\int_s \boldsymbol{\nabla} \times \mathbf{K} \cdot d\mathbf{s} = \int_{y=0}^{y=y_0} \int_{x=x_0}^{x=2x_0} 2\mathbf{a}_z \cdot \mathbf{a}_z \, dx \, dy = 2x_0 y_0.$$

(b) It is convenient to work this part of the problem in rectangular coordinates as well, so we start by converting $\mathbf{K}$ to the form $\mathbf{K} = -y\mathbf{a}_x + x\mathbf{a}_y$. We separate the contour c into its four segments, and evaluate the line integral as

$$\oint_c \mathbf{K} \cdot d\boldsymbol{\ell} = \oint_c \left(-y\mathbf{a}_x + x\mathbf{a}_y \right) \cdot \left(dx \, \mathbf{a}_x + dy \, \mathbf{a}_y \right)$$

$$= \int_{x_0}^{2x_0} (-y) \bigg|_{y=0} dx + \int_0^{y_0} x \bigg|_{x=2x_0} dy$$

$$+ \int_{2x_0}^{x_0} (-y) \bigg|_{y=y_0} dx + \int_{y_0}^0 x \bigg|_{x=x_0} dy.$$

Notice that we use the limits of integration for each side of the rectangle to reflect the counter-clockwise path around the rectangle. Evaluating these simple integrals, we find

$$\oint_c \mathbf{K} \cdot d\boldsymbol{\ell} = 0 + 2x_0 y_0 + (-x_0)(-y_0) - x_0 y_0 = 2x_0 y_0.$$

(c) Since the results of parts (a) and (b) agree, Stokes' Theorem is satisfied, as it must be.

Stokes' Theorem allows us a clearer local physical picture of the curl operator. We consider a region in space in which a vector field $\mathbf{E}(\mathbf{r})$ exists, and apply Stokes' Theorem to a small loop of area Δs, as illustrated in Fig. D.8. This process tells us that

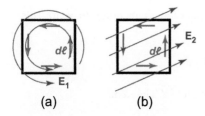

FIGURE D.8 A demonstration of the connection between the curl of a vector field **E** and the line integral **E** · $d\ell$ of that field around a small loop. In (a) the vector field **E**$_1$ is tightly circulating, and a component of **E**$_1$ lies parallel to each side of the rectangle. Therefore, all sides contribute to the path integral around the complete loop with the same sign, and we expect that this integral is large. Compare this to the vector field **E**$_2$ shown in (b), which is rather straight in this same space. Any contribution to **E**$_2$ · $d\ell$ on one side of the rectangle is exactly canceled by an equal but opposite contribution on the opposing side, and the path integral $\oint$ **E**$_2$ · $d\ell$ around the entire path is zero.

$$\oint_c \mathbf{E} \cdot d\ell = \int_s \boldsymbol{\nabla} \times \mathbf{E} \cdot d\mathbf{s} \simeq (\boldsymbol{\nabla} \times \mathbf{E})_n \, \Delta s,$$

where the subscript n indicates the component of the curl **E** normal to the surface Δs. Dividing by Δs gives us

$$(\boldsymbol{\nabla} \times \mathbf{E})_n = \frac{1}{\Delta s} \oint_c \mathbf{E} \cdot d\ell. \tag{D.18}$$

This expression is actually the formal general definition of the curl operator that we avoided earlier in this discussion, but we are now ready to use it for our interpretation of the curl. The path integral on the right side gives us a quantitative measure of the "circulation" of the field, or how much the field is rotating about some axis in the n-direction. If the vector field **E** circles around in a tight loop, like a whirlpool in a fluid or the vector field **E**$_1$ shown in the figure, then **E** · $d\ell$ for each side of the loop is the same sign, and the path integral around the entire loop is large. If, however, the field lines of **E** are straight and unchanging, as shown by **E**$_2$ in the figure, then the contributions to **E** · $d\ell$ from the different sides of the loop cancel one other, and the circulation is small. Thus the name "curl" is in fact very descriptive of this operation.

Along the same lines, a vector field **E** whose curl is zero is known as *irrotational*. As we show in Chapter 1, static electric fields have a zero curl. See that discussion for important properties of these fields.

PROBLEMS

The Gradient of a Scalar Field

PD-1 (a) Determine $\boldsymbol{\nabla}f(\mathbf{r})$ for the scalar field $f(\mathbf{r}) = x^2 - y^2 + 2xy + 2z^2$. (b) Determine the magnitude of $\boldsymbol{\nabla}f(\mathbf{r})$ at the point $(1, 2, 3)$. (c) Determine the point (x, y, z) at which $\boldsymbol{\nabla}f(\mathbf{r}) = 2\mathbf{a}_z$.

PD-2 Determine $\boldsymbol{\nabla}g(\mathbf{r})$ for the scalar field $g(\mathbf{r}) = \rho^2 z^3 - a_0^2 \cos^2 \phi$, where a_0 is a constant. Plot a two-dimensional slice of your result in the $z = 0$ plane.

PD-3 Determine $\boldsymbol{\nabla}h(\mathbf{r})$ for the scalar field $h(\mathbf{r}) = e^{-R^2/a^2}$. Plot a two-dimensional slice of your result in the $z = 0$ plane. On this same plot, sketch lines on which h is 0.7788, 0.3678, and 0.1054. What is the angle between these lines and the arrows representing $\boldsymbol{\nabla}h(\mathbf{r})$?

PD-4 Determine the unit vector normal to the surface defined by $xe^{-z/h} = 2$. Sketch the curve $xe^{-z/h} = 2$ for $h = 1$. Show the normal unit vectors at $x = 2$, 5.44, and 14.8.

The Divergence of a Vector Field

PD-5 For the following vector fields, determine their divergence. (a) $\mathbf{D}(x, y, z) = [\mathbf{a}_x 2yz + \mathbf{a}_y 2zx + \mathbf{a}_z 2xy]/(xyz)$; (b) $\mathbf{E}(x, y, z) = \mathbf{a}_x x + \mathbf{a}_y y$; (c) $\mathbf{F}(x, y, z) = \mathbf{a}_x 2yz + \mathbf{a}_y 2zx + \mathbf{a}_z 2xy$.

PD-6 For the following vector fields, determine their divergence. (a) $\mathbf{D}(\rho, \phi, z) = \mathbf{a}_\rho/\rho$; (b) $\mathbf{E}(\rho, \phi, z) = \mathbf{a}_\phi \rho + \mathbf{a}_z z \cos \phi$; (c) $\mathbf{F}(\rho, \phi, z) = \mathbf{a}_z z/\rho$.

PD-7 For the following vector fields, determine their divergence. (a) $\mathbf{D}(R, \theta, \phi) = \mathbf{a}_R/R^2 + \mathbf{a}_\theta \cos \theta \sin \phi/R$; (b) $\mathbf{E}(R, \theta, \phi) = \mathbf{a}_R R \cos \theta \cos \phi - \mathbf{a}_\theta R \cos \theta \sin \phi$; (c) $\mathbf{F}(R, \theta, \phi) = \mathbf{a}_\theta \cos \theta \sin \phi/R$.

The Divergence Theorem

PD-8 Consider the vector $\mathbf{A} = \mathbf{a}_x A_0 \sin(x/a)$. Determine the surface flux for this vector field through the surface of a cube centered at the origin with sides of length $2b$ with faces perpendicular to the x, y, and z axes. (b) Determine the volume integral of $\nabla \cdot \mathbf{A(r)}$ over this same cube. (c) Is the Divergence Theorem satisfied?

PD-9 Consider the vector $\mathbf{B} = \mathbf{a}_\rho B_0 \cos\phi$. Determine the surface flux for this vector field through the surface of a cylinder centered on the z-axis extending from $z = -b$ to b and of radius a. (b) Determine the volume integral of $\nabla \cdot \mathbf{B(r)}$ over this same cylinder. (c) Is the Divergence Theorem satisfied?

The Curl of a Vector Field

PD-10 For the following vector fields, determine their curl. (a) $\mathbf{D}(x,y,z) = [\mathbf{a}_x 2yz + \mathbf{a}_y 2zx + \mathbf{a}_z 2xy]/(xyz)$; (b) $\mathbf{E}(x,y,z) = \mathbf{a}_x x + \mathbf{a}_y y$; (c) $\mathbf{F}(x,y,z) = \mathbf{a}_x 2yz + \mathbf{a}_y 2zx + \mathbf{a}_z 2xy$.

PD-11 For the following vector fields, determine their curl. (a) $\mathbf{D}(\rho,\phi,z) = \mathbf{a}_\rho/\rho$; (b) $\mathbf{E}(\rho,\phi,z) = \mathbf{a}_\phi \rho + \mathbf{a}_z z \cos\phi$; (c) $\mathbf{F}(\rho,\phi,z) = \mathbf{a}_z z/\rho$.

PD-12 For $\mathbf{B} = \mathbf{a}_\phi C\rho^{-1}$ and $\mathbf{A} = -\mathbf{a}_z C \ln(\rho)$, verify that $\mathbf{B} = \nabla \times \mathbf{A}$.

PD-13 For the following vector fields, determine their curl. (a) $\mathbf{D}(R,\theta,\phi) = \mathbf{a}_R/R^2 + \mathbf{a}_\theta \cos\theta \sin\phi/R$; (b) $\mathbf{E}(R,\theta,\phi) = \mathbf{a}_R R \cos\theta \cos\phi - \mathbf{a}_\theta R \cos\theta \sin\phi$; (c) $\mathbf{F}(R,\theta,\phi) = \mathbf{a}_\theta \cos\theta \sin\phi/R$.

Stokes' Theorem

PD-14 Consider the vector $\mathbf{C} = \mathbf{a}_x C_0 \sin(y/a)$. Determine the path integral $\oint_c \mathbf{C} \cdot d\ell$ around a square with corners at (b,b), $(b,-b)$, $(-b,-b)$, and $(-b,b)$. (b) Determine the surface integral of $\nabla \times \mathbf{C(r)}$ over this same square. (c) Is Stokes' Theorem satisfied?

PD-15 Consider the vector $\mathbf{D} = \mathbf{a}_\rho D_0 \cos 2\phi + \mathbf{a}_\phi D_1 \sin\phi$. Determine the path integral $\mathbf{D} \cdot d\ell$ around a closed quarter-circle in the first quadrant of the x–y plane of radius a. (b) Determine the surface integral of $\nabla \times \mathbf{D(r)}$ over this same quarter-circle. (c) Is Stokes' Theorem satisfied?

APPENDIX E
Additional Vector Properties

E.1 Distributive and Derivative Product Properties

We're nearing the conclusion of our review of useful mathematical tools that we will need in our studies of electromagnetics. Before we are ready to move on, however, we present several additional properties of vector addition and multiplication. Some of these properties may seem obvious, but we include them (without proof) as a reference for later use. In the following properties, Φ and Ψ are arbitrary scalar fields, while $\mathbf{E}$ and $\mathbf{F}$ are vector fields.

Distributive Properties

$$\nabla(\Phi + \Psi) = \nabla\Phi + \nabla\Psi \tag{E.1}$$

$$\nabla \cdot (\mathbf{E} + \mathbf{F}) = \nabla \cdot \mathbf{E} + \nabla \cdot \mathbf{F} \tag{E.2}$$

$$\nabla \times (\mathbf{E} + \mathbf{F}) = \nabla \times \mathbf{E} + \nabla \times \mathbf{F} \tag{E.3}$$

Derivative Product Properties

$$\nabla(\Phi\Psi) = \Phi\nabla\Psi + \Psi\nabla\Phi \tag{E.4}$$

$$\nabla(\mathbf{E} \cdot \mathbf{F}) = (\mathbf{E} \cdot \nabla)\mathbf{F} + (\mathbf{F} \cdot \nabla)\mathbf{E} \tag{E.5}$$
$$+ \mathbf{E} \times (\nabla \times \mathbf{F}) + \mathbf{F} \times (\nabla \times \mathbf{E})$$

$$\nabla \cdot (\Phi\mathbf{E}) = \Phi\nabla \cdot \mathbf{E} + \mathbf{E} \cdot \nabla\Phi \tag{E.6}$$

$$\nabla \cdot (\mathbf{E} \times \mathbf{F}) = \mathbf{F} \cdot (\nabla \times \mathbf{E}) - \mathbf{E} \cdot (\nabla \times \mathbf{F}) \tag{E.7}$$

$$\nabla \times (\Phi\mathbf{E}) = \Phi(\nabla \times \mathbf{E}) + (\nabla\Phi) \times \mathbf{E} \tag{E.8}$$

$$\nabla \times (\mathbf{E} \times \mathbf{F}) = \mathbf{E}\nabla \cdot \mathbf{F} - \mathbf{F}\nabla \cdot \mathbf{E} \tag{E.9}$$
$$+ (\mathbf{F} \cdot \nabla)\mathbf{E} - (\mathbf{E} \cdot \nabla)\mathbf{F}$$

Each of these can be easily verified in a particular coordinate system by carrying out the operations explicitly. As a demonstration, we prove Eq. (E.6) in rectangular coordinates in Example E.1. Others can be shown in a similar manner.

Example E.1 Confirmation of Eq. (E.6) in Rectangular Coordinates

Show that Eq. (E.6) is valid in rectangular coordinates.

Solution:

We start by writing the vector $\mathbf{E}$ in terms of its x-, y-, and z-components,

$$\nabla \cdot (\Phi \mathbf{E}) = \nabla \cdot \left[\Phi \left(\mathbf{a}_x E_x + \mathbf{a}_y E_y + \mathbf{a}_z E_z \right) \right].$$

Then we find the x-, y-, and z-components of the product $\Phi \mathbf{E}$ by distributing Φ,

$$\nabla \cdot (\Phi \mathbf{E}) = \nabla \cdot \left[\mathbf{a}_x \Phi E_x + \mathbf{a}_y \Phi E_y + \mathbf{a}_z \Phi E_z \right].$$

Next, we use the definition Eq. (D.7) of the divergence in rectangular coordinates,

$$\nabla \cdot (\Phi \mathbf{E}) = \frac{\partial}{\partial x} (\Phi E_x) + \frac{\partial}{\partial y} \left(\Phi E_y \right) + \frac{\partial}{\partial z} (\Phi E_z).$$

Since Φ, E_x, E_y, and E_z can in general each be functions of x, y, and z, we use the product rule for differentiation to write

$$\nabla \cdot (\Phi \mathbf{E}) = \left(\frac{\partial \Phi}{\partial x} E_x + \Phi \frac{\partial E_x}{\partial x} \right) + \left(\frac{\partial \Phi}{\partial y} E_y + \Phi \frac{\partial E_y}{\partial y} \right) + \left(\frac{\partial \Phi}{\partial z} E_z + \Phi \frac{\partial E_z}{\partial z} \right),$$

which we rearrange to form

$$\nabla \cdot (\Phi \mathbf{E}) = \left(\frac{\partial \Phi}{\partial x} E_x + \frac{\partial \Phi}{\partial y} E_y + \frac{\partial \Phi}{\partial z} E_z \right) + \Phi \left(\frac{\partial E_x}{\partial x} + \frac{\partial E_y}{\partial y} + \frac{\partial E_z}{\partial z} \right).$$

Finally, using the definition of the gradient Eq. (D.1) and the dot product Eq. (A.1) for the first expression in parentheses, and the definition of the divergence Eq. (D.7) for the second, we find

$$\nabla \cdot (\Phi \mathbf{E}) = \nabla \Phi \cdot \mathbf{E} + \Phi \nabla \cdot \mathbf{E},$$

which is equivalent to Eq. (E.6).

Example E.2 Examples of Eq. (E.5)

Demonstrate that Eq. (E.5) is correct for vectors $\mathbf{E} = x E_0 \mathbf{a}_x$ and $\mathbf{F} = y F_0 \mathbf{a}_x - x F_0 \mathbf{a}_y$.

Solution:

We start by evaluating the left side of Eq. (E.5),

$$\nabla (\mathbf{E} \cdot \mathbf{F}) = \nabla \left(x E_0 y F_0 \right) = E_0 F_0 \left(\frac{\partial}{\partial x} (xy) \mathbf{a}_x + \frac{\partial}{\partial y} (xy) \mathbf{a}_y \right)$$

$$= E_0 F_0 \left(y \mathbf{a}_x + x \mathbf{a}_y \right).$$

Next, we examine the right side of Eq. (E.5), one term at a time. The first term is

$$(\mathbf{E} \cdot \nabla) \mathbf{F} = \left(E_x \frac{\partial}{\partial x} + E_y \frac{\partial}{\partial y} + E_z \frac{\partial}{\partial z} \right) \left(y F_0 \mathbf{a}_x - x F_0 \mathbf{a}_y \right)$$

$$= x E_0 \frac{\partial}{\partial x} \left(y F_0 \mathbf{a}_x - x F_0 \mathbf{a}_y \right) = -x E_0 F_0 \mathbf{a}_y. \tag{E.10}$$

Similarly, the second term is

$$(\mathbf{F} \cdot \mathbf{\nabla}) \mathbf{E} = \left(F_x \frac{\partial}{\partial x} + F_y \frac{\partial}{\partial y} + F_z \frac{\partial}{\partial z} \right)(x E_0 \mathbf{a}_x)$$

$$= \left(y F_0 \frac{\partial}{\partial x} - x F_0 \frac{\partial}{\partial y} \right)(x E_0 \mathbf{a}_x) = y E_0 F_0 \mathbf{a}_x. \tag{E.11}$$

The third and fourth terms involve cross products and curls. We use the curl of Eq. (D.13). The third term is

$$\mathbf{E} \times (\mathbf{\nabla} \times \mathbf{F}) = \mathbf{E} \times \begin{vmatrix} \mathbf{a}_x & \mathbf{a}_y & \mathbf{a}_z \\ \dfrac{\partial}{\partial x} & \dfrac{\partial}{\partial y} & \dfrac{\partial}{\partial z} \\ y F_0 & -x F_0 & 0 \end{vmatrix} = \mathbf{E} \times (-2 F_0) \mathbf{a}_z$$

$$= \begin{vmatrix} \mathbf{a}_x & \mathbf{a}_y & \mathbf{a}_z \\ x E_0 & 0 & 0 \\ 0 & 0 & (-2 F_0) \end{vmatrix} = 2 x E_0 F_0 \mathbf{a}_y. \tag{E.12}$$

The fourth term is

$$\mathbf{F} \times (\mathbf{\nabla} \times \mathbf{E}) = \mathbf{F} \times \begin{vmatrix} \mathbf{a}_x & \mathbf{a}_y & \mathbf{a}_z \\ \dfrac{\partial}{\partial x} & \dfrac{\partial}{\partial y} & \dfrac{\partial}{\partial z} \\ x E_0 & 0 & 0 \end{vmatrix} = 0. \tag{E.13}$$

Summing the four terms from the right side of Eq. (E.5) yields $E_0 F_0 \left(y \mathbf{a}_x + x \mathbf{a}_y \right)$, identical to the result from the left side.

E.2 Second-Order Derivatives

As we develop our notions of the properties of electric and magnetic fields, we will make extensive use of the spatial derivative operations (gradient, divergence, and curl) applied to various field quantities. In addition, we will see that we often want to use second-order derivatives of field quantities. These second-order derivatives can take several forms, as we introduce in this section. These include

$$\mathbf{\nabla} \cdot \mathbf{\nabla} \Phi, \tag{E.14}$$

$$\mathbf{\nabla} \times \mathbf{\nabla} \Phi, \tag{E.15}$$

$$\mathbf{\nabla} (\mathbf{\nabla} \cdot \mathbf{E}), \tag{E.16}$$

$$\mathbf{\nabla} \cdot \mathbf{\nabla} \times \mathbf{E}, \tag{E.17}$$

$$\mathbf{\nabla} \times \mathbf{\nabla} \times \mathbf{E}. \tag{E.18}$$

To convince yourself that this list is complete, remember that the gradient operator always acts on a scalar Φ and generates a vector, the divergence operator acts on a vector and generates a scalar, and the curl acts on a vector and generates a vector. Therefore, for example, we can take the divergence or curl of $\mathbf{\nabla} \Phi$, but it makes no sense to write the gradient of this quantity. Each of these second-order derivatives will play a special role at one point or another within our studies, and we will introduce them here.

E.2.1 $\nabla \cdot \nabla \Phi$

We encounter this operation when we develop Poisson's Equation, and its special case Laplace's Equation, in Chapter 2. This second-order derivative is usually abbreviated as

$$\nabla \cdot \nabla \Phi = \nabla^2 \Phi,$$

and is called the **Laplacian** of Φ. We can express the Laplacian in rectangular coordinates using Eqs. (D.7) and (D.1) as

$$\nabla^2 \Phi = \left(\mathbf{a}_x \frac{\partial}{\partial x} + \mathbf{a}_y \frac{\partial}{\partial y} + \mathbf{a}_z \frac{\partial}{\partial z} \right) \cdot \left(\mathbf{a}_x \frac{\partial \Phi}{\partial x} + \mathbf{a}_y \frac{\partial \Phi}{\partial y} + \mathbf{a}_z \frac{\partial \Phi}{\partial z} \right)$$
$$= \frac{\partial^2 \Phi}{\partial x^2} + \frac{\partial^2 \Phi}{\partial y^2} + \frac{\partial^2 \Phi}{\partial z^2}. \tag{E.19}$$

Example E.3 Determination of $\nabla^2 \Phi$ in Rectangular Coordinates

Consider the scalar field $\Phi(x, y, z) = (1/2)(x^2 - y^2 + 2xy)$. Determine $\nabla^2 \Phi$.

Solution:
We use Eq. (E.19):

$$\nabla^2 \Phi = \left(\frac{\partial^2}{\partial x^2} + \frac{\partial^2}{\partial y^2} + \frac{\partial^2}{\partial z^2} \right) \left\{ \frac{1}{2} \left(x^2 - y^2 + 2xy \right) \right\}$$
$$= \frac{\partial}{\partial x} \left\{ \frac{1}{2} \left(2x - 0 + 2y \right) \right\} + \frac{\partial}{\partial y} \left\{ \frac{1}{2} \left(0 - 2y + 2x \right) \right\}$$
$$= \frac{1}{2} \left(2 - 2 \right) = 0.$$

Similarly, in cylindrical coordinates the Laplacian is

$$\nabla^2 \Phi = \frac{1}{\rho} \frac{\partial}{\partial \rho} \left(\rho \frac{\partial \Phi}{\partial \rho} \right) + \frac{1}{\rho^2} \frac{\partial^2 \Phi}{\partial \phi^2} + \frac{\partial^2 \Phi}{\partial z^2}, \tag{E.20}$$

while in spherical coordinates it is

$$\nabla^2 \Phi = \frac{1}{R^2} \frac{\partial}{\partial R} \left(R^2 \frac{\partial \Phi}{\partial R} \right) + \frac{1}{R^2 \sin \theta} \frac{\partial}{\partial \theta} \left(\sin \theta \frac{\partial \Phi}{\partial \theta} \right) + \frac{1}{R^2 \sin^2 \theta} \frac{\partial^2 \Phi}{\partial \phi^2}. \tag{E.21}$$

Example E.4 Determination of $\nabla^2 \Phi$ in Spherical Coordinates

Consider the scalar field $\Phi(R, \theta, \phi) = (1/R^2) \cos \theta \sin \phi$. Determine $\nabla^2 \Phi$.

Solution:
Since we are given Φ in spherical coordinates, we use Eq. (E.21) to write

$$\nabla^2 \Phi = \frac{1}{R^2} \frac{\partial}{\partial R} \left[R^2 \frac{\partial}{\partial R} \left(\frac{\cos \theta \sin \phi}{R^2} \right) \right]$$
$$+ \frac{1}{R^2 \sin \theta} \frac{\partial}{\partial \theta} \left[\sin \theta \frac{\partial}{\partial \theta} \left(\frac{\cos \theta \sin \phi}{R^2} \right) \right] + \frac{1}{R^2 \sin^2 \theta} \frac{\partial^2}{\partial \phi^2} \left(\frac{\cos \theta \sin \phi}{R^2} \right).$$

After we perform the first differentiation of each term we find

$$\nabla^2 \Phi = \frac{1}{R^2} \frac{\partial}{\partial R} \left[R^2 \left(\frac{-2\cos\theta\sin\phi}{R^3} \right) \right]$$
$$+ \frac{1}{R^2 \sin\theta} \frac{\partial}{\partial \theta} \left[\sin\theta \left(\frac{-\sin\theta\sin\phi}{R^2} \right) \right] + \frac{1}{R^2 \sin^2\theta} \frac{\partial}{\partial \phi} \left[\frac{\cos\theta\cos\phi}{R^2} \right],$$

and differentiating each term again

$$\nabla^2 \Phi = \frac{1}{R^2} \left[\frac{2\cos\theta\sin\phi}{R^2} \right]$$
$$+ \frac{1}{R^2 \sin\theta} \left[\frac{-2\sin\theta\cos\theta\sin\phi}{R^2} \right] + \frac{1}{R^2 \sin^2\theta} \left[\frac{-\cos\theta\sin\phi}{R^2} \right].$$

Then we only need to simplify to find

$$\nabla^2 \Phi = \frac{2\cos\theta\sin\phi}{R^4} - \frac{2\cos\theta\sin\phi}{R^4} - \frac{\cos\theta\sin\phi}{R^4 \sin^2\theta}$$

$$= \frac{-\cos\theta\sin\phi}{R^4 \sin^2\theta}.$$

Note that the result for $\nabla^2 \Phi$ is a scalar field.

E.2.2 $\nabla \times \nabla\Phi$

This second derivative is very special, in that, for *any* scalar field $\Phi(\mathbf{r})$, the curl of the gradient of the scalar field must vanish, or

$$\nabla \times (\nabla\Phi) = 0. \tag{E.22}$$

We can understand this in terms of our description of the physical meaning of the curl and gradient operators. The vector function $\nabla\Phi$ always points "uphill" on the landscape of the function $\Phi(\mathbf{r})$, originating at local minima and terminating at local maxima. These vector lines cannot swirl around to form a loop. Equation (E.22) expresses this concept very compactly in vector terms. Let us now demonstrate this explicitly using notation in rectangular coordinates, and follow with a more general proof that involves Stokes' Theorem and integral calculus.

In rectangular coordinates, we can use Eq. (D.13) to write the curl of $\nabla\Phi$ as

$$\nabla \times (\nabla\Phi) = \begin{vmatrix} \mathbf{a}_x & \mathbf{a}_y & \mathbf{a}_z \\ \dfrac{\partial}{\partial x} & \dfrac{\partial}{\partial y} & \dfrac{\partial}{\partial z} \\ \dfrac{\partial\Phi}{\partial x} & \dfrac{\partial\Phi}{\partial y} & \dfrac{\partial\Phi}{\partial z} \end{vmatrix},$$

where we write the x-, y-, and z-components of $\nabla\Phi$ using Eq. (D.1). When we write out the determinant term-by-term, this expression becomes

$$\nabla \times (\nabla\Phi) = \mathbf{a}_x \left(\frac{\partial}{\partial y}\frac{\partial\Phi}{\partial z} - \frac{\partial}{\partial z}\frac{\partial\Phi}{\partial y} \right) + \mathbf{a}_y \left(\frac{\partial}{\partial z}\frac{\partial\Phi}{\partial x} - \frac{\partial}{\partial x}\frac{\partial\Phi}{\partial z} \right)$$
$$+ \mathbf{a}_z \left(\frac{\partial}{\partial x}\frac{\partial\Phi}{\partial y} - \frac{\partial}{\partial y}\frac{\partial\Phi}{\partial x} \right).$$

Notice that each component is the difference between two second-order derivatives of the function Φ. But for well-behaved (i.e. continuous, differentiable) scalar fields Φ, the order of differentiation is unimportant, and each component of $\boldsymbol{\nabla} \times (\boldsymbol{\nabla}\Phi)$ vanishes, leading to Eq. (E.22).

Our general proof of this identity starts by considering the geometry illustrated in Fig. D.6(a), in which we have a closed loop c bounding a surface s. We'll let the vector field $\mathbf{E}$ in that figure be $\boldsymbol{\nabla}\Phi$, and examine the surface integral,

$$\int_s \boldsymbol{\nabla} \times (\boldsymbol{\nabla}\Phi) \cdot d\mathbf{s}.$$

By Stokes' Theorem, this surface integral must be equivalent to the path integral of $\boldsymbol{\nabla}\Phi$ around the contour c, that is

$$\oint_c (\boldsymbol{\nabla}\Phi) \cdot d\boldsymbol{\ell}.$$

But $(\boldsymbol{\nabla}\Phi) \cdot d\boldsymbol{\ell}$ is just $d\Phi$, which we can see by writing out explicitly the dot product. In rectangular coordinates, for example, $\boldsymbol{\nabla}\Phi \cdot d\boldsymbol{\ell}$ is

$$\boldsymbol{\nabla}\Phi \cdot d\boldsymbol{\ell} = \left(\mathbf{a}_x \frac{\partial \Phi}{\partial x} + \mathbf{a}_y \frac{\partial \Phi}{\partial y} + \mathbf{a}_z \frac{\partial \Phi}{\partial z}\right) \cdot \left(\mathbf{a}_x dx + \mathbf{a}_y dy + \mathbf{a}_z dz\right)$$

$$= \frac{\partial \Phi}{\partial x} dx + \frac{\partial \Phi}{\partial y} dy + \frac{\partial \Phi}{\partial z} dz, \tag{E.23}$$

which, by the rules of partial differentiation, is just $d\Phi$. Then the line integral $\int_c d\Phi$ is equal to the difference between Φ evaluated at the two end points of c. Since c is a closed loop, the starting point and ending point for this integration are the same, and $\int_c d\Phi = 0$. Therefore, $\oint_s \boldsymbol{\nabla} \times (\boldsymbol{\nabla}\Phi) \cdot d\mathbf{s}$ must be zero, a result that is valid for any closed path c, and we conclude that the integrand itself must be 0, proving Eq. (E.22).

In Example E.5 we will demonstrate this identity by calculating $\boldsymbol{\nabla} \times (\boldsymbol{\nabla}\Phi)$ for a specific scalar field Φ and showing that this second-order derivative is indeed equal to zero.

Example E.5 Demonstration that $\boldsymbol{\nabla} \times (\boldsymbol{\nabla}\Phi) = 0$

Consider the scalar field $\Phi(x, y, z) = (1/2)(x^2 - y^2 + 2xy)$. Show that $\boldsymbol{\nabla} \times (\boldsymbol{\nabla}\Phi) = 0$.

Solution:
We start by finding $\boldsymbol{\nabla}\Phi$:

$$\boldsymbol{\nabla}\Phi = \mathbf{a}_x \frac{\partial \Phi}{\partial x} + \mathbf{a}_y \frac{\partial \Phi}{\partial y} + \mathbf{a}_z \frac{\partial \Phi}{\partial z} = \mathbf{a}_x(x+y) + \mathbf{a}_y(x-y).$$

The curl of this vector is

$$\boldsymbol{\nabla} \times (\boldsymbol{\nabla}\Phi) = \begin{vmatrix} \mathbf{a}_x & \mathbf{a}_y & \mathbf{a}_z \\ \dfrac{\partial}{\partial x} & \dfrac{\partial}{\partial y} & \dfrac{\partial}{\partial z} \\ (x+y) & (x-y) & 0 \end{vmatrix} = \mathbf{a}_z\left(\frac{\partial}{\partial x}(x-y) - \frac{\partial}{\partial y}(x+y)\right) = 0.$$

The converse of this identity is also true, and extremely important in electrostatics. That is, for any *irrotational* (i.e. curl-free) vector field, there exists a scalar field whose gradient is equal to the vector field. This property allows us to define a scalar electric potential, which we are assured must exist for static fields since $\boldsymbol{\nabla} \times \mathbf{E} = 0$. See Chapter 1, where we explore the properties of this electric potential. A proof that a scalar potential function exists is shown in SideNote E.1.

SideNote E.1 Existence of a Scalar Potential when $\nabla \times \mathbf{E} = 0$

We consider a vector field whose curl is zero, and define a quantity Φ_{BA} as the path integral of $\mathbf{E}$ between two points A and B. That is,

$$\Phi_{BA} = -\int_A^B \mathbf{E} \cdot d\boldsymbol{\ell}. \tag{E.24}$$

The minus sign in this definition is for convenience later. Its presence here does not affect in any way the development of this discussion. For a general field $\mathbf{E}$, the line integral Φ_{BA} depends not just on the end points A and B, but also on the path c taken between these points. But if $\nabla \times \mathbf{E} = 0$, we can show that Φ_{BA} depends *only* on the end points, and not at all on the path. To show this, we consider the two paths c_1 and c_2 shown in Fig. E.1 connecting A and B, and determine the difference between the Φ_{BA} for the two paths, $\Phi_{BA,c_1} - \Phi_{BA,c_2}$. Using Eq. (E.24), this difference is

$$\Phi_{BA,c_1} - \Phi_{BA,c_2} = -\int_{A,c_1}^B \mathbf{E} \cdot d\boldsymbol{\ell} + \int_{A,c_2}^B \mathbf{E} \cdot d\boldsymbol{\ell}.$$

But when we reverse the direction of the path, integrating from point A to point B instead of from point B to point A, this just changes the sign of the integral, or

$$\int_{A,c_2}^B \mathbf{E} \cdot d\boldsymbol{\ell} = -\int_{B,c_2}^A \mathbf{E} \cdot d\boldsymbol{\ell}.$$

So the difference $\Phi_{BA,c_1} - \Phi_{BA,c_2}$ that we seek is

$$\Phi_{BA,c_1} - \Phi_{BA,c_2} = -\int_{A,c_1}^B \mathbf{E} \cdot d\boldsymbol{\ell} - \int_{B,c_2}^A \mathbf{E} \cdot d\boldsymbol{\ell}.$$

The sum of these two integrals then represents integration along the path c_1 from A to B, and then returning to A along c_2, which forms a closed path. We can combine these two integrals to form

$$\Phi_{BA,c_1} - \Phi_{BA,c_2} = -\oint \mathbf{E} \cdot d\boldsymbol{\ell},$$

which must be zero when $\nabla \times \mathbf{E} = 0$, by Stokes' Theorem, Eq. (D.16). Therefore, Φ_{BA,c_1} and Φ_{BA,c_2} are equal, and we conclude that Φ_{BA} is independent of path for an irrotational field.

This independence of Φ_{BA} on the path taken from point A to point B is very important, for it allows us to assign a value $\Phi(x_A, y_A, z_A)$ at point A (where the coordinates of point A are (x_A, y_A, z_A), which we will represent by $\mathbf{r}_A$), and likewise a value $\Phi(x_B, y_B, z_B)$ at point B, represented as $\Phi(\mathbf{r}_B)$, such that the difference is $\Phi_{BA} = \Phi(\mathbf{r}_B) - \Phi(\mathbf{r}_A)$. In other words, $\Phi(\mathbf{r})$ is a scalar field, related to $\mathbf{E}$ through

$$\Phi(\mathbf{r}_B) - \Phi(\mathbf{r}_A) = -\int_A^B \mathbf{E} \cdot d\boldsymbol{\ell}, \tag{E.25}$$

for any two points A and B. This integral shows us how to find the change in $\Phi(\mathbf{r})$ between two points if we know $\mathbf{E}$. This is a very useful relationship, but we'd also like to go the other direction. That is, if we know $\Phi(\mathbf{r})$ in some region, can we use this to determine the vector field $\mathbf{E}$? The answer is "yes," which we show as follows. Let the

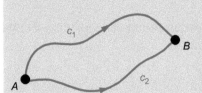

FIGURE E.1 Two paths c_1 and c_2 leading from point A to point B. We use this figure in our proof that for any irrotational vector field $\mathbf{E}$, there is a scalar field $\Phi(\mathbf{r})$ such $\mathbf{E} = -\nabla \Phi(\mathbf{r})$.

distance between points A and B become very small, such that the path integral of $\mathbf{E}$ can be approximated as

$$\Phi(\mathbf{r}_B) - \Phi(\mathbf{r}_A) \simeq -\mathbf{E} \cdot d\boldsymbol{\ell},$$

where $d\boldsymbol{\ell}$ is the incremental length pointing from A to B. In rectangular coordinates, we can use Eq. (C.2) for $d\boldsymbol{\ell}$, and write

$$\Phi(\mathbf{r}_B) - \Phi(\mathbf{r}_A) = -\mathbf{E} \cdot \left(\mathbf{a}_x\, dx + \mathbf{a}_y\, dy + \mathbf{a}_z\, dz\right)$$
$$= -E_x dx - E_y dy - E_z dz. \tag{E.26}$$

But we can write the left side of this equation as

$$\Phi(\mathbf{r}_B) - \Phi(\mathbf{r}_A) = \Delta\Phi = \frac{\partial\Phi}{\partial x}dx + \frac{\partial\Phi}{\partial y}dy + \frac{\partial\Phi}{\partial z}dz,$$

and comparison of each of the components of this equation with Eq. (E.26) gives us $E_x = -\partial\Phi/\partial x$, $E_y = -\partial\Phi/\partial y$, and $E_z = -\partial\Phi/\partial z$. Writing this in the short-hand vector notation, we have

$$\mathbf{E} = -\boldsymbol{\nabla}\Phi. \tag{E.27}$$

With this, we have completed our proof that, for any irrotational vector field $\mathbf{E}$, there is a scalar field Φ such that $\mathbf{E} = -\boldsymbol{\nabla}\Phi$.

In Chapter 1 we address this idea, and explore the properties of the function Φ further, but the importance of this property cannot be overemphasized, for the entire concept of the electric potential for static fields rests on this concept.

E.2.3 $\boldsymbol{\nabla}\left(\boldsymbol{\nabla}\cdot\mathbf{E}\right)$

We encounter this second-order derivative on occasion, particularly as we develop the wave equation, which helps us to understand the propagation of electromagnetic waves (freely propagating, or guided in waveguide structures or transmission lines). In that the divergence of the electric field $\mathbf{E}$ is zero in a charge-free region, and the divergence of the magnetic flux density $\mathbf{B}$ is always zero, we frequently find that, in problems involving electric and magnetic fields, this second-order derivative vanishes as well. This operator is related to $\boldsymbol{\nabla}\times\boldsymbol{\nabla}\times\mathbf{E}$, as we will discuss later in this section.

E.2.4 $\boldsymbol{\nabla}\cdot\boldsymbol{\nabla}\times\mathbf{A}$

Here we encounter a second case of a second-order differential that must be identically zero, this time for any vector field; that is, the divergence of the curl of any vector field $\mathbf{A}$ must be 0, or

$$\boldsymbol{\nabla}\cdot(\boldsymbol{\nabla}\times\mathbf{A}) = 0. \tag{E.28}$$

In words, this is a statement that there are no sources or sinks (i.e. terminal points) of $\boldsymbol{\nabla}\times\mathbf{A}$. As we saw earlier, $\boldsymbol{\nabla}\times\mathbf{A}$ always points in the direction of the axis of rotation of the vector $\mathbf{A}$. (Eq. (E.28) does not imply that $\mathbf{A}$ cannot circulate, which would contradict what we already know, but rather it implies that the lines of $\boldsymbol{\nabla}\times\mathbf{A}$ must close on themselves.) Let's follow the lead of our previous identity (i.e. $\boldsymbol{\nabla}\times(\boldsymbol{\nabla}\Phi) = 0$), and demonstrate the validity of this theorem by explicitly writing it out in rectangular coordinates, and then show a more general proof. In rectangular coordinates, we write the left-hand side of Eq. (E.28) as

$$\boldsymbol{\nabla} \cdot (\boldsymbol{\nabla} \times \mathbf{A}) = \frac{\partial}{\partial x} (\boldsymbol{\nabla} \times \mathbf{A})_x + \frac{\partial}{\partial y} (\boldsymbol{\nabla} \times \mathbf{A})_y + \frac{\partial}{\partial z} (\boldsymbol{\nabla} \times \mathbf{A})_z .$$

Then, writing out the different components of $\boldsymbol{\nabla} \times \mathbf{A}$, this becomes

$$\boldsymbol{\nabla} \cdot (\boldsymbol{\nabla} \times \mathbf{A}) = \frac{\partial}{\partial x} \left(\frac{\partial A_z}{\partial y} - \frac{\partial A_y}{\partial z} \right) + \frac{\partial}{\partial y} \left(\frac{\partial A_x}{\partial z} - \frac{\partial A_z}{\partial x} \right) + \frac{\partial}{\partial z} \left(\frac{\partial A_y}{\partial x} - \frac{\partial A_x}{\partial y} \right).$$

We can rearrange these terms, grouping by A_x, A_y, and A_z, to write

$$\boldsymbol{\nabla} \cdot (\boldsymbol{\nabla} \times \mathbf{A}) = \left(\frac{\partial}{\partial y} \frac{\partial}{\partial z} - \frac{\partial}{\partial z} \frac{\partial}{\partial y} \right) A_x + \left(\frac{\partial}{\partial z} \frac{\partial}{\partial x} - \frac{\partial}{\partial x} \frac{\partial}{\partial z} \right) A_y$$
$$+ \left(\frac{\partial}{\partial x} \frac{\partial}{\partial y} - \frac{\partial}{\partial y} \frac{\partial}{\partial x} \right) A_z.$$

Since the order of differentiation of each of the components of $\mathbf{A}$ is unimportant, each of these terms cancel, and we have demonstrated the validity Eq. (E.28).

We present a more general proof by considering a vector field $\mathbf{A}$ that is differentiable over a volume v, and examining the volume integral of $\boldsymbol{\nabla} \cdot (\boldsymbol{\nabla} \times \mathbf{A})$ over this volume:

$$\int_v \boldsymbol{\nabla} \cdot (\boldsymbol{\nabla} \times \mathbf{A}) \, dv.$$

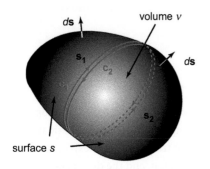

FIGURE E.2 The volume v is bounded by the closed surface s.

We illustrate the volume v in Fig. E.2. By the Divergence Theorem, this integral is equivalent to the surface integral,

$$\oint_s \boldsymbol{\nabla} \times \mathbf{A} \cdot d\mathbf{s},$$

where s is the closed surface that surrounds the volume v. Let us divide the surface s into two parts, which we label s_1 and s_2. In Fig. E.2, s_1 is the surface on the left side of the volume, extending up to the path labeled c_1, and s_2 is the surface on the right side of the volume, extending to the path labeled c_2. (These two paths actually lie on top of one another, but we label them separately since we need to consider their directions, as discussed below.) In terms of these two partial surfaces,

$$\int_v \boldsymbol{\nabla} \cdot (\boldsymbol{\nabla} \times \mathbf{A}) \, dv = \int_{s_1} \boldsymbol{\nabla} \times \mathbf{A} \cdot d\mathbf{s} + \int_{s_2} \boldsymbol{\nabla} \times \mathbf{A} \cdot d\mathbf{s}.$$

The two surface integrals in this equation can be reduced to path integrals using Stokes' Theorem, resulting in

$$\int_v \boldsymbol{\nabla} \cdot (\boldsymbol{\nabla} \times \mathbf{A}) \, dv = \oint_{c_1} \mathbf{A} \cdot d\boldsymbol{\ell} + \oint_{c_2} \mathbf{A} \cdot d\boldsymbol{\ell}.$$

Now we consider the directions of the paths c_1 and c_2, which must by strict convention be opposite one another, as we show in Fig. E.2. This statement follows from the direction of $d\mathbf{s}$, which always points outward from the volume v, and the right-hand rule for determining the direction of the path c from $d\mathbf{s}$, which we discussed earlier in the paragraph following Eq. (D.16). Since paths c_1 and c_2 are identical except for their directions, $\oint_{c_2} \mathbf{A} \cdot d\boldsymbol{\ell} = -\oint_{c_1} \mathbf{A} \cdot d\boldsymbol{\ell}$, and their sum is 0, leading us to the conclusion $\int_v \boldsymbol{\nabla} \cdot (\boldsymbol{\nabla} \times \mathbf{A}) \, dv = 0$. Since this argument is valid for any volume v, the integrand itself must be 0, thus proving Eq. (E.28).

Let's demonstrate this identity for a specific vector field.

Example E.6 Demonstration that $\nabla \cdot (\nabla \times \mathbf{F}) = 0$

Consider the vector field $\mathbf{F} = \mathbf{a}_\rho \rho \cos \phi + \mathbf{a}_\phi z$. Show that $\nabla \cdot (\nabla \times \mathbf{F}) = 0$.

Solution:
We start by using Eq. (D.14) to find $\nabla \times \mathbf{F}$ in cylindrical coordinates:

$$\nabla \times \mathbf{F} = \mathbf{a}_\rho \left(\frac{1}{\rho} \frac{\partial F_z}{\partial \phi} - \frac{\partial F_\phi}{\partial z} \right) + \mathbf{a}_\phi \left(\frac{\partial F_\rho}{\partial z} - \frac{\partial F_z}{\partial \rho} \right)$$
$$+ \mathbf{a}_z \frac{1}{\rho} \left(\frac{\partial (\rho F_\phi)}{\partial \rho} - \frac{\partial F_\rho}{\partial \phi} \right),$$

where $F_\rho = \rho \cos \phi$, $F_\phi = z$, and $F_z = 0$. The result is $\nabla \times \mathbf{F} = -\mathbf{a}_\rho + \mathbf{a}_z (z/\rho + \sin \phi)$. Then we use Eq. (D.8) to find $\nabla \cdot (\nabla \times \mathbf{F})$:

$$\nabla \cdot (\nabla \times \mathbf{F}) = \frac{1}{\rho} \frac{\partial}{\partial \rho} (\rho (\nabla \times \mathbf{F})_\rho) + \frac{1}{\rho} \frac{\partial}{\partial \phi} (\nabla \times \mathbf{F})_\phi + \frac{\partial}{\partial z} (\nabla \times \mathbf{F})_z.$$

We substitute the components of $(\nabla \times \mathbf{F})$ that we found above to write

$$\nabla \cdot (\nabla \times \mathbf{F}) = \frac{1}{\rho} \frac{\partial}{\partial \rho} (-\rho) + \frac{1}{\rho} \frac{\partial}{\partial \phi} (0) + \frac{\partial}{\partial z} \left(\frac{z}{\rho} + \sin \phi \right)$$
$$= \frac{-1}{\rho} + \frac{1}{\rho} = 0.$$

This shows that the divergence of the curl of this vector field is indeed equal to zero, as it must be.

The converse of this identity is also valid, and important in our development of magnetic flux densities in Chapter 4. Specifically, for any solenoidal vector field $\mathbf{B}$ (i.e. a field $\mathbf{B}$ such that $\nabla \cdot \mathbf{B} = 0$), there exists a vector field $\mathbf{A}$ such that $\mathbf{B} = \nabla \times \mathbf{A}$. In our study of magnetic fields this identity allows us to define a vector magnetic potential, analogous to the scalar electric potential Φ. This magnetic potential $\mathbf{A}$ possesses several properties that differ from the electric potential, which we explore in Chapter 6. We also discuss methods to determine this potential in that chapter. In SideNote E.2 we only show that this potential exists.

SideNote E.2 Existence of a Vector Potential when $\nabla \cdot \mathbf{B} = 0$

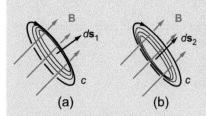

FIGURE E.3 (a) The surface s_1 is bounded by the contour c. We determine the flux of $\mathbf{B}$ through this surface. (b) The surface s_2, different from the surface s_1, is also bounded by the same curve c.

The proof of this identity is similar in some aspects to the proof in SideNote E.1 regarding the existence of the scalar potential for irrotational fields. We consider that the field $\mathbf{B}$ in Fig. E.3(a) is defined and differentiable in the region of space shown, and that we can determine the flux Φ of $\mathbf{B}$ passing through a surface s_1 bounded by a closed contour c. Then,

$$\Phi_{s_1} = \int_{s_1} \mathbf{B} \cdot d\mathbf{s}_1.$$

But s_1 is not the only surface bounded by the contour c. In fact, s_1 drawn in Fig. E.3(a) may not have been your first image of this surface when we mentioned a surface bounded by c. For many of us, our first image of this surface may have been a taut, flat surface stretched across c. But you should agree that s_1, which is bowed to the right in Fig. E.3(a), is equally valid, as is surface s_2, illustrated in Fig. E.3(b), which is bowed to the left. In fact, there are an infinite number of different such surfaces, and the flux Φ of **B** through any of these surfaces must be the same. To show this, we will consider the flux Φ_{s_2} through the surface s_2, and compute the difference:

$$\Phi_{s_1} - \Phi_{s_2} = \int_{s_1} \mathbf{B} \cdot d\mathbf{s} - \int_{s_2} \mathbf{B} \cdot d\mathbf{s}.$$

But the difference of these two surface integrals must be equivalent to the surface integral of $\mathbf{B} \cdot d\mathbf{s}$ over the entire *closed* surface comprised of s_1 and s_2 together. Remember that s_1 and s_2 are bounded by the same path c, and that direction of the surface element $d\mathbf{s}$ is defined by the right-hand rule. (See the discussion in the paragraph following Eq. (D.16) for this right-hand rule.) As shown in Figs. E.3(a) and (b), these surface elements each point up and to the right. But when we combine the two surfaces into one closed surface, we must reverse the direction of $d\mathbf{s}_2$ for surface s_2 since, by convention, $d\mathbf{s}$ always points *outward* from a closed volume. So

$$\Phi_{s_1} - \Phi_{s_2} = \oint_{s_1 + s_2} \mathbf{B} \cdot d\mathbf{s}.$$

But we can use the Divergence Theorem, Eq. (D.10), to write the surface integral on the right as the volume integral of the divergence of **B**, which must be zero since $\nabla \cdot \mathbf{B} = 0$. Therefore the two fluxes Φ_{s_1} and Φ_{s_2} must be the same. This may seem a bit strange, in that the only overlap shared by the two surfaces s_1 and s_2 is the contour c around the "rim" of the surface. Yet we have proven that the surface integrals of **B** over s_1 and s_2 must be identical. What is going on at the rim that causes this to be true? What is happening along this path? We draw our inspiration to answer this question from our proof that a scalar potential function must exist for an irrotational field, given in SideNote E.1. In that proof, we showed that, since the line integral $-\int_c \mathbf{E} \cdot d\boldsymbol{\ell}$ was independent of the path taken between the end points, but instead depended only on the end points $\mathbf{r}_A$ and $\mathbf{r}_B$ themselves, we could define a function Φ at points $\mathbf{r}_A$ and $\mathbf{r}_B$, such that $-\int_c \mathbf{E} \cdot d\boldsymbol{\ell}$ is equal to the difference $\Phi(\mathbf{r}_B) - \Phi(\mathbf{r}_A)$. In the present case, we see that when $\nabla \cdot \mathbf{B} = 0$, the flux of **B** through the surface s is independent of the surface, but rather depends only on the contour c that bounds s. We can infer from this that along this contour c, there is a function $A(\mathbf{r})$ such that $\Phi_s = \int_c A d\boldsymbol{\ell}$ if $A(\mathbf{r})$ is a scalar function, or $\Phi_s = \int_c \mathbf{A} \cdot d\boldsymbol{\ell}$, if $\mathbf{A}(\mathbf{r})$ is a vector function. We quickly realize that only the latter case makes sense, since the sign of the flux Φ_s tells us the direction of **B**, and only a vector $\mathbf{A}(\mathbf{r})$ can differentiate between the two directions of **B**. Furthermore, by Stokes' Theorem, Eq. (D.16), we identify $\nabla \times \mathbf{A}$ with the vector field **B**. So not only are the two fluxes Φ_{s_1} and Φ_{s_2} equal to one another, as we showed before, they can also be determined in terms of the path integral on c of this new vector **A**:

$$\Phi_{s_1} = \Phi_{s_2} = \oint_c \mathbf{A} \cdot d\boldsymbol{\ell},$$

where **B** and **A** are related through $\mathbf{B} = \nabla \times \mathbf{A}$.

We call **A** a potential, in that its spatial derivative (in this case, by spatial derivative we mean curl) provides us with the vector **B**, just as we saw earlier for the potential $\Phi(\mathbf{r})$ for an irrotational field. Unlike that potential, however, which was a scalar field, this potential function is a vector field. We will explore the properties of this potential function in detail when we study magnetic effects later.

E.2.5 $\nabla \times \nabla \times \mathbf{E}$

Through direct expansion of the curl operation, twice, we can show in a few lines that for any differentiable vector field $\mathbf{E}$,

$$\nabla \times \nabla \times \mathbf{E} = \nabla (\nabla \cdot \mathbf{E}) - \nabla^2 \mathbf{E}, \tag{E.29}$$

where $\nabla^2 \mathbf{E}$ is the Laplacian of the field $\mathbf{E}$. This derivation follows a similar path as we used earlier to demonstrate the BAC CAB rule for the vector triple product, Eq. (A.11). (When working through this, if you're so inclined, be careful to maintain the order of the ∇ operators and $\mathbf{E}$. In particular, since ∇ operates from the left, be sure to keep the components of $\mathbf{E}$ on the right.) Recall that we briefly discussed $\nabla (\nabla \cdot \mathbf{E})$ two sections earlier. The Laplacian of a vector is defined relatively simply in rectangular coordinates,

$$\nabla^2 \mathbf{E} = \mathbf{a}_x \nabla^2 E_x + \mathbf{a}_y \nabla^2 E_y + \mathbf{a}_z \nabla^2 E_z, \tag{E.30}$$

where we use the Laplacian of a scalar function defined earlier in Eq. (E.19).

We use Eq. (E.29) in Chapter 4 to help define a magnetic vector potential, and again in Chapter 8 when we develop a second-order differential equation, known as the **wave equation**, that the electric field $\mathbf{E}(\mathbf{r})$ and the magnetic field $\mathbf{H}(\mathbf{r})$ must obey, and which governs the propagation of electromagnetic waves. We defer further description to those chapters.

The form of $\nabla^2 \mathbf{E}$ is more complicated when written in cylindrical or spherical coordinates, but fortunately we will usually be able to limit our exposure to these latter forms. We include them here for completeness. In cylindrical coordinates, the Laplacian of $\mathbf{E}$ is

$$\nabla^2 \mathbf{E} = \mathbf{a}_\rho (\nabla^2 \mathbf{E})_\rho + \mathbf{a}_\phi (\nabla^2 \mathbf{E})_\phi + \mathbf{a}_z (\nabla^2 \mathbf{E})_z, \tag{E.31}$$

with the components of the Laplacian given by

$$
\begin{aligned}
(\nabla^2 \mathbf{E})_\rho &= \nabla^2 E_\rho - \frac{E_\rho}{\rho^2} - \frac{2}{\rho^2} \frac{\partial E_\phi}{\partial \phi}, \\
(\nabla^2 \mathbf{E})_\phi &= \nabla^2 E_\phi - \frac{E_\phi}{\rho^2} + \frac{2}{\rho^2} \frac{\partial E_\rho}{\partial \phi}, \\
(\nabla^2 \mathbf{E})_z &= \nabla^2 E_z.
\end{aligned}
\tag{E.32}
$$

The first term in the expression for each component is the scalar Laplacian, given in Eq. (E.20).

Finally, in spherical coordinates, we have

$$\nabla^2 \mathbf{E} = \mathbf{a}_R (\nabla^2 \mathbf{E})_R + \mathbf{a}_\theta (\nabla^2 \mathbf{E})_\theta + \mathbf{a}_\phi (\nabla^2 \mathbf{E})_\phi, \tag{E.33}$$

where the components of the Laplacian are

$$\left(\nabla^2 \mathbf{E}\right)_R = \nabla^2 E_R - \frac{2}{R^2}\left(E_R + \frac{\partial E_\theta}{\partial \theta} + \cot\theta\, E_\theta + \csc\theta\, \frac{\partial E_\phi}{\partial \phi}\right),$$

$$\left(\nabla^2 \mathbf{E}\right)_\theta = \nabla^2 E_\theta - \frac{1}{R^2}\left(\frac{E_\theta}{\sin^2\theta} - 2\frac{\partial E_R}{\partial \theta} + \frac{2\cos\theta}{\sin^2\theta}\frac{\partial E_\phi}{\partial \phi}\right), \tag{E.34}$$

$$\left(\nabla^2 \mathbf{E}\right)_\phi = \nabla^2 E_\phi - \frac{1}{R^2\sin\theta}\left(\frac{E_\phi}{\sin\theta} - 2\frac{\partial E_R}{\partial \phi} - 2\cot\theta\frac{\partial E_\theta}{\partial \phi}\right).$$

The first term in the expression for each component here is the scalar Laplacian in spherical coordinates, given in Eq. (E.21).

E.3 Completeness of Vector Fields

Finally, we close our review of the mathematical tools by discussing two important theorems related to the completeness of our knowledge of a vector field $\mathbf{A}(\mathbf{r})$ as expressed through $\nabla \cdot \mathbf{A}(\mathbf{r})$ and $\nabla \times \mathbf{A}(\mathbf{r})$. In the first, we are motivated by the following analog for scalar functions in one variable. You are probably familiar with the fact that, for a function $f(x)$, if we know the slope of this function, $f'(x) = df(x)/dx$, over some range of x, and if we know the value of $f(x_0)$ at one location x_0 within that range, then we can unambiguously determine the function $f(x)$ at any x within this range. For this one-dimensional case, we find $f(x)$ by direct integration, as in

$$f(x) = f(x_0) + \int_{x_0}^{x} f'(x')\, dx'.$$

Can this concept be extended to the case of vectors, which have three spatial components, each of which is a function of three spatial coordinates? The answer is yes, as expressed in the following theorem. That is, when we know the divergence and the curl of a vector field $\mathbf{A}(\mathbf{r})$ that is finite and continuous over some volume v in space, and when we also know the normal component of the vector $\mathbf{A}(\mathbf{r})$ at the boundary of this volume, defined by a surface s, then we can determine uniquely the vector $\mathbf{A}(\mathbf{r})$ within this space. We can prove this by presuming that $\mathbf{A}(\mathbf{r})$ is not unique, that there are at least two solutions $\mathbf{A}_1(\mathbf{r})$ and $\mathbf{A}_2(\mathbf{r})$, and then showing that the difference between these two vectors must necessarily be zero.

SideNote E.3 Vector Completeness Theorem

We start by expressing the divergence and curl of $\mathbf{A}(\mathbf{r})$ as

$$\nabla \cdot \mathbf{A}(\mathbf{r}) = \rho(\mathbf{r}) \tag{E.35}$$

and

$$\nabla \times \mathbf{A}(\mathbf{r}) = \mathbf{j}(\mathbf{r}), \tag{E.36}$$

where $\rho(\mathbf{r})$, a scalar charge density, and $\mathbf{j}(\mathbf{r})$, a vector current density, are sources of the field $\mathbf{A}(\mathbf{r})$, which is differentiable over the entire volume v. (This interpretation of $\rho(\mathbf{r})$ and $\mathbf{j}(\mathbf{r})$ as sources is not important now, but will become clear later, when we discuss electric and magnetic fields in more detail.) We assume that two fields $\mathbf{A}_1(\mathbf{r})$

and $A_2(r)$ each individually satisfy Eqs. (E.35) and (E.36), and satisfy the same normal boundary conditions.

Let's define the difference between vectors $A_1(r)$ and $A_2(r)$ as $\Delta A(r) = A_1(r) - A_2(r)$. If we can show that $\Delta A(r)$ must be zero within this volume v, then $A_1(r)$ and $A_2(r)$ must be the same vector, and there can be only one vector $A(r)$ in this region.

First, we use Eqs. (E.35) and (E.36) to write

$$\nabla \cdot \Delta A(r) = \nabla \cdot (A_1(r) - A_2(r)) = 0 \qquad (E.37)$$

and

$$\nabla \times \Delta A(r) = \nabla \times (A_1(r) - A_2(r)) = 0.$$

In words, the difference vector $\Delta A(r)$ is solenoidal and irrotational. Since $\nabla \times \Delta A(r) = 0$, we can define a potential $\phi(r)$, such that

$$\Delta A(r) = -\nabla \phi(r). \qquad (E.38)$$

Substituting this into Eq. (E.37) gives us

$$\nabla \cdot \Delta A(r) = 0 = -\nabla \cdot (\nabla \phi(r)) = -\nabla^2 \phi(r), \qquad (E.39)$$

where we recognize the Laplacian operator $\nabla^2 \phi(r) = \nabla \cdot \nabla \phi(r)$ from Section E.2. Equation (E.39) is known as Laplace's Equation, and we examine this in considerable detail in later chapters as we solve electrostatic boundary value problems. For now, however, we can examine only a limited solution relevant to the present proof.

To show that $\Delta A(r)$ must necessarily be zero, let us examine the integral

$$\int_v |\Delta A(r)|^2 \, dv,$$

where the integration is carried out over the entire volume v. Since the integrand $|\Delta A(r)|^2$ is non-negative, this integral must also be non-negative. But we can use Eq. (E.38) to write this as

$$\int_v |\Delta A(r)|^2 \, dv = \int_v \nabla \phi(r) \cdot \nabla \phi(r) \, dv. \qquad (E.40)$$

Now we use Eq. (E.6) for the divergence of the product of a scalar field and vector field, using $\nabla \phi(r)$ for E and $\phi(r)$ for Φ, which gives us

$$\nabla \cdot \{\phi(r) \nabla \phi(r)\} = \phi(r) \{\nabla \cdot \nabla \phi(r)\} + \nabla \phi(r) \cdot \nabla \phi(r).$$

But by Eq. (E.39), $\nabla \cdot \nabla \phi(r)$, which is equivalent to $\nabla^2 \phi(r)$, is 0, so we can write Eq. (E.40) as

$$\int_v |\Delta A(r)|^2 \, dv = \int_v \nabla \cdot \{\phi(r) \nabla \phi(r)\} \, dv.$$

We use the Divergence Theorem to change this volume integral to a surface integral,

$$\int_v |\Delta A(r)|^2 \, dv = \int_s \phi(r) \nabla \phi(r) \cdot ds = -\int_s \phi(r) \Delta A(r) \cdot ds.$$

Since the normal components of A_1 and A_2 are the same at the bounding surface, this surface integral must be zero, which gives us

$$\int_v |\Delta A(r)|^2 \, dv = 0.$$

Since the integrand $|\Delta A(\mathbf{r})|^2$ must be positive or zero, and since the integral over the entire volume is zero, it must be that $\Delta A(\mathbf{r}) = 0$, or $A_1(\mathbf{r}) = A_2(\mathbf{r})$ for all $\mathbf{r}$. Therefore we conclude that the vector $\Delta A(\mathbf{r})$ is uniquely determined, proving this theorem.

The second, and final, theorem that we discuss in this section is **Helmholtz's Theorem**. We will not prove this theorem here, as its proof depends on tools that we have not developed, so we simply state the theorem without proof. Helmholtz's Theorem tells us that we can always express any vector field $\mathbf{E}$ as the sum of two parts, one of which is solenoidal and the other of which is irrotational. Using the results of Section E.2, then, we can write the solenoidal component as the gradient of a scalar function $-\phi(\mathbf{r})$, and the irrotational component as the curl of a vector field $\mathbf{A}(\mathbf{r})$. We make extensive use of these *potential* functions.

PROBLEMS

Distributive and Derivative Product Properties

PE-1 Confirm Eq. (E.5) by explicitly writing out the terms of the operators in (a) rectangular coordinates; (b) cylindrical coordinates; (c) spherical coordinates.

PE-2 Confirm Eq. (E.7) by explicitly writing out the terms of the operators in (a) rectangular coordinates; (b) cylindrical coordinates; (c) spherical coordinates.

PE-3 Confirm Eq. (E.8) by explicitly writing out the terms of the operators in (a) rectangular coordinates; (b) cylindrical coordinates; (c) spherical coordinates.

Second Order Derivatives

PE-4 For the scalar field (a) $f(\mathbf{r}) = (x + y)/\sqrt{x^2 + y^2}$, find $\nabla f(\mathbf{r})$. Then confirm that $\nabla \times (\nabla f(\mathbf{r})) = 0$. (b) Follow the same instructions, using $g(\mathbf{r}) = \exp[-(x-x_0)/a] \times \exp[-(y-y_0)/b]$.

PE-5 (a) For the vector field $\mathbf{A}(\mathbf{r}) = \mathbf{a}_x y^2 - \mathbf{a}_y x^2 + \mathbf{a}_z xy$, find $\nabla \times \mathbf{A}(\mathbf{r})$. Then confirm that $\nabla \cdot (\nabla \times \mathbf{A}(\mathbf{r})) = 0$. (b) Follow the same instructions, using $\mathbf{B} = \mathbf{a}_\rho 6\rho \cos \phi - \mathbf{a}_\phi 6\rho \sin \phi$.

PE-6 For an arbitrary vector field $\mathbf{E}$, verify Eq. (E.29) in (a) rectangular coordinates, (b) cylindrical coordinates, and (c) spherical coordinates.

APPENDIX F
Summary of Vector Identities

F.1 Vector Identities

$\mathbf{A} + \mathbf{B} = \mathbf{B} + \mathbf{A}$ Appendix A

$\mathbf{A} + [\mathbf{B} + \mathbf{C}] = [\mathbf{A} + \mathbf{B}] + \mathbf{C}$ Appendix A

$b\mathbf{A} = b\left(A_x\mathbf{a}_x + A_y\mathbf{a}_y + A_z\mathbf{a}_z\right) = bA_x\mathbf{a}_x + bA_y\mathbf{a}_y + bA_z\mathbf{a}_z$ Appendix A

$\mathbf{A} \cdot (\mathbf{B} \times \mathbf{C}) = \mathbf{B} \cdot (\mathbf{C} \times \mathbf{A}) = \mathbf{C} \cdot (\mathbf{A} \times \mathbf{B})$ Eq. (A.10)

$\mathbf{A} \times (\mathbf{B} \times \mathbf{C}) = \mathbf{B}(\mathbf{A} \cdot \mathbf{C}) - \mathbf{C}(\mathbf{A} \cdot \mathbf{B})$ Eq. (A.11)

F.2 $d\boldsymbol{\ell}$, Gradient, Divergence, Curl, and Laplacian

Rectangular Coordinates

$$d\boldsymbol{\ell} = \mathbf{a}_x\, dx + \mathbf{a}_y\, dy + \mathbf{a}_z\, dz \qquad \text{Eq. (C.2)}$$

$$\nabla\psi = \mathbf{a}_x\, \frac{\partial\psi}{\partial x} + \mathbf{a}_y\, \frac{\partial\psi}{\partial y} + \mathbf{a}_z\, \frac{\partial\psi}{\partial z} \qquad \text{Eq. (D.1)}$$

$$\nabla \cdot \mathbf{A} = \frac{\partial A_x}{\partial x} + \frac{\partial A_y}{\partial y} + \frac{\partial A_z}{\partial z} \qquad \text{Eq. (D.7)}$$

$$\nabla \times \mathbf{A} = \begin{vmatrix} \mathbf{a}_x & \mathbf{a}_y & \mathbf{a}_z \\ \dfrac{\partial}{\partial x} & \dfrac{\partial}{\partial y} & \dfrac{\partial}{\partial z} \\ A_x & A_y & A_z \end{vmatrix} = \mathbf{a}_x\left(\frac{\partial A_z}{\partial y} - \frac{\partial A_y}{\partial z}\right) \qquad \text{Eq. (D.13)}$$

$$+ \mathbf{a}_y\left(\frac{\partial A_x}{\partial z} - \frac{\partial A_z}{\partial x}\right) + \mathbf{a}_z\left(\frac{\partial A_y}{\partial x} - \frac{\partial A_x}{\partial y}\right)$$

$$\nabla^2\Phi = \frac{\partial^2\Phi}{\partial x^2} + \frac{\partial^2\Phi}{\partial y^2} + \frac{\partial^2\Phi}{\partial z^2} \qquad \text{Eq. (E.19)}$$

$$\nabla^2\mathbf{E} = \mathbf{a}_x\nabla^2 E_x + \mathbf{a}_y\nabla^2 E_y + \mathbf{a}_z\nabla^2 E_z \qquad \text{Eq. (E.30)}$$

Cylindrical Coordinates

$$d\boldsymbol{\ell} = \mathbf{a}_\rho\, d\rho + \mathbf{a}_\phi \rho\, d\phi + \mathbf{a}_z\, dz \qquad \text{Eq. (C.3)}$$

$$\boldsymbol{\nabla}\psi = \mathbf{a}_\rho \frac{\partial \psi}{\partial \rho} + \mathbf{a}_\phi \frac{1}{\rho}\frac{\partial \psi}{\partial \phi} + \mathbf{a}_z \frac{\partial \psi}{\partial z} \qquad \text{Eq. (D.2)}$$

$$\boldsymbol{\nabla}\cdot\mathbf{A} = \frac{1}{\rho}\frac{\partial(\rho A_\rho)}{\partial \rho} + \frac{1}{\rho}\frac{\partial A_\phi}{\partial \phi} + \frac{\partial A_z}{\partial z} \qquad \text{Eq. (D.8)}$$

$$\boldsymbol{\nabla}\times\mathbf{A} = \frac{1}{\rho}\begin{vmatrix} \mathbf{a}_\rho & \rho\mathbf{a}_\phi & \mathbf{a}_z \\ \frac{\partial}{\partial \rho} & \frac{\partial}{\partial \phi} & \frac{\partial}{\partial z} \\ A_\rho & \rho A_\phi & A_z \end{vmatrix} = \mathbf{a}_\rho\left(\frac{1}{\rho}\frac{\partial A_z}{\partial \phi} - \frac{\partial A_\phi}{\partial z}\right) \qquad \text{Eq. (D.14)}$$

$$+ \mathbf{a}_\phi\left(\frac{\partial A_\rho}{\partial z} - \frac{\partial A_z}{\partial \rho}\right) + \mathbf{a}_z \frac{1}{\rho}\left(\frac{\partial(\rho A_\phi)}{\partial \rho} - \frac{\partial A_\rho}{\partial \phi}\right)$$

$$\boldsymbol{\nabla}^2\Phi = \frac{1}{\rho}\frac{\partial}{\partial \rho}\left(\rho\frac{\partial \Phi}{\partial \rho}\right) + \frac{1}{\rho^2}\frac{\partial^2 \Phi}{\partial \phi^2} + \frac{\partial^2 \Phi}{\partial z^2} \qquad \text{Eq. (E.20)}$$

$\boldsymbol{\nabla}^2\mathbf{E}$ See Eqs. (E.31) and (E.32).

Spherical Coordinates

$$d\boldsymbol{\ell} = \mathbf{a}_R\, dR + \mathbf{a}_\theta R\, d\theta + \mathbf{a}_\phi R\sin\theta\, d\phi \qquad \text{Eq. (C.4)}$$

$$\boldsymbol{\nabla}\psi = \mathbf{a}_R \frac{\partial \psi}{\partial R} + \mathbf{a}_\theta \frac{1}{R}\frac{\partial \psi}{\partial \theta} + \mathbf{a}_\phi \frac{1}{R\sin\theta}\frac{\partial \psi}{\partial \phi} \qquad \text{Eq. (D.3)}$$

$$\boldsymbol{\nabla}\cdot\mathbf{A} = \frac{1}{R^2}\frac{\partial(R^2 A_R)}{\partial R} + \frac{1}{R\sin\theta}\frac{\partial(A_\theta \sin\theta)}{\partial \theta} + \frac{1}{R\sin\theta}\frac{\partial A_\phi}{\partial \phi} \qquad \text{Eq. (D.9)}$$

$$\boldsymbol{\nabla}\times\mathbf{A} = \frac{1}{R^2\sin\theta}\begin{vmatrix} \mathbf{a}_R & R\mathbf{a}_\theta & R\sin\theta\mathbf{a}_\phi \\ \frac{\partial}{\partial R} & \frac{\partial}{\partial \theta} & \frac{\partial}{\partial \phi} \\ A_R & RA_\theta & R\sin\theta A_\phi \end{vmatrix} = \mathbf{a}_R \frac{1}{R\sin\theta}\left(\frac{\partial(\sin\theta A_\phi)}{\partial \theta} - \frac{\partial A_\theta}{\partial \phi}\right) \qquad \text{Eq. (D.15)}$$

$$+ \mathbf{a}_\theta \frac{1}{R}\left(\frac{1}{\sin\theta}\frac{\partial A_R}{\partial \phi} - \frac{\partial(RA_\phi)}{\partial R}\right) + \mathbf{a}_\phi \frac{1}{R}\left(\frac{\partial(RA_\theta)}{\partial R} - \frac{\partial A_R}{\partial \theta}\right)$$

$$\boldsymbol{\nabla}^2\Phi = \frac{1}{R^2}\frac{\partial}{\partial R}\left(R^2\frac{\partial \Phi}{\partial R}\right) + \frac{1}{R^2\sin\theta}\frac{\partial}{\partial \theta}\left(\sin\theta\frac{\partial \Phi}{\partial \theta}\right) + \frac{1}{R^2\sin^2\theta}\frac{\partial^2 \Phi}{\partial \phi^2} \qquad \text{Eq. (E.21)}$$

$\boldsymbol{\nabla}^2\mathbf{E}$ See Eqs. (E.33) and (E.34).

F.3 Differential Vector Identities

$$\boldsymbol{\nabla}(\Phi + \Psi) = \boldsymbol{\nabla}\Phi + \boldsymbol{\nabla}\Psi \qquad \text{Eq. (E.1)}$$

$$\boldsymbol{\nabla}\cdot(\mathbf{E} + \mathbf{F}) = \boldsymbol{\nabla}\cdot\mathbf{E} + \boldsymbol{\nabla}\cdot\mathbf{F} \qquad \text{Eq. (E.2)}$$

$$\boldsymbol{\nabla}\times(\mathbf{E} + \mathbf{F}) = \boldsymbol{\nabla}\times\mathbf{E} + \boldsymbol{\nabla}\times\mathbf{F} \qquad \text{Eq. (E.3)}$$

$$\nabla(\Phi\Psi) = \Phi\nabla\Psi + \Psi\nabla\Phi \qquad\qquad \text{Eq. (E.4)}$$

$$\nabla(\mathbf{E}\cdot\mathbf{F}) = (\mathbf{E}\cdot\nabla)\mathbf{F} + (\mathbf{F}\cdot\nabla)\mathbf{E} + \mathbf{E}\times(\nabla\times\mathbf{F}) + \mathbf{F}\times(\nabla\times\mathbf{E}) \qquad\qquad \text{Eq. (E.5)}$$

$$\nabla\cdot(\Phi\mathbf{E}) = \Phi\nabla\cdot\mathbf{E} + \mathbf{E}\cdot\nabla\Phi \qquad\qquad \text{Eq. (E.6)}$$

$$\nabla\cdot(\mathbf{E}\times\mathbf{F}) = \mathbf{F}\cdot(\nabla\times\mathbf{E}) - \mathbf{E}\cdot(\nabla\times\mathbf{F}) \qquad\qquad \text{Eq. (E.7)}$$

$$\nabla\times(\Phi\mathbf{E}) = \Phi\nabla\times\mathbf{E} + \nabla\Phi\times\mathbf{E} \qquad\qquad \text{Eq. (E.8)}$$

$$\nabla\times(\mathbf{E}\times\mathbf{F}) = \mathbf{E}\nabla\cdot\mathbf{F} - \mathbf{F}\nabla\cdot\mathbf{E} + (\mathbf{F}\cdot\nabla)\mathbf{E} - (\mathbf{E}\cdot\nabla)\mathbf{F} \qquad\qquad \text{Eq. (E.9)}$$

APPENDIX G
Integral Tables

$$\int x^n dx = \frac{x^{n+1}}{n+1}, n \neq -1 \tag{G.1}$$

$$\int \frac{dx}{x} = \ln x \tag{G.2}$$

$$\int x \ln(x) dx = \frac{x^2}{2} \ln(x) - \frac{x^2}{4} \tag{G.3}$$

$$\int \frac{dx}{(a+bx)^{1/2}} = \frac{2(a+bx)^{1/2}}{b} \tag{G.4}$$

$$\int \frac{dx}{(a+bx)^{3/2}} = -\frac{2}{b} [a+bx]^{-1/2} \tag{G.5}$$

$$\int \frac{xdx}{(a+bx)^{3/2}} = \frac{2}{b^2} \left[(a+bx)^{1/2} + a(a+bx)^{-1/2} \right] \tag{G.6}$$

$$\int \frac{dx}{[a^2+x^2]^{1/2}} = \ln \left(x + \sqrt{a^2+x^2} \right) \tag{G.7}$$

$$\int \frac{dx}{[a^2-x^2]^{1/2}} = \sin^{-1} \frac{x}{|a|} \tag{G.8}$$

$$\int \frac{xdx}{[a^2 \pm x^2]^{1/2}} = \pm\sqrt{a^2 \pm x^2} \tag{G.9}$$

$$\int \frac{dx}{a^2+x^2} = \frac{1}{a} \tan^{-1} \left(\frac{x}{a} \right) \tag{G.10}$$

$$\int \frac{xdx}{a^2+x^2} = \frac{1}{2} \ln \left(a^2+x^2 \right) \tag{G.11}$$

$$\int \frac{dx}{[a^2+x^2]^{3/2}} = \frac{1}{a^2} \frac{x}{[a^2+x^2]^{1/2}} \tag{G.12}$$

$$\int \frac{xdx}{[a^2 \pm x^2]^{3/2}} = \mp \frac{1}{[a^2 \pm x^2]^{1/2}} \tag{G.13}$$

$$\int \sqrt{(a^2 - x^2)^3}\,dx = \frac{1}{4}\left[x\sqrt{(a^2 - x^2)^3} + \frac{3a^2 x}{2}\sqrt{a^2 - x^2} + \frac{3a^4}{2}\sin^{-1}\frac{x}{|a|}\right] \qquad (G.14)$$

$$\int \frac{dx}{b + c\cos(ax)} = \frac{2}{a\sqrt{b^2 - c^2}}\tan^{-1}\left[\sqrt{\frac{b - c}{b + c}}\tan\left(\frac{ax}{2}\right)\right],\ \text{valid for } b > c \qquad (G.15)$$

$$\int \frac{\cos(x)dx}{b + c\cos(x)} = \frac{x}{c} - \frac{b}{c}\int \frac{dx}{b + c\cos(x)} \qquad (G.16)$$

$$\int \csc(ax)dx = -\frac{1}{a}\ln|\csc(ax) - \cot(ax)| \qquad (G.17)$$

APPENDIX H
Several Useful Taylor Expansions

$$e^x = 1 + x + \frac{x^2}{2!} + \frac{x^3}{3!} + \cdots, \text{ valid for } |x| < 1 \tag{H.1}$$

$$\sin x = x - \frac{x^3}{3!} + \frac{x^5}{5!} + \cdots \tag{H.2}$$

$$\cos x = 1 - \frac{x^2}{2!} + \frac{x^4}{4!} + \cdots \tag{H.3}$$

$$\tan x = x + \frac{1}{3}x^3 + \frac{2}{15}x^5 + \cdots \tag{H.4}$$

$$\frac{1}{1 \pm x} = 1 \mp x + x^2 \mp x^3 + \cdots \tag{H.5}$$

$$\sqrt{1 \pm x} = 1 \pm \frac{x}{2} - \frac{x^2}{8} \pm \frac{x^3}{16} + \cdots \tag{H.6}$$

$$\frac{1}{\sqrt{1 \pm x}} = 1 \mp \frac{x}{2} + \frac{3x^2}{8} \mp \frac{5x^3}{16} + \cdots \tag{H.7}$$

$$\ln(1 + x) = x - \frac{x^2}{2} + \frac{x^3}{3} - \frac{x^4}{4} + \cdots, \text{ valid for } |x| < 1 \tag{H.8}$$

$$\frac{1}{\sqrt{1 + x}} = 1 - \frac{1}{2}x + \frac{1 \cdot 3}{2 \cdot 4}x^2 - \frac{1 \cdot 3 \cdot 5}{2 \cdot 4 \cdot 6}x^3 + \cdots, \text{ valid for } |x| < 1 \tag{H.9}$$

$$(1 + x)^n = 1 + nx + \frac{n(n-1)}{2!}x^2 + \frac{n(n-1)(n-2)}{3!}x^3 \tag{H.10}$$
$$+ \cdots, \text{ valid for } |x| < 1$$

APPENDIX I
Review of Complex Numbers

Complex numbers are based on the number $\sqrt{-1}$, which we denote as j. (The symbol i is also often used for $\sqrt{-1}$, so you may see this notation in other texts. We stick with the symbol j, as this enjoys wider usage in engineering texts.) We can write any complex number z in terms of its real and imaginary parts,

$$z = x + jy,$$

where x is called the real part of z, $x = \Re(z)$, and y is called the imaginary part of z, $y = \Im(z)$. We can perform all of the common arithmetic operations with complex numbers that can be performed with real numbers. For example, we add two complex numbers $z_1 = a + jb$ and $z_2 = c + jd$ by adding the real parts to find the real part of the sum, and by adding imaginary parts to find the imaginary part of the sum. That is,

$$z_1 + z_2 = (a + c) + j(b + d).$$

Similarly, to find the difference of two complex numbers, we subtract the real parts and subtract the imaginary parts:

$$z_1 - z_2 = (a - c) + j(b - d).$$

We can think of j as just a number that multiplies b or d, and apply the distributive property to obtain these results.

It is often useful to visualize a complex number $z = x + jy$ on the complex plane, as shown in Fig. I.1. On this plane, we plot the real part of z along the horizontal axis, also known as the real axis, and the imaginary part of z along the vertical, or imaginary, axis. Addition of complex numbers, as discussed above, can be carried out graphically on this plane. We can represent z_1 as an arrow pointing from the origin to the point (a, b), and z_2 as an arrow pointing from the origin to the point (c, d). To add two numbers graphically, we move one of these arrows, let's say z_2, placing its tail at the head of the other, z_1. The sum of the complex numbers $z_1 + z_2$ is represented by an arrow from the tail of z_1 (at the origin) to the head of z_2. Addition of complex numbers is very similar to addition of two-dimensional vectors.

An alternative means of specifying a complex number uses its magnitude and phase (or angle). The magnitude of $z = x + jy$ is the length of the line from the origin to the point z, or

$$|z| = \sqrt{x^2 + y^2},$$

and the angle or phase of the complex number is

$$\alpha = \tan^{-1}(y/x).$$

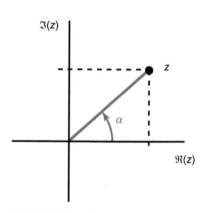

FIGURE I.1 A complex number $z = x + jy$ can be represented on the complex plane.

In terms of the magnitude and phase, we have $x = \mathfrak{R}(z) = |z| \cos \alpha$ and $y = \mathfrak{I}(z) = |z| \sin \alpha$. We then have two equivalent forms for expressing any complex number z. That is, in terms of its real and imaginary parts, or in terms of its magnitude $|z|$ and phase α. These two forms are perfectly equivalent, and we should choose the form that is more convenient for us as we encounter them in our analysis. For example, if we need to add or subtract two complex quantities, we need to express them in terms of their real and imaginary parts. We haven't yet discussed multiplication or division of complex numbers, but we will shortly see that these operations are simpler when the numbers are written as the magnitude and phase angle, although we can use either form.

Using the notation from the previous paragraph, we can write $z = x + jy = |z|(\cos \alpha + j \sin \alpha)$. The term inside the parentheses is, from Euler's Identity, equal to $e^{j\alpha}$, as can be shown by writing each of these functions using their Taylor expansions. This gives us the compact notation of

$$z = |z|e^{j\alpha},$$

a succinct expression of the complex number z in terms of its magnitude and phase.

In working with complex notation, a handy operation that we will use extensively is that of complex conjugation. The complex conjugate of a number $z = x + jy$ is $z^* = x - jy$; that is, the sign of the imaginary component is inverted. When plotted on the complex plane, as in Fig. I.1, the complex conjugate z^* is the reflection of z about the real axis. Similarly, a complex number written using the complex exponential follows the same prescription, $(e^{j\alpha})^* = e^{-j\alpha}$. Since $\cos \alpha$ is an even function, and $\sin \alpha$ is odd, we see that changing $j\alpha$ in the complex exponential to $-j\alpha$ changes the sign of the imaginary part, as we stated above.

Let's now take a look at multiplication and division of two complex numbers. We can multiply two complex numbers by multiplying each of the terms:

$$z_1 z_2 = (a + jb)(c + jd) = (ac + jad + jbc + j^2 bd).$$

Then we use $j^2 = -1$ and rearrange terms to find

$$z_1 z_2 = (ac - bd) + j(ad + bc).$$

This operation is often easier to carry out when the complex numbers are expressed in their magnitude and phase form, $z = |z|e^{j\alpha}$. Then the product is of the form

$$z_1 z_2 = |z_1|e^{j\alpha_1}|z_2|e^{j\alpha_2},$$

which, since the product of exponentials is found by adding the angles,

$$z_1 z_2 = |z_1||z_2|e^{j(\alpha_1 + \alpha_2)}.$$

In words, the magnitude of the product is the product of the magnitudes, and the angle of the product is the sum of the angles. A particularly useful product is that of a complex number with its conjugate, zz^*, which, whether carried out in real and imaginary parts,

$$zz^* = (x + jy)(x - jy) = x^2 + y^2,$$

or magnitude and phase form

$$zz^* = |z|e^{j\alpha}|z|e^{-j\alpha} = |z|^2,$$

we get the same result. That provides us a simple prescription for determining the magnitude of a complex number – that is

$$|z| = \sqrt{zz^*}.$$

Notice that the magnitude $e^{j\alpha}$, where α is any real phase, is simply 1, as can be shown directly by applying this rule.

Division is a little more complicated, but we can easily carry this out using the tools that we have already discussed. When written in terms of real and imaginary parts, we have

$$\frac{z_1}{z_2} = \frac{a + jb}{c + jd}.$$

We'd like to get rid of the imaginary term in the denominator. To do this, we multiply the top and bottom by z_2^*:

$$\frac{z_1}{z_2} = \frac{(a + jb)}{(c + jd)}\frac{(c - jd)}{(c - jd)}.$$

Multiplying through term by term, we find

$$\frac{z_1}{z_2} = \frac{(ac + bd) + j(-ad + bc)}{(c^2 + d^2)}.$$

Alternatively, division of two complex numbers when the complex numbers are expressed in magnitude and phase form is somewhat simpler:

$$\frac{z_1}{z_2} = \frac{|z_1|e^{j\alpha_1}}{|z_2|e^{j\alpha_2}} = \frac{|z_1|}{|z_2|}e^{j(\alpha_1 - \alpha_2)}.$$

In words, the magnitude of the ratio of two complex numbers is the ratio of the magnitudes, and the phase of the ratio is the difference of the phases.

One complex function that is of great interest to us in the present context is that of the complex exponential function for a harmonic wave, which through Euler's Identity we can write

$$e^{\pm j\omega t} = \cos(\omega t) \pm j\sin(\omega t). \tag{I.1}$$

The magnitude of $\exp(\pm j\omega t)$ is 1, and its angle, or phase, is constantly changing in time. We can picture $\exp(j\omega t)$ as sweeping around in the complex plane in the counter-clockwise direction at a constant rate ω, while $\exp(-j\omega t)$ sweeps in the clockwise direction. At $t = 0$, either of these functions points to the right – that is, along the positive real axis.

From Eq. (I.1) we can pick out the real and imaginary parts to write

$$\mathfrak{R}\left(e^{j\omega t}\right) = \cos(\omega t)$$

and

$$\mathfrak{I}\left(e^{j\omega t}\right) = \sin(\omega t).$$

We can invert these to express $\cos(\omega t)$ as

$$\cos(\omega t) = \frac{1}{2}\left(e^{j\omega t} + e^{-j\omega t}\right),$$

and $\sin(\omega t)$ as

$$\sin(\omega t) = \frac{-j}{2}\left(e^{j\omega t} - e^{-j\omega t}\right).$$

One of the primary motivations for using the complex exponential functions in place of sines and cosines is the simplification that they offer when determining derivatives or integrals with respect to time t. Since

$$\frac{d}{dt}e^{j(\omega t + \varphi)} = j\omega e^{j(\omega t + \varphi)},$$

the derivative of the complex exponential with respect to time is equivalent to *multiplying* the function by $j\omega$. The derivative of an exponential is still an exponential. Similarly, the indefinite time integral of this function is equivalent to the function divided by $j\omega$. In contrast, differentiation or integration of the cosine turns it into a sine function, and differentiation or integration of a sine gives us a cosine.

APPENDIX J

Hyperbolic Trigonometric Functions

We introduced the hyperbolic cosine function in Section 2.7 when discussing the charge density and capacitance for a pair of long, parallel wires, but only for real variable x. We encountered these functions again in Section 9.5, where we discussed the input impedance of transmission lines. In this context, the argument of the hyperbolic functions can be complex. We review the properties of these hyperbolic functions here, starting with a definition of the hyperbolic sine and hyperbolic cosine as

$$\sinh z = \frac{1}{2}\left(e^z - e^{-z}\right) \tag{J.1}$$

and

$$\cosh z = \frac{1}{2}\left(e^z + e^{-z}\right). \tag{J.2}$$

Notice the similarity between these functions and Euler's Identity for complex exponential functions! The argument z is in general complex, $z = x + jy$. We will also make use of the hyperbolic tangent function, defined as

$$\tanh z = \frac{\sinh z}{\cosh z} = \frac{e^z - e^{-z}}{e^z + e^{-z}}. \tag{J.3}$$

For real arguments, $y = 0$ and the hyperbolic functions are all real as well, since they only depend on exponentials of a real variable x. We show plots of these functions in Fig. J.1(a). Note that as $|x|$ gets large, $\cosh(x)$ and $\sinh(x)$ approach $\pm(1/2)e^{|x|}$ and $\tanh(x) \to \pm 1$. For small x, these hyperbolic functions are

$$\sinh(x) = x + \frac{x^3}{3!} + \frac{x^5}{5!} + \cdots,$$

$$\cosh(x) = 1 + \frac{x^2}{2!} + \frac{x^4}{4!} + \cdots,$$

and

$$\tanh(x) = x - \frac{x^3}{3} + \frac{2x^5}{15} - \cdots.$$

The inverse functions are defined analogous to inverse functions of regular sine, cosine, and tangent functions. For example, for $u = \cosh(x)$, then $x = \cosh^{-1}(u)$. We show plots of $\sinh^{-1}(u)$, $\cosh^{-1}(u)$, and $\tanh^{-1}(u)$ in Fig. J.1(b).

Two useful conversions of $\sinh^{-1}(x)$ and $\cosh^{-1}(x)$ to natural log functions are

$$\sinh^{-1}(x) = \ln\left(x + \sqrt{x^2 + 1}\right) \tag{J.4}$$

and

$$\cosh^{-1}(x) = \ln\left(x \pm \sqrt{x^2 - 1}\right). \tag{J.5}$$

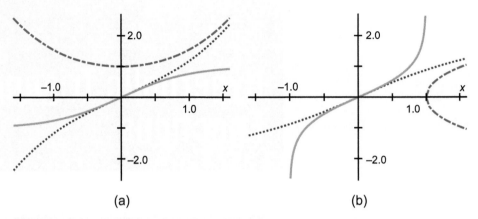

FIGURE J.1 (a) Plots of hyperbolic trigonometric functions: $\sinh(x)$ (blue, dotted); $\cosh(x)$ (red, dot-dashed); and $\tanh(x)$ (green, solid). (b) Plots of inverse hyperbolic functions: $\sinh^{-1}(x)$ (blue, dotted); $\cosh^{-1}(x)$ (red, dot-dashed); and $\tanh^{-1}(x)$ (green, solid).

The latter, of course, is only valid for $x \geq 1$, since cosh is never less than 1. To prove Eq. (J.4), let $u = \ln(x + \sqrt{x^2 + 1})$. Then,

$$e^u = x + \sqrt{x^2 + 1}$$

and

$$e^{-u} = \frac{1}{x + \sqrt{x^2 + 1}}.$$

We can simplify the expression for e^{-u} by multiplying the numerator and denominator by $x - \sqrt{x^2 + 1}$. This exponential then becomes

$$e^{-u} = \sqrt{x^2 + 1} - x.$$

Now let's combine e^u and e^{-u} as

$$\frac{1}{2}\left(e^u - e^{-u}\right) = \frac{1}{2}\left[\left(x + \sqrt{x^2 + 1}\right) - \left(\sqrt{x^2 + 1} - x\right)\right].$$

The left side is, of course, $\sinh(u)$, and the right side is simply x. Since $\sinh(u) = x$, then $u = \sinh^{-1}(x)$. But we started by defining $u = \ln(x + \sqrt{x^2 + 1})$. Therefore,

$$\sinh^{-1}(x) = \ln(x + \sqrt{x^2 + 1}),$$

and our proof is complete. The proof of Eq. (J.5) is similar, but we start by defining $u = x \pm \sqrt{x^2 - 1}$. We leave the remainder to the interested reader.

For lossless transmission lines, for which $\alpha = 0$, the complex propagation term γ simplifies to just $j\beta$, and the hyperbolic functions take the friendlier form

$$\sinh(j\beta z) = \frac{1}{2}\left(e^{j\beta z} - e^{-j\beta z}\right) = j\sin(\beta z),$$

$$\cosh(j\beta z) = \frac{1}{2}\left(e^{j\beta z} + e^{-j\beta z}\right) = \cos(\beta z),$$

and

$$\tanh j\beta z = \frac{e^{j\beta z} - e^{-j\beta z}}{e^{j\beta z} + e^{-j\beta z}} = \frac{2j\sin(\beta z)}{2\cos(\beta z)} = j\tan(\beta z).$$

Fundamental Constants and Physical Quantities

Fundamental Constants

Constant	Symbol	Value
Speed of light in vacuum	c	299 792 458 m/s
Fundamental unit of charge	e	$1.602\ 176\ 634 \times 10^{-19}$
Permittivity of free space	ε_0	$8.854\ 187\ 812\ 8\ (13) \times 10^{-12}$ F/m
Planck constant	h	$6.626\ 070\ 15 \times 10^{-34}$ Js
Boltzmann constant	k_B	$1.380\ 649 \times 10^{-23}$ J/K
Permeability of free space	μ_0	$1.256\ 637\ 062\ 12\ (19) \times 10^{-6}$ H/m $\approx 4\pi \times 10^{-7}$ H/m
Avogadro constant	N_A	$6.022\ 140\ 76 \times 10^{23}$ units/mol

Physical Quantities

Quantity	Symbol	Value
Frequency of cesium atomic clock reference		9 192 631 770 Hz
Mass of the electron	m_e	$9.109\ 383\ 7015\ (28) \times 10^{-31}$ kg
Mass of the proton	m_p	$1.672\ 621\ 923\ 69\ (51) \times 10^{-27}$ kg
Mass density of copper	ρ_{Cu}	8.94 gm/cm^3
Radius of the Earth	R_e	6378 km

The numbers within parentheses indicate the uncertainty of the quantities.

In addition, several physical parameters of materials are listed in tables, as follows.

Quantity	Table
Relative permittivities	1.1
Dielectric strengths	1.2
Conductivities	3.1
Relative permeability and magnetic susceptibility	4.1

APPENDIX L
Symbols and Unit Conversions

Variable	Definition	SI unit	First used in
α	Attenuation constant	m^{-1}	Sec. 8.6.2
$(\mathbf{a}_x, \mathbf{a}_y, \mathbf{a}_z)$	Unit vectors, rectangular components	Unitless	Appendix A, B.2
$(\mathbf{a}_\rho, \mathbf{a}_\phi, \mathbf{a}_z)$	Unit vectors, cylindrical components	Unitless	Sec. B.2
$(\mathbf{a}_R, \mathbf{a}_\theta, \mathbf{a}_\phi)$	Unit vectors, spherical components	Unitless	Sec. B.2
$\mathbf{A}$	Magnetic vector potential	Wb/m	Sec. 4.1
β	Phase constant	m^{-1}	Sec. 8.6.2
$\mathbf{B}$	Magnetic flux density	T (tesla) = Ns/Cm = Vs/m^2, or Wb/m^2	Sec. 4.1
$\tilde{\mathbf{B}}$	Magnetic flux density, phasor	T	Sec. 8.1.4
B_r	Residual magnetic flux density	T	Sec. 4.7, Fig. 4.41
c	Phase velocity in vacuum	m/s	Sec. 8.1
C	Capacitance	F (farad) = C/V = J/V^2	Sec. 2.5
C'	Capacitance per unit length	F/m	Sec. 2.5
δ	Skin depth	m	Sec. 8.6.4
$\mathbf{D}$	Displacement field	C/m^2	Sec. 1.8
$\tilde{\mathbf{D}}$	Displacement field, phasor	C/m^2	Sec. 8.1.4
ε	Permittivity of a material	F/m	Sec. 1.10, Table 1.1
ε_0	Permittivity of free space	F/m	Sec. 1.1
ε_r	Relative permittivity, or dielectric constant	Unitless	Sec. 1.10, Table 1.1
e	Proton charge	C (coulomb)	Sec. C.2
$\mathbf{E}$	Electric field	V/m = N/C	Sec. 1.1
$\tilde{\mathbf{E}}$	Electric field, phasor	V/m	Sec. 8.1.4
E_{max}	Dielectric strength	V/m	Sec. 1.10, Table 1.2
$\mathcal{E}$	Electromotive force, EMF	V	Sec. 5.1
Φ_E	Electric flux	Vm	Sec. 1.2
Φ_M	Magnetic flux	Wb (weber), Vs	Sec. 4.4
f	Cyclical frequency	Hz, cycles/s	Sec. 8.6.4
$\mathbf{F}$	Force	N (newton) = kg m/s^2	Sec. 1.1
γ	Propagation constant	m^{-1}	Sec. 8.6.2
Γ	Reflection coefficient	Unitless	Sec. 8.8
Γ_g	Reflection coefficient at source	Unitless	Sec. 9.8
Γ_L	Reflection coefficient at load	Unitless	Sec. 9.5

Variable	Definition	SI unit	First used in
G'	Conductance per unit length	S/m	Sec. 9.1
$G_D(\theta, \phi)$	Directive gain	Unitless	Sec. 11.1
η	Intrinsic impedance of a medium	Ω	Sec. 8.2
η_0	Intrinsic impedance of vacuum	Ω	Sec. 8.2
$\mathbf{H}$	Magnetic field	A/m	Sec. 4.7
$\tilde{\mathbf{H}}$	Magnetic field, phasor	A/m	Sec. 8.1.4
H_c	Coercive field	A/m	Sec. 4.7, Fig. 4.41
$\check{\mathit{I}}(t)$	Time-dependent current	A	Sec. 5.1
I	Current	A = C/s	Sec. C.3
$\tilde{I}$	Current, phasor	A	Sec. 9.2
$\mathbf{J}_d$	Displacement current density	A/m^2	Sec. 8.6.1
$\tilde{\mathbf{J}}_d$	Displacement current density, phasor	A/m^2	Sec. 8.6.1
$\mathbf{J}_s$	Surface current density	A/m	Sec. C.3
$\mathbf{J}_{s,eff}$	Effective surface current density	A/m	Sec. 4.6
$\mathbf{J}_v$	Volume current density	A/m^2	Sec. C.3
$\tilde{\mathbf{J}}_v$	Volume current density, phasor	A/m^2	Sec. 8.1.4
$\mathbf{J}_{v,eff}$	Effective volume current density	A/m^2	Sec. 4.6
k	Wave number	m^{-1}	Sec. 8.1
$\mathbf{k}$	Wave vector	m^{-1}	Sec. 8.2
$\boldsymbol{\kappa}$	Wave vector, complex	m^{-1}	Sec. 8.6
$\hat{\mathbf{k}}$	Unit vector parallel to $\mathbf{k}$	Unitless	Sec. 8.1.3
λ	Wavelength	m	Sec. 8.1
L	Inductance	H = Ω s = Vs/A = J/A^2	Secs. 5.1, 6.2
L'	Inductance per unit length	H/m	Sec. 6.2
μ	Material permeability	H/m	Sec. 4.7
μ_e	Electron mobility	m^2/Vs	Section 3.4
μ_0	Permeability of free space	H/m	Sec. 4.1
μ_r	Relative permeability	Unitless	Sec. 4.7, Table 4.1
$\mathbf{m}$	Magnetic dipole moment	A m^2	Sec. 4.5
$\mathbf{M}$	Magnetization field	A/m	Sec. 4.6
n	Turns density	m^{-1}	Sec. 4.1
n	Number density	m^{-3}	Sec. 3.6
N	Number of turns	Unitless	Sec. 4.3
ω	Radial frequency	Radians/s	Sec. 8.1
$\mathbf{p}$	Electric dipole moment	C m	Sec. 1.4.1
$\mathbf{P}$	Polarization field	C m^{-2}	Sec. 1.7
P	Transmitted power	W = J/s	Sec. 8.7
P_r	Radiated power	W	Sec. 11.1
P^{abs}	Power absorbed	W	Sec. 3.6
$\overline{P^{abs}_s}$	Time-averaged power absorbed per unit area	W m^{-2}	Sec. 8.8
P^{abs}_v	Power dissipation density	W m^{-3}	Sec. 3.6
$\overline{P^{abs}_v}$	Time-averaged power dissipation density	W m^{-3}	Sec. 8.6.1
$\mathcal{P}$	Poynting vector, power density	W m^{-2}	Sec. 8.7
$\overline{\mathcal{P}}$	Time-averaged Poynting vector, power density	W m^{-2}	Sec. 8.7
q, Q	Charge	C (coulomb)	Sec. C.2
q_e	Electron charge	C (coulomb)	Sec. C.2
Q_{enc}	Charge enclosed with a Gaussian surface	C (coulomb)	Sec. 1.8
(ρ, ϕ, z)	Cylindrical components	(m, , m)	Appendix B

ρ_ℓ	Linear charge density	C/m	Sec. C.2
ρ_s	Surface charge density	C/m^2	Sec. 1.1
$\rho_{s,eff}$	Effective surface charge density	C/m^2	Sec. 1.7
ρ_v	Volume charge density	C/m^3	Sec. C.4
$\tilde{\rho}_v$	Volume charge density, phasor	C/m^3	Sec. 8.1.4
$\rho_{v,eff}$	Effective volume charge density	C/m^3	Sec. 1.7
r	Position vector	m	Appendix A
r'	Position vector, source	m	Sections D.1, 1.1
(R, θ, ϕ)	Spherical components	(m, ,)	Appendix B
R	Resistance	Ω (ohm) = V/A	Sec. 3.4
R'	Resistance per unit length	Ω/m	Sec. 9.1
R_0	Characteristic resistance, lossless line	Ω	Sec. 9.5
R_r	Radiation resistance	Ω	Sec. 11.1
R_s	Surface resistance	Ω	Sec. 8.8
σ	Conductivity	S/m (siemens per meter)	Sec. 3.4, Table 3.1
S	Standing wave ratio	Unitless	Sec. 8.10
$\tan \delta$	Loss tangent	Unitless	Sec. 8.6.1, Table 8.2
τ	Transmission coefficient	Unitless	Sec. 8.8
T	Period	s	Sec. 8.1
T	Torque	N m	Sec. 6.3
u	Velocity	m/s	Section 3.1
u_p	Phase velocity	m/s	Sec. 8.1
$U(\theta, \phi)$	Radiation intensity	W	Sec. 11.1
$v(t)$	Tune-dependent voltage	V	Sec. 5.1
V	Electric potential	V (volts) = J/C	Sec. 1.4
$\tilde{V}$	Voltage, phasor	V	Sec. 9.2
W	Work	J (joule) = Nm = kg m^2/s^2	Sec. 1.4
w_e	Electric energy density	J/m^3	Sec. 2.1
W_e	Electric energy	J	Sec. 2.1
w_m	Energy density of a magnetic field	J/m^3	Sec. 6.1
W_m	Magnetic energy	J	Sec. 6.1
W'_m	Magnetic energy per unit length	J/m	Sec. 6.2
χ_e	Electric susceptibility	Unitless	Sec. 1.10
χ_m	Magnetic susceptibility	Unitless	Sec. 4.7
(x, y, z)	Rectangular components	(m, m, m)	Appendix A
X_s	Surface reactance	Ω	Sec. 8.8
Y_0	Characteristic admittance	S	Sec. 9.6
Y_{in}	Input admittance	S	Sec. 9.5
Z_0	Characteristic impedance	Ω	Sec. 9.2
Z_g	Generator impedance	Ω	Sec. 9.5
Z_{in}	Input impedance	Ω	Sec. 9.2
Z_s	Surface impedance	Ω	Sec. 8.8
$\langle \xi \rangle$	Ensemble average of quantity specified by ξ		Sec. 8.1.4
$\overline{\xi(t)}$	Time average of quantity specified by $\xi(t)$		Sec. 8.6.1

APPENDIX M

Answers to Selected Problems

Preface

P1-3 mJ

P1-4 μW

Chapter 1

Coulomb's Law

P1-1 (b) $8.155 \times 10^{-11} (-1, 0, -1)$ N

P1-4 (b) $(-1.05, +0.294, -0.098) \times 10^9$ V/m

P1-5 $\mathbf{a}_R \rho_v R / 3\varepsilon_0$

P1-12 $0; \mathbf{a}_R Q / 4\pi\varepsilon_0 R^2$

P1-15 For $z > 0$, $\rho_s / 2\varepsilon_0 \left(1 - z/\sqrt{a^2 + z^2}\right)$;

for $z < 0$, $-\rho_s / 2\varepsilon_0 \left(1 + z/\sqrt{a^2 + z^2}\right)$

P1-17 $\mathbf{a}_z \dfrac{\rho_\ell a z}{2\varepsilon_0 (a^2 + z^2)^{3/2}}$

Differential Properties of E

P1-19 (a) $\nabla \times \mathbf{E}_1 = 0$, so this can be a valid field; $\rho_v = 2\varepsilon_0 E_0 / R$, which could be possible except at the origin

P1-21 (c) $\nabla \times \mathbf{E}_3 = \dfrac{E_0}{R^3 \sin\theta} [\mathbf{a}_R \cos\theta + \mathbf{a}_\theta \sin\theta] \neq 0$, so this is not valid as an electrostatic field

Gauss' Law in Vacuum

P1-34 For $\rho < a$, $\mathbf{a}_\rho \rho_0 / \varepsilon_0$; for $\rho > a$, $\mathbf{a}_\rho \rho_0 a / \varepsilon_0 \rho$

P1-38 For $-a/2 < z < a/2$, $\mathbf{a}_z \rho_0 (z^2 - a^2/4)/2\varepsilon_0$; zero elsewhere

Maximum magnitude at $z = 0$, $-\rho_0 a^2 / 8\varepsilon_0$

Electric Potential

P1-41 (a) 2 μC/m³; (b) 56 kV; (c) 56 mJ

P1-44 (a) $-2x\mathbf{a}_x + 2y\mathbf{a}_y - 4z\mathbf{a}_z$; (b) 0; (c) -35.4 pC/m³, (d) 4 μJ

P1-47 For $R < a$, $V(R) = \rho_s a / \varepsilon_0$, $\mathbf{E}(R) = 0$;

for $R > a$, $V(R) = \rho_s a^2 / \varepsilon_0 R$, $\mathbf{E}(R) = \mathbf{a}_R \rho_s a^2 / \varepsilon_0 R^2$

P1-50 For $z > 0$, $V(z) = \rho_s (\sqrt{a^2 + z^2} - z)/2\varepsilon_0$, $E_z = \rho_s (1 - z/\sqrt{a^2 + z^2})/2\varepsilon_0$;

for $z < 0$, $V(z) = \rho_s (\sqrt{a^2 + z^2} + z)/2\varepsilon_0$, $E_z = -\rho_s (1 + z/\sqrt{a^2 + z^2})/2\varepsilon_0$

Perfectly Insulating Dielectric Materials

P1-62 $\rho_{s,eff} = P_0 b$ (outside surface), $-P_0 a$ (inside surface), zero (top and bottom);

$\rho_{v,eff} = -2P_0$; $E_\rho = -P_0 \rho / \varepsilon_0$

Gauss' Law Revisited

P1-66 For $a < R < b$, $\mathbf{D} = \mathbf{a}_R Q/4\pi R^2$, $\mathbf{E} = \mathbf{a}_R Q/4\pi\varepsilon_1 R^2$;

for $b < R < c$, $\mathbf{D} = \mathbf{a}_R Q/4\pi R^2$, $\mathbf{E} = \mathbf{a}_R Q/4\pi\varepsilon_2 R^2$;

$V_0 = \frac{Q}{4\pi}\left[\frac{1}{\varepsilon_2}\left(\frac{1}{b} - \frac{1}{c}\right) + \frac{1}{\varepsilon_1}\left(\frac{1}{a} - \frac{1}{b}\right)\right]$; $\rho_s = Q/4\pi a^2$ at $R = a$, $-Q/4\pi c^2$ at $R = c$

Boundary Conditions between Two Perfect Dielectrics

P1-70 $(\mathbf{a}_x 8/3 - \mathbf{a}_y 6)$ V/m; 56.3°; 66.0°

P1-73 $(\mathbf{a}_x 4.4 - \mathbf{a}_y 5.8)$ C/m²; 7.1°; 10.6°

Boundary Conditions for Perfect Dielectrics

P1-75 $\mathbf{D} = \mathbf{a}_x 5$ nC/m²; $\mathbf{E} = \mathbf{a}_x 188$ V/m; $\mathbf{P} = \mathbf{a}_x 10/3$ nC/m²; $\rho_{s,eff} = -10/3$ nC/m²

Chapter 2

Electrostatic Energy

P2-1 -1.25×10^{-19} J

Electrostatic Forces

P2-6 $\mathbf{a}_x 1.3 \times 10^{-6}$ N

Poisson's and Laplace's Equations

P2-11 For $a < R < b$, $V(R) = -C_1/R + C_2$, $\mathbf{E} = -\mathbf{a}_R C_1/R^2$; $\mathbf{D} = -\mathbf{a}_R \varepsilon_1 C_1/R^2$;

for $b < R < c$, $V(R) = -C_3/R + C_4$, $\mathbf{E} = -\mathbf{a}_R C_3/R^2$; $\mathbf{D} = -\mathbf{a}_R \varepsilon_2 C_3/R^2$;

where $C_1 = -V_0/\Delta$, $C_2 = -V_0\left(\frac{1}{b} - \frac{\varepsilon_1}{\varepsilon_2}\left(\frac{1}{b} - \frac{1}{c}\right)\right)/\Delta$, $C_3 = -\varepsilon_1 V_0/\varepsilon_2\Delta$,

$C_4 = -\varepsilon_1 V_0/c\varepsilon_2\Delta$, $\Delta = \left(\frac{1}{a} - \frac{1}{b}\right) + \frac{\varepsilon_1}{\varepsilon_2}\left(\frac{1}{b} - \frac{1}{c}\right)$;

$\rho_s = \frac{V_0}{a^2\Delta}$ at $R = a$, $\rho_s = -\frac{V_0}{c^2\Delta}$ at $R = c$

Capacitance

P2-18 $W_e = \frac{1}{2}\rho_s^2 S\left[\frac{d_1}{\varepsilon_1} + \frac{d_2}{\varepsilon_2}\right]$; $C = S\left[\frac{d_1}{\varepsilon_1} + \frac{d_2}{\varepsilon_2}\right]^{-1}$

P2-24 $C = \frac{2\pi L}{\ln(b/a)}\left(\frac{\varepsilon_1 + 2\varepsilon_2}{3}\right)$

Method of Images

P2-31 (a) $V = \frac{\rho_\ell}{2\pi\varepsilon_0}\ln(a/d)$; (c) $\rho_s = -\frac{\rho_\ell a}{2\pi}\left(\frac{d^2}{a^2} - 1\right)(d^2 - 2ad\cos\phi + a^2)^{-1}$; $-\rho_\ell$

Chapter 3

Resistance

P3-8 $\frac{1}{4\pi}\frac{3}{\sigma_1 + 2\sigma_2}\left(\frac{1}{a} - \frac{1}{c}\right)$

P3-10 (a) $-\mathbf{a}_y I/A$; (b) $\mathbf{J}_v(\sigma_1 + \sigma_2 y)^{-1}$; (c) $(I/A\sigma_2)\ln(1 + \sigma_2 d/\sigma_1)$;

(d) $(1/A\sigma_2)\ln(1 + \sigma_2 d/\sigma_1)$

Power Dissipation in Resistors

P3-11 $4\pi\sigma_1 V_0^2\left(\frac{1}{a} - \frac{1}{c}\right)^{-1}$

Boundary Conditions Revisited

P3-15 (a) $D_1 = Q/4\pi R^2$, $E_1 = Q/4\pi\varepsilon_1 R^2$, $J_{v,1} = \sigma_1 Q/4\pi\varepsilon_1 R^2$;

$D_2 = \varepsilon_2\sigma_1 Q/4\pi\varepsilon_1\sigma_2 R^2$, $E_2 = \sigma_1 Q/4\pi\varepsilon_1\sigma_2 R^2$, $J_{v,2} = \sigma_1 Q/4\pi\varepsilon_1 R^2$;

(b) $\frac{1}{4\pi}\left[\frac{1}{\sigma_1}\left(\frac{1}{a} - \frac{1}{b}\right) + \frac{1}{\sigma_2}\left(\frac{1}{b} - \frac{1}{c}\right)\right]$

Chapter 4

Biot–Savart Law

P4-3 $\mathbf{B} = \mathbf{a}_z(\mu_0 I a^2/2\pi)(z^2 + a^2/4)^{-1}(z^2 + a^2/2)^{-1/2}$

P4-5   For $x > d/2$, $\mathbf{B} = \mathbf{a}_y\mu_0 J_v d/2$;

       for $-d/2 < x < d/2$, $\mathbf{B} = \mathbf{a}_y\mu_0 J_v x$;

       for $x < -d/2$, $\mathbf{B} = -\mathbf{a}_y\mu_0 J_v d/2$

P4-7 $\mathbf{a}_\phi\mu_0 In \sin\alpha$

Vector Differential Properties of Magnetic Fields

P4-13   (c) $\nabla \cdot \mathbf{B}_3 = 0$, so this can be a valid field;

       $\mathbf{J}_v = \mu_0^{-1}\nabla \times \mathbf{B}_3 = \mathbf{a}_R\mu_0^{-1}B_0 \cot\theta/R^3$. This is pretty strange, but

       could exist, except at the origin where it becomes infinite.

Ampère's Law in Vacuum

P4-16   For $x > d/2$, $\mathbf{B} = \mathbf{a}_y\mu_0 J_v d/2$;

       for $-d/2 < x < d/2$, $\mathbf{B} = \mathbf{a}_y\mu_0 J_v x$;

       for $x < -d/2$, $\mathbf{B} = -\mathbf{a}_y\mu_0 J_v d/2$

Vector Magnetic Potential

P4-20 $\mathbf{A} = \mathbf{a}_z(\mu_0 I/4\pi)[\ln(y^2 + (x + d/2)^2) - \ln(y^2 + (x - d/2)^2)]$;

       $\mathbf{B} = (\mu_0 I/2\pi)\left[\dfrac{y\mathbf{a}_x-(x+d/2)\mathbf{a}_y}{y^2+(x+d/2)^2} - \dfrac{y\mathbf{a}_x-(x-d/2)\mathbf{a}_y}{y^2+(x-d/2)^2}\right]$

P4-21   For $\rho < a$, $\mathbf{A} = \mathbf{a}_\phi\mu_0 nI\rho/2$; for $\rho > a$, $\mathbf{A} = \mathbf{a}_\phi\mu_0 nIa^2/2\rho$

Magnetization

P4-23 $\mathbf{B} = \mathbf{a}_x 2\mu_0 M_0/3$

P4-24 $\mathbf{J}_{s,eff} = \mathbf{a}_\phi M_0$ on the rim, zero on the top and bottom surfaces;

       $\mathbf{J}_{v,eff} = 0$; $\mathbf{B} = \mathbf{a}_z\mu_0 M_0 ta^2/2(a^2 + z^2)^{3/2}$

Ampère's Law Revisited

P4-26   398 A/m; 0.50 T; 398 A/m; 0.5 mT

Boundary Conditions for Magnetic Fields

P4-29 $(\mathbf{a}_x 4 + \mathbf{a}_y 5 + \mathbf{a}_z 25)$ A/m; 32.6°; 14.4°

P4-31 $\mathbf{a}_x 10$ A/m; $\mathbf{a}_x 62.8$ µT

Chapter 5

P5-2 $\mathbf{E} = \mathbf{a}_\phi B_0\omega \sin\omega t\, \rho/2$; $\mathbf{J}_v = \mathbf{a}_\phi\sigma B_0\omega \sin\omega t\, \rho/2$;

       $P^{abs} = \pi\sigma B_0^2\omega^2 \sin^2\omega t\, a^3 d/6$; absorbed power decreases by a factor $N^{1/2}$.

P5-3   The potential of the outer ring is $\omega B_0(b^2 - a^2)/2$ higher than the inner ring;

       $I = \omega B_0(b^2 - a^2)/2R$ through the resistor from the outer ring to the inner;

       $P = \omega^2 B_0^2(b^2 - a^2)^2/4R$

P5-5   (c) $V(t) = (N\omega\mu_0 I_0 \sin\omega t\, a^2/2d)\left(1 + (a/2d)^2 + \cdots\right)$

P5-11 $V(t) = -\omega AB_0 x_0 \cos\omega t$; $i(t) = -\omega AB_0 x_0 \cos\omega t\, /R$;

       $P(t) = (\omega AB_0 x_0)^2 \cos^2\omega t\, /R$

Chapter 6

Magnetic Energy

P6-3   (a) 1.0 mJ/m³; 3.0 mJ/m³; (b) 0.5 µJ/m

Inductance

P6-10   0.77 µH/m

Magnetic Forces and Torques

P6-15 $W_m = -mB_0 \cos\theta$, where $mB_0 = 0.6$ mJ; $\tau_z = mB_0 \sin\theta$, restoring the loop so that **m** is parallel to **B**.

Chapter 7

Displacement Current Density

P7-1   (a) $\mathbf{a}_\rho \dfrac{Q}{2\pi L\rho}$; (b) $\mathbf{a}_\rho \dfrac{Q}{2\pi\varepsilon L\rho}$; (c) $\mathbf{a}_\rho \dfrac{1}{2\pi L\rho}\dfrac{dQ}{dt}$;

(d) $I_d = \dfrac{dQ}{dt}$; (e) $V(t) = \dfrac{Q}{2\pi\varepsilon L}\ln(b/a)$; (f) $C = \dfrac{2\pi\varepsilon L}{\ln(b/a)}$; (g) $C\dfrac{dV(t)}{dt} = I_d$

Relative Current Densities

P7-6 5.6×10^9 r/s, or 0.90 GHz

Chapter 8

Propagation in Lossless Media

P8-2 $(-\mathbf{a}_x E_{0y} + \mathbf{a}_y E_{0x})\sin(\omega t - kz)/\eta$

P8-4   (a) $\omega = 5.78 \times 10^6$ r/s; $\lambda = 326$ m; $k = 0.0193$ m^{-1}; $T = 1.09$ μs;

(b) $\omega = 5.78 \times 10^6$ r/s; $\lambda = 217$ m; $k = 0.0289$ m^{-1}; $T = 1.09$ μs

Uniform Plane Waves

P8-8   (a) wave propagates in +y-direction; (b) 477 MHz; (c) 0.314 m; (d) 4.0

P8-12   (a) $\mathbf{a}_y H_0 e^{-jkz}$ where $H_0 = 26.5$ mA/m; 0.01 m; 1.88×10^{11} r/s; 30.0 GHz;

(b) $\mathbf{a}_y H_0 e^{-jkz}$ where $H_0 = 39.8$ mA/m; 0.01 m; 1.26×10^{11} r/s; 20.0 GHz

Polarization

P8-17   (a) $E_L = -E_R = 7.07$ V/m, phase difference is π;

(b) induce π phase shift in E_R to make $E_L = E_R = 7.07$ V/m

Power Dissipation Density

P8-20   (a) 2.94 W/m^3; (b) 6.9 W/m^3; (c) 46.5 W/m^3

Wave Propagation in Absorbing Media

P8-22 $\beta = 141$ m^{-1}; $\alpha = 8.7$ m^{-1}; $\eta = (0.56 + j0.035)$ Ω; $u_p = 4.5 \times 10^7$ m/s; $\lambda = 0.0445$ m; $\Delta z = 0.080$ m

Non-conducting Absorbing Dielectric Media

P8-26 $(2.24 - j0.16)\varepsilon_0$

P8-27   (a) 1.87 mm^{-1}; 0.035 mm^{-1}; (b) 3.33 mm;

(c) 3.4×10^6 m/s; (d) $(8\,000 - j300)\varepsilon_0$

Wave Propagation in a Good Conductor

P8-29   (a) 6.8 GHz; (b) $E_0 = 38$ μV/m; (c) 4.6 μm;

(d) 4.3×10^4 m/s; (e) 14 A/m^2

Power Flow

P8-32   3.1 kV/m; 8.2 A/m

P8-34   (a) $\alpha = 5.3 \times 10^{-4}$ m^{-1}; (b) $\varepsilon_r = 9.0$

Reflection of Uniform Plane Waves, Normal Incidence

P8-35 (a) $-1/3$; $2/3$; (b) -92 V/m; 183 V/m;
(c) 11.1 W/m^2; 88.9 W/m^2; yes

P8-37 0.83×10^8 S/m; $174.5°$

Standing Waves

P8-45 (a) 3; (b) min at $z = 0$, since Γ is negative; (c) 5 V/m; (d) 9.0

Chapter 9

Lossless Transmission Line

P9-2 (a) 50.6 pF/m; (b) 0.220 µH/m; (c) 66.0 Ω; (d) 4.19 m^{-1};
(e) 1.73×10^8 m/s; (f) 0.15 A; (g) $\mathbf{a}_\rho$ 6.1 kV/m; $\mathbf{a}_\phi$ 15.9 A/m

Losses

P9-5 (a) $C' = 243$ pF/m; $L' = 0.183$ µH/m; $R' = 0.581$ Ω/m; $G' = 6.10$ mS/m;
(b) $Z_0 = (27.5 + j0.12)$ Ω; $\gamma = (0.096 + j16.8)$ m^{-1};
(c) $\lambda = 0.375$ m; $u_p = 1.5 \times 10^8$ m/s; (d) $I_0^+ = 182e^{-j0.25°}$ mA; (e) 7.2 m

Standing Waves

P9-8 $S = 3$; $V_0^+ = 5$V; $V_0^- = 2.5$ V; $\lambda = 5$ cm

Reflections at a Load

P9-11 $(51.0 - j3.0)$ Ω
P9-13 $(20.2 - j72.5)$ Ω
P9-15 450 Ω and 5.6 Ω
P9-19 (a) 1.5; (b) 1.31
P9-21 (a) 3; (b) $V_0^+ = 5.0$ V; $V_0^- = 2.5$ V; (c) 0.5; (d) 150 Ω; (e) 50 cm; (f) 9
P9-23 $V_0^+ = (-0.083 + j4.33)$ V; $V_0^- = (0.634 + j0.108)$ V; $V_L = (0.551 + j4.44)$ V;
$I_L = (-14.3 + j84.6)$ mA; $P_L = 0.368$ W

Impedance Matching

P9-24 86.6 Ω
P9-27 $\ell_L = 0.230\lambda$ and $\ell_{St} = 0.369\lambda$; or $\ell_L = 0.048\lambda$ and $\ell_{St} = 0.131\lambda$

Pulses

P9-38 $V_1^+ = 6.67$ V; $V_1^- = 0.606$ V; $V_2^+ = -0.0673$ V; $V_2^- = -0.0061$ V

Chapter 10

General Approach

P10-1 $u_p = \dfrac{c}{a+a\omega}$; $u_g = c\,(1 + 2a\omega)$

Parallel Plate Waveguide

P10-3 $\omega_c = 3.77 \times 10^{11}$ r/s

P10-4   (a) $\beta = 200$ m^{-1}; $\Lambda = 0.0325$ m

P10-5 $u_p = 3.14 \times 10^8$ m/s; $u_g = 2.86 \times 10^8$ m/s; $Z_{TM} = 360$ Ω; $Z_{TE} = 395$ Ω

P10-7   (a) $P = 1.76$ mW; (b) $P'_c = 10.2$ μW/m; (c) $\alpha_c = 0.00291$ m^{-1}

P10-11   (b) $a = 0.0173$ m

Rectangular Waveguide

P10-12 $f_c = 10$ GHz for TE$_{02}$

P10-13   (b) $\beta = 139$ m^{-1}; $\Lambda = 0.0454$ m

P10-16 $u_p = 6.91 \times 10^8$ m/s; $u_g = 1.30 \times 10^8$ m/s; $Z_{TM} = 164$ Ω; $Z_{TE} = 869$ Ω

P10-19   (a) $P = 1.06$ mW; (b) $P'_c = 30.5$ μW/m; (c) $\alpha_c = 0.0144$ m^{-1}

P10-24   (a) $b = 3$ cm, $a < b$; (b) $a = 1.67$ cm; (c) $b = 1.73$ cm

Circular Waveguide

P10-25 $\omega_c = 2.76 \times 10^{10}$ s^{-1} for TE$_{11}$; $\omega_c = 5.75 \times 10^{10}$ s^{-1} for TM$_{11}$

P10-26 $u_p = 3.34 \times 10^8$ m/s; $u_g = 2.69 \times 10^8$ m/s; $Z_{TE} = 420$ Ω for TE$_{11}$ mode

Resonant Cavity

P10-28 $\omega = 7.37 \times 10^{10}$ s^{-1} for TE$_{112}$

P10-29 $\omega = 7.37 \times 10^{10}$ s^{-1} for TM$_{112}$

P10-30 $Q = 1.1 \times 10^4$

Chapter 11

Elemental Dipole Antenna

P11-1   19.1 A

P11-2   (a) $R \gg 5$ cm; (b) 0.47 V/m; (c) 0.29 mW/m^2;
            (d) $P_r = 0.98$ W; $R_r = 0.078$ Ω; $P_L = 0.24$ W

Long Dipole Antenna

P11-5   (a) 16 A; (b) 1.48 m: (c) 12.2 μW/m^2

Antenna Arrays

P11-9   Max at $\phi = 180°$, 109.5°, 70.5°, and 0°

Appendix A

Vector Addition and Subtraction

PA-1   (a) $-3\mathbf{a}_x + 2\mathbf{a}_y + 4\mathbf{a}_z$

Scalar Product or Dot Product

PA-2   (d) -38

PA-3   (d) 17.8

PA-4   (a) 53.6°

PA-6 $(-0.536, -3, -3.964)$ and $(-3.964, -3, -0.536)$

PA-7   (d) $(-0.557, 0.371, 0.743)$

Vector Product or Cross Product

PA-8   (c) $(-28, -32, 10)$

Triple Products of Vectors

PA-12 $(17, -38, -74)$

Appendix B

Conversion between Coordinate Systems

PB-2   (a) $(-2.12, 2.12, -4)$; (b) $(5, 143.1°, 135°)$

Conversion of Sets of Unit Vectors

PB-6   (a) $\mathbf{a}_\theta = \mathbf{a}_x/2 + \mathbf{a}_y/2 + \mathbf{a}_z/\sqrt{2}$

Appendix C

Integration Along a Path

PC-1   3/2; 1

Integration Over a Surface

PC-5   46 m^2
PC-7   36 cm/hr
PC-8   (a) $2\pi E_0(1 - \cos\theta_0)$; (b) same

Appendix D

The Gradient of a Scalar Field

PD-2 $2\rho z^2 \mathbf{a}_\rho + \dfrac{1}{\rho}a_0^2 2\cos\phi\sin\phi\mathbf{a}_\phi + 3\rho^2 z^2 \mathbf{a}_z$

The Divergence of a Vector Field

PD-5   (b) 2
PD-6   (b) $\cos\phi$

The Divergence Theorem

PD-9   (a) 0; (b) 0; (c) yes

The Curl of a Vector Field

PD-11   (c) $\mathbf{a}_\phi z/\rho^2$
PD-13   (c) $-\mathbf{a}_R \cot\theta \cos\phi/R^2$

Stokes' Theorem

PD-14   (a) $-4bC_0 \sin(b/a)$; (b) same; (c) yes

Appendix E

Distributive and Derivative Product Properties

PE-4   (a) $\dfrac{[-\mathbf{a}_x y + \mathbf{a}_y x](x-y)}{(x^2+y^2)^{3/2}} = \mathbf{a}_\phi \dfrac{x-y}{x^2+y^2}$

Image Credits

5.12 Faraday disc generator: Emile Alglave, Public domain, via Wikimedia Commons

6.15 Magnetically levitated train: User Alex Needham (own photography) on en.wikipedia, Public domain, via Wikimedia Commons

6.17 Magnetic bases in the (a) "Off" and (b) "On" positions: Malyszkz, Public domain, via Wikimedia Commons

6.23 Senator George P. Wetmore, Rhode Island: Harris & Ewing, Inc., Public domain, via Wikimedia Commons

7.1 James Clerk Maxwell: Fergus of Greenock, Public domain, via Wikimedia Commons

8.3 (a) Heinrich Rudolf Hertz, (b) sketch of Hertz' spark gap radio transmitter and parabolic reflector: (a) Robert Krewaldt, Public domain, via Wikimedia Commons; (b) Heinrich Rudolf Hertz, Public domain, via Wikimedia Commons

8.4 (a) Two tall antennas on the surface of the Earth: Naval Postgraduate School, Public domain, via Wikimedia Commons

8.5 (a) The global constellation of satellites, (b) GPS III satellite: (a) In-fadel/iStock/Getty Images Plus, (b) USAF, Public domain, via Wikimedia Commons

8.6 (b) Dr. Gladys West: US Air Force, Public domain, via Wikimedia Commons

8.11 Electromagnetic spectrum: Philip Ronan, GNU Free Documentation License, https://commons.wikimedia.org/wiki/File:EM_spectrum.svg

8.12 A wireless router: Evan-Amos, Public domain, via Wikimedia Commons

8.16 (a) Wilhelm Röntgen, (b) the first medical X-ray image, (c) illustration of a Crookes X-ray tube: (a) Nobel foundation; (b) Wilhelm Röntgen, Public domain, via Wikimedia Commons; (c) Nandalal Sarkar/iStock/Getty Images Plus;

8.17 Detailed, all-sky picture of the infant universe: NASA/WMAP Science Team, Public domain, via Wikimedia Commons

8.36 Synthetic negative index material: Cynthia.L.Dreibelbis@nasa.gov, Public domain, via Wikimedia Commons

9.9 Slotted line for measurement of the standing wave pattern on a transmission line: Herbertweidner (Diskussion) Herbertweidner at German Wikipedia, Public domain, via Wikimedia Commons

10.2 Photonic crystal optical fiber: Public domain, via Wikimedia Commons

11.7 Schematic of a free electron laser (FEL): Dmitry Kovalchuk/iStock/Getty Images Plus

11.8 (a) AM radio broadcast antenna, (b) A Yagi-Uda antenna: (a) Chetvorno, Public domain, via Wikimedia Commons; (b) unknown author, Public domain, via Wikimedia Commons

11.14 (a) Cell phone tower: Lucas Ninno/Moment/Getty Images

Photo Credits

TN 0.1 Electromagnetics in the cinema: Yagi Studio/Stone/Getty Images

TN 1.1 Electrostatic air cleaners: lucia meler/Moment/Getty Images

TN 1.2 Electrostatic spraying: AlexanderLipko/iStock/Getty Images

TN 1.3 Photocopiers: Influx Productions/DigitalVision/Getty Images

TN 1.4 Faraday cages: VICTOR HABBICK VISIONS/SCIENCE PHOTO LIBRARY/-Science Photo Library/Getty Images

TN 1.5 Electric eels: Mark Newman/The Image Bank/Getty Images

TN 1.6 van de Graaff generators: Bettmann/Contributor/Bettmann/Getty Images

TN 8.1   Wave propagation in the atmosphere: MarioGuti/E+/Getty Images

TN 8.2   Satellite TV: seraficus/E+/Getty Images

TN 8.3   GPS: janiecbros/E+/Getty Images

TN 8.4   Atomic clock: Science & Society Picture Library/Contributor/SSPL/Getty Images

TN 8.5   Doppler traffic radar: Douglas Sacha/Moment Unreleased/Getty Images

TN 8.6   Doppler weather radar: Drew Angerer/ Getty Images News/Getty Images

TN 8.7   Doppler astronomy: AFP/Stringer/AFP/Getty Images

TN 8.8   Wi-Fi: Lee Sie Photography/Contributor/Moment Mobile/Getty Images

TN 8.9   Remote control: 2windspa/E+/Getty Images

TN 8.10 RFID: MirageC/Moment/Getty Images

TN 8.11 IR thermometers: dowell/Moment/Getty Images

TN 8.12 X-ray imaging: Universal History Archive/Universal Images Group/Getty Images

TN 8.13 Cosmic microwave background: Encyclopaedia Britannicaor/Universal Images Group/Getty Images

TN 8.14 Health effects of electric/magnetic fields: Peter Cade/Stone/Getty Images

TN 8.15 Absorptive polarisers: Jyoti Sangya/Moment/Getty Images

TN 8.16 CDs, DVDs, Blu-ray: EThamPhoto/The Image Bank/Getty Images

TN 8.17 LCD: Travelif/Photographer's Choice RF/Getty Images

TN 8.18 Underground utility sensing: Stocktrek Images/Stocktrek Images/Getty Images

TN 8.19 Thin film: studiocasper/iStock/Getty Images

TN 8.20 Multi-layer thin film coatings: Jackyenjoyphotography/Moment/Getty Images

TN 8.21 Negative refractive index materials: N-Photo Magazine/Future/Getty Images

TN 9.1   Slotted line: courtesy of the author

TN 10.1 Optical fibres: Roberto/iStock/Getty Images Plus

TN 10.2 Photonic crystals: abzee/E+/Getty Images

TN 10.3 Microwave ovens: Michael Haegele/The Image Bank/Getty Images

TN 11.1 Free electron lasers: ERIC PIERMONT/AFP/Getty Images

TN 11.2 Radio & TV broadcast antenna: Busà Photography/Moment/Getty Images

TN 11.3 Extremely low frequency communications: alxpin/E+/Getty Images

TN 11.4 Microwave relay stations: ollo/E+/Getty Images

TN 11.5 Smartphone antennae: xijian/E+/Getty Images

References

General

Cheng, David K. (1992) *Field and Wave Electromagnetics*, Addison-Wesley, 2nd edition.

Hayt, Jr., William H. and Buck, John A. (2012) *Engineering Electromagnetics*, McGraw Hill, 8th edition.

Jackson, John David (1999) *Classical Electrodynamics*, John Wiley & Sons, 3rd edition.

Lorrain, Paul, Corson, Dale, and Lorrain, François (1988) *Electromagnetic Fields and Waves*, W. H. Freeman and Company.

Reitz, John R. and Milford, Frederick J. (1967) *Foundations of Electromagnetic Theory*, Addison-Wesley, 2nd edition.

Sadiku, Matthew N. (2015) *Elements of Electromagnetics*, Oxford University Press, 6th edition.

Ulaby, Fawwaz T. (2005) *Electromagnetics for Engineers*, Pearson Prentice Hall.

Chapter 1

Charles-Augustin de Coulomb
https://en.wikipedia.org/wiki/Charles-Augustin_de_Coulomb
www.aps.org/publications/apsnews/201606/physicshistory.cfm#:~:text=By%20bringing%20a%20similarly%20charged,quantitative%20study%20of%20electric%20force

Copy Machines
www.explainthatstuff.com/photocopier.html
www.scientificamerican.com/article/how-does-a-photocopier-wo/
http://labman.phys.utk.edu/phys136core/modules/m5/electrostatic_devices.html

Corona Discharge
https://en.wikipedia.org/wiki/Corona_discharge#:~:text=A%20corona%20discharge%20is%20an,conductor%20carrying%20a%20high%20voltage.&text=Corona%20discharge%20from%20high%20voltage,waste%20of%20energy%20for%20utilities
www.sciencedirect.com/topics/engineering/corona-discharge

Electric Eel
https://en.wikipedia.org/wiki/Electric_eel#:~:text=The%20electric%20eel%20has%20three,organ%2C%20and%20the%20Sach's%20organ.&text=These%20organs%20are%20made%20of,adds%20to%20a%20potential%20difference

Electrostatic Air Cleaners
www.hunker.com/12167146/how-do-electrostatic-air-filters-work

www.modernalchemyair.com/technology/electrostatic/

Electrostatic Spraying
https://vandacoatings.co.uk/blog/how-does-electrostatic-spraying-work/

Faraday Cage
https://en.wikipedia.org/wiki/Faraday_cage

Henry Cavendish
www.britannica.com/biography/Henry-Cavendish/Experiments-with-electricity
www.famousscientists.org/henry-cavendish/
https://en.wikipedia.org/wiki/Coulomb%27s_law

Johann Carl Friedrich Gauss
https://en.wikipedia.org/wiki/Carl_Friedrich_Gauss

Lightning and Lightning Rods
www.weather.gov/safety/lightning-rods
https://science.howstuffworks.com/nature/natural-disasters/lightning7.htm
www.weather.gov/safety/lightning-science-electrification
www.weather.gov/media/pah/WeatherEducation/lightningsafety.pdf

Spark Plug
www.thoughtco.com/inventors-of-the-spark-plug-4074529

Tesla Coil and Nikola Tesla
https://en.wikipedia.org/wiki/Nikola_Tesla
www.smithsonianmag.com/history/the-rise-and-fall-of-nikola-tesla-and-his-tower-11074324/
www.history.com/topics/inventions/nikola-tesla
www.britannica.com/biography/Nikola-Tesla
https://en.wikipedia.org/wiki/Tesla_coil
www.livescience.com/46745-how-tesla-coil-works.html

van de Graaff Generator
https://science.howstuffworks.com/transport/engines-equipment/vdg.htm

Chapter 2

Batteries
https://en.wikipedia.org/wiki/Separator_(electricity)
www.explainthatstuff.com/batteries.html
www.science.org.au/curious/technology-future/batteries#:~:text=A%20battery%20is%20a%20device,be%20used%20to%20do%20work
https://web.mst.edu/~gbert/BATTERY/battery.html
https://chem.libretexts.org/Bookshelves/General_Chemistry/Map%3A_Chemistry_-_The_Central_Science_(Brown_et_al.)/20%3A_Electrochemistry/20.3%3A_Voltaic_Cells

Capacitance Manometer
www.mksinst.com/n/baratron-capacitance-manometers

Condenser Microphone
https://mynewmicrophone.com/the-complete-guide-to-electret-condenser-microphones/#What-Is-An-Electret-Condenser-Microphone?
www.youtube.com/watch?v=oqLhvg0I2BU
www.youtube.com/watch?v=Mxp3eCCQyas

MEMs, rf MEMs
www.everythingrf.com/community/what-are-rf-mems#:~:text=Radio%20Frequency%20Microelectromechanical%20Systems%20(RF,cost%2C%20size%2C%20and%20weight

Pierre-Simon Laplace
www.famousscientists.org/pierre-simon-laplace/
https://en.wikipedia.org/wiki/Pierre-Simon_Laplace

Semiconductor Junctions
https://en.wikipedia.org/wiki/P%E2%80%93n_junction
www.electronics-tutorials.ws/diode/diode_2.html

Siméon-Denis Poisson
www.britannica.com/biography/Simeon-Denis-Poisson

Supercapacitor
www.nationalgeographic.com/news/energy/2013/08/130821-supercapacitors/
https://en.wikipedia.org/wiki/Supercapacitor

Touch Screens
https://electronics.howstuffworks.com/iphone1.htm
www.youtube.com/watch?v=cFvh7qM6LdA
www.youtube.com/watch?v=wKuqNuzM1oM

Chapter 3

Bayard-Alpert Ionization Gauge
https://arunmicro.com/news/how-does-an-ion-gauge-work/

Cathode Ray Tube
https://computer.howstuffworks.com/monitor7.htm

Fluorescent Lights
https://home.howstuffworks.com/fluorescent-lamp2.htm
https://pwg.gsfc.nasa.gov/Education/wfluor.html
www.youtube.com/watch?v=JjPGGttPOe0

Georg Simon Ohm
www.code-electrical.com/ohmslaw.html#:~:text=In%201827%20Georg%20Simon%20Ohm,like%20water%20in%20a%20pipe.&text=In%20electrical%20terms%2C%20voltage%20is,by%20the%20letter%20%22R%22
www.britannica.com/biography/Georg-Ohm

Superconductors
www.sciencealert.com/superconductivity#:~:text=Superconductivity%20is%20a%20phenomenon%20whereby,efficiency%2C%20losing%20nothing%20to%20heat
https://home.cern/science/engineering/superconductivity

Vacuum Tubes
https://vacuumtubes.net/How_Vacuum_Tubes_Work.htm
https://circuitdigest.com/article/what-is-vacuum-tube-and-how-does-it-work

Chapter 4

André-Marie Ampère
https://en.wikipedia.org/wiki/Andr%C3%A9-Marie_Amp%C3%A8re
www.britannica.com/biography/Andre-Marie-Ampere
www.famousscientists.org/andre-marie-ampere/

Aurora Borealis
www.northernlightscentre.ca/northernlights.html
www.qrg.northwestern.edu/projects/vss/docs/space-environment/3-what-is-solar-wind.html#:~:text=The%20solar%20wind%20is%20a,It%20is%20made%20of%20plasma

Cyclotron
https://en.wikipedia.org/wiki/Cyclotron
www.britannica.com/technology/cyclotron
http://hyperphysics.phy-astr.gsu.edu/hbase/magnetic/cyclot.html

Earth's Magnetic Field
https://image.gsfc.nasa.gov/poetry/tour/AAmag.html#:~:text=The%20magnetic%20field%20of%20Earth,the%20crust%20and%20enters%20space
www.youtube.com/watch?v=51usJ74pPP8
https://geomag.nrcan.gc.ca/mag_fld/fld-en.php
https://en.wikipedia.org/wiki/Dynamo_theory

Félix Savart
https://mathshistory.st-andrews.ac.uk/Biographies/Savart/
https://en.wikipedia.org/wiki/F%C3%A9lix_Savart

Hall Effect Probe
http://hyperphysics.phy-astr.gsu.edu/hbase/magnetic/Hall.html
https://en.wikipedia.org/wiki/Hall_effect

Hans Christian Oersted
www.famousscientists.org/hans-christian-oersted/

Helmholtz Pair
www.universetoday.com/84140/helmholtz-coil/
https://en.wikipedia.org/wiki/Helmholtz_coil
http://hyperphysics.phy-astr.gsu.edu/hbase/magnetic/helmholtz.html

Ion Mass Spectrometer
http://hyperphysics.phy-astr.gsu.edu/hbase/magnetic/maspec.html
https://en.wikipedia.org/wiki/Mass_spectrometry

Ion Velocity Selector
http://hyperphysics.phy-astr.gsu.edu/hbase/magnetic/maspec.html
https://en.wikipedia.org/wiki/Wien_filter

Jean-Baptiste Biot
www.britannica.com/biography/Jean-Baptiste-Biot
https://en.wikipedia.org/wiki/Jean-Baptiste_Biot#:~:text=Jean%2DBaptiste%20Biot%20(%2F%CB%88,studied%20the%20polarization%20of%20light

Magnetic Memory
www.ukessays.com/essays/computer-science/understanding-how-magnetic-storage-devices-work-computer-science-essay.php#:~:text=The%20surfaces%20of%20disks%20and,can%20be%20stored%20on%20them.&text=The%20write%2Fread%20heads%20of,over%20the%20disk%20or%20tape
https://computer.howstuffworks.com/removable-storage2.htm
https://en.wikipedia.org/wiki/Magnetic_storage

Magnetic Resonance Imaging
www.nibib.nih.gov/science-education/science-topics/magnetic-resonance-imaging-mri#:~:text=How%20does%20MRI%20work%3F,-MRI%20of%20a&text=MRIs%20employ%20powerful%20magnets%20which,pull%20of%20the%20magnetic%20field
www.ncbi.nlm.nih.gov/pmc/articles/PMC1121941/

Tokamak
https://en.wikipedia.org/wiki/Tokamak#:~:text=A%20tokamak%20(%2F%CB%88to%CA%8A,produce%20controlled%20thermonuclear%20fusion%20power
www.energy.gov/science/doe-explainstokamaks
www.iter.org/mach/Tokamak

Chapter 5

Dynamic Microphone
www.youtube.com/watch?v=oqLhvg0I2BU
www.youtube.com/watch?v=Mxp3eCCQyas

Faraday Disc Generator
https://en.wikipedia.org/wiki/Homopolar_generator

Metal Detectors
www.explainthatstuff.com/metaldetectors.html
https://en.wikipedia.org/wiki/Metal_detector

Michael Faraday
www.sciencehistory.org/historical-profile/michael-faraday
www.famousscientists.org/michael-faraday/

Ruhmkorff Induction Coil
www.youtube.com/watch?v=iVbNGJX88gc
http://physics.kenyon.edu/EarlyApparatus/Electricity/Induction_Coil/Induction_Coil.html

Variable Autotransformer
https://en.wikipedia.org/wiki/Autotransformer

Wind Turbines
www.energy.gov/maps/how-does-wind-turbine-work

www.eia.gov/energyexplained/wind/electricity-generation-from-wind.php#:~:text=
Electricity%20generation%20with%20wind&text=In%202019%2C%20wind
%20turbines%20in,kilowatts)%20of%20electricity%20generation%20capacity

Wireless Charging Stations and Pads
www.computerworld.com/article/3235176/wireless-charging-explained-what-is-it-
and-how-does-it-work.html

Chapter 6

D.C. and A.C. Motors
www.explainthatstuff.com/induction-motors.html#:~:text=How%20does%20an
%20AC%20motor%20work%3F,-Unlike%20toys%20and&text=Unlike%20in%20a
%20DC%20motor,the%20outside%20of%20the%20motor

https://en.wikipedia.org/wiki/AC_motor

Electromagnets
https://en.wikipedia.org/wiki/Electromagnet#:~:text=So%20the%20maximum
%20strength%20of,around%201.6%20to%202%20T

https://nationalmaglab.org/education/magnet-academy/learn-the-basics/stories/
magnets-from-mini-to-mighty

https://home.cern/news/news/accelerators/demonstrator-magnet-produces-record-
magnet-field

https://nationalmaglab.org/about/maglab-dictionary/permanent-magnet#:~:text=This
%20would%20be%20your%20normal,50%20times%20stronger%20than%20that

Magnetic Base
https://en.wikipedia.org/wiki/Magnetic_switchable_device#:~:text=A%20magnetic
%20switchable%20device%20(often,be%20turned%20on%20or%20off.&text=A
%20round%20permanent%20magnet%20is,allow%20rotation%20of%20the
%20magnet

Magnetic Homing (birds)
www.nationalgeographic.com/news/2007/9/birds-can-see-earths-magnetic-field/

https://ssec.si.edu/stemvisions-blog/how-do-birds-navigate

Magnetic Levitation
https://en.wikipedia.org/wiki/Magnetic_levitation

http://hyperphysics.phy-astr.gsu.edu/hbase/Solids/maglev.html

Regenerative Brakes
https://electricvehiclesnews.com/History/historyearlyIII.htm

www.shorpy.com/node/5734?size=_original#caption

https://greeninginc.com/blog/new-tech/where-are-regenerative-brakes-headed/

Traffic Signal Sensors
www.fhwa.dot.gov/publications/research/operations/its/06108/02.cfm
www.cedengineering.com/userfiles/Detection%20For%20Traffic%20Signals.pdf
www.automatesystems.co.uk/how-traffic-light-sensors-work/#:~:text=Every
%20traffic%20light%20signal%20has,helps%20it%20direct%20traffic%20flow.&
text=These%20sensors%20use%20different%20technologies,light%20sensors
%20are%20induction%20loops

Chapter 7

Heinrich Hertz
www.aaas.org/heinrich-hertz-and-electromagnetic-radiation
www.famousscientists.org/heinrich-hertz/
https://ieeexplore.ieee.org/stamp/stamp.jsp?arnumber=8428525

James Clerk Maxwell
www.britannica.com/biography/James-Clerk-Maxwell
https://physicsworld.com/a/james-clerk-maxwell-a-force-for-physics/
https://mathshistory.st-andrews.ac.uk/Biographies/Maxwell/
https://en.wikipedia.org/wiki/James_Clerk_Maxwell

Chapter 8

Absorptive Polarizers
https://en.wikipedia.org/wiki/Polarizer#:~:text=A%20Polaroid%20polarizing
%20filter%20functions,plastic%20with%20an%20iodine%20doping
https://science.howstuffworks.com/innovation/everyday-innovations/sunglass6
.htm#:~:text=Polarized%20filters%20are%20most%20commonly,parallel
%20relation%20to%20one%20another

CDs, DVDs, and Blu-ray Discs
www.explainthatstuff.com/cdplayers.html
https://en.wikipedia.org/wiki/Compact_disc
Cope, John A. "The physics of the compact disc," *Phys. Educ.* 28, 15–21 (1993)
www.physik.uni-wuerzburg.de/fileadmin/physik-fpraktikum/_imported/fileadmin/
11999999/AFM/The_Physics_of_the_Compact_Disc.pdf

Cosmic Microwave Background
http://cosmology.berkeley.edu/Education/CosmologyEssays/The_Cosmic_Microwave_
Background.html
https://en.wikipedia.org/wiki/Discovery_of_cosmic_microwave_background_
radiation

Doppler Astronomy
https://en.wikipedia.org/wiki/Doppler_effect#Astronomy

Doppler Traffic Radar
https://copradar.com/chapts/chapt1/ch1d1.html

Doppler Weather Radar
www.weather.gov/jetstream/how
www.everythingweather.com/weather-radar/bands.shtml

Global Positioning System (GPS)
www.gps.gov/systems/gps/
www.afspc.af.mil/News/Article-Display/Article/1707464/mathematician-inducted-into-space-and-missiles-pioneers-hall-of-fame/
www.livescience.com/32660-how-does-an-atomic-clock-work.html#:~:text=When%20exposed%20to%20certain%20frequencies,precise%20way%20to%20count%20seconds

Health Effects of Electric and Magnetic Fields
www.niehs.nih.gov/health/topics/agents/emf/index.cfm#:~:text=IARC%20Classifies%20Radiofrequency%20Electromagnetic%20Fields,a%20malignant%20type%20of%20brain

Infrared Thermometer
www.instrumentchoice.com.au/news/how-do-infrared-thermometers-work

Liquid Crystal Displays
www.explainthatstuff.com/lcdtv.html
www.xenarc.com/lcd-technology.html
www.nobelprize.org/prizes/physics/1991/summary/
https://computer.howstuffworks.com/monitor5.htm

Radio Frequency Identification
www.harting.com/UK/en-gb/topics/4-uses-rfid-technology-industry
https://electronics.howstuffworks.com/gadgets/high-tech-gadgets/rfid.htm

Remote Control
https://electronics.howstuffworks.com/remote-control.htm

Satellite TV
https://electronics.howstuffworks.com/satellite-tv.htm
https://theconnectedhome.com/how-does-satellite-tv-work/
www.space.com/29222-geosynchronous-orbit.html
www.sciencelearn.org.nz/resources/270-satellite-communications#:~:text=At%20the%20satellite%2C%20the%20signal,Earth%20using%20a%20different%20frequency.&text=For%20television%20reception%2C%20the%20signal,sent%20to%20the%20television%20decoder

Sensing Instruments for Underground Utility Location
www.engineersupply.com/underground-utility-location-equipment-explained.aspx
https://www.globalspec.com/learnmore/building_construction/building_construction_tools_machines/underground_locating_equipment

Wave Propagation in the Atmosphere
https://en.wikipedia.org/wiki/Radio_propagation

www.sciencedirect.com/topics/earth-and-planetary-sciences/plasma-frequency#:~:
text=The%20plasma%20frequency%20is%20the,for%20conditions%20in
%20Earth's%20ionosphere
www.ieee.ca/millennium/radio/radio_differences.html

Wi-Fi
https://en.wikipedia.org/wiki/Wi-Fi

X-ray Imaging
https://science.howstuffworks.com/x-ray.htm
www.nobelprize.org/prizes/physics/1901/rontgen/biographical/
www.arpansa.gov.au/understanding-radiation/what-is-radiation/ionising-radiation/x-
ray#:~:text=X%2Drays%20are%20commonly%20produced,(braking%20radiation
%20or%20bremsstrahlung)

Chapter 10

Microwave Ovens
www1.lsbu.ac.uk/water/microwave_water.html
www.youtube.com/watch?v=kp33ZprO0Ck
www.nap.edu/read/2266/chapter/4#15

Optical Fibers
https://searchnetworking.techtarget.com/definition/fiber-optics-optical-fiber
www.corning.com/worldwide/en/innovation/the-glass-age/science-of-glass/how-it-
works-optical-fiber.html

Chapter 11

Extremely Low-Frequency Communications
https://fas.org/nuke/guide/usa/c3i/elf.htm

Microwave Relay Stations
https://en.wikipedia.org/wiki/Microwave_transmission
www.wired.com/2015/03/spencer-harding-the-long-lines/

Radio and Television Broadcast Antennas
https://en.wikipedia.org/wiki/Antenna_(radio)#:~:text=Thus%20the%20small
%20loop%20antenna,according%20to%20the%20receiver%20tuning
https://en.wikipedia.org/wiki/Circular_polarization

Smart Phone Antennas
www.antenna-theory.com/design/cellantenna.php

Index